MEDIEVAL IRELAND

AN ENCYCLOPEDIA

MEDIEVAL IRELAND
AN ENCYCLOPEDIA

Seán Duffy, Editor

Associate Editors
Ailbhe MacShamhráin
James Moynes

Routledge
New York and London

Published in 2005 by
Routledge
270 Madison Avenue
New York, NY 10016
www.routledge-ny.com

Published in Great Britain by
Routledge
2 Park Square
Milton Park, Abingdon
Oxon OX14 4RN U.K.
www.routledge.co.uk

Routledge is an imprint of the Taylor & Francis Group.
Printed in the United States of America on acid-free paper.

10 9 8 7 6 5 4 3 2 1

Library of Congress Cataloging-in-Publication Data
Medieval Ireland: an encyclopedia / Seán Duffy, editor ; associate
editors, Ailbhe MacShamhráin, James Moynes.
p. cm.
Includes bibliographical references and index.
ISBN 0-415-94052-4 (hb: alk. paper)
1. Ireland—History—1172–1603—Encyclopedia. 2. Ireland—History—To 1172—Encyclopedia.
3. Ireland—Civilization—Encyclopedia. 4. Civilization, Medieval—Encyclopedia.
I. Duffy, Seán. II. MacShamhráin, Ailbhe. III. Moynes, James.

DA933.M43 2005
941.503' 03—dc32 2004011295

Contents

Introduction

Medieval Ireland: An Encyclopedia presents the multiple facets of life in Ireland before and after the Anglo-Norman invasion of 1169, from the sixth to the sixteenth century. It provides reliable, scholarly information for the student, scholar, or general reader who wishes to learn more about this vivid period of history. The medieval period in Ireland was rich in culture, and *Medieval Ireland* provides information on such facets as architecture, art, craftsmanship, language, mythology, and religion. Further, many public figures of this time period in Ireland—ranging from kings to saints to poets—are portrayed throughout the text. The common life of the medieval Irish is covered in such topics as agriculture, coinage, law, clothing, villages, and games. Giving context to these subjects are the many outside influences that affected Irish civilization at this time. The Viking and Anglo-Norman invasions are discussed, as are the literary and cultural influences of many European countries. Thus the work is useful to people studying related topics, including Anglo-Saxon England, Carolingian Gaul, Norman England, and Viking-Age Scandinavia.

The fact that this reference work is dedicated solely to medieval Ireland—not prehistory and not contemporary affairs—gives it a value missing from other more chronologically broad-ranging works. If one is interested in just the medieval period, this volume is tailor-made. This encyclopedia is not just a history book, but as far as practicable it reaches beyond history to all recoverable areas of medieval Irish society. By confining the content to the thousand or so years that make up medieval Ireland, and exploring all aspects of that time, we have produced a unique volume. The major areas of knowledge on the subject of medieval Ireland are accessible within the covers of this book, as either separate essays or discussed within a broader context in another essay.

HOW TO USE THIS BOOK

Medieval Ireland: An Encyclopedia is arranged in an easy-to-use A to Z reference format that contains a series of 347 essays. The essays fall into two categories—biographies and thematic entries—and range from 250 to 2,500 words in length. Each essay provides an analysis of the topic at hand as well as directing the reader to explore the topic more thoroughly through **See Also** cross-references and a list of **References and Further Reading**. The ease of use is enhanced by a **thematic Table of Contents** at the beginning of the book, in addition to the standard **alphabetical Table of Contents**. Other features of the book include **illustrations** and **maps** that visually support the text. Readers will also find that the **Blind Entries** will lead them to articles that discuss a subject that is not covered as a separate entry but is essential to the study of medieval Ireland. Finally, the book contains a **detailed, analytical Index** that also helps the reader further navigate the work.

CONTENTS

More than 100 scholars have provided in-depth essays to *Medieval Ireland: An Encyclopedia*. The essays encompass all aspects of Irish society and culture throughout the one thousand or so years that make up medieval Ireland. The entries are categorized thematically, displaying the breadth of coverage dedicated to this important historical epoch in Irish history.

Archeology, Architecture, and Art

These entries provide an analysis and representation of the artistic contributions made to the history of Ireland. Topics include Abbeys, Castles, High Crosses, Iconography, and Sculpture.

Economy and Society (Government, Law, Military, Politics, and Religion)

Among these topics there are entries that explain the many aspects of society with regard to government (e.g., Parliament), economics (e.g., Coinage), law (e.g., Brehon Law), military (e.g., Armies), politics (e.g., Lordship of Ireland), and religion (e.g., Ecclesiastical Organization).

Developments and Periods

These essays review medieval Ireland by tracing the major events that affected this time period. Included are the Conversion to Christianity, Famine and Hunger, the Battle of Clontarf, Gaelic Revival, and the Renaissance.

Lineage

These essays provide the ancestry and ascension of the many families (e.g., Fitzgerald) and dynasties (e.g., Connachta) that flourished during medieval times in Ireland.

Manuscripts and Texts

These essays highlight the major texts (e.g., Book of Kells) preserved from this time period and discuss their influences, impact, and contribution to medieval Ireland.

Persons (Literary, Political, and Religious)

These essays give the known details of an individual's life as well as his significance to the study of this period. Included are political (e.g., Brian Boru), religious (e.g., St. Patrick), and literary (e.g., John Scottus Ériugena) figures.

Persons (Dynasties, Families, and Categorized by Group)

The essays on groups of people (such as children, pilgrims, and slaves) give perspective to the customs, laws, and functions of Irish society. The articles on dynasties (e.g., Connachta) and families (e.g., Fitgerald) delve into the roles and power of the Irish nobility.

Places

These articles describe the emergence and significance of the many towns, royal seats, and historical sites (e.g., Tara, Ailech, and Dublin) during the medieval period in Ireland.

Scholarship

There is extensive discussion of Irish scholarship throughout this work, and the essays present the topics of learning (e.g., Education, Language, and Sciences), literary genres (e.g., Hiberno-English), and literary influences (e.g., Scottish Literary Influence).

Acknowledgments

I am grateful to the New York-based staff of Routledge, in particular Marie-Claire Antoine, Mark O'Malley, and Mary Funchion (none of whom I have ever met, although all seem like old acquaintances at this stage), for their forbearance with me and with this project. Oh how, as deadlines have come and gone, they have learned what Mary, being Irish, knew—that we simply have no word that conveys quite the urgency of *mañana*. James Moynes and Ailbhe MacShamhráin have greatly assisted me in bringing the project to completion. James and I are both later medievalists, and he will, I know, join me in acknowledging our joint debt to Ailbhe, one of the most gifted early Irish medievalists of his generation and a friend and scholar of limitless generosity.

I have been astonished by the generosity also shown by so many scholars (again, many of whom I have yet to meet) in many countries across Europe, North America, and Australasia who have agreed to write for this book. Many did so at very short notice. Many are *the* leading authorities in their subjects in the world. They are extremely busy people who have little to gain on their career path by taking the time to contribute to such a volume, and, if truth be told, neither will it make much difference to their bank balance. They did so out of a commitment to the subject that is truly genuine. Their goal is simply to make a contribution to the study of medieval Ireland and to make their scholarship available to others in this generation and to come, and I am enormously thankful to every one of them. I hope that the finished product does justice to their efforts.

Seán Duffy

Publisher's Note

The Routledge Encyclopedias of the Middle Ages

Formerly the Garland Encyclopedias of the Middle Ages, this comprehensive series began in 1993 with the publication of *Medieval Scandinavia*. A major enterprise in medieval scholarship, the series brings the expertise of scholars specializing in myriad aspects of the medieval world together in a reference source accessible to students and the general public as well as to historians and scholars in related fields. Each volume focuses on a geographical area or theme important to medieval studies and is edited by a specialist in that field, who has called on a board of consulting or associate editors to establish the list of articles and review the articles. Each article is contributed by a scholar and, typically, is followed by a bibliography and cross-references to guide further research.

Routledge is proud to carry on the tradition established by the first volumes in this important series. As the series continues to grow, we hope that it will provide the most comprehensive and detailed view of the medieval world in all its aspects ever presented in encyclopedia form.

Vol. 1 *Medieval Scandinavia: An Encyclopedia*. Edited by Phillip Pulsiano.

Vol. 2 *Medieval France: An Encyclopedia.* Edited by William W. Kibler and Grover A. Zinn.

Vol. 3 *Medieval England: An Encyclopedia*. Edited by Paul E. Szarmach, M. Teresa Tavormin, and Joel T. Rosenthal.

Vol. 4 *Medieval Archaeology: An Encyclopedia*. Edited by Pamela Crabtree.

Vol. 5 *Trade, Travel, and Exploration in the Middle Ages*. Edited by John Block Friedman and Kristen Mossler Figg.

Vol. 6 *Medieval Germany: An Encyclopedia*. Edited by John M. Jeep.

Vol. 7 *Medieval Jewish Civilization: An Encyclopedia*. Edited by Norman Roth.

Vol. 8 *Medieval Iberia: An Encyclopedia*. Edited by Michael Gerli.

Vol. 9 *Medieval Italy: An Encyclopedia*. Edited by Christopher Kleinhenz.

Medieval Ireland: An Encyclopedia, edited by Seán Duffy, is Volume 10 in the series.

Contributors

T. B. Barry
Trinity College Dublin
Black Death; Manorialism; Motte-and-Baileys; Waterford

Rolf Baumgarten
Dublin Institute for Advanced Studies
Etymology

Jacqueline Borsje
Universiteit Utrecht
Witchcraft and Magic

Cormac Bourke
Ulster Museum, Belfast
Reliquaries

John Bradley
National University of Ireland, Maynooth
Bridges; Carolingian; Kildare; Kilkenny; Mills and Milling; Tara; Trim; Villages; Walled Towns

Paul Brand
All Souls College, Oxford
Common Law; Courts; Local Government; March Law; Modus Tenendi Parliamentum; Records, Administrative; Wills and Testaments

Dorothy-Ann Bray
McGill University, Montreal
Brigit; Íte; Mo-Ninne

Caoimhín Breatnach
University College Dublin
Historical Tales; Lismore, Book of; Rawlinson B502

Aidan Breen
National University of Ireland, Maynooth
Adomnán; Áed Mac Crimthainn; Blathmac; Classical Influence; Cogitosus; Dícuil; Grammatical Treatises; Literature, Hiberno-Latin; Marriage; Muirchú; Poetry, Hiberno-Latin Literature; Records, Ecclesiastical; Rhetoric; Sciences

Stephen Brown
University of Wisconsin-Madison
Education

Paul Byrne
Department of Finance, Dublin
Diarmait mac Cerbaill; Lóegaire Mac Neill; Mide; Niall Noígiallach; Uí Néill; Uí Néill, Northern; Uí Néill, Southern

Michael Byrnes
Dun Laoghaire, Co. Dublin, Ireland
Céli Dé; Feis; Máel-Ruain; Tribes; Tuarastal; Túath

Letitia Campbell
Trinity College Dublin
Cormac mac Cuilennáin; Éoganachta; Fedelmid mac Crimthainn; Mac Carthaig, Cormac

Howard Clarke
University College Dublin
Dublin; Fraternities and Guilds; Population; Urban Administration

Anne Connon
National University of Ireland, Galway
Aífe; Derbforgaill; Gormlaith (d. 1030); Queens; Sitriuc Silkenbeard; Uí Briúin

Peter Crooks
Trinity College Dublin
Anglo-Irish Relations; Factionalism; Feudalism; Lionel of Clarence; Mac Murchada, Diarmait; Marshal; Racial and Cultural Conflict; Society, Functioning of Anglo-Norman; Society, Grades of Anglo-Norman; Ulster, Earldom of; William of Windsor

John Reuben Davies
Derby, United Kingdom
Ecclesiastical Organization

James E. Doan
Nova Southeastern University
Ua Dálaigh

Charles Doherty
University College Dublin
Airgialla; Clientship; Derry; Érainn; Leth Cuinn and Leth Moga; Naval Warfare; Óenach

Clare Downham
University of Aberdeen
Cerball mac Dúngaile; Fine Gall; Viking Incursions

Paul Dryburgh
National Archives, London
Mortimer

Seán Duffy
Trinity College Dublin
Bruce, Edward; Courcy, John de; John; Limerick; Ua Briain, Muirchertach; Ua Néill, Domnall

David Edwards
University College Cork
Coshering; Coyne and Livery; Tánaiste; Ua Néill of Clandeboye

Steven Ellis
National University of Ireland, Galway
Fitzgerald, Gerald, 8th Earl; Lancastrian-Yorkist Ireland

Adrian Empey
The Church of Ireland Theological College, Dublin
Butler-Ormond

Nicholas Evans
Dublin Institute for Advanced Studies
Áedán mac Gabráin; Annals and Chronicles; Scotti/Scots; Scottish Influence

Maria FitzGerald
Dublin, Ireland
Clothing

Elizabeth FitzPatrick
National University of Ireland, Galway
Ailech; Emain Macha; Inauguration Sites; Promontory Forts; Ringforts; Roads, Routes

Fiona Fitzsimons
Eneclann Ltd., Dublin
Gossiprid

Marie-Therese Flanagan
Queen's University Belfast
Anglo-Norman Invasion; Henry II; Strongbow

John Gillingham
Brighton, Sussex Co., England
Giraldus Cambrensis

Angela Gleason
Dublin Institute for Advanced Studies
Entertainment; Games; Music; Poets/Men of Learning; Satire

Dianne Hall
The University of Melbourne
Nuns

Andrew Halpin
National Museum of Ireland
Coinage; Craftwork; Hoards; Weapons and Weaponry

Peter Harbison
Royal Irish Academy, Dublin
High Crosses; Iconography; Pilgrims and Pilgrimage; Sculpture

Beth Hartland
University of Durham
Clare, de; Geneville, Geoffrey de; March Areas; Valence, de; Verdon, de

Michael W. Herren
York University, Toronto
Biblical and Church Fathers Influence; Ériugena, John Scottus; Sedulius Scottus

Martin Holland
Trinity College Dublin
Cashel, Synod of I; Cashel, Synod of II; Church Reform, Twelfth Century; Gilla Pátraic, Bishop; Gille (Gilbert) of Limerick; Kells, Synod of; Malachy (Máel-M'áedóic); Ráith Bressail, Synod of

Poul Holm
University of Southern Denmark
Slaves

David Robert Howlett
University of Oxford
Airbertach Mac Cosse; Cathal mac Finguine; Cerball mac Muireccáin; Glendalough; Gormlaith (d. 948); Patrick; Pre-Christian Ireland; Máel-Muru Othna; Ua Tuathail (O'Toole), St. Lawrence

Benjamin Hudson
Pennsylvania State University
Clontarf, Battle of; Diarmait mac Máele-na-mbó; Marianus Scottus; Prophecies and Vaticinal; Ua Briain, Tairrdelbach

Bart Jaski
Universiteit Utrecht
Brian Boru; Kings and Kingship; Máel-Sechnaill I; Máel-Sechnaill II

Henry A. Jefferies
Thornhill College, Derry
Armagh; Cork; Mac Carthy; Munster; Ua Briain

Dorothy Johnston
University of Nottingham
Richard II

Eamon P. Kelly
National Museum of Ireland
Sheela-Na-Gig

Fergus Kelly
Dublin Institute for Advanced Studies
Agriculture; Law Schools, Learned Families; Law Texts

Stuart Kinsella
Christ Church Cathedral, Dublin
Christ Church Cathedral; St. Patrick's Cathedral

Brian Lacey
Discovery Programme, Dublin
Colum Cille

Colm Lennon
National University of Ireland, Maynooth
Renaissance

Angela Lucas
National University of Ireland, Maynooth
British Library Manuscript Harley 913; Hiberno-English Literature

Elizabeth Malcolm
University of Melbourne
Medicine

Conleth Manning
Department of the Environment, Dublin
Burials; Clonmacnoise; Ecclesiastical Settlements; Ecclesiastical Sites; Parish Churches, Cathedrals

Anthony McCormack
National University of Ireland, Maynooth
Desmond Geraldines (FitzGeralds of Desmond)

Darren McGettigan
University College Dublin
Mac Lochlainn; Mac Mahon; Mac Sweeney; Maguire; Ua Domnaill; Ó Ruairc

J. J. N. McGurk
Claremorris, Co. Mayo, Ireland
Scriptoria

Margaret McKearney
Howth, Co. Dublin, Ireland
FitzHenry, Meiler

Neil McLeod
Murdoch University, Perth, Western Australia
Brehon Law; Society, Grades of Gaelic

Tom McNeill
Queen's University Belfast
Archaeology; Castles; Downpatrick; Savage; Tower Houses

Bernard Meehan
Trinity College Dublin
Durrow, Book of; Kells, Book of; Manuscript Illumination

Karena Morton
Boyle, Co. Roscommon, Ireland
Wall Paintings

James Moynes
Dublin, Ireland
Lacy, de; Lacy, Hugh de; Maynooth; Pale; Plunkett

Evelyn Mullally
Queen's University Belfast
French Writing in Ireland

John L. Murphy
DeVry University
Jews in Ireland

Margaret Murphy
The Discovery Programme, Dublin
Central Government; Chief Governors; Cumin, John; Henry of London; Lordship of Ireland; Parliament

Kevin Murray
University College, Cork
Fenian Cycle

Muireann Ní Bhrolcháin
National University of Ireland, Maynooth
Leinster, Book of; Lyrics; Máel-Ísu Ua Brolcháin

Máire Ní Mhaonaigh
St. John's College, Cambridge
Cináed Ua hArtacáin; Cuán Ua Lothcháin; Flann Mac Lonáin; Flann Mainistreach; Lebor na hUidre; Muirchertach Mac Liacc; Scandinavian Influence; Uí Maine, Book of; Welsh Influence

Kenneth Nicholls
University College, Cork
Bermingham; Charters and Chartularies; Mac Donnell; Ua Catháin; Poer; Society, Functioning of Gaelic

Tomás Ó Cathasaigh
Harvard University
Echtrae

Eamon Ó Cíosáin
National University of Ireland, Maynooth
French Literary Influence

Colmán N. Ó Clabaigh
Glenstal Abbey, Co. Limerick, Ireland
Annals of the Four Masters; Mellifont; Military Orders; Papacy; Religious Orders

Cormac Ó Cléirigh
Dublin, Ireland
Fitzgerald

Dáibhí Ó Cróinín
National University of Ireland, Galway
Armagh, Book of; Christianity, Conversion to; Palladius; Paschal Controversy

Tadhg Ó Dúshláine
National University of Ireland, Maynooth
Foras Feasa ar Éirinn

Mícheál Ó Mainnín
Queen's University, Belfast
Mac Aodhagáin

Nollaig Ó Muraíle
Queen's University, Belfast
Dinnsenchas; Genealogy; Lecan, Book of; Mac Firbhisigh; Placenames

Pádraig Ó Néill
University of North Carolina
Áes Dána; Anglo-Saxon Literary Influence; Glosses; Languages; Metrics; Nicolas Mac Máel-Ísu; Osraige; Personal Names; Romance; Tírechán

Pádraig Ó Riain
University College, Cork
Ciarán; Hagiography and Martyrologies

Emmet O'Byrne
University College, Dublin
Armies; Mac Lochlainn, Muirchertach; MacMurrough; MacMurrough, Art; Military Service, Anglo-Norman; Military Service, Gaelic; Ua Conchobair, Ruairí; Ua Conchobair, Tairrdelbach

Elizabeth Okasha
University College, Cork
Inscriptions

Tadhg O'Keeffe
University College Dublin
Abbeys and Religious Houses; Architecture

Thomas O'Loughlin
University of Wales, Lampeter
Canon Law; Devotional and Liturgical; Moral and Religious Instruction; Penitentials

Timothy O'Neill
Dublin, Ireland
Leabhar Breac; Ports; Ships and Shipping; Trade

Aidan O'Sullivan
University College Dublin
Crannóga; Diet and Food; Fishing; Houses; Woodlands

Michael Richter
Universität Konstanz
Columbanus; National Identity; Peregrini

Michael Ryan
Chester Beatty Library, Dublin
Early Christian Art; Jewelry and Personal Ornament; Metalwork

Katharine Simms
Trinity College, Dublin
Bardic Schools/Learned Families; Duanairí; Gaelic Revival; Gaelicization; Ua Néill; Women

David Stifter
Universität Wien
Triads; Wisdom Texts

Michael Terry
University of Toledo
Poetry, Irish

Gregory Toner
University of Ulster, Coleraine
Invasion Myth; Mythological Cycle; Ulster Cycle

Mary Valante
Appalachian State University
Famine and Hunger

Freya Verstraten
Trinity College, Dublin
Burke; Connacht; Ua Conchobair

Katerine Walsh
Universität Innsbruck
FitzRalph, Richard

Keith Waters
Durham, United Kingdom
Fosterage; Gerald, Third Earl of Desmond; O Conchobhair-Fáilge

Nora White
National University of Ireland
Comperta; Fiachnae mac Báetáin

Dan M. Wiley
Hastings College, Nebraska
Aideda; Connachta; Dál Cais; Déisi; Laigin

Bernadette Williams
University College, Dublin
Clyn, Friar John; Hiberno-Norman (Latin)

Nicholas Williams
Baile Atha Cliath, Ireland
Mac Con Midhe, Giolla Brighde

Jonathan Wooding
University of Wales, Lampeter
Immrama

Alex Woolf
University of St. Andrews
Amlaíb Cuarán; Cruthni; Ulaid

Mark Zumbuhl
University of Glasgow
Leinster; Uí Chennselaig; Uí Dúnlainge

Alphabetical Entry List

Thematic Entry List

(1) Archaeology, Architecture, and Art

Abbeys
Archaeology
Architecture
Bridges
Burials
Castles
Christ Church Cathedral
Crannóga
Early Christian Art
Ecclesiastical Settlements
Ecclesiastical Sites
High Crosses
Hoards
Houses
Iconography
Inauguration Sites
Inscriptions
Jewelry and Personal Ornament
Manuscript Illumination
Metalwork
Motte-and-Baileys
Parish Churches and Cathedrals
Reliquaries
Ringforts
Sculpture
Sheela-Na-Gig
St. Patrick's Cathedral
Tower Houses
Villages
Wall Paintings
Walled Towns

(2) Dynasties

Airgialla
Connachta
Cruthni
Dál Cais
Déisi
Eóganachta
Érainn
Laigin
Osraige
Uí Briúin
Uí Chennselaig
Uí Dúnlainge
Uí Néill
Uí Néill, Northern
Uí Néill, Southern
Ulaid

(3) Developments and Periods

Anglo-Norman Invasion
Black Death
Christianity, Conversion to
Church Reform, Twelfth Century
Clontarf, Battle of
Gaelicization
Famine and Hunger
Gaelic Revival
Lancastrian-Yorkist Ireland
Racial and Cultural Conflict
Renaissance
Viking Incursions

(4) Families

Bermingham
Burke
Butler-Ormond
Clare, de
Desmond Geraldines (FitzGeralds of Desmond)
Fitzgerald
Lacy, de
Mac Aodhagáin
Mac Carthy
Mac Donnell
Mac Fhir Bhisigh
Mac Lochlainn
Mac Mahon
Mac Murrough
Mac Sweeney
Maguire
Marshal
Mortimer
Ó Conchobhair-Fáilge
Ó Ruairc
Plunkett
Poer
Savage
Ua Briain
Ua Catháin
Ua Conchobhair
Ua Dálaigh
Ua Domnaill
Ua Néill
Ua Néill, Clandeboye
Valence, de
Verdon, de

(5) Learning

Áes Dána
Bardic Schools, Learned Families
Duanairí
Education
Languages
Law Schools, Learned Families
Medicine
Sciences
Scriptoria

(6) Literature: genres

Ulster Cycle
Historical Tales
Invasion Myth
Hagiography and Martyrologies
Aideda
Comperta
Devotional and Liturgical
Dinnsenchas
Echtrai
Etymology
Genealogy
Glosses
Grammatical Treatises
Hiberno-English Literature
Hiberno-Latin Literature
Hiberno-Norman (Latin)
Immrama
Law Tracts
Lyrics
Metrics
Moral and Religious Instruction
Mythological Cycle
Penitentials
Prophecies and Vaticinal Literature
Rhetoric
Romance
Satire
Triads
Wisdom Texts
Poetry, Hiberno-Latin
Poetry, Irish

(7) Literature; influences

Anglo-Saxon
Biblical and Church Fathers
Carolingian
Classical
French
Scandinavian
Scottish
Welsh

(8) Manuscripts and texts

Annals of the Four Masters
Armagh, Book of
Durrow, Book of
Forus Feasa ar Éirinn
Kells, Book of
Leabhar Breac

Lebor na hUidre
Lecan, Book of
Leinster, Book of
Lismore, Book of
Uí Maine, Book of

(9) Music and Leisure

Entertainment
Games
Music

(10) Persons literary

Áed Mac Crimthainn
Airbertach Mac Cosse
Blathmac
Cináed Ua hArtacáin
Clyn, Friar John
Cogitosus
Cuán Ua Lothcháin
Dícuil
Ériugena (John Scottus)
Flann Mac Lonáin
Flann Mainistreach
Gerald, Third Earl of Desmond
Gilla Pátraic, Bishop
Giraldus Cambrensis
Mac Con Midhe, Giolla Brighde
Máel-Muru Othna
Máel-Ísu Ua Brolcháin
Marianus Scottus
Muirchertach Mac Liacc
Muirchú
Sedulius Scottus
Tírechán

(11) Persons political

Áedán mac Gabráin
Aífe
Amlaíb Cuarán
Brian Boru
Bruce, Edward
Cathal mac Finguine
Cerball mac Dúngaile
Cerball mac Muireccáin
Cormac mac Cuilennáin
Courcy, John de
Derbforgaill
Diarmait mac Cerbaill
Diarmait mac Máele-na-mbó
Fedelmid mac Crimthainn
Fiachnae mac Báetáin
Fitzgerald, Gerald
FitzHenry, Meiler
Geneville, Geoffrey de
Gormlaith (d. 948)
Gormlaith (d. 1030)
Henry II
John
Lacy, Hugh de
Lionel of Clarence
Lóegaire Mac Neill
Mac Carthaig, Cormac
Mac Lochlainn, Muirchertach
Mac Murchada, Diarmait
Mac Murrough, Art
Máel-Sechnaill I
Máel-Sechnaill II
Niall Noígiallach
Richard II
Sitriuc Silkenbeard
Strongbow
Ua Briain, Muirchertach
Ua Briain, Tairrdelbach
Ua Conchobair, Ruaidrí
Ua Conchobair, Tairrdelbach
Ua Néill, Domnall
William of Windsor

(12) Persons religious

Adomnán
Brigit
Ciarán
Colum Cille
Columbanus
Cumin, John
FitzRalph, Richard
Gille (Gilbert) of Limerick
Henry of London
Íte
Malachy (Máel-Máedóic)
Máel-Ruain
Mo-Ninne
Nicholas mac Máel-Ísu
Palladius

Patrick
Ua Tuathail (O'Toole), St. Lawrence

(13) Persons by category

Children
Fraternities and Guilds
Pilgrims and Pilgrimage
Poets/Men of Learning
Queens
Scotti/Scots
Slaves
Women

(14) Places

Ailech
Armagh
Emain Macha
Clonmacnoise
Connacht
Cork
Derry
Downpatrick
Dublin
Fine Gall
Glendalough
Kildare
Kilkenny
Leinster
Leth Cuinn and Leth Moga
Limerick
Maynooth
Mellifont
Mide
Munster
Pale, The
Placenames
Ports
Tara
Trim
Ulster, Earldom of
Waterford

(15) Records: historical

Annals and Chronicles
Charters and Chartularies
Records, Administrative
Records, Ecclesiastical
Wills and Testaments

(16) Government and administration

Central Government
Local Government
Chief Governors
Courts
Parliament
Urban Administration

(17) Economy and Society

Agriculture
Clothing
Coshering
Coinage
Coyne and Livery
Craftwork
Diet and Food
Fishing
Manorialism
Marriage
Mills and Milling
Personal Names
Population
Roads and Routes
Society, Grades of Gaelic
Society, Grades of Anglo-Norman
Society, Functioning of Gaelic
Society, Functioning of Anglo-Norman
Ships and Shipping
Trade
Woodlands

(18) Law

Brehon Law
Common Law
Canon Law
Clientship
Fosterage
Gossiprid
Law Texts
March Law

(19) Military

Armies
Weapons and Weaponry

Military Service, Gaelic
Military Service, Anglo-Norman
Naval Warfare

(20) Politics

Anglo-Irish Relations
Feis
Feudalism
Factionalism
Kings and Kingship
Lordship of Ireland
March Areas
Modus Tenendi Parliamentum
National Identity
Óenach
Oireacht
Tánaiste
Tribes
Tuarastal
Túath

(21) Religion

Cashel, Synod of I (1101)
Cashel, Synod of II (1172)
Céli Dé
Ecclesiastical Organization
Jews in Ireland
Kells, Synod of
Military Orders
Nuns
Papacy
Paschal Controversy
Pre-Christian Ireland
Ráith Bressail, Synod of
Religious Orders
Witchcraft and Magi

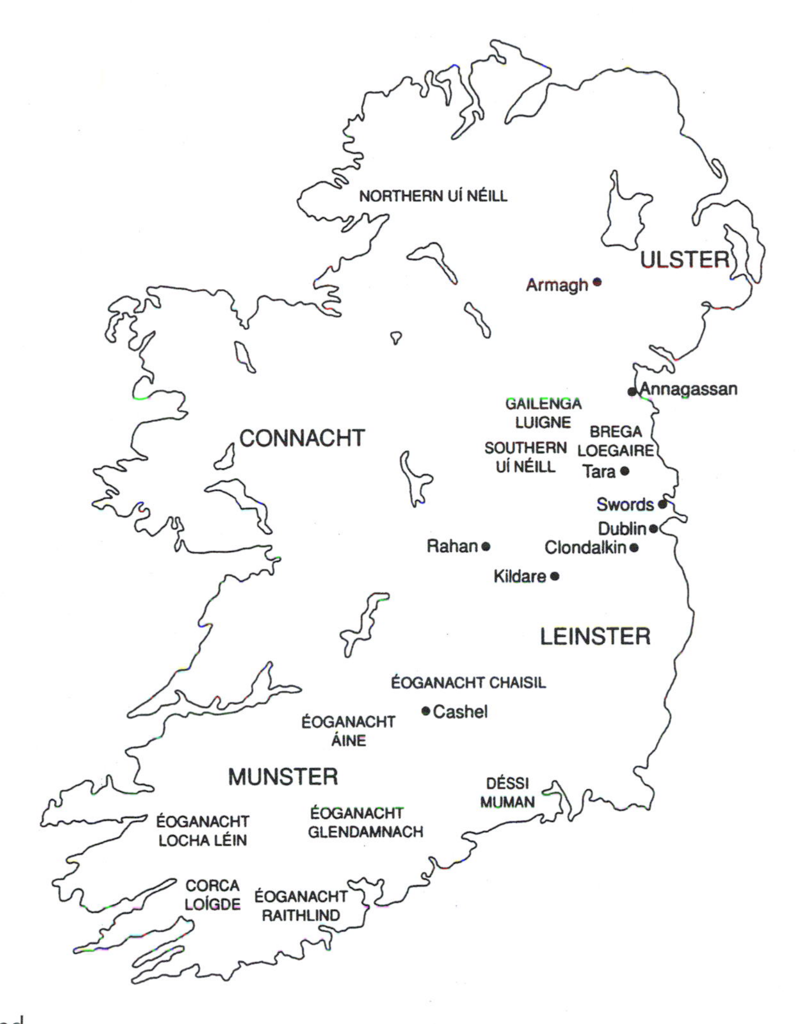

Early Medieval Ireland

ABBEYS AND RELIGIOUS HOUSES

Terminology is sometimes problematic in the study of medieval religious communal life and its material remains in Ireland, especially in the period before 1100. The erstwhile assumption of scholars that all ecclesiastical sites of the early Christian period, up to and including the age of Viking incursions, were monastic has given way in recent years to greater caution, driven by an increasing awareness of the complexity of the early Church's institutional and territorial structures, and of its provision of pastoral care to contemporary society. Strictly speaking, the designation "monastery" indicates the one-time presence of monks living in community according to a Rule, a code of behavior prescribed by one of the early church's intellectual heavyweights, and while many of the sites were certainly monastic by this measure, the organization and practice of religious life at many other sites—especially those small, archaeologically attested but barely documented, sites—simply remain unknown.

Claustral Planning

Religious foundations of the twelfth century and later generally have better documentary records, as well as higher levels of fabric-survival, so problems of interpretation and terminology are considerably less acute. Churches and associated building complexes designed for worship and habitation by religious communities are easily identified, and hence the adjective "monastic" can be used more confidently. Unlike pre-1100 foundations, most of these monasteries were claustrally planned. This claustral plan, which originated in continental Europe before A.D. 800 and first appeared in Ireland around 1140 (at Mellifont), comprised of a central square or rectangular cloister (*clustrum*, courtyard) with the key buildings arranged around it and fully enclosing it. The church was usually on the north side, the refectory (dining hall) was always on the side directly opposite, and the chapter house (a ground-floor room wherein the community assembled daily to discuss its business) and dormitory (a long first-floor room) were on the east side. The west side of the cloister comprised cellarage and additional habitation space; in Cistercian abbeys the *conversi*, lay brethren who undertook much of the manual work, were accommodated here. What made the claustral plan so attractive across the entire monastic landscape of high medieval Europe was its practical efficiency: Distances between parts of the monastery were maximized or minimized according to the relationships between the activities carried out in them. Moreover, its tightly regulated plan was a fitting metaphor for a monastic world that was itself highly regulated.

Abbeys, Priories, Friaries

Popular local tradition in Ireland, commonly abetted by ordinance survey maps, usually identifies twelfth-century and later monasteries as "abbeys," but is often incorrect in doing so. Less than a quarter of the 500-plus establishments of religious orders founded in post-1100 Ireland were genuinely abbeys, communities of male or female religious under the authority of abbots or abbesses. Of slightly lower grade were priories, communities of male or female religious presided over by priors or prioresses, officers of lower rank than abbots and abbesses; these were more numerous than abbeys, and constituted about one-third of that total. Friaries, communities of friars (literally, brothers) whose main work was preaching, make up most of the very significant remainder.

The abbeys and priories of twelfth- and early-thirteenth-century Ireland are mainly associated with

Moyne Abbey, Co. Mayo. © *Department of the Environment, Heritage and Local Government, Dublin.*

the Augustinian canons regular, priests living according to the Rule of St. Augustine, and, in the case of abbeys only, the Cistercian monks, followers of the Rule of St. Benedict. Monasteries of both groups survive in significant numbers in areas formerly under Gaelic-Irish and Anglo-Norman control. Priories of Augustinian canons regular occur more frequently in urban settings than the monastic houses of other orders, in part because of their willingness to engage in pastoral work, their modest space requirements, and their presence in Ireland at the time of colonization.

There were also other orders present in Ireland at this time, but they have left behind little above-ground archaeology. Premonstratensian canons, for example, had about a dozen houses in Ireland, but little remains of any of them. The sole house of Cluniac monks, founded by Tairrdelbach Ua Conchobair at Athlone circa 1150, is lost. Carthusian monks from England had one house, Kinalehin, founded circa 1252; dissolved ninety years later and then re-colonized shortly afterwards by Franciscan friars, the archaeological remains are mainly Franciscan, though elements of the Carthusian priory and fabric are still evident. There were also about seventy convents of nuns, mainly Augustinian canonesses. Of the few that survive, the nunnery of St. Catherine near Shanagolden stands out: Its church projects from the middle of the east side of the cloister, a very idiosyncratic arrangement.

Benedictine Houses

Benedictine monks were also present in Ireland, including some at Christ Church cathedral in the late 1000s, but they had surprisingly few houses there compared with contemporary England, where they enjoyed the patronage of the Normans. Malachy's energetic promotion of the Cistercians and Augustinians as the landscape of reformed monasticism in Ireland was taking shape was evidently to their cost. A couple of Benedictine houses, Cashel and Rosscarbery, were subject to *Schottenklöster* (Irish Benedictine monasteries in central Europe) but we know virtually nothing about their archaeology or architecture. Ireland's most substantial medieval Benedictine survival is at Fore, a late-twelfth-century foundation of the Anglo-Norman de Lacy family; the fabric of this claustrally planned monastery was altered considerably during the Middle Ages, but parts of the original church of circa 1200 remain.

Cistercian Houses

The Cistercian order, founded in 1098 in Burgundy, was a pan-European institution in the twelfth century, and its arrival in Ireland in 1142 is one of the key moments in the country's history. Fifteen Cistercian abbeys were founded in the thirty years before the Anglo-Norman invasion, and twice as many again were founded (by both Anglo-Norman and native Irish patrons) in the subsequent century. The last medieval foundation was at Hore, near Cashel, in 1272.

Cistercian architecture in Europe has a distinctively austere personality: The churches are generally simple cruciform buildings with flat-ended, rather than apsidal, presbyteries and transeptal chapels, and their interior and exterior wall surfaces tend to be unadorned. The Irish examples conform to this general pattern even though the two earliest foundations, Mellifont (founded 1142) and Baltinglass (founded 1148) have churches of slightly unusual plans.

Mellifont's construction was overseen by a monk of Clairvaux, Robert. Little of its original architecture remains. The principal surviving features at Mellifont are the late Romanesque lavabo, an elaborate structure for the collection and provision of water to monks about to enter the refectory, and the slightly later chapter house. Mellifont's community was originally composed of Irish and French brethren, but racial and cultural conflict between them persuaded the French to leave shortly after the foundation. Similar conflicts emerged and were sometimes resolved after armed conflict when the Anglo-Normans sought control of Ireland's Cistercian monasteries.

Augustinian Houses

Unlike the Cistercian Order, which entered Ireland in the company of monks from overseas, the Augustinian canons regular of pre-Anglo-Norman Ireland were simply indigenous religious who, in response to the twelfth-century Church reform, adopted the Rule of

St. Augustine as a way of life. Of the 120-odd monasteries of Augustinian canons founded in twelfth- and thirteenth-century Ireland, the number established before 1169 is uncertain; that number may be as high as one-third of the total, but the problem is that foundation dates are not as secure as those for Cistercian abbeys. Archaeology is of little help here, as there was no such thing as an "Augustinian style" of monastic architecture at any stage in the Middle Ages.

Anglo-Norman support for Augustinian canons manifested itself in continued patronage of existing houses and in the foundation of new houses. Some of these were very substantial: Athassel priory, for example, had one of the most extensive monastic complexes and one of the finest churches in medieval Ireland, while the now-destroyed St. Thomas's in Dublin, founded as a priory in 1177 and upgraded to an abbey fifteen years later, was one of Ireland's small number of mega-rich monastic houses. The claustral plan was widely employed in Augustinian houses founded by Anglo-Normans; there is no evidence of its use in Augustinian contexts prior to 1169 even though the Cistercians were using it from the 1140s.

Friars' Houses

Friars—Augustinian, Carmelite, Dominican, and Franciscan—first appeared in Ireland in the early thirteenth century, but most of the 200-odd friaries date from the period after 1350, and many of these had Gaelic-Irish patrons. Friary churches tend to be long and aisleless; large transepts were often added to their naves to increase the amount of space available for lay worship. Slender bell towers rising between the naves and choirs are perhaps the most distinctive features of friary churches.

Friaries were also claustrally planned, but their cloisters are generally much smaller than those in Cistercian abbeys or Augustinian priories, and are invariably to the north of the churches rather than to the south, which was the normal arrangement. The cloister ambulatories (or alleyways) themselves were sometimes unusual: Instead of timber lean-to roofs they often had stone-vaulted roofs which also supported the first-floor rooms of the claustral buildings. Consequently, while friary cloister courts often seem rather cramped, the dormitories often seem very spacious.

Beyond the Cloisters

Claustrally planned buildings constituted the functional and geographic inner cores of monastic possessions. Those possessions often included extensive lands with out-farms (called granges). The Cistercians were particularly adept at exploiting such lands.

A well-endowed monastery, whatever its affiliation, would normally have an enclosing precinct wall with a gatehouse; within the wall, a separate house for the abbot or prior; an infirmary (or infirmaries, as some houses had separate accommodation for monks, *conversi,* and the poor); a guest house; and gardens, orchards and dovecotes (*columbaria*) to provide for the refectory tables. Dovecotes are especially interesting. These small dome-roofed buildings of circular plan were frequently built very close to the churches, as at Ballybeg, Fore, and Kilcooley—Augustinian, Benedictine, and Cistercian foundations, respectively.

TADHG O'KEEFFE

References and Further Reading

Craig, Maurice. *The Architecture of Ireland from the Earliest Times to 1880*. London: Batsford, 1982.

Leask, Harold G. *Irish Churches and Monastic Buildings*. 3 vols. Dundalk: Dundalgan Press, 1955–1960.

Mooney, C. "Franciscan Architecture in Pre-Reformation Ireland.," *Journal of the Royal Society of Antiquaries of Ireland* 85 (1955): 133–173; 86 (1956): 129–169; 87 (1957): 1–38 and 103–124.

O'Keeffe, Tadhg. *An Anglo-Norman Monastery. Bridgetown Priory, County Cork, and the Architecture of the Augustinian Canons Regular in Medieval Ireland*, Kinsale: Gandon Editions/Cork County Council, 1999.

Stalley, Roger. *The Cistercian Monasteries of Ireland*. London and New York: Yale University Press, 1987.

See also **Architecture; Christ Church Cathedral; Church Reform, Twelfth Century; Ecclesiastical Sites; Religious Orders; St. Patrick's Cathedral; Ua Conchobair, Tairrdelbach**

ADOMNÁN MAC RÓNÁIN (*c.* 624–704)

Adomnán mac Rónáin was the ninth abbot of Iona (679–704) and biographer of Colum Cille, Iona's founding saint. According to the genealogies, he was the son of Rónán mac Tinne, one of the Cenél Conaill branch of the Uí Néill, and a kinsman of Colum Cille, his father being five generations descended from Colum Cille's grandfather, Fergus, son of Conall Gulban. His mother's name is given as Ronnat, one of the Cenél nÉnnae branch of the Northern Uí Néill, situated around what is now Raphoe in County Donegal. He is first mentioned in the *Annals of Ulster* in the year 687 as having been on a mission to Aldfrith, king of Northumbria, to obtain release of prisoners taken in a raid on Brega by his half-brother Ecgfrith in 685, whom he then escorted back to Ireland. On that occasion, he presented King Aldfrith, who was Irish on his mother's side, with a copy of his *De locis sanctis*, an account of a voyage to and journeys in the Holy Land and Jerusalem, purportedly taken from a narrative given him by Arculf, a Gaulish bishop, and supplemented by information in the volumes in the library at Iona. According to Bede's *Ecclesiastical History*, it was while he was in Northumbria that Adomnán adopted the "universal observance" of the church on the matter of the dating of Easter, having spent some time with Ceolfrith and the Anglian monks at Wearmouth or Jarrow, and having accepted their guidance on the matter. However that may be, Iona did not finally accede to the Roman Easter until 716. But it is nonetheless likely that he was anxious to effect a reconciliation of Iona with the English and the majority of the Irish churches. In 697, he journeyed again to Ireland to promulgate the *Cáin Adomnáin* at a synod in Birr (County Offaly), a piece of legislation intended to protect non-combatants in times of war, by a system of fines. The guarantor list attaching to it of ninety-one ecclesiastical and secular potentates from every part of Ireland, including three from Scotland, is a genuine, contemporary document. Adomnán continued as abbot of Iona until his death.

De locis sanctis shows a considerable knowledge of the works of Jerome and other patristic authors and makes reference to his consultation of *libri graecitatis* ("books of Greek words"). It subsequently formed the basis for a later work of Bede's on the holy places. The *Cáin Adomnáin* places particular proscription upon the abuse of women in war or raids and imposes heavy fines, payable in part to the Columban community and in part to the kin or lord of the injured or deceased party, upon those guilty of doing so and upon those guilty of the murder, injury, or molestation of women. It is a humane and innovative piece of legislation that reflects Adomnán's concerns with the preservation of peace and civil order and the protection of women, and is a milestone in Irish law.

Adomnán's major opus, his *Vita sancti Columbae*, written about 700, was based upon both written and oral tradition relating to the saint, some of it derived from some written memoranda of Cumméne Ailbe, abbot of Iona from 657 to 669, and some written notes (*paginae*), and partly from contemporary recollections of him. It displays a wide-ranging knowledge of the Bible and of other hagiographic and patristic texts. It is a remarkable account, written in an eloquent but not verbose Latin style, of the sanctity, prophecies, and *uirtutes* of a great Celtic saint, for whom Adomnán had considerable veneration. His desire to elevate Colum Cille to the status of a universal saint has given us one of the best and earliest pieces of hagiography to emerge from the Irish Church.

In addition, the few penitential canons ascribed to Adomnán are quite probably his. The text is certainly of eighth-century date at the latest, and there is an explicit reference in Canon 16 to one of the canons of the seventh-century text known as the Second Synod of Patrick (*c.* 26), dealing with a problematic case of remarriage after divorce. His awareness of the Romani provenance of this synod makes it very probable that the *Canones Adamnani* are of seventh-century composition.

The career of Adomnán is a remarkable achievement. He was singularly successful as a churchman, scholar, diplomat, and legislator, and his striving towards the unification of the Irish Church may have promoted that second flowering of scholarly and literary activity which characterizes the eighth century in Ireland.

Aidan Breen

References and Further Reading

Anderson, A. O. and M. O. Anderson, eds. *Adomnan's Life of Columba*. Oxford: Oxford Univerity Press, 1991.

Bieler, Ludwig, ed. *The Irish Penitentials. Scriptores Latini Hiberniae* 5. Dublin: Dublin Institute for Advanced Studies, 1963. Reprinted 1975.

Herbert, M. and P. Ó. Riain, eds. *Betha Adamnáin*. (The Irish Life of Adamnán). London: Irish Texts Society, 1988.

Herbert, M. *Iona, Kells and Derry*. Oxford: Clarendon, 1988.

Meehan, D., ed. *De locis sanctis*. (The Holy Land). *Scriptores Latini Hiberniae* 5. Dublin: Dublin Institute for Advanced Studies,1958.

Meyer, K. *Cáin Adamnáin*. (The Canon of Adamnán). Oxford, 1905.

Ní Dhonnchadha, M. "The guarantor list of *Cáin Adomnáin*, 697." *Peritia: Journal of the Medieval Academy of Ireland* 1 (1982): 178–215.

———. "The Lex Innocentium: Adomnan's Law for Women, Clerics and Youths, 697 A.D." In *Chattel, Servant or Citizen*, edited by M. O'Dowd and S. Wichert. Belfast: Institute of Irish Studies, 1994.

O'Loughlin, T., ed. *Adomnán at Birr, A.D. 697: Essays in commemoration of the law of the innocents*. Dublin: Four Courts Press, 2001.

Picard, J.-M. "The purpose of Adomnan's Vita Columbae." *Peritia: Journal of the Medieval Academy of Ireland* 1 (1982): 160–177.

See also **Aedán mac Gabráin; Annals and Chronicles; Biblical and Church Fathers; Brehon Law; Canon Law; Classical Influence; Collum Cille; Education; Hagiography and Martyrologies; Hiberno-Latin; Kells, Book of, Languages; Literature; Paschal Controversy; Patrick; Prophecies; Scriptoria; Women**

ÁED UA (OR MAC) CRIMTHAINN (*fl.* 1150–1160)

He was a descendant of an old ecclesiastical family from County Laois, who were the hereditary *comarbai* of Colum moccu Loígse, sixth-century founder of the monastery of Tír dá Glas (or Terryglass) in County Tipperary, and friend of Colum Cille. Áed was one of the principal compilers and scribes of the great twelfth-century literary-historical compendium, the Book of Leinster, also known as *Lebar na Núachongbála*, the Book of Oughavall, which was his family's ancestral home. Much of the writing of the manuscript may have been completed there. He signs himself on 32r (p. 313): "*Aed mac meic Crimthaind ro scrib in leborso ⁊ ra thinoil a llebraib imdaib*" (Áed Úa Crimthaind wrote this book and collected it from many books). He is not the finest scribe in the Book of Leinster, but he evidently played a key part in the compilation and redaction of the many texts which went into it. The identification of his hand in certain places is still a matter of some paleographic difference, particularly in those entries referring to events post-dating 1166.

He is also the recipient of the earliest Irish personal letter, written to him by Finn mac Gormáin, bishop of Kildare (d. 1160), copied into the tale known as *Cath Maige Mucrama* on 206v of the manuscript. It is the earliest vernacular example of the medieval *ars dictaminis*. It has the usual form of a rhetorical epistle, praising him for his learning as "chief historian of Leinster in wisdom and knowledge and book-lore, and science and learning." It requests that the tale *Cath Maige*, being dictated to his scribe by Finn, be completed by Áed, who apparently had access to a better or fuller copy. It concludes by asking him to send a copy of the *duanaire* of Mac Lonáin, "so that we may study the meanings of the poems that are in it." The letter also styles him *fer léigind* (man of learning) to the king of Leth Moga, perhaps Diarmait Mac Murchada. Although there is no independent evidence for the assertion, it is often claimed that the famous references to the exile of Diarmait Mac Murchada, "king of Leinster and the Foreigners" in 1166 and to his death in 1171, at the end of the prose regnal list of Leinster (f.39d)—*Saxain. iar sain miserabiliter regnant Amen*. (And, after that, the Saxons miserably reign)—indicate a close personal relationship between Áed and Diarmait. His span of scribal activity on the manuscript can be judged from the date of his first personal entry, which records the death of Domnall Ua Conchobair in 1161. Best identifies his last entry in the notice of the death in 1201 of Ruaidrí mac Con Ulad, but it is very doubtful that an individual who was *fer léigind* of Leth Moga in the middle of the twelfth century was still alive in 1201.

Aidan Breen

References and Further Reading

Best, R. I., et al., eds. *The Book of Leinster, formerly Lebar na Núachongbála*. Vol 1. Dublin: Dublin Institute for Advanced Studies, 1954.

Bhreathnach, E. "Two contributors to the Book of Leinster: Bishop Finn of Kildare and Gilla na Náem Úa Duinn." In *Ogma: Essays in Celtic studies in honour of Próinséas Ní Chatháin*, edited by M. Richter and J.-M. Picard. Dublin: Four Courts Press, 2002.

Forste-Grupp, S. L. "The Earliest Irish Personal Letter." *Proceedings of the Harvard Celtic Colloquium* 15 (1995): 1–11.

See also **Aés Dána; Anglo-Norman Invasion; Glendalough, Book of; Historical Tales; Laigin; Lebor na hUidre; Leinster; Manuscript Illumination; Poets, Men of Learning; Scriptoria; Uí Chennselaig**

ÁEDÁN MAC GABRÁIN (*fl. c.* 574–606)

The Irish king of Dál Riata in Scotland from about 574 to 606, Áedán mac Gabráin was a member of Cenél nGabráin and son of a previous king, Gabrán mac Domangairt. While many aspects of his reign are disputed (partly because the two main sources, Adomnán's "Life of St. Columba" and the Irish chronicles, often provide contradictory evidence), it is clear that Áedán was involved significantly in the politics of both Ireland and Britain.

There are no certain examples of Áedán being engaged in military activity in Ireland, but he was involved in Irish politics. With Columba active in Dál Riata at Iona and elsewhere during his reign, Áedán was involved with the saint's Cenél Conaill relatives in Ireland, meeting Áed mac Ainmirech, ruler of Cenél Conaill and the northern Uí Néill, at Druim Cett, although it is disputed whether the Irish chronicle date of 575 for this event or a date between 586 and 597 is correct. While the outcome of the meeting is unclear, it is likely that the political relationship between the Uí Néill and Dál Riata was discussed. In Adomnán's "Life of St. Columba," Columba (Colum Cille) ordains (reluctantly) Áedán as king, the first instance of this practice in Gaelic literature. An earlier version of this tale by Cumméne of Iona makes it explicit that Áedán and his successors should be on friendly terms with Cenél Conaill (and perhaps by implication enemies of the Cruthni and Ulaid), which was possibly a reflection of a treaty made between Áedán and Áed mac Ainmirech at Druim Cett, but also a retrospective explanation of the tribulations of Áedán's descendants after fighting Cenél Conaill in 637.

The Irish chronicles portray Áedán as militarily active in Britain, undertaking an expedition to the Orkneys in 580–581, fighting in either the Isle of Man or near the Firth of Forth in 582–583, and at the battle of Leithreid in 590. He was defeated by the Northumbrians in 600, probably the same event as the battle of Degsastan (dated to 603) described by Bede in his *Ecclesiastical History of the English People*. Bede states that Áedán attacked in response to Northumbrian encroachment on British territory, which is an indication of Dál Riata interest in the area south of the Clyde-Forth line at this time, an interest halted by this defeat. Adomnán also mentions a battle in Anglo-Saxon territory, perhaps the battle of Degsastan or one in 596, in which Áedán's son Domangart was killed, and a victorious battle against the *Miathi*, in lowland Scotland, where two other sons were killed.

In contrast to this checkered picture produced by the earlier sources, Áedán appears in a number of later Irish texts as a powerful ruler, being described as "king of Alba" in the eleventh-century *Liber Hymnorum* and as a conqueror of the Picts in the "Tripartite Life of Patrick." These depictions probably reflect the reinterpretation made in the tenth century that the kingship of Alba was the successor of Dál Riata.

NICHOLAS EVANS

References and Further Reading

Anderson, Marjorie O. *Kings and Kingship in Early Scotland.* Edinburgh and London: Scottish Academic Press,1973.

Bannerman, John. *Studies in the History of Dalriada.* Edinburgh and London: Scottish Academic Press, 1974.

Jaski, Bart. "Druim Cett Revisited." *Peritia: Journal of the Medieval Academy of Ireland* 12 (1998): 340–350.

Sharpe, Richard. *Adomnán of Iona: Life of St. Columba.* London: Penguin Books, 1995.

See also **Annals and Chronicles; Hagiography and Martyrologies; Kings and Kingship; Scots, Scotti**

ÁES DÁNA

Áes Dána (literally, "the people of skill, craft") is a collective term which identifies the practitioners of certain professions held in high esteem in medieval Ireland, while also distinguishing them from the farming community (*áes trebtha*). The *áes dána* comprised professions involving not only skills of artisanship, but also speech and knowledge. Examples of such people were doctors, lawyers, judges, harpists, and blacksmiths. Not so clear, however, is whether ecclesiastical scholars (typically called *scribae* and *sapientes*) belong in this broad category. In the hierarchical society that was early Ireland the *áes dána* enjoyed special status. Thus, for example, the law tracts stipulated stiffer penalties for offenses against such people and conferred greater weight on their sworn evidence. Likewise, early Irish literature attests to a prejudice in their favor over other classes, perhaps because that literature was composed by members of the *áes dána*. For example, an Old Irish proverb declares that "an art is better than an inheritance of land."

Within the *áes dána* itself there were hierarchies, to judge by the scale of honor-prices accorded them in native Irish law (Brehon Law). Lowest in status were artisans such as the turner and leatherworker; somewhat higher the chariot builder and the engraver; higher again the harpist; and at the high end as a group, such professions as blacksmith, carpenter, physician, and lawyer. The lawyer, in turn, could be either judge (*brithem*) or advocate (*aigne*). The former at the highest level of his profession served as the official judge of the people and the legal advisor of the king (*brithem túaithe*).

But preeminent among the *áes dána* was the profession of poetry (*filidecht*). Just as the word *dán* had the specialized meaning of "poem," so too those who composed poetry, the *filid* ("poets"), were regarded as the *áes dána* par excellence. Alone among the secular *áes dána* they enjoyed the privilege of *nemed*, a quasi-sacred status that put them on a par with the king and the bishop of the people. To them was entrusted the preservation and transmission of *senchas*, the body of knowledge, usually transmitted in verse, which comprised the traditional lore of the *túath*. It included such matter as the genealogies of the ruling family, *dindshenchas* (the lore of places), and the origin legends of the tribe. In early Irish literature they were often credited with the power of prophecy (the word *fili* is etymologically connected with "seer"), the *imbas forosna* (literally, "encompassing knowledge which illuminates"). Thus, at the beginning of the *Táin bó Cúailgne* ("the Cattle-Raid of Cooley") a woman *fili* named Fedelm prophesies disaster for Queen Medb's expedition into Ulster, declaring that she "sees" red on the army. This and certain other aspects of the *fili*'s functions may have been inherited from the druids, presumably another group of the *áes dána*, who died out after the introduction of Christianity.

The exalted status of the highest grade of *fili*, the *ollam*, depended in the first instance on acquisition of the necessary qualifications. An eighth-century Old-Irish legal tract, *Uraicecht na Ríar* ("the Primer of Rules"), discusses the training of the *fili*. It required many years of education during which the aspiring candidate moved successively through seven grades (and three sub-grades) of learning, probably on the analogy of the ecclesiastical grades. The distinction between one grade and the next was a matter of learning, not office. Additionally, the profession was hereditary: A *fili* had to be the son and grandson of a *fili*. Once he acquired his position, he was expected to behave in a manner appropriate to a *nemed* person. It was his duty to eulogize the king and, where necessary, to satirize injustices within the *túath*. By means of this role, he performed both a normative and corrective function which no one else (except perhaps the cleric) could dare undertake. Although not an entertainer as such, he was expected to be able to recite traditional tales when called upon by the king. His poems had to be competent in subject-matter, and technically without flaw. More broadly, as the repository of tribal *senchas* (which he had memorized), the *fili* was expected not only to conserve this lore in versified form but also to interpret it and make it relevant to his own time. In addition to the *fili*, there was another, inferior, type of poet, known as the *bard*. What primarily distinguished the two was the *bard*'s lack of professional training. He was someone with natural ability who had not studied in the poetic schools; he might, for example, perform compositions of the *fili*.

By the thirteenth century, control of the profession of *filidecht* had shifted to a group of literary families who trained candidates for the profession in what are commonly called the Bardic Schools. No doubt, the realignment was related to major ecclesiastical and political changes that occurred during the twelfth century: the demise of the older churches following ecclesiastical reforms and the introduction of the Continental religious orders; and the Anglo-Norman invasion. But how it was effected remains unclear; one suggestion is that the new learned families were the descendants of hereditary officials who maintained possession of monastic lands after the monasteries themselves, the original centers of learning, had disappeared. Their ability to adapt to the new political order meant that they maintained (and even enhanced) their special status by receiving patronage from the Gaelicized Norman lords.

Two other branches of the *áes dána* that thrived in the post-Norman period were law and medicine. Again, these were controlled by certain families, who trained suitable candidates in their schools and depended on the Gaelic and Gaelicized Norman aristocracy for patronage. For example, the Ua hIceadha (O'Hickey) family served as physicians to the Ua Briain rulers of Thomond, and the Ua Casaide (O'Cassidy) family to Mág Uidhir (Maguire) of Fermanagh. Since all of these professions, especially the poets, depended on the patronage of the ruling families, the collapse of the Gaelic order in the seventeenth century inevitably brought their demise.

PÁDRAIG Ó NÉILL

References and Further Reading

Greene, David. "The Professional Poets." In *Seven Centuries of Irish Learning, 1000–1700*, edited by Brian Ó Cuív. Cork: Mercier Press, 1971.

Kelly, Fergus. *A Guide to Early Irish Law*. Dublin: Dublin Institute for Advanced Studies, 1988.

See also **Bardic Schools/Learned Families; Brehon Law; Poetry, Irish; Society, Grades of Gaelic; Túath**

AGRICULTURE

The Old Irish law texts of the seventh–eighth centuries A.D. are the main written source of information on pre-Norman agriculture in Ireland, but valuable information is also provided by other categories of text in Irish and in Latin, particularly annals, penitentials, and saints' lives. In the period from the Anglo-Norman invasion until the end of the sixteenth century, the Irish annals continue to be an important source of information on agriculture as practiced in those parts of the country under Gaelic control. Information on agriculture in the rest of the country is provided by rentals, deeds, and other documents in Norman-French, English, and Latin. Interaction between Irish and Anglo-Norman farming practices is indicated by the borrowing of vocabulary in both directions. For example, the Irish word *speal*, "scythe," is probably of Middle English origin, indicating that large-scale hay-making was introduced after the Anglo-Norman invasion. Similarly, much of the farmwork on an Anglo-Norman manor was done by persons classed as *betagh* (Irish *bíattach*, "unfree tenant"), who were almost always Irish, and would no doubt have held to at least some of the agricultural practices of their forefathers.

Crops

The archaeological evidence indicates that cereals have been grown in Ireland since Neolithic times. It is clear, however, that the coming of Christianity in the fifth century A.D. with the subsequent establishment of monasteries brought various innovations in cereal-production from the Roman world. An eighth-century law text, *Bretha Déin Chécht*, lists seven types of cereal grown in Ireland, arranged in order of value. Predictably, the most highly valued cereal is bread-wheat (*cruithnecht*), though it can hardly have been much grown in the rather cool Irish climate. The second cereal on the list is rye (*secal*, from Latin *secale*), which is likely to have been more widely grown as it tolerates harsher conditions. Other cereals included in the list are *suillech*, which is perhaps to be identified as spelt wheat, and *ibdach*, probably two-row barley, as it was used to make beer. Next on the list is *rúadán*, a reddish wheat which is doubtless "emmer," and then *éornae*, "six-row barley." At the bottom of the list is the least prestigious cereal, *corcae*, "oats"—a twelfth-century legal commentary states that a sack of oats is worth only half a sack of barley. The law text on clientship, *Cáin Aicillne*, provides a description of the type of land which is suitable for the growing of barley, and stresses that it should be level, deepdraining and properly manured. Plowing was generally carried out in the spring, using a team of oxen. The usual term for such a team is *seisrech*, which contains the numeral *sé*, "six," so it is possible that all six oxen were yoked simultaneously. It was probably more usual, however, for four oxen to be used. They seem to have been yoked abreast and led by a front plowman (*cennairem*) walking backwards ahead of his team, while a rear plowman (*tónairem*) directed them from behind. The Old Irish law texts contain no mention of the coulter (*coltar*), but it is referred to in twelfth-century commentary. Harrowing was carried out by horses. After the young corn appeared, it was kept free from weeds, of which the most pernicious was darnel (*díthen*), which has poisonous seeds. The law texts also lay down heavy penalties on the owners of marauding livestock that damage growing corn. When ripe, the cornstalks were cut with a sickle, and the ears of corn collected in a reaping-basket. The ears were then threshed with a stick or flail (*súist*), and dried in a kiln (*áith*). The dried corn was stored in a barn (*saball*); a fragmentary law text on cats stresses that the cat should patrol the area around the barn to keep mice away. Writing in the late twelfth century, Giraldus Cambrensis refers to mice as a particular pest in Ireland.

Apart from cereals, other plants featured in the early Irish diet. The texts refer occasionally to peas and beans, and it is likely that both were normally kiln dried and stored for winter use. Another vegetable that is frequently mentioned in the Old Irish texts is *cainnenn*, which probably means "onion." It was clearly grown in fairly large quantities, as it formed part of the food-rent which a client paid to his lord. Smaller quantities of other vegetables were also grown, including *braisech* (cabbage), *foltchép* (chives), *borrlus* (leek?), *imus* (celery?), and *cerrbacán* (skirret?). The medico-legal text *Bretha Crólige* emphasizes the importance of vegetables in the diet of invalids. Apples and plums seem to have been grown on a small scale in the early period, but cultivated pears and cherries were evidently not introduced until after the Norman invasion. The main dye-plants were woad (*glaisen*) and madder (*roid*).

Cattle

Cattle occupied a position of central importance in early Irish society, and feature prominently throughout Old and Middle Irish literature. Fines, tributes, fees, and other payments were commonly expressed in terms of cattle, the standard unit being the milk cow (*bó mlicht*). Cattle were valued primarily for their milk and for milk-products such as butter and cheese. Beef was also consumed, and hides were used for making shoes, bags, belts, and the like. Early Irish cattle seem generally to have been small and black—much like modern Kerry cattle—but there are also references to red, brown, dun-colored, and white cattle. There is no mention of the provision of hay for livestock in documents from the pre-Norman period. The Old Irish law texts refer to the practice of keeping an area of "preserved grass" to nourish the cattle over the winter, and there is also mention of branches of holly and ivy being supplied as winter fodder. In the summer, it was clearly a frequent practice for cattle and other livestock to be driven off to hills or other rough ground where they grazed under the care of herdsmen. At night they were kept in a pen (*búaile*), whence the Anglicized term "booleying." This practice was regularly opposed by English officials. For example, in 1595, Edmund Spenser denounced the "Irish manner of keeping boolies in the summer upon the mountains and living after that savage sort."

Other Livestock

Sheep were kept primarily for their wool, but were also valued for their meat and milk. The Old Irish law texts assign greater value to white sheep than to dun-colored or black sheep. In the twelfth century, Giraldus Cambrensis commented on the prevalence of black sheep in Ireland, and it is likely that larger white-fleeced breeds were introduced after the Anglo-Norman invasion. There is less mention of goats in Irish sources, and in legal commentary their value is lower than that of sheep. The flesh of the pig was prized beyond that of any other animal, and roast pig was the traditional main dish at feasts. Pigs were commonly fattened up on acorns in the woods. The Old Irish law text *Críth Gablach* states that a well-to-do farmer should own a horse for riding as well as a workhorse. In about the late thirteenth century, the great plow-horse, which was originally developed on the Continent for military use, was introduced to Ireland, and by the end of the fifteenth century it seems that oxen were largely superseded for this purpose. The Old Irish law texts contain many references to hens, and there is occasional mention of ducks and geese. A separate law text is devoted to honeybees, which indicates that they were of considerable economic importance. Doves may have been reared for consumption in early Irish monasteries, but there seem to be no records of dovecotes in this country until after the Anglo-Norman invasion. The rabbit was an Anglo-Norman introduction, and elaborate warrens were constructed to house them. Fishponds—mainly for introduced species such as perch, carp, and pike—were a regular feature of the Anglo-Norman manor.

Farm Layout

For the early period, the law texts are an important source of information on the layout of the Irish farm, and much of what they tell us is confirmed by archaeology. The farmhouse was round and constructed of wattle packed with insulating material, and there was an adjoining out-house. The farmhouse was surrounded by an enclosed area (*les*) of approximately 100 feet in diameter, which contained structures such as the sheep pen, calf pen, pigsty, and hen coop. Outside the *les*, the typical farm had a vegetable garden, as well as a kiln for drying corn and a barn for storing it. The Old Irish law texts regularly distinguish between the infield (*faithche*), which refers to the better land around the farmhouse, and the outfield (*sechtarfhaithche*) farther away. The main law text on farming, *Bretha Comaithchesa*, provides detailed descriptions of what constitutes a proper field boundary, and distinguishes the stone wall, trench-and-bank, bare fence, and oak fence. The proper dimensions and method of construction are specified in the text. For example, the bare fence is constructed with posts and hazel rods, and is capped with an interwoven blackthorn crest—the medieval equivalent of barbed wire.

Farm Labor

It is clear from the Old Irish law texts that most of the work on the farm of a commoner was carried out by him and his family. However, plowing was often undertaken in cooperation by up to four farmers who pooled their resources of oxen and equipment, and plowed their lands in turn. Livestock were also regularly herded cooperatively, with animals belonging to a number of farmers looked after in a single herd. Higher up the social scale, the main work on the farm of a lord (*flaith*) was carried out by slaves or servants. In addition, lords were entitled to fixed amounts of labor from commoners with whom they had an agreement of clientship. These clients (*céili*) also supported the lord's household by the provision of an annual food-rent in return for the fief—usually of cattle—supplied by the lord. Some commoners were simultaneously clients of two or three lords. Monasteries functioned in a similar manner to lay lordships, and relied on the labor of church clients, as well as that of the monks. After the Anglo-Norman invasion, there is evidence that the independence of the commoner decreased, both in areas under Gaelic and under English control. The rent-paying *bíattach* became an unfree tenant bound to the land, without the option of transferring from one lord to another, or of serving more than one lord.

Land-Tenure and Control

The Old Irish law texts make a general distinction between a person's inherited share of kin-land (*fintiu*), and land which he has personally acquired. Naturally, he has greater legal entitlement to sell or bequeath acquired land, and can only dispose of kin-land with the agreement of the greater family unit (*fine*) and of his lord. A large amount of land was owned by the Church, and it is clear that many agricultural innovations are of monastic origin. For example, the use of the water mill, which revolutionized the processing of cereals, is likely to have spread from the monasteries. The law texts recognize the rights of adult dependents—wives or sons—to veto contracts made by a landowner which could damage the well-being of the farm. In general, it is clear that the early Irish farmer farmed so as to support his family and to produce a surplus to fulfill his obligations to his lord, king, and church. Old Irish texts provide little information on trade in agricultural produce, and it seems that any such trade was small-scale and local. The establishment of Norse towns on the eastern and southern coasts in the ninth and tenth centuries undoubtedly stimulated trade in foodstuffs and other commodities, and it is significant that the Irish word for market (*margadh*) is a borrowing from Old Norse. An eleventh-century poem in Irish refers to the sale of livestock at the fair (*óenach*) of Carmun, probably in the present County Kildare. After the Anglo-Norman invasion there was a flourishing export trade in wool and sheepskins, mainly to England.

FERGUS KELLY

References and Further Reading

Curtis, Edmund. "Rental of the Manor of Lisronagh, 1333, and Notes on 'Betagh' Tenure in Medieval Ireland." *Proceedings of the Royal Irish Academy* 43 C (1935–1937): 41–76.

Duignan, Michael. "Irish Agriculture in Early Historic Times." *Journal of the Royal Society of Antiquaries of Ireland* 74 (1944): 124–145.

Kelly, Fergus. *Early Irish Farming: A study based on the law-texts of the seventh and eighth centuries A.D.* Early Irish Law Series 4. Dublin: Dublin Institute for Advanced Studies, 1997. Reprinted 2000.

Lucas, A. T. *Cattle in Ancient Ireland.* Kilkenny: Boethius Press, 1989.

O'Loan, John. "Livestock in the Brehon Laws." *The Agricultural History Review* 7, no. 2 (1959): 65–74.

Otway-Ruthven, A. J. "The Organisation of Anglo-Irish Agriculture in the Middle Ages." *Journal of the Royal Society of Antiquaries of Ireland* 81 (1951): 1–13.

See also **Anglo-Norman Invasion; Annals and Chronicles; Brehon Law; Diet and Food; Famine and Hunger; Houses; Law Texts; Law Tracts; Mills and Milling; Penitentials; Slaves; Society, Functioning of Gaelic; Society, Functioning of Anglo-Norman; Society, Grades of Gaelic; Trade**

AIDEDA

In medieval Irish literary terminology, the word *aided* refers to a tale in prose or prosimetrum that relates the violent demise of a hero, king, or poet. Like the *Comperta*, *Echtrai*, and *Immrama*, it belongs to a system of nineteen tale-types or general topics, which medieval Irish scholars used as a means of classifying much of their narrative literature. Judging from the number of tales that survive, the *aided* must have been a popular tale-type. In fact, some thirty-five death-tales that contain the word in their title are extant, most of which are written in Old or Middle Irish (*c.* 650–1200). Almost half of these are historical tales, while much of the remainder belong to the Ulster Cycle. Yet the *aided* is not the only tale-type in which the violent deaths of kings and heroes can be narrated. Such stories often form integral episodes in other tale-types, especially the *catha* (battles), *togla* (destructions), and *oircne* (slaughters). Indeed, some of the most famous death-tales in the Irish language—stories like *Togail Bruidne Da Derga* (The Destruction of Da Derga's Hostel) and *Orgain Denna Ríg* (The Destruction of Dinn Ríg)—are not classified as *aideda* at all, but belong to these other tale-types. Regardless of the titles under which they survive, death-tales formed an important part of the Irish literary tradition.

Origins and Development

Comparative evidence from other Indo-European cultures, Greek and Indic in particular, suggests that the *aided*, as a tale-type, is ancient, and some Irish examples do preserve elements of demonstrable antiquity. However, since the 1950s, scholars have begun to change the way they view early Irish literature, the death-tales included. These stories are no longer regarded as the products of an age-old oral tradition, but as the products of the ecclesiastical scriptoria in which they were written. Studies, especially since the 1980s, have shown that the creators of these texts drew on a wide range of materials, both foreign and domestic. As a result, they were able to fashion narratives that at one time hearkened back to the pre-Christian past but at the same time addressed contemporary political and social concerns. The *aideda* themselves proved particularly adaptable in this respect, so much so that they continued to be composed and reworked from their textual beginnings in the eighth century right up until the eighteenth and nineteenth centuries, when death-tales like *Oidheadh Chloinne Lir, Oidheadh Chloinne Tuireann*, and *Oidheadh Chloinne Uisnigh* enjoyed widespread popularity.

The Mythology of Death

The heroes of Irish myth and legend do not wither away from disease or old age, but like their counterparts in other traditions, they die dramatic deaths that mark the culmination of their heroic biographies. Their deaths, like their births, take place in a well-ordered universe in which every event has its time and place, and in the *aideda*, the time and place of death are usually liminal. Heroes tend to die at transitional points in the seasonal calendar (like *Samain*) and at transitional points in the physical landscape (like fords). One of the more common liminal spaces in the Irish death-tales is the quasi-otherworld banquet hall, or *bruiden*. In this setting, the doomed hero partakes of the so-called fatal feast, often in the company of a strange woman thought to be a figure of death. Many a king in Irish literature from Conaire Mór to Diarmait mac Cerbaill meets his end in a *bruiden*. The hero can suddenly find himself in one of these liminal spaces through what appears to be happenstance, through his own actions (often this involves the violation of his *geissi*, or taboos), or through the complex interaction of human and supernatural agents bent on the hero's destruction. But however it comes about, once the hero enters this liminal space at the proper time, his demise is assured.

The Threefold Death

No aspect of the early Irish *aideda* has received more attention than the motif of the threefold death, in which the victim is killed by three different means in rapid succession, often wounding, drowning, and burning. Examples of this motif can be found in the literature and folklore of many countries, including Wales, France, and Estonia. Although its origins and development are obscure, some scholars believe that the motif may have its beginnings in a putative Indo-European tri-functional sacrifice, in which human victims were offered to a trio of divinities. Potential support for this theory comes from the archaeological record. Over the years, a number of prehistoric bodies have been unearthed from the bogs of northern and western Europe, some of which, like the Lindow Man from Cheshire, England, show signs of ritualistic threefold death.

But whatever its supposed origins, the motif of the threefold death in early Irish literature has little to do with paganism, much less human sacrifice. Two of the best examples of this motif are found in *Aided Diarmata meic Cerbaill* (The Death of Diarmait mac Cerbaill) and *Aided Muirchertaig meic Erca* (The Death of Muirchertach mac Erca). Just as in these stories, all the other instances of this motif in Irish sources are set in the early Christian period, specifically the sixth and seventh centuries, and center on conflicts between what would now be called Church and State. Furthermore, almost all the examples follow the same narrative pattern, which at its core consists of three main stages: (1) A crime is committed against the Church; (2) a prophecy that the offender will die a threefold death is pronounced; and (3) the prophecy is fulfilled as the offender dies in the manner foretold. Death, then, is seen as divine retribution for sins against God and his Church. Like the *aideda* set in the pagan past, death in these stories comes at the instigation of human and cosmic forces as a result of the hero's actions.

DAN M. WILEY

References and Further Reading

Bhreathnach, Máire. "The Sovereignty Goddess as Goddess of Death?" *Zeitschrift für celtische Philologie* 39 (1982): 243–260.

Green, Miranda J. *The World of the Druids*. London: Thames and Hudson Ltd, 1997.

Jackson, Kenneth. "The Motive of the Threefold Death in the Story of Suibne Geilt." In *Féil-sgríbhinn Éoin Mhic Néill*, edited by John Ryan. Dublin: Three Candles Press, 1940.

Mac Cana, Proinsias. *The Learned Tales of Medieval Ireland*. Dublin: Dublin Institute for Advanced Studies, 1980.

Melia, Daniel. "Remarks on the Structure and Composition of the Ulster Death Tales." *Studia Hibernica* 17–18 (1978): 36–57.

Meyer, Kuno. *The Death-Tales of the Ulster Heroes*. Dublin: Royal Irish Academy, 1906.

Ó Cathasaigh, Tomás. "The Threefold Death in Early Irish Sources." *Studia Celtica Japonica*, New Series 6 (1994): 53–75.

Radner, Joan Newlon. "The Significance of the Threefold Death in Celtic Tradition." In *Celtic Folklore and Christianity: Studies in Memory of William W. Heist*, edited by Patrick K. Ford. Santa Barbara, Calif: McNally and Loftin, 1983.

Rees, Alwyn, and Brinley Rees. *Celtic Heritage: Ancient Tradition in Ireland and Wales*. London: Thames and Hudson Ltd, 1961.

Sjoestedt, Marie-Louise. *Gods and Heroes of the Celts*, translated by Myles Dillon. London: Methuen & Co, Ltd, 1949.

Ward, Donald J. "The Threefold Death: An Indo-European Trifunctional Sacrifice?" In *Myth and Law among the Indo-Europeans*, edited by J. Puhvel. Berkeley: University of California Press, 1970.

Wiley, Dan M. "Stories About Diarmait mac Cerbaill from the Book of Lismore." *Emania* 19 (2002): 53–59.

See also **Diarmait mac Cerbaill; *Comperta*; *Echtrai*; Historical Tales; Pre-Christian Ireland; Prophecies and Vaticinal; Ulster Cycle**

AÍFE

In 1170, Aífe, daughter of Diarmait Mac Murchada, married Richard FitzGilbert de Clare, the Anglo-Norman baron better known as Strongbow. Their union fulfilled one half of the promise made by Mac Murchada in return for Strongbow's help in regaining his lost kingdom of Leinster. Strongbow's succession to that kingship upon Mac Murchada's death in 1171 fulfilled the other half.

Strongbow's succession has traditionally been seen as running contrary to both Irish and English practice. English law held that only in the absence of male heirs could a man succeed in right of his wife, but Mac Murchada had at least one son living in 1171. It has been suggested, however, that Mac Murchada may have regularized his marriage with Aífe's mother under canon law, thereby rendering Aífe his only legitimate offspring alive at that time. In terms of Irish tradition, it has been further suggested that a precedent of sorts for Strongbow's succession lay in the twelfth-century phenomenon of imposing dynasts upon thrones to which they had no ancestral claim. Marrying the daughter of one's predecessor was, moreover, a common characteristic of peaceful transfers in Irish dynastic control.

Styling herself "Countess of Ireland," Aífe issued charters concerning both her native Leinster and, following Strongbow's death in 1176, her English dower lands. The earl was initially succeeded by the couple's son, Gilbert, who died in 1185 while still a minor, leaving their daughter Isabella as sole heiress. In 1189, history repeated itself when William Marshal married Isabella, and succeeded to Leinster in right of his wife.

ANNE CONNON

References and Further Reading

Flanagan, Marie Therese. *Irish Society, Anglo-Norman Settlers, Angevin Kingship: Interactions in Ireland in the late 12th century*. Oxford: Clarendon Press, 1989.

See also **Anglo-Norman Invasion; Kings and Kingship; Leinster; Mac Murchada, Diarmait; Marriage; Strongbow; Tánaiste; Ua Tuathail (O'Toole), St. Lawrence**

AILECH

Ailech, or the Grianán of Ailech, was the *caput,* or principal royal seat of the early medieval Northern Uí Néill kings of Cenél nÉogain, until they moved their headquarters to Tulach Óc in the kingdom of Airgialla at the beginning of the eleventh century. The place-name Ailech was also used as the distinguishing sobriquet of the Northern Uí Néill dynasty. Ailech is popularly identified as a large multi-period fortification situated on Greenan Mountain at the southern end of the Inishowen Peninsula, County Donegal. However, Elagh, which is an Anglicized form of "Ailech," in nearby

Aerial view of Grianán of Ailech, Co. Donegal. © *Department of the Environment, Heritage and Local Government, Dublin.*

County Derry, could also have been the location of the historic Northern Uí Néill capital. The chronicles note the destruction of the Grianán of Ailech by the army of Muirchertach Ua Briain, king of Munster, in 1101. It was demolished in revenge for the destruction of the Uí Briain stronghold at Cenn Corad (Kincora), Killaloe, County Clare, which had been destroyed by Domnall Mac Lochlainn of the Northern Uí Néill in 1088.

The reputed site of Ailech on Greenan Mountain commands extensive views over Lough Foyle and Lough Swilly, and its lofty location, combined with the fact that it can be seen for a considerable distance, suggests that it was as much to be viewed as to view from. It consists of a triple-ramparted hillfort at the center of which lies an early-medieval *caiseal*, or stone fort. In addition, there are the vestiges of a mound or tumulus of possible Neolithic or Bronze Age date, the site of a ceremonial road approaching the fortification, and a holy well. The three earthen ramparts that enclose and predate the central *caiseal* appear to constitute a hillfort of the Late Bronze Age or Iron Age period. The *caiseal* was an early medieval addition to the hillfort and its construction perhaps signified the appropriation of Ailech as the headquarters of the Northern Uí Néill in the sixth century. Its present-day form is the result of significant rebuilding undertaken in the 1870s by Dr. Bernard, the Bishop of Derry. The *caiseal* is a very fine drystone structure with an elegant external batter. It has an internal diameter of circa 24 meters and rises internally in three terraces, with each tier accessible by means of inset staircases. The walls are about 4 meters thick and rise to a height of 5 meters. An entrance passageway, which is roofed with stone lintels, leads from the east into the interior. Additional stone passages run into the fort from the south and the northeast. Outside the *caiseal*, at a distance of 25 meters, one meets with the inner rampart of the hillfort, which survives as a heather-clad earthen bank. A low cairn of stones, possibly representing the mound or "tumulus" that George Petrie noted on his plan of Ailech (1835), is situated midway between the inner and middle ramparts of the hillfort. Both of these ramparts survive as quite eroded features that follow the contours of Greenan Mountain. A holy well dedicated to St. Patrick lies on the south side between the middle and external banks of the hillfort. Parallel breaches in the three ramparts of the hillfort, at the east, and a corresponding entrance in the *caiseal* wall, indicate the former presence of a ceremonial roadway that was apparently lined by stone settings, leading into the heart of the site. The road ran between two upstanding ledges of rock as it approached the summit of Greenan Mountain. The appropriation of such a multi-period site as a royal residence and as a place of king-making would have been in keeping with the typical exercise of royal authority and royal display of power by early-medieval Gaelic ruling families.

Ailech is the subject of three *dindshenchas* poems that account for the origin of the name, the deeds of the legendary heroes associated with it, and the blessing of the site by St. Patrick. According to the text *Vita Tripartita*, compiled circa 900, Patrick went to Ailech and blessed the fort, and left his flagstone (*lecc*) there, and prophesied that kings and ordained persons out of Ailech would have supremacy over Ireland. The flagstone was subsequently called Lecc Phátraic, and it was upon this that future kings of the Northern Uí Néill were to be inaugurated at Grianán of Ailech. A local tradition in Derry identifies Lecc Phátraic with "St. Columb's Stone," a large flagstone engraved with two shod footprints that lies in the garden of Belmont House near the city of Derry. This identification, however, cannot be supported.

The expansionist policy of the Northern Uí Néill saw them encroaching on the territory of the Airgialla as early as the tenth century. They specifically targeted Tulach Óc in Airgialla, which was colonized between 900 and 1000 by the Cenél mBinnig, a branch of the Ailech dynasty. By 1000 the ruling branch of the Cenél nEógain had established their royal headquarters at Tulach Óc. They had apparently set their sights on the kingdom of Airgialla as early as the ninth century. Their first success came in 827 when Niall Caille defeated the combined forces of Airgialla and Ulaid at the battle of Leth Cam. In the aftermath of Leth Cam the chronicles for this period reflect the hold that the Cenél nEógain had over Airgialla. The attraction of Tulach Óc for the kings of Ailech lay in the probability that it was the traditional inauguration site of the kings of Airgialla. To gain control of it would have struck at the very core and source of the kingship of Airgialla. That Tulach Óc was chosen as the preeminent inauguration place of Cenél nEógain in preference to Ailech or Armagh is evidence enough of the political importance attached to it. The first king of Ailech to be inaugurated there, and in a ceremony presided over by an ecclesiastic, was possibly Áed Ua Néill. He was installed as king of Cenél nEógain by Muirecán, *comarba* of Patrick, "in the presence of Patrick's community," while Muirecán was in Tír Eógain on visitation in 993.

ELIZABETH FITZPATRICK

References and Further Reading

Byrne, Francis J. *Irish Kings and High-Kings*. London: Batsford, 1973. Reprint, Dublin: Four Courts Press 2001.

Jaski, Bart. *Early Irish Kingship and Succession*. Dublin: Four Courts Press, 2000.

Lacy, Brian. *Archaeological Survey of County Donegal*. Lifford, Ireland: Donegal County Council, 1983.

See also **Derry; Earthworks; Inauguration Sites; Kings and Kingship**

AIRBERTACH MAC COSSE (d. 1016)

Allowing that he held the offices of *fer léigind* ("man of learning"; lector) and *airchinnech* (superior) of Ros Ailithir (Roscarberry, County Cork), Airbertach mac Cosse's reputation as a scholar, among present-day historians, rests mainly on four surviving works on the basis of which he has been viewed as a Latinist, a commentator on the Psalms, and a poet who utilized geographical and biblical themes. Little is known of his background. His genealogy is unknown, although it seems reasonable to conclude that his origins lay in Munster. He may have belonged to the minor population group of Uí Dobráin, which features among the subject peoples of Dál Messin Corb—a Laigin dynasty—and is associated with other lineages which had mid-Munster connections. He joined the community of Ros Ailithir, which included among its founding-fathers Fachna of Corco Loígde and Colmán *Ailithir* (the pilgrim). The reputation for Latin learning which the foundation enjoyed is reflected in the Old Irish Triads. During his time as *fer léigind*, in 990, the site was attacked by a Hiberno-Scandinavian force (probably from Waterford), which carried him off as hostage. He was ransomed on Inis Cathaig (Scattery Island, County Clare) by the powerful Munster high king, Brian Bóruma (Boru). Subsequently, he became superior of his community. He died in 1016.

The four surviving works most widely associated with Airbertach are found in the manuscript compilation Rawlinson B 502, at the Bodleian Library, Oxford, which, Ó Riain argues, is to be identified with the Book of Glendalough. These include a compound tract, written in 982, the principal concern of which is a study of the Psalms. One verse, seemingly an interpolation, refers to Airbertach as having translated some of the subject matter from Latin to Irish. There is also a lengthy poem, with a geographical theme, based on the writings of Orosius and Isodore of Seville, which is expressly ascribed to "in fer léigind Mac Coise." The two remaining poems, one dealing with the kings of Judah and the other with a battle in which the Israelites defeated the Midianites, are assigned to Airbertach because they are found in conjunction with the above-discussed compositions in Rawlinson B 502. Although it is not unreasonable to attribute these biblical poems to Airbertach, the possibility remains that they were produced by one of his students—or at least by another Ros Ailithir-based scholar.

Another work which features in the same manuscript compilation, the biblical opus known as "Saltair na Rann," may also, in the view of Gearóid MacEoin, have been composed by Airbertach. Widely regarded as one of the finest examples of Middle Irish verse, this apparently unfinished epic has been dated to 988 on the basis of a chronological passage which, it seems, formed part of the original poem. Various arguments advanced by MacEoin on this matter, including the suggestion that the poem's incomplete state reflects a suspension of Airbertach's work following his capture by the Norsemen in 990, have drawn opposition from James Carney and others.

AILBHE MACSHAMHRÁIN

References and Further Reading

Carney, James. "The Dating of Early Irish Verse Texts, 500–1100." *Éigse* 19 (1982–1983): 177–216.

MacEoin, Gearóid. "The Date and Authorship of Saltair na Rann." *Zeitschrift für celtische Philologie* 28 (1960–1961): 51–67.

———, (ed.). "A poem by Airbertach mac Cosse." *Ériu* 20 (1966): 112–139.

———. "Observations on Saltair na Rann." *Zeitschrift für celtische Philologie* 39 (1982): 1–28.

O'Leary, A. "The Identities of the Poet(s) Mac Coisi: A reinvestigation." *Cambridge Medieval Celtic Studies* 38 (Winter 1999): 53–71.

Ó Néill, P, ed. "Airbertach mac Cosse's poem on the Psalter." *Éigse* 17 (1977–1979): 19–46.

See also **Devotional and Liturgical literature; Poets, Men of Learning**

AIRGIALLA

Airgialla, "those who give hostages," was a collective name for a group of peoples around the Sperrin Mountains in the north of Ireland and in the midlands. They consisted of nine main tribal groups: Uí Maic Caírthinn, south of Lough Foyle; Uí Fiachrach of

Ardstraw; Uí Thuirtri east of the Sperrins (collectively known as Uí Maccu Úais); the Fir Chraíbe and the Fir Lí west of the Bann; the Airthir around Armagh; the Uí Chremthainn in Fermanagh, parts of Tyrone and Monaghan; the Uí Méith in Monaghan; and the Mugdorna, who also stretched into Meath. Branches of the Uí Moccu Úais were in Westmeath and Meath also. It is possible that the Déisi around Tara were Airgialla. Originally probably subject to the Ulaid (Ulstermen) they were gradually, from the sixth century onwards, brought under the control of the Uí Néill, especially by the Cenél nEógain who were expanding from their homeland in Inishowen across Lough Foyle and eastwards across Counties Derry and Tyrone.

Following the defeat of the Ulaid in the battle of Mag Roth (Moira, County Down) in 637 to 638, they enjoyed a degree of independence from both their former masters and the expanding Cenél nEógain. After the devastating defeat of the Ulaid at the battle of Fochairt in 735, the Cenél nEógain dominated the Airgiallan territories from the shores of Lough Foyle to the coast of Louth. The Airgialla provided military service for the Uí Néill and propaganda was produced explaining their evolving relationship with them. Defeated in the battle of Leth Cam beside Armagh in 827, they became vassals of the Cenél nEógain. It is very likely that it was Airgiallan patronage that helped Armagh rise to power during the seventh century to become the chief church in Ireland. The Airthir ("Easterners") had control of the offices in the church of Armagh. The Clann Sínaigh monopolized the abbacy from 996 until the twelfth century. The Uí Thuirtri migrated east of the Bann from 776 onwards and lost their link with the Airgialla after 919. The Airgiallan peoples in the midlands were absorbed by the various branches of the Southern Uí Néill. As the northern and southern branches of the Uí Néill drifted apart, two kingdoms emerged as a wedge between them. In Counties Leitrim and Cavan, the kingdom of Uí Briúin Bréifne (later O'Rourkes) was formed. Parallel with this kingdom to the north in Counties Fermanagh and Monaghan and parts of Louth, a consolidated kingdom of Airgialla emerged, and partly as a result of continuing pressure from the Cenél nEógain, who absorbed their northern borders, they moved toward the southeast. By the eleventh century the leading family was Ua Cerbaill (O'Carroll). Donnchad Ua Cerbaill pushed the southern boundaries of this kingdom to the Boyne in the twelfth century and had the seat of the diocese of Clogher transferred to his power center in Louth. When the Anglo-Normans conquered Louth, this area became known as "English Oriel" and this portion of O'Carroll's kingdom was transferred to the diocese of Armagh. The diocese of Clogher represents the medieval kingdom of Airgialla.

Charles Doherty

References and Further Reading

Bhreathnach, E., et al. "The Airgialla Charter Poem." In *Tara: Kingship and landscape*, edited by Edel Bhreathnach. Dublin: Royal Irish Academy, 2004.

Byrne, Francis John. *Irish Kings and High Kings*. London: Batsford, 1973.

Charles-Edwards, T. M. *Early Christian Ireland*. Cambridge: Cambridge University Press, 2000.

O'Brien, Michael A., ed. *Corpus genealogiarum Hiberniae*. Dublin: Dublin Institute for Advanced Studies, 1962.

See also **Uí Néill, Northern; Uí Néill, Southern; Ulaid**

AMLAÍB CUARÁN (*fl. c.* 940–981)

Amlaíb Cuarán (Ólafr Kvaran), the son of Sihtric Cáech (d. 927), belonged to the second generation of the Uí Ímair dynasty, which came to dominate the Hiberno-Norse world in the course of the tenth century. His father Sitriuc and uncle Ragnall had led the return of the Vikings to Ireland in 917 and, after eliminating rivals to Scandinavian leadership in the Irish sea world, they turned their attention to native dynasties. Sitriuc's refoundation of Dublin was secured by his victory over Niall Glúndub, king of Tara, at Islandbridge in 919. The following year, upon Ragnall's death, Sitriuc succeeded as senior member of the dynasty and moved his center of operations to Northumbria. It was there that he died in 927. Sitriuc's death led to the loss of most of Northumbria to Æthelstan of Wessex, although his brother Gofraid (927–934) and nephew Amlaíb mac Gofraid (934–941) continued to contest control with the West Saxons. Amlaíb Cuarán, a child when his father died, appears in the historical record as king of Northumbria on the death of his cousin Amlaíb mac Gofraid in 941. Two other members of the dynasty, "mac Ragnaill" and Blacaire mac Gofraid, the latter based in Dublin, ruled the Irish dominions. Amlaíb was expelled from Northumbria by Edmund of Wessex in 943, having first been forced to undergo baptism, and his whereabouts were unknown for two years.

In 945, the Annals of Ulster record that "Blacaire gave up Dublin and Amlaíb succeeded him." This notice is immediately followed by one in which Amlaíb and Congalach Cnogba, the new king of Tara, were engaged in military action against the Northern Uí Néill dynast Ruaidrí Ua Canannáin. Because Blacaire and Congalach were implacably hostile to one another, one can only suppose Amlaíb's accommodation in Ireland was arranged for him by the king of Tara. In 946 Amlaíb plundered

Mide, but in 947 he and Congalach were defeated in battle at Slane by Ua Canannáin. The following year, 948, Blacaire was back in Dublin, only to be slain by Congalach, and Amlaíb was back in Northumbria. It seems likely that the defeat at Slane had convinced the Dubliners that the alliance with Congalach was a mistake and that Amlaíb had been expelled for promoting it.

Amlaíb continued to rule in Northumbria until 952, when he was expelled by the populace. He then disappeared for about a decade while Blacaire's nephew Gofraid ruled in Dublin. This Gofraid mac Amlaíb died in 963, and the following year Amlaíb Cuarán returned to the Irish stage with a raid on Kildare. Amlaíb seems to have maintained his alliance with the family of Congalach, who had been slain by Gofraid in 956. Curiously the woman who succeeded as abbess of Kildare in 963, did not belong, like her predecessors, to the Fothairt dynasty, but was Congalach's daughter Muirenn. This seems more than coincidence. At this time Amlaíb had his own daughter, Ragnaillt, married to Domnall mac Congalaig. Amlaíb himself married Dúnlaith, the sister of Domnall ua Néill, king of Tara, and widow of Domnall mac Donnchada king of Mide (d. 952). At some point between about 966 and 970, Amlaíb married Gormflaith daughter of Máel Mórda, king of the Laigin, and his relations with Domnall ua Néill soured. The king of Tara targeted the monasteries that fell under Amlaíb's protection at Louth, Dromiskin, Monasterboice, and Dunleer. In 976, he also destroyed Skreen, the Columban church adjacent to Tara, which seems to have been patronized by Amlaíb. In the same year Domnall mac Congalaig, Amlaíb's son-in-law, died.

In 980, following the retirement into religion of Domnall ua Néill, Amlaíb fought a great battle at Tara against his own stepson Máel Sechnaill mac Domnaill, king of Mide. It is possible that Amlaíb was presenting his own claim to the kingship of Ireland, but he was defeated and Máel Sechnaill took the kingship. Amlaíb, after four decades at the heart of Insular politics retired to Iona where he died in penance the following year. One son, Ragnall, was killed in the battle of Tara, but a number of others, including Glún Iairn, Sitriuc, Aralt, Ímar, and Dubgall, survived their father and continued to play a significant role in Irish history. Amlaíb's career marks the process of nativization of the Vikings. His father's generation were pagan Scandinavians, but his own patronage of monasteries, retirement to Iona, and the Gaelic names borne by some of his children bear witness to the extent to which the Hiberno-Norse were now as much Irishmen as foreigners.

Alex Woolf

References and Further Reading

Downham, Clare. "The Chronology of the Last Scandinavian Kings of York." *Northern History* 40 (2003): 25–51.

Smyth, A. P. *Scandinavian York and Dublin*. Dublin: Templekieran Press, 1979.

Woolf, Alex. "Amlaíb Cuarán and the Gael, 941–81." In *Medieval Dublin III*, edited by Seán Duffy. Dublin: Four Courts, 2002.

See also **Cináed Ua hArtacáin; Viking Incursions**

ANGLO-IRISH RELATIONS

Anglo-Irish relations were given constitutional expression when King Henry II of England (1154–1189) came to Ireland in 1171 and took the formal submission of the Irish kings. Yet given the geographical proximity of Britain and Ireland, it is certain there had always been interactions between the peoples of the two islands. Ireland was not absorbed by the Roman Empire, despite the claim of the historian Tacitus that the governor of Roman Britain from 77–83 C.E., Agricola, contemplated an invasion. Contact with Roman Britain took the form of raiding and trading. In the early medieval period, Irish missionaries were influential in Britain, and political relations with Scotland and Wales were intimate. Dating Ireland's contact with England is more problematic. Unlike Ireland, the peoples that made up England were culturally diverse. The English kingdom was a comparatively recent invention, the very word *Engla-lnd* only appearing in the late tenth century. Before a certain point, therefore, it may be nonsensical to talk of "Anglo-Irish" relations. For a brief period in the tenth century, the Viking kings of Dublin were also kings of York. But although this is evidence of contact, it is questionable whether it should be dubbed "Anglo-Irish" relations.

On the other hand, it seems that the Viking fleets of Ireland were coveted by the Anglo-Saxon kings, and in the eleventh century Ireland's contacts with England come into focus. At the time of the Norman conquest of England, the sons of Harold Godwinsson sought refuge in Ireland from the Normans. It seems that the Norman kings of England aspired to control Ireland. According to his death notice in the *Anglo-Saxon Chronicle*, had William "the Conqueror" (1066–1087) lived two more years, "he would have conquered Ireland by his prudence and without any weapons." Giraldus Cambrensis records that the conqueror's son, William II "Rufus" (1087–1100) gazed from the coast of Wales towards Ireland and boasted that "For the conquest of the land, I will gather all the ships of my kingdom, and will make of them a bridge to cross over."

The interest was not all from predatory English kings. The archbishop of Canterbury, Lanfranc (d. 1089), claimed—partly on the basis of Bede's *Ecclesiastical History*—to be primate of all Britain, including Ireland. This claim was given some foundation when the bishop-elect of Dublin, Gille-Pátraic, went to him for consecration in 1074. Moreover, Lanfranc professed to be doing no more than following the practice of his predecessors. The Irish link with Canterbury brought with it relations with English monastic foundations such as St. Albans and Winchester.

These ecclesiastical contacts were supplemented in the eleventh and twelfth centuries by political connections. Ireland's long association with Welsh politics, including the fact that the founder of the Welsh ruling dynasty of Gwynedd, Gruffudd ap Cynan (d. 1137), was born in Dublin, inevitably brought it into contact with the Normans occupying the Welsh march. The king of Munster and high king, Muirchertach Ua Briain (d. 1119), had a Norman son-in-law in the lord of Pembroke, Arnulf de Montgomery. And in 1165, the year before Diarmait Mac Murchada was expelled from Leinster and sought military aid from King Henry II, the native Welsh chronicle reported that a fleet from Dublin (a town under Diarmait's control) came to Henry II's aid in his abortive campaign against the native Welsh.

England and the Lordship of Ireland

Ireland's connection with England was, therefore, long standing by the 1160s. But the Anglo-Norman invasion, and more particularly the expedition of Henry II of 1171–1172, brought England and Ireland into a formal relationship that has present-day ramifications. King Henry II became the "lord of Ireland" and the land of Ireland became vested in the English crown. There was large-scale peasant migration from England to settle the new acquisition, and with the settlers came English institutions, law, castles, and the introduction of a manorial economy.

It is wrong to imagine that Henry II was forced into this relationship with Ireland by the actions of Anglo-Norman adventurers led by Strongbow. No less than his predecessors, Henry II was happy to add Ireland to his empire. It has been suggested that the notorious papal privilege *Laudabiliter* (1155), which sanctioned an invasion of Ireland, was sought by the archbishops of Canterbury in order to regain primacy over the Irish Church; but if so, the archbishops required royal support. Moreover, there is evidence that, as early as 1155, Henry II was planning to make Ireland an appanage for his brother William. The Anglo-Normans who travelled to Ireland to aid Mac Murchada from 1167 did so with the consent of King Henry II. If they briefly believed they could act independently of the king of England, then Henry II's expedition of 1171–1172 stamped royal authority on Ireland.

One consequence of the invasion was that "Anglo-Irish relations" came to mean the connections between England and the English colonists in Ireland. The Gaelic population was rapidly eliminated from the equation. In the thirteenth century there were sparse contacts, such as when the king of Connacht, Feidlim Ua Conchobair (d. 1265), fought in the Welsh campaign of King Henry III in 1245. But Gaelic contact with the king of England was exceptional rather than commonplace.

Ireland's exact constitutional position in relation to England was initially ambiguous, and various plans were made for the lordship. In 1177 the lordship of Ireland was granted to the king's fourth son, John (d. 1216), the future king of England. It may be that Henry II intended that Ireland would descend as a kingdom in the cadet line of the English royal house, though probably remaining subject to the overlordship of the king of England. A crown was sent by the pope to make the Irish monarchy a reality, but the scheme was not put into effect. When John became king of England in 1199, Ireland once again became vested in the kingship of England. The constitutional position of Ireland was clarified in 1254. In that year King Henry III (1216–1272) granted Ireland to his eldest son Edward, the future king. However, Henry III stipulated that Ireland should never be alienated from the English crown. He retained the ultimate authority over Ireland for himself, and on Edward's succession in 1272 the two lands were once again reunited. This remained the situation until King Henry VIII adopted the title "King of Ireland" in 1541.

The key figure then in the relationship between Ireland and England was the English king. He was lord of Ireland and was required to protect his subjects there. Yet he was most notable for his absence. Henry II and his son John both visited Ireland. But after 1210, despite some good intentions, the only medieval king to visit the lordship personally was Richard II (1377–1399), who made two expeditions, in 1394–1395 and 1399. It is difficult to assess the impact of this absenteeism. Exhortations to the king to visit Ireland and remedy the colonists' ills became frequent from the fourteenth century. But it is unclear what—short of an aspirational renewed conquest—would have strengthened the lordship. There is a strange tendency among Irish historians to favor Kings John and Richard II, seemingly on the sole basis that they crossed the Irish Sea. In fact their expeditions were in many ways damaging and patently unrealistic.

Neglect of Ireland stemmed from the king's preoccupation with other enterprises, in Britain and in continental Europe. In the thirteenth century Ireland was exploited to fund Edward I's campaigns against Wales and Scotland. From the fourteenth century, however, amid the hardship provoked by the Bruce invasion, the Black Death and the Gaelic revival, Ireland ceased to be profitable. It was hoped that the expedition of Lionel of Clarence in the 1360s would rejuvenate the colony so that it could contribute to England's continental campaigns. This naive policy climaxed with Richard II's expeditions of the 1390s. It foundered when Richard II lost his crown to Henry Bolingbroke while in Ireland in 1399.

The later medieval period is complicated by the growth of a "middle nation" among the colonists in Ireland, sometimes called the "Anglo-Irish" by historians. This group referred to themselves as English and always insisted that they were loyal to the king. Yet their growing awareness of a discrete identity from England arguably altered the constitutional position of Ireland. The Irish parliament of 1460 declared that "the land of Ireland is, and at all times has been, corporate of itself . . . freed from any special burden of the law of the realm of England." It is still debated whether this declaration had any historical foundation. But, in a sense, that is irrelevant. The important point is that the voice of the Irish colony—the parliament—declared that Ireland was separate, not from the king, but from the kingdom of England.

The growing alienation of Ireland from England had become dangerous by the end of the medieval period, particularly after the Tudor dynasty won the crown in 1485. In 1487, in an act of extraordinary defiance, a boy called Lambert Simnel was crowned as King Edward VI at Christ Church Cathedral, Dublin. In 1494, a second pretender called Perkin Warbeck found support in Ireland. Yet more insidious were the conspiracies of Anglo-Irish lords and England's international enemies. Ireland was becoming a strategic liability. This fear that Ireland could be used as a "backdoor" into England—a fear that was realized several times in the modern era—came to be the predominant factor in English policy towards Ireland.

The administration of Henry VIII (1509–1544) recognized that the Irish problem had to be addressed. One response to the Kildare rebellion of 1534–1535 was the decision to change the constitutional position of the king. In 1541, King Henry VIII adopted the title "King of Ireland," rather than merely "lord," in an attempt to make the entire population amenable to English law and customs. The lordship of Ireland had at last become a kingdom. Ultimately the policy of accommodation faltered, and it became apparent to English administrators that the only solution was a renewed conquest and plantation of the country. The legacy of this policy is the embitterment that has characterized so much of Anglo-Irish relations into modern times.

Peter Crooks

References and Further Reading

Conway, Agnes. *Henry VII's Relations with Scotland and Ireland, 1485–1498; With a chapter on the acts of the Poynings Parliament, 1494–5 by Edmund Curtis.* Cambridge: Cambridge University Press, 1932.

Davies, R. R. *The First English Empire.* Oxford: Oxford University Press, 2000.

Ellis, Steven. "Henry VII and Ireland, 1491–1496." In *England and Ireland in the Later Middle Ages: Essays in honour of Jocelyn Otway-Ruthven,* edited by James Lydon. Dublin: Irish Academic Press, 1981.

Flanagan, Marie Therese. *Irish Society, Anglo-Norman Settlers, Angevin Kingship: Interactions in Ireland in the late 12th century.* Oxford: Clarendon Press, 1989.

Frame, Robin. *English Lordship in Ireland, 1318–1361.* Oxford: Clarendon Press, 1982.

———. "*Les Engleys Nées en Irlande*: The English Political Identity in Medieval Ireland." In *Britain and Ireland, 1170–1450.* London and Rio Grande, Ohio: The Hambeldon Press, 1998. First published in *Transactions of the Royal Historical Society*, 6th series, 3, (1993): 83–103.

Hudson, Ben. "William the Conqueror and Ireland." In *Irish Historical Studies* 29, no. 114 (1994):145–158.

Lydon, James F. "The Middle Nation." In *The English in Medieval Ireland: Proceedings of the first joint meeting of the Royal Irish Academy and the British Academy, Dublin, 1982,* edited by James F Lydon. Dublin: Royal Irish Academy, 1984.

———. "Ireland and the English crown, 1171–1541." *Irish Historical Studies* 29, no. 115 (1995): 281–294.

Richter, Michael. "The First Century of Anglo-Irish Relations." *History* 59 (1974): 195–210.

See also **Anglo-Norman Invasion; Bruce, Edward; Gaelic Revival; Henry II; John; Lionel of Clarence; Lordship of Ireland**

ANGLO-NORMAN INVASION

The commencement of the so-called Anglo-Norman invasion of Ireland is dated conventionally to 1169, although the first overseas mercenaries in fact arrived in the autumn of 1167 in the company of Diarmait Mac Murchada, the king of Leinster, who had been forced into exile in 1166 and had sought military assistance from Henry II, king of England, to recover his kingdom. The date 1169 derives from the near-contemporary account, the *Expugnatio Hibernica* (The Taking of Ireland), completed around 1189 by Gerald of Wales (Giraldus Cambrensis), an apologist for Anglo-Norman intervention who consistently exaggerated the role of his own relatives in that enterprise, the first of whom, his maternal uncle, Robert FitzStephen, arrived in May 1169. Although the term Anglo-Norman to describe the incomers enjoys wide currency, there is no scholarly consensus on its use; Norman, Cambro-Norman, and Anglo-French have also been used. All are anachronistic: Contemporary sources of both Irish and English provenance consistently described the incomers as *Saxain,* i.e., English. The earliest were adventurers from the South Wales area and of mixed ethnic background: hence the terms Cambro-Norman, and sometimes also Flemish, the latter referring more specifically to those drawn from the Rhos peninsula, where the English king, Henry I (1100–1135), had established a Flemish colony. A more identifiably English influx was already apparent by August 1170 when Richard FitzGilbert, lord of Strigoil, popularly known as Strongbow, arrived in Ireland. Although he was a landholder in South Wales, he also had extensive lands in England from where he drew some of his followers, whom he was to install as his tenants in Leinster following the death of Diarmait Mac Murchada in the spring of 1171. The English element was further reinforced by the personal intervention in Ireland in 1171 of Henry II. The use of the term "invasion" might also be debated, since the earliest incomers arrived as mercenaries in the employ of Diarmait Mac Murchada and invariably fought alongside Irish forces until Diarmait's death in 1171.

The major military expedition led by Henry II to Ireland in October 1171 marked a significant new phase in the English advance. Henry remained in Ireland for a six-month period, during which time he obliged Strongbow to acknowledge him as his overlord for Leinster. Henry also made a speculative grant of the kingdom of Meath to Hugh de Lacy, who had extensive landed interests in England, the Welsh borders, and Normandy. Moreover, Henry decided that the Irish port towns should be appropriated for his own use. He issued a charter granting the city of Dublin to his men of Bristol, which not only confirmed the established trading links between the two cities, but was also an early indication that he was ready to exploit the economic resources of the Hiberno-Norse east-coast towns. During his stay, Henry did not travel beyond Leinster nor deploy his army against Irish forces. A substantial number of Irish kings voluntarily offered their personal submission to him, while the Irish episcopate was also prepared to endorse his intervention in the expectation that greater political stability would be achieved and the bitter disputes that had characterized the pursuit of the office of high king during the twelfth century might be brought to an end.

As a consequence of Henry's personal intervention, a link between a part of Ireland and the English crown was inaugurated, the constitutional repercussions of which are still resonating. In 1175 the Treaty of Windsor was negotiated between Henry and Ruaidrí Ua Conchobair, king of Connacht, and claimant of the high kingship of Ireland. This divided Ireland into two spheres of influence, one under Henry, the other under Ruaidrí, with the latter acknowledging Henry as his overlord. The boundaries delimited by the treaty proved unstable, however, with individual English adventurers rapidly expanding beyond them, a notable instance being the intrusion into Ulaid (Ulster) in 1177 of the soldier of fortune, John de Courcy. In May 1177, Henry II modified the arrangements of the Treaty of Windsor by designating his youngest son, John, as lord of Ireland, with the intention that when he came of age he should personally assume control of the English colonists in Ireland. The king also made an additional series of speculative grants to actual and potential colonists in Munster. In 1185 John went to assume the lordship of Ireland in person, but retreated after a nine-month period, having failed to assert control over the English settlers there, and having suffered a series of military defeats at the hands of Irish kings. Nonetheless, John had made a further series of speculative grants to English members of his entourage, the most notable of whom was Theobald Walter, ancestor of the Butler earls of Ormond.

In 1199, all his brothers having died, John became king of England, an event that could not have been foreseen by Henry, and this was to forge a more direct administrative link between the English crown and the English-held areas of Ireland. Until 1204, John's lordship in Ireland was but one among an assemblage of diverse territories that stretched from the Anglo-Scottish border to the Pyrenees, each of which had its own customs and laws. However, King John's 1204 loss of the duchy of Normandy to Philip Augustus, king of France, altered the English crown's relations with Ireland. English-held Ireland may be said to have been transformed more directly into a colony. A mark of that relationship was that many of the institutions for the governance of England and English laws were transferred to Ireland, although these were to be applied only to English-controlled areas.

King John was said to have taken the laws of England with him on his second expedition to Ireland in 1210 during which he sought and largely succeeded in asserting control over his English subjects in Ireland, though much of his success was to be compromised subsequently by the baronial wars in England that culminated in the procural of *Magna Carta* from the king, to be followed by the eleven-year minority of his son Henry III. The loss of a substantial portion of its continental lands altered the character of the English crown's interest in Ireland. The English lordship of Ireland came to be seen as an annex of an England-centered sphere, while the English settlers in Ireland, the most noteworthy of whom still retained lands in England, sought to remain within the political orbit of the English royal court. In the early decades, a fairly rapid superimposition of English political overlordship had been established in the southeast, central, and northeast of Ireland which demarcated a zone of Anglicization, the visible impact of which is still evidenced on the landscape by the surviving mottes and baileys and stone castles that were erected.

A slower transformation followed of the social, economic, and ethnic landscape of significant parts of the country and the creation of communities that remained self-consciously English. English settlement was concentrated in the physically better-endowed lands of the south and east as well as in the port towns. Invasion and colonization are different if often sequential processes. Invasion typically involves the establishment of lordly or royal control and the imposition of a new aristocracy. Certainly, a new French-speaking aristocracy was installed in Ireland, the more important of whom continued to hold lands on both sides of the Irish Sea. Colonization involves settlement of the land and dispossession of the previous occupiers. Claims for a substantial peasant migration in the train of the new aristocracy have frequently been made, though it remains largely undocumented and impressionistic, and the numbers and density of actual settlers are very difficult to estimate. The establishment of so-called rural boroughs as a spur to colonization, where some of the tenants of a private lord were granted the privilege of holding their plots by the preferential legal and economic status of burgage tenure suggests that, in reality, there were difficulties in attracting settlers to Ireland. Even in the densest areas of English settlement, there were natural impediments to the process of colonization in the mountainous terrain, woodlands and bogland, and the Irish population survived on these less fertile lands retaining its essentially Gaelic character and remaining as pockets of colonial weakness. It proved difficult to maintain or give permanent effect to the colonizing impetus.

The high-point of English colonial initiative had been reached by the mid-thirteenth century, after which a combination of unfavorable political and economic circumstances ensured the so-called Gaelic revival. A steady colonial retreat occurred even in core regions such as the Wexford area, where the first settlers had established themselves, and where the town was exhibiting signs of urban decline already by the end of the thirteenth century. A critical turning-point in a process of de-colonization and loss of English governmental control was reached with the outbreak of plague in 1348. A distinctive "Anglo-Irish" political identity emerged out of the peculiar strains of perennial insecurity experienced by the colonial ruling elite in Ireland, coupled with a sense of its neglect, disregard and misunderstanding by the English crown, while culturally it formed an intermediate grouping characterized by varying degrees of Gaelicization or assimilation. Tensions between the English born in Ireland and the English of England who were sent recurrently as administrators remained constant. An English invasion there may have been in the twelfth century, but a conquest of Ireland was never achieved. In reality, the greater part of Ireland did not experience thoroughgoing Anglicization, and on the eve of the Tudor plantations English governmental control had shrunk to the defensive area known as the Pale, the colonial hinterland of Dublin.

MARIE THERESE FLANAGAN

References and Further Reading

Cosgrove, Art, ed. *A New History of Ireland*. Vol. 2, *Medieval Ireland, 1169–1534*. Oxford: Oxford Unversity Press, 1987. Revised edition, 1993.

Davies, R. R. *Domination and conquest: The experience of Ireland, Scotland and Wales, 1100–1330*. Cambridge: Cambridge University Press, 1990.

Duffy, S. *Ireland in the Middle Ages*. New York: St. Martin's Press, 1997.

Flanagan, Marie Therese. *Irish Society, Anglo-Norman Settlers, Angevin Kingship: Interactions in Ireland in the late 12th century*. Oxford: Clarendon Press, 1989.

Lydon, J. *The Lordship of Ireland in the Middle Ages*. Rev. ed. Dublin: Four Courts Press, 2003.

Smith, B. *Colonisation and Conquest in Medieval Ireland: The English in Louth, 1170–1330*. Cambridge: Cambridge University Press, 1999.

See also **Castles; Connacht; Courcy, John de; Giraldus Cambrensis; Henry II; John; Lacy, Hugh de; Mac Murchada, Diarmait; Mide; Motte-and-Baileys; Pale, The; Strongbow; Ua Conchobair, Ruaidrí; Ulaid**

Anglo-Normans: See also **Military Service, Anglo-Norman; Society, Functioning of; Society, Grades of Anglo-Norman**

ANGLO-SAXON LITERATURE, INFLUENCE OF

Despite geographical proximity and periods of close cultural ties, evidence of such influence on Irish literature is surprisingly scarce. Several reasons for this can be suggested at least for the seventh and early eighth centuries. During that period the Anglo-Saxons were much more likely to have been the recipients than the donors of influence. Ireland sent Christian missionaries to England in the seventh century who introduced Latin literacy and Irish script, while also providing hospitality for considerable numbers of Anglo-Saxon students who came to study in its schools of higher learning. Moreover, Anglo-Saxon literature in the vernacular was unlikely to have had much influence on its Irish counterpart, not only because of the language barrier but also because the English literary tradition was not well established until a full century after that of Ireland. An exception may be King Aldfrith of Northumbria (685–705), known as Flann Fína in Irish, to whom Irish literary tradition dubiously attributed several gnomic works in Irish.

The available evidence suggests that Anglo-Saxon literary influence—such as it was—was exercised through the medium of ecclesiastical Latin, a culture which both areas shared as part of their common Christian heritage. Verifiable instances of that influence are the Latin works of Anglo-Saxon England's greatest scholar, the Venerable Bede (d. 735). His commentaries on biblical exegesis, metrics, and computistics seem to have been known and studied in Ireland by the second half of the eighth century. Two manuscripts written by Irish scribes contain between them three of Bede's computistical works, *De rerum natura, De temporibus,* and *De temporum ratione*. Although copied in the first half of the ninth century and on the Continent, these manuscripts contain glosses which from their language (Old Irish) and phonology suggest that Bede was being studied in the Irish schools by the second half of the eighth century. Further evidence of Bede's influence on the Irish schools as a biblical scholar is found in the "Old-Irish Treatise on the Psalter," a commentary composed in Irish in the first half of the ninth century which attributes to him a comment on Psalm 1. Although no such work on the Psalms has been verified for Bede, the appeal to his authority and the use of the Irish form of his name (*Béid*) testifies to his high status in Ireland. Moreover, Bede's most famous work, the *Ecclesiastical History of the English People*, which was partially translated into Irish in the early tenth century, left its mark on medieval Irish annals and historiography.

Other influences can be traced to Anglo-Saxon England's continuing contacts throughout most of the eighth century with the Gaelic monastery of Iona, the center from which the Irish mission to Northumbria had been directed. The so-called Penitential of Theodore, composed in southern England in the late seventh century, is cited as an authority in the *Collectio canonum Hibernesis*, a collection of Irish ecclesiastical legislation co-authored in the early eighth century by Cú Chuimne of Iona. Likewise, the presence of a stratum of Anglo-Saxon saints in the early Irish martyrologies probably derived from a Northumbrian martyrology which passed to Iona and thence to Ireland during the eighth century.

PÁDRAIG Ó NÉILL

References and Further Reading

Dillon, Myles. "The Vienna Glosses on Bede." *Celtica* 3 (1956): 340–343.

Hull, Vernam. "The Middle Irish Version of Bede's *De locis sanctis*." *Zeitschrift für celtische Philologie* 17 (1927): 225–240.

Ireland, Colin. *Old Irish Wisdom Attributed to Aldfrith of Northumbria: An edition of Bríathra Flainn Fhína maic Ossu*. Tempe, Ariz.: Arizona Center for Medieval and Renaissance Studies, 1999.

Kenney, James F. *The Sources for the Early History of Ireland: Ecclesiastical*. New York: Columbia University Press, 1929.

Ní Chatháin, Próinséas. "Bede's *Ecclesiastical History* in Irish." *Peritia: Journal of the Medieval Academy of Ireland* 3 (1984): 115–130.

Ó Riain, Pádraig. *Anglo-Saxon Ireland: The evidence of the Martyrology of Tallaght*. H. M. Chadwick Memorial Lectures 3. Cambridge: Department of Anglo-Saxon, Norse, and Celtic, University of Cambridge, 1993.

See also **Glosses; Hagiography and Martyrologies**

ANNALS AND CHRONICLES

The Irish Chronicles, kept in Ireland throughout the medieval period, are a major source for Irish society and politics. They are largely annalistic in form, being divided into years, called "annals," rather than having other time-periods, such as reigns, as the main structural principle. They record the deaths of notable ecclesiastical and lay figures, battles, military campaigns, droughts, plagues, and unusual events, such as eclipses and miracles, but they very rarely provide evidence for life among the lower grades of society.

The style of the Irish chronicles is generally terse and factual, generally lacking the long descriptions, detailed accounts, statements of sympathy, animosity, or references to causation that are found in many chronicles from the rest of Europe. Initially predominantly written in Latin, but increasingly in Irish from the ninth century onwards, the vocabulary and syntax of the Irish chronicles is very formulaic and repetitive, producing a highly artificial chronicle style shared by scholars in a number of different centers. However, in the twelfth century, the chronicles do start to become more verbose to some extent, although the lack of a narrative thread between events continues to be an important feature. At roughly the same time Irish chronicles were adapted in texts such as the Fragmentary Annals of Ireland (probably compiled in the eleventh century for Osraige dynasts) and *Cogad Gáedel re Gallaib* (from the early twelfth century, portraying Munster as the savior of Ireland from the Scandinavians) to form non-annalistic narrative chronicles with clearer political messages.

It is quite likely that the chronicles' origins were in the practice of noting down events in the margins of Easter tables, although earlier continental chronicles may also have been influential. When such notes were subsequently copied without the Easter tables, "K" or "Kl," the same abbreviations for "Kalends (first) of January" used in Easter tables, were also employed in the chronicles to mark the beginning of each annal. The reasons for the subsequent maintenance of the chronicles afterwards are unclear, largely because the chroniclers themselves rarely give any indication of their motives. While a general interest in the past, common to all societies, is likely, the ordering of time according to Christian principles could also have been a factor, as Daniel McCarthy has recently shown in his studies of the Christian dating methods (such as A.D. dating) in the Irish chronicles. The high number of deaths recorded in the chronicles perhaps were designed to emphasize the transience of the earthly life. Political bias, mainly manifested through the selective inclusion and exclusion of certain events, could be another reason. Overall, it is likely that a combination of motivations were important, depending on the interests of the individual chronicler.

The Development of the Irish Chronicles before 1200

Attempts to reconstruct the development of the Irish chronicles have been complicated both by the tendency of the chroniclers to combine and rewrite chronicles, making it difficult to identify constituent sources, and by the lateness of the surviving manuscripts: They date from the late eleventh to the seventeenth century, usually centuries after the events they describe. Modern scholars have adopted varying methods for the identification of chronicle sources, using the frequency of references to particular places, local details, or the chronology of the chronicles to locate sources, producing different results. It is generally accepted that most of the Irish chronicles share a common source before C.E. 911 known as the "Chronicle of Ireland," but there is disagreement about whether events only recorded in one source were also part of this text or were derived from chronicles kept before C.E. 911.

It is likely that contemporary records of events found in the Chronicle of Ireland were kept as early as the late-sixth century for Scottish and Irish events, although the record is likely to have been subsequently altered to promote the powerful Uí Néill dynasty and St. Patrick. From about 660 to 740, it is clear that a chronicle was kept at Iona off the west coast of Scotland; this may have been the source for much of the Irish, as well as Scottish, chronicle records for this period. After 740, the Scottish element is greatly reduced, so it is unlikely that Iona continued to be a major source. From 740 to 911, constituent chronicles of the Chronicle of Ireland have been proposed for Armagh, Clonard in the midlands, and the area to the north of the river Liffey (called "Brega" and "Conaille" at the time). These views have been based on the interest the chronicles display in events in Armagh and the east midlands, although many events further to the west around the Shannon and Brosna rivers are also recorded. It is during the period from 731 to 911 that a number of non-Irish sources, including a "Book of Pontiffs" (from Rome), the "Chronicle of Marcellinus" (from sixth-century Constantinople), and early eighth-century works by the northern English monk Bede, were used by the Irish chroniclers to add Imperial and Papal events to the section from C.E. 431 to 720.

After C.E. 911, the Chronicle of Ireland was continued independently at different centers (although the Irishman Marianus Scottus also finished an unrelated chronicle in 1076 at a monastery in Mainz in Germany). The Annals of Ulster, found in a late-fifteenth-century manuscript, contains a continuation of the Chronicle of Ireland kept in Brega, Conaille or Armagh, but from the late tenth century it is clearly an "Armagh Chronicle," and from 1189 to the 1220s, this text was continued at Derry. The section from 1014 to the 1220s is also found in the Annals of Loch Cé, which survives in a sixteenth-century manuscript.

The other main continuation of the Chronicle of Ireland is found in a number of manuscripts called the "Clonmacnoise group," the main representatives of which are the fourteenth-century Annals of Tigernach (with a text which ends at 1178) and the seventeenth-century *Chronicum Scotorum* (which ends at 1150), but it is also found in the less-substantial Cottonian Annals, the Annals of Roscrea, the Annals of Clonmacnoise, and the Annals of the Four Masters. The high degree of interest in both the affairs of Clonmacnoise, and Brega to the east, in the decades immediately following C.E. 911 could be explained by the close links between the monasteries of Clonmacnoise and Clonard. However, the large number of detailed Clonmacnoise entries indicates that at least by the late eleventh century, if not before, the text had become a "Clonmacnoise Chronicle," with Clonmacnoise events from as early as perhaps the eighth century added to the Chronicle of Ireland.

At some point between C.E. 911 and about 1060, the version of the Chronicle of Ireland in this Clonmacnoise-group text was radically altered, by the addition of more material from Bede's *Chronica Maiora*, and from lists of kings of Ireland and Irish provinces. These sources were also added to a possibly preexisting section (called the "Irish World Chronicle") which covered the period from Creation to the coming of Palladius in C.E. 431. Combined with events from the Irish Invasion Myth, this not only made the chronicle more international in content, but also projected back concepts such as the "kingship of Ireland" and the provincial kingships into the prehistoric past, to provide a coherent account of Irish history.

The Annals of Inisfallen are found in the earliest Irish chronicle manuscript, produced in 1092 or shortly after in Munster. The text written then, which is closely related to that used by the compiler of *Cogad Gáedel re Gallaib*, was a compilation of a Munster chronicle source and at least one other chronicle, including a Clonmacnoise-group text. At some stage many entries were omitted, abbreviated, and rewritten, turning it into a Munster-orientated chronicle. After 1092, the chronicle was maintained by a number of scribes, as can be seen from the manuscript, probably in Munster at Lismore from 1092 to 1130, and at Inisfallen in the next surviving section from 1159 onwards (with gaps). Another chronicle, Mac Cárthaigh's Book, is closely related to the Annals of Inisfallen from the twelfth to fourteenth centuries, but it also contains material from South Ulster or Oriel and Giraldus Cambrensis's account of the Anglo-Norman invasion.

The Development of the Irish Chronicles after 1200

In the late medieval period the Irish chronicles continue to have complex interrelationships; often, different sections of the same manuscript were originally unrelated to each other, rather than being continuations of the same text. In the thirteenth and fourteenth centuries, a chronicle from northern Connacht forms the basis for a number of sets of annals, including the seventeenth-century Annals of Clonmacnoise, the Annals of Ulster, and the fifteenth-century Miscellaneous Annals from 1237–1249 and 1302–1314. The section of the Annals of Loch Cé from the early thirteenth century to 1316, and the fifteenth-century Annals of Connacht both contain the north Connacht chronicle, which had been altered by the learned Ua Máelchonaire family in the fifteenth century and the Ua Duibgeannáin historical family in the late fifteenth century or sixteenth century.

The common source of the Annals of Loch Cé and the Annals of Connacht also contains material from 1180 to about 1260 that, if not actually based on the Cottonian Annals, was derived from a text closely related to it. The Cottonian Annals, surviving in a thirteenth-century manuscript, contain an abbreviated version of the pre-Palladian Irish World Chronicle, the Chronicle of Ireland, and annals up to 1228 written at the Cistercian monastery of Boyle in northern Connacht. It was then continued to 1257, perhaps at the Premonstratensian monastery of Holy Trinity at Loch Cé.

In the fourteenth and fifteenth centuries another common source was used in the Annals of Ulster, the Annals of Connacht before 1428, and the section of the Annals of Loch Cé from 1413 to 1461. This source seems to have concentrated on northern Connacht and south Ulster, to be continued by the Mac Magnuis family at Clogher in the late fifteenth century and incorporated into the earliest manuscript of the Annals of Ulster, produced under the direction of Cathal Mac Magnuis in the late fifteenth century.

In the later Middle Ages there were also a number of annalistic chronicles more concerned with events in England and the Continent, which were written in Latin rather than in a mixture of Latin and Gaelic, often linked to the new Continental religious orders, and kept in English-controlled areas after 1169. The basis for most of these texts was a chronicle probably brought over from Winchcombe in England in the late eleventh century by Benedictine monks and maintained subsequently in Dublin at Christ Church. This chronicle was combined in the early thirteenth century at the Cistercian monastery of St. Mary's in Dublin with Irish Cistercian material, Giraldus Cambrensis's works, and English histories. It was a major source for the Annals of Multyfarnham, compiled in the late thirteenth century by the Franciscan friar Stephen Dexter; the Annals of Christ Church, produced in the early fourteenth century; and Penbridge's Annals, which also constitute a separate source from 1291 to 1370. The Fransiscan friar John Clyn, another continuator of this common source, produced a text at Kilkenny whose draft version was probably used by the Dublin friars who wrote the inappropriately named "Kilkenny Annals" at the same time in the early fourteenth century.

Use of the Irish Chronicles in Modern Scholarship

The Irish chronicles have been used by modern scholars as a prime source for accounts of the political and ecclesiastical centers in Ireland, mainly through turning the brief statements in the chronicles into historical narratives. Another approach has been to count the frequencies of certain types of entries, such as Viking raids and death-notices of types of ecclesiastics, to see trends. The degree to which such evidence reflected reality is debatable, depending on a detailed understanding of the chronicles themselves. However, many significant factors, such as the interests of the chroniclers, the contexts of the chronicles' composition, and how the texts were altered in later periods, still require further research, before the usefulness of the chronicles can be determined.

NICHOLAS EVANS

References and Further Reading

Anderson, Marjorie O. *Kings and Kingship in Early Scotland.* Edinburgh and London: Scottish Academic Press, 1973.

Charles-Edwards, Thomas. *Early Christian Ireland.* Cambridge: Cambridge University Press, 2000.

Etchingham, Colmán. *Viking Raids on Irish Church Settlements in the Ninth Century. A reconsideration of the Annals.* Maynooth, Ireland: The Cardinal Press, 1996.

Freeman, Martin A. *Annála Connacht: The Annals of Connacht (A.D. 1224–1544).* Dublin: Dublin Institute for Advanced Studies, 1944.

Grabowski, Kathryn, and David Dumville. *Chronicles and Annals of Medieval Ireland and Wales: The Clonmacnoise-Group Texts.* Studies in Celtic History 4. Rochester, N.Y. and Woodbridge, England: Boydell and Brewer, 1985.

Gwynn, A. "Some Unpublished Texts from the Black Book of Christ Church, Dublin." *Analecta Hibernica* 2 (1931): 310–329.

Hennessy, William M., ed. *Chronicum Scotorum: A Chronicle of Irish Affairs from the Earliest Times to A.D. 1135, with a supplement containing the events from 1141 to 1150.* London: Rolls Series, 1866.

———., ed. *The Annals of Loch Cé. A Chronicle of Irish Affairs from A.D. 1014 to A.D. 1590.* Vol. 1. London: Rolls Series, 1871.

Hughes, Kathleen. *Early Christian Ireland: Introduction to the Sources.* London: Hodder & Stoughton, 1972.

Mac Airt, Séan, ed. *The Annals of Inisfallen (MS Rawlinson B. 503).* Dublin: Dublin Institute for Advanced Studies, 1951.

———, and Gearóid Mac Niocaill, eds. *The Annals of Ulster (To A.D. 1131).* Dublin: Dublin Institute for Advanced Studies, 1983.

———. *The Medieval Irish Annals.* Medieval Irish History Series 3. Dublin: Dublin Historical Association 1975.

McCarthy, Daniel. "The Chronology of the Irish Annals." *Proceedings of the Royal Irish Academy* 98C (1998): 203–255.

O'Dwyer, B. W. "The Annals of Connacht and Loch Cé and the Monasteries of Boyle and Holy Trinity." *Proceedings of the Royal Irish Academy* 72C (1972): 83–102.

Smyth, A. P. "The Earliest Irish Annals: Their first contemporary entries, and the earliest centres of recording." *Proceedings of the Royal Irish Academy* 72C (1972): 1–48.

Stokes, Whitley. ed. *The Annals of Tigernach.* 3 vols. Vol. I reprinted from *Revue Celtique* 1895–1896 at Felinfach: Llanerch Publishers, 1993; Vol. II reprinted from *Revue Celtique* 1896–1897 at Felinfach: Llanerch Publishers, 1993.

See also **Anglo-Saxon Literary Influence; Annals of the Four Masters; Clyn, Friar; Giraldus Cambrensis; Invasion Myth; Mac Firbhisigh; Marianus Scotus; Ua Cléirigh; Religious Orders; Viking Incursions**

ANNALS OF THE FOUR MASTERS

The title given to the chief historical work of a small team of scribes and historians under the leadership of the Franciscan friar Mícheál (Tadhg) Ó Cléirigh, these annals were compiled in two stages between 1632 and 1636 in the "place of refuge" of the Donegal Franciscan community at Bundrowse on the Donegal/Leitrim border. Known to its compilers and patron as *Annála Ríoghachta Éireann* (The Annals of the Kingdom of Ireland), its more popular (if inaccurate) title *The Annals of the Four Masters* first appears in 1645 in the introduction to the *Acta Sanctorum Hiberniae* (Deeds of the Saints of Ireland) of the Franciscan hagiologist John Colgan, who adapted the phrase from a thirteenth-century commentary on the Franciscan rule. The *Annals* form part of the remarkable historical, doctrinal, catechetical, and hagiographical publishing program undertaken by the exiled Irish Franciscan community in Louvain (Belgium) in the first half of the seventeenth century. From their arrival in Louvain in 1607, the friars labored to produce Irish language material for their missionary work in Ireland and Scotland and for circulation among exiled Irish Catholics on the continent. In 1614, they acquired their own printing press and in 1617 moved to their permanent site at St. Anthony's College. Important publications included Bonaventure (Giolla Brighde) Ó hEoghasa's *An Teagasg Críosdaidhe* (Antwerp 1611, Louvain 1614), the first catechism to be printed in Irish; Flaithrí Ó Maoilchonaire's translation of a Catalan devotional text *Desiderius* (1616); and Aodh Mac Aingil's *Sgáthán Shacramuinte na hAithridhe* (Mirror of the Sacrament of Penance, 1618).

Like many of the friars involved in this program of scholarship, Mícheál Ua Cléirigh was a member of a hereditary learned family who had traditionally been professors of history to the Ó Domhnaill (Ua Domnaill) lords of Tír Conaill. Born circa 1590 and baptized Tadhg, he appears to have followed a military career in the Spanish Netherlands before joining the friars in 1623 in Louvain, where he was received as a lay brother and given the religious name Mícheál. His older brother Bernardine (Maolchonaire) had already joined the order. Friar Mícheál's skill as a scribe and historian was soon recognized, and in 1626 he was sent back to Ireland by the guardian of St. Anthony's, Hugh Ward, to gather whatever he could of the surviving ecclesiastical, political, and hagiographical material with a view to its publication. He spent eleven years traveling from his base in Donegal to various religious houses and lay schools throughout the country, transcribing saints' lives and martyrologies, compiling genealogies of the saints and kings of Ireland, and producing a new redaction of the *Lebor Gabála* (The Book of Invasions), which was the standard account of the early history of Ireland. To assist him, Ó Cléirigh assembled a small team of scribes from among his own kinsmen and members of other traditional learned families. His chief assistants were Fearfeasa Ó Maoilchonaire, Cúchoigcríche Ó Duibhgeannáin, and Cúchoigcríche

Ó Cléirigh, while Conaire Ó Cleirigh and Muiris Ó Maoilchonaire worked with him for shorter periods.

The *Annals of the Four Masters* represent a compilation or conflation of earlier annals and other historical sources. While a number of the sources listed by Ó Cléirigh in a preface to the *Annals* still survive, others have been lost, making the *Four Masters* the sole authority for much of the material they contain, particularly after 1500. As this source material came from a variety of scholarly traditions using different dating systems, the work of compilation and editing proved a major difficulty, and the chronology of the *Four Masters* is defective for large sections of the work. The content is also heavily weighted in favor of entries relating to the North of Ireland and to Connacht as the compilers do not seem to have been aware of the principal Munster source, the *Annals of Inisfallen*, or of any of the Anglo-Irish chronicles, including those compiled by fellow Franciscans.

Though steeped in the conventions of traditional Irish historiography, the *Annals of the Four Masters* differ significantly in scope and tone from earlier works. Earlier works represented the concerns of a particular monastic community or learned family while in theory, if not always in practice, the *Four Masters* concerned themselves with the whole of Ireland.

Bernadette Cunningham has demonstrated the extent to which the compilers were influenced by the ideals of Counter-Reformation Catholicism emanating from Louvain. Priority in each entry was given to ecclesiastical events of that year, such as the deaths of bishops or abbots. Details in earlier sources considered unedifying in a Counter-Reformation context are silently edited and events like the dissolution of the monasteries, or the destruction of the relics in the 1530s, which earlier annals saw as part of military campaigns, are presented by the *Four Masters* as the action of heretics. This confessional and controversial emphasis in their work did not however prevent the Franciscan scholars of Louvain and Rome from exchanging sources and information with Anglican antiquarians like Archbishop James Ussher of Armagh and Sir James Ware.

Two complete sets of the annals, each consisting of three volumes, were produced. One set was presented to Fearghal Ó Gadhra of Coolavin, who had sponsored the project, and the second set was to be forwarded to Louvain for publication, but only two volumes were sent, one of which has now disappeared. The five surviving volumes of the six originally produced are now housed in the libraries of Trinity College Dublin, The Royal Irish Academy, and University College, Dublin. The edition and translation of the *Annals* published by John O'Donovan and Eugene O'Curry in six volumes between 1848 and 1851, though not a critical one, is remarkable for O'Donovan's extensive scholarly apparatus and remains the most accessible and useful edition of the text.

Colmán N. Ó Clabaigh

References and Further Reading

Cunningham, Bernadette. "The Culture and Ideology of Irish Franciscan Historians at Louvain 1607–1650." In *Ideology and the Historian*, edited by Ciarán Brady. Historical Studies 17. Dublin: The Lilliput Press, 1991.

Giblin, Cathaldus. "The Annals of the Four Masters." In *Dún Mhuire Killiney 1945–1995: Léann agus Seanchas*, edited by Benignus Millett and Anthony Lynch. Dublin: The Lilliput Press, 1995.

Ó Cléirigh, Tomás. *Aodh Mac Aingil agus Scoil Nua-Ghaeilge i Lobháin*. Baile Átha Cliath (Dublin): An Gúm, 1935, Reprinted 1985.

Walsh, Paul. *The Four Masters and their Work*. Dublin: The Three Candles, 1944.

———. *Irish Leaders and Learning Through the Ages*. New edition, ed. Nollaig Ó Muraíle. Dublin: Four Courts Press, 2003.

See also **Annals and Chronicles; Bardic Schools, Learned Families; National Identity; O Cléirigh; Poets, Men of Learning; Records, Ecclesiastical; Religious Orders**

ARCHAEOLOGY

Archaeology is the study of the past through the medium of the physical remains of human activity, using three categories of evidence: sites, artifacts, and human effects on the natural environment. Often associated largely with the study of prehistoric societies, it has made a real contribution to the study of medieval Ireland. The strengths of archaeology lie in its independence from written documents, which emanate from particular groups in past society and reflect their interests, and on its study of long-term processes rather than events. In contrast to prehistoric archaeology, medieval archaeology is underpinned by working within a documented period, notably with fewer chronological problems and greater identification of past individuals and groups. The main handicap suffered is the destruction of evidence, both in the past and through modern development of land.

The study of medieval archaeology in Ireland has not been a story of even progress. In the years around 1900, and before, Irish scholars took their place with those of Britain and western Europe. Their study was based on above-ground sites and buildings; much of the work was aimed at establishing dates of such monuments as round towers, relating them to the Church of the 10th century and later, rather than exaggerated claims of antiquity. A major figure was Goddard Orpen, who showed that mottes were indeed early earthwork castles and not prehistoric sites. In the new Republic, archaeology was strongly supported but the nationalistic climate encouraged archaeologists to concentrate on the early-medieval (or "Early Christian" period; the name is in itself significant) sites but neglect the later part. The emphasis was on art historical analysis of artifacts rather than on excavation or sites in general; excavation techniques of the period were unable to examine timber structures, and the resources to undertake or analyze widespread field survey were lacking. In Northern Ireland, real achievements in research, excavation, and control came after 1950, with the establishment of the Archaeological Survey and the intensive study of County Down. Since the 1960s, there have been advances in the study on a number of fronts. Laboratory techniques have been systematically deployed, relating to chronology (radiocarbon or, most dramatically, tree-ring dating); the environment (pollen or animal and human bone studies); statistical tests for the analysis of site distributions; and analytic techniques of materials used in artifacts. Research now combines evidence from sites and artifacts, or field survey and excavation, while the involvement of the state in the salvaging of sites threatened by destruction through development has had a major impact on the volume of evidence recovered and potentially available for study.

The contribution, actual and potential, of archaeology to the study of medieval Ireland shows successes and weaknesses. The earliest medieval period, between the fifth and eighth centuries, is marked by an explosion in the volume of evidence, compared with the time before or, indeed, elsewhere in western Europe. Sites of the period survive in the thousands: secular ringforts and crannogs; Christian monasteries and lesser churches. The artifacts from the time include some of the most famous craft objects from Ireland: the Tara brooch, the Book of Kells, High Crosses, and the Ardagh chalice. They are clearly the product of wealth (a manuscript will need many calves to die for its parchment) and indicate that Ireland, especially through the Church, was closely in touch with Britain and Europe. The wealth aligns with the evidence of a rural environment with few trees and heavily managed by man for a farming economy based on agriculture and dairying. The archaeology focuses attention on the revolution which occurred to start the period's expansion and also the detailed management of land. All the sites relate to a hierarchy, such as is described in the documentary sources, but we do not know their exact relationship, nor do we understand the reasons behind the variations in the geographical distributions: why Leinster has few sites of any sort, why crannogs should be found mainly in the Midlands, or why ringfort densities can very widely across small distances.

To the political historian, the Vikings were military attackers in Ireland of the ninth and tenth centuries. In archaeology, however, they are much more associated with the foundation of major market towns, notably Dublin. Here was an organized urban site from the tenth century, very similar to York in its streets and economy based on crafts and trade; the major difference lies only in the material for the houses. The archaeologist can point to the survival of church sites near areas of Viking dominance to stress the possibility of coexistence between Scandinavians and Irish; no ringforts or crannogs show convincing signs of destruction. This peaceful emphasis may be as illusory as the picture of continual violence. The influx of silver from England and Europe through the towns in Ireland had to be paid for, probably by the export of slaves; it may be no coincidence that the tenth century appears to have seen an increase in the building of underground structures, *souterrains*, probably as refuges against raiders. At the same time, there also appears to have been a hiatus in church crafts, such as stone carving and the production of manuscripts, while metal artifacts reflected new styles brought in with the Vikings. Aggression and trade may not have been mutually exclusive.

After 1167, the seizure of large areas of land in Ireland by English lords was followed an explosion in the volume of archaeological evidence similar to that of the sixth through eighth centuries, with many new sites and artifacts, against a documentary background stressing political or military events. Archaeology stresses the English lords' agenda of spreading a market economy, with agriculture providing produce (especially grain) through centralized estates, to be sold through lesser towns and the great ports for lords' profit, to build new castles and church buildings. The new buildings reveal much of the lords' motivation: their commitment to stay in Ireland, their desire to reflect European contemporaries, and their stress on display rather than military conquest. The profits also involved merchants and farmers, stimulating the rise and importation of new crafts, such as pottery, often located in the towns founded as the engine of the economy. In the countryside, the lack of many truly nucleated settlements, unlike the villages found in many parts of England, question the assumption that the English lordships were based on wholesale immigration of peasant communities: Rather, it was through the organization of estates and the piecemeal arrival of individuals that the changes were effected. Modern archaeology has concentrated on those areas of high visibility, principally through excavation in advance of major development schemes in the large towns (Dublin, Waterford, Cork). These are well-studied sites; deploying resources on them has resulted in the relative neglect of other, less well-known areas. Principal among these is the world of the Gaelic Irish, but also the small towns and rural sites; no manor site has been excavated recently.

Two periods have been overlooked by archaeologists. The 150 years before 1200 have been lost, between the assumptions that life was a continuation of the fifth-through eighth-century world and that the incursion of English lords marked a fundamental change throughout Ireland. The potential indications that changes had occurred before 1150 have been neglected, other than those in the Church, where there is a combination of new sites (houses of the Continental Orders) and documentary accounts of reform. New forms of lordship may have caused new sites to manage the landscape, but these have not been sought. The period after the mid-fourteenth-century population collapse associated with the Black Death has been dominated by the documentary historians' picture of decline. The archaeological evidence of modest but real prosperity, implied by the widespread building of friaries or parish churches and tower houses, has not been fully deployed, while the difficulties of identifying pottery of the period has led to a serious underestimation of the vitality of towns in the period. Archaeology would stress the period as one in which the process of cultural fusion, started in the thirteenth century, between English and Irish and most obviously represented in the development of a distinctive Irish Late Gothic style of building, has been overshadowed by the documentary evidence for conflict. In both these cases the archaeology has suffered from the same problems. The context is dominated by political history, which stresses short-term and military events over the long-term processes that are the strength of archaeology. A second problem is the tyranny of the geographical fact that Ireland is an island, which leads to the assumption that it is a unity. Regional differences are downplayed in the face of the uniform literate culture of the upper classes. This is best seen in the assumption that the arrival of the English force in Wexford in 1169 would have changed the life of a Connacht peasant.

T. E. McNeill

References and Further Reading

Edwards, N. *The Archaeology of Early Medieval Ireland.* London: Batsford, 1990.

Mallory, J., and T. E. McNeill. *The Archaeology of Ulster.* Belfast: Institute of Irish Studies, 1991.

O'Conor, K. *The Archaeology of Medieval Rural Settlement in Ireland.* Dublin: Discovery Programme, 1998.

O'Keeffe, T. *Medieval Ireland: An Archaeology.* Stroud: England Tempus Publishing, 2001.

Hurley, M., and O. Scully. *Late Viking Age and Medieval Waterford: Excavations 1986–1992.* Waterford, Ireland: Waterford City Council, 1997.

See also **Burials; Castles; Craftwork; High Crosses; Houses; Inscriptions; Kells, Book of; Villages; Walled Towns**

ARCHERY

See **Weapons and Weaponry**

ARCHITECTURE

Early Medieval

Most of the architecture that survives from earlier-medieval (pre-twelfth-century) Ireland was ecclesiastical in nature, and most of the individual buildings that still stand to an appreciable height were churches. One of the enduring puzzles about Ireland's rich Christian civilization at this time is that these churches were buildings of almost willful simplicity; the skill and energy invested in manuscript illumination, the production of metalwork, and the carving of High Crosses were rarely deployed to provide an appropriately sumptuous architectural setting for worship.

The churches, most of which probably date from the tenth or eleventh centuries, are invariably single-cell structures of small size, with linteled west-end doors, limited fenestration, and no architectural sculpture; some of them have antae, pilaster-like projections of the side walls past the end walls, which most writers believe to have supported the end-timbers of the roofs.

There is also evidence for timber churches in early-medieval Ireland. Some of the written accounts, such as Cogitosus's seventh-century Life of St. Brigit, suggest carpentered and ornamented buildings of considerable

Bunratty Castle, Co. Clare. © *Department of the Environment, Heritage and Local Government, Dublin.*

Timahoe Round Tower, Co. Laois. © *Department of the Environment, Heritage and Local Government, Dublin.*

design sophistication, but we have no independent test of the accuracy of such accounts; significantly perhaps, the timber churches identified in archaeological excavation were simple post-built structures.

Round towers, or *cloigtheacha* ("bell-houses"), as the annalists described them, first appeared on church sites in the tenth century and continued to be built to the same basic design into the later 1100s. Distinctively tall, narrow, and elegant, they represent a triumph of the native mason's craft in the face of a difficult technical challenge.

Romanesque

The second half of the eleventh century saw the emergence in western Europe of Romanesque architecture, a complex stylistic movement with explicit formal and iconographical references to earlier Roman architecture. Irish church-builders were clearly aware of these developments, and by 1100 some of the characteristic elements of the tradition—round arches and barrel vaults—were beginning to appear there. In the early twelfth century a distinctively indigenous Romanesque tradition, borrowing heavily from the chevron-rich Anglo-Norman Romanesque, developed in Munster, eventually diffusing across the island by the end of the century. Churches are the only surviving representatives of this architectural tradition; given the importance of secular patronage in church-building, contemporary high-status residential architecture was also Romanesque. Cormac Mac Carthaig's eponymous church at Cashel (1127–1134) is the most substantial survival, but it may always have been an exceptional building. Most of the other churches of the period were of simple plan and unsophisticated superstructure, their portals and chancel arches being the only parts embellished with Romanesque motifs; key works include the portals at Killeshin (c. 1150) and Clonfert (c. 1180). The Cistercians independently introduced their own Burgundian version of Romanesque into Ireland in the mid-twelfth century.

The Anglo-Norman invasion did not mark the end of Romanesque building in Ireland. Rather, there was a late flowering of the style in Cistercian and Augustinian monastic churches founded in the lands of the Ua Conchobair kings and their subordinates to the west of the Shannon. Indeed, the Anglo-Normans themselves were more familiar with Romanesque than Gothic at the time of their arrival, since Gothic was only starting to take root in England and Wales in the late twelfth century. The Romanesque transepts of Christ Church Cathedral, for example, were built by their masons, while the halls and donjons in a number of their early castles (Adare, Ballyderown, and Trim, for example) also belong within the Romanesque tradition of their home territory.

Early Gothic

The first building projects in the Gothic style began as the twelfth century closed, and their patrons were Anglo-Norman. New Cistercian monasteries with English mother-houses, specifically the abbeys of Grey (started after 1193), Inch (started after *c.* 1200), and Duiske (started after 1204), played an important role in the dissemination of the style. But the critical buildings were probably the cathedral churches in Dublin (the nave of Christ Church; St. Patrick's in its entirety) and the now-lost cathedral in Waterford. Key elements of what is called "Early English" Gothic, such as pointed lancet windows and "stiff-leaf" capitals, were on display in these buildings during the early thirteenth century. The masons who worked on these projects were trained in the west of England, the region from which many Anglo-Norman settlers in Ireland had come.

Gothic church-building in the Early English style spread rapidly through the lordship in the first half of the thirteenth century, but the projects produced modest results. Relatively few of the new buildings were aisled or transeptal, or had any internal vaulting; the exceptions were either cathedral churches in prosperous sees (Newtown Trim, for example), major monastic churches (Athassel, for example), or parish churches associated with very powerful lords (New Ross, for example).

In the second half of the thirteenth century the new friaries of the mendicant orders provided opportunities for masons to practice their skills, and the general proposerity of the colony, at least in the third quarter of the century, provided favorable conditions for the building industry in general. Yet the period was marked by a rather unimaginative consolidation of the Early English style rather than any concerted attempt to keep pace with the increasingly elaborate Gothic work in contemporary England. But some building projects of the early fourteenth century (Athenry and Fethard friary churches, for example) featured traceried windows in the so-called Decorated style of contemporary English Gothic, and these works, few though they are, certainly undermine any assertion that Ireland was too war-torn in the early 1300s to have accommodated serious architectural endeavours.

Late Gothic

Levels of political patronage of architecture in the fifteenth century surpassed those of the thirteenth century. Projects of the era, ecclesiastical and secular, were also more widespread geographically, embracing areas that were under "Irish" and "English" political control. The architectural details of this late Gothic phase were derived largely from English stylistic traditions: Elements of the early-fourteenth-century Decorated style still remained, but were now augmented with elements from the so-called Perpendicular style, which was popular in contemporary England. Impulses from the latter tradition are especially evident in the Pale, not least in the three famous Plunkett family churches of Dunsany, Killeen, and Rathmore.

The fifteenth-century projects included additions to or partial rebuildings of many of the existing cathedrals, abbeys, priories, and friaries, as well as brand-new mendicant friaries of exceptional architectural merit in western Ireland (Rosserk and Moyne, for example). Patrons' investment in their own home comfort and outward display are represented by new tower-houses and other forms of castle, the doorways, windows, and battlemented parapets of which often parallel those in ecclesiastical buildings. Most of Ireland's medieval parish churches—the buildings most neglected by architectural historians—were substantially altered in the fifteenth and early sixteenth centuries, and very many seem to have been rebuilt.

TADHG O'KEEFFE

References and Further Reading

Craig, Maurice. *The Architecture of Ireland from the Earliest Times to 1880*. London: Batsford, 1982.

O'Keeffe, Tadhg. *Romanesque Ireland: Architecture and ideology in the 12th century*. Dublin: Four Courts Press, 2003.

Hourihane, Colum. *Gothic Art in Ireland 1169–1550: Enduring vitality*. Yale: Yale University Press, 2003.

Leask, Harold G. *Irish Churches and Monastic Buildings*. 3 vols. Dundalk: Dundalgan Press, 1955–1960.

Stalley, R. "Irish Gothic and English Fashion." In *The English in Medieval Ireland*, edited by James Lydon. Dublin: Royal Irish Academy, 1984.

See also **Abbeys and Religious Houses; Castles; Christ Church Cathedral; Church Reform, Twelfth Century; Ecclesiastical Sites; Mac Carthaig, Cormac; Parish Churches, Cathedrals; St. Patrick's Cathedral; Tower Houses; Walled Towns**

ARMAGH

Armagh (Ard Macha) became the ecclesiastical capital of Ireland in the middle ages, its status based on its supposed associations with Patrick.

Prehistory

Historians have been tempted to associate Armagh's emergence as Ireland's premier Christian center with a pre-Christian cultic legacy reflected by a sizable collection of stone carvings at Armagh, though the provenance of some of the stones is poorly documented and certainty about their origins is elusive. At nearby Emain Macha, named after the same goddess as Armagh, archaeologists excavated a major religious structure dating to circa 95 B.C.E.

Patrick

Annals claim that Patrick founded a church at Armagh in 444, but those annals were written retrospectively and are unreliable. In fact, apart from his *Confession* and his *Letter to Coroticus*, no documents survive from Ireland in Patrick's time, and neither composition associates him with Armagh. The *Book of Armagh*, written in 807, incorporates the earliest records to connect Patrick with Armagh: the *Book of the Angel*, written about 640 to 650; a catalogue of "Patrician" churches compiled by Bishop Tírechán circa 670; and a *Life of Patrick* composed by Muirchú maccu Machthéni in the 680s or 690s, though based upon earlier records.

Muirchú claimed that Patrick's church was not founded on the hilltop at Armagh, where the Church of Ireland cathedral stands, but lower down the hill at Templenaferta. Excavations at that site uncovered a series of burials dating from circa 420–685 C.E., establishing it as a very early Christian foundation. Excavations nearby uncovered evidence of a substantial ditch, which surrounded the hilltop in the fifth century. That suggests that the church at Templenaferta was founded beside a secular power center at Armagh in the early fifth century. However, there is no independent evidence to confirm the claim that Patrick ever founded a church at Armagh, though Armagh was certainly within his mission field.

Seventh-century records show that Patrick already enjoyed a national reputation. Cummian's letter of 633 referred to Patrick as "papa noster," recognizing him as the "father" of the Irish Church. Pope-elect John IV's letter of 640 addressed to leading churchmen in northern Ireland, with Bishop Tómméne of Armagh at the head of the list, points to Armagh being one of the leading ecclesiastical centers in the north of Ireland (if not *the* leading center) by that date. The works of Muirchú and Tírechán and the *Book of the Angel* amplified Patrick's existing reputation and bound it with the church at Armagh.

Wealth

Armagh's subsequent rise to national importance was associated with the cult of Patrick, but also on the great wealth the church leaders at Armagh were able to command, both in terms of land and the tributes drawn from dependent churches and monasteries in its far-flung *paruchia*. Armagh claimed jurisdiction over many supposedly Patrician foundations, and churches like Sletty that placed themselves under Armagh's protection. Political factors must have played a role in Armagh's accumulation of such wealth and power, though the process is obscured by lack of evidence.

Certainly Armagh was at the center of the kingdom of the Airthir, a branch of the Airgialla federation. The churchmen at Armagh, as one can see most blatantly in Muirchú's *Life*, looked to the Airthir and Uí Néill for patronage. The abbots of Armagh probably gained possession of their rich hinterland as discarded segments of the royal Airthir dynasty reprised themselves as ecclesiastical dynasties under the protection of the church. The successful courting of the Uí Néill meant that as they progressed towards a national hegemony (never fully realized), the church of Armagh's claims to national primacy were promoted in their train. The close tie between Armagh and the Uí Néill is symbolized by Áed Findliath, the king of Tara, having a house in Armagh in 870.

Armagh's monastery grew over time, as reflected in the growing number of church offices recorded in eighth- and ninth-century annals. It became a sizeable ecclesiastical settlement. A Viking raid on the city in 1020 destroyed the fort at Armagh and all the buildings in it, save the library, countless houses in and around Armagh, the great stone church on the hill and at least two lesser churches, and the students' accommodations and "much gold and silver, and other precious things."

Vikings

The first Viking raid on Armagh was recorded in 832, and they were frequent thereafter. The raiders came for slaves as well as precious religious objects and other portable wealth. A hoard lost by Vikings in the Blackwater River shows the high quality of metalwork being carried out at Armagh at the time. The church at Armagh survived repeated raids, apparently undiminished, though the loss of manuscripts and ecclesiastical treasures, not to mention lives and mundane goods, must have been considerable over the years.

Twelfth-Century Reforms

Armagh's claims to being Ireland's primatial see were formally acknowledged at the synod of Raith Breasail in 1111. The synod was part of the "twelfth-century reform" which sought to bring the Irish Church more closely in line with that elsewhere in Latin Christendom. The reform movement is closely associated with Malachy of Armagh, though Malachy faced tremendous opposition to his reforming efforts from the Clann Sínaig, the hereditary abbots of Armagh. The reformers eventually prevailed, though Popes Adrian IV and Alexander III directed Henry II of England to launch the Anglo-Norman invasion of Ireland in the third quarter of the twelfth century to complete the reforms.

Later Middle Ages

Armagh and most of its hinterland remained the possession of the archbishops in the later middle ages. Archbishop Máel Pátraic Ua Scannail (1261–1270) built Armagh's medieval cathedral, which survives in a heavily "restored" guise. It was described in 1553 as "one of the fairest churches in Ireland." Ua Scannail also founded a Franciscan friary in Armagh whose ruins can still be seen. An Augustinian priory built around the same time to house the reformed monastic community of Armagh was described as "the best building in Armagh" by Bishop Chiericati, the papal nuncio to the court of Henry VIII. There was also a small Céile Dé community in the city which survived into the sixteenth century. At the close of the middle ages, there was a convent at Templenaferta which boasted four carved panels in white alabaster of Italian cinquecento design.

Armagh remained a significant town throughout the later middle ages. English soldiers under Lord Deputy Sussex set fire to Armagh in 1557, though less than a quarter of the town was actually destroyed, which may reflect something of its size. However, by the end of the Tudor wars of conquest, Armagh had been all but completely wrecked. Bartlett's map of Armagh in 1601 shows extensive ruins of stone houses, as well as ecclesiastical buildings in Armagh by that time.

HENRY A. JEFFERIES

References and Further Reading

De Paor, Liam. "The Aggrandisement of Armagh." In *Historical Studies VIII*, edited by T. D. Williams. Dublin: Gill & Macmillan, 1971.

Jefferies, Henry A. *Priests and Prelates of Armagh in the Age of Reformations*. Dublin: Four Courts Press, 1997.

Ó Cróinín, Dáibhí. *Early Medieval Ireland, 400–1200*. London and New York: Longman, 1995.

———. "Saint Patrick." In *Armagh: History and Society*, edited by A. J. Hughes and W. Nolan. Dublin: Geography Publications, 2001.

Ramsey, G. "Artefacts, Archaeology and Armagh." In *Armagh: History and Society*, edited by A. J. Hughes and W. Nolan. Dublin: Geography Publications, 2001.

Sharpe, Richard. "St. Patrick and the See of Armagh." *Cambridge Medieval Celtic Studies* IV (1982): 33–59.

Stuart, James. *Historical Memoirs of the City of Armagh*. Newry: privately published, 1819.

See also **Airgialla; Church Reform, Twelfth Century; Ecclesiastical Settlements; Ecclesiastical Sites; Ecclesiastical Organization; Emain Macha; Houses; Malachy; Muirchú; Patrick; Pre-Christian Ireland; Religious Orders; Tírechán; Uí Néill; Vikings**

ARMAGH, BOOK OF

A vellum manuscript consisting originally of 222 folios (*c.* 195 x 145 mm; folios 1 and 41–44 are now missing) in three parts: The first contains a dossier of texts mostly in Latin but partly in Old Irish, comprising almost all the earliest biographical and historical materials relating to St. Patrick; the second part is the only complete copy of the New Testament surviving from the early Irish Church; the third contains the *Life of St. Martin of Tours* by Sulpicius Severus (in a unique recension). The manuscript is particularly important, both for its contents and because it can be dated. In a brilliant piece of detective work, Charles Graves, later Bishop of Limerick, in the mid-nineteenth century deciphered two partially erased colophons which revealed that the book had been written (with perhaps one or more assistants) by an Armagh scribe Ferdomnach (d. 846) at the behest of Torbach, *heres Patricii* (i.e., successor of Patrick and abbot of Armagh) in the year 807. The book was revered in the later Middle Ages as a relic because of a colophon on folio 24v which reads *Hucusque uolumen quod Patricius manu conscripsit sua* (Thus far the volume that Patrick wrote in his own hand); in later centuries, this text came to be called *Canóin Phátraic* (Patrick's Canon).

Leather satchel of the Book of Armagh. © *The Board of Trinity College Dublin.*

The Armagh collection of texts written by Patrick himself appears to have been incomplete at time of writing (807), if not before. Thus only his *Confessio* is copied (in a defective version) into the manuscript; Patrick's *Letter to the soldiers of Coroticus* is missing, apparently deliberately omitted. Copies of the two documents survive in continental transcripts, however, thus revealing the defective nature of the Armagh recension. Copies of the earliest surviving hagiographical works on Patrick, by Tírechán and Muirchú maccu Machtheni, appear also to have survived only in defective versions at Armagh, and it has even been suggested that they may, in fact, have been added at a later date to the rest of the collection, after their respective texts had been procured from elsewhere.

The gospel text in the Book of Armagh is classified as Vulgate with some Old Latin admixtures. One recent study has detected affinities between the Armagh text and that of the Echternach Gospels main text (Paris, BN lat. 9389). The other New Testament texts are of more or less equal purity (the Acts of the Apostles and the Apocalypse being especially so). At the end of Matthew's Gospel the scribe has added a collect for that saint's feast-day (on which day that particular page was written), while at the end of John, excerpts from Gregory the Great's *Moralia in Iob* are arranged in a geometrical design around the diamond-shaped closing words of the Gospel. At the end of the additions (*Additamenta*) to Tírechán's *Collectanea* of Patrician ecclesiastical sites, two groups of cryptic catchwords and abbreviations are inserted, neither having any connection with the Patrician material. The second of these groups (folio 53v.) consists of a number of allusions to Pope Gregory, with the full text of the *Hanc igitur* prayer of the Roman canon. Gregory was especially revered in the early Irish Church, as was Martin of Tours, and some authorities have claimed that the text of the *Vita Martini* in the Book of Armagh was transmitted to Ireland within half a century of Sulpicius Severus's death (410).

Whereas the Patrician section is undecorated, the four gospels are elaborated with large and small initials, often with bird- or animal-heads and spirals, and pen-and-ink drawings of the evangelist symbols, while some initials in the third (Martinian) section are also elaborated. Initials towards the end of the book, in the Acts of the Apostles, the Epistles, and the Life of St. Martin, are colored. On folio 170r. there is a diagrammatic representation of Jerusalem, the walls of which are decorated with interlace in the Irish style. The closing texts of John's Gospel and the opening of the Apocalypse are written in spectacular diamond-shaped patterns of exquisite calligraphy. There are some individual Latin words written in transliterated Greek.

The Book of Armagh is important also as the oldest dateable Irish manuscript containing continuous prose texts in the Irish language. Its value for linguists is therefore exceptionally high. Passages from the Patrician dossier also preserve the oldest surviving evidence for the use of charters to record land and property transactions in early Ireland. So also, Ferdomnach's scribal note *Scripsi hunc ut potui librum* ("I have written this book as best I could," folio 18v.) is likewise the earliest dateable example of Irish Latin hexameter verse.

Despite its substantial contents, the Book's small dimensions, and the arrangement of the gatherings, suggest that it was intended originally to be used as six separate booklets. Superficially akin to the well-known Irish class of "pocket gospels," it is unlikely that the gospels (or the entire New Testament sections) were used for liturgical or ceremonial purposes. A pair of wooden boards, apparently from an early binding, still survives, which may at some earlier date (perhaps already in the ninth century) have formed the cover of the New Testament or some other part of the manuscript. On the other hand, its exalted status was the reason why, in 1004, when Brian Boru visited Armagh in the course of a triumphal circuit around Ireland, he had his secretary (Calvus Perennis, alias Máel-Suthain) insert a note in the Book claiming Brian as *imperator Scottorum* ("emperor of the Irish").

The earliest datable reference to the Book of Armagh is preserved in the (seventeenth-century) *Annals of the Four Masters*, who record that in 937 a shrine or case (*cumdach*) was provided for the Book by Donnchad mac Flainn, King of Ireland. In the fifteenth century, the Book was provided with a leather carrying-satchel, which still survives. At some unknown date the hereditary office of "Steward of the Canon" was created to ensure the safekeeping of the manuscript. It seems to have passed from the possession of the keepers sometime after 1680 into the hands of Arthur Brownlow of Lurgan (County Armagh), and was eventually deposited by a member of the Brownlow family in the library of the Royal Irish Academy, in Dublin. In 1853 it was donated, after purchase, to the Library of Trinity College Dublin.

DÁIBHÍ Ó CRÓINÍN

References and Further Reading

Gwynn, John, ed. *Liber Ardmachanus. (The Book of Armagh).* Dublin: Royal Irish Academy, 1913.

Kenney, James F. *The Sources for the Early History of Ireland: Ecclesiastical.* New York: Columbia University Press, 1929.

Gwynn, E. J., ed. *Book of Armagh, the Patrician documents.* Dublin: Royal Irish Academy, 1937.

Alexander, J. J. G. *Insular manuscripts, 6th to the 9th century.* Survey of Manuscripts Illuminated in the British Isles 1, no. 53. London: Harvey Miller, 1978.

Bieler, Ludwig, ed. *The Patrician Texts in the Book of Armagh. Scriptores Latini Hiberniae* 10. Dublin: Dublin Institute for Advanced Studies, 1979.

See also **Annals of the Four Masters; Armagh; Brian Boru; Devotional and Liturgical Literature; Manuscript Illumination; Patrick; Reliquaries**

ARMIES

Armies in Ireland trace their origins to the legendary Fianna and their leader Finn mac Cumaill. From at least the eleventh century, the Irish kings maintained small permanent fighting forces later known as their *teaghlach* or *lucht tighe*—meaning "troops of the household." These were well-equipped and were divided into footmen and *marcshluag* (cavalry). Highly skilled professional soldiers, they were often given houses and lands among the king's mensal lands. It was clear that, from the reign of Brian Boru (d. 1014), Irish kings could take large forces of spearmen, swordsmen, archers, slingers, and horsemen on campaign, often combining them in operations with naval forces. To put such forces into the field, Irish kings must have developed an extensive support network to maintain, arm, and feed their troops on campaign. The size of these armies and the destructive scale of Irish warfare were aptly demonstrated in 1151 at the battle of Móin Mór, where seven thousand soldiers fell, if the annals are to be believed. What characterized Irish warfare during this period was the rapid mobility of armies. For example, Ruaidrí Ua Conchobair (d. 1198) developed large forces of highly mobile and well-armed horsemen—mainly drawn from the upper classes of his vassals. In comparison, Irish infantry forces seem mostly to have been lightly armed footmen. However, it is likely that the Irish elite soldiery had adopted Ostman-style chain mail armor; finds of armor-piercing arrowheads at Waterford show that some of its defenders wore chain mail. Moreover, Ruaidrí perhaps developed his permanent foot soldiers of his *teaghlach* or *lucht tighe* into a form of heavy infantry—similar to the household jarls of the Anglo-Saxon kings. Another major development in the composition of Irish armies was the growing dependence of Irish kings upon mercenaries later known as *ceithirne congbála* (retained bands). And from the early 1100s, Irish kings—such as Muirchertach Mac Lochlainn (d. 1166) were looking abroad—recruiting Hebridean-Norse forces and fleets from the Western Isles of Scotland to serve in Ireland. The military power of a great king such as Ruaidrí was maintained by the levy of Gaelic military service—illustrating the extent of a king's overlordship over his vassals. All the able-bodied population—apart from the learned and the clergy—were eligible for service. A king's principal military commander was the *marasgal* (marshal), an office whose origins lay probably in the earlier *dux luchta tige* (the head of the king's household). The marshal's principal duty was the organization of the king's army, particularly the levying and billeting of troops along with the fining of those who failed to render military service.

However, warfare and armies changed forever after the return in 1167 of Diarmait Mac Murchada (d. 1171) from Britain with English and Welsh mercenaries. The devastation of East Leinster by these forces demonstrated that they were vastly superior to their Irish opponents. Yet it would be a mistake to view Irish and English armies as uniracial. Other Irish kings soon followed Mac Murchada's example of building his forces around an English spine; Domnall Mac Gilla Pátraic of Osraige (d. 1185) hired Maurice de Prendergast in 1169 to resist Mac Murchada, and exemplified the fluid nature of military service, rendering feudal service to Richard de Clare (Strongbow, d. 1176). Further, Cathal Crobderg Ua Conchobair of Connacht (d. 1224) strengthened his forces in 1195 by employing the services of Gilbert de Angulo (d. 1212), demonstrating the hybrid nature of the forces in his pay.

On the other hand, English armies in Ireland were dependent upon military feudalism, whereby all royal tenants, both English and Irish, were obliged to render military service in the feudal host. Essentially, the arms of the feudal host were made up of knights, men at arms, footmen, archers, and hobelars (forces of lightly armed and mobile horsemen adapted to the conditions of Irish warfare). Throughout much of the thirteenth century, English armies continually demonstrated their superiority in pitched battles with the Irish. The major difference between the Irish and English armies of this time was the quality of their cavalry. In contrast with the lightly armed Irish horseman, the heavily armored English knight was mounted on a large horse known as a charger. The defeat at Athenry in September 1249 of Tairrdelbach Ua Conchobair (d. 1266), king of Connacht, showed that Irish forces could not resist the massed charge of English cavalry. This led to innovations to balance the military equilibrium. In 1259, Áed son of Feidlim Ua Conchobair (d. 1274), prince of Connacht, formed a marriage-alliance with the Hebridean-Norse king of the Western Isles. As part of his bride's dowry, he gained 160 fighting men known as *galloglass*—heavy infantry which fought in formations designed to counter English cavalry-charges.

The weakness of the Dublin government for much of the middle ages—combined with absence of a royal standing army—meant that English forces were to become increasingly hybrid. As time progressed, galloglass became a feature of English armies in Ireland. But the development of large private armies by the English magnates of Ireland was crucial to the survival of their power on the frontiers. Clearly, they were adopting Gaelic elements. In Ulster, the de Burgh earls adopted the *buannacht* (bonaght; the wages and provisions of a galloglass), which involved quartering galloglass throughout the earldom, while the earls levied the *tuarastal* (wages) of these elite soldiers upon the people. During the parliament of 1297, it emerged that English magnates often hired Irish troops, billeting them upon their own English tenants—prompting the outlawry of this practice. Other English magnates in Ireland billeted troops upon their tenants; it was reputed that James Fitzgerald (d. 1463), seventh earl of Desmond, first imposed *coinnmhead* (coyne; billeting) upon his earldom. During the early decades of the fifteenth century James Butler (d. 1452), fourth earl of Ormond, imposed forces of "kernety" and galloglass throughout his lands in Tipperary and Kilkenny—granting them the right to take a *cuid oidche* (cuddy; a night's portion of food, drink, and entertainment) from every freeholder's house. The change in the composition of private English armies was dramatically illustrated in the usage by Desmond and Ormond of kernety—a form of military police, traditionally only in the service of Irish lords, for arresting offenders and acting as guards of a lordship. That Ormond instituted this form was remarkable—but even more remarkable was the fact that his 120 kernety were drawn evenly from the Purcells and the Codys, families of English lineage. The rise of the Fitzgerald earls of Kildare from 1456 further displayed the hybrid nature of armies in Ireland. In 1474, Thomas Fitzgerald (d. 1478), seventh earl of Kildare, established a permanent fighting force, the "Fraternity of St. George," comprising 160 archers and 63 spearmen. However, the

Kildares' real military strength lay in their large forces of Mac Domnaill galloglass—forcing the Leinster Irish to recruit galloglass of their own. Such was the power of the Kildares that they were able to billet their galloglass upon the Pale, levying "coyne and livery" upon Englishmen for their maintenance.

From the late 1510s, the English government became convinced of the necessity of reform in Ireland and gradually royal armies returned. The collapse of the Kildare rebellion in 1535 created a countrywide political vacuum, so the Dublin government sought to extend royal jurisdiction throughout the country, demanding the dissolution of all private armies and the abolition of coyne and livery. There was vehement resistance—particularly from the Irish lords. Towards the end of the sixteenth century, Irish leaders such as Áed Ua Néill (d. 1616), second earl of Tyrone, and Fiach Ua Broin (d. 1597) emerged to revolutionize Irish armies and warfare by adopting foreign ideas, tactics, training, and formations. Tyrone trained a red-coated Ulster army to fight in the Spanish *tercio* formation, using both pike and musket. He won great victories at Clontribret in 1595 and at Yellow Ford three years later, but his defeat at Kinsale in 1601 effectively ended resistance from coordinated Irish forces. However, the allegiance owed to the great lords was still hard to destroy completely. Indeed, it took the armies of Oliver Cromwell (d. 1656), lord protector of England, during the late 1640s and 1650s to finally tear up the last roots of the private armies.

EMMETT O'BYRNE

References and Further Reading

Barry, Terry, et al., eds. *Colony and Frontier in Medieval Ireland*. London: Hambleton Press, 1995.

Bartlett, Robert and Angus McKay, eds. *Medieval Frontiers Societies*. Oxford: Oxford University Press, 1989.

Bartlett, Thomas and Keith Jeffrey, eds. *A Military History of Ireland*. Cambridge: Cambridge University Press, 1996.

Byrne, Francis. *Irish Kings and High-Kings*. London: Batsford, 1973.

Harbison, Peter. "Native Irish Arms and Armour in Medieval Gaelic Literature, 1170–1600." *Irish Sword* 12 (1975–1976): 174–180.

Lydon, James. "The Hobelar: An Irish Contribution to Medieval Warfare." *Irish Sword* 2 (1954–1956): 13–15.

———, ed. *Law and Disorder in the Thirteenth-Century Ireland*. Dublin: Four Courts Press, 1997.

———. *The Lordship of Ireland in the Middle Ages*. Dublin: Four Courts Press, 2003.

Morgan, Hiram. *Tyrone's Rebellion*. London: Boydell Press, 1999.

Nicholls, Kenneth. *Gaelic and Gaelicized Ireland*. Rev. ed. Dublin: Lilliput Press, 2003.

O'Byrne, Emmett. *War, Politics and the Irish of Leinster, 1156–1606*. Dublin: Four Courts Press, 2003.

Otway-Ruthven, Jocelyn. "Knight Service in Ireland." *Journal of the Royal Society of Antiquities* 79 (1959): 1–7.

———. "Royal Service in Ireland." *Journal of the Royal Society of Antiquities* 98 (1968): 37–39.

Simms, Katherine. "Warfare in Medieval Irish Lordships." *Irish Sword* 12 (1975–1976): 98–105.

———. *From Kings to Warlords*. London: Boydell Press, 1987.

See also **Brian Boru; Diarmait Mac Murchada; Mac Lochlainn, Muirchertach; Military Service, Anglo-Norman; Military Service, Gaelic; Military Orders; Strongbow; Ua Conchobair, Ruaidrí**

B

BARDIC SCHOOLS, LEARNED FAMILIES

Before the Twelfth-Century Church Reform

Although Julius Caesar mentions large schools run by druids for the youth of Celtic Gaul in the first century B.C.E., we know little or nothing about the education of poets and other men of learning in early Ireland before the eighth century C.E. Around this period, Liam Breatnach has argued, higher grades of poet, the *filid*, became differentiated from the oral "bards" by their literacy. They used written Old Irish texts to pursue studies of grammar, versification, genealogy and history that were closely modeled on the Latin curriculum of the church schools in early medieval Ireland. Almost every scholar of native learning recorded in the annals before 1200, whether poet (*fili*), expert in Irish traditional history (*senchae*), or judge of customary law (*brethem*), is identifiable as a cleric, or a teacher in a church school. However from the late tenth to the twelfth centuries, the annals also notice a few learned court poets, some of whose verses in praise of Irish kings still survive. A number of their surnames, Ua Cuill (Quill), Ua Sléibín (Slevin), and Mág Raith (Magrath) recur in the later Middle Ages, showing their descendants continued to practice the same hereditary art. During this transitional "Middle Irish" period, the distinction between literate *filid* and oral bards was lost. The best of the bards became literate, while *filid* lost their connection to the church schools after the twelfth-century reform of ecclesiastical organization in Ireland. New orders of Augustinian canons and Cistercian monks ran schools for their novices, which had no place for the study of Irish genealogies or customary law.

Secular Schools of the High Middle Ages

At the end of the twelfth century, Irish bardic verse developed a new standardized language based on contemporary Early Modern Irish speech, together with sets of elaborate metrical rules that presuppose a formal training for the new generation of court poets. They were dominated at this time by the Ua Dálaig (O'Daly) family, recorded in the twelfth century not only as local chieftains of Corkaree in modern County Westmeath, but also as gifted poets. Individual members were celebrated as "the best poet in Ireland" and even "chief poet of Ireland and Scotland." Two were court poets to the Mac Carthaig kings of south Munster in the mid-twelfth century, and two more, the famous Muiredach of Lissadell (fl. 1213) and the religious poet Donnchad Mór Ua Dálaig (d. 1244) are found in early-thirteenth-century Connacht. In a poem by Muiredach, he refers to himself as "Ua Dálaig of Meath," the head of his family, traveling with a little band of three companions, whose "master," or teacher, he is.

In the fourteenth century, we have evidence for fixed schools, each located at the home of a chief poet, using books in their studies. The first surviving Early Modern Irish textbook for poets, a tract on Metrical Faults, is preserved in a mid-fourteenth-century manuscript, National Library of Ireland G 2–3 (the "Ó Cianáin Miscellany"). Gofraid Finn Ua Dálaig (d. 1387), the most famous of the Munster branch of his family, is credited with two long poems which instruct students on meters and rhyme. He himself was trained in the school of the Mág Raith poets of north Munster. His works, and those of his fellow-pupil, Maelmuire Mág Raith, contain references to reading a book together with their teacher, or "fosterer." They use the Irish word for pupils, *daltae*, which means also "fosterchildren," and they make mention of the darkened beds on which student poets lay while composing their poems, and a process of *sgagad* (sifting), by which the best students were picked out and certified as fully-qualified poets.

Schools of History, or "Senchas"

The fourteenth century also saw a revival of the study of traditional Irish historical lore and genealogies, which involved transcribing Old Irish saga texts, historical tracts, and genealogies from twelfth-century manuscripts of the pre-reform church schools. This activity was led above all by Seaán Mór Ua Dubagáin (O'Dugan, d. 1372), court poet and historian to the chief William Ua Cellaig (O'Kelly) of Uí Maine in east County Galway. The Ua Dubagáin family reputedly functioned as archivists to the church settlement of Clonmacnoise. Seaán Ó Dubagáin was a scribe of early portions of the Book of Uí Maine, and teacher to Adam Ua Cianáin (O'Keenan, d. 1373), scribe of the "Ó Cianáin Miscellany." Both these manuscript compilations not only reproduce the genealogies of the main royal dynasties of early Ireland, but link the pedigrees of fourteenth-century Irish chiefs to their remote royal ancestors, or in some cases, alleged ancestors.

Another major manuscript of the late fourteenth century, the Book of Ballymote, is associated with the Ua Duibgennáin (O'Duignan) school of traditional historians or *seanchaide*. Coming from the area of County Leitrim, a member of this family, Fergal Muimnech, "the Munsterman," Ua Duibgennáin (d. 1357), crossed the Shannon to erect a church at the holy well of St. Lasair, of Kilronan, County Roscommon, in 1339, where he and his descendants remained as *erenaghs* (stewards) of the church lands there, and professional historians to the Mac Diarmada (Mac Dermot) chiefs of north Roscommon. This family also produced the now lost Annals of Kilronan, a year-by-year chronicle of Connacht affairs from which the sixteenth-century scribe Philip Ua Duibgennáin drew most of his entries for the still-extant Annals of Loch Cé, compiled for his patron Brian Mac Diarmada, chief of Moylurg. The sixteenth-century Annals of Connacht are largely drawn from closely related historical material compiled by the neighboring school of the Ua Mael Chonaire (O'Mulconry, Conroy) family of south Roscommon, recorded as poets and historians to the Ua Conchobair (O'Conor) kings of Connacht from at least the beginning of the thirteenth century.

Early Irish Texts Preserved by the Schools

As well as recording political events of their own day, genealogies of later medieval Gaelic rulers, and court poetry addressed to prominent aristocrats and ecclesiastical figures, the later medieval schools of bardic learning have preserved for us countless early Irish literary, historical, and legal texts, originally composed between about 700 and 1150, which would otherwise have been lost. The Mac Firbisig school of Lecan (County Sligo) could claim a continuous tradition since the early twelfth century, and accumulated an extensive family library. The chief source for other schools of historians concentrated around the Shannon basin may have been the dispersed library of eleventh- and twelfth-century manuscripts from the pre-reform school of Clonmacnoise. The best-known extant example of these is *Lebor na hUidre*, the Book of the Dun Cow, a collection of Old Irish sagas transcribed about 1100 C.E. This ancient manuscript was handed over to the Ua Conchobair chief of Sligo in 1359 as ransom for the son of Ua Sgingin, a member of a Connacht ecclesiastical family serving as court historian to the Ua Domnaill (O'Donnell) chief of Tír Conaill (County Donegal). The faded writing was restored and re-inked at Ua Conchobair's expense, but the manuscript was returned over a century later, as spoils of war to a victorious Ua Domnaill chieftain. When the Ua Sgingin historians in Tír Conaill died out in the fifteenth century, they were replaced by the Ua Cléirig (O'Clery) family of churchmen, poets and historians to the Ua Domnaill chiefs. The Uí Chléirig originally came from Ua Cellaig's territory of Uí Maine, where many of the old churchlands of Clonmacnoise lay.

Ulster Poets

From the same geographical area, soon after 1400, the Mac an Baird (Ward) family of Uí Maine, whose surname indicates that they were descended from "bards," the oral court poets of early Ireland, also entered the service of Ua Domnaill of Tír Conaill. By the sixteenth century they formed a major poetic school in Tír Conaill, and another branch had spread to County Monaghan, serving the Mac Mathgamna (MacMahon) chiefs there. Their best-known author was Fearghal Óg Mac an Bhaird of the Tír Conaill branch, whose work comments on the Nine Years War (1594–1603), the Flight of the Earls (1607), the Ulster plantation, and Counter-Reformation clerics in the Irish College at Louvain. Other famous northern poetic families were the Mac Con Mide (MacNamee) poets from Castlederg (County Tyrone), the Ua hUiginn school on the northern borders of Sligo, who addressed poems to the chieftains of Ulster and Connacht generally, the Ua hEodhusa (O'Hussey) poets of Fermanagh and the Ua Gním (Agnew) family of eastern Ulster. These latter were alleged to be descended from Scottish immigrants originally called Agnew. Fer Flatha Ua Gním's poems reflect the impact of the Ulster plantation on the old Gaelic families who had been his patrons.

Law Schools

Originally church schools played a key role in reducing the mass of inherited Irish customary law to written texts between the seventh and the ninth centuries, and then examining these texts in detail through glosses and commentaries. These secular studies were discontinued by the twelfth-century Continental monastic orders, and for a while in the later thirteenth century, Irish annals record no scholars of native law, whether clerics or laymen. Once again the first signs of revival appear in Uí Maine. The Mac Áeducáin (Egan, Keegan) family were lay landowners under the rule of the Ua Cellaig chiefs, who had become experts in Old Irish customary law by the beginning of the fourteenth century. In 1309, Gilla na Náem mac Duinnsléibe Meic Áeducáin was the first to be described as "*ollam* [professor] of Connacht in law," a "chief master of jurisprudence." In all, some forty members of this family were noted in the Irish annals, the overwhelming majority as experts in law. Their most famous school was that of Mac Áeducáin of Ormond in North Tipperary, but another was sited in Dún Daigre (Duniry, County Galway), and the family established separate branches, serving Anglo-Irish lords and Irish chieftains of Connacht, Meath, Longford, and north and south Munster. Their most famous manuscript, the early-fifteenth-century *Leabhar Breac* (Speckled Book) of Duniry, shows that their schools were not confined to copying, glossing, and commenting on the law tracts, since much of this book's contents consists of religious tracts and saints' lives from the pre-reform church schools of the twelfth century. In the Ua Briain (O'Brien) lordship of Thomond (County Clare), lawyers of the Mac Áeducáin family were rivalled by the almost equally prolific Mac Fhlannchada (MacClancy) law school, and the more-localized Ua Duib dá Boirenn (O'Davoren) school, serving the Ua Lochlainn (O'Loughlin) chiefs of the Burren, County Clare. To the scribes of this latter school we owe many surviving copies of Old Irish law tracts. Other less prominent law schools were those of the Ua Deoradáin (O'Doran) family in Leinster, Ua Breisléin (O'Breslin) in Fermanagh, and Mac Birrthagra (MacBerkery) in Eastern Ulster.

Medical Schools

Because later Irish annals concentrate on Connacht and Ulster, learned families from other parts of Ireland are often best known by the manuscripts they left behind. This is especially true of the medical families, many of whom were located in the south of Ireland, such as the Ua hIceda (O'Hickey), Ua Cuinn (Quin), Ua Laide, Mac an Lega (both anglicized as "Lee") physicians of Munster, the Ua Bolgaide (Bolger) family in Leinster, and the Ua Cenndubáin (Canavan) physicians of south Connacht. Better-documented by the annals were the Ua Siadail (O'Shiel), Ua Duinnsléibe (Dunlevy), and Ua Caiside (Cassidy) families of Longford, Donegal, and Fermanagh respectively. Medical schools were the exception to other centers of bardic learning in that their Irish medical tracts were translations of Latin textbooks from contemporary Continental schools of medicine, giving their patrons the benefit of the latest scientific advances, such as they were. Their pupils, however, shared the basic training in Irish spelling, grammar, and metrics which was common to all the bardic schools, and men from medical families often served as scribes, compiling learned anthologies of history, poetry, and law in the other schools.

The music of harp and tympanum (an instrument like a zither) was also studied in bardic schools, and we know the names of leading musicians' families, Ua Coinnecáin (Cunningham) and Mac Cerbaill (MacCarvill). However, no Irish musical notation has survived from the medieval period.

KATHARINE SIMMS

References and Further Reading

Breatnach, Liam, ed. *Uraicecht na Riar: The Poetic Grades in Early Irish Law*. Dublin: Dublin Institute for Advanced Studies, 1987.

Caerwyn Williams, John E. *The Court Poet in Medieval Ireland*. London: British Academy, 1971.

Carney, James. *The Irish Bardic Poet*. Dublin: The Dolmen Press, 1967.

———. "The Ó Cianáin Miscellany." *Ériu* 21 (1969): 122–147.

Henry, Françoise, and Geneviève Marsh-Micheli. "Manuscripts and Illuminations." In *New History of Ireland 2*, edited by Art Cosgrove, 780–815. Oxford: Oxford University Press, 1987.

MacCana, Proinsias, "The Rise of the Later Schools of Filidheacht." *Ériu* 25 (1974): 126–146.

Murphy, Gerard. "Bards and Filidh." *Éigse* 2 (1940): 200–207.

Ó Cuív, Brian, ed. *Seven Centuries of Irish Learning*. Dublin: Stationery Office, 1961.

Ó Muraíle, Nollaig. *A Celebrated Antiquary*. Maynooth, Ireland: An Sagart Press, 1996.

O'Rahilly, Thomas F. "Irish Poets, Historians and Judges in English Documents, 1538–1615." *Proceedings of the Royal Irish Academy* 36 (1921–1924): 86–120.

Simms, Katharine. "The Brehons of Later Medieval Ireland." In *Brehons, Serjeant and Attorneys*, edited by Daire Hogan and W. Nial Osborough, 51–76. Dublin: Four Courts Press, 1990.

———. "Literacy and the Irish Bards." In *Literacy in Medieval Celtic Societies*, edited by Huw Pryce, 238–258. Cambridge: Cambridge University Press, 1998.

Walsh, Paul. *Irish Men of Learning*. Edited by Colm O'Lochlainn. Dublin: Three Candles Press, 1947.

See also ***Áes Dána; Duanairí;*** **Education; Law Schools, Learned Families; Medicine; Scriptoria; Historical Tales; Dinnsenchas; Genealogy; Grammatical Treatises; Law Tracts; Lyrics; Metrics; Wisdom Texts; Poetry, Irish;**

Glendalough, Book of; Leabhar Breac; Lebor na hUidre; Lecan, Book of; Lecan, Yellow Book of; Leinster, Book of; Lismore, Book of; Uí Mhaine, Book of; Music; Mac Con Midhe, Giolla Brighde; Poets, Men of Learning; Clonmacnoise; Annals and Chronicles; Brehon Law.

BERMINGHAM

The medieval Irish lineage of Bermingham (in the sixteenth century, sometimes written Brimegham) was a branch of a knightly family resident at Birmingham in England. The first to appear in Ireland was Robert de Bermingham, to whom Earl Richard "Strongbow" granted the Irish kingdom of Offaly. Although Robert left only a daughter and heiress Eva, wife of Gerald fitz Maurice (FitzGerald), he seems to have divided Offaly with a brother, perhaps the William who occurs circa 1176, the latter receiving the northern part, known as Tethmoy (*Tuath Dhá Mhuigh*), in modern northeastern County Offaly. Tethmoy descended to Piers de Bermingham (d. 1254), the real founder of the family in Ireland, and from whom they derived the Irish surname of "Mac Feorais." Piers participated in the occupation of Connacht by Richard de Burgh after 1225, receiving the territories of Dunmore (County Galway) and Tireragh (County Sligo), while his second son Meilir received a separate enfeoffment of Athenry, where he founded a walled town and a Dominican friary. Piers's heir was his grandson, Piers fitz James de Bermingham of Tethmoy, celebrated by the colonists as a great warrior against the Gaelic Irish, but infamous in history for his treacherous massacre (1305) of the O'Connors of Offaly, who were his guests. His uncle Meilir married Basilie de Worcester, heiress of the great Tipperary baronies of Knockgraffon and Kiltinan, and these lands were to be the subject of complicated exchanges, difficult to disentangle, between their son and heir, Piers fitz Meilir of Athenry, and his cousins of Tethmoy. Another son of Meilir, William, was archbishop of Tuam from 1289 to 1314.

Piers of Tethmoy's son, John de Bermingham, as a reward for defeating Edward Bruce at Faughart in 1318, was created earl of Louth in 1319, with a grant of liberty authority over that county, and was chief governor of Ireland from 1321 to 1323. Resentment against his rule in Louth, with which he had no hereditary links, led to his massacre, along with his followers and many of his Connacht kinsmen, by the local gentry in 1329. His brother Sir William, who inherited his lands, was accused in 1331 of plotting with the first earl of Desmond and others to divide up Ireland between them and make Desmond king. He was imprisoned with his son Walter (d. 1350) by the new English governor Sir Anthony Lucy, and hanged in 1332. Walter was released, pardoned, and reinstated, becoming chief governor of Ireland (1346–1349), in which capacity he made a last, briefly successful attempt to reestablish the royal authority in Connacht. The direct Tethmoy line ended with his son, another Walter, in 1361. Most of Tethmoy was retaken by the O'Connors, while a collateral line of Berminghams, rejecting royal authority and the claims of the Preston family, heirs of Walter's sister, established an autonomous Gaelicized lordship in the adjacent lands of Carbury (County Kildare), which lasted until 1548. They figure through the fifteenth century alternately as ravaging Meath in company with the O'Connors and as allies of the English against them, while succession to the lordship passed in the Gaelic manner by "tanistry" between several lines.

Richard de Bermingham of Athenry (d. 1322), son of Piers fitz Meilir, was later remembered for his great defeat of the Gaelic Irish of Connacht at Athenry (1316); he was, however, married to a Gaelic wife, the mother of his successor Thomas (d. 1375). During the latter's time the Bermingham lands in Connacht, now concentrated in the hands of the Athenry family, suffered severe losses: Tireragh was recovered by its Gaelic lords, the O'Dowdas, while the O'Kellys—who inflicted a severe defeat on Thomas in 1372—subsequently occupied much of the Athenry territory, reducing the Bermingham lordship to Dunmore and a small area around Athenry. Knockgraffon and Kiltinan, after being held by a junior branch, reverted briefly to Thomas's son, the long-lived Walter (d. 1431), before being sold to the Butlers in 1410. Walter of Athenry, who served as sheriff of Connacht, was knighted by Richard II at Waterford in 1394, but thereafter the family's links with the English administration disappear. On the death in 1473 of Walter's son Thomas, the latter's son and namesake had to contend in turn with two cousins, one of whom—the son of a Richard who had died in 1438—succeeded in ousting him for a year. It is obvious that the lordship of Athenry in spite of being recognized as a peerage dignity by the Crown—perhaps a recognition of the former importance of the Bermingham name—was starting to pass by Gaelic modes of succession. After the death of the younger Thomas's son Meilir *Buidhe* (the yellow-haired) in 1529, the lordship passed to a Thomas whose descent is unclear, and then to Richard (a descendant of the Richard of 1438), who consolidated his position by killing his namesake, Meiler's son, and whose descendants continued as lords of Athenry.

In the thirteenth and fourteenth centuries, many other branches of the lineage existed in Connacht, Meath and Tipperary, many of whom later died out or were reduced to insignificance. In the fifteenth, a member of the Carbury branch acquired by marriage the hereditary chief sergeantship of Meath and lands in that county: his grandson Patrick Bermingham of Corbally (d. 1532) was Chief Justice of the King's Bench in Ireland, a post previously occupied (1474–1489) by a Philip Bermingham, perhaps from a branch that settled at Baldongan in County Dublin.

KENNETH NICHOLLS

References and Further Reading

Robin Frame. *English Lordship in Ireland, 1318–1361* Oxford: Clarendon, 1982.

Nicholls, Kenneth. *Gaelic and Gaelicised Ireland in the Middle Ages.* Dublin: Gill and MacMillan, 1972.

Orpen, Goddard Henry. *Ireland Under the Normans 1216–1333* 4 vols (Oxford University Press, 1911 and 1920).

Perros (Walton), Helen. "Crossing the Shannon Frontier: Connacht and the Anglo-Normans, 1170–1224.' In *Colony and Frontier in Medieval Ireland: Essays Presented to J. F. Lydon,* edited by T. B. Barry, et. al. London: Hambledon, 1995: 117–138.

Sweetman, H. S., ed. *Calendar of Documents Relating to Ireland.* 5 vols. London: Longman, 1875–1886.

Otway-Ruthven, A. J. *A History of Medieval Ireland. 2nd ed.* New York: Routledge Books, 1980.

See also **Anglo-Norman Invasion: Bruce, Edward; Burgh; Butler-Ormond; Factionalism; Gaelic Revival; Gaelicisation; Kildare; Leinster; Ua Conchobair, Failge; Racial and Cultural Conflict; Strongbow.**

BIBLICAL AND CHURCH FATHERS SCHOLARSHIP

Irish activity in biblical studies can properly be said to have begun with St. Patrick in the fifth century. The saint's writings rely heavily on scripture, particularly the Epistles of Paul, which Patrick cites effectively to illustrate his own situation as an exile. Apart from Patrick's writings, the fifth century remains dark, and there is very little literary evidence for most of the sixth. However, towards the beginning of the seventh century, signs reveal that the intensive study of the Bible had been in progress in Ireland for some time. The Old Irish poem *Amra Choluimb Chille*, believed by many to be an early-seventh-century work, credits St. Columba (Colum Cille, d. 597) not only with regular reading of scripture, but also with editing a copy of the Psalms. Also associated with Columba is the *Cathach* (Battler), a manuscript of the Psalms which, according to legend, was carried by the saint even into battle. This manuscript survives as "Dublin, Royal Irish Academy, s.n.," assigned variously to the late sixth or early seventh century. Jonas of Bobbio (seventh century) records that St. Columbanus (d. 615) wrote a commentary on the Psalms in his youth. This work, unfortunately, has not been recovered.

The period from the seventh to the ninth century marks the high point of Irish medieval biblical studies, encompassing not only the copying and glossing of biblical books, but also the writing of scriptural commentaries and at least one work of theology devoted to the Bible. Irish gospel books of the seventh century include the Codex Usserianus Primus (Dublin, Trinity College, MS 55 [A.4.15]), which contains an Old Latin text, and the Book of Durrow, which has a Vulgate text. Of somewhat later date (seventh- and eighth-century) are the Book of Mulling, the Book of Dimma, and Codex Usserianus Secundus (The Garland of Howth). The Irish had a predilection for gospel books, as shown by such famous later productions as the mac Regol (Rushworth) Gospels and the Book of Kells. Only the ninth-century Book of Armagh contains a complete New Testament. Apart from psalters, surviving copies of Old Testament books are rare, though countless citations from it prove that it was very well known.

The great gloss collections belong to the eighth and ninth centuries, though the survival of ancient forms of Irish words shows that glossing in the vernacular began very early. The two most famous collections are the Würzburg glosses on the Epistles of Paul: in Würzburg, Universitätsbibliothek, MS M.p.th.f.12 (dated to the end of the eighth century or beginning of the ninth); and the Milan glosses on the Psalms (in Milan, Biblioteca Ambrosiana, C. 301 inf., saec. VIII/IX, and Turin, Biblioteca Nazionale. F. IV 1 fasc. 5–6, saec. VIII/IX). The Würzburg glosses are in Latin and Irish, and belong to different periods, but taken together they reveal the range of Irish knowledge of patristic biblical commentaries. Pelagius is more heavily cited (though not all attributions are correct) than any other authority, but numerous other fathers are quoted or referred to as well: Origen (in Rufinus's translation), Hilary in the so-called Ambrosiaster commentary, Pseudo-Primasius (Pelagius in the edition of Cassiodorus), Jerome, Augustine, Gregory the Great, and Isidore.

The Milan gloss collection is based upon a Latin version of the Commentary on the Psalms (in Greek) by Theodore of Mopsuestia, who was branded as a heretic in the "Three Chapters Controversy." This commentary has been wrongly identified with the lost commentary on the Psalms by Columbanus. The fact that the commentary survives in two early Irish manuscripts in Irish hands is of considerable interest for the Irish role in the preservation of patristic texts, particularly texts of questionable orthodoxy. However, Theodore's Antiochene exegesis, though known and used in Ireland, never gained advantage over allegory.

While it is true that Irish scholars were intensely engaged in the study of the scriptures, they seem to have been more interested in preserving patristic authorities in florilegia or epitomes than in creating original commentaries themselves. Laidcenn of Clonfert-Mulloe in the seventh century wrote an epitome of Gregory the Great's *Moralia in Iob* (Ethics in Job). Sedulius Scottus in the ninth century compiled two florilegia on biblical texts: one on the epistles of Paul (using primarily Pelagius and Jerome), another on Matthew. The so-called Bibelwerk (*c.* 800) is also in the format of a florilegium. An exception is John Scottus Ériugena, who wrote a truly original commentary on the Gospel of John (as well as a famous homily on the same subject). Another original commentary, assuming it to be Irish and of the seventh century, is of contested authorship. This is the commentary on Mark which has been attributed to a certain Cummean. The work raises a number of theological and ecclesiological questions that were of interest to the Irish in the seventh century: the role of grace versus free will, the idea of a "first grace" (baptism) and a "second grace or mercy" (forgiveness through penance), and the inverse formula of the Eucharist (i.e., Christ's transfiguring himself into bread and wine, as opposed to a change of substance). The commentary on Mark stands out both for its theological interest and the fact that it is a line-by-line exegesis of a biblical text in the tradition of Jerome, Augustine, and Gregory.

Arguably the most interesting and challenging of all early Irish scriptural works of scholarship is the *De mirabilibus sacrae scripturae* (On the Miracles of Holy Scripture), composed in Ireland in 655 (as dated by a computistical formula). The author refers to himself as Augustine and claims to be addressing the monks of Carthage. The work addresses the question of miracles as presented in both Testaments as related to the scriptural claim that God rested after creation. "Augustine" ingeniously explains that there is a distinction between the "opus," which God perfected, and "labor," from which he need not cease. Nor does God's labor interfere with nature. For example, when Lot's wife was turned into a pillar of salt, God did not distort nature, for salt was already present in the human body (in the form of tears, for example). In his preface, "Augustine" pointedly attacks allegorical exegesis and states his preference for the letter of scripture.

Curiously, Pseudo-Augustine's work made only limited use of the real Augustine. The author shows a general familiarity with the great bishop's ideas, but rarely cites him verbatim. Such neglect of Augustinian texts is exhibited in other Irish works and compilations prior to the ninth century. The favored authorities of Irish exegesis were Jerome, Pelagius, Gregory the Great, and Isidore. In the earliest period (the time of Columba and Columbanus), the British saints Gildas and Uinniau were treated with special reverence (though more for their pronouncements on the monastic life than for scriptural exegesis).

The use of the Irish vernacular in scriptural scholarship is at least as old as the eighth century, and probably older. Not only is it prominent in the two great gloss collections mentioned above, it was also employed in what might be called "free-standing" works: the "Old Irish Treatise on the Psalter" (*c.* 800?) and the macaronic "Lambeth Commentary on the Beatitudes" (eighth century?). Irish was also employed by John Scottus Ériugena in his scholia to different books of the Old and New Testaments. The vernacular held a virtual monopoly in Ireland in the tenth and eleventh centuries, when Latin scholarship was in sharp decline. It was the language of numerous new works that found their inspiration as often in biblical apocrypha as in the canonical scriptures. In the ninth century, the Irish on the Continent were noted for their ability to employ Greek in scriptural study. This is shown especially in three manuscripts in Irish hands from the mid-ninth century that contain word-by-word interlinear Latin translations of the Greek texts: St. Gall 48 (Gospels), the Basel Psalter, and the Dresden Pauline Epistles (destroyed in World War II).

Much recent discussion of Irish biblical scholarship has centered on an article by Bernhard Bischoff that attributed a significant number of writings about the scriptures (some commentaries, others not) to Irish scholars active in the early Middle Ages. Bischoff assembled a set of criteria, consisting mostly of verbal formulae and the use of particular writers (especially Pelagius and Virgilius Maro Grammaticus) as indicators of Irish provenance. The validity of these criteria was much debated during the latter half of the twentieth century. However, when such *Merkmale* can be combined with other types of evidence, such as the use of scriptural lemmas of the Irish type, "Hibernian" spellings, or misreadings of established Irish manuscript abbreviations, they gain in validity.

MICHAEL W. HERREN

References and Further Reading

Bischoff, Bernhard. "Irische Schreiber im Karolingerreich." (Irish Authors in the Carolingian Empire.) In *Mittelalterliche Studien: Ausgewählte Aufsätze zur Schriftkunde und Literaturgeschichte* (Medival Studies: Selected Essays on Paleography and Literary History), edited by B. Bischoff, 3:39–54. 3 vols. Stuttgart: Anton Hiersemann, 1966–1981.

———. "Wendepunkte in der Geschichte der lateinischen Exegese im Frühmittelalter." (Turning Points in the History of Latin Exegesis in the Early Middle Ages.) In *Mittelalterliche Studien* (cited above), 1:205–274.

Gorman, Michael. "A Critique of Bischoff's Theory of Irish Exegesis." *The Journal of Medieval Latin* 7 (1997): 178–233.

Herren, Michael W., and Shirley Ann Brown. *Christ in Celtic Christianity: Britain and Ireland from the Fifth to the Tenth Century.* Studies in Celtic History 20. Woodbridge, Suffolk: The Boydell Press, 2002.

Kenney, J. F. *The Sources for the Early History of Ireland. Ecclesiastical: An Introduction and Guide.* Records of Civilization Sources and Studies 11. New York: Columbia University Press, 1929. Revised by L. Bieler, 1966.

Lapidge, Michael and Richard Sharpe. *A Bibliography of Celtic-Latin Literature, 400–1200.* Dublin: Royal Irish Academy, 1985.

McNamara, Martin. "Psalter Text and Psalter Study in the Early Irish Church." *Proceedings of the Royal Irish Academy* 73C, no. 7 (1973): 201–298.

McNamara, Martin. "The Irish Tradition of Biblical Exegesis, A.D. 550–800. In *Iohannes Scottus Eriugena: The Bible and Hermeneutics*, edited by Gerd Van Riel, et al., 25–54 Leuven, Belgium: University Press, 1996.

Wright, Charles D. "Bischoff's Theory of Irish Exegesis and the Genesis Commentary in Munich clm 6302: A Critique of a Critique." *The Journal of Medieval Latin* 10 (2000): 115–175.

See also **Colum Cille; Columbanus; Devotional Literature; Ériugena John Scottus; Glosses; Moral and Religious Instruction; Sedulius Scottus**

BISHOPS

See **Ecclesiastical Organization**

BLACK DEATH

Ireland, like most of Western Europe, suffered from the bubonic plague, or the "Black Death," in the years from 1348 to 1350. Unlike its nearest neighbor, England, the surviving contemporary sources for this catastrophic event are very limited. Even archaeological evidence is meager. Therefore, in order to understand the impact of this event on Ireland, we are forced to rely on parallel studies in other European countries that are better chronicled in the Middle Ages. However, even this assumption may not be wholly tenable in the light of recent research into the complexities of the pattern of medieval Irish settlement. Given the ease at which this disease spread among the population in Europe generally, Ireland's almost unique rural settlement pattern may have affected the plague's incidence to a greater extent than can be accurately gauged.

All that is known for certain is that the Black Death probably arrived in Ireland at the ports of Howth and Drogheda, both located north of Dublin, in August 1348, and spread to the capital, Dublin, soon afterwards. The Franciscan friar John Clyn in

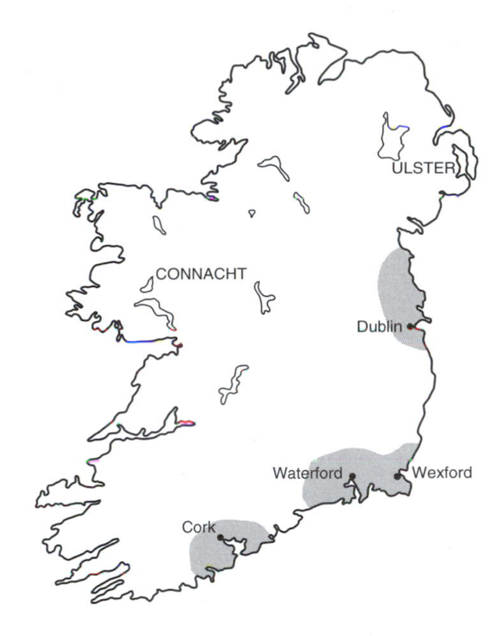

The Black Death in 1348.

his contemporary annals was able to accurately chart the progress of the disease within the houses of his Order, first in Dublin and then in Kilkenny. He also stated that the total number of citizens of Dublin who succumbed was 14,000, a very high number (and doubltless an exaggeration) out of a possible total urban population of not much more than 20,000 at that time.

The several surviving Gaelic-Irish chronicles all record that the plague had reached the west of Ireland by 1349, and seem therefore to be derived from the same original source, and this has led some scholars to suggest that the more dispersed Gaelic-Irish population in the west and northwest may have largely escaped from the pestilence. This, however, is predicated upon two more assumptions: firstly that the disease was, in fact, bubonic plague and secondly that all the Anglo-Irish population were living in nucleated settlements. There are some epistemological indicators of this plague that have suggested to researchers such as Twigg that the Black Death may have been an outbreak of anthrax, although most academic opinion does not support his theory. Secondly, in this period a significant proportion of the rural Anglo-Irish population within the colony was also living in dispersed settlements, such as the moated houses of isolated manors, and may therefore have escaped the worst effects of the plague.

Nevertheless, all the indications are that the urban population, which was crowded together in largely unsanitary conditions, suffered most from this disease, as happened elsewhere in Europe. The total population loss caused by this worst outbreak of plague in the Middle Ages would probably have been somewhere in the region of 25 percent to 33 precent, if contemporary European mortality models are taken into account. That it ceased to spread after about two years is probably as a result of the eventual death of all the carriers, infected rats, as well as the building-up of some kind of natural immunity over this period by the healthiest members of the Irish population. One should also note that although the Black Death as such ended at that point, the recurrence of plague was a regular phenomenon in succeeding generations.

TERRY BARRY

References and Further Reading

Butler, Richard, ed. *The Annals of Ireland by Friar John Clyn and Thady Dowling*. Dublin: Irish Archaeological Society, 1849.

Freeman, A. Martin, ed. *The Annals of Connacht (1224–1544)*. Dublin: Government Publication Office, 1944.

Kelly, Maria. *A History of the Black Death in Ireland*. Stroud, England: Tempus, 2001.

Twigg, Graham. *The Black Death: A Biological Reappraisal*. London: Batsford, 1984.

Ziegler, Philip. *The Black Death*. London: Collins, 1969.

See also **Agriculture; Annals and Chronicles; Clyn, Friar; Diet and Food; Dublin; Famine and Hunger; Medicine; Pilgrims; Population; FitzRalph, Richard; Villages, Walled Towns**

BLATHMAC (*fl.* EIGHTH CENTURY)

Blathmac was the son of Cú Brettan mac Congusso (d. 740), who was perhaps king of the Fir Roiss, a sept of the Airgialla, located in modern counties Louth and Monaghan. In the eighth-century saga of the battle of Allen (718), his father is represented as a combatant and ally of the king of Tara, Fergal mac Máele-dúin. He was a poet and author of devotional poems on the Passion of Christ and the Virgin Mary, which are found uniquely in MS G 50 in the National Library of Ireland. They apparently do not survive in their entirety: the first, *Tair cucom, a Maire bóid*, now has 149 stanzas, originally perhaps 150; the second, *A Maire, grian ar clainde,* is a continuation of the first. There may have been a third, thus comprising a triptych of 3 by 150. The language of the poems is Old Irish, of a form contemporary with the eighth-century glosses, and they were therefore probably composed between 750 and 770.

Blathmac's poetry shows familiarity with Scripture, patristic literature, and biblical apocrypha. It is especially interesting that it draws upon the apocryphal *Acts of Thomas* as a source for some incidents in the life of Christ. His praise of Mary is framed within a description of Christ's life, miracles, teaching, and passion. His metaphors and motifs are drawn from contemporary Irish social, legal, and religious institutions, and his poems therefore give us a valuable insight into the Zeitgeist of the period.

AIDAN BREEN

References and Further Reading

Byrne, F. J. *Irish Kings and High-Kings*. London: Batsford, 1973.

Carney, J. "Old Ireland and Her Poetry." In *Old Ireland*, edited by Robert McNally, 147–152. Dublin: Gill, 1965.

———. *The Poems of Blathmac Son of Cú Brettan, Together with the Irish Gospel of Thomas and a Poem on the Virgin Mary.* Dublin: Irish Texts Society, Educational Company of Ireland, 1966.

Good, J. "The Mariology of the Blathmac Poem." *Irish Ecclesiastical Record* 104 (1964): 1–7.

Lambkin, B. "The Structure of the Blathmac Poems." *Studia Celtica* 20–21 (1985–1986): 67–77.

———. "Blathmac and the Céile Dé: A Reappraisal." *Celtica* 23 (2000): 132–154.

O'Dwyer, P. *Devotion to Mary in Ireland, 700–1100*. Dublin: Editions Tailliura, 1976.

See also **Airgialla; Áes dána; Biblical and Church Fathers Influence; Languages; Poetry, Irish; Poets, Men of Learning**

BREHON LAW

"Brehon law" is a term used to describe the native Irish legal system. This system operated in Gaelic Ireland until the early seventeenth century. The phrase "Brehon law" comes from the Irish word for a "judge," which was Anglicized as "brehon." The Irish themselves generally referred to their law as *fénechas* (Irish jurisprudence). The term "Brehon law" is nevertheless an apt one. Irish law was "judge-made" law; its texts distill the legal rules and remedies developed over the centuries by highly trained professional jurists. It was an "organic" system that reflected the complexities of Irish society. This explains its richness and sophistication. Had lawmaking been the preserve of Ireland's many petty kings, it is likely that it would have been a more rudimentary affair, focused primarily on coercive rules and the accumulation of state revenues. King-made law would also have been fragmented and transitory: Ireland was made up of numerous petty kingdoms, arranged in turbulent alliances. However, Brehon law was the product of a learned class which transcended political boundaries. As a result, Brehon law was "national," in the sense that it was a cultural phenomenon of Ireland as a whole, with few (if indeed any) discernible regional variations.

This is not to say that the petty kings had no role in Irish law. As heads of their respective kingdoms they presided over their royal courts. But the king most likely pronounced the judgment recommended to him by his *brehon*. Kings were also empowered to pass emergency regulations in times of war and pestilence. These edicts were probably as narrowly focused as they were temporary, and no fragment of any of them survives.

On the other hand, a vast treasury of judge's law survives. The principal monument is the *Senchas Már*, "The Great Collection of Traditional Learning." This consisted of about fifty separate texts. Twenty-one of these survive more or less intact, and fragments of most of the others remain. Most of the texts deal with a discrete topic of law. For example, the first text in the *Senchas Már* is a tract "On the Four Divisions of Distraint" (*Di Chetharshlicht Athgabála*). Distraint was a process by which parties could force their opponents to court by impounding their cattle.

Specific Topics of Law

Relatively little is known about Irish court procedure. Notable features are the use of trained advocates, the prominence given to the evidence of eyewitnesses, and the right to appeal if a judge had made an error of law. There was no jury. Formal oaths setting out an allegation or denial had to be supported by a fixed number of "oath-helpers" of good reputation. If a party could not prove their case by bringing eyewitnesses, they could resort to an ordeal, such as casting lots, "trial by battle," or the ordeal of the cauldron. (The latter involved plunging one's hand into boiling water. The hands of the truthful were assumed to heal promptly.)

A person's legal status was dependent on his or her degree of wealth or professional training, and each grade in society had its own "honor-price." It is sometimes claimed that the status of women in early Irish law was considerably more advanced than that in comparable medieval cultures. But that claim finds little, if any, factual support in the laws. Most legal rights were dependent on the ownership of property, and most property was owned by men. Land was passed down through kin-groups which were agnatic, that is, reckoned through the male line. A man's land was inherited in equal shares by his sons. Only when a man had no sons would his land pass to his daughters, and then only for their lifetimes. Upon their deaths, the land was redistributed among their father's male relations.

Acts of violence were generally settled by a payment of compensation known as an *éric* fine. If a free person was murdered, the *éric* was equal to 21 cows, regardless of the victim's rank in society. In addition, each member of the victim's agnatic kin received a payment based on their own honor-price. There were separate payments for the kin-group of the victim's mother, and for the victim's foster-kin. (Many Irish children were brought up by foster-parents.) If this compensation was not paid, the members of any of these three separate kin-groupings could take vengeance against the offender's kin.

In cases of injury, a number of different fines were paid to the victim. The first component was again the *éric* fine, which varied with the nature of the injury. Injuries involving scarring or permanent damage incurred additional payments. On top of this, the injured person was entitled to a set fraction of their honor-price. Fines were halved if the injury was the result of mere negligence. The fact that Brehon law recognized a distinction between accident, negligence, and deliberate harm is a notable aspect of its sophistication.

Persons seriously injured through negligence were entitled to "sick-maintenance." In the early texts, the key feature of sick-maintenance was the *dingabáil* (removal) of the injured person from their home. They were taken away to be nursed and cared for in a suitable residence. They were entitled to be accompanied by a retinue appropriate to their status in society, and food of a defined standard had to be provided for the whole party. In addition, the offender had to pay for the fees charged by the physician. (Those fees are set out in the text *Bretha Déin Chécht*, The Judgments of Dian Cécht.) The offender also had to provide a substitute to perform the injured person's duties during their period of convalescence. In the case of intentional injuries (and in the case of some high-status persons), removal on sick-maintenance was replaced by a payment which varied with the status of the victim. In time this payment became the norm in cases of negligence as well, and the older institution of physical removal faded away.

After kinship, the most important legal institution appears to have been clientship. Members of the free farming classes would become the clients of noblemen, in return for grants of cattle. (These farmers generally farmed their own land.) In return, they owed their lords annual payments of food and fixed amounts of labor. Clientship agreements lasted for the life of the lord. When the client died, his heirs carried on the clientship agreement until the lord's death. Between nobles, too, there was a form of clientship which established hierarchies of homage and political support.

Contracts were supported by witnesses, whose job it was to remember the terms of the contracts, and by sureties who undertook to ensure those terms were fulfilled. Irish contract law required anyone selling property to disclose any hidden defects they knew about, and it gave both parties until sunset to back out of the contract. Contracts were generally unenforceable if made while drunk, or by people without full legal capacity.

The law of marriage is set out in a text called *Cáin Lánamna* (The Law of Close Relationships). The normal form of marriage was one between social equals following the betrothal of the woman by her kin in exchange for the payment of a bride-price. Wealthy men might also have a secondary wife, usually one from a lower class. *Cáin Lánamna* also gives details of a range of more-casual sexual relationships, including those where a man merely visits a mistress on a regular basis. Divorce was readily available.

The Authors of the Law Texts

The compilers who produced the *Senchas Már* claimed that its texts had been produced in the fifth century. Those texts were supposedly written by a legendary sage (Dubthach moccu Lugair), working under the supervision of St. Patrick. This story is now recognized as a typically anachronistic piece of medieval propaganda. Linguistic analysis has shown that the *Senchas Már* texts were originally composed between 650 and 750 C.E. On the other hand, it is quite likely that they were produced by eminent clerics working with members of the traditional learned classes (or indeed by clerics who had themselves been trained in Brehon law). We know, for instance, that some other law texts were the results of collaborative efforts by clerics, poets, and trained lawyers. This is true, for example, of *Bretha Nemed toísech* (the first collection of the Judgments of Privileged Persons), which was composed between 721 and 742 C.E. It is also most likely true in the case of *Cáin Fhuithirbe* (The Law of Fuithirbe), composed circa 680 C.E.

The important role played by the church in recording Brehon law has only been recognized relatively recently. For much of the twentieth century it was believed that the law texts were produced in traditional law schools. It was assumed that these schools had, somewhat reluctantly, adopted writing from the monks and had finally made written records of texts that had existed for some time as oral compositions. This view stressed the supposedly "archaic" nature of the texts, and valued them chiefly as repositories for long-obsolete elements of Indo-European law. The legal endorsement of polygamy was seen as evidence that they were not produced by monasteries.

However, by the seventh century the church had good reason to compile an authoritative account of secular law. It had managed to secure for itself a prominent place in the status-based hierarchy of Irish society, as well as significant property rights. The Church also had the appropriate facilities, in the form of scriptoria, as well as the necessary financial resources for the job. The role of the church in recording Brehon law was not, however, a legislative one; the church was not "creating" law in a vacuum. The law that was recorded was that of the existing secular society, and it shows all the hallmarks of a long development. So it is that the Brehon law tracts provide for murder to be "bought off" by the payment of an *éric* fine and sanction polygamy, despite the fact that these were institutions that the church considered to be less than ideal.

The church did, indeed, attempt to modify the Brehon law by legislation. But it did this primarily by harnessing the emergency power of the kings to make temporary laws. The church promulgated a number of special edicts which were guaranteed by local kings. These edicts include *Cáin Domnaig* (which introduced fines for breach of Sunday observance) and *Cáin Adomnáin* (the Law of Adomnán), also known as the *Lex Innocentum* (Law of the Innocents). *Cáin Adomnáin* has the rather undeserved reputation of introducing laws to protect women. In fact, it introduced a new layer of fines in the case of preexisting offences. These new fines went to the church rather than to the victim. Most, and probably all, of these promulgated church laws were concerned primarily with collecting revenue for the church. Very few of them survive. Only the two mentioned here are intact, and there is no copy of either among the Brehon law manuscripts. They were, however, well known to the lawyers and are referred to quite often in the Brehon law commentaries.

The Brehon law-texts also show the influence of the Irish poets, with many texts containing sections of highly alliterative prose and a good number containing significant portions of poetry. This poetic aspect was, until recently, considered to be good evidence for the "oral origin" of many of the law texts. But what it shows rather is that the texts are the polished product of a concerted effort by members of the learned elite.

This elite class produced authoritative law texts covering every area of law in the seventh and eighth centuries. Only one or two new texts were produced in the centuries that followed. The texts were functional works, used for studying the law and, no doubt, in preparing litigation. Many of the texts were later provided with an apparatus of learned glosses, which explain the terms of the main text. Lengthy "commentaries" on similar topics were also added as the law changed, and indeed as the Irish language itself changed. These glosses and commentaries date mainly from the twelfth to the sixteenth centuries. The surviving legal manuscripts also date from this period.

NEIL MCLEOD

References and Further Reading

Binchy, Daniel, ed. *Studies in Early Irish Law.* Dublin: Dublin Institute for Advanced Studies, 1936.

Breatnach, Liam. *Uraicecht na Ríar: The Poetic Grades in Early Irish Law.* Dublin: Dublin Institute for Advanced Studies, 1987.

———. "Lawyers in Early Ireland." In *Brehons, Serjeants & Attorneys*, edited by D. Hogan and W. N. Osborough, 1–13. Dublin: Irish Academic Press, 1990.

———. "Law." In *Progress in Medieval Irish Studies*, edited by K. McCone and K. Simms, 107–121. Maynooth, Ireland: The Department of Old Irish, Saint Patrick's College, 1996.

Hancock, W. N., et al., eds. *Ancient Laws of Ireland.* Vols 1–6. Dublin: Her Majesty's Stationery Office, 1865–1901.

Kelly, Fergus. *A Guide to Early Irish Law.* Dublin: Dublin Institute for Advanced Studies, 1988.

———. *Early Irish Farming.* Dublin: Dublin Institute for Advanced Studies, 1997.

———, and Charles-Edwards, Thomas. *Bechbretha. An Old-Irish Law-Tract on Bee-Keeping.* Dublin: Dublin Institute for Advanced Studies, 1983.

McLeod, Neil. *Early Irish Contract Law.* Sydney: University of Sydney, Centre for Celtic Studies,1992.

Patterson, Nerys. *Cattle Lords & Clansmen: The Social Structure of Early Ireland.* Notre Dame: University of Notre Dame Press, 1994.

Stacey, Robyn Chapman. *The Road to Judgment: From Custom to Court in Medieval Ireland and Wales.* Philadelphia: University of Pennsylvania Press, 1994.

See also **Canon Law; Common Law; Law Schools, Learned Families; Law Texts; March Law**

BRIAN BORU (926[?]–1014)

Brian Boru was arguably the most famous medieval Irish king, due to his achievement in becoming the undisputed king of Ireland and his death by the Norsemen at Clontarf in 1014. Later tradition turned him into the first true high king of the island and a heroic fighter for Ireland's freedom against the oppression of the heathen Vikings. Historians of the modern era have regarded him as an upstart from Munster who broke into the domination that the kings of Tara had enjoyed over Ireland for centuries. More true to the facts, Brian played a pivotal role in the transformation of the Irish political landscape in the tenth and eleventh centuries.

The Age of Brian Boru.

Career

According to many Irish annals, Brian was in the eighty-eighth year of his life when he was slain in 1014, and thus was born in 926 or 927. His birth is also recorded retrospectively in 923 or 942. His mother was Bé Bind, the daughter of Aurchad (d. 945), king of West Connacht. He may have been called Brian "Bóruma" from the territory of Bóruma near Killaloe in Thomond, in the heartlands of the Dál Cais. His epithet is also rendered "Bóraime," meaning "of the cattle-tribute," but this is probably a later interpretation. Brian was one of the twelve sons of Cennétig mac Lorcáin of the Dál Cais, who died as king of Thomond in 951. The Dál Cais profited from the weakness of the divided Eóganachta, especially after the death of Cellachán Caisil, king of Cashel, in 954. Afterwards, the kingship of Cashel was occupied by lesser men whose careers were cut short by violent death. This situation gave Mathgamain mac Cennétig the opportunity to extend his domination to the south. According to *Cogad Gáedel re Gallaib* (The War of the Irish against the Foreigners), a text which dates from the reign of Brian's great-grandson Muirchertach Ua Briain, he set up his camp near Cashel in 964. Mathgamain wanted to become king of Cashel in order to free Munster from its cruel Viking occupation. But stories about the subsequent liberation of Munster and the claim that the kingship of Cashel was the ancient birthright of the Dál Cais are simply propaganda to legitimize Mathgamain's coup. Contemporary annals recognize him as king of Cashel when he and his allies attacked Limerick in 967. In the years afterwards he subjugated his rivals for the kingship of Munster, whom he subsequently enlisted as his supporters. Mathgamain was treacherously killed by such new allies in 976, but within two years, the kings responsible were defeated and slain by his brother Brian. As the new king of Munster, Brian first consolidated his position at home, before starting a series of campaigns to obtain the hostages of the kings of Osraige, Leinster, and Connacht. This ensured him of the hostility of Máel-Sechnaill II, the new king of Tara, who retaliated by plundering Leinster and Connacht. The ruling dynasties of the two provinces had long been traditional allies of Clann Cholmáin of Mide (Meath). A period of more than fifteen years followed in which both kings tried to gain the upper hand in the two provinces, while occasionally raiding each other's territories. A direct confrontation was either avoided or did not give one side a decisive victory. In the long run Brian's tactics, stamina, and diplomacy paid off. He maintained a firm grip on the Munster kings, built a number of fortresses to defend his home territory, launched several campaigns at the same time, employed the Norse fleets of Limerick and Waterford along the Shannon and against Dublin, and turned former enemies into supportive allies.

Of Brian's sons, Murchad is most often mentioned as an army-leader in his service. The annals state that he was in the sixty-third year of his life when he was slain in 1014. Murchad was the son of Mór, who was the daughter of Eidin (d. 906), king of the Uí Fhiachrach Aidni of southern Connacht; Murchad also fathered Conchobar and Flann. Brian's other sons were Domnall (d. 1010 or 1011), who was the son of either Dub Coblach (d. 1009), daughter of Cathal (d. 1010), king of Connacht, or of the daughter of Carlus, king of Uí Áeda Odba in Mide, who is also recorded as the mother of Tadg (d. 1023). Brian was also married to Gormfhlaith (d. 1030), daughter of Murchad (d. 972), king of Leinster, and mother of Donnchad (d. 1065). Since Donnchad was an adult in 1014, this last relationship dates from before 997, when Brian and Máel-Sechnaill came to terms at a meeting near Clonfert. On this occasion they divided Ireland into two spheres of influence according to an old scheme: the north (Leth Cuinn) was given to Máel-Sechnaill, the south (Leth Moga) to Brian. Brian exchanged his hostages of Connacht with those of Leinster and Dublin which had been in Máel-Sechnaill's possession. Nominally, Brian was now overlord of Dublin, a major prize if he could tap its resources. Hence both kings made an expedition "and took the hostages of the foreigners to ensure good behaviour towards the Irish," as one annalist states. Yet Brian had to reckon with Máel Mórda, king of Uí Fáeláin in Leinster, and Sitriuc Silkenbeard, king of Dublin, who were, respectively, the brother and son of Brian's wife Gormfhlaith. Both had a long-standing row with Brian's allies, the Uí Dúnchada in Leinster and the Norsemen of Waterford. When they openly defied his overlordship, Brian gathered his forces, and routed them in the battle of Glenn Máma in 999. Dublin was plundered, and Sitriuc fled, but he found no asylum in the north. Upon his return he gave his submission, and it may be on this occasion that he married Brian's daughter Sláine. Dublin was now in Brian's hands, and this tilted the balance of power in his favor. In 1002, Brian managed to take the hostages of the men of Connacht and Mide after Máel-Sechnaill's pleas for help to the northern Uí Néill had been rebuffed. When the kings of Ailech and Ulaid slew each other in battle in 1004, Brian, accompanied by most of the Irish royalty and their hostages, brought an army to Armagh the next year. He left twenty ounces of gold on the altar of St. Patrick, and had his secretary add to the Book of Armagh a note in which he is proclaimed as *imperator Scottorum* (emperor of the Irish). This can be regarded as a claim that he ruled both the Irish and the Norse in Ireland, and may even imply suzerainty over the Gaels of Scotland, some of whom fought on his side. In 1006, Brian took his forces on a circuit through the territories of the northern Uí Néill and the Ulaid, acting as a lord would when visiting his clients. But his overlordship was not recognized, and it would take several other campaigns in 1010 and 1011 before Brian secured the hostages of all Leth Cuinn. Thus Brian achieved what no Munster king and few kings of Tara had been able to do, obtaining the submission of all the Irish overkings and Viking kings. It is symptomatic of the political relationships between the Irish kings that his success was shortlived.

Clontarf

In 1012, Flaithbertach ua Néill, king of Ailech and Brian's son-in-law, started to reassert his position as overking of the northern part of Ireland. The next year the Laigin and Dublin Norse revolted, and neither Brian nor Máel-Sechnaill was able to quell them at once. According to both Irish and Old-Icelandic saga-literature, Gormfhlaith played a decisive role in stirring her brother Mael Mórda to revolt, and in enlisting the support of the leader of the Vikings of the Irish Sea and the Orkneys. In 1014, Brian and Máel-Sechnaill raised camp near Dublin, accompanied only by the forces of Munster, southern Connacht, and Mide. Máel-Sechnaill retreated just before battle at Clontarf was joined, and Brian's forces merely won a Pyrrhic victory. Brian, Murchad, and his son

Tairrdelbach were slain, as well as many other Munster leaders. Dublin remained untouched, for after the battle young Donnchad led the Munster forces back home. Almost immediately strife broke out between the various contigents over the kingship of Munster. It set the pace for future struggles, which would keep the kings of that province occupied until the time of Tairrdelbach ua Briain. In later tradition, Brian and Murchad became the paragons of good kingship and bravery. The lists of those who were present at Clontarf swelled as allies of the Uí Briain wanted to include their forefathers among those present at the legendary battle. Brian, Murchad, and Clontarf hence entered the world of saga-literature and fiction, and ultimately became part of the "national" struggle of the Irish against foreign foes.

Achievements

For a brief period, Brian could by right claim to be the undisputed king of Ireland. Nevertheless, his domination was based on the usual terms by which an over king obtained the submission of other kings. Brian did not found a new institution or create a national monarchy of sorts, but he dealt a fatal blow to the kingship of Tara. Its wane in the tenth century, the fragmentation of the Uí Néill, and the weakness of the Eóganacht, paved the way for more vigorous dynasties on their fringes, such as the Dál Cais of Thomond and Uí Briúin Bréifne of eastern Connacht. Brian's rise accelerated the process in which new alliances were forged and the political map of Ireland was reshuffled. Additionally, it clearly showed that any able king could dominate large parts of Ireland. But it also underlined that it remained difficult to establish a lasting ascendancy even in one's own lifetime. Irish political relations remained largely personal and temporary, and Brian's power mainly rested on security at home, enlisting allies and former enemies in his campaigns, and the wearing-down of those who resisted his ambitions. This went hand in hand with the exploitation of the Norse ports for their economical and military resources, a strategical deployment of fleets, and unceasing campaigning. He exemplifies the development of a more "total" form of warfare, which gradually replaced seasonal campaigning and decisive pitched battles. Brian favored a prudent and careful approach in his actions. Ironically, the rare occasion that he engaged in a full-scale battle was to be his undoing, and it took the Dál Cais more than a generation to recuperate from their losses. Brian's career seems to belie his reputation as the one who established law and order in Ireland, so that women could travel alone without being harassed. There are also not many indications that he particularly stimulated learning and scholarship—although sources are slim on this topic. If anything in this realm, he stimulated the occupation of ecclesiastical centers in Munster by his relatives. For example, Brian's brother Marcán was superior of Terryglass, Inis Celtra, and Killaloe at his death in 1010. Brian was also keen to stay on good terms with Armagh. He recognized its supreme position in Ireland, and granted immunity to the churches of Patrick in 1012. It was also to Armagh that the dead king was taken, where the community of Patrick waked at his body for twelve nights in his honor. Brian had not been able to create a lasting overkingship in Ireland, but he established the domination of Munster by his descendants and relatives. The Ó Briain family would continue to rule Thomond for centuries thereafter.

Bart Jaski

References and Further Reading

MacShamhráin, Ailbhe. "The Battle of Glenn Máma, Dublin and the High-Kingship of Ireland: A Millennial Commemoration." In *Medieval Dublin II*, edited by Seán Duffy, 53–64. Dublin: Four Courts Press, 2001.

Newman, Roger Chatterton. *Brian Boru: King of Ireland*. Dublin: Anvil Books, 1983.

Ryan, John. "Brian Boruma, King of Ireland." In *North Munster Studies*, edited by Etienne Rynne, 355–374. Limerick: Thomond Archaeological Society, 1967.

See also **Armagh, Book of; Armies; Battle of Clontarf; Clientship; Dál Cais; Gormfhlaith; Kings and Kingship; Naval Warfare; Máel-Sechnaill; Muirchertach Mac Liacc; Ua Briain; Viking Incursions**

BRIDGES

There is no evidence for bridges in prehistoric Ireland. Fords were used as crossings and the earliest structures appear to have been interrupted causeways, built of stones and punctuated by gaps enabling the water to flow through. An example survives at Skeagh on the river Shannon between counties Leitrim and Roscommon, and it is probable that the causeways (*tóchair*) built at Athleague, Athlone, and Dublin in 1001 by Máel-Sechnaill II were of this form.

The earliest documentary evidence for bridge building occurs in Cogitosus's *Life of Brigit* (*c.* 650), which makes it clear that it was a prerogative of kings and the responsibility of the local community. The oldest known bridge, dendrochronologically dated to 804, spanned the river Shannon at Clonmacnoise. It was over 500 feet long, 10–12 feet across, and consisted of two parallel rows of oak trunks, set 16–20 feet apart, hammered into individual base plates of beams and planks, and driven to a depth of ten feet into the river clays. The form of the superstructure is unknown, but planks, post-and-wattle, and poles are mentioned in the early sources. A somewhat later timber plank bridge, probably of thirteenth-century date, is known from the river Cashen, County Kerry. It was carried on trestles fitted into sole plates that had been pegged into the riverbed. It had a span of 600 feet. The earliest stone bridges were of clapper form. A probable pre-twelfth-century example on the river Camoge at Knockainey, County Limerick, survived until the 1930s.

In the eleventh century, major bridges were built across the Shannon at Athlone, Athleague, and Killaloe. These were constructed not just to facilitate trade and communication, but also to permit the rapid deployment of armies. At least four successive bridges were built at Athlone in the course of the twelfth century. All follow the same pattern, built by the kings of Connacht to give them easy access into midland Ireland, and destroyed as quickly as possible by the kings of Mide to prevent such incursions. In Anglo-Norman Ireland the responsibility for maintaining bridges rested with the local community, which could rarely afford the costs involved. Accordingly, from the early thirteenth century, "pontage" grants were given to communities permitting them to levy tolls on commodities brought into the town for sale. The monies so collected were spent on building and maintaining the bridge. In the fourteenth century important bridges were built at Kilcullen, County Kildare (1319) and Leighlinbridge, County Carlow (1320). These had the effect of shifting settlements from the older ecclesiastical sites down to the bridging points, where they have remained to this day. Surviving medieval bridges, such as Adare, Askeaton, Slane, Trim, and Babes Bridge, County Meath, are characterized by pointed segmental arches, a width of between six and nineteen feet, arch spans of about twenty feet, prominent cutwaters, and high parapets that were sometimes battlemented. Ancillary structures such as gatehouses, water gates and slips were common in towns while chapels (Dublin) and houses (Baal's Bridge, Limerick, and Irishtown Bridge, Kilkenny) were also constructed on bridges.

John Bradley

References and Further Reading

O'Keeffe, Peter and Tom Simington. *Irish Stone Bridges: History and Heritage*. Dublin: Irish Academic Press, 1991.

O'Sullivan, Aidan. "The Clonmacnoise Bridge: An Early Medieval River Crossing in County Offaly." Archaeology Ireland Heritage Guide 11. Bray: Archaeology Ireland, 2000.

See also **Architecture**

BRIGIT (*c.* 452–*c.* 528)

The founder and patron saint of the monastery of Kildare, St. Brigit (also Brigid, Bríd, Bride, Bridget) is renowned as one of the three pillars of the early Christian Church in Ireland, along with Patrick and Colum Cille. According to later medieval tradition, her remains were buried with theirs at Downpatrick. She was also the patron saint of the Leinstermen and was said to protect them in battle. Her feast day is February 1.

No historical facts regarding Brigit and her works can be determined with any certainty; her very existence has been a matter of debate. All that is known about her is based on tradition, legend, and folklore, but a considerable number of documents relating to her have survived. These documents are among the earliest known hagiographical material in Ireland and include two extant Lives in Latin which date from the seventh century; a hymn to Brigit, attributed to Ultán of Árd-mBreccáin, may also date from the seventh century. Among the other documents are two subsequent Latin Lives of uncertain dates; a fragment of a Life in Old Irish, from around the late eighth or early ninth centuries; a Latin Life composed by Lawrence of Durham in the twelfth century; and a homiletic Life in Middle Irish contained in the Book of Lismore. Later hymns to Brigit also survive, and she appears prominently in the martyrologies.

Brigit's traditional genealogy makes her a member of a prominent family of the Fothairt; she was supposedly born at Faughart, near Dundalk, in County Louth. The author of one of her seventh-century Lives, Cogitosus, a monk of Kildare, relates that she was born to noble Christian parents; her father was Dubthach and her mother, Broicsech. Cogitosus describes the preeminence of the monastery in Ireland, as a community for both men and women and as an episcopal see ruled jointly by the abbess, Brigit, and her chosen bishop, Conláed (Conleth). Their tombs, according to Cogitosus, are placed on either side of the main altar in the church. Despite these details, Cogitosus's Life consists mostly of a series of miracle stories based on the traditions of the community at Kildare: Brigit tames wild animals, controls the weather, miraculously provides food, and even hangs her wet cloak on a sunbeam. The other seventh-century Life, an anonymous work known as the *Vita Prima* (because it is the first of Brigit's Lives in the *Acta Sanctorum* compiled by the Bollandists), uses the same sources as Cogitosus but contains a higher incidence of folkloristic material. In this Life, Brigit is the daughter of a nobleman, Dubthach, and a slavewoman whom Dubthach sells to a druid at his wife's urging. The slavewoman is set to work in the dairy, where she gives birth to Brigit on the threshold at dawn. This birth legend persisted in Brigit's tradition, as did her association with dairying and cattle. As an infant, Brigit refused to eat the druid's food; she would eat only the milk of a white, red-eared cow milked by a holy virgin. As a young girl, she too worked in the dairy and produced vast quantities of butter and cheese. Later, as abbess of Kildare, she entertained a group of bishops for whom she milked her cows three times in one day. In modern iconography, Brigit is often depicted with a cow.

Brigit was renowned for her charity and her hospitality. As a child, she gave away so much of her father's goods that he tried to sell her but was prevented by the local king, who was impressed by the girl's piety and virtue. As an abbess, she continually gave to the poor, even giving away the bishop's vestments; owing to her sanctity, she received perfect substitutes just in time for the celebration of the mass. At Easter, she miraculously provided ale for all of her churches from a small amount of malt. A poem attributed to Brigit ("St. Brigit's Alefeast"), from no later than the ninth century, expresses her desire to provide a lake of ale for Christ.

Brigit's hagiographers present her as a powerful and influential leader in both the ecclesiastical and secular communities. She receives bishops, including her contemporary, St. Patrick, and negotiates with local rulers. In the Old Irish Life, the anonymous author relates how, at Brigit's consecration as a nun, the presiding bishop mistakenly read over her the orders of a bishop instead. This incident has led to speculation that Brigit was a female bishop, but this idea cannot be supported; in the same text, Brigit must call upon her priest to perform some necessary sacerdotal functions. The abbess of Kildare, however, did hold a high status within the early Irish church, which may have included the honors and privileges held by a bishop, but historically the bishop of Kildare performed all the requisite episcopal functions.

St. Brigit is often associated with a pagan Irish goddess, also named Brigit, whose own traditions have influenced the saint's. The goddess Brigit appears to be the same as the pan-Celtic deity Brigantia, the tutelary goddess of the Brigantes. In Irish mythology, she was the daughter of the great god, the Dagda, and was the patron of smithying, healing, and poetry; she was also identified with a fire cult. A tenth-century text, Cormac's Glossary (*Sanas Cormaic*) calls the goddess whom the poets (the *filid*) worshipped; she had two sisters, also named Brigit, and from these all goddesses in Ireland were named Brigit. Other sources make her the wife of Bres, a mythological king; when their son, Rúadan, is killed, Brigit reportedly keened the first lament heard in Ireland. Giraldus Cambrensis (Gerald of Wales), in his *Topographia Hiberniae* (The Topography of Ireland, *c.* 1185), recounts that nineteen nuns at Kildare, each in turn, watched over a perpetual fire in St. Brigit's honor; on the twentieth night, the nuns left the fire to St. Brigit to tend. This fire never produced any ash and was kept within an enclosure that no man was permitted to enter. St. Brigit's feast day coincides with the pagan Celtic festival of Imbolc, a fertility celebration and one of the four great festivals of the Celtic year. St. Brigit, too, in her tradition is revered as the patron of smiths, healers, and poets. Based on these associations, some have considered the saint to be a euhemerized and Christianized version of the goddess, but the strict relationship is inconclusive. A revival of the cult of the goddess in the twentieth century generated further speculation regarding the saint; however, although Brigit the saint has many of the same attributes as the goddess Brigit, her overall tradition is within a Christian milieu.

St. Brigit became closely associated with the Virgin Mary. The renowned bishop Ibor, as related in the Old Irish Life, saw Brigit appear in a dream as Mary and prophesied her arrival. Her Middle Irish Life celebrates her as the "Queen of the South, the Mary of the Gael." In a later Scottish tradition, Brigit appears as the midwife of Christ.

Brigit's cult spread into Scotland and England, where she is often referred to as St. Bride, and into Wales, where she is known as St. Ffraid. Several dedications to her exist in place-names such as St. Bride's and Bridewell. Her cult also spread to continental Europe. Although her historicity remains a matter of debate, the veneration of St. Brigit continues to the present day.

DOROTHY ANN BRAY

References and Further Reading

Bray, Dorothy Ann. "The Image of St. Brigit in the Early Irish Church." *Etudes Celtiques* 24 (1987): 209–215.

———. "Saint Brigit and the Fire from Heaven." *Etudes Celtiques* 29 (1992): 105–113.

Connolly, Seán, ed. and trans. "*Vita Prima Sanctae Brigitae*." (The First *Life* of St. Brigit.) *Journal of the Royal Society of Antiquaries of Ireland* 119 (1989): 5–49.

———, and Jean-Michel Picard, eds. and trans. "Cogitosus: Life of Saint Brigit." *Journal of the Royal Society of Antiquaries of Ireland* 117 (1987): 5–27.

Davies, Oliver, trans. "Ultán's Hymn." In *Celtic Spirituality*, edited and translated by Oliver Davies, with Thomas O'Loughlin, 121. New York: Paulist Press, 1999.

Greene, David, ed. and trans. "St. Brigid's Alefeast." *Celtica* 2 (1954): 150–153.

Harrington, Christina. *Women in a Celtic Church: Ireland, 450–1150*. Oxford: Oxford University Press, 2002.

Kenney, James F. *The Sources for the Early History of Ireland: Ecclesiastical*. New York: Columbia University Press, 1929.

McKenna, Catherine. "Between Two Worlds: Saint Brigit and Pre-Christian Religion in the *Vita Prima*." In *Identifying the Celtic: CSANA Yearbook 2*, edited by Joseph F. Nagy, 66–74. Dublin: Four Courts Press, 2002.

Ó hAodha, Donncha, ed. and trans. *Bethu Brigte*. (Life of Brigit.) Dublin: Dublin Institute for Advanced Studies, 1978.

de Paor, Liam, ed. and trans. "Cogitosus's Life of St. Brigid the Virgin." In *Saint Patrick's World*, edited by Liam de Paor, 207–224. Dublin: Four Courts Press, 1993.

Sharpe, Richard. *Medieval Irish Saints' Lives*. Oxford: Clarendon Press, 1991.

Stokes, Whitley, ed. and trans. "Life of Brigit." In *Lives of the Saints from the Book of Lismore*, edited and translated by Whitley Stokes, 182–200. Oxford: Clarendon Press, 1890.

See also **Cogitosus; Colum Cille; Hagiography and Martyrologies; Íte; Kildare; Mo-Ninne; Nuns; Patrick; Pre-Christian Ireland**

BRITISH LIBRARY MANUSCRIPT HARLEY 913

London British Library Manuscript Harley 913 is most notable for containing, alongside material in Latin and French, seventeen English poems that are the earliest written examples of Hiberno-English, the English language in Ireland.

The manuscript is of parchment, dates from circa 1330 and contains forty-eight items over sixty-four folios. It is very small, measuring only 140 mm by 95 mm. The bulk of the manuscript appears to have been written by a single scribe, the size of the handwriting varying according to the demands of space available.

The codicology of the manuscript reveals a structure of five "Booklets," each complete in itself. Thirteen of the English poems, distributed among other items, are contained in Booklets two and three. Their titles are almost all post-medieval in date: *The Land of Cockaygne, Five Hateful Things, Satire, Song of Michael of Kildare, Sarmun, Fifteen Signs before Judgment, Seven Sins, Fall and Passion, Ten Commandments, Christ on the Cross, Lollai, Song of the Times,* and *Piers of Bermingham.* Here one finds those poems with Kildare associations, which have given the name "Kildare Poems" to the poems as a whole. Four further poems, *Elde, Erthe, Nego*, and *Repentance of Love,* are found among the items of Booklets four and five. All but three of the poems are unique to this manuscript: *Elde* and *Erthe* belong to a textual tradition outside Ireland; a (later) variation of *Lollai* is also found in England.

The manuscript shows signs of having been dismembered and assembled again incorrectly: *Seven Sins* begins on folio 48 and continues on folio 22r; *Elde* begins on folio 54r and continues on folio 62r. Evidence from London British Library Manuscript Lansdowne 418, a collection of Irish material made by Sir James Ware in 1608, indicates that eleven of its items came from "a small old book" called the "Book of Ross or Waterford"—known as "Harley 913"—in which only six of the items are still to be found. The other five, including a poem beginning "Yung men of Waterford," are no longer present, presumably lost when the manuscript was disarranged.

The seventeen Hiberno-English poems are unique linguistically. They are also unique among Middle English poems in exhibiting signs of Irish influence in their composition. The contents of the manuscript as a whole—powerfully homiletic, with some satirical pieces—suggest in their themes and images a strong Franciscan connection. The manuscript contains a list of Franciscan houses beginning with the provinces of Ireland. Kildare and Waterford, mentioned in the manuscript, had Franciscan houses, as had New Ross (mentioned in a notable French poem "The Entrenchment of New Ross"). Friar Michael, who claims authorship of one poem, says he is a "frere menour" (141). The subject of another poem, *Piers of Bermingham*, was buried in the Franciscan Priory in Kildare town. Also present are memorials of St. Francis and the Franciscan order. The satirical material, including such poems as *The Land of Cockaygne*, as well as Latin pieces such as the *Abbot of Gloucester's Feast* and *Missa de Potatoribus* (Mass of the Drinkers), while exposing mankind's wrongdoings to laughter rather than to homiletic censure, avoids any criticism of friars.

The manuscript's small size, taken with its contents and its well worn appearance, suggests that it was made to be a travelling preacher's "pocket-book." Such small books were produced in large numbers to meet the needs of frairs. Studies of Franciscan manuscripts indicate that Franciscans had a special liking for small portable books.

The early history of the manuscript can only be surmised, and what is known of its later movements contains significant gaps. Internal evidence suggests that to materials from Kildare were added materials from New Ross and finally from Waterford. The materials for the manuscript could have been assembled and copied in Waterford, probably at the Franciscan house. An inscription on folio 2 shows that in the sixteenth century the manuscript was owned by George Wyse, mayor (1561) and bailiff (1566) of Waterford. Perhaps it came to the family when Sir William Wyse, who was attached to the court of Henry VIII, acquired property in Waterford at the dissolution of the monasteries. In 1608, Ware made his above-mentioned transcriptions. In 1697, it was owned by John Moore, bishop of Norwich (1691–1701), as mentioned by Bernard in his *Catalogus*. In 1705, it was in the possession of Thomas Tanner, bishop of St. Asaph, who allowed George Hickes to print the *Land of Cockaygne* in his *Thesaurus*. Subsequently it was owned by Robert Harley, first earl of Oxford, with whose library it came to the British Museum in 1754.

ANGELA M. LUCAS

References and Further Reading

D'Avray, D. L. "Portable Vademecum Books Containing Franciscan and Dominican Texts." In *Manuscripts at Oxford: An Exhibition in Memory of Richard Hunt . . . on themes selected and described by some of his friends*, edited by A. C. de la Mare and B. C. Barker Benfield, 60–64. Oxford: Bodleian Library, 1980

Benskin, M. "The Hands of the Kildare Poems' Manuscript," *Irish University Review* 20 (1990): 163–193.

Bernard, E. *Catalogi Librorum Manuscriptorum Angliae et Hiberniae.* Oxford: Sheldonian Theatre, 1697.

Hickes, G. *Linguarum vett. Septentrionalium thesaurus.* Oxford: Sheldonian Theatre,1705.

Lucas, A. M. and P. J. Lucas. "Reconstructing a Disarranged Manuscript: The Case of MS Harley 913, a Medieval Hiberno-English Miscellany," *Scriptorium* 14 (1990): 286–299.

Lucas, Angela M. *Anglo-Irish Poems of the Middle Ages.* Blackrock, Ireland: Columba Press, 1995.

See also: **Bermingham; Cyn, Friar John; Education; French Literature, Influence of; Hiberno-English Literature; Hiberno-Norman (Latin); Kildare; Languages; Lyrics; Manuscript Illumination; Metrics; Moral and Religious Instruction; Poets/Men of Learning; Racial and Cultural Conflict; Satire; Waterford**

BRUCE, EDWARD (*c.* 1275–1318)

Edward Bruce, lord of Galloway (from 1308), earl of Carrick (from 1313), and king of Ireland (1315–1318), was a younger brother of Robert I of Scotland (1306–1329). He was heir-presumptive to the Scottish throne when he invaded Ireland in May 1315, and did so with King Robert's full support. It was alleged against members of the Anglo-Irish de Lacy family that they invited Edward (presumably to help recover the lordship of Mide [Meath] lost to the family through the female line), but one contemporary claimed that Edward was invited by a nobleman with whom he had been "educated," possibly a reference to fosterage as practiced in the Gaelic world and by the Bruces. The obvious candidate is Domnall Ua Néill of the Northern Uí Néill, who acknowledged his role in a letter to the pope in 1317, adding that he voluntarily ceded to Edward his own hereditary royal claim.

The assembly that met on April 26, 1315, at Ayr, facing the Antrim coast, was perhaps a muster for the fleet that sailed from there, landing on May 26, possibly at Larne, or Glendun farther north (Robert Bruce was there in July 1327), where Edward had a land-claim inherited from his great-grandfather Duncan of Carrick (in Galloway). The 6,000 troops landed without opposition, the Anglo-Irish government being taken unawares, with the chief governor in Munster and the earl of Ulster, Richard de Burgh, in Connacht. The earl's tenants—Sir Thomas de Mandeville, the Bissets, Logans, and the Savages—took to the field unsuccessfully against the Scots under Thomas Randolph, earl of Moray, and when the invaders marched on Carrickfergus, the town fell easily, although its heavily fortified castle required a prolonged siege. Probably while at Carrickfergus, up to twelve Irish kings came to Edward and, the annals record, he "took the hostages and lordship of the whole province of Ulster without opposition and they consented to his being proclaimed King of Ireland, and all the Gaels of Ireland agreed to grant him lordship and they called him King of Ireland."

King Edward Bruce now campaigned along the Six Mile Water, burning Rathmore near Antrim, before heading south through the Moiry Pass, where Mac Duilechain of Clanbrassil and Mac Artain of Iveagh apparently ambushed him. But on June 29, 1315, Bruce stormed the de Verdon stronghold of Dundalk, which he ransacked, including its Franciscan friary. The chief governor, Edmund Butler, assembled the feudal host, and de Burgh gathered his Connacht tenants (and Irish levies under Feidlim Ua Conchobair) and both converged south of Ardee around July 22. The Scots and Irish were ten miles away at Inniskeen, and after a skirmish near Louth Edward cautiously adopted Ua Néill's advice and retreated via Armagh to Coleraine, which they burned, sparing the Dominican friary but demolishing the bridge over the Bann. De Burgh alone pursued Bruce to Coleraine but was forced to withdraw to Antrim for lack of provisions (being also weakened by Ua Conchobair's return to Connacht). When the Scots crossed the Bann in pursuit, aided by the sea captain Thomas Dun, they defeated de Burgh in battle at Connor on September 1, and he withdrew humiliated to Connacht, the annals calling him "a wanderer up and down Ireland, with no power or lordship."

In mid-November Edward again marched south, and by November 30 was at Nobber, County Meath, where he left a garrison and advanced on Kells to challenge Roger Mortimer, lord of Trim, whose large but disloyal army fled the battlefield around December 6, Mortimer himself returning to England. Bruce burned Kells and turned west to Granard, County Longford, attacking the Tuit family manor and the Cistercians at Abbeylara (accused in Ua Néill's "Remonstrance" to the pope of spear-hunting the Irish by day and saying Vespers by evening). Edward raided the English settlements in Annaly, County Longford, and spent Christmas at Loughsewdy, *caput* of the de Verdons' half of Meath, before razing it to the ground. Meath manors (like Rathwire) still in de Lacy hands were untouched by Bruce, suggesting they had joined him, for which they were later convicted and dispossessed.

Edward next appeared in Tethmoy, County Offaly, home of the de Berminghams, adversaries of the de Lacys, apparently being waylaid in Clanmaliere by O'Dempsey, who remained loyal to the Dublin government. Having reached the lands of John FitzThomas (soon to be earl of Kildare, and second only to de Burgh among the resident Anglo-Irish baronage), Bruce attacked Rathangan castle and progressed to Kildare itself, but the garrison refused to surrender. Travelling to Castledermot, then north again via Athy and Reban, Edward was near the mound of Ardscull when the colonists assembled to deal with the threat. Led by Butler, John FitzThomas and his son, their cousin Maurice FitzThomas of the Munster Geraldines (later first earl of Desmond), and members of the Power and Roche families, the colonists faced the Scots in battle at Skerries near Ardscull on January 26. Although the government army exceeded Edward's, quarrels among its leaders handed victory to the Scots, despite heavier losses.

Bruce then retired into Laois, safe among Irish supporters and boggy terrain unsuitable for cavalry. The routed Anglo-Irish retired to Dublin and swore on February 4, 1316, to destroy the Scots on pain of death. Edward II's envoy, John de Hothum, wrote from Dublin urgently requesting £500 to replenish a treasury empty because of the war and the famine (felt throughout Europe) that had followed unusually bad weather and a disastrous harvest. Bruce too found Irish enthusiasm waning, as they blamed him for the desperate conditions that coincided with his occupation, and was unable to push home the advantage after Skerries. He burned FitzThomas's fortress at Lea, County Laois, and by February 14 was near another Geraldine castle at Geashill. But the government army was now assembling near Kildare. Bruce retreated, his army being reported at Fore, County Westmeath soon after, dying of hunger and exhaustion, arriving back in Ulster base by late February.

After reputedly holding a parliament in Ulster, Edward visited Scotland briefly in late March. Carrickfergus Castle still held out, despite Thomas Dun's sea blockade, although the garrison was reportedly reduced to cannibalism and, by September 1316, had surrendered (under terms that Edward, characteristically, honored). He also captured but re-lost Greencastle, County Down, and secured Northburgh Castle in Inishowen. Robert Bruce himself was rumoured to be in Ulster late that summer but cannot have been there long (if at all), since on September 30, Edward was at Cupar in Fife with Robert and the earl of Moray, where, styling himself "Edward, by the grace of God, king of Ireland," he approved his brother's grant to Moray of the Isle of Man. Edward perhaps had designs on Man himself and agreed to this in return for reinforcements in Ireland. Help was certainly needed, as the tide was turning against him. In October 1316, Edward II put a bounty of £100 on his head, and soon afterwards 300 Scots men-at-arms were killed in Ulster.

King Robert therefore set sail for Carrickfergus from Loch Ryan in Galloway, arriving about Christmas, the annals noting that he brought a great army of *galloglass* (Hebridean warriors) "to help his brother Edward and to expel the foreigners from Ireland." By late January 1317, they were on the move, allegedly numbering 20,000 by the time they reached Slane, County Meath in mid-February, ravaging the countryside as they went. The earl of Ulster was at Ratoath manor and possibly attempted to ambush the Scots, but he failed and fled to Dublin, taking refuge in St. Mary's Abbey. The citizens panicked and the mayor seized the de Burgh family and imprisoned them in Dublin Castle, suspecting collusion with the Scots: Earl Richard had a thirty-year association with the Bruces, and in 1302, Robert Bruce married Richard's daughter (now queen of Scotland), but they were no longer allies and the suspicions seem unfounded.

By February 23, King Robert was at Castleknock, and the Dubliners strengthened their defenses by dismantling the Dominican priory to fortify vulnerable stretches of the city walls near the bridge across the Liffey. They also fired the western suburbs to deny the Scots cover, and, although the conflagration raged beyond control and did enormous damage, the tactic worked. The Bruces did not risk a siege and, joined by the de Lacys, headed via Naas to Castledermot, where they burned the Franciscan friary. They proceeded through Gowran, County Kilkenny, reaching Callan by March 12. The Anglo-Irish were assembled in Kilkenny (led by the justiciar Edmund Butler, the second earl of Kildare, Maurice FitzThomas of Desmond, and Richard de Clare of Thomond), but dared not oppose the Scots in battle, who continued into Munster where they plundered Butler's town of Nenagh. The O'Briens had led Bruce to expect widespread support but, as with the O'Conors in Connacht, local rivalries intervened. So, having seized de Burgh's fortress at Castleconnell on the Shannon, putting Limerick within sight, the Scots proceeded no further. Butler led 1,000 men toward them in early April and, as hunger took its toll (the famine being even more severe than in 1316), Roger Mortimer, now king's lieutenant, landed at Youghal with fresh troops and began marching north. King Robert sensed the danger and began a retreat. His hungry and exhausted troops, having sheltered for a week in woods near Trim, struggled back to Ulster about May 1, whereupon Robert returned to Scotland, apparently not reappearing in Ireland for nearly a decade.

The fourteenth-century verse biography known as *The Bruce* records another gathering of the Irish at Carrickfergus, after which "every one of the Irish kings went home to their own parts, undertaking to be obedient in all things to Sir Edward, whom they called their king." Ua Néill then wrote to the pope on behalf of "the under kings and magnates of the same land and the Irish people" seeking papal support for the invasion. Nothing further is known of events for the next eighteen months until, after a bumper crop in that year's harvest, Bruce marched south again in October 1318. With the de Lacys in tow, anxious to recover their Meath lands and do down their occupiers, they headed for Dundalk, property of the de Verdons (who held half of Meath), when they were met by an opposing force on a hillside near Faughart on October 14. Their opponents were the de Verdons and their tenants, commanded by John de Bermingham of Tethmoy, an old antagonist and keeper of the half of Meath acquired by Roger Mortimer.

Although reinforcements were purportedly on their way from Scotland, Edward did not wait, and rushed to battle accompanied by Hebridean *galloglass* under Mac Domnaill and Mac Ruaidrí (both of whom were killed). Amid intense fighting, he himself was slain by a townsman of Drogheda, whose body was later found resting on that of the vanquished "king of Ireland." Contrary to local tradition, Bruce was not buried at Faughart, but was decapitated and his body quartered, one quarter, with his heart and hand, being sent to Dublin, the others to "other places." The victor, de Bermingham, brought Bruce's head to Edward II, who rewarded him with the new earldom of Louth. The collapse of Bruce's regime was joyously greeted by the Anglo-Irish, and probably went unlamented by the Irish too (one native obituarist certainly condemns him) because, after three years of war and famine, Edward inevitably found himself being blamed for events beyond his control. His claim to Ireland died with him, and was not resurrected by his heirs.

SEÁN DUFFY

References and Further Reading

Barbour, John. *The Bruce*. Edited by A. A. M. Duncan. Edinburgh: Canongate, 1997.

Duffy, Seán, ed. *Robert the Bruce's Irish Wars: The Invasions of Ireland, 1306–1329*. Stroud, England: Tempus, 2002.

Duncan, A. A. M. "The Scots' Invasion of Ireland, 1315." In *The British Isles, 1100–1500*, edited by R. R. Davies, 3–37. Edinburgh: John Donald Ltd., 1988.

Frame, Robin. "The Bruces in Ireland, 1315–1318." *Irish Historical Studies* 19 (1974).

McNamee, Colm. *The Wars of the Bruces: Scotland, England and Ireland, 1306–1328*. East Linton, Scotland: Tuckwell, 1996.

Clonmacnoise, Co. Offaly. © *Department of the Environment, Heritage and Local Government, Dublin.*

See also **Lacy, de; Mide (Meath); Ua Néill, Domnall; Uí Néill, Northern**

BURIALS

Knowledge of burial traditions in Ireland in the period immediately prior to the conversion to Christianity is limited. This period, normally called the Iron Age, extended from at least 500 B.C.E. to circa 400 C.E., and is characterized by a dearth of archaeological information, especially about settlement and, to a lesser extent, burial. Ring ditches or ring barrows (small circular ditched enclosures, with an external bank and often a central mound, in the case of barrows, or just a ditch, in the case of ring ditches) were used for burial throughout this period, mostly for burials that were cremated but also on occasion for inhumations. Inhumated burial, sometimes in cemeteries, appears to become more common towards the end of the Iron Age, possibly as a result of influence from Roman Britain, and many of these cemeteries continued to be used after the introduction of Christianity.

Early Medieval Period

There is evidence, both historical and archaeological, that ancestral burial grounds continued to be used for a few centuries after the introduction of Christianity. These burial grounds are mentioned in early Irish canons and a number of sites such as Knoxspark (County Sligo), Ballymacaward (County Donegal), and possibly Millockstown (County Louth), where there is no evidence of a church, appear to be examples of this type of cemetery. An extraordinary example is Cloghermore Cave, County Kerry, where pagan-style burial continued up to the ninth century, when the use of the cave for burial was taken over by Vikings.

In common with Christian practice elsewhere, early medieval burials in Ireland were extended inhumations, usually in cemeteries, aligned roughly east-west with the head to the west, and unaccompanied by grave goods. Usually the burials were quite shallow, with burial in a wooden coffin being the exception rather than the rule. Burial in a shroud appears to have been the norm, though there is no evidence for the use of pins to close the shroud. Stones or slabs were used in various ways in association with burials. Sometimes stones were placed at each side of the skull, or under it, forming a pillow. In some cases slabs were set on edge to line the sides and ends of the grave and in other cases slabs, serving as lintels, were placed on these. The latter are sometimes called lintel graves and good examples were found at Reask, County Kerry. In other cases slabs were set on the surface over graves to mark their location, and there is sometimes evidence for slabs set on end to act as head and foot stones as at High Island, County Galway. Slabs with crosses and sometimes an inscription, asking for a prayer for the deceased, are assumed to have been set on top of graves, though mostly they have been found out of context. The inscriptions, normally in Irish, usually take the form: *oroit* (pray; usually contracted to *or*) *do* (for) followed by the name of the person commemorated. It has been possible to date a small number of these slabs where the individual commemorated is also mentioned in the annals. The slabs date from at least the eighth century up until the twelfth. The largest collection of them is at Clonmacnoise, with over seven hundred complete or fragmentary examples, though not one of these has been found in place over a grave. Some of the latest in the series, dating from the eleventh or twelfth century, are inclined to be rectangular or trapezoidal in shape and of reasonably large size. Examples of this type of slab survive in settings over graves at Inis Cealtra (County Clare), and Glendalough, though no archaeological excavation of these graves has taken place.

Pillar stones inscribed with the ogham script (known as "ogham stones") were by no means always used to mark burials, but their frequent occurrence in early medieval cemeteries would suggest that some of them did so. The ogham script consists of notches and strokes carved on the angles of these stones, which date from about the fourth to the eighth century and are found mainly in South Munster. The language used is an early form of Irish, and the inscriptions commemorate individuals and their family affiliations.

Often burials took their alignment from an upstanding feature on the site such as a church and, if a later church was built on a different alignment, the burials generally followed suit. Rather than being buried in rows side-by-side, some of these cemeteries were laid out as string burials where the rows ran lengthways, the head of one burial following on from the feet of the previous one. This layout was noted in some of the earliest burials excavated at Clonmacnoise beneath the site of the Cross of the Scriptures.

Just as in ancestral burial grounds important ancestors may have had their graves marked out in a special way, the graves of founding saints or other holy persons came to be distinguished from the generality of graves in Christian cemeteries. Having a saint's grave or possession of the relics of a saint made a church a focus of pilgrimage, and sometimes the claim was made that burial in the same cemetery as the saint qualified the deceased for automatic entry to heaven. The remains of holy persons from the eighth century and later were often disinterred and placed in an outdoor stone shrine or in a metalwork reliquary within a church. A number of stone slab shrines are known from sites in the west of Ireland, particularly Kerry, and some had a hole in the end slab, through which the relics or the ground over the bones could be touched. At some important sites, such as Clonmacnoise and Ardmore (County Waterford), a small church was built over the supposed grave of the saint.

The main type of non-Christian burial found in Ireland during the medieval period is that of the Vikings or pagan Scandinavians. They first settled in Ireland in the ninth century and founded some towns and smaller trading posts and settlements. Their burials at this early stage were accompanied by grave goods such as swords and personal ornaments. Their most famous burial ground was at Kilmainham (Island Bridge), just west of Dublin. By at least the later tenth century they were Christianized and indistinguishable from the rest of the Irish in their burials.

Post-Norman Period

Some decades after the Anglo-Normans first came to Ireland, new types of grave memorials appear in the form of effigies carved in relief and coffin-shaped floor slabs. These mainly marked interments within the church and the effigies were usually the memorials of important individuals, usually bishops or lords. The effigies were placed in specially constructed niches in the side-walls, or as lids for free-standing sarcophagi, or in later times, as the tops of table or altar tombs. For less-exalted individuals, coffin-shaped floor slabs were used, with usually a floriated cross and sometimes an inscription carved on them. The occurrence of sarcophagi, carved out of a single stone, is confined mainly to the Leinster region, the area most heavily settled by the Anglo-Normans. Tomb inscriptions in French in the thirteenth and early fourteenth century are also found in this area, but Latin was the language most commonly used for tomb inscriptions up to the end of the sixteenth century. As a compromise between the effigy and the floor-slab with floriated cross, a figure or figures were sometimes incised on a flat slab and in other cases, only the head was carved, with or without the cross beneath it. It was only the wealthier classes who would have been commemorated in this way; the majority of the population continued to be buried in simple pits aligned east-west, with the head to the west, in cemeteries attached to the church. Ecclesiastics of all sorts were buried with their heads to the east, the theory being that they would face their flock when rising on the last day. From about the twelfth century, important ecclesiastics such as bishops were sometimes buried with metalwork or other items associated with their position, such as a chalice, which was excavated in a grave at Mellifont; a crozier at Cashel; and a ring and mitre at Ardfert. Scallop shells found with burials at Tuam indicate that these persons had made the pilgrimage to Compostella in Spain.

There is a lack of tomb sculpture, referred to by Hunt as the "hiatus," from 1350 to 1450, mainly due to the Black Death, which had a catastrophic effect on the colony, especially the towns, where many died as a result of the plague. When the carving of slabs and altar tombs became common again in the later fifteenth century, saints, especially the apostles, were carved on the side panels of the tombs and symbols of Christ's passion were carved on both tombs and floor slabs. A preoccupation with man's mortality led to the carving of effigies as cadavers in some cases; a fine example of this is the Rice monument in Waterford cathedral. Contemporary with the fine altar tombs of the late fifteenth and sixteenth centuries, especially in the regions around Dublin and Kilkenny, were slabs with seven-pointed crosses and long inscriptions in Gothic letters. These styles of memorials continued into the seventeenth century in the case of Catholics of both Gaelic and Old English origin, while new styles of commemoration of the Protestant New English appear from the mid sixteenth century.

Conleth Manning

References and Further Reading

Conway, Malachy. *Director's First Findings from Excavations in Cabinteely.* Glenageary: Margaret Gowen & County Ltd., 1999.

Fry, Susan L. *Burials in Medieval Ireland, 900–1500: A Review of the Written Sources.* Dublin: Four Courts Press, 1999.

Hunt, John. *Irish Medieval Figure Sculpture, 1200–1600.* 2 Vols. Dublin and London: Irish University Press, Sotheby Parke Bernet, 1974.

Hurley, Maurice F. and Sarah W.J. McCutcheon. "St. Peter's Church and Graveyard." In *Late Viking Age and Medieval Waterford: Excavations, 1986–1992*, edited by Maurice F. Hurley, et al., 190–227. Waterford: Waterford Corporation, 1997.

Lionard, Pádraig. "Early Irish Grave Slabs." *Proceedings of the Royal Irish Academy* 61C (1961): 95–169.

Macalister, R. A. S. *Corpus inscriptionum insularum Celticarum.* 2 Vols. Dublin: Stationery Office, 1945–1949.

Maher, Denise. *Medieval Grave Slabs of County Tipperary, 1200–1600 A.D.* Oxford: BAR British Series 262, 1997.

Manning, Conleth. "Archaeological Excavation of a Succession of Enclosures at Millockstown, County Louth." *Proceedings of the Royal Irish Academy* 86C (1986): 135–181.

Marshall, Jenny White, and Grellan D. Rourke. *High Island: An Irish Monastery in the Atlantic.* Dublin: Town House & Country House, 2000.

O'Brien, Elizabeth. "Pagan and Christian Burial in Ireland During the First Millennium A.D.: Continuity and Change." In *The Early Church in Wales and the West*, edited by Nancy Edwards and Alan Lane, 130–137. Oxford: Oxbow monograph 16, 1992.

———. "A Reconsideration of the Location and Context of Viking Burials at Kilmainham/Islandbridge, Dublin." In *Dublin and Beyond the Pale: Studies in Honour of Patrick Healy*, edited by Conleth Manning, 35–44. Bray: Wordwell, 1998.

See also **Altar-Tombs; Clonmacnoise; Ecclesiastical Sites; Ecclesiastical Settlements; Inscriptions; Parish Churches, Cathedrals; Reliquaries.**

BURGH

Lords of Connacht

The progenitor of the de Burghs (Burkes, Bourkes, de Búrca) in Ireland was William de Burgh, who is often given the epithet "the Conqueror." He is not to be (although he sometimes is) confused with one William fitz Adelm (or Audelin), who filled the offices of seneschal and deputy to Henry II. The origins of the de Burgh family lie in Norfolk. William came to Ireland with the Lord John in 1185 and obtained a grant of land in Munster very soon after. De Burgh erected the castle of Kilfeakle in 1192. He maintained friendly relations with Domnall Mór Ua Briain, whose daughter he married around 1193. The marriage-alliance strengthened his position in Munster substantially and he soon started colonization. John made William a speculative grant of Connacht in 1195. He also held lands from Theobald Walter (ancestor of the Butlers) and was granted more lands by John in 1199 and 1201. By the early thirteenth century he held extensive lands in what are now counties Tipperary and Limerick. The de Burghs also intermarried with other prominent Irish families, such as the Uí Chonchobair and Uí Chellaig, and rapidly Gaelicized.

The Augustinian priory of St. Edmund in Athassel, County Tipperary, was built by William de Burgh in about 1200. The de Burghs later increased its endowments, and several members of the family (Walter son of William [d. 1208]; Richard earl of Ulster [d. 1326]; and probably also Earl William [d. 1280]) were buried in the priory, including the founder.

After receiving the speculative grant from John, de Burgh interfered in Connacht affairs with the help of Ua Briain's forces. He came to the assistance of Cathal Carrach, grandson of Ruaidrí Ua Conchobair, who was opposing the claim to the kingship of Connacht of his kinsman, Cathal Crobderg son of Tairrdelbach Ua Conchobair. Cathal Carrach was proclaimed king, but de Burgh soon switched sides and in 1201 supported Cathal Crobderg. Their forces combined to kill Cathal Carrach, after which event Crobderg was inaugurated as king. Subsequent to another change of heart on de Burgh's side, he was summoned before the king of England. De Burgh then had his lands in Limerick and Tipperary re-granted, but lost his claim to Connacht. William died in 1205 and his lands were taken in custody until his heir Richard came of age in 1214. On September 13, 1215, Richard Mór ("the Great," or "Senior") obtained confirmation of the speculative grant his father had received. This grant, however, was not put into effect immediately, as Cathal Crobderg obtained a very similar grant confirming him in the possession of Connacht on the same day. Cathal Crobderg's son and successor Áed, however, forfeited the grant of Connacht, and so in 1227 Richard's grant was put into effect. Richard then became lord of Connacht, holding twenty-five of the thirty cantreds of which the province was comprised; his demesne lands were situated in what is now County Galway. The remaining five cantreds came to be known as the "King's Five Cantreds" and were leased to Ua Conchobair. Richard rebuilt Galway Castle in 1232, and four years later began building what became the center of his power, the castle of Loughrea. He was justiciar of Ireland from 1228 to 1232, and in this office he was nominally and briefly succeeded by his uncle Hubert de Burgh, earl of Kent and justiciar of England from 1215 to 1232. Hubert backed Richard in his efforts to increase his influence and wealth. When Hubert fell out of favor with the king, his nephew Richard was ordered to surrender his own lands. However, after supporting the king in his war against Earl Marshal in Leinster in 1234, he recovered his lands and proceeded with the conquest and subinfeudation of Connacht.

Lords of Connacht and Earls of Ulster

Richard died on an expedition to Poitou in the service of King Henry III in 1243, when his eldest son and heir Richard was not yet of age. The young Richard obtained seisin of his father's possessions in 1247, but died a year later. Another period of minority followed, and Richard's lands were given into the custody of Peter de Bermingham until Richard Mór's second son Walter came of age. In 1261, at the Battle of Callann, de Burgh, the justiciar, and the Fitzgeralds were defeated by MacCarthaig, after which the south of Munster was lost to government control. Meanwhile de Burgh's center of gravity shifted further to the north. Walter was given the title Earl of Ulster on July 15, 1263, and subsequently held sway over an enormous area. His award of the earldom reflected the feeling of the English government that the Anglo-Irish colony was under threat.

Walter founded the priory of St. Peter for Dominican friars in Lorrha (County Tipperary) in 1269. The de Burghs also built a Franciscan friary in Limerick in the thirteenth century. Walter died in 1271 and was succeeded by his son Richard "the Red Earl," who was still a minor. Richard was the son of Walter and his wife Avelina, daughter of the long-serving justiciar of Ireland John fitz Geoffrey. During Walter's lifetime, a civil war had broken out between the de Burghs and the Fitzgeralds, from whom the de Burghs had accumulated extensive lands in Connacht (in 1264 Walter de Burgh had seized two Geraldine castles). The Red Earl's great opponent was John FitzThomas, who was appointed first earl of Kildare in 1316. In 1294, FitzThomas imprisoned Richard, and John burnt the priory of Athassel in 1319. The dispute was, however, substantially resolved by the exchange of Geraldine lands in Connacht for lands elsewhere in Ireland. Richard married a distant relative, Margaret, the great-granddaughter of Hubert de Burgh. On several occasions he managed to depose an Ua Néill king and install his own favorite king of Cenél nEógain, from the newly formed Uí Néill branch called Clandeboye. He played a similar part in Connacht with the Uí Chonchobair.

Richard went on campaigns to Scotland in 1296 and 1303. In spite of the fact that his daughter

Elizabeth married Robert Bruce, Earl of Carrick (later, King Robert I of Scotland), Richard opposed Edward Bruce's army when it landed in Ireland in 1315. He was defeated in the same year in the battle of Connor where his cousin William Liath ("the Grey," d. 1324) was captured by the Scots (he was released a year later). Nevertheless, the earl's loyalties were questioned and he was apprehended by the citizens of Dublin in 1317. However, when Bruce was defeated, de Burgh was able to recover his territory. Richard was the most powerful nobleman in Ireland in his time. His paternal inheritance was enhanced through his mother Avelina's rights to estates in Munster. He briefly held the Isle of Man, which he restored to the king in 1290. He was appointed Lord Lieutenant of Ireland twice (1299–1300 and 1308). Around 1300, Richard founded St. Mary's priory for Carmelite friars in Loughrea (County Galway), built the castle of Ballymote (County Sligo), and possibly also started the building of spectacular Dunluce Castle (County Antrim). In 1305, he erected Northburgh castle in County Donegal, and about five years later he rebuilt Sligo castle, originally a Geraldine fortress. In 1326, he retired to Athassel abbey, where he died shortly afterwards.

William de Burgh and Richard de Bermingham won a victory at the Battle of Athenry in 1316, defeating King Feidlim Ua Conchobair, who fell in the battle. Paradoxically, it was after this Anglo-Irish victory that rural English landowners abandoned the area around Roscommon, with the exception of Sir David de Burgh, ancestor of the MacDavids of Clanconway. Until the seventeenth century, members of this branch held lands in the heart of the Ua Conchobair territory.

William "the Brown Earl," son of John de Burgh and Elizabeth de Clare, succeeded his grandfather Richard while still a minor. He was knighted in 1328 by King Edward III and was on that occasion given possession of his estates. Conflicts with the Geraldines and de Mandevilles of Ulster, as well as hostilities within the de Burgh family (see below) finally led to the murder of the earl in 1333. William's marriage to Maud of Lancaster produced one child, Elizabeth, who was still a baby when her father died. She later married Lionel, Duke of Clarence, a son of Edward III who was chief governor of Ireland from 1361–1364 and 1365–1366. The earldom of Ulster was passed on to the Mortimers through the marriage of their daughter Phillipa to Edmund Mortimer. Their great-grandson Richard, Duke of York, held the earldom in the first half of the fifteenth century, after which it was passed on to his son Edward IV and thus into the hands of the English crown.

Clan MacWilliam Burke and Clanrickard Burke

The de Burgh family split into several branches. Sir William Óg ("Young" or "Junior") de Burgh, son of Richard Mór was an antecedent of the Clan Mac William de Burgh. He was killed by Áed son of Feidlim Ua Conchobhair at the Battle of Ath-an-Chip in 1270. William's son Sir William Liath (d. 1324) was deputy justiciar of Ireland from 1308 to 1309. He founded Galway friary (on St. Stephen's Island) for Franciscan friars in 1296. William Liath's sons Walter and Edmund Albanach ("the Scot") were granted the custody of the late earl's lands in Connacht, Tipperary, and Limerick in 1326. When William "the Brown Earl" succeeded to the earldom he was at enmity with his kinsman Walter, who aspired to the kingship of Connacht. In 1332, Walter was captured and starved to death by William, who supported the descendants of Áed son of Cathal Crobderg in their claim to the kingship of Connacht. Walter's brother Sir Edmund Albanach (d. 1375) was the ancestor of the MacWilliam Íochtar of Lower (northern) Connacht, who held lands in Mayo. He led a longstanding feud against the Clanrickard Burkes or the MacWilliam Uachtar of Upper (southern) Connacht, who held lands in Galway. Richard deBurgh "an Fhorbhair" (d. 1343), the head of the Clanrickard, supported Sir Edmund (a son of Richard the Red Earl) against Edmund Albanach. The latter, however, managed to drown Sir Edmund in Loch Mask in 1338.

From this time, and all through the rest of the medieval period, these two great factions in the de Burghs of Connacht opposed each other. After the division in the Ua Conchobhair dynasty in 1384, at which the main line of the family split into Ua Conchobhair Donn and Ua Conchobhair Ruadh, each branch of the de Burghs supported one line of the Uí Chonchobhair. Ua Concobhair Donn was backed by Clanrickard (as well as Ua Conchobhair Sligigh), while Ua Conchobhair Ruadh was supported first of all by Sir Thomas, son of Sir Edmund Albanach, and later by his descendants.

Thomas and his rival Sir William (or Ulick) of Clanrickard alternated as official representatives of the Dublin government. When Thomas died in 1402, his branch of the family lost contact with the Dublin Administration, while the Clanrickards continued to provide sheriffs of Connacht from among their family. Until the end of the fifteenth century, the Lower MacWilliams were the stronger of the two factions.

Of the Clanrickard, Sir William de Burgh ("Uilleag an Fhíona") (d. 1423) was knighted by Richard II in 1395 in Waterford. Moreover, he was appointed one of the justices of Connacht in 1401.

Freya Verstraten

References and Further Reading

Blake, W. J. "William de Burgh: Progenitor of the Burkes in Ireland." *Journal of the Galway Archaeological and Historical Society* 7 (1911–1912).

Claffey, John A. "Richard de Burgh, Earl of Ulster, (c.1260–1326)." Ph.D. thesis, University College Galway, 1970.

Dudley Edwards, R. "Anglo-Norman Relations with Connacht, 1169–1224." *Irish Historical Studies* 1 (1938–1939): 135–153.

Frame, Robin. *Colonial Ireland, 1169–1369*. Dublin: Helicon Limited, 1981.

Gwynn, Aubrey and R. Neville Hadcock. *Medieval Religious Houses: Ireland; With an Appendix to Early Sites*. London: Longman, 1970.

Nicholls, Kenneth. *Gaelic and Gaelicised Ireland in the Middle Ages*. Dublin: Gill and MacMillan, 1972.

Ó Raghallaigh, Tomas "Seanchus Burcach." *Journal of the Galway Archaeological and Historical Society* 13 (1922–1928): 50–60 and 101–137; 14 (1928–1929): 30–51 and 142–167.

Orpen, Goddard Henry "Richard de Burgh and the Conquest of Connaught." *Journal of the Galway Archaeological and Historical Society* 7 (1911–1912): 129–147.

Orpen, Goddard Henry. *Ireland Under the Normans 1216–1333* 4 vols (Oxford University Press, 1911 and 1920).

———. "The Earldom of Ulster." *Journal of the Royal Society of Antiquaries Ireland* 43 (1913): 30–46 and 133–143; 44 (1914): 51–66; 45 (1915) 123–42.

Perros (Walton), Helen. "Crossing the Shannon Frontier: Connacht and the Anglo-Normans, 1170–1224." In *Colony and Frontier in Medieval Ireland: Essays Presented to J.F. Lydon*, edited by T. B. Barry, et al. London: Hambledon, 1995: 117–138.

Sweetman, H. S., ed. *Calendar of Documents Relating to Ireland*. 5 vols. London: Longman, 1875–1886.

Otway-Ruthven, A. J. *A History of Medieval Ireland*. 2nd ed. New York: Routledge Books, 1980.

See also **Bermingham; Connacht; Factionalism; Gaelicisation; Ua Conchobair; Ulster, Earldom of**

BUTLER-ORMOND

Origins

Theobald Walter, elder brother of Hubert, Archbishop of Canterbury from 1193–1205, was the ancestor of the Butler family in Ireland. His father, Hervey Walter, was a knight from Amounderness in Lancashire. The name Butler, soon to replace the family name, was derived from the honorific title of Butler of the household of John, Lord of Ireland and youngest son of Henry II. Theobald later assumed the hereditary title of Butler of Ireland, by virtue of which the family enjoyed the prise of wines entering Irish ports for several centuries. The reason for Theobald's rise to power in Ireland must be linked to the influence of a maternal aunt, wife of Ranulph de Glanville, justiciar of England from 1180 to 1189. It was this vital connection with the court of Henry II that opened up opportunities of advancement to the sons of a relatively obscure knight. Both Hubert and Theobald appear to have grown up in Ranulph's household. When Theobald set out for Ireland in 1185 as a member of John's household, he was accompanied by his uncle. Ranulph wasted no time in exploiting this position, with the result that shortly after the expedition landed at Waterford, he and Theobald jointly received a grant from John of extensive territory in the kingdom of Limerick (Thomond or North Munster). Theobald subsequently fell from favor when John became king in 1199. Two years later the kingdom of Limerick was granted to William de Braose, but Theobald's title to his lordship was secured by the timely intervention of Hubert, who headed the list of witnesses in a charter confirming his possessions, the de Braose grant notwithstanding.

The Butler Lordship

John's grant of five and a half "cantreds" (baronies) in the kingdom of Limerick for the service of twenty-two knights was speculative. The territory lay well beyond the limits of Anglo-Norman settlement in 1185. Apart from a desire to reward his followers for their military services, John seems to have intended the conquest of Munster, doubtless as a means of extending his lordship of Ireland and securing his demesnes in Munster. The grants to Theobald, William de Burgh, and Philip of Worcester included the modern county of Tipperary and some adjacent territories in County Limerick, County Clare, and County Offaly. Little is known of the progress of the conquest before Theobald died circa 1206, except that it was fiercely contended by Domnall Ua Briain, King of Limerick. However, it is possible to reconstruct both the outline and the organization of the lordship on the basis of later manorial surveys, which bear the imprint of an original plan that can confidently be attributed to Theobald on the basis of a grant of "the tuath of Kenelfenelgille" (the manor of Drum) to one of his vassals in the cantred of Eliogarty, probably between 1190 and 1200. This important deed reveals that before he died, the future shape of the settlement was already discernible. At the center of the cantred lay Theobald's chief manor (*caput*) of Thurles, from which radiated the fiefs of military tenants owing feudal services to their lord. This distinctively uniform scheme of settlement was repeated in all of the territories granted to Theobald. He organized his lordship in the kingdom of Limerick around four manorial centers or *capita*: Nenagh and Thurles (County Tipperary), Caherconlish (County Limerick), and Dunkerrin (County Offaly), forming a contiguous group of lordships. Theobald was also granted important fiefs in the lordship of Leinster by John during the minority of Isabelle, daughter and heiress of Strongbow. These he organized into lordships focused on three great manorial centers: Gowran (County Kilkenny), Tullow (County Carlow), and Arklow (County Wicklow). These enormous grants amounted to about 750,000 statute acres, placing Theobald, if not in quite the same category as de Courcy, de Lacy, or Strongbow, then certainly among the major tenants-in-chief of the crown, thereby laying the foundation of the future greatness of the family.

In the two centuries that followed, the heirs of Theobald acquired and lost other territories in Ireland, particularly in Uí Maine in Connacht, but many of them had no enduring value. In fact, the Butlers suffered major losses of territory at the hands of the O'Kennedys (Uí Chennéidig), O'Carrolls (Uí Cherbaill), and others in the course of the first half of the fourteenth century, particularly in northern County Tipperary and adjacent lands in County Offaly and northern County Kilkenny. As a consequence of these losses, much of the original heartland of the lordship was lost: Nenagh, the chief seat of the family, was reduced to a frontier outpost by the end of the fourteenth century. However, the absenteeism of the neighbouring Anglo-Irish lords, especially in the neighbouring county of Kilkenny, permitted the Butlers to compensate for their losses elsewhere. The purchase of Kilkenny Castle from the Despensers in 1391, replacing Nenagh as their chief seat, was only the final piece of a series of acquisitions in the county over the course the preceding century. Besides, the grant of the liberty of Tipperary to James Butler, first Earl of Ormond, in 1329, had the effect of extending the family's jurisdiction over the entire county, or at least what remained of it after the losses sustained in the north in the course of the same century. This shift in the territorial center of gravity was further reinforced in the course of the fifteenth century, when the demesnes of the earls of Ormond became concentrated in County Kilkenny, leaving County Tipperary largely in the hands of cadet branches.

The Earldom of Ormond

Although the Butlers were clearly important tenants-in-chief in the thirteenth century, they did not play a prominent political role. While Irish magnates did feature in the royal administration from time to time, the governorship was frequently controlled either by churchmen or by royal servants dispatched from England. However, as the political situation in the Irish colony deteriorated in the fourteenth century, the crown increasingly relied on the cheaper option of appointing Irish magnates to look after the troubled affairs of Ireland. Besides, the great lordships of Ulster, Mide (Meath), and Leinster were more often than not in the control of absentee lords, leaving the way open for those who remained, most notably the FitzGeralds and the Butlers. The first member of the Butler family to play a significant role was Edmund, who was lord deputy of Ireland 1304–1305 and 1312–1314, and chief governor (justiciar) 1315–1316, during the Bruce crisis. He was granted the earldom of Carrick in 1315, and was occasionally styled earl, but he was never created earl probably because he was unable to visit England before he died in 1321. It was his son James, made Earl of Ormond in 1328 during the Mortimer regency, who received the grant of the liberty of Tipperary for the term of his life, apparently as the price of his support. While the grant of the liberty, which made him palatine lord of Tipperary, was in some respects a de facto recognition that the county was increasingly a liability to the royal administration rather than a source of profit, it must also be seen as an honorific underpinning of the new title. The jurisdictional powers conveyed with the liberty were precisely the same as those exercised in the previous century by the lords of Ulster, Meath, and Leinster. It was in effect an official recognition that the Butlers had now achieved the rank formerly accorded only to the greatest Anglo-Irish magnates. Not least among the ironies of the new title was the fact that, during the lifetime of the first earl, Butler control of the cantred of Ormond began to disintegrate.

The apogee of the power and influence of the Butler earls in the medieval period coincided with the careers of James, third earl of Ormond from 1385 to 1405, and his son James, fourth earl of Ormond from 1411 to 1452. The third earl was justiciar of Ireland in 1384, and then deputy. He was subsequently justiciar in 1393, preparing the way for Richard II's first expedition to Ireland, and finally justiciar and later deputy in 1404–1405. As a fluent Irish speaker and influential magnate, he negotiated important submissions on behalf of the king. His son, the "White Earl," was undoubtedly the most influential Irish magnate in the first half of the fifteenth century. He was eight times chief governor of Ireland: lieutenant 1420–1422, 1425–1426, and 1442–1444; justiciar 1426–1427; and deputy 1407–1408, 1424,

1441–1442, and 1450–1452. Like his son, the fifth earl, he saw military service in continental Europe on several occasions, and was a frequent visitor to England. Within the Butler lordship he exercised firm control over the rivalries of the cadet branches, at the same time successfully managing the Irish septs on the frontiers.

Rivalry with the Fitzgeralds

Once the absentee lords were no longer serious rivals for power, it was in the nature of things that the remaining Anglo-Irish magnates would engage in the struggle for supremacy. In the fourteenth century a bitter feud arose between the Butlers and the FitzGerald earls of Desmond. During the minority of the second earl of Ormond, the earl of Desmond ravaged Ormond and Eliogarty in 1345, which seems to have instigated a devastating revolt by the O'Kennedys of Ormond and other Irish septs in the Butler lordship. The cause of these disputes is hard to determine, but they were probably provoked by territorial rivalries. On one memorable occasion in the chapel of Dublin castle in 1380 in the presence of Edmund, Earl of March, the celebrant, the bishop of Cloyne, began the preface in the mass with the words: "Eternal God, there are two in Munster that destroy both us and our property, to wit the earls of Ormond and Desmond, together with their bands of followers, whom in the end may the Lord destroy, through Jesus Christ our Lord." Such rivalries were further complicated by political alignment occasioned by the Wars of the Roses, which placed both the Desmond and Kildare FitzGeralds in the opposing Yorkist camp. The attainder and execution by the Yorkists of James, fifth Earl of Ormond, in 1461, left Ireland effectively in control of the FitzGerald earls until the succession of Piers Butler to the earldom in 1515.

Relations with the Native Irish

The Ormond deeds contain a number of fourteenth-century treaties between the earls and their Irish subjects, including three with the O'Kennedys of Ormond: 1336, 1356, and 1358. While the treaties reflect the changing balance of power between overlord and subject in the context of a Gaelic revival, they reveal some elements that were characteristic of the relationship reaching back to the invasion. Those elements included a system of judicial arbitration based on the compensatory provisions of Brehon law; an annual rent, sometimes expressed in monetary terms, but which almost certainly took the form of an ancient cattle rent reaching back into pre-Norman arrangements, and continuing into the sixteenth century; attendance at the earl's court in Nenagh; and military service in the form of cavalry and foot soldiers. It is unlikely that this arrangement survived in its judicial aspects into the fifteenth century. However, it is clear that even after the wars of the Gaelic recovery in the previous century, the Irish septs on the periphery of the lordship as often as not formed alliances with the Butlers, probably to secure protection from their rivals, or a consequence of internal power struggles. In this way, the third and fourth earls in particular anticipated the kind of Gaelic alliances that one associates with Gerald, the Great Earl of Kildare, in the later fifteenth century.

Cadet Branches

The emergence of powerful cadet branches was a notable feature of the later medieval period. The most important of these groups were the Butlers of Cahir, who traced their lineage from a liaison between the third earl and Catherine of Desmond. Their bitter rivals, the Butlers of Polestown (County Kilkenny), also traced their ancestry to the third earl, regarding themselves as next in line to the succession. Such rivalries were aggravated through family ties with the FitzGeralds, involving the Cahir Butlers with Desmond, and the Polestown Butlers with Kildare. The Butlers of Dunboyne, whose Tipperary base was the manor of Kiltinan, also became entangled in these rivalries. Repeated and only partially successful efforts were made by the earls in the course of the fifteenth and first half of the sixteenth centuries to contain such rivalries by a series of ordinances issued in assemblies composed of the inhabitants of the lordship.

Adrian Empey

References and Further Reading

Carte, Thomas. *An History of the Life of James, Duke of Ormond*. London: 1736.

Empey, C. A. "The Butler Lordship in Ireland, 1185–1515." Ph.D. diss., Trinity College Dublin, 1970.

———. "The Settlement of the Kingdom of Limerick." In *England and Ireland in the later Middle Ages*, edited by James Lydon, 1–25. Dublin: Irish Academic Press, 1981.

———. "The Norman Period, 1185–1500." In *Tipperary: History and Society*, edited by William Nolan and Thomas G. McGrath, 71–91. Dublin: Geography Publications, 1985.

———. "County Kilkenny in the Anglo-Norman Period." In *Kilkenny: History and Society*, edited by William Nolan and Kevin Whelan, 75–95. Dublin: Geography Publications, 1990.

———, and Katharine Simms. "The Ordinances of the White Earl and the Problem of Coign in the Later Middle Ages." *Proceedings of the Royal Irish Academy*, 75C, no. 8 (1975): 162–187.

See also **Anglo-Norman Invasion; Burgh, de; Clyn, Friar John; Coyne and Livery;John, King of England; Desmond Geraldines; Fitzgeralds; Gaelic Revival; Gaelicization; Kilkenny; Lancastrian-Yorkist Ireland; Leinster; Manorialism; Munster**

C

CANON LAW

Canon Law, both as the actual decrees of legislators (popes, bishops, synods) and as jurisprudence (collections of decrees, systems, and commentaries on law), saw itself throughout the Middle Ages as in continuity from the texts of Divine Law (i.e., the Christian Old Testament, read as "the Law and the prophets" [cf. Matthew 7:20]) and Christ as the new lawgiver. It drew out this Law into its details, and applied it to new situations. This connected dual focus, an ideal "then" and a specific "now," make it a source of unique value (and complexity) to the historian. In it we observe the concerns of a society (e.g., power structures, land-holding, status), how they managed problems, how they viewed social and legal ideas (e.g., their conception of Christianity), and their image of an ideal society.

In the insular context, Canon Law is found in three forms. First, in that specifically insular form, the penitentials: manuals prescribing penalties in reparation for individuals' sins. Second, in synodal legislation: both the acts of synods that took place or as legislation that is presented as having come from a synod (e.g., the First Synod of Patrick), and as "a law" on specific topics such as *Cáin Adomnáin* (the Law of Adomnán), which came from the Synod of Birr (697). Third, in *collectiones*: law books for those who applied the law in administration or a court situation. From Ireland we have, comparatively, an embarrassment of riches in all these forms from the early medieval period, and the earliest evidence for the interaction of Christian law with legal systems from outside the Greco-Roman world. Thus Brehon Law manifests the influences of Canon Law in its language, discussions of problems, and decisions, whereas Canon Law was adapted to Irish legal practices, expressed in canonical forms applying native principles on matters such as land-holding, divorce and inheritance, and procedures, and took over elements from that law to solve difficulties that had emerged in Canon Law in the fifth and sixth centuries on the Continent. For example, the problems of sins after baptism and public penance, which bedevilled Canon Law from the fourth to the seventh century, for example, in Spanish collections of law, were solved in Ireland by adopting the native notion of an honor price as a means of satisfying justice after an injury. The crime against God was processed analogously to a crime of an inferior against a superior in the native system, and this solution passed through Irish legal texts to the rest of the Latin Church (cf. Archbishop Theodore's judgment of the *Penitential of Finnian*). This need to integrate two legal corpora—native and canon—may be seen as a distinctive feature of Irish Canon Law. If the "lawyers" of both systems were not to be continually at loggerheads they had to be able to systematize the contents of their respective laws and develop jurisprudence for this process.

We see this occurring in the greatest product of Irish Canon Law: the *Collectio canonum hibernensis*. Compiled in Ireland in the late seventh to early eighth century, it is one of the earliest, and possibly the earliest, systematic presentations of Christian law in Latin. While earlier collections were arranged in a chronological format, the *Collectio* gathered specific legal problems and judgments together under general headings and attached to each judgment its basis as law. Thus, for example, all the laws relating to bishops (e.g. duties, powers, and selection) were gathered into one place (Bk 1: On Bishops), and arranged in a logical order with the particular "laws" that might inform judgment listed in order of precedence (e.g., *Lex dicit* ["the Old Testament says"], followed by *Synodus . . . dicit* [the Synod of . . . says], followed by *X ait* [the authority X has said]). In this way the differences between legal positions, for example, two conflicting laws from different synods, were overcome by presenting an encyclopedic, and supposedly consistent, picture of the law. The collectors had searched the Scriptures, the Christian laws available to them—from both Greek (mainly synods) and Latin (synods and decretals from popes) worlds—and all the major Latin (and some Greek) theological authors, and had abstracted any material that was relevant to their topics. Then by retaining all these items within a system they effectively created the belief that all these sources were in harmony—for every "judgment" (be it a verse of Scripture or an injunction from one of the Fathers) had its place—and agreed with their own insular legislative agenda.

The *Collectio* also reveals the insular situation, in that there are elements drawn from Irish law in such matters as inheritance and marriage, rules of evidence, and property. Since it takes from Irish law, and in other places outlines how Canon Law can function alongside native law (e.g., on divorce), it was necessary to develop a theory of the origins of legal systems and their respective competencies. The *Collectio* did this by developing a notion of a "natural law" from Paul (Romans 2:14), which could function alongside a "revealed [Christian] law." This notion of related, yet distinct, systems allowed Christian law to accommodate itself within social structures other than those of the world in which Christianity arose and where it forged its basic legal structures. It is this systematic and comprehensive arrangement, combined with the possibility of being integrated with other systems, that explains the popularity of the *Collectio* as a legal textbook and a model of imitation and excerption, from the eighth to the twelfth centuries, on the Continent. We know that the *Collectio* was being used in the Rhineland as a textbook in the first half of the eighth century, and that the Carolingians used it as a legal resource. Later it formed a base for many other legal collections, and some of its material and jurisprudence eventually found its way into the *Decretum* of Gratian (prior to 1159).

THOMAS O'LOUGHLIN

References and Further Reading

Brundage, James A. *Medieval Canon Law.* London: Longman, 1995.

Kuttner, Stephan G. *Harmony from Dissonance: An Interpretation of Medieval Canon Law.* Latrobe, Pennsylvania: The Archabbey Press, 1960.

O'Loughlin, Thomas. *Celtic Theology: Humanity, World and God in Early Irish Writings.* London: Continuum, 2000.

O'Loughlin, Thomas. *Adomnán at Birr, A.D. 697: Essays in Commemoration of the Law of the Innocents.* Dublin, Four Courts Press, 2001.

[Five articles on the *Collectio canonum hibernensis*]. *Peritia: Journal of the Medieval Academy of Ireland* 14 (2000): 1–110.

See also **Adomnán; Biblical and Church Fathers; Brehon Law; Christianity, Conversion to; Ecclesiastical Organization; Penitentials**

CAROLINGIAN (LINKS WITH IRELAND)

The Carolingian dynasty, named for its most famous son, Charles (Latin *Carolus*) the Great, or Charlemagne (*c.* 742–814), which ruled the greater part of Christian western Europe between 751 and 887, gave its name to a major cultural phase, the impact of which extended to Ireland.

The dynasty rose to prominence in the early seventh century and gradually reduced their titular Merovingian kings to figureheads. Pippin III (d. 768) deposed the last Merovingian ruler and, in 751, was proclaimed king of the Franks. Twenty years later the kingdom passed to Charlemagne, who embarked on a series of conquests that made him the undisputed master of the Christian West except for Britain, Ireland, parts of southern Italy, and northern Spain. On Christmas day 800, he was crowned emperor of the restored Roman Empire by Pope Leo III. Central to the maintenance of the empire was the education of an administrative elite, and some of the greatest scholars in Europe, including a number of Irish, were brought to the imperial capital at Aachen. The liberal arts were promoted, classical texts were copied and preserved, and a new script was devised—Carolingian minuscule—which was adopted in Ireland. Charlemagne was succeeded by his son Louis the Pious, on whose death in 840 the empire was divided into three, with the western section, including most of Gaul, going to Charles the Bald (823–877). Carolingian authority was further weakened by subsequent partitions and, although members of the dynasty ruled in France until 987, the last holder of the imperial title was Charlemagne's great-grandson, Charles the Fat, who was deposed in 887.

The Carolingian monastery of Echternach, which became a royal monastery in 751, was one likely means of access to the court for eighth-century Irishmen. In 767, Vergilius (d. 784) was appointed bishop of Salzburg, after successfully completing a diplomatic mission for Peppin III. Einhard, Charlemagne's biographer, commented that letters existed in which Irish kings praised the emperor and addressed him as "lord;" these probably related to the security of the pilgrim routes and the maintenance of hostels. In 772–774, Charlemagne directed that goods plundered from pilgrims be restored to the Irish church on the island of Honau, near Strasbourg. Irish scholars at the court of Charlemagne and Louis the Pious included Josephus Scottus (*fl.* 782–796), Dúngal (*fl.* 804–827), Clemens Scottus (*fl.* 815–831), and Dícuil. Irish influence reached its zenith at the court of Charles the Bald, where the circle included Murethach of Auxerre (*fl.* 840–850), Sedulius Scottus, Ériugena, and Martin Scottus (*fl.* 850–875), teacher of Greek at the court school in Laon. In 846, Charles the Bald confirmed the re-establishment of hospices for Irish pilgrims, while direct political contact is attested in 848 when Charles received an Irish embassy that presented him with gifts, requested safe passage to Rome for the "king of the Irish" (presumably Máel-Sechnaill I), asked for an alliance, and reported that their king had won a great victory over the Vikings. No political alliance was made, however, and with the waning of Carolingian power, the influence of Irishmen in France also declined.

John Bradley

References and Further Reading

Becher, Matthias. *Charlemagne*. New Haven and London: Yale University Press, 2003.

Kenney, J. F. *The Sources for the Early History of Ireland: Ecclesiastical*. New York: Columbia University Press, 1929. Reprint, Dublin: Padraic Ó Táilliúr, 1979 (pp. 517–604 in the 1979 edition).

McKitterick, Rosamond, ed. *The New Cambridge Medieval History, volume II: c.700-c.900*. Cambridge: University Press, 1995.

See also **Dícuil; Ériugena John Scottus; Peregrini; Pilgrims and Pilgrimage; Sedulius Scottus**

CASHEL, SYNOD OF I (1101)

In the year 1101, a synod was convened at Cashel by Muirchertach Ua Briain in his capacity as king of Ireland. Although many people are reported as attending, both cleric and lay, the only names we have, apart from that of Ua Briain, are his brother Diarmait and bishop Máel-Muire Ua Dúnáin. According to one source, Ua Dúnáin presided over the synod as papal legate; however, doubt has been cast upon the veracity of that claim. Unusually for a synod of this period, its decrees (or at least some of them) have survived; they are found in an Ua Briain genealogy and are believed to be genuine.

There is, however, a dispute over their interpretation. Some would see them reflecting the papal reforms then taking place elsewhere in the church. For example, the first decree is about *aithlaích* or *aithchléirig* (often translated as "ex-laymen," "ex-clerics," but also "laymen or clerics who are now penitents"); some historians translate this decree in a way that would suggest that they are being prohibited from purchasing a church. Because of this they see it as a prohibition on simony, a vice that the reform papacy of the time was very keen to stamp out, although they are unable to explain why it applies to the particular category of people in question. Interpreted differently, the decree is seen as a reaffirmation of a long-standing church rule that prohibited such people from taking possession of a church; it is thus a conservative rather than a reforming decree. Similar interpretations could be applied to another decree that prohibits laymen from becoming *airchinnig* (heads of ecclesiastical establishments). While expressing puzzlement as to why the prohibition is limited to the office of *airchinnech*, the decree is nevertheless seen to be particularly Gregorian in character. This is because popes and their legates were, around that time, busy on the continent seeking to free the church from the control of lay princes. This interpretation, however, assumes that the office of *airchinnech* had been taken over by laymen and that the synodsmen were now declaring the practice illegal. However, against this it is argued that the laymen who had taken over the office were in fact clerics, but without ecclesiastical orders, and that the church always forbade laymen from holding the office. The decree merely re-affirms this and is not therefore a reform.

Finally, there is the decree that defines what relationships are considered to be incestuous. Although this is accepted as being very limited in its scope, it is nevertheless seen to be an effort made to address perceived irregularities in Irish marriage practices. There had been many complaints, especially from non-Irish people, about these around the time of the synod. However, it can be argued that many of the foreign complaints were based upon the fact that there was a substantial difference between Irish and mainstream church laws on what was considered to be incestuous. Irish laws, it is argued, were based upon the Mosaic laws, and marriage was allowed between first cousins. Laws in the rest of the church at that time prohibited marriage between people who were related up to the seventh degree of relationship. The decree passed at Cashel confirms existing Irish law and is therefore a restatement of that, rather than being an effort to bring Irish laws into line with those that prevailed in the rest of the church.

This synod is seen as a reforming synod by virtue of the decrees it passed by those who interpret them as reforming decrees. However, there is another event that occurred at the synod, by virtue of which it is perhaps more entitled to carry that title. And it is this event in particular that the annalists picked out for mention, referring to it as unprecedented in Irish history: the grant of Cashel, the ancestral seat of the kings of Munster, by Muirchertach Ua Briain, as a gift to the Irish church forever. The significance of this became clear ten years later when Cashel was chosen at the synod of Ráith Bressail as the metropolitan see for the southern province in the new church structure planned there. But it was also significant in that it is the first indication we have of Ua Briain's changed strategy in relation to church reform at the beginning of the twelfth century. Henceforward, he would pursue a course that would see that reform carried out within a purely Irish context only, with no place in it for Canterbury.

MARTIN HOLLAND

References and Further Reading

Gwynn, A. *The Irish Church in the Eleventh and Twelfth Centuries*, edited by Gerard O' Brien. Dublin: Four Courts Press, 1992.

Hughes, K. *The church in early Irish society.* London: Methuen, 1966.

Gwynn, A. *The Ttwelfth-Century Reform,* A history of Irish catholicism II. Dublin & Sydney: Gill and Son, 1968.

Watt, J. *The Church in Medieval Ireland.* 2nd ed. Dublin: University College Dublin Press, 1998.

Holland, Martin. "Dublin and the reform of the Irish church in the eleventh and twelfth centuries." *Peritia: Journal of the Medieval Academy of Ireland* 14 (2000): 111–160.

See also **Church Reform, Twelfth Century; Ráith Bressail, Synod of; Ua Briain, Muirchertach**

CASHEL, SYNOD OF II (1172)

Twenty years after the synod of Kells had received papal confirmation for the new organizational structure for the church in Ireland, another synod took place at Cashel. This, however, was a synod of a different kind; it assembled at the request of King Henry II of England, shortly after his arrival in Ireland on October 17, 1171. He first went to Lismore, the see of Gilla-Críst Ua Connairche, then papal legate in Ireland. Afterward he proceeded via Cashel to Dublin, and thus had the opportunity to meet two archbishops, Donnchad Ua hUallacháin (Cashel) and Lorcán Ua Tuathail (Dublin). Through these contacts arrangements were put in place for a synod to meet at Cashel soon afterward.

We have to rely on non-Irish sources, in particular Giraldus Cambrensis, for information on the synod, as the Irish annals curiously ignore it. This may have a bearing on how it has come to be interpreted by historians. We are told that the papal legate, Gilla-Críst, presided, and the archbishops of Cashel, Dublin, and Tuam (Cadla Ua Dubthaig)—together with the bishops of their provinces, other bishops, and clergy—attended. Due to age and infirmity, the archbishop of Armagh, Gilla-Meic-Liac, was not present, but he later gave his consent to its decisions. Henry II apparently did not attend and was represented instead by Ralph, abbot of Buildwas; Ralph, archdeacon of Llandaff; Nicholas, the king's chaplain; and other royal officers.

The decrees enacted were:

- the laity to repudiate spouses related to them by kinship or marriage and to adopt lawful marriage contracts;
- infants to be catechised before the doors of the church and baptised in consecrated fonts in baptismal churches;
- parishioners to pay tithes of animals, crops, and other produce to their own church;
- church property to be free from all exactions made by secular magnates;
- clerics, whose kinsmen are obliged to contribute to composition payments, to be exempt from such contributions;
- all the faithful, in their final illness, to make a will in the presence of their confessor and neighbors in the manner specified;
- those who die after a good confession to receive their due in terms of masses, vigils, and manner of burial.

It was also decided that in Ireland, all matters relating to religion were to follow the observances of the English church. Some have interpreted this as referring to liturgical practices only; others see it as encompassing more, and therefore being much more fundamental, especially since it is claimed that the Irish bishops swore fealty to Henry at around this time. Given the absence of Irish sources it is difficult to be sure what the position of the Irish bishops was, either collectively or individually, although we know that they had communicated concerns about the state of the Irish church to Pope Alexander III. But from Henry's point of view, the synod was a diplomatic success. He was able,

St. Patrick's Rock, Cashel, Co. Tipperary. © *Department of the Environment, Heritage and Local Government, Dublin.*

through Archdeacon Ralph, to report to the pope on the assistance he had received from the bishops and to have, in return, papal instructions issued to the Irish bishops and kings, instructing them to support his rule in Ireland.

MARTIN HOLLAND

References and Further Reading

Gwynn, A. *The Irish Church in the Eleventh and Twelfth Centuries*, edited by Gerard O' Brien. Dublin: Four Courts Press, 1992.

Watt, J. A. *The Church and the Two Nations in Medieval Ireland.* Cambridge: 1970.

Flanagan, M. T. "Henry II, the council of Cashel and the Irish bishops." *Peritia: Journal of the Medieval Academy of Ireland* 10 (1996): 184–211.

See also **Giraldus Cambrensis; Henry II; Kells, Synod of**

CASTLES

Castles emerged with the new aristocratic society of Europe in the tenth century, providing lords with secure centers from which to control estates of land. From the start they were characterized by defensive features, both for use against attack and as a display of wealth and power. By the eleventh century, these defended residential power centers were being constructed to articulate the lands of kings, major aristocrats, and landed knights. As the competition for power developed, they grew more elaborate and required greater resources to construct; from the twelfth century they could also display an increasing elaboration of provision for ceremony and the life of a large hierarchical household. The study of castles leads us directly to the resources and priorities of their aristocratic or royal builders.

The system of lordship which required a castle was different from the traditions of early medieval Ireland. Castle lordship was based on a spatial organization and stability that saw the enduring control of land as the primary core of power, rather than personal relations between lord and man. An estate organized around a castle imposed its own order on the landscape and those who lived in it; the focus of the castle made a ready center for other activities. Possession of a castle was clear evidence of possession of the lordship, and it could therefore be transmitted more easily from a lord to his successor than traditional Irish lordship. Castles in Ireland show how the new order of feudal Europe differed from the early medieval polity. Because of the investment required to build castles, they are good guides to the real intentions of the lords who built them. Castle designs show the balance of lords' provision of accommodation, display, and defense and suggest their priorities. In Ireland, it has been suggested that castles may have been constructed before the arrival of English lords in 1169, just as in England before the Norman Conquest of 1066; in both cases this implies a change in traditional lordship. These are either remains of stronger fortification than was normal in early medieval Ireland (Downpatrick, Duneight, Dunamase, or Limerick) that could be attributed to the period, or else to references to sites called castles in contemporary documents. Unlike pre-1066 England, the sites are extensions of Irish royal power rather than the organization of lesser estates.

Between 1169 and the crisis of the mid-fourteenth century, castles are overwhelmingly associated with English lordships. From the start, castles were built either of stone and mortar or of earth and timber, or a combination of both. The choice of medium was dictated by resources (stone castles cost at least ten times as much) or the need for speed (earth and timber buildings could be erected in one year, not ten or twenty); one of the most enduring misconceptions is that earthwork castles were more primitive and earlier than stone ones. The early construction in stone is a clear sign of the new lords' commitment to remaining in Ireland. Trim has been extensively excavated, showing how the stone great tower was inserted into an earthen enclosure. The tower (three periods of construction: 1180s, 1195–1196 and 1203–1204) was a magnificent building, providing suites of rooms for the lord and members of his household. The approach to the castle from Dublin was marked by a gate tower of a French design unique to the British Isles. Defensively, however, the great tower was not particularly effective by the standards of the day. The great round tower at Nenagh looked, like the Trim gate, to Pembroke and France for its inspiration. Carrickfergus was much simpler, a tower for the lord's household and a small courtyard with the public hall, chapel, and kitchen. The bulk of castles of the late twelfth and early thirteenth centuries were of earth and timber, and the great majority of these apparently were mottes, erected where speed was important or where the lord was prepared or able to commit fewer resources. The distribution of mottes in Ireland is surprising, in that they are not evenly spread over the earlier English lordships. The conclusion from the study of the castles of the first two generations of English lordships is that they were built overwhelmingly by lords to celebrate their seizure of land, not to conquer it nor to hold it against potential or actual rebellions.

Most of the castles, and all of the most elaborate ones, put up before 1200 were built for aristocrats, not for the king. Neither Henry II nor Richard I seized much land or built significant castles to assert their power; only with John, who built strong castles at

Dublin and Limerick after 1205, did royal castles join the first rank. The weakness of the English king in terms of power on the ground lasted throughout the medieval period; power, land, and castles were indissolubly linked. Castles also changed the landscape of settlement. Major castles formed the centers of new towns (Carrickfergus and Carlingford were linked to new ports). Some, like Trim, were founded on monastic sites that could be easily developed as towns, perhaps because they were already centers of population. No evidence has yet been noted for the management of the rural landscape, through parks or routes, for castles in Ireland as it has in England, but this may be the result of the lack of looking rather than true absence.

The story of castles in the 150 years from around 1200 is one of steady development along preexisting lines—castles that usually reflected contemporary English (rarely French) practice, although there are signs of variations by Irish masons. Some of the major stone castles show the successive additions that resulted from their continuing positions as chief places of lordships: Trim (new hall range) or Carrickfergus and Dungarvan (twin-towered gate house). Probably because of the relative lack of resources, castles with additions are the exception; most lords seem to have been perforce content with the buildings they inherited. New lordships, of course, required new castles. The expansion of his family in north Connacht caused Richard de Burgh to construct Ballymote and Ballintober, while his ambitions in western Ulster produced Green Castle, County Donegal. The royal weakness in castle building continued after John's reign. Roscommon was a major castle, but the next largest project was at modest Roscrea. Limerick was left unfinished, and there was little work at Dublin. This is the period when the towered enclosure dominated by a grand gatehouse holds sway in European castles. In Ireland we see the prevalence of the twin-towered gatehouse: Castle Roche (1230s), Roscommon (1270s), and Ballymote (after 1299). Even if the model is grand, however, the scale of castles in Ireland tended to be more modest. Green Castle, in Donegal, reflects Edward I's great castles in north Wales, but at half scale. This is not a question of a lack of awareness of developments elsewhere (Roscommon foreshadows Edward's castle of Harlech ten years afterward), but of a lack of resources. The overall design is usually simpler, and economies are made in the accommodation of the households, but not in the lord's rooms. The defensive strength of castles in Ireland is similarly severely reduced in comparison with Europe; lords in Ireland do not seem to have anticipated much warfare. Some major buildings (Swords or Ballymoon), provide elaborate accommodation for the lord and his household, while remaining essentially undefended enclosures.

Lesser castles are elusive in the thirteenth century. The principal remains appear to be individual stone buildings. These are often interpreted as hall-houses, which implies that one building may combine hall, chamber, and stores, but there must have been other elements: kitchens, farm buildings, and so forth. Some buildings (Witches Castle, Castle Carra) are too small for halls, and it is more likely that they were chamber towers attached to wooden halls. Few display strong enclosures; in documentary accounts of manorial centers they may be described as surrounded merely by a hedge. By contrast, some borders in Ireland were marked by small, stone-built enclosures, lacking towers or gates, and apparently offering simple shelter for small (*c.* 20–50 men) bands stationed there to protect against raids. Until the mid-fourteenth century, there are few cases of Gaelic Irish lords constructing castles. A number of motte castles in mid-Ulster and some in Connacht, but only one or two stone castles, may be suggested as their work.

The fourteenth century saw a considerable change in castle building. The great castles seem confined to a few major lords, principally the great earls of late Anglo-Ireland: Kildare, Desmond, and Ormond. The finest surviving example is Askeaton castle, built for Desmond during the fifteenth century; a window in the great hall is very similar to one in the nearby friary founded in 1389. There is a great hall in the outer court; more-privileged visitors could penetrate to the inner court, where the earl and his immediate household were accommodated in a great tower with major state rooms and private chambers. Apart from being set on an island in the River Deel (which is not difficult to cross), the castle is weakly defended, without towers along the perimeter or (apparently) a gate house. Other castles such as Newcastle West (also Desmond), Adare (earl of Kildare), or Granny (Ormond) show the same pattern of fine domestic accommodation for the lord but provision for smaller households and weak defenses. Even the larger castles were slow to provide for the deployment of guns. Unlike the earlier castles, it is not easy to find close parallels in England or France for either the architectural details or the overall design.

The vast majority of castles from after the mid-fourteenth century—several thousand were built—were tower houses. As the name implies, the key to them is a stone tower, although in some cases at least a hall and other buildings of less-substantial material were attached. The accommodation is modest, suited for a family in the more modern sense rather than a large household. They have defensive features, but careful analysis often shows these to be (even more than is usual among castles) more gestures of display than effective defense, even against low-level violence. For example, there are flanking towers that do not cover the ground floor doors. In different regions of Ireland (Co. Limerick or Co. Down) they have been shown to be built to a common pattern, which contrasts with normal castle building, wherein each castle emphasizes its originality. Their builders seem to be stressing their adherence to a common group. Tower houses are common in Scotland and the north of England as well as Ireland, and seem to be associated with particular groups of people and their lifestyles. The core of these groups consisted of new gentry who prospered in the conditions of the weakened power of the great lords during the later fourteenth and fifteenth centuries. The Irish lords could associate themselves better with this sort of castle than the more elaborate earlier designs, and with them castle building first became a common feature among the Gaelic Irish. Tower houses also found favor among town merchants and rural priests. Tower houses are the key feature for the detection and understanding of settlement after the Black Death.

The final building of castles in Ireland came in the late sixteenth and early seventeenth centuries. As well as the continuing construction of tower houses, there were a number of more or less fortified houses constructed, to which contemporaries gave the name of castles, especially if they were combined with a strong enclosure or bawn. They often provided a display of gables and corner towers, and the defensive features of gun loops combined with elaborate machicolations and fake battlements. The most interesting set of these castles or strong houses is associated with the Plantation of Ulster, where the castles or houses (contemporaries use both words) use a variety of features derived from the Irish, English, and Lowland Scottish architectural repertoires. English planters tended to build in English style and Scots in Scottish style, but both used Irish workmen.

T. E. McNeill

References and Further Reading

McNeill, T. E. *Castles in Ireland.* Routledge, 1997.

Sweetman, P. D. *The Medieval Castles of Ireland.* Boydell, 1999.

See also **Architecture; Houses; Tower Houses; Anglo-Norman Invasion; Archaeology; Dublin; Feudalism; Kilkenny; Limerick; Manorialism; Maynooth; Motte-and-Baileys; Pale, the; Ringforts; Trim; Villages; Walled Towns; Weapons and Weaponry**

CATHAL MAC FINGUINE (d. 742)

Arguably the most powerful Munster king before Brian Boru, Cathal belonged to the Eóganacht Glennamnach dynasty. His father, Finguine (d. *c.* 695) son of Cú-cen-Máthair, is styled king of Munster in his obit, but little is known of his reign. Cathal's predecessor as overking was Cormac (d. 713) grandson of Máenach, of the Eóganachta of Cashel. The Munster regnal list, in claiming a twenty-nine-year reign for Cathal, implies that he succeeded his predecessor immediately. However, there are indications that he struggled to assert his authority at provincial level. In 715, Murchad son of Bran, the Uí Dúnlainge over king of Leinster, marched on Cashel. Further doubt is cast on the extent of Cathal's sway during these early years by the record of a rival, Eterscél son of Máel-umai of Eóganacht Áine, who is styled king of Cashel in his obit at 721. However, from that time onward, Cathal emerges not only as a strong over king of Munster, but as the dominant political force in Leth Moga, and as a serious threat to the political order that the Uí Néill dynasties strove to establish.

Cathal's marriage to Caillech (d. 731) daughter of Dúnchad Ard, a princeling of Uí Meicc Brócc, possibly reflects an early initiative—even before his accession to kingship—to forge alliances with dynasties to the east of Cashel. In any event, he had a daughter Tualaith, and a son (or, more likely, grandson) Artrí. Later, he secured a judicious marriage pact with the Uí Dúnlainge dynasty of Leinster, when his daughter Tualaith wed Dúnchad son of Murchad. It is not clear whether the marriage in question was arranged before or after 721, but in that year, as is widely noticed in the annals, Cathal joined forces with Murchad to plunder Brega (east Co. Meath and north Co. Dublin). Following this, the (admittedly partisan) *Annals of Inisfallen* make the dramatic claim that the Uí Néill king of Tara, Fergal son of Máel-dúin, submitted to Cathal. An appended text, which reckons Cathal and Brian Boru among five Munster kings who "ruled Ireland," resembles (eleventh- or twelfth-century) Ua Briain propaganda and fuels misgivings about the submission claim.

Yet, it is clear that Cathal overshadowed Leth Moga. Although it was the Uí Dúnlainge ruler, Murchad, and his subkings who defeated and slew Fergal mac Máele-dúin (722) in the crucial battle of Allen (Co. Kildare), Cathal presumed to intervene as kingmaker in Leinster following Murchad's death in 727. He supported his son-in-law, Dúnchad, against his brother, Fáelán. In the ensuing battle at Knockaulin (Co. Kildare), Cathal and his principal ally, the king of Osraige, were discomfited, Dúnchad was fatally wounded, and Fáelán seized the overkingship of Leinster. It is possible that Fáelán's subsequent marriage to Tualaith was intended as an affront to Cathal (the new Leinster king apparently opposed his designs), in which event her motives might well be questioned. Alternatively, it may have represented an attempt at settling differences. Either way, the years that followed witnessed strenuous efforts by Cathal to assert authority over Leinster. In 732, he was defeated by the king of Uí Chennselaig—but invaded Leinster again three years later. The *Annals of Ulster* maintain that he was repulsed with heavy losses, including his ally the king of Osraige, but Inisfallen insists that he secured victory over his recalcitrant son-in-law, Fáelán.

The case for Cathal, in parallel with his Leinster ventures, having challenged Uí Néill supremacy in Mide has drawn considerable debate. The *Annals of Ulster* at 733 records battles at the symbolic sites of Tailtiu (Teltown, Co. Meath) and Tlachtga (Hill of Ward, Co. Meath) involving one Cathal. Many (e.g., Binchy, Byrne, Jaski, Swift, Herbert) identify the protagonist as Cathal mac Finguine, but some (Ó Riain, Charles-Edwards) contest this, instead suggesting Cathal mac Áeda of Síl nÁeda Sláine. Those favoring Cathal mac Finguine note that he had previously invaded Brega; besides, the annal record at 733 contains no patronymic, which might be expected for a less well-known dynast. This annal entry aside, however, there are strong indications that Cathal not only claimed sway beyond Munster but was feared as a threat by the Uí Néill. The poem "Teist Cathail meic Finguine," in the Book of Leinster, styles him *Ardrí Temrach* (high-king of Tara), while the law tract "Bretha Nemed Toísech" (probably a Munster product dating to 721–742) makes reference to the *Feis* of Tara. Even more significant is the probability that the Uí Néill regnal poem "Baile Chuinn," in mentioning *fer fingalach* (a kin-slaying man)—a descendant of Corc (an Eóganachta dynast) who is "overlord of Munster of great princes in Tara"—refers to Cathal mac Finguine.

In 737, Cathal attended a *rígdál* (royal meeting) at Terryglass (Co. Tipperary) with Áed Allán son of Fergal, the new Uí Néill king of Tara. Presumably, the aim was to conclude a nonaggression pact, but the location of the meeting, bordering Leth Moga and Leth Cuinn, suggests mutual respect, with neither king summoned into the other's realm to betoken submission. The subsequent extension of the Rule of Patrick throughout Ireland—which recognized the ecclesiastical authority of Armagh in Munster—need not imply capitulation by Cathal, and may even have allowed him greater influence over Munster's Patrician foundations. However, it seems unlikely that Cathal—unless under duress—would have allowed Áed Allán to intervene in Leth Moga. Acknowledging that the annal record for 738 is somewhat disordered, it is possible to interpret Áed Allán's invasion of Leinster and defeat of its rulers—which seemingly drew no reaction from Cathal—as itself a response to the latter having taken the hostages of Leinster earlier in the year.

Cathal died in 742. Later assessments of his reign as a "milestone" ensured his place in Middle Irish literature, being featured in the historical tale *Cath Almaine* (the Battle of Allen) and in the satire *Aisling Meic Conglainne* as the king from whom the hero Mac Conglainne expels a demon of gluttony. Following his death, the succession record for Munster is confused; apparently, he was followed by ephemeral kings from other Eóganachta lineages. His son—or grandson—Airtrí (d. 821) regained the kingship in 796, and was ancestor of later Eóganacht Glennamnach kings.

AILBHE MACSHAMHRÁIN

References and Further Reading

The Annals of Inisfallen, ed and trans. Seán MacAirt. Dublin: Dublin Institute for Advanced Studies, 1944.

The Annals of Ulster to A.D. 1131. ed and trans. Seán MacAirt & Gearóid MacNiocaill. Dublin: Dublin Institute for Advanced Studies, 1983.

The Book of Leinster, vi, ed. Anne O'Sullivan Dublin: Dublin Institute for Advanced Studies, 1983, pp. 1376, 1379.

"The Ban Shenchus," ed. Margaret Dobbs, *Revue Celtique*, 47–9 (1930–1932), 185, 223;

E. Bhreathnach, F. J. Byrne, J. Carey & K. Murray, "Baile Chuinn Chétchathaig: edition." Edel Bhreathnach (ed.), *Tara: kingship and landscape*. Dublin: Royal Irish Academy, 2004, forthcoming.

Breatnach, Liam. "Canon law and secular law in early Ireland: the significance of *Bretha Nemed*," *Peritia*, 3 (1984), 439–59;

Byrne, Francis John. *Irish kings and high-kings*. London: Batsford, 1973; new ed. Dublin: Four Courts Press, 2001. xix, 150, 188–9, 203, 205f, 207–11.

Charles-Edwards, Thomas. *Early Christian Ireland*. Cambridge: Cambridge University Press, 2000. 280, 477–8.

Jaski, Bart. *Early Irish kingship and succession*. Dublin: Four Courts Press, 2000. 54, 219–20.

Mac Niocaill, Gearóid. *Ireland before the Vikings*. Dublin: Gill & Macmillan, 1972. 121, 123–4, 130.

MacShamhráin, Ailbhe & Byrne, Paul. "Prosopography I: kings named in Baile Chuinn Chétchathaig and the 'Airgialla Poem'." Edel Bhreathnach (ed.), *Tara: kingship and landscape*. Dublin: Royal Irish Academy, 2004, forthcoming.

Ní Chon Uladh, Póilín. "The Rígdál at Terryglass (737 A.D.)," *Tipperary Historical Journal*, (1999), 190–6;

Ó Corráin, Donnchadh. *Ireland before the Normans*. Dublin: Gill & Macmillan, 1972. 3, 23, 97, 101.

Ó Riain, Pádraig Introduction to P. Ó Riain (ed.), *Cath Almaine*. Dublin: Dublin Institute for Advanced Studies, 1978. xiii-xv

Swift, Catherine. "The local context of Óenach Tailten." *Ríocht na Midhe*, 11 (2000), 24–50, esp. 41.

See also **Eóganachta; Historical Tales; Leinster; Leth Moga; Mide; Munster; Osraige; Satire; Uí Dúnlainge; Uí Néill**

CÉITINN, SEATHRÚN

See **Forus Feasa ar Éirinn**

CÉLI DÉ

The *Céli Dé* (plural of *Céle Dé*, anglicized culdee, "servant of God") were a religious movement that emerged in *Leth Moga*, "the southern half of Ireland," in the mid-eighth century. It flourished under the leadership of Máel-Ruain, founder of the monastery of Tallaght, which became the center of the *Céli Dé* movement. In documents relating to the early Irish church the term *Céle Dé* was used prior to the ninth century to refer to religious persons in service to God, but thereafter came to mean an adherent to teachings of the new movement. The movement was characterized by intensified devotion to the ascetic spirit already present in Irish Christianity. Particular interest was placed on study and prayer and the desire to live as an anchorite. This was manifest by adherence to a strict code of practice that required its followers to recite the Psalter daily, to live a life of poverty, to practice charity to the poor and care for the sick, to practice mortification, to live a celibate life avoiding women, and to separate oneself from the world (especially on Sunday on which no work of any sort was to be done). In all of this, one was expected to avoid excess and live a life of moderation inspired by the love of God.

This life of religious asceticism was not in itself new to the Irish church, and a number of these practices can be identified in the early church. Although often called the "*Céli Dé* reform" by scholars, this classification has not been universally accepted. Those who see the *Céli Dé* as a reform movement point to a general drift by monasteries in the eighth century toward worldly concerns, especially characterized by their increased wealth and tendency for hereditary leadership. These scholars see the emergence of the *Céli Dé* as a reaction to a decline in earlier standards, and as a blueprint for reform. Hughes, for example, noted a rise in the number of anchorites cited in the annals and suggested that this was the result of a renewed vigor inspired by the *Céli Dé*. Etchingham, however, rejected this conclusion and suggested that the rise was due to a more complete annalistic record, and he argued that the *Céli Dé* were just a continuation of the anchorite tradition.

The difficulty in identifying the *Céli Dé* as a reform movement stems from the decentralized and fragmented style of its organization. Although the monastery of Tallaght was established as a center for the *Céli Dé* teachings, many of its leading adherents, such as Máel-Díthruib of Tír dá glass, were attached to older monastic settlements. Although the *Céli Dé* might make up the whole monastic body, as at Tallaght, they also might have a reduced presense as a distinct house attached to an older foundation such as Ros-cré or, as at Armagh, be a special group residing within the monastic enclosure. These followers were free to pursue the ascetic ideal as they felt was appropriate. One of the most famous examples of this relates to the consumption of alcohol. Máel-Ruain required strict abstinence at Tallaght, but Dub-Littir of Finglas, also a *Céli Dé* foundation, advocated relaxation of this practice during the feasts of Christmas, Easter, and Whitsun. Although the *Céli Dé* weren't uniform in their practice, it is clear that those who aspired to its teachings considered themselves to be different from members of the "old churches."

The earlier ascetics of the Irish church had been eremitic, and the *Céli Dé* also valued the anchorite tradition. One of the products of the movement was the introduction into the vernacular of nature poetry, characterized by internal and end rhyme, new meter, and alliteration. A number of other works have been credited to the *Céli Dé* movement. Among the most notable are the "rules," of which there are several. Some of the most prominent are *The Customs of Tallaght*, *The Rule of the Céli Dé*, and *The Rule of Fothad*. The number and slight variation of each bear testament to the diversity within the movement. The *Céli Dé* are also credited with producing the earliest extant Irish martyrology—the *Félire Óengusso*—as well as *The Martyrology of Tallaght*.

Despite the fervor of its adherents, the movement did not endure long. The decline of the *Céli Dé* lay in the autonomy that each house enjoyed. Because there was no single constitution or authority to protect its interests, each house was vulnerable to attack by outside forces, and by the tenth century the movement began to disappear.

MICHAEL BYRNES

References and Further Reading

Etchingham, Colmán. *Church Organisation in Ireland A.D. 650 to 1000*. Naas: Laigin Publications, 1999.

Gwynn, E. J., ed. and trans. "The Rule of Tallaght." *Hermathena* 44, 2nd supplemental volume, 1927.

Gwynn, E. J., and W. J. Purton, ed. and trans. "The Monastary of Tallaght." *Proceedings of the Royal Irish Academy* 29C, no. 5 (1911): 115–179.

Hughes, Kathleen. *The Church in Early Irish Society*. London: Methuen and Co. Ltd., 1966.

Kenney, James F. *The Sources for the Early History of Ireland: Ecclesiastical*. New York: Columbia University Press, 1929.

O'Dwyer, Peter. *Céli Dé*. Dublin: Carmelite Publications, 1977.

Stokes, W., ed. and trans. *The Martyrology of Oengus the Culdee: Félire Óengussa Céli Dé*. (Henry Bradshaw Society 29). London,1905. Reprint, Dublin, 1984.

See also **Devotional and Liturgical literature; Fedelmid mac Crimthainn; Ecclesiastical Organization; Hagiography and Martyrologies; Máel-Ruain; Moral and Religious Instruction; Penitentials; Poetry, Irish; Scriptoria**

CENTRAL GOVERNMENT

Central government refers to the bureaucratic machine that administered the medieval lordship of Ireland on behalf of the king of England. From 1171, the English king found it necessary to put in place a system that allowed him to govern his subjects in Ireland and preserve and develop his rights there from a distance. While recent research has identified considerable continuity in many features of settlement, economy, and society in Ireland before and after the conquest, in the sphere of government there was little recognizable administration with which the Anglo-Normans could link. Therefore, in most respects, the machinery of government had to be modeled on that found in England, where by the end of the twelfth century an organized administrative system was taking shape, which could be imposed on the newly conquered lands.

The system developed gradually in Ireland, probably originating in the lord John's visit of 1185. The king's place was taken by a chief governor who had the title of justiciar, or king's lieutenant. He was assisted by a number of professional administrators: the treasurer, chancellor, and escheator. The two principal departments or offices of the administration were the exchequer and the chancery.

The Exchequer

The exchequer was the first specialized department of government to appear in Ireland. This development mirrored events in England, where the exchequer had been the first branch of government to become detached from the king's itinerant household. From the time John became lord of Ireland, financial arrangements were put in place to collect revenue and pay expenses, and by 1200 this organization was being described as the exchequer. In its early days it appears that the justiciar was primarily responsible for rendering the accounts of the exchequer, but from at least 1217 the title of treasurer emerged as the designation of the chief clerk in charge of the exchequer. During the early thirteenth century the treasurer was second in importance only to the justiciar, but lost this position with the increase in the powers of the chancellor.

The treasurer was assisted by first one and later two chamberlains, and during the course of the thirteenth century a range of other minor officials appear that correspond to those of the English exchequer, such as the chancellor of the exchequer who was responsible for the exchequer seal, the chancellor's clerk, and the remembrancers who kept the records. The exchequer had its own court, which could determine financial disputes. The judges of this court were called barons of the exchequer and they are mentioned as early as 1207. They appear at this stage to have been prominent members of the administration. The emergence of professional barons, whose sole job it was to determine actions in the exchequer court, dates from the later thirteenth century.

The primary function of the exchequer was to receive and disburse the Irish revenues and to bring royal officials, particularly the sheriffs, to account. Each accountable minister was required to appear at Dublin twice a year, at Easter and Michaelmas. The exchequer was composed of two departments, which between them kept four main series of rolls. The receipt and issue rolls were produced by the lower exchequer and were records of income and expenditure. The upper exchequer produced the pipe rolls, which contained the audited accounts of the various officials who received and spent money on behalf of the crown. The memoranda rolls were also produced by the upper exchequer and contained records of correspondence, proceedings regarding accounts, and other miscellaneous material.

Very few original rolls now exist, and the workings of the exchequer must be reconstructed by means of calendars and transcripts.

Exchequer revenues were mainly derived from the profits of justice, the royal demesne (including the towns), escheats (see below), and royal service or scutage. They varied considerably throughout the medieval period, reaching their peak in the 1290s when, for a time, exchequer income was more than £9,000 annually. The revenues declined drastically in the early fourteenth century to an annual average of only £1,200 in the early years of the reign of Edward III. They rose slightly during the later fourteenth century, averaging about £2,000 per annum. However, in the fifteenth century, when the area under the control of the central government severely contracted, they shrank even more.

In contrast to England, where by the fourteenth century all government departments had ceased to move about with the royal household, in Ireland much more of the government remained itinerant. The exchequer was the only completely sedentary part of the administration. It sat in Dublin until 1361, when it was briefly and unsuccessfully moved to Carlow in an attempt to bring it closer to the towns and shires of the south.

The Chancery

The chancery was the secretariat of the Dublin government, which issued letters bearing the Irish great seal in the name of the king. These included writs of summons to parliament and copies of English statutes for circulation in Ireland. The chancery also drafted reports on the state of Ireland and messages to the king from the council or parliament. In Ireland, as in England, the chancery developed rather more slowly than some of the other departments. During the first fifty years of the lordship it appears that chancery business was conducted by the justiciar's household clerks, and letters bore the justiciar's own seal. The separate Irish chancery began in 1232, and the first Irish chancellor was Ralph Neville, bishop of Chichester and chancellor of England, who was granted the Irish chancery for life. He performed his functions through a deputy, and on his death in 1244, his deputy, then Robert Luttrell, was continued as chancellor of Ireland. From this date the office had a continuous independent existence, and the chancellor became the most senior member of the Irish council. The office was held by a mixture of professional administrators and ecclesiastics, such as the priors of the Hospital of St. John of Jerusalem (the Knights Hospitallers) in Ireland.

The chancellor was paid an annual fee and also received the fees paid for sealing letters with the great seal. He was expected to maintain a staff of clerks out of this amount. However, a constant criticism of the chancery was that it was under-staffed, with too few clerks, and those of doubtful qualifications. About 1285 it was alleged that there was only one chancery clerk, who knew little of its business. Several attempts were made at reform and reorganization, but they met with little success. English chancery clerks were sent over from time to time, particularly in the fourteenth century, in an attempt to raise standards, but the Irish chancery continued to be seen as rather unprofessional, and it did not develop specialist functions similar to those that developed in England. Similarly, the medieval Irish chancery rolls did not develop the elaborate subdivisions of the English chancery. It issued two basic kinds of letters, letters patent and letters close, which were differentiated by their wording and by the method of attaching the seal. Letters patent, which included appointments and grants and were intended to be shown to interested parties, had a seal hanging from a cord or strip of parchment, while the seal on letters close, which contained instructions and orders, was attached after the document had been rolled up and had to be broken in order to read it. The annually compiled patent and close rolls of the chancery contained copies of these letters. Most of this original documentation has not survived, but there have been attempts to reconstruct the rolls from a variety of substitute material.

The chancery had no fixed abode but traveled around the country, usually in the company of the justiciar who witnessed all the letters issued by this department. An attempt in 1395 to provide a permanent base in Dublin was unsuccessful.

The Escheator

The escheator was the official who was in charge of administering and accounting for the lands that came into the king's hands as a result of vacancies in bishoprics and religious houses, minority of the heirs of tenants-in-chief, and forfeitures. In Ireland this office assumed a much greater significance than its equivalent in England because of the importance the king attached to securing the feudal revenues of the colony, particularly those arising from vacant bishoprics and religious houses. The Irish escheator therefore, unlike his English counterpart, was among the most prominent of Irish ministers, holding a place next in importance after the chancellor and treasurer (in 1346, for example, the justiciar was instructed to act in all matters by the advice of the chancellor, treasurer, and escheator). The escheator was a leading member of the council and was paid an annual fee of £40. He could, however, earn a good deal more than this: in the middle of the fourteenth century there is a record of the escheator being paid 20 shillings for the execution of a single writ.

The first known, regularly appointed escheator was Geoffrey of St. John, appointed in 1250. Before that date it appears that the justiciar was responsible for accounting for escheats, although there is some mention of a specially appointed officer. After 1250 there is a regular succession of escheators, and the workings of the office become clear. The escheator took possession of land on behalf of the king and held inquisitions to determine the exact value of the lands and the identity and age of the next heir. If the land remained for a time in the king's hands it was administered by the escheator, either personally or by appointed custodians. The escheator's functions were spread all over the lordship and necessitated the employment of deputies or sub-escheators. These officials first appear in the records around 1270, but probably existed much earlier.

The escheator accounted for the accrued revenues of his office at the exchequer. In the thirteenth century the income of the escheatry was substantial, but it declined severely in the following centuries and so, too, did the importance of the office of escheator. He was gradually removed from his important position in the official hierarchy, and by the fifteenth century he had ceased to be a member of the council or an official on an equal footing with the chancellor and treasurer.

The central government of medieval Ireland, modeled on that of England with some modifications, can be seen to have operated more or less competently during the first centuries of the Anglo-Norman lordship. However, the power of this central authority could never be considered to be equivalent to that of royal government in England, and in the later Middle Ages the system started to break down as a result of a combination of circumstances. The records reveal frequent changes in personnel, as effective officials were promoted to positions in the English administration and corrupt and inefficient ones were ejected. There were many instances of peculation and malpractice, and in general, confidence in the agents of government was not high. The very low level of the annual finances meant that the government could not maintain a regular payment of fees and wages to its own officials. The surviving records reveal that even the fees of high officials were regularly in arrears. This increased the incidence of corruption as officials sought to redress the shortfall in expected income by peculation. The growing political and cultural fragmentation of Ireland piled on difficulties for a centralized system of government. As the maintenance of peace and stability became more costly and more problematic, so too did the potential of the administration to govern become correspondingly less effective and less efficient.

MARGARET MURPHY

References and Further Reading

Connolly, Philomena. *Medieval Record Sources*. Dublin: Four Courts Press, 2002.

Frame, Robin. *Colonial Ireland, 1169–1369*. Dublin: Helicon, 1981.

Lydon, James. *Ireland in the Later Middle Ages*. Dublin: Gill and MacMillan, 2003.

Otway-Ruthven, A. J. *A History of Medieval Ireland*, 3rd ed. New York: Barnes and Noble, 1993.

Otway-Ruthven, A. J. "The Medieval Irish Chancery." In *Album Helen Maud Cam* 2, 119–138. Louvain, 1960.

Richardson, H. G., and G. O. Sayles. *The Administration of Medieval Ireland, 1172–1377*. Dublin: Stationery Office for the Irish Manuscripts Commission, 1963.

See also **Chief Governors; Lordship of Ireland; Parliament; Anglo-Irish Relations; Coinage; Common Law; Courts; Dublin; Feudalism; Government, Local; Military Service, Anglo-Norman; *Modus Tenendi Parliamentum*; Records, Administrative; Urban Administration**

CERBALL MAC DÚNLAINGE (d. 888)

During the reign of Cerball mac Dúnlainge (842–88) the Osraige rose from relative obscurity to become a major player in Irish politics. The most lavish account of his deeds survives in an eleventh-century saga embedded in *The Fragmentary Annals of Ireland.* This saga appears to have been written under the sponsorship of Cerball's great-great grandson Donnchad mac Gilla-Pátraic, who ruled Osraige (1003–1039) and Leinster (1033–1039). The exaggerations and anachronisms found in the saga urge a degree of caution in its use as a historical source. Cerball is also mentioned in the Icelandic *Landnámabók* and later sagas. These demonstrate that a number of prominent Icelandic families claimed descent from Cerball as a figure of legend.

Cerball's kingdom, Osraige, was strategically placed between the heartlands of Munster and Leinster. At the beginning of Cerball's reign, Osraige owed allegiance to overkings of Munster. However, in the 850s and 860s the fortunes of Munster declined, a factor that can be seen to aid Cerball's advancement.

Nevertheless, Cerball also faced dangers from Viking incursions. He is most renowned for his victories over Vikings that are elaborated in *The Fragmentary Annals of Ireland.* Cerball's first battle against Vikings is reported in 846. He also allied with some Viking groups when it suited his policies. In the late 850s he joined forces with Ívarr, a king of the "Dark foreigners." In 859, they raided Southern Uí Néill, thus challenging the power of the Uí Néill overking Máel-Sechnaill mac Máele-Ruanaid. In consequence of this attack, a royal meeting was arranged at Rathugh (Co. Westmeath) in 859. Osraige was formally ceded from Munster control and placed under the authority of Máel-Sechnaill. The event caused Cerball to reject his alliance with Ívarr. Further hostilities against Vikings are recorded for the remainder of Cerball's reign, although a temporary alliance with one viking group is recorded in 864.

Cerball was able to deal effectively with the threats posed by other Irish kings. Cerball enjoyed good relations with the Loígis of Leinster (his sister Lann was initially married to the king of this population group). Nevertheless, Cerball engaged in hostilities against other kings in the province on at least three occasions. The marriages of Cerball's daughters to kings of Uí Cheinnselaig and Uí Dróna in Leinster may indicate attempts to reduce border warfare.

Cerball's relations with Munster fluctuated. In 864, he attacked the heartlands of the province. He later allied with Dúnchad mac Duibdábairenn, who became overking of Munster in 872. They plundered Connacht together in 871 and 873. His alliance with Dúnchad later collapsed, and Cerball campaigned in Munster in 878.

Cerball's relations with the powerful Uí Néill rulers seem to have been flexible and pragmatic. From 859, Cerball supported Máel-Sechnaill against his rival Áed, overking of Northern Uí Néill. Nevertheless, Cerball quickly joined sides with Áed following Máel-Sechnaill's death. Cerball's sister Lann assisted in securing these important alliances by marrying both kings in succession.

Cerball ruled for a total of forty-six years. His longevity, success, and the dramatic potential of events in his career encouraged the later development of legends about him.

CLARE DOWNHAM

References and Further Reading

Clarke, Howard B. et al., eds. *Ireland and Scandinavia in the Early Viking Age*. Dublin: Four Courts Press, 1998.

Ó Corráin, Donnchadh. "Nationality and Kingship in Pre-Norman Ireland." In *Nationality and the Pursuit of National Independence*, Historical Studies 11, edited by T. W. Moody, 1–35. Belfast: Blackstaff Press, 1978.

Ó Corráin, Donnchadh. "Viking Ireland: Afterthoughts." In H. B. Clarke *et al.* eds. *Ireland and Scandinavia in the Early Viking Age*, edited by H. B. Clarke et al., 442–445, 447. Dublin: Four Courts Press, 1998.

Radner, Joan Newlon, ed. and trans. *Fragmentary Annals of Ireland.* Dublin: Dublin Institute for Advanced Studies, 1978.

Radner, Joan N. "Writing History: Early Irish Historiography and the Significance of Form." *Celtica* 23 (1999): 312–325.

See also **Leinster; Máel-Sechnaill I; Munster; Osraige; Viking Incursions**

CERBALL MAC MUIRECCÁIN (d. 909)

The last strong Uí Dúnlainge king of the Laigin in the period before the Battle of Clontarf, Cerball mac Muireccáin belonged to the lineage of Uí Fáeláin. His father Muireccán, styled *rex Naiss & Airthir Liphi* (king of Naas and the eastern Liffey-Plain), was slain by the Vikings in 863. Of his brothers, Domnall, his predecessor-but-one in the kingship, was killed by his own retainers in 884, while Máelmórda, who fell fighting the Vikings at Cenn Fuait (perhaps Confey, Co. Kildare) in 917, was ancestor of most of the later Uí Fáeláin rulers.

It seems that, especially in the earlier years of his reign, Cerball struggled to enforce his authority against counter-claims by rival Uí Dúnlainge lineages—including Uí Muiredaig and Uí Dúnchada. The acknowledgement as *tánaiste* of Bran (d. 894), son of his immediate predecessor Muiredach of Uí Dúnchada, perhaps represented a concession to that lineage. Aside from intradynastic challenges, Cerball faced a protracted conflict with the neighboring kingdom of Osraige, ruled by the sons of Cerball mac Dúngaile. A poem in the *Book of Leinster*, "The Quarrel about the Loaf," which tells of contention between an old woman of Leinster and a Munster soldier over billeting rights, is a metaphoric account of a border dispute concerning Mag Dála, a plain in south County Laois. The poem preserves a catalogue of Cerball's subkings, and it is probably significant that all but one are from north Leinster dynasties. Uí Dúnlainge was apparently under considerable strain by the late 890s; a garbled entry in *AFM*, as viewed by Byrne, records the celebration by Diarmait son of Cerball mac Dúngaile of the *Óenach Carmain*, in effect a claim on the overkingship of Laigin. Confronted by such pressures, Cerball mac Muireccáin sought alliance with Clann Cholmáin, a powerful lineage of the Southern Uí Néill. He married Gormlaith daughter of Flann Sinna, king of Tara, but it is not stated that she was the mother of his son Cellach (sl. 924). His wife, according to a poem in the *Book of Leinster*, arranged the murder of Cellach Carmain and his wife Aillenn—dynastic rivals who perhaps belonged to Uí Muiredaig.

Clearly, Cerball did benefit from his alliance with the Uí Néill over king. In 902, with forces from the midland kingdom of Brega, he attacked Dublin and expelled its Hiberno-Scandinavian rulers. For a time, at least, one major threat to Uí Dúnlainge had been removed. Four years later, he joined his father-in-law, Flann Sinna, in a preemptive strike against Osraige and Munster. They pillaged their way from Gabrán (Gowran, Co. Kilkenny) across to Limerick. When the Munstermen retaliated in 908, Cerball supported Flann in blocking an invasion force at Belach Mugna (Ballaghmoon, Co. Kildare); the fatalities included Cormac mac Cuilennáin, king of Cashel, and Cellach son of Cerball mac Dúngaile, king of Osraige.

The accounts of Cerball's death are difficult to reconcile. Whether or not he sustained wounds at Belach Mugna, as claimed by a text in the *Book of Leinster* (which also alleges that he mistreated his wife, Gormlaith), he died the following year. A colorful story in the so-called *Fragmentary Annals* tells of a horse-riding accident at Kildare, whereby he fell backward onto his own spear, which was held by a servant. The *Book of Leinster* kinglist echoes the line concerning a fall onto a spear—perhaps a metaphor for assassination—however, the invariably staid *AU* merely records in somber tone that he "died of a sickness." There is a strong tradition that he was buried at Cell Corbbáin, probably located in the vicinity of Naas, County Kildare, where he is said to have maintained his court.

Certainly Cerball made a marked impression on the historical consciousness of Leinster. He is the subject of several praise-poems ascribed to Dallán mac Móre, reputedly his court poet, and of elegies attributed to Dallán and to Gormlaith. One poem, the "Song of Cerball's Sword," credits him with a strike against the Uí Néill royal site of Knowth, which appears fanciful in the light of the surviving record. It is doubtful whether, earlier in his reign, he would have had the resources to invade the Uí Néill realms or, in the later years reason to do so, given his alliances with the kings of Tara and Brega. However, his achievement in bringing relative stability to Uí Dúnlainge, in stalling Osraige expansion, and in removing—albeit temporarily—the threat from Scandinavian Dublin was presumably noted. Even *AU* styles him *rex optimus Laginentsium*—a most excellent king of Leinster.

Ailbhe MacShamhráin

References and Further Reading

Best, R. I., O. Bergin, and M. A. O'Brien, eds. *The Book of Leinster.* 6 vols. Dublin: Dublin Institute for Advanced Studies, 1954–1983, pp. (i) 182, 223–225; (iv) pp. 955–958.

Byrne, Francis John. "Historical Note on Cnogba." *R.I.A. Proc.* 66 C (1967): 387–388.

Byrne, Francis John. *Irish Kings and High-Kings*, new ed. Dublin: Four Courts Press, 2000, pp. 163–164.

MacShamhráin, Ailbhe. *Church and Polity in Pre-Norman Ireland*. Maynooth: An Sagart, 1996, pp. 78, 80, 84, 86, 137.

Meyer, Kuno, ed. "The song of the Sword of Cerball." *Gaelic Journal* 10 (1900): 613–616.

Ó Cróinín, Dáibhí. "Rewriting Irish Political History in the Tenth Century." In *Seanchas: Studies in Early and Medieval Irish Archaeology, History and Literature in Honour of Francis J. Byrne*, edited by Alfred P. Smyth. Dublin: Four Courts Press, 2000, pp. 212–224.

O'Nowlan, T. P., ed. "The Quarrel About the Loaf." *Ériú* 1 (1904): 128–137.

See also **Uí Dúnlainge; Laigin; *Tánaiste*; Osraige; Cerball mac Dúngaile; Leinster, Book of; *Óenach*; Uí Néill, Southern; Gormlaith; Cormac mac Cuilennáin; Dublin.**

CHARTERS AND CHARTULARIES

The study of Irish charters is made difficult by the massive loss of archives that began with the wars and land-confiscations of the seventeenth century and did not end with the destruction in 1922 of the Public Record Office in Dublin. The earliest Irish records recording land transactions were in the form of *notitiae*, often entered in religious manuscripts, such as the well-known eleventh- and twelfth-century examples in the Book of Kells. The *notitia* form continued to be used in Irish-language documents down to 1600. By the early twelfth century, however, Irish kings were making grants of lands to monasteries (and possibly to laymen), written in Latin and using the standard formulae of the European charter of the day. The Anglo-Norman invasion and settlement introduced a society in which the use of charters was universal. The only major Irish medieval secular charter collection that survives intact is that of the Butlers of Ormond, now in the National Library of Ireland. Most of the documents of ecclesiastical provenance in the collection have been printed in full: the remainder down to 1603, with omissions, are listed in a published *Calendar*, often highly inaccurate. Only portions of the Kildare archive are known to survive, while those of some minor families, the Dowdalls of County Louth and the Sarsfields and Lombards of Cork, have survived more or less intact. Portions of other archives also survive. Of the cities, only Dublin and Waterford have preserved medieval charters.

Completely spurious (forged) charters are rare in Ireland. A commoner form of forgery was the "improvement" of charters by the insertion of spurious clauses when they were presented (if the originals were in poor condition) for certification by bishops or others, or (in the case of municipal charters of privileges) for confirmation by successive English sovereigns. An extreme example is King John's charter to Waterford, where we can trace its growth in successive versions with the insertion of further, often wildly anachronistic, privileges. Domnall Mór Ua Briain's foundation charter of Clare Abbey is preserved only in an "improved" version of 1461. A number of Irish monastic chartularies survive, all of which (except the Great Register of St. Thomas, Dublin) have been published, as have a chartulary of the archbishops of Dublin (*Crede Mihi*), the chartulary of the episcopal see of Limerick (*The Black Book of Limerick*), and three surviving lay chartularies (*The Red Book of Ormond*, *The Red Book of the Earls of Kildare*, and *The Gormanston Register*). The other Dublin archiepiscopal chartulary, Archbishop Alen's *Register*, has been calendared in print. The Elizabethan chartulary of Sir Richard Shee, which contains much medieval material, remains in manuscript.

A form of record that became important in late medieval Ireland was the notarial instrument, since Irish notaries (appointed by papal and imperial authority) in the autonomous regions were not restricted, as were their counterparts in England, to purely ecclesiastical matters. A considerable number survive, often highly artistic in the "signs" by which they were authenticated.

Kenneth Nicholls

References and Further Reading

Dauvit Broun, *The Charters of Gaelic Scotland and Ireland in the early Central Middle Ages* (Cambridge, 1995).

Philomena Connolly, *Medieval record sources* (Dublin, 2002).

J. T. Gilbert (ed.), *Historic and municipal documents of Ireland, AD 1172-1320* (London, 1870).

J. T. Gilbert (ed.), *Facsimiles of the national manuscripts of Ireland*, 4 vols (Dublin, 1874-8).

Richard Hayes (ed.), *The manuscript sources for the history of Irish civilization*, 11 vols (Boston, Mass., 1965); *first supplement, 1965-1973*, 3 vols (Boston, Mass., 1979).

Gearóid Mac Niocaill (ed.), *Na búirgéisí xii-xv aois*, 2 vols (Dublin, 1964).

J. C. Walton, *The royal charters of Waterford* (Waterford, 1979).

Herbert Wood, *A guide to the records deposited in the Public Records of Ireland* (Dublin, 1919).

Dr Marie Therese Flanagan is preparing an edition of the charters of Irish kings.

See also **Feudalism; Hiberno-Norman (Latin); Kells, Book of; Law, Common; Records, Administrative; Records, Ecclesiastical; Scriptoria; Urban Administration; Wills and Testaments**

CHIEF GOVERNORS

The term "chief governor" has been used by historians from the eighteenth century on to describe those officials who, under various titles, occupied the position of principal officer in the central administration of the lordship of Ireland. From the end of the twelfth century up to the middle of the fourteenth the chief governor was usually styled "justiciar." In the later medieval period the title "lieutenant" or "deputy lieutenant" was the most common designation for the chief governor, and the title of justiciar was only applied to those individuals who held the post on a temporary or emergency basis. While the title lieutenant was clearly intended to convey a status higher than that of justiciar, it has been shown that virtually all the powers vested in the lieutenants had in fact been exercised by the justiciars.

Origins of the Office

An office equivalent to that of chief governor had developed in England in the twelfth century in response to the need to provide for a deputy or regent to act as permanent head of the administration during the king's frequent absences on the Continent. Such chief officers can be identified from the reign of Henry I, but it was not until the reign of Henry II that the distinct office of justiciar was instituted. During the period of the conquest of Ireland the justiciar of England was Richard de Lucy, who acted in this capacity until 1179 when he was replaced by Ranulf de Glanville. A succession of holders can be traced down to 1234. The office was revived by the baronial reformers in 1258 and continued to be filled until 1265.

There was therefore an office ready-made to be imported into Ireland, and the lists of chief governors of Ireland usually commence in 1172, headed by Hugh de Lacy. However, historians now argue convincingly both that the office of justiciar of Ireland did not exist under Henry II and that de Lacy cannot be seen as the person appointed to be the king's alter ego in the lordship. The earliest secure use of the title justiciar with reference to Ireland comes in 1185 with the appointment of John de Courcy, following the return of Prince John to England. It seems probable, therefore, that the office was created under the lordship of the king's sons, while Henry himself adopted a more ad hoc approach to the problem of governing Ireland in his absence, using a combination of royal clerks and local magnates with varying degrees of success. This policy was similar to that used in Brittany, another territory acquired by Henry II through conquest. De Lacy's purported elevation to justiciar in 1172 is based on the evidence of Howden's chronicle. However, there appears little justification from other evidence to attribute to him, at this stage, a role greater than that of custodian of Dublin. There is more substance to assigning the role of the king's principal agent in Ireland in these early years to William Fitz Audelin, a competent and trusted member of Henry II's household. Fitz Audelin's commission ended late in 1173 when Strongbow returned to Ireland and succeeded him as the king's principal agent. From 1173 until 1176 Strongbow can be seen operating as the king's chief representative, although he did not use any specific title to describe his role, nor is one attributed to him in royal records. When Strongbow died in 1176, Fitz Audelin was once again dispatched to Ireland. It was only after Fitz Audelin's permanent departure from Ireland in 1181 that Hugh de Lacy can be described as operating in the capacity of chief governor.

Appointment and Renumeration

Chief governors—justiciars and later, lieutenants—were appointed by the king. The earliest surviving instrument of appointment of a justiciar is that of Meiler FitzHenry in 1200. The justiciar was usually appointed for an indeterminate period of time "during the king's pleasure," but in the later fourteenth century the custom developed of appointing the lieutenant for a set term of years. The chief governor was advised on matters of policy and administrative business by a council made up of the chief ministers and some of the important resident magnates. The composition and meetings of the council became more formalized during the course of the thirteenth century. The council had no fixed center but, like the justiciar, was endlessly itinerant. In certain circumstances, when speedy action was required, the council could appoint a chief governor and receive royal confirmation later.

The justiciar's salary is first referred to in 1226, when Geoffrey de Marisco was granted £580 a year for the custody of Ireland. Two years later it was fixed at £500 a year for Richard de Burgh, and this remained the standard sum for the rest of the Middle Ages. This sum was expected to provide for the custody of castles and for the justiciar's own men-at-arms, while the royal service due to the king was to provide for other military expenses. In the course of the fourteenth century justiciars and lieutenants began to enter their office by way of indenture with the king. These indentures set out the military forces to be maintained by the chief governors and the payments they were to receive over and above the set fee. With the increasing disorder that characterized the second half of the fourteenth century, these payments increased exponentially. In 1369, William of Windsor was to have the considerable sum of £20,000 for military purposes.

Until the end of the thirteenth century the justiciar was required to render an account to the king of all receipts and issues during his period of office. Very few of these accounts have survived. The earliest is James de Audley's, rendered eight years after his death in 1272, and the fullest is that for John de Sandford, covering his tenure of the chief governership from 1288 to 1290.

Powers and Responsibilities

The authority vested in the chief governor was delegated to him by the crown, and he was at all times subject to the control of the lord of Ireland. However, with this proviso, his powers were extensive. As commander-in-chief of the army, head of the civil administration, and supreme judge, he decided all important matters of policy in Ireland. One of the most important powers exercised by the chief governors was the right to proclaim the royal service—to summon the chief magnates to war in person, or to pay a sum called scutage in order to provide someone in their place. He also appointed and dismissed officers of the administration and granted or withheld pardons for offenses committed against the crown. The chief governor also exercised the royal right of purveyance, which enabled him to seize food, goods, and transport for his household and for war, although he was expected to pay a fair price to the owners of the goods. The chief governors had the power to appoint deputies in their absences. During the thirteenth and fourteenth centuries the justiciars had normally remained resident in Ireland. However, many of the fifteenth-century lieutenants remained in England and acted almost entirely by deputy. Occasionally a resident chief governor could appoint a deputy to deal with a specific task or area. In the later fourteenth century the growing pressures of war and the disorganization of the administration frequently necessitated the appointment of deputy chief governors to act for certain districts. Thus, in 1376 Stephen de Valle was appointed to supervise Meath and Munster as the deputy of the chief governor, Maurice FitzThomas.

On his travels around the medieval lordship of Ireland, the justiciar was accompanied by his own law court, of which he was the chief justice. He was provided with professional legal services by the itinerant justices, and from at least as early as 1270 one of these justices became a full-time official of the justiciar's court. From 1324 there are references to two justices of the justiciar's bench. The work of the court was recorded on the justiciary rolls. After Richard II's visit to Ireland, the justiciar's court developed into the court of the King's Bench.

Some Prominent Chief Governors

It has been traditional to see John Wogan as one of the ablest and most energetic of chief governors, although this may be due to the fact that his tenure of office was lengthy and comparatively well-documented. He was appointed justiciar in October 1295 and held the office until June 1308, and again from May 1309 until August 1312. The Justiciary Rolls for the period of his office have survived in calendared form and bear testimony to the ceaselessly itinerant nature of the office and the multiplicity of tasks performed by Wogan. By the end of Wogan's tenure the area subject to direct royal government was substantially increased.

Another high-profile chief governor was Thomas de Rokeby, appointed in July 1349 to succeed Walter de Bermingham. Rokeby had made his reputation as a soldier and administrator during the Anglo-Scottish wars, and his appointment has been seen as marking a change in royal policy toward Ireland. Rokeby's attempts to recover and fortify land by a combination of warfare and collaboration with marcher and Gaelic lords were partly successful. In July 1350, a series of instructions from England required the justiciar to undertake a general overhaul of the Irish administration, involving a full examination of the rolls of the exchequer and other courts and an investigation of corrupt practices. This enquiry has left no trace of its results, but appears to have inspired the ordinances issued at a great council held in Kilkenny in November 1351. These ordinances display the usual preoccupation with provision for defense and were to be re-enacted in substance by the Statutes of Kilkenny in 1366. When he died in office in 1357, Rokeby was praised by the Dublin Annals for having fought the Irish well and paid for his provisions.

These same Annals some years earlier had displayed considerable animosity toward another justiciar, Ralph de Ufford, stating that when he died in April 1346 there was great rejoicing and the clergy celebrated Easter with special joy. Ufford, who was the grandson of Robert de Ufford, justiciar under Edward I, had married the widow of the earl of Ulster and had important interests in Ireland in her right. His tenure saw the swift dismissal of some long-serving officials of the administration, the temporary destruction of the power of the earls of Desmond and Kildare, and the distribution of forfeitures to members of his own retinue. These actions undoubtedly made him some enemies, although the Irish Council, writing to the king shortly after his death, spoke warmly of him.

The names of other chief governors are linked with scandals, usually financial. Geoffrey de Marisco, for example, who held the office from 1215 to 1221, was accused of systematically using the king's revenues in Ireland for his own purposes and of defying all attempts to control his actions and force him to render account. Similar accusations were made against Stephen de Fulborne, justiciar from 1281 to 1288, necessitating a wide-ranging investigation of his activities in 1284. It was said that he filled offices only on receipt of a substantial bribe and that he engaged in profiteering during the Welsh wars. Although great irregularities were found in his account, he was pardoned all arrears above £4,000 and continued in office until his death in 1288.

A significant number of the most prominent figures in the medieval history of the lordship and of England held the position of chief governor at some point in their career. Up until the middle of the fourteenth century most were magnates or ecclesiastics resident in Ireland, although from time to time English administrators were sent over. From the appointment of Lionel of Clarence as lieutenant in 1364, there was an increasing tendency to appoint English magnates who governed largely through deputies who were usually Anglo-Irish magnates. In the fifteenth century there was a renewed effort to recruit chief governors from among those who were powerful landowners in Ireland. By removing the obligation to account for revenue and increasing the level of control over the council, the office was made attractive to the great resident magnates, particularly the earls of Ormond, Desmond, and Kildare. These men were keen to exploit the office to augment their own power and interests, and the days when the chief governor could be seen as the king's alter ego were well and truly at an end.

MARGARET MURPHY

References and Further Reading

Ellis, Stephen. *Ireland in the Age of the Tudors, 1447–1603*. London: Longman, 1998.

Everard, Judith. "The 'Justiciarship' in Brittany and Ireland under Henry II." *Anglo-Norman Studies* XX (1998): 87–105.

Flanagan, Marie Therese. "Household Favourites: Angevin Royal Agents in Ireland Under Henry II and John." In *Seanchas: Studies in Early and Medieval Irish Archaeology, History and Literature in Honour of Francis J. Byrne*, edited by Alfred P. Smith, 357–380. Dublin: Four Courts Press, 1999.

Frame, Robin. *Colonial Ireland, 1169–1369*. Dublin: Helicon, 1981.

Lydon, James. *The Lordship of Ireland in the Middle Ages*, 2nd ed. Dublin: Four Courts Press, 2003.

Otway-Ruthven, A. J. "The Chief Governors of Medieval Ireland." *Journal of the Royal Society of Antiquaries of Ireland* 95 (1965): 227–236.

Otway-Ruthven, A. J. *A History of Medieval Ireland*, 3rd ed. New York: Barnes and Noble, 1993.

See also **Anglo-Irish Relations; Armies; Courts; Feudalism; Government, Central; Law, Common; Lionel of Clarence; Lordship of Ireland; Military Service, Anglo-Norman; Parliament; Records, Administrative; William of Windsor**

CHILDREN

The majority of information on childhood in the medieval period in Gaelic Ireland stems from the extant legal sources that mainly deal with fosterage and enable us to draw conclusions on childhood in general. Both parents bore responsibility for the rearing of the child. This practice could be influenced by the status of the couple involved and also by the circumstances of the conception (for example, whether permission had been granted by the woman's kin-group for the union to take place). A child was completely dependent on his parents (or guardian), and therefore bore no legal responsibility in his own right, nor could he undertake independent legal action in any capacity. The age of seven was a time of change for a child, with his honor price being tied closely to that of his parents from that point onward. Prior to the age of seven, the legal material suggests that the child carried the honor price of a cleric, which would have awarded him particular standing and protection, in theory. While a dependent and residing within the homestead, any action a child committed was the responsibility of, and/or compensated for, by the father.

Fosterage is one particular method of childrearing emphasized throughout the sources, and was common practice in medieval Ireland. This was a method of childrearing whereby adults, other than the biological parents, undertook to raise a child for a particular period of time. Fosterage is a well-established tradition when Ireland enters into the historic period. The foundation of this practice may be found in the Indo-European past, but as to its impetus and origins, little is known. The common terms applied to foster father (*aide*) and foster mother (*muime*) are terms of affection in Old-Irish. *Altram*, the term for fosterage, stems from the action of feeding and nourishing, the basic requirements of a dependent. The term *dalta* refers to the foster child. There is no distinction made in terminology between wet nurse and foster mother in Irish, as is the case in many other languages. The possibility of nursing as an optional first step in the fostering process finds support in the legal commentary, which notes three age groupings within fosterage: the first period, up to seven years of age, thus not dictating a set starting age; the second grouping, from seven to twelve years of age; and the third grouping, from twelve to seventeen years of age. The commencement age for fosterage could vary widely depending on the circumstances.

It is important to note that fosterage was a formal contract within the Irish tradition. The medieval Irish legal material notes two types of fosterage: one of affection and one for payment. The maternal kin-group, and more specifically the mother's brother, was a popular choice with whom to foster. Conditions under which this was to be conducted are specified. The placement of a child in another household may have been in part for the child's own protection, in a society that was polygynous in nature. The potential rivalry between a child and a nonbiological parent in the natal homestead, and more particularly among siblings and half-siblings, was the product of a society in which all male children had equal claim to patrimony.

The fosterage fee was calculated according to rank, and appears to involve a cattle payment to be returned with the child at the end of the fostering period. The child may also take particular goods with him when going into fosterage, such as items of clothing. The legal material notes that it cost more to foster a female child. There is not agreement, however, why this was the case.

At the core of fosterage was the education of the child, and by extension this permits us to see children and education in a wider context in medieval Ireland. The importance of education is emphasized through the imposition of a fine (two-thirds of the fosterage fee) if one of the required skills was not taught. The type of education a child received was intrinsically linked to his status. There was a strong pastoral emphasis to the education of children of the free-man grade (for example, boys were taught kiln-drying and wood-cutting; girls learned use of the quern, the kneading-trough, and herding of lambs). Thus, practical education is stressed. The children of higher noble grades were taught more varied pursuits (for example: the boys, board games and horse-riding; the girls, sewing, embroidery, and cutting-out). There are also references to play and toys, including particular games, which are mostly of a physical nature. The archaeological record, in addition to evidence for board games associated more with adults, has produced at least some toys. These include a wooden figure—possibly a doll—from the crannóg of Lagore (Co. Meath), a lead spinning-top from the rath of Ballycatteen (Co. Cork), and a model boat from a Hiberno-Scandinavian context at Dublin's Wood Quay.

It would appear that a professional family would educate a child within the tradition of his father's profession, for example, a poet. The church was also a vehicle where children were educated, possibly as part of the community of secular monastic tenants, or with the intention of taking religious orders. The church played an important role in the sphere of childrearing, taking responsibility for children who were abandoned, given as oblations, or placed in the monastic setting to pursue education (to later return to the secular life). The ties created between children and their superiors within the religious environment appear to mirror the secular bonds created between child and parent, or teacher and pupil in the professional sphere.

On the general upbringing of a child, we are informed that this was determined by the child's status. According to the legal commentary, the attire of the child reflected his status through color, the number of clothing changes, and accessories to the clothing (trimmings and brooches). Similarly, the quality of food indicated the status of the child. The selection of condiments and the basis of certain dishes (for example, based on water or new milk) varied according to rank.

The child's maternal and paternal kin-groups bore responsibility for arranging and paying for the fosterage period (what is referred to as co-fostering). Each kin-group provided half the fosterage fee. The maternal kin could protest against a fosterage placement. If the fosterage undertaken was one of affection, the foster father and mother were not liable for the crimes committed by the foster child. If it was fosterage for payment, the foster parents would be financially responsible, thus indicating the serious nature of this practice.

The age of the child when a crime was committed, the nature of the crime, and the number of prior offenses were taken into consideration when deciding what punishment was suitable. The foster father had to pay for the fines incurred, until he vocally "proclaimed" his foster son to his natural father. This step removed any financial responsibility for certain crimes, if the child was habitually criminal while under his charge. Discipline noted is chastisement as an initial measure, followed by fasting if the child repeats the act. Finally, restitution for the crime committed is prescribed within the legal sources. If the child was blemished in any way while in fosterage, the foster father forfeited two-thirds of the fosterage fee. Within the legal sphere, foster parents had the power of proof, judgement, and witness over foster children. This position was normally restricted to the natural kin.

Completion of fosterage was strictly regulated. Two errors noted are the premature returning of the child by the foster parents, or the premature taking back of the child by the biological parents. Either action resulted in compensation payment to the offended party. On completion of fosterage (around fourteen years of age for a girl and around seventeen for a boy), the *sét gertha* (*sét* of maintenance) was an important payment made to the foster child by the foster parents. This payment reflects the lifelong bond and obligation placed on the foster child to maintain the foster parents in later life, if necessary. Children were obliged to maintain their biological parents in similar circumstances. The provision of care for foster parents in times of poverty, and general maintenance in old age (*goire*), was expected. Such extensions of rights that we may associate with biological relationships reflect the standing of the foster kin vis-à-vis the natural kin.

Further benefits to all parties created through this method of childrearing include a source for military aid, legal support, and intensification of friendship between the foster parents and natural parents. The financial benefits in the form of compensation awarded to foster relatives when unlawful injury was inflicted on their fosterling or fellow-fosterling at any time of life, was a further factor in sustaining relationships. Much evidence attesting to this institution is found in the literary sources, particularly within the saga literature and bardic poetry. In general, there is consensus across the sources that a fosterage relationship should bring prosperity to both households involved in the process. Although the practice of fosterage was condemned by canon law in the Middle Ages, and although legislation prohibiting its practice was issued on several occasions in the late medieval period by the secular authorities, the range of short- and long-term benefits of fosterage played a large part in sustaining the power of the institution into the early modern period. It is a well-attested practice into the early modern period and affords an insight into childhood and upbringing in general.

BRÓNAGH NÍ CHONAILL

References and Further Reading

Charles-Edwards, T. M. *Early Irish and Welsh Kinship.* Oxford: Oxford University Press, 1993, pp 78–82.

Edwards, Nancy. *The Archaeology of Early Medieval Ireland.* London: Batsford, 1991.

Kelly, Fergus. *A Guide to Early Irish Law.* Dublin: Dublin Institute for Advanced Studies, 1988, pp 86–90.

See also **Education; Fosterage; Games; Society, Functioning of Anglo-Norman; Society, Functioning of Gaelic**

CHRIST CHURCH CATHEDRAL

Dublin's cathedral was dedicated to the Holy Trinity, but the name "Cristchirche" emerged in 1444. The cathedral was probably founded around 1028, the year the Hiberno-Norse king, Sitriuc "Silkenbeard," made a pilgrimage to Rome. Due to canon law irregularities in the organization of the Irish Church, Dublin was to become a suffragan diocese of Canterbury from at least 1074, following the consecration of its second bishop, Gilla Pátraic. With Muirchertach Ua Briain as secular ruler of Dublin, together they laid the foundations of what would become the twelfth-century church reform.

Gilla Pátraic introduced the first of the religious orders to Holy Trinity: Benedictine monks, who remained until their expulsion around 1100 by Bishop Samuel. It was during his episcopate that the Dublin diocese was subsumed into Glendalough under the 1111 synod of Ráith Bressail, and not until the synod of Kells in 1152, under Bishop Gréne, did Dublin diocese and Holy Trinity cathedral achieve archiepiscopal and metropolitan status, respectively. Lorcán Ua Tuathail, brother-in-law of Diarmait Mac Murchada, succeeded in 1162 and established a priory of Augustinian canons regular at the cathedral.

Following the Anglo-Norman invasion of Dublin in 1170, property granted to the cathedral priory by former Irish and Hiberno-Norse kings was confirmed by Henry II and his son John. This was an estimated 10,500 acres arranged under the manorial system, including Grangegorman, Clonkeen, Glasnevin, and Balscadden. Both Lorcán (d. 1180) and Richard de Clare "Strongbow" (d. 1176) predeceased the Romanesque rebuilding of the cathedral traditionally attributed to them. Building work by English West Country masons began in the mid-1180s under the first Anglo-Norman archbishop, John Cumin. In 1216, under his successor Henry of London, Holy Trinity became the diocesan cathedral for Glendalough following its unification with Dublin. By 1220, St. Patrick's cathedral had been founded by Henry, and the remainder of the century saw an architectural and constitutional jostling for supremacy between Holy Trinity's regular and St. Patrick's secular chapter. A Gothic nave (1230s), partially extant, and an extension to the chancel (1280s) were built at Holy Trinity. However, the constitutional wrangling ceased only when, in 1300, both signed a *composicio pacis* acknowledging both as diocesan cathedrals, but Holy Trinity as the elder. Surviving fire in 1283 and the fall of the steeple in 1316, Holy Trinity was an accustomed venue for the Irish parliament, which often met, as in 1450, in the common hall. The belfry was rebuilt by 1330, and by 1337 to 1342, the surviving account roll gives a glimpse of the priory's administration, the records of which are unusually plentiful for an Irish medieval institution. Despite recurrent outbreaks of plague from 1348, the next decades saw a choir extension built by Archbishop John de St. Paul and the acquisition of an English illuminated psalter by Prior Stephen de Derby. In 1366, the Kilkenny statutes disqualified the native Irish from membership of the chapter.

Richard II knighted four Irish kings in the cathedral in 1395, while the coronation of the Yorkist pretender Lambert Simnel as King Edward VI took place at Christ Church in 1497.

St. Augustine's rule defined the priory's religious life, enhanced by liturgical manuscripts such as the martyrology, psalter, and a book of obits. These were used in chantry chapels such as St. Lo (1332) and St. Laurence O'Toole (1485) and were enhanced by a choir of four choristers in 1480, whose education by a music master was confirmed by Prior David Wynchester in 1493. The most elaborate chapel was a perpendicular gothic chantry dedicated to St. Mary, built in 1512 by Gerald Fitzgerald, eighth earl of Kildare, who was buried there the next year.

Holy Trinity also led the moral and religious instruction of Dubliners. In 1477, the archbishop of

Armagh and papal nuncio, Octavian de Palatio, preached there in support of a crusade against the Turks. The earliest known morality play from Ireland, *The Pride of Life,* survived in the fourteenth-century priory account roll, and in 1528 the priors of Holy Trinity, Kilmainham, and All Hallows attended performances of passions at Hoggen Green.

Commerce coexisted with this world, with many guilds or fraternities having chapels at Holy Trinity, such as the merchants' Trinity guild (1451) or the guild of St. Edmund, asked in 1466 to provide bows and arrows for the defense of the city. Shops soon emerged from crypt cellars. The "utestale[s]" mentioned in 1423 had oaken beams and stone roofs by 1466. Internally, maintenance was continuous. Four windows were newly glazed in 1430 in St. Mary's chapel, a structure with a complex building history, little of which survives. South of it lay the long quire where, in 1461, the east window blew in, destroying numerous relics but notably excluding the *Baculus Ihesu* (staff of Jesus). The priory's earnest protection of visiting pilgrims' "immunities," as in 1493, can be attributed to the lucrative supply of income that they provided. The cathedral's relics were publicly burned in 1538 by Archbishop George Brown. Christ Church was the sole religious house to survive the dissolution of the monasteries in 1539, abandoning its monasticism for a secular constitution based on St. Patrick's. Henry VIII confirmed Prior Castle, alias Paynswick, as first dean in 1541, and by 1544, three prebendal parishes were established: St. Michael's, St. Michan's, and St. John's. If not the Reformation, then certainly the fall of the roof and south wall of the nave in 1562, partially destroying the Strongbow monument, signalled the end of the medieval period.

STUART KINSELLA

References and Further Reading

Milne, Kenneth, ed. *Christ Church Cathedral Dublin: A History.* Dublin: Four Courts Press, 2000.

See also **Abbeys; Architecture; Church Reform, Twelfth Century; Cumin, John; Dublin; Education; Fraternities and Guilds; Gilla Pátraic, Bishop; Henry of London; Moral and Religious Instruction; Music; Parish Churches and Cathedrals; Pilgrims and Pilgrimages; Records, Ecclesiastical; Religious Orders; Sculpture; St. Patrick's Cathedral; Scriptoria; Ua Tuathail (O' Toole), St. Laurence**

CHRISTIANITY, CONVERSION TO

The year 431 marks the date of the official introduction of Christianity to Ireland. In that year (according to Prosper of Aquitaine, *Chronicle*, s.a.) Pope Celestine I dispatched the newly ordained Palladius as "first bishop to the Irish believing in Christ" (*primus episcopus ad Scottos in Christum credentes*). Prosper appears to allude again to the mission of Palladius in his polemical tract *Contra Collatorem* (written in the later 430s in defense of Celestine against his detractors), when he refers to Celestine's having made Britain (*Romana insula*, the Roman island) Catholic, while making Ireland (*barbara insula*, the barbarous island) Christian. Prosper was here referring to an earlier episode, in 429, when Celestine dispatched Germanus, bishop of Auxerre, to Britain in order to combat a recent recrudescence of the heresy known as Pelagianism. That mission (again according to Prosper) had been undertaken at the instigation of a deacon named Palladius, who is undoubtedly identical with the man of that name sent to Ireland in 431. It is generally assumed that the mission to Ireland in 431 followed on from the one to Britain in 429.

Native tradition, however, associates the beginnings of Irish Christianity with Patrick, not Palladius, who was subsequently written out of Irish history. Because Palladius disappears from the historical record in Ireland (and elsewhere) after 431, Irish historians were forced to fill the perceived void in the historical narrative by dating Patrick's arrival immediately afterward, in 432. Patrick, a Briton by birth and upbringing, was captured at age sixteen by Irish pirates in a raid on his family's estate (*uillula*), "along with many thousands of others" (as he says himself), and was brought to Ireland as a slave. His account of that episode, and of the events that unfolded because of it, has survived in his famous *Confessio*, which is a unique testimony to the experiences of a Roman citizen snatched from his home by alien marauders, and who lived to tell the tale. The *Confessio* and the only other writing of Patrick's to survive, his letter addressed to the soldiers of Coroticus, offer unique insights into the everyday experiences of a man in the front line of missionary activity beyond the frontiers of the Roman Empire. Unfortunately, we do not know the dates of Patrick's mission in Ireland. In fact, we have no dates at all for the saint, for the simple reason that he offers none, and no other reliable contemporary source exists that might fill that gap.

Modern scholars are unanimous that Patrick's two surviving writings reveal an individual of genuine spiritual greatness. Historians have been troubled, however, by the fact that Patrick nowhere in his writings refers to Palladius or anyone else involved in missionary activity in Ireland, but constantly reiterates the claim that he has gone "where no man has gone before." It is not at all impossible, therefore, that Patrick came to Ireland *before* Palladius, rather than after him, perhaps in the late fourth century or in the generation before Palladius was dispatched by Pope Celestine to the "Irish believing in Christ." That would perhaps offer the most satisfactory explanation for Patrick's otherwise inexplicable silence about the work of others before him on the Christian mission in Ireland. An earlier missionary period for Patrick would also account for the presence in Ireland of Christians before 431, those "Irish believing in Christ" to whom Palladius was sent as first bishop. Certain expressions in Patrick's writings would seem to add weight to this surmise, since he appears to be writing at a time when the Roman presence is still all-pervasive in his native Britain. On the other hand, the more "traditional" dating of his career (arrival in 432, death in 461 or 493), runs up against the difficulty that the Roman legions had long since departed the "Saxon shore" and left Britain a prey to Anglo-Saxon invaders. Since Patrick makes no mention of these cataclysmic events, it seems reasonable to infer that his silence on the subject is due to the fact that he had left his native home long before the Anglo-Saxon occupation of Britain.

Palladius's mission left nothing like the same impression on the Irish historical mind as Patrick's did, and yet there are occasional traces of a transitional period during which Christianity was still finding its feet, not yet securely established as the "national" religion. In fact, that was probably not to be the case until the late sixth or early seventh century, at the earliest. The first phase of missionary activity is represented, for example, by a remarkable survival: a list of the days of the week in a mixture of Irish and Latin, a witness to the first faltering attempts by Irish Christians to adapt to the new concepts introduced by the Roman religion. This phase of conversion is evident also in the fact that the earliest Christian vocabulary used by Irish converts simply recycled the terminology of the older native beliefs. Thus the Irish terms for "God," "belief," "faith," "grace," and so on are all words used to express similar concepts in the pre-Christian religion. We know next to nothing about the progress of Christianity in Ireland in the fifth century, and Patrick himself refers only once (and that disparagingly) to native Irish practices of sun-worship "and other abominations," but he does not elaborate. In time, of course, the newer religion was to replace the earlier one entirely, but not before the latter had left an indelible mark on the Irish Christian mind. How much of the new Irish Christian religion was due to the activities of Palladius and his continental comrades, and how much to Patrick and the efforts of later British clergy, is difficult to judge. The evidence, such as it is, seems to indicate that the British influence in the longer term was the stronger of the two.

No document from the Palladian mission has survived, whereas Patrick's two writings became the foundation for a body of legends that turned the humble Briton into an all-powerful, conquering Christian warrior who wiped out paganism and converted the Irish people singlehanded. In the process of this reinvention, however, the true character of the man was sacrificed for the purpose of creating a mythological figure whose "heroic" deeds formed the basis for outlandish claims made by Irish churchmen in the centuries after him. When Patrick emerges into the light of history again in 632, in the famous Paschal letter of Cummian, he is there referred to as *sanctus Patricius* (the holy Patrick) *papa noster* (our father)—the earliest indication we have that Patrick enjoyed a special status in the Irish Church by that time.

DÁIBHÍ Ó CRÓINÍN

References and Further Reading

Binchy, D. A. "Patrick and His Biographers, Ancient and Modern." *Studia Hibernica* 2 (1962): 7–173.

Ó Cróinín, D. *Early Medieval Ireland.* London: Longman, 1995, pp. 14–40.

Thompson, E. A. *Who Was Saint Patrick?* Woodbridge: Boydell, 1985.

See also **Armagh; Armagh, Book of; Art, Early Christian; Biblical and Church Fathers Scholarship; Classical Influence; Ecclesiastical Organization; Ecclesiastical Settlements; Ecclesiastical Sites; Inscriptions; Muirchú; Palladius; Patrick; Pre-Christian Ireland; Tírechán**

CHURCH REFORM, TWELFTH CENTURY

Apart from the Hiberno-Norse towns of Dublin and Waterford, the church in Ireland lacked a permanent diocesan structure in the eleventh century. The reason for this is largely to be found in the fact that Ireland was never part of the Roman Empire and thus lacked the administrative structure upon which the western church elsewhere based its organization. The circumstances surrounding the foundation of the diocese of Dublin early in the century are obscure, but Dublin would later play an important role in the events surrounding the introduction of a new diocesan system for the church throughout the whole country.

Canterbury and the Irish Church

Shortly after the Norman conquest of England in 1066 a controversy arose between Canterbury and York over how the primacy of the Church of England should be interpreted. From documentation associated with this we get our first evidence that the new archbishop of Canterbury, Lanfranc, believed that Ireland as well as York was subject to the primacy of Canterbury. Two years later, in 1074, a vacancy occurred in the see of Dublin and its new bishop, Gilla-Pátraic (or Patrick), was consecrated in London by Lanfranc. From evidence associated with this we get an insight into how Lanfranc planned to exercise his claimed primacy over the Irish church: He would do so through the agency of the see of Dublin. It would be the metropolitan see for the whole island of Ireland and owe allegiance directly to Canterbury; its bishops would be consecrated by the archbishop of Canterbury and profess obedience to him.

After his consecration, Bishop Patrick passed on letters from Lanfranc to the king of Dublin and to Tairrdelbach Ua Briain, then the most powerful king in Ireland. In these, Lanfranc exhorted the kings to act against certain abuses that he had heard occurred in Ireland. In his letter to Ua Briain, however, he urged him to convene an assembly of religious men to eradicate what he calls "evil customs" from Ireland. And it would appear that Ua Briain responded; a synod, held in Dublin in 1080, was apparently convened by him. Thus began a level of co-operation between Ua Briain, together with certain Irish bishops, and Canterbury—cooperation that continued after his son, Muirchertach Ua Briain, succeeded him in 1086. King and bishops took part in the election of successive bishops of Dublin, and of a bishop for the newly erected see of Waterford in 1096—all of whom openly professed their obedience to the archbishop of Canterbury, to whom the electors sent them for consecration. It is not clear, however, whether they understood the exact nature of Canterbury's enterprise in Ireland. In any case, their cooperation was about to come to an end, as became apparent when the first synod of Cashel met in 1101.

Muirchertach Ua Briain Exchanges Canterbury for Armagh

Evidence for the change in Ua Briain's attitude to Canterbury is found in a specific action he took at that synod. He granted Cashel, the seat from ancient times of the kings of Munster, to the church forever—not to some local church, but to the whole of the church in Ireland. The significance of this became clear ten years later, when Cashel was chosen as the metropolitan see of the southern province in a new diocesan structure set out at the synod of Ráith Bressail. Already in 1101, therefore, Ua Briain had a vision of this new structure that had an important role for Cashel, but none for Canterbury.

Major problems, however, stood in the way of its realization. The first of these was Armagh, the most prestigious ecclesiastical establishment in Ireland; it would have to be included in whatever new structure was introduced. But the ecclesiastical organization there at this time was traditional; the man with the highest level of authority was the abbot, usually referred to as the *comarbae* (heir) of Patrick. Although a cleric, he had no ecclesiastical orders and was married. In addition, he belonged to a family that had controlled the office since the middle of the tenth century. This obviously presented a problem for Ua Briain, since Armagh, in the new church structure then being envisaged, would become a metropolitan see ruled ultimately by an archbishop. The change required there was of such major proportions that strong leverage was needed; that leverage was Dublin and the role that was mapped out for it in Ireland by Canterbury. Should this be made a reality, Dublin would usurp a position in the Irish church that Armagh clearly saw as belonging to it. An opportunity to apply this leverage was available to Ua Briain when he visited Armagh in 1103, and it would appear that he was successful, as subsequent events suggest. It is likely that nothing could be done as long as the then-incumbent of the abbacy lived, but the swift action of his successor, Cellach, would indicate that a decision had been made to go along with the plans for reform as envisaged by Ua Briain. Within six weeks of his predecessor's death in 1105 Cellach assumed ecclesiastical orders, and in the following year, perhaps only a few months later, he was consecrated a bishop, significantly while on a visit to the Ua Briain territory, Munster.

With Armagh won over to reform, Ua Briain was now faced with the problem of finding an ecclesiastic who would carry the project to its next stage. Although there is no direct evidence to link him with the man so chosen (Gille or Gilbert of Limerick), events surrounding the selection clearly point to Ua Briain. The most significant of these is Gille's appointment to the bishopric of a new Hiberno-Norse diocese in Limerick, the town in which Ua Briain then had his headquarters. Furthermore, his selection followed a pattern with which Ua Briain and his father before him were familiar. Just like bishops of Dublin and Waterford before him, in the selection of whom the Uí Briain kings were directly involved, Gille had been a monk in a Benedictine abbey in England. However, unlike them, he was not sent to the archbishop of Canterbury for consecration, thus reflecting the changed attitude of Ua Briain we have noted as existing since 1101. After his appointment, he set out to prepare the clergy for the upcoming changes. He wrote a short tract, *De statu ecclesiae* (Concerning the Constitution of the Church), in which he set out the organizational structure of the whole western church, from layman to pope, and a short description of their various functions or duties. This tract and an accompanying letter—in which he expresses the wish that the diverse practices that he says exist in Ireland would yield to a single, uniform one in conformity with Rome—are extant in twelfth- or thirteenth-century copies. Given that the tract is mainly concerned with church structure, we can already see what the main preoccupation of the reformers was at this point. This would become even clearer a few years later when the synod of Ráith Bressail met, over which Gille presided as papal legate.

The Introduction of a Diocesan System to the Irish Church

At this synod a scheme was prepared for the introduction of a new hierarchical structure into the church. Following a plan believed to have been set out originally for the English church, whereby there would be two provinces, each with a metropolitan and twelve suffragans, it was decided that Ireland would be divided into two metropolitan provinces, one at Armagh, the other at Cashel—Armagh holding the primacy. The sees and boundaries were set out for all the dioceses; however, while twelve suffragans were assigned to the Armagh province, Cashel only got eleven. Most significantly of all, there was no mention of Dublin, but it seems fairly clear that in assigning only eleven suffragans to Cashel, room was being left for Dublin's subsequent inclusion. However, considerable effort would be needed to get it to join in, given the nature of its relationship with Canterbury. Evidence that such an effort was being made may be seen in the action of Cellach of Armagh, who took over the Dublin see after the death of its bishop, Samuel, in 1121. There was resistance to this in Dublin, and a subdeacon called Gréine was quickly elected and sent to the archbishop of Canterbury for consecration. He failed, however, to gain possession of the see on his return, although he did so some years later. Dublin now stood apart from the newly organized Irish church.

In 1129, Cellach died, and his chosen successor was Malachy (Máel-Máedóic). This represented a further break in tradition at Armagh in that, unlike Cellach, Malachy did not belong to the family that had provided abbots since the middle of the tenth century. There was strong resistance to Malachy's appointment from this family. However, since it was essential that Armagh, the seat of the primate, be retained within the fold of the reformers, Malachy had to be installed there. This explains why secular forces, particularly those in Munster who favored reform, took such an active part in Malachy's installation. However, his position there remained difficult, and a man who was acceptable both to the reformers and to local secular rulers—Gilla-Meic-Liac—was chosen in his stead, and Malachy resigned. He now pursued the interests of reform on a larger stage.

Papal Approval for the New Structure

Although the new diocesan system had been set out in 1111, papal approval for the two incumbents of its archbishoprics—by the granting of *pallia*—had not been sought, as far as is known, before Malachy did so in 1140. Although unsuccessful in this bid to get the *pallia*, Malachy's journey to Rome was not in vain. It brought him to Clairvaux (France) and to Arrouaise (Flanders) and resulted in the introduction of the Cistercian order and the rule of the canons of Arrouaise into Ireland. In addition, Pope Innocent II appointed Malachy as his legate in Ireland in place of the ailing Gille and told him to re-apply for the *pallia* after he had gained the agreement of all in Ireland. The obstacle here was Dublin, and this was Malachy's main task on his return home. Little in detail is known about how Malachy now pursued this task, but it can be inferred that agreement was reached with Dublin at some point thereafter and was approved at a synod held in 1148 at Inis Pádraig. The agreement involved the recognition of Dublin as a metropolitan see gaining suffragans that had previously been part of the province of Cashel, as set out at Ráith Bressail. Also approved there was a new province of Connacht (carved out from that of Armagh), with its metropolitan see at Tuam; this reflected the current status of Tairrdelbach Ua Conchobair, now king of Ireland as well as king of Connacht.

Immediately after the synod, Malachy went to meet the pope to get his approval for the synod's decisions, but he died on the way at Clairvaux. The request was transmitted by others, and this time it was successful. Pope Eugenius III sent his legate, Cardinal John Paparo, to Ireland bearing *pallia* for the four new archbishops. After some difficulties put in his way by King Stephen of England, motivated perhaps by a desire to prevent Canterbury's interests in Ireland being jeopardized, he eventually arrived in Ireland. He convened a synod in March 1152 that met at two locations, Kells and Mellifont. As well as the enactment of decrees, the consecration of archbishops and bishops, and arrangements regarding what dioceses should belong to the various metropolitans, Cardinal Paparo formally presented *pallia* on behalf of the pope to the four new archbishops at the synod. With this, Ireland had come into line with the rest of the western church. It now had a hierarchy of bishops within a canonically constituted, territorially defined diocesan system by which the church would henceforward be administered.

MARTIN HOLLAND

References and Further Reading

Gwynn, A. *The Irish Church in the Eleventh and Twelfth Centuries*, edited by Gerard O' Brien. Dublin: Four Courts Press, 1992.

Gwynn, A. *The Twelfth-Century Reform, A History of Irish Catholicism* II. Dublin & Sydney: Gill and Son, 1968.

Hughes, K. *The church in early Irish society.* London: Methuen, 1966.

Holland, Martin. "Dublin and the Reform of the Irish Church in the Eleventh and Twelfth Centuries." *Peritia: Journal of the Medieval Academy of Ireland* 14 (2000): 111–160.

Watt, J. *The church in medieval Ireland*, 2nd ed. Dublin: University College Dublin Press, 1998.

See also **Cashel, Synod of I (1101); Ecclesiastical Organization; Gille (Gilbert) of Limerick; Kells, Synod of; Malachy (Máel-Máedóic); Ráith Bressail, Synod of; Religious Orders**

CIARÁN

The name Ciarán, from *ciar* (dark, black), is assigned by the list of homonymous Irish saints to over twenty reputedly distinct saints, of whom only the first two, Ciarán mac int Shaír and Ciarán mac Laigne, respectively of Clonmacnoise (Offaly) and Saigir (Seirkieran, Offaly), attained prominence. As well as being used of the eponymous ancestor of the Ciarraige tribal group (hence *Ciarraí*, "Kerry"), the name was also attached, without the diminutive ending *-án*, to the female patron of Cell Chéire, Kilkeary (north Tipperary).

St. Ciarán of Clonmacnoise

The Clonmacnoise saint was the better known of the two Offaly bearers of the name, mainly due to the outstanding role of his church as a center of cultural and politico-ecclesiastical activity. Several, as yet unevaluated, Lives were written for him, in Latin and Irish. The Latin text is preserved in three recensions, the Irish in one only. Ancestrally attached to the Latharna (now Larne, Antrim), the saint is said to have been born on the plain of Connacht in Roscommon. This probably reflects both the position of his church, Clonmacnoise, on the boundary of Connacht and the power of the kings of that province during the second half of the twelfth century, when the Life (in its present form) may have been written. Another association of the saint in the Life is with Finnian, whose church of Clonard rivaled Clonmacnoise for primacy among the churches of the kingdom of Meath. Each became a diocesan center, one of West and the other of East Meath. Significantly, a prophecy of Ciarán's authority over the whole of the northern half of Ireland is placed in Finnian's mouth. Moreover, Ciarán is assigned a critical role in the foundation story of a house of nuns at Clonard—probably St. Mary's, a house of Augustinian nuns founded there by Murchad Ua Máel Shechlainn, king of Meath. A niece of Murchad, Agnes, became abbess there, and his daughter, Derbfhorgaill, rebuilt the Nuns' Church at Clonmacnoise in 1167. Short of a full investigation of the Life, these events supply a probable *terminus post quem*. That its author was an Augustinian canon is reflected *inter alia* by the choice of Ciarán's first teacher, Diarmait, patron of Inchcleraun, a house of canons on Lough Ree. A legendary dun cow associated with Ciarán, *Odar Chiaráin*, is commemorated in the name of the earliest surviving vernacular manuscript, *Lebor na hUidre* (Book of the Dun Cow), which was compiled at Clonmacnoise in the late eleventh century. Ciarán's feast fell on September 9.

Ciarán of Saigir

Although less well-known in Ireland than his Clonmacnoise namesake, Ciarán of Saigir exercised wider influence abroad through his (groundless) assimilation to the cult of Perran of Perranzabuloe in Cornwall, which led to the adaptation of one version of his Life for use as a Life of Perran's. There are three Latin and two Irish versions of his Life. These, as we now have them, were probably written in the late twelfth or early thirteenth century, against the background of a church which, although within the diocese of Ossory (Osraige), was physically separated from it by a stretch of land belonging to the diocese of Killaloe. The arrival of the Anglo-Normans led to the replacement of Seirkieran as a diocesan center by Kilkenny. Previously, however, a house of Augustinian canons was founded here, and their attitude may be reflected in the Life's attribution to the saint of *paruchia* (jurisdiction) over all of Ossory. Until superseded by Canice of Kilkenny, Ciarán is reputed to have been Ossory's chief saint. He is also said, spuriously, to have brought Christianity to Ireland before Patrick. His mother allegedly belonged to the Corca Loígde of the Clonakilty area, who were said to have provided several early kings of Ossory. This would explain the presence of a church of his on the island of Cape Clear. His feast falls on March 5.

PÁDRAIG Ó RIAIN

References and Further Reading

Doble, G. H., *Saint Perran, Saint Keverne, and Saint Kerrian.* Shipston-on-Stour: Cornish Saints Series No. 29, 1931.

Kenney, J., *The Sources for the Early History of Ireland (Ecclesiastical).* New York: Cornell University Press, 1929, pp. 316–317, 378–380.

Plummer, C., ed. *Vitae Sanctorum Hiberniae.* 2 vols. Oxford: Clarendon Press, 1910, (I) pp. xlviii–liv, 200–233.

Plummer, C., ed. *Bethada Náem nÉrenn. Lives of Irish Saints.* 2 vols. Oxford: Clarendon Press, 1922, (I) pp. xxv–vii, 103–124.

Sharpe, R. "Quatuor Sanctissimi Episcopi: Irish Saints Before St Patrick." In *Sages, Saints and Storytellers. Celtic Studies in Honour of Professor James Carney*, edited by D. Ó Corráin, L. Breatnach, and K. McCone, 376–399., Maynooth, 1989.

Stokes, W., ed. *Lives of the Saints from the Book of Lismore.* Oxford, Clarendon Press, 1890, pp. 117–134.

See also **Clonmacnois; Connacht; Ecclesiastical Organization; Hagiography and Martyrologies; Lebor na hUidre; Mide; Osraige**

CINÁED UA hARTACÁIN

Cináed ua hArtacáin (d. 975) was an accomplished, prolific Mide poet who was closely connected with the Síl nÁeda Sláine ruler, Congalach mac Maíle Mithig (d. 956), the most powerful king in Ireland in his day. Significantly, among his other patrons was the Hiberno-Norse king of Dublin, Amlaíb Cúarán (d. 981), who was associated with Skreen, County Meath. A *dinnshenchas* poem on Achall (the Hill of Skreen) names Cináed as author and cites Amlaíb as dedicatee, from whom the poet received *ech d'echaib ána Aichle* (a horse of the noble horses of Achall) as payment for his composition. We may speculate that the Norse king was also the recipient of Cináed's work on Benn Étair (Howth, Co. Dublin), since he is described as having assumed the kingship of that territory in the Achall poem. Five other poems in this genre attributed to Cináed survive, all concerned with places in his immediate vicinity. Two focus on Tara, a third on Brug na Bóinne (the Boyne Valley), and a further pair of poems describe what may well be literary locations, but which are perceived nonetheless as being within the same general region. Thus, Ochan, burial place of King Níall Noígíallach (Níall of the Nine Hostages), is Ochan Mide, and Ráth Ésa (the fort of Ésa or Étaín) recounts how a king of Tara, Eochaid Airem, successfully retrieved his wife and daughter from Midir of the Otherworld, a tale told in more detail in *Tochmarc Étaíne* (The Wooing of Étaín).

These *dinnshenchas* compositions all attest to Cináed's considerable command of narrative tradition, which is in even greater evidence in his best-known work, *Fíanna bátar i nEmain* (Champions Who Dwelt in Emain [Navan Fort]). This detailed composition constitutes a virtual compendium of the *aideda* (death-tales) of Ireland's premier heroes and kings and functions as an important index to stories already in existence, in some form, in Cináed's time. That it was valued is indicated by its reworking in the twelfth-century by Finn, bishop of Kildare, who added a number of verses to the copy he transcribed into the *Book of Leinster* making references to more recent notable destructions, including those at the battles of Clontarf (1014) and Móin Mór (1151). Not surprisingly, stanzas by Cináed also preface one version of the downfall of King Conchobar mac Nessa. Furthermore, his composition on the Boyne is incorporated into *Senchas na Relec* (Burial Ground Lore) preserved in *Lebor na hUidre*, and he is said to be the author of verse contained in *Lebor Gabála Érenn* (The Book of the Taking of Ireland, commonly known as The Book of Invasions). This fame may not simply be due to the inherent interest of the subject matter of his work. A skilled metrical craftsman who, in his own words, *rofitir rind-chert cech raind* (knows the rule of rhyme for every verse), Cináed must also have been admired for his technical accomplishments. These are particularly well displayed in his poems on Tara, which employ elaborate, intricate meters with accuracy and precision. It is with some justification, therefore, that he is termed *príméces* (primary poet) of Leth Cuinn (the northern half of the country) and of Ireland in his obituary notices in contemporary chronicles.

Máire Ní Mhaonaigh

References and Further Reading

Best, R. I., and Bergin, Osborn, eds. *Lebor na Huidre: Book of the Dun Cow.* Dublin: Royal Irish Academy, 1929, pp. 129–132.

Gwynn, Edward, ed. and trans. *The Metrical Dindshenchas.* 5 vols. Todd Lecture Series 8–12, 8: 6–13, 28–37, 46–53; 9: 2–9, 10–17, 36–41; 10: 104–109. Dublin: Royal Irish Academy, 1903–1935.

Meyer, Kuno, ed. and trans. *The Death-tales of the Ulster Heroes.* Todd Lecture Series 14, 18. Dublin: Royal Irish Academy, 1906.

Meyer, Kuno, ed. "Mitteilungen aus irischen Handschriften: Cináed húa Artagáin .cc." *Zeitschrift für celtische Philologie* 12 (1918): 358–359.

Stokes, Whitley, ed. and trans. "On the Deaths of Some Irish Heroes." *Revue Celtique* 23 (1902): 302–322; 27 (1906): 202–203.

See also ***Áes Dána; Aideda;*** **Amlaíb Cúarán; Dinnshenchas; Dublin; Emain Macha;** ***Lebor na hUidre*****; Leinster, Book of; Mide (Meath); Niall Noígiallach; Poetry, Irish; Poets/Men of Learning; Tara; Uí Néill, Southern**

CLARE, de

The history of the de Clare family in Ireland covers the period from the Anglo-Norman invasion to the mid-fourteenth century. The de Clare earls of Gloucester and Hertford held extensive estates in England and Wales throughout this period, and in Normandy until 1204. Their importance in an English context can be judged from the fact that the head of the family in the late thirteenth century, Earl Gilbert the Red, was considered to be the leading magnate of his day and married Joan of Acre, one of the daughters of Edward I. The family fortunes in Ireland, however, were started by a member of a junior branch of the de Clare earls of Hertford—Richard de Clare (d. 1176), known as Strongbow, the lord of Striguil and the dissatisfied claimant to the earldom of Pembroke. Failure in the male line caused Strongbow's lordship of Leinster to pass to his son-in-law, William Marshal. In 1245, the last of William Marshal's childless sons died, causing the great lordship of Leinster to be divided between William the elder's five daughters and their heirs. The lordship of Kilkenny descended via Isabel, his third daughter, to her son Richard de

Clare, Earl of Gloucester and Hertford. It was thus via inheritance, and not conquest, that the main line of the de Clare family came to hold land in Ireland.

Both Earl Richard and his son, Gilbert the Red Earl, visited their lordship of Kilkenny. Earl Richard's brief trip in 1253, however, was probably made out of pique at the fact that Henry III would not allow Richard to accompany him abroad. Richard seems to have been much more interested in his lands in the march of Wales; perhaps understandably, as they were worth some £2,000 to £2,500 *per annum*, whereas his lands in Ireland were valued at £350 in around 1247. Earl Gilbert the Red's sojourn in Ireland in 1293 to 1294 was a more prestigious affair, accompanied as he was by his wife, Countess Joan. The earl's purpose in traveling to Ireland in 1293 was a double one, and at least partly prompted by the king. The earl first undertook to effect "the pacification of Kilkenny" following disturbances by the native Irish of Leinster. He was later present (probably as the king's unofficial representative) at the suit between John FitzThomas, the lord of Offaly, and William de Vescy, the justiciar and lord of Kildare who had been accused of speaking treasonably against the king. Gilbert may have been more interested in his lands in Kilkenny than his father was; he certainly took the opportunity afforded by this trip to investigate, and attempt to expand, his rights in the lordship of Kilkenny. This interest was probably focused on optimizing the revenues available to be sent to the earl in England.

Gilbert the Red was the last of the earls of Gloucester to visit his Irish lands. Nevertheless, his death in 1295 did not represent the end of direct de Clare involvement in Ireland. In 1276, Thomas, one of Gilbert's younger brothers, had received a speculative grant of Thomond from Edward I, to be conquered from the Uí Briain. Thomas, a favoured household knight, sought to follow in the footsteps of his famous predecessor, Strongbow, in carving out an hereditary patrimony in Ireland with the sword, and has hence been regarded as a "throwback" to the late twelfth century. Thomas made a name for himself, posthumously at least, by betraying his erstwhile ally, Brian Ruad, as recorded in *Caithréim Thoirdhealbhaigh*. Whether this attempt at conquest could have worked is open to debate; certainly it was not helped by Thomas's early death from disease in 1287 or the minority (and then absenteeism) of his son Gilbert. Effective de Clare lordship in Thomond ended at the battle of Dysert O'Dea in 1318, when Thomas's second son, Richard, was killed. De Clare lordship in Kilkenny had already ended by this point with the death of the last de Clare earl of Gloucester at the battle of Bannockburn in 1314.

The de Clare lands in both Thomond and Kilkenny passed into the hands of heiresses, but this was not the end of the family's involvement in Ireland. One-third of Kilkenny passed to Gilbert's youngest sister, Elizabeth, who also inherited Clare in England, whereby she became lady of Clare. In addition to her inheritance in Kilkenny, Elizabeth continued to hold dower lands from her marriages to John de Burgh (d. 1313) and Theobald de Verdon (d. 1316) in Ulster, Connacht, Munster, and Meath until her death in 1360.

BETH HARTLAND

References and Further Reading

Altschul, Michael. *A Baronial Family in Medieval England: The Clares, 1217–1314.* Baltimore: John Hopkins Press, 1965.

Davies, R. R. *Lordship and Society in the March of Wales 1282–1400.* Oxford: Oxford University Press, 1978.

Dictionary of National Biography. Oxford: Oxford University Press, 1917 –

Frame, Robin. *Colonial Ireland, 1169–1369.* Dublin: Helicon, 1981.

Frame, Robin. *English Lordship in Ireland, 1318–1361* Oxford: Clarendon Press, 1982.

Lloyd, Simon. "Crusader Knights and the Land Market in the Thirteenth Century." *Thirteenth Century England* 2 (1987): 119–136.

Simms, K. "The Battle of Dysert O'Dea and the Gaelic Resurgence in Thomond." *Dal g Cais* 5 (1979): 59–66.

See also **Anglo-Norman Invasion; Fitzgerald; Gaelic Revival; Kilkenny; Leinster; Marshal; Munster; O Briain; Strongbow; Verdon, de**

CLASSICAL INFLUENCE

The question of knowledge in medieval Ireland of Classical Greek and Latin literature is controversial and complex. A great deal of research remains to be done into the lines of transmission of the Greek material that is to be found in Hiberno-Latin, given that the general orthodoxy is that Greek had disappeared from Western Europe by the end of the fifth century. The great bulk of Hiberno-Latin literature that shows an extensive knowledge of classical Greek and Latin literature survives in manuscripts written outside of Ireland. Much of it was written by emigrés like Columbanus, Johannes Scottus Ériugena, Sedulius Scottus, and Martin of Laon, so that there remains a doubt about the actual extent of knowledge of classical literature within Ireland. It cannot be denied that Irish expatriate scholars such as these contributed a great deal to the dissemination and use of Greek literature in Carolingian Europe, but few would accept that their knowledge of Greek had a foundation in their native schools. However, there is no doubt that there was some knowledge among the Irish of classical, as well as patristic, Greek and Latin literature, but we cannot be certain in many cases whether it was acquired on the continent or is a product of the Irish schools. Archaeological evidence and incidental references in the literature of the Graeco-Roman world also indicate that the Irish were well aware of the existence of their Mediterranean neighbors and their languages and culture.

Current dogma on the status of classical influence on medieval Irish scholarship varies from extreme scepticism to mild optimism. The conventional way to assess the knowledge of Greek is to examine the written evidences of its transcription into Greek letters or in Latin transliteration. The pitiful state of the preservation of Irish manuscripts has left us with little enough remains of Greek material per se. But in an overwhelmingly Latin-speaking ecclesiastical environment, there is likely to have been a limited requirement for the preservation of extensive passages of Greek, except for specific liturgical or other purposes. The Anglo-Saxon church was introduced to Greek learning by the arrival at Canterbury of Theodore and Hadrian, who, Bede tells us, were as well-conversant with Greek as Latin and left behind them a generation of students who were proficient in both languages. The glosses on the Pentateuch first discovered by Bischoff, and attributed by him to Theodore, contain only the bare minimum of Greek vocabulary, and then largely in Latin transliteration. But it has been shown that they preserve a substantial knowledge of Greek patristic literature. What we have is a knowledge of Greek dressed up in Latin form for speakers of Latin. This understanding must also be applied to Hiberno-Latin literature.

It is generally thought that Ireland would have had very little contact with the few sources of spoken Greek or of Greek literature remaining in Italy. But there is increasing evidence of knowledge of Greek patristic and liturgical material in the early period after conversion to Christianity, and this is not unrelated to the question of classical influence. Where did Ailerán of Clonard acquire his knowledge of Greek *onomastica sacra*, or the manuscript known as *Liber Commonei* find its Greek liturgical material? Some knowledge of Greek can therefore be clearly discerned from the few remaining fragments of biblical or liturgical Greek copied into some manuscripts, chiefly for pious ostentation, as well as from the Fahan Mura inscription discussed by Macalister and largely ignored since.

It has been shown recently that the Irish had an acquaintance with classical clausular structure and meter, and some may have had a basic reading knowledge of Greek itself. Allusions to classical mythology are scattered throughout Hiberno-Latin literature in works composed at home, such as the *Liber hymnorum*, and in hagiography, biblical exegesis, and grammatical and scholastic texts. Some influence from the late classical world in these sources is therefore certain.

Regarding classical Latin literature, Virgil was certainly well known to them, as were several other authors including Petronius and Lucretius, in the transmission of whose works the Irish played a part. The oldest manuscript of the *Scholia Bernensia* on Virgil's *Eclogues* and *Georgics* came through Irish hands. Many emigré Irish scholars played an important role in the preservation of fragments or texts of classical Latin authors otherwise little known in the Middle Ages. The poems attributed to Columbanus contain the earliest allusions to and use of a range of classical literature. Jonas of Bobbio's biography of Columbanus says that in his youth he had received some grounding in liberal arts and grammar. However, the authorship of five poems formerly attributed to him has been disputed, primarily because of their implications for some knowledge of classical literature in the early Irish schools, and the poems have been attributed to a later Columbanus of Saint Trond. From the early seventh century at the latest, the Irish became acquainted with the works of the late classical grammarians, such as Donatus, Priscian, and many lesser-known grammatical works, but the earliest Hiberno-Latin literature to show a knowledge of Aristotelian philosophy or the works of Boethius comes from the period of the Carolingian renaissance, and therefore cannot be attributed with certainty to the Irish schools. Another emigré, Dicuil, wrote several works on geography, computus, grammar, and astronomy that show a knowledge of Pliny, Solinus, and Ptolemy, who were also known in the Irish schools.

In the later medieval period, perhaps from the tenth century onward, vernacular adaptations and translations of Greek and Latin classics were being produced. One could instance the *Togáil Troí*, a Middle-Irish version of the fictitious account given by the fifth-century author called Dares Phrygius. It is the earliest vernacular translation of an admittedly pseudoclassical work of literature. The fourteenth-century tale called *Merugud Uilix Maic Leirtis* (The Wanderings of Ulysses, Son of Laertes) is an adaptation of Homer's *Odyssey*, which is not without literary merit. There are echoes of Homeric and other classical literature elsewhere in Middle and Early Modern Irish, which are not all due to coincidence or to some putative common Indo-European inheritance. There are some free versions of Latin epics also, such as Statius' *Thebaid* (*Togáil na Tébe*), Virgil's *Aeneid*, and Lucan's *Civil War* (*In Cath Catharda*), all dating from the fourteenth to the fifteenth century.

Aidan Breen

References and Further Reading

Bieler, L. "The Island of Scholars." *Revue du Moyen Age Latin* vii (1952): 213–234.

Bolgar, R. R. *The Classical Heritage and its Beneficiaries.* Cambridge: 1954.

Bolgar, R. R., ed., *Classical Influences on European Culture.* Cambridge: 1971.

Daintree, D. "The Transmission of Virgil and Virgil Scholia in Medieval Ireland." *Romanobarbarica* 16 (1999): 33–47.

Harris, J. R. *Adaptations of Roman Epic in Medieval Ireland.* Lewiston, NY.: Edwin Mellen, 1998.

Herren, M. "Classical and Secular Learning Among the Irish Before the Carolingian Renaissance." *Studi Medievali*, 3rd ser. 18 (1977): 815–880.

Hofman, R. *The Sankt Gall Priscian Commentary.* 2 vols. Münster: Nedus, 1996.

Holtz, L. "La Redécouverte de Virgile aux VIIIe et IXe Siécles d'aprés les Manuscrits Conservés." *Lectures Médiévales de Virgile, Collection de l'école Fraçnaise de Rome* 80 (Rome, 1985): 9–30.

Howlett, D. R. *The Celtic Latin Tradition of Biblical Style.* Dublin: 1995.

Ó Cuív, B. "Medieval Irish Scholars and Classical Latin Literature." *PRIA* 81C (1981): 239–248.

Stanford, W. B. *Ireland and the Classical Tradition.* Dublin: 1976.

Walker, G. S. M., ed. *Sancti Columbani Opera.* Dublin: 1957.

See also **Biblical and Church Fathers; Columbanus; Dícuil; Ériugena, John Scottus; Hiberno-Latin; Literature; Sedulius Scottus**

CLIENTSHIP

This is *céilsine* in early Irish law. It is the word used to describe the relationship between a *céile* (companion, fellow) and his lord, *flaith. Céile* is cognate with Latin *cliens* (client, dependent). It could be translated "vassal," but this is avoided since "vassal" is so closely associated with later feudalism. Thomas Charles-Edwards has recently highlighted the difference between the Irish and Continental systems as "Frankish lordship worked through land, Irish lordship through capital." The capital was primarily (though not exclusively) livestock, the "fief," that a lord granted to a client. It was a contractual personal relationship that bound lord and man together. Clientship permeated the entire social and political fabric of Ireland in both secular and ecclesiastical society.

At the top level of society, clientship among kings and nobility was primarily political. At the bottom were slaves, and between the slaves and the farm-owning class were the semifree, *bothach*, *fuidir*, and *senchléithe*. The *bothach* was a cottager and the *senchléithe* were serfs, tied to the soil. The *fuidri* would seem to have dropped into the semifree class as a result of crime or inability to sustain themselves as independent farmers. They were the clients of the classes above and provided the labor force on farms and in households. The free independent farmers were the most important class in relation to clientship, for they formed the backbone of the economy.

There were two forms of clientship: *sóer chéilsine* ("free clientship," in the sense of legally independent) and *dóer chéilsine* ("base clientship," in which part of the contracting party's legal independence was absorbed by the lord). Both involved free commoners. In each the *céile* received a "fief" of stock from a lord, for a period of contract of seven years. The *sóer chéile* accepted three cows. He paid heavy interest each year, but at the end of the period the fief became his absolute property. This person was a wealthy farmer. Each party could opt out of the contract without penalties. The other duties of the *sóer chéile* were those of *manchuine*, "personal service" of attendance upon his lord, and *urérge* (homage) to him. He helped his lord pursue the feud and took part in mourning his death. He formed part of his lord's *dám*, his "company," when on public business, and was a member of his war band. It was a position of prestige. If for some reason the *sóer chéile* failed to maintain his contract, he could drop to the level of *dóer chéile*.

The *dóer chéile* was given a more generous "fief," and his interest was less per year. He faced heavy penalties should he withdraw from the contract. *Dóer chéilsine* was originally called *giallnae*, derived from *giall* (hostage). The *dóer chéile* was given an extra payment, *séoit taurchluideo* (chattels of subjection), which was equal to his honor price. This, with the earlier name *giallnae*, may suggest that this form of clientship was originally applied to defeated peoples. In return for this payment the lord assumes some legal responsibility for him, hence his dependent legal status. The lord received compensation for injury done to his client, for example, for homicide or theft (one-third of all payments were due to him). The *dóer chéile* paid interest to his lord in livestock and also in foods of various kinds—both meat and cereals—and candles. His inferior status is revealed in the labor services he was required to supply: taking part in his lord's harvest and also in digging an extra rampart around his lord's fort. He had also to take part in the military hosting. Lords collected their rent for the most part in guesting upon their base clients during the *aimser chue*, the period of winter visitation between January 1 and the beginning of Lent. The number of guests and the quality of food were regulated.

It would appear that there were differences in clientship between the northern and southern parts of Ireland. In the north the *dóer chéile*'s contract lasted until the death of his lord. If the client died first, his heirs had to maintain the contract. On the lord's death, the original grant or their offspring remained with the client after the seven-year period had been properly completed. This rule applied in Munster too, as long as the contracting parties were of close rank. The greater the difference in rank, the longer the client and his heirs had to serve the lord before they took possession of the "fief." There were other differences, too. In general it would seem that the southern *dóer chéile* were worse off, although it is possible that the duty of hospitality was not heavy in the south. The organization of clientship within the vast estates of the Church would seem to have been rather similar to that in the secular world. As well as the geographical differences in clientship, there are hints of changes taking place over time. Base, as well as free clients, became part of the lord's *dám* (company). There would seem to have been a general increase in labor services. This may account for the references to tenants absconding from Church estates.

The main description above is drawn from the law tracts of the seventh and eighth centuries. Changes had been taking place, but the rate of change was greatly accelerated following the raids by the Norse in the ninth century. Between the tenth and twelfth centuries society became more militaristic. Only the most important kings had any real political control, and their emerging lordships had a feudal-like structure. Minor kings became their officials. A system of taxation emerged that was based on areas of assessment. Much of this development is imperfectly understood. Against this background there is some evidence of a leveling downward within the ranks of the freemen to produce a mass of peasant rent-payers in the course of time. These were the *bíataigh* (*betagii*), or "betaghs," as cited in later English documents. They were literally the "food-providers" of their lords.

CHARLES DOHERTY

Clonmacnoise Round Tower, Co. Offaly. © *Department of the Environment, Heritage and Local Government, Dublin.*

References and Further Reading

Charles-Edwards, T. M. *Early Christian Ireland.* Cambridge: Cambridge University Press, 2000.

Kelly, Fergus. *A Guide to Early Irish Law.* Dublin: Dublin Institute for Advanced Studies, 1988.

Patterson, Nerys. *Cattle Lords and Clansmen. The Social Structure of Early Ireland*, 2nd ed. Notre Dame and London: University of Notre Dame Press, 1994.

See also **Brehon Law; Kings and Kingship; Law Texts; Society, Grades of Gaelic; Society, Functioning of Gaelic**

CLONMACNOISE

Clonmacnoise was one of the most important early medieval ecclesiastical sites in Ireland, and today has a number of ruined churches and a fine collection of high crosses and cross-slabs.

It was founded by St. Ciarán in the 540s, and being on a crossing point of two routeways—the north-to-south-flowing river Shannon itself, on whose east bank it lies, and a major land route running east to west—it soon became an important center of population, trade, and craftsmanship, as well as religion. In time it became an important center of learning, and especially the keeping of annals. It gradually became the center of an affiliation of monasteries and other churches, and in this way grew to have considerable power and influence. Being situated on the boundary of two provinces—Mide, or Meath, and Connacht—it was sometimes under the sway of the kings of Connacht, and at times of the kings of Mide.

It was raided on many occasions by Irish enemies, Vikings, and in the years around 1200 by the Anglo-Normans. By this time its influence was on the wane as a result of a number of factors. The church reformers of the twelfth century had sounded the death knell of the old monastic/ecclesiastical system by establishing territorial dioceses and introducing continental monastic orders. Clonmacnoise became the seat of a bishop, but soon lost territory to the powerful Anglo-Norman–dominated diocese of Meath and became one of the smallest and poorest dioceses in Ireland. Also, with the defeat of the Gaelic kingdoms, it lost its important royal patrons and went into serious decline from the early thirteenth century on. The settlement also declined, and by the early sixteenth century there were only a few thatched houses around the churchyard.

Most of the churches are in the old walled graveyard. The largest is the cathedral, incorporating much of the *daimliag* (stone church) built in 909. In subsequent building phases a new west doorway was inserted around 1200, the south wall was rebuilt to the north of its original line in the late thirteenth century, and a new north doorway, windows, and vaulting over the east end were added in the 1450s. The smallest church, Temple Ciarán, also dates to around 900, and like the cathedral has antae (projections of the side walls beyond the end walls). It was regarded as the burial place of St. Ciarán. The freestanding round tower was the bell tower for the establishment, and the annals record that it was completed in 1124. It stands today to a little more than half its original height.

There are two important Romanesque churches: the Nuns' Church (1167), a nave-and-chancel building to the east of the main site, with an ornate chancel arch and west doorway; and Temple Finghin, a fine nave-and-chancel church with an attached round tower. The period around 1200 saw further building activity in the Transitional style, especially Temple Connor, Temple Melaghlin, and the now much-ruined castle, built by the chief governor of Ireland in 1214.

CONLETH MANNING

References and Further Reading

Kehnel, Annette. *Clonmacnois—The Church and Lands of St. Ciarán.* Münster: Lit Verlag, 1997.

King, Heather A., ed. *Clonmacnoise Studies Volume 1: Seminar papers 1994.* Dublin: Dúchas, 1998.

———, ed. *Clonmacnoise Studies Volume 2: Seminar papers 1998.* Dublin: Dúchas, 2003.

Macalister, R. A. S. *The Memorial Slabs of Clonmacnoise, Kings County.* Dublin: The University Press for the Royal Society of Antiquaries of Ireland, 1909.

Manning, Conleth. *Clonmacnoise.* Dublin: Office of Public Works, 1994. Second edition Dublin: Dúchas, 1998.

See also **Architecture; Ciarán; Early Christian Art; Ecclesiastical Settlements; Ecclesiastical Sites; High Crosses; Iconography; Inscriptions; Parish Churches, Cathedrals; Scriptoria; Sculpture**

CLONTARF, BATTLE OF

The Battle of Clontarf took place on Good Friday, April 23, 1014, and is named from a field east of the Viking fortress of Dublin called *cluain tarbh* (the bulls' meadow). High King Brian Boru (Bóruma) mac Cennétig fought an alliance of Sitriuc Silkenbeard of Dublin, Máelmórda mac Murchada of Leinster, and Sigurd Hlodvisson of the Orkneys. Clontarf was one of the most famous battles in Ireland, becoming a standard chronological marker in Irish historical writings. Medieval Irish historians saw Clontarf as the battle that broke Viking power in Ireland, but later historians place it within the context of Irish political maneuvering.

The events leading to the battle of Clontarf began in 1013. Brian was an outsider in the Irish political order, and his ascendancy was challenged by the elites he had displaced. A rebellion began among the northern Uí Néill in 1012, and it was joined by the kings of Dublin and Leinster in 1013. Legend blames Brian's wife Gormfhlaith for precipitating the conflict by encouraging either her son Sitriuc (according to the Norse) or her brother Máelmórda (according to the Irish) to rebel. Brian led an army to Dublin in the autumn of 1013, but was unable to reduce the town, and his forces retired late in the year. At the same time, Sitriuc recruited Sigurd of the Orkneys, supposedly with the promise of the kingship of Ireland.

In April 1014, Brian returned to Dublin, to fight what would be the battle of Clontarf. His troops were drawn from Munster and southern Connacht. Brian approached the town from the west, and awaited reinforcements on the *faithce* (green space), probably Oxmantown Green on the north bank of the Liffey. Brian's ally was the southern Uí Néill king, and previous high king, Máel Sechnaill II, who was also the former husband of Gormfhlaith. On the day of the battle, however, Máel Sechnaill and his troops appear to have stood aloof from the fight. The Viking-Leinster army had drawn troops from the Orkneys, Hebrides, Isle of Man, northern England, and, possibly, Normandy. They assembled at Howth and camped at *Mag nElta*, where Clontarf is situated. In a diversionary tactic, Brian's son Donnchad (Sitriuc's maternal half-brother) took troops to ravage the lands south of Dublin.

The battle of Clontarf began on Friday morning. Brian, who was in his seventies, observed the battle from a ridge and his son Murchad led the troops. The Viking-Leinster coalition was commanded jointly by Máelmórda and Sigurd. Sitriuc's presence is uncertain; the Norse sagas claim that he led troops, but some Irish records claimed that he remained in Dublin. Both sides divided their troops into battle groups and used banners to identify them. Brian's mercenary troops fought in one unit under the command of a *mórmáer* (great-steward) from Scotland named Domnall mac Eimhin of Mar. Reports of the battle emphasize the combatants' different methods. The Irish did not wear armor, although the commanders seem to have had helmets. They used shields for protection. Wealthy warriors used swords and Viking axes, while others used spears and knives. The Vikings had body armor made of iron, possibly an early version of chain mail, which eliminated the need for shields. For fighting they used spears and swords. Only a few appear to have used axes, but they did use bows and arrows.

The hostilities raged for hours. The advantage was initially with the Viking-Leinster forces, but passed to Brian's troops by the afternoon. The fighting spread over a wide area west of Howth. Much of the combat took place near or among the wharves and piers along the Liffey, which had been built to accommodate the large trading vessels that called at Dublin. A grandson of Brian named Tairrdelbach drowned after he was knocked unconscious under a weir. The outcome was decided late in the afternoon when the cohort from Dublin was broken at *drochat Dubgaill* (Dubgall's Bridge), probably at the site of "Old Dublin Bridge" from Bridgefoot Street to Oxmantown. After Sigurd and Máelmórda were slain, the Viking forces tried to escape to their ships along the Liffey, but their retreat turned into a slaughter. They had beached their boats above the high-water mark, but an unusually high tide floated the ships into the middle of the channel.

Disaster struck the Irish as well. Brian's son Murchad was slain in the battle, and Brian was cut down by escaping Vikings led by Brodor of York. The tract *Cogad Gáedel re Gallaib* (War of the Irish against the Vikings) has a story that the Vikings were passing by Brian, believing him to be a priest, when a mercenary previously in his service recognized him. The chronicler Marianus Scottus claims that Brian was at prayer when he was slain. Although Brian's troops held the field, with Brian and Murchad dead they were unable to proceed further. They were too decimated by the slaughter to storm the fortress of Dublin. The survivors waited on the battlefield for two days until Donnchad returned on Easter Sunday. Their return home was hindered by fighting within their own ranks together with opposition from Brian's subjects, who now rose in rebellion.

Who won the battle of Clontarf? The insular records claim it as an Irish victory, even though Brian's objective, the capture of Dublin, was not achieved. A contemporary Viking poem, however, flatly states that it was a victory for the Vikings. From what is now known, perhaps the most fair assessment is that the battle was a stalemate that exhausted both sides.

The battle of Clontarf demonstrated the military power of the Irish. Fighting an evenly matched opponent, Brian's troops held the field against an international force led by, in the case of Sigurd, one of the premier warriors of the northern world. Stories about Clontarf circulated throughout Europe, from Iceland to Francia. Brian passed into legend as the great hero-king of the Irish. Nevertheless, the victory at Clontarf failed to unify the Irish, and ambitions towards national monarchy would be temporarily obscured by factionalism and dynastic rivalries.

BENJAMIN HUDSON

References and Further Reading

Goedheer, A .J. *Irish and Norse Accounts about the Battle of Clontarf.* Haarlem: H.D. Tjeenk Willink & Zoon N.V., 1938.

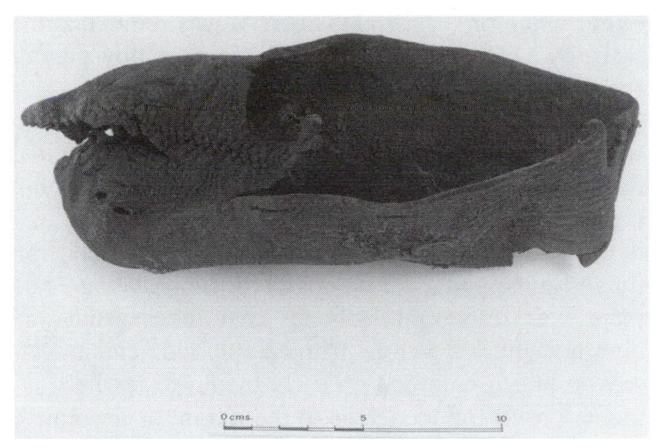

Leather shoe, Beleevna More, Co. Tyrone. *Photograph reproduced with the kind permission of the Trustees of the National Museums and Galleries of Northern Ireland.*

Ó Corráin, Donncha. *Ireland Before the Normans.* Dublin: Gill and Macmillan, 1972.

Ryan, John. "The Battle of Clontarf." *Journal of the Royal Society of Antiquaries of Ireland* 68 (1938): 1–50.

Todd, J. H., ed. *Cogadh Gaedhel re Galliabh.* (*The War of the Gaedhil with the Gaill.*) London: Rolls Series, 1867.

See also **Amlaíb Cuarán; Brian Boru; Dublin; Dál Cais; Fine Gall; Gormlaith (d: 1030); Laigin; Leinster; Máel Sechnaill II, Scandinavian Influence; Ua Briain; Uí Néill, Southern; Viking Incursions; Weapons and Weaponry**

CLOTHING

Our picture of clothing in medieval Ireland is derived primarily from figurative scenes—particularly those from carved stone crosses and effigies and from illuminated manuscripts—but also from descriptions in contemporary literary sources. Additionally, there is a body of surviving textiles: a small assemblage dating to the early medieval period and more extensive collections recorded from medieval urban excavations in Dublin, Waterford, and Cork. While most of the extant textiles do not compose entire garments, they do provide useful supplementary information.

Society in Ireland throughout the medieval period was hierarchical in nature, with clearly defined social grades. In this context, clothing had a primary functional role to protect the wearer from extremes of climate, but could also act as a signal of the wearer's status or cultural origins. The Irish law tracts made some attempt to regulate dress styles by imposing restrictions on the number of colors in garments worn by various ranks. The prevailing style of dress in the early medieval period comprised a *léine* (tunic) worn under a *brat* (cloak). The *léine* was an ankle-length, sleeveless garment worn next to the skin and made of either white or *gel* (bright) linen. It was secured at the waist by a belt and could be hitched up to allow greater freedom of movement.

The *brat* was rectangular in shape and made from wool, and was sometimes large enough to wrap around the body five times. It could be brightly colored, with ornate decorative borders. The archaeological evidence suggests that the use of dyestuffs extracted from the red-dyeing madder plant and the blue-dyeing woad plant was important from at least the seventh and eighth centuries, while fringed, plaited, and tablet-woven braids recorded on early medieval textile fragments provide evidence as to the nature of decorative borders. The *brat* was secured on the breast by a bronze, silver, or iron brooch or pin, depending on the individual's social status and wealth.

Figurative art also suggests that *truibhas* (trousers) were worn by horsemen and others engaged in outdoor activities. A series of small figures wearing knee-length *triubhas* are recorded in the text of the Book of Kells. The wearing of the *léine* and *brat* secured with a penannular brooch is recorded on the Cross of the Scriptures at Clonmacnois, County Offaly, in a scene interpreted as the laying of the church foundation post by Abbot Colmán and King Flann around 910.

There are few women depicted in the figurative art of the early medieval period, but descriptions in the myths and sagas indicate that the *brat* and *léine* were worn by both sexes. From the early ninth century onward women covered their heads with a veil or headdress.

The introduction to Old Irish before 900 of a number of Old Norse loan words—such as *skyrta*, which became *scuird* (shirt, tunic, cloak), and *brok*, which became *bróg* (hose, trousers, [and later] shoes)—suggests that the Viking incursions had an impact on dress. In particular, the Vikings may have introduced the short tunic and trousers outfit, as well as the *ionar*, a form of tunic worn over the *léine*. The Scandinavians are also generally credited with the introduction of silk cloth into Ireland through their increased trading connections.

The Anglo-Norman Invasion of 1169 and the establishment by Henry II of a stronghold in the Dublin region in 1171 introduced a new aristocracy to Ireland, who followed the fashions of London and Europe. The contrast in dress and appearance between the recently arrived Anglo-Normans and the Gaelic Irish is highlighted in the descriptions and illustrations of Giraldus Cambrensis in his *Topographia Hiberniae*. The *léine*, *brat*, and *ionar* continued to be worn by the Gaelic Irish in the medieval period; the *brat* came to be called the "Irish mantle" and the *léine* the "saffron shirt." Other garments of importance included a short-hooded cloak called a *cochall* and a poncho-type cloak of colored and patterned cloth called a *fallaing*, as well as woollen *truibhas* with feet and soles. Contemporary Anglo-Normans are shown wearing tunics of mid- to lower-calf length with Magyar-style sleeves, belted at the waist with a white sash from which a scabbard was suspended, along with a traditional mantle or cloak. The contrast in dress styles was probably most apparent during the initial colonization period, while the following centuries saw considerable mutual cultural influence, as evidenced by various statutes and laws that sought to discourage Anglo-Norman descendants from adopting Gaelic modes of dress and appearance.

Anglo-Norman men and women wore an underdress, or kirtle, and an overgown, or surcoat. The kirtle was round-necked with tight-fitting sleeves and was secured at the waist or hips with a girdle (for women) or sash (for men), from which personal objects such as keys or scabbards were suspended. The surcoat could be sleeved or sleeveless, with deep armholes and with vertical slits called fitchets that provided access to objects suspended from the girdle. Both male and female versions of the surcoats had a slit at the neck, which during the twelfth and thirteenth centuries was commonly secured by a ring brooch. This dress fastener was introduced by the Anglo-Normans, but a number of ring brooches were recovered from Gaelic Irish Crannóga. In winter, a mantle or fur-lined cape was also worn. The Irish mantle appears to have been adopted by both communities, and came to be an important trade item.

In the mid-fourteenth century a closer-fitting outfit emerged for Anglo-Norman men, consisting of a knee-length garment called a *gipon* (later doublet) worn with hose. The wearing of a doublet by Noah in the *Book of Ballymote*, which dates to 1400, would suggest that this was also adopted by the Gaelic Irish. A gown with buttons on the sleeves and bodice and a full knife-pleated skirt, seen on the double effigy at Knocktopher, County Kilkenny, is interpreted as an Irish adaptation of the Anglo-Norman *houppelande*, and a garment of similar type was recovered from a bog in Moy, County Clare. The wearing of hoods with long, pointed extensions—called *liripipes* by the Irish—represents an expression of mutual cultural influence.

As Ireland fell under increasingly direct English rule during the sixteenth century, the ascendancy redoubled its efforts to supplant Gaelic traditions and customs. In terms of clothing, this manifested itself as a growing struggle between the increasingly sober styles of London and the relative flamboyance of indigenous medieval dress.

MARIA FITZGERALD

References and Further Reading

Barber, J. W. "Early Christian Footwear." *Journal of the Cork Historical and Archaeological Society* 36 (1981): 103–106.

Deevy, M. *Medieval Ring Brooches in Ireland*. Bray: Wordwell, 1998.
Dunlevy, M. *Dress in Ireland*. London: Batsford, 1989.
FitzGerald, M. "Dress Styles in Early Ireland (*c*. 5th–*c*. 12th A.D.)" MA Thesis, University College Dublin, 1991.
FitzGerald, M. "Textile Production in Prehistoric and Early Medieval Ireland." Ph.D. Thesis, Manchester Metropolitan University, 2000.
Lucas, A. T. "Footwear in Early Ireland." *Journal of the County Louth Archaeological Society* 13 (1956): 309–394.
McClintock, H. F. *Old Irish and Highland Dress*. Dundalk: Dundalgan Press, 1950.

See also **Craftwork; High Crosses; Jewelry and Personal Ornament; Kells, Book of; Society, Grades of Anglo-Norman; Society, Grades of Gaelic; Viking Incursions**

CLYN, FRIAR JOHN (d. 1349?)

John Clyn was an Anglo-Irish Franciscan friar and the author of *Annals of Ireland by Friar John Clyn*, written in Kilkenny and covering the period from the "beginning of the world" to 1349. According to the seventeenth-century antiquarian James Ussher, Clyn was born in Leinster and held the degree of doctor. The surname Clyn is not common in Ireland, but there is a townland a few miles from Kilkenny called Clinstown. From the annals, we learn that Clyn became the first guardian of the friary of Carrickbeg (Carrick-on-Suir) in 1336, when the earl of Ormond presented the property to the Franciscans. Clyn was present in Kilkenny friary in 1348 during the Black Death, when he identified himself as the author of the annals. The annals are famous for a dramatic first-hand account of the Black Death in Ireland in 1349. A very rough seventeenth-century transcript claims that Clyn was also guardian of the Franciscan friary of Kilkenny. Clyn's original manuscript is no longer extant; Sir Richard Shee, sovereign (mayor) of Kilkenny, possessed the manuscript in 1543, and by 1631 it had been acquired by David Rothe, bishop of Ossory. Four main seventeenth-century transcripts survive, and they state that the annals were copied from the community book of the Franciscans of Kilkenny. There is scant reference to Franciscan affairs, but as the annals reportedly were part of the community book of the Franciscans of Kilkenny, there would have been no need for such information in the annals. The annals consist of very brief entries, with years often repeated and out of sequence, until 1333. All four transcripts agree that in 1333 a new section of the annals commenced. Clyn's main interest is in the military society of the area surrounding Kilkenny in a troubled period of Anglo-Irish history. Internal evidence suggests that Clyn was familiar with military society and displayed a great interest in knighthood, noting who was knighted by whom. Clyn respected a certain code of conduct, which led him to express displeasure at actions, perpetrated by either the native Irish or the Anglo-Irish, that were contrary to the highest standards of knighthood. Clyn has sometimes been considered as hostile to the Irish, and indeed during this troubled period it was only to be expected that they should receive censure, but Clyn is remarkable for his criticism of the troublesome members of the Anglo-Irish nation also. Clyn is particularly dismayed by treachery or betrayal, in any form and by either nation. On balance, Clyn only refers to the Irish nation in relation to its effect on the Anglo-Irish nation. Clyn exhibited a particular familiarity with the local Mac Gillapatrick family. Among the Anglo-Irish, it is the de la Frene family that occasions most interest. The dominant personality in Clyn's annals is Fulk de la Frene, whose knighting by the earl of Ormond Clyn reports in 1335. Fulk emerges, in Clyn's annals, as a strong military man, and this is reflected by the reports of his victories over the Irish and his success in expelling Anglo-Irish troublemakers. The longest entry in the annals is for 1348, which describes the horrors of the Black Death, an event that the writer regarded as truly catastrophic and apocalyptic. Clyn's account of the plague opens with pilgrimages to the local St. Mullins Well; these were, he tells us, inspired by fear of the plague. His entry includes the number of people who died in Dublin from August to Christmas, the number who had died in the Franciscan friaries of Drogheda and Dublin from the beginning of the plague to Christmas, and the information that the plague was at its height in Kilkenny during Lent. Although Clyn enters the number of Dominicans who died in Kilkenny, he makes no mention of Franciscan deaths, but this information could have been entered in another section of the community book. Clyn also includes an account of the plague in Avignon and a lengthy account of an apocalyptic vision given to a monk at the Cistercian monastery at Tripoli in 1347. It is with great sorrow, and a great eulogy, that Clyn reports, in his last entry, the death of Fulk in 1349. The seventeenth-century transcripts suggest that Clyn died of the plague. Another possibility is that Clyn was moved to a different friary as part of a possible redistribution necessary after the decimation of some friaries. A third possibility is simply that Clyn ceased to write once his friend, and perhaps patron, Fulk de la Frene, had died.

Bernadette Williams

References and Further Reading

Butler, Richard, ed. *Annals of Ireland by Friar John Clyn and Thady Dowling, Together with the Annals of Ross*. Dublin: Ir. Arch Soc, 1849.

Williams, Bernadette, "The Latin Franciscan Anglo-Irish Annals of Medieval Ireland." Ph.D. dissertation, University of Dublin, Trinity College, 1991.

———."The Annals of Friar John Clyn—Provenance and Bias." *Archivum Hibernicum* 47 (1993): 65–77.

See also **Anglo-Irish Annals; Black Death; Kilkenny; Leinster**

COARBS

See **Church Reform, Twelfth Century; Ecclesiastical Organization**

COGITOSUS (*fl. c.* 650)

Cogitosus (Ua hAédo) was an Irish monk and author. It is likely that the unusual nom-de-plume Cogitosus is a translation of the rare Irish name Toimtenach. He may be the same Toimtenach of Mainister Emín (Monasterevin) mentioned in the genealogies. He was author of a Latin Life of St. Brigit of Kildare (d. 525) written not much later than 650, and therefore the earliest extant piece of hagiography in Hiberno-Latin. It is written in an unpretentious Latin style, his aim throughout being to emphasize the presence of God's power in Brigit, manifested through her miracles, her great faith in God, and her charity toward the poor.

In the epilogue he addresses himself as "the blameworthy descendant of Aed." The Áed to whom he claims relationship is probably Áed Dub, bishop and abbot of Kildare (d. 639), a member of the Uí Dúnlainge dynasty of the northern Laigin. He states that he was "compelled in the name of obedience" by the community of Kildare to write a Life of their foundress. We know from other sources that Brigit's church was at Kildare, though Cogitosus nowhere mentions it in the text, nor does he tell us that he was a member of it. Muirchú moccu Machthéni claimed Cogitosus as his spiritual father and the first hagiographer among the Irish in the prologue to his Life of Patrick.

It is probable, judging from similarities between the material in Cogitosus and the later lives of Brigit, especially Vita I, that Cogitosus drew from existing written material that had preserved some traditions of her life and miracles. He states that Kildare claimed to be "the head of almost all the Irish churches with supremacy over all the monasteries of the Irish and its *paruchia* extends over the whole land of Ireland, reaching from sea to sea" (Prol. 4). It was a double foundation, with one monastery for monks, including some priests, with a prior over them, and another for nuns, ruled by an abbess. Its first bishop was Conláed, asked by Brigit to become bishop so "that he might govern the church with her in the office of bishop and that her churches might not lack in priestly orders" (Prol. 5). Kildare's importance through her contacts abroad is shown by Conláed having obtained his episcopal vestments from overseas. Cogitosus' description of the basilica at Kildare is a unique seventh-century eyewitness account of the structure and furniture of an Irish church. It had three chapels—containing painted pictures, an ornate altar, and the sarcophagi of Brigit and Conláed, "adorned with a refined profusion of gold, silver, gems and precious stones with gold and silver chandeliers hanging from above" (§ 32), all under one roof—that served as a place of worship for both communities and for laity and pilgrims together. We are told that Kildare was one of the greatest centers of pilgrimage in Ireland, "a vast metropolitan city and the safest city of refuge in the whole land of the Irish for all fugitives, and the treasures of kings are kept there" (32.9).

AIDAN BREEN

References and Further Reading

Connolly, S., and J.-M. Picard. "Cogitosus' Life of Brigit." (Trans.) *Journal of the Royal Society of Antiquaries of Ireland* 117 (1987): 5–27.

Esposito, M. "On the Earliest Latin Life of St. Brigit of Kildare." *Proceedings of the Royal Irish Academy* 30 C (1912): 307–326.

———."Notes on Latin Learning and Literature in Medieval Ireland 1.6: Cogitosus. (*c.* 620–680)." *Hermathena* 20, no. 45 (1926–1930): 225–260, at 251–257.

Mc Cone, K. "Brigit in the Seventh Century: A Saint with Three Lives." *Peritia: Journal of the Medieval Academy of Ireland* 1 (1982): 107–145.

Ó Briain, F. "Brigitana." *Zeitschrift für Celtische Philologie* 36 (1976): 112–137.

Sharpe, R. "Vitae S Brigitae: The Oldest Texts." *Peritia: Journal of the Medieval Academy of Ireland* 1 (1982): 811–106.

See also **Brigit; Ecclesiastical Organization; Ecclesiastical Settlements; Hagiography and Martyrologies; Hiberno-Latin; Patrick; Uí Dúnlainge**

COINAGE

The Earlier Middle Ages

It is an academic cliché that Ireland was a coinless society throughout the first millennium A.D. Small numbers of coins from the Roman world circulated in Ireland in the early centuries A.D., but were not used as currency and were probably kept as curiosities or for their bullion value. The cliché still holds good for most of the first millennium, but its applicability in the ninth and tenth centuries is increasingly questioned. Proper coin usage

Henry III penny. © *Courtesy of John Stafford-Langan.*

in Ireland begins with the Vikings, culminating in the first minting of Irish coins, in Dublin around 997. The main evidence for Hiberno-Norse coin usage in the ninth and tenth centuries consists of hoards of coins, deposited for safekeeping. These coins, mainly Anglo-Saxon but including coins from further afield, presumably reached Ireland through Viking activity. Most Viking-age coin hoards, however, occur in areas that would have been under Irish, rather than Scandinavian, control—notably the powerful midland kingdom of Mide (Meath). Analysis of the occurrence and distribution of these hoards suggests that most were deposited by Irish, rather than Hiberno-Norse, hoarders.

This, in turn, raises obvious questions about the use of coins among the Irish, at least in those areas that have produced the preponderance of hoard evidence. Gerreits and Kenny question the assumption that the Irish did not use coinage, as such, even after Dublin began minting its own coinage. Kenny suggests that contact with the Hiberno-Norse may have created a "heightened awareness of coins and coin usage," especially in the kingdoms bordering Dublin—Mide, Brega, and north Leinster. It is still assumed that in most cases Dublin, or another of the Scandinavian port towns, was the point of entry or production of the coins, which then passed into Irish hands through trade, or as tribute or booty. How this coinage was used by the Irish remains to be fully explored. It should be remembered, however, that even outside of Ireland coins were used at this date only for a restricted range of functions, such as major trading transactions, payment of taxes or tribute, or payment for military service. It was probably not until the thirteenth century and later that coinage was in sufficiently common supply to be used for ordinary, daily transactions.

Hoard evidence suggests a marked increase in the amount of coin circulating in Dublin at the end of the tenth century, paving the way for the first Irish coinage, minted in Dublin from *circa* 997 under the authority of Sitriuc III. This coinage was a direct (and relatively good) copy of the contemporary English silver penny of Aethelred II. Hiberno-Norse coinage continued to be minted until the mid-twelfth century. Although continuing to imitate English issues, it quickly deteriorated in quality, culminating in the bracteates of the twelfth century—discs of silver so thin that they could be struck only from one side. Inscriptions become unintelligible and eventually disappear altogether, so that it is impossible to be certain where, when, or by whom the later coinage was minted. Production of this series seems to have ceased before 1170, but even this is uncertain.

The Later Middle Ages

After the English invasion of Ireland the volume of coinage in circulation, and its usage, gradually increased as part of wider economic changes. The first Irish coinage of the new dispensation was issued under the authority of John, as lord of Ireland, from the late 1180s. Besides Dublin, mints operated in Waterford, Limerick, Kilkenny, and in Carrickfergus and Downpatrick, where John de Courcy briefly issued coinage in his own name at the end of the twelfth century. These early mints struck silver halfpennies and farthings; it was not until after John became king (1199) that pennies were minted. These coins were minted to the full English standard, and the pennies, in particular, seem to have circulated freely in England and beyond. Indeed, it has been suggested that the real purpose of the large-scale minting of the thirteenth century was to provide a convenient mechanism for exporting silver from Ireland, to help pay for English military expeditions elsewhere. The Irish production of halfpennies and farthings was unusual, if not unique, and these coins probably circulated mainly within Ireland.

There is little evidence for minting in Ireland between around 1210 and 1250. In 1251 to 1254, Henry III resumed the minting of silver pennies, to the full English standard, in Dublin. This prolific issue was particularly widely circulated and frequently turns up in European contexts. No halfpennies or farthings were minted until after 1279, however. The need for smaller denominations was made up, partly by cutting pennies into halves and quarters, and by the use of unofficial base-metal coinage—such as the hoard of over 2,000 pewter tokens found in a late-thirteenth-century pit excavated at Winetavern Street, Dublin, and clearly intended for use in the taverns there. In 1279, Edward I reformed the coinages of England and Ireland, and large-scale minting of good-quality silver pennies, halfpennies, and farthings resumed in Dublin, Waterford, and (in 1295) in Cork.

Minting ceased again after *circa* 1302, and for the following century and a half very little coinage was produced. This was a consequence of the fourteenth-century

European economic depression, but also reflected the substantial outflow of silver from Ireland in the thirteenth century. In the absence of new coinage, old, foreign, debased, and forged coins circulated widely. Minting was revived under Edward IV (1461–1483) when the first attempt was made—at the insistence of the Anglo-Irish parliament—to develop a distinctive Irish coinage to a lower standard (i.e., containing less silver), which was less likely to flow out of the country. The first base-metal coinage—farthings and half-farthings of copper and copper alloys—was also introduced. Minting extended to towns such as Drogheda and Trim and continued until *circa* 1500, after which Ireland's coinage tended to be minted in England. The first use of the characteristic harp on Irish coinage was under Henry VIII in 1534, and the first Irish shillings were minted under Edward VI (1547–1553). Henry VIII (1509–1547), perennially short of money for his campaigns, also began a serious debasement of the Irish coinage toward the end of his reign. It is estimated that in 1535 Irish pennies typically contained over 90 percent silver, while by 1560 this had fallen as low as 25 percent.

ANDY HALPIN

References and Further Reading

Doherty, C. "Exchange and Trade in Early Medieval Ireland." *Journal of the Royal Society of Antiquaries of Ireland* 110 (1980): 67–89.

Dolley, M. *The Hiberno-Norse Coins in the British Museum*. London: British Museum, 1966.

Gerreits, M. "Money Among the Irish: Coin Hoards in Viking Age Ireland." *Journal of the Royal Society of Antiquaries of Ireland* 115 (1985): 121–139.

Kenny, M. "The Geographical Distribution of Irish Viking-Age Coin Hoards." *Proceedings of the Royal Irish Academy* 87C (1987): 507–525.

O'Sullivan, W. "The Earliest Irish Coinage." *Journal of the Royal Society of Antiquaries of Ireland* 79 (1949): 190–235.

O'Sullivan, W. *The Earliest Anglo-Irish Coinage*. Dublin: National Museum of Ireland, 1964.

Seaby, P., ed. *Coins and Tokens of Ireland*. London: B. A. Seaby, 1970.

See also **Hoards; Sitriuc Silkenbeard; Trade**

COLUM CILLE

Colum Cille (and other variations), the name by which he became known throughout Ireland and Gaelic Scotland, was revered as one of the three patron saints of medieval Ireland, along with Patrick and Brigit. He is also known by the Latin name, Columba. Almost every aspect of Colum Cille's life became heavily mythologized, but around 700 his hagiographical Life was written in Latin by Adomnán, which, together with other sources, can be used to reconstruct some of the details of the saint's actual life. The annals assign various dates to the main events of his life, suggesting that he was born around 520 (or 523) and left Ireland—ultimately to found the monastery of Iona—in 562 (or shortly afterward). His obituary is dated to 597 by the *Annals of Ulster*, which has been widely accepted, although Dr. Daniel McCarthy has argued that the chronology of the *Annals of Tigernach*, which would place his death at 593, may be more accurate. Legends say his birth occurred on December 7, the day St. Buite of Monasterboice died, and that this fell on a Thursday. There were strong traditions, especially in Scotland, associating the saint with Thursday.

Colum Cille belonged to the Cenél Conaill, part of the northern Uí Néill. Legends claimed that he could have become high king. He certainly was extremely influential in the highest aristocratic and royal circles of his time, both in Ireland and northern Britain. Tradition points to a number of sites in the Gartan area of Donegal, said to be the locations of his birth and other events in his early life. Eithne, his mother, among other possibilities is claimed to have belonged to the Corbraige of Fanad, north of Gartan. She too was venerated as a saint. Her alleged grave is pointed out on Eileach an Naoimh, in the Garvellachs, in the Inner Hebrides. Colum Cille's father, Fedelmid, was said to be a great-grandson of Niall Noígiallach. The saint had a brother, Iogen, and three sisters: Cuimne, Sinech, and Mincoleth, who was mother of the sons of Enan (after whom Kilmacrennan, Co. Donegal, is named). Several other members of his family became monks and priests, and some were also commemorated as minor saints. Legends claim that his original name was Crimthann, meaning something like "fox" or "deceitful one." It is not certain that the name Colum Cille was ever used when he was alive, although it occurs in very early texts about him. Almost certainly his "Christian" name was the Latin *Columba* (Dove). The Irish "Colum Cille" (Dove of the Church) may itself be part of the growth of his cult.

While he was still a deacon, Colum Cille spent time in Leinster studying with an "old master," Gemmán. He is also said to have studied sacred scripture with a bishop Uinniau, identified by Pádraig Ó Riain as St. Finbarr of Movilla, County Down. Although there are many legends that purport to tell us about Colum Cille's life as a young cleric, it is only with his departure for Iona in 562 (at about forty-one years of age) that we have any reliable information about him. He is credited with founding Derry (his first and thus most beloved church) as early as 546, but it is clear that this date is too early, and the name of an alternative founder is recorded elsewhere. He is said to have founded many other monasteries in Ireland (such as Moone,

Co. Kildare, and Swords, Co. Dublin), but it is certain that they too were established much later. One such place was Kells, County Meath, associated with the Book of Kells. Although traditions claim that Colum Cille founded this monastery, we know that it was established about 804–807. Likewise, the famous manuscript was sometimes later, erroneously, attributed to the saint. The one monastery that we can be certain that Colum Cille did found in Ireland, probably in the 580s on one of his return trips from Iona, was Durrow in County Offaly. Other places in Ireland that have strong traditional connections with the saint include: Glencolumbkille, in County Donegal, and Tory Island, nine miles off the coast of the same county. In the 1850s, William Reeves listed thirty-seven churches in Ireland, and fifty-three in Scotland, that had dedications to Colum Cille; most of these, however, originated later than the life of the saint.

Adomnán links Colum Cille's departure from Ireland with the battle of Cúl Dreimne in 561. Cúl Dreimne was near Drumcliff, County Sligo, where later there was a monastery dedicated to St. Colum Cille. The annals say that victory was gained for the northern Uí Néill on that ocasion, "through the prayer of Colum Cille," over Diarmait mac Cerbaill. The facts about this battle became enshrouded in legends that suggest that Colum Cille himself had been responsible for it, and that it was as penance for this that he went into exile. Modern scholars and the earliest sources available, however, would suggest that Colum Cille's exile was voluntary; Adomnán called him a "pilgrim for Christ." This sort of practice, exile for the love of God, became known in the Irish church as "white martyrdom." A list, drawn up about the early eighth century, claims that he left Ireland accompanied by twelve companions.

Iona

Colum Cille eventually established his most important monastery (*c.* 562/3) on Iona, a small island off the Isle of Mull, off the west coast of Scotland. This was to become one of the most influential ecclesiastical centers in western Christendom. A rectangular earthwork enclosing about eight acres, built, apparently, by earlier settlers, was re-occupied by the Irish monks; within it they built their monastery. We can get some impression of what that monastery was like from Adomnán's Life. The great restored Benedictine Abbey that dominates the island now was begun in the early thirteenth century, but there are some earlier buildings, including one that is pointed out as the burial place of the saint, despite another tradition that he was reburied at Downpatrick. Among the significant early monuments on Iona are the three great High Crosses, dedicated respectively to St. Oran, St. John, and St. Martin, that were carved, in that order, between about 750 and 800.

Many other churches and monasteries in Ireland, Scotland, and northern England were founded from Iona, that is, the federation known as the *Familia Columbae*; some of these, although not all those claimed, were founded within the lifetime of the saint. Colum Cille did travel away from Iona, up the Great Glen of Scotland and even back to Ireland. He attended a significant meeting in Ireland around 590, between his relative, the important Donegal king Áed mac Ainmerech, and the king of an Irish colony in Scotland, Áedán mac Gabráin. The Convention of Druimm Cete, as this meeting came to be known, and at which Colum Cille is said in legend to have saved the poets from expulsion from Ireland, gave rise to a whole host of legends. Colum Cille is, himself, remembered as a poet. Three Latin poems attributed to him are possibly genuine, but there are many others in Irish, which later propagandist poets put into his voice. The apparently contemporary manuscript of the Psalms, known as the Cathach, is also claimed as his work, but this has neither been substantiated nor disproved.

Colum Cille died on Iona, probably on June 9 (his feast day) in 593 or 597. Very shortly after his death, the long poem known as the *Amra Choluimb Chille* (Elegy of Colum Cille) was written, allegedly by the famous poet Dallán Forgail. Other similar poems praising Colum Cille were written in the seventh century, and we know that a *liber de virtutibus sancti Columbae* (book on the virtues of saint Columba) was compiled in the mid-seventh century, although only one paragraph survives as a quotation in a version of Adomnán's Life. In the later twelfth century another Life of Colum Cille was written in Irish, probably in Derry. This took the form of a homily for preaching on his feast day. The text is structured, mainly, as an account of the saint's alleged journey around Ireland, founding churches and monasteries.

The most elaborate Life of Colum Cille was written just over a thousand years after his birth, in 1532. This was prepared for Maghnus Ua Domhnaill who, in 1537, became chieftain of Tír Conaill (most of Co. Donegal). Throughout the Middle Ages, many other works, both hagiographical and secular, were also written about, or referred to, Colum Cille, each of which developed the fabulous legends about him and moved his fictional character ever more distantly from that of his true identity. Professor Pádraig Ó Riain has also argued that several other early Irish saintly characters, such as St. Cainnech of Achad Bó, are really aliases of Colum Cille.

In England, Durham Cathedral, as inheritor of some of the traditions of the seventh-century monastery on Lindisfarne Island off the coast of

Northumberland, preserved aspects of the cult of Colum Cille down to the later Middle Ages, including some of the important early texts about him. His cult was also brought to continental Europe by various means.

BRIAN LACEY

References and Further Reading

Anderson, M. O. *Adomnán's Life of Columba*, revised ed. Oxford: Clarendon Press, 1991.

Bourke, Cormac, ed. *Studies in the Cult of Saint Columba.* Dublin: Four Courts Press, 1997.

Clancy, Thomas Owen, and Márkus, Gilbert. *Iona: The Earliest Poetry of a Celtic Monastery.* Edinburgh: Edinburgh University Press, 1995.

Herbert, Máire. *Iona, Kells and Derry: The History and Hagiography of the Monastic Familia of Columba.* Dublin: Four Courts Press, 1996.

Lacey, Brian. *Colum Cille and the Columban Tradition*. Dublin: Four Courts Press, 1997.

Lacey, Brian. *Manus O'Donnell: The Life of Colum Cille*. Dublin: Four Courts Press, 1998.

Lacey, Brian. "Columba, Founder of the Monastery of Derry?—'Mihi manet Incertus'." *Journal of the Royal Society of Antiquaries of Ireland* 128 (1998): 35–47.

MacDonald, Aidan. "Aspects of the Monastery and Monastic Life in Adomnán's Life of Columba." *Peritia: Journal of the Medieval Academy of Ireland* 3(1984): 271–302.

Ó Riain, Pádraig. "Cainnech *alias* Colum Cille, Patron of Ossory." In *Folia Gadelica*, edited by R.A. Breatnach, 20–35. Cork: Cork University Press, 1983.

Picard, Jean-Michel. "Adomnán's *Vita Columbae* and the Cult of Colum Cille in Continental Europe." *Proceedings of the Royal Irish Academy* 98C (1998): 1–23.

Reeves, William. *The Life of St. Columba, Founder of Hy; Written by Adomnan.* Dublin: The Irish Archaeological and Celtic Society, 1857.

Sharpe, Richard. *Adomnán of Iona: Life of St. Columba.* London: Penguin Books, 1994.

See also **Adomnán; Ailech; Annals and Chronicles; Brigit; Conversion to Christianity; Durrow, Book of; Hagiography and Martyrologies; Manuscript Illumination; Paschal Controversy; Patrick; Poetry, Irish; Poetry, Latin; Scriptoria**

COLUMBANUS (*c.* 540–615)

Columbanus was born in Leinster and died in Bobbio. He is the earliest Irish Latin author known by name. He was a scholar, *peregrinus* (see *peregrinatio*), and abbot. Columbanus can be approached by way of his own writings (imperfectly preserved) and by the earliest hagiographical account of him by Jonas of Bobbio. All of his extant writings, with one exception (the hymn *Precamur patrem*, preserved in the Antiphonary of Bangor, a manuscript that was written in Bangor before 600 and came to Bobbio a century later) originated after his departure from Ireland in around 590. Jonas, the author of his hagiographical account, wrote a generation after Columbanus's death. He had no personal knowledge of Columbanus; his account of the years in Ireland and in Italy is exceedingly brief. Jonas appears to have drawn the bulk of his information from monks in the Burgundian monasteries.

Columbanus "the dove" (a favorite monk's name in Ireland) received his first Latin education in Leinster. His writings show that Latin education of a high standard was then available in Ireland. Jonas mentions his teacher Sinilis, probably identical with a renowned expert in computistics. Columbanus encountered a woman hermit who encouraged him to become a *peregrinus*. However, before leaving Ireland, he spent some years in the monastic community of Bangor under its founder abbot Comgall. Jonas writes that Columbanus composed in Ireland a commentary on the Psalms, now lost, and hymns, one of which has survived.

The Latin of Columbanus's writings is of exceptionally high quality; he acquired it in Ireland and had little opportunity to further it in Gaul. He is believed to have been familiar with some of the classical Latin poets, but also with the poetry of Venantius Fortunatus.

Columbanus left for the continent around 590, together with twelve monks (*imitatio Christi*) determined not to come back. Eventually he settled in the Vosges mountains, then part of the Merovingian kingdom of Burgundy, and established monasteries at Annegray, Luxeuil, and Fontaines. The foundation of more than one monastery (of which Luxeuil took pride of place) became necessary when he attracted a great number of followers from among the native aristocracy. It cannot be established how his highly ascetic life proved so attractive. He only worked as a monastic leader and never seriously entertained the idea of missionary work. It is likely that the thirteen *Instructiones*, which provide the best insight into his monastic ideals, were written in Gaul. From Luxeuil he wrote to Pope Gregory I, and among other issues he requested a copy of Gregory's *Regula Pastoralis*, of which he wrote with respect. This shows that he was aware of what was current in the Christian world of his days.

For most of the time Columbanus enjoyed the support of the regional monarchy, a fact obscured by the eventual clash. However, he did not want the supervision of the local bishop, which was the norm on the continent but was unknown in Ireland. He was summoned to a synod in around 602 to justify himself, but refused to attend and instead wrote a letter in which he explained his position (Ep. 2; Walker, 1970). Another important point of disagreement would appear to have been the Irish Easter that his monks observed; in this field he defended the Irish position vigorously, especially against the bishop of Rome. He also had brought with him the Irish system of penance and has left his own penitential.

In 610, he fell out with the Burgundian royal court, especially with Queen Brunhilde, by his criticism of the moral laxity of the king. He was told to return to Ireland, but in any case to leave the kingdom. It was specifically ruled that he was not allowed to take any of his Frankish monks with him (Jonas I. 20). This is the most plausible context for his writing of two monastic rules, whereas previously he would have ruled by his personal authority, and perhaps also his penitential. After his departure, dissent seems to have surfaced at Luxeuil over the harsh style of life. Attempts to ship Columbanus back to Ireland failed miraculously, and this is the reason why he ended up in Northern Italy after a brief spell in the area of Lake Constance. It would appear that he had a sufficient group of non-Frankish monks with him to establish a viable community. This would mean that recruits from Ireland had joined him after his departure. There is no other evidence of his continued contacts with Ireland after his departure. One of his Irish companions on the way to Italy, with whom he fell out, was Gallus, who in 612 stayed behind in the Alps (the monastery named after him was established in 720).

In the Lombard kingdom, Columbanus quickly received the protection of King Agilulf, who encouraged him to preach against Arianism, the heretical variety of Christianity that was observed by a great number of Lombards. A treatise that he wrote on this topic has not been preserved. There was more religious strife in Italy (the "Three Chapters" controversy). Columbanus wrote again to the pope and demanded that he settle that issue. These two works can be firmly placed into his last years.

It was under royal protection (much like earlier on in Burgundy) that Columbanus established around 613 the monastery of Bobbio (in the Appennines south of Pavia, diocese of Piacenza), centered around a preexisting, then ruined, church dedicated to St. Peter. A charter by King Agilulf for this foundation is not preserved in the original but would appear to be, in essence, reliable. Bobbio was the first royal Lombard monastic foundation; it was to enjoy royal favour over the next centuries There Columbanus died, barely two years later. He was well-remembered and honored in Italy, more so than in his native country. Bobbio became the most important monastery in northern Italy. From this monastery a considerable number of manuscripts have been preserved that contain material in Old Irish and were written in Irish script. The historical context of this phenomenon is as yet unclear. Bobbio was also the first monastery in the West to receive a papal exemption (in 628), harking back to Columbanus's refusal to bow to episcopal supervision. To this day, Bobbio is called in Italy "the Monte Cassino of the north."

MICHAEL RICHTER

References and Further Reading

Krusch, B., ed. *Jonae Vitae sanctorum Columbani, Vedastis, Iohannis*. Hannover, Leipzig, 1905.

M. Lapidge, ed. *Columbanus. Studies on the Latin writings*. Woodbridge, 1997.

Richter, M. *Ireland and Her Neighbours in the Seventh Century*. Dublin and New York, 1999.

Walker, G. S. M., ed. *Sancti Columbani Opera*. Dublin, 1957. Reprint 1970.

See also **Classical Influence; Hiberno-Latin**

COMMON LAW

The Common Law is the body of legal rules that was developed by the royal courts in England from the last quarter of the twelfth century onward, and applied by them within England on a nationwide basis. No authoritative written summary of these rules was compiled, though many of them were embodied in the written judgments of the courts, and they were also summarized and discussed in medieval legal treatises. The Common Law was so called to distinguish its rules from rules of purely local application. The term Common Law is also used to refer in a general way to the legal institutions that are or were characteristic of English law. In the later Middle Ages, these include the use of courts of a particular general type; the initiation of civil litigation by writs of a limited range of standard types issued by the king's chancery; the initiation of most criminal proceedings through indictment at the king's suit; the use of jury trial for fact finding in civil and criminal proceedings; and the existence of a bifurcated lay legal profession with a recognized professional expertise but no connection with the law faculties of the universities.

During the first half-century of English invasion and settlement in Ireland there is comparatively little evidence about the legal customs followed in those parts of Ireland that were controlled by the new settlers. The settlers seem to have made use of some of the characteristic remedies of the early English Common Law and some of its characteristic modes of proof. There is also evidence for the introduction of some of the characteristic institutions of English land law, such the widow's right to dower. During King John's visit to Ireland in 1210, he drew up a charter that is known from later references to have established the general principle that the English Common Law was to be applied in the courts of the lordship of Ireland. Soon afterward the king sent a register of writs, containing the standard types of writ then available from the chancery in England for the initation of litigation, with instructions authorizing the justiciar to issue such writs in Ireland. Further orders were given in 1234 and 1236 for making available in Ireland particular forms of writ, and in 1246 the general principle stated that all writs "of common right" should be available in Ireland. Mandates were also sent from England to explain specific legal rules and procedures and instruct that they be applied in the courts of the lordship. These provide evidence of a determination that the Irish Common Law remain close to its English model. Such evidence ceases in the second half of the thirteenth century. This does not, however, mean that the courts of the lordship were now left free to develop their own distinctive law and custom. It is in this period that, for the first time, cases started being removed by writs of error from Irish courts to the court of King's Bench in England, where judgments were upheld or overthrown on the basis of English legal rules. From 1236 onward, legislation enacted for England was also sent to Ireland, with orders that it be applied there. This ensured that the Irish Common Law was not left behind at a time when the English Common Law was being drastically remodeled by legislation. The last time English legislation was simply sent to Ireland with instructions for its application was in 1411. The same effect, however, was achieved (albeit with prior Irish agreement) from the fourteenth century onward by the adoption or reenactment by the Irish parliament of specific items of existing English legislation. In 1494 to 1495, Poynings' Law, enacted by the Irish parliament, authorized the adoption of all general, public legislation enacted in England prior to that date.

By 1300, there had also come to be a legal profession in Ireland that bore a close resemblance to its English counterpart: a small group of professional serjeants (specialists in pleading for clients in the courtroom) practiced in the main royal courts of the lordship. There was also a separate group of professional attorneys, whose main responsibilities were in the preliminary stages of litigation and in briefing the serjeants, but the surviving evidence does not reveal how large a group this was. Professional lawyers also practiced in the city courts in Dublin and in county courts. Law students traveled to England from Ireland to learn the law from at least 1287 onward, and from the 1340s they attended the Inns of Court in London to do so. However, more elementary legal education was also available in Ireland, probably in Dublin. The Irish legal profession failed to establish the monopoly over the main judicial appointments that the English legal profession had secured by around 1340. This was mainly because English lawyers (with no previous Irish connections) continued to be appointed to serve as justices in Ireland. The appointment of English justices to serve in the Irish courts and the education of the Irish legal elite in England must also have played a role in ensuring that the Irish Common Law continued to bear a close resemblance to its English cousin.

Even in the thirteenth century there was some development of a distinctive Irish custom within the Common Law—at least in part from an accommodation with native Irish law, for example in allowing, in the case of homicides against native Irishmen, the payment of compensation (*éraic*) rather than imposition of the death penalty. The existence of an independent Irish parliament with unfettered freedom (prior to 1494–1495) to enact its own legislation also allowed the development of a body of statutory law modifying the Common Law, quite distinct from that of England. The earliest such legislation now known dates from 1278.

Paul Brand

References and Further Reading

Brand, Paul. "The Early History of the Legal Profession in the Lordship of Ireland, 1250–1350" and "Ireland and the Literature of the Early Common Law." In Brand, P. *The Making of the Common Law.* London and Rio Grande: The Hambledon Press, 1992.

Donaldson, A.G. *Some Comparative Aspects of Irish Law.* Durham, N.C. and Cambridge: Cambridge University Press, 1957

Hand, Geoffrey J. *English Law in Ireland, 1290–1324.* Cambridge: Cambridge University Press, 1967.

See also **Anglo-Norman-Invasion; Chief Governors; Courts; Education; Feudalism; Government, Central; Government, Local; John; March Law; Parliament; Records, Administrative; Society, Functioning of Anglo-Norman; Urban Administration**

COMPERTA

Comperta ("birth tales," plural of Old Irish *compert* [conception/birth]) are among the classes of tales found in Irish narrative literature, and they deal almost exclusively with the events surrounding the conception and birth of a hero. The native classification was according to tale type, usually the first word of the title. Examples of the classes *Comperta, Imrama* (Voyages), and *Echtrai* (Expeditions) are: *Compert Con Culainn* (Birth of Cú Chulainn), *Imraim Brain* (Voyage of Bran), and *Echtrae Chonnlai* (Expedition of Connlae). Two lists of the various classes of tales survive in Irish manuscripts and are referred to as List A and List B, of which List B includes the class *Comperta.* These lists contain the stories that a medieval Irish poet would have been expected to be able to narrate.

The genre of the *Compert*, or birth tale, appears to be quite ancient. It is present not only in Celtic tradition, but also in mythologies worldwide (e.g., the birth of Jesus and also Hercules in Greek myth) and is a major element in the life of the hero. It would seem that each episode in the heroic biography (e.g., *Compert*, *Aided* [violent death], etc.) corresponds with the different stages of the ritual life cycle, and so these tales are of a symbolic rather than a factual nature. The heroic biography emphasises the conception and birth of the hero, which is consistently of an extraordinary nature and is sometimes found incorporated into another story, such as the larger story of the hero's life. For example, the story of the birth of Cormac mac Airt has come down to us in two tales, which include not only the conception and birth of Cormac, but also his life story, containing many of the features of the heroic biography. However, sometimes the conception and birth of the hero appears as a tale in its own right. The most prominent of these in Irish literature are *Compert Con Culainn*, *Compert Conchobuir*, and *Compert Mongáin*.

Two versions of *Compert Con Culainn* have come down to us from an original probably composed in the eighth century. In what is likely the older version, Dechtine, daughter of Conchobor, king of Ulster, adopts a child who is the son of the god Lug. The child dies and Lug appears to Dechtine in a dream, telling her that she is pregnant by him and that she would give birth to a boy whom she was to call Sétantae. She subsequently marries and aborts the fetus. She again becomes pregnant, this time by her husband, and gives birth to a boy whom she calls Sétantae. It is this child who is later renamed Cú Chulainn.

There are also two versions of *Compert Conchobuir* (the same Conchobor, king of Ulster, who appears above as uncle of Cú Chulainn). Again, there was a probable eighth-century original of this tale. The earliest version tells how one day the druid Cathbad comes upon Nessa, princess of Ulster. In answer to her question regarding what the hour was lucky for, Cathbad declares "Begetting a king upon a queen." Nessa becomes pregnant by Cathbad at her own request and carries the child for three years and three months. Although Conchobor's father is Cathbad, he is known as Conchobor mac Nessa.

Also a likely eighth-century composition, *Compert Mongáin* opens with Fíachnae mac Báetáin, the king of Ulster, leaving for Scotland to fight alongside his friend, Aedán mac Gabráin, against the Saxons. While Fíachnae is away, a noble-looking man visits his wife. He convinces her that Fíachnae is in mortal danger and that he will help her husband if she will sleep with him and bear him a famous son, Mongán. She sleeps with the stranger and he keeps his promise. Fíachnae returns safely and his wife bears a son, known as Mongán mac Fíachnai, although he was the son of the god Manannán mac Lir, who was, in fact, the stranger who came to her.

Undoubtedly *Comperta* were written for many other Irish heroes but have been lost. Under the heading of *Comperta* in List B mentioned above, five tales appear. Of these only two have survived: *Compert Con Culainn* and *Compert Conchobuir.* Another, *Compert Cormaic Uí Chuinn* (Birth of Cormac grandson of Conn), does not survive, but is found, as is stated above, incorporated into two other tales. The other two, *Compert Conaill Chernaig* and *Compert Cheltchar maic Uithechair*, do not exist in the extant literature. Furthermore, the tale *Compert Mongáin* does not appear at all, showing that this is hardly an exhaustive list.

NORA WHITE

References and Further Reading

Carney, James. *Studies in Irish Literature and History.* Dublin: Dublin Institute for Advanced Studies, 1955.

Gantz, Jeffrey. *Early Irish Myths and Sagas.* Harmondsworth: Penguin, 1981.

Mac Cana, Proinsias. *The Learned Tales of Medieval Ireland.* Dublin: Dublin Institute for Advanced Studies, 1980.

Meyer, Kuno. *The Voyage of Bran.* London: Llanerch, 1895.

Ó Cathasaigh, Tomás. *The Heroic Biography of Cormac mac Airt.* Dublin: Dublin Institute for Advanced Studies, 1977.

Rank, Otto. *The Myth of the Birth of the Hero.* New York: Vintage Books, 1964.

See also **Áes Dána; Aideda; Echtrai; Immrama; Poets/Men of Learning; Ulster Cycle**

CONNACHT

Early History

Connacht is provided with natural borders by the river Shannon and Loch Ree in the east and the Curliew Mountains on the northeast. The Ulster Cycle has the legendary warrior-queen Medb and her husband Ailill rule the province from their royal seat at Cruachain (Rathcroghan), the capital of Connacht, a complex of ringforts, mounds, and earthworks. It is located in the traditional heartland of the later kings of Connacht, the Uí Chonchobair, who employed the nearby prehistoric burial cairn of Carnfree as a royal inauguration site.

The name Connacht is derived from the Connachta dynasty, which according to tradition takes its name from Conn Cétchathach ("of the Hundred Battles"), legendary king of Ireland. According to the genealogies, from him were descended the brothers Niall Noígiallach ("of the Nine Hostages"), Brión, Fiachra, and Ailill, progenitors of the Uí Néill, Uí Briúin, Uí Fiachrach, and Uí Aililla, respectively. The Uí Néill allegedly originated in Connacht, but migrated into the midlands and then the north of Ireland. The other three dynasties remained in Connacht.

Between the fifth and the eighth centuries the Uí Fiachrach was the most prominent Connacht dynasty. However, it split into two branches, with the Uí Fiachrach Aidni settling in the south and the Uí Fiachrach Muaide in the northwest of the province. Rivalry within Uí Fiachrach led to its weakening, and from the third quarter of the eighth century none of their kings became kings of Connacht. In the eighth century the Uí Aililla was also in decline. The vacuum left by this dynasty opened the way for the Uí Maine, a quite powerful dynasty, although unrelated to the Connachta. The Uí Maine, of which the Uí Ceallaigh was later the dominant branch, settled in the southeast.

Of the various Connacht dynasties, the Uí Briúin emerged as the strongest. This dynasty split into the Uí Briúin Ai, Uí Briúin Seóla, and Uí Briúin Bréifne. The former stayed in the original Uí Briúin territory around the traditional royal seat in Connacht. They again splintered, and one of their branches developed into the Síl Muiredaig, from whom sprang the Uí Chonchobair kings of Connacht. Due to the dynasty's later significance, early regnal lists of Connacht have undergone extensive revision to give the Uí Briúin more distinction. The Uí Briúin Seóla were forced into lands centred on Loch Corrib, and the Uí Briúin Bréifne found a new home in what are now approximately Counties Leitrim and Cavan. Throughout the medieval period Bréifne was regarded as being part of the province also. Another area that was considered to be part of Connacht, though only in the early Middle Ages, was that portion of Thomond (literally North Munster) that is now County Clare. According to tradition, the region was conquered in the fifth century by Munster kings; however, the hegemony of the Connacht king Guaire in the seventh century seems to have reached into Thomond. Clonmacnoise, founded in the sixth century, became the richest and most prestigious of the ecclesiastical centers in Connacht's sphere of influence, though it suffered from many Viking incursions and technically, sited just east of the river Shannon, it lay outside the province. It became the burial place of the kings of Connacht.

The Anglo-Norman Era

Although Connacht enjoyed prominence in ancient times, it exerted no great influence beyond its own borders again until the twelfth century, with the rise of the Uí Chonchobair. From Connacht's most powerful sept emerged Tairrdelbach Mór Ua Conchobair (1088–1156), whose remarkable career culminated in his occupying the high kingship of Ireland. His son and successor Ruaidrí ruled at the time of the Anglo-Norman invasion and was Ireland's last high king. A clear mark of Tairrdelbach's eminence was his success in procuring a pallium for Tuam, the archbishopric of Connacht, comprising the dioceses of Clonfert, Killala, Achonry, Annaghdown, Mayo, Roscommon, Kilmacduagh, and Elphin.

Foremost among the subjects of the Uí Chonchobair were the Meic Diarmata of Moylurg, who played an essential part in the inauguration of the king of Connacht. Another princely family within the Síl Muiredaig was the Meic Donnchada of Tír Ailillo. The Uí Flaithbertaigh, the dominant branch of the Uí Briúin Seóla dynasty, were contenders for the kingship of Connacht in the eleventh century. They were unable to procure a supreme uncontested position, however, and were pushed into Iar-Connacht (Connacht west of Loch Corrib). Umall was occupied by the Uí Máille, neighbours of the Conmaicne Mara, who have left a trace of their prominence in the name Connemara. Of the Uí Briúin Bréifne, the Uí Ruairc gained prominence in the west, and the Uí Ragallaig in the east of Bréifne.

The province is not the most suitable for agriculture due to the presence of extensive bogs, rocky outcrops, and forests. Because of these factors, and also owing to its location remote from the governmental center of gravity in the Anglo-Norman era, colonization was not as thorough in Connacht as it was, for example, in Leinster. Early in the thirteenth century, the king of England granted the king of Connacht a part of the province as hereditary land, the "King's Five Cantreds" (cantreds were territorial units, many of which later became baronies). This area comprised roughly County Roscommon, with small parts of Counties Galway and Sligo. The remaining twenty-five cantreds were granted to the de Burghs (Burkes). In the second half of the thirteenth century the area under control of the Uí Chonchobair was further reduced to a mere three cantreds. The de Burghs held demesne lands southeast of Galway and subinfeudated the greater part of Connacht to families such as Bermingham, Fitzgerald, and Costello (formerly de Angelo, Nangle). By the end of the medieval period, the de Burgh family had split into the MacWilliam de Burghs, with lands in the northwest, and the ClanRicard, with lands in the south of the province extending into Thomond, while Bermingham and Costello held lands in mid-Connacht. The main line of the Uí Chonchobair split into the Uí Conchobair Ruad and Uí Chonchobair Donn in 1384. Another branch, the Uí Chonchobair Sligigh, settled in Sligo.

Freya Verstraten

References and Further Reading

Byrne, Francis John. *Irish Kings and High Kings*, 2nd ed. Dublin: Four Courts Press, 2001.

Dillon, Myles. "The Inauguration of O'Conor." In *Medieval studies. Presented to Aubrey Gwynn*, edited by J. A. Watt, J. B. Morrall, and F. X. Martin, 186–202. Dublin: Three Candles, 1961.

Knox, H. T. "Occupation of Connaught by the Anglo-Normans after A.D. 1237." *Journal of the Royal Society of Antiquaries Ireland* 32 (1902): 132–138, 393–406; and 33 (1903): 58–74.

Ó Corráin, Donncha. *Ireland Before the Normans (The Gill History of Ireland, Vol. 2)*. Dublin: Gill and MacMillan, 1972.

Walsh, Paul. "Christian Kings of Connacht." *Journal of the Galway Archaeological and Historical Society* 17 (1937): 124–143.

Walton, Helen. "The English in Connacht, 1171–1333." Unpublished Ph.D. dissertation, Trinity College Dublin, 1980.

See also **Anglo-Norman Invasion; Bermingham; Burke; Clonmacnoise; Connachta; Gaelicization; Leth Cuinn and Leth Moga; Niall Noígiallach; Ruaidrí Ua Conchobair; Tairrdelbach Ua Conchobair; Ua Conchobair; Uí Briúin; Uí Mhaine, Book of; Uí Néill; Ulster, Earldom of**

CONNACHTA

Connachta is the collective name for the dynasties that dominated the province of Connacht and claimed descent from a mythic figure named Conn Cétchathach. In the early historical period, the name applied only to the dynasties of Uí Fíachrach, Uí Briúin, and Uí Ailello, which shared a common fifth-century ancestor and were known collectively as *na téora Connachta* (the Three Connachta). Aside from legends and some scattered references in the early annals, very little is known about their origins and activities before the eighth century.

The Connachta of the prehistoric period are much celebrated in Irish myth and legend. Their perennial feuds with the Ulaid, which may have a basis in fact, provided the background for the Ulster Cycle, while some of their more famous kings, like Cormac mac Airt, became enshrined in the early Irish historical tales. Legend has it that they had lived in Ireland for centuries and had controlled much of the north, an area known as Leth Cuinn, from their ancient capitol at Cruachu on Mag nAí in County Roscommon. Some legends also connect them with Tara. Although their ultimate origins are not known, one story suggests that they came to Ireland from Spain in the distant past, under the leadership of Tuathal Techtmar, the grandfather of Conn.

Out of these legendary beginnings, the historical Connacht dynasties emerged in the course of the fifth and sixth centuries A.D. They are said to originate with three brothers named Fíachra, Brïon, and Ailill, whose father, Eochaid Mugmedón, was king. Their activities are closely connected with those of their half-brother, Níall Noígíallach, who became the progenitor of their collateral kin, the Uí Néill. In the fifth century, it is claimed, all four siblings together with their families struck out in different directions, possibly from Mag nAí, and began the conquest of what was to become their historical homelands. Whatever the truth behind these events, the historical Connachta and Uí Néill did share a sense of common kinship. It is likely too that they originally recognized a joint over king, presumably the king of Tara, since there is little evidence for a separate provincial kingship of Connacht during this period. Exactly when the two groups finally parted ways is not clear, but it must have happened after the death of Ailill Molt (d. 482), apparently the last of the Connachta featured in the Tara kinglists.

During the fifth and sixth centuries, the descendants of Fíachra, Brïon, and Ailill gained control of the best lands in Connacht and asserted their suzerainty over the local populations. These people included groups such as the Conmaicne, Partraige, Greccraige, Cíarraige, Luigni, and Gailenga. For some time, though, they failed to gain ascendancy over the Uí Maine, a powerful kingdom in the southeast of the province. What evidence there is for this period suggests that the Uí Fíachrach were dominant, although they faced fierce competition from the Uí Briúin. The former had split early on into two main lines, the one controlling the northern coasts of the province, the other the southern border. The northern line consisted of four main septs that were collectively known as Uí Fíachrach in Tuaisceirt. These included the Uí Fíachrach Múaide on the river Moy; the Uí Fíachrach Muirsce in the north of County Sligo; the Uí Amolngada, the family of Tírechán, in the north of County Mayo; and the Fir Cherai in west-central Mayo. The southern line lived along the Munster border and was known as Uí Fíachrach Aidne. They reached the height of their power in the seventh century under Guaire Aidne (d. 663), who later became a celebrated figure in Irish legend. But despite their early prominence, both the northern and southern lines were on the decline in the eighth century, and after the death of Donn Cothaid in 773, the Uí Fíachrach never again produced an over king of Connacht.

During this same period, the Uí Ailello enjoyed local autonomy in their lands north of Mag nAí, though they never played a major role in Connacht politics. In the eighth century, they ran into constant conflict with their subject peoples, the Luigni and Greccraige, and possibly also the expanding Uí Briúin. They were eventually wiped out at the battle of Ard Maicc Rimi in 792.

With the extinction of the Uí Ailello and the decline of the Uí Fíachrach, the kingship of Connacht from the late eighth century on became the sole prerogative of the Uí Briúin. In later centuries, their royal families would play a major role in Irish politics.

DAN M. WILEY

References and Further Reading

Byrne, Francis John. *Irish Kings and High-Kings*, 2nd ed. Dublin: Four Courts Press, 2001.

Charles-Edwards, T. M. *Early Christian Ireland*. Cambridge: Cambridge University Press, 2000.

Mac Niocaill, Gearóid. *Ireland Before the Vikings*. Dublin: Gill and Macmillan Ltd, 1972.

Ó Corráin, Donncha. *Ireland Before the Normans*. Dubin: Gill and Macmillan Ltd, 1972.

Ó Cróinín, Dáibhí. *Early Medieval Ireland 400–1200*. New York: Longman Group Ltd, 1995.

O'Rahilly, Thomas F. *Early Irish History and Mythology*. Dublin: Dublin Institute for Advanced Studies, 1946.

See also **Anglo-Norman Invasion; Connacht; Historical Tales; Leth Cuinn; Niall Noígiallach; O Ruairc; Tírechán; Ua Conchobair, Ruaidrí; Ua Conchobair, Tairrdelbach; Uí Briúin; Uí Néill; Ulster Cycle.**

CORK

The history of Cork may be traced to the foundation of a monastery, reputedly by St. Finbarr, around 606. Already by the late seventh century it was one of the foremost ecclesiastical settlements in Ireland. Its abbot was numbered among the leading Irish churchmen at the synod of Birr (697). Its law school had a national reputation. A contemporary law tract accorded its abbot the same status as the king of Munster. Nonetheless, early Christian Cork was the location of a monastery and not of a town.

Drishane Castle, Co. Cork. © *Department of the Environment, Heritage and Local Government, Dublin.*

Cork experienced its first recorded Viking incursion in 820. In 846 there was a Viking fortress at Cork, possibly the "castle" destroyed in 865. A Viking town may have developed at Cork from the early tenth century, though it has left little impression in historical records and has so far eluded systematic archaeological study. Following the Meic Carthaig revolt against Muirchertach Ua Briain in 1118, Cork became the capital of the kingdom of Desmond (South Munster). The Meic Carthaig constructed a major castle at Shandon, immediately north of Cork. It seems that the town grew under Meic Carthaig auspices, with a mixed population of Irish and Scandinavians.

According to contemporary Irish annals, contradicting Giraldus Cambrensis, Cork was seized and sacked by English knights in 1177. Cork was granted the status of a royal borough by Prince John around 1189. From the early thirteenth century the two central islands of Cork, an area of 14.5 hectares, were enclosed with great stone walls. It was necessary to raise the level of the ground within the walled town considerably to prevent periodic flooding. The channel between the islands was transformed into docks shielded by a fortified ship-gate, as reflected in Cork's coat of arms.

Cork's trade increased tremendously in volume, first with Bristol and then with southwestern France. Its population grew commensurately, with many people employed in handling traded goods and in processing animal hides and foodstuffs. By the early fourteenth century there was considerable suburban development at Shandon and Fayth, subordinate boroughs on the north and south banks of the Lee. Archaeologists have found evidence of housing of improved quality and greater density.

The fourteenth century proved calamitous for Cork. Climatic deterioration, and the breakdown of law and order in its rural hinterland, reduced the volumes of agricultural surplus available for processing and trading in Cork. These economic difficulties were compounded by the Black Death, which exacted a heavy toll on the city. For a century after the plague there is very little trace in the archaeological record of Cork's overseas trade.

The recovery of the English economy from the mid-fifteenth century, and the increased order imposed on Cork's hinterland by the MacCarthys and the earls of Desmond, facilitated a modest revival in Cork's economic fortunes. Much of the increased wealth, however, was concentrated in the hands of a small mercantile oligarchy. They monopolized control of the corporation and they built impressive tower houses for themselves within the city walls, or nearby. The examination of skeletal remains from St. Mary's Dominican priory shows that for most medieval Corkonians, life was short and very hard. Nonetheless, the advent of the Tudors seemed to promise a continued amelioration in Cork's fortunes.

HENRY A. JEFFERIES

References and Further Reading

Jefferies, Henry A., and Gerard O'Brien. *History of Cork*. Dublin: Four Courts Press, 2004.

O' Flanagan, P., and C. G. Buttimer ed. *Cork: History and Society*. Dublin: Geography Publications, 1993.

See also **Anglo-Norman Invasion; Ecclesiastical Settlements; Eóganachta; Mac Carthaig, Cormac; Trade; Walled Towns**

CORMAC MAC CUILENNÁIN (836–908)

Cormac mac Cuilennáin was a member of the Eóganachta Chaisil branch of the Eóganachta, though like Fedelmid mac Crimthainn (d. 847), no ancestor of his had been king of Munster since Óengus mac Nad Fraoích, grandson of the legendary founder of the Eóganachta dynasties, Corc of Cashel, direct ancestor of the most successful eastern Eóganachta branches, whose death is mentioned in the annals around 489. Cormac mac Cuilennáin became king of Munster in 902 and may have been a compromise candidate in the absence of strong opposition from the main branches of the dynasty. From his early years he is reputed to have been of a scholarly and pious nature, and may have been ordained priest and bishop, though this cannot be verified. Although of an ascetic nature, he is said to have been betrothed or even married to Gormlaith, daughter of Flann Sinna mac Máele-Sechnaill (southern Uí Néill king of Tara from 879 to 916), but rejected her because of his wish to remain celibate. It may be noted that celibacy was not a requirement for high office in the Church at that time. Gormlaith was then married to Cerball mac Muireccáin, king of Leinster, who is said in a bardic poem to have treated her badly, and later very happily married to Niall Glundub, northern Uí Néill king of Tara who was killed by the Vikings at Islandbridge, Dublin, in 919. Serial marriage was not unusual among women of noble birth in medieval Ireland, in a society that sanctioned divorce and used marriage as a means of cementing alliances.

Cormac mac Cuilennáin is credited with several scholarly works, among them genealogical tracts for the whole of Ireland in which Éber son of Míl, ancestor of the Eóganachta, comes first instead of Éremon, ancestor of the Uí Néill. The most famous extant work assigned to him is the *Sanas Chormaic*, a glossary containing etymologies and explanations of over 1,400 obsolete and difficult Irish words that may have been part of the lost *Psalter of Cashel*, a compilation of origin tales, genealogies, and tribal histories, part of which may be found in the MS Laud Misc 610 (now in the Bodleian Library). The fragment that has survived contains Munster origin tales and tribal histories, as well as the aforementioned genealogical tracts.

As the Uí Néill continued to threaten the sovereignty of Munster, Cormac was the last Eóganachta king of Munster to challenge northern hegemony. In 906 the southern Uí Néill king of Tara, Flann Sinna, assisted by the king of Leinster, Cerball mac Muirecháin, led his forces into Munster and was met and defeated by the Munstermen under Cormac, at Mag Léna (modern Tullamore, Co. Offaly). In 908, Flann Sinna, once again with the crucial assistance of Cerball, king of Leinster, and Cathal, king of Connacht, returned to the attack, as Cormac—instigated, according to an eleventh century text, by Flaithbertach mac Inmainén of the Múscraige, Abbot of Inis Cathaig—claimed tribute from Leinster and is said to have signified his intention of assuming the position of high king. The text was written in the interests of the Osraige, neighbours and former vassals of the Eóganachta, and there is unlikely to be much truth in it. It is more likely that Cormac's intentions were to discourage the Uí Néill from further attacks on Munster. In a battle fought at Belach Mugna near Leighlinbridge in County Carlow, the Munstermen suffered a complete defeat and Cormac was killed in the battle. He was succeeded by Flaithbertach mac Inmainén, the last king of Munster to be a cleric. The practice of elevating clerics to the kingship is unique to the south of Ireland, although not to Munster, as it occurred in south Leinster also. With the death of Cormac mac Cuilennáin, the decline of the Eóganachta overkingship, which had begun in the previous century, became more pronounced, and it was replaced by the Dál Cais, in the person of Brian Boru, in 978.

LETITIA CAMPBELL

References and Further Reading

Byrne, F. J. *Irish Kings and Highkings*. London: Batsford, 1972. Reprint with additional notes and corrigenda, Dublin: Four Courts Press, 2001.

Meyer, Kuno. "The Laud Genealogies and Tribal Histories." *Zeitscrift f-ur Celtische Philologie* viii (1911): 291–338.

Ó Corráin, Donnchadh. *Ireland Before the Normans*. Gill and Macmillan, 1972.

Ó Cróinín, Dáibhí. "Re-writing Irish Political History in the Tenth Century." In *Seanchas: studies in early and medieval Irish archaeology, history and literature in honour of Francis J. Byrne*, edited by A. P. Smyth, 212–224. Dublin: Four Courts Press, 2000.

Radner, J. N., ed. *Fragmentary Annals of Ireland*. Dublin: Dublin Institute for Advanced Studies, 1978, pp.150–167.
O'Donovan, John, trans. and anno., and Whitley Stokes, ed. *Sanas Chormaic*, or *Cormac's Glossary*. Calcutta: Privately Published, 1868.

See also **Cerball mac Muireccáin; Eóganachta; Gormlaith (d. 948); Uí Néill, Northern; Uí Néill, Southern**

COSHERING

Coshering, or coshery, was a late medieval Anglo-Irish term derived from the Irish word *cóisir*. In the Old Irish period it was known as *cáe* or *cóe*. Some scholars think that, as it was applied in the fifteenth and sixteenth centuries, coshering/*cóisir* may have been adapted from the French word *causerie*.

Coshering was a type of obligatory hospitality demanded by a lord of his subjects toward the maintenance of his retainers and followers. Usually it took the form of a banquet lasting two days and two nights held at or near the time of a major religious festival, particularly Christmas and Easter, but also Whitsuntide and Michaelmas. By the 1400s it was used in Anglo-Irish as well as Gaelic Irish territories. It bore certain similarities to coyne and livery, with which it was sometimes confused by observers, but unlike coyne it was levied exclusively on the wealthier subjects of a lordship—that is, on those rich enough to provide a proper feast.

Writing in the 1580s, the Dublin commentator Richard Stanihurst claimed coshering was an occasion of great merriment and celebration, and he wrote colorfully of the activities of the bards, harpers, jesters, professional gamblers, and storytellers who attended these feasts. However, it had its ugly side, too. Often a lord could be attended by as many as a hundred followers (not just armed retainers, but friends and allies), and his demands for food and drink could be burdensome to his host. The English government abolished the practice early in the seventeenth century.

David Edwards

References and Further Reading

Binchy, D.A. "Aimser Chue." In *Féil-sgríbhinn Éoin Mhic Néill: Essays and Studies Presented to Professor Eoin MacNeill*, edited by J. Ryan. Dublin: Sign of the Three Candles, 1940.
Hore, Herbert, and James Graves, eds. *Social State of Southeast Ireland in the Sixteenth Century*. Dublin: Royal Historical and Archaeological Association of Ireland, 1870.
Nicholls, Kenneth. *Gaelic and Gaelicized Ireland in the Middle Ages*, 2nd ed. Dublin: Lilliput Press, 2003.
Simms, Katharine. "Guesting and Feasting in Gaelic Ireland." *Journal of the Royal Society of Antiquaries of Ireland* 108 (1978): 67–100.

See also **Armies; Coyne and Livery; Entertainment; Gaelicization; Military Service, Gaelic; Society, Functioning of Gaelic**

COURCY, JOHN DE

John de Courcy (d. 1219?), known as "Prince of Ulster," was from a family originating in Courcy-sur-Dives in Calvados who held Stoke Courcy (Stogursey) in Somerset. He was probably a brother of William de Courcy III (d. 1171), lord of Stogursey (both had a brother Jordan), and son of William de Courcy II (*fl. c.* 1125). The latter's wife, Avice de Rumilly, was daughter of William Meschin of Copeland in Cumbria, and John succeeded to a fraction of his estate (at Middleton Cheney, Northamptonshire), which suggests his illegitimacy.

Nothing is known of de Courcy's early life, but he was possibly reared in northwest England whence many of his Ulster tenants hailed. The "Song of Dermot" claims Henry II granted Ulster to John "if by force he could conquer it," but no evidence exists of Irish involvement until 1176 when, arriving with the king's deputy William fitz Audelin, he joined the Dublin garrison. Giraldus records him growing impatient, assembling 22 knights and 300 others, marching north in late January 1177, and invading Ulaid (modern Antrim and Down)—against fitz Audelin's wishes, maintains Roger of Howden. About February 1, John reached Downpatrick, forced Ruaidrí Mac Duinn Sléibe, king of Ulaid, to flee, and built a castle. Unsuccessful mediation was attempted by the papal legate, Cardinal Vivian, who arrived from the Isle of Man having solemnized King Gudrødr's marriage to a daughter of Mac Lochlainn of Cenél nEógain. John married Gudrødr's daughter, Affreca (in 1180, according to the unreliable "Dublin annals of Inisfallen"). After early setbacks, he slaughtered the Ulaid at Down on June 24, but was defeated twice in 1178: by the Fir Lí (from the lower Bann), and by the Airgialla and Cú Ulad Mac Duinn Sléibe, who killed 450 Englishmen.

In 1179, John initiated a grand program of ecclesiastical patronage, founding new abbeys and priories, and subjecting unreformed monasteries to new orders with mother-houses predominantly in Cumbria. In 1185, he "discovered" at Down the bodies of saints Patrick, Brigit, and Colum Cille, and had them formally reburied. He kept a book of Colum Cille's prophecies (in Irish), believing they forecast his conquest. He altered the dedication of Down cathedral from the Holy Trinity to St. Patrick (for which, according to the *Book of Howth,* God later took vengeance), commissioned a Life of Patrick by Jocelin of Furness, minted halfpence bearing the saint's name—and it is possible that Patrick de Courcy, later lord of Kinsale, was his (illegitimate) son.

Though archaeological evidence such as mottes and stone castles is considerable, documentary records of John's rule are meager. He quickly won the support of Irish clerics and some Irish rulers. Howden says Irishmen aided his invasion, and "Mac Carthaigh's Book" has Irishmen wasting Ulaid with him in 1179. Niall Mac Mathgamna of Airgialla plundered Louth with him in 1196. When, in 1197, his brother Jordan was killed by an Irish adherent, he ravaged the northwest with support of Irishmen, and of Gallowaymen under Duncan of Carrick (then rewarded with Ulster lands).

After the failed 1185 expedition of John, lord of Ireland, de Courcy became chief governor and restored order to the lordship. When Lord John rebelled against King Richard in 1193 to 1194, de Courcy remained loyal, joined Walter de Lacy of Mide (Meath) against John's allies, and aided Cathal Crobderg Ua Conchobair against William de Burgh. In 1200, Cathal fled to Ulster, but when de Courcy and Hugh II de Lacy invaded Connacht in 1201, de Courcy was captured and brought to Dublin to swear allegiance to John, now king. He and the de Lacys later became enemies, and Hugh de Lacy's Meath tenants defeated de Courcy in battles at Down in 1203 and 1204. He gave hostages to King John, went to England in 1205, and had his English lands restored, but Ulster was awarded to Hugh de Lacy, along with the title "earl" that was never held by de Courcy. He rebelled, went to the Isle of Man, and was given a fleet to invade Ulster by his wife's brother, King Rögnvaldr, but he failed to capture Dundrum Castle and took refuge in Tír nEógain with Áed Ua Néill. In November of 1207 he returned to England, only reappearing in 1210 to help King John overthrow the now disgraced de Lacy. However, de Courcy never regained Ulster and possibly survived as a royal pensioner. The justiciar was ordered in 1213 to provide land for his wife Affreca, and was ordered to secure her dower lands on September 22, 1219, suggesting that John had recently died. He possibly was buried, as he wished, at Canons Ashby, Northamptonshire (although Affreca was buried in Grey Abbey, County Down). No legitimate children are known.

Seán Duffy

References and Further Reading

Duffy, Seán. "The First Ulster Plantation: John de Courcy and the Men of Cumbria." In *Colony and Frontier in Medieval Ireland:Eessays Presented to J. F. Lydon*, edited by Terry Barry, Robin Frame, and Katharine Simms, 1–27. London: The Hambledon Press, 1995.

MacNeill, T. E. *Anglo-Norman Ulster: The History and Archaeology of a Medieval Barony.* Edinburgh, John Donald Ltd., 1980.

Orpen, Goddard Henry. "The Earldom of Ulster." *Journal of the Royal Society of Antiquaries of Ireland* 43 (1913): 30–46, 133–143; 44 (1914): 51–66; 45 (1915): 123–142; 50 (1920): 166–177; 51 (1921): 68–76.

See also **Airgialla; John; Ulaid; Ulster, Earldom of**

COURTS

Within the later medieval lordship of Ireland there existed five different types of court: royal courts, communal courts, town courts, private courts, and ecclesiastical courts. Royal courts functioned in much the same general way as their counterparts in England, in accordance with a model established in the last quarter of the twelfth century. They were run by small groups of full-time justices appointed by or in the name of the king. They required specific written royal authorization for most of the business they heard, and they kept a full written record of that business. Although there are references from the early thirteenth century onward to a king's court in Ireland, it does not seem to have been constituted on the classic English royal court model. It is only in 1221 that we first find a royal court run by a group of justices and holding sessions (*assise*), both in Dublin and elsewhere in individual counties in the lordship. By the middle of the thirteenth century its sessions in Dublin were coming to be described as sessions of the Dublin Bench, and those sessions had a distinctive countrywide civil jurisdiction of their own. The Dublin Bench only, however, became a fully independent court with its own separate group of justices and meeting on a regular daily basis during four terms each year (on the model of its Westminster namesake) in the 1270s. The irregular sessions held by the same group of royal justices (and from the 1270s onward a separate group of royal justices) outside Dublin (and from time to time in Dublin itself for County Dublin) resembled sessions of the General Eyre in England, with the same mixture of civil and criminal jurisdiction and responsibility for conducting local inquiries. These general county visitations, however, ceased at around the same time as their English counterparts, after the first quarter of the fourteenth century. By then the most urgent civil business was being heard (as in England) on a much more frequent basis by assize justices, and the more urgent criminal business by justices of jail delivery: from 1310 onward the same justices seem to have been commissioned for both kinds of business. The earliest evidence of a separate justiciar's court comes only from the second half of the thirteenth century, and the first evidence of that court meeting on a regular basis only from 1282. Prior to the fifteenth century it seems to have traveled around the lordship with the chief governor, but then (like its English counterpart, the court of King's Bench) came to be stationary in a single place (Dublin). When Richard II visited Ireland in 1395 it became for a while his own itinerant court, and thereafter it retained the name of King's Bench. One other royal court emerged in Ireland in the last quarter of the fifteenth century: the Irish court of chancery. Like its English counterpart (which had emerged almost a century earlier), this was a court of equity with only a single judge (the Irish chancellor), and it heard business on the basis of bills submitted to it.

The communal courts of the lordship were local courts serving specific areas of the lordship. Like their English counterparts, the running of these courts was shared by local officials who presided over their proceedings, and by local landowners with an obligation to attend the court on a regular basis as one of its "suitors" and who were responsible for making judgments in them. They also shared with their English counterparts the practice of meeting only on an occasional basis and for fixed intervals of time. At the upper level of these communal courts were the county courts, presided over by sheriffs, which served the individual counties of the lordship, whether in royal hands or part of the greater liberties. These possessed a mainly civil jurisdiction, but were also the venue for all outlawries. At the more local level were the courts of the individual cantreds (later baronies), which corresponded to the English hundred, or wapentake, courts. These seem to have possessed a minor civil jurisdiction, but it was also at these courts that the local sheriff held twice each year the sheriff's tourn, to enquire into various misdemeanours and more serious criminal offences committed locally. Towns within the lordship also generally had their own courts, often described as hundred courts, and which came under the control of the town authorities. They normally met rather more often than other types of communal court and were the main courts for the enforcement of town custom. They also claimed an exclusive jurisdiction over civil litigation in the town.

The lordship possessed various different types of private court. In the earliest days of the lordship the great private courts of the lords of Leinster and Meath enjoyed a virtual independence. Between 1200 and 1208 these courts came under attack from the Crown, and the outcome was that henceforth major criminal pleas in these liberties were reserved to the lord of Ireland; that it became possible to appeal from the liberty courts to the king; and that lands belonging to the church within the liberties henceforward fell directly under royal jurisdiction. By the middle of the thirteenth century the major liberties possessed what seem to have been their own versions of the Dublin Bench, with a wide civil jurisdiction and run by justices appointed by the lord of the liberty. They also possessed private county courts that functioned like their royal counterparts, but whose profits went to the lord of the liberty. There also existed in the feudalized parts of Ireland manorial courts, with a civil and disciplinary jurisdiction roughly equivalent to that of their English counterparts.

Separate from this variegated network of lay courts were the courts run by the Catholic Church in Ireland as part of its European network of ecclesiastical courts at a diocesan and provincial level. The law enforced and applied in these courts was the Church's canon law, but with some local modifications. They possessed jurisdiction over matrimonial and testamentary matters, over various types of moral offense, and over the internal running of the Church, except where the lay courts managed to make good their claim to determine certain matters, such as those involving property rights.

Paul Brand

References and Further Reading

Brand, Paul. "The Birth and Development of a Colonial Judiciary: The Judges of the Lordship of Ireland, 1210–1377." In *Explorations in Law and History: Irish Legal History Discourses, 1988–1994*, edited by W. N. Osborough, 1–48. Blackrock, County Dublin: Irish Academic Press, 1995.

Hand, Geoffrey J. *English Law in Ireland, 1290–1324*. Cambridge: Cambridge University Press, 1967.

See also **Brehon Law; Canon Law; Common Law; Law Texts; March Law**

COYNE AND LIVERY

As it is usually understood, coyne and livery was the single most important tax in later medieval Ireland. It comprised the key element in the system of tributes and exactions used in the native lordships whereby the lords and chieftains required their subjects to give free entertainment (food, lodging, etc.) to their servants and followers. Often used as a tax to meet the maintenance of a lord's army, the extent to which it could be imposed determined the military strength of a lordship; conversely, a strong military lord could impose it as often as he liked, once he had the troops to enforce it. Oppression and racketeering were the bedfellows of coyne and livery.

It is important to avoid giving too precise a definition of the tax. The term "coyne and livery" is a hybrid one, used by English writers to describe a range of taxes in use in the Gaelic and gaelicized lordships, all of which revolved around the taking of free entertainment in some form or other. For instance, the cuddy (*coid oidche*) was specifically the taking of a night's entertainment, but was sometimes dubbed coyne and livery by English observers. Similarly, the bishops in

Gaelic areas levied "noctials," something very like coyne and livery: essentially the compulsory hospitality demanded of their tenants and clergy.

The word "coyne," sometimes rendered "coign," dates to post-Norman times and is derived from the Gaelic words for billeting and quartering, *coinnem* and *coinnmed*. An earlier term, *congbáil*, used in the Brehon Law texts and meaning entertainment or maintenance, may provide its pre-Norman root. *Meiconia*, or *miconia*, was most likely a Hiberno-Latin derivate of *coinnmed,* in use by the late fifteenth century. *Sorthan*, or "sorren," was an equivalent term used in parts of Munster. "Livery" referred to the provision of free food and bedding for a lord's cavalry ("horsemen") and fodder and stabling for his horses and their grooms, known as "horseboys."

Already endemic across Gaelic Ireland, coyne and livery spread into the lands of the "Englishry" in the course of the fourteenth century, as the military authority of the royal administration in Dublin declined and responsibility for the defense of outlying colonial regions was increasingly devolved to local lords and landowners. The chronology of its adoption is difficult to determine exactly. Even in the late thirteenth century the marcher lords of the colony were increasingly predisposed to the forced requisition of food and lodging by their private armies through purveyance. As the fourteenth century progressed and military arrangements evolved, at some point purveyance began to merge with coyne and livery, which began to replace it—presumably because it was more flexible and better-suited to the defense of ethnically mixed lordships populated by Gaelic as well as Anglo-Irish inhabitants (already the rank and file of the private armies comprised mainly Gaelic soldiers). By the early fifteenth century, coyne was in general use in all of the main Anglo-Irish territories in the south and east of the country. Its widespread imposition was probably the main reason for the growing regional dominance of the Butler earls of Ormond and the Fitzgerald earls of Kildare and Desmond.

Coyne was notoriously oppressive. For lesser landowners and tenants in many parts of the country, it was understandably difficult to refuse hospitality to a lord and his men. The line between hospitality given voluntarily and hospitality taken compulsorily was a thin one, and, of course, repeated usage in time created the legal basis for its becoming a customary exaction. In some areas, particularly the Ormond territories in Kilkenny and Tipperary, the lord's right to coyne and livery was usually controlled to some extent by his greater need to retain the support of the local gentry and to govern by consensus. Such constraints did not always apply elsewhere. Coyne was an especially heavy burden in the Desmond lordship, imposed as often as once a fortnight on the earls' subjects, many of whom were reduced to subsistence levels of existence as a result. Especially onerous was the requirement of some lords that coyne and livery be offered not just to them and their troops, but "without limitation" to their friends and followers also. Efforts to regularize coyne by transmuting its exaction into a money charge sometimes backfired, if the soldiers themselves were allowed to collect what was due. Despite such problems, however, the imposition of coyne and livery brought more advantages than disadvantages for the lords. In particular, its arbitrary nature meant that additional forces could be hired and maintained at short notice.

In the sixteenth century, as the English crown reasserted its power, measures were taken to abolish coyne and livery across the island. Gradually it was abandoned, beginning with the Ormond lordship in the 1560s, and by the early seventeenth century it had entirely disappeared.

David Edwards

References and Further Reading

Edwards, David. *The Ormond Lordship in County Kilkenny, 1515–1642: The Rise and Fall of Butler Feudal Power.* Dublin: 2003.

Empey, C. A., and Katharine Simms. "The Ordinances of the White Earl and the Problem of Coign in the Later Middle Ages." PRIA 75 C (1975), 161–187.

Hore, Herbert, and James Graves. The Social State of South-East Ireland in the *Sixteenth Century.* Dublin: RHAAI, 1870.

Nicholls, Kenneth. *Gaelic and Gaelicized Ireland in the Middle Ages*, 2nd ed. Dublin: Lilliput Press, 2003.

Nicholls, Kenneth. "Gaelic Society and Economy in the High Middle Ages." In *New History of Ireland, II: Medieval Ireland, 1169–1534*, edited by Art Cosgrove. Oxford: Clarendon Press, 1987.

Simms, Katharine. "Guesting and Feasting in Gaelic Ireland." Journal of the Royal Society of Antiquaries of Ireland 108 (1978), 67–100.

See also **Armies; Coshering; Entertainment; Gaelicization; Military Service, Gaelic; Society, Functioning of Gaelic**

CRAFTWORK

As in any preindustrial society, a wide range of crafts provided the necessities of life in medieval Ireland. Such crafts can be studied through documentary and archaeological evidence—ideally through a synthesis of both—but are particularly prominent in archaeological work. Excavation of rural (secular and ecclesiastical) and urban sites has produced vast amounts of evidence for all aspects of crafts, from gathering and processing of raw materials to the use and disposal of finished products. Major advances in understanding have been made, especially in terms of technical processes and patterns of distribution and trade. Less fully understood are the organization and scale of crafts and the role of professional specialists, as opposed to traditional, domestic craftworking. Towns are increasingly prominent in this research. Areas and buildings associated with specific crafts, including amber-working, woodworking, and comb making, have been identified in excavations in Dublin and Waterford. Unfortunately, because this evidence tends to be relatively early (no later than the thirteenth century), it cannot readily be reconciled with late-medieval documentary evidence for craft guilds and craft areas. As the mass of data produced by urban excavation is digested and analyzed, however, further progress should be made on these and other issues.

Woodworking and Carpentry

Wood was perhaps the most important natural resource in medieval Ireland, used to construct buildings and machinery and as the raw material for a wide range of artifacts. Problems of preservation have deprived us of the bulk of early medieval wooden structures and artifacts, but Ireland is relatively fortunate in its abundant bogs and wetland sites (notably crannogs), which provide important glimpses of what once existed. Wooden structures are least likely to survive, but horizontal mills provide evidence of high-quality carpentry techniques as early as the seventh century. Unexpectedly, a ringfort (normally a "dry" site where organic preservation would not be expected) at Deer Park Farms, County Antrim, preserved in waterlogged deposits substantial remains of post-and-wattle walling of a series of houses dating from around 700. Such chance survivals point to the existence in early medieval Ireland of well-established woodworking and carpentry traditions (also attested in documentary sources) and complement the mass of later evidence for buildings in Dublin and Waterford.

Shipbuilding, a specialized form of carpentry, was clearly practiced in several towns. Danish scholars have demonstrated that the Skuldelev 2 ship, perhaps the finest known Viking warship, was almost certainly made in Dublin in the 1040s. McGrail's study of Dublin ship timbers provides strong evidence for a continuing shipbuilding tradition, capable of producing even large ships, in the late twelfth and early thirteenth centuries. Documentary evidence exists for shipbuilding in thirteenth-century Waterford, Limerick, and Drogheda, and these industries may well have functioned over extended periods.

The almost endless range of artifacts carved from wood, most of them strictly utilitarian, includes exceptional objects like the gaming board from Ballinderry crannog, County Westmeath, and the decorated objects from Dublin, published by Lang. The scarcity of pottery

Ceramic cooking pot, Derrymagowan, Co. Armagh. *Photograph reproduced with the kind permission of the Trustees of the National Museums and Galleries of Northern Ireland.*

on most early medieval sites underlines the importance of wood as the raw material for bowls, buckets, barrels, and other vessels. Coopering, a highly skilled craft, is attested in the eighth century at Moynagh Lough crannog, County Meath, and by later vessels such as the oak butter churn from Lissue ringfort, County Antrim, and the yew bucket from Ballinderry crannog, County Westmeath. A number of exceptional stave-built buckets with decorative metal bindings and fittings are thought to be high-status vessels for both secular and ecclesiastical use. Carving of single-piece vessels (from solid blocks of wood) continued from prehistory and was enhanced in the early medieval period by the technique of lathe turning. Little evidence survives for lathes per se, but the products of turning—both finished vessels and waste cores—are known from many sites, including the early crannogs of Lagore and Moynagh Lough. Later sites, especially in Dublin, Waterford, and Cork, have produced abundant evidence for coopering and turning; indeed, large amounts of turning waste at High Street, Dublin, were interpreted as evidence for a lathe turning workshop in the vicinity. Woodworkers were clearly familiar with and exploited the different properties of various species, as seen, for instance, in the consistent use of ash wood for turned bowls and plates.

Stone Carving and Masonry

Stone carving was practiced in many different contexts and for different purposes, but the production of architectural stone was always a major area of activity. Prior to the twelfth century Irish stone buildings (mainly churches) display little in the way of complex carving, but high crosses and decorated grave slabs are evidence of highly skilled (and presumably professional) craftsmen. The introduction of Romanesque and Gothic architectural styles produced a marked increase in the quality and prominence of carved stone as architectural detailing. Some of this, particularly in the twelfth and thirteenth centuries, is attributed to foreign masons brought to Ireland for the purpose, as the output of native masons was generally old-fashioned and often of limited technical scope, largely because economic conditions provided little basis for well-established schools of masonry or sculpture.

Leather-Working

The overwhelming economic importance of animal husbandry ensured, as a by-product, the ready availability of many raw materials, particularly leather, bone, and wool. Leather was used primarily for footwear, but also for many other objects. Lucas' seminal study of footwear requires updating, because since its appearance, vast quantities of footwear have been recovered from urban contexts that appear to differ significantly from the rural material, being more similar to broader European traditions. At High Street, Dublin, Ó Ríordáin excavated an enormous deposit of leather fragments, the waste of generations of cobblers or shoemakers active on that part of the site in the twelfth and thirteenth centuries. Leather sheaths and scabbards, some highly decorated, are also common finds in urban contexts, but one cannot, as yet, definitively state that they were produced in the towns where they are found. Much rarer, but even more spectacular, are the decorated satchels produced (almost certainly in a rural, Gaelic milieu) to contain important manuscripts, such as the *Book of Armagh*, and shrines, such as the *Breac Maodhog*.

Bone- and Antler-Working

Evidence for bone- and antler-working is widespread, where conditions favor survival of the material. Bone would have been especially common, but antler was preferable for many processes. The craft often involved considerable skill, but relatively simple technology—often little more than a sharp knife. Combs were the most complex product, often involving the careful assembly of several individual components into a finely carved and beautifully decorated unit. There is good evidence from Dublin and Waterford for specialist comb-makers in the Hiberno-Norse and Anglo-Norman periods, but a well-known reference to a comb-maker in the tenth-century ecclesiastical settlement of Kildare reminds us that such craftsmen were not confined to towns.

Pottery

Pottery, although extremely important to the archaeologist, was neither a prestigious nor a particularly profitable craft. Pottery was not widely produced in early medieval Ireland; indeed, many areas seem to have been virtually aceramic, apart from small quantities imported from France and the Mediterranean. In the northeast, however, coarse hand-made pottery, evocatively but misleadingly known as "souterrain ware," was produced from the seventh to eighth centuries until the twelfth to thirteenth centuries. The vessels are typically flat-bottomed, bucket-shaped pots, unglazed and with little decoration. Many were clearly cooking pots, while others may have been storage vessels. While it occurs commonly in east Ulster (especially Cos Down and Antrim), there is no evidence for large-scale or commercial production of souterrain ware. Its production may, indeed, have been entirely a domestic craft. Ireland's first commercial pottery industry was probably established in Dublin in the wake of the English conquest. Similarities with Ham Green pottery, from Bristol, suggest that the Dublin industry was established, soon after 1171, by Bristol potters. Produced in a range of vessels such as glazed jugs, cooking pots, dripping pans, bowls, skillets, and even money boxes, Dublin pottery was initially hand-made, but from the early thirteenth century was wheel-thrown. An extremely productive industry, it continued to function to the end of the medieval period and beyond. Local pottery industries were also established in many other parts of the country in the wake of the English conquest. Most were probably based in towns such as Cork, Waterford, and Kilkenny, but this cannot be established with certainty until actual production sites are discovered. To date, only Downpatrick and Carrickfergus (and possibly Drogheda) have produced actual kilns of this period. Medieval potters also made floor and roof tiles for churches and other important buildings.

Textile Crafts

Textile production was clearly a widespread craft, and few excavated medieval sites, whether rural or urban, have failed to produce some evidence—even if only in the form of the ubiquitous spindle whorl. Spinning wool into thread was clearly a widespread domestic craft. Weaving the thread into cloth was, in theory, more specialized, but evidence for looms (mainly in the form of loom weights and accessories such as pin beaters and weaver's swords) is surprisingly common. It, too, may have been a relatively common domestic craft until the later Middle Ages, when specialist weavers undoubtedly operated in many towns and imported cloth was more readily available. Processes for finishing textiles, such as fulling and dyeing, tend to leave less physical evidence, and their organization is still very imperfectly understood. Excavations at an ecclesiastical settlement on the remote island of Inishkea North, County Mayo, however, revealed what appears to have been an early medieval workshop producing dyes from locally collected dog whelk molluscs.

Crafts in Imported Materials

Amber and jet or lignite (used for beads, pendants, bracelets, and other ornaments) do not occur naturally in Ireland. Because Irish amber is clearly imported, probably from the Baltic (although small amounts occur along the east coast of Britain), one might assume that such ornaments were imported in a finished state. There is evidence for amber-working, however, in Dublin at least, including a tenth-century property plot at Fishamble Street where amber-working clearly continued for generations. Similarly, there is evidence for working of walrus ivory, another imported material. Glass- and enamel-working was largely an adjunct of decorative metalworking and may have been mainly carried out by metalworkers. The main items not related to metalwork were glass beads and bracelets, many of which were undoubtedly produced in Ireland, but the organization of the craft is poorly understood. Whether glass vessels or window glass were ever manufactured in medieval Ireland remains uncertain.

ANDY HALPIN

References and Further Reading

Dunlevy, M. "A classification of early Irish combs." *Proceedings of the Royal Irish Academy* 88C (1988): 341–422.

Edwards, Nancy. *The Archaeology of Early Medieval Ireland*, chap. 5. London: Batsford, 1990.

Halpin, A. *The Port of Medieval Dublin*. Dublin: Four Courts Press, 2000.

Hencken, H. O'Neill. "Lagore Crannog: An Irish Royal Residence of the Seventh to Tenth Century A.D." *Proceedings of the Royal Irish Academy* 53C (1950): 1–248.

Hurley, M. F., O. M. B. Scully, and S. W. J. McCutcheon, eds. *Late Viking Age and Medieval Waterford: Excavations 1986–1992*. Waterford: Waterford Corporation, 1997.

Lang, J. T. *Viking-Age Decorated Wood: A Study of Its Ornament and Style*. Dublin: Royal Irish Academy (National Museum of Ireland Medieval Dublin Excavations 1962– 1981, Ser. B, vol. 1), 1988.

Lucas, A. T. "Footwear in Ireland." *Journal of the County Louth Archaeological Society* 13 (1956): 309–394.

McCutcheon, C. "Medieval Pottery in Dublin: New Names and Some Dates." In *Medieval Dublin* I, edited by S. Duffy, 117–125. Dublin: Four Courts Press, 2000.

McGrail, S. *Medieval Boat and Ship Timbers from Dublin*. Dublin: Royal Irish Academy (National Museum of Ireland Medieval Dublin Excavations 1962–1981, Ser. B, vol. 3), 1993.

Ó Ríordáin, A. B. "The High Street Excavations." In *Proceedings of the Seventh Viking Congress, Dublin 1973*, edited by B. Almqvist and D. Greene, 135–140. Dublin: Royal Irish Academy/Viking Society for Northern Research, 1976.

Wallace, P. F. "Carpentry in Ireland A.D. 900–1300—The Wood Quay Evidence." In *Woodworking Techniques Before A.D. 1500*, edited by S. McGrail, 263–299. Oxford: British Archaeological Reports, 1982.

See also **Crannóga/Crannogs; Dublin; Ecclesiastical Sites; Fraternities and Guilds; Houses; Jewelry and Personal Ornament; Metalwork; Mills and Milling; Ringforts; Ships and Shipping; Trade; Waterford**

CRANNÓGA/CRANNOGS

In the Middle Ages, people built and lived on small artificial islands in lakes, constructed of stone, earth, and timber. In the early medieval period, these islands were often referred to in saints' Lives, annals, and sagas using the words *inis* or *oileán*, perhaps signifying that people made little distinction between such places and natural islets. By the mid-thirteenth century, the word *crannóg*—the word used today—began to be used in the annals. Scholars have typically defined crannogs as islands built of stone, earth, timber, and organic materials, usually circular or oval in plan and enclosed within a surrounding palisade of planks, posts, or stone walls. However, a broader definition would include those crannogs without palisades, as well as other deliberately enhanced natural islands, rocky outcrops, and mounds and rock platforms along lakeshore edges.

History of Research

Since the nineteenth century, crannogs have been the focus of much antiquarian and archaeological investigation in Ireland and Scotland. In Ireland, these include the pioneering crannog surveys of W. F. Wakeman in the northwest in the 1870s. In 1886, W. G. Wood-Martin published his significant and influential synthesis, *The Lake Dwellings of Ireland*. In the 1930s and 1940s, there were important archaeological crannog excavations by the Harvard Archaeological Expedition at Ballinderry No. 1 (Co. Westmeath), Ballinderry No. 2 (Co. Offaly), and at Lagore (Co. Meath). In recent decades, regional and local surveys by the Archaeological Survey of Ireland in the Republic, by the Environment and Heritage Service in Northern Ireland, and by other scholars have revealed a diversity of size, morphology, siting, and location. Recent archaeological excavations, particularly at Moynagh Lough (Co. Meath) and Sroove (Co. Sligo), have also revealed evidence for houses, pathways, fences, pits, working areas, and the debris of domestic life and industrial production.

Origins and Chronology

The origins and chronology of crannogs have largely been reconstructed through the results of excavations, artifactual studies, and radiocarbon and dendrochronological dating. It is now evident that crannogs were being built and occupied in the Late Bronze Age and possibly into the early Iron Age, when they appear to have variously functioned as defended lake dwellings, metalworking platforms, and as places for cult activities such as the deposition of metalwork into water. However, it is also clear that the most intensive phase of crannog building, occupation, and abandonment in Ireland lies within the early Middle Ages (*c.* 400–1000). There appears to have been an explosion of crannog building in the late sixth and early seventh centuries A.D., probably due to both social change and political upheaval. It is also evident that many early medieval crannogs were occupied over several hundred years, although this was not always continuous. It is also clear that many crannogs were built and reoccupied in the later Middle Ages, when the Irish annals indicate that they were being used as lordly strongholds, prisons, hospitals, ammunition stores, and as places to keep silver and gold plate. They were also used as Gaelic Irish strongholds in Ulster in the sixteenth and seventeenth centuries, when they are described in English commentaries and depicted in the pictorial maps of Richard Bartlett, completed around 1602. There is also evidence that some crannogs were used as seasonal dwellings or refuges for the poor, and as hideouts for outlaws, in the eighteenth and nineteenth centuries.

Geographical Distribution

The geographical distribution of Irish crannogs is now broadly understood, and it is likely that there are at least 1,200 sites (although undoubtedly many remain undiscovered along marshy and wooded lakeshores). Crannogs are widely distributed across the lakelands of the midlands, northwest, west, and north of Ireland. They are particularly concentrated in the northwest drumlin lakes of Cavan, Monaghan, Leitrim, and Roscommon. Crannogs are known in every county of Ulster, in a belt stretching from Fermanagh, through south Tyrone and Armagh, to mid-Down, with particular concentrations in Monaghan and Cavan. Crannogs are more dispersed across the west and northeast, although concentrations can be identified, such as those around Castlebar Lough (Co. Mayo) and Lough Gara (Co. Sligo). Other regions, such as Westmeath, have smaller numbers, but a few crannogs have been identified further south and east. Crannogs tend to be found on smaller lakes, being apparently infrequent on the very large midland lakes of the River Shannon (e.g., Lough Ree) and River Erne.

Siting and Morphology

Recent archaeological studies indicate that crannogs vary significantly in local siting, morphology, and construction. For example, in Westmeath, most crannogs are actually found in quite shallow water, often being connected to the nearby shoreline by narrow stone causeways. However, other crannogs can also be found in much deeper water, up to 5 to 6 meters in depth, situated at some distance from the shoreline (e.g., 60–100 m). It is also evident that crannogs often vary in size, from relatively large cairns 18 to 25 meters in diameter by 3 to 4 meters in height (e.g., Croinis, Lough Ennell), to smaller mounds 8 to 10 meters in diameter and 1 to 2 meters in height (as at Sroove, Co. Sligo). Recent archaeological surveys on some lakes, such as Lough Gara (Co. Sligo) and Lough Derravarragh and Lough Ennell (Co. Westmeath), indicate that they are often together in groups, with several smaller platforms on the shoreline overlooking a larger, impressive crannog in the water. This may reflect a sequence of crannog occupation across time, or the expression of social and ideological relationships between lords and their tenants in the Middle Ages.

Crannogs also produce evidence, from both archaeological survey and excavation, for a wide range of other structures, such as houses, outdoor working spaces, middens of bone and discarded objects, defined entrances, jetties, pathways, and stone causeways. Crannogs have also produced large assemblages of artifacts, both as a result of archaeological excavation and as discoveries made both accidentally or by design (e.g., by treasure hunters in the 1980s). These material assemblages have included items of clothing (shoes, textiles), personal adornment (brooches, pins, rings), weaponry (swords, spearheads, axes, shields), and domestic equipment (knives, chisels, axes).

Interpreting the Social and Cultural Role of Crannogs

Traditionally, scholars have interpreted the social and economic role of medieval crannogs in terms of power, defensiveness, and social display. Thence, they have often been seen as island strongholds or isolated refuges occupied at times of danger (i.e., as might be suggested by their occasional remoteness and difficulty of access). There is certainly plenty of historical evidence that many were attacked and burned during raids and warfare, and the evidence for weaponry and the impressive scale of their timber and roundwood palisades does suggest a military or fortress role for some. In the north midlands, most crannogs are situated on modern barony boundaries, suggesting that they were formerly situated at the edges and frontiers of early medieval territories. Both archaeology and historical sources also suggest that at least some medieval crannogs were high-status or even royal sites used for feasting, as redistribution centers for the patronage of crafts and industry, and for projecting through their size and architecture the power and wealth of their owners. Early medieval crannogs such as Lagore (Co. Meath) and Island MacHugh (Co. Tyrone) certainly could be interpreted as the island residences of kings or nobles, perhaps being used as summer lodges, defensive strongholds, and as places for management of public gatherings and assemblies. Some early medieval crannogs have also been associated with the patronage and control of craft production (typically metalworking). For instance, Moynagh Lough (Co. Meath), a probable lordly crannog, particularly during its mid-eighth-century occupation phase, was clearly a place where various specialist craft workers resided and worked. In contrast, Bofeenaun crannog on Lough More (Co. Mayo) appears to have been an isolated site devoted to the processing of iron ore, and may have been the forge of a blacksmith intent on preserving the secrets of his craft.

On the other hand, it is clear from archaeological surveys that most crannogs were essentially small island or lakeshore dwellings, occupied at various times by different people, not necessarily of high social status. Recent archaeological excavations at Sroove, on Lough Gara (Co. Sligo) have suggested that some small, early medieval crannogs were the habitations of social groups or households who had little wealth or political power. Indeed, several crannogs have produced relatively modest material assemblages and could be interpreted as the island homesteads of the "middle classes," with most activity focused on cattle herding and arable crop production along the lakeshore. Others may have been used periodically, seasonally, or for particular specific tasks. In other words, different types of crannogs were built, used, and occupied by various social classes in medieval Ireland.

Moreover, while there is commonly an image of early medieval crannogs as secular dwellings, it is also likely that many were used by the church, given the significant role of the church in the early medieval settlement landscape. The discovery of early medieval ecclesiastical metalwork (handbells, crosses, and bookshrines) on some midlands crannogs (occasionally in proximity to actual church sites and monasteries) suggests their use as shrine islands for storage of relics, or perhaps even as island hermitages (akin to the small hermitages occasionally found off monastic islands on the Atlantic coast).

Conclusions

In conclusion, despite being a subject of interest for over a hundred years, there is much that remains to be discovered and interpreted about crannogs in medieval Ireland. Recent studies have adopted multidisciplinary approaches, further exploring the social, cultural, and ideological perception of islands and crannogs among medieval communities, and thence how they were used to negotiate social identities of power, gender, and kinship. It is also likely that future projects will stress the importance of understanding medieval crannogs within their wider social and cultural landscapes, in both regional and local terms.

AIDAN O'SULLIVAN

References and Further Reading

Bradley, John. "Excavations at Moynagh Lough, Co. Meath." *Journal of the Royal Society of Antiquaries of Ireland* 111 (1991): 5–26.

Fredengren, Christina. *Crannogs: A Study of People's Interaction with Lakes, with Particular Reference to Lough Gara in the North-West of Ireland.* Bray: Wordwell Books, 2002.

Kelly, E. P. "Observations on Irish Lake-Dwellings." In *Studies in Insular Art and Archaeology*, edited by Catherine Karkov and Robert Farrell, 81–98. (American Early Medieval Studies 1) Cornell: 1991.

O'Sullivan, Aidan. *The Archaeology of Lake Settlement in Ireland.* Dublin: Royal Irish Academy/The Discovery Program, 1998.

O'Sullivan, Aidan. *Crannogs: Lake-Dwellings in Early Ireland.* Dublin: Town House, 2000.

O'Sullivan, Aidan. "Crannogs in Late Medieval Gaelic Ireland, c.1350–c.1650." In *Gaelic Ireland: Land, Lordship and Settlement, c. 1250–c.1650*, edited by P. J. Duffy, D. Edwards, and E. Fitzpatrick, 397–417. Dublin: Four Courts Press, 2001.

O'Sullivan, Aidan. "Crannogs—Places of Resistance in the Contested Landscapes of Early Modern Ireland." In *Contested Landscapes: Landscapes of Movement and Exile*, edited by Barbara Bender and Margot Winer, 87–101. Oxford: Berg, 2001.

See also **Agriculture; Archaeology; Craftwork; Ecclesiastical Settlements; Houses; Ringforts**

CRUTHNI

The term "Cruthni" was applied to a number of early Irish population groups by the writers of Old and Middle Irish, although it is unclear whether it represents a self-description by these groups or whether it was an externally generated label. The word is an Irish cognate of Welsh Prydain and earlier Celtic Pretani, from which the modern English terms Britain and Britons derive. The equivalence of British "p" with Irish "kw," later "c," is well established, and the modern English forms in "b" derive from Latin forms themselves, based on a mishearing of the British pronunciation. Cruthni thus, in some sense, means "Britons," although by the Middle Ages the Irish distinguished between the descendants of the Romano-British population living south of the river Forth (in Scotland), whom they called Bretain (Britons), and the unromanized tribes to the north, whom they called Cruthni and who were known in Latin as *Picti*, the Picts. It is incorrect, however, to designate the Irish Cruthni as "Picts," for they were never so-called by Irish Latinists such as Adomnán, who while using *Picti* for those in northern Britain, simply Latinized "Cruthni" when referring to the Irish.

Groups of Cruthni were found across Ireland, but only three groups were of any significance: the Sogain, a major subject people of the Connachta whose seven tribes were scattered across Connacht proper and Mide; the Loígis (whose name survives as modern Laois) in western Leinster; and finally a large group of *túatha* occupying most of Antrim and some of the neighboring districts in the northeast. It is with this last group that we shall principally be concerned. If the term "Cruthni" does imply an origin in Britain (the alternative view would be that the island takes its name from the people) then there is no reason to suppose any particular connection between the various Cruthnian groups within Ireland. Their ancestors may have made the crossing at different periods and from different places in Britain. Since the migrations are not recorded in the annals it is fairly certain that any such migrations happened before the fifth century A.D., perhaps long before.

While the Sogain, the Loígis, and the smaller groups of Cruthni scattered throughout Ireland were for the most part loyal vassals within established over kingdoms, the northeasterners were a major force in their own right. Their territory in the mid-sixth century seems to have covered most of the land between Lough Foyle and the Lagan. They were bordered on the southeast by the Ulaid, on the southwest by the Airgialla, and on the west by the Northern Uí Néill. They were at this stage divided into many *túatha*. At the battle of Móin Daire Lotheir in 563 two of the Cruthian *túatha* under Baetán mac Cuinn, aided by the Uí Néill, fought against their over king Áed Brecc. Áed, together with seven of his allied kings, was slain, and at least one further Cruthnian, Eochaid Láeb, is said to have escaped the battle. This puts at least eleven kings of the Cruthni at this battle, giving some sense of their extent. Two years later Áed Brecc's successor as Cruthnian over king, Áed Dub, slew Diarmait mac Cerbaill, king of Tara. In 637 the Cruthni were defeated by Domnall mac Áeda, king of Tara, at Mag Rath (Moira, County Down), and the battle became a central point of saga. These events of the decades around 600 give some indication of the importance of the Cruthni in the early period.

By the eighth century a single kingdom had emerged to dominate the Cruthni, that of the Dál nAraide of Mag Line (Antrim), and its name gradually drove that of the Cruthni from the record. In the later ninth and early tenth centuries Dál nAraide even began to exert their authority over the Ulaid, and they rewrote their own origin legends to make it appear that they had always been Ulaid.

Modern scholars have always stressed that there is no evidence of British speech or other cultural traits among the Irish Cruthni, but it must be conceded that we have no early texts emanating from their province and that the record of their personal names, and so forth, may have been normalized by the chroniclers, just as they were able to produce Gaelic forms of the names of Pictish kings in Britain whose own preferred spelling forms are preserved elsewhere. This said, had any migration occurred before the fifth century it is quite likely that British and Irish Celtic would have been close enough that convergence of dialects rather than language shift would have been required to bring the two tongues together.

A further problem regards the relationship between the Cruthni and Dál Riata. By the seventh century the kingdom of Dál Riata had emerged, extending from the valley of the Bush in Antrim as far as Mull and the adjacent mainland in Scotland, with ecclesiastical centers at Armoy, Kingarth, and Iona and royal centers at Dunseverick, Dunadd, and Dunollie. In Ireland, Dál Riata was completely enclosed by Cruthnian *túatha*, while in Scotland it appears to have been the beachhead from which Gaelic language and culture was eventually to spread to the whole region. At the end of the sixth century Áedán mac Gabráin, king of Dál Riata, seems to have displaced Áed Dub as over king of the Cruthni and become a dominant figure in the north of both Britain and Ireland. Later genealogists gave Dál Riata a distinct descent from the Cruthni, making them either exiles from the Dingle peninsula or a branch of the Ulaid, but the close geographical and political relationship between the two groups begs some questions. It is also odd that the Irish population group described as "Britons" should be such close neighbors of a population group in Britain perceived as Irish by their neighbors. Possibly both groups represent the two halves of a people who bridged the gap between Britain and Ireland, in cultural as well as geographical terms.

In religious terms, the most important Cruthnian saint was Comgall, founder of Bangor and friend of Colum Cille. His foundation at Bangor lay on the borderlands between the Ulaid and the Cruthni.

ALEX WOOLF

References and Further Reading

Mallory, J. P., and T. E. MacNeill. *The Archaeology of Ulster.* Belfast: The Institute of Irish Studies, 1991, pp. 176–178.

O'Rahilly, T. F. *Early Irish History and Mythology.* Dublin: D.I.A.S., 1946, pp. 341–352.

See also **Áedán mac Gabráin; Colum Cille; Connachta; Leinster; Mide; Túatha**

CÚÁN ÚA LOTHCHÁIN

Cúán úa Lothcháin (d. 1024) was a professional poet inextricably linked to Tara and its rulers, in particular Máel Sechnaill mac Domnaill, who predeceased him by two years. Indeed Pádraig Ó Riain has argued that it is to Cúán we owe the literary revival of Tara as a symbolic seat of kingship designed to advance the standing of his northern employer in the face of stiff opposition from the latter's Munster contemporary, Brian Boru. This is seen most clearly perhaps in his *dinnshenchas* poem on *Temair toga na tulach/foatá Ériu indradach* (Tara noblest of hills, under which is Ireland of the battles) and in his description of Tailtiu, a site closely associated with Tara. The latter, written in support of Máel Sechnaill, who in a conscious declaration of power celebrated the famous *óenach* (gathering) there in 1006 after a gap of almost eighty years, endorses the monarch as *oen-milid na hEorapa* (the sole champion of Europe), *ordan íarthair domuin duind* (the glory of the noble western world), and, most significantly, as the rightful king of Tara whose rule bestowed peace and plenty on his subjects. Máel Sechnaill is also directly addressed in a *dinnshenchas* poem on the Boyne which may be by our poet; a second version of the river's origin is recounted in another composition that Cúán almost certainly wrote, though the surviving ascription is only partly legible. Moreover, he employed the tragic story of Eochaid Feidlech and his three sons to underline the legitimacy of Máel Sechnaill's rule in his poem on Druim Criaich, near Tara. In other compositions he ventured outside his favored territory, addressing such far-flung places as the River Shannon and Carn Furbaide (Granard, Co. Longford). Nonetheless, Tara remained his chief focus, being accorded pride of place in one recension of *Dinnshenchas Érenn* (The Place-Name Lore of Ireland), which Cúán may have authored, according to Tomás Ó Concheanainn.

Promotion of Tara is also a feature of his other work. Thus, his tract on royal prohibitions concerns itself with the king of Tara in the first instance. Similarly, the narrative poem *Temair Breg, baile na fían* (Tara of Brega, homestead of the champions), which may have been written by him, furnishes Níall Noígíallach (Níall of the Nine Hostages) and his descendants, of whom Máel Sechnail was one, with a patent for the all-important kingship of Tara, serving as the poetic counterpart to the corresponding prose version, *Echtra mac nEchach Mugmedóin* (The Adventure of the Sons of Eochaid Mugmedón). Moreover, such was his fame that he was later cited as an authority by his patron's rivals, Brian Boru's twelfth-century descendants claiming his imprimatur for their long-held right to the kingship of Cashel. On him were also fathered a number of poems, including one on the three famous trees of Ireland, whose author describes himself as Cúán *ó Caeindruim* (from Cáendruim). In reality, it was from Tethba in Mide that Cúán hailed, and it was here too that he was murdered by local inhabitants in 1024. Our poet had the last word, however, as the annalist recounted: *Brenait a n-aenuair in lucht ro marb. Firt filed innsein* (The party that killed him became putrid within the hour. That was a poet's miracle).

MÁIRE NÍ MHAONAIGH

References and Further Reading

Dillon, Myles, ed. and trans. "The Taboos of the Kings of Ireland." *Proceedings of the Royal Irish Academy* 54 C (1951–1952): 1–36.

Gwynn, Edward, ed. and trans. *The Metrical Dindshenchas.* 5 vols., Todd Lecture Series, 8–12, 8: 14–27; 10: 26–33, 292–297; 11: 30–35, 42–57, 146–162. Dublin: Royal Irish Academy, 1903–1935.

Joynt, Maud, ed. and trans. "Echtra mac Echdach Mugmedóin." *Ériu* 4 (1908–1910): 91–111.

Mac Airt, Seán, and Gearóid Mac Niocaill, eds. and trans. *The Annals of Ulster (To A.D. 1131), Part I: Text and Translation.* Dublin: Dublin Institute for Advanced Studies, 1983, pp. 462–463.

Meyer, Kuno, ed. "Mitteilungen aus Irischen Handschriften." *Zeitschrift für Celtische Philologie* 5 (1905): 21–23.

Ó Concheanainn, Tomás. "A Pious Redactor of Dinnshenchas Érenn." *Ériu* 33 (1982): 85–98.

Ó Riain, Pádraig. "The Psalter of Cashel: A Provisional List of Contents." *Éigse: A Journal of Irish Studies* 23 (1989): 107–130.

See also **Dinnshenchas; Metrics; Tara**

CUMIN, JOHN (d. 1212)

John Cumin, archbishop of Dublin, was born probably in the 1130s into a minor Somerset family. He spent his early years in this county before entering into royal service at a young age. By the 1160s he had become a trusted royal official and earned the powerful patronage of King Henry II. He served the king in the judiciary, the chamber, and as a negotiator on a number of important diplomatic missions. He acquired deacon's orders and in 1166 was appointed to the archdeaconry of Bath. He remained loyal to the king during the dispute with the archbishop of Canterbury, Thomas Becket, a loyalty that resulted in his excommunication by the archbishop. After Henry II had weathered the Becket controversy, he set about rewarding those clerks who had remained faithful to him. When the strategically vital see of Dublin became vacant on the death of Lorcán Ua Tuathail (Laurence O'Toole) in 1180, the king secured the election of his trusted servant, John Cumin.

Cumin was consecrated archbishop by Pope Lucius III at the papal court in Velletri in February 1182, and then returned to join the royal court. His first securely recorded visit to his diocese was in the autumn of 1184, when he was dispatched to Ireland to prepare for the Lord John's imminent visit. He remained in Ireland after John's return to England to hold a provincial council in Dublin early in 1186, but soon returned to England, where he remained until after the death of Henry II in 1189 and the coronation of Richard I the following year. The death of his patron has been seen as marking a watershed in Cumin's career—the point at which he diverted his energies away from royal politics and toward ecclesiastical administration. He returned to his diocese and maintained more or less constant residence until 1196. The realignment of his priorities had repercussions, and he became embroiled in a dispute with the Irish chief governor, Hamo de Valognes, over the temporalities of his see. This resulted in his exile from Dublin for a period of nine years. In 1205, King John, faced with the threat of papal excommunication and interdict, agreed to restore the archbishop's full liberties and temporal possessions, and John Cumin returned to Dublin. The remaining six years of his episcopate passed without apparent incident, and he died in October, 1212, at an advanced age.

While it is clear that John Cumin's qualifications for office were of a decidedly secular nature, he sought at the outset of his episcopacy to combine his involvement with the royal court with an active ministry. His willingness to address the particular problems of the Irish church can be seen in the nature of the legislation approved by the provincial council that he summoned to meet in Dublin in 1186. The canons promulgated by the council are believed to be largely Cumin's own work, and they displayed a particular concern with the proper administration of the sacraments. This was also the subject of the opening sermon of the council, preached by the archbishop himself. Furthermore, the archbishop legislated to enforce clerical celibacy and regularize marriage practices, and in many ways allied himself with the aims of the native reform party in Ireland.

Cumin asserted on a number of occasions that his church was in dire need of reform and that the people he was sent to govern were in need of instruction and civilizing. His stated aim in 1192, when he raised the church of St. Patrick's to collegiate status and instituted a group of thirteen clerks to serve there, was to improve the educational status of the people of his diocese. It is not clear whether the archbishop intended that St. Patrick's be elevated to cathedral status, his later absence from his diocese making it difficult to assess his motives. He made no provision for officials in the college, but did grant the canons similar liberties and privileges to those enjoyed by the secular canons of Salisbury cathedral.

From the beginning of his episcopate, Cumin was concerned with the consolidation and defining of the temporal possessions of the Dublin diocese. These were extensive, and after the amalgamation with the diocese of Glendalough they included the manors of St. Sepulchre, Swords, Finglas, Clondalkin, Tallaght, Shankill, Ballymore, and Castlekevin, along with the land of Coillacht—an extensive wooded area extending from the upper Dodder to Tallaght. The archbishop exercised jurisdiction in these manors through his seneschals and bailiffs. Cumin was keen to exploit the commercial potential of his possessions. In 1193, he established an annual eight-day fair at Swords at the feast of St. Colum Cille, and at some date before 1199 he was granted a Saturday market at Ballymore, County Wicklow.

The archbishop was generous with gifts of lands and tithes, and in his benefactions he showed a particular favor for houses of nuns. He founded the priory of Grace Dieu, County Dublin, transferring nuns from nearby Lusk. He endowed the convent with several churches, including the valuable church of St. Audoen inside the walls of Dublin. He also granted a carucate of land to the nuns of Timolin, County Kildare. He was similarly generous to members of his own family, and used his position and power to assist the careers and fortunes of his three nephews—Gilbert and Walter Cumin and Geoffrey de Marisco.

It was while defending his temporal possessions that the archbishop fell foul of the justiciar Hamo de Valognes. The initial cause of the dispute is unclear, but appears to have concerned the nature of royal forest rights in lands newly acquired by the Dublin church. John Cumin excommunicated Valognes and several of his retinue, and in 1197 left for England, having placed Dublin under interdict. It is obvious that Cumin did not have the same standing with Henry II's sons as he had had with that monarch, as he spent several years seeking redress from first Richard and then John. His cause was championed by Pope Innocent III, who eventually brokered a settlement by which the archbishop was restored to his full liberties and temporal possessions. Cumin's biographers have noted the irony that, given his stance some decades earlier during the Becket dispute, he should find himself, like Becket, exiled from his church and dependent on Rome for help. Giraldus Cambrensis appears to make a direct reference to Cumin's exile from Dublin when he remarks that the archbishop would have made outstanding improvements to the condition of his church had he not been prevented by the secular powers from so doing.

John Cumin played a large part in introducing the principal elements of Anglo-Norman ecclesiastical administration into the Dublin province. However, he appears to have approached the task with a certain amount of circumspection, and in many of his actions displayed a willingness to interact with Irish ecclesiastics. His diplomatic skill, honed as an advocate for Henry II, was fully manifest in 1192 in his establishment of the secular college of St. Patrick's, an action that might have been expected to arouse both the fears of the existing cathedral chapter at the priory of Holy Trinity (Christ Church) and the hostility of the Irish. Yet, when the new college was consecrated on St. Patrick's Day, the ecclesiastical procession set out from Holy Trinity and was led by the two most senior Irish ecclesiastics, Archbishops Mattheus of Cashel and Eugenius of Armagh.

John Cumin held the archbishopric of Dublin for thirty-one years. When he died in 1212, "old and full of days" according to the annalist of St. Mary's Abbey, Dublin, he was laid to rest in the church of Holy Trinity.

MARGARET MURPHY

References and Further Reading

Gwynn, Aubrey. "Archbishop John Cumin." *Reportorium Novum* 1 (1955–1956): 285–310.

MacShamhrain, Ailbhe. "The Emergence of the Metropolitan See: Dublin, 1111–1216." In *History of the Catholic Diocese of Dublin*, edited by James Kelly and Dáire Keogh, 51–71. Dublin: Four Courts Press, 2000.

Murphy, Margaret. "Balancing the Concerns of Church and State: The Archbishops of Dublin, 1181–1228." In *Colony and Frontier in Medieval Ireland: Essays Presented to J. F. Lydon*, edited by Terry Barry, Robin Frame, and Katharine Simms, 41–56. London: The Hambledon Press, 1995.

Murphy, Margaret. "Archbishops and Anglicisation: Dublin, 1181–1271." In *History of the Catholic Diocese of Dublin*, edited by James Kelly and Dáire Keogh, 72–91. Dublin: Four Courts Press, 2000.

Sheehy, Maurice. *When the Normans Came to Ireland*, 2nd ed. Cork: Mercier Press, 1998.

Watt, John. *The Church in Medieval Ireland*, 2nd ed. Dublin: University College Dublin Press, 1998.

See also **Chief Governors; Christ Church Cathedral, Dublin; Giraldus Cambrensis; Glendalough; Henry II; Henry of London; John; Nuns; Ua Tuathail, St. Lorcán; St. Patrick's Cathedral**

D

DÁL CAIS

Dál Cais was the name of the Munster people based in eastern County Clare that rose to prominence in the latter half of the tenth century and produced a number of powerful kings, including Brian Boru. Although they claimed kinship with the Eóganachta, who had dominated the province since the dawn of history, the Dál Cáis actually belonged to the larger population of Munster Déisi, who were ethnically Érainn. This Déisi population originally formed a loose conglomerate stretching from southern Waterford into Limerick, but by the eighth century, they had divided into two separate groups—the Déisi Muman of Waterford and southern Tipperary, and the western Déisi of Limerick. In the latter territory were the Déis Deiscirt, who were eventually eclipsed by their neighbors, and the Déis Tuaiscirt, who later changed their name to Dál Cais. Although legend has it that they conquered their lands in eastern County Clare from Connacht in the fifth century, historical sources suggest that they did not actually gain possession of this territory until the early eighth.

It is not known exactly when the Déis Tuaiscirt adopted the name Dál Cais, but it is first used of them in the *Annals of Inisfallen* under the year 934, in an entry recording the obit of their king, Rebachán mac Mothlai. His death is also important for marking a major transition in internal Dál Cais politics. For some time prior, the kingship had been controlled by Rebachán's sept, the Clann Óengusso, but at his death, the office was seized for the first time by their rivals, the Uí Thairdelbaig. Under this new leadership, the Dál Cais began a program of expansion that would soon make them one of the most powerful kingdoms in Ireland.

This expansion began in earnest with the rule of Cennétig mac Lorcáin, who had extended his sway over much of north Munster by the time of his death in 951. Building on his father's success, Mathgamain mac Cennétig extended Dál Cais rule even further by seizing the kingship of Cashel and thus putting an end to centuries of Eóganacht rule in Munster. More importantly, though, Mathgamain also gained control of the Norse settlements of Waterford and Limerick, the resources of which were needed to sustain Dál Cais expansion. In 976, Mathgamain was succeeded by his brother, Brian Boru, who went on to become the most powerful king in Ireland. However, the deaths of Brian and a number of his important kinsmen at the Battle of Clontarf in 1014 crippled Dál Cais. By the time they reemerged in the late eleventh century, the ruling family of Uí Thairdelbaig had adopted the surname Ua Briain (O'Brien), and they enjoyed a considerable revival under Tairrdelbach (d. 1086) and, later, Muirchertach Ua Briain. But with the latter's death in 1119, the days of the great Dál Cais kings came to an end. After the Anglo-Norman Invasion, the O'Briens were essentially restricted to their lands in Thomond, where they retained some power throughout the later Middle Ages.

DAN M. WILEY

References and Further Reading

Byrne, Francis John. *Irish Kings and High-Kings*, 2nd ed. Dublin: Four Courts Press, 2001.

Ó Corráin, Donncha. *Ireland before the Normans*. Dublin: Gill and Macmillan Ltd, 1972.

Ó Cróinín, Dáibhí. *Early Medieval Ireland 400–1200*. New York: Longman Group Ltd, 1995.

See also **Brian Boru; Déisi; Eóganachta; Érainn; Munster; Ó Briain; Ua Briain, Muirchertach; Ua Briain, Tairrdelbach**

DANCE

See **Music**

DÉISI

As a proper noun, the word Déisi (sg. Déis) means "subject peoples" and was the name borne in the historical period by two important Érainn populations, the one in Brega, the other in Munster. The former were known as both the Déisi Breg and the Déisi Temro (the subject peoples of Tara) because they occupied lands just south of the ancient site. In the eighth century, their kingdom was eclipsed by the expanding Southern Uí Néill dynasty of Síl nÁedo Sláine, though with the decline of their overlords, they were able to regain their independence in the eleventh. Their revival, however, did not outlast the subsequent Anglo-Norman Invasion.

Larger were the Déisi populations in Munster, which originally formed a single, if discontinuous, conglomerate stretching from the extreme southeast to the north of the province. In the fourth or fifth century, a branch of the Déisi from the Waterford area established a colony in Dyfed (southwest Wales), where they retained power until the tenth century. Later, their Irish counterparts split into two main divisions about the beginning of the eighth century: The Déisi Muman lived in County Waterford and southern Tipperary and the western Déisi in eastern Limerick. In the latter territory, the most important people were the Déis Tuaiscirt, who conquered east County Clare. By the early tenth century, they adopted the name Dál Cais and subsequently became the most powerful kingdom in Ireland, under their ruler Brian Boru (d. 1014).

DAN M. WILEY

References and Further Reading

Byrne, Francis John. *Irish Kings and High-Kings*, 2nd ed. Dublin: Four Courts Press, 2001.

Jackson, Kenneth. *Language and History in Early Britain*. Edinburgh: Edinburgh University Press, 1953.

Ó Corráin, Donncha. *Ireland before the Normans*. Dublin: Gill and Macmillan Ltd, 1972.

See also **Dál Cais; Érainn; Munster; Southern Uí Néill**

DERBFORGAILL

Derbforgaill (1108–1193), daughter of Murchad Ua Máelsechlainn, king of Mide, and wife of Tigernán Ua Ruairc, king of Bréifne, owes her place in history mainly to her abduction at the hands of Diarmait Mac Murchada, king of Leinster. Following an internal Leinster rebellion against Mac Murchada more than a decade after the abduction, Ua Ruairc seized the chance to avenge his insulted honor and marched on Mac Murchada, expelling the Leinster king across the Irish Sea. Mac Murchada's subsequent recourse to foreign military aid in regaining his kingdom brought about the Anglo-Norman invasion. Surveying this chain of events, contemporary and later observers laid the blame for the invasion at Derbforgaill's feet, dubbing her the Irish Helen of Troy. Different sources have attributed varying motivations for the abduction, including revenge and overweening lust. Given the circumstances surrounding the kidnapping, however, which occurred in Mide in 1152 following Tigernán's temporary deposition as king of Bréifne by Diarmait and Tairrdelbach Ua Conchobair, it is likely that political motivations were at least partly at play. Mac Murchada and Ua Ruairc were competitors in the territorial dismemberment that took place in Derbforgaill's homeland of Mide following that kingdom's twelfth-century collapse as a major power. In addition to representing a dramatic undermining of Ua Ruairc's authority, it has accordingly been suggested that Derbforgaill's seizure may have symbolized Mac Murchada's pretensions toward Mide. Some sources report that Derbforgaill's brother Máel-Sechnaill colluded with Mac Murchada in arranging the abduction; possibly Máel-Sechnaill, who had newly come into power in the eastern portion of Mide, felt his best chances for survival lay in an alliance with Mac Murchada. Derbforgaill's own role in the kidnapping has been a further matter of dispute, with some sources portraying her as an innocent victim led to the kidnapping site by her brother. Others, no doubt influenced by the report that Derbforgaill was accompanied into captivity by all her cattle and her wealth, accuse her of having been complicit in the affair. Complict or not, however, through the intervention of Tairrdelbach Ua Conchobair, Derbforgaill, her cattle, and her wealth returned to Ua Ruairc within the year.

That wealth must have been fairly considerable, for in 1157 Derbforgaill is recorded as donating a large sum of gold to the newly consecrated monastery of Mellifont in Drogheda. While Mellifont had links to Ua Ruairc, Derbforgaill was also a generous patron of churches associated with her own family. In 1167, she finished the Nuns Church at Clonmacnoise, a foundation linked to the Arroasian convent at Clonard where her cousin Agnes was abbess. Derbforgaill retired into religious life in Mellifont in 1186, dying there seven years later at the age of eighty-five. Although she is the single most historically documented woman in pre-Norman Ireland, there is no record of Derbforgaill's having had any children. The fact that Tigernán had a son called Máel-Sechnaill, a name with strong family links to Derbforgaill not hitherto seen in the Ua Ruairc genealogies, may, however, indicate she had at least one child.

ANNE CONNON

References and Further Reading

Byrne, Francis John. *Irish Kings and High-Kings*. London: Batsford, 1973.

Doan, J. "Sovereignty Aspects in the Roles of Women in Medieval Irish and Welsh Society." *Proceedings of the Harvard Celtic Colloquium* 5 (1985): 87–102.
Flanagan, Marie Therese. *Irish Society, Anglo-Norman Settlers, Angevin Kingship: Interactions in Ireland in the Late Twelfth Century.* Oxford: Clarendon Press, 1989.
Martin, F. X., and Scott, A. B. *Expugnatio Hibernica: The Conquest of Ireland.* Dublin: Royal Irish Academy, 1978 (notes 5, 6, p. 286).

See also **Anglo-Norman Invasion; Clonmacnois; Mac Murchada, Diarmait; Mellifont; Mide; Queens**

DERRY

An early ecclesiastical settlement and modern city, Derry is situated on an island in the river Foyle, although the old rivercourse to the west (the Bogside) had become a swamp by the early historic period. *Daire* means "oak wood"—its oaks were "sacred" throughout its history. The earliest name form was Daire Calgach (the oak wood of Calgach [person unknown]). By the early twelfth century it is called Daire Coluim Cille, named after St. Colum Cille (d. 597), the traditional founder of the church. However, it has been strongly argued by Lacy that its foundation was by a Fiachrach mac Ciaráin, either alone or as joint founder with Colum Cille.

The place was mentioned by Adomnán (d. 704) in his "life" of Colum Cille. It had a church, a graveyard, and a harbour. It was a place of refuge. The reference to a scribe in *Annals of Ulster* 720, Caech Scuili, indicates that it had a scriptorium. It was mainly staffed by the Cenél Conaill, Colum Cille's own dynasty. Their rivals, the Cenél nEógain, controlled the surrounding area from 789 onward and made Derry their capital during the eleventh and twelfth centuries. Cenél nEógain success against the Vikings prevented it from becoming an international trading town, but by the late twelfth century it had urban characteristics and had become head of the Columban familia. It had at least three churches and a round tower. Famous abbots were Gilla-Meic-Liag (d. 1174) who succeeded St. Malachy as head of the Irish church at Armagh in 1137, and Flaithbertach Ua Brolcháin (d. 1175), under whose direction the town was reshaped. Its school produced the lost Book of Derry, and the lost Gospel of Martin. Much of the literature about Colum Cille was produced here during this period as part of the propaganda associating his name with the site. Annals (now incorporated in the *Annals of Ulster*) were written there in the late twelfth to early thirteenth century, allowing us a glimpse of secular events.

As a result of the church reform of the twelfth century the monastery adopted the rule of Canons Regular of St. Augustine by about 1220. Soon after 1224, a Dominican priory was founded. For political reasons the seat of the Cenél nEógain diocese was at Ráith Luraig in Maghera, County Derry, although some bishops may have lived in Derry. However, in 1254 the Tempull Mór became the cathedral of the diocese of Derry, despite opposition from the Cenél Conaill. Each side had an erenagh (lay head of church) family, Mac Lochlainn (Cenél nEógain) and O'Deery (Cenél Conaill), living in Derry until 1609. The archbishop of Armagh, John Colton, made a famous visitation on 10 October, 1397. The Anglo-Normans came to Derry in 1197, but despite the possibility of establishing a planned colonial town at various times during the following centuries, this did not happen, and the town was largely uninfluenced by them. By the mid-sixteenth century the site had become strategically important to the English, who now gained control and built its (still intact) stone walls between 1613 and 1618.

CHARLES DOHERTY

References and Further Reading

Adomnán of Iona. *Life of St. Columba.* Translated by Richard Sharpe. London: Penguin Books, 1995.
Herbert, Máire. *Iona, Kells and Derry.* Chap. 9. Oxford: Oxford University Press, 1988. Reprint, Dublin: Four Courts Press, 1996.
Lacey, Brian. "Columba, Founder of the Monastery of Derry?—'Mihi manet Incertus.'" *Journal of the Royal Society of Antiquaries of Ireland* 128 (1998): 35–47.
O'Brien, Gerard, ed. *Derry and Londonderry: History and Society.* Dublin: Geography Publications, 1999.

See also **Colum Cille; Church Reform, Twelfth Century; Ecclesiastical Organization; Ecclesiastical Settlements; Ecclesiastical Sites; Scriptoria; Uí Néill, Northern**

DESERTED VILLAGES

See **Villages**

DEVOTIONAL AND LITURGICAL LITERATURE

Christianity, until the age of print, cannot be described as a "religion of the book," as its belief structures did not focus on a book as the vehicle or content of its revelation. However, from its earliest times as a sect within Second-Temple Judaism (before 71 C.E.) it (1) elaborated itself in terms of a regular liturgy (e.g., Acts 2:42), (2) presented itself as the fulfillment of prophesies contained in the Scriptures (e.g., Acts 17:2), and (3) sought to regulate its teaching through the use of written records which, in time, became the repository of its memory (e.g., the notion of Four Gospels or of a New Testament), to the extent that (4) its teaching was conceived as the interpretation and representation of those records (i.e., exegesis). In all these activities books were essential, and because liturgy touches every Christian, and almost always involves reading the religion's scriptures, books (along with the appropriate skills of reading, writing, and book production) need to be widely available for even the basic functioning of the religion. Indeed, it was probably Christianity's need for a plentiful supply of books that led to the dominance of the codex form—pages bound on the left margin—over the roll form preferred in antiquity. In this process the needs of the liturgy was the single most important, constant, driving force, both for the reproduction of existing books and for the writing of new texts.

Evidence from Ireland

Turning to Ireland, assuming that by the late fifth century Christianity had established a firm foothold, we are faced with some surprises. First, written literature in Ireland appeared with the arrival of Christianity, but for the first time in the West the language of Christian literature, Latin, was not already part of the culture. Therefore, Ireland from the start related to Christian literature in a manner different from the way that literature was received elsewhere (e.g., Gaul or Spain) at the time, and in a way that anticipated the way Christian literature would be received by the Germanic peoples. In Ireland, written literature was primarily in a foreign language that had to be acquired by formal study, and which, as a consequence, required the need to develop the writing of the vernacular language in order to provide books in that language to support the pastoral needs of the new religion. As a result, although Irish was always the spoken language of the island in the early Middle Ages, most of the written materials were in Latin, and much of the literature—whether in Latin or Irish—was linked in one way or another to the Church and its needs. From this perspective we can view most of the materials catalogued in Kenney's *Sources* as "liturgical" or "devotional." Second, Christianity was present in Ireland for at least a century before we have any hard evidence (Patrick's writings aside) of texts written in Ireland (the penitential of Finnian and possibly—if he wrote while in Ireland—Columbanus) or a surviving manuscript (the *Cathach* of Colum Cille, from around 600). And third, even allowing for the fact that liturgical and devotional books never survive in quantity (they are used until no longer fit for use, and an outdated book had no library value unless it came to be regarded as a relic), we have only a few remnants from Ireland. Therefore, any attempt to recreate what the liturgy was like in Ireland at any point before the twelfth century is a far more difficult task than for almost any other area of the Latin West. For instance, only one *libellus missae* (booklet [of the text] of a mass) survives (the *Stowe Missal*—incorrectly categorized as a *missale* [missal] in the nineteenth century—from around 800), despite the fact that it represents a basic liturgical book of which every church would have had several at all times.

It is impossible to characterize the liturgy in Ireland in the early Middle Ages in any way that specifies it from that of the western churches in general. But it is worth noting the following points. First, the earliest books came, presumably, from sub-Roman Britain, and via Britain from Gaul, yet where we can note similarities with texts from elsewhere these seem to show more contacts with Spain than elsewhere. Second, we should expect the survival of earlier forms on the periphery of Europe, given that Ireland is at the end of traveling routes, but this expectation is not fulfilled by the evidence. Liturgical and devotional materials from Ireland seem to evolve broadly in parallel with elsewhere, probably as a result of the movement of clerics who had a professional, and very frequently intense, interest in these matters (e.g., the author of the *Navigatio sancti Brendani*). Indeed, in some cases the earliest extant references for particular liturgical actions come from Ireland (e.g., Muirchú's description of the Easter vigil fire). Third, from our viewpoint, there is a danger of equating the liturgy, or Christian devotional practice, with its supporting written products (or even with those products that have survived). But the Christian cult was essentially the repetition of known activities within a fixed pattern of time (be that the weekly gathering for the Eucharist or the annual gathering at a local saint's well); the books merely supplied the fixed spoken texts that were one element in the overall activity. Moreover, while later liturgical books often supply the "stage directions" (rubrics) for the whole event, this is not true for the early medieval period. Hence, what books have survived have to be read in conjunction with all the other parts of the liturgy that have survived: for example, sacred vessels (e.g., patens and chalices), references to liturgy in canon law, architectural evidence for the size of church buildings, images of liturgy either literary or graphic, other objects (e.g., high crosses) that indicate cult usage, and activities such as pilgrimage or tours with relics at times of pestilence.

Fourth, the post-medieval distinction between devotion and liturgy makes little sense in the period prior to 1100; the former can be seen as an outgrowth from liturgy and as the supporting culture of liturgy. In any study of the religious culture of the period, one can only properly examine a cult artifact—be it a text or an object—by locating it in relation to the main liturgical structures. Fifth, extant devotional books often tell us more incidentally about actual devotions than simply what devotional material they contain. For example, a vernacular litany is a devotional artifact in itself, but saints' lives may indicate how they were used, or describe how the liturgy or devotional practices were perceived. Sixth, the twelfth-century changes in religious life marks a watershed in devotion in Ireland, as elsewhere in Europe; from then, devotion became more distinct from liturgy, and Irish materials are simply regional variants on what is happening elsewhere.

The Range of Liturgical Books

We can classify the books relating to liturgy thus:

1. Books needed for the Eucharist: (a) books containing formularies either (i) for specific days that include variables (e.g., collects) and invariables (e.g., the eucharistic prayer)—these are known as *libelli missae*—or (ii) collections of formularies that cover the liturgical year in whole or part: missals; and (b) lectionaries containing readings for this liturgy—gospel books, as such, may or may not be suitable for use at the liturgy.
2. Books needed for the Office: (a) Psalters; (b) Antiphonaries, which contain additional material for the Office and collects for it; and (c) hymnals, which may also contain other prayers (both in Latin and Irish) and vernacular hymns.
3. Other formularies: (a) liturgies for visiting the sick; (b) books of blessings; (c) texts relating to penance, such as the penitentials; and (d) books containing other prayers that cannot be held in the memory, for example, litanies. Blessings, penitential texts, and other prayers existed in both Latin and Irish.
4. Books relating to the liturgical year: (a) calendars—an essential text for the liturgy; (b) martyrologies, which formed a part of the daily monastic liturgy; and (c) works of computistics for determining the movable feasts (concern with the paschal controversy is a species of liturgical thought).

The Range of Devotional Books

The largest category of devotional books (and unlike the core liturgical texts, these books are found in both languages) is saints' Lives. Their primary purpose was edification and they were read in public, often as part of the liturgy. Next in importance are directions for monastic living in the form of both "rules" and guides (e.g., the *Apgitir Crábaid* [alphabet of devotion] from, possibly, the eighth century). Last, there are allegories that describe ideal Christian communities, most famously the monastic allegory of the *Navigatio sancti Brendani*.

To assess the extent of what survives from the early medieval period one should consult Lapidge and Sharpe's *Bibliography* for the Latin material, and Kenney for the Irish material, using the categories cited above.

THOMAS O'LOUGHLIN

References and Further Reading

Curran, Michael. *The Antiphonary of Bangor and the Early Irish Monastic Liturgy.* Dublin: Irish Academic Press, 1984.

Kenney, James F. *The Sources for the Early History of Ireland: Ecclesiastical—An Introduction and Guide.* New York: Columbia University Press, 1929. Reprint, Dublin: Pádraig Ó Táilliúir, 1979.

Lapidge, Michael, and Richard Sharpe. *A Bibliography of Celtic-Latin Literature 400–1200.* Dublin: Royal Irish Academy, 1985.

O'Loughlin, Thomas. *Celtic Theology: Humanity, World and God in Early Irish Writings.* London: Continuum, 2000.

O'Loughlin, Thomas. "The Praxis and Explanations of Eucharistic Fraction in the Ninth Century: the Insular Evidence." *Jahrbuch für Liturgiewissenschaft* 45 (2003): 1–20.

Palazzo, Eric. *A History of Liturgical Books: From the Beginning to the Thirteenth Century.* Collegeville, Minn.: The Liturgical Press, 1993.

Warren, Frederick Edward. *The Liturgy and Ritual of the Celtic Church.* 2nd ed., with new introduction by Jane Stevenson. Woodbridge: The Boydell Press, 1987.

See also **Biblical and Church Fathers; Conversion to Christianity; Hagiography and Martyrologies; Languages; Metalwork; Moral and Religious Instruction; Reliquaries; Scriptoria**

DIARMAIT MAC CERBAILL

Firm historical information about Diarmait is scarce. The available annals, which may include some near contemporary material, but which were certainly augmented in subsequent centuries, suggest that his career was not a particularly successful one. Diarmait's prominence in Irish history derives primarily from the fact that he was said to be the father of Colmán Már and Áed Sláine, the putative progenitors of the Clann Cholmáin Máir and Síl nÁedo Sláine. These were dynastic groups that rose to prominence in the seventh and eighth centuries, respectively, and were dominant among the southern Uí Néill (based in the Irish midlands), as well as assuming over kingship of the Uí Néill on many occasions. In consequence of his status as a common Uí Néill ancestral figure, Diarmait served as a suitable emblematic figure for later mythmakers to convey particular messages about their own times—messages that frequently dealt with the tensions between church and state. Possibly the earliest writer to adapt Diarmait to his own ends was St. Colum Cille's biographer, Adomnán, writing at Iona around 700, who claimed that Diarmait was *totius Scotiae regnatorem Deo auctore ordinatum* (ordained by God as ruler of all Ireland).

There is no indication as to when, or where, Diarmait was born. The genealogical tradition recounts that Diarmait was a son of Fergus Cerrbél and grandson of Conall Cremthainne, son of Niall Noígiallach. The Banshenchas claims that his mother was Corbach, daughter of Maine of the Laigin. Diarmait was purportedly married to Mugain, of the Eóganachta of Munster, and to Muirenn Máel, of the Partraige of Connacht. The rivalry between these two women is recounted in a tenth-century text *Geinemain Áedha Sláine* (the conception of Áed Sláine). According to the *Geinemain*, Mugain, who had been made barren by God because of her hostility to Muirenn, was blessed by St. Finnian of Moville (presumably an error for Finnian of Clonard), in consequence of which she became pregnant with Áed Sláine. Diarmait is said to have had two additional wives, both from Conmaicne in Connacht: Eithne, who was mother of Colmán Már, and Brea, mother of Colmán Bec. The twelfth-century *Accalam na Senórach* alone mentions another wife of Diarmait: Bé Binn from Scotland.

Diarmait became king of Tara in 545 following the death of Tuathal Máelgarb, said to have been a grandson of Coirpre, son of Niall Noígiallach. According to later legend, shortly before his accession to the kingship, Diarmait was in hiding from Tuathal at Clonmacnoise and assisted St. Ciarán in the building of the first church at that foundation (a tradition that is also recorded on a panel of the early tenth-century *Cross of the Scriptures* at Clonmacnoise). On this occasion, Ciarán predicted that Diarmait would be "king of Ireland" on the following day, and this came to pass when Tuathal Máelgarb was killed that night. This legend presumably owes its origins to a period, before the tenth century, when the relationship between Clonmacnoise and the Clann Cholmáin was especially close. We know that the mid-ninth-century was such a period: Ruaidrí mac Donnchada, who died in 838, was *tánaise abb* (i.e., designated to succeed as abbot) of both Clonmacnoise and Clonard, and was abbot of other, unspecified, churches. He was a son of Donnchad Midi, the Clann Cholmáin over king of the Uí Néill. The tale *Aided Diarmata* (the violent death of Diarmait), reveals a more strained relationship between Clann Cholmáin and Clonmacnoise. It recounts how Ciarán, angry at the fact that Diarmait had slain an enemy of his, one Flann, on lands that he had just given to the Saint, declared that Diarmait would suffer the same triple death as Flann, namely, through wounding, drowning, and burning.

In 558 or 560 (the annals give both dates), Diarmait celebrated the *Feis Temro*. Some of the collections of annals state that this was the last celebration of the *Feis*. The *Feis Temro* was a celebration held once during the reign of a king, and was, according to D.A. Binchy "a pagan fertility rite, with a quasi-divine king at its centre."

Later legend explained the abandonment of Tara as follows. Diarmait had violated the sanctuary given by St. Ruadán of Lorrha to Áed Guaire, the king of Uí Maine in Connacht, when he seized Áed and took him prisoner to Tara. Ruadán, supported by St. Brendan, went to Tara to demand the release of Áed and fasted against the king to force his hand. Ruadán also placed a curse on Tara so that it would henceforth remain deserted. The king, finally recognizing the superior power of the clergy, relented and agreed to release Áed. This tale may date from as late as the eleventh century, when the issue of sanctuary rights bedeviled the relationship between Church and State.

In contrast to the positive light in which Adomnán portrayed Diarmait, the annals recount that a dispute between Diarmait and Colum Cille (the precise nature of which is unclear) was the cause of the battle of Cúil Dreimne (near Benbulben in County Sligo) in 561, when an army comprising the Cenél nEógain, Cenél Conaill, and the Connachta, which, reputedly, had the full support of Colum Cille, routed Diarmait. Diarmait, however, is said to have relied on the support of the Druids.

Diarmait was defeated at Cúil Uinsen in Tethbae (approximately equivalent to modern County Longford) by Áed mac Brennáin, the king of Tethbae, in 562 or 563, and he fled from the battlefield.

Diarmait was killed (allegedly at Mag Line, in County Antrim) in 564 or 571 by Áed Dub mac Suibni, of the Cruithni, who was said to have been his former foster child. The *Aided Diarmata* relates that Diarmait's demise took the form of the "triple death" foretold previously. Áed Dub's responsibility for Diarmait's death may be historically based, as he is condemned by Adomnán for this deed. Some of the annals recount that Diarmait's head was buried at Clonmacnoise, while his body was buried at Connor (Co. Antrim).

PAUL BYRNE

References and Further Reading

Anderson, A. O., and M. O. Anderson, eds. *Adomnan's Life of Columba*. Edinburgh: Nelson, 1961.

Binchy, D. A. "The Fair of Tailtiu and the Feast of Tara." *Ériu* 18 (1958): 113–138.

Byrne, Francis John. *Irish Kings and High-Kings*. London: Batsford, 1973.

Mac Airt, Seán, and Gearóid Mac Niocaill, eds. *The Annals of Ulster (to A.D. 1131)*. Dublin: Dublin Institute for Advanced Studies, 1983.

McCone, Kim. *Pagan Past and Christian Present in Early Irish Literature*. Maynooth: An Sagart, 1990.

O'Grady, Standish H, ed. *Silva Gadelica: A Collection of Tales in Irish*. London: Williams and Norgate, 1892.

Ó hÓgáin, Dáithí. *Myth, Legend & Romance, an Encyclopaedia of the Irish Folk Tradition*. New York: Prentice Hall, 1991.

Ryan, Rev. John. *Clonmacnois: A Historical Summary*. Dublin: Stationery Office, 1973.

See also **Adomnán; Clonmacnoise; Colum Cille; Uí Néill; Uí Neill, Southern**

DIARMAIT MAC MÁELE-NA-MBÓ (Reigned 1036–1072)

Diarmait mac Máele-na-mbó (died February 7, 1072 at Odba, Co. Meath) was king of the southeastern kingdom of Uí Chennselaig and overlord of the province of Leinster (1042–1072). Some medieval historians claimed that he was high king of Ireland, with opposition (1063–1072). Diarmait's father Donnchad mac Diarmata (fl. 1003) was more familiarly known as *Máel na mbó* (the Cattle Rustler) and his mother was Aífe, daughter of Gilla Pátraic mac Donnchada, king of Osraige. Diarmait was one of the most powerful Irish rulers of the eleventh century, as well as the most internationally oriented of his contemporaries. For almost a quarter century he controlled the wealthy southeastern quarter of Ireland, while engaging in adventures in England, the Isle of Man, and Wales. The kingdom of Uí Chennselaig was located northwest of Wexford, and by the eleventh century its capital was at Ferns. Until the early eighth century it had provided kings for the province of Leinster, but had been in obscurity for the following three centuries. All this changed in the course of Diarmait's career. He became king of Uí Chennselaig in 1036, after blinding a rival named Ruaidrí mac Taidc. Diarmait understood that the success of a prince depended very much on the resources he commanded. Toward that end, he began to bring the Viking towns of eastern Ireland under his control. In 1037, Diarmait raided the Viking settlement at Waterford.

Diarmait's uncle Donnchad mac Gilla Pátraic, the king of Osraige, had been recognized as king of Leinster in 1036, but after his death in 1039 his family was unable to maintain their hold on the province. Diarmait began attacks on the new king, Murchad mac Dúnlainge of the northern Leinster dynasty of Uí Muiredaig, and in 1040 he raided the churches of Moone, Castledermot, and Dunmanogue (all in Co. Kildare) in Murchad's territory. His successes emboldened Diarmait to attempt to extend his authority throughout Leinster. With his brother Domnall Remar ("the Fat"), Diarmait raided the neighboring kingdom of Uí Bairrche. The attempt was premature, however, and the brothers were defeated in a battle at Kilmolappogue (Co. Carlow), where Domnall was slain.

These activities alarmed Diarmait's neighbors, especially Donnchad mac Briain, the king of Munster and claimant to the Irish high kingship. He was also the father of Diarmait's wife Derbforgaill (d. 1080), and their son was Murchad (d. 1070). In 1041, Donnchad burned Ferns, and in revenge Diarmait attacked Killeshin (Co. Laois). Murchad, the king of Leinster, was slain in 1042, and sometime thereafter Diarmait was recognized as the provincial king of Leinster. Nevertheless, Donnchad forced Diarmait to surrender hostages (the customary sign of submission) in 1048. Diarmait promptly demonstrated that his power remained undiminished by leading a raid against the Munster kingdom of Déisi and then, in alliance with Niall mac Eochaid of Ulaid, overrunning Mide and destroying its churches. The following year they were raiding Brega.

Shortly after mid-century, there occurred two events that changed Diarmait from a typical Irish prince to a player on the international stage. In September 1051, Diarmait provided refuge for two Anglo-Saxon nobles named Harold and Leofwine Godwinson, who had sailed from Bristol to Ireland. The brothers had been forced to flee England as part of a power struggle between their father and King Edward "the Confessor." They remained in Ireland for nine months before returning to Britain in June 1052, with a fleet supplied by Diarmait. With the successful reinstatement of the Godwinsons in September, Diarmait now had powerful friends in Britain. This led to increased trade between southwest Britain and southeast Ireland. One aspect was the slave trade that would be denounced by Bishop Wulfstan II of Worcester.

An even more important event occurred in 1052, when Diarmait captured the important Viking town of Dublin. The little that is known of this triumph suggests that it was intended to be a mere raid of Fine Gall, but it escalated into the capture of the town after its king, Echmarcach Rögnvaldsson (Mac Ragnaill), fled. Diarmait placed his son Murchad in Dublin. Now he controlled one of the wealthiest towns in the northern Atlantic, which also had one of the most powerful fleets. Using Dublin as a base, Diarmait raided the lands to the north and west. A favored target was Mide, whose rulers, the Southern Uí Néill dynasty of Clann Cholmáin, had been politically ineffective since the first quarter of the eleventh century. In 1053, his army attacked Mide, and would raid it three more times: 1059, 1068, and 1072. Despite his increasing power, he was again forced to submit to Donnchad mac Briain, who raided Fine Gall in 1053.

With wealth from Dublin, and an ally in his nephew Gilla Pátraic, the king of Osraige, Diarmait attacked Donnchad in 1054, raiding Emly and Duntrileague. He sent a naval expedition to raid Scattery Island in 1057. His campaigns received reinforcement in 1058, when Diarmait made an alliance with his foster son, and Donnchad's nephew, Tairrdelbach Ua Briain. The allies raided Limerick, and Diarmait defeated Donnchad in a battle fought at the Galtee Mountains. The next year Diarmait returned to Munster in order to destroy Donnchad's strongholds. Now it was Donnchad's turn to submit to Diarmait. A decisive battle was fought in 1062 at Cleghile (Co. Tipperary), where Diarmait and Tairrdelbach defeated Donnchad's army. At the height of his power, Diarmait used his new position as king-maker to indulge in political assassination in 1066, when he and Tairrdelbach paid the Connacht prince Áed Ua Conchobar thirty ounces of gold to kill a rival. They invaded the province in the next year.

Since 1052, Diarmait had become increasingly involved in adventures outside Ireland, either directly or through assistance to foreign lords. The English nobleman Ælfgar, the earl of Mercia, fled to Ireland, almost certainly to Diarmait, in 1055 and was supplied with a fleet. In 1058, Diarmait allied with the Norwegians on an expedition into western England. This was led by the Norwegian heir-apparent Magnus, son of the Norwegian king Harald "Hard-Counsel." Only troops from Dublin were present for that enterprise, but they were joined at the last minute by Ælfgar and the Welsh prince Gruffudd ap Llywelyn. That venture might have suggested other endeavors to Diarmait. In 1061, Murchad led a fleet to the Isle of Man, where he defeated his father's old foe Echmarcach and collected taxes from the kingdom of the Isles. The Welsh prince Gruffudd ap Llywelyn might have fled to Diarmait's court after he had been forced from Wales in 1064 by Harold Godwinson; within a generation there were circulating stories that he had been slain in Ireland in circumstances of treachery. After his friend Harold Godwinson was defeated and slain in October 1066 at the battle of Hastings, Diarmait's court became one of the centers for the Anglo-Saxon resistance. He gave refuge to Harold's sons and supplied fleets for their raids on England in 1068 and 1069.

Diarmait's final years saw the crumbling of his empire. His sons Murchad and Glúniairn died in 1070; Murchad died on March 21 of a plague that ravaged Dublin. Elderly and without the assistance of his sons, Diarmait could not maintain order even in Leinster. Tairrdelbach Ua Briain was forced to come to his foster-father's aid in 1070 and 1071. One center of unrest was Diarmait's own family. His grandson Domnall (the son of Murchad) and his nephew Donnchad (the son of Domnall Remar) competed for supremacy, and Tairrdelbach was forced to intervene in order to prevent open warfare. In 1071, Diarmait made his final visit to Tairrdelbach, leaving his blessing on Munster. Leinster affairs were quiet enough in 1072 that Diarmait felt able to lead an expedition into Meath. This was his last miscalculation, for in the battle of Odba, fought on February 7, he was defeated and executed by Conchobar Ua Máelshechnaill. In his obituary the so-called *Annals of Tigernach* describe Diarmait as king of the Welsh, the Isles, Dublin, and the southern half of Ireland.

Diarmait's career shows that early Irish kings were not deliberately insular and provincial. He was an ardent student of his opponents' tactics and methods, using Dublin's fleet as astutely as any Viking prince. Diarmait clearly understood how vital were the Viking trading centers for political supremacy in Ireland. His rise to power directly matched the acquisition of economic resources in Dublin. These Viking towns were also international ports, and they provided the avenue for Diarmait into the wider world of Irish Sea power politics through alliances with English, Norwegian, and Welsh princes.

This is not to ignore Diarmait's importance for the revival of his family's fortunes in Leinster and Ireland. After Diarmait, Uí Chennselaig was the dominating force in the province and one of the most powerful families of eastern Ireland. Within Ireland, Diarmait assembled an impressive network of alliances that went from the Ulaid in the north to Osraige in the south, across to Dál Cais and Connacht in the west. Ironically, it was not his grandiose empire building outside Ireland that brought about Diarmait's end, but a typical and petty conflict with his neighbors.

BENJAMIN HUDSON

References and Further Reading

Ó Corráin, Donncha. *Ireland before the Normans*. Dublin: Gill & Macmillan, 1972.

Ó Corráin, Donnchad. "The Career of Diarmait mac Máel na mBó, King of Leinster," *Journal of the Old Wexford Society* 3 (1970–71): 27–35, and 4 (1972–73): 7–24.

Ryan, John. "Pre-Norman Dublin." *Journal of the Royal Society of Antiquaries in Ireland* 79 (1949): 73–88.

Stokes, Whitley. "The Annals of Tigernach: The Fourth Fragment, A.D. 973–1088." *Revue Celtique* 17 (1896): 336–420.

See also **Dublin; Fine Gall; Laigin; Leinster; Leth Cuinn and Leth Moga; Mac Murchada; Marianus Scottus; Ua Briain, Tairrdelbach; Uí Chennselaig**

DICUIL

Dicuil (*c.* 760–post 825) was an Irish scholar-exile at the courts of Charles the Great and Louis the Pious, and an important author of several works on geography, computus, grammar, and astronomy. The only details known of his life are what can be garnered from incidental references in his works. He was teacher at the Palace School of Louis the Pious in about 815. The date of Dicuil's death is not known.

His first work, *Liber de astronomia*, is a verse-computus written between 814 and 816 in four books, to which a fifth book was later added. In 818, he wrote the *Epistula censuum*, a verse treatise on weights and measures. He also made a copy of Priscian's *Partitiones XII Versuum Aeneidos Principalium*, which he summarized in twenty-seven hexameters appended to it. He also wrote an *Epistola de questionibus decim artis grammaticae*, which no longer survives.

In 825, he wrote two treatises, *De prima syllaba*, a tract on prosody, and *Liber de mensura orbis terrae*, a treatise on geography and unquestionably his most important work. Dicuil used a wide range of sources, directly or indirectly, for this treatise. Some of these are now lost or only partly preserved, such as the *Cosmographia* of Julius Caesar in the recension of Julius Honorius, as well as some derivative of the emperor Agrippa's map of the world, probably that known as the *Diuisio* (or *Mensuratio*) *orbis* of emperor Theodosius. Among his other sources are Pliny the Elder, Solinus, Isidore of Seville, and Caelius Sedulius. He had clearly spent some time in the islands north of Britain and Ireland. As he says himself: "Near the island Britannia are many islands, some large, some small, and some medium-sized. Some are in her sea south and some in the sea to her west, but they abound mostly to the north-west and north. Among these I have lived in some, and have visited others; some I have only glimpsed, while others I have read about." (*Liber de mensura orbis terrae* VII § 6)

Dicuil tells us that he was present when a monk, who had returned from a pilgrimage to the Holy Land sometime before 767, was received by Suibne on Iona. This "master Suibne" to whom Dicuil refers was most likely Suibne, abbot of Iona from 767 to 772. Dicuil had acquired his geographical knowledge of the islands around Britain from some time spent as a monk on Iona, and from first-hand oral accounts of the voyages of Irish hermit-monks in the eighth century to the islands north of Britain, to the Orkneys, Shetlands, and Faroes, and to Iceland (*Thule*), where they sojourned from February to August. Some of the islands—perhaps the Faroes—had been occupied by Irishmen "for nearly a hundred years." His description of the eastern Mediterranean, including Egypt and Palestine, is largely derived from written sources, though he also refers to oral information communicated from a traveler to those parts, a "brother Fidelis," from whom he also got one of the earliest descriptions in Western vernacular literature of a Nile crocodile!

Aidan Breen

References and Further Reading

Bermann, Werner. "Dicuils De Mensura Orbis Terrae." In *Science in Western and Eastern Civilization in Carolingian Times*, edited by P. L. Butzer and D. Lohrmann, 527–537. Basel, 1993.

Esposito, M. "An Unpublished Astronomical Treatise by an Irish Monk Dicuil." *R.I.A. Proc.* 26C (1907): 378–446 (with addenda and corrigenda by the editor in *Z.C.P.* 8 (1910): 506–07).

Gautier-Dalché, Patrick. "Tradition et Renouvellement dans la Représentation d'Espace Geographie au IX Siecle." *Studi Medievali* ser. 3, 24.1 (1983): 121–165.

Howlett, D. R. *The Celtic Latin Tradition of Biblical Style*. Dublin, 1995, pp.124–129. idem: "Dicuill on the Islands of the North." *Peritia* 13 (1999): 127–34.

Manitius, M. "Micons v. St-Riquier 'de primis syllabis,'" *Münchener Museum* 1 (1912).

Strecker, K. "Studien zu karolingischen Dichtern: Zu Micons Schrift de prima syllaba," *Neues Archiv* 43 (1920): 477–87.

van de Vyver, A. "Dicuil et Micon," *Revue Belge de Philologie et d'Histoire* 14 (1935): 25–47 (with an edition of *Liber censuum*).

Tierney, J. J. (ed.), with contribution by L. Bieler, *Dicuili Liber de Mensura Orbis Terrae*. SLH 6 (1967).

Smyth, A. *Kings, Warlords and Holymen,* Edinburgh University Press, 1984, 167–69.

Wooding, J. M. "Monastic Voyaging and the *Navigatio*." In Smyth, A., ed., *The Otherworld Voyage in Early Irish Literature. An Anthology of Criticism*. Dublin: Four Courts Press, 2000, pp. 226–245, at p. 238 ff.

For further bibliography, see Lapidge, Michael, and Richard Sharpe. *A Bibliography of Celtic-Latin Literature 400–1200*. Dublin: Royal Irish Academy, 1985, nos. 660–664.

See also **Classical literature, Influence of; *Eachtrai*; Ériugena (John Scottus); Grammatical Treatises; Hiberno-Latin Literature; Poetry, Hiberno-Latin; Sciences; Sedulius Scottus**

DIET AND FOOD

In the Middle Ages, the production of food was a significant aspect of most people's lives, involving endless labor in the sowing and harvesting of crops and the management of cattle, sheep and other animals. It also involved work in the preparation of foods both for immediate consumption and for long-term storage. However, food was also immensely important in social and ideological terms, being used to perform and express identities of social rank, gender, and ethnicity. Food—its production, preparation and exchange—provided the basis of most social and economic relationships between people. It was also the means by which households extended hospitality to kin and strangers, and Simms, Kelly, and O'Sullivan have all discussed the elaborate customs and traditions that evolved around the display, consumption, and use of it (Kelly 1997, 321; Simms 1978; C. M. O'Sullivan, 2004).

Early Medieval Cereals and Vegetables

In the early medieval period (A.D. 400–1200), historical and archaeological evidence indicates that bread and

Wooden churn, Lissue, Co. Antrim. *Photograph reproduced with the kind permission of the Trustees of the National Museums and Galleries of Northern Ireland.*

milk were the basic foodstuffs consumed and that these were supplemented for proteins, minerals, and flavoring by meat, vegetables, and fruit (Lucas 1960; Ó Corráin 1972, 51–61; Kelly 1997, 316–59). Early Irish laws indicate that the range of cereals grown and eaten included oats, barley, wheat, and rye, used for making bread, porridges, cakes, and beer. Different grains were accorded different status, and according to early Irish laws (typically seventh to eighth century A.D.) wheaten bread was a high-status food (Sexton 1998). There is abundant archaeological evidence for drying of cereal grain in corn-drying kilns and the grinding of grain in both domestic rotary querns and horizontal mills. Vegetables for soups were grown in small gardens around the dwelling, and included *cainnenn* (probably onions), celery, and possibly parsnips or carrots, peas, beans and kale. Wild garlic and herbs may also have been gathered in the woods, along with apples (which were grown in orchards), wild berries, and nuts.

Early Medieval Milk and Meats

Between the seventh and the tenth century A.D. (and after), cattle were primarily kept to provide milk and all its products: cream, butter, curds, and cheeses, as well as thickened, soured, and skimmed milk drinks, all referred to in old Irish as *bánbíd* (white foods). As argued by McCormick, faunal analyses of cattle bones from the large middens found on early medieval crannogs such as Moynagh Lough and Lagore (Co. Meath) and Sroove (Co. Sligo) also indicate that cattle herds were carefully managed for dairying (McCormick 1987). Rennet from calves and sheep was used in making cheese, while butter was clearly made in large amounts. Wooden buckets, tubs, and churns recovered from early medieval Crannogs also indicate the preparation and storage of such produce, while tubs of "bog butter" may have been placed in bogs for preservation.

However, meat was also important and evidently eaten by both rich and poor (to judge from the ubiquitous amounts of animal bone found on settlement sites). There is a strong sense, though, that meat was more commonly consumed by the prosperous members of society. Beef was eaten in large amounts, typically being from the unwanted, slaughtered male calves and aged milch cows. Pigs were the source of fresh pork and salt bacon, sausages, and black puddings. Sheep were kept for mutton, lamb meat, and milk. Wild animals that were hunted and trapped (mostly for sport by the nobility) included deer, wild boar, and badger. It is also evident that Ireland's relatively restricted range of freshwater fish species (e.g, salmon, trout, and eels) were caught in fishweirs. In coastal regions, shellfish (limpets, periwinkles, oysters, mussels, cockles, and scallops) were gathered on rocky foreshores, for both food and industrial purposes. The shells were frequently discarded in large middens, perhaps adjacent to unenclosed coastal settlements. Seals and wildfowl may have been occasionally hunted, while stranded porpoises and whales may also have been used when the opportunity arose. Edible seaweeds, such as dulse, were also gathered for food. Some potential foods were regarded as taboo. Therefore, carrion and dog were avoided, while the church banned the eating of horse meat (although there is archaeological evidence for its occasional consumption).

The feast (*fled*) was an important institution in early Irish society, being held, for example, during seasonal festivals or to commemorate a royal inauguration. At an early medieval feast, the distribution of different cuts of meat was probably made on the basis of social rank (McCormick 2002). Early Irish historical sources (e.g., laws, wisdom texts, narrative literature) also suggest that social ranking had a profound influence on the foods that people generally ate, with the nobility eating more meats, honey, onions, and wheat. Wine was also imported by Gaulish and Frankish traders, while more exotic spices and condiments may also have been brought into the island in glass and pottery vessels. If the early Irish diet was balanced and healthy, there were also periods of famine and hunger (particularly at stages in the sixth and seventh centuries), and the occasional long winters would have led to food supplies running out.

Hiberno-Norse Towns

In Hiberno-Norse Dublin in the tenth and eleventh century A.D., archaeological and palaeobotanical evidence (including analysis of fecal fill of cesspits) suggests that the townspeople would have been self-sufficient in some ways, raising pigs and goats and growing their own vegetables within their own properties. The surrounding rural landscape would have been the main source of cattle meat and dairy products, wheat and barley, as well as various gathered fruits, hazelnuts, berries (e.g., sloes, rowan berries, bilberries), and mosses. Marine mollusc shells such as periwinkles and mussels indicate the consumption of foods gathered from the foreshore. According to Geraghty, faunal analyses suggest that cattle found in the town were all steers; no calves were present, suggesting that herds were being specifically driven into the town for slaughtering for beef (Geraghty 1996, 67). Some imported foods included plums, walnuts, and of course, wine. Despite this, there is some skeletal evidence for seasonal shortages of food and malnutrition, while it is likely that the proximity of wells to cesspits led to stomach ailments and intestinal parasites (Geraghty 1996, 68).

Gaelic Irish and Anglo-Norman Diet and Food Traditions

By the later Middle Ages, it is possible that there were regional and cultural variations in diet and food consumption. Oats, dairy produce, salted meats, and animal fats may have been primarily consumed by the Gaelic Irish, while the diet of people in the Anglo-Norman towns and neighboring regions may have been dominated by wheat, meats, fish (particularly salted and smoked herring), and fowl. However, in reality there may have been a more complex ethnic and cultural blending of dietary traditions, with spices, wines, and rich foods being consumed by social elites, while most people ate dairy produce and cereals. Meat consumption appears to have been dominated by cattle, and animals were slaughtered at a mature age when their hides and horns could also be used. On the other hand, sheep, pigs, and goats were also highly important. In archaeological excavations in Hiberno-Norse and later medieval Waterford, massive amounts of sheep bones were uncovered (McCormick 1997). The Anglo-Norman manorial economy also led to the introduction of rabbits into Ireland, and these were probably kept in warrens, while doves were kept in dovecots for an extra delicacy on the table. Fish and shellfish were also consumed. Medieval fishweirs found on Strangford Lough and on the Shannon estuary indicate the catching of salmon, eels, and trout (among other fish) in the twelfth and thirteenth century A.D. (O'Sullivan 2001; McErlean and O'Sullivan 2002).

In the Anglo-Norman manorial economy, tillage and arable crops were a significant aspect of the agricultural organization of the landscape. Cereal crops were threshed, dried in kilns, and brought to water mills for grinding. Processed grain was used for preparation of bread, stews, and pottages, as well as for making alcohol. Ale, rich in calories and vitamins, was brewed professionally and in the home, and was consumed (apparently in large quantities) in both aristocratic and peasant households (O'Keeffe 2000, 68). However, there were also periods of hunger and famine, particularly in the early fourteenth century, when bad weather and warfare combined to wreak havoc on the Irish population.

In the sixteenth century, cattle continued to be of major social and economic importance to Gaelic Irish, particularly in the north and west where a mobile cattle-herding system emerged, well-suited to a time of political instability and warfare. Dairy products such as milk, butter, cheeses, whey, and curds dominated diets. Oats were also of some importance, being used for porridges and for making dry oaten cakes. Cattle were occasionally bled for food, the blood being mixed with butter and meal to make puddings.

AIDAN O'SULLIVAN

References and Further Reading

Geraghty, S. "Viking Dublin: Botanical Evidence from Fishamble Street." National *Museum of Ireland Medieval Dublin Excavations*, 1962–81, Ser. C, vol. 2. Dublin: Royal Irish Academy, 1996.

Kelly, F. *Early Irish Farming*. Dublin: Dublin Institute for Advanced Studies, 1997.

Lucas, A. T. "Irish Food Before the Potato." *Gwerin* 3 (1960): 1–36.

McCormick, F. "Stockrearing in Early Christian Ireland." Unpublished Ph.D. thesis, The Queen's University of Belfast, 1987.

McCormick, F. "The Animal Bones." In *Late Viking Age and Medieval Waterford: Excavations 1986–1992*, edited by M. F. Hurley and O. M. B. Scully, 819–852. Waterford: Waterford Corporation, 1997.

McCormick, F. "The Distribution of Meat in a Hierarchical Society: The Irish Evidence." In *Consuming Passions and Patterns of Consumption*, edited by P. Miracle and N. Miller, 25–32. Cambridge: McDonald Institute for Archaeological Research, 2002.

McErlean, T., and A. O'Sullivan. "Foreshore Tidal Fishtraps." In *Strangford Lough: An Archaeological Survey of Its Maritime Cultural Landscape*, edited by T. McErlean, R. McConkey, and W. Forsythe, 144–180. Belfast: Blackstaff Press, 2002.

Simms, K. "Guesting and Feasting in Gaelic Ireland." *Journal of the Royal Society of Antiquaries of Ireland* 108 (1978): 67–100.

O'Neill, T. *Merchants and Mariners in Medieval Ireland*. Blackrock, Co. Dublin: Irish Academic Press, 1987.

Ó Corráin, D. *Ireland Before the Normans*. Dublin: Gill and McMillan, 1972.

O'Keeffe, T. *Medieval Ireland: An archaeology*. Stroud: Tempus, 2000.

O'Sullivan, A. *Foragers, Farmers and Fishers in a Coastal Landscape: An Intertidal Archaeological Survey of the Shannon Estuary*. Dublin: Discovery Programme Monographs 4, Royal Irish Academy, 2001.

O'Sullivan, C. M. *Hospitality in Medieval Ireland, 900–1500*. Dublin: Four Courts Press, 2004.

Sexton, R. "Porridges, Gruels and Breads: The Cereal Foodstuffs of Early Medieval Ireland." In *Early Medieval Munster: Archaeology, History and Society*, edited by M. A. Monk and J. Sheehan, 76–86. Cork: Cork University Press, 1998.

Sexton, R. *A Little History of Irish Food*. Dublin: Gill and McMillan, 2001.

See also **Agriculture; *Crannóga*; Famine and Hunger; Fishing; Manorialism; Mills and Milling; Trade**

DINNSHENCHAS

The term *dinnshenchas* (lore of prominent or famous places) denotes a popular genre of early Irish literature that purported to explain the origin of well-known Irish placenames. Material coming under this general heading pervades almost all aspects of that literature; it forms a significant part of such great literary works as *Táin Bó Cúalnge* and, even more notably, *Acallam na Senórach*, as well as cropping up in hagiographical texts. In such works, some well-known place name is mentioned and a question posed as to how it got that name. There may also be reference to another name (usually quite fanciful) by which the place was reputedly known prior to the incident said to have given rise to the present toponym. The explanation is usually given in terms of pseudoetymology, or the invention of a suitable eponym. It may take the form of what has been termed an "elaborate legendary anecdote" relating to a fictitious, and often mythological, individual and some imaginative incident in which he or she was reputedly involved (e.g., a river or lake named from a legendary princess said to have drowned there). While such purported "explanations" may be enjoyed as entertaining stories, they rarely if ever shed any worthwhile light on the true origins or meaning of the names, and therefore have little in common with the results of modern scholarly study of placenames.

The collection specifically known as *Dinnshenchas Érenn* (the *dinnshenchas* of Ireland) is a large body of toponymic lore put together in the eleventh and twelfth centuries. It occurs as both poetry and prose in a number of late medieval Irish manuscripts, beginning with the twelfth-century *Book of Leinster*. The relationship between the various recensions is quite complex and has led to a marked diversity of views among leading scholars. The following account is particularly indebted to the work of Tomás Ó Concheanainn, who follows Rudolf Thurneysen in recognizing three recensions.

Recension A, the metrical version, consists of a series of 107 poems. In most cases, a single place is the subject of a single poem, but some places (e.g., Benn Étair, Druim Fíngin, and Mag Femin) may be treated of in more than one poem. This recension occurs uniquely—albeit broken into four sections—in the *Book of Leinster*; between the sections other material is interposed, including elements of Recension B. (The provincial division of the poems is as follows: 40 relating to Leinster, 26 to Connacht, 16 to Munster, 10 each to Meath and Ulster, and 5 uncertain or involving more than one province.)

Recension B, the prose version, comprises about one hundred separate items and survives—albeit incomplete— in three manuscripts: the *Book of Leinster* and two from the sixteenth century—Rawlinson B 506, in the Bodleian Library, Oxford, and Gaelic MS XVI in the National Library of Scotland, Edinburgh. (The latter two can be designated, respectively, Bd and Ed.)

Recension C combines both prose and verse, 176 items in all, with the legend relating to a particular place name given first in prose and then as a poem. The recension occurs in several manuscripts from the fourteenth to the sixteenth centuries, most notably in the books of Ballymote, Lecan, and Uí Mhaine (all of which were either complete or being compiled by the closing years of the fourteenth century), and also in a manuscript preserved in the municipal library in Rennes, Brittany. The prose material from the Rennes manuscript was edited by Whitley Stokes. (The recension occurs in as many as nine other manuscripts, some of them from as late as the eighteenth century, while scattered extracts are found in several additional manuscripts.)

Edward Gwynn—to whom we are indebted for his five-volume edition of the *Metrical Dinnshenchas*—thought in terms of just two recensions, A and B together and C, constituting his "First" and "Second" Recensions, respectively. The traditional scholarly understanding was that the prose and metrical versions were essentially separate, and Gwynn's view was that the prose had been "put together largely by making abstracts of the corresponding poems . . . the prose [being] usually no more than a brief extract from the poem." According to the same scholar, Recension C (his "Second Recension") was the result of a late twelfth-century "Reviser" joining the poems of A to the prose of B, and supplying further prose and verse equivalents as necessary. Tomás Ó Concheanainn, however, takes an entirely different view, seeing Recension C as the earliest one, with B as "an abridged recension made from the prose of C," and A as "an anthology . . . extracted from an early text of C." He also suggests the date 1079 as a superior limit to be assigned to the formation of the extant versions of C, and he identifies the anonymous twelfth-century man of learning whom Gwynn dubbed the "Reviser" as "in fact, the original redactor of *Dinnshenchas Érenn*, though by no means the author of all its components."

While most of the poems in the *Dinnshenchas* are anonymous, a number are ascribed to seven well-known poets of the ninth to eleventh centuries, including Máel-Mura of Othain (Fahan, Co. Donegal), who died in 887, Cormac mac Cuilennáin, learned king-bishop of Cashel (d. 908), Cináed ua hArtacáin (d. 974), Mac Liac (d. 1016), Cúán ua Lóthcháin (d. 1024), Eochaid Eolach ua Céirín (*fl. c.* 1050), and the celebrated Flann Mainistrech (of Monasterboice, Co. Louth), who died in 1056—Cináed being said to have composed no fewer than seven poems, and Cúán four. Nine other poems are said to be the work of another seven poets of whom little appears to be known other than their names, while there are also fictitious attributions to Colum Cille, Finn mac Cumaill, and others. A long poem, "Éire iarthar talman torthig," preserved at the end of a copy of the *Dinnshenchas* in the Book of Uí Mhaine, is attributed to Gilla-na-Náem Ua Duinn, of the monastery of Inis Clothrann on Lough Ree, and is said to have been composed in 1166. It summarizes the *Dinnshenchas* legends of some 97 places—giving one quatrain to each place. Gwynn also compiled a list of 46 early Irish texts, most of them well-known—including *Táin Bó Cúalnge, Fled Bricrenn, Táin Bó Fraích, Tochmarc Étaíne,* the *Vita Tripartita* of St. Patrick, the *Banshenchas,* and others—that he suggests were the source of various articles in the *Dinnsenchas*. All of this reflects the fact that *dinnshenchas* was an essential part of the body of learning that a medieval Irish poet or literary practitioner was expected to master.

Nollaig Ó Muraíle

References and Further Reading

Bowen, Charles. "A Historical Inventory of the *Dindshenchas*." *Studia Celtica* 10/11 (1975–76): 113–37.

Gwynn, Edward, ed. *The Metrical Dindshenchas*. Todd Lecture Series VIII, IX, X, XI, and XII. Dublin: Royal Irish Academy, 1903, 1906, 1913, 1924, 1935. Reprinted by School of Celtic Studies, Dublin Institute for Advanced Studies, 1991.

———. "The Dindshenchas in the Book of Uí Maine." *Ériu* 10 (1926–28): 68–91.

Ó Concheanainn, Tomás. "An Dinnsheanchas Próis." Ph.D. thesis, National University of Ireland, Galway, 1977.

———. "The three forms of *Dinnshenchas Érenn*." *The Journal of Celtic Studies*, vol. III, no. 1 (1981): 88–131.

Ó Murchadha, Diarmuid, and Kevin Murray. "Place-names." In *The Heritage of Ireland: Natural, Man-made and Cultural Heritage: Conservation and Interpretation, Business and Administration*, edited by Neil Buttimer, Colin Rynne, and Helen Guerin, 146–155. Cork: The Collins Press, 2001.

Stokes, Whitley, ed. "The Bodleian Dinnshenchas." Folklore 3 (1892): 467–516.

———, ed. "The Edinburgh Dinnshenchas." *Folklore* 4 (1893): 471–497.

———, ed. "The Prose Tales in the Rennes Dindshenchas." *Revue Celtique* 15 (1894): 272–236, 418–484; 16 (1895): 31–83.

Thurneysen, Rudolf. *Die irische Helden- und Königsage bis zum siebzehnten Jahrhundert*. 36–46. Halle, 1921.

See also **Cináed ua hArtacáin; Cormac mac Cuilennáin; Cúán ua Lothcháin; Etymology; Flann Mainistrech; Lecan, Book of; Leinster, Book of; Máel-Muru; Placenames; Uí Maine, Book of**

DOMNALL MIDI

See **Uí Néill, Southern**

DOWNPATRICK

Downpatrick is situated at the main entry point into the Lecale peninsula in south-east County Down, on a ridge overlooking the Quoile river and marshes. It has had at least four names. Before the eleventh century Dún Lethglaise was most common. Dún dá Lethglas becomes more general after the eleventh century, often used in a secular context, until it gives way in the thirteenth century to simply Dún, Dunum, or Down. The "Patrick" element does not seem to have been added until the seventeenth century. An ecclesiastical settlement existed at the site by the eighth century, and continued into the twelfth century, established on the hill at the western end of the ridge. Excavation in the 1980s showed that what was claimed in the 1950s to be a Bronze Age hillfort was in fact the monastic vallum. From the early eleventh century, coincidental with the introduction of the name Dún dá Lethglas, references in annals link it to the ruling dynasty of Ulaid. These culminate in 1177 when John de Courcy, in his seizure of the kingdom, made straight for Downpatrick and captured it, expelling the king. It has been conjectured that the strong fort at the edge of the marsh on the northern promontory, the so-called English Mount, might be the site of a twelfth-century royal center.

Down, with Connor, was named the site of the see of Ulaid at the Synod of Ráith Bressail in 1111, the first well-known Bishop being the future St. Malachy, elected in 1124. His career at Down was a complex story of secular involvement and hostility. Initially he was more centered on Bangor, but his two successors, also named Malachy, established Down firmly as the diocesan center. As the site of a cathedral and a royal center it also served as the focus for further monastic patronage under the first Malachy, notably that of Erenagh, less than three miles away, founded in 1127, the earliest house in Ireland of a Continental Order (Savigny). Annalistic references also indicate a subsidiary, potentially urban settlement attached.

Bishop Malachy III successfully negotiated the crisis of the conquest by John de Courcy in 1177. The mound within the English Mount enclosure is best explained as an unfinished motte, and there is no evidence of a castle of the earl of Ulster in Down; the bishop was the lord of the town. It developed the characteristics of an English town: trade, with some evidence of use as a port and a mint under John de Courcy; industry, with a pottery kiln of the earlier thirteenth century; and institutions, such as a mayor by 1260. It continued to be strongly ecclesiastical in nature, with a friary and other churches founded around it. How far it was physically enclosed is unclear. In the fifteenth century it was probably overshadowed by the trading success of the port of Ardglass, but remained the administrative head of the area. The century after 1550 treated it harshly, with the burning (and failure to rebuild) of the cathedral and the dissolution of other monasteries.

T. E. McNeill

References and Further Reading

Buchanan, R. H. and A. Wilson. Downpatrick. *Irish Historic Towns Atlas*, no. 8. Dublin: Royal Irish Academy, 1997.

McCorry, M. *The Medieval Pottery Kiln at Downpatrick*. Oxford: British Archaeological Reports (British series), 2001.

McNeill, T. E. *Anglo-Norman Ulster*. Edinburgh: John Donald, 1980.

Wilson, A. M. *Saint Patrick's Town. A History of Downpatrick and the Barony of Lecale*. Belfast: The Isabella Press, 1995.

See also **Courcey, John de; Ecclesiastical Settlements; Ecclesiastical Sites; Patrick; Ulaid; Ulster, Earldom of; Walled Towns**

DOWRY

See **Marriage**

DUANAIRÍ

The earliest reference to a *duanaire* as an anthology of bardic poetry, instead of the older meaning of "songster" or "poet," comes in a note by Find Ua Gormáin, Bishop of Kildare from 1148 to 1160, to the scribe of the *Book of Leinster,* asking for the *duanaire* of Flann mac Lonáin (d. 896, or 918). Scattered texts of poems by Flann still survive, but the anthology the bishop mentioned seems lost. Surviving anthologies by particular poets or groups of poets include the fifteenth-century Ua hUiginn *duanaire*, which was bound up in the collection of fragments styled "the so-called Yellow Book of Lecan," and Trinity College MS 1363, containing poems by a certain Seifín (no surname) and his two sons, and composed in the looser metrical style of *bruilingeacht* associated with less-highly trained professional poets. Both collections may have served as textbooks for the bardic schools, since extracts from the poems found in them are quoted in later medieval grammatical treatises used in the training of professional poets.

However, our earliest surviving *duanaire* is the fourteenth-century *Book of Magauran*, praising the rulers of a barony in northwest County Cavan, with their wives and other relatives. Such anthologies of especially treasured poems composed for a chief, or his family, are the commonest class of *duanaire*. Since these texts were chosen for their personal associations rather than their aesthetic merit, they give a better impression of the range of bardic poetry available to medieval aristocrats, from simple *ógláchas* meters used by half-educated local bards, to the elaborate and expensive *dán díreach* meters of the top practitioners. Such a collection can also show interesting historical developments in a single chief's career. Published aristocratic *duanairí* include the *Book of O'Hara*, the *Leabhar Branach*, the *Duanaire Mhéig Uidhir*, and a number of fragmentary collections edited by James Carney as *Poems on the Butlers* and *Poems on the O'Reillys*. Important unpublished collections include poems on the Roche family, in the fifteenth-century Book of Fermoy, and poems to the seventeenth-century Theobald Viscount Dillon and his descendants in Royal Irish Academy MS 744 (A.v.2).

A further type of *duanaire* consisted of miscellaneous poems collected for their literary merit. One of the earliest surviving miscellanies is the Nugent MS, National Library G 992, produced by the Ua Cobhthaigh school of Meath poets in the late sixteenth century. It contains not only Ua Cobhthaigh poems, but religious and secular works by admired authors from the thirteenth and fourteenth centuries, normally ones cited in the grammatical treatises. By the seventeenth century, Irish aristocrats themselves had become more highly educated. The Nugents acquired the Ua Cobhthaigh volume, and other nobles commissioned similar miscellaneous anthologies for their leisure reading. The largest such collection is the *Book of O'Conor Don*, originally compiled in the Netherlands around 1631 for the exiled Captain Somhairle MacDonnell from Antrim. It is these seventeenth-century miscellaneous anthologies that have preserved for us the bulk of the bardic poetry that survives today.

Katharine Simms

References and Further Reading

McKenna, Lambert, ed. *Aithdioghluim Dána, A Miscellany of Irish Bardic Poetry, Historical and Religious, Including the Historical Poems of the Duanaire in the Yellow Book of Lecan* (2 vols). London: Irish Texts Society, 1939, 1940.

McKenna, Lambert, ed. *The Book of Magauran*. Dublin: Dublin Institute for Advanced Studies, 1947.

McManus, Damian. "The Irish Grammatical and Syntactical Tracts: A Concordance of Duplicated and Identified Citations." *Ériu* 48 (1997): 83–102.

Ó Cuív, Brian. *The Irish Bardic Duanaire or Poem-Book*. Dublin: Dublin Institute for Advanced Studies, 1974.

See also **Bardic Schools, Learned Families; Flann Mac Lonáin; Gaelicization; Gerald, 3rd Earl of Desmond; Grammatical Treatises; Leinster, Book of; Metrics; O'Neill, Clandeboye; Poetry, Irish**

DUBLIN

Origins

As a settlement, Dublin belongs to a common category in medieval Europe, commanding the lowest fordable, and later bridgeable, site on a major river. This attribute is reflected in the earlier of two Irish placename forms—Áth Cliath (ford of hurdle-work)—first recorded reliably in the sixth century C.E. The settlement itself was located on a low east-west ridge overlooking the south bank of the River Liffey, where three long-distance routeways (Irish *slighte*) converged. The proximity of a magnificent bay facing the neighboring island of Britain would have made Áth Cliath a major focal point of communications by land and by sea. Economically, it may have functioned as a trading place as well as an agrarian and fishing community. A second pre-urban nucleus was situated near the tidal "black pool" (Irish *linn duib*) in the River Poddle, a small tributary of the Liffey. Cumulatively there is strong evidence for the existence of an ecclesiastical community of some importance founded no later than the early seventh century. As an inversion compound, Duiblinn gave rise ultimately to the international name of the medieval and modern city. A fourth long-distance overland route coming from the south-west along The Coombe (Irish *com*, "valley") forded the Poddle near the site of St. Patrick's Cathedral and terminated at the southern entrance to the postulated ecclesiastical enclosure. It should be emphasized that this construct of the dual origins of Dublin is essentially hypothetical, but a wide range of documentary and topographical evidence is broadly consistent within itself and in conformity with this interpretation.

The Early Town

A third nucleus of settlement, which came to be known as Dyflinn in Norse speech, evolved by stages after the initial Viking takeover of 841. The annalists refer to a naval encampment (Irish *longphort*) in the ninth century and to a stronghold (Irish *dún*) in the tenth. In between there was a period of enforced exile, at least for the Norse leadership, in the years from 902 to 917. For much of the Viking period Dublin appears to have functioned economically as an emporium, with a strong emphasis on the slave trade. Warrior-merchants conducted occasional raids inland on horseback; their war leaders were regarded as kings, most of whom belonged to the dynasty established by Ívarr the Boneless (d. 873). After the Viking recapture of Dublin in 917 there must have been a good deal of social interaction with the local Irish population in the hinterland of the main settlement. This we know from the house-types—not typically Scandinavian—and from the relatively small number of pagan burials (between eighty and ninety) that have been discovered so far, the majority upstream at Kilmainham and Islandbridge. After decades of political turbulence and economic uncertainty, it appears from both the archaeological and the documentary evidence that the settlement achieved a greater measure of permanency during the long reign of Amlaíb Cuarán (945–980). The first defensive embankments at Wood Quay have been dated to around 950: the *dún* of Áth Cliath had come to represent the town of Dyflinn.

King Amlaíb (Norse *Óláfr*) took an Irish wife, may have understood the Irish language, and is said to have been converted to Christianity. Thereafter a mixed culture—part Irish, part Norse—characterized the Hiberno-Norse inhabitants of Dublin for many decades. Norse culture was represented by two public monuments: the Thingmót, or assembly mound, and the Long Stone, or megalith, marking the taking of Dublin. It was also represented by burial mounds (Norse *haugar*) that presumably denoted the graves of pagan kings, while the archaeological record includes decorated wood, graffiti, ringed pins, runic inscriptions, and models and timbers of ships that are indicative of a strong Scandinavian cultural identity. After around 980, however, the townspeople were drawn politically into the Irish system of kingship; from 1052 onward their overlord was usually an Irish high king or would-be high king.

Increasing social acculturation took on a dramatic spiritual dimension around 1030 with the construction of the first cathedral of the Holy Trinity, commonly called Christ Church. More or less coincidentally, the defended area of the town was doubled to about 12 hectares, and its population in the mid-eleventh century has been estimated at 4,500. A major sign of economic prosperity is the conversion of the earth and timber defenses to stone around the end of that century. By 1112, and probably earlier, there was a permanent bridge across the Liffey. In 1170, there were, besides the cathedral, seven parish churches inside the walls and about the same number outside. There were, in addition, two suburban monasteries and a third on the northern bank of the principal river. Thus the Hiberno-Norse legacy was a fully-developed town in all essentials, except that of chartered status.

High Medieval Expansion

The immediate outcome of the Anglo-Norman takeover of Dublin in 1170 to 1172 was that Dublin acquired a foreign overlord, the king of England, who granted it to the merchants of Bristol in southwestern England. For two decades Dublin was in their charge, and redevelopment was rapid. Examples are the construction of the first castle, probably earthwork, in the southeastern angle of the existing walls; the rebuilding of the western gate as Newgate; and the rerouting of the Poddle in artificial channels along both sides of Patrick Street. Not far away, St. Patrick's was rebuilt as a collegiate church and training center for priests, dedicated on the saint's feast day in 1192. A few weeks later, an independent charter of urban liberties was granted by John, lord of Ireland, after which Dublin developed into a normally loyal, English-orientated city. Its royal status was reinforced by the construction of a powerful, courtyard type of stone castle in the second and third decades of the thirteenth century. Another vast undertaking was a second transformation of St. Patrick's, this time as a cathedral dedicated in 1254. Thus Dublin came to contain both the biggest royal castle and the biggest church in medieval Ireland. Most dramatic of all, however, was a large-scale program of land reclamation from the Liffey. Starting in the last years of the twelfth century, a series of wooden revetments were constructed, behind which various materials were deposited. Around 1260 a stone quay wall was built at Wood Quay, while access to this new land was gained by means of openings in the old Hiberno-Norse wall.

The principal opening was probably King's Gate, in Winetavern Street, just inside which stood the guild hall of the merchants of Dublin. In an upper room of this building the city council held regular meetings and court sessions, reflecting the close linkage between wealth acquired through trade and the political power so characteristic of medieval cities. From 1229 onward the council met under the presidency of a mayor—a rare and much sought-after privilege in this period. The city's main commercial axis extended from Newgate to the pillory southeast of Christ Church Cathedral, and thence northward to the river via Fishamble Street. Part of this street alignment had the expressive name of Bothe Street (modern Christchurch Place), reflecting the practice of erecting booths as market stalls in the roadway itself. Other market spaces were to be found in the suburbs, for by the end of the thirteenth century about three-quarters of the city's population were living outside the defensive walls. Indeed, suburbs of different types extended in all directions. Besides the second cathedral, these populous districts contained parish churches and all but one of the city's religious houses. There were also a number of public spaces, the largest of which were St. Stephen's Green on the south side and Oxmantown Green on the north. Immediately outside the western city wall and ditch lay Fair Green, the venue for international fairs held each summer and lasting for a fortnight. In association with the monks of St. Thomas's Abbey, an elaborate fresh water supply system had been installed by 1245, the main aqueduct following the course of the street markets as far as the pillory. Thus, by the end of the thirteenth century the Anglo-Norman city had far outstripped the Hiberno-Norse town in physical size and in institutional sophistication.

Late Medieval Crises and Stagnation

The history of Dublin in the fourteenth and fifteenth centuries is typical of that of countless European cities and towns: a succession of crises followed by economic stagnation, yet accompanied by the preservation and even elaboration of municipal life. The first disaster of this period was an accidental fire (a common hazard, in practice) in the northern suburb of Oxmantown in 1304, when part of St. Mary's Abbey was burnt as well. Much more serious, however, was the deliberate firing of the western suburb on the mayor's instructions in 1317, in order to deprive the threatening Scottish army, led by Edward and Robert Bruce, of cover for a protracted siege. The fires may have got out of control, for St. Patrick's Cathedral was damaged and other buildings were destroyed. So extensive was the devastation that the citizens sought, and received, financial compensation from the English government. Best known among the disasters, of course, is the plague pandemic (commonly called the Black Death) that reached Dublin in 1348. The late-thirteenth century population of the city has been estimated at 11,000 (some scholars favor a higher figure), so by analogy with other cities of comparable size, it may have fallen to around 6,000 by the end of the century. Another problem seems to have been caused by natural silting of the Liffey; merchants are said to have been avoiding the city directly, preferring to land their goods at Dalkey in particular. A concealed reason for this practice may have been a desire to evade customs payments, for the English crown reacted by obliging traders using the out-ports to pay duty at the same rate as at Dublin itself.

The walled enclosure survived the 1317 crisis intact, but its maintenance would have become a massive burden on the citizens, whose numbers and resources were reduced after the initial outbreak of plague. Indications of the ruinous nature of sections of walling and of the need for rebuilding occur from 1427 onward. In addition to the main defensive circuit, extramural gateways had to be built so as to offer at least minimal protection, especially at night-time. Referred to in 1351 as the king's chief castle in Ireland, Dublin Castle is said to have required major repairs in 1358, yet by 1380 it was close to complete dereliction, to the extent that official meetings could no longer be held, nor records be stored there.

At a time of growing Anglo-Irish insecurity and nervousness, the citizens were increasingly obliged to defend themselves by participating in military expeditions into the mountains to the south. Their victory at Little Bray early in the fifteenth century earned them the gratitude of King Henry IV and the gift of the great civic sword that still survives. As the English colonial grip on Ireland slackened, attempts were made to expel Irish residents in the 1450s unless they had been living in the city for at least twelve years. The depleted suburbs may have been home to significant numbers of people of Irish or part-Irish descent, despite the overwhelmingly English veneer of the municipal records. English influences continued to be strong, however, as we can see in the language of the assembly rolls from the early 1450s and in cultural imports such as the annual Corpus Christi procession.

A Springboard for Modern Times

The 1530s brought the Middle Ages in Dublin to a dramatic close. The revolt of Thomas Fitzgerald (commonly known as Silken Thomas) had important consequences locally as well as nationally. Although the insurgents had cannon and made vigorous assaults on the castle and on Newgate, the constable and the citizens between them conducted a successful defense. As in 1317, compensation was sought from the English crown and, conscious of new technology, it took the practical form of six small cannon—one for each of the main gates—along with a supply of gunpowder. The rebels left their mark on Dublin in a quite different way by murdering a political enemy, John Alen, the archbishop. The task of his successor George Brown, an appointee of King Henry VIII, was to bring elements of religious reform to his archdiocese. Sacred relics were destroyed or dispersed, monasteries and hospitals were closed down, and Christ Church Cathedral was secularized. The definitive ruination of the medieval city had begun. As on other occasions of this kind, however, opportunities arose and were seized upon. Some monastic buildings were converted to other uses; others were demolished and their materials recycled; and the site of one—All Saints' Priory—was awarded to the citizens as a corporate body and was eventually adapted for a college of Dublin University (Trinity College). Much else remained the same, as John Speed's map of 1610 indicates, until the more radical developments of the late-seventeenth century.

H. B. Clarke

References and Further Reading

Bradley, John, ed. *Viking Dublin Exposed: The Wood Quay Saga*. Dublin: The O'Brien Press, 1984.

Clarke, H. B. "Gaelic, Viking and Hiberno-Norse Dublin." In *Dublin through the Ages*, edited by Art Cosgrove, 4–24. Dublin, College Press, 1988.

———, ed. *Medieval Dublin*, 2 vols. Dublin: Four Courts Press, 1990.

———. *Dublin, Part I, to 1610*. Irish Historic Towns Atlas, no. 11. Dublin: Royal Irish Academy, 2002.

———. *Dublin c. 840 to c. 1540: The Medieval Town in the Modern City*, 2nd ed. Dublin: Royal Irish Academy, 2002.

———. *The Four Parts of the City: High Life and Low Life in the Suburbs of Medieval Dublin*. Dublin: Dublin City Public Libraries, 2003.

Clarke, H. B., Sarah Dent, and Ruth Johnson. *Dublinia, the Story of Medieval Dublin*. Dublin: The O'Brien Press, 2002.

Connolly, Philomena, and Geoffrey Martin, eds. *The Dublin Guild Merchant Roll, c. 1190–1265*. Dublin: Dublin Corporation, 1992.

Curriculum Development Unit. *Viking and Medieval Dublin*, rev. ed. Dublin: O'Brien Press Educational, 1988.

Duffy, Seán, ed. *Medieval Dublin I: Proceedings of the Friends of Medieval Dublin Symposium 1999*. Dublin: Four Courts Press, 2000.

———. *Medieval Dublin II: Proceedings of the Friends of Medieval Dublin Symposium 2000*. Dublin: Four Courts Press, 2001.

———. *Medieval Dublin III: Proceedings of the Friends of Medieval Dublin Symposium 2001*. Dublin: Four Courts Press, 2002.

———. *Medieval Dublin IV: Proceedings of the Friends of Medieval Dublin Symposium 2002*. Dublin: Four Courts Press, 2003.

———. *Medieval Dublin V: Proceedings of the Friends of Medieval Dublin Symposium 2003*. Dublin: Four Courts Press, 2004.

Lennon, Colm, and James Murray eds. *The Dublin City Franchise Roll, 1468–1512*. Dublin: Dublin Corporation, 1998.

Lydon, James. "The Medieval City." In *Dublin through the Ages*, edited by Art Cosgrove Dublin: College Press, 1988., pp. 25–45.

Milne, Kenneth, ed. *Christ Church Cathedral, Dublin: A History*. Dublin: Four Courts Press, 2000.

Wallace, P. F. *The Viking-Age Buildings of Dublin* (2 parts). Dublin: Royal Irish Academy, 1992.

See also **Amlaíb Cuarán; Battle of Clontarf; Bruce, Edward; Christ Church Cathedral; Cumin, John; Fine Gall; Henry of London; St. Patrick's Cathedral; Sitriuc Silkenbeard; Walled Towns.**

DÚNS

See **Promotory Forts; Ringforts**

DURROW, BOOK OF

The Book of Durrow (Trinity College Dublin MS 57) contains a Latin copy of the four Gospels, in a version close to the Vulgate text compiled in the fourth century by St. Jerome. The Gospels are preceded by "etymologies," mainly of Hebrew names; "canon tables," or concordances of gospel passages common to two or more of the evangelists; summaries of the gospel narratives, known as *Breves causae*; and *Argumenta*, prefaces characterizing the evangelists. The first gospel text (Matthew) begins on folio 22r. The manuscript contains 248 folios, now measuring 245 × 145 mm, and is written in the script known as Irish majuscule.

The Book of Durrow is perhaps the earliest example of a fully decorated insular gospel book. The decoration is greatly influenced by metalwork motifs. The gospel texts are prefaced with an Evangelist symbol, a carpet page (containing only abstract decoration), and an elaboration of the opening words of the gospel. In St. Jerome's scheme, the Man symbolized Matthew, the Lion symbolized Mark, the Calf stood for Luke, and the Eagle for John. In a reversion to a pre-Vulgate order, John's Gospel in the Book of Durrow was prefaced by the Lion, and Mark's Gospel by the Eagle. There is no extant carpet page for Matthew's Gospel. This may be lost, or it may have been the present folio 3v. There is doubt about the location of several pages, the consequence of the volume having been broken down to single leaves prior to its repair and rebinding in 1954. The present final leaf, folio 248r, a carpet page with "lattice-work" decoration, was at one time placed earlier in the volume, while the symbol of the Man, folio 21v, was formerly the last leaf.

The book's date and place of origin have aroused considerable academic debate. It was probably produced early in the eighth century. It takes its name from the monastery of Durrow, County Offaly, one of several foundations by St. Colum Cille (*c.* 521–597), whose principal house was on Iona. His name occurs in an early inscription on folio 247v, and for a long time the manuscript was regarded as a relic of the founder. In the seventeenth century, the custodian of the manuscript was reported as dipping it into water, and giving the water to sick cattle as a cure. The book is first located with certainty in Durrow in the period between 877 and 916, when Flann Sinna son of Máel Sechnaill I, king of Ireland, placed it in a *cumdach* (shrine). The shrine has been lost since 1689. Damage at the beginning and end of the book and at its edges

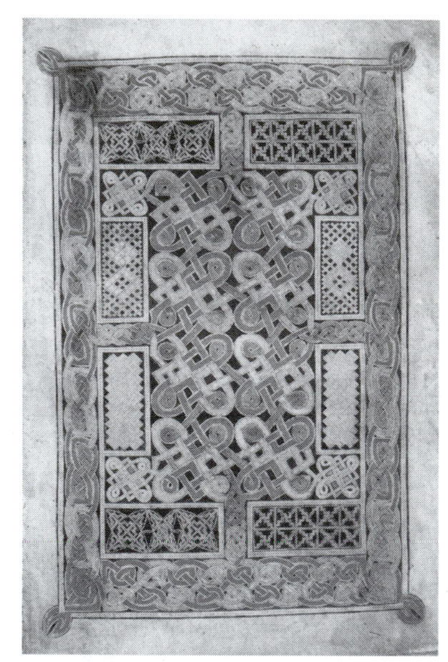

Folio 125v from the Book of Durrow. © *The Board of Trinity College Dublin.*

resulted from the ill-fitting nature of the shrine. Claims have been made that the manuscript originated on Iona, under the sponsorship of the scholarly abbot Adomnán in the period from 682 to 686, or in Northumbria.

After the dissolution of the monastery of Durrow in the mid-sixteenth century, the custody of the book passed to Henry Jones, who presented it, and the Book of Kells, to Trinity College Dublin while he was bishop of Meath between 1661 and 1682.

BERNARD MEEHAN

References and Further Reading

Evangeliorum Quattuor Codex Durmachensis, a facsimile of the Book of Durrow in 2 Volumes. Lausanne: Olten, 1960. With introductory matter, "Editor's Introduction," by A A Luce; "The Paleography of the Book of Durrow," by L. Bieler; "The Art of the Book of Durrow," by P. Meyer; "The Text of Codex Durmachensis Collated With the Text of Codex Amiatinus, Together With an Inventory or Summary Description of the Contents of Each Page of the Manuscript," by G. O. Simms.

Meehan, Bernard. *The Book of Durrow. A Medieval Masterpiece at Trinity College Dublin*. Dublin: Town House, 1996.

Werner, Martin. "The Cross-Carpet Page in the Book of Durrow: The Cult of the True Cross, Adomnan, and Iona." *The Art Bulletin* 72 (1990): 174–223.

Werner, Martin. "The Book of Durrow and the Question of Programme." *Anglo-Saxon England* 26 (Cambridge University Press, 1997): 23–39.

See also **Colum Cille; Early Christian Art; High Crosses; Iconography; Kells, Book of; Máel Sechnaill I; Manuscript Illumination; Metalwork; Scriptoria; Southern Uí Néill**

E

ECHTRAI

Echtrae (expedition, adventure) is an early Irish tale type that tells of the hero's journey to another world. In some of these tales, the hero remains in the Otherworld. One of the earliest of them is *Echtrae Chonnlai* (Connlae's Expedition), which probably dates to the eighth century. Connlae, a son of Conn Cétchathach, is approached by an unnamed woman. She says she has come from "the land of the living, where there is neither death nor sin," and invites Connlae to go with her. He is torn between love of his people and desire for the woman, but he eventually departs with her and is never seen again. In *Echtrae Nerai*, the hero leaves the court of Ailill and Medb at Cruachain (Rathcroghan in County Roscommon) and goes in search of a drink for a hanged man. When he returns, Nerae follows some warriors into the cave at Cruachain and finds himself in a *síd*, or Otherworld abode. He secretly marries a woman there; she warns Nerae that the Connachta must assail the *síd*, for otherwise Cruachain will be destroyed. When the Connachta have accomplished this, Nerae returns to the *síd* and stays there forever.

Sometimes a mortal is summoned by an Otherworld being to secure the defeat of an Otherworld adversary. In *Echtrae Laegairi* (Laegaire's Expedition), Fiachna mac Rétach, king of a *síd*, comes to Connacht and offers gold and silver to anyone who will help him win back his wife from Goll, king of the fort of Mag Mell (The Plain of Delights). Laegaire, son of the king of Connacht, answers the call with fifty men, and he vanquishes Goll in battle. Loegaire marries Fiachna's daughter, Dér Gréine, and remains in the *síd*.

In *Echtrae Fergusa Maic Leiti* (The Expedition of Fergus mac Leiti), the hero's conquest of a monster costs him his life. Fergus mac Leiti, king of Ulster, goes to the seashore. He falls asleep and sprites (*lúchorpáin*, literally little-bodied ones) come and bear him out to sea. Fergus demands from them the power to pass under water. They grant him this but forbid him to go under Loch Rudraige (Dundrum Bay in County Down). In defiance of this prohibition, Fergus dives into Loch Rudraige, and he sees a monster so fearful that his face is disfigured at the sight of it. A king with such a blemish is no longer eligible to rule, but the elders of Ulster contrive to conceal it from the people and even from the king himself. One day, however, Fergus discovers the truth. He goes to Loch Rudraige, dives in, and fights the monster for a day and a night. He emerges triumphant with the head of the monster, and then he dies.

In some stories the hero returns from the Otherworld with treasures that are used for the benefit of his people. In *Echtrae Cormaic maic Airt* (The Expedition of Cormac mac Airt), the king of Tara returns with a cup that enables him to distinguish truth from falsehood.

There is an overtly dynastic message in *Echtrae mac n-Echach Muigmedóin* (The Expedition of the Sons of Echu Mugmedón), a tale in Middle Irish that shows how Niall Noígiallach acquired the kingship of Ireland for himself and his descendants, the Uí Néill. The crucial episode in the tale is a hunting expedition undertaken by Niall and his four half-brothers. Having lost their way, they cook and eat their quarry, and then each of the brothers in turn sets out in search of drinking water. They find a well that is guarded by a hideous hag (*caillech*) who demands a kiss in return for water. Of Niall's brothers, all but one decline to approach the hag; Fiachra gives her a passing kiss but gets no drink in return. Niall not only kisses the hag but also makes love to her. She is immediately transformed into a beautiful young woman who identifies herself as "The Sovereignty of Ireland" and decrees that Niall and his descendants will be kings of Ireland forever. As exceptions to this she mentions two descendants of Fiachra who will receive the kingship as a reward for the kiss that Fiachra gave her. As a piece of propaganda for the Uí Néill, the story of the transformed hag would be most effective if only Niall enjoyed her favor, but it is modified here because its author is constrained by the known tradition that Niall was succeeded as king by Fiachra's son Dath Í and that Dath Í's son Ailill Molt was also king.

Tomás Ó Cathasaigh

References and Further Reading

Carney, James. *Studies in Irish Literature and History.* Dublin: Dublin Institute for Advanced Studies, 1955.

Mac Cana, Proinsias. *The Learned Tales of Medieval Ireland.* Dublin: Dublin Institute for Advanced Studies, 1980.

McCone, Kim. *Echtrae Chonnlai and the Beginnings of Vernacular Narrative Writing in Ireland.* Maynooth: Maynooth Monographs, 2001.

Ó Cathasaigh, Tomás. *The Heroic Biography of Cormac mac Airt.* Dublin: Dublin Institute for Advanced Studies, 1977.

Rees, Alwyn, and Brinley Rees. *Celtic Heritage.* London: Thames and Hudson, 1961.

See also ***Aideds*****; Connachta;** ***Immrama*****; Uí Néill; Ulster Cycle**

EARLY CHRISTIAN ART

Because the term "Early Christian" has been used by archaeologists to denote the period from roughly 400 A.D. to 1169 A.D., confusion has arisen about what constitutes Irish Early Christian Art—the term has been applied indiscriminately to buildings, metalwork, manuscripts, and even weapons of different periods. By transferring the term from Early Christian Europe, where it has a definite meaning, the specifically Christian aspects have been confused and the transformations that took place in Irish society and in the Irish church over a period of almost eight hundred years have been glossed over.

Irish Christian art is richly allusive—partly abstract, partly figurative, and subtly symbolic in style. In its decorative elements it is often playful and sophisticated. Although it is sometimes difficult to see, it reflects contemporary European traditions fairly closely and was not an isolated and impoverished tradition. It was, however, one of the last barbaric arts of Western Europe, and its animal ornament has often misled commentators to confuse its pagan origins with a strong underground paganism when all the patterns had become naturalized in a Christian milieu.

The Missionary Period: Later Fourth to Sixth Centuries A.D.

Irish art at the time of the arrival of the first formal Christian missions was best expressed in metalwork—the art form typically associated with aristocratic, warrior societies. The style seems to have combined a strong provincial Roman influence from Britain with elements of survival of the La Tène style of the Iron Age to create a late flowering of curvilinear abstract ornament that has been termed Ultimate La Tène. We can only speculate on the nature of earliest Christian art, as dating evidence does not support any very early attributions of sculpture with Christian symbols. The missionaries must have brought with them codices, altar vessels, and vestments—all the requirements for the practice of the liturgy—not to mention novel ideas about buildings for public worship and the disposal of the dead. None of this survives; all we have are echoes in later literature. The seventh-century writer, Tírechán, states that square patens and book covers made by St. Patrick's bronzesmith, Assicus, were to be seen in Elphin and Armagh. Armagh preserved relics of SS. Peter and Paul said to have been brought by Patrick himself. A mid-seventh century description of the church of Kildare describes a sophisticated wooden structure with the tombs of its founding saints lying before the altar. Over them hung crowns. The church was also decorated with images, presumably paintings. A penannular brooch from Arthurstown, County Carlow, carries a simple cross motif on its terminals. It is arguably of fifth- or sixth-century date. The appearance of Christian symbols on personal ornaments is part of a widespread Christian practice to afford protection to the wearer.

Simple crosses on pillar stones are still, often without clear justification, dated to the sixth or seventh century, but likely early examples are to be seen at Reask, County Kerry. One Reask pillar has incised scrollwork and a cross of arcs in a circle, while another fragment has a pair of birds flanking a cross, a fundamental Christian motif. A tall inscribed pillar at Kilnasaggart, County Armagh, can be confidently dated, but it is much later. An inscription on it records a certain Ternoc, perhaps he whose death is recorded in 714 or 716 A.D. The pillar is likely to be the earliest firmly dated stone sculpture in Ireland.

The Seventh Century

The seventh century saw a progressive enrichment of the repertoire of Irish craftsmen as Irish missionary activity in northern Britain (begun in the sixth century by St. Colum Cille) and mainland Europe brought Ireland into contact with the wider world. Our earliest surviving manuscripts, the *Cathach* (a psalter associated in legend with St. Colum Cille) and the fragmentary gospel book, *Codex Usserianus Primus,* date to the early seventh century. They already show distinctive traits of script and ornament in embryo but clearly also echo Mediterranean influences, in particular the large-dot-outlined red cross with pendant alpha and omega in the Ussher manuscript and the dolphin-like beast in the Cathach.

At least two ambitious metal reliquaries may be dated to the seventh century—both are made of engraved tinned bronze plates with reserved designs of Ultimate La Tène scrollwork. In addition to a contemporary interest in hagiography, they clearly signal the burgeoning cult of native saints. One is preserved in the Irish monastery of Bobbio, Italy; the other was found at Clonmore, County Armagh. Both are so-called house-shaped shrines. Of the two, the Clonmore shrine is the more highly decorated, with a fine composition that recalls some of the spiral scrollwork of the *Book of Durrow.* The lyre-shaped patterns of the Bobbio shrine are simpler and may have been executed abroad in an Irish idiom, conceivably as early as the early seventh century when St. Columbanus founded his monastery there.

It is with the *Book of Durrow* that we see clearly the emergence of the insular Christian style. The place of origin of the manuscript is disputed, but its broad historical context is well understood. It was produced in one of the monasteries of the Columban family of churches in a milieu where Irish, Anglo-Saxon, and Pictish influences were apparent. With its rich carpet pages of spiral ornament, one page of animals of Germanic inspiration, and its interlace patterns that ultimately hark back to Mediterranean sources, *Durrow* is an eclectic work that heralds the great flowering of the arts in the following century. We cannot be sure of the date of *Durrow.* Opinion has ranged from the early seventh to the late eighth centuries A.D., with the balance favoring a date in the last quarter of that century, but it is by no means a firmly established conjecture.

By the end of the seventh century, a rich polychrome style had grown up in Ireland and parts of northern Britain. Metalworkers probably led the way by adding some of the colorful effects of Germanic jewelery and, above all, the adoption of a radically modified version of Germanic, especially Anglo-Saxon, animal ornament. The summa of this style is to be found in a series of metalwork objects and the *Book of Lindisfarne*, a gospel book that was probably created around the year 700 A.D. at the monastery of that name. The metal objects are primarily from eastern Ireland: the Tara Brooch, the Donore door furniture, and the Hunterston Brooch, which, although found in Ayrshire, is of an Irish type. The style is typified by a brilliant and witty use of animal patterns. The beasts are long bodied, sinuously interlaced, and, while entirely fabulous, are provided with convincing anatomical detail. The gospels are self-evidently Christian and carry portrait pages that clearly derive from Mediterranean prototypes. The Tara and Hunterston Brooches are something new—hybrids of the native penannular form and sumptuous Anglo-Saxon disk brooches. Their ornament appears to have marked Christian overtones recalling the three genera of beasts of Genesis and showing symmetrically opposed beasts in a manner reminiscent of the placement of animals in homage on either side of a cross or chalice in Merovingian and Ravennate sculpture. On the reverse of the Tara Brooch there is a frieze of birds that echoes processions of fowl on the Canon Table arcades of gospel manuscripts. The brooches are clear expressions of the pervasiveness of Christian ideas. The Donore door fittings reflect the style of the Tara Brooch closely. Included in the find is a lion-head door handle that clearly belongs in an ecclesiastical context and is an Irish interpretation of an antique form.

The Eighth and Ninth Centuries

The emergence by about 700 A.D. of the polychrome style set the tone for Irish metalwork and manuscript painting for the next 150 years or so and also provided the stone sculptor with much of his decorative repertoire. The great surviving achievements of that period are the remarkable altar vessels from Ardagh, County Limerick, and Derrynaflan, County Tipperary, a great array of shrines and reliquaries, and some manuscripts of undoubted Irish origin.

The silver chalice from Ardagh and the paten from Derrynaflan have much in common; they both, like the Tara Brooch, are examples of the high-polychrome metalwork style. The Ardagh Chalice largely conceals its Christian symbolism, but it carries two medallions on its bowl that contain prominent crosses of arcs. It also bears the names of the Apostles in fine incised lines in a sea of dot punctulations. The Derrynaflan Paten carries scenes of Christian import in filigree, including one—a stag and snakes—recalling a tale in the *Physiologus*. The Derrynaflan Chalice is less colorfully ornamented, but its filigree is of great interest because it shows elements of common Christian iconography—griffons, birds, beasts, and quadrupeds, probably lions—that are widespread in early medieval European sculpture and metalwork as part of the Tree of Life and related motifs. These patterns are not mindless "ornament" but are carefully contrived symbols appropriate to the vessels that carried the elements of the Eucharist.

The house-shaped shrine that made its appearance in the seventh century is the most numerous of the surviving pieces. Complete shrines from Ireland, Scotland, Italy, and Norway survive, along with many fragments. It is likely that all of these shrines contained corporeal relics.

The greatest manuscript of the period, *The Book of Kells*, was, arguably, created on Iona and only brought to Ireland in the tenth century. It nevertheless captures in full the essence of the style in its sophisticated use of animal symbolism, interlace, curvilinear ornament and portraiture, and illustration. The provenance and date of the manuscript are hotly contested which, if nothing else, indicates the close community of culture that existed between Ireland and the peoples of what is now Scotland. Smaller manuscripts—the so-called pocket gospel books—were produced in Ireland. The Stowe Missal—a copy of the Gospel of St. John bound together with a service of the mass and a commentary on its spiritual meaning—was made about the year 800 A.D. Altogether simpler than the high style of painting, it has fine script and workmanlike ornament. It was considered a venerable relic and was later enshrined. The *Book of Armagh* (early ninth century) is remarkable for its elegant script and line drawings.

One characteristic Irish phenomenon is the book-shrine—a box-shaped reliquary, sometimes of metal-covered wood but also sometimes of entirely metal construction—that seems to have derived from the practice of the early Roman liturgy of keeping the book of mass readings in a sealed box. The earliest surviving example is the one found in dismantled condition near a crannog in Lough Kinale that bore a bronze cross on its front. Fine examples that were preserved by hereditary keepers include the Shrine of the Cathach made in the later eleventh century; the Shrine of the Stowe Missal made in the later eleventh century; and the Soiscél Molaise, restored around 1000 A.D., which bears a fine motif of a cross and evangelist symbols.

An intriguing reliquary is the belt-shrine found in a bog at Moylough, County Sligo. Dating to the eighth century, it contains the leather belt preserved in its hinged segments. It carries a large imitation buckle and counterplate in a style reminiscent of the large buckles of seventh-century burials in Burgundy. Other applied ornaments of glass, enamel, and stamped silver mimic the belt stiffeners of the prototypical continental belts. Could this have been the belt of an Irish holy person who had traveled abroad, or could it have been made to enshrine the belt of someone from the continent who had come to Ireland and who had been venerated as holy?

The appearance of freestanding high crosses is somewhat mysterious. Older theories proposed a very gradual development of the form, largely in isolation over a period of a couple of centuries. The evidence of inscriptions and a newer approach to understanding regional relationships all suggest that the phenomenon of the high cross developed rather rapidly during the ninth century and that massive external influences account for much of the iconography and perhaps even for the form of the wheeled cross itself. The native contribution is to be seen in the ornamental detail where the familiar themes of spiral and trumpet scroll, interlace, animal interlace, and other patterns already well established in metalwork make their appearance. The Irish sculptors owed a great debt to contemporary Italian carvings and to inspiration in other media. The figurative scenes that dominate many ninth- and tenth-century crosses have their roots in early Christian funerary sculpture, manuscript, wall paintings, and ivory and even wooden objects. Some crosses covered in ornament derived from metalwork prototypes seem to be local transpositions to a native idiom of the idea of the *crux gemmata* (the jeweled cross), the triumphant instrument of the Redemption that was erected on Golgotha and remembered in later Christian art such as the great apse mosaic in St. Apollinare in Classe, Ravenna. It is one of the cherished myths of Irish Christian art that the development of ambitious sculpture in immoveable stone represents a response by Irish churches tired of losing their portable treasures to the Viking onslaught. On the contrary, the creation of these great works of art was a sign of confidence on the part of patrons and artists, and the sculpture is often associated with powerful and wealthy religious foundations or their dependencies.

Carved stone grave markers gradually increase in sophistication during the eighth and ninth centuries with elaborate cross-forms and ornament. A particularly fine series extending to the twelfth century is preserved at Clonmacnoise, County Offaly. An elegant, probably eighth century, slab at Tullylease, County Cork, with a large cross similar to an example in the Lindisfarne Gospels calls for a prayer for Berechtuin, probably the founder of the monastery. These are all simple monuments with essentially two-dimensional sculpture. At Fahan, County Donegal, is a massive upright cross-slab with a gabled top and a high-relief cross. It has a Greek doxology inscribed on it. The cross-slab is broadly analogous to similar monuments in Pictland and an eighth or ninth century date for it seems plausible.

Later Developments

In the eleventh and twelfth centuries, Irish art underwent a revival when important reliquaries were created (the Shrines of the Cathach, Stowe Missal, St. Patrick's Bell, and St. Manchan's Shrine) or repaired and other new works of importance—crosiers and objects such as the Cross of Cong—were created. These new pieces hark back to earlier styles in what appears to have been a conscious attempt at revival of old glories, but they also incorporate the Scandinavian animal ornaments. In sculpture, a new style of high cross appears with a large relief figure of Christ on the front. Decoration in the Romanesque style is carved on the new churches of stone that appear on many sites. The revival associated with the reform of the Irish church and the emergence of a new and formidable type of kingship was short-lived—it lingered only briefly after the Anglo-Norman Invasion.

MICHAEL RYAN

References and Further Reading

Edwards, Nancy. *The Archaeology of Early Medieval Ireland.* London: Batsford, 1990.

Harbison, Peter. *Pilgrimage in Ireland.* London: Barrie & Jenkins, 1991.

———. *The Golden Age of Irish Art: The Medieval Achievement 600–1200.* London: Thames and Hudson, 1999.

Henry, Francoise. *Irish Art in the Early Christian Period to* A.D. *800.* London: Methuen, 1965.

———. *Irish Art during the Viking Invasions 800–1020* A.D. London: Methuen, 1967.

———. *Irish Art in the Romanesque Period 1020–1200.* London: Methuen, 1970.

Hourihane, Colum, ed. *From Ireland Coming: Irish Art from the Early Christian to the Late Gothic Period and its European Context.* Princeton: Princeton University Press, 2001.

Neuman de Vegvar, C. "Romanitas and Realpolitik in Cogitosus's Description of the Church of St. Brigit, Kildare." In *The Cross Goes North,* edited by Martin Carver. York: The Boydell Press, 2002.

Ryan, Michael. *Studies in Medieval Irish Metalwork.* London: Pindar Press, 2000.

Ryan, Michael, ed. *Treasures of Ireland.* Dublin: Royal Irish Academy, 1983.

Youngs, S., ed. *The Work of Angels: Masterpieces of Celtic Metalwork 6th-9th Centuries* A.D. London: British Museum, 1989.

See also **Armagh, Book of; Durrow, Book of; Early Christian Sites; High Crosses; Iconography; Inscriptions; Jewelery and Personal Ornament; Kells, Book of; Manuscript Illumination; Metalwork; Reliquaries; Scuplture**

ECCLESIASTICAL ORGANIZATION

The Early Middle Ages

Christianity had reached Ireland by the beginning of the fifth century. We may infer from linguistic evidence that the missionaries of the new faith came mainly from the Romano-British Church, rather than Gaul (a notable quantity of Irish loanwords derive from a British dialect of Latin). By 431, the number of Christians in Ireland warranted episcopal oversight, and Pope Celestine I commissioned a deacon named Palladius to be first bishop "to the Irish believing in Christ." The other (and by the seventh century) more famous fifth-century apostle of the Irish, Patrick, came from Britain; he is a less shadowy figure than Palladius, having left us some of his own writings. Although he does not provide a very clear picture of the organization of the Irish Church, Patrick does tell us that, as an Irish bishop, he was responsible to a synod of British bishops. Therefore Ireland had no metropolitan bishop.

In the Gaul of Palladius's time or the Britain of Patrick's, a bishop presided over a *ciuitas* (or city), and a metropolitan bishop over a province; in Ireland, however, there were no such cities or provinces. Because Ireland had not been part of the Empire, it lacked the imperial administrative framework upon which ecclesiastical organization in the Latin West was based. However, evidence for the earliest phase of ecclesiastical development suggests that the principles by which the Church was organized in the imperial provinces were nevertheless adapted to Ireland's peculiar circumstances. Our most important source for the period after the conversion is a collection of early Irish canons called the *Synod of the Bishops* (*Synodus episcoporum*) that may date from the late fifth century. This text shows a bishop had authority over a *plebs* or *parochia*, that the clergy were subject to him, and that no one could perform any function within the *plebs* without his permission. The word *plebs* here represents the Irish *túath* (or small kingdom); the Latin and Irish words both mean "people." This, for ecclesiastical purposes, was the counterpart to the *ciuitas* in the Empire; the *túath* was, therefore, the earliest and fundamental unit of episcopal government in Ireland, once the Church had become widely established.

In the middle of the sixth century, many of the great ecclesiastical centers of the Irish Church were founded: Bangor, Clonard, Clonmacnoise, and Iona are among the most famous. These principal houses had daughter foundations that remained attached to the mother church with varying degrees of closeness. The practice in Iona, and probably the other large communities, was for the abbot to appoint *praepositi* (or priors) to supervise the daughter houses; the abbot would also make visitations to the subordinate houses. In the first half of the seventh century, a papal letter on the paschal question shows that bishops were now at least sharing power with the heads of these greater foundations.

The principal ecclesiastical settlements in early medieval Ireland were known as *ciuitates* and were composed of more than monks. They included both those whose vocation was monastic prayer and those on behalf of whom the monks proper offered their prayers; a professional military element might also be included within the wider community. The largest of the *ciuitates* began to take on the appearance of urban centers, with dense populations, a variety of crafts, and a delimited boundary with special legal status. Thus, places such as Armagh, Clonmacnoise, Cork, and Kildare became towns under the jurisdiction of the head of the church, the erenagh/*princeps*, who was often not necessarily in even minor orders.

The concept of the erenagh (Old Irish *airchinnech*, meaning chief or head), who was the governor of temporalities, is an extraordinary feature of the Irish Church; we find him described either explicitly as erenagh/*princeps*, or sometimes as abbot, but performing the same function as the erenagh/*princeps*. The Irish ecclesiastical *princeps* is attested in the *Synod of the Bishops* and is, therefore, an early feature of Church organization rather than the result of degeneracy as lay ecclesiastical rulers were, say, in Carolingian churches. Many of the churches, with their estates that were ruled by an erenagh/*princeps*, were controlled by ecclesiastical dynasties that acted like secular magnates and, in some cases, were minor branches of the secular ruling dynasties.

The most important churches, however, were still defined by their episcopal status, and bishops remained at the center of ecclesiastical organization; bishops feature prominently in the Irish annals. In fact, from the seventh century to the tenth, the evidence seems to point to a large degree of continuity in Irish ecclesiastical organization. We see not a Church dominated by abbots, as was once thought, but the authority of bishops, abbots, and *coarbs* existing side by side in an apparently complicated ecclesiastical structure. We find that the three types of authority—that of bishop, abbot, and *coarb*—might be exercised by one person alone, by separate individuals, or be combined in different permutations. So, for example, we have Bishop Crunnmáel, abbot of Cell Mór Enir (*Annals of Ulster*, 770.12). The *coarb* (Old Irish *comarba*; Latin *heres*), or heir/successor of the founding saint, is found as early as the seventh century in the *Liber Angeli* in relation to Patrick. Indeed, Irish churchmen sometimes described the pope as "*coarb* of Peter." Therefore, from the earliest period for which we can discern Irish Church organization in any detail, the coarbial aspect is part of the structure of authority, together with the episcopal and the abbatial. This coexistence and combination of offices seems to be the most distinctive feature of early Irish ecclesiastical authority.

The great churches—the cult centers of saints—might have had both a *paruchia* and a *familia*. By the middle of the twentieth century, scholarship had come to understand the *paruchia* as a group of daughter monasteries controlled by the abbot of the mother house in an ecclesiastical structure dominated by monastic government. The links between the mother house and subordinate churches might extend beyond the boundaries of any one *túath*, and the authority of the abbots of the greater monasteries could thus overshadow that of the local bishops. However, the earlier sense of *paruchia* (or *parochia* as it was written in Britain and Gaul) as the territory subject to, but distinct from, the episcopal church is closer to the sense that seems to be understood in Irish canon law. Thus, scholarship is now tending to view the *paruchia* as a bishop's zone of pastoral jurisdiction, in principle territorially cohesive. However, the person who presided over the *paruchia* need not always have been a bishop and may have been a nonclerical *princeps*/erenagh, assisted by clerical ministers. The term *paruchia* can be used to mean both a basic sphere of jurisdiction (the *paruchia* of a particular church) and an extended one, comprising smaller units, some of which might be contiguous and others not (such as the *paruchia* of Armagh).

The *familia* of a saint comprised not only the people who belonged to his principal church but also those of the dependent churches and dependent kindreds. In this respect, the term corresponds to the restricted and extended senses in which *paruchia* is used. The occurrence in the sources of the terminology of *plebs* and *túath*, in connection with ecclesiastical jurisdiction, also emphasizes that the fundamental aspects of ecclesiastical jurisdiction were territory and community.

Latin and vernacular prescriptive texts—the vernacular laws, the *Collectio Canonum Hibernensis*, and the *Ríagail Phátraic* (Rule of Patrick) —and hagiography all bear witness to an ecclesiastical hierarchy in the early eighth and ninth centuries, which is in principle episcopal with a parallel between superior episcopal jurisdiction and over kingship. The concept of hierarchy is adapted to that of nonclerical church rulers, with the rank of an archbishop or metropolitan credited to the nonclerical governor of a church of great eminence.

Entries in the Irish annals seem to refer to bishops who enjoyed superior jurisdiction over spheres greater than a basic episcopal diocese (e.g., Óengus of the Ulstermen, who died in 665), and an eighth-century legal tract (*Uraicecht Becc*) refers to "a supreme noble bishop" who is equal in status to the king of a whole province such as Munster. Thus, territorial bishoprics and an episcopal hierarchy were realities in the early medieval period, but spheres of jurisdiction were unstable and probably altered in response to ecclesiastical and political change.

The Nature of Irish Monasticism

The appearance of the monastic life among Patrick's converts was the culmination of his mission: Patrick considered celibacy to be the highest form of religious life. As well as the usual male and female celibates, there were also widows and married people who had taken a vow of sexual abstinence; he was particularly concerned with the monastic vocation of women. Patrick's ambitions for the celibate life might make the emergence of great monasteries in sixth-century Ireland easier to understand. Palladius, too, had monastic connections.

Patrick perceived the monastic vocation to be one that may be lived in the world, outside a monastic enclosure, and this unorganized approach to monasticism may explain some of the odd features of the Irish Church, such as the wide extension of monastic vocabulary, which makes it difficult to distinguish religious houses from secular churches or the head of a monastery from the head of any other independent church. In the seventh century, Tírechán talks frequently of "monks of Patrick," many of whom were female. The smaller churches to which they belonged were often nunneries combined with a male pastoral clergy. In some cases, these nunneries could be episcopal churches, too, so Patrick's nuns could provide the bases from which bishops worked. The cenobitic type of monasticism, which formed the heart of all these communities, is something that is attested throughout the early Middle Ages.

The other type of monasticism, that of the hermit, took two forms. There was the unregulated or solitary holy man, usually poorly educated, who tended to be viewed with disapproval; his counterpart was the authorized anchorite, a well-educated person, conventually trained, who was highly regarded. This second variety usually lived with others who were devoted to high levels of mortification, within or near an ecclesiastical community. Their dwelling was known as a *dísert* (literally, a desert); by the ninth century, this term was being applied to enclosures of female religious as well as to some prominent churches. There was often interchange between the two communities; the cenobitic monks might spend temporary periods in the eremitic life, and the anchorite was sometimes called to take on ecclesiastical office.

Church Reform

Until the twelfth century, the Irish Church was so organized that it lacked the jurisdiction of a metropolitan archbishop; moreover, it had failed to embrace developments that in the rest of Europe had seen a separation between pastoral and monastic churches two centuries earlier. In the ninth and tenth centuries, major churches, as repositories of wealth and property, were the natural targets for Viking attacks but suffered no permanent damage. The establishment of Viking settlements, especially Dublin, provided an opportunity for change. In the eleventh century, the Scandinavian settlers became Christian, and their churches sought links with the English Church. This drew Ireland to the attention of the reforming archbishops of Canterbury, Lanfranc and Anselm. The first bishoprics to follow the Roman model—a compact territorial diocese ruled from an episcopal see—appeared in the Scandinavian kingdoms on the coast: Dublin first, in 1074; Waterford in 1096; and Limerick sometime later. The influence of Canterbury on the reform of the Irish Church was marked. Lanfranc certainly considered the Irish Church to be subordinate to the English; Goscelin of Canterbury presented St. Augustine as the primate of England, Scotland, and Ireland in his hagiographical works; and four successive bishops of Dublin were consecrated by archbishops of Canterbury, as was Malchus (a monk of Gloucester Abbey) as bishop of Waterford.

The main problems for the reformers were that the ecclesiastical rulers—the erenaghs—had too much power, bishops too little, and the Church was not organized into territorial dioceses along the Roman model. Three national synods were held, at Cashel (1101), Ráith Bressail (1111), and Kells-Mellifont (1152), which established diocesan organization and absorbed Dublin, the first of the reformed territorial bishoprics, into a national Church under the primacy of Armagh.

The eventual consequence of Anglo-Norman invasion in 1169, and the subsequent colonization of eastern Ireland, was the division of the Church into English and Irish factions. This division was effected by the Second Synod of Cashel (1172), the first ecclesiastical council to be controlled by the English. Not only did it promulgate reforming decrees concerning such matters as the payment of tithes, freedom of the Church from lay control, and clerical privileges but it also resolved that the Church in Ireland should adopt the practices of the English Church in all matters. From now on the system of ecclesiastical appointments, ecclesiastical courts, clerical privilege, and so on, would obtain in Ireland as they did in England. The diocesan structure that was created in the twelfth century survived largely unaltered through the Middle Ages. In the cathedrals, the old monastic chapters became canonical chapters and, in general, the Irish Church more closely resembled the Church in England and continental Europe. Many of the older churches, however, came to lose their status, and though most adopted the rule of St. Augustine, the thirteenth century saw a rapid decline, as hereditary coarbs and erenaghs now lived off estates that had once supported great churches. The new dioceses, inadequately endowed with assets taken from churches, did little better, and the once-great institutions of Clonmacnoise and Glendalough were too poor to survive as episcopal sees. The reformers' moral program, the imposition of clerical celibacy, and the enforcement of canonical marriage also largely failed.

During the thirteenth century, the dioceses were subdivided into parishes; this process occurred more extensively in the English colony, where arrangements for the support of parish clergy differed from those that continued to exist in Gaelic Ireland. Factional considerations influenced episcopal nominations, so that the bishops of Ireland were divided along lines of nationality. There were attempts in the thirteenth century, which were opposed by the papacy, to exclude Irishmen from the episcopate. The provinces of Armagh and Tuam were governed almost exclusively by Irish bishops, while Dublin was the preserve of

Englishmen, and Cashel had a mixture of Irish and Englishmen. Another divisive issue concerned the primacy of Armagh; Dublin was ultimately successful in withholding its recognition of Armagh's primacy, while Cashel unsuccessfully attempted a similar policy.

An accompanying feature of Church reform was enthusiasm for the new religious orders, and the Cistercians flourished in the twelfth century under the guidance of St. Malachy, who is the most important figure in the reorganization of Irish monasticism. By the fourteenth and fifteenth centuries, however, the Cistercian order was largely decadent in Ireland. In contrast, the mendicant orders came to thrive and the Augustinians remained active. Such religious communities were unaffected by the dissolution of the monasteries under Henry VIII, and this circumstance provided an environment in which Roman Catholicism was able to survive the Reformation in the West of Ireland.

John Reuben Davies

References and Further Reading

Charles-Edwards, T. M. *Early Christian Ireland.* New York: Cambridge University Press, 2000.

Corish, Patrick J., ed. *The Christian Mission (A History of Irish Catholicism, Vol 1:3).* Dublin: Gill and Macmillan, 1972.

Etchingham, Colmán. *Church Organisation in Ireland A.D. 650 to 1000.* Maynooth: Laigin Publications, 1999.

Gwynn, Aubrey. In *The Irish Church in the Eleventh and Twelfth Centuries,* edited by Gerard O'Brien. Dublin: Four Courts Press, 1992.

Hughes, Kathleen. *The Church in Early Irish Society.* London: Methuen, 1966.

Kenney, James F. *Sources for the Early History of Ireland:* Vol 1. *Ecclesiastical.* New York: Columbia University Press, 1929.

Mooney, Canice. *The Church in Gaelic Ireland: Thirteenth to Fifteenth Centuries.* Dublin: Gill and Macmillan, 1969.

Reeves, William. *Ecclesiastical Antiquities of Down, Connor and Dromore.* Dublin: Hodges and Smith, 1847.

Sharpe, Richard. "Churches and Communities in Early Medieval Ireland." In *Pastoral Care before the Parish,* edited by John Blair and Richard Sharpe. Leicester: Leicester University Press, 1992.

———. "Some Problems Concerning the Organization of the Church in Early Medieval Ireland'. *Peritia: Journal of the Medieval Academy of Ireland* 3 (1984): 230–270.

Sheehy, M. P. *When the Normans Came to Ireland.* Cork: Mercier Press, 1975.

Watt, J. A. *The Church and the Two Nations in Medieval Ireland.* Cambridge: Cambridge University Press, 1970.

———. *The Church in Medieval Ireland (The Gill History of Ireland, Vol 5).* Dublin: Gill and Macmillan, 1972.

See also **Canon Law; Church Reform, Twelfth Century; Christianity, Conversion to; Ecclesiastical Settlements; Palladius; Patrick; Racial and Cultural Conflict; Religious Orders; Túath**

ECCLESIASTICAL SETTLEMENTS

In early medieval Ireland, larger ecclesiastical settlements were the main centers of population and it is a subject of debate whether the more important of these should or should not be classed as towns. In terms of scale alone, settlements such as Armagh, Clonmacnoise and Kells, County Meath, were certainly large enough by the eleventh/twelfth centuries to qualify as towns, even if they served a different primary function.

Calling many of these sites monasteries gives the modern reader a wrong impression of a single-sex community entirely devoted to religion and subsistence farming and living within an enclosure that has only monastic buildings. Many of these sites did have monks and nuns and the buildings associated with them, but there was generally also a large dependent community with ecclesiastical families, servants, monastic tenants, craftsmen and other workers, traders, and so forth. Large examples of these sites could be seen as monastic towns with the principal ecclesiastical elements at the core. Also, many of these sites did not have any monastic community and were proprietorial or dependent churches, but they could still have an associated settlement.

Historical Evidence

Various buildings associated with the ecclesiastical core such as the *tech mór* (great house), *proindtech* (refectory), *cucann* (kitchen), the abbot's house, and the guest house or enclosure are mentioned in historical sources. Because all of these buildings as well as the huts or cells of the monks were built of perishable materials, they have not survived. Even when churches were commonly built of mortared stone in the eleventh and twelfth centuries, the other buildings continued to be built of timber or post and wattle. On the west coast and islands, some smaller ecclesiastical sites and eremitical monasteries were built of dry stone from about the ninth century (mostly they were built of timber in the earlier period), and remains of cells and other structures survive as well as the church or oratory. However, the interpretation of the buildings is difficult, even when excavated, and the validity of comparing these small sites with the large ecclesiastical sites is open to question. The gate of the enclosure, sometimes called *doras na cille*, is mentioned on occasion, but only one gate structure survives: the masonry gatehouse at Glendalough, with its two arches and antae (see Ecclesiastical Sites). The area to the west of the main church appears to have been the *platea*, or main open space and gathering place within the ecclesiastical core. The main high crosses and round tower are usually within or on the edge of this area, and patterns have been discerned in the layout of these features along with the shrine or small church built over the saint's grave. There is a need for research excavation at some suitable site to throw light on the layout of the timber buildings apart from churches in the ecclesiastical core.

Incidental references to ecclesiastical settlements give some indication of their size and layout and the types of people living in the areas around the ecclesiastical core. A story told in the *Fragmentary Irish Annals* relating to the year 909 describes a king of Leinster entering Kildare on horseback eastward along the "street of the stone steps" when a comb maker, setting out his antlers at his workshop, caused the horse to rear up, resulting in the king being thrown backward onto his own spear, which was being carried by his servant. Armagh had distinct sectors, each called a *trian* (third): *Trian Saxan*, *Trian Masain*, and *Trian Mór*, meaning the English, middle, and great third, respectively. An indication of the number of houses on these sites is given by some references to Clonmacnoise, where in a raid in 1179, 105 houses were burned and in 1205 the abbot's enclosure and forty-seven houses near it were destroyed. The construction of stone-paved roadways at Clonmacnoise, from one part of the settlement to another, is referred to on three occasions in the annals.

Archaeological Evidence

Archaeological evidence for the more domestic buildings directly associated with the ecclesiastical establishment is very scarce. These buildings would have been built of perishable materials, and at virtually all of these sites, the ecclesiastical core has been used for burial for many centuries. The scale of subsequent burial activity is likely to have seriously disturbed the archaeological stratigraphy of the core area and, with burial continuing in many cases, it may be inappropriate even to contemplate excavation.

Nendrum, County Down, was excavated extensively in the 1920s and claims were made that the structures found within the middle enclosure were the domestic buildings, scriptorium, school house, and so forth of the monastery. However, the excavation was not of a high quality and little confidence can be placed in the sequence, dating, or identification of structures and features.

There is good evidence for large enclosures surrounding ecclesiastical settlements. These were first highlighted as a result of aerial photographs taken by J. K. S. St. Joseph in the 1960s. Leo Swan pioneered the study of these enclosures and, through his own aerial work, greatly increased the number of known sites. In size these enclosures average 90 to 120 meters in diameter, with some measuring as large as 400 meters. The enclosing element was usually a bank and external ditch, but some had a large dry-built wall. At the larger sites there is sometimes evidence for two or three concentric enclosures, with the inner enclosure containing the main ecclesiastical core and the more secular dependent settlement being confined to the outer enclosure or enclosures. Occasionally there is evidence for further suburban settlement along approach roads or attached to churches just beyond the outer enclosure.

In some cases, such as Kiltiernan, County Galway, and Moyne in the parish of Shrule, County Mayo, there is further clear evidence for radial divisions within the enclosure. In the case of Moyne, excavation showed that the most visible divisions were late medieval in date. This serves as a warning about interpreting the early history of these sites from the visible surface remains alone. To understand their early development, archaeological excavation is essential.

Many of the more important sites continue as urban centers today such as Armagh, Kells, Tuam, and Kildare, and the enclosures are reflected in the modern street patterns. Most excavation that has taken place within these sites has been limited in extent and is often not very informative regarding the layout of the settlement. A notable exception has been Heather King's excavations in the new graveyard at Clonmacnoise. Here, at each side of a roadway, platforms for circular houses and other structures were found as well as corn-drying kilns, a boat slip, and evidence for metal working and comb making. The best-preserved evidence in this part of the site dated from about the eighth century. Evidence for small circular houses has also been found at smaller sites such as Kilpatrick, County Westmeath, and evidence for metal working and comb making is commonly found on these sites.

The Economy

Mills were an important feature of the economy of these settlements, and most significant sites would have had a mill on or relatively close to the site. For example, up until the nineteenth century, there was a mill still functioning close to the main settlement at Clonmacnoise on the southeast side. It is likely that it was on the site of the early medieval mill. Considerable effort was expended at St Mullin's, County Carlow, to dig a millstream to power a mill on the ridge close to the churches. As early as the twelfth century this millstream was regarded as the work of the saint, and wading through its waters became part of the pilgrimage at St Mullin's. In the case of Nendrum on a small island in Strangford Lough, there was no stream that could be harnessed and a tide mill was constructed. Excavations have uncovered a sequence of three mills here, the first built in C.E. 619 and the third in 789. Structural oak timbers, which survived in the waterlogged conditions, provided the precise dendrochronological (tree-ring) dates. One of the most extraordinary mill sites in the country is that located close to the small monastic site on High Island off County Galway, where a small lake served as the millpond as was also the case at nearby Inishbofin where Colman, famed for his part in the Paschal Controversy, settled with his followers from Britain after the Synod of Whitby (664 C.E.).

Barns for the storage of corn must also have been an essential feature on these sites, along with the corn-drying kilns that have been found during excavation and would have been essential for drying the grain prior to milling. Smaller amounts of grain could have been ground with rotary quern stones (hand mills), which are commonly found. Animal bones, resulting from the eating of meat, are usually found in the occupation deposits and samples studied from Clonmacnoise show parallels with urban rather than rural sites. These settlements were also production, trade, and market centers.

CONLETH MANNING

References and Further Reading

Bradley, John. "The monastic Town of Clonmacnoise." In *Clonmacnoise Studies Volume 1: Seminar Papers 1994*, edited by Heather A. King. Dublin: Dúchas The Heritage Service, 1998.

Doherty, Charles. "The Monastic Town in Early Medieval Ireland." In *The Comparative History of Urban Origins in non-Roman Europe*, edited by Harold B. Clarke and Anngret Simms. Oxford: British Archaeological Reports, 1985.

McErlean, Thomas. "Tidal Power in the Seventh and Eighth Centuries A.D." *Archaeology Ireland* 15, No. 2 (Summer 2001): 10–14.

Manning, Conleth. *Early Irish Monasteries*. Dublin: Country House, 1995.

Norman, E. R. and J. K. R. St. Joseph. *The Early Development of Irish Society: The Evidence of Aerial Photography.* Cambridge: The University Press, 1969.

Swan, Leo. "Enclosed Ecclesiastical Sites and their Relevance to Settlement Patterns of the First Millennium A.D." In *Landscape Archaeology in Ireland, BAR British Series 116*, edited by Terence Reeves-Smyth and Fred Hammond. Oxford: British Archaeological Reports, 1983.

———. "Monastic Proto-towns in Early Medieval Ireland: The Evidence of Aerial Photography, Plan Analysis and Survey." In *The Comparative History of Urban Origins in non-Roman Europe*, edited by Harold B. Clarke and Anngret Simms. Oxford: British Archaeological Reports, 1985.

See also **Archaeology; Burials; Craftwork; Ecclesiastical Organization; High Crosses; Houses; Parish Churches, Cathedrals**

ECCLESIASTICAL SITES

The old model for understanding the organization of the early Irish church was that the system introduced by the early missionaries such as Patrick was based on bishops, but that after a couple of centuries monasteries became dominant and remained so up until the twelfth-century reform. In recent years this model has been challenged and it is now argued that bishops retained considerable authority within the church right through this period.

The large monastic sites were certainly powerful entities but were not just monasteries in the modern understanding of the word. There were monks, priests, and ecclesiastics of one sort or another, but there was also a large dependent lay population consisting of estate workers, servants, monastic tenants, craftsmen, and traders as well as important ecclesiastical families. Also, it is clear that not all churches were monastic and that there were different classes of ecclesiastical sites such as bishop's churches, churches associated with families or tuatha, dependent churches, hermitages, and so forth.

Most early ecclesiastical sites appear to have had a large circular or oval enclosure around them. Most of these sites are marked today by a modern cemetery, usually taking up only a small part of the original enclosure, and a church, usually ruined and dating from the later medieval period. In the case of very important sites, remains of Romanesque or earlier churches can survive.

Churches

Very little is known about the earliest churches constructed in Ireland except that they were built of perishable materials: either timber, post and wattle, or clay.

Reask Stone, Co. Kerry. © *Department of the Environment, Heritage and Local Government, Dublin.*

The earliest biographers of St. Patrick, writing in the late seventh century, refer to churches built of clay. The *Life of St. Brigit* by Cogitosus, also late seventh century, described a large timber church at Kildare with a screen down the center dividing male from female and another cutting off the east end where the shrines of Brigit and Conláed were displayed. The earliest Irish word used for a church in the annals, and that only from 762, is *dairthech* (oak house), and this seems to have signified a timber-framed oak church. The earliest reference to a stone church was at Armagh in 789, where the unique term used for it—*oratorium lapideum*—is indicative of the novelty of the structure at the time. This church, built at a time when Armagh was successfully promulgating the cult of St. Patrick and his preeminence as the converter of the Irish, may have been the first large mortared-stone building in the country, setting a trend that was only gradually followed by other important ecclesiastical sites such as Clonmacnoise, where the *damliac* (now the cathedral) was built in 909.

The building of stone churches became more frequent at major sites during the tenth century and became virtually the norm by the eleventh. Most of the earliest surviving churches, especially those of tenth-century date, have a feature known as antae, where the side walls are continued for a short distance beyond the end or gable walls. These features appear to be uniquely Irish and served to support the barge boards or end rafters of the roof, which was carried over the gables. Antae appear to have been replaced during the eleventh century by east- and west-projecting corbels at the corners, which served the same purpose of carrying the roof over the gables and supporting the end rafters.

Early masonry churches were invariably plain rectangular structures with a west doorway, usually linteled, and normally two windows, one in the center of the east wall and one in the south wall. The gables were steeply pitched and, in some cases, the masonry was characterized by the use of large thin stones placed on edge. The largest surviving example, and coincidentally the earliest exactly dated one, is Clonmacnoise Cathedral, which originally measured 18.8 by 10.7 meters internally. Some very small examples, such as Temple Ciarán at Clonmacnoise and St. Declan's Oratory at Ardmore, appear to have served as shrines over the grave of the founding saint.

Small dry-stone rectangular churches with corbeled roofs like the best-preserved example at Gallarus, County Kerry, are a local style of building largely confined to west Kerry. They probably date from about the ninth to tenth century and certainly do not go back to the beginnings of the early medieval period. Examples excavated at Church Island (Valentia) and Reask, County Kerry, have been shown to be late in the sequence of activity on the site, and the post holes of a wooden church were found beneath the Church Island building. Likewise, post holes for earlier wooden churches have been found beneath stone churches at Ardagh, County Longford, and at Carnsore, County Wexford.

Apart from the carving of a cross over the doorway or on the underside of the lintel in a handful of cases, these early masonry churches do not have any surviving decoration. This changed in the twelfth century with the arrival of Romanesque architecture, when the arches and sides of doorways, chancel arches, and sometimes windows were lavishly decorated with carvings. Around the same time the fashion of building the chancel as a separately roofed smaller unit attached to the nave and of building stone roofs on small churches developed. A feature of Hiberno-Romanesque architecture is that in most cases the only change to the form of the building was the addition of a chancel, decorated chancel arch, and decorated doorway. Very few examples have any further elaboration, and Cormac's Chapel at Cashel is unique with its elaborate blind arcading internally and externally combined with vaulted ceilings, stone roofs, corbel tables, attached square towers, and tympana on its doorways. It is at the same time the finest and the most atypical Romanesque church in Ireland.

Round Towers

Free-standing bell towers of circular plan, known as round towers, are a feature of the more important ecclesiastical sites and were built between the tenth and twelfth centuries. Known as a *cloigtheach* (bell house) in Irish, these remarkable structures can be up to 33 meters in height, with a doorway placed well above ground level in most cases. Offsets or corbels in the inner face of the wall indicate the former locations of wooden floors, which must have been connected by ladders. There were usually four floors between the entrance floor and the belfry level, and these had each a window facing in a different direction. The belfry level itself had a number of windows, usually four. It was thought in the past that handbells would have been rung from the top floor, but Stalley has recently suggested that more conventional bells may have been hung at the top floor and been operated by long ropes. The roof was a cone of mortared stone, although many surviving examples have later roofs or are severely truncated, being particularly susceptible to lightning strike. The earliest reference in the annals to a *cloigtheach* is to one at Slane, County Meath, when it was burned by the Hiberno-Norse of Dublin in 950. This tower does not survive. The only example with a building date from the annals is that at Clonmacnoise, which was completed in 1224. The latest examples such as Timahoe (County Laois), Ardmore (County Waterford), and Devinish (County Fermanagh) have original

Romanesque features indicating a date around the middle of the twelfth century.

Cemeteries, Holy Wells and Bullauns

Burial was an important function of ecclesiastical sites from an early date, but it is important to understand that for many lay people the use of nonecclesiastical ancestral burial grounds continued for some centuries after the introduction of Christianity. The earliest cemeteries attached to churches may have been used only for priests and monks and lay people directly associated with the church or ecclesiastical settlement.

An enigmatic feature of many ecclesiastical sites is a bullaun stone. Usually either sandstone or granite, these large or small boulders have a ground-out hollow or hollows in them. There is great uncertainty as to how or why these hollows or basins were formed. Theories vary between practical uses such as grinding grain or metal ore with smaller stones or pestles to ritual or devotional use of turning stones within the hollows. There are a large number of bullauns at certain sites such as Glendalough.

Holy wells are certain natural springs usually associated with a saint and resorted to for their supposed curative powers for particular or general ailments. They are often close to but seldom within ecclesiastical sites and many have been enhanced with kerbs, steps, and well houses in stonework and even concrete over the years. They often play or played an important part in the local Pattern (patron) Day or pilgrimage. There are strong indications in some cases that their sanctity goes back to pre-Christian times.

Sometimes the former existence of an early medieval ecclesiastical site can be deduced from clues such as a place name with "kill" in it, usually indicating a church or church property; a large oval enclosure; a holy well or other traditional association with a saint; a bullaun stone; and so forth. The combination of a number of these clues tends to increase the certainty of the identification.

CONLETH MANNING

References and Further Reading

Hamlin, Ann. "The Study of Early Irish Churches." In *Ireland and Europe: The Early Church*, edited by Próinséas Ní Chatháin and Michael Richter. Stuttgart: Klett-Cotta, 1984.

Manning, Conleth. "Clonmacnoise Cathedral." In *Clonmacnoise Studies Volume 1: Seminar Papers 1994*, edited by Heather A. King. Dublin: Dúchas The Heritage Service, 1998.

———. "References to Church Buildings in the Annals." In *Seanchas: Studies in Early and Medieval Irish Archaeology, History and Literature in Honour of Francis J. Byrne*, edited by Alfred P. Smyth. Dublin: Four Courts Press, 2000.

Ó Carragáin, Tomás. "A Landscape Converted: Archaeology and Early Church Organisation on Iveragh and Dingle, Ireland." In *The Cross Goes North*, edited by Martin Carver. Woodbridge: The Boydell Press, 2002.

Stalley, Roger. "Sex, Symbol and Myth: Some Observations on the Irish Round Towers." In *From Ireland Coming: Irish Art from the Early Christian to the Late Gothic Period and its European Context*, edited by Colum Hourihane. Princeton: Princeton University Press, 2001.

Swan, Leo. "Enclosed Ecclesiastical Sites and their Relevance to Settlement Patterns of the First Millennium A.D." In *Landscape Archaeology in Ireland*, edited by Terence Reeves-Smyth and Fred Hammond. Oxford: BAT British Series 116, 1983.

See also **Architecture; Ciaran; Columcille; Early Christian Art; Ecclesiastical Settlements; High Crosses; Inscriptions; Parish Churches, Cathedrals**

EDUCATION

Between the fifth and the sixteenth centuries, the two main institutions of Irish education were the monastic schools, founded by Christian missionaries in the early fifth century, and the bardic schools, originating in ancient Celtic times. It has been argued that the bardic schools predate the arrival of Christians in Ireland, existed parallel to or were integrated with the monastic system and, after the decline of monastic schools in the tenth and eleventh centuries, continued to thrive up until the seventeenth century. Between the fifth and the ninth centuries, however, the Christian monastic schools modeled on a classical Latin curriculum were the predominate educational form. What we know of education in medieval Ireland comes from manuscripts produced in Christian monasteries, accounts of the lives of the Irish saints, glosses, the Brehon Laws, and translations of Latin texts produced in monasteries. Most of these accounts describe a rigorous and dynamic monastic life that created learned scholars who influenced Irish and European culture, religion, and scholarship.

The Ancient *Filid*

When Christian missionaries arrived in Ireland, they encountered an already thriving educational system under the tutelage of the ancient learned class, the *filid* (scholars). Among their many social roles, the *filid* were educators and taught natural and moral philosophy. Three tracts of the Brehon Laws, the *Senchus Már*, the *Crith Gablach* (the branched purchase), and the *Uraicecht Becc* (the small primer), provide a sharper picture of the schools created by the *filid*. The school itself involved a regular course of training with seven ascending grades: *Fochluc*, *Mac fuirmid*, *Dos*, *Cana*, *Cli*, *Anruth*, and *Ollamh*. The whole course lasted twelve years, and each year was assigned a specific curriculum. To complete the course, the student progressed through instruction mainly in grammar, philosophy, and poetry. The *filid* enjoyed a very high social status, producing works of history, topography, romance and heroic tales, narrative, lyric and elegiac poetry, law tracts, folklore, epigrams, and songs. Of course, all of the productions by the *filid* were oral. It took the arrival of Christianity and the monastic schools to write down and record these accomplishments.

Monastic Schools

The fifth to seventh centuries witnessed a rapid increase in monasteries throughout Ireland, established by Christian missionaries of whom the Latinists Palladius and St. Patrick are the most familiar. Attached to many of these monasteries were monastic schools where monks instructed students in the Latin ecclesiastical tradition. Reading and writing Latin and the rigorous study of the Latin bible constituted the main conduits to knowledge. In these schools, monks studied Christian authors, the Scriptures, ecclesiastical rules, theology, canon law, and ritual. The seventh-century biography of Columbanus, written by Jonas, reveals that Columbanus as a youth received instruction in "liberal letters," grammar, and religious doctrine. By "liberal letters," scholars assume the presence, to a greater or lesser degree, of a classical curriculum comprised of the trivium of grammar, rhetoric, and logic and the quadrivium of arithmetic, geometry, astronomy, and music. Although the stern Patrician tradition would always maintain a foothold in monastic education, monastic schools are usefully characterized by their catholic interests in Roman knowledge. By the seventh century, monks studied and transcribed the works of Virgil, Horace, Marital, Juvenal, Claudian, Statius, and Ausonius. Using these Christian and non-Christian writings, the Irish monks labored to instruct their students about the pursuit of wisdom within a wider theological frame.

In general, early Christian monasteries were made up of either a small community dedicated to living a religious life or a tiny church where a single cleric served the local lay and religious residents. By the seventh century, the clergy lived a much more communal life in monastic settlements. Some of the larger monastic communities included Clonmacnoise (supposedly founded by St. Ciarán), Iona (founded by St. Colum Cille), Monasterboice (founded by St. Buite), and Glendalough (founded by Cóemgen [Kevin]).

Writing

A monastic education was synonymous with writing. Transcription of manuscripts by scholars in the monastic schools everywhere flourished and writing became the means through which Irish monks communicated their learning to Europe. Scholars used quill pens and ink made from charcoal to write on parchment or vellum and the skins of goats, sheep, or calves for works intended for preservation. Long, thin wooden tablets covered in wax and etched on by an iron style were used as practice boards for impermanent notes. Many of the early monasteries boasted a *sciptorium* (a school for penmanship), but as the monasteries grew in size, it is reasonable to assume that in the larger monasteries the *scriptorium* was a separate building where the *scriba* (scribe) worked and where finished texts were stored. The Brehon Laws enumerate seven degrees of religious learning within the monastic schools: *Fealmac* (a boy after reading his psalms), *Freisneidhed* (an interrogator), *Fursaintid* (an illustrator), *Sruth do Aill* (a stream from a cliff), *Saí* (professor), *Anruth* (a noble stream), and *Rosaí* (great professor). Whether or not the monastic school actually adhered to these categories is not known, but scholars contend that these designations provide insight into the pedagogical organization within the early monastic schools.

Certainly Latin was the predominate language taught in monastic schools. However, scholars still wonder: How did Latin arrive in a country that had limited contact with the Roman Empire? Undoubtedly, Christian missionaries brought Latin to Ireland, but the successful absorption of Latin into Irish scholarship necessitated instructional manuals and books for learning and teaching the language. Ó Cróinín plots a trajectory using two grammar textbooks, the *Ars Asporii* and the *Anonymus ad Cuimnanum*, from an educational system focused on studying Christian texts in the early sixth century to the celebration and serious study of classical texts by the eighth century. The *Ars Asporii*, an adaptation of the Roman grammarian Donatus's *Ars Minor*, provides a rudimentary grammar guide for beginning Latin students within the context of Christian devotion. The *Anonymus ad Cuimnanum*, produced two centuries later, renders a much more subtle and complex pedagogy; it treats Latin grammar as an autonomous subject and shows the influx of the grammarians Charisius, Consentius, Diomedes, and Probus into Irish thinking. By the eighth century, copies of many books transcribed in Irish scriptoria reached monastic libraries throughout Europe. Much of ancient saga literature owes its preservation to the monastic schools. In addition, Greek and Hebrew eventually found their way into the curricula. Although there is considerable debate about when Greek entered the monastic schools, scholars agree that by the ninth century Greek was known and studied, as we see in the works of early Irish hymnodists who often refer to Greek myths in their compositions and, more directly, in the writings of John Scottus Eriugena (an Irish scholar who worked in the Court of Charles II), especially his *De Divisione Natura*, which attempted to reconcile Neoplatonic ideas of emanation with Christian doctrine related to creation.

In addition, there can be no doubt that the monastic schools also taught Irish. As early as 600, Irish appeared side-by-side with Latin in the form of glosses—remarks, comments, and additions written by the copyist in the margins of the manuscript—that often explicate the Latin text in the Irish language. The *Auraicept na nÉces* (the Instruction of the Poets or Scholars), a treatise on Irish grammar, appeared in the middle of the seventh century and continued to be worked on by monastic and lay authors until the eleventh century. The *Auraicept* outlines, among other topics, the origins of Gaelic; the Latin and Irish treatment of semi-vowels; the seven elements of speech in Irish; and the alphabets of Hebrew, Greek, and Latin.

During these centuries, students and scholars traveled to Ireland to benefit from Irish monastic education. Monasteries generally welcomed foreign students and lay students (those who were not intended for the church but a civil or military life), and there are a few records suggesting that women might have studied in the monasteries. In addition, evidence for the influence of monastic education on the *filid* and the native secular schools is found in the absorption of Latin into secular instruction and the writing of the Brehon Laws in the eighth century. At the same time, Irish scholars such as the famed St. Colum Cille and St. Columbanus fanned across Europe and founded monastic schools.

There have been two primary (and opposed) opinions about this period of scholarship in Irish history. The first, popularized by such writers as Douglas Hyde, suggests that Irish monastic schools from the sixth to the end of the ninth century preserved knowledge during the Dark Ages by devoting their cultural and religious institutions to scholarship and combated illiteracy and ignorance with the two-handed engine of classical texts from Greece and Rome and Christian religious texts. The second opinion argues the other extreme—that classical knowledge in ancient Ireland was limited, and the famed scholars in Europe such as John Scottus Eriugena acquired their classical knowledge in exile. Between these two claims resides the majority of scholarly opinion that might be summarized as follows: Even though there may have been variations in the quality and standards of education as well as the number of classical texts available to Irish scholars, what remains clear is that the texts read and transcribed, the skill of the Irish monks in writing and instruction, and their influence over the intellectual life of Europe were all formidable.

Bardic Schools

When Viking raids began in the last decade of the eighth century, monastic life was permanently impacted. The Vikings targeted monasteries because of their riches and encountered little resistance to their plundering. Between 775 and 1071 C.E., Glendalough itself was pillaged on numerous occasions and destroyed by fire at least nine times. Devastation, however, was not the only order of the day. Since the Vikings also settled in many parts of Ireland, their culture intermingled with the Irish. Evidence for the increasing internationalization of Irish learning is found in twelfth-century translations of *The Aeneid*, *The Pharsalia*, and *The Thebais*. In addition, many important native histories were written during this time, including *Cogad Gáedel re Gallaib* (the War of the Irish against the Foreigners), written in Munster in the early twelfth century, and *Lebor na hUidre* (the Book of the Dun Cow), a twelfth-century manuscript traditionally associated with Clonmacnoise.

In the late eleventh and twelfth centuries, with the monastic schools weakened by Viking attacks, the bardic schools where the *filid* trained began to thrive again and were to some degree comparable to monastic centers of learning. Each bardic school was generally associated with a poetic family such as Ua Dálaigh in Cork and Ua hUiginn in Sligo. Students studied languages, metrics, mythology, history, genealogy, dinnshenchas, and, predictably, Latin, learning their lessons orally from the Latin and Irish manuscripts. Also, the *filid* class began to develop new forms of poetry, another indication that Irish intellectual life remained vital in the later Middle Ages. This structure of education, where an elite family would cultivate learning, might also have been true for the legal, medical, and musical professions.

By the thirteenth and fourteenth centuries, the continental monastic orders of the Cistercians, Benedictines, Dominicans, Franciscans, and Augustinians had moved into Ireland and superceded the older Irish monasteries. These orders introduced a pervading movement in education toward Aristotelianism, which emphasized logic over the literary, historical, or mythological study of classical works. Aristotle continued to dominate Irish education well into the seventeenth century while the rest of Europe "rediscovered" the classics of Rome and Greece during the Renaissance. Despite the influx of European scholars and educators, Ireland still had no university and many students traveled to England or other parts of Europe for advanced studies; it would take until 1591 for a viable university to be established in Ireland in the form of Trinity College Dublin. In sum, from the fourteenth century on, a series of ordinances attempted to suppress the Irish language and

Irish customs, restrictions that were increasingly successful in changing the shape of Irish education so that, in the first half of the sixteenth century, the monasteries were dissolved and replaced by grammar schools and Jesuit schools.

MATTHEW BROWN

References and Further Reading

Atkinson, Norman. *Irish Education: A History of Educational Institutions*. Dublin: Allan Figgis, 1969.

Dowling, P. J. *A History of Irish Education: A Study in Conflicting Loyalties*. Cork: Ireland Mercier Press, 1971.

Hanson, W. G. *The Early Monastic Schools of Ireland*. Cambridge: W. Heffer & Sons, Ltd., 1927.

McGrath, Fergal. *Education in Ancient and Medieval Ireland*. Dublin: Studies "Special Publications," 1979.

Ó Cróinín, Dáibhí. *Early Medieval Ireland 400–1200*. London: Longman, 1995.

Richter, Michael. *Medieval Ireland: The Enduring Tradition*. London: MacMillan Education, 1983.

See also **Bardic Schools/Learned Families; Brehon Law; Classical Influence; Clonmacnoise; Colum Cille; Columbanus; Conversion to Christianity; Ecclesiastical Settlements; Ériugena, John Scottus; Glendalough, Book of; Glosses; Languages; Manuscript Illumination; Moral and Religious Instruction; Pre-Christian Ireland; Religious Orders; Scriptoria; Viking Incursions**

EISCIR RIATA

See **Roads and Routes**

EMAIN MACHA

Emain Macha, the pseudo-historical capital of Ulster and the principal setting of the *Táin Bó Cúailnge* (The Cattle-Raid of Cooley), lies three miles west of the ecclesiastical city of Armagh. It derives its name from the goddess Macha, who is also immortalized in the place name Ard Macha or Armagh, which translates as the "heights of Macha." Pseudo-historical texts claim that Emain Macha was established as the center of the Ulaid between the seventh and fourth centuries B.C.E. and that the power of its dynastic rulers declined in the fourth or fifth century C.E. Emain Macha comprises a concentration of forty-six prehistoric monuments, central to which is Navan Fort. The fort is a large circular earthen enclosure, 286 meters in diameter, consisting of a broad, deep ditch and external rampart. A ring ditch and an impressive mound 6 meters high, both of which were excavated by D. M. Waterman between 1963 and 1971, lie within it. The most exciting discovery within the mound was a very large multiring timber structure of radial plan, 40 meters in diameter. The large central post of that structure produced a tree-ring date of late 95 B.C.E. or early 94 B.C.E., after which it was covered by a large cairn of limestone blocks. Additional elements in the Emain Macha complex include the sites of two possible passage tombs of fourth millennium B.C.E. date that lie to the north of Navan Fort, and a small natural lake called Loughnashade is situated to the northeast. A multivallate hill fort known as Haughey's Fort, occupied in the period 1300–900 B.C.E., and an artificial pond called the King's Stables are located approximately 1,000 meters west of Navan Fort. The development of the Navan complex appears to have begun about the thirteenth century B.C.E. when Haughey's Fort and the King's Stables were constructed, while Navan Fort apparently became the new focus of ritual activity from about the tenth century B.C.E. onward.

It has been suggested that the mound within Navan Fort may have been purpose-built for kingly inauguration, but there is no evidence, as yet, that a sense of royalty and an established custom of inaugurating kings on mounds actually prevailed in the Irish late prehistoric period. Emain Macha is, however, frequently evoked as an ideal kingship center during the later medieval period. The Uí Néill kings of the fourteenth century, for instance, closely identified themselves with the heroes of the Ulster Cycle and particularly with Conchobar mac Nessa and his abode at Emain Macha. In an attempt to physically attach himself and his dynasty to the ancient seat of the legendary kings of the Ulaid, Niall Óg Ua Néill had a temporary house built there to entertain poets and learned men in 1387. The perception of Emain Macha as the most desirable inauguration site for Ulster royalty is also evoked in later medieval bardic poetry. The thirteenth-century poet Gilla Brigde Mac Con Mide, in his *aisling* (dream or vision) on the desired inauguration of Roalbh Mac Mathgamna as chief of Airgialla, uses Emain Macha as the setting for the inauguration. In the dream he sees Roalbh made chief by the "poet bands of the world" who are arranged in order upon the mound within Navan Fort.

ELIZABETH FITZPATRICK

References and Further Reading

Aitchison, Nicholas B. *Armagh and the Royal Centres in Early Medieval Ireland*. Woodbridge: Boydell and Brewer, 1994.

Lynn, Chris J. *Excavations at Navan Fort 1961–71 by D. M. Waterman*. Belfast: Stationary Office Publications, 1993.

Simms, Katharine. "Propaganda Use of the Táin in the Later Middles Ages." *Celtica* 15 (1983): 142–149.

Waddell, John. *The Prehistoric Archaeology of Ireland*. Galway: University Press, 1998.

See also **Armagh; Inauguration Sites; Kings and Kingship; Poetry, Irish; Poets, Learned Men; Uí Néill; Ulaid; Ulster Cycle**

ENGLISH LITERATURE, INFLUENCE OF

See **British Library Manuscript Harley 913; Eachtrai; French Literature, Influence of; Hiberno-English Literature; Romance**

ENTERTAINMENT

Medieval Ireland features a wide variety of entertainment, professional entertainers, and performers. Most prominent is an array of performing fools. Several early Irish terms exist for these performers. Foremost as a performing fool was the *drúth*. The term is related to the term druid, although the two figures are distinct. The *drúth* offered various kinds of entertainment, most prominently physical and vocal antics best associated with the medieval jester. Impersonating and mocking the congenital fool, also known as *drúth*, was also featured. Several descriptions of the *drúth* include comments suggesting the performing fool was indeed mentally deficient. The professional *drúth* is often described in colorful motley clothing, with long shaggy hair. This semblance was clearly an important part of his trade. According to several Law Tracts, damage to his clothing or hair demanded compensation. A common figure of the saga texts, the *drúth* is most closely associated with royal and other high-ranking members of society. A professional fool was often part of a retinue, receiving both payment and protection from his patron.

A further entertainer of medieval Ireland was the *fuirseoire*, best described as a jester or buffoon. Like the *drúth*, the *fuirseoire* is most often connected with the royal court. Likely a paid professional, the *fuirseoire* entertained through mimicry, contortions, and elements of fright. Judging from descriptions his performance may also have included singing and other vocal antics. In several descriptions he performs to accompanied music. Named alongside the *drúth* and lapdog in a trio expected at a royal banquet, the *fuirseoire* seems to have been a common professional performer, likely maintained by his employer as part of a retinue.

Several entertainers of medieval Ireland are difficult to identify. The *creccaire*, attested alongside other performers, seems to have been a type of mummer or perhaps scarifier, eliciting fear from his audience through physical and vocal antics. He is described in one text as making a green branding upon his eyes, possibly referring to a disguise or commonly accepted and recognizable tattoo. The *creccaire* was clearly not very highly regarded in society and in one text is described receiving a "crooked bone" as his portion at a feast. A further unsavory, although clearly popular, entertainer was the *braigetoir*. This performer is the early Irish representation of the widespread and apparently popular medieval entertainer the "farter." An early Law Tract offers the clearest, most unequivocal definition, stating that these performers render their craft "out of their backsides."

Beyond various types of fools and jesters, further popular Irish entertainment was performed by magicians, jugglers, featsters, and acrobats. Several entertainers are described as conjuring magic, while others are noted for particularly athletic and acrobatic performances. Juggling was a popular entertainment, often taking place in alehouses. As noted in a Law Tract, jugglers were culpable for damage or injury caused by errant throws. Culpability and compensation for their errors depended on several factors, including the shape of the juggled objects and the distance of the audience from the performance. According to several descriptions, juggling and similar feats were often accompanied by music.

Music was a popular entertainment of medieval Ireland. Expected at assemblies, festivals, and banquets, music played an important role in medieval Irish entertainment. Music was a standard accompaniment to any occasion and an entertainment available to all classes of society. While music was often the result of amateur improvisational sessions, professional musicians held a relatively high status and were well paid, either as retinue to wealthy patrons or as traveling performers. Most prevalent among professional musicians were harpists, timpánists, and pipers. Vocalists also provided entertainment, although lack of mention in the sources suggests performances were largely spontaneous and probably not professional. By the late medieval period, English influence brought about critical change in Ireland's music tradition. The introduction of sheet music counteracted traditional oral transmission, while classical voice training moved away from the conventional rhythmic, lilting tones, and melodies that continue to characterize Irish music of today.

Various sporting events also provided popular entertainment in medieval Ireland. Field games attracted large audiences, evidenced by legislation providing protection and compensation in case of injury for both participant and spectator. Horse and chariot racing were also popular. Races at seasonal fairs and festivals were often the high point of such gatherings, drawing eager crowds. Presumably to ensure enjoyment and prevent conflict, strict guidelines at the fair at Carman describe bans on arguments, warfare, politics, and judgments while the races were held. Particularly associated with the Liffey Valley, horse racing and breeding in medieval Ireland enjoy a continuum to the present day, reflected in the prominence of Kildare and environs in modern racing.

Angela Gleason

References and Further Reading

Simms, Katharine. "Guesting and Feasting in Gaelic Ireland." *Jnl. Roy. Soc. Antiq. Ire.* 108 (1978): 67–100.

Kelly, Fergus. *A Guide to Early Irish Law.* Dublin: Dublin Institute for Advanced Studies, 1988.

Ramsey, Greer. "A Breath of Fresh Air: Rectal Music in Gaelic Ireland." *Archaeology Ireland,* 16 no. 1 (Spring 2002): 22–23.

See also **Law Tracts; Music; Society Functioning of, Gaelic**

EÓGANACHTA

The Eóganachta emerged at the beginning of recorded history to become the dominant dynasty in the south of Ireland, the ancient kingdom of Munster, at approximately the same time as the Uí Néill in the north of the country. The name Eóganachta may indicate descent from a divine or human ancestor connected with the yew tree, suggesting a parallel with the Gaulish tribe, the Eburones, "yew people." The yew tree is said to have been regarded as sacred by the Eóganacht. The importance of this tree held right into the historic period, the ending of Eóganacht power is said to have been symbolized by the destruction of an ancient yew tree at the Eóganacht monastery of Emly by the Dál Cais. The genealogies claim that they descended from the mythical Eógan Mór (also known as Mug Nuadat), son of Ailill Ólum. It may have been that the Eóganachta were colonists who returned or were driven from their conquests in Britain, which could account for the Latin borrowing for the name of their capital, Caisel (modern Cashel, County Tipperary, from *castellum*). Around the beginning of the fifth century, the rulers of some Irish kingdoms in North Wales were expelled by the original inhabitants, which may or may not be coincidental, although traditionally Munster colonists in Britain were said to have been the Uí Liatháin and the Waterford Déisi. The favored-ally status of the Déisi with the Eóganachta in historical times may well stem from such colonial activities.

A remarkable feature of the Eóganacht kingship was its association with Caisel and Christianity. Unlike the capitals of the other provinces such as Tara, Emain Macha, or Cruachain, Caisel had not been a ritual center in prehistoric times and was associated in historic times only with the Eóganachta. Tírechán's seventh-century biography of Patrick tells us that Patrick himself baptized the sons of Nadfraích *super Petram Cothrigi* at Caisel. This is an obvious invention but would appear to be an early effort to tie Munster in with Armagh and the supremacy of the Uí Néill. The story of Conall Corc and his sons and their acquisition of the kingship of Munster has some parallels with the tales that attach to Niall Noígiallach and his rise to power in the north of Ireland. The legends of Corc of Caisel included both pagan elements, found in a seventh-century text, *Conall Corc and the Corcu Loígde*, and Christian elements, in the eighth-century texts, *The Exile of Conall Corc* and *The Finding of Caisel*. The *literati* of the Eóganachta liked to depict the kingship of Munster as a benevolent place, more peaceful than the Uí Néill kingship in the north.

Six main branches may be identified: Eóganacht Áine, Eóganacht Chaisil, Eóganacht Glendamnach, Eóganacht Airthir Chliach, Eóganacht Locha Léin, and Eóganacht Raithlind. Others such as Eóganacht Arran; Eóganacht Ruis Argait, also called Ninussa; and the Uí Fidgenti in Limerick and Uí Liatháin in Cork, who are included in the tract *The Expulsion of the Déisi* as one of the Eóganachta dynasties, may have been segments of the main branches or grafted onto the ruling stem at a later date in the case of the last two named. Ultimately all of the Eóganachta were said to have been descended from Eógan Mór, son of Ailill Ólum. However, Eóganacht Raithlind (Uí Eachach Muman) and Eóganacht Locha Léin may also have been later grafts onto the main Eóganacht stock and rarely figured in the kingship of Caisel, which remained with a few exceptions in the grasp of the eastern Eóganachta, the descendants of Óengus mac Nad Fraích maic Cuirc of Caisel, who consisted of Eóganacht Chaisil, Eóganacht Glendamnach, and Eóganacht Airthir Chliach. Eóganacht Caisel were settled around Caisel itself; Eóganacht Glendamnach around Glanworth in north Cork; Eóganacht Áine at Knockaney in County Limerick; Eóganacht Locha Léin around Loch Léin in Killarney; and Eóganacht Raithlind in the Lee and Bride valleys to the west of Cork Harbor.

The Eóganachta had a complicated relationship with their vassal kingdoms. There seems no doubt that the Déisi of Waterford and the Múscraige (in at least six widely separated *túatha* from north Tipperary to west Cork) acted as facilitators for the Eóganachta and were treated as favored allies, and extant texts show that the kingship of Caisel/Munster was based on mutual obligations between the king and his subkings and vassals. While the text may be aspirational rather than the letter of the law, it shows that the king of Caisel was expected to give compensatory gifts in order of precedence in return for the services, such as hosting, tribute, and so forth that he got from his vassal states.

The most notable kings of Munster were Cathal mac Finguine of Eóganacht Glendamnach (d. 742) and Fedelmid mac Crimthainn (d. 847), both of whom were notable warriors who went on the offensive against the ever-increasing power of the Uí Néill kings of Tara. The Eóganacht ruled Munster for over 500 years until the rise of the Dál Cais in the tenth century, who legitimated Brian Boru's usurpation of the kingship of Munster in 978 by grafting an ancestor of their own, Cormac Cas, onto the Eóganacht genealogy as a son of Ailill Ólum, brother of the eponymous ancestor of the Eóganachta, Eógan Mór. The Dál Cais version alleged that the Eóganacht had ignored a decree of Ailill Ólum to have the kingdom alternate between the descendants of his two sons. Needless to say, this theory of alternation only lasted in Brian's lifetime, as the Dál Cais had no intention of alternating with a weakened Eóganachta. The Eóganachta staged a comeback in the first quarter of the twelfth century in the person of Cormac Mac Carthaig and went on, even after the Anglo-Norman Invasion, to rule parts of Cork and Kerry until the fall of the Gaelic Order in the seventeenth century. The MacCarthys, O'Sullivans, O'Donoghues, O'Keeffes, Kirbys, Moriartys, and many other well-known Munster families claim descent from these rulers of early medieval Munster.

LETITIA CAMPBELL

References and Further Reading

Byrne, F. J. *Irish Kings and High Kings*. London: 1972; reprint with additional notes and corrigenda, Dublin: Four Courts Press, 2000.

Dillon, Myles. "The Story of the Finding of Cashel." *Ériu 16* (1952): 63.

Hull, Vernan. "Conall Corc and the Corcu Loígde." *Proceedings of the Modern Languages Association of America 62* (1947): 887–909.

———. "The Exile of Conall Corc." *Proceedings of the Modern Languages Association of America 56* (1941): 937–950. (Éoganacht origin legend with introduction, text, translation and notes.)

Mac Niocaill, Gearóid. *Ireland before the Vikings*. Dublin: Gill and Macmillan, 1972.

Meyer, Kuno. "The Laud Genealogies and Tribal Histories." (Zeitschrift fur Celtische Philologie viii 1911): 291–338: *De Bunad Imthechta Eóganachta*, 312–315; Munster version of *Lebor Gabála Érenn*.

Ó Corráin, Donnchadh. *Ireland before the Normans*. Dublin: Gill and Macmillan, 1972.

See also **Kings and Kingship; Mac Carthaig, Cormac; Munster**

ÉRAINN

This is the name of one of the ancient peoples of Ireland. It is first attested in the *Geography* of Claudius Ptolemaeus of Alexandria (*c.* 150 A.D.) as *Ivernioi*. Ptolemy also records the "town" *Ivernis*, "the Fertile Place," from which is derived the name of the island, *Ivernia*, and the people, *Ivernioi*. Ptolemy's *Ivernioi* inhabited the southwest of Ireland. According to the genealogists the Érainn are found in other parts of Ireland as well. Genealogical theory changed over time so the status of the Érainn and their relationships with other peoples evolved in accordance with the evolving political landscape. The main groups classed as Érainn were the Corcu Loígde, in historical times located in southwest County Cork, the Múscraige of Cork and Tipperary, the Corcu Duibne of Kerry, the Corcu Baiscinn of west Clare, the Dál Riata of north Antrim, and the Dál Fiatach (Ulaid) of County Down. The genealogists considered the Érainn, the Laigin, and the Cruthin as being distinct races. In the historical period the Ulaid (Ulstermen) were the most prominent of the Érainn and they, together with the Laigin (Leinstermen), were regarded as "free races." By the eighth century the Éoganachta of Munster and the Connachta (in particular the Uí Néill) had come to dominate the island and they made up the third "free race," the Féni. In time the Érainn were brought within the circle of the Féni as a relative of "Míl."

It is clear that the Érainn had been politically important in the proto-historic period, although in the historical period many of them had been reduced to servile or politically subordinate status. In the saga literature the ancestor of many of the Érainn, Conaire, was depicted as the just and beneficent king of Tara. Lugaid mac Con of the Corcu Loígde was said to have been king of Tara and was succeeded by Cormac mac Airt (ancestor of the Uí Néill). The Corcu Loígde (Loigodewa, "the people of the Calf Goddess") were the most important of the Munster Érainn. Genealogical theory claimed that they shared power with the more recent Eóganachta. Early tradition suggests that the Osraige, a major people between Munster and Leinster, had been ruled by or were in alliance with the Corcu Loígde. Indeed they may have been closely related. This association was disrupted during the sixth century, however, when the Eóganachta rose to dominate Munster with the help of the Uí Néill. It is likely that the Corcu Loígde had been dominant in Munster, if not beyond the province, before the rise of the Eóganachta and for this reason had been given the status of most-exalted vassals of their new masters. By the twelfth century the Corcu Loígde still retained an element of prestige when the core of their territory became the diocese of Ross. St. Ciarán of Saigir, patron of the Osraige, was one of their kin. Their lord during the later Middle Ages was O'Driscoll, whose wealth was based upon the sea, trading in wine with Gascony.

CHARLES DOHERTY

References and Further Reading

Byrne, Francis John. *Irish Kings and High Kings*, new ed. Dublin: Four Courts Press, 2001.

Charles-Edwards, T. M. *Early Christian Ireland*. Cambridge: Cambridge University Press, 2000.

De Bernardo Stempel, Patrizia. "Ptolemy's Celtic Italy and Ireland: A Linguistic Analysis." In *Ptolemy. Towards a Linguistic Atlas of the Earliest Celtic Place-names of Europe*, edited by David N. Parsons and Patrick Sims-Williams. Aberystwyth: CMCS Publications, Department of Welsh, University of Wales, 2000.

Ó Corráin, Donnchadh. "Corcu Loígde: Land and families." In *Cork: History and Society. Interdisciplinary Essays on the History of an Irish County*, edited by Patrick O'Flanagan and Cornelius G. Buttimer. Dublin: Geography Publications, 1993.

See also **Éoganachta; Munster; Osraige; Uí Néill; Ulaid**

ÉRIUGENA, JOHN SCOTTUS (*fl.* 848–870)

John Scottus was born some time in the first quarter of the ninth century and died in or after 870. Neither birth nor death date is known. That he was born in Ireland is proved by the epithet he gave himself: "Eriugena" (born in Ireland). His documented activity embraces the period approximately 848 to 870, beginning with his role in the predestination controversy and ending with his last datable poem, addressed to King Charles the Bald. It is not known when he came to the continent, but what we know of his career took place in the western kingdom of the Carolingian Empire, mostly in what is now northeastern France. Centers such as Compiègne, Saint-Denis, Soissons, Laon, and Reims figure in his itinerary, but it is very difficult to know where he was at any particular time. Legend holds that John left Charles's kingdom and became a teacher in the court of King Alfred in Wessex, but this is widely discounted.

Other certain facts of his life are likewise few. Despite his being arguably the most outstanding theologian of his time, he held no position of ecclesiastical authority. According to contemporary evidence (Prudentius of Troyes, *De praedestinatione*, PL cxv, 1043A), he was "*nullis ecclesiasticis gradibus insignitus*." For a brief period he was a teacher in Charles's court. However, he may have lost his position at the court when he came under attack for his views on predestination, which were condemned at two councils (Valence in January 855 and Langres in May 859). It is not known if he suffered any penalty for his views, although this possibility is suggested by the relatively long silence between 851, when he published his own *De praedestinatione*, and 858, when his first securely dated poem was written. Finally, there is some evidence to suggest that John not only wrote and taught but also practiced medicine.

More, perhaps, is known of John's literary contacts, including some distinguished pupils. Contemporaries in John's circle include the Irish scholar Martin of Laon (Martin Hiberniensis) and Wulfad, abbot of Saint-Médard and later bishop of Bourges, to whom John dedicated his *magnum opus*, the *Periphyseon*. He was also on friendly terms with Hincmar, bishop of Laon. Among his students are counted Heiric of Auxerre; Wicbald, bishop of Auxerre; and, possibly, Hucbald of Saint-Amand. John's students may also have included some of his countrymen, as the numerous Irish glosses in his biblical scholia suggest. Unfortunately, it is not known to what extent (if at all) John was in contact with his famous fellow Irishman Sedulius or members of his circle.

John's education can be reconstructed only from a study of his sources. It is difficult to tell what learning he brought with him from Ireland and what he acquired on the Continent, although certain stylistic mannerisms betray his Irish early education. An aspect of his education that affected practically all of his writings was his study of Greek. How and where he acquired a working knowledge of the Greek language, which had all but disappeared from Western Europe after the sixth century C.E., is a mystery. However, John employed Greek, with increasing skill, over the course of his career and in different contexts. These include translations of Greek patristic works, citations of Greek authors in his own writings, the use of Greek for constructing etymologies, and—most impressively—the *graeca* found in his poems. These exhibit Greek elements ranging from a word or a phrase to whole lines. Several poems are written entirely in Greek, although there are some imperfections.

Even for a time noted for the collection of manuscripts and the expansion of libraries John's reading in both secular and religious literature would be considered impressive. Of the Roman classics, John had a deep knowledge not only of Vergil but he also had an acquaintance with Lucretius, Cicero, Pliny the Elder, and possibly even Petronius. Like most of his contemporaries he accessed the Roman (and some Greek) classics through intermediaries, principally Macrobius, Martianus Capella, Calcidius, Priscian, and Isidore. His knowledge of Christian Latin writers is noteworthy. He cited the poets Paulinus of Nola and Avitus in his own poems, which also contain echoes of Iuvencus, Corripus, and Venantius Fortunatus. A thorough grounding in the Latin fathers would have been expected in John's day, and John was no exception. He read Augustine widely and deeply, knew Ambrose well, and cited other well-known Latin patristic authorities such as Jerome. However, the most impressive aspect of John's reading is the Greek fathers. In addition to those fathers who could be known from ubiquitous Latin translations—Origen and Basil—John knew and used several Greek fathers in the original Greek. He published translations of the corpus of works of Pseudo-Dionysius, the *Ambigua* and the *Quaestiones ad Thalassium* of Maximus the Confessor, the *De opificio hominis* of Gregory of Nyssa, and the *Solutiones ad Chosroem* by Prisicianus Lydus, which was doubtless the source of John's citations of Aristotle. He apparently also made his own translations of the passages cited from Epiphanius's *Ancoratus*.

Although John Scottus is admired today principally for his two works, the *De praedestinatione* and the *Periphyseon*, his activity as a scholar was highly diverse, comprising learning aids, commentaries and scholia collections, translations, and original compositions. John moved easily between secular and religious material. Grammar (in the broadest sense of the word) appears to have been his foundational discipline. Among the learning aids that can be linked to John is an edition of Macrobius's treatise *De differentiis et societatibus graeci latinique uerbi*. It is also likely that John contributed to the paradigms of Greek words found in the famous codex *Laon 444*, edited by his colleague Martin. Another manuscript, British Library, MS Harley 2688, also contains Greek paradigms and a word list transmitted under the name "IWANHC."

To list John's activity as a scholiast and commentator is not easy, since some scholia appear in manuscripts mixed with scholia by other writers and because some attributions are disputed. Four fully fledged commentaries—one scriptural, one patristic, and two secular—can be attributed to John with certainty. The first is his commentary on the Gospel of John, on whose prologue Eriugena also wrote a homily. While the commentary survives in a single manuscript with corrections and additions in his own hand, the homily appears in fifty-four manuscripts; clearly the latter work was the most influential of all John's compositions during the Middle Ages. His commentary on the *Celestial Hierarchy* of Pseudo-Dionysius was based on his own translation of the work but shows a deeper grasp of Greek. The third commentary, the *Annotationes in Marcianum* (Capellam), surviving in two chief recensions, is usually thought to be an early work. It was surely intended to be a complete commentary on one of the most-studied authors of the early Middle Ages, but large parts of it consist of only brief scholia. It is remarkable in its day for its classical erudition and offers insights into Eriugena's exegetical methods that have not been fully explored. The fourth commentary is on Priscian's *Institutiones* and is not yet edited. Other commentaries have been assigned to John, namely one on Boethius's *opuscula sacra* and another on the poet Prudentius.

Various scholia collections must also be considered. The recent demonstration that the so-called *Glossae divinae historiae* were indeed written by John contributes to the appreciation of our author as a pedagogue. The scholia, consisting primarily of "hard words" drawn from both the Old and New Testaments, are explained (usually but not always correctly) by "easier" Latin, by Irish, or, less helpfully, by Greek! The collection can be studied with profit for clues to John's Irish education. Other scholia sets also represent recent discoveries. It is now recognized that the copy of Priscian, Leiden, Universiteitsbibliotheek, MS B.P.L. 67, written by the Irishman Dubthach in 838, contains numerous glosses and scholia in the hand of Eriugena (now firmly identified as i-1). Of special interest regarding John's classical enthusiasms is the fact that he recopied Priscian's citations of Homer and other ancient Greek authors in the margins of several folios of this manuscript and attempted to translate them into Latin. John also left some scholia to the "philosophical" sections of Book 6 of Vergil's *Aeneid*. These appear in an Irish miscellany of scholia to classical Latin poets: Bern, Burgerbibliothek, MS 363. John is also credited (not wholly implausibly) with a life of Vergil.

John's translations of Greek patristic works and of Priscianus Lydus have been mentioned above. To these might be added a translation of the gospels in St. Gall, Stiftsbibliothek, MS 48, which some scholars assign to the circle of Sedulius. In any case, the interlinear Latin translation of the Greek text (both were written in an Irish hand) bears witness to Irish biblical scholarship in the mid-ninth century. The translation, which cannot be classified by a siglum, was specially composed to serve as a crib to the Greek text to enable the most industrious scholars of the day to appreciate the original.

Of John's free-standing compositions—the *carmina* (poems), *De praedestinatione*, and *Periphyseon*—the *carmina* are exceptionally valuable for what they tell us about John's enthusiasm, friendships, and attitudes, his passion for Greek, his affection and respect for his king, even his undisguised anti-Jewish attitudes. They also record events of the day: the civil war of 858, Charles the Bald's donation to the abbey of St. Denis (867), and the king's plan to construct a major church dedicated to St. Mary at Compiègne (870).

The *De praedestinatione* was commissioned by Hincmar, archbishop of Reims, to combat the imputed heresy of Godescalc (Gottschalk of Orbais), who argued that God predestined not only the elect but also the damned and denied any notion of free will, even the efficacy of the sacraments, since they could offer no help to those already condemned. Eriugena's refutation of Godescalc apparently went too far in the direction of asserting free will, for it inspired attacks by Prudentius of Troyes and Florus of Lyon and led to the official condemnations mentioned above. Interestingly, Eriugena's work reveals his deep knowledge of Augustine, for John used passages from Augustine to refute Godescalc's claims for double predestination!

For modern readers the *Periphyseon*, or *De divisione naturae* (On the Division of Nature) is by far Eriugena's most important work, one that indisputably established John's reputation as one of the most outstanding thinkers of the Middle Ages. This long work, arranged as a dialogue between master and pupil, was the product of several revisions; indeed, John's rethinking of his composition can be seen in the pages of Reims, Bibliothèque municipale 875, which contain numerous corrections and additions in his own hand. The work shows the influence of John's dialectical training and a Neoplatonic cast of thought mediated by Augustine, Plato's *Timaeus*, and especially the writings of Pseudo-Dionysius. John divides "nature" into four categories: nature that creates and is not created, nature that is created and creates, nature that is created and does not create, and nature that does not create and is not created. Categories 1 and 4 refer to God as the origin of all things and their end, respectively. Category 2 embraces the forms, or divine ideas, that are coeternal with God, yet dependent upon him. These, in turn, create the intelligible world (Category 3). In the end, all nature will return to God. This "return" will not be an annihilation of individuality but a reunion in a spiritual state in the ultimate source of all things. As matter will no longer exist, the elect and the damned will not be dispatched to a physical heaven or hell but will subsist in God, with each group saved or damned according to their respective consciences. While such radical notions, rooted as they were in the theology of the Christian Orient, must have been abhorrent to John's contemporaries (indeed, his writings were often labeled heretical), they do not cease to command the admiration of modern readers for their intelligence and originality.

MICHAEL HERREN

References and Further Reading

Beierwaltes, Werner, ed. *Eriugena Redivivus: Zur Wirkungsgeschichte seines Denkens im Mittelalter und im Übergang zur Neuzeit.* Heidelberg: Universitätsverlag Carl Winter, 1987.

Brennan, Mary. *A Guide to Eriugenian Studies: A Survey of Publications 1930–1987.* Fribourg: Éditions Universitaires, 1989.

———. "Materials for the Biography of Johannes Scottus Eriugena." *Studi Medievali* ser. 3a, 27.1 (1986): 413–460.

Cappuyns, Maïeul. *Jean Scot Érigène, sa vie, son oeuvre, sa pensée.* Louvain: Abbaye du Mont César, 1933.

Contreni, John, and Padraig P. Ó Néill. *Glossae divinae historiae: The Biblical Glosses of John Scottus Eriugena.* Florence: SISMEL Edizioni del Galluzzo, 1997.

Ganz, David. "The Debate on Predestination." In *Charles the Bald: Court and Kingdom,* edited by Margaret Gibson and Janet Nelson. Oxford: B.A.R. International Series 101, 1981.

Herren, Michael W. *Iohannis Scotti Eriugenae Carmina.* Scriptores Latini Hiberniae 12. Dublin: Institute for Advanced Studies, 1993. [text and English translation]

Jeauneau, Édouard. *Études Érigéniennes.* Paris: Études Augustiniennes, 1987.

———. *Maximi Confessoris Ambigua ad Iohannem iuxta Iohannis Scotti Eriugenae interpretationem.* Corpus Christianorum Series Graeca 18. Turnholt: Brepols, 1988.

———. *Iohannis Scotti Eriugenae Periphyseon.* Corpus Christianorum Continuatio Medievalis 161, 163, 164. Turnhout: Brepols, 1996. [Latin text with additions removed]

Jeauneau, Édouard, and Paul Dutton. *The Autograph of Eriugena.* Turnhout: Brepols, 1996.

Lapidge, Michael, and Richard Sharpe. *A Bibliography of Celtic-Latin Literature 400–1200.* Dublin: Royal Irish Academy, 1985.

Marenbon, John. *From the Circle of Alcuin to the School of Auxerre: Logic, Philosophy and Theology in the Early Middle Ages.* Cambridge: Cambridge University Press, 1981.

Moran, Dermot. *The Philosophy of John Scottus Eriugena: A Study of Idealism in the Middle Ages.* Cambridge: Cambridge University Press, 1989.

Sheldon-Williams, I. P., Ludwig Bieler, Édouard Jeauneau, and John O'Meara. *Johannes Scottus, Periphyseon.* 4 vols. Scriptores Latini Hiberniae 7, 9, 11, 13. Dublin: Institute for Advanced Studies, 1968–1995.

See also **Biblical and Church Fathers Scholarship; Classical Influence; Education; Glosses; Poetry, Hiberno-Latin**

ETYMOLOGY

In contrast to modern linguistic etymology, which studies the origin of words (combining expression and meaning), their interrelationship, and their historical changes, medieval etymology is ontological insofar as it assumes that the relationship between the signifier (word, name) and the signified (thing, person) is not arbitrary but that the investigation of the former will throw light on the nature of the latter. ("Etymology" < *etymologia*, is from Greek *etymon* "the true, original thing" + *-logia* = "the science of origin"). In modern times, medieval etymology has been much ridiculed because of its "unscientific" approach, especially its way of using or adapting morphologically similar and semantically suitable words, its method of dividing words into (sometimes dual language) components (Early Irish *bélra n-etarscartha* "the science, literally language, of separation"), and the absence of the postulate of uniqueness. Medieval etymology is theoretically well founded.

It has been claimed that the Bible with its numerous etymologies and etymological origin tales—see, for instance, the double explanation of the names of the sons of Jacob or the origin tale of *Passover*—was the model for the medieval Irish scholars. However, the stimulus for systematic etymological research and application by Irish scholars from the seventh century onward came from Isidore (Bishop of Seville, † 636). His *Etymologiae* (also called *Origines*), an encyclopedic collection of heterogeneous materials arranged according to subjects, was at the same time their methodological guide and their practical model. Isidore's philosophy is briefly: *Omnis enim rei inspectio etymologia cognita planior est.* (The investigation of every thing is clearer once the etymology is known.) (*Etym.* I 19,2; see also I 7,1.)

In Irish texts etymologies are found as part of interlinear and marginal glosses or they are integrated into the text as, for instance, the explanations of technical terms at the beginning of the legal text *Críth Gablach*. Etymology is either implicit or identified as such. There are general collections (alphabetically arranged like book X "De vocabulis" of the *Etymologiae*) such as *Cormac's Glossary* and specialized collections such as the legal *O'Davoren's Glossary*, which includes valuable quotations, or the more elaborately explained names (epithets) of famous persons in *Cóir Anmann* (The Correctness of Names). The self-contained etymological origin tales of the names of places, in prose and verse versions, are called *Dinnsenchas* (The Lore of Famous Places). In the prehistoric ("synthetic") part of the Irish history book *An Lebar Gabála* (The Book of Invasions; Invasion Myth) whole sections originate in this type of etymology—see, for instance, the explanation of *Scotti* (i.e., the Irish) from *Scythi* (otherwise also from Pharaoh's daughter *Scotta*).

In double-barreled place names the element after generics like *dún* (fort) or *sliab* (mountain) was usually interpreted as the name of a person. Thus in the Ulster epic *Táin Bó Cuailnge* at a place called *Áth Buide* (presumably, The Yellow Ford) Cú Chulainn killed an adversary by the name of *Buide*; therefore the ford was called *Áth Buide* (Buide's Ford), as is explicitly stated. In *Betha Senáin* (*The Life of St. Senáin*) the name of his island, *Inis Chathaig* (Scattery Island), is elaborately explained through the presence and activities of a monster called *Cathach*, which he expelled.

The presence of etymology in all types of Irish texts and, particularly, its contribution to the growth of medieval Irish literature deserve further investigation.

ROLF BAUMGARTEN

References and Further Reading

Arbuthnot, Sharon. "Short Cuts to Etymology: Placenames in *Cóir Anmann*." *Ériu: Founded as the Journal of the School of Irish Learning* 50 (1999): 79–86.

Baumgarten, Rolf. "Placenames, Etymology, and the Structure of *Fianaigecht*." In *The Heroic Process—Form, Function and Fantasy in Folk Epic: Proceedings of the International Folk Epic Conference, University College Dublin, 2–6 September 1985*, edited by Bo Almqvist et al., 1–24. Dún Laoghaire (County Dublin): Glendale Press, 1987 [also in *Béaloideas: The Journal of the Folklore of Ireland Society* 54–55 (1986–1987): 1–24].

———. "Etymological Aetiology in Irish Tradition." *Ériu* 41 (1990): 115–122.

———. "Creative Medieval Etymology and Irish hagiography (Lasair, Columba, Senán)." *Ériu* 54 (2004): [forthcoming].

Herren, Michael. "On the Earliest Irish Acquaintance with Isidore of Seville." In *Visigothic Spain: New Approaches*, edited by Edward James. Oxford: Clarendon Press, 1980.

Hillgarth, Jocelyn N. "Ireland and Spain in the Seventh Century." *Peritia: Journal of the Medieval Academy of Ireland* 3 (1984): 1–16.

Lindsay, Wallace M. *Etymologiarum sive originum libri XX Isidori Hispalensis episcopi (Scriptorum Classicorum Bibliotheca Oxoniensis).* 2 vols. Oxford: 1911. Republished, with Spanish translation and notes, by José Oroz Reta and Manuel-A. Marcos Casquero, *San Isidoro de Sevilla: Etimologías—Edición Bilingüe; Introducción General* by Manuel C. Díaz y Díaz (*Biblioteca de Autores Cristianos, Vols 433–434*). 2 vols. Madrid: 1982–1983.

Russell, Paul. "The Sounds of a Silence: The Growth of Cormac's Glossary." *Cambridge Medieval Celtic Studies* 15 (1988): 1–30.

See also **Biblical and Church Fathers Scholarship; Dinnsenchas; Glosses; Invasion Myth**

F

FACTIONALISM

Within the colonial community of medieval Ireland, factionalism is a theme traditionally associated with the later Middle Ages. Factionalism was, however, endemic from the outset of English involvement in Ireland. The 1169 invasion was essentially an initiative of an acquisitive Anglo-Norman nobility, and lust for land immediately caused rivalries to develop within the new colonial baronage. In this scramble for power, the Anglo-Normans manipulated the factious Gaelic political system, supporting competing Gaelic lords in an attempt to undermine their own Anglo-Norman rivals. These struggles at times amounted to civil war. What is interesting is that, far from attempting to assuage such violence, it was unofficial royal policy to promote rivalries in Ireland. An example presents itself at the first instance of royal intervention in Ireland, the expedition of Henry II in 1171–1172. Concerned by the independence afforded by Strongbow's Leinster power base, the king counterbalanced him with a grant of Mide (Meath) to Hugh I de Lacy. Such practices inevitably caused friction and ultimately violence within the Anglo-Norman community in Ireland.

We should note that to excite conflict in this manner was contrary to the contemporary concept of the king as the provider of justice and the arbitrator, rather than promoter, of disputes. It was, moreover, extraordinary in practical terms, since the king's natural instrument of government in remote lordships was the nobility. Yet it was precisely this group that royal policy undermined. An explanation for this behavior is found in the English crown's preoccupation with affairs on the continent, which meant that it could not afford to allow the baronage of its new acquisition in Ireland too great a measure of independent power.

Examples of the king exploiting personal rivalries in order to curb the power of nobles in Ireland are manifold in the first century of the lordship's history. King John's license to Hugh II de Lacy to oust John de Courcy from Ulster in 1205 is perhaps the supreme case, but the practice was perpetuated by seemingly less capricious monarchs. The assassination of Richard Marshal by members of the Irish baronage in 1234 was particularly sinister because it took place under the king's peace. Henry III was almost certainly complicit in the murder, reputedly offering Marshal's Leinster lands to those who could "bring him, dead or alive, before the king." It could be objected that the barons in Ireland were merely fulfilling their obligation to oppose an enemy of the king. However, given that those involved in the assassination had a history of conflict with the Marshal family, it is difficult to escape the conclusion that Henry III deliberately exploited rivalries in Ireland to eliminate a political opponent.

Although the disturbances in Ireland typically had local causes, there was a tendency for them to become immersed in the factional conflict that intermittently engulfed England. At the very least the barons of Ireland saw such moments of confusion as an opportunity to indulge their local ambitions. During the Barons's Wars in England in the 1260s, for instance, a dispute emerged between the Geraldine leader, Maurice fitzMaurice, and Walter de Burgh, recently made earl of Ulster. In December 1264, a number of the earl's supporters—including the royal chief governor—were taken captive by the Geraldines, who may have acted at the instigation of the anti-royalist barons in England led by Simon de Montfort. This family rivalry re-emerged in the next generation. In 1294, Richard de Burgh, Walter's son, was imprisoned by the then leader of the Geraldines, John fitzThomas, baron of Offaly. In both cases, however, it seems likely that the real issue was domestic and stemmed from de Burgh's theoretical overlordship of Geraldine lands. John fitzThomas's jealous protection of his rights was not untypical. In 1294, a protracted legal dispute between him and the chief governor, William de Vescy, led to mutual slandering and a challenge to decide the case by wager of battle. John fitzThomas lost by default when he failed to appear in court.

The fourteenth and fifteenth centuries were a period of great instability in English politics, and parties in England frequently attempted to gain support in Ireland. For instance, in the confusion following Edward II's deposition in 1327, Roger Mortimer attempted to curry favor in Ireland by creating the earldoms of Ormond and Desmond in 1328–1329. In doing so, he was playing to one side of a factional dispute that had plagued Ireland throughout the 1320s. After Mortimer was overthrown in 1330, Edward III was highly suspicious of the Anglo-Irish nobility. They had come to expect that they could act independently of royal authority with impunity, and a number of decades were to pass before the king came to rely on them again as instruments of government. Meanwhile, fissures were growing between the Anglo-Irish and the officials born in England who were imposed on the Irish administration. A crisis was reached in 1341, and the problem recurred throughout the medieval period as the "English born in Ireland" increasingly emphasized their autonomy.

Modern historians have stressed that factionalism was one of the great weaknesses of the colony and contributed to its inexorable decline, but it is possible to exaggerate this development. Royal authority certainly diminished in the late Middle Ages, but factionalism was hardly the principal cause. Admittedly the development in its place of strong local lordships centered around the earldoms of Kildare, Ormond, and Desmond led to intense competition for control of the office of chief governor. From 1414, there was a prolonged struggle for political power between the earl of Ormond and the Talbot family. When the Wars of the Roses gripped England from the 1450s, the opposing houses of Lancaster and York became identified in Ireland with the earls of Ormond and Kildare, respectively. This Geraldine-Butler feud continued into the early modern era. However, the intricacies of these various conflicts have only casually been studied, and we should be cautious about attributing to them the "decline" of English lordship in Ireland. The great lineages survived in frontier conditions by employing unorthodox but expedient methods. Their private armies and networks of power admittedly could be used for destructive purposes, both against each other and against the administration. However, although they were technically illegal, it was arguably these methods that ensured their survival and contributed to the endurance rather than decline of English control over much of Ireland.

PETER CROOKS

References and Further Reading

Empey, C. A., and Katharine Simms. "The Ordinances of the White Earl and the Problem of Coign in the Later Middle Ages." *Proceedings of the Royal Irish Academy* 75 section C (1975): 161–187.

Frame, Robin. *English Lordship in Ireland, 1318–1361*. Oxford: Clarendon Press, 1982.

———. "Ireland and the Barons' Wars." In *Ireland and Britain, 1170–1450*, edited by Robin Frame. London: The Hambledon Press, 1995. First published in *Thirteenth Century England I*, edited by P. R. Coss and S. D. Lloyd. Woodbridge: Boydell and Brewer Ltd., 1992.

Griffith, Margaret. "The Talbot-Ormond Struggle for Control of the Anglo-Irish Government." *Irish Historical Studies* 2 (1940–1941): 376–397.

See also **Anglo-Norman Invasion; Chief Governors; Courcey, John de; Desmond; Fitzgerald; Henry II; John; Kildare; Lacy, de; Lacy, Hugh de; Lancastrian-Yorkist Ireland; Leinster; Marshal; Mide; Mortimer; Ormond; Racial and Cultural Conflicts; Strongbow; Ulster, Earldom of**

FAIRS

See **Agriculture; *Feis*; Manorialism; Óenach; Trade**

FAMINE AND HUNGER

Sources from the early Christian period contain little specific information on the subject of hunger and famine. While the best evidence for specific periods of hunger comes from the various sets of Irish annals, the evidence is only as reliable as the sources themselves. Saints' Lives describe miraculous cures of diseases that affected both cattle and people, but they almost never refer to feeding the hungry. By the time of the Viking Age and after, sources that describe not only the facts of famine and hunger but also the causes and long-term effects are far more common. It is clear that since bread and dairy products were medieval Ireland's food staples, anything that negatively affected grain crops or reduced the milk production in cows had the potential to cause hunger and famine. At the worst times, diseases then struck the weakened population.

Weather conditions were a primary cause of famines. Prolonged droughts, although rare in Ireland, reduced the grain crops. More common were periods with too much rain. In the spring this could drown young crops; later in the year too little sun could stunt crop growth. A cold spring meant the planting must wait.

A prolonged and cold winter meant that even the cattle had too little to eat and that alternative (i.e., wild) foods were not even available. The tale *Erchoitmed Ingine Gulidi* contains a particularly devastating description of a farm on the verge of starvation due to too much cold. A rainy or windy autumn could cause damage to crops still in the fields, leaving many insufficiently prepared for winter.

Weather itself was not the only cause of famine, however. Diseases among the cattle or grain crops, which often followed unusual weather, could likewise create times of hunger and famine. For example, the *Annals of Ulster* record for the year 900 "A rainy year . . . Great scarcity affected the cattle." The first Irish Life of Cóemgen clearly demonstrates that protection from bad weather meant protection for the cattle: "For however great the frost and snow on every side of it [the fort], it never penetrates within. And beasts and cattle in time of cold and snow habitually find grass there." Cattle diseases are recorded more often than crop diseases, but even so a scarcity of grain is recorded a number of times.

There is little evidence that war alone caused famine in medieval Ireland, but prolonged periods of fighting could contribute to an already difficult situation. Probably the most tragic example of this occurred between 1315 and 1318. Famines are reported from all over northern Europe for these years. When coupled with Edward Bruce's invasion of 1315–1318, things became even worse for the Irish. The cost of wheat and oats rose sharply; although when crops finally improved the prices dropped back down.

It is clear that most famines primarily affected the young, the elderly, and others already in a weakened state. A. T. Lucas has shown that in times of trouble, wild plants including nettles, water cress, and sorrel, were eaten when they were available. Even tree bark would be eaten if the situation were bad enough. As Fergus Kelly has pointed out, killing off cattle for food was a last, desperate resort, since that only led to food supply problems in the future.

It was during these most difficult times that large numbers of refugees were known to flee to other regions, often wherever their king had allies. However, even if they did manage to arrive in an unaffected area, there was rarely much help to be found. Monasteries helped when they could, and some people permanently attached themselves to monasteries as base tenants due to exactly these circumstances. However, monasteries could not do much to help if they themselves were stricken.

In general, periods of hunger and famine in medieval Ireland were most often brief and localized. The most devastating medieval famines were those that affected not only all of Ireland but other nearby lands as well. For example, a famine that the Chronicum Scottorum reports as ended in 1004 is also described in sources such as the Anglo-Saxon Chronicle. However, while weather patterns that negatively affected all of northern Europe would also impact Ireland, such occurrences were rare until the "Little Ice Age" of the fourteenth century. The Black Death, of course, followed and devastated an already weakened population. Even so, the fifteenth century saw a distinct improvement in weather and therefore food production all over northern Europe, and continued outbreaks of the Black Death were never again as devastating as the first.

MARY VALANTE

References and Further Reading

Crawford, E. M., ed. *Famine: The Irish Experience 900–1900: Subsistence Crises and Famines in Ireland*. Edinburgh: John Donald, 1989.

Kelly, F. *Early Irish Farming*. Dublin: Dublin Institute for Advanced Studies, 1997.

Lucas, A. T. "Nettles and Charlock as Famine Food." *Breifne: Journal of Cumann Seanchais Bhreifne* 1 (1959): 137–146.

———. "The Sacred Trees of Ireland." *Journal of the Cork Historical and Archaeological Society* 68 (1963): 16–54.

See also **Agriculture; Annals and Chronicles; Black Death; Diet and Food**

FEDELMID MAC CRIMTHAINN (*c.* 770 TO 847)

Over king of Munster from 820 to 847, Fedelmid mac Crimthainn's birth year is given in the annals as 770, although the date may have been a later interpolated entry. He was a member of the Eóganacht Chaisil branch of the Eóganachta, although he was not from a dominant segment. The last of his ancestors to hold the kingship of Munster had been Fingen mac Áedo (+ 619), and Fedelmid's accession was unusual in a time when kingly succession was determined by relationship to a recent king, that is a son, grandson, or brother. He may have been a compromise choice at a time when Munster had been under attack by the Uí Néill kingship of Tara and needed a strong warrior as its king. He was closely associated with the Céle Dé church reform movement that began in the eighth century in Ireland and had much in common with the Carolingian reform associated with Benedict of Aniane. The Céli Dé were ascetics who disapproved strongly of the worldly state of the church in Ireland, particularly the great monasteries that were patronized by the great kings and nobles of Ireland, who often stored their wealth in stone buildings provided for that purpose. As monasteries/ecclesiastical settlements became more associated with secular interests they became vulnerable to attack, but Fedelmid was the first great Irish king to introduce the practice. It is possible that his attacks on ecclesiastical settlements such as Clonmacnoise were inspired by a puritan zeal to cleanse the old and sinful ways of the unreformed church. However, it is more likely that such attacks were part of his attempts to replace the Uí Néill as kings of Tara.

In 823, Fedelmid proclaimed the law of Patrick in Munster, with Artri mac Conchobair his favored candidate for bishop of Armagh. In the same year he commenced a war of attrition against ecclesiastical settlements in Uí Néill or border territories when he burned the monastic site of Gallen (County Offaly). In 826, he burned Delbna Bethra (in western Mide) with an army from Munster. In 827, he met the Uí Néill king of Tara, Conchobar, at Birr for what the annals called a "royal conference." Any agreement made at this meeting was short-lived as in 830, Fedelmid was recorded as inflicting a defeat on the Connachta and the Uí Néill. In 831, he plundered the territory of Conchobar near Slane, and the latter replied by plundering the Liffey plain. In 823, he burned the church lands of Clonmacnoise "to the very door of the church" and put to death members of the community. He did likewise at Durrow. He burned the *termon* (sanctuary lands) of Clonmacnoise in 832 and plundered the surrounding land in Delbna Bethra three times in that same year. His first reverse came from Cathal, son of Ailell, king of Uí Maine, at Mag nAí (County Roscommon) in 835.

In 835, Fergus, son of Bodbchad, king of Carraic Brachaide, was killed by Munstermen and in 836, Dúnlang, son of Cathúsach, abbot of Cork, is recorded as having died without communion in Cashel of the kings. The *Annals of Inisfallen* do not record the death of Dúnlang but mention the entry of Fedelmid into the abbacy of Cork the same year. The foundation of Cork had been involved in several battles with other ecclesiastical settlements in that period. He attacked Uí Maine in Connacht in 837. In 838, the annals report "a great royal conference in Cluain Chonaire Tomar, between Fedelmid and Niall [Caille]," that is, at Cloncurry in County Kildare. Whatever the outcome of his meeting with Niall, Fedelmid attacked Mide and Brega—the lands of the Southern Uí Néill—the following year. He also ravaged the kingdom of Delbna Bethra as well as the neighboring Uí Néill kingdom of Cenél Fiachrach (Fir Chell in the annals). The southern annals record that in 840, he harried the north from Birr to Tara and, in 841, Fedelmid led an army to Carman. The king of Tara, Niall Caille, marched against him and defeated the forces of Munster at Mag Óchtar. Following mac Crimthainn's defeat at the hands of Niall Caille, the Uí Néill were again firmly in place as the dominant kingship in Ireland. However, he attacked Clonmacnoise again in 846.

He died in 847 and is described in the annals as "king of Munster, a scribe and anchorite and the best of the Irish." When Fedelmid mac Crimthainn succeeded to the kingship of Munster it was in a weakened state, as his predecessor according to the regnal lists, Tnúthgal mac Donngail of Eóganacht Glenamain, is not mentioned in the annals. In fact, the annals call Artrí mac Cathail, also of Eóganacht Glenamain, who died in 821, the year after Fedelmid became king, king of Munster. The decline of the Eóganacht Locha Léin, whose over kingdom of Iarmumu collapsed when the subject peoples of west Munster transferred their direct allegiance to Cashel around the end of the eighth century, made Fedelmid's position in Munster secure and allowed him to move beyond the province in order to take on the ever-increasing threat of the Uí Néill.

Fedelmid mac Crimthainn was an ecclesiastic, although it is not certain that he was a bishop. The phenomenon of kings who were also ecclesiastics seems to be unique to the south of Ireland, and it is possible that Fedelmid may have inaugurated the tradition of ecclesiastical kingship. His career can best be explained in the light of increasing aggression from the Uí Néill toward Munster and his membership of the Céli Dé, which colored his view toward the great monasteries. It could be said of him that he was the last great Eóganacht king of Cashel, as the Eóganachta dynasties began to go into decline from that time. At a time when Ireland was under attack by the Vikings it is remarkable to note that Fedelmid never struck a blow against them. Fedelmid mac Crimthainn was later revered as a saint and his feast day was celebrated on August 28, according to the Martyrology of Donegal.

LETITIA CAMPBELL

References and Further Reading

Byrne, F. J. *Irish Kings and High Kings*. London: Batsford, 1973.

Charles-Edwards, T. M. *Early Christian Ireland*. Cambridge: Cambridge University Press, 2000.

Hennessy, William, ed. *Chronicum Scotorum*. Rolls Series, London: H. M. Stationary Office, 1866.

Hughes, Kathleen. *The Church in Early Irish Society*. London: Methuen, 1966.

Mac Airt, Seán, ed. *Annals of Inisfallen*. Dublin: Dublin Institute for Advanced Studies, 1952.

Mac Airt, Seán, and Gearóid Mac Niocaill, ed. *Annals of Ulster to A.D. 1131*. Dublin: Dublin Institute for Advanced Studies, 1983.

Murphy, Denis, ed. *Annals of Clonmacnoise*. Dublin: Royal Society of Antiquaries of Ireland, 1896.
Ó Corráin, Donnchad. *Ireland before the Normans*. Dublin: Gill & Macmillan, 1972.
O'Donovan, John, ed. *Annals of the Kingdom of Ireland*. Dublin: Hodges, Smith & Co., 1854.
O'Dwyer, Peter. *Céle Dé*. Dublin: Editions Táilliúra, 1981.

See also **Céli Dé; Clonmacnoise; Ecclesiastical Settlements; Eóganachta; Munster; Uí Néill, Southern**

FEIS

Feis, (plural *feisi*, *fesa*) (spending the night, a feast), traditionally translated "feast" as in *feis Temro* (the feast of Tara), is etymologically the verbal noun of the Old Irish verb *fo-aid* (to spend the night, to sleep), hence the formula *feis la mnaí* means "to sleep with a woman" or "to marry a woman." *Feis* is also a component of the term *banais* (wedding, marriage feast), which is sometimes used in its place.

A *feis* was originally a ritual celebrating the sovereignty of a king, held once in his reign, although not necessarily at the beginning. This ceremony was conceived of as a sacral marriage of the king to the goddess of the territory. Mac Cana believes that the goddess represented the land and people as well as the judicial and spiritual realm of the territory. Through his marriage with the goddess the king became (or was confirmed as) the temporal ruler of her territory. If the king were a just ruler the land would flourish, be fertile, and the people prosper. This concept is embodied in the ideas of *fír flaithemon* (truth of a ruler). Such a territory could be as small as the single *túath* or as large as the kingship of Tara.

Historical accounts of *feisi* are not very numerous. The most famous *feis*, the *feis Temro*, occurs only three times in the *Annals*, the last in 560 during the reign of Diarmait mac Cerbaill. Between the twelfth and fourteenth centuries the term *feis* re-emerges in the historical documents, this time in much greater detail. The most famous account is of the inauguration of the Cenél Conaill king, as related by Giraldus Cambrensis (iii §25). He alleges that the people gathered together at the inauguration site where the successor to the kingship sexually embraced a white mare, which was then slaughtered and cooked into a broth. The king was then bathed in this broth and he and the people both drank of it. Unsurprisingly, this account has been regarded as propaganda, painting the Irish as a pagan, barbaric people. His description, however, has not been entirely regarded as fabrication. Ritual horse sacrifice as part of inauguration rites of a king was a noted feature of Indo-European societies. Likewise, Byrne has suggested that in the eighth century there seems to have been a confused tradition linking the broth bath with the inauguration ceremony. Additionally, the public inauguration site mentioned by Giraldus was an important feature of every *túath*. Another account in the *Annals of Connacht* describes the inauguration of Fedlimid Ua Conchobhair in 1310 as having been a public ceremony at which he married the province of Connacht and his kingship was proclaimed.

Although rare in the early historical sources, there are continuous references to *feisi* in the literature throughout the early and medieval Irish periods. It is from the idealized tradition of Irish literature that the *feis* ritual has been understood, particularly the role of the sovereignty goddess figures such as Medb, Eithne, and Ériu. Given that the *feis* had a pagan origin and involved the marriage of a pagan goddess with the king, it is surprising that it survived into the early Christian period. Its survival has been credited to the conservative nature of Irish society, in particular its learned classes, although this is disputed. Although it seems that *feis Temro* was no longer held after the reign of Diarmait mac Cerbaill, other *feisi* seem to have persisted at the local level. There does not appear to have been a standard ceremony, but rather its form and content seem to have varied from region to region. There were a number of basic characteristics in the inauguration: the granting of the rod of sovereignty, the holding of a race, a procession symbolizing the regions under the king's rule, the singing of praise poetry, and the drinking of some sort of liquor.

By the later Middle Ages the symbolic marriage had disappeared from the inauguration ceremony. In the early Irish concept of kingship, the king was married to the *túath* and became its representative and chief defender but never the owner of the territory. In the later Middle Ages the concept of kingship had changed. Where previously the king had been the representative of the *túath*, he was now a lord and at his inauguration the land passed into his possession with the people acting more as tenants. Under this sort of altered governance, the purpose behind the marriage ritual was lost from the ceremony.

MICHAEL BYRNES

References and Further Reading

Binchy, D. A. "The Fair of Tailtiu and the Feast of Tara." *Ériu* 18 (1958): 113–138.
Dillon, Myles. "The Consecration of Irish Kings." *Celtica* 10 (1973): 1–8.
Freeman, A. M., ed. and trans. *Annála Connacht. The Annals of Connacht A.D. 1224–1544*. Dublin: Dublin Institute for Advanced Studies, 1944.
Jaski, Bart. *Early Irish Kingship and Succession*. Dublin: Four Courts Press, 2000.

Mac Cana, Proinsias. "Women in Irish Mythology." *The Crane Bag* Vol. 4 No. 1 (1980): 7–11.

O'Meara, J. J., ed. "Giraldus Cambrensis, *Topographia Hibernie*. Text of the first recension." *Proceedings of the Royal Irish Academy* 52C no. 4 (1948–1950): 113–178.

Simms, Katharine. *From Kings to Warlords: The Changing Political Structure of Gaelic Ireland in the later Middle Ages*. Woodbridge: The Boydell Press, 1987.

See also **Diarmait mac Cerbaill; Giraldus Cambrensis; Inauguration Sites; Kings and Kingship; Túatha**

FENIAN CYCLE

Fíanaigecht (later spelling, *Fiannaíocht*) (Fenian Cycle) refers to the stories centered on the legendary character Finn mac Cumaill, his *fían* (warrior band), his son Oisín, and his grandson Oscar. From the earliest literary attestation in the seventh century among the Laigin, cultivation of this material spread and became associated in the Old Irish period with places as far apart as south Tipperary, west Cork, the Midlands, and east Ulster. Classified by modern scholars as one of the four medieval Irish literary cycles (along with the Ulster cycle, the cycle of Historical Tales [or cycles of the Kings] and the Mythological cycle), it emerged from fragmentarily documented beginnings to become the dominant literary genre of the post-Norman period in Ireland.

The warrior band, an institution with Indo-European roots, was an integral part of medieval Irish society, occupying an important position on its boundaries. Some scholars have argued, however, that the *fían*'s existence on the margins of society contributed to the early literary neglect of *Fíanaigecht* material by Christian redactors and scribes who wished to discourage warrior bands and associated practices. The lack of relevance of this material to a society obsessed with history and genealogy is another reason cited for its initial lack of cultivation. Its rise in popularity in the post-Norman period has been attributed to a lessening of church opposition to the genre and its adaptability to changing literary tastes.

Similar to the figure of Arthur in Britain, the cult of Finn grew from its localized beginnings to spread throughout Ireland and the rest of the Gaelic-speaking world. During this process, particularly under the influence of the synthetic historians in the tenth and eleventh centuries, a position was found for Finn and his *fían* in the historical and literary record. They were often portrayed as the standing army of King Cormac mac Airt, defending Ireland in the third century against foreign invasion, often from a base at Cnoc Ailinne (Knockaulin, County Kildare) in Laigin territory.

Fenian lays and ballads began to be composed at least as early as the eleventh century, and these became the dominant literary form of the tradition from the late medieval period onward. The two most important extant ballad collections are those preserved in the sixteenth-century Scottish manuscript, the *Book of the Dean of Lismore*, and the seventeenth-century *Duanaire Finn*, compiled among the Irish exiles in Ostend, Belgium. The fame of Fenian balladry had spread all over Europe by the nineteenth century, thanks to James Macpherson. He published three works in the 1760s that purported to be translations of epic poems written by Finn's son, Oisín (in Macpherson's spelling Ossian). These "translations" were, in the main, creations of his own imagination, although they were partly based on genuine ballad tradition. From Macpherson's "Ossian" the term "Ossianic" emerged, a word that is still occasionally used to refer to the cycle as a whole and to the ballad tradition in particular.

Prose material was also extensively cultivated and includes the Middle Irish texts *Tochmarc Ailbe* and *Macgnímrada Find*, the later *Feis Tighe Chonáin*, and the very well known *Tóraigheacht Dhiarmada agus Ghráinne*, a classic example of the love triangle. The central text in *Fíanaigecht* tradition, however, is *Acallam na Senórach* (The Colloquy of the Ancients), recently translated by Dooley and Roe. This long tale that focuses, *inter alia*, on the accommodation reached between the native and Christian traditions, features the encounters between St. Patrick and the last surviving Fenian warriors, most notably Caílte. In this frame-tale, the journey of saint and warrior around fifth-century Ireland is recounted, and the different moral codes of the *fían* and the church are compared, contrasted, and ultimately harmonized. It is a veritable treasure trove of *Fíanaigecht* material, described by Murphy as "a reservoir into which a brilliant late-twelfth-century innovator had diverted several streams of tradition which previously had normally flowed in separate channels." The single largest medieval Irish text, it was written in approximately 1200 in prose interspersed with poetry, the "prosimetrum" form so favored in Irish tradition.

The human and mythic characteristics of Finn mac Cumaill are very fully documented in *Fíanaigecht* tradition. These range from comparisons with Lug and with the Welsh Gwynn ap Nudd; to Finn's role of seer in the early literature; to the magical and supernatural environment that surrounds Finn, his *fían*, and his family in the later material. His attractiveness to an audience is ensured, however, by his continual presentation as a character with all too human qualities and failings. We see him as an unsympathetic and isolated youth, a successful and a spurned lover, a wise and learned poet, a jealous and embittered old man, and a strong and vigorous hunter/warrior; thus he reflects many aspects of the human condition. This complexity of character, coupled with repeated evocation of the beauties of nature, the clever use of *dinnshenchas* (place-name lore), and the recurring presence of magic and the supernatural have combined to place *Fíanaigecht* at the very heart of Irish culture, as is still the case in the modern folk tradition.

KEVIN MURRAY

References and Further Reading

Almqvist, Bo, Séamas Ó Catháin, and Pádraig Ó Héalaí, ed. *Fiannaíocht: Essays on the Fenian Tradition of Ireland and Scotland* (An Cumann le Béaloideas Éireann, 1987) [= *Béaloideas* 54-5 (1986–1987)].

Carey, John, ed. *Duanaire Finn: Reassessments*. Subsidiary Series xiii. London: Irish Texts Society, 2003.

Dooley, Ann, and Harry Roe. *Tales of the Elders of Ireland*. Oxford: Oxford University Press, 1999.

Meyer, Kuno. *Fianaigecht*. Todd Lecture Series xiv. Dublin: Royal Irish Academy, 1910.

MacNeill, Eoin, and Gerard Murphy. *Duanaire Finn*, 3 vols, vii, xxviii, xliii. London: Irish Texts Society, 1908–1953.

Murphy, Gerard. *The Ossianic Lore and Romantic Tales of Medieval Ireland*. Cork: Mercier Press, 1955.

Nagy, Joseph Falaky. *The Wisdom of the Outlaw: The Boyhood Deeds of Finn in Gaelic Narrative Tradition*. Berkeley: University of California Press, 1985.

Ó Fiannachta, Pádraig, ed. *An Fhiannaíocht*, Léachtaí Cholm Cille xxv. Maigh Nuad: An Sagart, 1995.

Ó hÓgáin, Dáithí. *Fionn mac Cumhaill: Images of the Gaelic Hero*. Dublin: Gill & Macmillan, 1988.

See also **Historical Tales; Laigin; Mythological Cycle; Ulster Cycle**

FERMORY, BOOK OF

See **Duanairí; Gerald, Third Earl of Desmond; Gaelicization; Poetry, Irish**

FEUDALISM

Feudalism is a term used by many historians to describe the operation of medieval society. The word derives from the Latin *feodem*, which can be translated as fief or fee, the unit of land granted by a lord to a subordinate (vassal) in return for aid and military service. No such term was in use in the Middle Ages, the concept of feudal tenure being devised in the sixteenth century by French legal historians. The term gained currency in the eighteenth-century Enlightenment, particularly with the political ideas of Baron de Montequieu (1689–1755), who described it in terms of a breakdown of royal authority and a resultant "feudal anarchy." The classic twentieth-century description is that of Marc Bloch (1886–1944), whose hugely influential *Feudal Society* was published in French in 1939–1940 and appeared in English in 1961. Bloch's formulation was extremely broad, but it was so by necessity since it was an attempt to distill centuries of European civilization into a few brief lines:

> A subject peasantry; widespread use of the service tenement (i.e., the fief) instead of a salary . . . ; the supremacy of a class of specialized warriors; ties of obedience and protection which bind man to man . . . ; fragmentation of authority—leading inevitably to disorder; and in the midst of all this, the survival of other forms of association, family and State . . . —such then, seem to be the fundamental features of European feudalism.

Since the 1970s, feudalism has come under sustained attack from historians. It is depicted, correctly, as a construct that postdates the medieval period. This in itself is not necessarily a problem; worse is the fact that the term is so all-encompassing that it seems to have almost no utility. Some have argued that it can still be useful if it is only used to describe the legal relationship between lord and vassal and the services that were owed in return for tenure of land. This, too, is much disputed. The mercurial nature of feudalism is well demonstrated by the contrast between traditional interpretations in mainland Europe and in Britain. The distinctive feature of feudalism for European historians is the fragmentation of royal authority—the so-called feudal anarchy of Montesquieu—whereas for British historians its essence is strong royal power and a precisely calculated hierarchy of land holders, the "feudal pyramid" familiar from schooldays.

Feudalism in Ireland

Feudalism is commonly thought of as coming to England with the Norman conquest of 1066 and being extended to Ireland with the Anglo-Norman invasion of the late 1160s. It is increasingly apparent, however, that early medieval Gaelic society was not so isolated from the European mainstream. In the eleventh and twelfth centuries continued contact with the continent ensured that Irish kings acted like feudal lords, albeit under the broad definition. By the time of the Anglo-Norman invasion, Ireland was moving toward a strong central kingship, and it is likely that this would increasingly have conformed to feudal fashions.

With the influx of Anglo-Normans came the precociously developed institutions and system of government of England. Ireland provided a clean slate for settlement (much as England had in 1066), which meant that the hierarchy of land tenure could be precisely defined, exceptionally so by European standards. If these institutions are thought of as feudalism in the narrow sense, then it arrived with the Anglo-Normans.

The entire story of the invasion is described in the sources in expressly feudal terms. When Diarmait Mac Murchada applied to King Henry II of England for military aid to reconquer his kingdom of Leinster, the chronicler Giraldus Cambrensis states that he did homage and fealty to the English king, the feudal ceremony in which one became the vassal of a lord, in return for which the lord was obliged to provide protection. Diarmait's own exhortation for men to come to his aid includes the promise: "Whoever shall wish for soil or sod, Richly shall I enfeoff them."

This undertaking is found in the French poem known as the *Song of Dermot and the Earl*, composed shortly after the invasion. It therefore gives us an Anglo-Norman rather than an Irish view of Diarmait's intentions. It is clear, however, that the poem's Francophone audience comprehended the settlement of Ireland in what we would call feudal terminology. To enfeoff, or to invest someone with a piece of land—a *feodem* (fief)—is the core of the feudal relationship.

At the pinnacle of the hierarchy was the king of England, who, as lord of Ireland, apportioned lands to his "tenants-in-chief," those who held directly of the crown. The military service that the holder owed was specified: Leinster was held for one hundred knights' fees, Meath for fifty, and so on. The lord of each area then set about subdividing his territory among his followers, a process known as "subinfeudation." The intention was to provide an army that could rapidly be summoned by the king or, more usually, the chief governor. Except for the fact that the church in Ireland did not have to provide military service for its lands, feudal obligations in Ireland were much the same as in England. The king had the right to take what are termed the "feudal incidents" of wardship, marriage, and relief. His tenants owed forty days of personal knight service annually, although in practice this was soon commuted to a money payment called scutage (shield money), which in Ireland was known as royal service. Scutage survived longer and was levied more frequently in Ireland than in England; there were, for instance, nine scutages in the troubled decade 1269–1279. With tenure of land came jurisdiction and the right to hold a court, although the king was the ultimate provider of justice and serious criminal cases were reserved to his judges.

In the first century after the invasion, some Irish lords made genuine efforts to adapt to the invaders' institutions in the hope that it would give them security of tenure. Cathal Crobderg Ua Conchobair (d. 1224), king of Connacht, sought and eventually received a charter for Connacht, which he held "during his good service." His son Feidlim (d. 1265) served Henry III on his campaign in Wales in 1245. The invaders were, however, prepared to use any pretext to expropriate the Gaelic lords. Equally it is likely that the wide kin groups of Gaelic Ireland did not support such feudal notions as primogeniture. The feudal experiment of the Ua Conchobair kings proved in the end to be a negative experience and one that bred acrimony and distrust.

Bastard Feudalism

It should be remembered that Ireland's incorporation into the feudal system occurred extremely late in the development of feudalism as a whole. Almost from its inception, therefore, the new lordship displayed signs of what historians have labeled "bastard feudalism." If any term has generated more debate than feudalism among historians of recent decades, it is surely its supposedly illegitimate successor. Bastard feudalism, so-named in the late nineteenth century, refers to a relationship between lord and man based on money pensions and a written contract rather than land. An older school of historians dated this "perversion" of the feudal system to the reign of King Edward III and the start of the Hundred Years War in the fourteenth century. They blamed it for disorder, violence, and, most extravagantly, for the Wars of the Roses (1455–1485). The Irish evidence supports more recent scholarship that has pushed the chronology further back until we must question if there was ever a purely feudal age. It seems likely that in Ireland, from the moment of English involvement, the clear feudal hierarchy was supplemented by less well-defined expediencies. The lord of Leinster, William Marshal (d. 1219), brought his bastard feudal affinity with him to Ireland in the early 1200s. Indeed, a society like Ireland, where warfare was endemic, was ideally suited to such developments. Lords on the frontiers required their own private armies if they were to hold on to their conquests. Edward I exploited this to the full in the late 1290s when he contracted armies from Ireland to serve in his attempted conquest of Scotland. So valued were these levies that the "Red earl" of Ulster was able to negotiate with the king for the highest pay awarded any earl in the campaign.

A related factor immediately apparent in Ireland and traditionally associated with the "decline" of feudalism is the growth of liberty jurisdiction, under which lords were given powers akin to those of the king within a specific region that had its own administration and courts. At the time of the initial invasion, Leinster, Meath, and Ulster were all created as liberties, and from the fourteenth century the earls of Kildare, Ormond, and Desmond all personally controlled liberties. These liberties are traditionally seen as encouraging factionalism between "over-mighty" subjects. However, given the weakness of royal government in Ireland, it was quite possibly the creation of paid private armies and the power conferred by liberties that ensured the endurance of English control over much of the country.

These are subjects that have been largely neglected since A. J. Otway-Ruthven first examined feudal institutions in Ireland in the 1950s, and the time is ripe to bring current research on late medieval society into an Irish context. The importance of doing so may lie in the fact that so-called bastard feudal connections straddle the two cultures of medieval Ireland. The Gaelic lords may have rejected what we think of as classic feudalism, but the student of bastard feudalism would find the private armies of Gaelic lords, particularly the mercenary galloglass who dominate the military history of the late medieval lordship, surprisingly familiar. From the second half of the fourteenth century onward, the records of the earls of Kildare and Ormond are littered with agreements of retinue between Anglo-Irish and Gaelic lords. Testament to the success of the system is its endurance to the end of the medieval period. The power network of the earls of Kildare—which criss-crossed the cultural frontier—was a challenge to the authority of the Tudor monarchy and provoked a show of unprecedented strength by Henry VIII to bring it down.

PETER CROOKS

References and Further Reading

Bloch, Marc. *Feudal Society*, trans. by L. A. Manyon. London: Routledge and K. Paul, 1961.

Carpenter, David, P. R. Coss, and David Crouch. "Debate: Bastard Feudalism Revised." *Past and Present,* 131, (1991): 165–203.

Ellis, Steven. "Taxation and Defence in Late Medieval Ireland: The Survival of Scutage." *Journal of the Royal Society of Antiquaries* 107 (1977): 5–29.

———. "The Destruction of the Liberties: Some Further Evidence." *Bulletin of the Institute of Historical Research* 54, no. 130 (1981): 150–161.

Frame, Robin. "Military Service in the Lordship of Ireland, 1290–1360: Institutions and Society on the Anglo-Gaelic frontier." In *Medieval Frontier Societies*, edited by Robert Bartlett and Angus MacKay. Oxford: Clarendon Press, 1989.

Hicks, Michael. *Bastard Feudalism.* Essex: Longman, 1995.

Otway-Ruthven, A. J. "Knight Service in Ireland." *Journal of the Royal Society of Antiquaries* 89 (1959): 1–15.

———. "Royal Service in Ireland." *Journal of the Royal Society of Antiquaries* 98 (1968): 37–46.

Reynolds, Susan. *Fiefs and Vassals: The Medieval Evidence Reconsidered.* Oxford: University Press, 1994.

See also **Anglo-Norman Invasion; Chief Governors; Factionalism; Kings and Kingship**

FÍACHNAE MAC BÁETÁIN

Fíachnae mac Báetáin, also known as Fíachnae Lurgan, was a member of the Dál nAraidi, the main dynasty of the people known as *Cruithni.* The Dál nAraidi kings, including Fíachnae, resided at *Ráith Mór* in *Mag Line* (Moylinny), east of Antrim town. Along with another *Cruthni* dynasty they supplied a number of over kings of Ulster between the sixth and the tenth centuries. However, most kings of Ulster were supplied by the "true Ulaid," of which the Dál Fiatach in eastern Down were the ruling dynasty. Fíachnae mac Báetáin's main enemy in Ulster was Fíachnae mac Demmáin of the Dál Fiatach dynasty. After winning many victories over his rival, Fíachnae mac Báetáin was finally defeated and slain by him in 626.

According to the *Annals of the Four Masters*, Fíachnae, whose rule was to span more than three decades, killed Áed Dub mac Suibni and succeeded him in the Ulster kingship in 588. In 597, in one of his many recorded victories in the *Annals of Ulster*, Fíachnae was active as far south as *Slíab Cuae* in Munster. In the *Book of Leinster* under *Ríg Ulad* (kings of Ulster) Fíachnae mac Báetáin can be found between Áed Dub and Fíachnae mac Demmáin as well as under the heading *Rig Dail Araide* (kings of Dál nAraidi).

We may deduce from this that Fíachnae was an important and powerful king, but is it possible that he achieved the ultimate title, that of King of Tara? Fíachnae is not included in Middle-Irish Tara king-lists, while several suspect Uí Neill dynasts are. The earlier *Baile Chuinn* (Conn's vision) king list, on the other hand, omits many of these Uí Neill kings from the late sixth and early seventh centuries. In their place it includes some unidentified names, such as *Féchno.* F. J. Byrne suggests that "it is not altogether impossible that he [Fíachnae mac Báetáin] is the mysterious Féchno who appears as Diarmait mac Cerbaill's successor to the high-kingship of Tara in the Baile Chuind." Some middle-Irish sagas would certainly seem to agree with this suggestion. One such tale actually states that Fíachnae was king of Ireland and Scotland. It may be concluded that Fíachnae mac Báetáin was a powerful Ulster king who extended his influence across the sea and into the midlands and south of Ireland to such an extent that he may at least have been a contender for the kingship of Tara.

In addition to the historical record, Fíachnae appears in the literature as father of the enigmatic Mongán mac Fíachnai. In the tale *Compert Mongáin*, Fíachnae goes to the aid of his ally Áedán mac Gabráin, king of Dál Ríata in Scotland, in a war against the Saxons. In his absence the god Manannán mac Lir visits his wife and Mongán is conceived. The fact that a cycle of tales was composed depicting Fíachnae's son Mongán as an extraordinary figure may be due to some degree to the extent of Fíachnae's power combined with his alleged relationship with the poets of the time.

NORA WHITE

References and Further Reading

Byrne, F. J. *Irish kings and High Kings.* (2nd ed.) Dublin: Four Courts Press, 2001.

Marstrander, Carl. "How Fiachna mac Baedáin Obtained the Kingdom of Scotland" *Ériu* 5 (1911): 113–119.

Meyer, Kuno. *The Voyage of Bran.* London: Llanerch, 1895.

See also **Áedán mac Gabráin; Comperta; Cruthni**

FINE GALL

Fine Gall (literally, kindred of the foreigners) was the name given to a stretch of territory north of the River Liffey that was ruled by the Scandinavians of Dublin. It thus developed after the foundation of the *longport* in 841, at the height of the Viking incursions. Today the name "Fingal" still applies to the area north of the city from the River Tolka to the Devlin River near Gormanstown.

Several place names reflect the Viking history of the area. The names Howth, the Skerries, Ireland's Eye, Lambay, and Holmpatrick found along the coast north of Dublin contain Norse place-name elements. While it is likely that Vikings settled in the district, archaeological evidence (for example, from the excavations at Feltrim Hill in North Dublin) indicates that an Irish population continued to flourish under Viking control. It is also clear that there was a high level of interaction between Gaelic and Scandinavian culture in the area. To date, however, archaeological indicators to Scandinavian settlement north of the Liffey are few.

It is uncertain when the Vikings seized the lands of *Fine Gall.* The temporary occupation of islands off Brega by Vikings in 852 may represent the beginnings of this expansion. Nevertheless, the conquest may have proceeded slowly. The name *Fine Gall* appears in *The Annals of the Four Masters* under the year 868 [= 866], but this seems to reflect a later gloss added to the chronicle record (cf. *Chronicum Scotorum*). More reliably, the name appears in Irish chronicles under the year 1013. On this occasion the Uí Néill over king Máel-Sechnaill II raided *Fine Gall*, including Drinan and Howth. Other references to this name are found from the eleventh century onward.

There is evidence that Dublin's northern hinterland once reached beyond the boundaries of modern Fingal. In the 970s, Vikings temporarily held sway over parts of Meath. By 1052, the northern boundary of their power had retracted to the Devlin River, which is the present boundary of Fingal. Vikings also subdued territory south and west of Dublin. This included all or part of the barony of Rathdown in the south. Indeed, the south county is, so far, the only part of Dublin's hinterland in which archaeological traces of Viking houses have been found—at Brownsbarn (near Clondalkin) and at Cherrywood (near Shankill). The latter has been identified as a longhouse, typical of the ninth century. To the west, the place name Leixlip (Old-Norse *laxhlaup*, salmon leap) may indicate the extent of Dublin's influence. *The Annals of Ulster* suggest that in 938 their influence extended as far as Áth Truisten, near Mullaghmast, in County Kildare. It may be significant that a "hog's back" tomb—that resembles a type from Scandinavian Northumberland—has been found in the area. Nevertheless, their territory shrank with the decline of Viking power from the late tenth century. The boundaries of modern County Dublin may thus represent the final stage in the long-term evolution of the port's territorial power.

Fine Gall was a distinct part of Dublin's wider hinterland. The whole territory was named *Dyflinaskiri* (Dublin shire) in Icelandic sagas. This may be equated with *crích Gall* mentioned in Irish sources. This hinterland was significant as a source of food, building materials, fuel, and other day-to-day goods for Dublin. *Fine Gall* in particular was prized for its agricultural fertility, and the area was later dubbed "the breadbasket of Dublin." In the late tenth and early eleventh centuries there is reference to another important resource, The Wood of Tórir, located near Clontarf. This was probably a source of fuel, building materials, game, nuts, and fodder for pigs for the Dublin market. *Fine Gall* also included the islands of Lambay and Ireland's Eye. These served at different times in Dublin's history as military outposts, trading posts, or refuges. For example, in 902, some of Dublin's inhabitants fled to Ireland's Eye when the port was sacked by troops from Brega and Leinster.

The political significance of *Fine Gall* for Dublin meant that it was often preyed upon by those seeking to win control over the town. At least fifteen attacks are recorded from 962 until 1162. In spite of these dangers a large number of ecclesiastical settlements appear to have flourished in *Fine Gall.* The most prominent were Lusk and Swords. Lusk was founded before the Vikings came to Ireland, but

Swords is first recorded in 994 and it may have received patronage from the kings of Dublin. Each site has a fine round tower that predates the Anglo-Norman invasion.

After the invasion, Henry II retained the lands around Dublin, including *Fine Gall*. These lands were then granted to colonists, including Hugh de Lacy and Almeric St. Lawrence, while large areas were confirmed in the possession of the church. Some Gaelic and Hiberno-Scandinavian landholders seem to have remained in the area despite the influx of new settlers. Nevertheless, the strategic significance and agricultural fertility of *Fine Gall* made it a core area of English colonization. Anglophone culture persisted there for the rest of the Middle Ages. A large number of castles were built from the late twelfth century, including those at Malahide, Swords, Howth, and Dunsoghly. These reflect both the politically disturbed conditions of the region and its wealth. From its foundation *Fine Gall* was closely linked with the fortunes of Dublin and has been an area characterized by the cultural diversity of its inhabitants.

CLARE DOWNHAM

References and Further Reading

Ball, F. Elrington. *A History of the County of Dublin*, 6 vols. Dublin: Alex, Thom & Co., 1902–1920.

Bhreatnach, Edel. "Columban Churches in Brega and Leinster: Relations with the Norse and the Anglo-Normans." *Journal of the Royal Society of Antiquaries of Ireland* 129 (1999): 5–18.

Bradley, John. "The Interpretation of Scandinavian Settlement in Ireland." In *Settlement and Society in Medieval Ireland: Studies Presented to F. X. Martin, O. S. A.* edited by John Bradley. Kilkenny: Boethius Press, 1988.

Clarke, Howard. "Christian Cults and Cult Centres in Hiberno-Norse Dublin and its Hinterland." In *The Island of St. Patrick*, edited by Ailbhe MacShamhráin. Dublin: Four Courts Press, 2004.

Hennessy, William, ed. *Chronicum Scotorum*. Rolls Series, London: 1866.

Mac Airt, Seán, and Gearóid Mac Niocaill, ed. *Annals of Ulster to A.D. 1131*. Dublin: Dublin Institute for Advanced Studies, 1983.

O'Donovan, John, ed. *Annals of the Four Masters—Otherwise Annals of the Kingdom of Ireland*. Dublin: Hodges, Smith & Co., 1854.

Ó Néill, John. "A Norse Settlement in Rural County Dublin." *Archaeology Ireland*, 13 no. 4 (Winter 1999): 8–10.

Simms, Anngret, and Patricia Fagan. "Villages in County Dublin: Their Origins and Inheritance." In *Dublin City and County: From Prehistory to Present, Studies in Honour of J. H. Andrews*, edited by F. H. A. Aalen and Kevin Whelan. Dublin: Geography Publications, 1992.

Valante, Mary. "Dublin's Economic Relations with Hinterland and Periphery in the later Viking Age." In *Medieval Dublin*, I, edited by Seán Duffy. Dublin: Four Courts Press, 2000.

See also **Anglo-Norman Invasion; Castles; Dublin; Ecclesiastical Settlements; Máel-Sechnaill II; Viking Incursions**

FIREARMS AND ORDNANCE

See **Weapons and Weaponry**

FISHING

In the Middle Ages, fishing was an important source of food, livelihood, and income in Irish coastal and estuarine landscapes and the ownership, regulation, and use of fisheries was often a significant aspect of social and economic relationships. Fish was of great importance in the medieval diet, as religious custom forbade meat consumption during Lent, Advent, and after Pentecost as well as on holy days and the eves of Christian celebrations. Moreover, in aristocratic and ecclesiastical households, some species of fish were regarded as delicacies and were often maintained in fishponds. Through the medieval period, both sea fish and freshwater fish were caught by boats, nets, and traps for local consumption or were preserved or transported in barrels to settlements elsewhere, occasionally across large distances.

In the early Middle Ages, it is likely that fishing was a small-scale, subsistence activity intended primarily to produce food for the domestic table, with the surplus distributed in local markets. Early Irish laws, dating to the seventh and eighth century A.D., regulated the use of fishweirs for catching salmon, trout, and eels (the range of Irish native species being quite limited). Sea fishing was probably less important at the time (Kelly 1997). By the tenth and the eleventh century A.D, the growth of urban populations, improved methods of salting and smoking preservation, and the development of Atlantic sea fisheries would have led to fishing becoming a much more significant source of wealth and power. It is likely that by the twelfth century and thirteenth century (if not earlier in many locations) most estuarine and riverine fish weirs would have been taken into the hands of monastic houses, bishops, and manorial lords (Hutchinson 1994; O'Neill 1987; Childs and Kowaleski 2000). Irish salmon and eels were particularly valuable sources of income from these freshwater fishweirs.

Recent coastal archaeological surveys have identified spectacular evidence for medieval fish weirs on the Shannon estuary, County Clare and County Limerick (O'Sullivan 2001), and on Strangford Lough, County Down (McErlean and O'Sullivan 2002). These were artificial barriers of stone or wood built to deflect fish into an opening where they could be trapped in nets or baskets. Most were ebb-weirs, catching fish on a falling tide, and were typically V-shaped stone or wooden structures with post-and-wattle fences and baskets of varying size and construction. Local and regional differences in size, location, building materials, and trapping mechanisms indicate the role of local tradition and practice in the work of fishing communities.

On the Shannon estuary, intertidal archaeological surveys have revealed evidence for several medieval wooden fishtraps, dated to between the fifth and the thirteenth century A.D (O'Sullivan 2001). The Shannon estuary fish weirs tended to be small structures, hidden away within the narrow, deep creeks that dissect the estuary's vast expanses of soft, impenetrable muds. Despite being relatively small, they could have literally sieved the water of all fish moving around with the tides. They were oriented to catch fish on either the flooding or ebbing tide and could in season have taken large catches of salmon, sea trout, lampreys, shad, flounder, and eels (the latter in October through November).

The earliest known fish weir is a small post-and-wattle fence (*c.* 8 meters in length) on the Fergus estuary, County Clare (a tributary of the Shannon estuary), dated to 447–630 A.D (O'Sullivan 1993–1994). Early medieval fish weirs have also been located on the mudflats of the Deel estuary, County Limerick, which flows into the upper Shannon estuary. These weirs provide intriguing evidence for local continuity of size, form, and location and appear to have essentially replaced each other between the eleventh and the late twelfth century A.D. Medieval fish weirs are also known from the Shannon estuary mudflats at Bunratty, County Clare, dating to between the eleventh and the thirteenth century A.D. At Bunratty 4, a complex V-shaped structure had at least three phases of use at the site, with several post-and-wattle fences repaired over a period of time, probably 20–30 years. It has been radiocarbon dated to A.D 1018–1159, indicating its possible use by a Gaelic Irish community at Bunratty prior to Anglo-Norman colonization. Probably the most spectacularly preserved medieval fish weir in Ireland is the site of Bunratty 6. This had two converging post-and-wattle fences (22 meters in length) of hazel, ash, and willow braced against the ebbing tide by diagonally placed poles. These fences led to a rectangular wooden structure on which was placed a massive woven basket trap (4.2 meters in length, 80 centimeters in diameter) dated to A.D 1164–1279. The Bunratty 6 fish weir was probably used by the population of the Anglo-Norman borough at Bunratty, one of the most important medieval settlements and ports in the region. Intriguingly, there is a strong continuity in fish-weir style and construction across time, indicating perhaps that Gaelic Irish *betaghs* were supplying the Anglo-Norman borough with fish for its domestic tables, fairs and markets (O'Sullivan 2003).

Archaeological surveys on Strangford Lough, County Down, have also revealed evidence for medieval fish weirs, mostly concentrated in Grey Abbey Bay in the northeast end of the Lough. At least fifteen wooden and stone-built structures have been recorded and the wooden traps in particular have been radiocarbon dated to between the eighth and thirteenth centuries A.D. (MacErlean and O'Sullivan 2002). Strangford Lough probably had a range of fish species, including salmon, sea trout, plaice, flounder, mackerel, cod, grey mullet, and skate, with large numbers of eels in the abundant kelp growth.

The Strangford Lough wooden fish weir fences measure between 40 meters and 200 meters in length. At the "eye" of the converging post-and-wattle fences, baskets or nets were probably hung on rectangular structures. The earliest fish weir at Chapel Island, radiocarbon dated to A.D. 711–889, may have been owned by the early medieval monastery of Nendrum, County Down, across the lough. In Grey Abbey Bay, 1.5 kilometers to the east, three wooden traps and four stone traps have been recorded. At South Island, a large V-shaped wooden trap has provided two separate radiocarbon dates of 1023–1161 A.D. and 1250–1273 A.D. Similar V-shaped wooden traps found elsewhere in the bay have produced radiocarbon dates of 1037–1188 A.D. and 1046–1218 A.D. The Strangford Lough stone-built fish weirs are broadly similar in size, form, and orientation. They typically measure between 50 meters and 300 meters in length, 1.1 meter in width, and they probably stood between 0.5 meter to 1 meter in height. The stone fish weirs are variously V-shaped, sickle-shaped, and tick-shaped in plan, mainly depending on the nature of the local foreshore.

The massive physical scale and form of the Strangford Lough weirs probably indicate a local response to the broad, sandy beaches of the lough, although it is also clear that these were intended to literally harvest all of the fish out of this part of the lough. The Strangford Lough structures were clearly in use in the bay throughout the Middle Ages. Some of the large wooden and stone weirs may have been the property of the Cistercian community of Grey Abbey, founded in 1193 A.D.

The later Middle Ages see an expansion in Ireland's offshore fisheries. By the early thirteenth century, Irish fleets from ports along the east and south coast (e.g., Ardglass, Drogheda, Dublin, Wicklow, Arklow, and Waterford) were operating in the herring fisheries of the Irish Sea and were exporting fresh, salted, and smoked fish (particularly herring and hake) in large amounts to

Bristol, Chester, and the west coast of England. The herring fisheries off Ardglass and Carlingford were also attracting hundreds of ships from Wales, southwest England, and Spain. During the fourteenth and fifteenth centuries, the rich fishing grounds off the south and west coasts, warmed by climatic change, were being harvested by foreign fishing fleets, particularly those of England, Wales, Scotland, Brittany, Gascony, and Iberia. Among these saltwater fish, herring was the most important export, along with cod, hake, pollock, whiting, and ling.

Although the Gaelic Irish population may not have fished these grounds themselves, they used them as an important source of income. Foreign fishing fleets operated out of sheltering havens under the control of Gaelic Irish lords (e.g., the O'Driscolls and O'Sullivan Beares of west Cork), who profited by victualing the fleets, by issuing licenses for fishing, and by charging for the use of their harbors and foreshores (Breen 2001). By the mid-fifteenth century, the English government, concerned at loss of customs revenue through illegal exports, attempted to restrict foreign fisheries off Baltimore, to little initial effect. In the early sixteenth century, foreign fishing fleets operating off Ireland were required to land a portion of their catch in Ireland. Nevertheless, by the late sixteenth century, reputedly 600 Spanish ships were fishing off Ireland.

AIDAN O'SULLIVAN

References and Further Reading

Breen, C. "The Maritime Cultural Landscape in Medieval Gaelic Ireland." In *Gaelic Ireland c.1250–c.1650*, edited by P. J. Duffy, D. Edwards, and E. Fitzpatrick. Dublin: Four Courts Press, 2001.

Childs, W., and M. Kowaleski. "Fishing and Fisheries in the Middle Ages." In *England's Sea Fisheries: The Commercial Sea Fisheries of England and Wales since 1300*, edited by D. J. Starkey, C. Reid, and N. Ashcroft. London: Chatham Publishing, 2000.

Hutchinson, G. *Medieval Ships and Shipping*. London: Leicester University Press, 1994.

Kelly, F. *Early Irish Farming*. Dublin: Dublin Institute for Advanced Studies, 1997.

McErlean, T., and A. O'Sullivan. "Foreshore Tidal Fishtraps" In *Strangford Lough: An Archaeological Survey of its Maritime Cultural Landscape*, edited by T. McErlean, R. McConkey, and W. Forsythe.Belfast: Blackstaff Press, 2002.

O'Neill, T. *Merchants and Mariners in Medieval Ireland*. Blackrock, County Dublin: Irish Academic Press, 1987.

O'Sullivan, A. "An Early Historic Period Fishweir on the Upper Fergus Estuary, Co. Clare." *North Munster Antiquarian Journal* 35 (1993–4): 52–61.

O'Sullivan, A. *Foragers, Farmers and Fishers in a Coastal Landscape: An Intertidal Archaeological Survey of the Shannon Estuary*. Discovery Programme Monographs 4. Dublin: Royal Irish Academy, 2001.

O'Sullivan, A. "A Day in the Life of a Medieval Fisherman . . . and of Intertidal Archaeologists." In *Lost and Found: Discovering Ireland's Past*, edited by J. Fenwick. Dublin: Wordwell, 2003.

See also **Diet and Food; Manorialism; Religious Orders; Ships and Shipping**

FITZGERALD

Barons of Offaly to 1316

The Anglo-Norman family known as the Fitzgeralds or Geraldines of Kildare emerged from relatively modest beginnings, being descended from Henry I's castellan Gerald of Windsor and Nest, daughter of the Welsh prince Rhys ap Tewdwr. The family came to Ireland with Richard "Strongbow" de Clare, lord of Strigoil in 1169, when the pioneering exploits of Maurice fitzGerald (d. 1176) earned him the reward of a grant of land in the form of the middle cantred of Offelan in County Kildare.

Maurice's middle son Gerald fitzMaurice (d. 1204) became the first Geraldine baron of Offaly through marriage to Eva de Bermingham, thereby gaining the important centers of Lea and Rathangan. He also acquired the manors of Maynooth and Rathmore in County Kildare from his elder brother William, baron of Naas. Finally, he took possession of Croom in County Limerick through his participation in the Anglo-Norman invasion of Thomond during the 1180s and 1190s. By his death, he had gained possession of the manors and estates that subsequently formed the core of the family's landed interests.

The Fitzgeralds became prominent in the colony's affairs under the leadership of Maurice fitzGerald, second baron of Offaly (d. 1257). First, fitzGerald expanded the family's holdings in Limerick by acquiring the manors of Adare and Grene. More importantly, he held office as chief governor from 1232 to 1245. Notwithstanding his involvement in the death of his feudal lord Richard Marshal, lord of Leinster, in April 1234, he quickly earned the trust of Henry III. He used his authority as chief governor to summon the feudal host of the lordship to participate in the invasion of Connacht in 1234–1237 led by Richard de Burgh and Hugh de Lacy, earl of Ulster. In return, de Burgh granted him the manors of Ardrahan and Kilcolgan in County Galway, while de Lacy granted him estates in Mayo and Sligo as well as claims to lands in Fermanagh and Donegal.

On his death, the family's holdings were divided between his grandson and heir Maurice fitzGerald, third baron of Offaly (d. 1268), who inherited the core estates in Kildare and Limerick, and his younger son Maurice fitzMaurice (d. 1286) who inherited the estates and claims in Connacht and Ulster.

In 1264–1265, the Fitzgeralds' rivalry with the de Burghs as to which lineage would dominate the northwest led to the outbreak of a bitter civil war that caused widespread devastation throughout Ireland and during which the two Geraldine magnates imprisoned the chief governor. However, a truce was established between the two lineages without settling the dispute and both Fitzgerald lords escaped punishment by fighting in the royalist cause in England in 1265–1266. In 1266, the third baron cemented his family's high status by marrying the king's niece Agnes de Valence (d. 1310).

However, fitzGerald drowned in 1268, sending the family's fortunes into near terminal decline. The Fitzgeralds were forced to endure a lengthy minority until 1285, while de Valence kept control of the family's Limerick properties for the rest of her life. More ominously, from 1272 onward, the Irish dynasties of the midlands in general, and the Ua Conchobair Failge dynasty in particular, became hostile to the settlers, and by 1284, Lea Castle had been burned. Maurice fitzMaurice died without male heirs in 1286 and in 1287, the fourth baron, Gerald fitzMaurice, died childless at the age of twenty-two. However, just before he died, contrary to customary law, he transferred the property and lordship to his cousin John fitzThomas fitzMaurice (d. 1316).

Although fitzThomas appears to have been the sole surviving male representative of the family, he effected a decisive reversal in the family's fortunes. His primary goal appears to have been the reunification of the second baron's legacy. First, the government helped him to temporarily pacify the midlands Irish dynasties. Second, he persuaded Maurice fitzMaurice's heiresses to bequeath him their properties and claims in Connacht and Ulster. On his return home from visiting King Edward in 1292, however, the unresolved question of supremacy in the northwest brought fitzThomas into conflict with both the chief governor William de Vescy and Richard de Burgh, earl of Ulster, after quarreling over the appointment of a king of Connacht in 1293. FitzThomas subsequently accused de Vescy of treason and succeeded in having him removed from office. In 1294, fitzThomas captured the earl of Ulster and attacked his supporters in Meath, Kildare, and Connacht in an explosion of lawlessness known as the "time of disturbance." King Edward intervened and summoned fitzThomas to Westminster in disgrace and ultimately stripped him of his family's Connacht and Ulster properties as a punishment.

FitzThomas redeemed himself with the king by repeated military service, twice in Scotland and once in Flanders. Although he was never entrusted with the office of chief governor, he played a leading role in the increasingly strenuous efforts to pacify the Irish midland dynasties. When Edward Bruce invaded Ireland in 1315, the aged fitzThomas remained loyal to Edward II and was one of the leaders of an army that was defeated by the Scots in January 1316. Immediately afterward, he traveled to England to confer with the king, who created him earl of Kildare in May 1316, four months before his death.

Earls of Kildare to 1534

For the remainder of the fourteenth century, fitzThomas's heirs played prominent roles in the lordship's affairs while consolidating their power in Kildare. The second earl Thomas fitzJohn (d. 1328) was granted the liberty of Kildare in 1318 and twice served as chief governor. Initially, the position of his younger son, the fourth earl Maurice fitzThomas (d. 1390), was threatened by the suspicions harbored by Edward III toward the Anglo-Irish magnates. However, after campaigning in France in 1347, the earl gained the king's trust and subsequently often served as chief governor in short spells. He also devoted much energy to the defense of the Kildare marches, both through military action and forging alliances with midlands Irish dynasts such as An Sinnach (Ua Catharnaig) and Még Eochagáin.

The first half of the fifteenth century saw a sharp decline in the Fitzgeralds' influence. The fifth earl, Gerald fitzMaurice (d. 1432), was principally known as an adherent of his son-in-law, the earl of Ormond, and was arrested in 1418 on charges of plotting against John Talbot, Lord Furnival, on Ormond's behalf. Overall, however, his tenure was marked by a major increase in the power of the Ua Conchobair Failge dynasty, which captured Rathangan within a year of his death.

The recognition of Thomas "fitzMaurice" fitzGerald (d. 1478) as seventh earl in 1456 represents a major turning point in the family's fortunes. The Yorkist seventh earl's success was closely connected to his lengthy period in office as chief governor during which he dovetailed the successful defense of the Pale with the advancement of his personal interests by subduing his Ua Conchobhair Failge neighbors and recovering large tracts of territory in the Kildare marches. Essentially, he pioneered the methods and tactics later used by his son Gerald, the eighth earl (d. 1513), to establish the "Kildare ascendancy."

Upon the death of the eighth earl, his son, Gerald fitzGerald (d. 1534), inherited both the earldom and the chief governorship. At first, the ninth earl continued to govern the lordship using the methods so successfully employed by his father, dispensing gifts and protection to his adherents in return for their service in peace and war. However, from 1515 the "Kildare system" encountered increasing levels of opposition within the lordship, from both his estranged brother-in-law Piers Ruad Butler, ably assisted by his wife Margaret and from sources within the Pale, whose opposition was based upon the negative effects upon the Pale of the introduction of March customs such as coyne and livery.

Kildare's position was progressively undermined at court until he was replaced in 1520 by the earl of Surrey. The Fitzgeralds' response of attempting to render the lordship ungovernable was temporarily successful and the earl was reappointed in 1524. However, under the strain of closer royal scrutiny and constant complaints from Ireland, the system broke down. Following a series of replacements and reappointments, the earl was summoned to England in 1534. There, facing replacement again, the dying earl ordered his son Thomas, lord of Offaly (d. 1537), to launch the rebellion that finally destroyed the family's dominance in Ireland.

General Observations

When the Kildare Geraldines' experiences are taken together, some common familial characteristics may be discerned. They expressed their religious piety conventionally through the foundation of religious establishments such as the Dominican house at Sligo or the Augustinian house at Adare. They showed a keen sense of self-awareness, as exemplified by their continued use of patronymics. In general, the Fitzgeralds did not express great interest in learning and their literary patronage does not compare with that of their Butler or Desmond peers. For all of their familiarity with native Irish customs and culture, they showed a clear preference for marriage with individuals of English descent or preferably of English birth. Overall, their success can be explained in terms of their military qualities, their cultivation of personal relations with the king, their ability to operate readily in both Irish and Anglo-Irish society, and, above all, their ruthless opportunism.

CORMAC Ó CLÉIRIGH

References and Further Reading

Bryan, Donough. *Gerald Fitzgerald the Great Earl of Kildare.* Dublin: The Talbot Press, 1933.

Ellis, Steven. *Reform and Revival: English Government in Ireland, 1470–1534.* Woodbridge: 1986.

Mac Niocaill, Gearóid, ed. *Crown Surveys of Lands 1540–41, with the Kildare Rental.* Dublin: 1992.

Ó Cléirigh, Cormac. "The Problems of Defence: A Regional Case-study." In *Law and Disorder in Thirteenth-century Ireland: The Dublin Parliament of 1297*, edited by James Lydon. Dublin: 1997.

Orpen, G. H. "The Fitzgeralds, Barons of Offaly." *Journal of the Royal Society of Antiquaries of Ireland* 44 (1914): 99-113.

Otway-Ruthven, A. J. "The Medieval County of Kildare." *Irish Historical Studies* 11, no. 43 (1959): 181–199.

See also **Anglo-Norman Invasion; Bermingham; Burke; Desmond; Gaelic Revival; Lacy de; Lancastrian-Yorkist Ireland; Lordship of Ireland; March Areas; March Law; Marshal; Ormond; Racial and Cultural Conflict; Valence, de**

FITZGERALD, GERALD (*c.* 1456–1513)

Gerald Fitzgerald, the eighth earl of Kildare, chief governor of Ireland (1478, 1479–1492, 1496–1513), was the eldest of four sons and two daughters of Thomas, the seventh earl, and his wife Joan, daughter of James, the sixth earl of Desmond. He was, according to Tudor commentators, unlearned: "rudely brought up according to the usage of the country" but "a mightie man of stature" (Carew) and "a warrior incomparable" (Stanyhurst). In 1472, he had command of twenty-four spearmen for defense of the English Pale, and in March 1478 was elected justiciar after his father's death. The young earl contested Edward IV's decision to appoint Henry Lord Grey as deputy lieutenant a few months later, in a characteristic demonstration of Geraldine power. Grey's reluctance to serve without local support prompted the king to summon the leading lords and officials to court. In the resultant settlement Kildare was given charge of a more broadly based administration, with detailed instructions about preserving good rule and the king's interests. The earl generally observed the spirit of this settlement until Edward's death.

The eighth earl's career marked the height of the family fortunes during the Kildare Ascendancy (1470–1534). Kildare was in many ways a typical early Tudor ruling magnate whose chief recommendation to successive English kings was his ability to rule the marches and protect their basic interests at little cost through the deployment of an extensive *manraed.* Hitherto secondary figures, the earls owed their rise to the crisis of lordship that followed the eclipse of leading magnates of the previous generation—Richard, duke of York; John Talbot, earl of Waterford; and James Butler, fourth earl of Ormond. Kildare power was deliberately built up by successive kings through continued reliance on the earl as governor, grants of land, and eventually marriage into the royal family, so reflecting the earl's enhanced status in Tudor circles.

His second wife, Elizabeth St. John, was Henry VII's first cousin, and his son and heir, Lord Gerald, married Elizabeth Zouche, also the king's relative. Yet royal favor also recognized the earl's organizational and military abilities. As deputy, Kildare built up a standing force of three hundred kerne, galloglass, and horsemen, and he reorganized the English Pale's southern defenses around his principal castles: Maynooth, Rathangan, Portlester, Lea, Kildare, Athy, Kilkea, Castledermot, Rathvilly, and Powerscourt. Marcher defense was also strengthened by matching his numerous children—one son and six daughters with his first wife, Alison FitzEustace, daughter of Lord Portlester; seven sons with his second wife—with prominent English and Gaelic families. Almost all the Gaelic lords whose lands bordered the Pale also paid the earl "black rents." From a Gaelic perspective his dealings with the border chieftaincies differed little from relations between a Gaelic overlord and his vassal chiefs. Kildare's court included a Gaelic entourage, and he spoke and wrote in Gaelic as occasion demanded.

Although admired by historians of the Irish Free State era as a champion of home rule and exemplar of a growing Anglo-Gaelic cultural rapprochement, Kildare was no "Anglo-Irish separatist." Certainly, he exploited the monarchy's renewed weakness after Edward IV's death, exacting better terms from Richard III and intruding as chancellor his brother, Thomas, against the king's wishes. Yet the corollary was his loyal support for the Yorkist cause in the years following. This went far beyond most English magnates and long clouded his relations with Henry Tudor. In 1487, Kildare backed the Yorkist pretender, Lambert Simnel, had him crowned Edward VI in Christ Church Cathedral, Dublin, and recruited four thousand Gaelic kerne commanded by his brother to invade England. The Yorkist army was heavily defeated in a three-hour battle at Stoke-by-Newark in which Thomas Fitzgerald was killed. Kildare held out for a few months but eventually submitted. Pardoned in 1488, the deputy and council refused to give better security for their conduct, threatening to "become Irish every of them," and in 1490 Kildare also evaded a summons to court (Harris). Yet, when in 1491–1492 another Yorkist pretender, Perkin Warbeck, landed at Cork, attracting support from the earl of Desmond, Kildare's cousin, Henry VII, responded much more energetically. He dispatched Sir James Ormond with two hundred troops and dismissed Kildare.

Henry's attempt to build up Ormond as a counterweight to Kildare resurrected the old feud between the two houses, precipitating serious disturbances. With Warbeck still at large, the leading nobles and officials were bound over for their conduct, Kildare in one thousand marks, and summoned to court. Sir Edward Poynings was appointed deputy with 653 troops to hold Ireland and carry out administrative reforms. Kildare actively supported Poynings, encouraging Ulster lords to submit, but in February 1495, he was arrested on charges of plotting with Irish enemies against the deputy, attainted by the Irish parliament, and shipped to England. Thereupon, the Geraldines rose in rebellion, led by Kildare's brother James, who seized Carlow castle. Once Poynings had broken Warbeck's blockade of Waterford, however, resistance collapsed, and the king became anxious for a settlement so as to reduce costs. By 1496, the English parliament had reversed Kildare's attainder, he had married the king's cousin, and a formal investigation of his contacts with the Ulster lords had cleared him of treason. Accordingly, following undertakings given before the king's council in August, he was reappointed deputy. Kildare's son, Lord Gerald, remained at court as pledge for his conduct.

Thereafter relations between king and earl remained harmonious. These were years of comparative peace, prosperity, and strong government in the lordship. Kildare made regular progresses throughout Ireland: he visited outlying towns that seldom saw the deputy, including Carrickfergus in 1503, Galway in 1504, and Limerick in 1510. In 1512, he captured Belfast and Larne castles. In 1503, he also visited court for his son's marriage, after which Lord Gerald returned to Ireland as treasurer. In 1504, the largest engagement of the period, at Knockdoe near Galway, saw the Pale levies of English bills and bows and Kildare's Gaelic clients defeat Ua Briain and Clanrickard Burke in a rare pitched battle. The king rewarded Kildare with election to the Order of the Garter. Periodic expeditions against Ua Briain reflected the extended horizons of English rule, but not all were successful. In 1510, he broke down the latter's bridge over the Shannon but suffered heavy losses. On Henry VII's death in 1509, Kildare was elected justiciar according to custom but soon was reappointed deputy by the young Henry VIII. Age was catching up with the old earl, too. He was wounded in 1511 while campaigning in the midlands. Two years later, he was seriously wounded on campaign—shot while watering his horse near Kilkea. He retired slowly to Kildare and died on September 3. His body was brought to Dublin and buried in Christ Church in the chapel he had built two years earlier.

According to the *Annals of Ulster*, he exceeded all the English in power and fame by keeping better justice and law, building more castles for the English, conquering more territory and razing more castles of the

Irish, but nonetheless being generous to the Gaelic *literati*. His servant, Philip Flattisbury of Johnstown, wrote that Kildare surpassed all previous governors in defeating Irish enemies and reducing them to the king's peace, recolonizing and rebuilding towns long destroyed, and constructing castles and bridges "to the great profit and defense of the English." He was succeeded as governor and earl by Lord Gerald.

STEVEN G. ELLIS

References and Further Reading

Brewer, J. S., and Bullen, W. (ed) *Calendar of Carew Manuscripts preserved in the Library at Lambeth*, 6 vols, London: H. M. Stationary Office, 1867–73, vol. 5 (1871), p. 180.

Bryan, D. *The Great Earl of Kildare: Gerald FitzGerald 1456–1513*. Dublin: The Talbot Press, 1933.

Ellis, S. G. *Tudor Frontiers and Noble Power: The Making of the British State*. Oxford: Clarendon Press, 1995.

Ellis, S. G. *Ireland in the Age of the Tudors 1447–1603: English Expansion and the End of Gaelic Rule*. London: Longman, 1998.

Harris, W. ed. "The Voyage of Sir Richard Edgecombe into Ireland in the Year 1488." *Hibernica* 2 vols. Dublin: privately published, 1747–1750.

Kildare, Marquis of. *The Earls of Kildare and their Ancestors from 1057 to 1773*. (3rd ed.) Dublin: privately published, 1858.

Mac Niocaill, G. *The Red Book of the Earls of Kildare*. Dublin: Irish Manuscripts Commission, 1964.

Mac Niocaill, G., ed. *Crown Surveys of lands 1540–41, with the Kildare Rental Begun in 1518*. Dublin: Irish Manuscripts Commission, 1992.

Miller, Liam and Power, Eileen (ed) *Holinshed's Irish Chronicle, 1577*. Dublin: Irish Academic Press, 1979, pp. 82–83.

Quinn, D. B. In *A New History of Ireland. II Medieval Ireland 1169–1534*. Oxford: Clarendon Press, 1987.

Sayles, G. O. "The Vindication of the Earl of Kildare from Treason in 1496." *Irish Historical Studies*, vii (1951).

See also **Chief Governors; Kildare**

FITZHENRY, MEILER

Meiler Fitzhenry was the son of Henry, the natural son of King Henry 1 by Nesta ap Rhys, daughter of Rhys ap Tewdwr, prince of South Wales. In 1157, Meiler succeeded to his father's possessions in the central and northeastern parts of modern Pembrokeshire. In 1169, he accompanied his uncle Robert fitzStephen to Ireland where he established his reputation as one of the premier knights. FitzHenry was appointed chief governor of Ireland by Richard I, and his position was later reaffirmed by King John who in 1204 ordered him to build a castle in Dublin to serve as a court and treasury. During his justiciarship Meiler clashed with both the clergy and his fellow barons. William de Braose was the first of the magnates to clash with Meiler, a dispute that John solved in 1200 by recalling Meiler to court to accompany him on circuit in England and in Normandy. In 1201 and 1202, vacancies arose in the bishoprics of Armagh and Tuam and, at Meiler's prompting, an illegal election was held that recommended colonists for the positions. Early in 1203, William de Burgh, constable of Limerick City, set out for Connacht to unseat Cathal Crobderg Ua Conchobair, who, unwilling to challenge de Burgh on the field, was saved by the intervention of the justiciar. As a result of de Burgh's expulsion, de Braose was appointed in July of that year to succeed him as constable of Limerick. William almost at once transferred the lucrative post to his son-in-law Walter de Lacy. Shortly afterward John ordered de Lacy to surrender Limerick to the justiciar, but this action did not meet with the approval of the de Lacy-de Braose axis and subsequently disturbances broke out in Meath.

Concurrently a conflict was in progress with John de Courcy, lord of Ulster. De Courcy was eventually overthrown and in 1205 Hugh de Lacy was installed as earl of Ulster. Meiler's most serious adversary was William Marshal, with whom he was in dispute over land. Meiler was a tenant-in-chief of two fiefs in Kerry and Cork, granted to him by John around the time he was confirmed as justiciar. The bulk of his lands were held by the Marshal in his capacity as lord of Leinster. A long-running dispute followed; eventually, in 1208, John summoned both opponents and several other barons of Leinster to discuss the friction in his lordship. However, while both were in London, Meiler's forces were defeated by an opposition reinforced by his previous ally, Hugh de Lacy; no longer a tenable power, he was replaced as justiciar. FitzHenry remained a powerful baron even after he ceased to be justiciar. Married to a niece of Hugh de Lacy, he had one son whom he outlived. When Meiler died in 1220 he was interred in the Augustinian monastery of Great Connell that he had founded in 1202. His grave is marked by perhaps the earliest example of an Anglo-Norman headstone in Ireland, bearing this inscription: "Conduntur tumulo Meyleri Nobilis ossa, Indomitus domitor totius gentis Hiberniae."

MARGARET MCKEARNEY

References and Further Reading

Curtis, E. *A History of Medieval Ireland*. (2nd. ed.) London: Metheun, 1938.

Flanagan, Marie-Therese. *Irish Society, Anglo-Norman Settlers, Angevin Kingship*. Oxford: Clarendon Press, 1989.

Gilbert, J. T. *History of the Viceroys of Ireland*. Dublin: James Duffy, 1865.

Lodge, John, enlarged by Mervyn Archdall. *The Peerage of Ireland, or a Genealogical History of the Present Nobility of that Kingdom*. Dublin: James Moore, 1789.

Orphen, G. H., ed. *The Song of Dermot and the Earl*. London: 1892.

Scott, A. B. and F. X. Martin, eds. *Gerald of Wales, Expugnatio Hibernica: The Conquest of Ireland*. Dublin: 1978.

See also **Anglo-Norman Invasion; Chief Governors; Courcy, John de; Henry II; John, King of England; Lacy, de; Marshal; Strongbow**

FITZRALPH, RICHARD

Richard (Ar[d]machanus) FitzRalph, theologian, Archbishop of Armagh, was born shortly before 1300 into a prosperous Anglo-Norman family in Dundalk, County Louth, and died at the papal court in Avignon around November 10 to 20, 1360. From approximately 1315 he studied arts and theology at Oxford, graduating with an M.A. in 1325 and a D. Theol. in 1331. At Oxford FitzRalph acquired skills in logic and metaphysics, impressive knowledge of the Bible, and a high level of competence as a theologian and preacher. From this period date his *Quaestio biblica* and his Commentary on the Sentences of Peter Lombard, which survives in revised form. He was the most important secular theologian to lecture on the Sentences in the later 1320s and was prepared to present both sides of an argument without taking a personal decision.

FitzRalph gained the patronage of John Grandisson, bishop of Exeter (1327–1369), and spent a year at the university of Paris as mentor of Grandisson's nephew, John de Northwode. FitzRalph owed early ecclesiastical preferrment to Grandisson's support and acquired a number of benefices in the diocese of Exeter and, possibly, also a canonry in Armagh.

In 1332, FitzRalph was elected chancellor of the university of Oxford, and his term of office was overshadowed by strife between the student population and the townspeople as well as between the northern and southern nations within the university community. This resulted in the "Stamford Schism" and the brief establishment of an alternative university at Stamford in Lincolnshire. The matter was brought before the pope in Avignon, where FitzRalph represented the university. This was the first of four lengthy visits to Avignon, where papal patronage and curial contacts were to play an important part in his subsequent career. At Avignon he gained a high reputation as a preacher, and on December 17, 1335, he became dean of Lichfield by papal provision.

FitzRalph's second and longest stay in Avignon, 1337–1344, occasioned the work that guaranteed his subsequent renown in ecclesiastical circles. His *Summa de Quaestionibus Armenorum* arose out of lengthy debates with representatives of the orthodox churches, who were seeking papal support against the Turkish threat. Here FitzRalph discussed questions of papal primacy and ecclesiastical authority that were taken up by participants at the councils of Basle (1431–1438) and Ferrara-Florence (1439–1440), then striving to unite the oriental churches with Rome. The *Summa* documents FitzRalph's approach to the Bible and his emphasis on scriptural proof, *sola scriptura*. It also reveals the beginning of his preoccupation with dominion and its dependency on grace, which was further developed by John Wyclif.

On the death of Archbishop David Mág Oireachtaigh in 1346, the cathedral chapter of Armagh immediately elected FitzRalph as successor, and he received papal confirmation on July 31, 1346. Early in 1347, he did homage to King Edward III and received the temporalities of his see before being consecrated bishop by Grandisson in Exeter cathedral on July 8, 1347. He traveled to Ireland early in 1348, where his first recorded sermon was preached in Dundalk on April 24, 1348. In his early sermons in Ireland FitzRalph invited comparison between Christ's coming to the Jews and the archbishop's return as pastor to his own people, the citizens of Dundalk and Drogheda. He was pastoral minded, concerned with reform and visitation, and defended vigorously the primatial rights of his see against the archbishop of Dublin, but he spent much of his episcopate outside Ireland. During his longest sojourn in Avignon as dean of Lichfield he had acquired the status of an "Irish expert" at the curia, and he returned there again in 1349 on diocesan business. Preaching in Avignon in August 1349 he painted a dramatic picture of Irish society, maintaining that violence was conditioned by the cultural clash between the two nations and lamenting the Irish reputation for theft and dishonesty.

FitzRalph promoted interest in the cult of St. Patrick, above all by giving publicity to the pilgrimage of the Hungarian knight, George Grissaphan, to St. Patrick's Purgatory in Lough Derg (County Donegal, Diocese of Clogher). The visions allegedly experienced there, *Visiones Georgii*, had a wide continental circulation in Latin and in several vernaculars. Propaganda for St. Patrick's Purgatory was disseminated from Avignon, presumably with the help of FitzRalph's nephew and representative there, Richard Radulphi, and pilgrims were attracted from all over Europe.

FitzRalph's attitude to the friars, whom he had initially respected, altered radically on becoming archbishop. Now he identified the cause of tension between the two nations with the ubiquitous presence of the friars in confessional and pulpit, where he regarded them as a disruption of parochial authority. He began to examine the biblical and legal foundations, and consequent justification, of their professsion and made the first clear statement of his criticism while preaching before Pope Clement VI on July 5, 1350. He subsequently developed his arguments on the poverty question, which he published in the treatise *De Pauperie Salvatoris* (On the Poverty of the Savior). With this text he returned to London on routine business in the summer of 1356, where its circulation caused the mendicant controversy to become acute. FitzRalph's friend, Richard Kilvington, dean of St Paul's cathedral, allowed the archbishop to defend himself in a series of sermons preached during the winter and spring of 1356–1357 at St Paul's Cross, the most prominent pulpit in London.

These represent the basis of his case laid before Pope Innocent VI in Avignon on November 8, 1357. Here he also dealt with his critics in the eighth book of *De Pauperie Salvatoris*, while the case between him and the friars dragged on inconclusively. After FitzRalph's death in November 1360, followed by that of several other participants a year later, the matter passed into oblivion.

FitzRalph's papers were preserved, presumably initially by Kilvington, and in approximately 1370 his remains were returned to Ireland. They were interred in the church of St. Nicholas, Dundalk, where the local cult of "St. Richard of Dundalk" led to calls for his canonization. With the support of several Irish bishops, a commission was convened in Rome to investigate the matter. The examination of his writings exposed similarities to the teachings of John Wyclif with regard to dominion and scriptural proof. The friars pointed to their enemy as the source of Wycliffite heresy, while Lollard sources referred to him as *noster sanctus Armachanus* (our holy Armachanus), with appropriate damage to FitzRalph's postumus reputation at the papal curia.

KATHERINE WALSH

References and Further Reading

Coleman, Janet. "FitzRalph's Antimendicant 'Proposicio' (1350) and the Politics of the Papal Court at Avignon." *Journal of Ecclesiastical History* 35 (1984): 376–390.

Courtenay, William J. *Schools & Scholars in Fourteenth-Century England*. Princeton, N.J.: Princeton University Press, 1987.

Dawson, James D. "Richard FitzRalph and the Fourteenth-Century Poverty Controversies." *Journal of Ecclesiastical History* 34 (1983): 315–344.

Dolan, Terence P. "Langland and FitzRalph: Two Solutions to the Mendicant Problem." *The Yearbook of Langland Studies* 2 (1988): 35–45.

———. "English and Latin Versions of FitzRalph's Sermons." In *Latin and Vernacular. Studies in Late-Medieval Texts and Manuscripts*, edited by Alexander J. Minnis. (York Manuscript Conferences: Proceedings Series 1) Woodbridge, Suffolk: 1989.

Dunne, Michael. "A Fourteenth-Century Example of an *Introitus Sententiarum* at Oxford: Richard FitzRalph's Inaugural Speech in Praise of the *Sentences* of Peter Lombard." *Mediaeval Studies* 63 (2001): 1–29.

Genest, Jean-Francois. "Contingence et révélation des futurs: la *Quaestio biblica* de Richard FitzRalph." In *Lectionum varietates. Hommage à Paul Vignaux (1904–1987),* edited by Jean Jolivet, Zénon Kaluza, and Alain de Libera. (Etudes de philosophie médiévale 65). Paris: Vrin, 1991.

Haren, Michael J. "Bishop Gynwell of Lincoln, Two Avignonese Statutes and Archbishop FitzRalph of Armagh's Suit at the Roman Curia against the Friars." *Archivum Historiae Pontificiae* 31 (1993): 275–292.

Haren, Michael, and Yolande de Pontfarcy, ed. *The Medieval Pilgrimage to St. Patrick's Purgatory. Lough Derg and the Medieval Tradition*. Enniskillen: Clogher Historical Society, 1988.

Radulphi, Ricardus. In *Summa Domini Richardi Radulphi Archiepiscopi Armacani . . . in Questionibus Armenorum*, edited by Johannis Sudoris, with an appendix containing four London sermons. Paris: privately published, 1511.

———. *Defensio Curatorum*. In Goldast von Haiminsfeld, Melchior, *Monarchia Sancti Romani Imperii*, Vol. 2. edited by Melchior Goldast von Haiminsfeld. Frankfurt,: 1612; and in *Fasciculus rerum expetendarum et fugiendarum*, Vol. 2. edited by Edward Brown. London: privately published, 1690.

———. *De Pauperie Salvatoris*, lib. I–IV. In John Wyclif, *De Dominio Divino*, edited by Reginald L. Poole. London: Wyclif Society, 1890.

Tachau, Katherine H. *Vision and Certitude in the Age of Ockham. Optics, Epistemology and the Foundations of Semantics 1250–1345*. (Studien und Texte zur Geistesgeschichte des Mittelalters 22). Leiden–New York–Kobenhavn–Köln: E. J. Brill, 1988.

Walsh, Katherine. *A Fourteenth-Century Scholar and Primate. Richard FitzRalph in Oxford, Avignon and Armagh*. Oxford: Clarendon Press, 1981, with index of manuscripts and extensive bibliography.

———. "Preaching, Pastoral Care, and 'sola scriptura' in Later Medieval Ireland: Richard FitzRalph and the Use of the Bible." In *The Bible in the Medieval World. Essays in Memory of Beryl Smalley*, edited by Katherine Walsh and Diana Wood. (Studies in Church History, Subsidia 4). Oxford: Basil Blackwell, 1985.

———. "Richard FitzRalph of Armagh († 1360). Professor–Prelate–'Saint'." *County Louth Archaeological and Historical Journal* 22 (1990): 111–124.

——— "Die Rezeption der Schriften des Richard FitzRalph (Armachanus) in lollardisch-hussitischen Kreisen." In *Das Publikum politischer Theorie im 14. Jahrhundert*, edited by Jürgen Miethke. (Schriften des Historischen Kollegs, Kolloquien 21). München: Oldenbourg, 1992.

———. "Der Becket der irischen Kirche: Der 'Armachanus' Richard FitzRalph von Armagh († 1360). Professor–Kirchenfürst–'Heiliger'." *Innsbrucker Historische Studien* 20/21 (1999): 1–58.

See also **Armagh; Black Death; Canon Law; Education; Moral and Religious Instruction; Pilgrims and Pilgrimages; Racial and Cultural Conflict; Religious Orders; St. Patrick**

FLANN MAC LONÁIN (d. 896)

Flann mac Lonáin was an elusive poet whose considerable fame ensured that verse ranging from an elegy of Écnechán mac Dálaig who died five years after Flann to Early Modern Irish eulogies of the Dál Cais were attributed to him. Of the historical figure himself, however, little is known; indeed Colm Ó Lochlainn accorded him phantom status. Whatever his precise guise, he appears to have been primarily associated with the North Munster territory to which the Uí Briain laid claim, as indicated by passing remarks in a poem put into his mouth addressing a giant, Fidbadach mac Feda Rúscaig, allegedly Óengus, son of the Dagda, in disguise. He is also said to be the posthumous author of a *dinnshenchas* poem on Slíab nEchtga (the Aughty mountains, County Clare) in which he describes himself as *file féig* (a keen poet). Nonetheless, the prose tale preceding this poem in one manuscript terms him *ollam* (chief poet) of Connacht and his genealogy in the paternal line similarly links him with the famous sixth-century king of that territory, Guaire Aidne. As far as his poetic talent is concerned, however, the same genealogy attributes it in no uncertain terms to his northern mother, Laitheóc Láidhech, claiming *ar dúthchus a máthar do dhechaidh sidhe re héicsi* (it was because of his mother's inheritance that he took up poetry). His career was short-lived; he was murdered in the territory of the Déisi, according to the *Annals of the Four Masters*, which describes him as *Uirghil shil Scota* (the Virgil of the Irish race) on his death.

Máire Ní Mhaonaigh

References and Further Reading

Bergin, O. J. "A Story of Flann mac Lonáin." In *Anecdota from Irish Manuscripts*, 5 vols., edited by O. J. Bergin, R. I. Best, Kuno Meyer, and J. G. O'Keeffe. Halle: Max Niemeyer, 1907–1913.

Gwynn, Edward, ed. and trans. *The Metrical Dindshenchas*, 5 vols. Todd Lecture Series. Dublin: Royal Irish Academy, 1903–1935.

Meyer, Kuno, ed. "Mitteilungen aus irischen Handschriften." *Zeitschrift für celtische Philologie* 8 (1910): 109–110.

Ó Donnchadha, Tadhg. *An Leabhar Muimhneach maraon le Suim Aguisíní*. Dublin: Oifig Díolta Foillseacháin Rialtais, n.d.

O'Donovan, John, ed. and trans. *Annála Ríoghachta Éireann: The Annals of the Kingdom of Ireland by the Four Masters From the Earliest Period to the Year 1619*. (2nd ed.), 7 vols. Dublin: Hodges, Smith, and Co., 1856.

Ó Lochlainn, Colm. "Poets on the Battle of Clontarf." *Éigse: A Journal of Irish Studies* 3 (1941–1942); 208–18; 4 (1943–1944): 33–47.

See also ***Áes Dána;*** **Dinnshenchas; Metrics; Poetry, Irish**

FLANN MAINISTRECH

Flann Mainistrech, son of Echthigern, acquired his epithet "monastic" from his association with the monastery of Monasterboice, County Louth, to whose famous school he was attached. His ties with the place were enhanced by the fact that his dynasty, Cíannachta Breg, had for long been associated with it. Indeed a number of his immediate ancestors served as church officials there, as did his son, Echthigern, who died as *airchinnech Mainistrech* (superior of Monasterboice) in 1067, a mere eleven years after his father. Both within his own family and in the wider world at large, however, Flann is set apart by his immense learning, as the relatively large corpus of his extant work amply attests.

Historical Scholarship

Much of this oeuvre can be described as historical scholarship for which he was accorded the title in *senchaid* (the historian) by admiring contemporaries. That a northern focus can be detected in some of his compositions is not surprising, best exemplified perhaps in the collection of seven poems on Uí Néill dynasties attributed to him in the *Book of Leinster*. Five of these deal with Cenél nÉogain, four of which may in fact be part of a continuous poem, as Eoin Mac Néill has claimed on the basis of their common meter, linking alliteration, and *dúnad* (closure) in which the final word of the last unit echoes the opening word of the first. Nonetheless, each of the self-contained sections has its own specific emphasis. Beginning with an explanation of the name of the family's main citadel, Ailech, Flann follows this with a versified list of its most famous kings.

Thematically related stanzas in a different meter intervene before the poet reverts to *snédbairdne* to recount his subject's notable victories and finally to glorify other significant exploits after which he signs himself *Flann fer légind ó Mainistir* (Flann, scholar, from Monasterboice). Regnal lists of the neighboring dynasties of Mide and Brega complete the series, which is paralleled by a companion set of seven interconnected works dealing with world kingship contained in a variety of manuscripts. Together, "the two treatises jointly form a metrical counterpart of the Annalistic prose material," in Seán Mac Airt's words, and he relates their composition to Flann's teaching at Monasterboice. If so, his curriculum was heavily influenced by the Eusebian view of world history found in contemporary chronicles that must have furnished the poet with his most important source material. Thus, in line with this, Flann describes a succession of dynasties in turn—Assyrian, Mede, Persian, Greek, Macedonian, Babylonian, and Roman—in accordance with his stated aim *deigríg domuin do thuirim* (to enumerate the good kings of the world), a task that he acknowledged as *ní soraid, ní snéid-shuilig* (not easy, not readily contrived). Part of the difficulty certainly involved metrical constraints, which Flann skilfully surmounted by recourse to eloquent chevilles. The result was a taut long list of considerable breadth, though supplying little more than the length of the reigns of various monarchs. Relative rather than absolute chronology underlies his two important metrical lists of pre-Christian and Christian kings of Tara in which his main preoccupation lay in recounting the manner of their deaths. The ambitious scope of this linked pair of poems, encompassing legendary rulers from Eochaid Feidlech to Nath Í and historical monarchs down to Máel Sechnaill mac Domnaill (d. 1022), respectively, mark it out as the earliest national king list, as Peter Smith has noted. In actual fact, however, Flann skilfully elaborated and advanced the work of learned predecessors, propelled by the intellectual currents of his own time. Among those termed "synthetic historians" by Mac Néill, his work can be read in terms of the gradual formulation of a doctrine of all-Ireland history to fit an established Christian framework. Indeed Flann's importance in this regard can be seen in the incorporation of a number of his poems into the eleventh-century national origin legend, *Lebor Gabála Érenn* (*The Book of the Taking of Ireland*, commonly known as *The Book of Invasions*).

Other Learned Activities

If Flann's extensive historical scholarship was appreciated by his contemporaries and their immediate descendants, so too were his other academic activities. These included considerable manuscript work to judge from a colophon in *Lebor na hUidre* claiming that our poet, together with a colleague, gathered texts from a selection of choice codices in Armagh and Monasterboice, including the now lost *Lebor Buide* (*Yellow Book*) and *In Lebor Gerr* (*The Short Book*) whose theft and removal overseas were lamented by the twelfth-century interpolator. Specifically mentioned is *Senchas na Relec* (Burial Ground Lore), of which *Aided Nath Í* (The Death-tale of Nath Í) is deemed to form part. We may note that the poetic version of *Genemain Áeda Sláine* (The Birth-tale of Áed Sláine), which in conjunction with its prose telling follows *Senchas na Relec* in the same manuscript, is also attributed to Flann. Moreover, he is said to have composed it *do chumnigud in gnima sin ocus día thaiscid hi cumni do chách* (to commemorate that event [Áed's miraculous birth] and to keep it in remembrance for everyone), an aim that may conceivably underlie his compilatory work. In fact, Áed's Uí Néill pedigree may also have attracted Flann, whose authorship is supported by a similar attribution in the *Book of Leinster*. That his subsequent fame made him an attractive advocate for Áed's Brega descendants, however, should be borne in mind. He is also cited in one version of *Aided Chonchobuir* (The Death-tale of Conchobar [mac Nessa]) as author of two stanzas, one of which unsurprisingly comments on an ancestor of his own dynasty, Tadc son of Cían. His connection with a poetic rendering of De Excidio Troiae (On the Destruction of Troy) is more difficult to assess. Nonetheless, his mastery of Irish and Latin coupled with his obvious intellectual range suggest that he would have had both the skill and the opportunity to rework the original composition by Dares Phrygius, or, alternatively, an existing vernacular prose adaptation. Be that as it may, a sufficient quantity of scholarship has survived of which his authorship is not in doubt to justify the accolade he was accorded on his death in 1056: *airdfer leighinn ocus sui senchusa Erenn* (eminent scholar and master of the historical lore of Ireland).

Máire Ní Mhaonaigh

References and Further Reading

Best, R. I., and Osborn Bergin, ed. *Lebor na Huidre: Book of the Dun Cow*. Dublin: Royal Irish Academy, 1929.

Best, R. I., Osborn Bergin, M. A. O'Brien, and Anne O'Sullivan, eds. *The Book of Leinster, Formerly Lebar na Núachongbála*. 6 vols. Dublin: Dublin Institute for Advanced Studies, 1954–1983.

Dobbs, Margaret E. "The Pedigree and Family of Flann Mainistrech." *Journal of the County Louth Archaeological Society* 5:3 (1923): 149–153.

Mac Airt, Seán, ed. and trans. "Middle Irish Poems on World-Kingship." *Études celtiques* 6 (1953–1954): 255–280; 7 (1955–1956): 18–41; 8 (1958–1959): 98–119, 284–97.

Mac Néill, Eoin, ed. "Poems by Flann Mainistrech on the Dynasties of Ailech, Mide and Brega." *Archivum Hibernicum* 2 (1913): 37–99.

Smith, Peter J. "Early Irish Historical Verse: The Evolution of a Genre." In *Ireland and Europe in the Early Middle Ages, Texts and Transmission: Irland und Europa im früheren Mittelalter: Texte und Überlieferung*, edited by Próinséas Ní Chatháin and Michael Richter. Dublin: Four Courts Press, 2002.

See also **Áes *Dána*; Ailech; Classical Literature, Influence of; Education; Invasion Myth; *Lebor na hUidre; Leinster, Book of;* Mide; Poetry, Irish; Uí Néill**

FORAS FEASA AR ÉIRINN

Foras Feasa ar Éirinn (The Foundation for the History of Ireland) is by definition a monumental task set for himself by one who had the means, the training, and the education to do so. Of aristocratic Anglo-Norman stock, the author, Seathrún Céitinn (Geoffrey Keating), was educated at home and on the continent, and both his outlook and works are very much the products of his background and time.

His family held extensive land holdings in the vicinity of Cahir, and some of Céitinn's poetic output, panegyrics, and elegies, on the Butlers of Cahir point to his having been educated by the Mac Craith and other noted Munster poetic families such as Mac Bruideadha. The young Céitinn was skilled in native lore and language by the time he left to further his education in the post-Tridentine seminaries of Reims and Bordeaux. His formation in the *ratio studiorum* developed by the Jesuits did much to form his methodology and style, and the evidence of his prose work point to extensive knowledge of classical, theological, and contemporary scholarship and a rhetorical mastery of homiletics and Christian apologetics.

Céitinn's early work in poetry and prose shows a passionate concern for the welfare of the homeland. The poem "Óm sgeol ar ArdMhagh Fáil, ní chodlaim oíche" gives vent in biblical terms to his anger at the devastation of Ireland after the defeat at Kinsale in 1601; the later dramatic lyric, "A bhean lán de stuaim," suggests perhaps a vocational crisis. While his early religious prose works show a deep concern to adapt the best of contemporary liturgical and devotional works for the use of the faithful, the later *Trí Bior-Ghaoithe an Bháis* shows a preoccupation with the theme of death, influenced perhaps by his experiences after his return to the home mission around 1610. On completion in 1631 he would have turned to his magnum opus, the *Foras Feasa*, completed around 1634, in the compilation of which the author has access to many printed sources and traveled extensively to examine valuable manuscripts, such as the Psalter of Cashel, in the possession of the learned family of Ó Maolchonaire of Clare. The work is not a chronicle but a synthetic, sympathetic interpretation of the story of Ireland from the beginning to the coming of the Anglo-Normans, divided into two books dealing with the periods before and after the coming of Christianity; a division that Bernadette Cunningham points out (*The World of Geoffrey Keating*) mirrors that of the Bible. Like the Bible, too, Céitinn's history is a compendium of mythology, topography, hagiography, and chronology. He is the first to use the word "béaloideas," (I, 48) now meaning "folklore," to describe the oral record and tradition of the people, influenced, perhaps by the developments in ecclesiastical historical methodology. The contemporary Louvain school of Irish history uses the more restrictive "béalphroceapta" to describe the traditional teaching of the church. Céitinn makes a spirited defense of his sources, which shows his highly developed critical sense: "If I make statements here concerning Niall Naoighiallach which the reader has not heard hitherto, let him know that I have song or story to prove every statement I advance here." His defense too of the account of the pre-Christian king Connor's empathy with the passion of Christ shows his knowledge and critical use of Christian apologetics: "And if anyone should deem it strange that Bacrach or any other druid, being Pagan, should foretell the death of Christ, how was it more fitting for the Sybils, who were Pagans, to have foretold Christ before His birth than for Bacrach or any of his kind? Hence the story is not to be thus discredited."

From the outset the work was enormously popular and copiously copied down to the nineteenth century; soon after its completion it was translated into English and John Lynch published a Latin translation at St. Malo in 1660. For all that it had its detractors from the outset. Bishop John Roche, in a letter to Luke Wadding in 1631, is dismissive of Céitinn as a historian: "One Dr. Keating laboureth much in compiling Irish notes towards a history in Irish. The man is very studious, and yet I fear that if his work come to light it will need an amendment of ill-warranted narrations: he could help you to many curiosities of which you can make better use than himself." The criticism has continued: Donnchadh Ó Corráin, contending that the author's post-Tridentine zeal for reform has considerable influence on his selective historical approach, dismissed his critical assessment of the story of the King with the horse's ears ('I think this part of the story is a romantic tale rather than history') as no more than an assessment any schoolboy would be capable of. It should be noted, however, that Céitinn's inclusion of the tale here may have something to do with the moral of this international folktale "that truth will out," in keeping with that sense of poetic justice that informs his renderings of other tales, such as "The Story of Deirdre" and "The Death of Conraoi." Breandán Ó Buachalla contends that the popularity of the *Foras Feasa* has more to do with its style than its contents, but the contention that Céitinn is "the father of Irish prose" has been contested by Cainneach Ó Maonaigh, who illustrates, successfully, that Aodh Mac Aingil was master of a more pithy, poetic style. Scholars as diverse as Aodh de Blácam, Caerwyn Williams, and Declan Kiberd properly identify and stress the importance of Céitinn's stated aim: "I set forth to write the history of Ireland . . . because I deemed it was not fitting that a country so honourable as Ireland, and races so noble as those who have inhabited it, should go into oblivion, without mention or narration being left of them." In that, Céitinn reveals himself as the successor of the bardic chroniclers and the precursor of that epic and record of a people on the verge of extension in Tomás Ó Criomhthain's autobiography, *An tOileánach.*

TADHG Ó DÚSHLÁINE

References and Further Reading

Cunningham, Bernadette. *The World of Geoffrey Keating.* Dublin: Four Courts Press, 2000.

Ní Mhurchú, M., and Breathnach, D. *1560–1781 Beathaisnéis.* Baile Átha Cliath: Clóchomhar, 2001.

Ó Buachalla, Breandán. *Annála Ríoghachta Éireann* agus *Foras Feasa ar Éirinn:* an comhthéacs comhaimseartha, *Studia Hibernica*, Nos. 22–23 (1982–1983): 59–105.

Ó Corráin, Donnchadh. "Seathrún Céitinn. c. 1580–c. 1644: an cúlra stairiúil." In *Dúchas*. Dublin: Coischéim, 1986.

See also **Áes Dána; Anglo-Norman Invasion; *Annals of the Four Masters*; Bardic Schools, Learned Families; Conversion to Christianity; Dinnsenchas; Education; Gaelicization; Genealogy; Giraldus Cambrensis; Invasion Myth; Moral and Religious Instruction; Mythological Cycle; Poetry, Irish; Poets, Men of Learning; Renaissance; Scots, Scotti**

FORESTS

See **Woodlands**

FOSTERAGE

Fosterage was the medieval Irish custom by which the parents of a child would send him or her to be raised and educated by another family. Two main categories of fosterage are discernible: fosterage for affection or fosterage for a fee. In cases of fosterage for a fee (higher for girls because they were considered less beneficial to the foster parents), costs were determined by the social standing of the child's father.

The purpose of fosterage was to cultivate closer ties between the two families. It could be used to strengthen marriage ties through fosterage with the child's maternal line or to form or reinforce ties through fosterage with allies or vassals. Its effectiveness in this capacity was due to the strong bond that often developed between the child and his foster parents and siblings. This bond was reflected in the Law Tracts that show that the child had obligations to support his foster parents in their old age, and should a fostered child be murdered, his foster family had a right to part of his honor price. Even the right and responsibility to avenge the murder of a foster son was extended to the foster family.

The age at which fosterage began varied widely; it could begin as early as infancy or as late as age ten. The age at which fosterage concluded seems to have been more formalized. Although there are indications that fosterage for both sexes could be considered complete at age fourteen or seventeen, it has been suggested that the most common custom was that girls remained in fosterage until age fourteen when they could marry and boys remained in fosterage until age seventeen, the age of maturity. During the period of fosterage, the foster parents were responsible for raising, educating, and maintaining the child in a manner appropriate to the social standing of the child's father; for example, a king's son was to be taught martial skills but a boy of lower rank was to learn the skills necessary for farming and animal husbandry. Even the child's diet reflected his rank—gruel and buttermilk being the daily staple of commoner children, while nobles enjoyed luxuries such as wheaten porridge and honey.

Following the Anglo-Norman Invasion, the Anglo-Normans adopted Irish customs such as fosterage and gossiprid to establish alliances with the Irish. By the fourteenth century, the adoption of these Irish customs had become a point of concern for the royal government because of the divided loyalties they engendered and because they were seen as one of the causes of Gaelicization. Laws, such as the Statutes of Kilkenny (1366), were passed outlawing the practice, but these laws appear to have had only limited effect, and they were undermined by the royal government's willingness to grant exemptions.

Fosterage continued to be practiced throughout the Middle Ages, but by the end of the sixteenth century the term referred to an even wider range of relationships, including purely financial arrangements wherein the "foster family" did not actually take custody of the child but rather paid a yearly sum to the child and fulfilled the traditional financial obligations of fosterage.

KEITH A. WATERS

References and Further Reading

Fitzsimons, Fiona. "Fosterage and Gossiprid in Late Medieval Ireland: Some New Evidence." In *Gaelic Ireland, c. 1250–c. 1650*, edited by Patrick J. Duffy, David Edwards, and Elizabeth FitzPatrick. Dublin: Four Courts Press, 2001.

Kelly, Fergus. *A Guide to Early Irish Law.* Dublin: Dublin Institute for Advanced Studies, 1988.

Nicholls, Kenneth. *Gaelic and Gaelicised Ireland in the Middle Ages*. Dublin: Gill and MacMillan, 1972.

See also **Brehon Law; Children; Gaelicization; Gossiprid; Society, Functioning of Anglo-Norman; Society, Functioning of Gaelic**

FOUR MASTERS

See **Annals of the Four Masters**

FRATERNITIES AND GUILDS

Fraternities and guilds were essentially urban phenomena, reflecting the strong tendency for medieval townspeople to form themselves into religious and social associations in order to defend and to promote common interests within a competitive and densely populated environment. Lay fraternities (sometimes called confraternities) were designed for men and women whose married state made it impossible for them to be members of the (male) First Order or (female) Second Order of the Franciscan movement. Guilds, on the other hand, were more purely secular organizations to start with. Merchants, many of them itinerant, were among the first to form such associations, but in the course of time craftworkers followed suit. After the initial outbreak of the Black Death in 1348, guilds acquired a more pronounced religious identity. They adopted a patron saint and held a procession on the appropriate feast day. Many guilds established a chantry chapel in a local church and supported the regular singing of masses by one or more priests. In addition, some late medieval guilds were more or less purely religious associations, with the result that fraternities and guilds overlapped institutionally to some degree.

The oldest guild in Ireland was Dublin's Guild Merchant. The city's landmark charter of urban liberties, granted in 1192, may have been requested by this guild, whose remarkable membership roll containing about 8,400 names extends from approximately 1190 to 1265. To start with, it was a general guild with a wide range of resident and nonresident members, who paid an entry fee that was eventually standardized at nine shillings. Altogether at least fourteen Irish towns—mainly the largest ones—came to have a guild merchant. The chief concern of these organizations was the installation of a local trading monopoly, to the disadvantage of all "foreign" (external) merchants. Craft guilds were exclusive organizations representing specialized groups. Almost all the surviving evidence dates from the fifteenth century or later, although some craft guilds may have originated earlier. They established and maintained standards of workmanship, requiring new recruits to execute a "masterpiece." Such guilds were usually governed by one or two masters assisted by two wardens. These officials were entitled to investigate offences committed by guild members, to examine apprentices and to arrest those who ran away, and often to regulate prices and wages. Craft guilds also fulfilled charitable and social functions, lending practical assistance to members in times of personal difficulty, providing funeral expenses and support for widows, and funding elementary schools. The religious guilds of Dublin and its hinterland were organized along similar lines, their primary function being to maintain a chantry.

In Ireland fraternities of laymen and laywomen took the form of the Franciscan Third Order Secular, starting in the middle of the thirteenth century in places that already had a First Order friary. An early example was instituted at Kilkenny in 1347 for the purpose of repairing the friars' church and building a steeple. Members lived in their own homes but were bound by vows with regard to religious instruction and practice, sexual abstinence, fasting, personal dress, and the performance of charitable works. A wife had to have her husband's consent before joining. It is possible that the number of lay fraternities in Ireland, as elsewhere, increased after 1348 as the plague pandemic was countered by more outward expressions of personal devotion.

H.B. CLARKE

References and Further Reading

Clark, Mary, and Raymond Refaussé, ed. *Directory of Historic Dublin Guilds*. Dublin: Dublin Public Libraries, 1993.

Connolly, Philomena, and Geoffrey Martin, ed. *The Dublin Guild Merchant Roll, c. 1190–1265*. Dublin: Dublin Corporation, 1992.

Gross, Charles. *The Gild Merchant: a Contribution to British Municipal History*. 2 vols. Oxford: Clarendon Press, 1890.

Ronan, M. V. "Religious Customs of Dublin Medieval Gilds." *Irish Ecclesiastical Record* 5th series, 26 (1925): 225–247, 364–385.

Webb, J. J. *The Guilds of Dublin*. Dublin: Sign of the Three Candles, 1929.

See also **Black Death; Parish Churches, Cathedrals; Religious Orders; Urban Administration**

FRENCH LITERATURE, INFLUENCE OF

The Anglo-Norman invasion and the twelfth-century humanist revival marked a turning point in Ireland's literary relations with continental Europe in the late medieval period. Anglo-Norman French in Ireland is attested by verse texts, legal and administrative records, and loan words absorbed into Gaelic. Some compositions in Irish indicate what could be called French influence, but much of the material involved is common to *latinitas* ("Latinity," Western European culture of the period in various languages). Direct transmission from medieval French sources probably occurred, but mediation *via* Latin or Middle English versions is also attested. Brian Ua Corcráin, author of the neo-Arthurian tale *Eachtra Mhacaomh an Iolair*

(The Tale of the Eagle Youth) stated that he "heard the bones of this story from a nobleman who said he had heard it told in French" and that he adapted it, adding short verse passages. Such lack of precision illustrates the difficulty of establishing sources for such texts in Gaelic, whether Irish or (later) Scottish. The Irish Hercules, *Stair Ercuil ocus a bhás*, is a Gaelic adaptation of an English version of Raoul Lefèvre's *Recueils des Histoires de Troies* (1464, French). Similarly, the Irish version of the travels of Sir John Mandeville (original in French) was translated in approximately 1475 from an English version. Two Irish Charlemagne tales derive not from a French *chanson de geste* but from a Latin chronicle.

Tales drawn from the Arthurian cycle are relatively few and of a late date compared to other European languages. The incomplete translation of the Quest for the Holy Grail (the Cistercian *La Queste del Saint Graal*), entitled *Lorgaireacht an tSoidhigh Naomhtha* by its editor, dates from the fourteenth or fifteenth century. It is the only direct version of an Arthurian tale in Irish, remaining close to the original(s)—details indicate that the author drew on more than one original, as it differs in places from Malory's *Tale of the Sankgreal* and also from the French. *Eachtra an Amadáin Mhóir* (The Story of the Great Fool) is a variation on the story of Perceval. It may derive from or be a response to the French originals and also contains elements of the tale of Gawain and the Green Knight. Such motifs were no doubt easily adapted, given their resemblance to some Ulster Cycle tales. The modern debate on the Irish origins of medieval French Arthurian myths is ongoing, but no awareness of such a connection surfaces in the medieval Gaelic material. Gawain appears in Gaelic tales and narrative poems (or "lays") as Sir Bhalbhuaidh or Uallabh, in *Eachtra an Mhadra Mhaoil* (The Tale of the Crop-eared Dog), in the Hebridean story *Sir Uallabh O Còrn*, and the lay *Am Bròn Binn* (The Melodious Sorrow). The fifteenth-century tale *Céilí Iosgaide Léithe* (Grey Thigh's Visit) is set in the framework of King Arthur and the Round Table and features a King of Gascony. Burlesque humor is an element in many of the above tales. Determining French or English origins is difficult as many of the surviving manuscripts and versions are post medieval. The relation between manuscript and oral versions has been the subject of scholarly debate since Alan Bruford's major study *Gaelic Folktales and Medieval Romances* (1966).

Other material includes *Eachtra Uilliam* (the French *Guillaume de Palerme*), translated from a sixteenth-century English prose version, and a variation on *Orlando Furioso* set in the Arthurian framework.

Late medieval Irish love poetry and love songs were influenced by French courtly poetry, transmitted by Anglo-Norman settlers according to Seán Ó Tuama's study (1962) classifying Irish folk songs under French categories. However, the concept of "amour courtois" used by Ó Tuama dates from the nineteenth-century work of Gaston Paris, whose interpretation has been revised by subsequent studies. Conclusive textual proof that the folk songs contain specifically French motifs as opposed to English or international elements is lacking. Ó Tuama conceded this but maintained the French hypothesis in his 1988 work on élite poetry, the *Dánta Grádha* (Love Poems). However, this corpus of texts is by predominantly post-medieval authors, with two exceptions, the poetry of one of whom, the Anglo-Norman third earl of Desmond, Gearóid Iarla (*c.* 1360), shows no clear French characteristics. Mícheál Mac Craith (1989) has demonstrated that many of the *Dánta Grádha* are not love poems in the proper sense and has traced some poems to English models. A further instance of possible French influence is the story of "the prince who never slept," found only in the Old French lay *Tydorel* (*c.* 1220) and in oral tales collected in Irish-speaking districts. The questions of which direction the tale moved in or whether it descends from a common Celtic archetype are unresolved.

ÉAMON Ó CIOSÁIN

References and Further Reading

Caerwyn Williams, J. E., and Patrick K. Ford. *The Irish Literary Tradition*. Cardiff: University of Wales Press, 1992; also in Irish version, Caerwyn Williams, J. E., and Máirín agus Ní Mhuiríosa., *Traidisiún Liteartha na nGael*. Dublin: An Clóchomhar, 1979.

Gowans, Linda. *Bibliography of Gaelic Arthurian Literature*. http://www.lib.rochester.edu/camelot/acpbibs/gowans.htm

Lorgaireacht an tSoidhigh Naomhtha, edited by S. Falconer. Dublin: Dublin Institute for Advanced Studies, 1953.

Mac Craith, Mícheál. *Lorg na hIasachta ar na Dánta Grá*. Dublin: An Clóchomhar, 1989.

Ó Doibhlin, B. "La France dans la littérature gaélique." In *The Irish-French Connection*, edited by Liam Swords. Paris: The Irish College, 1978.

Ó Tuama, Seán. *An Grá in Amhráin na nDaoine*. Dublin: An Clóchomhar, 1962.

Ó Tuama, Seán. *An Grá I bhFilíocht na nUaisle*. Dublin: An Clóchomhar, 1988.

See also **French Writing; Giraldus Cambrensis; Records, Administrative**

FRENCH WRITING IN IRELAND

The earliest surviving French writing in Ireland dates from the end of the twelfth century. In England, the victory of William the Conqueror in 1066 had brought the Norman dialect of French to the ruling classes, and over the next century a distinctive Anglo-Norman dialect of French evolved. This is the French that came to Ireland with the coming of the Normans. Unfortunately, the surviving corpus of French in Ireland is too small to show that a distinctively Hiberno-Norman dialect of French can be said to have evolved in its turn.

Anglo-Norman writing is more noted for factual record than for imaginative fantasy, and the two earliest surviving items bear this out. The first is a fragmentary chronicle of some three and a half thousand lines, recounting how Strongbow came to the aid of Diarmait Mac Murchada and the subsequent political activities of King Henry II in Ireland. The anonymous author identifies with the *Engleis*, the Anglo-Norman allies of the king of Leinster, and is therefore hostile to all the other Irishmen who opposed Diarmait. He describes himself as obtaining his information orally from a certain Morice Regan, Diarmait's interpreter. This is possibly the only case on record from medieval Ireland of a French speaker in contact with an Irish speaker. The author may well have been a French-speaking Welsh Norman like many of the invaders themselves or a second-generation Irish Norman. The fragment begins with the abduction of Derbforgaill and breaks off at the siege of Limerick in 1175. The author gives a detailed account of names and events, and his chronicle is a primary source for the history of Ireland in the twelfth century.

In the late thirteenth century the Dominican Jofroi of Waterford cowrote a French adaptation of the pseudo-Aristotelian *Secretum Secretorum.* However, he has no other surviving connection with Ireland beyond his name. It appears from an allusion in his work that he was based in Paris and he wrote in an eastern French dialect.

The second surviving French work definitely produced in Ireland is on a very unusual theme. It records in verse the Walling of New Ross in 1265. This anonymous poem of some two hundred lines, perhaps the work of an itinerant Franciscan, celebrates the communal effort of the various trades of the town and even a contingent of women to build a defensive wall around it. The immediate reason for the building of the wall is the fear inspired by the conflict between Maurice fitzGerald and Walter de Burgh, but the expressed objective appears to be a desire to defend this colonial enclave so that no "Ires en Irland" would dare attack it.

The poem survives in MS BL Harley 913, which dates from approximately 1330. In the same manuscript there are two short rhetorical poems in French by Thomas Fitzgerald, first earl of Desmond. One begins "*Soule su, simple e saunz solas*," the other "*Folie fet qe en force s'afie*." They are entitled 'proverbs' but they are in fact literary plays on moral commonplaces, an unexpected side to the earl's preoccupations.

Less hostile to the native Irish than Harley 913 is the composite manuscript, Cambridge, Corpus Christi College MS 405. It was probably made approximately 1327 and contains many items of Irish interest in Latin. It also includes two Irish works in Anglo-Norman verse. The first is a summary of world geography based on Honorius of Autun, rendered into French rhyming couplets by the otherwise-unknown Perot de Garbalei or Garbally. The second is by Adam of Ross, possibly a Cistercian, who composed a verse version of the legendary infernal vision of St. Paul.

Also in the fourteenth century, Richard Ledrede, the English Franciscan bishop of Ossory, made an attempt to impose the sacred on the secular by composing Latin hymns to the airs of French songs popular in Kilkenny in his day. The first line of some of these songs are thus preserved along with Ledrede's Latin works in the Red Book of Ossory.

As in England, French was frequently used in Ireland for legal and administrative purposes. Formal letters and charters occasionally appear in French. A notable case is that of the Statutes of Kilkenny in 1366. The purpose of the Statutes was to halt the way the English colonists were "going native" and adopting Irish language and customs. Both the English and the Irish within the "land of peace" were forbidden to speak "la lang Irroies." Curiously, by this date there is no allusion to the use of French itself. It is treated like Latin, an essentially written language used for administrative purposes.

EVELYN MULLALLY

References and Further Reading

Colledge o.s.a., Edmund, ed. *The Latin Poems of Richard Ledrede, O.F.M., Bishop of Ossory 1317–1360.* Toronto: Pontifical Institute of Mediaeval Studies, 1974.

Dean, Ruth J., and Maureen Boulton. *Anglo-Norman Literature: A Guide to Texts and Manuscripts.* London: Anglo-Norman Text Society, 1999.

Hardiman, James. *Laws—A Statute of the Fortieth Year of King Edward III: Enacted in a Parliament Held in Kilkenny,* A.D. *1367* [Edition and translation of the Statutes of Kilkenny]. Dublin: Irish Archaeological Society, 1843.

Lucas, Angela M. *Anglo-Norman Poems of the Middle Ages.* Dublin: Columba Press, 1995. [MS Harley 913 is described on pages 14–26.]

Monfrin, J. "La Place du *Secret des Secrets* dans la litterature franaise medievale." In *Pseudo-Aristotle: The Secret of Secrets; sources and influences*, edited by W. F. Ryan and C. Schmitt. London: Warburg Institute, 1982. [There is no published edition of Jofroi's work.]

Mullally, Evelyn. "Hiberno-Norman Literature and its Public". In *Settlement and Society in Medieval Ireland: Studies Presented to F.X. Martin, o.s.a*, edited by John Bradley. Ireland: Boethius Press, 1988.

Mullally, Evelyn, ed. and trans. *The Deeds of the Normans in Ireland: La Geste des Engleis en Yrlande* [formerly known as *The Song of Dermot and the Earl*]. Dublin: Four Courts Press, 2002.

Shields, H. "The Walling of New Ross—A Thirteenth-century Poem in French." *Long Room* 13 (1975–1976): 24–33.

Sinclair, K. V. "Anglo-Norman at Waterford: The Mute Testimony of MS Cambridge Corpus Christi College 405." In *Medieval French Textual Studies in memory of T. B. W. Reid*, edited by Ian Short. London: Anglo-Norman Text Society, 1984.

See also **Anglo-Norman Invasion; Annals and Chronicles; Education; English Literary Influence; Fitzgeralds of Desmond; Gerald, Third Earl of Desmond; Hiberno-English; Hiberno-Norman Latin; Languages; Parliament; Poetry, French; Records, Administrative; Romance**

G

GAELIC REVIVAL

The political revival of the Gaelic communities in the later Middle Ages had a number of stages. Those Gaelic rulers who retained some territory after the Anglo-Norman invasion relied at first on the protection of the English king to keep the aggression of the Anglo-Irish barons within bounds. This policy failed during the long minority of King Henry III, when several of his council of regency, such as Earl William the Marshal and the chancellor Hubert de Burgh, were closely connected to the barons in Ireland. Their encouragement led to a renewed westward expansion, the conquest of Connacht by Richard de Burgh, and attempts by the FitzGerald lord of Sligo to conquer Donegal, and by the Fitzgeralds of Desmond, or south Munster, to expand into the southwest at the expense of the MacCarthaig lords.

Attempts to Revive the High Kingship of Ireland

These pressures led about the middle of the thirteenth century to a general movement among the younger generation of Gaelic princes to withdraw their allegiance from King Henry and his son, the Lord Edward, and to take up arms to recover lost territory. Important local victories were won by Gofraid Ua Domnaill: at Credran (1257) against the Fitzgerald lords of Sligo; and by Fingin and Cormac Mac Carthaig at Callan (1261) and Mangerton (1262) respectively, against the Fitzgeralds of Desmond, which halted the momentum of conquest. Brian Ua Néill, the king of Tír Eogain in mid-Ulster, hatched a more ambitious plan for an alliance with the heirs of Ua Conchobair, king of Connacht, and of Ua Briain, king of Thomond or North Munster, in support of Brian's own claim to be king of all the Irish of Ireland. This hope perished with the defeat and death of Ua Néill at the battle of Downpatrick (1260). In 1263, the king of Norway, Haakon Haakonson, came with a fleet to assert his lordship of the western Isles of Scotland. He was asked to extend his expedition to Ireland and accept the kingship of the Irish, but this came to nothing, as Haakon refused and died shortly afterwards. The final attempt to put forward a single king over Ireland as an alternative to the English king's lordship came between 1315 and 1318, when Edward Bruce invaded with a Scottish army and claimed the kingship of Ireland, supported by Domnall Ua Néill and other Gaelic chiefs.

Decline of the Colony in the Fourteenth Century

After the collapse of this ambitious attempt at countrywide resistance, with the defeat and death of Edward Bruce near Dundalk in 1318, the more effective recovery of Gaelic power took place at a regional level during the fourteenth century. It happened as much through the weakness of the Anglo-Irish colony as any added strength on the part of the Irish. The fourteenth century saw an extended decline in weather conditions across northern Europe, leading to bad harvests and famines. In Ireland this took greatest effect in the cereal-growing regions of the south and east, where the English colonists were concentrated. Similarly the plague known as the Black Death, which swept across Europe from 1347 to 1349, entered Ireland through the major seaports, which were inhabited by the Anglo-Irish; the Gaelic communities, engaged in pastoral farming and living dispersed in rural settlements, were less severely affected.

Poverty and depopulation in the Anglo-Irish colony led to a fall in financial profits for aristocratic landowners and less taxation revenue for the English government. England was in any case focused for most of the fourteenth century on its Hundred Years War with France. At both private and public level there was less investment in military defense against the incursions of the neighboring Gaelic chieftains, and military retinues maintained by the earls and barons were supported by billeting mercenary soldiers in the houses of the tenant farmers. Anglo-Irish peasants and townspeople were faced—on the one hand—with increased attacks from the Irish, and the unpleasant burden of billeted soldiers and added taxation for their upkeep, and—on the other hand—with the attraction of farms and jobs that had become vacant in England in the aftermath of the plague. They emigrated back over the Irish sea in considerable numbers, while some others, left farming on the frontiers of an Irish chieftain's domains, bought immunity by submitting and paying tribute to their powerful Irish neighbor rather than to an absentee English landlord.

Military Recovery

The armies of the Irish chieftains over the same period became increasingly professional. Instead of relying on musters of their own subjects, chiefs employed bands of "kernes" (*ceithirne*, *ceatharnaigh*; light-armed native Irish mercenaries) and "galloglasses" (*gallóglaigh*, troops of heavy-armored Scots from the Western Isles). The first galloglasses arrived in the mid-thirteenth century, but their numbers were reinforced by political exiles from Scotland after the Bruce wars. They too were billeted on peasant farmers in the Gaelic lordships, an exaction known as "coyne and livery." The chieftains themselves, with their families and household guards, formed the cavalry, wearing suits of mail and helmets, and armed with long spears. A series of major Irish victories in the fourteenth century demonstrated their effectiveness: in 1318, at Dysert O'Dea, where the death of Lord Richard de Clare and the subsequent failure of his heirs ensured lasting independence for the Ua Briain lordship of Thomond; in 1346, when Brian Mór Mac Mathgamna (MacMahon) of Monaghan defeated the Anglo-Irish of Louth, killing four hundred of them; or in 1374, when Niall Mór Ua Néill defeated and killed the Seneschal of Ulster at Downpatrick. However, real territorial gains for the Irish chiefs came from a gradual war of attrition on the borders of the colony, resulting in considerable expansion for Ua Conchobair Failge (O'Conor Faly) along the southern borders of Meath and Kildare, for Ua Broin and Ua Tuathail (O'Byrne and O'Toole) in Wicklow, for Mac Murchada Caemánach (MacMurrough Kavanagh) in Wexford and Carlow, and for Ua Cerbaill (O'Carroll) in Tipperary. In Ulster, the murder of Earl William de Burgh in 1333, and the absenteeism of his heirs, led to virtual independence for the chiefs there, but in Connacht and Desmond, or south Munster, the Anglo-Irish Burkes and Fitzgeralds respectively dominated the local chiefs, although the English government itself had little control in those areas.

KATHARINE SIMMS

References and Further Reading

Frame, Robin, "Two Kings in Leinster." In *Colony and Frontier in Medieval Ireland*, edited by Terence B. Barry, et al., 155–175. London and Rio Grande: Hambledon Press, 1995.

Hayes-McCoy, Gerard A. *Irish Battles*. London: Longman, 1969.

Lydon, James F. "Lordship and Crown: Llywelyn of Wales and O'Connor of Connacht." In *The British Isles, 1100–1500*, edited by Rees R. Davies, 148–163. Edinburgh: Edinburgh University Press, 1988.

MacNeill, Eoin. *Phases of Irish History*. Dublin: Talbot Press, 1919.

Ó Murchadha, Diarmaid. "The Battle of Callann, A.D. 1261." In *Journal of the Cork Historical and Archaeological Society* 66 (1961):105–115.

Simms, Katharine. "Late-Medieval Tír Eoghain: The Kingdom of 'the Great Ó Néill'." In *Tyrone: History and Society*, edited by Charles Dillon and Henry A. Jefferies, 127–162. Irish County History Series, Dublin: Geography Publications, 2000.

———. "Gaelic warfare in the middle ages." In *A Military History of Ireland*, edited by Thomas Bartlett and Keith Jeffery, 99–115. Cambridge: Cambridge University Press, 1996.

See also **Armies; Black Death; Bruce, Edward; Burke; Clare, de; Coyne and Livery; Desmond; Famine and hunger; Fitzgerald; Gaelicization; Mac Murchada, Art Caomhánach; Military service, Gaelic; Mortimer; Ua Domnaill; Ua Néill, Ua Néill, Domnall; Weapons and Weaponry.**

GAELICIZATION

Gaelicization is a rather controversial concept. Nationalist historians used to cite the Latin tag *Hiberniores ipsis Hibernis*, "more Irish than the Irish themselves," to convey that many originally English families who settled Ireland in the Middle Ages came to speak Irish, wear Irish costume, defy the orders of the English kings or their representatives, and often allied with Irish chieftains to make war on their Anglo-Irish neighbors. Art Cosgrove has since demonstrated that this Latin phrase did not belong to the medieval period and was of uncertain authorship. Constitutional historians such as Steven Ellis and Robin Frame have pointed out that the regional independence and feuding tendencies of the Anglo-Irish frontier barons were not specifically Irish, but could be found in many other societies across Europe where central government was weak and distant.

Legislation Against Gaelicization

Nevertheless, statutes of Anglo-Irish parliaments from 1297 onward contain repeated complaints that certain Englishmen within the lordship of Ireland had become "degenerate" (in Latin, *degeneres*), that is, that they abandoned the characteristics of their own people and adopted those of their Gaelic Irish neighbors. Such laws focus on language and dress; alliance with Gaelic Irish families through intermarriage, fosterage, and gossiprid; patronage of Irish poets and musicians; and the exaction of forced hospitality by Anglo-Irish magnates, for the support of their household retinues and troops, from Anglo-Irish neighbors who were not legally their tenants. This was a custom based on the Gaelic lord's prerogative of "guesting" (*coinnmheadh*) at his subjects' expense, known to the Anglo-Irish as "coyne and livery" or "coigny."

Only the famous Statutes of Kilkenny, drawn up in 1366 during the viceroyalty of Prince Lionel of Clarence, attacked the Irish language itself, and decreed that landowners of English descent who could speak only Irish should be forced to learn and use the English language on pain of forfeiting their estates. This was exceptional, and may be related to the simultaneous promotion of the English language as against the use of French in England at the height of the Hundred Years War. Normally, parliamentary legislation dealt only with aspects of Gaelicization perceived as threatening the peace of the colony. Even after 1366, Anglo-Irish nobles were permitted to intermarry with the families of Gaelic magnates, and send their children to be fostered with them if they obtained royal license to do so, and the connection was officially considered to promote, rather than threaten, the precarious peace between what were called the "English lieges of our lord the king" and the "wild Irish." Similarly, barons were permitted to exact "coyne," or the billeting of their armed retinues, from tenants living on their own estates. Irish harpers and musicians were to be excluded from Anglo-Irish banquets because they might act as spies. Irish chaplains who knew no English were not to officiate in parishes occupied by the colonists, because they could not hear confessions and minister to their flock adequately. Wearing Irish dress was said in the thirteenth century to expose an Englishman raiding his neighbor's lands to increased risk of being killed in mistake for an Irishman, since the penalty for killing an Englishman was death, while killing an Irishman incurred only a financial penalty, in line with native Irish law. In the late fifteenth century, merchants speaking Irish and wearing Irish dress in the small market towns of Meath were seen by parliament as symptoms of economic decline in that area.

The Anglo-Irish Nobility and Bardic Poetry

Parliamentary legislation chiefly expressed the view of the English-speaking burgesses of the towns, together with knights of the shire from those Irish counties closest to the site of a particular parliament, whether held in Dublin, Kilkenny, or Trim, and the administrators on the King's Council, many of them English born. Another primary source for the study of this subject is Irish bardic poetry commissioned and, in some cases, actually composed by the frontier barons themselves. Some thirty-eight Irish poems are ascribed to the third earl of Desmond, Gerald "the Rhymer" FitzGerald (d. 1398), mostly about love or personal matters, but some expressing his close friendship for the Mac Carthaig (Mac Carthy) lords of Muskerry, and alleging, however disingenuously, that he attacks his Irish friends only from fear that otherwise he would be imprisoned in London by the King of the Saxons. A bardic poem to Edmund Butler, sixth Lord Dunboyne (*fl.* 1445), declares that his right to rule his estates derives from his royal Irish descent through his Ua Briain (O'Brien) mother as well as his Butler father. These poems partly support the caution expressed by Ellis and Frame, in that they contain no rejection of the English king's authority during the medieval period, but with the Reformation and Tudor reconquest of the sixteenth century, the tone changes. A poem to the rebel James fitz Maurice FitzGerald (d. 1578) describes the Geraldines as descendants of Greeks, who will ally with the Irish to resist the English forces. Poems by another rebel, William Nugent, younger brother of the baron of Slane, express the cultural alienation he felt while a student at Oxford in the later sixteenth century, and the large collection of poems addressed to Theobald Viscount Dillon and his seventeenth-century descendants frequently repeats the preposterous fiction that the Anglo-Irish Dillons are descended in direct male line from the ancient Irish dynasty of the Uí Néill.

Family Structure

Marriage between families of English and Irish descent affected family structure, with legal implications. The first generation of Anglo-Norman barons had promptly intermarried with Irish royal dynasties, and as Seán Duffy has pointed out, in the thirteenth century, when the influence of the English colony was at its height, these family connections could result in the Anglicization of the Irish nobles rather than the Gaelicization of the newcomers. However as the colony grew poorer and more neglected by central government, cross-cultural influence swung the other way, and is seen in the acquisition of noble Irish concubines by the barons, whose children were acknowledged as family members with certain rights of inheritance. Thus where aristocratic families in other parts of Europe often died out for lack of heirs, the Anglo-Irish earls and barons multiplied into small armies of Geraldines, Burkes, and Butlers, their ranks swelled by numerous bastards and adherents. In a number of cases the marcher lords avoided the strict rules of English primogeniture, and elected leaders from among the wider kindred when direct heirs failed, leading to confrontations with the Crown in the case of the fourteenth-century de Burghs (Burkes) and the fifteenth-century FitzGerald earls of Desmond.

Gerald FitzGerald, eighth Earl of Kildare, while negotiating in 1488 with Henry VII's envoy to be pardoned for his support of the Yorkist pretender Lambert Simnel, is said to have objected to a particular clause inserted into the text of his submission, threatening that he and his fellow-conspirators would "sooner turn Irish every one" than agree to that condition. Edmund Curtis saw this episode as supporting his claim that there was a movement for Anglo-Irish "Home Rule" in the late fifteenth century. What is undeniable is that the threat implies that the rebellious Anglo-Irish did not perceive themselves as having turned "Irish" yet. Gaelicization had its limits.

KATHARINE SIMMS

References and Further Reading

Cosgrove, Art. "Hiberniores ipsis Hibernis." (More Irish than the Irish.) In *Studies in Irish History, Presented to R. Dudley Edwards*, edited by Art Cosgrove and Donal McCartney, 1–14. Dublin: University College Dublin, 1979.

Curtis, Edmund. *A History of Medieval Ireland from 1110 to 1513*. First edition. Dublin: Maunsel and Roberts, 1923.

Duffy, Seán."The Problem of Degeneracy." In *Law and Disorder in Thirteenth-Century Ireland: The Dublin Parliament of 1297*, edited by James Lydon, 87–106. Dublin: Four Courts Press, 1997.

Ellis, Steven G. "Nationalist Historiography and the English and Gaelic Worlds in the Late Middle Ages." In *Irish Historical Studies* 25 (1986): 1–18.

Frame, Robin. "Les Engleys nées en Irlande: The English Political Identity in Medieval Ireland." *Transactions of the Royal Historical Society* 6, no. 3 (1993): 83–103. Reprinted in Robin Frame, *Ireland and Britain, 1170–1450*, 131–150. London and Rio Grande: Hambledon Press, 1998.

Lydon, James, ed. *The English in Medieval Ireland*. Dublin. Royal Irish Academy, 1984.

Nicholls, Kenneth. *Gaelic and Gaelicised Ireland in the Middle Ages*. The Gill History of Ireland 4. Dublin: Gill and Macmillan, 1972.

Simms, Katharine. "Bards and Barons: The Anglo-Irish Aristocracy and the Native Culture." In *Medieval Frontier Societies*, edited by Robert Bartlett and Angus MacKay, 177–197. Oxford: Clarendon Press, 1989.

See also **Parliament; Fosterage; Gossiprid; Poets, Men of Learning; Coyne and Livery; Lionel of Clarence; FitzGerald; FitzGerald, Gerald, third Earl of Desmond; FitzGerald, Gerald, eighth Earl of Kildare; Burke; Butler**

GAMBLING

See **Games**

GAMES

The sources of medieval Ireland reveal a variety of games. Field games, particularly stick and ball games (mentioned in the tract *Mellbretha*, "sport judgements"), seem to have been quite popular in medieval Ireland. References to the games, including clues to their equipment and strategy, are described in literary sources dating back to the seventh century. The richest and most informative descriptions of early field games are found in the Ulster Cycle saga "The Cattle Raid of Cooley" concerning the hero Cú Chulainn. Scenes from the saga literature generally describe games in which numerous participants vie for one or several balls. Goals are scored by either driving or throwing the ball(s) through a hoop or across a border. Evidence for the skills and strategy needed in stick and ball games is hinted at in a fourteenth-century saga. In the tale, a skillful foreigner keeps a ball aloft from one end of a strand to the other, catching it occasionally with hands, knees, shoulders and feet.

Hurling, a popular contemporary game, is first mentioned under that name in the medieval period. The earliest testimony to hurling (*horlinge*) is found in a statute issued at Kilkenny in 1366, describing a game played with clubs and ball along the ground. The statute, one of many designed to suppress native custom and activity, outlaws *horlinge* as a distraction from more constructive pursuits such as archery and military training.

The earliest physical representations of what *may* be an early playing stick (*cammán*) are found on the tenth-century high crosses at Kells and Monasterboice; there is a more clear-cut, although late, depiction on a fifteenth-century grave-slab from County Donegal, showing a sword alongside a thin playing-stick with curved end, above which a ball lies. The image suggests the subject was known as both a superior soldier and sportsman; early Irish literature portrays field games as favorite pursuits of warriors.

Field games were generally held on the greens of a fort or enclosure. Early Law Tracts describe penalties for injuries to participants and damage to structures while playing on public greens. Evidence from saga tales suggests that assemblies and fairs were the most common settings for field games, often with spectators present. Field games were clearly violent affairs and injuries were common. Descriptions of both injuries and penalties are common. Field games are also commonly described as an appropriate means of settling quarrels and disputes, in many cases ending in injury or death.

Apart from field sports, board games were also a common pursuit. Archaeological evidence for board games generally belongs to one of two contexts, pre-Christian settlements and Viking Age settlements of the tenth to twelfth centuries. The best known surviving evidence is the Ballinderry game board, a well-preserved wooden board roughly 25 cm square. Forty-nine holes are bored into the main panel in a 7 by 7 arrangement. Several games have been suggested for the board though none is certain. Boards are often described in the literature as intricately carved and adorned with precious gems. Clearly valuable personal possessions, they are described as gifts, tributes, and spoils of war.

Three specific board games are mentioned in the early literature, *fidchell*, *brandub*, and *buanbach*. *Fidchell* survives in Modern Irish as the word for chess. Chess, however, did not reach Ireland until at least the twelfth century, perhaps not until the thirteenth. As references to *fidchell* appear as early as the eighth century, *fidchell* predates chess's introduction to Ireland and is clearly a separate game. A similar misinterpretation occurs with *brandub*, consistently translated "backgammon." Little is known of the board game *brandub* and references to it are most often found in association with *fidchell*. It is likely the games were played on the same board.

Fidchell, literally "wood sense," is cognate to the Welsh board game *gwyddbwyll* and likely represents the same game or family of games. *Fidchell* is by far the best attested of Ireland's early board games. In references from the saga-literature play at the *fidchell*-board often lasts several games, particularly when a stake is involved. Unfortunately, little is known of the strategy and arrangement of the early board games. Occasional hints concerning the physical layout and the movements of pieces are found for *fidchell*, describing it as a chase-game whereby a principal piece is surrounded by defenders. This allows a tentative understanding, though one which is far from complete. Comparisons to the contemporary games of "fox and geese" and variants of Scandinavian *tafl* (table) games have been suggested.

Skill at *fidchell* and other board games is directly associated with military skill. The games and their playing pieces are frequently used as metaphors for battles and soldiers. Kings and heroes are frequently described playing *fidchell*. The saga hero Cú Chulainn and his charioteer Láeg are keen players. Queens are occasional participants, though their participation is generally related to reveal their military skill. Children are also portrayed as eager and well-trained players. An early Law Tract lists instruction in the playing of *fidchell* and *brandub* as two skills which must be taught to a foster-son. The *fidchellach*, or "*fidchell*-player," is a common figure of the early literature. He was an on-call opponent for kings and lords, and in one text is described as the "household pet."

Apart from board games, dice games, and other games of chance were played in early Ireland. Dice of various shapes and sizes are found in Iron Age and pre-Christian contexts, generally in burials. Dice in later contexts, particularly Viking settlements, are occasionally found in burials but also and more generally in living areas.

The early literature provides little specific information or background to gaming and gambling. "Bone-players" are mentioned in an early poem concerning the annual fair at Carman, and "bone-games" appear in a list of boys' feats in a late saga tale. Dice are often found in archaeological sites alongside gaming-pieces, gaming-counters, glass and stone beads, and so forth. Several burials, usually of young males having met violent ends, include scattered dice and glass or bone gaming counters, suggesting the fate of crooked or perhaps unlucky participants.

Dice and games of chance were likely common pastimes, played from an early age. Literary accounts also describe the casting of lots, or *crannchor* (literally "wood-throw"). Beyond simple gaming and gambling, legal and hereditary issues were often solved through the casting of lots. Gaming and gambling seem to have become increasingly popular in late-medieval Ireland, with several games and gaming pieces introduced by the English.

Angela Gleason

References and Further Reading

Binchy, D. A., (ed.). "*Mellbretha.*" *Celtica* 8 (1968): 144–154.
Kelly, Fergus. *A Guide to Early Irish law.* Dublin: Dublin Institute for Advanced Studies, 1988.
Edwards, Nancy. *The Archaeology of Early Medieval Ireland.* London: Batsford, 1990.
O'Sullivan, Aidan. "Warriors, Legends and Heroes: The Archaeology of Hurling." *Archaeology Ireland* 12, no. 3 (Autumn 1998): 32–34.

See also **Children; Entertainment; Law Tracts; Ulster Cycle**

GENEALOGY

Genealogical texts, written in Irish and detailing the descent of the chief families of Gaelic (and later Anglo-Norman) Ireland, are an important source for the history of Ireland from early medieval to early modern times.

It has been claimed that the body of medieval Irish genealogies is the largest of its kind for any country in Europe—"unique . . . in its chronological extent and its astonishing detail." The collections preserved in two manuscripts, Bodleian MS Rawlinson B 502 and the Book of Leinster (from the earlier and later twelfth century respectively), contain the names of some 12,000 persons (mainly men, and from the upper echelons of society), many of whom were historical figures living between the sixth and twelfth centuries. They share over 3,300 personal names and belong to numerous tribes, dynasties or family-groups. (By the early tenth century some had begun to bear surnames.) There is mention of thousands of further individuals in several surviving genealogical collections from the post-Norman period—the greatest of all, Dubhaltach Mac Fhir Bhisigh's mid-seventeenth-century Book of Genealogies, lists about 30,500 individuals sharing more than 6,600 personal names.

The genealogies relate to invasion myth—claiming to trace the ancestry of virtually all the Gaelic people of Ireland (and of Scotland) back to one or other of the sons of Míl Espáinne (Irish for *Miles Hispaniae*, "soldier of Spain"): most of the main dynasties (apart from those of Munster and east Ulster) were supposedly descended from his son Éremón. Various subject peoples are traced to certain of the reputed pre-Gaelic inhabitants of Ireland, such as Fir Bolg. The genealogical scheme as a whole is made to complement and support the body of origin legends that were, by the later eleventh century, brought together to produce *Lebor Gabála Érenn* (The Book of the Taking of Ireland). The entire scheme, in turn, is linked into, and indeed modeled on, the genealogical scheme that underlies the Old Testament—Míl's descent being traced back via Japheth son of Noah to Adam.

Irish genealogical texts are chiefly of two kinds: (1) single-line pedigrees that trace an individual's ancestry back through the paternal line; (2) *cróeba coibnesa*, "branches of relationship" (or *cróebscaíled*, "ramification"), that detail the side-branches of a family down through the generations. With the assistance of the latter, it may be possible to construct a detailed genealogical table for an entire sept or extended family. Genealogical texts may also contain various incidental materials, both prose and poetry, such as origin legends and chunks of family history.

The oldest genealogical texts we possess are a series of archaic poems detailing the genealogies of Leinster kings, and which, according to some authorities, may reflect a period as early as the fifth century. Certainly some genealogical texts have roots that can be traced back to the early seventh, or even late sixth, century, and some early non-genealogical texts, such as Tírechán's late-seventh-century hagiographical account of St Patrick (in Latin), also include brief scraps of genealogical lore. The existing body of medieval Irish genealogies is thought to represent a revision made (probably in Armagh) about A.D. 1100 of a text from a lost Munster manuscript known as the Psalter of Cashel, thought to date from about a century earlier (although the Psalter is traditionally attributed to the learned king-bishop Cormac mac Cuilennáin, who was slain in 908).

Although members of the Irish learned classes may have been been able to commit to memory quite lengthy pedigrees, the early genealogical texts betray their non-oral origins by extensive use of Latin. Although recensions dating from the twelfth century onwards are almost entirely in Irish, many Latin formulae (*a quo*, *ut supra*, *ut dixit*, etc.) continued to be used in early modern genealogical texts. (Some interesting Irish genealogical matter written in English may be found in various compilations from the sixteenth and seventeenth centuries, such as the collections by Sir George Carew and, from the early eighteenth century, Roger O'Ferrall's celebrated, and still unpublished, *Linea Antiqua*.)

Since genealogies were used in early Ireland to bolster political and territorial claims, the forging of pedigrees to reflect changing political relationships and circumstances was something of a minor industry. A particular pedigree, therefore, may be an entirely accurate record of a line of descent, or it may be a complete fabrication, or (more probably) a mixture of both.

Women are generally mentioned only incidentally in the largely patrilineal and male-dominated secular genealogies, although they fare rather better in the early Irish saints' genealogies and, of course, even more so in the genealogical work known as the *Banshenchas*, or "Lore of Famous Women." This eleventh- and twelfth-century work—occurring as both prose and verse—purports to trace the descents and marriage-alliances of well-known women from Irish mythology and, following the coming of Christianity, from the Meath and Leinster royal dynasties.

There is also a substantial body of genealogies of hundreds of early Irish saints, but these have been characterised as "generally fictional," their purpose being generally "to conceal rather than lay bare the saint's true origins." Nevertheless, they may be of considerable value for the light they can shed, for example, on the growth and spread of a saint's cult.

The recording and updating of genealogies were disrupted, along with other aspects of native learning, by the twelfth-century church reform and advent of continental religious orders and, soon after, by the Anglo-Norman invasion of 1169. The first post-Norman genealogical manuscripts now extant date from the mid-fourteenth century. They include the Ua Cianáin manuscript (National Library of Ireland MS G 2) penned, perhaps in Fermanagh, in the 1340s, and the east Connacht manuscript (TCD 1298 [H.2.7]). From the end of that century, we have the Book of Uí Mhaine and one of two great north Connacht codices, the Book of Ballymote; slightly later is the Book of Lecan, and from later in the fifteenth century, Laud MS 610 and the *Leabhar Donn.* Of the later collections the greatest of all is Dubhaltach Mac Fhir Bhisigh's Book of Genealogies, produced chiefly in Galway in the mid-seventeeth century, but the tradition of compiling genealogical collections continued at least into the following century.

Irish genealogical manuscripts from the early sixteenth century onwards began to recognize new political realities, by including the pedigrees of some of the leading Anglo-Irish families. This often reflects the degree of gaelicization undergone by such families. Some Norman families (such as the Plunkets, Powers, Bennetts and Dillons) went further and had themselves assigned a pseudo-Gaelic ancestry.

The genealogies represent a very important, though often neglected, source for Ireland's earlier history. When used in conjunction with the annals, they can be used to cross-check, or flesh out, material in the latter. While the pre-Norman genealogical recensions deserve a great deal of further study, the later collections—from the mid-fourteenth century onward—have, until now, scarcely been studied at all, let alone edited and made available in print.

Nollaig Ó Muraíle

References and Further Reading

Dobbs, Margaret E. "The Ban-shenchus." *Revue Celtique* 47 (1930): 284–339; 48 (1931), 163–234; 49 (1932), 437–489.

Kelleher, John V. "The Pre-Norman Irish Genealogies." *Irish Historical Studies* 16 (1968): 138–153.

Ní Bhrolcháin, Muireann. "*An Bansheanchas*." In *Na Mná sa Litríocht* (*Léachtaí Cholm Cille* XII), edited by Pádraig Ó Fiannachta, 5–29. Maynooth, Ireland: An Sagart, 1982.

Nicholls, Kenneth. "The Irish Genealogies: Their Value and Defects." *The Irish Genealogist* 2 (1975): 256–261.

———. "Genealogy." In *The Heritage of Ireland: Natural, Man-made, and Cultural Heritage: Conservation and Interpretation, Business and Administration*, edited by Neil Buttimer, et al., 156–161. Cork: The Collins Press, 2000.

O'Brien, Michael A., ed. *Corpus Genealogiarum Hiberniae*, I. Dublin: Dublin Institute for Advanced Studies, 1962; reprinted 1968 and subsequently with new introduction by John V. Kelleher.

Ó Corráin, Donnchadh. "Irish Origin Legends and Genealogy: Recurrent Etiologies." In *History and Heroic Tale: A Symposium*, edited by T. Nyberg, et al., 51–96. Odense, Denmark: 1985.

———. "Creating the Rast: The Early Irish Genealogical Tradition." Carroll Lecture 1992. *Peritia* 12 (1998): 177–208.

O'Donovan, John, ed. *The Genealogies, Tribes and Customs of Hy-Fiachrach, Commonly called O'Dowda's Country . . . from the Book of Lecan, . . . and from the Genealogical Manuscript of Duald Mac Firbis . . .* Dublin: Irish Archaeological Society, 1844.

Ó Muraíle, Nollaig. *The Celebrated Antiquary: Dubhaltach Mac Fhirbhisigh (c. 1600–1671), His Lineage, Life and Learning.* Maynooth, Ireland: An Sagart, 1996; revised and reprinted 2002.

———, ed. *Leabhar Mór na nGenealach: The Great Book of Irish Genealogies, compiled 1645–1666 by Dubhaltach Mac Fhirbhisigh.* 5 vols. Dublin: De Búrca, 2003-2004.

Ó Riain, Pádraig. *Corpus Genealogiarum Sanctorum Hiberniae*. (The Genealogies of the Irish Saints.) Dublin: Dublin Institute for Advanced Studies, 1985.

Pender, Séamus, ed. "The O Clery Book of Genealogies." *Analecta Hibernica* 18 (1951), xi–xxxiii, 1–198.

See also **Hagiography; Gaelicization; invasion Myth; Lecan, Book of; Leinster, Book of; Mac Fhir Bhisigh family; personal names; society, functioning of Gaelic; Uí Maine, Book of**

GENEVILLE, GEOFFREY DE (*c.* 1226–1314)

The career of Geoffrey de Geneville, lord of Vaucouleurs in Champagne, who came to hold land in England, Wales, and Ireland, is a late example of the "aristocratic diaspora" typical of the high Middle Ages, when nobles moved across Europe in search of better fortunes. Geoffrey's fortunes were secured at the English court by the intervention of Peter de Savoy, the uncle of Queen Eleanor and the husband of Geoffrey's stepsister, Agnes de Faucigny. Peter obtained the marriage of Matilda de Lacy, coheiress of Walter de Lacy, lord of Meath, for de Geneville in 1252, whereby he became lord of Ludlow in the Welsh March and lord of Trim in Ireland.

Geoffrey's importance in an Irish and British context stemmed not only from his landholdings but also from his loyal service to both Henry III and Edward I. It was his loyalty throughout the baronial rebellion of the 1250s and 1260s that secured him a place in the favor of the future Edward I. He was of particular importance in Ireland during the disturbances caused by the war between the de Burgh and the FitzGerald families. Following the capture of the justiciar in December 1264, Geoffrey, who was already a member of the council in Ireland, assumed control of the government and had secured reconciliation between the warring parties by April 1265, thereby creating a stable enough situation for troops to be safely dispatched to

England in time for the battle of Evesham. It was also to Geoffrey's castle of Ludlow that Edward fled on escaping from Montfortian captivity in May 1265. Later, Geoffrey's loyalty during another constitutional crisis, and his experience as an assistant to the marshal of the army in Wales in 1282, led to his appointment as marshal of the king's army for the 1297 expedition to Flanders.

In 1270, Geoffrey made one of the most astute moves of his career in accompanying Lord Edward on crusade. This shared experience secured the bonds of his relationship with the future king of England, and it is no coincidence that de Geneville was named as the new justiciar of Ireland in 1273. De Geneville has been criticised as justiciar largely on account of his failure to solve the problems caused by the native Irish in the Wicklow mountains. In defending himself against such criticism, however, Geoffrey would probably have referred to the difficulty of the job and the "secret opposition" which he faced. The monks of Roscommon, at least, gave a favorable report of him as justiciar noting that he was "a man of great condition and discretion."

Geoffrey continued to serve Edward I after his resignation as justiciar. In addition to his service in the king's armies in Wales and Flanders, he was entrusted as an envoy of the English king. His first diplomatic engagement had been in 1267, during the negotiations with Llywelyn ap Gruffudd of Wales; after 1280, his commissions usually took him to the continent, where he was employed in attempts to secure a general European peace (1280–1283, 1290–1291) and a final peace between England and France (1297–1301). During 1280 to 1281, at least, his employment as an envoy at Paris afforded him the opportunity to winter on his estate at Vaucouleurs. He returned from his last and arduous diplomatic mission to Rome in April 1301, at the age of seventy-five. Thereafter Geoffrey may not have left Ireland until his death on October 21, 1314.

In 1252, it was far from clear that Geoffrey would concentrate his career as a magnate in Ireland. Indeed, the first concrete evidence of de Geneville visiting Ireland dates from 1262. Nevertheless, it was to his Dominican Priory of Trim, and not the college that he founded at Vaucouleurs, that Geoffrey chose to retire in 1308, and where he was ultimately buried. Long before this date, Geoffrey had unburdened himself of his lands in England and Wales (to his son Peter in 1283) and Vaucouleurs (to his second son Walter in 1294). Trim was not necessarily the obvious choice to retain in his own hands.

Between 1279 and 1307, Geoffrey was engaged in an ongoing battle with the Dublin administration over the extent of his jurisdiction within his liberty of Trim, which was seized by the government in 1293 and 1302. Geoffrey's long service to the Crown, however, was repaid with support from Edward I, who generally responded very favorably to Geoffrey's lengthy petitions. On Geoffrey's retirement, these well-defended rights passed to his granddaughter Joan and her husband Roger Mortimer.

BETH HARTLAND

References and Further Reading

Bartlett, Robert. *The Making of Europe: Conquest, Colonization, and Cultural Change, 950–1350.* London: 1993.

Cokayne, G. E. *Complete Peerage of England, Scotland and Ireland, Great Britain and the United Kingdom,* edited by V. Gibbs, et al. Vol. 2, *Eardley of Spalding to Goojerat.* London: St. Catherine's Press, 1926.

Delaborde, H.-F. "Un Frère de Joinville au Service de l'Angleterre." (A Brother of Geneville in the Service of England.) *Bibliothèque Nationale De l'Ecole des Chartes* 54 (1893): 334–343.

Hand, G. J. *English Law in Ireland, 1290–1324.* Cambridge: University of Cambridge Press, 1967.

Hartland, Beth. "Vaucouleurs, Ludlow, and Trim: The Role of Ireland in the Career of Geoffrey de Geneville (c. 1226–1314)." *Irish Historical Studies* 33 (2001): 457–477.

Watson, G. W. "The Families of Lacy, Geneva, Joinville, and La Marche." *The Genealogist* 22 (1905): 1–16.

See also **Dublin; FitzGerald; Lacy, de; Trim**

GERALD, THIRD EARL OF DESMOND (*c.* 1338–1398)

Gerald, third earl of Desmond (also known as Gerald FitzMaurice) was the third son of Maurice FitzThomas, first earl of Desmond (d. 1356) and Avaline, daughter of Nicholas FitzMaurice, Lord of Kerry. At the time of the first earl's death in 1356, Gerald's eldest brother Maurice succeeded to the earldom but died in 1358. Maurice FitzMaurice's heir was his brother Nicholas, who was judged to be of unsound mind. Owing to the importance of the earls of Desmond for the stability of southwest Munster at the time, the king waived his right to the keeping of the lands of the mentally unsound and, in 1359, granted Gerald the custody of the earldom (which included lands in Limerick, Kerry, Waterford, Tipperary, and Cork) as well as the liberty of Kerry.

Gerald married Eleanor (d. 1392), the daughter of James Butler, Earl of Ormond (d. 1382). However, this marriage alliance did not prevent a violent feud from erupting between the two families during the 1380s and 1390s, probably as the result of land disputes and conflicts of interest in Munster. Gerald also came into conflict with Ua Briain of Thomond. This conflict raged on and off throughout the 1370s and early 1380s, but by 1388, Gerald had established a closer relationship with Ua Briain. This new relationship was cemented by the fosterage of Gerald's third son James (d. 1463) with Conchobhar Ua Briain (d. 1426).

Gerald served as the chief governor of Ireland from February 1367 until June 1369 and refused a second, temporary appointment in 1382, but otherwise his involvement in the government of the lordship of Ireland was largely restricted to holding offices in Munster: he received several judicial commissions (1363 and 1382); he was appointed keeper of the peace in Cork, Limerick, and Kerry as well as chief keeper of the peace for that region (various appointments 1387–1391); and he received an unusual appointment as the chief governor's deputy in Munster (1386).

Despite his long and active political career, he is best known for his poetry. A number of poems written in the Irish vernacular attributed to him survive; most have been published, but without translation. His skill and importance as a poet have often been overestimated by historians. Gerald lacked the training of a true bardic poet, so his poems imitate the bardic style rather than achieving the full bardic form. This should not be taken to suggest his poems lack literary merit: they show significant talent as well as a substantial knowledge of Irish mythological cycles and historic tales. His poetry also offers historical insights including information concerning Gerald's capture and imprisonment by Brian Sreamhach Ua Briain (d. 1400) in 1370 as well as a close relationship with Dairmait Mac Carthaigh (d. 1381). This relationship has led to speculation that Gerald himself may have been fostered with the Mac Carthaigh Mór.

Gerald was succeeded as earl by his eldest son John (d. 1399).

KEITH A. WATERS

References and Further Reading

Mac Niocaill, Gearóid. "Duanaire Ghearóid Iarla." *Studia Hibernica* 3 (1963): 7–59.

Carney, James. "Literature in Irish, 1169–1534." In *A New History of Ireland*, edited by Art Cosgrove, 2:688–706. Oxford: Clarendon Press, 1987.

Simms, Katharine. "Bards and Barons: The Anglo-Irish Aristocracy and the Native Culture." In *Medieval Frontier Societies*, edited by Robert Bartlett and Angus MacKay, 177–197. Oxford: Clarendon Press, 1989.

Otway-Ruthven, A. J. *A History of Medieval Ireland*. New York: Barnes and Noble, 1993.

Gibbs, Vicary. *The Complete Peerage*. Vol. iv. London: St. Catherine Press, 1916.

See also **Chief Governors; Gaelicization**

GILLA-NA-NÁEM UA DUINN

See **Dinnshenchas; Placenames**

GILLA-PÁTRAIC, BISHOP

Nothing is known about the date or place of Gilla-Pátraic's birth. All that is known about him before his consecration as bishop of Dublin is that he had been a priest and Benedictine monk under Wulfstan, abbot and bishop of Worcester in England. Neither is it clear how he was chosen to succeed Bishop Dúnán, the first bishop of Dublin. According to a church document, a copy of which still exists among the archives of Canterbury, he was elected by the clergy and people of Dublin, but it is possible that Archbishop Lanfranc of Canterbury may have been involved in choosing him. This man, in fact, consecrated him bishop at St Paul's in London in 1074 and exacted a profession of obedience from him to both himself and his successors at Canterbury.

Gilla-Pátraic and Reform of the Irish Church

Two years before, in 1072, there is evidence in a letter written by Lanfranc to the pope that he considered Ireland to be part of the area over which Canterbury exercised primacy. The death of Bishop Dúnán presented him with an opportunity to put this claim to primacy over the Irish church into practice. His plan was to make Dublin the metropolitan see for all Ireland under the primacy of Canterbury. It was the only option he had; it was impracticable to incorporate the whole of Ireland into the metropolitan see of Canterbury.

Thus a process was set in train that would ultimately lead to the introduction of a new organizational structure to the Irish church in the following century. In this Gilla-Pátraic played a crucial role, for it would appear that it was he, backed by Archbishop Lanfranc, who engaged the interest of the most powerful king in Ireland at the time, Tairrdelbach Ua Briain, in the matter of church reform. This can be seen by the presence of Tairrdelbach at a synod held in Dublin in 1080, which may have been a response to the urgings of Lanfranc as expressed in a letter brought back by Gilla-Pátraic after his consecration; the synod was probably facilitated by Gilla-Pátraic. It can also be seen by his presence at the selection of Gilla-Pátraic's successor, Donngus, in 1085 and his dispatch to Lanfranc for consecration.

But perhaps more important still would have been the influence Gilla-Pátraic most likely brought to bear on the young Muirchertach Ua Briain, the son of king Tairrdelbach, whom the latter installed as king of Dublin the year after Gilla-Pátraic took possession of his see in the same town. Muirchertach initially continued his father's policy of cooperation with Canterbury's activity in Ireland after he succeeded his father as king of Munster and aspirant king of Ireland in 1086, although it is not clear whether either king was aware of its motivation.

Later, however, Muirchertach adopted a different position, deciding that the Irish church should be organized within an Irish context only, that is, independent of Canterbury. When this policy was put into action at the synod of Ráith Bressail in the year 1111, the Dublin diocese remained outside the new hierarchical structure agreed there. However, provision was made for Dublin to ultimately join and cut its ties with Canterbury. Efforts were made afterwards to bring this about, but it was not until 1148, at the synod of Inis Pátraic, that agreement was reached. Dublin was, finally, fully integrated into the Irish church structure; it had, however, to give up its aspirations, first striven for by Gilla-Pátraic, to be the metropolitan see for the whole island of Ireland. However, it did retain metropolitan status, but with a smaller province and under a different primate, the archbishop of Armagh.

The Writings of Gilla-Pátraic

There is nothing about church reform in Gilla-Pátraic's writings and, indeed, nothing about Ireland apart from the poem *De mirabilibus Hiberniae*, probably the earliest of them. It is a versified translation of an unknown Old Irish text, and closely resembles an Irish prose version of the "Wonders of Ireland" in the Book of Ballymote (*c.* 1400) and less so a shorter version in the Book of Uí Mhaine. It differs widely from a version in the Norse *Speculum Regale* and in *Topographia Hiberniae* by Giraldus Cambrensis.

The doctrinal poem *Constet quantus honos humane conditionis* probably written at Worcester, is concerned with the belief that man is made in God's image; its metres are so varied that an early twelfth century copy treats it as five short poems. The poem *Ad amicum de caduca vita* is, as its name suggests, a meditation on the transience of life, which he sent to a friend. The long allegorical poem *Mentis in excessu* carries numerous glosses to help the reader interpret its moral teaching.

The charming short poem *Perge carina* was written to accompany a copy of his prose work *Liber de tribus habitaculis animae*, which was being sent from Dublin to old friends in Worcester. The latter, perhaps because of its subject matter (heaven, hell, and people in the world), was the most popular of all his writings. More than a hundred manuscript copies are known to exist, none of them Irish: the earliest was written quite soon after his death in 1084. However, in the twelfth century it was often attributed to Caesarius of Arles, less frequently to Eusebius of Emesa, and, in later centuries, to Augustine. Transmission of his poetry is poorer: apart from the earliest witnesses (twelfth century Cistercian manuscripts), all other extant texts are anonymous and are scarce. Both poetry and prose were transmitted predominantly by English scriptoria.

Gilla-Pátraic's writing style is simple and straightforward, influenced by that of Virgil and of Servius's commentary on the *Aeneid*; the influence of Boethius and of Paulinus of Nola can also be detected as can contact with "Hisperic" latinity. Among the doctrinal sources used were the works of St. Augustine and, perhaps, Saints Benedict and Gregory the Great, and the *Collationes* of Cassian. Gilla-Pátraic's writings give an important insight into the activity of the monastic school of Worcester in the time of St. Wulfstan.

MARTIN HOLLAND

References and Further Reading

Gwynn, A., ed and trans. *The Writings of Bishop Patrick*. Scriptores Latini Hiberniae 1. Dublin: Dublin Institute for Advanced Studies, 1955.

Boutémy, A. "Le recueil poétique du manuscrit Additional du British Museum" (The Poetry Collection of the British Museum manuscript Additional). *Latomus* 2 (1938): 30–52 and 37–40.

Cross, J. E. "*De signis et prodigiis*" In *Versus S. Patricii episcopi de mirabilibus Hibernie*." *Proceedings of the Royal Irish Academy*, 71C (1971): 247–254.

Gwynn, A. *The Irish Church in the Eleventh and Twelfth centuries*. Edited by Gerard O'Brien. Dublin: Four Courts Press, 1992.

Holland, Martin. "Dublin and the Reform of the Irish Church in the Eleventh and Twelfth Centuries." *Peritia: Journal of the Medieval Academy of Ireland* 14 (2000): 111–160.

Holland, Martin. "The Synod of Dublin in 1080." In *Medieval Dublin III*, edited by Seán Duffy, 81–94. Dublin: Four Courts Press, 2002.

See also **Church Reform, Twelfth Century; Ecclesiastical Organization; Hiberno-Latin Literature; Moral and Religious Instructional Literature; Raith Bressail, Synod of; Ua Briain, Muirchertach; Ua Briain, Tairrdelbach**

GILLE (GILBERT) OF LIMERICK

Apart from Malachy, Gille is the most important ecclesiastic who took part in the twelfth-century Church Reform in Ireland. Despite this, relatively little is known about him. It is not known where or when he was born, and it is not even sure that he was Irish, although it is most likely that he was. As well as that, there is a problem with his name. Geoffrey Keating, using sources relating to the synod of Ráith Bressail, called him Giolla Easpuig; this together with the English translation of his Latin name, Gilbert, have been commonly used. However, contemporary sources suggest that his name was, in fact, Gille.

Apart from a dubious reference to his being abbot of the ancient monastery of Bangor, the only thing known for certain about his life before he became bishop of Limerick is that he had once associated with Anselm, archbishop of Canterbury, while at Rouen in Normandy. This is known from a later exchange of letters they wrote sometime after August 1107. The same letters reveal that Gille was not consecrated by the archbishop of Canterbury, a fact that fits well with our understanding of the choice of Gille as bishop of Limerick. He had been chosen for that position by king Muirchertach Ua Briain, in order to take charge of reforming the church within an Irish context and separate from Canterbury. Some time after his appointment, he wrote a tract on the constitution of the church, *De statu ecclesiae*, and sent it, with an accompanying letter, to "the bishops and priests of the whole of Ireland." In the letter he deplored the diversity of religious practices that he said existed in Ireland, and he called for unity of practice in conformity with the rules of the Roman church. In order to help achieve this, he said that a church structure was required in which all members would find their place; he then placed a sketch of this structure at the start of his tract. He used it to explain the relationship between the different levels in the structure; for example, he said that an archbishop may have between three and twenty bishops within his province. After that he proceeded to give the duties and function of the people at each level, from layman to pope.

In sending this tract to the bishops and priests of Ireland, Gille was preparing them for the changes that were being contemplated and which would be revealed at the synod of Ráith Bressail (1111). Gille presided over this synod as papal legate; the first, according to St. Bernard of Clairvaux, "to function as legate of the apostolic see throughout the whole of Ireland." Unfortunately we know nothing about the circumstances surrounding his appointment by Pope Paschal II. One thing is clear however; the pope would not have appointed him without being assured of his worthiness. It is, therefore, a tribute to the character of the man that he would be entrusted with such a signal honor.

Before the synod of Ráith Bressail began, Gille already had established his new diocese of Limerick and its cathedral, St Mary's. This is clear from the documents associated with that synod. The enactments of the synod itself were revolutionary; a whole new church structure, similar but not identical to the one he had outlined in his tract, was to be introduced into the Irish church as a replacement for the existing one, with its ancient traditions. This is a measure of the task which Gille faced. Thereafter, however, very little is found about him in the sources. He visited Westminster in 1115 and took part in a consecration there; he also performed some episcopal duties at the abbey of St. Albans in England. But, much more importantly, he took decisive action at a time that was crucial to the continued survival of the new reform structure. When Cellach, the bishop of Armagh, died in 1129, there was an attempt made by conservative forces there to reverse the church's commitment to reform. Gille strongly urged Malachy, the successor chosen by reformers, to take on these forces. An assembly of bishops and secular princes was called to add force to his urgings, and Armagh was ultimately successfully kept within the reform camp. Malachy would eventually succeed Gille as papal legate in 1140, Gille being at that time elderly and frail. His death in 1145 is the only time that he gets a mention in the Irish annals and even then, only in one.

MARTIN HOLLAND

References and Further Reading

Fleming, J. *Gille of Limerick (c. 1070–1145): Architect of a Medieval Church.* Dublin: Four Courts Press, 2001.

Gwynn, A. *The Twelfth-century Reform.* History of Irish Catholicism 2. Dublin and Sydney: Gill and Son, 1968.

———. *The Irish Church in the Eleventh and Twelfth centuries.* Edited by Gerard O'Brien. Dublin: Four Courts Press, 1992.

Holland, Martin. "Dublin and the Reform of the Irish Church in the Eleventh and Twelfth centuries." *Peritia: Journal of the Medieval Academy of Ireland* 14 (2000): 111–160.

Hughes, K. *The Church in Early Irish society.* London: Methuen, 1966.

Watt, J. *The Church in Medieval Ireland.* 2nd edition. Dublin: University College Dublin Press, 1998.

See also **Canon Law; Cashel, Synod of I (1101); Church Reform, Twelfth Century; Dál Cais; Ecclesiastical Organization; Limerick; Malachy (Máel-Máedóic); Muirchertach; Raith Bressail, Synod of; Ua Briain**

GIRALDUS CAMBRENSIS (GERALD DE BARRI)

Gerald de Barri (Giraldus Cambrensis) was the first foreigner to write a book about Ireland—indeed, in the late 1180s he wrote two in swift succession, the *Topographia Hibernie* (Topography of Ireland) and the *Expugnatio Hibernica* (Invasion of Ireland), both from the standpoint of a hostile outsider. The brilliance of these two books (the most popular of all his many works) elaborated and established an idea that was already beginning to take root in intellectual circles in Europe and especially in England, the idea that the Irish were an inferior and barbarous people. So influential did Gerald's expression of this idea become that in the seventeenth century, John Lynch was moved to write:

> The wild dreams of Giraldus have been taken up by a herd of scribblers . . . I find the calumnies of which

> he is the author published in the language and writings of every nation, no new geography, no history of the world, no work on the manners and customs of different nations appearing in which his calumnious charges against the Irish are not chronicled as undoubted facts . . . and all these repeated again and again until the heart sickens at the sight.

Gerald was born circa 1146 at Manorbier on the coast of Dyfed, a place he described as "by far the most beautiful spot in all Wales." From Manorbier Castle, he wrote, "you can see ships scudding before the east wind on their way to Ireland." His father was the Anglo-Norman lord of Manorbier, William de Barri; his mother was Angharad, daughter of Gerald of Windsor, the first Norman castellan of Pembroke, and of a celebrated Welsh princess, Nest. In the 1190s he described the men of his family as marcher lords "winning south Wales for the English" and his own descent as "one part Trojan [i.e., Welsh] and three parts English and Norman."

In *De rebus a se gestis*, the autobiography he wrote when in his sixties, he recalled building sand churches on Manorbier beach while his elder brothers built sand castles, towns, and palaces. His father called him "my bishop" and sent him to school, first with his uncle David FitzGerald, bishop of St. David's (1148–1176), and then to St. Peter's abbey at Gloucester. Between 1165 and 1179, he spent a dozen years at Paris, receiving the best education that the finest schools in Europe could offer. He studied the liberal arts, especially rhetoric, then canon and civil law, and made a start on theology. He also gave lectures on rhetoric and law, later claiming that his eloquence, and the pleasure of listening to the voice of a handsome man, made him a highly successful teacher.

From 1174 to 1176, he interrupted his studies, returned home, and—though only his own account is available—made heroic efforts to reform the Welsh church, in particular to enforce both the payment of tithes and the celibacy of the clergy. An archdeacon of Brecon who kept a mistress was deposed, and at the instance of the archbishop of Canterbury, his uncle gave him the archdeaconry; in later life he usually referred to himself as "the archdeacon." According to Gerald, had it not been for Henry II's refusal to countenance a bishop in Wales who had Welsh connections, he would have succeeded his uncle at St David's in 1176. However it seems to have been those same Welsh connections that led Henry to take him into service circa 1184 as a royal clerk. He remained in government service for about ten years, and received an annual fee from the exchequer from 1191 to 1202.

He made three trips to Ireland. He went there first in February 1183 with his brother Philip, who had just been granted three cantreds (two as yet unconquered from the Irish) by their uncle, Robert FitzStephen. He went again when Henry II assigned him to the expedition that landed at Waterford on April 26, 1185, under his son John's command. Gerald was highly critical of John's conduct in Ireland, in part because the king's son disregarded the advice of Gerald's kinsmen, the Geraldines. Despite this John offered him, at least according to Gerald, a choice of Irish bishoprics: Leighlin, Ferns, or even the two combined. But at this stage of his life, only an English bishopric would do. After John's departure, Gerald stayed on until the early summer of 1186. In his own view, he won great fame by accusing the Irish clergy of drunkenness and neglect of their pastoral duties in a sermon he preached at a Lenten Council at Dublin in 1186. More importantly, he continued to collect material and began to draft his two Irish works. The *Topographia* he divided into three books: in the first, he described Ireland's situation, climate, flora, and fauna; in the second, he dealt with marvels and miracles; in the third, he covered Irish history from its mythical beginnings until the moment that he called the coming of the English (*adventus Anglorum*). It was here that he made explicit his view of the Irish as a barbarous, primitive, and savage people, Christian in name only. In the *Expugnatio* he composed a narrative of events from the 1160s to the 1180s, a chronicle in which his own kinsmen, the Geraldines, were the conquering heroes, fighting to bring civilization to a benighted land.

As soon as the *Topographia* was finished, he set about publicizing it. Not content with the conventionally sycophantic—dedicating it to King Henry and praising him as "our Alexander of the West"—he also put on a one-man literary festival at Oxford in 1188. He staged readings of its three parts over three days, and paid for three book launch parties. It was, he boasted, a magnificent and expensive achievement, the like of which had not been seen since antiquity. When ordered by the king to accompany Archbishop Baldwin of Canterbury in preaching the crusade in Wales in 1188, Gerald took the opportunity to give Baldwin a copy of the book and make sure he read it. He completed the *Expugnatio* in 1189 and dedicated it to Richard I, offering his services as a historian to the new king. Richard, however, preferred to employ him as an expert in Welsh affairs. Still in government service, he received and rejected offers of Welsh bishoprics, while writing two more remarkable and innovative books, *Itinerarium Kambriae* (a narrative of the Welsh preaching tour) and an ethnographic monograph, *Descriptio Kambriae* (The Description of Wales). In both he criticized barbarous Welsh mores, but also found more to praise than he had in Ireland.

Another book written at this time, his *Life of Geoffrey, Archbishop of York*, was in effect propaganda on behalf of John's rebellion against Richard, and the failure of that rebellion meant the end of Gerald's career as a courtier-cleric. Not even by dedicating the first edition of *The Description of Wales* (*c.* 1194) to Hubert Walter (the king's choice as justiciar and archbishop of Canterbury) could he ward off Hubert's anger. For a few years, Gerald led a quiet life at Lincoln pursuing his theological studies and writing saints' lives. In 1199, however, he not only accepted election as bishop of St. David's, he also revived its old claim to be the archbishopric of Wales—an assault on Canterbury's rights over the Welsh churches, rights which he himself had previously upheld, notably in 1175 and 1188.

This fight for a form of Welsh independence won him the support of some of the Welsh princes, for a while at least, and it involved him in several journeys to the papal court. Whereas as researcher and author he may have won Pope Innocent III's admiration, he was no match for Hubert Walter's political skills and financial resources. By 1203, the cause was lost, and Gerald once again retired from the fray, disillusioned with Welsh princes, and announced (again) that he preferred literary immortality to worldly success. Not even the offer of the archbishopric of Cashel, made, Gerald claims, by his cousin Meiler FitzHenry, justiciar of Ireland, during the course of a third visit to Ireland to see his friends and relatives, could tempt the sixty-year-old to take up high office. But from that time on, Gerald emphasized the Welsh side of his ancestry and insisted that throughout his life his enemies had used his Welshness to bring him down. It was this that led to him being identified as Giraldus Cambrensis, or Gerald of Wales. But this is not how he had identified himself in his earliest works. In the *Topographia* he wrote "we English." In a famous passage in the *Expugnatio*, he put a speech into the mouth of his uncle, Maurice FitzGerald. Besieged in Dublin in 1171, Maurice tells his followers that they can expect help from no one, "for just as we are English to the Irish, so we are Irish to the English."

During the last twenty years of his life, mostly at Lincoln, he continued to write, especially about the St. David's case, and to produce new editions of his earlier works. When Prince Louis of France brought an army to England in 1216 and 1217, Gerald denounced the tyranny of the kings of England, extolling the liberty that people enjoyed under Capetian rule. But with the defeat and withdrawal of Louis's troops, this too ended in disappointment. Gerald was dead by 1223, but his keen eye and his fine Latin style had won for him the immortality he craved, above all thanks to his four Irish and Welsh books, ironically the ones written while worldly ambition kept him busy in the service of English kings.

JOHN GILLINGHAM

References and Further Reading

Gerald of Wales. *The History and Topography of Ireland*. Translated by J. J. O'Meara. Harmondsworth: Penguin Classics, 1982.

———. *Expugnatio Hibernica* (The Conquest of Ireland). Translated by A. B. Scott. Dublin: Royal Irish Academy, 1978.

Bartlett, Robert. *Gerald of Wales, 1146–1223*. Oxford: Clarendon, 1982.

See also **Anglo-Norman Invasion; Annals and Chronicles; Fitzgerald**

GLENDALOUGH

An ecclesiastical settlement had developed at Glendalough (Glenn-dá-Locha, valley of the two lakes), County Wicklow, before the mid-seventh century, as shown by the obits recorded for bishops Colmán (660) and Dairchell (678), both of whom were probably also abbots. The foundation is ascribed to Cóemgen (St. Kevin; d. 618), who is genealogically linked to Dál Messin Corb, a proto-historical dynasty of the Laigin, and who is the subject of Latin and Irish "Lives," but about whom little of historical worth is known. The earliest settlement was at the upper lake, where the foundations of a beehive hut survive; terracing may be traced on the adjacent hillside. Located here are the churches of Templenaskellig and Reefert (*Ríg-fert*), the burial ground of Leinster kings. Expansion towards the lower lake was apparently underway by the eighth century, and was facilitated by the dynasty of Uí Máil, the influence of which is discernible in the record of abbatial succession. However, before 800 C.E., as the wealth of the settlement increased and its network of dependencies expanded, Glendalough had attracted the rulers of Uí Dúnlainge, whose role in its affairs is clearly reflected in hagiographical tradition. By the eleventh century, the ecclesiastical center and its dependencies were dominated by Uí Muiredaig, a branch of Uí Dúnlainge, whose most distinguished churchman was St. Lorcán Ua Tuathail (d. 1180).

Meanwhile, Glendalough was attacked by Vikings in 834 and 836—later coming under pressure from the Scandinavian kings of Dublin. By the eleventh century, however, it seems that a peaceful Hiberno-Scandinavian presence had been established there, as attested by finds of a coin hoard and of a grave-slab carved by a stone-mason named Gutnodar. The settlement was, by this time, well developed commercially. Twelfth-century annals mention a watermill, while a market cross (which formerly stood in a flat open space beside the river) may date to the same period. Most of the surviving ecclesiastical remains, especially those in the lower valley, certainly date to this time and owe much to Uí Muiredaig patronage. Near the cathedral are the churches of saints Cóemgen and Ciarán, the "Priest's

House" (perhaps a repository for relics) and a fine example of a round tower. West of the main settlement lies the "Lady Church," a foundation for women religious, and to the east is St. Saviour's, founded for Augustinian canons by St. Lorcán. Glendalough was chosen as an episcopal see at the reforming Synod of Ráith Bresail in 1111, but following the Anglo-Norman invasion, pressure from the Dublin-based English administration saw the diocese united with Dublin in 1216. Commercial activity was maintained at Glendalough; there is archaeological evidence of ironworks in the thirteenth and fourteenth centuries. Increasingly perceived by the English, once the Gaelic revival gathered pace, as a haven for "Leinster rebels," the settlement was burned in 1398. Occupation continued at the site, and the Leinster Irish nobility strove, with varying degrees of success, to revive the bishopric, but Glendalough gradually faded from the historical record during the fifteenth century.

AILBHE MACSHAMHRÁIN

References and Further Reading

Plummer, Charles, ed. "*Vita Sancti Coemgeni*" (Life of St. Kevin). In *Vitae Sanctorum Hiberniae*, 1:234–257. Oxford: Clarendon Press, 1910.

———, ed. "Vie et miracles de S Laurent, archeveque de Dublin" (The Life and Miracles of St. Lawrence, Archbiship of Dublin). *Analecta Bollandiana* 33 (1914): 121–186.

———, ed. "Betha Caimgin" (Life of St. Kevin). In *Bethada Nóem nÉrenn* (Live of Irish Saints), 1:125–167 and 2:121–161. Oxford: Clarendon Press, 1922.

Long, Harry. "Three Settlements in Gaelic Wicklow, 1169–1600: Rathgall, Ballincor, and Glendalough." In *Wicklow: History and Society*, edited by K. Hannigan and W. Nolan, 248–256. Dublin: Geography Publications, 1994.

MacShamhráin, Ailbhe S. "Prosopographica Glindalechensis: The Monastic Church of Glendalough and its Community, Sixth to Twelfth Centuries." *Journal of the Royal Society of Antiquaries of Ireland* 119 (1989): 79–97.

———. *Church and Polity in Pre-Norman Ireland: The Case of Glendalough*. Maynooth, Ireland: An Sagart, 1996.

See also **Church Reform, Twelfth Century; Ecclesiastical Organization; Ecclesiastical Settlements; Ecclesiastical Sites**

GLENDALOUGH, BOOK OF

See **Rawlinson B502**

GLOSSES

A gloss, in its simplest form, is an explanation of a difficult word (lemma). Typically, it is entered close to its lemma, between the lines or on the margins of the manuscript, and in a subordinate script. The practice of glossing arose from the need to elucidate difficult words in commonly used texts, and from the fact that most of these texts were written in a foreign language, Latin. However, glosses also occur in certain vernacular texts that contain technical vocabulary, notably the Old-Irish law tracts.

Glosses were composed in Latin, Irish, or a mixture of both. Glossing in the vernacular had already taken hold in the seventh century as is evident from a scattering of Old-Irish glosses in the so-called Ussher Gospels (Dublin, Trinity College, MS 55) and from an archaic stratum of glosses in a ninth-century copy of Priscian's *Institutiones Grammaticae* (St. Gall, Stiftsbibliothek, MS 904). During the eighth and first half of the ninth century the glossing of Latin texts with vernacular words was widely practiced by the Irish, as is evident from three manuscripts of Irish origin which between them contain over 15,000 Old-Irish glosses (as well as numerous Latin glosses): a copy of the Pauline Epistles, a commentary on the Psalms, and the text of Priscian's grammar mentioned above.

The surviving glosses range in complexity from simple calques on individual words to complex interpretations of biblical passages. They serve such common functions as: supplying information about the grammatical properties of a lemma; clarifying its meaning with illustrations; highlighting its relationship with other words in the immediate context; and offering commentary or interpretation. Another type, the so-called syntactical gloss, consists of symbols (combinations of dots or letters) attached to Latin words of the text, which effectively rearrange the Latin word order to conform to that of the vernacular. Although one naturally thinks of glosses as designed to help students, some may have served the teacher. For example, the Ussher Gospels contain glosses that provide merely the opening words of excerpts from St. Jerome's commentaries, suggesting they may have been intended to jog the teacher's memory.

Once the preserve of linguists and lexicographers, the study of glosses has shifted from language to content, from printed editions to manuscript contexts, and from vernacular words in isolation to the interaction of vernacular and Latin glosses. This new approach brings glossography into the mainstream of literary evidence. Thus, glosses can testify to the use of rare or unusual literary sources in Ireland, such as Pelagius's Commentary on the Pauline Epistles and Chromatius's treatise on St. Matthew's Gospel. Secondly, glosses identify the kinds of words that the Irish found difficult or interesting in a Latin work. Thirdly, glosses offer important insights into the methodologies employed by Irish scholars, notably, their recourse to etymology to explain difficult words, their use of grammar to expound biblical passages, and their fondness for juxtaposing conflicting interpretations. This tradition of glossing continued well into the twelfth century, though after the ninth century it was expressed mainly in Latin.

PÁDRAIG Ó NÉILL

References and Further Reading

Contreni, John J., and Pádraig P. Ó Néill , eds. *Glossae Divinae Historiae: The Biblical Glosses of John Scottus Eriugena.* Florence: SISMEL, 1997.

Draak, M. "Construe Marks in Hiberno-Latin Manuscripts." *Mededelingen der Koninklijke Nederlandse Akademie van Wetenschappen, afd. Letterkunde*. new series 20 (1947): 261–282.

Hofman, Rijcklof. *The Sankt Gall Priscian Commentary, Part 1.* 2 vols. Münster: Nodus Publications, 1996.

Stokes, Whitley, and John Strachan. *Thesaurus Palaeohibernicus: A Collection of Old-Irish Glosses, Scholia, Prose and Verse*. 2 vols. Cambridge: Cambridge University Press, 1901–1903. Reprinted with supplement, Dublin: Dublin Institute for Advanced Studies, 1975.

See also **Biblical and Church Fathers Scholarship; Etymology; Grammatical Treatises; Hiberno-Latin**

GORMLAITH (d. 948)

Daughter of a southern Uí Néill king of Tara, it is alleged that she was successively queen-consort of Munster, Leinster, and Tara, and also a poetess. Her part as thrice-married queen has prompted much discussion in relation to sovereignty symbolism. Historically, there is perhaps a stronger case, as Ó Cróinín argues, for viewing her as party to dynastic intrigues in early-tenth-century Leinster. The political priorities of her father, Flann Sinna (d. 916) of Clann Cholmáin, make her role in a marriage-alliance with the Uí Fáeláin dynasty of Leinster understandable. More difficult to justify is the assertion of the Middle Irish poem "Éirigh [a] ingen an rígh" that she was previously married to the bishop-king of Cashel, Cormac mac Cuilennáin. The latter, it is stressed, was celibate—making their marriage merely a symbolic union. Record of Cormac's death in 908—he was killed in the battle of Belach Mugna—implies that her marriage to the victor of that battle, Cerball (d. 909) mac Muireccáin, Uí Fáeláin over king of Leinster, lasted no more than a year. A text in the Book of Leinster, which claims that Cerball spent this year recovering from wounds sustained at Belach Mugna, portrays him as a violent bully who mocked the memory of Cormac and treated Gormlaith so badly that, at least once, she felt the need to return to her father. She subsequently married Niall Glúndubh, the Cenél nÉogain king of Tara, who fell at the battle of Islandbridge in 919.

Gormlaith's reputation as a poetess was enlarged by early-modern tradition, related in the *Annals of Clonmacnoise* and repeated by several modern commentators, that she was left in want by her royal husbands and became a wandering rhymer, reliant on the support of common folk. It may be noted that her obit in the more-sober *Annals of Ulster* says nothing of this. Leaving aside very late ascriptions to Gormlaith of miscellaneous verses, which range in date of composition, Middle Irish sources assign to her laments for Cerball and Niall but, perhaps significantly, not Cormac. Difficulties relating to this marriage leave it probable that it is a fiction—created when memory of Gormlaith became assimilated to the "sovereignty goddess" who had three husbands. In contrast, the case for accepting as historical her marriage to Cerball is strengthened by a *dindshenchas* poem in the Book of Leinster, which also presents a different view of their relationship, implying that she was involved in intrigue on his behalf. She is blamed for the deaths of Cellach Carmain, who was an Uí Muiredaig dynast, and his wife Aillenn—apparently rivals of her husband. This circumstance, along with the fact that Cerball had the support of Flann Sinna at Belach Mugna, fits well with a Clann Cholmáin–Uí Fáeláin alliance in the years prior to that battle. Gormlaith outlived her last husband by almost thirty years, which suggests that she reached quite an advanced age. Record of her death in penitence suggests that she ended her days in a convent.

AILBHE MACSHAMHRÁIN

References and Further Reading

Bergin, Osborn, ed. "Poems Attributed to Gormlaith," In *Miscellany Presented to Kuno Meyer*, edited by Osborn Bergin and Carl Marstrander, 343–369. Halle a.S.: M. Niemeyer, 1912.

Best, R. I., et al., eds. *The Book of Leinster* Vols. 1 and 4. Dublin: Dublin Institute for Advanced Studies, 1954–1983.

Ní Dhonnchadha, Máirín, ed. "Éirigh [a] ingen an rígh; (Bibliotheque Royale, Brussels, MS 20978-9, f. 54v)." In *Seanchas: Studies in Early and Medieval Irish Archaeology, History, and Literature in Honour of Francis J. Byrne*, edited by Alfred P. Smyth, 234–237. Dublin: Four Courts Press, 2000.

———. "On Gormlaith, Daughter of Flann Sinna." In *Seanchas*, edited by A. P. Smyth, 225–233. Dublin: Four Courts Press, 2000.

Ó Cróinín, Dáibhí. "Rewriting Irish Political History in the Tenth Century." In *Seanchas*, edited by A. P. Smyth, 212–225. Dublin: Four Courts Press, 2000.

Trindade, W. Ann. "Irish Gormlaith as a Soverignty Figure." *Études Celtiques* 23 (1986): 143–156.

See also **Uí Néill, southern; Cerball mac Muireccáin; Cormac mac Cuilennáin**

GORMLAITH (d. 1030)

This Gormlaith was the daughter of Murchad mac Finn of the Uí Fáeláin branch of Uí Dúnlainge, king of Leinster from 966 to 972. The last of three royal women of that name to become the center of a considerable body of literary material, Gormlaith daughter of Murchad was famous in Irish tradition as the most ambitious and aggressive of historical Irish queens. Her reputation is indicated by a short tract in the genealogies of the Book of Leinster describing a vision wherein the daughter of an unnamed king of Connacht (possibly Tadg mac Cathail of Uí Briúin Aí) slept with the king of Leinster, subsequently bearing him a son—Máelmórda—who took the kingship of Leinster, and a daughter—Gormlaith—who took the kingship of Ireland.

While Gormlaith did not literally take the kingship of Ireland, at least one of her husbands did. A poem in the Leinster genealogies describing Gormlaith as taking "a leap at Dublin, a leap at Tara, a leap at Cashel of the goblets overall" indicates that marriage was seen to be her route to power. Commentary following the poem explains the leap at Dublin as her union with Amlaíb Cuarán, the Norse king of Dublin, by whom she bore a son, Sitriuc Silkenbeard, another king of Dublin. The leap at Cashel, meanwhile, represents her marriage to Brian Boru, Dál Cais king of Munster and, later, of Ireland. Gormlaith's son by Brian, Donnchad, was also king of Munster and a contender for the kingship of Ireland. The "leap at Tara" is more problematic. The commentary asserts that after Amlaíb, Gormlaith married Máel-Sechnaill II, the southern Uí Néill king of Tara, and later sources state that she was the mother of Máel-Sechnaill's son Conchobar, king of Tara. The eleventh-century king of Tara by the name of Conchobar, however, was not Máel-Sechnaill's son, but his grandson. Furthermore, while the sources most closely contemporaneous with Gormlaith mention Donnchad and Amlaíb, they make no reference to Conchobar or to her marriage with Máel-Sechnaill. Although such a coupling would have been plausible, possibly the "fact" of their union arose as a later addition to the literary tradition surrounding the queen.

The most vivid aspect of this literary tradition is the central role it ascribes to Gormlaith in instigating the Battle of Clontarf. The twelfth-century *Cogad Gáedel re Gallaib* depicts Gormlaith as inciting her brother Máelmórda to rebel against her former husband, Brian Boru, while the thirteenth-century *Brennu-Njáls* saga portrays the queen as a beautiful, but wicked, Machivellian manipulator, instructing her son Sitriuc to gain the support of the Vikings against Brian at all costs. No doubt this image of Gormlaith has been exaggerated for dramatic and thematic effect; however her portrayal in the later sources makes it clear that the reputation Gormlaith daughter of Murchad left to posterity at her death in 1030 was that of a strong character, at-home in the political sphere, and adept at using ties of blood and marriage in the service of her goals.

Anne Connon

References and Further Reading

Ní Dhonnchadha, Máirín. "On Gormfhlaith Daughter of Flann Sinna and the Lure of the Sovereignty Goddess." In *Seanchas: Studies in the Early and Medieval Irish Archaeology, History, and Literature in Honour of Francis J. Byrne*, edited by A. P. Smyth, 225–237 Dublin: Four Courts Press, 2000.

———. "Tales of Three Gormlaiths in Medieval Irish Literature." *Ériu* 52 (2002): 1–24.

Trindade, Ann. "Irish Gormflaith as Sovereignty Figure." *Études Celtiques* 23 (1986): 143–156.

See also **Amlaíb Cuarán; Battle of Clontarf; Brian Boru; Kings and Kingship; Máel-Sechnaill II; Queens; Uí Dúnlainge**

GOSSIPRID

In medieval Ireland, as in continental Europe, kinship bonds provided the framework of society. These familial bonds could be adapted to "secure" political relationships between an overlord and his clients. From the fourteenth to early seventeen centuries, political alliances in Ireland were most frequently underpinned by marriage, fostering, and gossiprid. Political marriage was common throughout Europe, and fostering likewise had obvious European parallels. However, the social custom of gossiprid appears to have been unique to Ireland.

Gossiprid was a pledge of fraternal association between a lord, who by the arrangement gained service, and his client(s), who received protection and patronage. It was a phenomenon of a bastard feudal system, as it raised the demands of personal lordship above those of the central government. There were four methods by which gossiprid could be practiced. First, the client could take a voluntary oath to complete a specific agreement on behalf of his lord. This form emphasized the personal relationship between the overlord and the individual client. All other forms of gossiprid were contracted with varying degrees of formality, to emphasize the communal relationship by which an overlord and his adherents created a political faction or affinity. Second, the clients could receive gifts or salaries from their overlord, and undertook to serve him, and to assist his followers and allies. Third, the clients pledged service to the lord and his following by the symbolic breaking of bread, and would again receive a "gift." And fourth, the most formal type of gossiprid was agreed by all contracting parties receiving the sacrament of communion, in pledge of their adherence to the faction.

Fiona Fitzsimons

References and Further Reading

Fitzsimons, Fiona. "Fosterage and Gossiprid in Late Medieval Ireland: Some New Evidence." In *Gaelic Ireland: Land, Lordship and Settlement, c. 1250–c. 1650*, edited by P. J. Duffy, et al. Dublin: Four Courts Press, 2001.

"H. C.'s Tract." Unpublished manuscript. National Archives, London, S.P. 63/203/119.

Nicholls, Kenneth. *Gaelic and Gaelicized Ireland*. 2nd edition. Dublin: Lilliput Press, 2003.

GRAMMATICAL TREATISES

Because Ireland never formed part of the Roman Empire, it would have been the first task of the early missionaries to teach the Latin language and its grammar. The invention of ogham, a cipher system based upon a form of the Latin alphabet, is evidence that there already existed in Ireland a knowledge of Latin sometime before the conversion of the island to Christianity. The profusion of commentaries and glosses upon the grammarians of the Classical and late Antique periods attests that the medieval Irish studied Latin with great earnestness, much as modern students would approach the study of a foreign language that was essential to their career advancement. Moreover, they compared their own complex language, Old Irish, to that of Latin. As the Celticist Maartje Draak has said:

> That method [of teaching Latin grammar] was thorough and on a considerable level—the more so if we take into account that it was an achievement by (and in the context of) an alien culture. The Irish teachers were interested, they were intellectually stimulated, but they were not over-awed by the Latin language. (Draak).

Having completed their elementary studies with Donatus, Irish students of Latin would have graduated to Priscian's great *Institutiones grammaticae*, the most extensive grammatical analysis surviving from late antiquity, of which four ninth-century Irish manuscripts have survived. But not everyone could have had access to or mastery of these original sources, so the Irish soon began to compose grammars, chiefly extracts compiled from ancient authorities, strengthening and illustrating specific points of grammar, and follow their sources in the identification and definition of the eight parts of speech. At least five such texts pre-date 700. Perhaps the earliest of these is the *Ars Asporii*, an adaptation of Donatus's *Ars maior*, but set against a Christian grammatical tradition that relied upon Scriptural sources and examples and used a monastic vocabulary, the *Anonymus ad Cuimnanum*, which may have been addressed to Cumméne (d. 669), abbot of Iona, and the *Ars Ambrosiana*. Malsachanus is the seventh- or eighth-century author of a grammatical tract on the verb-participle. Nothing is known of him, but his work is of importance for its extensive use of a seventh-century Irish tract on the verb, and his use of earlier Classical Latin grammarians such as Donatus, Consentius, and Eutyches. These texts drew upon a very wide range of sources, some of which are now unknown.

The mass of Hiberno-Latin grammatical material can best be summarized under a few important authorial headings, since most of the Irish schoolmasters composed (or compiled) Latin grammars for their students. Clemens Scottus, who flourished about 800, was teacher at the Palace School at Aachen and is the author of a Latin grammar based upon an ancient commentary on the grammar of Donatus, set in dialogue form between master and pupil. It opens with an excursus on philosophy and some discussion on the classification of the sciences, which largely follows Isidore of Seville. The bulk of the work is taken up with a discussion of the eight parts of speech. Clemens's rigorously organized and competent grammar is closely related to another Irish-Latin grammar known as *Donatus Ortigraphus*. Fragments preserved in Würzburg containing some grammatical material may also have been compiled by Clemens, and it is possible that he brought with him to Würzburg the famous glossed Pauline codex (M.p.th.f.12).

The ninth-century Irish scholar Cruindmael was the author of a metrical grammar, *Ars metrica*, which in its treatment and source-usage belongs to a well-defined group of Irish grammatical tracts. It used many sources, including the older classical grammars by Donatus, Servius, Pompeius, and later authors. Sedulius Scottus and his scholarly circle flourished in the mid-ninth century. His writings are numerous and include commentaries on Classical Latin grammarians, including Eutyches. The St. Gall copy of Priscian's *Institutiones*, written within the circle of Sedulius, is heavily glossed in Old Irish and Latin.

Martin of Laon (819–875), wrote extensively on grammar and composed a commentary on Martianus Capella's *De nuptiis Philologiae et Mercurii*, a standard textbook on the liberal arts in the Middle Ages. Martin had a competent knowledge of Virgil and other Latin poets and has preserved some fragments of the lost commentary on Virgil by Aelius Donatus. Muiredach, of Auxerre and Metz, is the ninth-century author of a commentary on Donatus's *Ars maior*. Israel Scottus (*c.* 900–968 or 969) was the Irish or Breton author of a versified grammar of the Latin verb and noun, an exposition of Donatus's grammar, and glosses on Porphyry's *Isagoge*.

AIDAN BREEN

References and Further Reading

Bischoff, B., and B. Löfstedt. *Anonymus ad Cuimnanum. Expossitio Latinitatis*. (Corpus Christianorum Series Latina 133D). Turnhout: Brepols, 1992, xii–xiii.

Chittenden, J. "*Donatus Ortigraphus Ars grammatica*." *Corpus Christianorum Continuatio Medievalis* 40D (1982). Law, V. *Insular Latin Grammarians* 1982.

Contreni, John J. *Codex Laudunensis 468: a ninth-century guide to Virgil, Sedulius and the Liberal Arts* (Armarium Codicum Insignium 3) Turnhout: Brepols, 1984.

Draak, M. "The Higher Teaching of Latin Grammar in Ireland During the Ninth Century" (Mededelingen der Koninklijke Nederlandse Akademie van Wetenschappen 30, no. 4). Amsterdam: North-Holland, 1967.

Holtz, L. "Sur trois commentaires irlandais de l 'Art majeur' de Donat au IXe siècle" (On Three Ninth-Century Irish Commentaries on Donatus's *Ars maior*). *Revue d'histoire des textes* 2 (1972): 45–73.

———. "Grammairiens irlandais au temps de Jean Scot: quelques aspects de leur pédagogie" (Irish Grammarians contemporary with John Scotus: Several Aspects of their Pedagogy.) In *Jean Scot Érigène et l'histoire de la philosophie*, 69–78. Paris: Centre National de la Recherche Scientifique, 1977.

———. *Donat et la tradition de l'enseignement grammatical* (Donatus and the Tradition of Teaching Grammar). Paris: Centre National de la Recherche Scientifique, 1981.

Jeudy, C. "Israel le grammairien et la tradition manuscrite du commentaire de Rémi d'Auxerre à l'*Ars Minor* de Donat" (Israel the Grammarian and the Manuscript Tradition of the Rémi d'Auxerre's Commentary on Donatus's *Ars minor*). *Studi Medievali* 3rd series, 18 (1977): 751–814.

Kenney, J. F. *The Sources for the Early History of Ireland: Ecclesiastical*. Vol. 1. New York: Columbia, 1929. Reprint, Dublin: Four Courts Press, 1993.

Lapidge, M. "Israel the Grammarian in Anglo-Saxon England." In *From Athens to Chartres: Neoplatonism and Medieval Thought*, edited by H. J. Westra, 97–114. Leiden: Brill, 1992.

———, and R. Sharpe. *A Bibliography of Celtic-Latin Literature, 400–1200*. Dublin: Royal Irish Academy, 1985.

Law, V. "Malsachanus Reconsidered: A Fresh Look at a Hiberno-Latin Grammarian." *Cambridge Medieval Celtic Studies* 1 (1981): 83–93.

Löfstedt, B. *Der hibernolateinische Grammatiker Malsachanus* (The Hiberno-Latin Grammar of Malsachanus). Uppsala: Uppsala Universitet, 1965

Puckett, M. "*Clementis qui dicitur 'Ars grammatica*': A Critical Edition." Ph.D. diss. Los Angeles: 1978. Supersedes *Clementis Ars grammatica*, J. Tolkiehn. 1928.

———. "Eine wenig beachtete hibernolateinische Grammatik." In *Irland und die Christenheit*, 272–276. Stuttgart: 1987.

See also **Biblical Church Fathers Influences; Classical Influence; Education; Eriugena; Etymology; Glosses; Hiberno-Latin; Inscriptions; Languages; Law Tracts; Metrics; Scriptoria; Sedulius Scottus**

HAGIOGRAPHY AND MARTYROLOGIES

A Chronology of Irish Hagiography

Hagiography (sacred writing, Lives of saints) is divided into two categories, one literary, the other liturgical. The former is mainly represented by the Lives of the saints, the latter by the records of their feasts in calendars and martyrologies. The chronology of the production of both categories in Ireland is erratic in character. Two martyrologies date to the early ninth century; four or five, to the late twelfth; numerous copies, to the early fifteenth; and one final record of the feasts of the Irish saints, to the 1620s. Similarly, the four Latin Lives of the period 650 to 700—two of Patrick, one each of Brigit and Colum Cille—were followed during the period up to the beginning of the Anglo-Norman invasion in 1169 by scarcely more than six such compositions— including Brigit's *Vita Prima* (First Life), and three vernacular biographies of Brigit, Patrick, and Adomnán. During the fifty or so years that followed, the bulk of the surviving record of the Irish saints, liturgical and literary, was composed. Then, during the fourteenth century, collections began to be made of Latin and vernacular Lives. A final phase, mostly devoted to copying, collecting, and publishing earlier works, began about 1580 and continued until about 1650.

Possibly cutting across the chronological pattern outlined above are the so-called O'Donohue Lives, of mainly midland saints, preserved in the fourteenth-century *Codex Salmanticensis*. Dates as early as the eighth century have been proposed for the Lives of this collection, which would imply the existence in Ireland of collections of Lives of local saints long before anywhere else. Judgment on the age of these Lives must, therefore, be reserved until much more work has been done on them.

Early Latin Lives

Periods of cultural tension often coincided with surges in hagiographical activity. Late seventh-century rivalry between the three great Irish churches, Armagh, Kildare, and Iona, was one of the factors that led to the composition of Lives for Patrick, Brigit and Colum Cille (Columba). Cogitosus's Life of Brigit ascribed to Kildare "supremacy over all the monasteries of the Irish . . . from sea to sea." Patrick's two seventh-century Lives, by Muirchú and Tírechán, attributed to the saint triumphal journeys to the midlands and west of Ireland, which aggrandized both Armagh and the hereditary ecclesiastical families to which these authors belonged. Neither author was interested in southern Irish churches. Only Tírechán made a token gesture towards Munster. The Life of Colum Cille (d. 597), composed by his successor on Iona, Adomnán (d. 704), has been described as the most sophisticated of the seventh-century Lives. Its concern with Iona's influence is evident in the visits to many other churches ascribed to Colum Cille, including Clonmacnoise and Terryglass, where the saint was subsequently localized. This Life was also designed to instruct, as when the then ongoing Paschal Controversy caused Adomnán to delay the date of the saint's death to avoid a clash with the "Easter festival of joy."

Brigit's *Vita Prima* is considered by some to predate that of Cogitosus. However, such incidental details as mention of anchorites and *serui Dei*, "servants of God," both reflective of the *Céli Dé* movement of the late eighth century, point to a later date.

Early Vernacular Lives

The decline of the *Céli Dé* movement in the early ninth century coincided not only with a period of intense Viking depredations but also an upsurge in hagiographical activity. Between about 850 and 950, three vernacular Lives were written—of Brigit, Patrick, and Adomnán—on behalf of Ireland's principal churches, Kildare, Armagh, and Iona's successor, Kells. The earliest, *Bethu Brigte* (Life of Brigit), assigned to its saint (who was the abbess of Kildare) a unique status equal to that of a bishop. The Tripartite Life composed for Patrick before 900 expanded greatly the itinerary attributed to the saint in his earlier Lives, especially with regard to Munster, where Armagh's influence had increased. *Betha Adamnáin* (Life of Adomnán) commented on church-state relations in the midlands about 950, from the point of view of the Columban authorities in Kells.

A date in the late eleventh century is indicated for the second vernacular Life of Brigit, the so-called Middle-Irish Life, by the presence of extracts in the manuscripts containing a *Liber Hymnorum* (Book of Hymns), which date to about 1100. While the Middle-Irish Life of Patrick may belong in the same period, that of Colum Cille has been dated to the late twelfth century.

Twelfth-Century Lives

Ireland witnessed a major ecclesiastical reorganization in the early twelfth century. However, despite its root-and-branch nature, during its first fifty or so years this reorganization failed to stimulate hagiographical activity. Neither Life nor calendar nor martyrology is preserved in the great manuscripts of the period 1050 to 1150, notably *Lebor na hUidre* and the Book of Glendalough.

Paradoxically, Irish hagiography was then being compiled abroad, at Lagny, near Paris, where a Life of Fursa was prefaced with an account of his upbringing in Ireland, based on oral witness, and at Clairvaux, where Bernard drew on information from the saint's companions to write a Life of Malachy. In England, a Life of Brigit was written by Laurence of Durham in the 1140s, and Geoffrey wrote a Life of Modwenna (Moninne) in the early twelfth century at Burton-on-Trent.

In Ireland, the turning point came in the 1160s, against the background of the Anglo-Norman invasion and the ensuing collision of cultural traditions. These events spawned numerous saints' Lives in Latin, the only language shared by Irish and Anglo-Normans. The early-thirteenth-century Life of Abbán, which used an English church (Abingdon) to make its point, was clearly directed at an English audience.

Collections of Saints' Lives

The earliest known collection of Irish saints' Lives was made at Regensburg (Germany) in the late twelfth century for inclusion in the Great Austrian Legendary, now preserved in various Austrian libraries (hence the name). Later, as the revival in learning reached Ireland from the Continent in the course of the fourteenth century, collections of saints' Lives commenced. The *Codex Salmanticensis*, which was possibly compiled at Clogher (Tyrone) in the early to mid-fourteenth century, was followed by collections made for houses of Austin canons on Saints' Island, Westmeath (Rawlinson B 485), shortly before 1400; and Abbeyderg, Longford (Rawlinson B 505), shortly after. Two Franciscan collections were made during the fifteenth century, one probably at Kilkenny (Marsh's Library MS Z 3. 1. 5.), the other (Trinity College MS 175) in south Leinster.

The late fourteenth century also witnessed the production of some vernacular Lives, notably in south Munster and Connacht. The Lives of Molaga and Finnchú of northeast Cork, and Lasair of Kilronan (Roscommon), probably date to this period. Collections of vernacular Lives are extant from the mid-fifteenth century. In the early sixteenth century, vernacular Lives were composed in northwest Ulster, including Manus O'Donnell's well-known Life of Colum Cille.

Hagiography in the Period 1580–1650

A new interest in Irish saints' Lives began with the publication at Antwerp in 1587 of *De vita S. Patricii Hyberniae* (On the Life of St. Patrick of Ireland) by Richard Stanyhurst. This new phase was distinctive in many ways, most notably through the role played by the mainly Jesuit and Franciscan Irish colleges on the Continent. The Franciscan scheme for the publication of Ireland's ecclesiastical remains, which was based in St. Antony's College, Louvain, involved such outstanding scholars as John Colgan, Míchéal Ó Cléirigh, and the Jesuit Stephen White. The scheme ensured the survival of numerous texts that would otherwise have perished. On the other hand, the survival of the main collections of Latin Lives was due to the endeavors of such Anglo-Irish scholars as Archbishop James Ussher and Sir James Ware. Despite the fact that much of this activity was directed towards the compilation of new histories of regional or national Christianity, and the illumination of the great religious disputes of the day, both groups occasionally exchanged materials.

The Liturgical Tradition

Since celebration of the saint's feast-day necessarily dates to soon after the subject's death, liturgical writings concerning feast-days are often regarded as the more authentically historic of the two strands of hagiography. Two types of written record were involved, calendar and martyrology; the former (selective) record comprised the feasts of a single church or closely linked group of churches, the latter (fuller) record included saints commemorated throughout Christendom. The earliest surviving record of this kind, the *Depositio Martyrum* of 354 A.D., was a calendar of feasts celebrated in Roman churches. The earliest martyrology, spuriously named Hieronymian after Jerome (d. *c.* 420), was compiled in the early seventh century, possibly at Luxeuil. Its earliest manuscript, the Martyrology of St. Willibrord of Echternach, which dates to shortly after 700, was perhaps written by an Irish scribe. All later martyrologies are based on the *nuda nomina* (bare names) of the Hieronymian lists, including the so-called historical versions, which added biographical details. Bede (d. *c.* 735) compiled the first historical martyrology, and his work was augmented in the ninth century, most notably by Ado of Vienne and Usuard of Paris. One of the earliest copies of Bede, the ninth-century St. Gall MS 451, reveals considerable Irish influence. Historical and Hieronymian martyrologies continued to be copied throughout the Middle Ages. The now standard Roman Martyrology was first drawn up on the instructions of Gregory XIII (d. 1585).

The Irish Martyrological Tradition

The Irish tradition is taken to begin with the compilation at Tallaght of two texts, one prose (Martyrology of Tallaght), the other verse (Martyrology of Óengus). The latter is a metrical version of the former, and both are dateable to about 830. The prose text, an abbreviated version of the Hieronymian martyrology, contains many features pointing to a Northumbrian provenance, very probably in the monastery of Lindisfarne. On the way from Northumbria to Tallaght, the martyrology passed through Iona and Bangor, where it received its first and second layers of Irish additions. However, the greater part of the Irish additions were made at Tallaght. These include a number of clerics involved in the *Céli Dé* movement. Not long after its completion, the prose martyrology appears to have been regarded as a relic. When a new copy was made shortly after 1150, probably at Terryglass, for inclusion in the Book of Leinster, further names were added from a copy of the Martyrology of Ado. A copy of Ado had reached Dublin in the early eleventh century but did not go into circulation before 1150. Now preserved in a thirteenth-century copy made at Christ Church, Dublin, Ado also served as a source of the Martyrology of Gorman, the Commentary on Óengus, and the Drummond Martyrology, all of which date to around the beginning of the Anglo-Norman invasion. The earliest, Gorman, while drawing Irish saints mainly from the Martyrology of Tallaght, otherwise made extensive use of a copy of the Martyrology of Usuard. Other martyrologies were compiled at this time at Lismullin near Tara (Martyrology of Turin) and Lismore (Martyrology of Cashel). The latter text, no longer extant, was devoted exclusively to Irish saints. Preserved in the same late-twelfth-century manuscript as the Martyrology of Turin is an Irish version of the metrical Martyrology of York.

The churches located in the English sphere of influence used copies of Usuard or Ado, usually of English provenance, but often containing Irish feasts. A copy of Usuard made at the Youghal Franciscan friary shortly before 1500 is now preserved in Berlin (Staatsbibliothek, MS Theol. Lat. Fol. 703). Following the revival of learning in the second half of the fourteenth century, several new copies were made *inter Hibernos* (among the Irish) of the Martyrology of Óengus. The latest native martyrology of note was that of Donegal, which Míchéal Ó Cléirigh and at least one other collaborator prepared in the 1620s. It is almost exclusively devoted to Irish saints.

Calendars

The early ninth-century Karlsruhe calendar (Landesbibliothek MS Cod. Aug. CLXVII) is the only surviving pre-Anglo-Norman text of this kind. Originating in Ireland (Clonmacnoise or Glendalough), it was later brought to Reichenau in Germany. Numerous (mostly unedited) calendars survive from churches of the areas under English influence, notably Dublin and Meath. The earliest calendar of this type from a church *inter Hibernos* forms part of a poem composed in the late fourteenth century.

PÁDRAIG Ó RIAIN

References and Further Reading

Aigrain, R. *L'hagiographie, ses sources, ses méthodes, son histoire*. Paris: Bloud et Gay, 1953.

Bieler, L. "The Celtic Hagiographer." *Studia Patristica* 5 (1962): 243–265.

Bieler, L., ed. *Four Latin Lives of St. Patrick*. Dublin: Institute for Advanced Studies, 1971.

———. *The Patrician Texts in the Book of Armagh*. Dublin: Institute for Advanced Studies, 1979.

Binchy, D. A. "Patrick and his Biographers: Ancient and Modern." *Studia Hibernica* 2 (1962): 7–173.

Charles-Edwards, T. M. *Early Christian Ireland*. Cambridge: University Press, 2000.

Dubois, J. *Les martyrologes du moyen âge latin* (Latin Martyrologies of the Middle Ages). Turnhout: Brepols, 1978.

Hennig, J. "Studies in the Latin Texts of the *Martyrology of Tallaght*, of *Félire Oengusso*, and of *Félire Húi Gormain*." *Proceedings of the Royal Irish Academy* 69C, no. 4 (1970): 45–112.

Herbert M., and P. Ó Riain, eds. *Betha Adamnáin: The Irish Life of Adamnán*. London: Irish Texts Society, 1988.
Herbert, M. *Iona, Kells, and Derry: the History and Hagiography of the Monastic Familia of Columba*. Oxford, University Press, 1988.
Hughes, K. *Early Christian Ireland: Introduction to the Sources*. London: Hodder and Stoughton, 1972.
Kenney, J. *The Sources for the Early History of Ireland: Ecclesiastical*. New York: Cornell University Press, 1929.
McCone, K. "Brigit in the Seventh Century: a Saint with Three Lives?" *Peritia* 1 (1982): 107–145.
Ó Riain, P. "St. Abbán: The Genesis of an Irish Saint's Life." In *Proceedings of the Seventh International Congress of Celtic Studies, Oxford, 1983*, edited by D. E. Evans et al., 159–170. Oxford: University Press, 1986.
———. "The Tallaght Martyrologies, Redated." *Cambridge Medieval Celtic Studies* 20 (1990): 21–38.
———. "A Northumbrian Phase in the Formation of the Hieronymian Martyrology: the Evidence of the Martyrology of Tallaght." *Analecta Bollandiana* 120 (2002): 311–363.
———, ed. *Four Irish Martyrologies: Drummond, Turin, Cashel, York*. London, Henry Bradshaw Society, 2003.
Plummer, C., ed. *Vitae Sanctorum Hiberniae. Lives of Irish Saints*, vol 1. Oxford: University Press, 1910.
———, ed. *Bethada Náem nÉrenn. Lives of Irish Saints*, vol. 2. Oxford: University Press, 1922.
Schneiders, M. "The Irish Calendar in the Karlsruhe Bede." *Archiv für Liturgiewissenschaft* (Liturgical Studies Archive) 31 (1989): 33–78.
Sharpe, R. "*Vitae S Brigidae*: the Oldest Texts." *Peritia* 1 (1982): 81–106.
———. *Medieval Irish Saints' Lives. An Introduction to Vitae Sanctorum Hiberniae*. Oxford: Clarendon Press, 1991.

See also **Adomnán; Anglo-Norman Invasion; Brigit; Céle Dé; Cogitosus; Colum Cille; Devotional and Liturgical Literature; Glendalough, Book of; Leinster, Book of; Patrick**

HEIRS AND HEIRESSES

See **Marriage; Society, Functioning of**

HENRY II

The accession of Henry II as king of England in 1154 resulted in not only a change of dynasty from Norman to Angevin, but also a greatly expanded assemblage of lordships and lands. To the cross-channel kingdom of England and Normandy that had resulted from the Norman conquest of England in 1066, was now added the county of Anjou (whence the dynastic name Angevin), where Henry succeeded his father; at the same time, in right of his wife, Eleanor, whom he had married in 1151, he also ruled Aquitaine and its associated lands. Henry's dominions thus stretched from the Scottish border on its most northerly frontier to the Pyrenees on its southernmost frontier, and his kingdom had a more obviously European dimension and encompassed a greater diversity of customary laws and practices than the more-narrowly integrated Anglo-Norman state that had been fashioned between 1066 and 1154.

Shortly after Henry's accession as king, a royal council at Winchester, attended by the ecclesiastical and lay magnates of England, debated a prospective conquest of Ireland, following which papal endorsement was sought from the English-born pope, Adrian IV, who in 1155 issued a papal privilege known as *Laudabiliter* ("Laudably," from the first word of the text) which authorized Henry to enter Ireland in order to advance the Christian faith among a people "still untaught and barbarous." The envoy sent to the papal court was John of Salisbury, secretary to Theobald, archbishop of Canterbury, and a personal friend of Pope Adrian. In his *Memoirs of the papal court*, John recounted how he secured the papal privilege from Adrian and how, while in Rome, he had taken the opportunity to inspect the papal archives for information on the status of the Irish church. He also incidentally revealed his animosity towards the Roman papal legate, Cardinal John Paparo, who in 1152 had presided over a church synod at Kells and given papal endorsement to an island-wide diocesan framework for the Irish church under the primacy of the archbishop of Armagh. A number of other contemporary English and Norman chroniclers also commented adversely on this decree of the Synod of Kells. Their dissatisfaction derived from the circumstance that from the late eleventh century onwards archbishops of Canterbury had consecrated a series of bishops for the Hiberno-Norse towns of Dublin, Waterford, and Limerick. The primacy of Armagh endorsed by Paparo at the Synod of Kells however precluded any further such consecrations. It may be inferred—both from the emphasis on Christian reformation and the circumstance that it was John of Salisbury who traveled to Rome to petition the pope—that a conquest of Ireland was mooted by the church of Canterbury in 1154 and probably as a possible means of recovering its former influence in the Hiberno-Norse towns. Henry may at first have been prepared to countenance the proposal but, in the event, he chose not to act on *Laudabiliter*. Pope Adrian was not willing to reverse the decision on the independence of the Irish church, while the justification offered for papal authorization of a conquest of Ireland was a claim to jurisdiction over islands, which the king may not have found particularly palatable—Britain, after all, was also an island. In any case, circumstances in 1155 were not particularly favorable for an overseas military enterprise. Henry had succeeded to a kingdom war-weary after nineteen years of civil strife, a factionalized aristocracy, and a greatly depleted exchequer;

within his own family, he faced opposition from his brother, William, who sought to challenge Henry's retention of the lordship of Anjou and its attachment to the kingship of England.

In 1166, Diarmait Mac Murchada, the exiled king of Leinster, traveled to the court of Henry II in Aquitaine and solicited military aid to assist him in recovering his kingdom. In 1165, Henry had hired the Dublin fleet for a military campaign in north Wales, almost certainly with Diarmait's consent as overlord of Dublin. Henry responded initially by authorizing Diarmait to raise mercenary troops within his dominions for deployment in Ireland. In the autumn of 1171, however, Henry decided to lead a major expedition to Ireland himself where he remained until Easter 1172. Some Anglo-Norman chroniclers claimed that he removed to Ireland in order to avoid the censure of papal legates in the wake of the murder of Thomas Becket, archbishop of Canterbury. This may indeed have been a consideration, but Henry's main aim was to assert control over those of his Anglo-Norman subjects who had gone to Ireland, and notably over the most important of them, the disaffected Richard FitzGilbert, lord of Strigoil and Earl of Pembroke (1148–1154) also known as Strongbow, whose lands in England, Wales, and Normandy were held directly from Henry II, and whom Henry had deprived of the earldom of Pembroke in 1154. Henry obliged Strongbow to acknowledge him as overlord of his newly acquired lands in Leinster, from which, however, Henry excepted the Hiberno-Norse ports of Dublin, Waterford, and Wexford, reserving them for his own use. Retention of the ports may have been crucial initially for reasons of security and control of traffic between the two countries, but Henry's charter to "his men of Bristol," granting them the city of Dublin, issued during his stay in Ireland, affords an early indication that Ireland also presented opportunities for economic entrepreneurship and profiteering by the king.

On the eve of his departure, Henry made a speculative grant of the kingdom of Mide (Meath) to a prominent Anglo-Norman magnate, Hugh de Lacy, who had accompanied him to Ireland. While this may be interpreted as aimed at securing political stabilization by providing a counterweight to Strongbow as lord of Leinster and by removing Meath, the most volatile and heavily contested area in twelfth-century Ireland, from Irish control, it was also another early indication that Ireland could be exploited as a potential source of patronage by the English crown. During his six-month stay, Henry had not ventured outside Leinster, nor did he engage in any military conflict with the Irish. A significant number of Irish kings traveled to his Christmas court at Dublin and voluntarily offered him recognition. Ruaidrí Ua Conchobair, king of Connacht and claimant to the high kingship, however, remained aloof, although in 1175 he negotiated a treaty by proxy at Windsor with Henry whereby he agreed to acknowledge Henry's overlordship of Meath and Leinster, in return for which Henry acknowledged Ruaidrí as over-king of the remainder of Ireland.

Following Strongbow's untimely death in April 1176, leaving a three-year-old son as his heir, responsibility for the administration of the lordship of Leinster fell on Henry II until Strongbow's heir came of age. In May 1177, at a royal council at Oxford, Henry divided Leinster into three administrative areas and signaled his intention to make more formal arrangements for Angevin lordship in Ireland by designating his youngest son, John, as lord of Ireland, and requiring the principal Anglo-Norman landholders in Ireland to swear fealty to John. At the same time he made a series of speculative grants in Munster. To Robert FitzStephen and Miles de Cogan, who were already actively involved in conquest there, he assigned the kingdom of Desmond, while Thomond was offered to Philip de Braose, who, however, proved incapable of implementing the grant which then lapsed.

In 1185, Henry judged that it was time for John to assume the Angevin lordship of Ireland in person and provided him with a team of experienced administrators and substantial resources, presumably with the intention that John would take up long-term residence in the country. Henry had already assigned regional lordships to his other sons as a means of consolidating links between the constituent parts of the Angevin dominions. Within eight months, John had returned to England, having failed to intensify Angevin lordship in Ireland. On the evidence both of Anglo-Norman commentators and of the Irish annals, John's chief failure lay in relation to Hugh de Lacy, then the most prominent Anglo-Norman in Ireland, who was rumored to aspire to the kingship of all Ireland, and to have dissuaded Irish rulers from acknowledging John. Henry was seriously displeased with this outcome and was contemplating sending John back to Ireland when news reached the English court of Hugh de Lacy's assassination, at which the king is said to have rejoiced. Henry's last executive decision in relation to his Irish lordship was to arrange, shortly before he died, the marriage between Isabella, daughter and heir of Strongbow, to William Marshal, who thereby succeeded to the lordship of Leinster in right of his wife.

Although Henry primarily had reacted to the activities of his subjects rather than determined the Anglo-Norman advance in Ireland; nonetheless, he was more than willing to avail of such opportunities for the dispensing of royal patronage as it afforded him. In his attitude toward the Irish, he may have been more tolerant of difference than were his own subjects and subsequent English kings, accustomed as he was to regional diversity within his many lordships through which he traveled constantly.

M. T. Flanagan

References and Further Reading

Duffy, Seán. *Ireland in the Middle Ages*. London: Macmillan, 1997.

Flanagan, Marie Therese. *Irish Society, Anglo-Norman settlers, Angevin Kingship: Interactions in Ireland in the Late Twelfth Century*. Oxford: Oxford University Press, 1989. (Reprint, 1998.)

See also **Anglo-Irish Relations; Anglo-Norman Invasion; Church Reform, Twelfth Century; Feudalism; Kings and Kingships; Lordship of Ireland; Mac Murchada; Papacy; Uí Chennselaig**

HENRY OF LONDON (d. 1228)

Henry of London, archbishop of Dublin (1212–1228) and Chief Governor of Ireland, was born in London, one of five sons of Bartholomew Blund, alderman of that city. He is sometimes called *magister* in the records, which suggests he had some university education, but it is not known where he earned this title. His early patron was Hugh de Nonant, bishop of Coventry, a trusted ally of Count, later King, John. Following his patron's example, Henry also attached himself to the count's retinue; once John became king, his successful career in the royal administration began. Henry was typical of the royal *curiales* who operated in a variety of capacities in Angevin governmental structure. He served in the judiciary, and as an official of the chamber; he supervised both the transport of the royal treasure and the organization of household supplies. He also undertook a number of diplomatic missions, including, significantly, at least two visits to Ireland. Henry proved his loyalty to the king by remaining steadfastly by his side from 1208 to 1214, when England and Wales lay under papal interdict. He was rewarded with an impressive collection of benefices, prebends, and titles, but two attempts on the part of King John to have his faithful clerk appointed to an English bishopric were unsuccessful. In 1212, however, he was appointed, apparently unopposed, to the archbishopric of Dublin when it fell vacant on the death of John Cumin.

From the start of his episcopate it was clear that Henry was expected to perform a dual role in Ireland, and this was underlined in 1213 by his appointment as justiciar. He held the office of justiciar from 1213 to 1215 with considerable success in both military and diplomatic spheres. He played a major role in the rebuilding of Dublin castle as well as supervising the construction of several other castles in strategic parts of the colony. He appears to have ensured the loyalty of the English barons in Ireland to King John during the crisis faced by that monarch in England and fittingly, was present at Runnymede in June 1215, where he was one of the chief witnesses to the Magna Carta.

Henry was relieved of the office of justiciar in 1215 as he was about to embark for Rome to attend the Fourth Lateran Council. He returned to Dublin in 1217 as papal legate and in this capacity presided over a general synod of Irish clergy. His appointment as papal legate was terminated abruptly in 1220, when he was associated with the discriminatory policy that sought to exclude Irish clerks from holding ecclesiastical offices. During this period he also became engaged in a number of disputes with both royal and municipal officials around the exercise of his jurisidictional rights in the Dublin area.

The conflicts were not serious enough to place the loyalty of the archbishop in doubt, and in 1221 he was again appointed justiciar of Ireland. This second term of office coincided with the rebellion of Hugh de Lacy, from whom Henry was forced to purchase a truce when he threatened Dublin in 1224. Soon after this he was replaced in office by William Marshall II, whereupon he crossed over to England and spent some time there and on the Continent. After his return to Ireland in 1226, his last recorded administrative duty was to examine Marshall's account as justiciar.

It has been said of Henry that he was more noted for his administrative expertise than his pastoral care and that his more overt spiritual actions frequently masked thinly disguised political aims. While attending the Lateran Council in 1216, he succeeded in obtaining papal confirmation for the unification of the diocese of Glendalough to Dublin and, with the establishment of the archiepiscopal manor of Castlekevin on lands previously held by that diocese, an important base was secured from which to subdue the Irish of Wicklow. Furthermore, the concern the archbishop expressed to the pope in 1215 for the religious in his diocese, who lived in scattered cells without proper discipline and guidance, was used as justification for the granting of the Irish Abbey of St. Saviour's of Glendalough to All Hallows Priory in Dublin. In completing the work of his predecessor Cumin by raising St. Patrick's to cathedral status and instituting the offices of dean, chancellor, treasurer, and precentor, he insured the means of rewarding those ecclesiastical civil servants so important to the administration.

The precise nature of the archbishop's association with the 1217 mandate excluding Irish clerks from episcopal offices is the subject of debate. Some have seen him as the principal architect of this policy, while others point to evidence of his willingness to work with

Irish churchmen and attribute the prime role to Geofrey de Marisco or William Marshall. During the years of Henry's episcopate there was a noticeable increase in the number of Englishmen appointed to Irish bishoprics. While Henry clearly would have approved of this trend, the extent of his involvement in securing the election of these individuals cannot be systematically assessed.

Although he was responsible for raising St. Patrick's as a rival cathedral to Christ Church, he did not neglect the older institution, but continued to regard it as very much his cathedral church. He progressed the building work on the cathedral church of Holy Trinity, completing most of the nave with stone and sculptors brought over from England. In 1220, Henry granted rents to the prior and convent for the building of a new entrance, and in return, the community undertook to celebrate his obit forever.

Henry of London was a generous benefactor to religious houses in and around Dublin and displayed a particular favor to institutions which cared for the sick and the poor. He made gifts to St. Thomas' Abbey to help with the care of paupers and to the hospital of St. John the Baptist. He also founded a hew hospital, St. James at the Steyne, which was intended to cater specifically for the needs of poor pilgrims who planned to visit the shrine of St. James at Compostella. The evidence of this practical piety goes some way towards mitigating the "faceless bureaucrat" tag which might otherwise be applied to him.

MARGARET MURPHY

References and Further Reading

Gwynn, Aubrey. "Henry of London, Archbishop of Dublin: A study in Anglo-Norman Statecraft." *Studies*, 38 (1949): 297–306, 389–402.

Murphy, Margaret. "Balancing the Concerns of Church and State: The Archbishops of Dublin, 1181–1228." In *Colony and Frontier in Medieval Ireland: Essays Presented to J. F. Lydon,* edited by Terry Barry, et al., 41–56. London: The Hambledon Press, 1995.

———. "Archbishops and Anglicization: Dublin, 1181–1271." In *History of the Catholic Diocese of Dublin*, edited by James Kelly and Dáire Keogh, 72–91. Dublin: Four Courts Press, 2000.

Turner, Ralph. *Men Raised from the Dust: Administrative Service and Upward Mobility in Angevin England*. Philadelphia: University of Pennsylvania Press, 1988.

Watt, John. *The Church in Medieval Ireland*. 2nd. ed. Dublin: University College Dublin Press, 1998.

See also **Chief Governors; St. Patrick's Cathedral**

HIBERNO-ENGLISH LITERATURE

Hiberno-English, or Irish English, is the name given to the English dialect that developed in Ireland after the invasion in 1169 by French-speaking Anglo-Normans, who had English speakers (as well as Welsh and Flemings) among their followers. The extent to which English spread subsequently among the population of Ireland is hard to ascertain. English-speaking followers of the Anglo-Norman military leaders established themselves in the towns of the east and southeast of Ireland, and the language continued in use in the area of greatest English influence, the Pale. Medieval Hiberno-English dates from the twelfth to the fifteenth centuries.

The written evidence for English in medieval Ireland is small and thus enormously important in relation to its volume. London British Library Manuscript Harley 913, dating from circa 1330, is the earliest and most notable repository of medieval Hiberno-English, containing amongst its forty-eight items seventeen poems written in the English of medieval Ireland. These poems include the *Land of Cockayne*, and the *Song* of Friar Michael of Kildare. There is no firm evidence to support the theory that all seventeen poems were composed by Friar Michael, and internal evidence does not justify naming Kildare as their place of origin. Nevertheless, though probably written down in Waterford, these Hiberno-English poems in MS Harley 913 are often called the "Kildare Poems." Several of the poems show strong evidence of Irish influence, and as a collection they display a capacity for both piety and satire.

Some Features of Medieval Hiberno-English

It is a general feature of the earliest medieval Hiberno-English that, while its individual linguistic features will nearly all be found in various Middle English dialects of England, especially those of the southwest and the west midlands, collectively they do not match any single dialect of Middle English. The evidence overall would suggest that the earliest settlers probably came from counties around the Severn estuary, and fits the historical fact that the Anglo-Norman invasion, planned by Richard de Clare (Strongbow), Earl of Pembroke, was largely launched from Bristol. For example:

1. The earliest texts show loss of final *-e* in words such as *tak*, *ber*. This loss occurred earliest in the Middle English dialect of the north of England, and its occurrence in other Middle English dialects is later than its earliest occurrence in Hiberno-English. Though spellings with the final *-e* are also evidenced (*take*, *bere*), it appears from rhymes that in all cases it was silent.
2. The spelling *s* for *sh* in unaccented syllables in words like *Englis* (English), *Iris* (Irish), *worsip* (worship), is also a feature of the northern dialect of Middle English.

3. The spelling *u/v* for initial *f* found in some words such as *uoxe* (fox), *uadir,* (father) is a feature of Middle English dialects of the south and south-west of England.
4. The present participle ending of verbs, *-end/-ind,* as in *glowind* (glowing), is a feature of Middle English dialects of the North and West of England.
5. The use of special letter forms ʒ (yogh) for *y* or *gh*, and þ (thorn) for *th* is common to all Middle English writing at this time, but some confusion between the two symbols is found in *neiþ* for *neiʒ* (nigh) in *Christ on the Cross*, and *touþ* for *touʒ* (tough) in *Song of the Times* in MS Harley 913.
6. The present tense ending of verbs are those of southern dialects of Middle English: singular endings: *-e*, *-ist/-est*, *-iþ/-eþ*; plural endings: *-iþ/-eþ*. The occasional final *d* or *t* instead of *þ/th*, as in *makit* for *makiþ*, and the use of *the* for *de* in the name Piers *the* Bermingham indicate a failure to distinguish between the sounds /t/ and /æ/, still a notable feature in Irish English today.

The poems of MS Harley 913 also contain some Irish words in their vocabulary including: *eri*, probably a borrowing from Irish *éraic* (compensation); *corrin*, from Irish *coirín* (can or tankard); *ketherin*, from Irish *ceithearn* (band of soldiers); *tromchery*, from Irish *trom chroí* (liver); *capil*, (horse) is later found in England in other fourteenth-century authors including Chaucer (and may have been borrowed from Irish into English via Norse), but has its earliest occurrence in English in *The Land of Cokaygne*.

Some Other Sources of Medieval Hiberno-English

The drama now called *The Pride of Life* was written down some time in the first half of the fifteenth century by two scribes in the spare spaces of an Account Roll of the priory of Holy Trinity Dublin (otherwise Christ Church Cathedral). Its composition may date from the second half of the fourteenth century. The original was destroyed in the 1922 fire at the Public Record Office in Dublin. The play is the earliest known example in English of a Morality Play, and there is no reason to suppose that it was not of local composition.

The slates of Smarmore, County Louth, discovered in 1959–1962, inscribed in English and Latin and probably intended for schoolroom use, provide limited but valuable evidence of the English language in Ireland in the fifteenth century. Though there are difficulties in identifying and dating the handwriting because of the unusual materials involved, musical notation on four of the slates dates them to the second quarter of the fifteenth century. The English on these slates seems in general to conform with what is known about the English language in Ireland at that period.

A complete list of texts containing medieval Hiberno-English, including town records and legal documents, is to be found in McIntosh and Samuels's "Prolegomena." In its medieval form, the English language in Ireland was virtually overcome by the spread of Irish among the original invaders as shown as early as the Statutes of Kilkenny. Two rural areas, however, Forth and Bargy in South Wexford, and Fingal in North Dublin, preserved until modern times the Hiberno-English of the earliest texts. The medieval period of Hiberno-English ends with the coming of new varieties of English, including the emerging "Standard English," with large numbers of "planters" in the sixteenth and seventeenth centuries. However, with firmly established medieval roots, Hiberno-English has the longest recorded history of extra-territorial English in the world.

ANGELA M. LUCAS

References and Further Reading

Bliss, Alan J. "The Inscribed Slates at Smarmore." *Proceedings of the Royal Irish Academy* 64 (1965): 33–64.

Britton, Derek, and Alan J. Fletcher. "Medieval Hiberno-English Inscriptions on the Inscribed Slates of Smarmore: Some Reconsiderations and Additions." *Irish University Review* 20 (1990): 55–72.

Dolan, T. P., and Diarmuid Ó Muirithe,. *The Dialect of Forth and Bargy, County Wexford*. Dublin: Four Courts Press, 1996. Revised edition of Poole, *A Glossary of the Old Dialect of Forth and Bargy.*

Henry, P. L. "A Linguistic Survey of Ireland: Preliminary report," *Lochlann* 1 (1958): 49–208.

Hickey, Raymond. "The Beginnings of Irish English." *Folia Linguistica Historica* 14 (1993): 213–238.

———. "A Lost Middle English Dialect." In *Historical Dialectology*, edited by Jacek Fisiak, 235–272. Berlin, New York, and Amsterdam: Mouton de Gruyter, 1988.

Hogan, Jeremiah J. *The English Language in Ireland*. Dublin: Educational Company of Ireland, 1927. Reprint, Maryland: McGrath, 1970.

Jordan, Richard. *Handbook of Middle English Grammar: Phonology.* Translated and revised by E. J. Crook. The Hague: Mouton, 1974.

Lucas, Angela M., ed. *Anglo-Irish Poems of the Middle Ages*. Blackrock: Columba Press, 1995.

McIntosh, Angus, and Samuels, M. L. "Prolegomena to a Study of Medieval Anglo-Irish." *Medium Ævum* 37 (1968): 1–11.

McIntosh, Angus, et al. *A Linguistic Atlas of Mediaeval English*. Aberdeen: Aberdeen University Press, 1986.

Mills, James, ed. *Account Roll of the Priory of the Holy Trinity, Dublin, 1337–1346*. Dublin: Royal Society of Antiquaries of Ireland, Extra vol. 1890–1891. Reprint, with introduction by James Lydon and Alan J. Fletcher, in *A History of Christ Church Dublin*. Vol. 2. Dublin: Four Courts Press, 1996.

See also **Anglo-Norman Invasion; Dublin; Education; Gaelicization; Hiberno-Norman (Latin) Literature; Kildare; Languages; Strongbow; Waterford**

HIBERNO-LATIN LITERATURE

Hiberno-Latin literature is the name given to a vast body of literature written in Ireland or by Irishmen abroad between the fifth and twelfth centuries. In some cases, this category includes material with a Welsh, Scottish, or Western European background. But if it can be shown that this material appears in manuscripts exhibiting insular paleographical features, or otherwise has content, style, or language characteristic of texts of known Hiberno-Latin provenance, then that also may be included in the category. The great bulk of Hiberno-Latin literature has not been preserved, either in Ireland or in Irish manuscripts; so that it has been rightly said by Mario Esposito, one of the pioneers in the field, that "a just appreciation of the nature and extent of Latin learning in medieval Ireland can only be obtained by a critical study of the Latin literature produced either in that country, or by Irishmen who had emigrated to Britain and the continent" (Esposito 1929). The survival of these texts in Anglo-Saxon or Continental manuscripts, often in copies made centuries after their original composition, testifies to the influence of Hiberno-Latin literature throughout Europe in the Middle Ages. Controversy still surrounds the authorship, provenance, and date of some of them.

The study of rhetoric, as part of the first stage of the monastic curriculum of grammar, rhetoric and dialectic, led to the cultivation of various kinds of composition—more elaborate, ornate composition for the epistolary style, of which the earliest examples are the letters of Columbanus, and for the rhetorical introductions to commentaries, treatises, and hagiographical compositions. A plainer Latin style was used for texts which were not meant to be read as literature, but as legal or instructional documents, such as monastic rules, penitentials, and canon texts.

The great period of literary activity began in the seventh century, from which about fifty original works survive. But the greatest period of productivity was among the Irish *peregrini* (those living abroad) in the eighth and ninth centuries, so that the quantity of literature from the entire period is consequently too numerous to be listed. Many texts are still in manuscript or have yet to be properly edited and studied. Only a handful of works survive from the sixth century.

Grammar

No grammar textbooks survive from before the mid-seventh century, but some from the later period show evidence of having been based upon seventh century originals. Perhaps the earliest are the *Ars Asporii*, a Christian adaptation of Donatus's *Ars Minor*, the *Anonymus ad Cuimnanum* and the commentary on Virgil compiled perhaps by Adomnán. Numerous other grammar texts, glossaries, and short tracts also survive.

Hagiography

The earliest Latin life of an Irish saint is possibly Jonas of Bobbio's Life of Columbanus, which, although written under the influence of the Irish educational system on the continent circa 639 to 643, is not Hiberno-Latin. The earliest Irish hagiographical composition is Cogitosus' *Vita Brigitae*, the Life of Brigit of Kildare, written circa 650. The next are Muirchú and Tírechán's lives of Patrick, dating between 661 and 700. Adomnán's *Vita Columbae* was completed circa 700, but perhaps as early as 692. It is based in part upon living tradition transmitted to Adomnán by people who had known Columba (Colum Cille). Cogitosus' Life of Brigit is fantastic and is little more than a catalogue of miracles and stories of the marvellous, whereas Muirchú's Life of Patrick is an attempt to form a consecutive narrative out of the disparate traditions relating to Patrick. There are fragments also of an early life of Brigit called the *Vita Prima*, which may have been written by Ailerán of Clonard (d. 665), a biblical scholar. Some of the later Irish *Vitae* may be based on earlier material.

Monastic Rules

The earliest now surviving is that of Columbanus of Luxeuil (d. 615), a very strict Rule modeled upon that written for the early foundation of Bangor, County Down, by Comgall (d. 602). The Rule of Comgall itself does not survive, but it is listed among other Irish monastic rules in a ninth-century catalogue of manuscripts from the medieval library of Fulda. Some documents from the British church may be even earlier than these. The anonymous Rule known as *Regula cuiusdam patris ad monachos* (The rule of a certain father for his monks) was written on the Continent sometime in the late seventh century by an Irishman. Both it and the lengthy *Regula Magistri*, as has been recently suggested, are Continental-Irish adaptations of the Columbanian rule to the milder rule of Benedict of Nursia.

Penitentials

Penitentials are booklets prescribing certain penances for various categories of sins, both for monks and laymen. The penitentials for laymen were both for monastic tenants (*manaigh*) or ordinary lay persons being ministered to by monastic clergy. The earliest penitential is that of Finnian of Clonard, which dates to the first half of the sixth century. Columbanus' penitential is next and then the great penitential of Cummian the Tall, dating from the mid-seventh century. There are many others, including a short collection of canons attributed to Adomnán, and also a number in Old Irish penitentials based upon Latin originals.

Canon Law

Canon law texts include the so-called First and the Second Synods of Patrick. The First Synod is a very early document ascribed to Patrick, Auxilius, and Isserninus. Though it survives in the form in which it was used during the Romani reform, it may be based upon an original from the fifth or sixth century. The Second Synod is an interesting collection of decisions upon various matters, specifically of the Romani reform movement of the seventh century. The early-eighth-century compilation of Irish canon law known as the *Collectio canonum Hibernensis*, put together for the use of an Irish church unified after the divisions of the Paschal controversy, is one of the earliest systematic canon collections. Other miscellaneous canonical documents also survive.

Theological Literature

There are several interesting theological and Scriptural treatises surviving from before 700, including the earliest treatises on the Catholic Epistles from the Latin church, and a commentary on Mark. The earliest datable text, composed in 655, is *De mirabilibus Sacrae Scripturae* (Wonders of the Holy Scripures), by an author using the name of Augustinus. In the naively rationalistic spirit of the Middle Ages, it attempts to give a physical explanation for the miracles in the Bible—for example, the sun standing still at Joshua's command, Moses and the Israelites crossing the Red Sea dry-shod, and an interesting explanation of tidal flow. The most widely diffused text, of which there are many hundreds of manuscripts and vernacular translations, is a description of the twelve sources of moral evil in the world and their remedy, *De XII abusivis*, written between 630 and 660. Some of the other interesting compositions of the period are Adomnán's tract on the Holy Places of Palestine (*De locis sanctis*), the Pseudo-Isidorian theological-cosmological treatise *De ordine creaturarum* and Dicuil's *De mensura orbis terrae*.

Epistolography

Quite a number of letters, both open and private, survive. These include the famous letters of Columbanus; the letter written in 632 and 633 to abbot Segéné of Iona, probably by Cummian the Tall, relating to the Easter question; and that known as Colmán's Letter to Feradach on the textual emendation of the poet Caelius Sedulius and other texts.

The Computus

The calculation of the Easter term, or computus, was of great importance throughout Christendom and of particular interest to the insular churches from earliest times. The foundation of the insular computus was the tract known as *De ratione Paschali*, for long described as "an insular forgery," but now known to be a fourth-century Latin translation of a treatise written by Anatolius, the third-century bishop of Laodicea. Other important tracts are the seventh-century *De ratione computandi* and Cummian's epistle. A considerable amount of this material survives only in manuscript. Bede and the Anglo-Saxons drew much of their computational knowledge from the Irish.

Liturgical

A great mass of hymns and other liturgical pieces survive from the earliest period up to the eleventh century. One of the earliest manuscripts is the Antiphonary of Bangor, written from 680 to 691. There are some palimpsested Continental codices and fragments and later Irish martyrologies, missals, sacramentaries, hymnals, and so forth. They show the eclectic range of sources from which the Irish and Welsh churches drew their liturgy—from Rome, Gaul, Moorish Spain, and Antioch.

Scholastic Texts

There are also some scholastic texts, including the pseudo-grammatical treatises and letters of Virgilius Maro, who may have been Irish, and pieces written in an extravagant Latin style known as Hisperic, such as the *Hisperica Famina*, and some amulet poems or *loricae*.

Charters

There are no surviving seventh-century charters, but the evidence of the Patrician dossier, which refers to early church and monastic foundation documents, indicates that there might once have been. Unlike Francia and Anglo-Saxon England, Ireland has only about twelve pre-twelfth-century charters, nine of which were copied into the Book of Kells.

Poetry

There exists a very large body of poetry in Latin—lyrical, liturgical, hagiographical, and technical—some of which may date to the fifth century, which cannot easily be summarized in content or character. Much of it is skilfully and beautifully composed and still rewards study.

Aidan Breen

References and Further Reading

Bieler, L. *The Irish Penitentials*. Dublin: Dublin Institute for Advanced Studies, 1963.

Dunn, M. *The Emergence of Monasticism: From the Desert Fathers to the early Middle Ages*. Oxford: Blackwell, 2000.

Esposito, M. "Notes on Latin Learning and Literature in Mediaeval Ireland." *Hermathena* 20, no. 45 (1929).

Flint, V. "The Career of H.A. Some Fresh Evidence." *Revue Benedictine* 82 (1972): 63–86.

Garrigues, M. O. "L'oeuvre d'Honorius Augustodunensis: Inventaire critique." *Abhandlungen der Braunschweigischen Wissentschaftlichen Gesellschaft* 38 (1986): 7–136; 39 (1987): 123–228.

Kenney, J. F. *The Sources for the Early History of Ireland: I Ecclesiastical*. New York: Columbia University Press, 1966 [1929]; Reprint Dublin: Four Courts Press, 1992.

Lapidge, M., and R. Sharpe. *A Bibliography of Celtic-Latin Literature, 400–1200*. Dublin: Royal Irish Academy, 1985.

McCarthy, D. P., and A. Breen. *The Ante-Nicene Christian Pasch: De Ratione Paschali: The Paschal Tract of Anatolius, Bishop of Laodicea*. Dublin: Four Courts Press, 2003.

Ó Cróinín, D. *Early Medieval Ireland 400–1200*. (London and New York: Longman, 1995).

Picard, J.- M. "Structural Patterns in Early Hiberno-Latin Hagiography." *Peritia* 4 (1985): 67–82.

Reynolds, R. "Further Evidence for the Irish Origin of Honorius Augustodunensis." *Vivarium* 7 (1969): 1–7.

Sharpe, R. "An Irish Textual Critic and the Carmen Paschale of Sedulius: Colmán's Letter to Feradach." *Journal of Medieval Latin* 2 (1992): 44–54.

Walsh, M., and D. Ó Cróinín, eds. *Cummian's Letter "De controversia paschali," Together with a Related Irish Computistical Tract, "De ratione conputandi."* Toronto: Pontifical Institute of Mediæval Studies, 1988.

See also **Biblical and Church Fathers; Canon Law; Charters and Chartularies; Columbanus; Devotional and Liturgical; Hagiography and Martyrologies; Penitentials; Sciences**

HIBERNO-NORMAN (LATIN)

In Anglo-Norman Ireland, as in all of medieval Europe, Latin, apart from being the language of the liturgy, was the principal language of record-keeping and written communication. Almost all ecclesiastical records were in Latin and from an early date Anglo-Irish clerics produced Latin chronicles and annals. Clerics used their knowledge of Latin, initially gained in the service of the liturgy, for secular administrative purposes. Thus, the great majority of records of central and local administration were rendered in Latin. Latin was taught, resulting in a great variation of expertise, and this is reflected in surviving manuscripts. Individual writers produced errors, which often crept into common usage. Latin changed principally, however, because the people who wrote it spoke their own vernacular language in their daily lives. The construction and arrangement of their own languages influenced their Latin and therefore, inevitably, local variations arose. Local vernacular words were absorbed into Latin, reflecting the environment in which they were produced. New words were also created to describe items for which there was no known Latin word; developments in farming, weaponry and other areas of scientific advancement demanded words not available in Classical Latin. In wills or inventories, when new terminology was needed, the Latin-trained clerk, seeking an appropriate word, would often simply Latinize the word in common parlance. In 1186, when describing the death of Hugh de Lacy, the scribe states that he was killed with an iron tool, "namely a pykays." The Latin used in Ireland was essentially that in use in England—in effect Anglo-Norman Latin—but it was also influenced by the origin of the individual clerk that produced it. He might be a newcomer from England or Anglo-Irish by birth. In either case, the scribe had the problem of dealing with the Irish language and usually made a valiant attempt to produce phonetically an approximation of the Irish name. A similar difficulty arose with regard to place names. The scribes in the administration in Dublin complained that they found difficulty in recording Irish names in documents. Irish names were rendered in the closest approximation to the English vernacular and then sometimes Latinized; for example, instead of the patronymics Ua or Mac, the word *filius* might be followed by the genitive of the surname. The Anglo-Irish author of the Annals of Multyfarnham, writing in the 1270s, had difficulty with Irish names and his attempt to reproduce the names in phonetic form gave rise to Macohelan for Maelsechlainn Mac Cochláin and Makemahon for Eachmharcach Ua hAnluain. With standard documents, like wills and land transfers, there were well established formulae, and scribes could hardly go wrong. Formulaic Latin was also used to record other legal transactions and court depositions. Very idiosyncratic Latin and peculiar spellings can sometimes be found in witness reports, which may be simple mistakes or may represent local pronunciations; a name can be found spelled differently on the same page—this is easily understandable if the subject matter was being dictated in a local accent. Knowledge of early pronunciation, spelling, syntax, and vocabulary of the local vernacular is useful in determining the word intended by the scribe.

That a more-than-adequate education in Latin was available in Ireland is evidenced by the *Itinerarium Symonis Semeonis ab Hybernia ad Terram Sanctam.* Latin books of sermons have survived, although the best-known, the *Liber Exemplorum*, was in fact written by an English Franciscan living in Ireland about 1275. The actual sermon would be rendered in the vernacular for preaching to the laity. A thirteenth-century Latin manuscript collection of sermons in Trinity College, Dublin, includes a few short sentences in Middle English and has occasional interjections such as *loc wel* and *lo lac wel* that call attention to certain passages. There was a greater understanding of Latin among the general public than might be imagined. It can be assumed that the layperson knew at least enough Latin to protect his ownership of property, and many knew more. During the Alice Kyteler witchcraft trial of 1324, William Outlaw, Alice's son, was sufficiently expert in Latin to be able to forge a writ against Bishop Ledrede; during the same trial, the bishop insisted that English, French, and Latin, be used in order to ensure that his case was understood by all.

BERNADETTE WILLIAMS

References and Further Reading

Emmison, F. G. *Archives and Local History*. London: 1966.
Gooder, Eileen. *Latin for Local History: An Introduction*. London: 1961.
Latham, R. E. *Revised Medieval Latin Word List from British and Irish Sources*. London: 1994.
Lewis and Short. *A Latin Dictionary*. Oxford: 1984.
Mantello, F. A. C., and A. G. Rigg, editors. *Medieval Latin: An Introduction and Biographical Guide*. Washington: 1966.
Trice Martin, Charles. *The Record Interpreter*. Sussex: 1982.
Thoyts, E. E. *How to Read Old Documents*. London: 1980.

See also **Anglo-Irish; Annals; Geraldus Cambrensis; Hugh de Lacy; Mellifont; Placenames; Charters and Chartularies; Wills and Testaments; Records, Administrative; Records, Ecclesiastical; Courts; Agriculture; Personal names; Weapons and weaponry;**

HIGH CROSSES

The Irish words for "High Cross" are first encountered in the year 957, when *The Annals of the Four Masters* mention the burning of Clonmacnoise "from the High Cross to the Shannon" (ó chrois aird co Sionnaind), and the same source refers to the "Cross of the Scriptures" (ó chrois na screaptra) also at Clonmacnoise, under the year 1060. Other rare references to crosses are neither descriptive nor helpful in identifying them with any surviving examples, of which there are—depending upon definition—more than two hundred. In the absence of historical sources to provide explanations, only observation, comparison, and art historical research can shed light on the nature, date, development, and purpose of these crosses—topics that have evinced much debate for more than a century, while still leaving many questions unanswered.

East face of the North Cross, Castledermot, Co. Kildare. © *Department of the Environment, Heritage and Local Government, Dublin.*

The crosses earn their description by reaching a height of more than 21 feet, including base (e.g., the Tall Cross at Monasterboice), and have the shape of a Latin cross, with arms usually more than two-thirds the way up the shaft. The bases, normally in the form of a truncated pyramid, are always separate, while the actual crosses can be made up of one or two separate blocks, sometimes with an additional roof stone on top. The preferred material was that which was available locally—limestone west of the Shannon, granite in the Barrow valley and County Down, and sandstone (the easiest to carve) in the remainder of the north, east, and midlands. Munster has few crosses, and Cork and Limerick have none. With the possible exception of the Drumbane grit that went into the making of St. Patrick's Cross on the Rock of Cashel, no quarry has been reliably proven to have been the source of the stone of any one particular cross.

The awesome height of the crosses show how the early Irish monks began to appreciate the monumental power of stone in the period after 800, when the Viking raids caused much else that remained in the monasteries to be reduced to ashes. Almost all crosses are found on ecclesiastical sites, though ignorance about the location of contemporary wooden churches and domestic buildings makes it difficult to realize how the crosses related to space and structures within the ecclesiastical compound—a problem further exacerbated by the probable change of location of at least some of the crosses in recent centuries. The idealized plan of St. Mullins in Carlow, found in the Book of Mulling, would suggest that some crosses stood near the entrance to the enclosure; others may have marked boundaries, and apparently none were grave-markers. The majority were presumably erected in order to bring those who stood and knelt beside them closer to God, and crosses carved with biblical scenes would have had the dual purpose of being able to explain the sacred scriptures pictorially and inducing feelings of piety in the beholders. However, inscriptions (even ones poorly preserved) found at the bottom of some cross-shafts and most legible to those kneeling in front of them show that the crosses were not purely religious in character, but also had a political dimension. These inscriptions include the names of two High Kings, father and son, Máel-Sechnaill I and Flann Sinna (whose reigns span the years 846 to 916) and another, Tairrdelbach (Turlough) Ua Conchobair (1119–1156), showing that the church and secular rulers had cooperated in erecting the crosses whose inscriptions helped to glorify the political dynasts. One name that recurs on crosses is that of Colman, who could conceivably have been the master sculptor of some of them, though we cannot say whether the carvers of the crosses were monks, or masons of a traveling workshop.

The crosses, however, are unlikely to have been developed in the stone form that we see today. It is likely that they were preceded by crosses made of other materials—a wooden core covered, at least partially, in bronze, is made extremely likely by crosses such as that at Dromiskin, County Louth, where decorative squares or bosses are almost certainly modeled on bronze originals. On the North Cross at Castledermot, County Kildare, the mason has even copied the heads of the nails (used to attach the spiral-ornamented bronze sheet to the wooden core) of the model on which the stone cross was based, and surviving pieces of bronze in Irish, British, and Norwegian museums may have formed part of such "prototype" crosses. We should thus envisage existing crosses in the landscape as the final and most lasting step in the development of the High Cross, which would have started in wood and bronze (and possibly other materials as well)—and probably on a much smaller scale before they reached the monumentalized stone form that we see today.

The Irish probably copied the idea from Britain, where stone crosses had become popular in the eighth century. Chronologically, the Irish crosses can be separated into a ninth- and tenth-century group and a twelfth-century group, while leaving open the possibility that the dating of each group could be expanded somewhat. Françoise Henry envisaged the development of the first group taking place in Donegal, where she saw the Fahan stele and the Carndonagh cross as emerging in the seventh and eighth centuries from upright cross-decorated standing stones, and then equipped with independent arms and figure carving. She supported her argument by pointing out that the Fahan stone bore a Greek inscription "Glory and Honor to the Father . . ." that used a formula approved by the Council of Toledo in 633. But the dating of these two crosses is still a matter of debate, and both can be seen as being closer to developments in Scotland than in the rest of Ireland.

Late twentieth-century thought would instead prefer to see the High Crosses evolving from upright pillars, not in Donegal but at Clonmacnoise, where they are decorated with horsemen, interlace of the ribbon and human varieties, spirals, and lions, emerging probably at a time when the papal throne was occupied by one bearing their name, Leo III (796–816). Among the half-dozen monuments of this group clustered around Clonmacnoise, there are some which were certainly crosses and not pillars, as can be seen at Twyford, where an inscription refers to a Tuathgal who may—but is certainly not proven to—have been the same as an abbot of Clonmacnoise of that name who died in 811. The same preference for animals—presumably having a symbolic meaning for those who carved them—is found with what is probably a roughly contemporary group of granite crosses at Finglas, County Dublin, and, more particularly at Moone, County Kildare. At Moone, at first on a cross with a hole at the centre, and later on the more-famous tall cross sculpted (perhaps with the other two) by the same master, we find a profusion of animals, some of which have a clearly Italian ancestry.

But the base of the tall cross at Moone introduces us here for the first time to a series of Old and New Testament scenes designed to relate Adam and Eve to the Crucifixion, and to show how the Lord shields the innocent from danger, while also featuring the desert hermits Paul and Anthony. This pair also appears on other granite crosses in the area, particularly those at Castledermot, County Kildare, a foundation of the Céli Dé, members of a reform movement whose return to ascetical monastic rule may be symbolized by Paul and Anthony, and whose interest in reading the Bible might illuminate the appearance of scriptural scenes on crosses in the Barrow Valley. But the emergence of the "classical" High Cross—sandstone, with large ring, and with much of the surface covered with a much greater variety of biblical images than are present at Moone and Castledermot—is to be found in the east, the midlands, and the north of Ireland at much the same time or marginally later. The question of when that time was has been a matter of considerable debate, but it must have started during, if not before, the reign of the aforementioned Clann Cholmáin king Máel-Sechnaill I (846–962), and continued during the reign of his son Flann (879–916). The iconography would plead for a date in the period between 830 and 880, but Conleth Manning would see one of the finest midland crosses—the Cross of the Scriptures at Clonmacnoise—to be contemporary with the building of the Cathedral there in the first decade of the tenth century. Kells, outside of Clann Cholmáin territory, a new foundation from Iona which had wide international contacts, may have been in the vanguard of the development of the classical Irish High Crosses as well as the hub for the dissemination of those more-extensive iconographic schemes, which—along with the more-naturalistic sculpture that portrayed them—was probably introduced from France (with little sign of intermediary stations in England). The similarity in composition of High Cross panels to those on Continental church frescoes intimates similar functions (with the crosses being Ireland's open-air response, in the country's absence of large, well-lit churches suitable for frescoes), but also encourages the notion that the High Crosses were originally painted (though not a trace of pigment remains), which would make sense in a scene like the *Mocking of Christ* on Muiredach's Cross at Monasterboice, where the color of the Savior's cloak (scarlet or purple) would have been important in transmitting the Bible's message.

Each cross illustrates a different selection of scriptural events and there are sufficient remains of the Broken Cross at Kells to demonstrate how the rare subjects on both sides of the cross were chosen by someone steeped in the Bible who was able to correlate the Old and New Testament scenes with one another, as was done with the frescoes facing one another on the north and south walls of Roman basilicas. The symbolic use of water in a number of the scenes chosen for the Broken Cross was doubtless intentionally designed to impart the message of the healing power of baptism, and demonstrates that scene selection on each cross was anything but random and was intended to impart a particular idea of church teachings. The Northern crosses are much stricter in dividing Old and New Testaments, usually placing each on a different side of the cross.

One group of crosses in the Tipperary–Kilkenny border area, centered on Ahenny, was long thought to be of crucial importance in the development of Irish High Crosses in the eighth century, but recent research is suggesting that these very intricate crosses—copied from models with complicated wooden carpentry techniques and covered with bronze plaques and bosses — are roughly contemporary with the midland Scriptural crosses, and were erected either by Máel-Sechnaill I or his brother-in-law Cerball mac Dúngaile, king of Ossory.

This earlier batch of High Crosses represents an important Irish contribution to European sculpture and forms the largest treasury figure sculpture with biblical iconography anywhere in Europe during the last quarter of the first Christian millennium. The richness and variety of their narrative sculptural scenes far surpasses anything known from Britain or from what is found on the precious Carolingian ivories on the Continent. In addition to the crosses with biblical sculpture, which number over eighty, there are about fifty with purely geometrical ornament and a further twenty with bosses—some decorated with interlace akin to that on a tenth-century wooden example found in the Wood Quay excavations in Dublin. Others bear no decoration, which makes them difficult to date, but they can be classed as High Crosses simply because of their height.

The earlier group of crosses wanes in the course of the tenth century, and in the twelfth century a new type of cross emerges. On these crosses, scriptural scenes are confined to Adam and Eve and the Crucifixion, and the main feature is the figure of the more-triumphant Christ with outstretched arms represented in high relief above or back-to-back with an episcopal figure in almost equally high relief, possibly symbolic of the new diocesan organization instituted by the twelfth-century reform movement in Ireland. These crosses are found in all four Irish provinces. Two at Tuam, County Galway, bear inscriptions that again demonstrate king and church cooperating to erect High Crosses. It is eminently possible that some of these late crosses, numbering over twenty, were erected to attract and impress pilgrims, whose activity was most popular throughout Europe in the twelfth century.

PETER HARBISON

References and Further Reading

Cronin, Rhoda. "Late High Crosses in Munster: Tradition and Novelty in Twelfth-Century Irish Art." In *Early Medieval Munster: Archaeology, History and Society*, edited by Michael A. Monk. and John Sheehan, 138–146. Cork: Cork University Press, 1998.

Harbison, Peter. *The High Crosses of Ireland*. 3 vols. Bonn: Romisch-Germaniches Zentralmuseum Mainz, 1992.

———. "A High Cross Base from the Rock of Cashel and a Historical Reconsideration of the 'Ahenny Group' of Crosses." *Proceedings of the Royal Irish Academy* 93C (1993): 1–20.

———. "The Extent of Royal Patronage on Irish High Crosses." *Studia Celtica Japonica* 6 (1994): 77–105.

———. "The Holed High Cross at Moone." *Journal of the County Kildare Archaeological Society* XVIII (Part IV), 1998-9, 493-512.

———. "The Otherness of Irish Art in the Twelfth Century." In *From Ireland Coming: Irish Art from the Early Christian to the Late Gothic Period and its European Context*, edited by Colum Hourihane, 103–120. Princeton: Princeton University Press, 2001.

Kelly, Dorothy. "The Heart of the Matter: Models for Irish High Crosses." *Journal of the Royal Society of Antiquaries of Ireland* 121 (1991): 105–145.

Manning, Conleth. *Clonmacnoise*. Dublin: 1994.

Ó Floinn, Raghnall. "Patrons and Politics: Art, Artefact, and Methodology." In *Pattern and Purpose in Insular Art*, edited by Mark Redknap, et al., 1–14. Oxford: 2001.

See also **Cerball mac Dúngaile; Church Reform, Twelfth Century; Clonmacnois; Early Christian Art; Ecclesiastical Sites; Iconography; Inscriptions; Máel-Sechnaill I; Sculpture**

HISTORICAL TALES

Description

Historical tales are narratives usually concerned with kingship, dynastic conflicts, and battles, in which the glories of one royal dynasty or another are recorded. They are frequently assigned to the genre known as the "Cycle of the Kings," a classification based primarily on the fact that the main thematic concern of such tales is with royal personages. The significance of many of these tales is not so much in the historical matters which they purport to recount but in the motivation of authors writing at some removal from the events narrated and whose main purpose was to recast earlier events, usually for the benefit of some local dynasty. The time and place of composition of such tales, therefore, are crucial factors in seeking to understand the motives of their authors. It must also be borne in mind that historical accuracy was not the paramount concern in the composition of such texts and it is not uncommon to find deliberate distortion of earlier historical records. Historical tales are thus often a more accurate reflection of events occurring at their time of composition rather than of events recounted in the narratives themselves.

Some of these tales recount events in historical battles. One example is *Cath Almaine* (The Battle of Allen), which is based on a battle fought in 722 at the Hill of Allen, County Kildare. There are two recensions of this tale. The composition of the earlier recension would seem to represent a tenth-century re-working of previously existing sources. The protagonists in this tale are the Northern Uí Néill under Fergal mac Maíle Dúin, and the men of Leinster under Murchad mac Brain. The tale celebrates the victory of the Leinstermen and was probably written for one of Murchad's descendants. Tales concerned with the Battle of Mag Rath (Moira, in modern Co. Down), which was fought in the year 637, have also been written. *Cath Maige Rath* (The Battle of Mag Rath) is preserved in two recensions, the earlier of which has been dated to the tenth century. The first recension purports to set out the cause and outcome of the conflict in which the protagonists are Domnall mac Áeda, over king of the Uí Néill and his foster-son Congal Cáech, of the Ulaid. Congal is eventually killed at Mag Rath. Another tale concerned with the battle of Mag Rath is *Fled Dúin na nGéd* (The Feast of Dún na nGéd), dateable, it would seem, from about the late eleventh to the mid-twelfth century. In this tale, the Ulaid ruler, who is named Congal Cláen, declares himself publicly slighted by Domnall and vows revenge. Congal subsequently seeks aid from his Scottish and British kinsmen and the narrative ends with a brief account of the battle of Mag Rath in which Congal and his allies are defeated by Domnall. The author of *Fled Dúin na nGéd* displays familiarity with *Cath Maige Rath* and has rewritten the earlier narrative with various embellishments of a learned nature for a different purpose. Characters in the narrative stand as surrogates for royal contemporaries of the author. Domnall mac Áeda of the text would seem to stand as surrogate for Domnall Mac Lochlainn of Cenél nEógain, who sought to regain traditional Uí Néill hegemony in Ireland at the close of the eleventh century. Domnall mac Áeda's relationship with the Ulaid as depicted in *Fled Dúin na nGéd* would seem to have been intended to reflect Domnall Mac Lochlainn's relationship with the Ulaid in the late eleventh century. The compiler's sympathies lie not with the aggrieved foster-son, but with his foster-father and with the viewpoint of stable authority, with the particular position of Domnall Mac Lochlainn in regard to the Ulaid. *Fled Dúin na nGéd* is one of many narratives, therefore, in which the present is represented in terms of the past, that is, contemporary concerns of the author are communicated by allusion to past events.

Tales of Battles with Foreign Invaders

Historical tales belonging to this category were also written for propaganda purposes. One such tale is *Cogad Gáedel re Gallaib* (The War of the Irish against the Foreigners). This text provides an account of the Scandinavian invasions of Ireland in the ninth and tenth centuries and the resistance to them by the Dál Cais, culminating in the victory of that dynasty under their leader Brian Boru, at the Battle of Clontarf in 1014. The narrative, however, was seemingly compiled in the early part of the twelfth century at the behest of Brian Boru's great-grandson, Muirchertach Ua Briain. The purpose of the narrative was not to provide an accurate historical record of events in ninth- and tenth-century Ireland but rather to enhance the position of the later Muirchertach and to legitimize the political power achieved by his dynasty by highlighting the remarkable deeds achieved by his ancestors. Another twelfth-century dynastic propaganda text concerned with resistance against the invading Vikings is *Caithréim Chellacháin Chaisil* (The Victorious Career of Cellachán of Cashel) which appears to have been written between 1127 and 1134. In this instance the kings of another dynasty, the Eóganachta, are glorified and their ancestor, Cellachán, is portrayed as defender of Ireland. It has been remarked upon by Donnchadh Ó Corráin (1974) that themes of the great and just king, of the sainted royal ancestor, and of patriotism as they were developed in Irish historical tales such as *Cogad Gáedel re Gallaibh* and *Caithréim Chellacháin Chaisil* are also present in the European literature of the age. Geoffrey of Monmouth's *History of the Kings of Britain*, a narrative held to boost the ideal image of English kingship, was completed within a decade of *Caithréim Chellacháin Chaisil*. Ó Corráin also suggests that *Cogad Gáedel re Gallaibh* may have had Asser's *Life of Alfred* as its model.

Post-Norman Period

Historical tales continued to be used for propaganda purposes in the post-Norman period. A notable example is *Caithréim Thoirdhealbhaigh* (The Victorious Career of Turlough). The narrative is concerned not only with the Tairdelbach Ua Briain (d. 1306) of the title but also with events after his death. In style and method it is closely modeled on its twelfth-century predecessor, *Cogad Gáedel re Gallaibh*. The narrative deals with the civil wars among the Uí Briain (O'Briens) of Thomond (North Munster) in the late thirteenth and early fourteenth centuries and appears to have been written about the middle of the fourteenth century. The two branches of the Uí Briain in conflict with one another are Clann Taidc (supported by the Anglo-Irish de Burghs) and Clann Briain Ruaid (supported by their Anglo-Irish allies, the de Clares). Clann Taidc are eulogized from start to finish. It has been argued that *Caithréim Thoirdhealbhaigh* was commissioned by a king from among Clann Taidc to discredit Clann Briain Ruaid and to prove that the descendants of Tadc Ua Briain were the authentic heirs of Brian Boru and thus of purer stock. It has also been pointed out that although the kings of Clann Taidc are portrayed in the mold of Brian Boru, the narrative betrays a sympathy for the foreigners; such acceptance is alien to the traditions of Gaelic dynastic propaganda in the pre-Norman period. This may reflect intimate ties between the Clann Taidc branch of the Uí Briain and Anglo-Irish magnates which developed from the 1270s on. Several marriages, for instance, served to cement the political alliance of Clann Taidc and the de Burghs. It has been remarked upon that although the de Burghs are occasionally criticized in *Caithréim Thoirdhealbhaigh*, they are more often described as "kingly" and "of English origin but now Irish-natured." *Caithréim Thoirdhealbhaigh* can be viewed as marking a significant development in Irish historical tales in that it reflects changing circumstances in post-Norman Ireland. It is an example of political propaganda in which, following an earlier tradition, in which heroes continue to be depicted as the epitome of all that was most praiseworthy in a Gaelic king, but also one in which the changed reality of ties with the Anglo-Irish lordship could also be reflected and, in this case, approved of.

CAOIMHÍN BREATNACH

References and Further Reading

Herbert, Máire. "*Fled Dúin na nGéd*: A Reappraisal." *Cambridge Medieval Celtic Studies* 18 (Winter 1989): 75–87.

Nic Ghiollamhaith, Aoife. "Dynastic Warfare and Historical Writing in North Munster." *Cambridge Medieval Celtic Studies* 2 (Winter 1981): 73–89.

Ní Mhaonaigh, Máire. "*Cogad Gáedel re Gallaib* and the Annals: A Comparison." *Ériu* 47 (1996): 101–126.

Ó Corráin, Donnchadh. "*Caithréim Chellacháin Chaisil*: History or Propaganda?" *Ériu* 25 (1974): 1–69.

Ó Riain, Pádraig. *Cath Almaine*. Dublin: Dublin Institute for Advanced Studies, 1978.

See also **Gaelicization; Ó Briain; Viking Incursions**

HOARDS

In archaeological terms, a hoard may be defined as a group of artifacts found together, usually not associated with any known archaeological feature, site or monument. Hoards are thus distinguished from assemblages of artifacts found in the excavation of settlement sites, graves or burial chambers (although there are rare examples of hoards found in the excavation of such sites). While some hoards may reflect nothing more than accidental loss, most are seen as the result of deliberate deposition, for which a variety of motives have been suggested. Hoards are an important feature of the archaeology of prehistoric (i.e., pre-Christian) Ireland, especially in the Bronze Age and Iron Age. Many of these hoards have been interpreted as ritual or votive offerings—that is, as a deliberate offering to deities or spirits without any intention of retrieval. As Ireland became a Christianized society, however, such practices disappeared and throughout the medieval period the purpose of hoard deposition, almost without exception, was to conceal the objects for safekeeping, with the intention of retrieving them. In a society without banks or safety deposits, and where most buildings were all-too-easily destructible, the concealment of valuables in the ground was often the only means of safekeeping in times of insecurity. In most cases, presumably, such hidden goods were subsequently retrieved by the owners, but evidently there were many cases where this did not happen and the objects remained concealed until accidentally found at a much later date.

It is in the nature of such hoards that they tend to contain valuable and important objects—precisely the types of object that are rarely encountered in the excavation of settlement sites. This underlines the importance of hoards in the study of medieval Irish society. A case in point is the hoard of handles, plates, and frames from Donore, County Meath. These highly decorated pieces, dating to the early eighth century, are thought to be door fittings from a church, or perhaps a portable shrine, that were concealed for safekeeping at an unknown date. Had they been left in their original location, it is most unlikely that they would have survived to the present.

The most important and spectacular hoards known from medieval Ireland are two groups of early medieval ecclesiastical metalwork, found at Ardagh, County Limerick, and at Derrynaflan, County Tipperary. Both hoards are best-known for their magnificent silver chalices, but each contained a disparate assemblage of eighth- and ninth-century objects that had clearly been concealed for security in the face of some threat. The Derrynaflan hoard is thought to have been deposited during the ninth century and the Ardagh hoard was probably deposited slightly later, perhaps in the early tenth century. It seems reasonable to suggest that the Derrynaflan hoard—composed entirely of liturgical objects, such as a paten and wine strainer, in addition to the silver chalice—represents the altar service of the early medieval monastic church at Derrynaflan. The Ardagh hoard also consists, at least in part, of liturgical vessels—including a bronze chalice, smaller and simpler than the silver chalices—but it was found on a secular site (a ringfort) and its origins are more difficult to reconstruct.

The Viking period also witnessed the appearance of a new and distinctive type of hoard—silver hoards, composed of combinations of coins, ingots, ornaments, and hack-silver (cut-up fragments of ingots or ornaments). Over 130 silver hoards, datable from the ninth to twelfth centuries, are known from Ireland and are an invaluable source of information on the Viking and Hiberno-Norse periods. What the hoards reveal, above all, is that hitherto unprecedented levels of wealth, in the form of silver, were reaching Ireland in this period—clearly as a result of Scandinavian activity. The hoards also chart, in outline, the progression of the Irish economy from an entirely coinless system, through a bullion economy, to a faltering coin-based economy. Analysis by Sheehan and Graham-Campbell has revealed that hoards of the ninth and early tenth centuries are almost entirely coinless—consisting of objects being kept for their bullion value, in an economy where silver bullion, rather than formal coinage, was the main medium of exchange. In the tenth century, mixed hoards—containing some coinage—become more common, and after about 940 A.D., hoards composed exclusively of coins begin to predominate. These comprise predominantly Anglo-Saxon coins, in the tenth century, and Hiberno-Norse coins in the eleventh century. A series of extremely large hoards of late-tenth-century Anglo-Saxon coins, discovered in late twentieth-century excavations in Dublin, is seen as part of the process leading to the minting of the first Irish coinage, which took place in Dublin circa 997. Although coin hoards are common, however, they represent a relatively insignificant amount of silver, in bullion terms, and Sheehan argues that most silver was imported into Ireland in the period from 850–950, before the use of coinage became common.

Analysis of the distribution of Viking-Age silver hoards reveals that while the coinless hoards are relatively evenly spread over much of the country (although with a discernable concentration in the Midlands), mixed hoards and coin hoards display a strong concentration on the east coast and east Midlands. This is taken as evidence for the central role of the Viking coastal settlements—particularly Dublin—in the dispersal of this silver within Ireland. The distribution also makes it clear, however, that much of the wealth represented by the hoards ended up in Irish hands, presumably, in the main, as a result of trade with the Vikings and Hiberno-Norse. In the later Middle Ages, hoarding of objects other than coins becomes so rare as to be effectively nonexistent. Coins continued to be hoarded, however, especially in times of insecurity, and coin hoards continue to be an important source of information.

ANDY HALPIN

References and Further Reading

Gerreits, M. "Money Among the Irish: Coin Hoards in Viking Age Ireland." *Journal of the Royal Society of Antiquaries of Ireland* 115 (1985): 121–139.

Graham-Campbell, J. "The Viking-age silver hoards of Ireland." In *Proceedings of the Seventh Viking Congress, Dublin 1973*, edited by B. Almqvist and D. Greene, 31–74. Dublin: Royal Irish Academy/Viking Society for Northern Research, 1976.

Kenny, M. "The Geographical Distribution of Irish Viking-Age Coin Hoards." *Proceedings of the Royal Irish Academy* 87C (1987): 507–525.

Ó Floinn, R. "The Archaeology of the Early Viking Age in Ireland." In *Ireland and Scandinavia in the Early Viking Age*, edited by H. B. Clarke, et al., 131–165. Dublin: Four Courts Press, 1998.

Ryan, M. F. *The Derrynaflan Hoard I: A Preliminary Account*. Dublin: National Museum of Ireland, 1983.

Sheehan, J. "Early Viking Age Silver Hoards from Ireland and their Scandinavian Elements." In *Ireland and Scandinavia in the Early Viking Age*, edited by H. B. Clarke, et al., 166–202. Dublin: Four Courts Press, 1998.

Sheehan, J. "Ireland's Early Viking Age Silver Hoards: Components, Structure, and Classification." In *Vikings in the West*, edited by S. Stummann Hansen and K. Randsborg. Copenhagen: Acta Archaeologica 71, 2000.

Wallace, P. F. and Ó Floinn, R., eds. *Treasures of the National Museum of Ireland: Irish Antiquities*. Dublin: Gill and Macmillan, 2002.

See also **Coinage; Jewelry and Personal Ornament**

HOSTAGES

See **Brehon Law; Kings and Kingship**

HOUSES

Introduction

Within the walls and under the sheltering roof of Middle Age houses, people slept, worked, prepared and ate food, gathered for social occasions and extended hospitality to others. The house could potentially be seen then as the main venue for the performance of personal and collective social identities. Indeed, archaeologists often see the house not merely as a backdrop for human action, but as a space through which social identities of social rank, gender, and kinship are ordered, produced, and reproduced over time, with doors, hearths, walls, and beds all constraining and enabling movement and daily practice.

Archaeological and Historical Evidence

There is a range of archaeological and historical evidence of early medieval Irish houses (between the seventh and the ninth century A.D. in particular). Studies have established a good understanding of their architectural development, location, shape, size, building materials, and internal features (Lynn 1994). At least 300 wooden and stone early medieval houses are known, mostly from ringforts, crannogs, and ecclesiastical enclosures, but also from Hiberno-Norse towns (e.g., Dublin, Waterford, Cork). Early medieval historical texts (e.g., early Irish laws, saints' lives, and narrative literature) also usefully describe house architecture. The eighth-century law tract *Críth Gablach* provides detailed discussion of the size of houses, construction details, and the types of domestic equipment used within them, all closely linked with ideas of social class and rank. The narrative literature also provides descriptions of fantastic houses that clearly owe more to the imagination than to real-life dwellings, but these do indicate the social and symbolic importance of houses, doorways, hearths, and internal arrangements. In the later medieval period, there is also good historical evidence, but relatively few archaeological excavations of houses have been carried out. It is also worth stating that most late medieval castles and tower-houses should also be seen as houses, places for domestic residence and daily activity.

Early Medieval Round Houses (500–800 A.D.)

The structures at the beginning of the early medieval period were usually round houses (usually found at the center of ringforts) constructed of post-and-wattle or stone walls, with wooden poles for roof joists, and thatched roofs of reed, turf, or straw. Most early medieval roundhouses were fairly small, typically 4 to 5 meters in diameter, with some houses slightly larger, at 6 to 10 meters in diameter. The internal floor space was typically 45 square meters, comprising a single small room. It is likely that house size was closely related to social rank, and both customary practice and law forbade an individual from building larger than a certain size. In the early Middle Ages, if people required more domestic space, they built a second house and attached it to the larger house to create a figure-of-eight shape (as at Dressogagh, Co. Armagh, and Deer Park Farms, Co. Antrim). This may have been the *cuile*, or back-house, referred to in the law tracts, which was possibly used as a kitchen or sleeping area.

Early Medieval Rectangular Houses (800–1000 A.D.)

Lynn's (1994) studies also show that there is a significant architectural change from the use of round houses to rectangular houses after about 800 A.D. At the end of the early medieval period (ninth to tenth century A.D.), rectangular houses built in stone or turf were common, and roundhouses became rare. On most settlement sites where there is clear chronological evidence, round houses can be seen to have been physically replaced by rectangular structures, as at Leacanabuile, County Cork (Ó Ríordáin and Foy 1941). The reasons for this transition in architectural styles are still unclear. It is possible that it reflects significant changes in early medieval Irish society by 800 A.D., with the emergence of semi-feudal socioeconomic relationships and changing concepts of land ownership and household size. The ownership and use of a rectangular house, which could more easily be divided up into compartments and sections, may have gone hand-in-hand with new ideas about personal status and concepts of private and public space. Rectangular houses (typically 6 to 8 meters in length) were usually simply constructed, with low stone, earth, or turf walls, and internal wooden poles to support thatched roofs. In terms of location, they tend to be found closer to entrances and towards the sides of ringfort enclosures.

The Social and Symbolic Organization of Early Medieval Houses

Experimental archaeology has shown that wooden roundhouses, if carefully maintained, could have lasted as long as fifty or sixty years (i.e., the lifetime of an individual). It is probable that early medieval houses had lifecycles that were related in a practical and metaphorical sense to those of their inhabitants. It is interesting then that some early medieval houses were deliberately rebuilt on the location of earlier dwellings (e.g., at Leacanabuile, Co. Cork, Dressogagh, Co. Armagh, and Deer Park Farms, Co. Antrim). This could be interpreted as an attempt to establish a historical continuity and symbolic link with the previous household. There is also archaeological evidence for the formal deposition of objects in the ground (i.e., in pits, the house floor and in wall slots of the "old" house) at the end of the life of one house and perhaps the beginning of the next. These deposits of broken rotary millstones and plow parts, items associated with agricultural labor and the domestic preparation of food, may have been intended to mark the "death" of the house (and perhaps a person associated with it).

In most early medieval roundhouses, doorways typically face east and southeast. This orientation is typically interpreted in terms of the practical shelter provided from prevailing wet, southwesterly winds. It is also possible that doorways were customarily oriented towards the sunrise, for long-standing symbolic or cosmological reasons. It is also clear that the entrance would have been oriented to enable the household to watch visitors entering the enclosure, as most doors point towards the ringfort entrance. The hearth or the fireplace would also have been of huge symbolic and social importance, being literally at the center of the house and most indoor domestic activities. Hearths seemed to have served as the constant around which domestic life moved. They were frequently defined in some way. Usually, long stones are set on edge to create a rectangle or square, with rounded stones placed at the corners. There is also sometimes evidence for wooden structures beside the hearth, probably serving to suspend cooking vessels or roasting meat. There is frequent evidence for the cleaning of hearths, and for their reconstruction across long periods of time (i.e., with hearths re-built at slightly different locations, on five to six occasions).

Houses in Hiberno-Norse Dublin and Waterford (*c.* 950–1200 A.D.)

Archaeological excavations in Dublin, Waterford, and Wexford have also provided much evidence about houses between the tenth and the thirteenth centuries A.D. In Hiberno-Norse Dublin, houses were usually located on the front end of long, narrow plots, which originally seem to have stretched from the street frontage to the town defenses. Each house was entered from the street, with a back or side exit into a plot out the back, where there may have been vegetable gardens, pigpens, workshops, and storehouses. Each house also would have had a latrine pit out the back. Studies of the paleofecal material in the pits allows understanding of peoples' diets, stomach ailments, and other health problems.

There were several different types of houses in use in Hiberno-Norse Dublin and late Viking-Age Waterford. Wallace's (1992) Dublin Type 1 houses were the most common (comprising about seventy percent of all houses in Dublin). The origins of the Dublin Type 1 house are still a matter of debate. It may have evolved in Ireland before the tenth century, or it may be an Irish version of the rectangular farmsteads found in Norse settlements in the Earldom of Orkney. These houses were sub-rectangular in plan, with double entrances, aisled partitions, and internal roof supports. They typically measured 7.5 m by 5.5 m; with walls up to 1.25 m high. The walls were of post-and-wattle, typically of ash, hazel, and willow. The roofs were supported on four main posts arranged in a rectangle within the floor area. The floors of the houses were covered with laid clay or post-and-wattle, and paleoenvironmental studies of floor deposits of dung, hair, mosses, food remains, ash, and brushwood have revealed much of living conditions and practices. The houses may have been rebuilt every 15 to 20 years.

There were usually two opposed doors, located in the end walls, one giving access to the street, the other to buildings at the rear of the plot. Internally, the floor space was divided into thirds, with the central strip, sometimes paved or graveled, being the broadest. A rectangular stone-lined fireplace was located in the center. Along the side walls, low benches were used both for sitting and sleeping. Sometimes corner areas near the doors were partitioned off to form a private space.

By the mid-twelfth century (in Waterford) and slightly later in Hiberno-Norse Dublin, there is a shift towards the use of rectangular houses constructed on sill-beams with earth-fast roof supports, or to houses built of stone walls. By the mid-thirteenth century, fully-framed timber houses emerge.

Houses in Late Medieval Ireland (1200–1500 A.D.)

In the late Middle Ages, there is also a wide range of archeological and historical evidence for houses of varying social status, architecture, and function. For example, both Anglo-Norman masonry castles and later English and Gaelic Irish tower-houses (variously dating to between the fourteenth and the seventeenth centuries) effectively served as impressive domestic residences for the upper social classes, both in towns and in the rural landscape. Tower-houses were entered from a doorway above a basement, with upper floors lit by windows and fireplaces. These different floors variously functioned as public spaces for receiving guests, eating, and daily living, or as private rooms for bedchambers. They also often had smaller buildings beside them, for feasting, storage, granaries, stables, and administration. While there has been a tendency to view these structures as primarily defensive or military in function, more recent studies have suggested that they be interpreted in terms of estate administration and domestic life (O'Conor 1998) and as venues for the performance of gendered, ranked, and ethnic identities (O'Keeffe 2001).

The houses of the lower social classes in the late Middle Ages have generally proven more difficult to identify, largely due to a lack of archeological excavations. Anglo-Norman rectangular stone and earthen houses dating to the thirteenth and fourteenth century have been excavated within manorial villages at sites like Caherguillamore, and Bourchier's Castle, County Limerick; Jerpoint Church, County Kilkenny; and Piperstown, County Louth (O'Conor 1998). These houses were typically small, rectangular, two-roomed structures, with one end serving as a cattle byre and the other as a living area. Houses of the Gaelic Irish peasantry have proven even more elusive and have often been thought of as insubstantial structures. Late medieval and early modern historical documents and maps describe or depict Gaelic Irish "creats" as small circular houses of wattle, clay, earth, and branches that may have been quickly disassembled. There is also evidence that the Gaelic Irish occupied small sub-rectangular houses built of stone and earthen walls, with cruck-trussed roofs, occasionally using them for summer booleying in the uplands or as ordinary dwellings.

Conclusions

In conclusion, houses remain as key artifacts of medieval societies in Ireland. Studies are still needed to properly establish the character and development of houses in the later Middle Ages. It is also likely that future scholarship will move on from questions of style and architectural development and inspired by sociology and anthropology, to address how houses were used in the construction and negotiation of social identities of ethnicity, power, gender, and kinship across the medieval period.

AIDAN O'SULLIVAN

References and Further Reading

Collins, A. E. P. "Excavations at Dressogagh Rath, County Armagh." In *Ulster Journal of Archaeology* 29 (1996): 117–129.

Hurley, M. F. and O. M. B. Scully. *Late Viking Age and Medieval Waterford: Excavations 1986–1992* Waterford: Waterford Corporation, 2000.

Lynn, C. J. "Early Christian Period Domestic Structures: A Change from Round to Rectangular Plans?" In *Irish Archaeological Research Forum* 5 (1978): 29–45.

———. "Deer Park Farms" In *Current Archaeology*, 113 (1987): 193–198.

———. "Early Medieval Houses." In *The Illustrated Archaeology of Ireland*, edited by Michael Ryan, 126–131. Dublin: Country House, 1991.

———. "Houses in Rural Ireland, A.D. 500–1000." In *Ulster Journal of Archaeology* 57 (1994): 1–94.

O'Conor, K. D. *The Archaeology of Medieval Rural Settlement in Ireland*. Dublin: Royal Irish Academy, 1998.

O'Keeffe, T. *Medieval Ireland: An Archaeology*. Stroud: Tempus, 2000.

———. "Concepts of Castle and the Construction of Identity in Medieval and Post-Medieval Ireland." *Irish Geography* 34, no. 1 (2001): 69–88.

Ó Ríordáin, S. P. and J. B. Foy. "The Excavation of Leacanabuile Stone Fort, Near Caherciveen, County Kerry." In *Journal of the Cork Historical and Archaeological Society* 46 (1941): 85–91.

Wallace, P. F. "Archaeology and the Emergence of Dublin as the Principal Town of Ireland." In *Settlement and society in medieval Ireland*, edited by John Bradley, 123–160, Kilkenny: Boethius, 1988.

———. "The Archaeological Identity of the Hiberno-Norse Town." *Journal of the Royal Society of Antiquaries of Ireland* 122 (1992): 35–65.

———. *The Viking Age buildings of Dublin*. National Museum of Ireland Medieval Dublin Excavations, 1962–1981, series A, vol. 1. Dublin: Royal Irish Academy, 1992.

See also **Architecture; Dublin; Law Tracts; Manorialism; Ringforts; Society, Grades of Anglo-Norman; Society, Grades of Gaelic; Tower Houses; Walled Towns; Villages; Waterford**

I

ICONOGRAPHY

Iconographical subjects in medieval Ireland change according to the period and the medium in which they appear, with a surprisingly small selection of biblical scenes in illuminated manuscripts. If we take the Books of Durrow and Kells to be "Irish," then we can include them here as demonstrating the use of evangelist symbols, which are also found in the Book of Armagh. Kells also illustrates St. John, the *Enthroned Christ* and the icon-like *Virgin and Child*. The smaller gospel books have figures of the Evangelists, but narrative iconography is rare among the early manuscripts. *The Temptation* and what is normally taken to be *The Arrest of Christ* occur in the Book of Kells and the codex numbered O.IV.20 in the Biblioteca Nazionale in Turin has *The Ascension* and a *Second Coming of Christ*. This latter scene is also encountered in Ms. 51 in the Stiftsbibliothek in St. Gall, which has evangelist figures and also depicts a *Crucifixion* of a very stylized kind—a subject that is not encountered again in Irish manuscripts until circa 1408–1411 in the *Leabhar Breac* in the Royal Irish Academy, where the marginally earlier Book of Ballymote features *Noah's Ark*.

All of these biblical scenes recur on the earlier group of Irish High Crosses which, en bloc, form the most extensive repository of Old and New Testament iconography in northwestern Europe during the first millennium. The Book of Genesis is a rich source for their pictorial material—*Adam and Eve*, *Cain and Abel*, *Noah's Ark*, *The Sacrifice of Isaac*, as well as certain Joseph scenes which, however, have been open to other interpretations. On the Broken Cross at Kells, Exodus provides subjects connected with the Israelites escaping from bondage in Egypt. The figure of David gets extensive coverage in various manifestations, and from the later books of the Old Testament we find, among others, *The Three Hebrew Children in the Fiery Furnace*, and *Daniel*. The Moone Cross shows Daniel surrounded by seven lions, a number found only in the apocryphal *Bel and the Dragon*, and New Testament scenes of *The Washing of the Christ Child* at Kells (and possibly others from an *Early Life of the Virgin* cycle at Duleek) confirm pictorially what we know already from literature, that the Irish monks had a good knowledge of the Apocrypha.

The New Testament is well-represented, but selectively so, on the High Crosses, with repetitious *Childhood of Christ* and *Passion* scenes, but only rare appearances of Christ's public life (apart from *The Baptism*) and, inexplicably, no representation of *The Nativity*. Not unexpectedly, it is the Passion and post-Passion scenes which predominate on the crosses, where *The Crucifixion* nearly always occupies the west face of the cross—often at the center of a ring, which may have cosmic symbolism. Back-to-back with it on the iconographically-richer crosses is *The Last Judgment*, or some variation of the figure of *Christ in Glory*. The only non-biblical figures to make occasional appearances on the crosses are the desert hermits *Paul and Anthony*.

While some of the biblical carvings display Irish characteristics such as the penannular (open-ring) brooch worn by Christ on Muiredach's Cross at Monasterboice, or the predominance of the piercing of the crucified Christ's left side, it is clear that the biblical iconography on the High Crosses is not Irish in origin, though opinions differ as to when and from whence it came to Ireland. The ultimate European source is early Christian Rome, where panels in pictorial cycles correspond to those on the High Crosses. Stalley would see at least some of the iconography coming to Ireland as early as the seventh century—from the Continent via England. Harbison, however, would prefer to see most, if not all, of it coming more directly from the Continent—and largely in the ninth century. The choice of biblical subjects, as well as their arrangement and compositional details, on the High Crosses when compared to those on fresco cycles in Italy and Central Europe (e.g., Mustair in Switzerland) suggest that fresco painters and stone carvers were ultimately deriving their inspiration from common sources (perhaps pattern books?), and trying to achieve similar aims of enlivening the sacred scriptures with pictorial cycles while at the same time trying to induce thoughts of piety in the beholders. Byzantine manuscripts (e.g., Parisinus Graecus 510 in the Bibliotheque Nationale in Paris) and other eastern examples may be explained through a Roman "crucible."

The Crucifixion with many subsidiary figures is featured on lintels of twelfth-century churches, but with few exceptions, it and other biblical material are absent from twelfth-century crosses, which show instead a more-triumphant Christ in high relief with outstretched arms—and the figure of a bishop, at least in some cases probably an embodiment of Hildebrandine church reform, rather than national saints or local abbots, as is sometimes thought. *The Crucifixion* is also reproduced on Romanesque plaques and later medieval metal work, including shrines which also feature *The Virgin*, *Saints Peter and Paul*, *John the Baptist*, *Catherine of Alexandria*, and possibly *Saints Patrick*, *Brigid*, and *Colum Cille* (*Columba*).

The greatest corpus of later medieval iconography is to be found on fifteenth- and sixteenth-century tombs, particularly in Leinster. Here, we frequently find *Apostles* as "Weepers," but other subjects include *The Crucifixion*, *The Trinity*, *Passion* scenes (most notably at Ennis, where they copy English alabasters), *Ecce Homo*, *Christ's Pity*, *The Magi*, and *The Virgin*. In addition we find international saints such as *Michael*, *Gabriel*, *John the Baptist*, *Margaret of Antioch*, *Catherine of Alexandria*, and *Thomas of Canterbury*, as well as the Irish saints *Brigid* and *Patrick*. More unusual are *St. Appolonia*, and Kilconnell's two French saints, *Denis* and *Louis of Toulouse*. *Saints Dominic and Francis* appear both on tombs and on architectural sculpture (e.g., Clonmacnois), where symbolic subjects such as the *Pelican Vulning* occur occasionally. Reformation zealots left few wooden statues of the Gothic period, but sculptures of *The Virgin*, with or without Child, are among the most common survivors. There is a group of *God the Father*, *Christ on Calvary*, and *John the Baptist*, formerly in Fethard, County Tipperary, and now in the National Museum in Dublin, where carvings of local saints are also preserved. The inspiration for these later medieval sculptures is generally foreign, often English.

Peter Harbison

References and Further Reading

Alexander, J. J. G. *Insular Manuscripts, 6th to 9th Century*. London: Harvey Miller, 1978.

Harbison, Peter. "Earlier Carolingian Narrative Iconography: Ivories, Manuscripts, Frescoes and Irish High Crosses." *Jahrbuch des romisch-germanischen Zentralmuseums, Mainz* 31 (1984): 455–471.

———. *The High Crosses of Ireland*. 3 vols. Bonn: R. Habelt, 1992.

———. *Irish High Crosses with the Figure Sculpture Explained*. Drogheda, Ireland: Boyne Valley Honey Company, 1994.

———. "The Biblical Iconography of Irish Romanesque Architectural Sculpture." In *From the Isles of the North: Early Medieval Art in Ireland and Britain*, edited by Cormac Bourke, 271-80. Belfast: Stationary Office Books, 1995.

———. *The Crucifixion in Irish Art*. Harrisburg, Penn.: Thomas More Press, and Dublin: Columba Press, 2000.

———. *From Genesis to Judgement: Biblical Iconography on Irish High Crosses*, Dublin: 2002.

Hunt, John. *Irish Medieval Figure Sculpture*. 2 vols, London and Dublin: Irish Academic Press, 1974.

MacLeod, Catriona. "Some Mediaeval Wooden Figure Sculptures in Ireland." *Journal of the Royal Society of Antiquaries of Ireland* 76 (1946): 155–170.

———. "Some Late Mediaeval Wood Sculptures in Ireland." *Journal of the Royal Society of Antiquaries of Ireland* 77 (1947): 53–62.

Ó Floinn, Raghnall. *Irish Shrines & Reliquaries of the Middle Ages*. Dublin: Country House and the National Msuseum of Ireland, 1994.

Stalley, Roger. "European Art and the Irish High Crosses." *Proceedings of the Royal Irish Academy* 90C (1990): 135–158.

See also **Armagh, Book of; Church Reform, Twelfth Century; Durrow, Book of; Early Christian Art; High Crosses; Illuminated Manuscripts; Kells, Book of; Leabhar Breac; Metalwork; Sculpture**

IDLEMEN

See **Military Service**

IMMRAMA

The *immrama* (rowings about; voyages) make up a genre which exemplifies the spirituality of early medieval Irish self-exile and monastic pilgrimage. There are four extant *immrama*, variously made up of prose and poetry or a mixture thereof: *Immram Brain Maic Febuil* (Voyage of Bran son of Febal), *Immram curaig Maíle Dúin* (Voyage of Máel-dúin's curach), *Immram Snédgusa ocus Maic Riagla* (Voyage of Snédgus and Mac Riagla), and *Immram curaig Úa Corra* (Voyage of the Uí Chorra's curach). The surviving versions of the tales range very widely in date. *Immram Brain Maic Febuil*, on linguistic grounds datable to the eighth century, stands very early in the development of narrative literature in the Irish language. *Immram curaig Úa Corra*, in its extant form, is dated near to the end of the middle ages, though there can be little doubt that a much earlier version of this story existed: the Uí Chorra are commemorated in the "Litany of Pilgrim Saints" (*c.* 800 C.E.) and the tale itself is referred to in medieval lists of titles of Irish tales. The older of the versions of *Immram curaig Maíle Dúin* (prose) and *Immram Snédgusa ocus Maic Riagla* (poetry) date from around the ninth and tenth centuries, respectively. The medieval tale lists also imply the past existence of at least one other *immram*, which is now lost, concerning Muirchertach Mac Erc—whose surviving "death-tale" (*aided*) includes a dream-voyage episode.

Ocean voyages feature in tales of several genres, for example tales of exile (*loingsea*) and otherworldly excursions (*echtrai*). What distinguishes the *immrama* from these genres is that in the *immram*, the voyage becomes the central motif, and the islands and marine phenomena encountered are the principal measure of progression in the narrative. The hero of the *immram* is drawn into a prolonged, often seemingly aimless voyage of exile, encountering perilous creatures and situations.

If past scholarship tended to identify "otherworld" genres in general with a pre-Christian cosmology, recent research has tended to note the largely Christian context of the *immrama*. *Immram Snédgusa ocus Maic Riagla* is a tale of two monks of Iona who send themselves into voluntary exile; Máel-dúin is the child of a (violated) nun; the Uí Chorra go into voluntary exile as atonement for their looting of churches—a crime for which texts such as the *Vita Patricii* (Life of Patrick) by Muirchú and *Cáin Adomnáin* (Law of Adomnán) describe exile as a specific penance. Even *Immram Brain Maic Febuil*, the earliest and least overtly Christian of the *immrama*, makes reference to the Biblical Fall and to birds who sing the monastic hours. References to inundated lands and the *Tír inna mBan* (Land of Women, a land of sinless pleasure) in *Immram Brain maic Febuil*, and also in *Immram curaig Maíle Dúin*, have been controversially held to represent pre-Christian religious conceptions; but we should note that even these putatively "native" motifs are accommodated within an undoubtedly Christian cosmology. All the *immrama* should be regarded as ecclesiastical in the greater part of their setting and causality.

Recent studies have, moreover, demonstrated that the *immrama* owe much to a Hiberno-Latin tradition which achieves its most developed form in the Latin prose tale *Nauigatio sancti Brendani abbatis* (the voyage of St. Brendan the abbot), written circa 800 C.E. This account of a fantastic voyage by Brendan to the *terra repromissionis sanctorum* (Promised Land of the Saints) expands upon an earlier sub-genre of voyage narratives found in Latin saints' lives such as the *Vita Columbae* (Life of Columba) of Adomnán, *Vita Albei* (Life of Ailbe), *Vita Fintani seu Munnu* (Life of Fintan or Munnu) and the *Vita Brendani* (Life of Brendan). The voyage episodes in these Latin saints' lives also exhibit many basic similarities with the *immrama*, to the extent of sharing some of the same locations and episodes. Saints Ailbe and Brendan themselves are also referred to in some of the *immrama*. An especial point of similarity lies in the initiating motifs of the Latin voyages and those of the *immrama*. Many of the Latin and Irish voyage narratives present some or all of their voyagers as going against the advice of mentors or spiritual directors, including stock characters who are supernumerary to the inital makeup of the crew and who for this reason bring judgement upon themselves. Likewise, both Latin and Irish voyage narratives appear to present parables concerning the balance between the personal desire for pilgrimage and the requirement to provide leadership. A distinctive expression of the monastic vocation through self-exile and pilgrimage (*peregrinatio*) presented problems for the early Irish monks in balancing the desire for *peregrinatio* with responsibilities towards their communities—as noted in theological writings from the time of Gildas and Columbanus onward. In the secular *immrama* the aspirations of lay heroes present allegories of monastic ideals—much as the aspirations of heroes of some later French romances seem to present secular, heroic, endeavor in terms of monastic ideals of purification and perfection.

Outside of this specific religious context, the *immrama* may be seen to have the timeless appeal of all travellers' tales that depict journeys upon the margins of the known world and which occasionally venture across the threshold of the "otherworld." While—particularly in the face of futile attempts to "retrace" these largely imaginary voyages—we should keep in mind that the *immrama*'s seeming evocation of the life lived on the sea is mostly a product of literary creation and subtle narrative transitions, we should nevertheless also observe that the earliest *immrama* and Latin voyage tales emerge in a period (*c.* 700–800) when Irish *peregrini* are described by Dícuil as exploring deserted islands in the Atlantic. Accounts of voyages to islands such as the Faroes (*c.* 730) and Iceland (*c.* 795) appear to have contributed to some scenes in the *immrama*; though many more islands in the tale are of mythic origin or simple invention.

Though the *immrama* were little known outside of Ireland until the modern period, their Latin counterpart, *Nauigatio sancti Brendani*, was Ireland's most popular contribution to medieval European literature, inspiring imitations in many European languages.

JONATHAN M. WOODING

References and Further Reading

Burgess, Glyn, and Clara Strijbosch. *The Legend of St. Brendan: A Critical Bibliography*, Dublin: Royal Irish Academy, 2000.

McCone, Kim. *Echtrae Chonnlai and the Beginnings of Vernacular Narrative Writing in Ireland*. Maynooth, Ireland: McCone, 2001.

Mac Mathúna, Séamus. *Immram Brain: Bran's Journey to the Land of the Women*. Tübingen, Germany: Niermeyer 1985.

———. "The Structure and Transmission of Early Irish Voyage Literature." In *Text und Zeittiefe*, edited by H. L .C. Tristram, 313–357. Script Oralia 58. Tübingen, Germany: Niermeyer, 1994.

Meyer, Kuno, trans. *Immram Brain*. (The Voyage of Bran.) *The Celtic Christianity e-Library*. http://www.lamp.ac.uk/celtic/BranEng.htm.

Ó hAodha, Donncha, trans. *Immram Snédgusa ocus Meic Riagla* (The Voyage of Snédgus and Mac Riagla). *The Celtic Christianity e-Library*. http://www.lamp.ac.uk/celtic/SnedgusaVerse.htm. First published as "The Poetic Version of the Voyage of Snédgus and Mac Ríagla." In *Dán do Oide: Essays in Memory of Conn R. Ó Cléirigh*, edited by A. Ahlqvist and V. Capková, 419–429 Dublin: 1997.

Stokes, Whitley, trans. *Immram Ua Chorra* (The Voyage of the Uí Chorra.) Translated 1893. *The Celtic Christianity e-Library*. http://www.lamp.ac.uk/celtic/UaCh.htm .

———, trans. *Immram curaig Maíle Dúin* (The Voyage of Máel Dúin's Curragh.) *The Celtic Christianity e-Library*. http://www.lamp.ac.uk/celtic/MaelDuin.htm. First published in *Revue Celtique* 9 (1888): 447–495; 10 (1889): 50–95.

Van Hamel, Anton, ed. *Immrama*. Medieval and Modern Irish Series 10. Dublin: Dublin Institute for Advanced Studies, 1941; reprinted 2004.

Wooding, Jonathan M., ed. *The Otherworld Voyage in Early Irish Literature: An Anthology of Criticism*. Dublin: Four Courts Press, 2000.

See also **Aided; Echtrae; Hagiography and Martyrologies; Pilgrims and Pilgrimages**

INAUGURATION SITES

The kings and chiefs of medieval Gaelic ruling families were ritually inaugurated at specially appointed open-air assembly sites within their respective territories. Thirty inauguration venues have been identified on the Irish landscape. Common to each of them is their setting on low-lying but far-seeing hills. A panoramic view of the territory over which the royal candidate was about to rule was fundamental to the Gaelic inauguration ceremony. The land that constituted a ruler's dominion was considered his betrothed, and the ceremony that conferred legitimate right to rule was accordingly portrayed as a marital feast or *banais ríghe* (literally, king's wedding feast). The place-names of inauguration sites tend to allude to royalty, to a sept name, or to a hilltop monument and the topography of the site. For instance the place of inauguration of Ua Dochartaigh in the sixteenth century was Ard na dTaoiseach (Height of the Chietains; Inishowen, Co. Donegal) while Carn Uí Eadhra (Lavagh, Co. Sligo) derived its name from that of the sept of Uí Eadhra (O'Hara). In particular, the words *ard* (height), *cnoc* (hill), *mullach* (top), *tulach* (hill), *lec* (flagstone), *carraig* (rock), *carn* (heap, pile or cairn) and *cruachan* (heap, pile, hill) are recurrent in the place-names of inauguration sites.

The range of archaeological monuments identified with certainty as inauguration places includes hilltop enclosures, more popularly earthen mounds, and less frequently natural places, ringforts, and churches. Sacred trees (*bileda*), stone chairs, inauguration stones and stone basins are also associated with some sites. Irish dynasties tended to appropriate existing prehistoric ceremonial landscapes for assembly and inauguration. The expedient purpose behind this was to visibly attach the pedigree of a royal candidate to an illustrious past, whether that took the form of an alleged burial place of an eponymous ancestor of the sept or a legendary heroic figure, or an ancient landscape associated with renowned events.

Mounds define thirteen of the thirty known inauguration venues. Their lack of homogeneity confirms their diverse origins. Some of them appear to be reused, unaltered prehistoric sepulchral monuments. Others show modifications, such as a flattened summit or an upper tier, that could have been the direct result of the adaptation of an existing prehistoric mound for inauguration ceremonies, and still others may have been wholly new additions to earlier ceremonial landscapes. The small summit diameters of some of them suggest that they were essentially throne mounds accommodating no more than the official inaugurator and the royal candidate who sat in a stone chair on the summit, or stood there, placing his foot on a stone. The idea of the enthroned chief raised upon a mound above the assembly is conveyed in a stylized and retrospective illustration of the performance of the rite of the single shoe during the inauguration of Ua Néill at Tulach Óg (Cookstown, Co. Tyrone), on an unsigned map of Ulster by Richard Bartlett or a copyist dated circa 1602. The main body of evidence for the use of mounds in the inauguration of Gaelic royalty lies in the annals, prose tracts, and bardic poetry from the twelfth century, but more particularly from the fourteenth century onward. Among those documented are Magh Adhair (Toonagh, Co. Clare), the inauguration venue of the Dál Cais dynasty and their Uí Briain successors; Carn Fraoich (Carns, Co. Roscommon) where the Uí Chonchobair chiefs of Síol Muiredaig received their office; the inauguration place of the Meig Uidhir at Sciath Ghabra (Cornashee, Co. Fermanagh); Carn Amalgaid (Killala, Co. Mayo) and Carn Ingine Bhriain (possibly Aughris, Co. Sligo) used as the respective pre- and post-Norman election sites of the Uí Dubda (O'Dowds); and Cnoc Buadha (Rathugh, Co. Westmeath) where the Southern Uí Néill king Máel-Sechnaill I held a *rígdál* (royal meeting) in 859 C.E. and where the Meig Eochagáin chiefs of Cenél Fiachach were inaugurated.

The use of the word *lec* in the place-names of some inauguration sites such as Mullach Leac (Leck, Co. Monaghan) and Leac Mhíchil (Ballydoogan, Co. Westmeath) hints at the presence of inauguration "furniture," whether in the form of an unadorned flagstone or, more ambiguously, a footprint stone or stone "chair." The pillar stone at Tara called the Lia Fáil is the only alleged inauguration stone mentioned prior to the fifteenth century. Being upright, its interpretation as an inauguration stone in the medieval sense is untenable. It features as a potent literary device and symbol of kingship in late medieval bardic poetry and prose texts, where it is variously called Leac Luigdech, Lec na nGíall, and Lec na Ríogh. In the Irish sagas and saints' Lives, additional *leaca* are linked with Irish kingship ritual. Among those mentioned, but never described, are Lec na nGíall, at Emain Macha, and Lec Phátraic, ordained by Patrick for the making of future kings at Grianán of Ailech. The act of standing upon a *lec* evidently formed part of the procedure of legitimizing the authority of a king or chief-elect. The stone itself played an integral role in the candidate's empowerment, and was at times attributed a particular potency, something of which may lie in the taboo of the king not being permitted to touch the mortal earth in his royal condition. Open-air stone inauguration chairs were used by both the Uí Néill of Tír Eógain at Tulach Óg and the Uí Néill of Clann Áeda Buide at Castlereagh (Co. Down), and possibly also by the Meic Matgamna of Airgialla and the Clann Uilliam Uachtair branch of the gaelicized Burkes. The Tulach Óg chair was illustrated by Richard Bartlett prior to its destruction by Lord Deputy Mountjoy in 1602. The cartographer shows a crude stone object composed of four individual pieces. It consists of a cumbersome base that may have been Lec na Ríogh ("Flagstone of the Kings") mentioned in the chronicles in 1432, to which the back and sides were later added. This was possibly done in the fourteenth century when Uí Néill dynasts invented the title *Rex Ultonie* (King of Ulster) for themselves. The only known surviving inauguration chair is that of the Clann Áeda Buide, which is a chair-shaped monolith housed in the Ulster Museum, Belfast. It may have been modeled on the Tulach Óg chair in the fifteenth century, when Clann Áeda Buide extended their dominion into south Antrim and north Down.

ELIZABETH FITZPATRICK

References and Further Reading

FitzPatrick, Elizabeth. "An Tulach Tinóil: Gathering-sites and Meeting-Culture in Gaelic Lordships." *History Ireland* 9, no. 1 (2001): 22–26.

———. "Assembly and Inauguration Places of the Burkes in Late Medieval Connacht." In *Gaelic Ireland, c. 1250–c. 1650: Land, Lordship and Settlement*, edited by Patrick J. Duffy, David Edwards and Elizabeth FitzPatrick, 357–374. Dublin: Four Courts Press, 2001.

———. "*Leaca* and Gaelic inauguration ritual in medieval Ireland." In *The Stone of Destiny*, edited by Richard Welander et al., 108–121. Edinburgh: Historic Scotland, 2003.

See also **Archaeology; Dál Cais; Emain Macha; Kings and Kingship; Máel Sechnaill I; Society, Functioning of (Gaelic); Tara**

INHERITANCE

See **Marriage; Society, Functioning of**

INSCRIPTIONS

Most inscriptions from medieval Ireland are found on stones, many of which are also carved, but there are also a few inscriptions on metalwork and other portable objects. Although there is some inscribed material of Roman provenance from Ireland (notably the coin hoard from Ballinrees, Co. Derry), the vast majority of the inscriptions date from early Christian times. In many cases these inscriptions are the earliest evidence for the history and culture of early Christian Ireland, but, unfortunately, not all the inscribed texts are complete and many are weathered. The inscriptions can be divided into two groups, depending on the script used: ogham and Roman alphabet. Inscriptions using ogham script are in the Irish language; Roman alphabet inscriptions can occur in Latin but are more commonly in Irish.

Ogham Inscriptions

In its standard form, the ogham alphabet consists of twenty characters, set in four groups of five. Each character is formed by the use of a varying number of strokes or notches oriented in different ways with reference to a stem line. The ogham alphabet was a deliberate creation, based on the alphabetic principle of one symbol for one sound. It is most likely to have been invented by Irish speakers in the south of Ireland, probably in the fourth century C.E. Its inventors were clearly familiar with the Latin alphabet and with at least some Latin grammar, but the ogham alphabet was developed for writing short epigraphic texts in the Irish language.

There are over 330 stones known from Ireland containing ogham inscriptions, with over one third of these found in County Kerry. From Ireland, the use of ogham spread into Wales, Cornwall, the Isle of Man, and Pictish Scotland. The date range of the Irish ogham inscriptions is from the fifth or sixth centuries to the seventh century. However the tradition continued longer in some of the places to which ogham spread. In particular, most of the Pictish inscriptions date from the period of the seventh to ninth centuries. After the seventh century in Ireland, ogham declined in use as an epigraphic script, but some scholastic knowledge remained and can be seen in the manuscript record and in the occasional inscribed stone or portable object.

Typically ogham-inscribed stones contain the text incised on the angles of the stone, starting at the bottom left and reading up the left side, along the top and down the right side. Word separation is not indicated. The texts are usually short and almost all contain at least one personal name in the genitive, dependent on an unexpressed word, probably meaning "stone." Many texts also contain a patronymic or an indication of sept or tribal affinity. The most typical text is of the form *X maqi Y*, "[stone] of X, son of Y." Irish ogham stones, unlike some of those from Wales and Cornwall, rarely contain a Roman-alphabet text inscribed on the same stone. An example of an ogham inscription is the stone from Ballinvoher, County Kerry, now in the National Museuem of Ireland (Macalister 1945). The text on this stone reads *Coimagni maqi Vitalin* "[stone of] Cóemán, son of Vitalinus."

Roman-Alphabet Inscriptions

Early medieval stone inscriptions that use the Roman alphabet have a date range of the sixth to the twelfth centuries. They thus first appear a little later than the earliest ogham stones but continue for a longer period. Inscribed portable objects are also recorded. From the twelfth and thirteenth centuries, there are a small number of inscriptions using Gothic or Lombardic script. Later medieval inscriptions using the standard Roman alphabet are recorded from the twelfth century onward. Since much of this material has not been compiled, numbers are hard to estimate. However, in Munster, Okasha and Forsyth (2001) recorded 129 inscriptions dating from the sixth to the twelfth centuries.

The majority of the medieval Roman-alphabet inscriptions are incised on stone in a form of insular script known as "half-uncial," although a few texts using decorative capitals are known. Most of the texts are set horizontally and word separation is rare. Most texts contain a personal name and many take the form of a request for prayer for the individual named. A typical text is of the form *oróit do X* (a prayer for X), with the word *oróit* abbreviated.

An example of such a stone is a large cross-slab from Lismore, County Waterford, probably dating from the ninth century (Okasha and Forsyth 2001). The face of the stone contains an incised Latin cross in a rectangular base. The text is set in two lines, the first reading upwards with the letters facing right, and the second reading horizontally above the cross, with the letters inverted with respect to it. The text reads *ór do donnchad*, for *oróit do Donnchad* (a prayer for Donnchad), but Donnchad has not been identified.

An example of an inscription on a piece of metalwork is the well-known eighth-century silver chalice from Ardagh, County Limerick (Ryan 1983). A girdle of gold filigree and glass studs encircles the chalice near the top. Immediately below this is an inscription in ornate capital letters, the letters standing out against a stippled background. The text is in Latin and consists of the names of eleven apostles and St. Paul.

Conclusion

Medieval Irish inscriptions are among the earliest written records and are therefore of the greatest importance in a study of the history and culture of Ireland. They are also linguistically important, as examples of early Irish, and furnish much information about early Irish names and nomenclature. Many are now well protected inside churches and museums, but some of those that are still standing outside are in need of care and preservation.

ELISABETH OKASHA

References and Further Reading

Macalister, R. A. S. *Corpus Inscriptionum Insularum Celticarum.* 2 vols. Dublin: Stationery Office, 1945 and 1949.

McManus, Damian. *A Guide to Ogam.* Maynooth, Ireland: An Sagart, St. Patrick's College, 1991.

Okasha, Elisabeth, and Katherine Forsyth. *Early Christian Inscriptions of Munster: A Corpus of the Inscribed Stones.* Cork: Cork University Press, 2001.

Ryan, Michael, ed. *Treasures of Ireland: Irish Art, 3000 B.C.–1500 A.D.* Dublin: Royal Irish Academy, 1983.

Sims-Williams, Patrick. *The Celtic Inscriptions of Britain: Phonology and Chronology, c. 400–1200.* Publications of the Philological Society 37. Oxford: Blackwell Publishing, 2002.

Swift, Catherine. *Ogam Stones and the Earliest Irish Christians.* Maynooth, Ireland: Department of Old and Middle Irish, St. Patrick's College, 1997.

See also **Early Christian Art; High Crosses; Metalwork; Sculpture**

INVASION MYTH

Irish scholars in the early Middle Ages had a keen interest in the origin of the Irish people, and stories concerning successive invasions of the country were already in circulation in manuscripts in the first half of the ninth century, if not earlier. The most fully developed and best-known account of the invasions of Ireland is *Lebor Gabála Érenn* (The Book of the Taking of Ireland), also popularly known as the Book of Invasions. It was written in the late eleventh century by an anonymous scholar whose aim was to create a comprehensive history of Ireland from Creation down to his own time. It is the culmination of centuries of development, and bears all the hallmarks of a compilation. However, it soon became the canonical account of Ireland's early history and was frequently copied and redacted over the following centuries. So great was its authority that various other related texts, such as *Cath Maige Tuired* (The Battle of Mag Tuired) and *Scél Tuáin meic Cairill* (The Story of Tuán mac Cairill) were altered to accommodate it.

Lebor Gabála Érenn

Lebor Gabála Érenn depicts six successive invasions beginning with the arrival of Cessair, the daughter of Bith son of Noah. However, all her followers perish in the Flood, except for Fintan mac Bóchra, who survives in many forms to relate the history of Ireland to future generations. The second invasion is led by Partholón who fights a battle against a demonic race (the Fomoiri) from over the sea. His forces are finally wiped out by plague. There then follows a third invasion led by Nemed. After Nemed's death, his people are oppressed by the Fomoiri until they rise up and attack their masters. Only thirty survive, some of whom go to Greece, the rest to the north of the world, and these survivors supply the next two invasions. The first of these (the fourth invasion in the overall scheme) return under the names of Fir Bolg, Gailióin, and Fir Domnann. These are the first invaders to be reflected in the names of historical tribes: Fir Bolg is a collective name applied elsewhere to the subject tribes (*aithechthúatha*) and connected to the continental Celtic tribal name Belgae; the Gailióin were later known as the Laigin, who give name to the modern province of Leinster; and the Fir Domnann are found in Connacht (known as the "Irrus Domnann") and in Celtic Britain (the "Dumnonii"). The group that went to the north of the world became skilled in the magic arts and came to be called Túatha Dé Danann (tribes of the goddess Danu/Danann). They arrive in Ireland and demand the kingship from the Fir Bolg (the name is here used collectively for all the previous invaders). This demand gives rise to the first battle of Mag Tuired in which the Fir Bolg are defeated, and is later followed by the second battle of Mag Tuired.

The sixth and final invasion is led by the sons of Míl Espáine, the ancestors of the dominant peoples of medieval Ireland who styled themselves Goídil (Gaels). The sons of Míl defeat the Túatha Dé Danann in battle and proceed to Tara where they encounter three goddesses (Banba, Fótla, and Ériu), each of whom wins a promise to have the land named after them. The Goídil are duped into returning to their ships by the Túatha Dé Danann, who then create a wind which blows them out to sea. The poet Aimirgin calms the wind so that the sons of Míl can land, and the Túatha Dé Danann are subsequently defeated in the battle of Tailtiu (Teltown, Co. Meath).

Although this text undoubtedly contains native elements, its extant structure and content is rooted firmly within Christian biblical tradition. The opening section is provided by the biblical account of creation, and the Great Flood is said to be the cause of the obliteration of Cessair's people. The name of the leader of the second invasion, Partholón, is clearly borrowed from Latin Bartholomeus, whose name is explained in Latin sources as "the son of he who stays the waters," that is, a survivor of the Great Flood. The story of the Gaels is particularly closely linked to biblical narrative and Latin learning. They are traced back to Japhet son of Noah. The first in their line, Fénius Farsaid, was present at Babel when the languages of the world were rendered incomprehensible to each other. His offspring were in Egypt at the same time as the Israelites, and Fénius's son married Scota, a pharoah's daughter. Her name is Latin for "Irishwoman" and her son, Goídel Glas, gave name to the Gaels (Goídil) and their language (Goídelc) which he created. Like the Israelites, they were later persecuted by the Egyptians, and were forced to wander the earth until they eventually reached Spain. They are led in their wanderings by Míl Espáine, whose name is derived from Latin *miles Hispaniae* (soldier of Spain) and reflects the belief that the Latin name for Ireland, *Hibernia*, was derived from Iberia. Like Moses, he leads his people on an extraordinary journey out of captivity but dies before they reach the land in which it has been prophesied that they will settle.

Composition of Lebor Gabála

The ambitious aim of the compiler of *Lebor Gabála Érenn* was to synthesize previously separate traditions and to create a continuous history of Ireland from the beginning of the world down to his own time. Most significantly, he merged two previously separate accounts: one dealing with the history of the Gaels and the other with the successive invasions of Cessair, Partholón, etc. The latter was inserted into the middle of the former, and the compiler was forced to draw the attention of his readers to the change in subject matter. He wrote mainly in prose but included a large number of preexisting poems, most of which were composed by four men: Eochaid úa Flainn (d. 1004), Flann Mainistreach (d. 1056), Tanaide (d. 1075?), and Gilla-Cóemáin mac Gilla-Shamthainne (*fl.* 1072). The prose often summarizes the poems and it is clear that the author is citing the poems as authorities.

Lebor Gabála is concerned with origins. As we have seen, the last three invasions supply the vassal tribes and the dominant septs of medieval Ireland, as well as the characters who appear elsewhere as gods, but are here usually portrayed as earthly magicians (the Túatha Dé Danann). The contemporary geography of Ireland is explained by the actions of successive waves of settlers. Each invader, save the antediluvian descendants of Cessair and the Fir Bolg, builds great earthworks and clears plains, and during their time lakes burst forth. We are told that Partholón cleared four plains and that seven lakes appeared, and that during the time of Nemed four lakes were formed and twelve plains cleared. The subsequent invasions, however, begin to introduce social institutions. During the fourth invasion, the country is divided into five provinces (*cóiced*, "fifth") and the Fir Bolg introduced the notion of kingship and its sacred character. After the defeat of the Túatha Dé Danann in the second battle of Mag Tuired, they deprived the Gaels of corn and milk. As a result, the country was divided into two with the Túatha Dé Danann retiring to the fairy mounds and hills while the Gaels inhabited the surface. Thus, the origins of Ireland's peoples, physical geography, and social institutions are explored and set within an historical framework provided by the Bible.

Lebor Gabála was an immediate success and many copies and revised editions were made in great codices such as the Book of Leinster, the Book of Ballymote and the Book of Lecan. Interest in the text was stimulated by the brief revival of native Irish history in the seventeenth century. The Franciscan historian Mícheál Ó Cléirigh produced his own version in 1631, and Geoffrey Keating included a version in his *Foras Feasa ar Éirinn* (*c.* 1633–1638).

Other Accounts of Invasions

Lebor Gabála Érenn stands at the end of a long tradition of invasion myths. The earliest continuous account of the peopling of Ireland is contained in the *Historia Brittonum* (History of the Britons), which was written in Wales in between 829 and 830 C.E. It is clearly derived from Irish sources and tells of various invasions of Ireland from Spain. The first invader is Partholomus, recognizable as Partholón in *Lebor Gabála*, but there is no mention of the prediluvian Cessair. He is followed by Nimeth *filius* Agnominis (Nemed in *Lebor Gabála*), the three sons of *mils Hispaniae* (Míl Espáine), and a certain Damhoctor (Irish *dám ochtair*, "company of eight"). The arrival of the Túatha Dé Danann is not included among the incursions, and the Fir Bolg are represented only by a later invader called Builc. This is followed by a separate account attributed to "the most learned of the Irish," which tells of the arrival of the Irish. The latter are said to be descendants of a Scythian noble who was banished from Egypt following the drowning of the pharoah's men in the Red Sea. After a sustained period in Africa, he settled in Spain, where his descendants remained until they finally moved to Ireland.

The Story of Tuán Mac Cairill

The *Story of Tuán mac Cairill* appears to have been written in the ninth century, but was revised several times under the influence of *Lebor Gabála*. Tuán is pressed by the Ulster cleric Finnia to recount the history of Ireland. Tuán explains that Ireland had not been settled before the Flood, and that he was the sole survivor of the invasion led by his father, the son of Agnoman. As time passed he took on various shapes: a stag, a boar, a hawk, and a salmon, until he was reborn to the wife of Cairell, king of Ulster. During all this time he observed the invasions of Ireland from his hiding place in the wilderness. The first of these was led by another son of Agnoman, Nemed; the invaders remained for a long time but eventually died out. They were succeeded by the Fir Domnann and the Fir Bolg, who were later ousted by the Gailióin and the Túatha Dé and Andé (tribes of gods and idols?). These in turn were defeated by the sons of Míl.

Origin Legends

A large body of legend in prose and verse is concerned with the origins and migrations of tribes and dynasties in the early medieval period. They generally relate how people came into possession of their lands, how they assumed (or lost) kingship, and their relationship with other population groups. Some can be dated as early as the seventh century and are closely related to genealogies alongside which they often appear. *Cath Crinna* (the battle of Crinna) tells how Cormac mac Airt defeated the Ulaid with the help of Tadg mac Céin and drove them from the Boyne, as a result of which Tadg was rewarded with all the lands he could encircle in his chariot. This legend attempts to explain the political landscape of the eighth century, when the land around the Boyne was occupied by the descendants of Tadg (the Cianachta), who were vassals of the Uí Néill (descendants of Cormac). The story known as "The Expulsion of the Déisi" tells how the Déisi were expelled from Tara, wandered through Leinster, and were eventually granted a homeland on the borders of Munster in reward for driving out the Osraige. This story creates a connection between the Déisi of Brega in the Midlands and the Déisi of Munster, but it almost certainly lacks any historical foundation: *déisi* simply means "vassals" and could have been applied independently to different subject tribes in different parts of the country. Indeed, inconsistencies between variant accounts of the story appear to reflect the changing fortunes and relations of different branches of the ruling dynasty. In stories such as this contemporary reality and relationships are projected into the past, illustrating the folly of attempts to use origin legends to reconstruct the history of pre-Christian Ireland.

GREGORY TONER

References and Further Reading

Carey, John. "Scél Tuáin meic Chairill." *Ériu* 35 (1984): 93–111.

———. "Origin and Development of the Cesair Legend." *Éigse* 22 (1987): 37–48.

———. *A New Introduction to Lebor Gabála Érenn*. London: Irish Texts Society, 1993.

———. *The Irish National Origin Legend: Synthetic Pseudohistory*. Quiggin Pamphlets on the Sources of Medieval Gaelic History. Cambridge: Department of Anglo-Saxon, Norse, and Celtic, 1994.

———. "Native Elements in Irish Pseudo-history." In *Cultural Identity and Cultural Integration: Ireland and Europe in the Early Middle Ages*, edited by Doris Edel, 45–60. Dublin and Portland: Four Courts Press, 1995.

Koch, J. T., and John Carey, eds. *The Celtic Heroic Age: Literary Sources for Ancient Celtic Europe and Early Ireland and Wales*. Malden, Mass.: Celtic Studies Publications 1994. 2nd edition, 1995.

Macalister, R. A. S. *Lebor Gabála Érenn*. 5 vols. London: Irish Texts Society, 1938–1956.

McCone, Kim. *Pagan Past and Christian Present*. Maynooth, Ireland: An Sagart, 1990.

Ó Corráin, Donncha. "Irish Origin Legends and Genealogy: Recurrent Etiologies." In *History and Heroic Tale: A Symposium*, edited by Tore Nyberg et al., 51–96. Odense, Denmark: 1985.

———. "Historical Need and Literary Narrative." In *Proceedings of the Seventh International Congress of Celtic Studies*, edited by D. Ellis Evans, 141–158. Oxford: Oxford University Press, 1986.

Scowcroft, R. M. "*Leabhar Gabhála*, Part I: The Growth of the Text." *Ériu* 38 (1987): 81–142.

———. "*Leabhar Gabhála*, Part II: The Growth of the Tradition." *Ériu* 39 (1988): 1–66.

See also **Biblical and Church Fathers; Dinnsenchas; Forus Feasa ar Éirinn, Genealogy; Mythological Cycle; Scriptoria**

ÍTE (d. 570 OR 577?)

After Brigit, Íte was one of the most prominent female saints in the Irish Church. Íte was the founder and abbess of the monastery of Killeedy, County Limerick. Her feast day is January 15. Her name also appears records her original name as Deirdre; a fourteenth-century Life by John of Tynemouth gives Derithea. She was the patron saint (*matrona*) of the Uí Conaill Gabra, who occupied the western part of present-day County Limerick. The main church at Clúain-chredail became Íte's monastery, Cell-Íte, now Killeedy. Three recensions of her Life remain extant, but none of the present forms can be dated earlier than the twelfth century.

Most of her life and works is legendary; few historical details can be determined. According to Íte's genealogy, she was a member of the royal family of the Déisi and was born near present-day Waterford. After she was consecrated as a nun, she migrated to Clúain-chredail, where she founded her own monastery. The date of her foundation is unknown, but she was present at Killeedy by 546. Although Killeedy was founded as a monastery for women, by the ninth century it had become a monastery for men.

Killeedy apparently supported a school for young boys. Íte is traditionally known for fostering young boys, among them St. Brendan the Navigator; she has been called the "foster-mother of the saints of Ireland." According to one recension of her Life, St. Brendan asked her the three things which pleased God and the three things which displeased him; Íte replied, "True faith in God with a pure heart, a simple life together with holiness, generosity together with charity" are pleasing, but that "a mouth detesting men, holding fast in the heart an inclination to evil, and smugness in wealth" are displeasing. Íte appears in the traditions of St. Brendan, offering him advice and guidance for his voyages. The ninth-century Martyrology of Óengus contains the anecdote that Íte asked to have the infant Jesus to nurse; the text records the poem "Ísucán" (Little Jesus), which is attributed to her but is of a later date.

Íte's Lives and traditions also depict a prophet, a healer, and an ascetic whose fasting is so rigorous that an angel warns her to desist. As part of her ascetic practices, according to the Martyrology of Óengus, she allowed a stag-beetle to eat at her side. In fear, her nuns killed it; Íte then prophesied that no nun would succeed her. Íte seems to have been especially devoted to the Trinity; as a young girl, she received a vision in which an angel gave her three precious stones, signifying the Trinity. Another holy woman once asked her why she was esteemed more by God than any other holy virgin; Íte replied that she lived in constant prayer and devotion to the Trinity. Íte's reputation was such that high-ranking clerics and rulers sought her out. She is called a "second Brigit" for her virtues. Her cult and fame spread beyond Ireland; she is mentioned in a poem on Irish saints by the English scholar Alcuin and appears in English martyrologies.

DOROTHY ANN BRAY

References and Further Reading

Bray, Dorothy Ann. "*Secunda Brigida*: Saint Ita of Killeedy and Brigidine Tradition." In *Celtic Languages and Celtic Peoples: Proceedings of the Second North American Congress of Celtic Studies*, edited by Cyril F. Byrne, et al., 27–38. Halifax, Nova Scotia: D'Arcy Mcgee Chair of Irish Studies, 1992.

Gwynn, Aubrey and R. Neville Hadcock. *Medieval Religious Houses: Ireland*. Dublin: Irish Academic Press, 1970.

Harrington, Christina. *Women in a Celtic Church: Ireland, 450–1150*. Oxford: Oxford University Press, 2002.

Kenney, James F. *The Sources for the Early History of Ireland: Ecclesiastical*. New York: Columbia University Press, 1929.

Plummer, Charles, ed. "Vita sancte Ite Virginis." (Life of the Sainted Virgin Ite.) In *Vitae Sanctorum Hiberniae* (Lives of the Irish Saints), edited by Charles Plummer, 2:116–130. Oxford: Clarendon Press, 1910.

Sharpe, Richard. *Medieval Irish Saints' Lives*. Oxford: Clarendon Press, 1991.

Stokes, Whitley, ed. and trans. *The Martyrology of Oengus*. London: Henry Bradshaw Society, 1905.

See also **Brigit; Hagiography and Martyrologies; Nuns**

J

JEWELRY AND PERSONAL ORNAMENT

The history of Irish early-medieval metalworking is best understood by examining the development of personal ornament. At the beginning of the Christian era, the well-to-do Irish wore cloak fastenings which derived—like other aspects of their costume—from late Roman Britain where there were Irish settlers and where ties of intermarriage ensured the presence of a strong British influence in Ireland that is reflected in language and writing as well as in changes in religion and economy. The basic brooch type was the zoomorphic penannular brooch ("zoomorphic" because the ring ends in stylized animal heads, "pennanular" because the ring is incomplete). The brooch was equipped with a free-swiveling pin and functioned by skewering the cloth of the cloak and pressing the ring down so that the pin passed through the gap. The ring was then rotated so that the pin lay on top of the ring and was pulled tight against it by the drag of the cloth. The terminals were often raised with respect to the ring to ensure that the pin did not slip back and pass between the terminals. As time went on the terminals became enlarged and were used as a field for the display of ornament.

Bronze penannular brooch, Coleraine, Co. Derry. *Photograph reproduced with the kind permission of the Trustees of the National Museums and Galleries of Northern Ireland.*

Stick pins of bronze, but sometimes of silver, were also manufactured—the most celebrated were the hand pins, so-called because their heads resembled a hand with fingers bent and pointing forwards. These had their origin also in later Roman Britain but some examples with fine enameled ornament and millefiori decorations were almost certainly manufactured as late as the earlier seventh century. Millefiori consists of fine rods of colored glass fused together so that when cut into platelets, patterns show in the cross-section.

The penannular brooch seems to have been the dominant type of high status ornament until the seventh century and in broad terms we can see a development in which ornament made of fine reserved lines of bronze is seen against a background of red enamel. Some brooches bear ornament that clearly derives from motifs on provincial late Roman military equipment. Others, such as one from Athlone that probably dates to the later sixth or earlier seventh century, show the development of a style that harks back to and partly reinvents a version of the La Tène style (Ultimate La Tène) of later prehistoric times, a repertoire that includes trumpet-scroll spirals with occasional bird-head terminals. Dating is difficult because of the lack of known contexts for many examples; some are probably later in date.

An exceptional brooch found at Ballinderry Crannog No. 2, County Offaly, heralds the major changes that took place in Irish metalwork at the beginning of the seventh century. This large brooch is made of tinned bronze and instead of fine line-engraved ornament, its terminals are filled with enamel in which are floated platelets of millefiori glass. The ring bears fine-cast lines, like simulated wire binding, and the entire decorative scheme is uncannily like that on a large hanging bowl of Celtic style found in the great Anglo-Saxon ship burial of Sutton Hoo. Not alone does this help to place the brooch chronologically in the earlier seventh century, but it also shows clearly how widely-separated workshops could influence one another. The seventh century saw extensive contacts between Ireland, Pictland, Anglo-Saxon England and the kingdom of the Franks and Italy. All of these played a part in the development of the complex and beautiful polychrome style of early medieval Ireland that had emerged by the century's end.

An experimental piece of gold filigree from Lagore Crannog, County Meath, shows at some point in the seventh century an Irish craftsman attempting to approximate elaborate filigree effects common on Anglo-Saxon work. Towards the end of the seventh century a change in Irish personal ornament takes place with the appearance of a kind of brooch that is often referred to as pseudo-penannular or more simply as the "Tara" type. The fashion is best represented by the two finest and probably earliest of the series—the so-called Tara Brooch (from Bettystown, Co. Meath) and the Hunterston Brooch found in Ayrshire in Scotland. These brooches have closed terminals but their ornament is laid out in panels that clearly reflect the penannular tradition. By closing the terminals, a large semicircular plate is created for the display of ornament. The pin head is an elaborate construction which mimics the ornament and form of the terminals. With the pin unable to pass through the terminals, the brooch cannot function any longer as an effective dress fastener and so a supplementary pin or a thong must have been employed to prevent the brooch from falling out. The majority of the brooches of this class are made of silver and are now recognized as being insignia of status which have their remote origins in the Roman and Byzantine practice of demonstrating rank by wearing large fibulae.

The ornamental possibilities were seized upon by the best craftsmen who had at their disposal not only a new range of techniques but also a new hybrid art style that combined animal ornament of Germanic origin with scrollwork in the Ultimate La Tène tradition, with plain interlace from the Mediterranean world—probably Italy—and Christian iconographical themes although these are very subtle. The Tara and Hunterston Brooch stand very close to the style of the Lindisfarne Gospels and are probably to be dated to the late seventh or very early eighth century.

The pseudo-penannular brooch remained fashionable in Ireland for the following two centuries—a corresponding tradition of penannular brooches but with similar elaborate ornament emerged in Pictland. The Pictish brooches are further distinguished from their Irish analogues by having simple loop-pinheads. Most examples are less accomplished than the masterpieces of Tara and Hunterston. In the ninth century, simplified animal patterns, often in openwork, along the margins and a generally plainer style (represented by two brooches in the Ardagh Hoard and such single finds as the examples from Loughmore, and Roscrea, Co. Tipperary, and Killamery, Co. Kilkenny) prefigure developments in brooch design which gained ground when Viking trade had made silver more abundant.

The penannular form may have remained in use throughout the period but there is no clear evidence of this. However, in the ninth century a new type of silver penannular developed in Ireland—the "bossed penannular brooch," so-called because its terminals are often embellished with silver bosses, sometimes connected by incised bands. These are often large brooches and fragments of them have been found as hacksilver in Viking hoards of the early tenth century. Their origins have been contested—their decoration was originally thought to have been derived from Scandinavian oval brooches worn in pairs by women. These have bosses connected by lines, and like some rare Irish examples have inset openwork ornament. The evidence suggests, however, that these penannular brooches are of local origin with some influence from silver Anglo-Saxon disc brooches of the ninth century.

Another form of silver penannular that arose in Ireland about that time is the "ball brooch" and its variant the "thistle brooch." These are brooches that have simple terminals and pin heads reduced to a large sphere in the case of the ball brooch, and to a sphere with a flaring projection rather like a partly-opened thistle flower on the thistle brooch. The thistle brooch is more widely distributed in the Viking lands, but its origin in the Irish brooch tradition is clear. The balls of many of the brooches are "brambled"—that is, grooved to give an appearance not unlike the surface of a blackberry fruit. The appearance of brambling on ninth century pseudo-penannulars and the Irish trait of making the pin head reflect the terminals locate the origin of these brooches neatly. Hoard evidence again shows the emergence of the type in the later ninth century. One of the four brooches found in the Ardagh Hoard was a ball brooch. One splendid Irish ninth-century penannular brooch from Loughan, County Derry (once known as the Dál Riada Brooch) is made of gold. It is decorated in gold filigree, stamped foil, brambled bosses, and openwork marginal animal ornament. Its pin however is modeled on the Pictish style. It typifies the experimentation of the creative workshops of the period.

Personal ornaments were not just for those of very high status. Throughout the period, simple pins, with closed free-swiveling rings, were used as cloak fasteners—they were not able to function as penannulars and perhaps a thong or cord was used to supplement the fastening. These "ringed pins" had more elaborate versions in which the ring was decorated or a small decorative circular head was sometimes substituted for it—in which case they are referred to as "ring brooches." Some of these are very elaborate—an especially fine large example is the Westness Brooch from Orkney, the decoration of which is not far short of the quality of the finest pseudo-penannulars. Simple ringed pins remained popular in a number of variants as late as the eleventh century and they are particularly well represented in the Viking-age levels of Dublin.

A further variant is the brooch with a hinged tab connecting a large pendant head to a pin. The best known of these are the "kite brooches" so-called from the shape of their heads which were often the field for fine ornament including sometimes, filigree in a style which owes much to Viking traditions. A very fine example was excavated from Viking-age deposits in Waterford and humbler versions have been found in Viking Dublin.

The fashion for wearing very sumptuous personal ornaments seems to have died out during the tenth century but the reasons for this are not clear.

MICHAEL RYAN

References and Further Reading

Graham-Campbell, James. "Two Groups of Ninth-century Irish Brooches." *Journal of the Royal Society of Antiquaries of Ireland* 102 (1972). 113–128.

——— "Bossed Penannular Brooches: A Review." *Medieval Archaeology* 19 (1975), 33–47.

Henry, Francoise. *Irish Art in the Early Christian Period to AD 800*. London, 1965.

——— *Irish Art during the Viking Invasions 800–1020 A.D.* London, 1967.

Nieke, Margaret R. "Penannular and Related Brooches: Secular Ornament or Symbol in Action." In *The Age of Migrating Ideas: Early Medieval Art in Northern Britain and Ireland*, edited by R. Michael Spearman and John Higgett, 128–134. Edinburgh, 1993.

Smith, Reginald A. "Irish Brooches of Five Centuries." *Archaeologia* 65 (1914), 223–250.

Stevenson R.B. K. "The Hunterston Brooch and its Significance." *Medieval Archaeology* 18 (1974): 16–42.

Whitfield, Niamh. "The 'Tara' Brooch: an Irish Emblem of Status in its European Context." (ed.) In *From Ireland Coming: Irish Art from the Early Christian to the Late Gothic Period and its European Context*, edited by Colum Hourihane, 210–247. Princeton, 2001.

Youngs, S. ed. *The Work of Angels: Masterpieces of Celtic Metalwork 6th-9th Centuries AD*. London, 1989.

See also **Early Christian Art; Hoards; Metalwork; Viking Incursions**

JEWS IN IRELAND

In history and in legend, connections between the Irish and the Jews exist. Chronicles linking the two peoples document contact more on fanciful invention than solid foundation. Medieval Ireland had little contact with real Jewish settlers. But their reputation preceded their actual arrival through centuries of Christian supposition.

Legend

Efforts to reconcile biblical and native traditions, while endeavoring to explain the settlement of Ireland, resulted in fabricated accounts of exploits by alleged Jewish ancestors—as in *Lebor Gabála Érenn*. This source lists Cessair, daughter of Bith, a son of Noah, among Ireland's first immigrants. Spurned by Noah, she arrived on her own ark with fifty women and three men. *The Book of Druim Snechta* counters this by promoting Banba as an escapee from the flood. Magog's son and Japhet's grandson Aithechda was held to be the distant progenitor of the Túatha Dé Danann. Their name was tied to the tribe of Dan. Pedigrees for Leinster and Munster's kings stretched back to Éremon and Éber, sons of Míl and, earlier, to the Patriarchs. The *Senchas Már* legal compilation claimed its predecessor as Mosaic law. Exodus inspired tales that Scota, Pharaoh's daughter, fled after defending Moses. After landing in Ireland, this widow of Míl set up Jacob's stone pillow from Bethel as the *Lia Fáil* ("stone of destiny"). Having fallen in the battle of Slieve Mis, her grave lies in "Scotia's Glen," Glanaskagheen in Kerry.

The ten lost tribes and the story of the *Lia Fáil* merge Torah with Tara. Another royal refugee, Tea-Tephi, after the fall of the First Temple in Jerusalem, reached the Hill of Tara. She married Eochaid, king of Ulster. Alternatively, King Heremon married Tara, daughter of Judah's last king, Zedekiah. The mountain Kippure near Dublin; scapegoats and Puck Fair; the names Eber and Hebrew; Hibernia and Iberia; Iveragh and *éver yam* (Hebrew: "a region across the sea"): all have been suggested as Irish–Hebrew proof-texts.

History

Trade along the Mediterranean over the Atlantic to the British Isles may have involved Hebrew merchants. The first recorded encounter happened much later.

The *Annals of Inisfallen* record that in 1079, *"Coicer Iudaide do thichtain dar muir & aisceda leo de Tairrdelbach, & a n-dichor doridisi dar muir."* (Five Jews came over sea with gifts to Tairrdelbach, and they were sent back over sea.) This brief entry can be debated. Its brevity may allude to an inhospitable reception given the Jews, and the harsh reaction by Munster's king of Thomond, Tairrdelbach Ua Briain, which led to the hasty departure of the five visitors. Louis Hyman suggests that they "pleaded to secure for their co-religionists the right of entry." The Jews may have come from England or Normandy. The inclusion of "over sea" can indicate that the sea voyage was brief, involving their passage over only one body of water. Stanley Siev opts for Rouen. England lacked a large Jewish presence in the period following the Norman Conquest; Rouen possessed a Jewish merchant class engaged in Northern trade. Limerick, a Thomond stronghold, presented a likely point of arrival and contact between the upriver Norman town and the Shannon. As the five were not taken captive, killed, or despoiled of their goods, Siev argues that the Irish king accepted their gifts and recognized the influence of the visiting trade representatives.

The same *Annals* for 1080 note that "Ua Cinn Fhaelad, king of the Déisi, went to Jerusalem." This may bolster a favorable reception given the earlier delegation, as this king embarked as a pilgrim to the Holy Land, not long before the First Crusade. However, Irish pilgrimages predate considerably the Jewish visit of 1079.

Regardless of the degree of hospitality given these pioneers, Jews did establish an Irish presence at a later date. On July 28, 1232, Henry III granted to Peter des Rivall (or Rivaux) not only control of the Irish Royal Exchequer but "Custody of the king's Judaism in Ireland," adding the provision that "all Jews in Ireland shall be intentive and respondent to Peter in all things touching the king." Letters to the Irish Jews repeated this appointment. Calendar entries between 1171 and 1225 offer scattered mentions of Jews but lack their residences, possibly indicating an English habitation.

As early as 1169 Josce, a Jewish lender from Gloucester, had advanced funds to two Anglo-Norman mercenaries who landed in Ireland to aid Diarmait Mac Murchadha against Ruairí Ua Conchobair; this transaction—which was punished—occurred well ahead of Strongbow's Anglo-Norman invasion. Monetary deals by Jews were hindered in Ireland as well. Prohibition of land transfers (*in Judaismo ponere*) to the Jews in Dublin occurs in its White Book for 1241. Deportation to Ireland was threatened for any Jew that opposed the royal levies raised by Henry III for his war against the Welsh in 1244. Aaron, Benjamin's son, was born in Colchester but was recorded in the Exchequer Rolls as *"Aaron de Hibernia, Judaeus."* Jailed at Bristol Castle, he was tried in 1283 for selling plate made out of parings from royal coinage.

The last citation in the Calendar of Documents occurs for Jews in Ireland at Easter 1286. The 1290 royal banishment from the realm of all Jews seems to have applied to those in Ireland. Surnames of "Jew" and "Abraham" do appear over the next two centuries, but these are not of Jewish origin.

After later royal expulsions of Jews, from Spain and Portugal, Ireland did offer refuge at the end of the medieval era at least temporarily, perhaps for those in transit to Jewish communities in London or in Bristol, where a Marrano or crypto-Jewish colony already existed. About 1492, Petrus Fernandes, a physician, was born. He practiced throughout the Continent, but died by 1540. That year, Thomas Fernandes of Viana in Portugal, facing accusations of being "New Christian," testified that he was the son of the late Master Fernandes, born in Ireland.

JOHN L. MURPHY

References and Further Reading

Buckley, Anthony D. "Uses of History Among Ulster Protestants." In *The Poet's Place. Ulster literature and society. Essays in honour of John Hewitt, 1907–1987*, edited by Gerald Dawe and John Wilson Foster, 259–71. Belfast: Institute of Irish Studies, 1991.

Hyman, Louis. *The Jews of Ireland from Earliest Times to the Year 1910*. Shannon: Irish University Press, 1972.

CELT: The Corpus of Electronic Texts, s.v. *"Annals of Inisfallen,"* http://www.ucc.ie/celt/online.

Ó hÓgáin, Dáithí. "Cessair." "Mil." "Tara." *Myth, Legend, and Romance. An Encyclopaedia of the Irish Folk Tradition*. New York: Prentice Hall Press, 1991.

Siev, Stanley. *The Celts and the Hebrews*. Dublin: The Irish Jewish Museum, and Shannon: The Centre for International Co-Operation, 1993.

See also **Anglo-Norman Invasion; Annals and Chronicles; Biblical and Church Fathers Scholarship; Invasion Myth; Pilgrims and Pilgrimage; Racial and Cultural Conflict; Records, Administrative; Ua Briain, Tairrdelbach; Trade**

JOHN (1167–1216), KING OF ENGLAND

John was the fourth son of Henry II, was lord of Ireland (from 1177), earl of Mortain (from 1189), and king of England (1199–1216). John's early years coincided with the Anglo-Norman invasion of Ireland, his father's establishment of the Lordship, the 1175 treaty of Windsor with Ruaidrí Ua Conchobair, and the collapse of that settlement following the death in 1176 of the most powerful invading baron, Strongbow, then

Henry's chief governor. The reliable chronicler Roger of Howden alone reports that the power vacuum was filled in 1176 when "The lord king of England, the father, gave Ireland to his son John." Other accounts record this occurring at Oxford in 1177, which Howden also reports, noting that Henry made John king of Ireland "having a grant and confirmation thereof from Alexander the Supreme Pontiff," which suggests some prior preparation. Howden concludes his account by stating that "after the king, at Oxford, had divided the lands of Ireland and their services, he made all those to whom he had entrusted their custody do homage to himself *and* his son John, and take oaths of allegiance and fealty to *them*."

Understandably, given his youth, the evidence for John's involvement with his new lordship in these years is nonexistent. It was only when worries started to mount in the early 1180s that Hugh de Lacy, lord of Mide (Meath) (who had recently married Ruaidrí Ua Conchobair's daughter) intended to make himself king in Ireland, that it became urgent to send John there. In the winter of 1184–1185, Lacy was recalled and Archbishop John Cumin of Dublin was sent ahead to prepare the way. The annals record for 1185 that "the son of the king of England came to Ireland with sixty ships to assume its kingship," and Howden writes that, at Windsor on March 31, Henry "dubbed his son John a knight, and immediately afterwards sent him to Ireland, appointing him king," while the Chester annals record that John "started for Ireland, to be crowned king there." He did not, however, possess a crown as Pope Lucius III refused Henry's request and it was only late in 1185 that his successor, Urban III, "confirmed it by his bull, and as proof of his assent and confirmation, sent him a crown made of peacocks' feathers, embroidered with gold." By this point, however, John had returned from Ireland in ignominy and the crown was never worn.

Giraldus Cambrensis accompanied John to Ireland in the same ship, having been sent by his father to record, as duly emerged, the history of the colony to date, and the new beginning that was anticipated. But despite Henry's careful and costly preparations, Giraldus claims the expedition "came to nothing and was totally unsuccessful." According to his uncorroborated and (as one of the Geraldine pioneers in Ireland) not unbiased testimony, the first mistake was made almost the moment John disembarked at Waterford, where "the Irish of those parts, men of some note, who had hitherto been loyal to the English and peacefully disposed" came in peace and accepted him as their lord. But, in a famous incident, they were mocked by John's youthful entourage, being pulled about by their beards, and consequently "made for the court of the king of Limerick [Ua Briain]. They gave him, and also the prince of Cork [Mac Carthaig], and Ruaidrí of Connacht, a full account of all their experiences at the king's son's court . . . They held out no hope of mature counsels or stable government in that quarter, and no hope of any security for the Irish." They deduced that greater injustices would follow and therefore plotted to resist, and "to guard the privileges of their ancient freedom even at the risk of their own lives."

This evidence is bolstered by charters issued during John's visit, including a grant to Theobald Walter (ancestor of the Irish Butlers) and Ranulf Glanville (John's former guardian, justiciar of England) of lands that later developed into the earldom of Ormond. Other lands in County Tipperary were assigned to William de Burgh (brother of Hubert, later justiciar of England). In addition, John built castles at Lismore, Ardfinnan, and Tibberaghny, provoking opposition from Domnall Mór Ua Briain, who, having voluntarily submitted to Henry II in 1171, now (because of these speculative grants on his kingdom's borders) attacked Ardfinnan and Tibberaghny. Also, Diarmait Mac Carthaig was treacherously killed in 1185 parleying with Theobald Walter's men at Cork. Instead of a triumphal procession through his new lordship, Howden observes that John "lost most of his army in numerous conflicts with the Irish," failure to pay his troops led to widespread desertion, and, after less than eight months, he returned to England penniless.

Apart from being ill-behaved and ill-advised, the expedition was undoubtedly spoiled by de Lacy, the annals observing that John "returned to his father complaining of Hugh de Lacy, who controlled Ireland for the king of England before his arrival, and did not allow the Irish kings to send him tribute or hostages." Hugh's death in 1186 cleared any obstacles in John's way. "When King Henry heard of it," say the Chester annals, "he prepared to send his son John once more into Ireland." Around Christmas 1186 papal envoys arrived proposing to crown John in Ireland, but when John was at Chester awaiting a favourable wind for the voyage news arrived of the death of his brother, Geoffrey of Brittany, whereupon Henry recalled John. Ireland remained therefore a kingdom without a king; its would-be ruler, as long as his father lived, preferred the style *filius Domini Regis* (even in Irish charters) to *Dominus Hibernie*, which was not always employed.

After 1185, John's powers of lordship were heavily circumscribed, and only after his father's death could he adopt a more interventionist approach. At Chinon in Touraine, where Henry died in July 1189, John granted Hubert Walter "all *my* vill of Lusk" (Co. Dublin). The witnesses included Bertram de Verdon and Gilbert Pipard, who both soon received grants of

Airgialla and Airthir, covering much of County Louth and southeast Ulster. In 1194 John made his most sweeping grant yet when de Burgh was allocated all Connacht. John's generosity was linked to his rebellion against his brother, Richard I, his land-grabbing henchmen being opposed by established loyalist barons like the de Lacys and John de Courcy. After his return to England and John's restoration, Richard intervened on behalf of Walter de Lacy to secure his succession to Meath and gave Leinster to Strongbow's son-in-law, William Marshal, which John had tried to prevent.

But when Richard died in April 1199 the lord of Ireland became king of England. John's new freedom of maneuver produced another spate of land grants (the most noteworthy his revival in 1201 of the claim of the Welsh Marcher baron William de Briouze to Limerick), castle construction, and westward colonization that had no regard for the sensitivities of the indigenous rulers. These years saw intense (and confusing) jostling for power among the barons as they rushed to breach the Shannon frontier, and warfare broke out between competing factions, each sponsoring rival O'Conors. These civil wars persisted throughout the first decade of John's reign. He encouraged Hugh II de Lacy to oust de Courcy in 1204, rewarding him in 1205 with a grant of Ulster as an earldom (the first in Ireland), but had fallen out with de Lacy (and Briouze) in turn by 1208.

John was sufficiently worried to make preparations for an Irish expedition, but it did not materialize until 1210. Anxious to bring his troublesome barons to heel, John began with a display of generosity to the native kings, who willingly accepted him as lord, Donnchad Cairbrech Ua Briain being knighted and receiving a charter for an (albeit petty) estate. However, the contemporary *Histoire des Ducs de Normandie* describes John quarrelling with Cathal Crobderg Ua Conchobair; the latter, having marched with John to Ulster to capture Carrickfergus from the de Lacy and Briouze factions, refused to hand over his heir as hostage (presumably because of John's treacherous reputation), whereupon John seized four of Cathal's sub-kings and officers, whom he brought back to England with him. Also, John entered negotiations at Carrickfergus with the most powerful northern king, Áed Méith Ua Néill, but the annals are clear about the outcome: "Messengers came to him [Ua Néill], to his house, to seek hostages, and he said: 'Depart, O foreigners, I will give you no hostages at all.' The foreigners departed and he gave no hostages to the king."

Thus, whatever the successes elsewhere of John's 1210 campaign (he dealt effectively with his Anglo-Norman opponents and is said to have brought Ireland's law and government into line with the English model, establishing an exchequer at Dublin), he failed to produce a settlement with his Irish subject-kings. The breakdown in relations was followed by a government backlash, the king instructing John de Gray, bishop of Norwich, to protect vital Shannon crossings with castles at Clonmacnoise and Athlone. Connacht was twice invaded by the colonists rival members of the Ua Conchobair dynasty, and Ua Néill too suffered, an English army going northwards in 1211, although it was routed by an alliance of northern kings. De Gray himself went north in 1212 and built castles at Cáel Uisce on the Erne and Clones, County Monaghan, launching raids into the heart of Ua Néill's kingdom while Thomas of Galloway's fleet attacked Derry to the rear.

Yet de Gray was defeated, little progress was made in undermining northern resistance, and John contemplated another Ireland expedition in 1213, although this became impossible when his baronial crisis struck. Most Irish barons (like the Welsh Marchers) remained loyal during the emergency, and John sought to win Irish support, taking Cathal Crobderg into his protection and ordering Henry of London, archbishop of Dublin, to buy scarlet cloth for robes for the Irish kings. But the Irish generally took advantage of the barons' preoccupation with English affairs and staged a recovery. In 1214, Ua Néill defeated the English in Ulster, demolishing Cáel Uisce and Clones and razing the port of Carlingford. In 1214–1215 Cormac Ua Máel Sechnaill attacked the castles of Meath and Offaly, John instructing his justiciar in July 1215 to ensure that the barons immediately fortified their lands in the marches.

About February 1216, Pope Innocent III ordered his legate in Ireland to "put down conspiracies against the king throughout the kingdom of Ireland," while another papal letter of the same date orders the punishment of clerics "who communicate with those excommunicated for insurrection against the king." John died in October 1316, but in the following January the papal legate was instructed "to take measures to preserve to [the new] King Henry [III] the fealty of his subjects in Ireland, and to recall those who have opposed him," another mandate of April 1217 urging him "to fulfill his office faithfully and prudently in bringing about a peace between the Irish and the king." This suggests that the Irish and the new king were at war, but that was the legacy left by King John. It is little surprise that nearly two centuries passed before another English king visited his Irish lordship.

Seán Duffy

References and Further Reading

Duffy, Seán. "King John's Expedition to Ireland, 1210: The Evidence Reconsidered." *Irish Historical Studies* 30 (1996–1997): 1–24.

——— "John and Ireland: The Origins of England's Irish Problem."In *King John. New interpretations*, edited by S.D. Church, 221–245. Woodbridge: 1999.

Warren, W.L. "The Historian as 'Private Eye.'" In *Historical Studies*, ix, edited by J.G. Barry, 1–18. Belfast: 1974.

——— "John in Ireland, 1185." In *Essays presented to Michael Roberts*, edited by J. Bosy and P. Jupp, 11–23. Belfast: 1976.

——— "King John and Ireland." In *England and Ireland in the Later Middle Ages: Essays in Honour of Jocelyn Otway-Ruthven*, edited by J.F. Lydon, 26–42. Dublin: 1981.

See also **Anglo-Norman Invasion; Henry II; Lordship of Ireland**

K

KEATING, GEOFFREY

See **Forus Feasa ar Éirinn**

KELLS, BOOK OF

The Book of Kells (Trinity College Dublin MS 58) contains the four Gospels in a Latin text based on the Vulgate text that St. Jerome completed in 384 A.D., intermixed with readings from the earlier Old Latin translation. The Gospels are prefaced by etymologies, mainly of Hebrew names (only one page survives); canon tables, or concordances of gospel passages common to two or more of the evangelists, compiled in the fourth century by Eusebius of Cesarea; summaries of the gospel narratives (*Breves causae*); and prefaces characterizing the evangelists (*Argumenta*). The first gospel text (Matthew) begins on folio 29r. The Book is written on vellum (prepared calfskin) in a bold and expert version of the script known as insular majuscule. It contains 340 folios, numbered 1–339. The number 36 was used twice, while 335 and 336 were bound and numbered in reverse order when they were foliated by J. H. Todd, Trinity College Librarian from 1852–1869. The folios now measure approximately 330 by 255 mm, but they were severely trimmed, and their edges gilded, in the course of rebinding in the nineteenth century. Originally a single volume, the Book of Kells has been bound for conservation reasons in four volumes since 1953.

Folio 34r from the Book of Kells. *The Board of Trinity College Dublin.*

The manuscript's celebrity derives largely from the impact of its lavish decoration. There are full pages of decoration for the canon tables (folios 1v–6r); symbols of the evangelists Matthew (the Man), Mark (the Lion), Luke (the Calf), and John (the Eagle) (folios 1r, 27v, 129v, 187v, 290v); the opening words of the Gospels: *Liber generationis* (Mt. 1.1) on folio 29r; *Initium euangelii iesu christi* (Mk. 1.1) on 130r; *Quoniam* (Lk. 1.1) on 188r; and *In principio erat uerbum {et} uerbum* (Jn. 1.1) on 292r; the Virgin and Child surrounded by angels (7v); a portrait of Christ (32v); complex narrative scenes, the earliest to survive in gospel manuscripts, representing the arrest of Christ (114r) and his temptation by the Devil (202v); a "carpet" page, made up wholly of decoration, depicting a double-armed cross with eight roundels embedded in a frame (33r). The Chi Rho page (34r), introducing Matthew's account of the nativity, is the single most famous page in medieval art. Other passages are emphasized through decoration on folios 8r (the opening of the *Breves causae* of Matthew); 13r (the beginning of the *Breves causae* of Mark); 12r, 15v, 16v, and 18r (the opening words of the *Argumenta* of the four Gospels); 19v (the words *ZACHA*[*riae*] at the opening of the *Breves causae* of Luke; 114v (the opening of Mt. 26.31, *Tunc dicit illis ihs omnes uos scan*[*dalum*]; 124r (Mt. 27.38, *Tunc crucifixerant xpi cum eo duos latrones*); 183r (Mk. 15.25, *Erat autem hora tercia*); 200r–202r (Lk 3.22–38); 203r (Lk. 4.1, *Ihs autem plenus spiritus sancto*); 285r (Lk. 24.1, *Una autem sabbati . . .*). There is a portrait of Matthew (28v) and John (291v), but no portrait of Mark or Luke survives. These were probably executed, like other major pages of the manuscript, on single leaves, so that the transcription of the text could continue without interruption, but they are presumed to have become detached and lost. In all, around thirty folios went missing in the medieval and early modern periods.

The extent, variety, and artistry of the decoration of the text pages are incomparable. Abstract decoration and images of plant, animal, and human ornament enliven and punctuate the text, with the aim of glorifying Jesus' life and message, keeping his attributes and symbols constantly in the eye of the reader. There are repeated images of the face of Jesus; the cross; the eucharist (grapes, chalices, communion hosts); and symbols of resurrection (the lion, the peacock, the snake). Certain images allude to the text: the word *dicit* (he said) is frequently composed of animals whose paws point at their mouths. Other images, such as those of men pulling each other's beards (on, for instance, folios 34r or 253v), present difficulties of interpretation.

The transcription of the text itself was remarkably careless, in many cases due to eyeskip, with letters and whole words omitted. Text already copied on one page (folio 218v) was repeated on folio 219r, with the words on 218v elegantly expunged by the addition of red crosses. Such carelessness, taken together with the sumptuousness of the book, have led to the conclusion that it was designed for ceremonial use on special liturgical occasions, such as Easter, rather than for daily services.

Three artists seem to have produced the major decorated pages. One of them, whose work can be seen on folios 33r and 34r, was capable of ornament of such extraordinary fineness and delicacy that his skills have been likened to those of a goldsmith. Four major scribes copied the text. Each displayed characteristics and stylistic traits while working within a scriptorium style. One, for example, was responsible only for text, and was in the habit of leaving the decoration of letters at the beginning of verses to an artist. Another scribe, who may have been the last in date, was fond of using bright colors—red, purple, yellow—for the text, and of filling blank spaces with the unnecessary repetition of certain passages. The extent to which there was an identity between scribe and artist, and the extent to which the original program of decoration was followed, are among key unanswered questions about the manuscript. There are clear indications that the manuscript was left uncompleted.

A wide range of pigments was employed. The most notable was a blue pigment derived from lapis lazuli. This was available in the Middle Ages from only one source, a mine in the Badakshan district of Afghanistan. Other blues were made from indigo or woad, native to northern Europe. Orpiment (yellow arsenic sulphide) was used to produce a vibrant yellow pigment; it was highly toxic and had to be used with care. Reds came from red lead or from organic sources that are difficult to identify. A copper green, reacting with damp, was responsible for perforating the vellum on a number of folios. Whites came from white lead or from chalk. The artists employed the technique of adding as many as three pigments on top of a base layer. The relief effect they achieved was largely lost when the leaves were wetted for flattening in the nineteenth century, and the full splendor of the manuscript in the Middle Ages can be judged only partially.

The date and place of origin of the Book of Kells have attracted a great deal of scholarly controversy. The majority opinion now tends to attribute it to the scriptorium of Iona (Argyllshire), but conflicting claims have located it in Northumbria or in Pictland. A monastery founded around 561 by St. Colum Cille on Iona, an island off Mull in western Scotland, became the principal house of a large monastic confederation. In 806, following a Viking raid on the island that left sixty-eight of the community dead, the Columban monks took refuge in a new monastery at Kells, County Meath, and for many years the two monasteries were governed as a single community. It must have been close to the year 800 that the Book of Kells was written, though there is no way of knowing if the book was produced wholly at Iona or at Kells, or partially at each location.

The manuscript seldom comes to view in the historical record. The *Annals of Ulster*, describing it as "the chief treasure of the western world," record that it was stolen in 1006 for its ornamental *cumdach* (shrine). Although the shrine has been missing since then, the book itself was recovered "two months and twenty nights" later under a sod. This episode probably accounts for the loss of leaves and text at the beginning and end of the manuscript. It remained at Kells throughout the Middle Ages, venerated as the great gospel book of Colum Cille, a relic of the saint, as indicated by a poem added in the fifteenth century to folio 289v. In the late eleventh and twelfth centuries, blank pages and spaces on folios 5v–7r and 27r were used to record property transactions relating to the monastery at Kells. In 1090, it was reported by the *Annals of Tigernach* that relics of Colum Cille were "brought" (this probably means "returned") to Kells from Donegal. These relics included "the two gospels," one of them probably the Book of Kells, the other perhaps the Book of Durrow. Following the rebellion of 1641 the church at Kells lay in ruins, and around 1653 the book was sent to Dublin by the governor of Kells, Charles Lambart, earl of Cavan, in the interests of its safety. A few years later it reached Trinity College, the single constituent college of the University of Dublin, through the agency of Henry Jones, a former scoutmaster general to Cromwell's army in Ireland and vice chancellor of the university, when he became bishop of Meath in 1661. Since the middle of the nineteenth century, it has been on display in the Old Library at Trinity College, and now attracts in excess of 500,000 visitors each year.

BERNARD MEEHAN

References and Further Reading

Bourke, Cormac. "The work of angels?" *The Innes Review* 50, no. 1 (spring 1999): 76–79.

Farr, Carol. *The Book of Kells. Its function and audience.* London: British Library, 1997.

Fox, Peter, ed. *The Book of Kells, MS 58, Trinity College Library Dublin: Commentary.* Luzern: Facsimile Verlag, 1990.

Henry, Françoise. *The Book of Kells: Reproductions From the Manuscript in Trinity College Dublin.* (With a study of the manuscript by Françoise Henry.) London: Thames and Hudson, 1974.

Henderson, George. *From Durrow to Kells. The Insular Gospel-books 650–800.* London: Thames and Hudson, 1987.

Meehan, Bernard. *The Book of Kells. An Illustrated Introduction to the Manuscript in Trinity College Dublin.* London: Thames and Hudson, 1994.

See also **Durrow, Book of; Manuscript Illumination; Reliquaries; Scriptoria**

KELLS, SYNOD OF

The synod of Kells in 1152 marks a very important stage in the Church Reform of the twelfth century in Ireland: a new administrative structure that had first been introduced at the synod of Ráith Bressail in 1111 finally received papal approval. However, it was not precisely the same structure; some changes were brought about in the meantime. The most important of these saw the number of metropolitan sees increase from the two that had been planned at Ráith Bressail (Armagh and Cashel) to four (Dublin and Tuam were now added).

Malachy Seeks Papal Approval

In 1139 to 1140, Malachy, probably acting on behalf of the elderly papal legate in Ireland, Gille, had gone to Rome to seek papal approval for the decisions which had been made at Ráith Bressail. Pope Innocent II, however, being aware of some unresolved problems, declined to give his approval; instead he made Malachy his legate and sent him back to Ireland to negotiate a settlement of outstanding disagreements. When that was successfully completed, he informed Malachy, a request for papal approval should be made again, obviously with the expectation that it would be granted. Little is known about the detail of the work done by Malachy on his return to Ireland, but it is clear that he had two problems to overcome. Dublin, which had remained aloof from the new diocesan arrangement, would have to be encouraged to join up and renounce its tie with Canterbury, and the ambitions of the king of Connacht (the reigning high king of Ireland), Tairrdelbach Ua Conchobair, would have to be accommodated. Solutions were worked out, and in 1148 a synod, which met on St. Patrick's Island, approved of them and sent Malachy to the pope to get his approval—this time with appropriate backing. Although Malachy, dying en route, never in fact got to meet the pope, the request was conveyed to the pope and it was successful. As a result the pope sent his legate, Cardinal John Paparo, to Ireland, carrying with him four *pallia* (the symbols of papal approval).

Papal legate, Cardinal Paparo, Goes to Ireland

On his first attempt to get to Ireland, in 1150, Paparo was refused a safe conduct through England by King Stephen unless he pledged himself to do nothing in Ireland that would injure England's interests there. The cardinal refused and returned indignantly to Rome. There has been speculation about the reason for Stephen's refusal—perhaps a dispute between him and the pope over the jurisdiction of papal legates or a concern about Malachy's relationship with King David of Scotland. It seems more likely, however, that it was an attempt by Stephen to prevent Paparo from bringing papal confirmation for an arrangement in Ireland that would see Canterbury's claims in Ireland finally extinguished. One chronicler states specifically that Paparo's action in Ireland was contrary to the dignity of the church of Canterbury.

With Paparo back in Rome, a delegation was sent by Irish kings and bishops asking that he be dispatched. And so, in the summer of 1151, he set out again for Ireland; this time his journey was facilitated by King David of Scotland. He was accompanied by Gilla Críst Ua Connairche, first abbot of Mellifont, now bishop of Lismore and permanent papal legate in Ireland (he had been a fellow monk with the current pope, Eugenius III, at Clairvaux), who may have been one of the delegation who had been sent to Rome. The cardinal arrived in Ireland at some time in October of 1151. Apart from a week he spent in Armagh, very little is known about his activities before the convening of the synod in the following March; approximately four months of his time is, therefore, unaccounted for. It is probable that he visited church leaders and lay magnates in preparation for the synod; perhaps he needed to check that the general agreement claimed for the new diocesan arrangement existed.

The Synod Meets

The synod met in March of 1152. There has been some confusion about its actual location. The annals say Drogheda, but Geoffrey Keating, quoting an old book that is no longer extant, gives Kells as the location. Putting both sources together, it is believed that there were two separate sessions of the synod. The first was held at Kells and concluded by March 6; it reconvened at Mellifont, near Drogheda, around Sunday, March 9, and concluded on Palm Sunday (March 23). It is not known how the business of the synod was divided between the two sessions, but it is likely that episcopal consecrations took place at Mellifont and that the four *pallia* were distributed to the archbishops there at the last sitting on Palm Sunday. Although the abbey church there was not consecrated until 1157, the building must have been sufficiently advanced in 1152 to allow these ceremonies to proceed.

According to the *Annals of the Four Masters* the synod was convened by the bishops of Ireland, along with the coarb of Patrick (i.e., the bishop of Armagh) and Cardinal Paparo, and was attended by 3000 ecclesiastics, both monks and canons. It does not tell us who the bishops were; fortunately Keating transcribed their names. Although he says there were twenty-two bishops and five bishops-elect present, he names only twenty bishops and two bishop's vicars: Gilla Críst Ua Connairche, bishop of Lismore and legate of the pope in Ireland; Gilla Mac Liac, coarb of Patrick and primate of Ireland (archbishop of Armagh); Domnall Ua Lonngargáin, archbishop of Munster (Cashel); Gréine, bishop of Áth Cliath (archbishop of Dublin); Gilla na Náem Laignech, bishop of Glendalough; Dúngal Ua Cáellaide, bishop of Leighlin; Tostius, bishop of Port Láirge (Waterford); Domnall Ua Fogartaig, vicar-general to the bishop of Osraige (Kilkenny); Fionn mac (Máel Muire Mac) Cianáin, bishop of Kildare; Gilla in Choimded Ua hArdmaíl, vicar to the bishop of Emly; Gilla Áeda Ua Maigin, bishop of Cork; Mac Ronain (Máel Brénainn Ua (Mac) Rónáin), bishop of Ciarraige (Ardfert); Torgestius, bishop of Limerick; Muirchertach Ua Máel Uidir, bishop of Clonmacnoise; Máel Ísu Ua Connachtáin, bishop of East Connacht (Elphin); (Máel Ruanaid) Ua Ruadáin, bishop of Luigne (Achonry); Mac Raith Ua Móráin, bishop of Conmaicne (Ardagh); Étrú Ua Miadacháin, bishop of Clonard (Meath); Tuathal Ua Connachtaig, bishop of Uí Briúin (Kilmore); Muiredach Ua Cobthaig, bishop of Cenél nEógain (Derry); Máel Pátraic Ua Bánáin, bishop of Dál nAraidne (Connor); and Máel Ísu mac in Chléirig Chuirr, bishop of Ulaid (Down). The two bishop's vicars are described as bishops in another source. They may, of course, have been consecrated as bishops at the synod, in which case the list of names given by Keating would tally with the number of bishops he said were present.

A notable absentee from this list is the archbishop of Tuam; the see had been vacant since the death of Muiredach Ua Dubthaig in 1150. But we are told the archbishop was given the *pallium* by Paparo at the synod; it is possible, therefore, that Áed Ua hOissín was one of the unnamed bishops-elect present and that he was consecrated before receiving it. Alternatively, the omission from the list may be an error, as his name is included in a list found in another source. The bishops of a number of other dioceses are also missing from the list: Raphoe, Louth, Duleek, Clonfert, Killala, Mayo, Kilmacduagh, Ferns, Killaloe, Scattery, Kilfenora, Roscrea, Cloyne, and Ross. Apart from conveying papal approval for the four archbishoprics, Paparo also had the task of setting out the dioceses. They were: Connor, Down, Louth (Clogher), Clonard, Kells, Ardagh, Raphoe, Derry, and Duleek (Armagh province); Killaloe, Limerick, Scattery, Kilfenora, Emly, Roscrea, Waterford, Lismore, Cloyne, Cork, Ross, and Ardfert, together with Ardmore and Mungret who claimed episcopal status (Cashel province); Glendalough, Ferns, Kilkenny (Ossory), Leighlin, and Kildare (Dublin province); Mayo, Killala, Roscommon, Clonfert, Achonry, and Kilmacduagh (Tuam province). It was later claimed that a decision was made by Cardinal Paparo that some small dioceses should be allowed to continue to exist until the incumbent died, at which time they would become rural deaneries.

The synod also passed decrees that, according to a contemporary chronicler, were preserved in Ireland and in papal archives. They are, however, no longer extant. We have, therefore, to depend upon the scant evidence in the annals and in what Keating transcribed for knowledge of their contents. The *Annals of the Four Masters* report that those present "established some rules thereat i.e. to put away concubines and mistresses from men; not to demand payment for anointing or baptizing (though it is not good not to give such, if it were in a person's power); not to take [simoniacal] payment for church property; and to receive tithes punctually." According to Keating's transcription the synod "entirely rooted out and condemned simony and usury, and commanded by Apostolic authority the payment of tithes." He wrote elsewhere that it also "(put) down robbery and rape and bad morals and evils of every kind besides."

On March 23, 1152, the day after the synod closed, Cardinal Paparo set sail for Rome, calling on the king of Scotland on his way.

MARTIN HOLLAND

References and Further Reading

Gwynn, A. *The Irish Church in the Eleventh and Twelfth Centuries*. Edited by Gerard O' Brien. Dublin: Four Courts Press, 1992.

Gwynn, A. *The Twelfth-Century Reform, A History of Irish Catholicism II*. Dublin & Sydney: Gill and Son, 1968.

Holland, Martin. "Dublin and the Reform of the Irish Church in the Eleventh and Twelfth Centuries." *Peritia: Journal of the Medieval Academy of Ireland* 14 (2000): 111–160.

Hughes, K. *The Church in Early Irish Society*. London: Methuen, 1966.

Lawlor, H. J. "A fresh Authority for the Synod of Kells." *Proceedings of the Royal Irish Academy* (C) 36 (1922): 16–22.

Watt, J. *The Church in Medieval Ireland*, 2nd ed., Dublin: University College Dublin Press, 1998.

See also **Annals of the Four Masters; Church Reform, Twelfth Century; Gille (Gilbert) of Limerick; Malachy (Máel-M'áedóic); Raith Bressail, Synod of**

KILDARE

Kildare is a cathedral town and county in eastern Ireland. The place name is derived from the Irish *Cell dara* (church of the oak tree), a feature interpreted as the survival of a pagan oak grove into Christian times. The presence of a nunnery associated with a perpetual fire, first described in the 1180s, has been regarded as the continuation of a pre-Christian tradition similar to that of the vestal virgins at Rome—although such views are often contested. Kildare sits on a hill rising above the Curragh, a sacred landscape since Early Bronze Age times (*c.* 2400–*c.* 1600 B.C.E.), and it is likely that a pre-Christian ritual site preceded the cathedral. The date at which a Christian ecclesiastical settlement was established is unknown, although it is assumed that it occurred in the fifth or sixth century. The site has been associated with Brigit from early times. The first securely dated bishop is Áed Dub mac Colmáin, who died in 639, by which time a cathedral evidently existed. This was an exceptional building and is described by Cogitosus in his *Life of Brigit*, written around 650, as a *basilica*, that is, a church with important relics. It is probably the same building as the *dairthech* (oak church) referred to in 762, and it may have stood until 1020, when the ecclesiastical complex was remodeled. Kildare was the preeminent church site of Leinster in the early medieval period, and as such it was a target of attack by both Irish and Vikings. Between 710 and 1155 it was burned or plundered on at least thirty-eight occasions. Its ecclesiastical importance was confirmed in 1111 by its designation as an episcopal see at the synod of Ráith Bressail.

In the early 1170s Strongbow used Kildare as a base, and by 1176 it was the principal manor of his north Leinster lordship. The castle was probably established at this time, although the first documentary evidence does not occur until around 1185. Kildare prospered during the thirteenth century. A new cathedral, traditionally ascribed to Ralph of Bristol (bishop, 1223–1232), was constructed; the Franciscan friary was founded around 1254 to 1260; a Carmelite friary was established around 1290; and the church of St. Mary Magdalen, with its associated hospital, was in existence by 1307. The town functioned primarily as a marketplace and collection point for the agricultural produce of the region, which was conveyed from there to Dublin. In 1248, after the death of the last of William Marshal's male heirs, Kildare passed into the hands of William de Vescy. During the 1290s the town was threatened by Gaelic-Irish and Anglo-Norman lords. In 1295 it was captured and the castle ransacked by Calbach Ua Conchobair Fáilge. Two years later the de Vescys surrendered their interest to the crown, and in 1316 both castle and town were granted to John FitzThomas FitzGerald, who was created earl of Kildare as a reward for his loyalty during the Bruce invasion. Kildare, with an estimated population of between 1000 and 1500, was never very large, but after the Black Death it shrank considerably, and by the late Middle Ages it was little more than a village.

JOHN BRADLEY

References and Further Reading

Andrews, J. H. *Kildare* (*Irish Historic Towns Atlas no. 1*). Dublin: Royal Irish Academy, 1986.

Gillespie, Raymond, ed. *St. Brigid's Cathedral, Kildare, a History*. Naas: Kildare Archaeological Society, 2001.

Harrington, Christina. *Women in a Celtic Church: Ireland 450–1150*. Oxford: Oxford University Press, 2002. (See esp. pp. 63–67).

See also **Brigit; Church Reform; Cogitosus; Ecclesiastical Settlement; Ecclesiastical Sites; FitzGerald; pre-Christian Ireland**

KILKENNY

Kilkenny is the name of a cathedral town, county (from *c.* 1207), lordship, and liberty (1247–*c.* 1402) in southeast Ireland. The town, which straddles the River Nore, derives its name from the Irish *Cell Chainnigh* (church of [St.] Canice). An earlier Christian settlement, the *martartech Mag Roigne* (relic house of Rogen's plain), was established in the fifth century, and although it

Kilkenny Castle from the Rose Garden. © *Department of the Environment, Heritage and Local Government, Dublin.*

continued through the Middle Ages as a church dedicated to St. Patrick, it was eclipsed in importance during the seventh and eighth centuries by the newer church of Canice. In 1111, at the synod of Ráith Bressail, St. Canice's became a cathedral. By the middle of the twelfth century, Kilkenny was one of the principal residences of the Mac Gilla Pátraic kings of Osraige and, when the Anglo-Normans arrived in 1169, it was the largest and most important inland settlement in the southeast.

Although an Anglo-Norman castle existed by 1173, when it was burnt by Domnall Mór Ua Briain, king of Thomond, it was not until the 1190s that an enduring Anglo-Norman settlement was established. William Marshal was a key figure in this regard. He came to Kilkenny in 1207 and granted its first charter. He obtained land from the bishop to enlarge the town, and he founded the Augustinian priory of St. John, which dominated the eastern bank of the town until the Dissolution. Marshal's most enduring contribution, however, was the construction of a stone castle of quadrangular plan with massive, circular, corner towers. It functioned as the administrative center of the lordship of Kilkenny, passing in succession from the Marshals to the de Clares in 1248 and to the Dispensers in 1317, before being sold in 1391 to James Butler, third earl of Ormond, whose descendants lived there until 1936.

Kilkenny was a twin town throughout the Middle Ages. The pre-Norman settlement, known as Irishtown, remained a separate borough with the bishop of Ossory as its lord. It was dominated by St. Canice's Cathedral, a Gothic structure initiated by the diocese's first Anglo-Norman bishop, Hugh de Rous (1202–1218). The Anglo-Norman town, known as Hightown or Englishtown, was laid out along a single main street linking the castle with the cathedral and was given its own parish church dedicated to St. Mary. About 1225 a Dominican priory was founded, and a Franciscan house was added between 1231 and 1234. In 1231 an urban administration was established with a sovereign (Latin *superior*) as its head. Over the succeeding centuries this body obtained market rights and jurisdictional privileges for the town that enabled it to surpass Irishtown in wealth and influence. The hinterland is excellent corn-growing country, and there were at least six mills in Kilkenny from the early thirteenth century. It was also an important center of cloth production, brewing, and iron-working. At its maximum in the late thirteenth century it is estimated that the combined towns had a population of about 4500. After 1300, the numbers declined. Kilkenny was devastated by the Black Death in 1348 and 1349, the effects of which were vividly described by the local Franciscan chronicler, John Clyn. Suburbs were abandoned, extramural chapels were demolished, and some burgages remained waste into the first quarter of the fifteenth century.

Kilkenny was a major venue for meetings of the king's council and parliament, one or the other of which convened there on at least thirty-four occasions between 1277 and 1425. The most famous (or infamous) gathering was the parliament of 1366 presided over by Lionel, duke of Clarence, which promulgated the statute of Kilkenny. The urban culture of the thirteenth and fourteenth centuries was vehemently hostile to the native Irish, viewing them as "natural enemies." This attitude evidently relaxed during the fifteenth century, when the *Liber primus Kilkenniensis*, the oldest town book, records burgesses and craftsmen with Gaelic surnames. A key factor in the process of Gaelicization was the purchase of Kilkenny by the earl of Ormond in 1391. This broke the link with English-based lords and introduced a family that had built up its power base by the skilful management of the Irish in the march of Tipperary. The fifteenth century was a period of urban consolidation characterized by subtle social and economic changes, reflected topographically by redevelopment and reconstruction. After 1425 there is evidence of a demand for building space within the walls and, after 1460, both major bridges and almost all the town gates, mural towers, churches, and religious houses were rebuilt. These developments coincide with the emergence of an oligarchy of about fifteen families that dominated the town into the early modern period.

John Bradley

References and Further Reading

Bradley, John. *Kilkenny* (*Irish Historic Towns Atlas 10*). Royal Irish Academy: Dublin. 2000.

See also **Clare, de; Lionel of Clarence; Marshal; Ormond-Butler; Osraige; Parliament; Urban Administration**

KINGS AND KINGSHIP

Medieval Ireland was marked by the existence of dozens of kingdoms, each ruled by a king who in the early medieval period was technically the highest nobleman in the *túath*. Most kings were subject to over kings, who were the policy-makers of the time. They based their authority over other lords and kings on ties of blood relationship and alliance. The integrity of such alliances partially depended on the power and personal qualities of the over king. The ruling kindreds of the Irish kingdoms were often caught between the forces of internal division and outward stability. The rule of inheritance and succession stimulated competition among relatives and expansion by the kindred's branches. Yet it also gave the kindred as a whole a measure of stability and flexibility, as the kindred hardly ever died out in the male line. Several royal dynasties remained in control of an area for many centuries.

Historical Roots

The historical roots of Irish kingship are still debated. It has been argued that pagan sacral kings, who ruled over tribes, were replaced by aristocratic kings, who ruled over kindreds in the period of the coming of Christianity and the rise of expansionist dynasties. The most ancient collective names are those only found in the plural (such as Laigin and Ulaid), and

Ireland circa 1100.

names ending in *-r(a)ige*, from *-rigion* (kingdom), such as Cíarraige and Osraige. These are held to express a tribal feeling, since they are connected to matters such as human characteristics, totem animals, or deities. Yet such "tribes" may well have been ruled by certain families, as they were among the continental Celts in the first centuries B.C. and A.D. This impression is sustained by names ending in the collective *-ne* (such as Conaill(n)e, Conmaicne), or containing the element *moccu* (seed) or the related formula (MAQ(Q)I) MUCOI found on ogam-stones. These names appear in connection with a personal name, either an ancestral deity or a human forefather. They may point to the existence of aristocratic families within small communities, at least from the fifth century B.C. onward. The rise of the aristocracy is difficult to date, but its development may have caused the demise of sacral kingship, as it did in ancient Greece and Rome. The ideology of sacral kingship remained a feature in the exercise of aristocratic kingship in the medieval period. A sacral king was regarded as the mediator between the kingdom and the supernatural world. This bond was forged by a sacred marriage between the king and the goddess of the territory, who was thus rejuvenated. A good king enjoyed divine favor; a bad king risked divine wrath by tempests, diseases, and criminal offspring. Hence it was expected that he ruled wisely, did not break the "ruler's truth" (*fír flaithemon*) or his "taboos" (*gessi*), and remained unblemished. Aspects of sacral kingship were continued in the medieval period in inauguration rituals and in political ideology, where they were appropriately Christianized and applied to all secular and ecclesiastical rulers.

Royal Duties

At around the eighth century there were probably over one hundred territories that were ruled by a *rí túaithe* (king of a people or territory). Although the title *rí* means literally "king," the holder was essentially the highest nobleman of the *túath*. He held the main nobility of the *túath* in clientship; they owed him tribute and support in exchange for protection and representation. Together with the bishop and the master-poet, the king had the highest status in the territory. A person's status was expressed by his honor price, which determined his legal rights and entitlements. This hierarchical aspect of early Irish society was balanced by an egalitarian approach to responsibilities. Anyone who neglected to fulfill his duties or acted contrary to his status risked losing his honor price if he did not make amends. Serious or structural abuse could incur permanent loss of honor price, and hence loss of authority. A king's power was thus not absolute, but sensitive to his public behavior and deeds. It followed that any responsible position had to be filled by the most suitable person. Hence the nobility and royal kindred chose the candidate who was considered best qualified to carry out the royal functions. These functions included representing the people in external matters, such as dealing with other kings in times of war and peace, and maintaining internal order, including acting as judge in serious matters. As a leader of the people, the king hosted a yearly assembly (*óenach*), had a council (*airecht*; later *oireacht*) with members of the secular and ecclesiastical elite, and conferred with other kings at a meeting (*dál*). He had a number of servitors to support him in his office, such as a steward, messenger, judge, and champion.

Succession

According to theory, the headship of a royal or noble kindred was due the most suitable person in regard to descent, age, and abilities. When the head of a kindred died, and he had no other near relatives, his oldest son succeeded him. The land of the father was divided among his legitimate sons in equal shares. The oldest son received the extra share that was attached to the headship of the kindred, and had the right to represent his brothers in external affairs. After him, the other sons succeeded according to age. The oldest son was normally considered the most experienced candidate, as long as he was the son of a betrothed wife or concubine, and fit to take the burden of lordship in regard to his physical, mental, economic and political qualifications. If not, a more suitable junior candidate could be chosen instead. If two candidates were equally qualified, they would have to cast lots. In practice, such matters were often resolved by internal struggle or by negotiation, by which a senior candidate could relinquish his claims in exchange for certain privileges. No candidate had an absolute right to the succession, not even the *tánaise ríg*. Daughters had no permanent right to kin-land, and heiresses could not pass on kin-land to their offspring. Hence, outsiders could not take the headship of a family that had died out in the male line by marrying an heiress, as became common in medieval Europe. When a lineage died out, their land reverted to their male next-of-kin. This catered to stability within the Irish dynasties in the long run, but division of the kin-land and collateral succession often resulted in temporary fragmentation of the kindred's assets and political power.

Dynastic Kingship

In theory, the descendants of the sons of a lord alternated in the headship of the kindred, as long as they were duly qualified. In practice, those who—for whatever reason—were passed over for the succession were often unable to attract sufficient clients to maintain noble status for several generations. Their descendants became commoners and clients of their more fortunate relatives. This fate could be avoided by joining the ranks of the poets or clerics, or by competing successfully for power. In order to relieve internal pressure and extend the domination of the kindred, a ruler could install brothers or sons as rulers over neighboring client-peoples. The new noble or royal branches thus created remained part of the same kindred, and nominally subject to an over king as their common head. The over kingship was often contested by the leaders of the most powerful branches of the kindred, and this often led to destructive succession struggles. An over king who was disobeyed raided the territory of his errant subkings, in order to drive off their cattle as tribute or to take their hostages as guarantees for future obedience. Internal warfare could weaken the kindred as a whole, with the succession erratically being taken by this branch or that. Usually, one or two branches came out on top and subjugated all others. Yet within a few generations the winning branch would itself be split up into rival lineages, and the whole cycle would start anew. This process remained typical for Irish dynastic kingship until the end of the Gaelic order in the decades around 1600.

Over Kingships

The importance of blood relationship for claims of submission and tribute is reflected in the Irish political nomenclature. The ruling dynasties are all named after a legendary or historical ancestor, whose name is preceded by a term expressing kinship, such as Corco (seed), Dál (division), Clann (children), Cenél (kindred), Síl (seed), and Uí (grandsons or descendants). All those who recognized the same ancestor were politically tied together. Certain dynasties were, by mutual consent or a procured relationship, held to be related. This is reflected in the Old-Irish word *cairdes*, which means "kinship" and by extension, "friendship." A powerful over king could claim that others were his relatives, and thus claim authority over them. Genealogical bonds expressed political bonds, hence the importance of the recording of genealogy in the medieval sources. The law tracts of around 700 recognize a hierarchy of kings of a *túath*, kings of several *túatha*, and the provincial kings. The provincial king ruled not only a powerful dynasty but also a defined territory that he habitually dominated, named a *cóiced* (literally "fifth"). A king of Ireland only existed on a theoretical basis, as no dynasty had been able to rule Ireland permanently.

Political Structure

Already before the eighth century the over kingships had begun to dissolve the *túath* as the basic sociopolitical unit. Most of the Irish petty kings were subject to an over king, and many were hardly independent rulers. The power of the over kings over their dynasties and neighboring kings increased in time, and about a dozen were of major consequence. The Uí Néill ruled in Mide, Brega, and The North (*In Túasceirt*); the Uí Briúin and Uí Fhiachrach in Connacht; the Uí Meic Uais and Uí Chremthainn in Airgialla; the Dál Fiatach and Dál nAraidi in Ulster; the Uí Dúnlainge and Uí Chennselaig in Leinster; and the Éoganachta in Munster. Until the tenth century the over kings of the Uí Néill and the Eóganachta dominated Ireland, and claimed suzerainty over Leth Cuinn and Leth Moga, respectively. This division of Ireland is named after Conn Cétchatach, the legendary forefather of the Connachta, Uí Néill, and Airgialla, and his alleged contemporary Mug Nuadat, ancestor of the Eóganachta. The kings of Tara came to overpower the kings of Ulster and Leinster as well. Hence Máel-Muru Othna (d. 887) attaches the Laigin and Ulaid (Dál Fiatach) to those who shared a common ancestor with the Uí Néill in his poem on the Irish invasion myth. A few kings of Tara, from Máel-Sechnaill I (ruled 846–862) onward, took hostages of the kings of Cashel and claimed to be kings of Ireland. Internal rivalry and losses against the Vikings were among the factors by which the Eóganachta and Uí Néill fell apart in the tenth century.

Later Developments

The career of Brian Boru (d. 1014) marked the end of the domination of the Éoganachta and Uí Néill. This gave other dynasties the opportunity to rise to power. Notable kings were now given the honorary title "high king" (*ard-rí*), a term subsequently used to denote the kings of Tara of old. This gave rise to the anachronistic notion of a high kingship of Ireland. In the new political order that ensued the leading families were Mac Murchada (Uí Chennselaig) in Leinster, Mac Carthaig (Éoganacht Caisil) in Desmond, Ua Briain (Dál Cais) in Thomond, Ua Conchobair (Uí Briúin Ái) in Connacht, Ua Ruairc (Uí Briúin Bréifne) in the northern Midlands, and Ua Domnaill (Cenél Conaill), Ua Néill, and Mac Lochlainn (Cenél nÉogain) in the North. Apart from Mac Lochlainn, they remained powerful from around 1150 to 1600, which testifies to the resilience of the main Irish dynasties. These families also had the tendency to extend their domination by planting branches on neighboring territories. After the Anglo-Norman invasion there was an increasing development toward the exercise of lordship among feudal lines, but on the whole Gaelic tendencies persevered. These included the donation of *tuarastal* and the impositions of coshering and coyne and livery. Internal rivalry, raiding, hostage-taking, and fluctuations in alliances and power remained characteristic for the Gaelic lordships. This hampered the implementation of the English surrender-and-regrant policy in the decades around 1600, by which the Irish kings and lords were recreated as English earls and barons, with the promise to follow English law and custom. In the end, the Irish royal families died out, lost power, or their chiefs went abroad, and few managed to keep up their noble stature.

BART JASKI

References and Further Reading

Byrne, Francis John. *Irish Kings and High Kings*, revised ed. Dublin: Four Courts Press, 2001.

Jaski, Bart. *Early Irish Kingship and Succession*. Dublin: Four Courts Press, 2000.

Simms, Katharine. *From Kings to Warlords. The Changing Political Structures of Gaelic Ireland in the Later Middle Ages*. Woodbridge: Boydell Press, 1987.

See also **Ailech; *Feis*; Feudalism; Fosterage; Genealogy; Marriage; Niall Noígiallach; Óenach; Tara**

KITCHENS

See **Castles; Ecclesiastical Sites**

KNIGHTS AND KNIGHTHOOD

See **Society, Grades of Anglo-Norman**

L

LACY, DE

The first member of the de Lacy family to arrive in Ireland was Hugh de Lacy, of the Hereford branch of the family, who accompanied Henry II on his expedition of 1171–1172 and received a grant of the entire kingdom of Mide (Meath), possibly as a check on the territorial ambitions of Strongbow and probably also to provide a buffer between the land of the unsubmissive high king, Ruaidrí Ua Conchobair, and the capital of the new colony at Dublin, custody of which was entrusted to Hugh. Assassinated in 1186, Hugh left two sons by his first marriage: Walter (d. 1241) and Hugh II de Lacy (d. 1242). William Gorm de Lacy, son of Hugh I's second marriage to "Rose" Ua Conchobair, was a close associate of his half brothers. As lord of Meath and earl of Ulster, respectively, Walter and Hugh were among the most powerful men in Ireland, but their relationship with King John was not an easy one, and they suffered forfeiture of their lands more than once.

Walter may have been a minor at the time of his father's death, as he did not gain possession of his full estate until 1194. In the late 1190s, Walter spent time on campaign in France, and Hugh acted on his behalf in Meath. In 1195, the brothers assisted John de Courcy in a war against the English of Leinster, and Walter's lands were escheated as punishment. But in 1199, Walter, having been fined 2,100 marks, regained the king's favor, and when John turned against de Courcy he used Hugh to bring about the downfall of his former ally.

In 1203, Hugh drove de Courcy out of Down and the following year was granted de Courcy's Ulster lands, in addition to lands in Connacht. The following year he was titled Earl of Ulster, the first earldom created in Ireland. The brothers combined to foil de Courcy's attempt to reenter Ireland this year, and Hugh spent this period campaigning in Ulster. But their relationship with the king deteriorated as they quarreled with his justiciar Meiler fitz Henry, and when William de Braose, Walter's father-in-law, fled to Ireland from John's wrath, the king crossed the Irish sea to humble the de Lacys.

The brothers fled to Scotland, and then to France. Walter regained the king's favor by 1215; his lands were returned, and the following year he was appointed sheriff of Hereford. Several years of loyal service in France and England followed. But Hugh was left in the cold for some years, and his lands were entrusted to Walter during this time. After spending time in France on the so-called Albigensian crusade, Hugh returned to England in 1221. Refused permission to go to Ireland, he crossed over illegally and entered an alliance with Áed Ua Néill, king of Cenél nEógain. The following year he conspired with Llywelyn ab Iorwerth, prince of north Wales, in a failed campaign against William Marshal, and in 1224 was again with Ua Néill in war against Áed Ua Conchobair.

During this rising Hugh was supported by his half brother, William Gorm. The rebellious pair was captured by William Marshal, and Hugh was deported to England; he was not reinstated until 1227. In 1228, Walter and Hugh were summoned to serve in France. Walter commanded a division in the 1230 invasion of Connacht. Both brothers supported the justiciar's struggle against Richard Marshal, were present at his defeat in battle on the Curragh in 1234, and were on campaign in Connacht again the following year.

William Gorm, who had fought for the king in France in 1230, was killed in Bréifne in 1233. Walter's health declined in the late 1230s, but Hugh's turbulent spirit remained. After the death of his son-in-law Alan, lord of Galloway in Scotland, Hugh supported Alan's illegitimate son Thomas in a failed rebellion against the Scottish king. In 1238, he was temporarily driven out of Ulster by the Mac Lochlainn family; one of his sons was killed in the fighting before he regained his position. Hugh died at Carrickfergus late in 1242, a year after his brother Walter.

Walter left no surviving male heirs, and the lordship of Meath was partitioned between his surviving granddaughters: Matilda, who married Geoffrey de Geneville, and Margaret, who married John de Verdon. A cadet branch, the de Lacys of Rathwire (possibly descended from a brother of the first Hugh) carried on the name in much-reduced circumstances. In 1309, a Hugh de Lacy was constable of Rinndown Castle in County Roscommon, but he must have resented the loss of the de Lacy patrimony, and with his brothers Walter and Amaury took advantage of Edward Bruce's invasion to conspire against the de Verdons. Accused of treason in 1315, they incredibly managed to convince a jury of their innocence, but in 1317 were forced to flee with the Scots army to Carrickfergus. The following year a number of the family died with Bruce at Faughart. The survivors fled to Scotland and their lands were confiscated, although a partial recovery was secured during Edward III's reign.

References and Further Reading

Duffy, Seán (ed). *Robert the Bruce's Irish Wars: The Invasions of Ireland 1306–1329* (2002).

Graham, B.J. "Anglo-Norman Settlement in County Meath." *Proceedings of the Royal Irish Academy 75* (1975): C, 223–248.

Otway-Ruthven, J. "The Partition of the de Verdon Lands in Ireland in 1332." In *J.R.S.A.I. 48* (1967):

Potterton, Michael. "The Archaeology and History of Medieval Trim, County Meath." Ph.D. diss., The National University of Ireland, Maynooth, 2003.

Wightman, W.E. *The Lacy Family in England and Normandy 1066–1194*. 1966.

See also **Courcy, John de; Geneville, Geoffrey de; John; Lacy, Hugh de; Marshal; Verdon, de**

LACY, HUGH DE

Hugh de Lacy (d. 1186) was born into the Hereford branch of the Lacy family, powerful landholders on the Welsh marches. He succeeded his father as Fourth Baron Lacy some time in the early 1160s, and campaigned in north Wales in the late 1160s. In October 1171, Hugh accompanied King Henry II to Ireland and was granted the former kingdom of Mide (for fifty knights' service), as well as custody of Dublin city and castle; it was intended that his authority would balance the growing power of Richard de Clare (Strongbow) and probably also curb the capacity of the high king, Ruaidrí Ua Conchobair, to resist English settlement.

Following Henry's return to England, Hugh set about consolidating his new lordship. At a meeting with Tigernán Ua Ruairc, king of Bréifne, at the Hill of Ward in Co. Meath a violent quarrel broke out in which Ua Ruairc was slain. Each side accused the other of bad faith over the incident, but the removal of Ua Ruairc could not have hurt Hugh's position in Meath. He returned to England before Christmas, and the following year was on campaign in Normandy in defense of King Henry during the rebellions of his sons in alliance with the king of France.

On his return to Ireland he was very active in securing his lands, building fortifications, and bringing in settlers from his English and Welsh estates. Such was his energy and ability that in 1177 he was appointed justiciar of Ireland. He chose Trim as his chief manor in Meath, where an earthen ringwork castle was soon succeeded by a stone keep. The "Song of Dermot and the Earl" gives a list of the chief men who settled as Hugh's vassals. One of them, Gilbert de Nugent, married Hugh's sister Roesia. Hugh also acquired a reputation for fair dealing with the native Irish, and encouraged them to remain as tenants under his lordship.

Hugh's first wife, Rose de Monmouth, had died by 1180, and in this year he married a daughter of Ruaidrí Ua Conchobair. But this marriage (undertaken without King Henry's permission) to the daughter of the last high king gave some the impression that Hugh had regal ambitions of his own. Such accusations were no doubt exaggerated, but Hugh was deprived of Dublin and recalled to England in 1181. He managed to reassure Henry of his loyalty, but when he was reinstated as justiciar the following year a royal clerk, Robert of Shrewsbury, was appointed to oversee his activities. When Henry's young son John came to Ireland as the colony's new lord, some commentators blamed Hugh for sabotaging the expedition, claiming he would not let the Irish pay tribute to John. Again the rumors against Hugh were probably overstated, but his power in Ireland by this time was unrivalled and he may well have felt reluctant to hand over his hard-won authority to a young and untested lord.

Hugh was beheaded in 1186 at Durrow by an agent of An tSionnach Ua Catharnaig; English chroniclers recorded that the English king was overjoyed at his death. But although it aroused royal jealousy, Hugh's success was instrumental in firmly establishing the new colony. His chief heirs from his first marriage were Walter, who succeeded him as lord of Meath, and Hugh, who subsequently overthrew John de Courcy and was belted earl of Ulster by King John. From his marriage to Ua Conchobair's daughter he had a son, William Gorm de Lacy. A postmortem dispute over the burial rights to Hugh's body (not recovered from the Irish until 1195) led to his head being buried in

Thomas's abbey, Dublin, beside his wife, and his decapitated corpse being solemnly interred in Bective abbey, Co. Meath, a dispute only resolved in 1205, in favor of St. Thomas's, where all of the remains were then reunited.

References and Further Reading

Brady, J. "Anglo-Norman Meath." *Riocht na Midhe* 2 (1961): 38–45.

Bartlett, Robert. "Colonial Aristocracies in the High Middle Ages." In *Medieval Frontier Societies*, edited by R. Bartlett and A. MacKay. Oxford: 1989.

Graham, B. J. "Anglo-Norman Settlement in County Meath," *Proceedings of the Royal Irish Academy* 75 (1975): 223–248.

Otway-Ruthven, A. J. "The Partition of the de Verdon Lands in Ireland in 1332." In *J.R.S.A.I.* 48 (1967): .

Potterton, Michael. "The Archaeology and History of Medieval Trim, County Meath." Ph.D. diss., The National University of Ireland, Maynooth, 2003.

Walsh, Paul. *Irish Leaders and Learning Through the Ages.* Edited by Nollaig Ó Muraíle. Dublin: , 2003

Wightman, W. E. *The Lacy Family in England and Normandy* 1066–1194. 1966.

See also **Henry II; John; Mide; Strongbow**

LAIGIN

Originally an ethnic term, the word "Laigin" refers to the people who dominated the southeast of Ireland and gave their name to the province of Leinster (*Cóiced Laigen*). Medieval sources closely associate the Laigin or "Leinstermen" with the Gáileóin and Domnainn, to whom they were probably related, but regard them as ethnically distinct from the Ulaid, Connachta, and other provincial powers. Although reliable information about their origins and rise to power is lacking, medieval legend suggests that the Laigin came to Ireland from either Britain or Gaul under the leadership of their ancestor Labraid Loingsech and seized control of the province of Leinster some time in the third or fourth century B.C.

By the dawn of the historical period, the Laigin had split into many separate dynasties and spread throughout the province. The most powerful of these—dynasties like the Uí Garrchon, Uí Enechglais, and Uí Failgi—lived in the north and vied for control of the Liffey valley, an area that encompassed Naas, Kildare, and Dún Ailinne, the symbolic center of their provincial kingship. During the fifth and sixth centuries, these dynasties were engaged in territorial wars with the expanding Uí Néill, to whom they ultimately lost possession of Tara and its environs. Conflicts with this dynasty would become a recurrent feature of Laginian history, particularly in the eighth and ninth centuries as the Uí Néill attempted to assert their suzerainty over successive kings of Leinster. By the mid-eighth century, these conflicts in conjunction with numerous internal feuds significantly weakened many of the old northern dynasties such that a new Laginian power structure arose, one dominated by the Uí Dúnlainge in the north and the Uí Chennselaig in the south.

From 738 to 1042, the Uí Dúnlainge ruled the Liffey valley and maintained an exclusive hold on the provincial kingship. However, by the early decades of the eleventh century, internal disputes, conflicts with the Vikings, and invasions by successive kings of Osraige had crippled the Uí Dúnlainge septs, allowing Uí Chennselaig to seize power. Before that time, the latter had enjoyed a measure of independence from Uí Dúnlainge in their new homeland around Ferns, though they were wracked by dynastic strife. But with the Uí Dúnlainge weakened, they seized the Laigin kingship in 1042 and dominated the province from that point until the Anglo-Norman Invasion. The person responsible for their rise to power was Diarmait mac Máele-na-mbó, who was one of the most powerful kings in Ireland at his death in 1072. In the early twelfth century, the ruling family of Uí Chennselaig adopted the surname Mac Murchada (Mac Murrough), and one of the first kings to bear it was Diarmait Mac Murchada (d. 1171), whose arrangement with Henry II made possible the Anglo-Norman Invasion. Once the English got control of Leinster, the Mac Murchada family would enjoy only occasional bouts of power, as they did under Art Mac Murchada in the late fourteenth century, but it was not until the late sixteenth century that they were completely brought to heel.

DAN M. WILEY

References and Further Reading

Byrne, Francis John. *Irish Kings and High-Kings*. Second Edition. Dublin: Four Courts Press, 2001.

Duffy, Seán. *Ireland in the Middle Ages*. London: Macmillan; New York: St. Martin's Press, 1997.

Mac Niocaill, Gearóid. *Ireland Before the Vikings*. Dublin: Gill and Macmillan Ltd, 1972.

Ó Corráin, Donncha. *Ireland Before the Normans*. Dublin: Gill and Macmillan Ltd, 1972.

Ó Cróinín, Dáibhí. *Early Medieval Ireland 400–1200*. London & New York: Longman Group Ltd, 1995.

O'Rahilly, Thomas F. *Early Irish History and Mythology*. Dublin: Dublin Institute for Advanced Studies, 1946.

Ó hUiginn, Ruairí. "The Literature of the Laigin." *Emania* 7 (1990): 5–9.

Smyth, Alfred P. *Celtic Leinster: Towards an Historical Geography of Early Irish Civilization A.D. 500–1600*. Dublin: Irish Academic Press Ltd, 1982.

See also **Anglo-Norman Invasion; Connachta; Diarmait mac Máele-na-mbó; Leinster; Mac Murchada; Uí Chennselaig; Uí Dúnlainge; Uí Néill; Ulaid; Viking Incursions**

LANCASTRIAN-YORKIST IRELAND

The deposition in September 1399 of King Richard II in favor of Henry of Lancaster, crowned King Henry IV, spelled the end of a period during which king and council had devoted rather more attention to events in Ireland than had traditionally been the case. The appointment of Lionel of Clarence as lieutenant in 1361 had inaugurated a relatively sustained effort to strengthen the English position in Ireland, culminating in Richard II's two personal expeditions to Ireland with powerful armies royal. The results, however, fell far short of expectations. The new Lancastrian regime, moreover, had more pressing commitments elsewhere and more slender resources with which to discharge them. Initially, an attempt was made with the appointment as lieutenant in 1401 of Henry IV's second son, Thomas of Lancaster, to maintain a substantial garrison to repress the Irish: Lancaster was promised 12,000 marks annually to maintain his estate, but actually received less than half this sum. By 1413, when Sir John Stanley was appointed lieutenant, the governor's normal salary had been reduced to a more manageable £2,000 a year. Reports from royal officers in Ireland descended into graphic detail to explain to the king the dire consequences for the defense of the Englishry of this shortage of money and *manræd*—the weakness of the marches, growing raids by the Irish, destruction and rebellion all around. Yet, for king and court—and the English political nation more generally—events in Ireland, bad as they seemed, were simply not a priority.

Ireland and the English Monarchy

The fact was that the good rule and defense of "the king's loyal English lieges" in Ireland had to be seen in the context of commitments elsewhere. Most important was the defense of the realm, threatened by invasion from Scots enemies to the north and a protracted uprising among "the mere Welsh" (1400–1415) which briefly (1403) attracted support from France and Brittany and also from the dissident earl of Northumberland. By the time internal dissension had been stamped out, the Hundred Years War with France had recommenced with sweeping English successes—the conquest of Normandy and large stretches of northern France, the occupation of Paris, and finally in 1422 the glittering prize of the French crown. Naturally, the exploitation of military victories on French battlefields took priority over petty raiding in the bogs of Ireland. The lordship's military resources were again tapped to consolidate these new conquests. In 1419, the prior of Kilmainham led a force of 700 men to serve under Henry V at the siege of Rouen. Ireland's premier earl, James Butler of Ormond, also participated, as he did in campaigns there from 1415 to 1416 and in 1430. Then from 1435, when the War turned sour, France remained a priority for different reasons, swallowing up scarce resources to shore up the crumbling English position.

The result was that little could be spared for Ireland, which ranked a bad fourth—after France and Scotland—in the regime's priorities. Lord Treasurer Cromwell's statement of royal income and expenditure presented to the English parliament in 1433 gives some insight into the overall position. Cromwell estimated the king's ordinary annual income (excluding taxation) at £64,800, but projected ordinary expenditure (excluding the French war, which was supposedly self-sufficient) at £80,700. Grants of taxation would hopefully make up the difference, but substantial debts had also accumulated, amounting to £168,400. The government's finances had probably deteriorated during Henry VI's minority (1422–1437), but only peace or sweeping military success could stabilize the position. In these circumstances, nothing much could be expected for or from the Irish theater of operations: Cromwell estimated the king's revenues there at £2,340 (a decidedly optimistic estimate), with expenditure at £5,026, thus leaving a deficit of almost £2,700 to be made good from England. By comparison, the financial deficit for the defense of Calais (costing almost £12,000) exceeded £9,000; that of Gascony (costing over £4,100) ran to £3,300; and defending the Anglo-Scottish frontier cost a further £4,800. Overall, the outlying territories provided a series of strategic posts and buttresses to defend the English mainland at a cost of £20,000.

The lordship's primary significance in all of this was the string of royal port towns stretching south from Carrickfergus and around to Galway, which facilitated English naval control of the Irish and Celtic Seas and denied the island to any continental prince. Some of these port towns, notably Carrickfergus and Galway themselves, were effectively English military outposts in Gaelic Ireland, but others had extensive English hinterlands—the eastern coastal plain around Dublin, Drogheda, and Dundalk, and the Barrow-Nore-Suir river basin, the two densest areas of medieval English settlement. Here more fertile land had permitted the introduction of English manorialism and mixed farming with nucleated villages and market towns. Militarily, these formed a series of strong points along comparatively stable marches that were a good deal easier to defend from the perennial Gaelic raids than the more thinly populated pastoral regions where the marches were fluid and shifting. Also strategically important was the king's highway down the Barrow valley connecting these two regions; but this was swept by Gaelic raids both from the midlands and the Leinster mountains. Yet what happened in the purely Gaelic parts—"the land of war" inhabited by "the wild Irish" living in their woods and bogs—was of little concern to the government.

What the beleaguered Lancastrian government aimed to do was to conduct a holding operation while addressing more pressing problems elsewhere. Successive governors could expect an annual salary of, at most, 4,000 marks—notably John Talbot, Lord Furnivall (later earl of Shrewsbury and Waterford), lieutenant from 1414 to 1420 and from 1445 to 1447; James Butler, earl of Ormond, a regular choice for two- or three-year periods between 1420 and his death in 1452; and Richard duke of York, lieutenant from 1447 to 1460. This salary, normally payable from the English exchequer, was intended to offset the deficit on the Irish revenues, so allowing governors to maintain an adequate force for defense, commonly 300 or 400 archers. Given that Waterford, Ormond, and York were the lordship's leading landowners, each with an extensive *manrœd*, these arrangements should in theory have been more than adequate. Yet, for various reasons, the reality was far different. Increased reliance on local landowners at this time promoted faction: the classic illustration was the escalating feud between Talbot and Butler. Besides encouraging Gaelic raids, the feud left the Dublin administration virtually paralyzed in the early 1440s. Yet outside governors without a local following obviously needed greater support. Escalating feuds between provincial magnates also epitomized the regime's collapse elsewhere at this time—for want of impartial justice by the feeble Henry VI. Another indication of incipient collapse was the worsening financial situation. Already during Talbot's first lieutenancy, the English exchequer's failure to maintain the payments agreed in his indenture forced the lieutenant to resort to coign and livery to maintain his troops—that is, to billet them on the country and to purvey supplies for his household without payment. Later lieutenants generally received less of what was owed; deputies appointed during their long absences commanded still smaller resources; and at £500 the salary allowed to a justiciar elected by the council to fill a casual vacancy was modest indeed. In short, a governor with even 2,000 marks a year to maintain a small retinue was very much the exception, and little was available in Ireland by way of taxation—700 marks per subsidy, £300 from a scutage.

The Crisis of Lordship and the Descent to Civil War

The result was that intensive royal government on the lowland English model, supervised by the central courts, the governor, and council, was increasingly restricted to "the four obedient shires" around Dublin, the region later called "the English Pale," which supplied most of the king's revenues. Here, the government encouraged the construction of towers and dikes in the marches to facilitate defense and inhibit cattle rustling. Elsewhere, however, apart from the predominantly self-governing royal towns, defense and good rule increasingly devolved on the region's ruling magnate—notably, the earls of Desmond in the southwest, and Ormond in south Leinster, whose private armies of kerne and galloglass were maintained by coign and livery in the Gaelic manner. Central supervision was intermittent, in part because the Barrow valley was now passable only with an armed escort. Occasionally, when affairs around Dublin permitted, the governor might make a progress southwards, perhaps mounting a short campaign against "Irish enemies" and holding brief judicial sessions. Yet, conditions in the Barrow valley worsened markedly following the death in 1432 of the leading lord there, the earl of Kildare. Ormond had married Kildare's daughter and secured most of the estates, but with no resident earl to defend them, outlying estates were overrun and key castles like Tullow and Castledermot were destroyed, thus undermining the whole march.

By the late 1440s, the English position was everywhere collapsing. The arrival as lieutenant in 1449 of the king's heir apparent, Richard duke of York, briefly gave new heart to the Englishry: wholesale submissions by Gaelic chiefs prompted the rash prediction that within twelve months "the wildest Irishman in Ireland shall be swore English." Then news arrived of the final English collapse in Normandy: York demanded immediate support, "for I had liever be dead" than have it chronicled "that Ireland was lost by my negligence." Soon after, rebellion broke out in England, and York departed, leaving Ormond as his deputy. Yet Ormond's death in 1452 was followed a year later by Waterford's death in distant Gascony in the final English collapse there, and suddenly Ireland's leading landowners were all absentees. Ormond and Waterford's successors never visited their Irish estates. In the ensuing crisis of lordship, the king recognized as earl of Kildare Thomas FitzMaurice of the Geraldines, grandnephew of the last earl, in a bid to strengthen the southern marches of "the four shires." Yet by then English politics were sliding towards civil war: York built up strong support, retaining the earls of Desmond and Kildare, although the absentee earl of Ormond sided with the court party.

War began in earnest in 1459. The lordship's potential as a retreat and recruiting ground for attempts on the throne was first appreciated by the Yorkists, the duke himself fleeing there after the rout of Ludford, while Warwick and York's son, the earl of March, retired to Calais. During the winter of 1459–1460, Warwick visited York in Waterford to coordinate a two-pronged invasion of England, and the army raised for

York's attempt on the throne the following autumn drew solid support from the English of Ireland. Although York was killed soon after, his son claimed the throne as King Edward IV, and his ensuing victory at Towton, the greatest battle of the Wars of the Roses, left the Yorkists in control. In 1462, the defeated Lancastrians tried to emulate York's strategy in reverse: a Lancastrian invasion of the lordship led by Ormond's brother coincided with risings in the midlands and Meath on behalf of Henry VI. The Lancastrians briefly secured control of the Ormond heartland around Kilkenny and Tipperary, capturing Waterford city, but elsewhere there was little support for the feeble Henry VI, and the risings collapsed following the rebel defeat by Desmond at Pilltown.

Thereafter, the lordship remained solidly Yorkist. In 1470–1471, divisions within the Yorkist camp permitted Henry VI's short-lived "readepcion." An Irish echo of this saw Kildare briefly heading a nominally-Lancastrian administration as deputy to Edward IV's renegade brother, Lord Lieutenant Clarence. Yet this time Edward had no need of Irish support to recover the throne, and following news of Barnet and Tewkesbury Edward was promptly proclaimed. These events were in marked contrast, however, to the aftermath of Richard III's defeat at Bosworth in 1485. Edward's death in 1483 had precipitated new splits among the Yorkists, eventually allowing Henry Tudor to seize the throne. Yet the Irish administration now headed by Kildare's son, the young 8th earl, exhibited a marked reluctance to proclaim Henry VII, even going so far as to convene a parliament in Dublin in Richard III's name fully two months after his death. It was to be another ten years before Tudor rule was fully accepted in English Ireland. In 1487, traditional loyalties remained sufficiently strong for the Yorkists to recover Ireland, crown an English king in Dublin, and then invade England with an army that was probably the largest raised there in 150 years. The Yorkist cause only finally expired at the siege of Waterford in 1495 when Desmond's army and "King Richard" IV's navy were dispersed by Sir Edward Poynings' artillery.

It is not difficult to explain the lordship's enthusiastic support for the Yorkists. We may discount York's supposed concession of legislative independence in a Home Rule parliament in 1460; this underlined royal weakness at a time when the Englishry craved closer ties with the court, not less. The key factor was the close relationship between the Yorkist leadership and the 7th earl of Kildare. York's retainer and deputy, Kildare was thereafter consistently favored by Edward IV, himself an experienced marcher lord and a far better judge of character than the saintly Henry VI. Kildare's long tenure of the governorship and other indications of Edward's favor, such as grants of land, enabled the earl to recover, restore, and extend his wasted ancestral possessions, expelling the Irish and fortifying the territory with towers and castles. His son, the 8th earl, continued this strategy, thereby also restoring the English position in Counties. Kildare and Carlow. This is not to say that relations between king and earl were invariably harmonious; there were clashes in 1468, 1478, and 1483. Yet, at bottom, Edward IV recognized (as the later Tudors did not) that good rule in a marcher society rested on reliable and resident marcher lords. Following the collapse of Lancastrian France, ruling magnates were the key to the English recovery in the remaining borderlands.

STEVEN ELLIS

References and Further Reading

Cosgrove, A., ed. "A New History of Ireland." II Medieval Ireland. Oxford: Clarendon, 1987.

Ellis, S. G. *Ireland in the Age of the Tudors 1447–1603: English Expansion and the End of Gaelic Rule*. London: Longman, 1998.

Ellis, S. G. *Tudor Frontiers and Noble Power: The Making of the British State*. Oxford: Clarendon, 1995.

Ellis, S. G. *Reform and Revival: English Government in Ireland, 1470–1534*. London: The Royal Historical Society, 1986.

Empey, C. A., and K. Simms. "The Ordinances of the White Earl and the Problem of Coign in the Later Middle Ages." In *Proceedings of the Royal Irish Academy*, 75 (1975): sect. C.

Goodman, A. *The Wars of the Roses: Military Activity and English Society, 1452–1497*. London: Routledge, 1981.

Gorman, V. "Richard, Duke of York, and the Development of an Irish Faction." In *Proceedings of the Royal Irish Academy*, 85 (1985): sect. C, 169–79.

Griffiths, R. A. "The English Realm and Dominions and the King's Subjects in the Later Middle Ages." *In Aspects of Government and Society in Later Medieval England: Essays in Honour of J. R. Lander*, edited by J. Rowe. Toronto: 1986

Griffiths, R. A. *The Reign of King Henry VI*. Berkeley and Los Angeles: University of California Press, 1981.

Johnson, P. A. *Duke Richard of York 1411–1460*. Oxford: Clarendon, 1988.

Lydon, J. F. *The Lordship of Ireland in the Middle Ages*. 2nd ed. Dublin: Four Courts, 2003.

Matthew, E. "The Financing of the Lordship of Ireland Under Henry V and Henry VI." In *Property and Politics: Essays in Later Medieval English History*. Edited by A. J. Pollard. Gloucester: Alan Sutton, 1984.

Otway-Ruthven, A. J. *A History of Medieval Ireland*. 2nd ed., London: Benn, 1980.

Pugh, T. B. "Richard Plantagenet (1411–1460), Duke of York as King's Lieutenant in France and Ireland." In *Aspects of Government and Society in Later Medieval England: Essays in Honour of J. R. Lander.* Edited by J. Rowe. Toronto: 1986.

Storey, R. L. *The End of the House of Lancaster*. 2nd ed. Gloucester: Alan Sutton, 1986.

Thompson, M. W. *The Decline of the Castle*. Cambridge: Cambridge University Press, 1987.

See also **Dublin; Kildare; Kilkenny; Lionel of Clarence; Manorialism; Pale, The; Parliament; Richard II**

LANGUAGES

Early medieval Ireland was host to a number of languages, two of which stand out: Irish and Latin. The first enjoyed preeminence as the language of the Irish people and their Gaelic culture, while the second commanded prestige as the language of the Church and ecclesiastical learning.

Irish

Irish belongs to the Goidelic branch of Celtic, itself an Indo-European language. Goidelic was introduced into Ireland by Celtic-speaking people, whose period of arrival remains highly uncertain, with proposed dates ranging from 1800 to 350 B.C.E. A date close in the second half of the first millennium B.C.E. seems plausible if one accepts the frequent claim in early Irish literature that Ireland comprised different ethnic strata of which the Goidelic speakers was the most recent. No doubt other languages were being spoken when they arrived, including perhaps non–Indo-European languages and other Celtic languages. The Goidels were able to impose a cultural hegemony on Ireland such that by the time of the earliest written records their language enjoyed a complete monopoly. That state of affairs lasted, despite Scandinavian and Anglo-Norman invasions, until the sixteenth century.

Irish, along with the other members of the Goidelic group (Scottish Gaelic and Manx), is traditionally labeled as Q-Celtic in contradistinction to P-Celtic, a terminology based on the phonological criterion that the former preserves the sound *kw*, which became *p* in the latter; thus, Irish *cenn* ("head") versus Gaulish *penno-*, Welsh *penn*. The importance of this yardstick in distinguishing Goidelic from Brittonic, the branch of Celtic spoken in Britain (now represented mainly by Welsh), has been overplayed. A more telling difference between the two branches is that in Goidelic accented words are stressed on the first syllable while in Brittonic the stress falls on the penultimate syllable.

The earliest evidence about Irish is found in inscriptions written in Ogam, an alphabet specifically invented for that language. It consists of twenty symbols in the form of notches (for the vowels) and strokes (for the consonants) carved on the adjoining faces (and their intersection) of a stone pillar. Despite its curious form, Ogam is based on the Roman alphabet and was already in existence by the fourth century C.E. Its origins may be sought in Roman Britain or even colonies of Christian missionaries coming from that province to Ireland. Although limited in subject matter (personal names) and linguistic forms (nouns in set formulas such as "X son of Y"), the surviving Ogam inscriptions provide invaluable information about the Irish language. Not only do they preserve the linguistic state of Irish in the fifth and sixth centuries, they demonstrate by contrast how much the language changed over the following two centuries. For example, contrast the Ogam name *CATTUBUTTAS* with its Old Irish counterpart, *Cathboth*. The former presents a form of Irish that had a fairly simple phonemic system and was still highly inflected in the manner of Latin, while the latter shows loss of the unstressed internal vowel and final syllable as well as a new phonemic distinction based on a dual quality of consonants.

The earliest conventionally written records of Irish date from the seventh century (perhaps even the late sixth century). They are written on parchment in the Roman alphabet, albeit a modified form marked by curious spelling features, such as the representation of the sounds /b/, /d/, and /g/, in certain well-defined environments, by the letters *p*, *t*, and *c*, respectively. This and other spelling peculiarities are explained by the theory that in composing their alphabet the Irish used as a model the Latin alphabet as it was pronounced by British speakers. Thus, British speakers of the sixth century pronouncing the Latin name Tacitus would likely have rendered it by /Tagidus/, reflecting sound changes that had occurred in their vernacular. But since they were unable (or unwilling) to change traditional Latin spellings to reflect such pronunciations, they established new equivalences between Latin symbols (in this case internal *c* and *t*) and local pronunciation. It seems likely that the Irish inherited these distinctly British treatments of certain Latin letters when they appropriated them for writing their own vernacular. Thus, Old Irish *sacart* ("priest") represents /sagṡRd/. The process of developing this new Irish alphabet probably took place in an ecclesiastical (most likely monastic) environment among British clergy working as missionaries and teachers in Ireland, perhaps in the sixth century.

For the early medieval period, modern scholars distinguish two major stages of Irish: Old and Middle. Old Irish is broadly defined as the stage of the language between 600 and 900 C.E.; Middle Irish between 900 and 1200 C.E. A further division of Old Irish distinguishes between "Archaic" and "Classical." The former, a relatively new field of research, refers to the oldest written records of Irish dating from the seventh century, which include the Cambrai Homily, glosses on the Pauline Epistles, and the Irish personal and place-names embedded in Latin documents such as Tírechán's memoir of St Patrick. Classical Old Irish has been reconstructed from a large body of glosses entered in the margins and between the lines of Latin texts such as the Pauline Epistles and Priscian's Grammar in manuscripts dated between circa 750 and 850 C.E.

Linguistically, Old Irish is characterized by a high degree of morphological complexity especially in its verbal system. Verbs can be either simple (*berid*, "she brings") or compounded with one or more prepositions (*as-beir*, "she says"). Each verb, whether simple or compound, has parallel independent and dependent forms, with special accommodations for infixing personal pronouns. For example, the compound verb *ad-cí* ("she sees") has a dependent form, *ní accai* ("she does not see"); with the infixed personal pronoun *m* ("me"), it becomes *atom-chí* ("she sees me"), and with the corresponding dependent, it becomes *ním-accai* ("she does not see me"). In the noun system, Old Irish preserved the three grammatical genders (masculine, feminine, and neuter) and the three numbers (singular, plural, and dual) of Indo-European, though it had a simplified paradigm of the noun (and adjective), consisting of five cases.

Unique to Goidelic (and Irish) is the phonemic distinction between nonpalatal (broad) and palatal quality for all consonants. For example, *túath* ("territory") with broad *-th* indicates a nominative singular, whereas with palatal *-th* it is accusative or dative; the latter is spelled *túaith* where the glide vowel *i* marks the palatalization. Another unusual feature of Old Irish is the melding of prepositions with a following personal pronoun; for example, the preposition *co* ("to"), when followed by the first person pronoun, has the form *cuccum* ("to me"). Perhaps most remarkable about Classical Old Irish is the consistency of its spelling and grammar and the apparent absence of dialect forms. These characteristics suggest that it was a standardized, literary language somewhat removed from ordinary speech.

By contrast Middle Irish seems chaotic. The complex verbal system of Old Irish is in the process of breaking down; the infixed pronouns disappear to be replaced by independent pronouns, and in the nouns and adjectives the neuter gender and the dual form gradually die out. Phonologically, all unstressed final syllables became "schwa" (/ɢ/), which meant that inflections based on distinguishing final vowels were confused and ultimately lost; for example, singular *céile* ("a companion") could no longer be distinguished from plural *céili*. This linguistic turmoil is also reflected in the spelling confusion of Middle Irish sources as scribes vacillate between the standardized spelling criteria of Old Irish and the realities of contemporary speech. Although the shift from Old to Middle Irish is dated approximately to 900 C.E., the process may have already begun in the Old Irish period, as suggested by spellings in the glosses that deviate from the classical norm, perhaps reflecting the influence of contemporary spoken language. The touchstone of Middle Irish is a work known as *Saltair na Rann*, a versified summary of biblical history that is generally thought to have been composed in 988. Although relatively neglected by comparison with Old Irish, Middle Irish deserves closer study not only because of its intrinsic importance but also because much of the vernacular literature of the Old Irish period has been preserved only in Middle Irish copies.

Latin

Although some Latin may have been spoken in Ireland as a result of trade contacts with the Roman Empire, its real impact was felt with the arrival of Christian missionaries (including Patrick) in the fifth century. They probably came from sub-Roman Britain, which meant that Ireland received Latin as pronounced in the British manner (see above). As the official language of the new religion, enshrined in its liturgy and its Bible, Latin had to be learned by Irish converts who aspired to ecclesiastical orders and the monastic life. But because it was a totally foreign language, the Irish had to learn it from scratch, with the result that they became remarkably efficient at mastering its grammar. Irish writers of Latin such as Columbanus, Adomnán, and Eriugena bear witness over several centuries to the continued excellence of the Latin taught in the Irish schools.

The flowering of Hiberno-Latin scholarship took place during the seventh century and first half of the eighth century. Thereafter, it appears that Latin was gradually displaced by Irish as the language of ecclesiastical learning. The Viking invasions of the ninth century may have contributed to this process by disrupting the monastic schools and encouraging the exodus to the Continent of scholars such as Eriugena and Sedulius Scottus. The Céli Dé movement, which became very influential in the early ninth century, may also have contributed by encouraging use of the vernacular in religious writings, perhaps because many of its adherents were not versed in Latin.

The most enduring witness to the influence of Latin is the body of words that Irish borrowed from it. Predictably, many of them are overtly religious in character, such as *cásc* (<Lat. *Pascha*, "Easter"), and *peccath* (<Lat. *peccatum*, "sin"), though some denote mundane aspects of daily life, for example, *muilenn* (<Lat. *molina*, "a mill"), and *scúap* (<Lat *scopa*, "a brush"). The traditional view holds that all these words divide neatly into two strata: an earlier group of borrowings consequent on Patrick's mission in the mid fifth century, and a later, larger stratum resulting from close ties with British monasticism in the sixth century. That view is being challenged with the plausible hypothesis that these Latin loanwords represent a continuum of borrowing during the fifth and sixth centuries.

Greek

The issue of knowledge and use of Greek in Ireland is problematic. Certainly, the Irish knew the Greek alphabet and the numerical significance of its symbols; they also knew individual Greek words that they culled from patristic writings, especially Jerome. Irish sources offer occasional glimpses of a more substantial knowledge of Greek, such as the Greek text of the Lord's Prayer and a seventh-century inscription on a stone at Fahan Mura. But the optimistic portrayal of Ireland as a haven of Greek learning is not supported by the surviving evidence.

Other Languages

Along with Latin, British missionaries brought their vernacular to Ireland; to them may be attributed many of the Welsh loanwords in Irish. Another language from Britain represented in Ireland, especially during the seventh and eighth centuries, was Old English, spoken by colonies of Anglo-Saxons. But judging by the paucity of Old English loanwords in Irish, their influence was slight. The most influential of the foreign vernaculars introduced into Ireland was Old Norse, which came with Scandinavian settlers in the early ninth century. It continued to be spoken until the late twelfth century when they were absorbed into the general Gaelic population. However important their depredations may have seemed to the monastic chroniclers, the Scandinavian speakers in Ireland must have always constituted a small body by comparison with the Irish-speaking population. Old Norse influence is evident in loanwords. The vast majority of them can be traced to the dialect of Old Norse spoken in southwest Norway, indicating that among the Scandinavians in Ireland the Norwegians (rather than the Danes) exercised a greater influence. Place-names are evident, some borrowed directly, such as Waterford (*Vethrafjörthr*) and Limerick (*Hlymrekr*), others consisting of Irish elements combined in a Norse way, for example, Gaultier<*Gall+tír* (=Irish, *Tír na nGall*) and Dublin<*Dub+linn* (=Irish, *Linn dub*). Predictably, most of the loanwords relate to areas of life that were unfamiliar to the Irish and for which Norse offered an abundance of lexicon, especially words dealing with ships and seamanship, trade, and coinage.

Pádraig Ó Néill

References and Further Reading

Greene, David. "Archaic Irish." In *Indogermanisch und Keltisch*. Edited by Karl Horst Schmidt. Wiesbaden: Dr. Ludwig Reichert Verlag, 1977.

McCone, Kim. "The Würzburg and Milan Glosses: Our Earliest Sources of Middle Irish." *Ériu* 36 (1985): 85–106.

———. *Towards a Relative Chronology of Ancient and Medieval Celtic Sound Change*. Maynooth, Ireland: Department of Old Irish, St. Patrick's College, 1996.

McCone, Kim. "Prehistoric, Old and Middle Irish." In *Progress in Medieval Irish Studies*. Edited by Kim McCone and Katharine Simms. Maynooth, Ireland: Department of Old Irish, St. Patrick's College, 1996.

McManus, Damian. "A Chronology of the Latin Loan-Words in Early Irish." *Ériu* 34 (1983): 21–71.

——— *A Guide to Ogam*. Maynooth, Ireland: An Sagart, St. Patrick's College, 1991.

Ó Cuív, Brian. *A View of the Irish Language*. Dublin: Stationery Office, 1969.

O'Rahilly, Cecile. *Ireland and Wales: Their Historical and Literary Relations*. London: Longmans, Green and Co., 1924.

Sommerfelt, Alf. "The Norse Influence on Irish and Scottish Gaelic." In *The Impact of the Scandinavian Invasions on the Celtic-speaking Peoples c. 800–1100 A.D.* Edited by Brian Ó Cuív. Dublin: The Dublin Institute of Advanced Studies, 1975.

Thurneysen, Rudolf. *A Grammar of Old Irish*. Dublin: The Dublin Institute for Advanced Studies, 1946.

See also **Adomnán; Anglo-Saxon Literary Influence; Céli Dé; Classical Literary Influence; Columbanus; Eriugena; Glosses; Hiberno-Latin Literature; Inscriptions; Placenames; Scandinavian Literary Influence; Sedulius Scotus; Tírechán**

LATRINES

See **Castles**

LAW SCHOOLS, LEARNED FAMILIES

It is clear from references in the ninth-century wisdom text *The Triads of Ireland* that the monasteries of Cork, Cloyne, and Slane were centers of legal learning. No precise information has survived regarding the location where individual law texts were written. There is evidence, however, that the main body of law texts—written in the seventh and eighth centuries—came from two main legal traditions, one based in Munster and the other in the northern Midlands and southern Ulster.

The pre-Norman annals contain references to fifteen persons described as *iudex* or *brithem*, "judge," of whom all but four are recorded as having held ecclesiastical office. For example, the Annals of Ulster record the death in 802 of Ailill son of Cormac, abbot of Slane, who is described as *iudex optimus*, "an excellent judge." In 806 the same annals record the death of Connmach, judge of the Uí Briúin of Connacht: he was evidently a layman. The annals of this period provide no clues as to the operation of the law schools. It is clear from the legal manuscripts, however, that the work of interpreting the Old Irish law texts began as early as the ninth century. The earliest practice seems to have been for glosses to be written between the lines of the text, and to consist largely of explanations of words which might be unfamiliar on account of linguistic change or because they belonged to the specialized legal vocabulary. Most law texts were also provided with commentaries, which expand upon the original text. In the earlier legal manuscripts the commentary is generally fitted into the margins of the page, whereas in later manuscripts of the fifteenth and sixteenth centuries the commentary is given a place in the body of the page.

After the Norman invasion of 1169, clerical involvement in Irish law diminished, and the law increasingly became the preserve of laymen from a small number of legal families. From the evidence of the annals and of the surviving legal manuscripts, it is clear that the MacEgan (*Mac Aodhagáin*) family was the most active and influential of these. There are more references to MacEgans than to any other legal family in the annals, and most surviving legal manuscripts have a MacEgan connection. They had schools in Ormond (Co. Tipperary) and at Duniry, Park, and other locations in County Galway. As well as being academic lawyers, the MacEgans were widely involved in legal practice. Between the fourteenth and sixteenth centuries they are known to have acted as lawyers for most of the ruling families of western and central Ireland. Their patrons included old Gaelic families such as Mac Carthy More, O'Connor Roe, and O'Conor Don, as well as Anglo-Norman lords such as Blake, Butler, and Barrett. The most prominent member of the MacEgan family in the surviving documents is Giolla na Naomh Mac Aodhagáin, whose death in battle in 1309 is recorded in the Annals of Connacht, where he is described as "chief legal expert of Connacht and a well-versed general master in every other art." Three works are attributed to Giolla na Naomh. The first is "An address to a student of law," a poem of twenty-five stanzas that summarizes the educational needs of a law student. He stresses the importance of legal precedents as a basis for right judgement, and recommends the careful study of law texts and wisdom texts. Another poem attributed to Giolla na Naomh deals with the law relating to distraint (*athgabál*). The longest surviving text attributed to Giolla na Naomh is a general treatise on Irish law. It is primarily based on the Old Irish law texts and their associated glosses and commentaries. In addition, there is a significant Anglo-Norman element, which illustrates the degree to which Irish law schools had by this period been influenced by English Common Law. Thus the treatise uses terminology of Anglo-Norman origin such as *baránta*, "guarantor," and *fínné*, "jury."

Other prominent legal families were the MacClancies (*Mac Fhlannchadha*) of Munster and the O'Dorans (*Ó Deoráin*) of Leinster. One of the most important of all the surviving legal manuscripts, now called *Egerton 88*, was the product of a minor legal family, the O'Davorens (*Ó Duibhdábhoireann*) of County Clare. This manuscript was compiled by Domhnall O'Davoren and his pupils between 1564 and 1569, and contains a variety of legal material, much of it not preserved elsewhere. The abundant marginal comments are also of great interest, as they provide insight into the life and general atmosphere of a sixteenth-century law school.

The Elizabethan wars, culminating in the Flight of the Earls in 1607, brought about the end of the Irish law schools, as the lords who formerly employed the legal families were dispossessed or adopted English law.

FERGUS KELLY

References and Further Reading

Kelly, Fergus. *A Guide to Early Irish Law* (Early Irish Law Series, vol. 3). Dublin: Dublin Institute for Advanced Studies, 1988, repr. 2001.

———. "Giollana Naomh: a Thirteenth-century Legal Innovator." In *Mysteries and Solutions in Irish Legal History.* Edited by Desmond Greer and Norma Dawson. Dublin: Four Courts Press, 2001.

Ní Dhonnchadha, Máirín. "An Address to a Student of Law." In *Sages, Saints and Storytellers: Celtic Studies in Honour of Professor James Carney* (Maynooth Monographs vol 2). Edited by Donnchadh Ó Corráin, Liam Breatnach, and Kim McCone. Maynooth: An Sagart, 1989.

Simms, Katharine. "The Brehons in Later Medieval Ireland". In *Brehons, Serjeants, and Attorneys: Studies in the History of the Irish Legal Profession.* Edited by Daire Hogan and W. N. Osborough. Dublin: Irish Academic Press, 1990.

See also **Brehon Law; Common Law; Law Texts; Mac Aodhagáin; Wisdom Texts**

LAW TEXTS

Most of our knowledge of early Irish or Brehon law comes from the Old Irish law texts, mainly composed in the seventh and eighth centuries A.D. Some of these texts have survived in a complete form in later manuscripts (generally of the fourteenth to sixteenth centuries), but many are to be found only in fragments.

Senchas Már

The best preserved collection of early Irish law texts is that of the *Senchas Már*, "great tradition," which is likely to have been organized as a unit about A.D. 800. The texts in this collection are all anonymous, and it is not known where or by whom it was put together. However, most of the place-names and personal names cited in the texts relate to the northern Midlands and southern Ulster, so it is probable that the material derived from this area. It may have been assembled in a monastic law school, such as that at Slane, County Meath.

Originally, the *Senchas Már* consisted of about fifty law texts, arranged in three groups. There seems to be no particular logic in the order in which the texts have been placed, though some texts dealing with similar topics are found together. For example, the text on the law relating to cats, *Catshlechta*, is followed by the text on dogs, *Conshlechta*. The First Third (*trian toísech*) of the collection commences with an introduction in which there is a general discussion of the legal topics that are covered, as well as a description of the role Saint Patrick was believed to have played in the codification of Irish law. The second text, *Di Chetharshlicht Athgabálae*, deals at length with distraint (*athgabál*), the formal seizure of another's property to enforce a legal claim against him. It is followed by three fragmentary texts: *Di Gnímaib Gíall* ("On the Acts of Hostages"), *Cáin Íarraith* ("The Law of the Fosterage Fee"), and *Cáin Shóerraith* ("The Law of the Free Fief"). The last of these deals with the institution of free clientship, and is followed by the nearly complete *Cáin Aicillne*, "The Law of Base Clientship." The next text, *Cáin Lánamna*, "The Law of Couples," has survived in its entirety, and is concerned mainly with marriage and divorce. The last text in the First Third is entitled *Córus Bésgnai*, "The Arrangement of Customary Behavior," approximately half of which survives. It discusses the nature of Irish law, the maintenance of order in society, and the relationship between the Church and the laity. It repeats material from the Introduction on the dissolution of contracts, and on Saint Patrick's involvement with Irish law.

The Middle Third (*trian medónach*) is the best preserved of the three sections of the *Senchas Már.* It contains sixteen texts, of which thirteen have been preserved in their entirety; considerable portions of the remaining three texts have also survived. The first text of the Middle Third is entitled *Na Sechtae*, "The Heptads," and is of special value to the student of early Irish law, as it covers a wide range of subjects, arranging the material in groups of seven, for example, the seven churches that may be destroyed with impunity, seven kings who are not entitled to honor-price, and seven women who have sole responsibility for rearing their offspring. The next text, *Bretha Comaithchesa*, "The Judgements of Neighborhood," deals with trespass by domestic animals, fencing obligations, and so on. Two specialized treatments of the law of neighborhood also occur in the Middle Third. These are *Bechbretha*, "Bee-judgements," which includes a discussion of trespass by honey-bees, and *Coibnes Uisci Thairidne*, "Kinship of Conducted Water," which provides rules for bringing water for a mill across a neighbor's land. The final text in the Middle Third is the partially preserved *Bretha im Gata*, "Judgements Concerning Thefts."

The Last Third (*trian déidenach*) is the least complete section of the *Senchas Már*, and there is still a good deal of uncertainty as to its original complement. In his study "On the Original Extent of the *Senchas Már*" Liam Breatnach lists twenty-three texts in the Last Third, and it is probable that the original number was higher. For example, there is evidence that the text on trapping deer, *Osbretha*—of which only a few fragments accompanied by later commentary survive—belonged here. Likewise, the Last Third may have contained *Bretha Luchtaine* and *Bretha Goibnenn*, texts on the law relating to carpenters and blacksmiths, respectively. No material that can be assigned to these texts has so far been identified, but Breatnach provides evidence that the associated *Bretha Creidine*, on the law relating to coppersmiths, belonged in this section. Only three texts belonging to the Last Third are complete. These are the short text on sick-maintenance (*othras*) and the longer medico-legal texts *Bretha Crólige*, "Judgements of Blood-lying," and *Bretha Déin Chécht*, "Judgements of Dían Cécht (a Legendary Physician)."

Other Legal Traditions

Another less clearly defined group of law texts has Munster associations, and includes *Bretha Nemed toísech*, *Bretha Nemed déidenach*, and *Cáin Fhuithirbe.* It seems that the wisdom text *Audacht Morainn* also belongs in this tradition, as it has verbal correspondences with both *Bretha Nemed* texts. Binchy suggested that the text on status *Uraicecht Becc*, "Small Primer," likewise comes from a Munster tradition, as it refers to the preeminence of the king of Munster, as well as to the monasteries of Cork and Emly. Other law texts—such as the invaluable excursus on status *Críth Gablach*, "Branched Purchase"—have no known connection with the *Senchas Már* collection or with the Munster group of texts. Another important text that stands apart from the rest is *Gúbretha Caratniad*, "The False Judgements of Caratnia," which gives fifty-one exceptions to the general principles of early Irish law.

Origin of the Texts

The linguistic evidence indicates that the essential features of the early Irish legal system go back at least as far as the Common Celtic period (*c.* 1000 B.C.). Thus, there are many correspondences between Irish, Welsh, and Breton legal vocabulary. For example, Old Irish *macc*, "surety," is cognate with Old Breton and Medieval Welsh *mach* of the same meaning. Similarly, Old Irish *díles*, "immune from legal process," is cognate with Old Breton *diles* and Medieval Welsh *dilys* of the same meaning. Correspondences of this type indicate that such basic legal concepts were recognized long before the coming of Christianity to Ireland in the fifth century A.D. Nonetheless, there is no doubt that the impact of Christian learning on early Irish law was immense. The introduction of Latin letters revolutionized the transmission of legal material, and allowed for legal topics to be treated in detail, whereas previously only the salient points could be passed on by word of mouth.

There is strong evidence that the law texts were written in monastic scriptoria, as the legal manuscripts use the same spelling system, script, punctuation, abbreviations, and illuminated capitals as are found in manuscripts of monastic origin. In addition, many of the law texts show the influence of the Latin grammarians in their use of the question-and-answer technique, and of etymological explanations of legal terms and other words. There are also strong Christian influences to be observed in the content of the law texts. In the text on status *Críth Gablach*, it is stated the king should rise up before the bishop "on account of the Faith," and many other texts make special reference to the privileged position in society of the Church and its clergy. There are also frequent references to Biblical principles and personalities, and some direct quotations and adaptations from Canon law. On the basis of this evidence, some scholars have held that most or all of the authors of the law texts were clerics. On the other hand, doubt has been expressed that clerics were responsible for law texts such as *Cáin Lánamna* and *Bretha Crólige*, in which concubinage and divorce have explicit legal status.

Style and Content of the Law Texts

The style employed in the law texts varies considerably. The majority of them are in prose, but some—particularly those associated with the Munster tradition—are largely in verse. The manner in which the information is presented is similarly variable. Texts such as the Heptads and Gúbretha Caratniad cover a wide range of legal issues, but most deal with a single topic, often quite specialized. Thus, the long text Bretha im Fhuillema Gell deals solely with the interest payable for pledges given by a person on behalf of another. The technical information present in such detailed treatments renders them of great interest to the social and economic historian, as well as to the student of law. For example, the medico-legal texts Bretha Crólige and Bretha Déin Chécht supply a great deal of information on early Irish medical practice. In general, it can be said that the authors of the law texts display an intelligent and humane attitude towards legal problems, and a deep concern that justice should be done. However, the hierarchical and inegalitarian nature of early Irish society is reflected throughout these documents. Disappointingly, there is hardly any case law, so it is difficult to know how the principles of Irish law were applied in practice. From the ninth century, it seems that very few further law texts were composed, and thereafter the energies of the law schools were mainly devoted to the work of copying and interpreting the existing texts through the provision of explanatory glosses and commentaries.

FERGUS KELLY

References and Further Reading

Binchy, D. A. *Críth Gablach* (Mediaeval and Modern Irish Series, vol. 11). Dublin: Dublin Institute for Advanced Studies, 1941, repr. 1970.

Breatnach, Liam. "Canon Law and Secular Law in Early Ireland: the Significance of *Bretha Nemed*." *Peritia: Journal of the Medieval Academy of Ireland* 3 (1984): 439–459.

———. "On the Original Extent of the *Senchas Már*." *Ériu* 47 (1996): 1–43.

Charles-Edwards, Thomas, and Fergus Kelly. *Bechbretha: an Old Irish Law-tract on Bee-keeping* (Early Irish Law Series, vol. 1). Dublin: Dublin Institute for Advanced Studies, 1983.

Hancock, H. N. et al. *Ancient Laws of Ireland* vols. 1–6. Dublin: Her Majesty's Stationary Office, 1865–1901.

Kelly, Fergus. *A Guide to Early Irish Law* (Early Irish Law Series, vol. 3). Dublin: Dublin Institute for Advanced Studies, 1988, repr. 2001.

———. "Texts and Transmissions: The Law-texts." In *Ireland and Europe in the Early Middle Ages*. Edited by Próinséas Ní Chatháin and Michael Richter. Dublin: Four Courts Press, 2002.

Ó Corráin, Donnchadh, Liam Breatnach, and Aidan Breen. "The Laws of the Irish." *Peritia: Journal of the Medieval Academy of Ireland* 3 (1984): 382–438.

See also **Brehon Law; Canon Law; Common Law; Ecclesiastical Organization; Law Schools, Learned Families; Medicine; Patrick; Scriptoria; Society, Grades of Gaelic; Wisdom Texts**

LEABHAR BREAC

Leabhar Breac (*The Speckled Book*) is in the library of the Royal Irish Academy, Dublin (Cat. No. 1230). The vellum manuscript has always been associated with the learned family of Mac Aodhagáin. It was also known as *Leabhar Mór Dúna Doighre*, as it was in the possession of a branch of that family who lived in Duniry near Portumna in the sixteenth century. The manuscript was compiled in the early fifteenth century, before 1411, from sources in the midlands bordering the river Shannon. Lorrha, in Co. Tipperary, Clonsost, in County Offaly, and Clonmacnoise are named by the scribe as places where he copied texts. This beautifully written, well preserved book is almost certainly the work of Murchadh Ó Cuindlis, a professional scribe who also worked for the Mac Firbhisigh family and whose hand has been identified in the *Yellow Book of Lecan* and the *Book of Lecan*.

The *Leabhar Breac* consists almost entirely of religious texts except for "Cormac's Glossary" (compiled by or for Cormac mac Cuilennáin) and the "Histories of Philip of Macedon and His Son Alexander the Great." One section is comprised of stories paraphrased from the Bible, combined with legends and poems such as the "Lament of the Mothers of Bethlehem" and a version of the "Legend of the True Cross." Another large portion has accounts of the sufferings of Christ, the apostles, and the martyrs. There are lives of St. Patrick, St. Brigit, St. Colum Cille, and St. Martin in the form of homilies and a version of the "Martyrology of Oengus." One of the most important texts is the witty "Vision of Mac Conglinne," a comical satire of monastic life and scholarship; the only complete copy of it survives in *Leabhar Breac*. The manuscript is written almost entirely in Irish with occasional passages in Latin.

The manuscript is one of the largest in terms of its page size (40.5 cm × 28 cm) to survive from the period and is also unusual in that it was written throughout by one scribe. Among the marginal jottings are notes that give useful information about the length of time it took to write certain sections. It has been calculated by Tomás Ó Concheannain that Ó Cuindlis wrote the thirty-five pages (pp. 141–175) in about six weeks, hence roughly one double column page per day. A note mentions that he only managed to copy a single column another day, but this was a complicated transcription of a fifty-two–line poem with interlinear glosses. The scribe mentions incidental details of daily life, such as the wonderful singing of a robin and the straying of the cat; in another part, the author describes warfare in the area as Lorrha is plundered by local magnate Murchad Ua Madagáin. But in common with scribes in every age it is the weather, particularly the cold, that is noted most frequently. On page 17, he mentions a snowfall on the first of March; later he remarks on the coincidence of his writing a homily on St. Patrick on the eve of his feast day (March 17), while the cold weather was again a problem some days later: "twenty nights from today till Easter Monday, and I am cold and weary without fire or covering" (Ó Longáin 33).

The decoration in the *Leabhar Breac* is confined to a series of colored capitals introducing various sections. In style these closely resemble the ribbon and wire type initials found in twelfth century Irish manuscripts and which the scribe collected most probably from different exemplars on his travels. In addition there is an unusual and large drawing of the Menorah candelabrium illustrating the "Story of the Children of Israel" and a drawing of the crucifixion, which, stylistically, is contemporary with the manuscript.

According to marginal jottings, the Mac Aodhagáin family of Duniry, County Galway, had the manuscript in their collection in the second half of the sixteenth century until 1595 at least. In 1629, it was in the nearby Franciscan friary of Kinalehin when Br Micheál Ó Cléirigh copied a saint's life from it, and the book remained in the area until the end of the seventeenth century. In the eighteenth century it was in the possession of the family of Conchur Ó Dálaigh (O'Daly) near Mitchelstown, County Cork, who loaned it to Bishop John O'Brien of Cloyne for use in the compilation of his Irish Dictionary. In 1789, the O'Dalys sold the manuscript to the Royal Irish Academy for £3. 13s 8d. In 1876, the Academy published a lithographic facsimile of *Leabhar Breac* based on the transcript of Joseph O'Longan. The manuscript was restored and rebound in 1973, and from 2003 a digitized copy may be viewed online.

TIMOTHY O'NEILL

References and Further Reading

O Concheanainn, Tomas. "The scribe of the Leabhar Breac" *Ériu*, 24 (1973), 64-79.

Ó Longáin, Joseph. *Leabhar Breac. A Lithographic Facsimile by Joseph Ó Longáin*. Edited by J. J. Gilbert. Dublin: Royal Irish Academy, 1876.

See also **Devotional and Liturgical Literature; Hagiography and Martyrologies; Mac Aodhagáin; Manuscript Illumination; Scriptoria**

LEBOR NA HUIDRE

Lebor na hUidre (The Book of the Dun Cow) is the earliest extant vernacular Irish manuscript. The fragmentary nature of seventeen of its thirty-seven texts indicates that it has not come down to us in its complete form, its surviving sixty-seven folios representing approximately half the original codex, according to Tomás Ó Concheanainn. Notwithstanding this, it constitutes a veritable treasure trove of Old and Middle Irish literature, both secular and religious, although the original order in which the texts appeared can no longer be determined. A copy of the longest medieval Irish narrative, *Táin Bó Cúailnge* ("The Cattle Raid of Cúailnge"), is found among its leaves, together with two of its *remscéla* (fore-tales). That tale's premier hero, Cú Chulainn, also features in other compositions therein, including those describing his birth and resurrection. The activities of the latter's Ulaid colleagues are similarly recounted, most notably in *Fled Bricrenn* ("Bricriu's Feast") and *Mesca Ulad* ("The Intoxication of the Ulaid"), as are those of royal personages. *Togail Bruidne Da Derga* ("The Destruction of Da Derga's Hostel") provides a literary biography of the prehistoric king Conaire Mór; other narratives focus on pivotal events in a particular monarch's reign. Among these is the otherworld journey of fair Connlae, son of King Conn Cétchathach, which forms one of a group of texts that emphasizes the supernatural in all its guises. Its Christian dimension is highlighted in a story relating the prophetic revelation of another of Conn's sons, Art, which finds thematic resonance in *Comthoth Láegaire co cretim* ("The Conversion of Láegaire to the Faith"). These are complemented by religious texts including *Dá Brón Flatha Nime* ("The Two Sorrows of the Kingdom of Heaven") and the homiletic tracts, *Scéla Laí Brátha* ("The Tidings of Doomsday") and *Scéla na hEsérgi* ("The Tidings of the Resurrection"), for which the manuscript constitutes our sole witness.

Scribes

This pair of homilies, along with a handful of other texts, are in the hand of the latest of the trio of scribes connected with the codex whose homiletic interest has earned him the designation "H." A thorough reviser, H inserted his new material either in sections of the manuscript that he had previously erased or on leaves intercalated precisely for that purpose. His work is also to be detected in the many interpolations in texts originally written by Scribe A, who began the manuscript, or by his more prolific successor, M, so named because of his identification with Máel-Muire, the author of the codex's two *probationes pennae*. Since this man can be located in time and place as the Máel-Muire mac Célechair who was killed by marauders in Clonmacnoise in 1106, the approximate date and provenance of the compilation were long considered secure. An analysis of the pen trials in question, however, has led Ó Concheanainn to posit that they were in fact written by H, whose language does not appear to be significantly later than that of the original scribes. Certain linguistic features do indeed suggest that the interpolator could well have been active about the turn of the twelfth century, but the scanty palaeographical evidence is difficult to evaluate, as Ó Concheanainn admits. In any event, it seems that the principal scribe and prodigious editor, one of whom was Máel-Muire, may even have been contemporaries, the reviser remolding the manuscript considerably, in accordance with both his own scholarly tastes and the various recensions of texts he himself had to hand.

Provenance

In the case of two thematically related tracts, *Aided Nath Í* ("The Death of Nath Í") and *Senchas na Relec* ("Burial Ground Lore"), H in fact provides an indication of their ultimate origin. In *Lebor na hUidre*'s sole colophon, he attributes their compilation to the eleventh-century scholars, Flann Mainistrech and Eochaid úa Cerín, who drew on a range of manuscripts both in Armagh and in Flann's Louth monastery of Monasterboice. Moreover, both H and the manuscript's main scribe cite *Cín Dromma Snechta* (*The Book of Drumsnat*) as a source in the case of four tales; six further tales thought to have been contained in this lost eighth-century manuscript are also preserved in our Book. As far as much of its base material is concerned, therefore, *Lebor na hUidre*'s associations are not with Máel-Muire's home monastery of Clonmacnoise but with the southeast Ulster/northeast Leinster area where the scribe's family originated. Furthermore, a similar geographical focus can be detected in many of the noninterpolated narratives, as Ó Concheanainn has shown. Accordingly, the likelihood is that the codex first took form at some distance from the monastic community whose patron's dun cow was later to give it his name. It is tempting to speculate, with Ó Concheanainn, that it was at Clonmacnoise that H undertook his dramatic alterations, in whose library he would have had access to alternative texts, though it must be admitted that his additional material also displays a northeastern bias in part. Notwithstanding this, our earliest record of the Book places it in Sligo in the mid-fourteenth century, having been acquired by the Uí Chonchobhair as ransom from the Uí Dhomhnaill of Donegal, who repossessed the work a century later. It was still in Ulster more than one hundred and fifty years after that, as Mícheál Ó Cléirigh's transcription of *Fís Adamnáin* ("Adamnán's Vision") from it in 1628 attests. Nonetheless, its sojourn in Connacht was a significant one, close textual connections between it and manuscripts of western provenance indicating that it was extensively drawn on by a range of scribes. Moreover, its influence can be detected at an earlier period also if the twelfth-century redactors of the *Book of Leinster* employed H's texts as exemplars, as

Ó Concheanainn has claimed. In truth, however, the early history of the manuscript is hazy and the textual evidence from which it must be constructed both complex and contested. Nor are we afforded more than brief glimpses of its fate in the later medieval period. It was in the nineteenth century that it finally came to rest in the Royal Irish Academy where it is still housed today.

MÁIRE NÍ MHAONAIGH

References and Further Reading

Best, R. I. "Notes on the Script of Lebor na hUidre." *Ériu* 6 (1912): 161–74.

Best, R. I. and Osborn Bergin, eds. *Lebor na hUidre: Book of the Dun Cow.* Dublin: Royal Irish Academy, 1929.

Ó Concheanainn, Tomás. "The Reviser of Leabhar na hUidhre." *Éigse: A Journal of Irish Studies* 15 (1973–1974): 277–288.

Ó Concheanainn, Tomás. "LL and the Date of the Reviser of LU." *Éigse: A Journal of Irish Studies* 20 (1984): 212–225.

Ó Concheanainn, Tomás. "Textual and Historical Associations of Leabhar na hUidhre." *Éigse: A Journal of Irish Studies* 29 (1996): 65–120.

Ó Concheanainn, Tomás. "Leabhar na hUidhre: Further Textual Associations." *Éigse: A Journal of Irish Studies* 30 (1997): 27–91.

Oskamp, H.P.A. "Notes on the History of Lebor na hUidre." *Proceedings of the Royal Irish Academy* 65 C (1966–1967): 117–137.

See also **Glendalough, Book of; Leinster, Book of; Manuscript Illumination; Scriptoria**

LECAN, BOOK OF

One of the great codices of late medieval Irish learning produced by the learned family of Mac Fhir Bhisigh in north Connacht, its full title is the *Great Book of Mac Fhir Bhisigh of Leacán*. It was compiled by Giolla Íosa (son of Donnchadh Mór) Mac Fhir Bhisigh, who was also the principal scribe—almost 260 of the manuscript's surviving 311 folios are in his hand. (A small number of the book's original vellum folios—perhaps a dozen or so—have been lost over the centuries.) Giolla Íosa informs us that he was writing the book "for himself and for his son after him," while three colophons pinpoint the time of writing—two refer to "the autumn Mac Donnchaid was killed" and a third to "the winter after Mac Donnchaid['s death]." Scholars have interpreted this information differently. Eugene O'Curry thought it indicated the year 1417, while Paul Walsh suggested that it reflected the death in 1416 of "Mac Donnchaid . . . chief of Tirerrill, in the present county Sligo." But, as Tomás Ó Concheanainn has pointed out, the only death of a MacDonagh chieftain that suits the context is that of Tomaltach mac Taidhg, king of Corann and Tír Oilealla since 1383, who was slain in a dispute in north Connacht in mid-August (early autumn in the medieval Irish view), 1397. (It was in this Tomaltach's house in Ballymote that part of the codex known as the *Book of Ballymote* was written circa 1391; Tomaltach, incidentally, was a second cousin of Giolla Íosa's wife, Caithirfhíona.) The suggestion that the manuscript was being written by 1397 is corroborated by some of the terminal dates in the valuable corpus of genealogies preserved in the *Book of Lecan*; these indicate that work was in progress during the period from 1397 to 1403.

Giolla Íosa's earliest assistant in writing the manuscript was Murchadh Ó Cuinnlis, apparently a native of east Galway. Evidently a pupil or apprentice of Giolla Íosa's—he refers to him as his *aidi* (master or teacher)—he appears to have left the Clann Fhir Bhisigh school at Lackan, County Sligo (whence the name of the book), by 1398, for from 1398 to 1399 he was in present-day County Tipperary, penning "an excellent manuscript" (Ó Concheanainn's description) that is now part of the composite volume, the *Yellow Book of Lecan*. A decade later—as Ó Concheanainn has shown—he was at work on "the largest Irish vellum manuscript by one scribe," the compendium of medieval Irish ecclesiastical material known as the *Leabhar Breac*.

In 1418, a later scribal assistant, Ádhamh Ó Cuirnín, penned some 23 folios of the manuscript for Giolla Íosa. (This same scribe has been recognized in recent times as having also written, circa 1425, a manuscript in the National Library of Scotland—known as the "Broad Book" of John Beaton, from its owner in 1700.) A third scribal assistant is unnamed, but he has been convincingly identified—on the basis of strong circumstantial evidence—as Giolla Íosa's (only?) son, Tomás Cam. Among the items he penned is the lengthy poem (of some 900 lines)—replete with genealogical and topographical detail—that his father composed as an inauguration ode for the local chieftain Tadhg Riabhach Ó Dubhda, who succeeded his brother, Domhnall, early in 1417. The poem contains a great deal of genealogical and topographical detail that mirrors that found in the fascinating prose survey of much of Counties Mayo and Sligo, which is preserved in the *Book of Lecan*, and is very probably also the work of Giolla Íosa. It may be noted that the author himself makes a number of textual interventions—in what seems a somewhat infirm hand—throughout the poem.

Unlike the other great Clann Fhir Bhisigh manuscript, the principal component (which I have dubbed *Leabhar Giolla Íosa*) of the so-called *Yellow Book of Lecan*, whose contents are almost wholly literary, most of the contents of the *Book of Lecan* have a historical or quasi-historical slant. The volume opens (fols. 1–13v, 16v–21v) with two slightly differing copies of the B-version of *Lebar Gabála Érenn* (*The Book of the Taking of Ireland*)—alternatively styled "Redaction 2" and *Míniugad*—and it closes (264–311) with the C-version—or "Redaction 3"—of the same work. The genealogies of the saints of Ireland (34–51) are followed (53–138v) by an important recension of the medieval Irish secular genealogies, with some related miscellaneous materials scattered throughout the volume (176–183, 213v, 215–229v). (Broadly similar versions of both the saints' and secular genealogies may be found in the other great north Connacht codex, the *Book of Ballymote*.) Other important texts in the book include *Sex Aetates Mundi* (22–26), the *Lebor Bretnach* (an Irish version of the *Historia Brittonum* by Nennius; 139–145), *Auraicept na nÉces* ("The Poets' Primer;" 151–162v), *Cóir Anmann* ("The Fitness of Names;" 173–175), *Lebor na Cert* ("The Book of Rights;" 194–202v), the *Banshenchus* ("History and Genealogies of Famous Women;" 203–212), and Version C of the *Dinnsenchas* ("Lore of Famous Places;" 231–263v). In addition, there are numerous shorter prose texts as well as many poems—about twenty of 20 quatrains and upwards (one running to 305 qq, another to 181). (Except in the genealogical portion of the manuscript, most pages are laid out in double columns of 51 lines each.)

The *Book of Lecan* seems to have remained in the hands of Clann Fhir Bhisigh until the early 17th century, but by October 1612 had come into the hands of Henry Perse, secretary to the Lord Deputy, Sir Arthur Chichester. It later passed into the possession of the scholarly James Ussher, Protestant archbishop of Armagh, from whose library in Drogheda it was lent in 1636 to Conall Mageoghegan of Lismoyny, County Westmeath, translator of the *Annals of Clonmacnoise*. Around this time it was also drawn upon as a source by Brother Michéal Ó Cléirigh, and it may have been in north Tipperary for a period after that. In 1640, Ussher left Ireland for England, never to return, and his library followed some time later. On his death in 1656, his daughter offered his valuable collection of books and manuscripts for sale. The Cromwellian government, wishing to prevent it from going to foreign purchasers, decided that it should return to Ireland, to form the nucleus of the "second college" being planned for Dublin. The latter proved abortive and, following the Restoration, Charles II bestowed the collection on Trinity College. There it was consulted in 1665 by none other than Dubhaltach Mac Fhir Bhisigh, kinsman of the compiler. It remained in TCD until the outbreak of the Williamite War in 1688. By 1702, it was noted as being missing from Trinity College and the following year it turned up in France. Some time subsequent to that it came into the possession of the Irish College in Paris. In 1787, through the intercession of Colonel Charles Vallancey, the rector of the Irish College, Abbé Charles Kearney, presented it to the newly founded Royal Irish Academy, Dublin, where it has remained ever since. (A small portion—nine folios, 142–150—has been part of another manuscript—1319 or H.2.17—in Trinity College, Dublin, since 1688.) In 1937, the Irish Manuscripts' Commission issued a facsimile edition of the manuscript, with a detailed introduction by Kathleen Mulchrone.

NOLLAIG Ó MURAÍLE, IRISH AND CELTIC STUDIES,
QUEEN'S UNIVERSITY BELFAST

References and Further Reading

The Book of Lecan: Leabhar Mór Mhic Fhir Bhisigh Leacain. Collotype facsimile with introduction and indexes by Kathleen Mulchrone. The Irish Manuscripts Commission, Dublin, 1937.

MacSwiney of Mashanaglass, Marquis. "Notes on the history of the Book of Lecan." *Proceedings of the Royal Irish Academy* 38 C (1928), 31–50.

Tomás Ó Concheanainn. "Scríobhaithe Leacáin Mhic Fhir Bhisigh." *Celtica* 19 (1987), 141–75.

———. "*Lebar Gabála* in the Book of Lecan." In *"A Miracle of Learning" – Studies in Manuscripts and Irish Learning: Essays in honour of William O'Sullivan*, edited by Toby Barnard, Dáibhí Ó Cróinín, and Katharine Simms, 68–90. Aldershot, etc.: Ashgate Publishing, 1998.

———. "A medieval Irish historiographer: Giolla Íosa Mac Fhir Bhisigh", in *Seanchas: Studies in Early and Medieval Irish Archaeology, History and Literature in Honour of Francis J. Byrne* (ed. Alfred P. Smyth), pp 387-95. Four Courts Press, Dublin, 2000.

O'Donovan, John, ed. *The Genealogies, Tribes and Customs of Hy-Fiachrach, commonly called O'Dowda's Country . . . from the Book of Lecan, . . . and from the Genealogical Manuscript of Duald Mac Firbis . . .* Dublin: Irish Archaeological Society, 1844.

Ó Muraíle, Nollaig. *The Celebrated Antiquary: Dubhaltach Mac Fhirbhisigh (c. 1600–1671), His Lineage, Life and Learning.* Maynooth: An Sagart, 1996; revised reprint, 2002.

———, ed. *Leabhar Mór na nGenealach: The Great Book of Irish Genealogies, compiled (1645–66) by Dubhaltach Mac Fhirbhisigh.* 5 vols. Dublin: De Búrca, 2003–2004.

Walsh, Paul. "The great Book of Lecan." In *Irish Men of Learning: Studies by Father Paul Walsh*, edited by Colm O Lochlainn), 102–88. Dublin: Three Candles Press, 1947.

———. "The Book of Lecan." *Journal of Galway Archaeological and Historical Society* 18 (1939), 94–5 (Reprinted in *Irish Leaders and Learning Through the Ages: Paul Walsh—Essays collected, edited and introduced by Nollaig Ó Muraíle*, 499–501. Dublin: Four Courts Press, 2003.

LECAN, YELLOW BOOK OF

See **Genealogy; Mac Firbhisigh**

LEINSTER

Leinster, or *Cóiced Laigen* (the Fifth of the Laigin), is one of the ancient provinces of Ireland. The dominant inhabitants were the Laigin, who believed they came from Gaul in prehistoric times. There were links across the Irish Sea in the early centuries C.E., for the Lleyn peninsula in Wales takes its name from the Laigin. Traditions preserved in historical tales and the annals show that the Laigin controlled a vast territory before the sixth century, including much of Brega and Mide; this land was ultimately lost to the Uí Néill, but there were Laigin kings of Tara before the reign of Niall Noígiallach. By the seventh century, Leinster's boundaries extended from the valley of the Liffey westwards to the Slieve Bloom Mountains, then southward around the highlands of Osraige and down the Barrow valley to the sea. It was divided into north Leinster, *Laigin Tuathgabair*, and south Leinster, *Laigin Desgabair*, and kings in each area enjoyed a considerable degree of independence. The main settlement was in the valleys of the Liffey, Barrow, and Slaney, and in the plains of Wexford and Kildare. In the latter area is the hillfort of Dún Ailinne (Knockaulin), a site comparable to Tara or Emain Macha, and an important centre of the early Leinster kingship.

From the seventh century to the twelfth, the leading dynasties were Uí Dúnlainge in the north and Uí Chennselaig in the south, but other Laigin groups had been paramount in Leinster beforehand. Uí Garrchon were settled in the Liffey valley, and at least two of their kings of Leinster fought against the Uí Neill around the end of the fifth century. Uí Enechglaiss also provided early provincial kings. These dynasties suffered as a result of Uí Neill expansion and were ousted from their fertile lands by Uí Dúnlainge, resettling east of the Wicklow Mountains. Uí Dúnlainge then strove for power with the Uí Máil. A few Uí Máil dynasts succeeded to the provincial kingship, the last being Cellach Cualann (d. 715). Subsequently, Uí Máil were deprived of both the Leinster kingship and the fertile lowlands by Uí Dúnlainge. Also significant were Uí Failge, who occupied the boggy lands at the headwaters of the Barrow. They had previously ruled a larger territory, but also suffered from Uí Néill encroachments. The Loígis were settled southeast of Slieve Bloom and were totally unrelated to the Laigin. They had the status of favored vassals of the Leinster kings, principally for their role as defenders of this border area. A further non-Laigin people were the Fothairt, who were scattered throughout Leinster, and to whom St. Brigit belonged.

In south Leinster Uí Chennselaig were dominant by the eighth century, but here too were earlier Laigin dynasties. Uí Bairrche occupied lands in Carlow, but were divided by Uí Chennselaig expansion so that one branch remained in the middle reaches of the Barrow valley, and another moved southwards to the Wexford coast. Also in south Leinster were groups of Fothairt, associated with Uí Bairrche, and a people called the Benntraige.

After 738, Uí Dúnlainge excluded all other peoples from the Leinster kingship. Their ascendancy was gradually eroded by the Uí Néill, who regularly tried to gain the submission of Leinster's kings. The depredations of Viking incursions also had a destabilizing effect on the Uí Dúnlainge hegemony; its decline was accelerated by the interference of the kings of Munster and Osraige. The establishment of the Viking settlement at Dublin in 841 was of undoubted significance, creating new maritime links, and providing a center of wealth on Leinster's doorstep. By the eleventh century, Dublin and its hinterland, Fine Gall, were closely linked to Leinster, and Irish kings strove for dominance over the rulers of the city and control of its resources. Meanwhile, Leinster had become a significant factor in the struggle for high kingship between Uí Néill and the Dál Cais. Domination by their kings and the kings of Osraige fatally undermined Uí Dúnlainge authority, and in 1042 Diarmait mac Máele-na-mBó of Uí Chennselaig took the Leinster kingship. Though ultimately unsuccessful in his challenge for the high kingship, he achieved more than any previous king of Leinster. Control of Dublin had been a key factor in his success, and this lesson was not lost on his contemporaries. His descendants, the Meic Murchada (Mac Murroughs) retained the provincial kingship and played an important role in later interprovincial struggles. A pivotal role was played by Diarmait Mac Murchada, whose expulsion from and return to Ireland led to the Anglo-Norman Invasion. Subsequently, a considerable number of colonists entered Leinster, notably the de Clares and Fitzgeralds, and established lordships in Kildare, Carlow, and Wexford.

The Mac Murroughs retained a degree of power in south Leinster, though hemmed in by the English. This changed with the career of Art Mór Mac Murchada Caomhánach from the 1370s to 1416, who created a solid kingdom in Carlow and northern Wexford, and Mac Murroughs succeeded to the title "king of Leinster" down to the sixteenth century. Uí Dúnlainge, represented principally by the families of Ua Broin (O'Byrne) and Ua Tuathail (O'Toole) were driven by the invaders from their lands into the Wicklow Mountains, thus suffering the fate they had inflicted on Uí Garrchon and Uí Máil centuries earlier. However, both families were able to create enduring Gaelic lordships in the highland fastnesses. The O'Tooles were often enemies of the English crown, but ultimately in 1541 Tairrdelbach Ua Tuathail submitted to Henry VIII of England. The O'Byrnes had lands in the southern Wicklow Mountains and in *Críoch Branach* (O'Byrne Territory) on the coast. They too fought against the English and other Gaelic lords, but by the later sixteenth century the coastal areas were largely in the orbit of the colonial administration. One branch of the family, the lords of *Críoch Raghnall* (Raghnall's Territory) in the Wicklow Mountains, resisted the English to the end of the sixteenth century, notably in the person of Fiach Ua Brain, whose death in 1597 heralded the end of Gaelic Leinster.

MARK ZUMBUHL

References and Further Reading

Byrne, Francis J. *Irish Kings and High-kings.* London: B.T. Batsford, 1973.

O'Byrne, Emmett. *War, Politics and the Irish of Leinster, 1156-1606.* Dublin: Four Courts Press, 2003.

Smyth, Alfred P. *Celtic Leinster. Towards an Historical Geography of Early Irish Civilization* A.D. *500-1600.* Dublin: Irish Academic Press, 1982.

See also **Anglo-Norman invasion; Diarmait mac máele-na-mbó; Dublin; historical tales; Laigin; MacMurchada, Diarmait; MacMurchada (MacMurrough) family; Mide; Uí Chennselaig; Uí Dúnlainge; Uí Néill; Viking incursions**

LEINSTER, BOOK OF

History

The *Book of Leinster* is one of the foremost manuscripts of the twelfth century following *Lebor na hUidhre*. Its modern name derives from its large content of Leinster-based texts, genealogies, and saga material, but the title *Leabhar Laighneach* is usually reserved for the collection of Leinster genealogies. It is now housed in the library at Trinity College, Dublin, Ireland, with the shelf number H 2 18, 1339. It has mistakenly been called *The Book of Glendalough*, but it is recognized as *Leabhar na Nuachongbála* (the book of the New Foundation), identified as the townland *Nuachongbáil*, Oughavall, near Stradbally in County Laois.

The territory had belonged to the Uí Chrimthainn, a member of whose family was the principal scribe of the manuscript. The patron may have been Diarmait Mac Murchada, who had one of his strongholds in *Dún Másc* close to *An Nuachongbáil*. In the twelfth century, the Uí Chrimthainn had become an ecclesiastical family and the land passed into the control of the Uí Mhórda (O'Moores). *Dún Másc* passed from Diarmait to his daughter's husband, Strongbow. From him the land passed to his daughter Isabel and then as part of her dowry to the Marshal Earls of Pembroke and to their descendants. Meiler fitz Henry exchanged some of his land in Kildare for property in Laois and reestablished an Augustinian monastery, and gave the monastery all the churches on his estate. Oughavall passed into the ownership of the priory.

The manuscript reappears in the fourteenth century at Oughavall and may have been kept in the vicarage meanwhile. There is an entry in the margin that says the book was in the possession of Calbach Ó Mórdha in 1583 and that it is on loan to Seán Ó Ceirín. It is noted elsewhere that it was in the possession of Ruaidhrí Ó Mórdha, Calbach's son, at a later date. There are two further connections with the Ó Mórdha family, a panygeric to the Clann Domnaill that included the Ó Mórdha family and a faded note referring to Conall son of David Ó Mórdha restoring the castle at *Dún Másc*.

If the manuscript was held at the vicarage, this would explain the later Anglo-Norman additions by scribes who used Latin and English script. These date from the early fourteenth century to those of poems written in the fifteenth century and include Pope Adrian's *Laudabiliter*, the papal bull that sanctioned the Anglo-Norman invasion two hundred years before.

W. O'Sullivan dates the foliation and rebinding to this period and maintains that the newly found manuscript was a highly prized possession. The castle at *Dún Másc* was burnt in 1324 and rebuilt by the Ó Mórdha family, and there is a praise poem to a Melaghlin Ó Mórdha, who died in 1502, on one of the pages.

It was loaned to various scholars, and when the binding disintegrated, separate parts were borrowed by antiquarians. The Franciscans had a section in their church in Donegal, but the Ó Mórdha family kept the main manuscript, which they took with them to Ballyna in County Kildare when they lost their lands in County Laois. Sir James Ware made a note of its existence there. In 1700, the Welsh archaeologist Edward Lhuyd bought the manuscript during a tour of Ireland. He collected many Irish manuscripts, including the *Yellow Book of Lecan*, and he made a habit of binding them together badly. But he did little to interfere with the *Book of Leinster*, apart from making notes on the foliation.

The book was bought by Sir Thomas Saunders Sebright after Lhuyd's death, and it was presented to Trinity College in 1782 by Sebright's son; it eventually reached the College in 1786. There was no effort to collate or bind the manuscript until 1841, when O'Curry was given the task of providing an index and rearranging the leaves. At this time it was boxed for the first time and referred to as H.2.18.

O'Sullivan says that the material is carefully divided into different sections, but that there are certain irregular sections, leading him to opine that this was not intended as a single manuscript. It now contains 410 folios, 310 in the first 5 volumes of the diplomatic edition by R. I. Best, O. Bergin, and M. A. O'Brien; the Anglo-Norman section is published separately by A. O'Sullivan as volume 6. Finally, the Ó Longáin family produced a lithographic copy in 1880.

Content

The manuscript contains an extensive collection of seminal texts, including *An Leabhar Gabhála* (a large collection of genealogies), *Cogad Gáedel re Gallaib*, *Sanas Chormaic*, *Tecosca Chormaic*, the metrical *Banshenchas*, and the metrical *Dinnshenchas*. There are over one hundred prose texts, including many famous Heroic Cycle sagas: *Táin Bó Cúailnge*, *Scéla Muicce meic Dathó*, *Aided Cheltchair*, *Aided Chonchobair*, *Aided Meidbe*, *Do Fallsigud Tána Bó Cúailnge*, *Loinges Mac nUislenn*, *Mesca Ulad*, *Scéla Chonchobair*, *Talland Étair*, *Brislech Mór Maige Muirthemne*, *Táin Bó Flidais*, and *Táin Bó Fraích*. There are some specifically Leinster sagas such as *Fingal Rónáin*, *Orgain Dind Ríg*, *Esnada Tige Buchet*, and the *Bóroma*. The translation of *Togail Traí* appears along with *Cath Maige Muccrama* and the wisdom text *Audacht Morainn*. The large collection of metrical material includes many that refer specifically to Leinster—*Fianna bátar in Emain*, *Cúiced Lagen na Lecht Ríg*, and *Temair Breg*—but others are general historical poems such as *Hériu ardinis na ríg*, *Can a mbunadas na nGael*, and the *Banshenchas*. Flann Mainistrech has thirteen poems in the book, by far the largest collection. Other poets, such as Cináed Ua hArtacáin, Dallán mac Móre, and Gilla-Coemáin, have only three poems. There are also poems relating to the Fenian cycle, including *Oenach indiu luid in rí*.

The material is not purely secular, however. The ecclesiastical content includes the genealogies of the saints, along with lists of Irish bishops; the mothers, sisters, and daughters of Irish saints; the martyrology of Tallaght; and a collection of stories about Mo-Ling. Some scholars have noted the untrustworthy nature of some of the texts, particularly E. J. Gwynn and T. Ó Concheanainn in reference to the *Dinnshenchas*.

Scribes and Decoration

One scribe claims the manuscript as his own, signing his name on page 313 as *Áed mac meic Crimthaind ro scríb in leborso 7 ra thinóil a llebraib imdaib* (Áed son of the son of Crimthainn wrote this book and he collected it from many volumes).

Further evidence for the involvement of Áed mac Crimthainn as a scribe is found in the short letter written to him on the margins of fol. 206 by Finn bishop of Kildare. He describes Áed as the "foremost historian of Leinster for his wisdom and learning and knowledge of books and intelligence and scholarship and let the end of this little story be written for me . . . "

This Finn has been recognized as Finn Ua Gormáin of Kildare, who died in 1160 and was himself a poet, but E. Bhreathnach identified a Finn Ua Cíanáin as a possible candidate, and his collaboration in the manuscript might explain the inclusion of the poem, *Clanna Ralge Ruis in Ríg*, that praises the Uí Fáilge above all other families in Leinster.

Áed makes two historical observations. On page 49, he mentions the death of Domnall son of Congalach Ua Conchobhair-Fáilge in 1161, and he also records his name and manner of his death in the list of Uí Fáilge kings on page 40 d 38. Secondly, he refers to the banishment of Diarmait Mac Murchadha in the year 1166. The writing of the text did not begin until after 1151, the year of the Battle of Móin Mór, which is mentioned by Bishop Finn in the additions that he makes to the poem of Cináed Ua hArtacáin. According to O'Sullivan there were additions made to the writing through 1189, when Cathal Cróbderg took the kingship of Connacht, and in 1198, when Ruairí Ua Conchobair died.

The leaves of the book have been numbered on different occasions. O'Curry made two attempts, one at the bottom of the page and another on the top, and this is used virtually throughout the manuscript. The first set of numbers may indicate the order in which O'Curry found the book, and the second set of foliation is still used as the pagination. There is also evidence that he attempted a second reading of the original pagination but failed to use this system.

R. I. Best recognized only one hand in the manuscript and identified the scribe as Áed mac Crimthainn, but O'Sullivan distinguishes four main hands: A, F, T, and U. They come from the same school, which explains the similarities that led Best to his conclusion. The main scribe is A, referring to Áed mac Crimthainn; F stands for the hand who wrote as Bishop Finn in the letter to Áed; T stands for the style used in both *Togail Traí* and the *Táin Bó Cuailnge*; finally, U stands for the scribe using the uncial *a* in the medial position. He also identifies two lesser hands: M, who wrote *Mesca Ulad*, and S for the scribe whom Áed employed to copy pages ccvii-ccxvi at Bishop Finn's request.

Áed's hand is formal and rounded with little contrast between thick and thin strokes, but the best hand is that of F, which may have been written by the Bishop himself or more likely by his scribe. Despite the fact that his hand has been re-inked, it is clearly a fine hand with a good contrast between thick and thin strokes. T also writes the Osraige genealogies; he fills blanks and writes some bridge section. His hand—being somewhat ragged and uneven, reminiscent of a scholar rather than a professional—is not as impressive as that of either A or F. U displays a variety of styles and it may indeed contain more than one hand. It is distinguished by the use of the uncial *a* and the *d* with a vertical tail that is unknown in the other hands of the manuscript.

The quality of the vellum varies; although originally white, it has acquired a brown stain over the centuries. Some leaves are formed from joining together various pieces of leaf. Most leaves are pricked and ruled on the recto side.

The decoration employed is usually in keeping with the traditions of the period, including mosaic filling, huge animals with snakes, and animal-headed letters. The colors employed are red and yellow along with green and purple. But the unique feature is the appearance of human heads hanging from initial letters. The most famous of these amusing human drawings is in the Banquet Hall of Tara. The scribes' abilities differ in this respect, as well: F draws the best animals, and A produces the same designs but of a much lower standard. The best design is at the beginning of the *Leabhar Gabhála*. U seldom attempts decoration and the quality is somewhat dull.

As a result of the manuscript's history, the binding has suffered great damage and at times it is impossible to identify where the original sewing took place. The pages do not always match in size; they vary by as much as a centimeter. The book was kept in loose parts when the binding finally collapsed, and O'Sullivan assumes that the last binding was carried out in the middle of the fourteenth century. At some point a knife was used to cut through the leaves, and thongs were passed through the holes to keep them together.

The *Book of Leinster* is the last of the large manuscripts produced by the unreformed Irish church and it became one of the sources for the large number of new manuscripts that were being produced for lay patrons from the thirteenth century onward. The earliest reference to *Leabhar na Nuachongbála* in this context is in the *Yellow Book of Lecan* col. 896 (Fasc 185 a 33) and it is also mentioned in the *Book of Ballymote* 263 a 19 and again in the *Book of Lecan* 42 d 14.

MUIREANN NÍ BHROLCHÁIN

References and Further Reading

Best, Richard Irving, O'Brien, Michael and Bergin, Osborn. *The Book of Leinster* (Diplomatic Edition, vols. 1–5). Dublin: DIAS, 1954, 1956, 1957, 1965, 1967.

Bhreathnach, E. "Two contributors to the Book of Leinster: Bishop Finn of Kildare and Gilla na Náem Úa Duinn." In *Ogma: essays in Celtic Studies in honour of Próinséas Ní Chatháin*, edited by Michael Richter and Jean-Michel Picard, 105–11. Dublin: Four Courts Press, 2002.

Gwynn, E.J., (ed.) *The Metrical Dindshenchas* i, Todd Lecture Series xiii (1903), ii Todd Lecture Series ix (1906), iii Todd Lecture Series x (1913), iv Todd Lecture Series (1924), v Todd Lecture Series, xii Todd Lecture Series (1935), (1937) 50-91.

Gwynn, A. "Some notes on the history of the Book of Leinster." *Celtica* 5 (1960): 8–12.

Ó Concheanainn, T. "LL and the date of the reviser of LU." *Éigse* 20 (1984): 212–225.

O'Sullivan Anne. *The Book of Leinster* (Diplomatic Edition, vol. vi). Dublin: DIAS, 1983.

O Sullivan, W. "On the scripts and make-up of the Book of Leinster." *Celtica* 7 (1966): 1–31.

O'Sullivan, A. "Leabhar na hUidhre: further textual associations." *Éigse* 30 (1997): 27–91.

See also **Áed mac Crimthainn; Anglo-Norman; Glendalough; Lebor na hUidre; Lecan, Book of; Mac Murchada, Diarmait; Manuscript Illumination**

LETH CUINN AND LETH MOGA

Traditionally, Ireland was divided into two halves in common with many other societies. The boundary (prehistoric in origin) ran along the Eiscir Riada, a natural gravel ridge (laid down at the end of the last glaciation) running in a line roughly between Dublin and Galway. The cosmographic structure of Ireland is elaborated in the mythological literature, particularly in *Lebor Gabála Érenn* (*The Book of the Taking of Ireland*)—twelfth century in its present form but preserving materials going back to at least the eighth century. The earliest division in the remote past was said to be between the sons of Míl (immediate ancestor of the inhabitants of the island), Éremón and Éber. By the eighth century, however, the theory that was to endure throughout the historical period had emerged. The northern half was known as *Leth Cuinn* (Conn's Half) and the southern half as *Leth Moga* (The Half of Mug). Conn (from whom the Connachta / Uí Néill) was the eponymous ancestor of the dynasties of the northern half. Mug or *Mug Nuadat* (the slave of Nuadu), also called Eógan, was the eponymous ancestor of the Eóganachta, the main dynasty of Munster. Since Conn means "head" or "chief" and *mug* means "slave," this also reflects the international pattern of binary opposition in such schemas. The theory of the two halves was elaborated in the eighth century and reflects the political equilibrium of that period with the Uí Néill dominating Tara in the northern half and the Eóganachta dominating Cashel in the southern half.

Leth Cuinn

Leth Cuinn, "Conn's Half," the northern half of Ireland. The Connachta derive their name from Conn (*Dál Cuinn*, "Division of Conn") and from them is named the province of Connacht in the west of Ireland. The Uí Néill emerged from the Connachta as an independent group by the late seventh century. The *Ulaid* (Ulstermen), who would seem to have exercised overlordship over the northern half in the protohistoric period, suffered massive defeats in the battles of Moira, County Down, in 637 and Faughart, County Louth, in 735. They never recovered. Their place was taken by the Cenél nÉogain who gradually moved eastward across the Foyle from Inishowen and dominated all of the north by the mid ninth century, making the Airgialla their vassals and dominating the Dál Riada of northeast Antrim, the Dál nAraide (Cruithin) of mid Antrim, the Dál Fiatach of Down, and many other groups that had formerly been under the Ulaid. In Connacht the Uí Briúin of County Roscommon came to dominate the province supplanting the Uí Fiachrach and Uí Maine, as well as a multitude of earlier populations. The southern Uí Néill had conquered the midlands from the Shannon to the sea absorbing many older population groups. The clan Cholmáin became the dominant group among them. Both branches of the Uí Néill sought to dominate Leth Moga—indeed it was essential if they were to claim the high kingship—although major campaigns against Munster were rare before the eighth century. Although Leinster was within the cultural area of Leth Moga in the seventh century, by the eighth century it was dominated by the Uí Néill. Máel-Sechnaill mac Máele-Ruanaid, king of the southern Uí Néill, was the first high king with real power. He campaigned in Munster and was the first high king to reach the south coast in 858. The position of Osraige, lying between Leinster and Munster, was ambiguous, but in 859 it was alienated to Leth Cuinn. There were Munster challenges to the Uí Néill: Cathal mac Finguinne in the eighth century, Fedelmid mac Cremthainn in the ninth century, Cormac mac Cuilennáin in the tenth century, and Brian Bóruma in the late tenth century to early eleventh century. Of these Brian was the most successful. In the post-Viking period, the emergence of Dublin as an international trading port and the feudalization of Irish society brought about a gradual reorientation of politics in Ireland. The old pattern of north-south conflict was no longer as important as the competition to control the wealth of Dublin. Medieval kingdoms were emerging in both halves of the country, led by powerful families such as Ua Briain in Munster, Ua Conchobhair in Connacht, Mac Lochlainn and Ua Neill in Ulster, and Mac Murrough in Leinster. Many of the major battles of the twelfth century took place in the midlands with control of Dublin as the ultimate goal. As a result of this pressure, the Uí Néill kingdoms of the midlands were greatly weakened. The arrival of the Anglo-Normans in the late twelfth century was the final blow to the ancient divisions since they conquered and settled large areas in both halves of Ireland.

Leth Moga

Leth Moga, "Mug's Half," "The Servant's Half," from *Mug Nuadat*, "Servant of Nuadu," otherwise known as Eógan, eponymous ancestor of the Eóganachta, the main dynasty of Munster. The capital of Leth Moga was Caisel (Cashel in Tipperary), a borrowing of the Latin *castellum* (castle, fortress), and tradition claims that it was founded in the fifth century. This may be read against the background of Irish settlement within the borders of the Roman Empire in Wales and Cornwall. Cashel was distinct from other tribal capitals in that, despite underlying pagan elements, its foundation story ("The Finding of Cashel") is strongly Christian and many, but not all, of its kings were also clerics. The Eóganachta were spread throughout Munster—the main branches were the Eóganachta Áine, East Limerick; Eóganachta Glendamnach, North Cork; and the Eóganachta Chaisil, Tipperary. The kingship of Munster circulated among the various branches of the Eóganachta until it was virtually monopolized by the Eóganachta Chaisil from the mid ninth century onward. The wide settlement of the Eóganachta across Munster assured control of their many subject peoples. Several Munster kings campaigned against the Uí Néill. Cathal mac Finguine harassed the midlands and Leinster from 733 until 738, but although the Uí Néill accepted his strength within Munster, he was not allowed control of Leinster. Fedelmid mac Crimthainn, king (820–847) and abbot, provided a serious challenge to the Uí Néill invading the midlands and Connacht in the 830s and 840s. He was no respecter of churches sacking both Clonmacnois and Kildare, as well as many others. In 840, he camped at Tara, indicating his ambition to achieve political dominance, and became involved in Armagh politics toward the same end. He attempted to hold the Óenach Carman in 841 to demonstrate his overlordship of Leinster, but suffered a massive defeat. Cormac mac Cuilennáin, king and bishop of Cashel (902–908) again attempted to win political supremacy, but was defeated at Belach Mugna in Leinster in 908. With Cormac's death the power of the Eóganachta collapsed. Their place was taken by an obscure group that rose to power through their ability to control the portages on the lower

Shannon—the Dál gCáis. From them came Brian Bórumha mac Cennétig (Brian Boru) who was arguably the first effective high king of Ireland from Munster. Brian dominated Munster and soon made his claim for the kingship of Ireland. His defeat at the battle of Clontarf beside the Norse town of Dublin highlights the reorientation of Irish politics at this time. By supplanting the Eóganachta as the traditional kings of Munster and the Uí Néill as the traditional occupants of the high-kingship, Brian introduced a new era in Irish politics. His descendants were to play a leading role in national affairs of the eleventh and twelfth centuries, controlling not only Dublin but also the Isle of Man for a period. They had international agreements and marriage alliances, corresponded with the Pope and Norman bishops—leading to the twelfth century church reform—and introduced Romanesque architecture. For them the concept of Leth Moga was no longer of major significance.

CHARLES DOHERTY

References and Further Reading

Francis John Byrne, *Irish Kings and High-Kings*, London, 1973; T.M. Charles-Edwards, *Early Christian Ireland*, Cambridge, 2000.

See also **Genealogy; Invasion Myth**

LIA FAIL

See **Inauguration Sites**

LIGHTHOUSES

See **Ships and Shipping**

LIMERICK

The Shannon River (Σηνοσ - Senos) features on Ptolemy's map of Ireland (*c.* 150 A.D.), which also locates a tribe near Limerick, the Gangani, not subsequently known. The later settlement developed on King's Island—then known as Inis Sibtond—the Érainn dynasty having a branch known as Érainn of Inis Sibtond, who may have controlled it about the beginning of the historical period. The battle at Luimnech (*c.* 575 A.D.) may refer to a location in Connacht, the earliest references to the northMunster Luimnech occuring in a law tract (*c.* 700 A.D.), where the estuary itself is intended; also, an early saint's life speaks of an island "in that sea called Luimnech," and a ninth-century tale mentions Loch Luimnig (*loch* here meaning "estuary").

By the sixth or seventh century, the subject peoples of Munster were known as *Déisi* (vassals). The land of *In Déis Tuaiscirt* (the northern vassalry) straddled the Shannon near Limerick. A branch, the Uí Caisin, included Luimnech among its lands. A parallel offshoot, Uí Tairdelbaig, established the kingdom of Dál Cais across the Shannon (from which Brian Boru descended). The introduction of Christianity to the Limerick area is associated with Uí Tairdelbaig, the city's patron saint, Mainchín son of Setna (St. Munchin), being a member of the dynasty, said to have been granted land on Inis Sibtond by Ferdomnach of Dál Cais to found a church (possibly the site of the modern St. Munchin's church).

Viking raiding parties used the Shannon from the 830s, attacking churches along its route and, by mid-century, had established a settlement at Limerick, building a fortress on Inis Sibtond. It was a very strategic site, protected from the west by the Shannon and elsewhere by the Abbey River. The Viking rulers of Inis Sibtond had access to the very interior of Ireland, making Limerick, after Dublin, perhaps the most important commercial center in the country. Tomrar son of Elge, "Jarl of the Foreigners," based himself there in 922 to ravage the Shannon valley. By this time the Norse kingship of Limerick, drawn from a Hebridean dynasty, had emerged as a regional power, challenging Dublin for supremacy of the Irish Scandinavians.

As Dál Cais strengthened its position in north Munster around the mid-tenth century, the potential of Limerick was recognized. In 967, Brian Boru's brother Mathgamain slaughtered the Limerick Norse in battle, burning their ships, plundering Inis Sibtond and its fortress (*dún*), and the Norse king Ívar was temporarily expelled. In 972, the Norsemen were driven out of Inis Sibtond and the *dún* was set on fire. The subjugation of the Limerick Norse was completed by Brian Boru in 977, when he slew Ívar and his sons on Scattery Island, after which the Dál Cais controlled Limerick and maintained a fleet on the estuary. Brian may have selected Inis Sibtond as one of his bases, if it is the *Inis Gaill Duibh* (Island of the Black Foreigner) where he built a stronghold (*daingen*) in 1012.

In 1016, the Dál Cais royal poet Mac Liag died at Inis Gaill Duibh and the city featured in the struggle in the 1050s between Donnchad son of Brian Boru and his nephew Tairdelbach Ua Briain (d. 1086). Later, Tairdelbach and his son Muirchertach (d. 1119) made the city their capital and summoned provincial kings there to make submission. Muirchertach refortified the island defences in 1101 by demolishing the Grianán of Ailech, royal site of the northern Uí Néill, and commanding his army "to carry with them, from Ailech to Limerick, a stone for every sack of provisions which they had" (AFM). Like his great-grandfather, Muirchertach had a Shannon fleet, probably based in Limerick, and used it to maintain suzerainty over the rival ports of Dublin and Waterford.

Under Tairdelbach and Muirchertach, Limerick emerged as a major center of the church reform movement. In 1111, its first properly consecrated bishop, Gille (Gilbert), presided as papal legate over the Synod of Ráith Bressail, which drew up a fixed territorial diocesan scheme. Gille tells us that he spent some time, and perhaps studied, in Rouen, where he met the future St. Anselm of Canterbury. When he wrote from Limerick to Anselm about 1107, he sent him twenty-five pearls, presumably obtained from local oyster harvesters. Trade and fisheries were Limerick's staples, and the wealth its masters could obtain from these is suggested by the goods which Tairdelbach Ua Briain (d. 1167), nephew of Muirchertach, escaped with from the town in 1151: besides the drinking-horn of Brian Boru, he made off with "ten score ounces of gold and sixty beautiful jewels" (AFM).

At the time of the Anglo-Norman invasion, Domnall Mór Ua Briain, son of Tairdelbach (d. 1167) was king of Thomond and ruler of Limerick. He rebuilt St. Mary's cathedral in the 1170s and may have introduced Continental religious orders to Limerick, being credited by Sir James Ware as founder of St. Peter's priory for Augustinian nuns just outside the city walls. Domnall Mór defeated the Anglo-Normans at Thurles in 1174, but they advanced on Limerick in 1175 in alliance with the high king, Ruaidrí Ua Conchobair. Giraldus Cambrensis describes how the attackers "found that the river was swift flowing and deep, and formed an intervening obstacle which they could not cross." The "Song of Dermot" also records that the city "was surrounded by a river, a wall, and a dyke (*fosse*), so that no man could pass over without a ship or a bridge, neither in winter nor in summer, except by a difficult ford."

Nevertheless, the Anglo-Normans took the town, but within two years were forced to abandon it again to Domnall Mór Ua Briain. Although nominally granted by Henry II to Philip de Briouze in 1177, the Anglo-Normans did not regain Limerick until after Domnall's death in 1194, infighting among the Uí Briain facilitating their return. William de Burgh held Limerick prior to 1203, briefly detaining Domnall's son as prisoner there, but King John revived the de Briouze interest, granting the lordship of Limerick to William de Briouze (though it was subsequently withdrawn). John may have ordered the construction of the castle that still bears his name, although no evidence to that effect exists (and he never stayed there), and, in order to penetrate west of the Shannon, the castle was followed by construction of Thomond Bridge.

There was no bridge where the later Baal's (or Ball's) Bridge stood at the time of the Anglo-Norman arrival, but it was built soon afterward, and became a prominent landmark. A grant of King John to Thomas fitz Maurice mentions "a burgage near the bridge on the left, at the entrance of the vill towards the north, within the walls of Limerick." In 1340, Edward III ordered funding for a bridge, possibly that which survived until its replacement in 1830. The origin of the name is unknown; one theory is that *baal* comes from the Irish *maol* (bald), and applied to bridges lacking parapets. Speed's map of 1610 calls it "The thye bridge," presumably because it linked the "English" and "Irish" towns on either side of the Abbey River.

The King's Island site comprised "King John's Castle" and a walled enclave surrounding it, which in the later Middle Ages became known as Englishtown, the rest of the island being less settled, the castle constable having grazing rights while the citizens also used it for recreation. The adjacent fisheries were highly prized and consequently controversial, especially the competing claims to a share in their profits. The *Black Book of Limerick* records an inquisition (1200–1201) by a jury of 36 inhabitants (12 Irishmen, 12 members of Limerick's old Norse community, and 12 new English residents) that found that the archbishop was entitled to "half of the fishery of Curragour, and the land of the mill on the water near the walls of the city, and altogether a tenth of all the fish which are caught by the fishermen of that city." Upriver from Curragour was the salmon fishery of Laxweir, which, as its Norse name indicates, existed since Viking times.

The mill recorded in 1200–1201 is probably that marked on the map of Limerick drawn circa 1590 (TCD MS 1209/58), named Thomas Arthur's Mill from one of the city's leading merchant families. The map has another mill called Queen's Mill, which may also date from King John's reign, when the bishop was compensated "for the damage done to him by the construction of the King's mills and fisheries at Limerick."

The priory of SS Mary and Edward, for Augustinian "Crutched Friars," apparently existed by 1216, and the Knights Templars and Hospitallers both had houses there, while the Franciscans were introduced by the de Burgh family circa 1267. There was a hospital of St. Mary, a poor-hospital of St. Laurence, and also a leper hospital in the city. Donnchad Cairbreach Ua Briain (d. 1242), a younger son of Domnall Mór, although not ruling Limerick, is said by Ware to have founded St. Saviour's Dominican priory, where he was buried in 1242.

One of his successors, Brian Ruad (d. 1277), apparently reasserted lordship, and the English hosted "to Limerick against Ua Briain" in 1271 (AI). His grandson

Tadc Luimnig (d. 1317, AI) may have been born in the city. Fighting around Limerick in 1313 appears to have involved rival Uí Briain factions, but the city was in English hands when in 1370 it was sacked and burned by the Irish of Thomond. In 1466, Tadg Ua Briain of Thomond placed on a formal basis the "black rent" long claimed by his dynasty from the city. Generations of neglect by the Dublin-based administration of the Lordship of Ireland—its energies concentrated on defense of the Pale—had left Limerick in a greatly weakened state by the end of the medieval period. Although the city's merchants fortified Irishtown in walls stronger than Englishtown on King's Island, and St. Mary's cathedral thrived on their patronage, the deficiency of royal government saw even its once great castle reduced to what contemporaries described as a ruin.

SEÁN DUFFY

References and Further Reading

John Begley. *The Diocese of Limerick, Ancient and Medieval.* Dublin, 1906.

Judith Hill. *The Building of Limerick.* Cork & Dublin, 1991.

Maurice Lenihan. *Limerick; its History and Antiquities.* Cork, 1866.

James McCaffrey (ed.). *The Black Book of Limerick.* Dublin, 1907.

T.J. Westropp. *The Antiquities of Limerick and its Neighbourhood.* Dublin, 1916.

See also **Church Reform, Twelfth Century; Dál Cais; Gille (Gilbert) of Limerick; Munster**

LIONEL OF CLARENCE

Lionel of Clarence (b. Antwerp, 1338; d. Alba in Italy, 1368), earl of Ulster, lord of Connacht, duke of Clarence. Lionel was the third son of King Edward III and inherited enormous lands in Ireland through his marriage to Elizabeth de Burgh, heiress of William de Burgh (d. 1333), the "Brown Earl" of Ulster. In 1360, a break in the war between France and England enabled Edward III to assent to requests from Ireland for a royal prince to be sent to remedy the "utter devastation, ruin and misery" of the lordship of Ireland.

Lionel was appointed king's lieutenant in 1361, and he arrived in Ireland that September. He was resident in Ireland, except for a visit to England in 1364, until 1366. His appointment heralded a policy of military intervention, heavily funded from England, that was to culminate in the two expeditions of King Richard II in the 1390s. The intention was to fulfill the king's duty to protect his subjects in Ireland and to revive the colony as a source of profit to the crown.

Lionel brought with him a large army, including many absentee landholders. Although Lionel held notional titles to Connacht and Ulster, these areas were barely in communication with the central government, and Lionel made little effort to reconquer them. Instead, his military campaigns were concentrated on Leinster, the midlands, and the southwest. It was of particular importance to secure Leinster because it was planned to move the Exchequer from Dublin to Carlow in an effort to make it easier for royal officers to make payments. Warfare in Ireland was rarely decisive, however, necessitating a constant presence in the form of garrisons, known as wards, posted across the country. This policy was expensive, and when funds from England disappeared from 1364, there were widespread desertions. From 1366, with the departure of Lionel and his army, conditions rapidly returned to the pre-expeditionary situation.

Lionel's arrival in Ireland was one of the few occasions on which anything approximating to a court life appeared in Ireland. Lionel renovated Dublin castle and ordered preparations to be made for sports and tournaments. These were aspects of royal government to which the remote residents of Ireland did not usually have access. Nonetheless, Lionel antagonized the Anglo-Irish by appointing officials born in England to implement his administrative reforms and by attempting to exact a subsidy from the Irish parliament. Lionel managed to compromise on these issues, but they reemerged after 1369 during the chief governorship of William of Windsor.

Except for a modest financial recovery, most of Lionel's achievements did not survive his departure from Ireland. His most enduring legacy was the Statutes of Kilkenny of 1366. These statutes codified much of the existing legislation dating back to 1297 that aimed at curbing Gaelicization. While the statutes were racially exclusionary, they were not a direct attack on Gaelic culture. They were defensive in tone, and their simplistic racial distinctions were designed to contain what was perceived as a principal cause of decline. Modern historians have stressed the extent to which the statutes merely summed up previous legislation. Yet the statutes of 1366 were unusually comprehensive and were reissued several times during the late medieval period. One of the paradoxes of late medieval Irish history is that among those who reissued the statutes were many who knowingly and wilfully transgressed their provisions.

Lionel left Ireland in November 1366; he is reputed to have sworn never to return. A marriage had been arranged for him to the daughter of the Visconti of Milan, and he died in Albain Piedmont, in 1368. His Irish estates descended through his daughter Phillipa to the Mortimer earls of March.

PETER CROOKS

References and Further Reading

Curtis, Edmund. "The viceroyalty of Lionel, duke of Clarence in Ireland, 1361–1367." *Journal of the Royal Society of Antiquaries of Ireland* 47 (1917) 165–81; 48 (1918) 65–73.

Connolly, Philomena. "The financing of English expeditions to Ireland, 1361–1376." In *England and Ireland in the Later Middle Ages: Essays in honour of Jocelyn Otway-Ruthven*, edited by James Lydon, 104–21. Dublin: Irish Academic Press, 1981.

See also **Central Government; Connacht; Gaelicisation; Leinster; Lordship of Ireland; Mortimer; Parliament; Richard II; Ulster, Earldom of; William of Windsor**

LISMORE, BOOK OF

Description

This is a fifteenth-century vellum codex also known as *Leabhar Mhic Cárthaigh Riabhaigh* (The Book of Mac Cárthaigh Riabhach). It is now called the *Book of Lismore* because it came to light during the course of structural alterations in Lismore Castle, County Waterford, in 1814. Unfortunately, the manuscript suffered at the hands of local "scholars" in Cork at that time, and many leaves were abstracted from it. It now consists of 198 folios. The writing (with the exception of the recto page of what is now folio 116) is in two columns. There are a number of eighteenth-century manuscripts that—it is believed—derive their contents in part, either directly or indirectly, from the *Book of Lismore*. It has also been suggested that some other texts found in one or more of these manuscripts may have been contained in the missing sections of the earlier codex. In 1930, the manuscript was transferred to Chatsworth, the Derbyshire seat of the Duke of Devonshire (the owner of Lismore Castle), and has remained there in private keeping. In 1950, a collotype facsimile edition of the manuscript was published by the Irish Manuscripts Commission with a descriptive introduction and indexes by R. A. S. Macalister. It should be pointed out, however, that the introduction suffers from some deficiencies and contains a number of errors. Some of the leaves abstracted from the manuscript at the beginning of the nineteenth century may have contained information about the patrons for whom it was compiled, the scribes, and the date (or dates) of compilation. Little information of this kind has survived in the manuscript in its extant form. A full examination of the hands of the manuscript has also yet to be undertaken. The fact that the original manuscript is in private keeping has meant that comparatively few scholars have had an opportunity to examine it. An important description of the manuscript; its history, foliation, and pagination; its scribes, contents, and missing leaves; and finally its binding has been made by Brian Ó Cuív.

Among the known patrons of one of the scribes, Aonghas Ó Callanáin, was Fínghin Mac Cárthaigh Riabhach of Cairbre in County Cork. Ó Callanáin, however, was not the chief scribe. Scribal notes on folios 2r, 7v, 11r, and 17r refer to a *lánamhna* (married couple), for whom texts on those leaves were written. Some scholars have identified this couple as the aforementioned Fínghin (d. 1505), lord of Cairbre, and his wife Caitlín (d. 1506), a daughter of Thomas Fitzgerald (d. 1468), eighth earl of Desmond. This identification is by no means certain, however, and the couple may be another husband and wife, possibly Fínghin's father and mother. The manuscript may have been written for this earlier couple and added to during Fínghin's time. The *Book of Lismore* was not prepared for the library of a monastery or of a professional scholar. It is one of a number of fifteenth-century composite volumes that were compiled for lay patrons. The contents of these manuscripts reflect the varied interests of the members of the Gaelic and Anglo-Irish nobility at the time. One other such manuscript is the *Book of Fermoy*, a manuscript comprising several sections written in different periods and containing a wide diversity of material. It was written mainly in the fifteenth century for the Anglo-Irish Roches of Fermoy, County Cork. It has been suggested that some parts of both the *Book of Fermoy* and the *Book of Lismore* were written by the same scribe.

Contents

Among the contents of the *Book of Lismore* are many texts of religious interest. There are saints' lives, including those of Brigit, Colum Cille, and Patrick. There are apocryphal texts, including a copy of *An Tenga Bithnua* (*The Evernew Tongue*), the title of a dialogue between the Hebrew sages and the spirit of the apostle Philip, who is called "Evernew Tongue" because when he was preaching to the heathen his tongue was nine times cut out and nine times miraculously restored. There is also a medieval account of Antichrist. The manuscript contains various other texts (both prose and poetry), of which the following is a selection: There is a copy of *Lebor na Cert* (*The Book of Rights*), which contains, among other material, a collection of poems on the stipends and tributes of the kingdoms of Ireland. It has been dated to the twelfth century. There are copies of *Caithréim Chellacháin Chaisil* ("The Triumph of Cellachán of Cashel"), one of the historical tales of Irish literature and of

Acallam na Senórach (*Colloquy of the Ancients*), an important collection of material relating to the legendary Fionn mac Cumhaill and his band of warriors, believed to have been written in the late twelfth century. A further tale is *Tromdám Guaire* ("The Oppressive Company of Guaire"), the oppressive company in question being Senchán Torpéist and his retinue of poets who visit Guaire, king of Connacht, and make unreasonable demands upon him. The tale is a satire on certain aspects of the role of the poet in medieval Ireland. The *Book of Lismore* also contains a number of Irish translations of foreign sources, including the only extant copy of *Leabhar Ser Marco Polo* (*The Book of Sir Marco Polo*), a translation of the Latin version of Marco Polo's *Il Milione*, probably written between 1320 and 1325. There is also a copy of *Gabháltas Séarlais Mhóir* (*The Conquest of Charlemagne*), believed to have been translated from Latin, possibly about 1400, and of *Stair na Lombardach* ("The History of the Lombards"), probably a fifteenth-century translation of a chapter ("*De S. Pelagio papa*") from *Legenda Aurea*, compiled by Jacobus de Voragine between 1260 and 1270.

CAOIMHÍN BREATNACH

References and Further Reading

The Book of Lismore. Facsimile with introduction by R.A.S. Macalister. Dublin: Irish Manuscripts Comission, 1950.

Stokes, Whitley (ed.), *Lives of the saints from the Book of Lismore*. Oxford: Clarendon Press, 1890.

Ó Cuív, Brian. "Observations on the Book of Lismore." *Proceedings of the Royal Irish Academy* 83C (1983): 269-92.

See also **Hagiography and Martyrologies; Hiberno-Norman (Latin); Historical Tales; Mac Carthy; Satire**

LOCAL GOVERNMENT

The main unit of local government in the later medieval Lordship of Ireland was one imported from England and with a long prior history in that country—the county. The first counties were probably created in the final years of the twelfth century. By the beginning of the second decade of the thirteenth century, separate counties of Dublin, Munster, Cork, and Waterford had come into existence in those areas reserved to the Crown at the time of the initial Anglo-Norman invasion of Ireland or during the later expansion of the Lordship. Further counties were created during the course of the thirteenth century. A separate county of Uriel or Louth was probably created in 1227 at the time that Hugh de Lacy recovered the liberty of Ulster. County Kerry was carved out of either the existing county of Munster or out of County Cork, probably in the 1220s. County Limerick had probably been carved out of the older county of Munster by the 1230s, and by the 1250s the remainder of that county had come to be called County Tipperary rather than Munster. Connacht, too, had its own sheriff by 1236, reflecting the progress of conquest in the west and the creation of a county there. It too subsequently had a separate county (of Roscommon) carved out of it, perhaps in 1288. It was not until 1297 that a separate county of Meath was established, coterminous with the original liberty of Meath, but with a sheriff directly responsible for only the de Verdon portion of that liberty. All these were royal counties, with sheriffs who were directly answerable to the Dublin administration. There were also private sheriffs within the greater liberties who were immediately answerable to the lords of these liberties and their stewards (or seneschals). The large liberty of Leinster had been divided into four separate administrative units from the late twelfth century on. A separate sheriff of County Kildare is first mentioned in 1224, before the partition of the liberty itself between coheirs. References to the other counties seem to come only after the division (to Co. Wexford in 1249; to County Carlow in 1254; to County Kilkenny in 1255), but the division itself probably followed the preexisting division into separate counties. The liberty of Ulster was also divided into a number of separate counties. In the fourteenth century there also emerged within each of the liberties counties consisting of lands belonging to the church (cross-lands) in the liberty that were exempt for this reason from the control of the lord of the liberty and directly subject to the king's rule. These sheriffs of the cross-lands also came to play a rule in acting in the counties within the liberties when the steward of the liberty failed to do so. The names of some of the counties were derived from those of preexisting native Irish administrative and political units, either provinces or kingdoms (Munster, Meath, Connacht, Uriel). Others were named after specific towns that formed the core of the counties concerned and constituted their administrative centers (Dublin, Cork, Waterford, Carlow, Kildare, Roscommon, Wexford, Kilkenny, Tipperary).

The county's main administrative official was the sheriff, who was chosen by the local county court. Governmental orders were transmitted from Dublin or from England to the sheriff for local execution within his county, and he was normally required to report back on what had been done or why it had not been done. The sheriff was also responsible for collecting moneys owed to the king within his county and transmitting them to the Exchequer in Dublin or spending the money locally and accounting for that when he next came to render his accounts in Dublin. The process of the king's courts was also dependent on him.

He was responsible for ensuring that defendants were summoned to court, the execution of court process against them and the enforcement of judgments within his county, and reporting back on what he had done. The sheriff was also the presiding officer of the county court held in the county he served and was responsible for executing its process and judgments as well. Sheriffs were assisted in the execution of their duties by a staff of under-sheriffs and clerks appointed by and answerable to them, and also by a chief sergeant and his subordinates, who were generally responsible for the local execution of royal mandates. There were also at least two coroners in each county, whose primary responsibility was to make enquiries into all suspicious deaths, but who might be required to act in place of the sheriff if he failed to execute any of his functions. For the counties within the liberties, however, communication from the central authorities was through the stewards, who were the main administrative officials of the liberties concerned and who then transmitted any necessary orders to the sheriffs.

Each county also possessed a county court. Like its English counterpart, this had a significant civil jurisdiction and sole power to proclaim the outlawry of a fugitive from justice. It was also a place for the choice of representatives for the county at the Irish parliament and also for the choice of sheriffs and coroners. The primary location for the proclamation of newly enacted legislation, and other matters the Dublin administration wished to draw to wider attention, was also the county court. In addition, the county court was a locus for wider decision-making on such matters as the imposition of local taxation to help pay the costs of local military activity.

The main administrative unit below the level of the county was the *cantred*. The term is related to one used in Wales and is etymologically equivalent to the English term "hundred," the term used for a similar sub-county administrative unit. They were also often based on preexisting areas, and generally coincided with the basic area of ecclesiastical administration above the parish and below the archdeaconry, the rural deanery. They were significant units for the purposes of taxation, law enforcement (the sheriff held a sheriff's tourn in each *cantred* twice a year), and general administration. But already before the end of the Middle Ages the term "barony" was coming to be used in place of *cantred* for these units.

The larger cities and towns of the later medieval lordship generally enjoyed a substantial degree of autonomy and were governed and administered by their own elected officials (mayors or bailiffs) and their councils. This autonomy was generally granted them by royal charter, and the charter also generally confirmed some of the distinctive customs that were observed in the town. Regular reissuing of these charters allowed regular updating of their powers and of the city's custom. However autonomous, they remained ultimately under the control of the Dublin administration.

How effective this structure of local government was at any stage in the later Middle Ages is more problematic. The late thirteenth century was probably the period when the control of the Dublin administration reached its maximum extent, but even then there were areas of Gaelic lordship within the existing counties in which the Dublin administration and its local agents were relatively ineffective. Thereafter there was a steady decline in its control and also therefore in the effective reach of the colony's local government structures. By the late fifteenth century, the area most firmly under its control was the area of the Pale, but some local government structures also survived outside that area, not just in major towns but also in some rural areas, often in discontinuous islands of settled governmental structure that had managed to survive the wider decline in the lordship's fortunes.

PAUL BRAND

References and Further Reading

Ellis, Steven. *Reform and revival: English government in Ireland, 1470-1536*. Woodbridge; Boydell Press.1986.

McGrath, Gerard. 'The Shiring of Ireland and the 1297 Parliament' in *Law and Disorder in Thirteenth-Century Ireland. The Dublin Parliament of 1297*, edited by James Lydon, Dublin: Four Courts Press,1997.

See also **Courts; Parliament**

LÓEGAIRE MAC NÉILL

Supposed king of Tara, son of Niall Noígiallach, and progenitor of Cenél Lóeguiri, a dynastic group which, according to their own genealogical tradition, were powerful in Ireland during the sixth and seventh centuries, ruling territories which extended from Loch Erne to the church of Rathlihen, north of the Sliabh Bloom Mountains. The dynasty's main power base seems to have been near the church of Trim in modern county Meath. Very little can be said with certainty about Lóegaire because all accounts of his activities considerably postdate his lifetime.

According to the annals and other sources, Lóegaire's floruit was in and around the second third of the fifth century. However, there are some indications that he may, in fact, have lived as early as the fourth century. The fifth century chronology for Lóegaire may have originated in the church of Ardbraccan, in Cenél Lóegaire, which seems to be the source of the Lóegaire episode in Tírechán's late seventh century collection of lore about St. Patrick. That church wished to associate the founder of the local dynasty with the saint, the date of whose arrival in Ireland had been set at 432.

Lóegaire plays a prominent role in the seventh century Patrician hagiography, where his supposed encounter with Patrick at Tara is a central element of the narrative. According to Muirchú, Lóegaire converted to Christianity following his encounter with St. Patrick. Tírechán, on the other hand, says that he refused to accept the Christian faith, because his father, Niall, would not allow this; he had ordained that Lóegaire should be buried, fully armed, in the ridges of Tara, facing the graves of Uí Dúnlainge of Leinster—traditional enemies of the Uí Néill—at Mullaghmast in County Kildare. This hostility is reflected in the annals—written long after Lóegaire's lifetime—which recount that Lóegaire routed the Laigin (Leinstermen) in the year 453. Fortunes were reversed five years later at the battle of Áth Dara when Lóegaire suffered a defeat at the hands of the same enemy; he was taken prisoner and released only when he gave the elements as sureties that he would cease to levy the *Bóruma Laigen* ("the cattle tribute of the Laigin"). A tract on the *Bóruma* recounts that Lóegaire broke his promise and the elements, accordingly, passed judgement on him and brought about his death. The account of his death in the annals—at the year 462—refers to this legend.

Presumably because of his association with St. Patrick, the pseudohistorical prologue to the *Senchas Már* (the major collection of Brehon Law tracts) claims that Lóegaire called a convention of the men of Ireland to reform the traditional laws in accordance with Christianity.

The *Bansshenchas* names two wives of Lóegaire as Angas, daughter of Ailill Tassach of the Éoganachta of Munster, and Muirecht, daughter of Eochaid Munremar, an ancestor figure of the Dál Riata of Antrim and Scotland. Genealogical accounts dating from different periods ascribe between twelve and fifteen sons to Lóegaire.

References and Further Reading

Mac Eoin, Gearóid, "The Mysterious Death of Loegaire mac Néill." *Studia Hibernica* 8 (1968): 21–48.

Ó hÓgáin, Daithí. *Myth, Legend & Romance, an Encyclopaedia of the Irish Folk Tradition*. New York: Prentice Hall, 1991.

Byrne, Paul. "Certain southern Uí Néill kingdoms (sixth to eleventh century)." Ph.D. diss., University College Dublin, 2000.

Byrne, Francis John. *Irish Kings and High-Kings*. London: Batsford, 1973.

See also **Adomnan; Brehon Law; Conversion to Christianity; Niall Noígiallach; Patrick; Tara; Trim; Uí Néill; Uí Néill, Southern**

LORDSHIP OF IRELAND

Although never in contemporary usage, modern historians use the term "lordship of Ireland" in acknowledgement of the fact that the king of England from 1171 to 1541 bore the title "lord of Ireland" (*dominus Hibernie*). The term sometimes carries a more restricted connotation, denoting that part of medieval Ireland over which the king of England exercised effective power.

The term was first given expression in 1155 in the papal privilege of Pope Adrian IV (1154–1159) known as *Laudabiliter.* This letter, which was addressed to Henry II, authorized a conquest of Ireland with the aim of reforming the Irish church. Henry contemplated at that point bestowing Ireland, so the chronicler Robert of Torigni claims, on his youngest brother William, but nothing came of the proposal, and he himself did not intervene in Ireland until 1171–1172, when papal support for his actions was used as an important legitimizing force. The role of Adrian IV in granting the lordship to Henry II was subsequently cited; for example, in 1317 in the Remonstrance of the Irish princes (see below). The submission of the Irish kings and bishops to Henry during his visit to Ireland was regarded as a public acceptance by them of his lordship, and in the words of Giraldus Cambrensis, Ireland was made subject to the English crown "as if through a perpetual indenture and an indissoluble chain."

In 1177, at a council in Oxford, Henry granted Ireland to his youngest son, John, who was the first to use the title "lord of Ireland." Henry's intention was that John should become King of Ireland, and plans were set in motion to obtain a crown from Rome for him. This can be seen as part of Henry's wider strategy to hold together the scattered lands of the Angevin Empire by entrusting them to the government of different sons. Had the future turned out as Henry had envisaged it, Ireland would have descended in a cadet line of the Plantagenet house. Instead a sequence of deaths resulted in John being made King of England in 1199. His accession to the throne was a significant moment in Irish history, and from that date the title *dominus Hibernie* became permanently part of the royal style for the rest of the Middle Ages, interestingly inserted immediately after "king of England" and before "duke of Normandy and Aquitaine and count of Anjou." Even so, official records in the first quarter of the thirteenth century sometimes contain references to the "kingdom of Ireland." During John's reign there was a greater degree of royal involvement in Ireland than at any other time during the medieval period, and many historians regard him as the real creator of the medieval lordship.

In 1254, Henry III endowed the future Edward I with wide territories that included the lordship of Ireland. However the lands were given to him on condition that they never be separated from the crown of England but remain "wholly to the kings of England for ever." This marked a decided change from the grant to John in 1177. While the land of Ireland could be granted to another person, the lordship remained separate and inalienably held by the crown. The principle was established that the lordship was vested in the English crown, not in any one king or royal line, and this was the constitutional principle that underpinned Anglo-Irish relations throughout the later medieval period.

Lordship of Ireland implied control over the whole territory of Ireland, but the reality was very different. It has been said that medieval Ireland was not so much a lordship as a patchwork of lordships, a reference to the fragmented geography of power that pertained throughout the land. Under Henry II and John, royal authority had been acknowledged by the submission of both Anglo-Norman magnates and Gaelic kings to their feudal lordship. However, by the time of Edward I it was only considered necessary to obtain the allegiance of the English lords who were theoretically in control of the whole island. Furthermore, the exclusion of the Irish as a race from the common law, the means by which a subject obtained the protection of his king, had the effect of denying to many of the inhabitants of Ireland the benefits of lordship.

The Remonstrance addressed to Pope John XXII in 1317 in the name of the Irish kings, magnates, and people complained that the Irish no longer held their lands directly of the crown nor benefited from the protection of a powerful overlord. Therefore, it was claimed, they were vindicated in their withdrawal of obedience from Edward II. A serious attempt was made during the reign of Richard II to reestablish the lordship of the English crown over the whole island. In 1385, Richard had briefly granted to his favorite Robert de Vere (who would bear the titles "marquis of Dublin" and "duke of Ireland") the lordship and lands of Ireland almost as an independent palatinate, and all writs ran in de Vere's name, his arms replacing those of the king in Ireland, although the experiment lapsed shortly afterward. The renewed submissions taken from Gaelic leaders by Richard during his own 1394–1395 expedition to Ireland (the first by a lord of Ireland since 1210) were a significant confirmation of royal lordship and the benefits of personally discharging the obligations that entailed.

Richard II failed to achieve his ideal of uniting all the inhabitants of Ireland under his lordship, and thereafter royal intervention in Ireland was limited in scope and interest. However, although Ireland was frequently ignored and neglected by England, at no point did the king ever consider relinquishing lordship in Ireland or abrogating his obligation as a lord to protect his subjects there. Moreover, at no point was the position of the king as lord of Ireland seriously threatened, not even by the separatist tendencies that were given expression in the challenge to the constitutional position of Ireland in the parliament of 1460.

Not long after this date, shortly after the Geraldine ascendancy began, the Irish parliament can be found reminding the king that Ireland was "one of the members of his most noble crown, and eldest member thereof." In 1541, at another parliament—this one held in Dublin—a bill was presented that stated that Henry VIII and his heirs "should from thenceforth be named and called king of the realm of Ireland." The bill was apparently passed without the slightest opposition. Thus, the medieval lordship of Ireland and the constitutional principle that had governed Anglo-Irish relations from 1171 was brought to an end.

MARGARET MURPHY

References and Further Reading

Davies, R.R. "Lordship or Colony." In *The English in Medieval Ireland*, edited by James Lydon, 142–60. Dublin: Royal Irish Academy, 1984.

Duffy, Sean. *Ireland in the Middle Ages*. Dublin: MacMillan Press, 1997.

Frame, Robin. *Colonial Ireland, 1169–1369*. Dublin: Helicon, 1981.

Lydon, James. "Ireland and the English Crown." *Irish historical Studies*, 29, no. 115 (1995), 281–94.

Lydon, James, *The Lordship of Ireland in the Middle Ages*, 2nd ed. Dublin: Four Courts Press, 2003.

See also **Anglo-Irish Relations; Henry II; John; Parliament; Richard II**

LYRICS

The corpus of early Irish poetry contains a tiny selection of lyrics, usually anonymous, ascribed to fictitious authors or to famous individuals by the use of masks. There are no great epic poems, but the sagas are frequently composed in a combination of prose and poetry; these contain most early lyrics. A few exceptional sagas are presented in meter alone. Some lyrics appear on the margins of manuscripts and as verses illustrative of unusual meters in the metrical tracts. The eleventh century religious poet Máel-Ísu Ua Brolcháin is one of the few names that appear in this period. His compositions included the bilingual *Deus, meus*.

The earliest pieces are in *rosc*, a style of meter without rhyme, rhythm, or stanzas, depending on linking and internal alliteration. The earliest example is the eulogy to Colm Cille, composed by Eochaid Dallán Forgaill circa 600. It begins:

God, God, may I beg of him
before I go to face Him
Through the chariots of battle.
God of heaven, may He not leave me
in the path where there's screaming
From the weight of oppression.

Over a period of time, possibly influenced by the new Latin poetry, there developed the new meters based on syllabic count that lasted until the seventeenth century. These contain stanzas, alliteration, consonance, and rhyme that are impossible to illustrate in translation.

Prosimetrum, the combination of prose and poetry, is unusually common in early Irish sagas. Although some sagas contain no poetry, most prose tales include poems that appear at points of high emotion. The poem by Líadain from the saga *The meeting of Líadain and Cuirithir*, where she bemoans hurting her lover, Cuirithir, by entering a convent, for example:

Without pleasure
the deed that I have done;
the one loved I have vexed (tormented.)
. . .

I am Líadan;
I loved Cuirithir,
It is as true as is said.
. . .

A roar of fire
has split my heart;
for certain, without him it will not live

Many of the early lyrics are characterized by a love of nature and an appreciation of birds and of animals:

The little bird
that has whistled
from the end of a bill
bright-yellow.

There is also a strong tendency for the use of masks throughout, and nearly a total absence of personal, emotional poetry. In the poem about his cat, *White Pangur*, the persona of the poet is at its most immediate here, and the voice feels modern, individual, and self-reflective:

I and white Pangur
practice each of us his special art:
his mind is set on hunting,
my mind on my special craft.

It is usual, at times, for a mouse to stick
in his net, as a result of warlike battlings.
for my part, into my net falls some
difficult rule of hard meaning.

These poets use older, preestablished masks that depend on the audience recognizing the character, for example, Finn Mac Cumaill:

I have tidings for you:
the stag bells;
winter pours;
summer is gone;

Wind is high and cold;
the sun is low;
its course is short;
the sea runs strongly . . .

Many masks are female; they give a male poet the power to express emotions that might otherwise be seen as female fragility. One historical female poet, Úallach daughter of Muinechán, appears in the annals, and the following poem, the *Caillech Bérre*, may have been composed by a woman.

Ebb-tide to me as to the sea;
old age causes me to be sallow;
although I may grieve thereat,
It comes to its food joyfully.

I am the Old Woman of Beare, from Dursey;
I used to wear a smock that was always new.
Today I am become so thin that I would not
Wear out even a cast-off smock.

Religious poetry also uses masks, for example, Colm Cille:

My hand is weary with writing;
my sharp great point is not thick;
my slender-beaked pen juts forth a
beetle-hued draught of bright blue ink.

The early lyrical hymns remain anonymous:

Shame to my thoughts
how they stray from me!
I dread great danger from it
on the day of lasting doom.

Finally, there are those stray stanzas, personal, funny, and touching:

I do not know
who Etan will sleep with,
but I do know that blond Etan
Will not sleep alone.

He's my heart,
a grove of nuts,
he's my boy,
here's a kiss for him.

Bitter is the wind tonight,
it tosses the sea's white hair;
I do not fear the wild warriors from Norway,
Who course on a quiet sea.

MUIREANN NÍ BHROLCHÁIN

References and Further Reading

Greene, D. and O'Connor, F. *A golden treasury of Irish poetry: A.D. 600 to 1200*. London: MacMillan, 1967.

Murphy, G., *Early Irish Lyrics*. Oxford: Oxford University Press 1965; reprint, Dublin: Four Courts Press, 1998.

Clancy, T.O. "Women poets in early medieval Ireland: stating the case." In *The fragility of her sex? Medieval women in their European context*, edited by C.E. Meek, C.E. and M.K. Simms, 43–72. Dublin: Four Courts Press, 1996.

Dillon, M. "Irish poetry." In *Early Irish Literature*, 149-98. Chicago: Chicago University, 1948, reprint Dublin: Four Courts Press, 1994.

Jackson, K. *Studies in early Celtic nature poetry*. Cambridge: Cambridge University Press, 1935; reprint, Wales: Llanerch publishers, 1995.

Mac Cana, P. "Notes on the combination of prose and verse in early Irish narrative." In *Early Irish literature–media and communication, Scriptoralia 10*, edited by Stephen N. Tranter and H.L.C. Tristram, 115–47. Tübingen: Gunter Narr, 1989.

Tymoczko, M. "A poetry of masks: the poet's persona in early Celtic poetry." In *A Celtic Florilegium: studies in memory of Brendan O Hehir*, edited by Kathryn A. Klar, Eve E. Sweetser, and Claire Thomas, 187–209. Lawrence, Massachusetts: Celtic Studies Publications, 1996.

See also: **Máel-Ísu Ua Brolcháin; Metrics; Poetry Irish; Women**

M

MAC AODHAGÁIN (Mac EGAIN)

Although members of the Clann Aodhagáin are on record as poets, clerics, and other professionals—note, for example, the Mac Aodhagáin contribution to the early seventeenth-century poetic contention *Iomarbhágh na bhFileadh*—the family is best known as the most influential of all the hereditary legal families of late medieval Ireland. The family produced both academic and practicing lawyers, in the latter case acting for the most prominent of the ruling families of Connacht and the adjacent midlands between the fourteenth and sixteenth centuries, as well as the Meic Carthaig (MacCarthys) of Desmond. Clann Aodhagáin was widely dispersed with important seats at *Baile Mhic Aodhagáin*, "Mac Egan's homestead" (anglicized Ballymacegan), at the northern end of Lough Derg in Ormond (north Tipperary); *Dún Daighre* (Duniry), between Loughrea and Portumna in southeast Galway; and *Páirc* (Park) near Dunmore in northeast Galway.

The most famous manuscript associated with Clann Aodhagáin is the fifteenth-century *Leabhar Breac*, a great collection of primarily religious material transcribed by Murchad Riabhach Ua Cuinnlis while employed in various Mac Aodhagáin scriptoria in Ormond in or around 1398. It is also known as *Leabhar Mór Dúna Daighre* on account of its being held at Duniry in the sixteenth century. The fourteenth-century Book of Ballymote may also be connected with Clann Aodhagáin, as sections of the manuscript were transcribed in Munster in the home of Domnall Mac Aodhagáin (sl. 1413), who appears to have served as tutor to the scribes Magnus Ua Duibhgeannáin, Solamh Ua Droma, and Robeartus Mac Síthigh. The majority of manuscripts with which members of the family can be connected are primarily legal compilations; indeed, it has been noted that most of our extant legal manuscripts have some connection with Clann Aodhagáin. The oldest surviving manuscript containing mostly legal material, the early fourteenth-century manuscript H.2.15A now held in Trinity College, Dublin, includes some commentary by Áed Mac Aodhagáin (sl. 1359) who worked on the manuscript in 1350 and who notes that it was formerly in the possession of his father, Conchobar. Another member of the family, Gilla na Náem mac Duinn Sléibe (sl. 1309), is likely to have been the author of a fourteenth-century legal manual (a copy of which is found in TCD H.3.18)—which has been described as "basically a précis of Old Irish law-texts"—as well as two items of verse: one a summary of the law text *Di Chethars [h]licht Athgabála*, the other offering advice to a student of law. An important legal manuscript with Mac Aodhagáin associations is the sixteenth-century Egerton 88 in the British Library, much of which is thought to have been written at the family law-schools in Galway. The influence of the Clann Aodhagáin law-schools persisted into the seventeenth century. Brother Mícheál Ua Cléirigh is thought to have visited the Ballymacegan school conducted by Flann (sl. *c.* 1643) and his elder brother Baothghalach Ruadh Mac Aodhagáin (*fl. c.* 1628), on three occasions. The other great antiquary of the time, Dubhaltach Óg Mac Fhirbhisigh is known to have visited Ballymacegan in 1643.

MÍCHEÁL Ó MAINNÍN

References and Further Reading

Blake, Martin J. "Two Irish Brehon Scripts: With Notes on the MacEgan Family." *Journal of the Galway Archaeological and Historical Society* 6 (1909): 1–8.

Costello, T. B. "The Ancient Law School of Park." *Journal of the Galway Archaeological and Historical Society* 19 (1940): 89–100.

Egan, Joseph J. and Mary Joan Egan. *The Birds of the Forest of Wisdom: History of Clan Egan*. Ann Arbor, Mich. Irish American Caltural Institute, 1979.

Kelly, Fergus. *A Guide to Early Irish Law*. Dublin: Dublin Institute for Advanced Studies, 1988.

McKenna, Lambert. *Iomarbhágh na bhFileadh: The Contention of the Bards*. 2 vols. London: Irish Texts Society, 1918.

MacHale, Conor. *Annals of the Clan Egan: An Account of the Mac Egan Bardic Family of Brehon Lawyers*. Enniscrone: published by the author, 1990.

Ní Dhonnchadha, Máirín. "An Address to a Student of Law." In *Sages, Saints and Storytellers: Celtic Studies in Honour of Professor James Carney*, edited by Donnchadh Ó Corráin, Liam Breatnach, and Kim McCone, pp. 159–177. Maynooth: An Sagart, 1989.

Ní Maol-Chróin, Caitilín. "Geinealaigh Clainne hAodhagáin A.D. 1400–1500, Ollamhain i bhFéineachus is i bhFilidheacht." In *Measgra i gCuimhne Mhichíl Uí Chléirigh*, edited by Sylvester O'Brien, pp. 132–139. Dublin: Assisi Press, 1944.

Ó Muraíle, Nollaig. *The Celebrated Antiquary: Dubhaltach Mac Fhirbhisigh (c. 1600–1671), His Lineage, Life and Learning*. Maynooth: An Sagart, 1996.

See also **Law Schools; Learned Families; Law Tracts**

MAC CARTHAIG, CORMAC (*fl.* 1138)

King of Munster 1123–1138, Cormac Mac Carthaig was the second son of Muiredach mac Carthaig of Éoganacht Chaisil, a branch of the Éoganachta group of dynasties who ruled Munster from the early medieval period until Brian Boru and the Dál Cais from north Munster supplanted them in 978. In 1070, the Uí Briain wrested the Éoganacht ancestral lands at Caisel (Cashel, Co. Tipperary) from them, and in Muiredach mac Carthaig's reign as king of Éoganacht Chaisil, they migrated westward and occupied territory in north Cork around Duhallow. The family took the surname of Mac Carthaig from Cormac's grandfather, Carthach (d. 1045). Up to 1114, Muirchertach Ua Briain was the undisputed king of Munster and de facto king of Ireland. This situation changed after a serious illness, and he retired from active life, dying in 1119. Cormac's elder brother Tadc became king of Éoganacht Chaisil around 1116 and began building support for the Mac Carthaig position, gaining support first from within Éoganacht Chaisil and then from other Éoganacht branches. By 1118, he had control over south Munster, and when the Uí Briain tried to reassert Dál Cais control, Tadc met them at Glanmire in County Cork and won a decisive battle. Tairrdelbach Ua Conchobair, king of Connacht, was at this time making a bid for the high kingship of Ireland and decided that it was in his best interests to keep Munster weak. He made a treaty with Tadc and Cormac Mac Carthaigh at Glanmire in 1118, formally recognizing Tadc as king of Desmond (south Munster), and the sons of Diarmait Ua Briain were given charge of Thomond (north Munster). From then on Ua Conchobair, king of Connacht, was the leading dynast in Ireland and divided Munster in 1121, 1122, and 1123. The Mac Carthaig brothers, Tadc and Cormac, led an expedition into Osraige in 1120 with some success as its king, Mac Gilla Pátraic, submitted at first. However the Uí Briain intervened and captured the Osraige hostages and handed them over to Conchobar Ua Conchobair who launched a series of attacks on Desmond in 1121 and 1123, destroying up to seventy churches. By early 1122, Tadc was forced to submit and was deposed the following year after a serious illness. His brother Cormac succeeded him and by all accounts was an inspiring political leader as well as an outstanding patron of the twelfth-century church reforms. In 1124, he became the first Éoganacht king of all Munster for 150 years. Between 1128 and 1131, he commissioned *Caithréim Chellacháin Chaisil*, ostensibly a biography of a tenth-century ancestor of the Mac Carthaig, Cellachán Chaisil, but in reality a propaganda tract on his family's behalf. From 1124, Mac Carthaig was one of the leaders of an alliance against Ua Conchobair. In 1125, he seized the kingship of Limerick. He challenged Ua Conchobair in 1126 but was defeated, following which his subkings deposed him and gave the kingship to his brother Donnchad. Cormac retired to the monastery of Lismore in 1127, and later that year Donnchad Mac Carthaig, with the chief subkings of Munster, submitted to Ua Conchobair who divided Munster again between Donnchad Mac Carthaig and Conchobar Ua Briain. This caused an extraordinary and unique reaction in Munster. Conchobar and Tairrdelbach Ua Briain went to Lismore and offered Cormac the kingship of Munster, the first time since the Dál Cais had seized the kingship of Munster from the Éoganachta in the tenth century that they had recognized an Éoganacht king, possibly seeing in Cormac the only leader capable of defeating Ua Conchobair. The alliance between Cormac and the Uí Briain was successful and produced the main opposition to Ua Conchobair's pretensions to the high kingship. With allies from Connacht and Mide, Mac Carthaig attacked Ua Conchobair, and in 1133 the Treaty of Abhaill Cathearne was concluded, the conditions of which marked the collapse of Ua Conchobair's supremacy and at which he agreed to confine his ambitions to Connacht. In 1134, the church built by "Cormac son of Muiredach mac Carthaig," now known as "Cormac's Chapel" on the Rock of Cashel in County Tipperary was consecrated. The annals record that Mac Carthaig became king of Osraige as well as Munster in 1136. The alliance between Mac Carthaig and the Uí Briain broke down in 1133, and a bitter struggle developed between them. In 1138, Cormac was murdered at the behest of Tairrdelbach Ua Briain, who seized the kingship of Munster.

LETITIA CAMPBELL

References and Further Reading

Bugge, Alexander, ed. *Caithréim Ceallacháin Caisil, the Victorious Career of Ceallacháin of Cashel*. Oslo: J. Chek. Gundersen Bogtrykkeri, 1905.

Mac Airt, Seán, ed. *The Annals of Inisfallen*. Dublin: Dublin Institute for Advanced Studies, 1951.

Ó hInnse, Séamus, ed. *Miscellaneous Irish Annals*. Dublin: Dublin Institute for Advanced Studies, 1947.

Ó Corráin, Donnchadh. *Ireland before the Normans*. Dublin: Gill and Macmillan, 1972.

———. "*Caithréim Chellacháin Chaisil*: History or propaganda." *Ériu* 25 (1974): 1–69.

See also **Brian Boru; Dál Cais; Éoganachta; Munster**

MAC CARTHAIG (Mac CARTHY)

The eponymous ancestor of the MacCarthys, Carthach, was a grandson of Donnchad mac Cellacháin, the last king of Munster drawn from the Éoganachta of Cashel. On Donnchad's death, the kingship of Munster was seized by Mathgamain, king of the Dál Cais. The Dál Cais, and particularly the descendants of Brian Boru, undermined the political power of the Éoganachta of Cashel and, by the 1070s, had even wrested possession of Cashel itself from them. Carthach himself was not a king, but a son of his, Muiredach mac Carthaig (d. 1092), did become the king of the Éoganachta of Cashel, though it seems likely that the territory he ruled over lay somewhere between Emly and Muskerry. Muiredach's brother and successor was killed soon afterward, probably by Cellachán Ua Cellacháin (O'Callaghan), who was killed in turn in 1115. It was at that point that Tadc Mac Carthaig, who may have been responsible for Cellachán's death, became the king of the Éoganachta of Cashel. Tadc and his brother Cormac were the founders of the MacCarthaig dynasty whose members would dominate southern Munster for almost five centuries.

Tadc MacCarthaig, as king of the Éoganachta of Cashel, may have possessed more prestige than power, but his regal pedigree allowed him to pose as the leader of a great rebellion of south Munster dynasties against Muirchertach Ua Briain, the ailing king of Munster, in 1118. The MacCarthaig-led rebellion succeeded in dividing Munster in two, a division that was confirmed by the next high king of Ireland, Tairrdelbach Ua Conchobair, king of Connacht.

Tadc MacCarthaig (c. 1123), king of Desmond (South Munster), aspired to being the king of all of Munster, but his ambitions were thwarted by Tairrdelbach Ua Conchobair, who did not wish to see a strong Munster on his southern flank. Tadc was succeeded by his brother Cormac who is now best remembered for his chapel at Cashel with its striking Romanesque architecture. Cormac MacCarthaig chaffed under the restraints imposed by Tairrdelbach Ua Conchobair, and he led several coalitions of provinces against the high king. Late in 1126, though, Cormac was defeated by Ua Conchobair, and the nobles of Desmond deposed Cormac as their king to escape Ua Conchobair's wrath. MacCarthaig retired to the religious community at Lismore but in February 1127 was persuaded by Conchobar Ua Briain, king of Thomond, to become the king of a united Munster and to resume the war against the high king. MacCarthaig commissioned the composition of *Caithréim Cheallacháin Chaisil*, a propaganda tract that glorified his rule of Munster by eulogizing his ancestor, Ceallachán (d. 954), the last great king of the Éoganachta of Cashel to be the king of Munster.

Cormac MacCarthaig proved to be an inspirational leader and, after a long and hard-fought war, finally overthrew the high kingship of Tairrdelbach Ua Conchobair in 1133. However, its goal achieved, his coalition duly disintegrated. In 1138, Tairrdelbach Ua Briain had Cormac assassinated and became king of Munster. It was not until a second rebellion in 1151, led by Cormac's son Diarmait MacCarthaig, that the kingdom of Desmond was revived.

Diarmait MacCarthaig was an effective ruler who aspired to be the king of Munster. Relations between the MeicCarthaig and Uí Briain were hostile. In 1171, Henry II came to Ireland to impose his authority over the Anglo-Norman adventurers who had invaded the country and recently defeated Ruaidrí Ua Conchobair, king of Ireland, in battle near Dublin. Diarmait took the opportunity of Henry's visit to seek an alliance with the English against Domnall Ua Briain, king of Thomond—but Ua Briain then submitted to Henry II to negate any advantage MacCarthaig had hoped for. These submissions of the Munster kings may have been instrumental in persuading Henry II to retain an interest in Ireland.

In 1177, Anglo-Norman adventurers led by Robert fitz Stephen and Miles de Cogan invaded the kingdom of Desmond and occupied the Hiberno-Norse town of Cork. MacCarthaig resisted, but the intervention of the Uí Briain, seeking to exploit Desmond's difficulties, undermined MacCarthy's position and forced him to submit to fitz Stephen and de Cogan. MacCarthaig ceded seven cantreds to the knights, as well as Cork and the cantred of Kerrycurrihy to Henry II, and promised to pay a tribute on his remaining twenty-four cantreds. In 1182, Diarmait MacCarthaig led a great assault against the English colonists, but failed to oust them from Desmond. In 1185, he was killed by some knights while parleying with them at Kilbane, west of Cork.

Domnall MacCarthaig, Diarmait's son and successor, maintained the war against the English and even asserted his overlordship over Cork at one point. However, once he died in 1206, there was a succession dispute among the MeicCarthaig that the English exploited to extend their conquests across almost all of the area of modern County Cork and to hold it down with a string of castles built as far west as Bantry and possibly beyond. Domnall's son, Diarmait (d. 1229), succeeded in reigning as king of Desmond, though he actually ruled a very diminished kingdom. Diarmait's brother, Cormac Finn MacCarthaig, reigned until his death in 1247 and was succeeded by another brother, Domnall Got. Domnall Got MacCarthaig had established a lordship for himself on O'Mahony territory in southwest Cork, the basis of the lordship of MacCarthaig Reagh (Riabhach) of Carberry. Domnall Got was killed by John fitz Thomas Fitzgerald, the leading colonist in Kerry in 1257.

Finín MacCarthaig, Domnall Got's son and successor, battled with success against the O'Mahonys, who wanted their land back, and against the English lords who dominated southern Munster. After he destroyed a number of English castles in 1261, the English mounted a major campaign against him. However, at the Battle of Callan, near Kenmare, on July 24, 1261, Finín routed the English decisively. He overthrew a series of castles along the south coast and drove the colonists eastward toward Cork and Kinsale. He was killed in battle at Rinrone late in 1261 while pressing home his advantage over the colonists.

In 1262, Domnall Ruad MacCarthaig became the king of Desmond and reigned until 1303. In 1280, he captured the castle at Dunloe. He established Feidlim, a grandson of King Diarmait who died in 1229, on the borderlands north of Killarney and recognized the lordship of Domnall Óc MacCarthaig, a son of Domnall Got, over Carberry. It became a feature of MacCarthaig power that ambitious collateral branches of the former royal lineage were established on borderlands to absorb their energies at the expense of their neighbors.

The main line of the royal lineage of Desmond took on the name "Mac Cartaig (MacCarthy) Mór" during the course of the fourteenth century, and the title of king was abandoned. The MeicCarthaig Mór had their powerbase in the south of what is now County Kerry—hence Kerry's popular designation as "The kingdom." They exercised an overlordship over the lesser MacCarthaig lordships in modern County Cork, but in the later Middle Ages that overlordship grew weaker over the increasingly powerful lordships of MacCarthaig Reagh of Carberry, MacDonogh (Mac Donnchada) MacCarthaig of Duhallow and MacCarthaig of Muskerry.

Cormac MacCarthaig (d. 1359), grandnephew of Domnall Ruad MacCarthaig, while king of Desmond adopted a policy of cooperating with the English crown in order to bolster his position vis-à-vis the earls of Desmond. MacCarthaig received a royal grant of extensive lands in Muskerry and Coshmang. These lands were entrusted to collateral branches of the former royal lineage of Desmond who pushed the boundaries of MacCarthaig power ever eastward. The MeicCarthyaig Mór succeeded in passing power from father to son from the time of Cormac, son of Domnall Ruad, through five subsequent generations to 1508. This kept the dynasty united and strong into the sixteenth century. In 1565, Domnall MacCarthy Mór (d. 1596) was made the first earl of Clancare, ensuring the dynasty remained powerful into early modern times.

Through the course of the later Middle Ages the MacCarthys Reagh came to dominate virtually all of the south of modern County Cork, pushing the English colony into coastal enclaves in the baronies of Ibawn and Courceys. By the late fifteenth century, MacCarthaig Mór exercised very little authority over MacCarthaig Reagh. The MacDonogh MeicCarthaig of Duhallow came to dominate the upper Blackwater valley in the latter Middle Ages, at the expense of the Barrys. They were subject to some continued overlordship by MacCarthaig Mór into the early sixteenth century. The MeicCarthaig of Muskerry proved to be the most dynamic of the collateral branches of the former royal dynasty. They tended to seek legitimation from the English crown for their land acquisitions as they expanded closer to Cork city. The most renowned of the MeicCarthaig lords of Muskerry was Cormac mac Taidc (d. 1495), founder of Kilcrea Friary and builder of the famous Blarney Castle. He pushed the boundaries of MacCarthaig power to Carrigrohane, just west of Cork. The castle at Blarney, built probably in the 1480s, allowed the MeicCarthaig of Muskerry to overawe the citizens of Cork and exact from them an annual "black rent" (a financial tribute). For a brief period from 1535 to 1536, Cormac Óc MacCarthaig, lord of Muskerry, enjoyed control of Kerrycurrihy, on the shores of Cork harbor. That barony reverted to the earl of Desmond on Cormac Óc's death, but MacCarthaig expansion around Cork continued piecemeal until late in the sixteenth century. Most of the former kingdom of Desmond was once more under the control of MeicCarthaig.

Henry A. Jefferies

References and Further Reading

Jefferies, Henry A. "Desmond: The Early Years and the Reign of Cormac MacCarthy." *Cork Historical and Archaeological Society* 88 (1983).

———. "Desmond before the Norman Invasion." *J.C.H.A.S.* 89 (1984).

Nicholls, Kenneth W. "The Development of Lordship in County Cork, 1300–1600." In *Cork: History and Society*, edited by P. O'Flanagan and C. G. Buttimer. Dublin: Geography Publications, 1993.

See also **Éoganachta; Mac Carthaig, Cormac; Munster; Ua Briain, Muirchertach; Ua Conchobair, Tairrdelbach**

MAC CON MIDHE, GIOLLA BRIGHDE (C. 1210–C. 1272)

Born in or near Ardstraw, County Tyrone, he is sometimes confused with an earlier poet, Giolla Brighde Albanach. Giolla Brighde Mac Con Midhe was hereditary poet to the O'Gormleys, who inhabited the area east of the Foyle between Derry and Strabane. The place, "Lerga Mic an Midhe" (< lerga Mhic Con Midhe "hillside pasture of Mac Con Midhe"), cited in a seventeenth-century tract as being near Strabane, probably contains the name of the poet's kindred. Although only one poem by Giolla Brighde for an O'Gormley survives, we have numerous poems by him for various chieftains of the O'Donnells: Domnall Mór (d. 1241) and his sons (d. 1247), Gofraid (d. 1258), Domnall Óg (d. 1282) and Aodh. Giolla Brighde also composed poems for Pádraig Ua hAnluain, lord of Orior (d. 1243), Áed Ua Conchobair, king of Connacht (d. 1247) and Roalbh Mac Mathgamna of Oriel (fl. 1270). He is perhaps best known for his poem "Aoidhe mo chroidhe ceann Briain" ("Brian's head is the care of my heart") for Brian Ua Néill and his allies who died fighting the Anglo-Normans in 1260 at the disastrous battle of Downpatrick. In this lament of over sixty quatrains, Giolla Brighde poignantly observes how unequal was the battle, since the Irish wore thin cloth but the Anglo-Normans steel armour:

> Leatrom ro chuirsead an cath
> Goill agus Gaoidhil Teamhrach:
> léinte caolshróill fá Chloinn gCuinn
> is Goill 'na n-aonbróin iaruinn.
>
> [Unequal combat did they join,
> the Foreigners and the Irish of Tara:
> there were shirts of thin satin about the Sons of Conn
> and the Foreigners were a single phalanx of iron.]

Giolla Brighde also composed some fine religious poetry, most notably the poem "Déan oram trócaire, a Thríonnóid" ("Have mercy on me, O Trinity"), in which he begs God to grant him children in place of those who have died. "A theachtaire thig ón Róimh" ("O messenger who comes from Rome") is a well-known poem ascribed to Giolla Brighde, which defends poets and poetry against attacks by the church:

> Dá mbáití an dán, a dhaoine,
> gan seanchas, gan seanlaoidhe,
> go bráth acht athair gach fhir
> rachaidh cách gan a chluinsin.
>
> [If poetry were suppressed, O people,
> so there was neither history nor ancient lays,
> every man for ever would die unheard of
> except for the name of his father.]

The vocabulary and style of this poem are somewhat different from the rest of Giolla Brighde's work, and he may possibly not have been the author.

Giolla Brighde's death is mentioned in a poem by Brian Ruadh Mac Con Midhe (+1452), "Lenfat mo cheart ar Cloinn Dálaigh" ("I will claim my right from the sons of Dálach"). The poem tells how Domnall Óg Ua Domnaill, to whom Giolla Brighde had addressed several poems, plundered the O'Gormleys. The latter made no attempt at resistance because they were burying Giolla Brighde. On hearing why they had not withstood him, Ua Domnaill returned all the booty he had taken out of respect for the dead poet.

NICHOLAS WILLIAMS

References and Further Reading

Williams, Nicholas. The Poems of Giolla Brighde Mac Con Midhe. Dublin: Irish Texts Society, 1980.

MAC DOMNAILL (MACDONNELL)

The Irish dynasty of Mac Domnaill or MacDonnells (with the exception of the minor lineage of the same name in Monaghan and Fermanagh) were descended from the great Scottish house of MacDonald of the Isles, the ultimately dominant branch of the descendants of Somerled (d. 1164) of Argyll, who established his power in the Western Highlands and Islands at the expense of the Scandinavian kings of Man and the Isles. Although their later genealogical traditions were confused, it is clear from the earlier sources that all the Irish galloglass lineages were descended from Alexander (d. 1299?), elder brother of Óengus Óg (d. c. 1330) who was ancestor of the later Lords of the Isles (and of the Macdonnells of the Glens, see below). Alexander's sons and grandsons were serving as galloglass in Ulster from the 1340s at least, and in 1366 Alexander's son Ragnall, defeated in his struggle for the lordship of the Isles against his cousin John of Islay, came to Ireland and was immediately engaged as a mercenary in the fratricidal struggles of the Uí Néill (O'Neills). By 1373, "MacDounayll, captain of the Scots dwelling in Ulster" was already a major figure in the Ulster political scene. This might have been Ragnall's son John *maol* ("the bald") who submitted to Richard II along with Ua Néill in 1395, styling himself "captain of his nation and constable of the Irish of Ulster." In his letter to the king he complains of his kinsman, Domnall (Donald, lord of the Isles), who had driven him from his own land into Ireland. Modern historians have confused this John *maol* with Domnall's brother John *mór* (see below). John *maol* was the ancestor of the later house of Mac Domnaill gallóglach, constables to Ua Néill, who held extensive territories in Tyrone and Armagh. The numerous Meic Domnaill galloglass in Connacht (where they appear by the 1360s) and Leinster were descended from Ragnall's brother Somairle and the latter's son Marcus (d. 1397). From the beginning, they (with their kinsfolk the MacDowells or meic Dubgaill) functioned as the galloglass arm of the faction headed by Mac Uilliam Íchtarach and Ua Conchobair Ruad, as their rivals the Mac Sweeneys (Meic Suibhne) were of that led by Mac Uilliam Uachtarach and Ua Conchobair Donn. From Marcus's son Tairrdelbach descended the numerous MacDonnells of County Mayo. A grandson of Toirdelbach went to Leinster to serve the "Great Earl" of Kildare as constable of galloglass, and was ancestor of the MacDonnells of Leinster, who after the fall of the Kildares in 1534 passed as galloglass into the service of the English crown.

The other group of MacDonnells in Ireland were the descendants of John *Mór* (d. 1422) of Duniveg in Islay, second son of John of Islay by his wife Margaret Stewart, daughter of King Robert II of Scotland. As has been noted, this John has been confused with his namesake the constable of Ulster. John Mór married Margery Bisset, heiress of the Glens of Antrim, and styles himself "lord of Dunevage and the Glynnis" in his agreement of 1403 with King Henry IV. His descendants, although entering into Irish marriages, seem to have been more interested in the affairs of the Isles than in their Irish lands until John *Mór's* grandson Sir John, having proclaimed himself lord of the Isles, was captured and executed with his sons by King James IV in 1496. His grandson, Alexander Mac Eoin Catháneig (d. 1536), also seems to have sought the lordship of the Isles. He must be distinguished from his uncle and successor, Alexander *Carrach* (d. after 1542), who may in fact have been ruling the Glens during his nephew's lifetime. By the sixteenth century, MacDonnells of this stock were spreading into other parts of eastern Ulster, and in Elizabethan times they were able to take over the entire MacQuillin territory in north Antrim. The earldom of Antrim was created for Sir Randal (Ragnall) MacDonnell in 1620.

KENNETH NICHOLLS

References and Further Reading

Nicholls, K. W. "Anglo-French Ireland and after." *Peritia* 1 (1982): 386–8.

Schlegel, Donald M. "The MacDonalds of Tyrone and Armagh." *Seanchas Ardmacha* 10, no. 1 (1980–1981): 193–219.

Kingson, Simon. *Ulster and the Isles in the Fifteenth Century:The Lordship of the Clann Domhnaill of Antrim*. Dublin: Four Courts Press, 2004.

Nicholls, Kenneth. "Notes on the Genealogy of Clann Eoin Mhóir." *West Highland Notes and Queries*, ser. 2, no. 8 (1991): 11–24.

MAC FHIR BHISIGH

This family—one of the foremost of the hereditary learned families that were a feature of late medieval Gaelic Ireland—produced significant manuscript collections of medieval Irish learning and literature. Based around Killala Bay in north Connacht, most notably at Lackan (alias Lecan or Leacán Meic Fhir Bhisigh), County Sligo, the family may have had an ecclesiastical background: a son of the family's eponymous ancestor is associated in his obituary (1138) with the great monastery of Cong.

In the period 1279–1414, the Irish annals record the deaths of seven members of the family, all of them described as noted men of learning. But their foremost scholar during this early period, Giolla-Íosa, son of Donnchadh Mór, is strangely missing from the annalistic record—although there is record of the death, by drowning, of his wife in 1412.

Giolla-Íosa penned, circa 1392, the principal portion of the composite volume known as the Yellow Book of Lecan; this manuscript of ninety-nine folios contains copies of some of the most important treasures of early Irish literature, most notably an almost complete copy of the early recension of the great Ulster Cycle tale, *Táin Bó Cúalnge*. By 1397, Mac Fir Bhisigh was compiling the codex known as the Book of Lecan. While Giolla-Íosa had the assistance of three other scribes at various times over the next two decades, most of the manuscript is in his hand. Another (much smaller) manuscript of his—now in the National Library of Scotland—is extant, and he is also reputed to have begun compiling, before the year 1397, a collection of annals of which most is now lost. Another work of his is a lengthy poem composed as an inauguration ode for the local chieftain Tadhg Riabhach Ó Dubhda, who succeeded his brother, Domhnall, in 1417. (Two other long historical poems attributed to him—and as yet unedited—are preserved in the Book of Lecan (one is of 94 qq. and the other of 60 qq.). The date of Giolla-Íosa's death—like that of his birth—is unknown. The last contemporary reference to him occurs in 1418.

A second great scholar belonging to Clann Fhir Bhisigh was the seventeenth-century genealogist and scribe, Dubhaltach Óg. He was not a descendant of Giolla-Íosa Mór—they belonged to different branches of the family, both based at Lackan—but his grandfather, also named Dubhaltach, was an accomplished scribe. Dubhaltach was probably born at Lackan, the eldest of four brothers. His birth date is not recorded (but probably circa 1600), and virtually nothing is known of his early life. He may have received some schooling in Galway, and more traditional training from the learned family of Mac Aodhagáin, at Ballymacegan, County Tipperary. We know that, in addition to his native Irish—of which he had an unrivalled mastery for his time—he had a good knowledge of English and Latin, as well as some Greek.

In May 1643, at Ballymacegan, Dubhaltach copied a glossary called *Dúil Laithne* ("Book of Latin"). In that year he transcribed for the Galway scholar Dr. John Lynch a collection of Leinster historical materials now known as Fragmentary Annals of Ireland. Probably still at Ballymacegan, he copied a valuable early legal tract, *Bretha Neimheadh Déidheanach*, and an important collection of early annals, *Chronicum Scotorum*. Settled in Galway by early 1645, he copied an ancient historico-genealogical text, *Senchas Síl Ír*, from the Book of Ó Dubhagáin (alias Book of Uí Mhaine); that copy now forms part of Mac Fhir Bhisigh's great Book of Genealogies.

In 1647, Dubhaltach translated from English into Irish some tracts on the Rule of Saint Clare for the Poor Clare nuns in Galway. By early 1649, he was working on his monumental *Leabhar Genealach*, or Book of Genealogies, a compendium of Irish genealogical lore from the medieval and early modern periods collected from many sources, some of them now lost. By the close of 1650, he had completed the main text of the manuscript, including a general index. The work was executed during a very disturbed period of Galway's history: 1649 to 1650 (for example, the bubonic plague killed some 3700 of the inhabitants), and shortly after its completion Sir Charles Coote's Cromwellian army began a nine-month siege of the city.

In 1653, at an unknown location, Mac Fhir Bhisigh added hagiographical material to the Book of Genealogies from the early-fifteenth century *Leabhar Breac*. Back in his home area in April 1656 to witness his hereditary lord, Dathí Óg Ua Dubhda (David O'Dowda), wed the latter's cousin, Dorothy O'Dowd, it may well have been he who drafted the interesting "Marriage Articles" (in English). That same year, Dubhaltach compiled a work on early Irish authors which survives in a later copy by him. In October 1657, in Sligo town, he copied into the Book of Genealogies an interesting early text from a source no longer extant. In 1662, he was mentioned in print for the first and only time in his lifetime, in the book, published in France, *Cambrensis Eversus*, by his friend John Lynch. In the early 1660s, too, he was listed as liable to pay hearth-tax on a dwelling in Castletown, not far from his native Lackan.

In 1664, Mac Fhir Bhisigh added significant material to the Book of Genealogies from unknown sources. By the end of 1665, he had reached Dublin, where he was soon employed by the Anglo-Irish historian and antiquary Sir James Ware, whom he furnished with English translations of small portions of the Annals of Inisfallen and of Tigernach and a section of the now-lost Annals of Lecan covering the years 1443 to 1468. He also wrote a tract in English on early Irish bishops, drawing on various documents (few now extant) from the archives of Clann Fhir Bhisigh. Back in County Sligo in spring 1666, Dubhaltach compiled a catalogue in Irish of early Irish bishops and then undertook an abridged version of the Book of Genealogies. His original copy of the abridgement is lost, so we cannot tell if he ever finished it; both of the earliest (early eighteenth-century) copies appear incomplete. Mac Fhir Bhisigh was in Dublin at the time of Ware's death on December 1 but then returned to Connacht. Seeking patronage from Sir Dermot O'Shaughnessy in County Galway, he composed a poem in his honor, with unknown results. He may have sought support from the Marquess of Antrim, in Larne, County Antrim, and left several important manuscripts in the hands of the local learned family of Ua Gnímh. Back in his home-area, he was stabbed to death at Doonflin, in January 1671, by one Thomas Crofton in circumstances that remain unclear.

Mac Fhir Bhisigh left a substantial scholarly legacy. He was one of the last traditionally trained members of a hereditary learned family, and by his diligence as a copyist, compiler, and translator he ensured the survival of several important sources of medieval and early modern Irish history.

NOLLAIG Ó MURAÍLE

References and Further Reading

Ó Concheanainn, Tomás. "Scríobhaithe Leacáin Mhic Fhir Bhisigh." *Celtica* 19 (1987): 141–175.

———. "A Medieval Irish Historiographer: Giolla Íosa Mac Fhir Bhisigh." In *Seanchas: Studies in Early and Medieval Irish Archaeology, History and Literature in Honour of Francis J. Byrne*, edited by Alfred P. Smyth, pp. 387–395. Dublin: Four Courts Press, 2000.

O'Donovan, John, ed. *The Genealogies, Tribes and Customs of Hy-Fiachrach, commonly Called O'Dowda's Country . . . from the Book of Lecan, . . . and from the Genealogical Manuscript of Duald Mac Firbis. . . .* Dublin: Irish Archaeological Society, 1844.

Ó Muraíle, Nollaig. *The Celebrated Antiquary: Dubhaltach Mac Fhirbhisigh (c. 1600–1671), His Lineage, Life and Learning.* Maynooth: An Sagart, 1996. Reprint, 2002.

———, ed. *Leabhar Mór na nGenealach: The Great Book of Irish Genealogies, compiled (1645–66) by Dubhaltach Mac Fhirbhisigh.* 5 vols. Dublin: De Búrca, 2003–2004.

Walsh, Paul. "The Learned Family of Mac Firbhisigh." In *Irish Men of Learning: Studies by Father Paul Walsh*, edited by Colm O Lochlainn, pp. 80–101. Dublin: Three Candles Press, 1947.

See also **Annals and Chronicles; Bardic Schools, Learned Families; Genealogies; Lecan, Book of; Lecan, Yellow Book of; Mac Aodhagáin**

MAC LOCHLAINN

A leading family of the Cenél nEógain branch of the northern Uí Néill dynasty, the Meic Lochlainn (Mac Loughlin) were descended from Lochlann mac Máelsechnaill, king of Inishowen, who died in 1023. There is some confusion among the medieval genealogists in regard to the ancestry of the Mic Lochlainn, due to a deliberately forged pedigree drawn up during the reign of the high king, Domnall Mac Lochlainn (d. 1121). In reality, the Meic Lochlann were descended from Domnall Dabaill, son of the Cenél nEógain king, Áed Findliath, whose other son was Niall Glúndub, ancestor of the Ua Néill dynasty of southern Tír nEógain. The Meic Lochlainn, who were known as the Clann Domhnaill, were a great warrior family, who suppressed their rivals, the Ua Néills, and then usurped their genealogy. However, they were greatly disadvantaged in wider Irish politics by their distance from the beneficial influence of the Norse towns in southern Ireland.

Domnall Mac Lochlann, (d. 1121), king of Aileach and high king of Ireland for twenty years, foiled the attempts by the Munster high king, Muirchertach Ua Briain, to subdue the Cenél nEógain. By using the good offices of the abbot of Armagh, Domnall continually made peace with Ua Briain from 1099 to 1113. In 1110, Domnall raided Connacht and seized three thousand prisoners and many thousands of cattle. Domnall died in 1121, aged seventy-three, being called "the most distinguished of the Irish for personal form, family, sense, prowess, prosperity and happiness, for bestowing of jewels and food upon the mighty and the needy." His son Niall Mac Lochlainn succeeded him as king of Tír nEógain.

Domhnall's grandson, Muirchertach Mac Lochlainn, was high king of Ireland from 1156 to 1166. A powerful king, Muirchertach counted men such as Diarmait Mac Murchada, king of Leinster, among his vassals, and made a policy of dividing rival kingdoms such as Mide, out among subservient claimants. In 1150, Muirchertach granted twenty cattle and a five-ounce gold ring to the abbot of Derry. In 1154, he hired a Norse fleet from the Hebrides and the Isle of Man to oppose the fleet of Tairrdelbach Ua Conchobair, king of Connacht. Muirchertach's flotilla, however, was defeated in a naval battle off Inishowen. In the same year, Mac Lochlainn obtained the submission of the Norse of Dublin and granted them *tuarastal* (a ceremonial gift to seal a vassal's submission) of 1,200 cattle. In 1157, Muirchertach attended the synod of Mellifont granting "seven score cows, and three score ounces of gold, to God and to the clergy" as well as an entire town-land near Drogheda. In 1159, Muirchertach defeated Ruadrí Ua Conchobair in a battle at Ardee and in 1162 led an army against the Norse of Dublin, who submitted, paying Mac Lochlainn "six score ounces of gold." Muirchertach was a ruthless king. In 1160, he had the influential Domnall Ua Gairmledaig, lord of Cinél Móen, assassinated. However, in 1166 Muirchertach made a fatal mistake when he blinded Eochaid Mac Duinnsléibe, king of Ulaid. This blinding outraged the north of Ireland, and Mac Lochlainn was abandoned by most of his army. He was defeated and killed by Mac Duinnsleibe's foster-father and guarantor, Donnchad Ua Cerbaill, king of Airghialla. In his obit, Mac Lochlainn was called "the chief lamp of the valour, chivalry, hospitality, and prowess of the west of the world in his time."

Following Muirchertach Mac Lochlainn's death, the Ua Neills emerged again as a force to be reckoned with in Tír nEógain and took the kingship from their Mac Lochlainn rivals. In 1167, the new high king, Ruairí Ua Conchobair divided Tír nEógain in two, north and south of the mountain, between Muirchertach's son, Niall Mac Lochlainn, and Áed "An Macaoimh Tóinleasc" Ua Néill. After the Anglo-Norman invasion of Ulster, the Meic Lochlainn assisted the Ulaid against John de Courcey. In 1196, Muirchertach Mac Lochlainn, king of Tír nEógain, was noted as a "destroyer of the cities and castles of the English." He was slain in that year by an Ua Catháin, a member of a new rising dynasty in County Derry. In 1215, Áed Mac Lochlainn was killed by the English.

In the early thirteenth century, the Meic Lochlainn began to occupy the ecclesiastical center of Derry but were becoming very unpopular among the Cénel nEógain. In the 1230s, Domnall Mac Lochlainn became very powerful. In 1235, he killed Domnall Ua Néill, the king of Tír nEógain, and assumed the kingship himself. In 1238, Domnall instigated a Gaelic Irish uprising against Hugh de Lacy, east of the Bann, and in 1239 he was victorious in the battle of Carnteel, fought near Dungannon, against some Ua Néill and Ua Gairmledaig rivals. However, Domnall was crushingly defeated at the battle of Caimeirge (a site traditionally said to be near Maghera in Co. Derry), by Brian Ua Néill and Máelsechnaill Ua Domnaill, king of Tír Conaill. Domnall and ten other Meic Lochlainn of his *derbhfine* (close family) were killed. The battle of Caimeirge proved to be decisive in the struggle for power in Tír nEógain between the Meic Lochlainn and the Ua Néills. Very unusual for Gaelic Ireland, the Mac Lochlainn family was totally eclipsed and never again threatened Ua Néill hegemony of Tír nEógain. After the battle, Brian Ua Néill married Mac Lochlainn's daughter, Cecilia, and a Mac Lochlainn chieftain, Diarmaid Mac Lochlainn, was killed at Brian Ua Néill's great defeat at Down Patrick in 1260.

Nevertheless, the Mac Lochlainns did survive in a very reduced condition in the peninsula of Inishowen. They remained as chieftains but failed to retain the lordship, even of Inishowen, which was taken over first by the Earls of Ulster and then by the powerful Cenél Conaill dynasty of Ua Dochartaigh. In 1375, Sean Mac Lochlainn, "Chief of his own tribe" died, and in 1510 "Mac Lochlainn, Uaithne" died. By the end of the sixteenth century, the Mac Lochlainn chieftains still survived, tributary to Ua Dochartaigh, and held two castles on the shore of Lough Foyle. In 1601, Hugh Carrogh Mac Lochlainn "chief of his sept," held Carrickmaquigley Castle (Red Castle) and Brian Óg Mac Lochlainn held Garnigall Castle (White Castle).

DARREN MCGETTIGAN

References and Further Reading

Byrne, Francis. *Irish Kings and High-Kings*. London, 1973.

Officio Rotulorum Cancellariae Hiberniae Asservatarum, Repertorium. Vol. 2, app. 5, *Inquisitionum*. Donegal.

Ó Corráin, Donncha. *Ireland before the Normans*. Dublin, 1972.

———. "Prehistoric and Early Christian Ireland." In *The Oxford Illustrated History of Ireland*, edited by R. F. Foster. Oxford, 1989.

O'Donovan, John, ed. and trans. *Annala Rioghachta Eireann, Annals of the Kingdom of Ireland*. Dublin, 1856.

Simms, Katharine. *From Kings to Warlords*. Bury St. Edmunds, 1987.

———. "Tír Eoghain 'North of the Mountain.'" In *Derry History and Society*, edited by Gerard O'Brien. Dublin, 1999.

The names of all the chief places of strength in O'Doherty's country called Inishowen, as well castles as forts, 12 April 1601, SP 63/208PT2/22-25.

See also **Airgialla; Connacht; Courcey, John de; Derry; Downpatrick; Dublin; Lacy, Hugh de; Leinster; Mac Lochlainn; Mellifont; Mide; Munster; Ua Néill; Tuarastal; Ua Briain; Ua Conchobair, Ruairí; Ua Conchobair, Tairrdelbach; Uí Néill, Northern; Ulaid; Ulster, Earldom of**

MAC LOCHLAINN, MUIRCHERTACH (C. 1110–1166)

Muirchertach Mac Lochlainn, son of Niall Mac Lochlainn (1091–1119), prince of Ailech, was a powerful high king of Ireland. The only event of note known of his early life was the killing of his father by the Ua Gairmledaig dynasty of Cenél Móen on 28 December 1119. Thereafter nothing is known of Muirchertach due to the prominence of his grandfather Domnall Mac Lochlainn (d. 1121) and his uncle Conchobar Mac Lochlainn (sl. 1136). Muirchertach first appears after the killing of Conchobar in 1136 by Mathgamain Ua Dubda (sl. 1139), lord of Clann Laithbhertaig, and the men of Mag nItha. Although Muirchertach succeeded his uncle as over king of Northern Ireland, he faced challenges, killing Gillamurra Ua hÓgain that year. In 1139, the king of Ulaid invaded Muirchertach's home kingdom of Tír nEógain to Tullaghoge, plundering the churches in the surrounding plains. Smarting still, Muirchertach then killed Mathgamain Ua Dubda and the chief men of his territory. In 1142, he defeated the Uí Dongaile at Feara Droma, but was severely wounded during that battle and weakened politically. For in 1143, Domnall Ua Gairmledaig (sl. 1160) expelled Muirchertach from Tír nEógain and assumed its kingship. During 1145, Muirchertach returned from what is now County Donegal with the Cenél Conaill and defeated Ua Gairmledaig, but failed to depose him. And it took a second expedition to separate Ua Gairmledaig from the kingship.

Muirchertach's return to power had countrywide significance, leading to the hostages of Leinster's being sent "to his house." His problems with Ulaid reappeared during 1147 when its king Cú Ulad Ua Duinnsléibe plundered Fernmag (the modern barony of Farney, County Monaghan). In response, Muirchertach and his ally Donnchad Ua Cerbaill of Airgialla (sl. 1167) attacked Ulaid and defeated Ua Duinnsléibe on 29 June, forcing him to surrender hostages. In 1148, Muirchertach and Ua Cerbaill invaded Ulaid again, dividing it between four lords. Ua Cerbaill regretted the removal of Cú Ulad. And with Tigernán Ua Ruairc of Bréifne (sl. 1172), he restored Cú Ulad in defiance of Muirchertach, prompting the latter to expel Cú Ulad and replace him with Donnchad Ua Duinnsléibe. Before Archbishop Gilla mac Liag (Gelasius) of Armagh (d. 1173), the situation was resolved temporarily when Ua Cerbaill and the Ulaid gave Muirchertach hostages. Muirchertach's tightening grip over Ulster was displayed later that year when the Cenél Conaill also gave him hostages, while Ua Gairmledaig was banished to Connacht. Yet Ulaid remained a problem. In 1149 Cú Ulad deposed Donnchad Ua Duinnsléibe, Muirchertach's protégé, from the kingship of Ulaid. Muirchertach immediately invaded, but Ua Cerbaill intervened and gave his own son as a hostage. Cú Ulad remained recalcitrant, provoking Muirchertach, and after more devastation Cú Ulad submitted, giving up his own son as a hostage.

The northern king then began his challenge for the high kingship with a royal progress in autumn 1149, taking hostages from Ua Ruairc, Murchad Ua Máelsechlainn of Mide (d. 1153), the men of Tethbae and the Conmaicní. Muirchertach and Ua Cerbaill then traveled through Leinster to Dublin, taking the submission of Diarmait Mac Murchada of Leinster (d. 1171) before making peace between the Leinster king and the Dublin Ostmen. In 1150, Mac Lochlainn's bid for the high kingship received church recognition.

In that year Archbishop Gelasius of Armagh (d. 1173) visited Tír nEógain to receive full tribute from its churches. On that occasion Muirchertach bestowed a gift of twenty cows upon the archbishop. During the same year, Bishop Flaithbertach Ua Brolcháin of Derry (viv. 1175) also visited Tír nEógain and got a full tribute from its churches. On that occasion, Muirchertach was more magnanimous. Besides a gift of twenty cows, Mac Lochlainn gave Ua Brolcháin a gold ring, a horse, and his own battle dress. And to emphasize his rising power, Muirchertach made a royal journey to Inis Mochta (near Slane, Co. Meath) to meet Ua Cerbaill and Ua Ruairc. While there, Tairrdelbach Ua Conchobair of Connacht (d. 1156) sent him hostages without compulsion. Muirchertach on this occasion banished Ua Máelsechlainn, dividing Mide (Meath) between Ua Conchobair, Ua Ruairc, and Ua Cerbaill. This settlement of Mide did not go well among the Uí Máelsechlainn. Such was the opposition to the rule of Ua Cerbaill and Ua Ruairc that Muirchertach had to crush another rising in Mide.

In 1151, Muirchertach felt threatened by Ua Conchobair's reassertion of his suzerainty over Munster after the battle of Moin Mór. Ua Conchobair's victory over Tairrdelbach Ua Briain (d. 1167) compelled the northern king to attack Connacht, forcing the overstretched Ua Conchobair and Mac Murchada to render hostages. During 1152, Muirchertach fell out with Ua Cerbaill over the latter's feud with Archbishop Gelasius of Armagh. Typically, he seized the opportunity to pose as the defender of the church and deposed his ally. With Ua Cerbaill out of the way, Muirchertach made a peace with Ua Conchobair near Ballyshannon. There Muirchertach seems not to have objected to Ua Conchobair's intended invasion of Munster, for afterward Ua Conchobair and Diarmait Mac Murchada met him at another conference at Rathkenny in County Meath, dividing Mide between Ua Máelsechlainn and his son Máelsechlainn Ua Máelsechlainn (d. 1155). They also attacked Ua Ruairc and forced him to give up his overlordship over Conmaicne before replacing him with a kinsman as king of Bréifne. In 1153, he felt compelled to aid Ua Briain after Ua Conchobair had banished him from Munster and divided it between Tadc Ua Briain (d. 1154) and Diarmait Mac Carthaig (sl. 1185). In Mide, Ua Conchobair and Tadc Ua Briain attempted to halt the northern army. Muirchertach, however, brushed Tadc aside and routed the Leinster cavalry. Ua Conchobair then retreated across the Shannon, but Muirchertach and Ua Ruairc inflicted a heavy defeat on his son Ruaidrí Ua Conchobair (d. 1198) at Fardrum in west Meath. Victorious, Muirchertach took hostages of Máelsechlainn Ua Máelsechlainn, confirming Mide to him and gave him the Leinster subkingdoms of Uí Failge and Uí Fáeláin. He did not forget to reward Ua Ruairc, restoring him to Bréifne, and gave him Conmaicne, but he made sure to take his hostages before leaving for home. A sign of his power at this time was his ability to billet the exiled army of the ill Ua Briain throughout Ulster and Mide. But once Ua Briain recovered, he returned with Muirchertach's help to Thomond and reassumed his kingship.

Ua Conchobair was determined not to allow Mac Lochlainn get the better of him. In 1154, he plundered the coastline of Tír Conaill and Inishowen. But Connacht's maritime dominance was challenged by Muirchertach's hired fleets from the Hebrides and Man. Although the Connacht fleet was victorious, Muirchertach proved stronger on land, plundering east Connacht before compelling Bréifne to recognize Ua Ruairc's kingship. That done, he took the submission of the Dublin Ostmen and gave them *tuarastal* of 1200 cows. Upon his homecoming, he banished the son (Cú Ulad) of Deorad Ua Flainn, king of Uí Thuirtri, to Connacht for blinding his own son. The death of Máelsechlainn Ua Máelsechlainn in 1155 brought Muirchertach south again to take the hostages of the men of Tethbae before giving Mide to Donnchad Ua Máelsechlainn (sl. 1160). Ua Conchobair and his son Ruaidrí displayed their resistance to Muirchertach's settlement of Mide by building another bridge at Athlone and sacking Cullentragh Castle. Early in the year 1156, the king of Connacht made further inroads into Muirchertach's client base, taking hostages of Ua Briain before making a peace with Ua Ruairc. But Ua Conchobair was too old to go another round with Muirchertach and died that May, leaving the kingship of Connacht to Ruaidrí.

The news of Ruaidrí's accession possibly fanned an Ulaid revolt against Muirchertach. After subduing Ulaid, Muirchertach marched south and took hostages from Mac Murchada before plundering Osraige. Inevitably, Ruaidrí competed with Muirchertach for dominance over the midland kingdoms and Mide. Accordingly, the whole region was transformed into an arena where Ua Conchobair and Mac Lochlainn clients struggled for their respective kingships. That said, Muirchertach sometimes had difficulty in controlling his own clients. In 1157, Donnchad Ua Máelsechlainn of Mide killed Cú Ulad Ua Caíndealbháin of Laeghaire despite Muirchertach's protection. At the consecration of the Cistercian monastery of Mellifont, Mac Lochlainn posed as its patron, granting it land, riches, and cows. At the same time, he also presided over a convention of the clergy that excommunicated Donnchad Ua Máelsechlainn for his crime. Muirchertach banished Donnchad from Mide and gave the kingship to Diarmait Ua Máelsechlainn (sl. 1169). He then began a countrywide circuit by taking hostages of Mac Murchada and attacking the midland kingdoms of

Uí Failge, Loígis, and Osraige, forcing their kings to flee to Connacht. Diarmait Mac Carthaig of Desmond (sl. 1185) quickly gave him hostages, allowing Muirchertach to successfully besiege the Ostman town of Limerick and banish Ua Briain. Before he marched home, he divided Munster between Mac Carthaig and Conchobar Ua Briain (bld. 1158). During Muirchertach's absence in Munster, Ruaidrí attacked the north, burning Incheny near Strabane, plundering parts of County Derry and later reversed the settlement of Munster. There was more trouble in the north during 1158. Then the Cenél Conaill revolted, forcing Muirchertach to lead a hosting of Ulaid and Airgialla to waste Fanad in County Donegal. In the south, Ruaidrí again threatened, invading Leinster and taking hostages of Loígis and Osraige. Moreover, Ruaidrí and his fleet attacked Tír nEógain before ransacking Tethbae in Mide. His settlement of Mide was also under pressure from Donnchad Ua Máelsechlainn. Although defeated by Diarmait Ua Máelsechlainn and Ua Ruairc, Donnchad significantly took refuge in Ruaidrí's Connacht.

During 1159, Muirchertach changed tack in Mide, banishing Diarmait Ua Máelsechlainn and restoring Donnchad, causing Ua Ruairc to join Ruaidrí. The Connacht king now invaded Mide and pushed into Ua Cerbaill's Airgialla. Somewhere close to Ardee, Muirchertach inflicted a massive defeat on the Connacht army, pursuing Ruaidrí to the Shannon and wasting Bréifne. After rewarding Donnchad, he returned home to assemble another army to invade Connacht. Although Muirchertach sacked the Ua Conchobair capital at Dunmore, Ruaidrí would not submit, leaving the high king no alternative but to withdraw. He then expelled Ua Ruairc from Mide and billeted troops there presumably to protect Donnchad's kingship. Pragmatically, Ua Ruairc sued for peace. The high king proved generous, confirming Bréifne to him. He also confirmed Leinster to Mac Murchada, but expelled Mac Fáeláin, Ruaidrí's principal supporter, to Connacht. In spite of such success, Muirchertach faced the reemergence of discord among the subkings of Tír nEógain during 1160. The principal rebels were Domnall Ua Gairmledaig, Muirchertach's old enemy, and Áed Ua Néill (sl. 1177). Although they enjoyed success, Muirchertach defeated them near the modern Newtownstuart. He dealt ruthlessly with Ua Gairmledaig, invading Cenél Móen and having him assassinated, then dispatching his head to Armagh. However, Mac Lochlainn's difficulties and the killing of Donnchad Ua Máelsechlainn encouraged a resurgent Ruaidrí to invade Mide, culminating in his giving its kingship to Diarmait Ua Maelsechlainn. At Assaroe, Muirchertach met the Connacht king, but they could not agree to a peace. Muirchertach then marched into Mide to take its hostages and those of Bréifne, only to be confronted by the combined armies of Ruaidrí, Diarmait Ua Máelsechlainn, and Ua Ruairc. Prudently, Muirchertach avoided battle, allowing Ruaidrí to tighten his hold on Munster.

Ruaidrí continued to erode the high king's authority in 1161. He and Ua Ruairc went into Leinster and took the hostages of Uí Fáeláin and Uí Failge, leaving Fáelán Mac Fáeláin and Muirchertach Ua Conchobair Failge respectively as kings. Muirchertach could not bridle this usurpation and plundered Bréifne and west Mide. There Muirchertach took the submissions of the Dublin Ostmen and Mac Murchada, confirming Leinster to the latter. Ruaidrí now decided to recognize the high king as his overlord. Given that the high king did not interfere with Ruaidrí's clients in Uí Fáeláin and Uí Failge, the Connacht king's submission was far from unconditional. On the plain of Tethbae, Ruaidrí gave Muirchertach four hostages for Uí Briúin, Conmaicne, Munster and Mide, although the annals of Clonmacnoise say he gave twelve. In return the high king confirmed Connacht to him. Moreover, Muirchertach granted half of Mide to Ruaidrí before confirming the other half to Diarmait Ua Máelsechlainn. At this time, Muirchertach was king of Ireland without opposition. Later that year he gave a further demonstration of the effectiveness of his high kingship, presiding over a great convention of all the laity and the clergy of Ireland at Dervor in Meath. During 1162, he brought an army, including a Connacht contingent, against the Dublin Ostmen to "take vengeance upon them for his wife and for her violation." Although he plundered Fingal, he could not reduce Dublin before his return home, leaving Mac Murchada and the Meathmen to prosecute the siege to a successful end. During 1163, the Connachtmen repudiated his high kingship, killing the bodyguard of his son Niall Mac Lochlainn (sl. 1176) as he feasted in Connacht. That said, Muirchertach still received a tribute of five score ounces of gold from the men of west Mide. And in 1164 he and Archbishop Gilla mac Liag of Armagh prevented the acceptance by Bishop Ua Brolcháin of Derry of the abbacy of Iona. On the other hand, he with Ua Brolcháin began to build the great church of Derry that year. But he was distracted from this work when the Fer Manach and Uí Fiachrach Arda Sratha attacked Tír nEógain, quelling them only by killing the latter's leader.

The trouble in Ulster during 1164 was nothing compared to that of 1165 to 1166. Then Muirchertach's old problem with Ulaid reemerged with a vengeance. Early that year he deposed Domnall son of Cú Ulad Ua Duinnsléibe as king of Ulaid, replacing him with his brother Eochaid Ua Duinnsléibe. This Eochaid was also Muirchertach's "gossip" and the foster-son of

Ua Cerbaill. After becoming king of Ulaid, Eochaid too revolted, provoking the high king. Eochaid was duly banished, allowing Muirchertach to give the kingship to Donnsléibe Ua Duinnsléibe. Undeterred, Eochaid attempted to take back his kingship, but was expelled by the Ulaid and captured by Ua Cerbaill. The latter brought a repentant Eochaid before Muirchertach at Armagh and asked for him to be restored. The high king agreed, but charged Eochaid a heavy penalty in hostages, jewels, and land, granting the land to Ua Cerbaill and the church. In 1166, Muirchertach's world and high kingship unwound amid serious rebellions in Armagh, Derry, and parts of Tír nEógain. After killing one of the probable perpetuators, Áed Ua Máelfabaill, lord of Carraig Brachaide in northwest Inishowen, he then spent Easter with Eochaid. But he became suspicious of Eochaid and arrested him after a feast. Although Eochaid was under the protection of Archbishop Gilla mac Liag and Ua Cerbaill, Muirchertach ordered him and three prominent nobles of the Dál Riata blinded. The punishment meted out to Eochaid cost Mac Lochlainn dear, as his allies deserted him. An outraged Ua Cerbaill revolted and appealed to Ruaidrí Ua Conchobair, recognizing him as high king at Drogheda. At the head of large army, Ua Ruairc and Ua Cerbaill invaded Tír nEógain to hunt Muirchertach down. They found him with a small force near the woods of Uí Echach in the Fews of Armagh, defying them. During the ensuing conflict, he was beheaded by a soldier of Airgialla. He was buried at Armagh (a snub to Bishop Ua Brolcháin of Derry) and was survived by five sons.

EMMET O'BYRNE

References and Further Reading

Hennessy, W., and B. Mac Carthy. ed. *The Annals of Ulster*. 4 vols. Dublin: 1887–1901.

———. *The Annals of Loch Cé*. 2 vols. London, 1871.

O'Donovan, John, ed. *The Annals of the Four Masters*. 7 vols. Dublin: 1851.

Ó hInnse, Séamus, ed. *Miscellaneous Irish Annals*. Dublin: Irish Manuscript Commission, 1947.

Mac Airt, Seán. *The Annals of Inisfallen*. Dublin: Dublin Institute for Advanced Studies, 1951.

Murphy, Denis. *The Annals of Clonmacnoise*. Dublin: Royal Irish Academy, 1896.

Stokes, Whitley, ed. "The Annals of Tigernach." *Rev. Celt* 16–18 (1895–1897).

Hogan, James. "The Irish Law of Kingship, with Special Reference to Ailech and Cenél nEógain." *Proceedings of the Royal Irish Academy*, vol. 40, sec. C (1932): 186–254.

Ua Ceallaigh, Séamus. *Gleanings from Ulster History*. Cork: Cork University Press, 1951. Reprint, Draperstown: Ballinascreen Historical Society, 1994.

Simms, Katharine. "Tír nEógain North of the Mountain." In *Derry & Londonderry: History and Society*, edited by Gerard O'Brien, pp. 149–174. Irish County History Series. Dublin: Geography Publications, 1999.

Byrne, F. J. *Irish Kings and High-Kings*. London: B. T. Batsford, 1973. Reprint, Four Courts Press: Dublin, 2001.

Ó Corráin, Donncha. *Ireland before the Normans*. Dublin, 1972.

See also **Mac Murchada, Diarmait; Ua Conchobair, Ruaidrí; Ua Conchobair; Tairrdelbach**

MAC MAHON

The Mac Mahon (*Mac Mathgamna*) family were Gaelic lords of Airgialla (Oriel), approximating to what is now County. Monaghan, during the late medieval period, on the marchlands of Gaelic Ulster and the northern Pale. They are first mentioned in the annals in 1181, and were related to the previous Ua Cerbaill ruling dynasty of Airgialla who surrendered the lordship to the Mac Mahons in the late twelfth century. The Mac Mahons were a border people, who came alternately under pressure from Ua Néill, lord of Tír nEógain, and the Anglo-Irish of Louth. Throughout the entire medieval period the Mac Mahon lordship was also a relatively poor region. There were no important religious houses or Mac Mahon castles, although during the fifteenth century the Mac Mahon lords did become noted patrons of bardic poets.

The first powerful Mac Mahon ruler was Niall, who ruled Airgialla in the late twelfth and early thirteenth centuries. Niall employed mercenary bandits and moved the Mac Mahons from Farney into Cremorne. He fought both English and Irish enemies, being an ally of John de Courcy. His most notable exploit was the killing of Éicnecán Ua Domnaill, lord of Tír Conaill in 1207, who was killed while on a raid into Fermanagh. The Mac Mahons came under severe pressure from the English of Louth in the 1250s and 1260s, which greatly weakened the family. They then came under the overlordship of the earls of Ulster and the de Verdons of Louth, who imposed claims for military service on them. As the fourteenth century progressed, the dynasty became more powerful, and by the end of that century various branches of the Mac Mahon family ruled the territories of Farney, Cremorne, Dartry, and Monaghan, with the Mac Kenna family of Truagh under the overlordship of the ruling Mac Mahon. The mid-fourteenth-century lord, Brian Mór, was the most powerful ruler of Airgialla. In 1346, he won a great victory over the English of Louth when he killed over 300 English soldiers. In 1365, Brian "assumed the lordship of Airgialla," but became embroiled in a war with Ua Néill of Tír nEógain for drowning Somhairle MacDomnaill, Ua Néill's galloglass constable. Brian Mór's chief fortress was at Rath-Tulach in the barony of Monaghan. He was slain in 1372 by one of his own bodyguards. Brian Mór levied black rents on the English of Louth, which included tributes of fine clothes, silver, and malt, and was referred to as "undisputed high king of Airgialla, a man who held sway [over the territory extending] from the Boyne to Derry, and from Gleann righe [Newry] to Bearamhan in Breifne." However, in 1368, even Brian Mór had to yield half of Airgialla to Niall Mór Ua Néill as *éiric* (payment; compensation; legal fine, especially for violation of honour or manslaughter) for killing Ua Néill's galloglass constable, and in 1370 Ua Néill killed "very great numbers of Mac Mahon's people." In the late fifteenth century, Áed Óc (1485–1496) was a powerful figure. In 1486, he burned 28 townlands belonging to the English in Airgialla and in 1494 inflicted a sharp defeat on an English force, killing 60 gentlemen. Having become blind, he died in 1496. His brother Magnus was noted for displaying the severed heads of his English enemies on the palisade around his bawn at Lurgan.

By the sixteenth century there were three main branches of the Mac Mahon family in Airgialla, the branches of Monaghan, Dartry, and Farney, all descended from Ruaidrí who died in 1446. From 1513, the Monaghan branch of the dynasty monopolized the chieftaincy. English interference in the Mac Mahon lordship became very serious as the sixteenth century progressed. In 1576, Walter Devereux, earl of Essex, was granted the barony of Farney, and in 1590, the Mac Mahon chieftain, Hugh Roe, was executed by Lord Deputy William Fitzwilliam, who divided Airgialla up among the chief lords and freeholders. During the Nine Years' War, Brian Mac Hugh Óg Mac Mahon of the Dartry branch, was a prominent leader, of note for betraying the confederate cause on the eve of the battle of Kinsale for a bottle of whiskey.

It is important to note that during the medieval period there was an important and completely separate Mac Mahon family, lords of Corcu-Baiscinn, in Thomond, descended from the high king, Muirchertach Ua Briain (d. 1119). In 1404, they are referred to as Mac Carthy's "chief maritime officer." The chieftain, Tadhg Caech Mac Mahon, lord of west Corc-Baiscinn, (1595–1602), was prominent during the Nine Years' War on the Confederate side. He was expelled from his lordship by the earl of Thomond and fled to Red Hugh O'Donnell, the lord of Tír Conaill. Tadhg Caech was shot dead in 1602, it being stated that "There was no triocha-chead in Ireland of which this Tadhg was not worthy to have been lord, for [dexterity of] hand, for bounteousness, for purchase of wine, horses, and literary works."

Darren McGettigan

References and Further Reading

Duffy, P. J. "The Territorial Organisation of Gaelic Landownership and its Transformation in County Monaghan, 1591–1640." *Irish Geography* (1981): 1–26.

MacDuinnshleibhe, Peadar. "The Legal Murder of Aodh Rua McMahon, 1590." *Clogher Record* (1955): 39–52.

O'Donovan, John., ed. and trans. *Annala Rioghachta Eireann, Annals of the Kingdom of Ireland*. Dublin, 1856.

Pender, S. "A Tract on MacMahon's Prerogatives." *Études Celtiques* (1936): 248–260.

Simms, Katharine. "Gaelic Lordships in Ulster in the Later Middle Ages." PhD diss., Trinity College Dublin, 1976.

Smith, Brendan. *Colonisation and Conquest in Medieval Ireland, The English in Louth, 1170–1330*. Cambridge, 1999.

See also **Airgialla; Anglo-Norman; Courcy, John de; Derry; Military Service; Muirchertach Mac Carthy; Pale; Ua Briain; Ua Domnaill; Ua Néill; Ulster, Earldom of; Verdon, de**

MAC MURCHADA, DIARMAIT

Diarmait Mac Murchada (b. 1110; d. Ferns, 1171), king of Leinster, was famous as the king who appealed for military aid to King Henry II of England (1154–1189) and thereby precipitated the Anglo-Norman invasion.

Mac Murchada is certainly one of the most maligned historical figures in what is sometimes termed the "Irish national memory." There, when he is remembered at all, it is as a traitor to Ireland, responsible for the oppression of his own race and for postponing by eight centuries the emergence of a national state. It has been with some vigour that Irish historians have taken on the task of revising this view, and although a scholarly biography has yet to be published on Mac Murchada, they have been generally successful. Leaving to one side the problems with teleological history that links twentieth-century problems with twelfth-century events, Mac Murchada's "treasonous" actions have become comprehensible, even natural, when he is studied in his own context.

Background and Early Career

We have an unusually full knowledge of Diarmait Mac Murchada because we can supplement Gaelic sources for his career, such as the annals and the *Book of Leinster*, with two Anglo-Norman texts documenting the invasion of Ireland: the *Expugnatio Hibernica* ("The Conquest of Ireland") by Giraldus Cambrensis, and the metrical history in French known as the *Song of Dermot and the Earl*. Nonetheless, his early career remains relatively obscure. Giraldus included a description of Diarmait in his work on the conquest of Ireland:

> Diarmait was tall and well built, a brave and warlike man among his people, whose voice was hoarse as a result of constantly having been in the din of battle.

It is not hard to believe that, by the time the first Anglo-Norman adventurers met Diarmait in the late 1160s, a career spent striving to maintain his position had hoarsened his voice.

The homeland of the Meic Murchada dynasty was Uí Chennselaig, a region in south Leinster with its center at Ferns in modern County Wexford. Leinster was traditionally ruled by the Uí Dúnlainge dynasties of north Leinster. Following the battle of Clontarf in 1014, political instability within the north Leinster dynasties, and their rivalry with the Meic Gilla Pátraic of Osraige in south Leinster, enabled a particularly able king of Uí Chennselaig to seize the kingship of Leinster in 1042. This was Diarmait mac Máel-na-mBó (d. 1072). He went on to take the kingship of Dublin in 1052 and to lay claim, admittedly with opposition, to the kingship of Ireland. From the death of Diarmait mac Máel-na-mBó until the coming of the Anglo-Normans, the kingship of Leinster remained in the hands of the Uí Chennselaig—an extraordinary feat for a small and previously unimportant kingdom from south Leinster.

Diarmait Mac Murchada was a great-grandson of this king, and he is said to have succeeded his brother Énna as king of Leinster in 1126. He can, in truth, have been little more than king of his homeland of Uí Chennselaig at first. The intervening Leinster kings had not retained the power that mac Máel-na-mBó had attained, and Diarmait—who succeeded aged only about fifteen—was by no means secure. He was threatened by dynastic, provincial, and interprovincial enemies. There were certainly others among the Uí Chennselaig who could have put aside the claim of a youth like Diarmait, and the northern dynasties that had lost the Leinster kingship less than a century before were typically hostile. In terms of external enemies, the threat to Diarmait is obvious from the first reference to him in the annals. In 1126, they report that the king of Connacht, Tairrdelbach Ua Conchobair, marched into Leinster and deposed "the son of Mac Murchada"—an inauspicious start to a career. In order to succeed, then, Diarmait was going to have to fight.

Diarmait's early career was spent consolidating his position in Uí Chennselaig and then asserting his power over Leinster. Tairrdelbach Ua Conchobhair had supported the claim to the kingship of Leinster of one of the Uí Fáeláin, a north-Leinster dynasty based in modern Kildare. Domestic trouble in Connacht in the 1130s, however, weakened Ua Conchobair's influence in Leinster affairs and allowed Diarmait to come to prominence. He did this spectacularly by perpetrating a notorious outrage on the abbess of Kildare. Kildare, with its shrine to St. Brigit, was Leinster's foremost monastic institution, and the king of Leinster traditionally held the right to appoint the abbess. In 1132, the incumbent was an Uí Fáeláin appointee. Diarmait wished to make way for his own candidate and had the abbess's suitability destroyed with a ruthless expedient. As the annals put it: "The nun herself was taken prisoner and put into a man's bed." His hold on Leinster was similarly maintained with severity. In 1141, the north-Leinster dynasties rose against Diarmait. He crushed the rebellion and had seventeen dynasts from Uí Dúnlainge families and an unspecified number of lesser nobles killed or blinded. This was an atrocity even for Diarmait's contemporaries. A similar action in 1166 precipitated the fall of the king of Ireland, Muirchertach Mac Lochlainn. The twelfth century was not an exclusively violent time. The annals, although replete with military hostings and depredations, also report a great number of peace conferences mediated by the church and reminiscent of the peace movements on the European continent a century earlier. Diarmait had established himself as a ruthless ruler. It was a point that impressed Giraldus Cambrensis, who remarks, "He [Diarmait] preferred to be feared by all rather than loved."

Interprovincial Politics and International Contacts

By 1142, with his rivals in north Leinster devastated, Diarmait was strong enough to become involved in interprovincial politics. His policy was expedient: he cooperated with whomever would best serve his interests. It is not the case that throughout his career he harbored a grudge against the kings of Connacht and sought his political and military allies in the north. Indeed, in the early 1140s he struck up an alliance with Tairrdelbach Ua Conchobair based on their mutual enmity for the Uí Briain of Munster. It is true that Tairrdelbach's son, Ruaidrí Ua Conchobair, later became a bitter enemy of Diarmait and was instrumental in his downfall in 1166. But this was in the future. Until Tairrdelbach's death in 1156, Diarmait's interests were best served by not crossing the Connacht king. Diarmait's concerns were twofold: to secure control over Osraige as a buffer between himself and the Uí Briain of Munster, and to exert influence in the affairs of his northern neighbor Mide. In both these matters Tairrdelbach Ua Conchobair was to help him.

That Diarmait was not merely ruthless but also politically adept is shown by his actions in 1151. In that year, he and Tairrdelbach Ua Conchobair inflicted a crippling defeat on Tairrdelbach Ua Briain at the battle of Móin Mór. This victory gave Diarmait enough power to intervene in Osraige and appoint kings favourable to him. There was, however, a new power growing in the north of the country in the form of Muirchertach Mac Lochlainn, king of Cenél nEógain. Mac Lochlainn and Diarmait, although they were allies later, were not initially well disposed toward each other. This was particularly true after Diarmait's joint action with Mac Lochlainn's principle rival—Tairrdelbach Ua Conchobhair—in destroying Ua Briain at Móin Mór. The same year, following his victory in Munster, Diarmait sent Mac Lochlainn the hostages of Leinster, seemingly of his own free will. In was a shrewd political move designed to avoid the enmity of Mac Lochlainn without preventing Diarmait from acting with Ua Conchobair when it suited him. In this way, Diarmait maximized the chances of securing his interests.

In 1152, with his control over Osraige newly secured, Diarmait involved himself in the affairs of Mide. In concert with Ua Conchobair and Mac Lochlainn, he attacked Tigernán Ua Ruairc (d. 1172), the king of Bréifne, and notoriously kidnapped Ua Ruairc's wife, Derbforgaill. Both Gaelic and Anglo-Norman sources report this tale with relish, but they vary on the question of motive. *The Annals of Clonmacnoise* report that Diarmait wished "to satisfie his insatiable, carnall and adulterous lust." *The Song of Dermot and the Earl*, however, portrays the unfortunate Derbforgaill as a pawn in Diarmait's power game with Ua Ruairc:

> Dermot, king of Leinster,
> Whom this lady loved so much,
> Made pretence to her of loving,
> While he did not love her at all,
> But only wished to the utmost of his power [to be avenged on Ua Ruairc].

It was supposedly in retaliation for this that Ua Ruairc insisted on Diarmait's expulsion from Ireland, which led directly to the appeal to King Henry II. In fact, Diarmait's flight from Ireland came some thirteen years after the kidnapping. Moreover, in 1166, Ua Ruairc was an ally of Ruaidrí Ua Conchobair, whose father Tairrdelbach had been allied with Diarmait in the attack on Ua Ruairc in 1152. Neither of these facts has, however, prevented the two events' being directly connected in popular imagination. The historian F.J. Byrne snubbed both this interpretation and Derbforgaill with the memorable comment that "[she] may have been fair, but was certainly forty." Instead, he attributed Ua Ruairc's hostility to his long-standing rivalry with Diarmait over Mide. Nonetheless, there can be little doubt that the Derbforgaill affair added a personal edge to an already acrimonious relationship.

The death of Tairrdelbach Ua Briain in 1156 altered the political situation in Ireland. Muirchertach Mac Lochlainn was now without serious rival, and Diarmait threw his lot in with him; Tigernán Ua Ruairc was soon associated with the new king of Connacht, Ruaidrí Ua Conchobair. It was at this time that Diarmait began to take a serious interest in the Ostman cities of Leinster. His great-grandfather, Diarmait mac Máel-na-mBó, had set a precedent of taking the kingship of Dublin in 1052, but it was some time since more than a nominal submission had been wrung from what was emerging as the capital of Ireland. In 1162, aided by Mac Lochlainn, Diarmait forced Dublin to submit and according to the annals "obtained a great sway over them, such as was not obtained for a long time." It was his connection with the foreigners or "Gall" of Dublin, and not his appeal for Anglo-Norman aid, that won for Diarmait the nickname "Diarmait na nGall."

Diarmait already had a long association with Dublin. He had founded the Augustinian nunnery of St. Mary de Hogges there in 1146, and sometime after 1161 established the priory of All Hallows on the site now occupied by Trinity College, Dublin. This relationship is instructive in terms of assessing his subsequent appeal to King Henry II. Through trade and its coveted fleet, Dublin had a centuries-old relationship with Wales and England. Previous kings of Leinster, in claiming authority over the city, were thereby brought into this transmarine network [see *Anglo-Irish relations*]. But one does not have to dig so far into the past for an association. In 1165, the native Welsh chronicle reports that Henry II hired a fleet from Dublin to fight in his abortive Welsh campaign of that year. Diarmait, in control of Dublin, surely had knowledge of this, possibly indicating a connection with Henry II dating from only one year prior to Diarmait's flight from Ireland in 1166.

Diarmait's power was now bound up with Mac Lochlainn, and when the latter fell in 1166 Diarmait's enemies, Ua Conchobair and Ua Ruairc, rapidly moved against him. The men of north Leinster and the Ostmen of Dublin took the opportunity to rebel, and Diarmait was forced to retreat to his heartland of Uí Chennselaig. He then took the decision to sail to Bristol and seek out Henry II [*see Anglo-Norman Invasion*]. He was back in Ireland with a small group of Anglo-Norman adventurers by 1167, and Ruaidrí Ua Conchobair, now king of Ireland, allowed him to retain his homeland of Uí Chennselaig. Diarmait was, however, set on greater things and had promised his daughter in marriage and the succession to Leinster to the earl of Pembroke, Richard de Clare (Strongbow), who arrived in Ireland in 1170. By the time Diarmait died at his capital of Ferns around May 1171, his Anglo-Norman forces—although not yet entirely secure—had destabilized the political situation in Ireland, causing other Irish kings to go into rebellion and shattering the power of Ruaidrí Ua Conchobair. They were successful enough to bring Henry II to Ireland late in 1171. With that royal expedition, the history of the English lordship of Ireland began.

Assessment

Historians debate the importance of Diarmait Mac Murchada. Arguing counterfactually, they question whether, even had he never appealed to Henry II, the situation would have been very different. Sooner or later, it has been said, an English king would have turned to a conquest of Ireland. Diarmait was merely a facilitator. Perhaps, but conquest did not necessarily have to take the form it did in Ireland. The "Normanizing" kings of Scotland, notably David I (1124–1153), show that Anglo-Norman culture could become influential by subtle infiltration as well as by invasion. It is therefore still open to question whether Diarmait's submission to Henry II made a full conquest of Ireland inevitable.

Another theme that has been stressed is that Diarmait's actions were not so extraordinary. In twelfth-century Ireland, kings were willing to adopt new methods to achieve and sustain their power. We should remember that requests for foreign aid were not exceptional and were naturally directed to the military source closest to hand. Ulster, for instance, had intimate contacts with the western isles of Scotland. As recently as 1154, Muirchertach Mac Lochlainn had hired a fleet from the isles led by one Mac Scelling to counter the naval power of Tairrdelbach Ua Conchobair. Mac Murchada's contacts in Leinster lay east and south; so it was that in 1166 he set sail for Bristol.

These views have won general acceptance; the only danger is that, as revisions turn into threadbare commonplaces, the significance of Diarmait will be explained away. He was—even by the standards of his time—a ruthless and manipulative ruler, and he would have had a reputation as such without any invasion. Recourse to foreign aid may have been natural step for him. But that should not dilute the fact that the Anglo-Norman invasion was the single greatest watershed in Irish history after the conversion of Ireland to Christianity, and Diarmait Mac Murchada was central to it.

PETER CROOKS

References and Further Reading

Byrne, F. J. *Irish Kings and High-Kings*. London: B. T. Batsford, 1973. Reprint, Four Courts Press: Dublin, 2001.

Davies, R. R. *Domination and Conquest: The Experience of Ireland, Scotland and Wales 1100–1300*. Cambridge: Cambridge University Press, 1990.

Flanagan, Marie Therese. *Irish Society, Anglo-Norman Settlers, Angevin Kingship: Interactions in Ireland in the late 12th Century*. Oxford: Clarendon Press, 1989.

Gillingham, John. "The English Invasion of Ireland." In *The English in the Twelfth Century: Imperialism, National Identity and Political Values*. Woodbridge: Boydell and Brewer, 2000.

Giraldus Cambrensis. *Expugnatio Hibernica: the Conquest of Ireland*. Edited with translation by A. B. Scott and F. X. Martin. Royal Irish Academy: Dublin, 1978.

Martin, F. X. "'No Hero in the House': Diarmait Mac Murchada and the Coming of the Normans to Ireland." O'Donnell lecture 19, 1975. Dublin, 1976.

Ó Corráin, Donnchadh. "The Education of Diarmait MacMurchada." *Ériu* 28 (1977): 71–81.

———. "Diarmait MacMurrough (1110–71) and the Coming of the Anglo-French." In *Worsted in the Game: Losers in Irish History*, edited by Ciarán Brady. Lilliput Press: Dublin, 1989.

Ó Cuiv, Brian. "Diarmaid na nGall." *Éigse* 16 (1975): 136–44.

Orpen, G. H., ed. *The Song of Dermot and the Earl*. Oxford: Clarendon Press, 1892.

See also **Anglo-Norman Invasion; Diarmait mac Máele-na-mbó; Henry II; Leinster; Mac Lochlainn, Muirchertach; Ua Conchobair, Ruairí; Ua Conchobair, Tairrdelbach**

MACMURROUGH

a family line of the Uí Chennselaig that held the kingship of Leinster from the middle of the 1040s to 1603. In the 1040s the Uí Chennselaig, under the leadership of Diarmait mac Máel na mBó (sl. 1072), emerged to claim the Leinster kingship and challenge for the high kingship. It was from Diarmait's son Murchadh that the later MacMurroughs traced their descent. The death of Murchad in 1070 and the death of Diarmait mac Máel na mBó at the battle of Odba in 1072 were considerable blows, leaving the kingdom of Leinster vulnerable to Thomond and Connacht. Thereafter the descendants of Diarmait's brother Domnall Remar (sl. 1041) tried to monopolize the Leinster kingship. It was not until 1114, when Donnchad mac Murchada (sl. 1115) defeated his cousin Máel mórda, that the MacMurroughs firmly established themselves over Uí Chennselaig. This Donnchad, however, had to share the Leinster kingship with the powerful Conchobar Ua Conchobiar (O'Connor Faly) of Offaly (sl. 1115). Their joint reign was brief, as Domnall Ua Briain and the Dublin Ostmen routed the pair—burying Donnchad with a dog in the floor of the Ostman assembly house. The kingship thereafter passed to the short-lived Diarmait son of Énna MacMurrough (d. 1117) before Donnchad's son Énna MacMurrough (d. 1126) was elected provincial king. Upon his death in 1126, it appears that he was succeeded by his younger brother—the famous Diarmait Mac Murchada (d. 1171). During his early reign, Diarmait's kingship was disputed by Domnall Mac Fáeláin of Uí Fáeláin (sl. 1141)—a supporter of Tairrdelbach Ua Conchobair, high king of Ireland. Mac Fáeláin's opposition to Diarmait was finally ended in 1141, allowing the latter to rule uninterrupted until 1166. In that year Diarmait was driven from his kingdom by his enemies—allies of Ruaidrí Ua Conchobair, high king of Ireland. Diarmait returned to Ireland in 1167, eventually establishing himself with English and Welsh help over Leinster and Dublin—threatening the O'Connor high kingship. Upon his death in May 1171, Diarmait's kingdom passed to his son-in-law—Richard de Clare (Strongbow) (d. 1176). This angered some MacMurroughs, culminating in Diarmait's brother Murchad MacMurrough (sl. 1172) with other Leinster nobles waging war on the English. De Clare did much to pacify the MacMurroughs, granting Uí Chennselaig to Muirchertach MacMurrough (d. 1192) and appointed Domnall Cáemánach (Kavanagh) (sl. 1175) as seneschal of the pleas of the Irish of Leinster. Trouble flared again from 1173 to 1174 with Domnall's followers routing some of de Clare's forces. But the 1175 killing of Domnall Cáemánach—described as king of Leinster—may have contributed to a gradual settling down of relations. Thereafter, leading MacMurroughs served as officers to the Marshal heirs of de Clare, attending upon English hostings to Ulster and Connacht in 1196 and 1225. Their good relationship with the English was evidenced in 1219. Then the MacMurroughs were among the "five bloods" to be enfranchised with common law by Henry III of England (d. 1272). Disaster struck the dynasty in 1225 when four prominent members were killed on an English expedition to Connacht—perhaps accounting for the subsequent long silence in the records. In the 1270s, the MacMurroughs under the leadership of two brothers—Muirchertach and Art—reappeared. This time the MacMurroughs were less well disposed toward the English, assuming in 1274 the leadership of an Irish rebellion raging in East Leinster from 1269. The assassination of these brothers at Arklow in July 1282 quietened the dynasty until the emergence of Muiris MacMurrough (d. c. 1314) in the middle of the 1290s. Under this Muiris and a series of later leaders, they sought to tack before the political winds. This they did with some success, fighting the English or serving them against the Leinster Irish whenever the occasion suited their purpose. The MacMurroughs achieved their greatest success under Art Mór (d. 1416/1417)—the greatest of the kings of medieval Leinster. Under his leadership, the dynasty enjoyed good relations with the Leinster nobility and successfully defied the second expedition of Richard II of England. After the death of Art in 1416/1417, the MacMurroughs declined rapidly, two rival branches emerging, one descended from Art's son Donnchad (d. 1478), the other from Art's other son Gerald (d. 1431). During the late 1440s Donnchad finally made peace with Domnall Riabhach (d. 1476), son of his brother Gerald, agreeing that the latter would be his successor. Under the kingship of Domnall Riabhach and through alliance with the Butlers of Ormond, MacMurrough fortunes revived. However, Domnall Riabhach's successor from the rival branch, Murchad Ballach, grandson of Donnchad (d. 1511/1512), was faced with the rise of the power of Gerald Fitzgerald, eighth earl of Kildare (d. 1513). On 19 August 1504, Murchad and Kildare's enemies were defeated by the earl at Knockdoe, County Galway. After this, Murchad Ballach accepted Kildare suzerainty until his death in 1511/1512, as did his successor, Art Buide Kavanagh (d. 1517), son of Domnall Riabhach. In this period, most of the important MacMurroughs began to side with Piers Butler (d. 1539), later ninth earl of Ormond, against the Kildares. After the death in 1531 of Art Buide's brother Muiris MacMurrough, king of Leinster, Gerald Fitzgerald (d. 1534), ninth earl of Kildare, secured the election to the Leinster kingship of his cousin and rival Cathaoir "MacInnycross" MacMurrough (d. c. 1544). While Cathaoir fought for Kildare throughout the Fitzgerald rebellion of 1534–1535, his power was curtailed after the failure of the rebellion. The MacMurroughs survived, of course, but their leaders were generally taken henceforth from among the descendants of Gerald son of Art Mór. The last of them to bear the title king of Leinster, Domnall Spáinneach Kavanagh, leader of Sliocht Airt Buide, finally submitted in April 1602 following the collapse of Gaelic power at the battle of Kinsale.

EMMETT O'BYRNE

References and Further Reading

Barry, Terry, Robin Frame, and Katharine Simms, ed. *Colony and Frontier in Medieval Ireland*. London: Hambleton Press, 1995.

Byrne, Francis. *Irish Kings and High-Kings*. London: B. T. Batsford.

Colfer, William. *Arrogrant Trepass*. Enniscorthy: Duffry Press, 2002.

Hore, Herbert and James Graves, ed. *Social State of the Southern and Eastern Counties of Ireland in the Sixteenth Century*. Dublin, 1870.

Lydon, James. *The Lordship of Ireland in the Middle Ages*. Dublin: Four Courts Press, 2003.

Morgan, Hiram. *Tyrone's Rebellion*. London: Boydell Press, 1999.

Nicholls, Kenneth. "The Kavanaghs, 1400–1700." *Irish Genealogist*, vol. , no. 4 (1977): 435–436.

———. *Gaelic and Gaelicized Ireland*. Dublin: Lilliput Press, 2003.

O'Byrne, Emmett. *War, Politics and the Irish of Leinster 1156–1606*. Dublin: Four Courts Press, 2003.

O'Corrain, Donnchadh. "The Ui Cheinnselaig Kingship of Leinster 1072–1126." *Journal of Old Wexford Society* 6 (1976–1977): 48–52.

Otway-Ruthven, Jocelyn. *A History of Medieval Ireland*. New York: Barnes & Noble, 1993.

Simms, Katherine. *From Kings to Warlords*. London: Boydell Press, 1987.

See also **Butler-Ormond; Clare, de; Diarmait mac Máele-na-mbó; Kings and Kingship; Leinster; MacMurrough, Art; Mac Murchada, Diarmait; Strongbow; Uí Chennselaig**

MACMURROUGH, ART (C. 1357–1416/1417),

son of Art MacMurrough (Mac Murchada) the elder (d. 1362), king of Leinster, and a daughter of Philip O'Byrne (Ua Broin) (viv. 1334). Art's father died a hostage at Trim Castle, County Meath in 1362, whereupon the kingship reverted to a collateral branch of the family, but it was recovered by Art's uncle Donnchad in 1369. When he was killed in 1375, the rival line retook the kingship in the person of another Art MacMurrough (d. 1414), son of Diarmait Láimhdhearg, who was recognized as king by the English in 1377. Art, though, did not recognize his rival's position and proclaimed himself king of Leinster in 1377. Throughout 1377, Art raided the counties of Wexford, Carlow, Kilkenny, and Kildare demanding an annual fee of 80 marks. Upon his coming to peace in January 1378, the English administration met Art's demands in full, as well as giving him £40 in compensation for the killing of Donnchadh.

Besides being a great soldier, Art was an astute politician, fostering links with other Irish dynasties. One of his preferred methods was the traditional ploy of marriage—giving kinswomen in marriage to Irish leaders, creating alliances stretching from Leinster to the Shannon. In north Munster and the midlands, he focused his charms upon the O'Connor Falys of Offaly, O'Dempseys of Clanmaliere, O'Dunnes of Iregan, O'Mores of Laois, O'Carrolls of Ely, the MacGillapatricks of North Ossory, and the O'Briens of Arra to the horror of the Butler earls of Ormond, while he developed relations with the O'Byrnes and the O'Tooles of East Leinster. By 1384 Art's importance among the wider Irish aristocracy was also evident, as he formed an alliance with Brian Sreamach O'Brien of Thomond. Behind Art's diplomatic front lay a determination to support his allies. In 1386/1387, the English of Ossory pressed the MacGillapatricks of North Ossory—leading Art to intervene and rout the settlers utterly. An incident from 1386–1388 also shows how active Art was among the midland Irish during this period, as the bishop of Meath earned £214 13s. 6d. for campaigning against Art and Tadhg O'Carroll.

Yet it would be a mistake to see Art as an outright enemy of the English of Ireland, for both Irish and English lived under his kingship. In 1384 he was compensated for an attack upon his tenants during a parley, while he and Gerald O'Byrne earned £48 14s. circa 1386 for fighting the Leinster Irish. His power over Carlow was further demonstrated in 1389, when he received 10 marks from the English for the killing of some followers. Art's close relationship with the English of East Leinster was dramatically illustrated in 1390. He had married Elizabeth de Veel, heiress to the Kildare barony of Norragh. However, the Statutes of Kilkenny of 1366 prohibited mixed-race marriages, meaning that Elizabeth's lands were forfeit and were granted to John Drayton in 1390. In 1391 Art petitioned unsuccessfully to have this decision reversed. In retaliation, Carlow was destroyed by Art in 1391/1392, leading a host that included O'Ryans, O'Nolans, and O'Carrolls. During 1392, Art and the O'Byrnes, O'Tooles, and O'Mores pillaged the counties of Carlow and Kildare as far as Naas, while the townsfolk of Castledermot paid him 84 marks to go away. Shortly afterward Norragh's revenues were restored to Elizabeth.

In October 1394, Richard II of England landed at Waterford in an expedition intended to arrest the colony's decline and to force Art's submission. In response, Art plundered New Ross, but Richard proved too powerful, forcing him to submit by October 30. As a result of the charges of James Butler, third earl of Ormond, Art was briefly imprisoned, but was released for other hostages. Near Tullow in January 1395, Art and Gerald O'Byrne promised to evacuate Leinster and become royal mercenaries and conquer fresh territories. At a later meeting near Carlow, both men pledged to forfeit 20,000 marks each if these promises were broken. In return Richard restored Norragh to Elizabeth, allowing Art to encourage the Leinstermen's acceptance of the agreement. Richard, elated by his success, brought Art to Dublin and knighting him there in March.

After Richard's departure for England in May, the agreement slowly crumbled. Some of the Irish were not serious about leaving their ancestral lands, while Ormond provoked conflict with them. Also, some English attempted to kidnap Art in Dublin, although he escaped. The Leinster nobility remained at peace until the rising of the O'Tooles during the summer of 1396 and that of the O'Byrnes in early 1397. Much of the Irish anger was directed at Ormond whose ambitions were supported by Richard's heir, the Lord Lieutenant Roger Mortimer. For much of 1398, though Art remained outwardly loyal, he approved of the attacks upon the English. The struggle culminated when the O'Byrnes and the O'Tooles, along with a contingent of Art's troops, killed Mortimer at Kellistown, County Carlow in July 1398. The killing of Mortimer brought Art and the Dublin government into direct conflict, leading him to attack the English of Leinster and Meath.

On hearing of Mortimer's death, Richard brought a second expedition to Ireland—revoking the agreement with Art and granting Norragh to the duke of Surrey in May 1399. The English king landed at Waterford on June 1, but he could not corner Art, pursuing him into the Leinster mountains. Richard's decision was a disaster, as Art ceaselessly harried the English. At a failed parley, Art—described as a tall handsome man with a stern countenance—told Richard's envoy that he would never submit. The landing in England of Henry Bolingbroke forced Richard to return home in July 1399. Despite Richard's departure, Art promised his wife never to rest until Norragh was restored to her—traveling to Munster in August to aid Maurice Fitzgerald, fifth earl of Desmond, against Ormond. There he recruited mercenaries and encouraged his ally Tadhg O'Carroll to harry the lands of Ormond. Upon his return to Leinster, Art set about reenforcing his kingship over the eastern part of the province—attacking the English of Wexford in 1401.

At this stage his fame was such that there exists a strong possibility that the letter from the Welsh leader Owain Glyndwr (d. c. 1416), intercepted at Waterford in November 1401, urging the Irish kings to join him in a struggle against the English, was intended for Art. He also resumed his routine of extracting black rents from the English of Leinster—receiving 10 marks for his defense of New Ross. In 1405, he again flexed his military muscle to force the government to pay him his fee of 80 marks, ravaging Castledermot, Wexford, and Carlow until it was paid. The next year saw the government try to clip his wings. The Lord Lieutenant Thomas of Lancaster, James Butler, fourth earl of Ormond, Thomas Fitzgerald, sixth earl of Desmond, and Prior Thomas Butler of Kilmainham campaigned into Art's territory to loosen his grip there. While they failed in their ultimate objective, this did not prevent them from trying again. During late August or early September 1407, the government attacked the MacMurroughs, fighting an inconclusive battle. A considerable reverse to Art's ambitions was the killing of Tadhg O'Carroll by the English at Callan on September 9. The pressure on Art possibly encouraged the English of Wexford to resist his demands for protection money. In June 1408, Art punished them—devastating the cantreds of Forth and Bargy in southern Wexford. During the following year, Art pressed the Dublin government, laying charges against the Wexford English for non-payment of his fee. In response the administration authorized the payment to Art of his 80 marks. In spite of this, Art's struggle with the English of Wexford intensified. MacMurrough power was further demonstrated in 1413, when he destroyed the town of Wexford.

Arguably this was the high point of MacMurrough power in Leinster. In his last years, Art declined and his sons lacked his ability. This new MacMurrough weakness became clear in 1414, when the English of Wexford burnt Idrone (Co. Carlow) and captured Art's second son, Gerald MacMurrough. Although Art's eldest son, Donnchad MacMurrough, rescued his brother, it was a sign of decline. Significantly during this period Art and the Butlers became allies as evidenced by the marriage of Donnchadto Aveline Butler, the half sister of Ormond. This may have been a reaction to the arrival in Ireland during 1414 of the Butlers' enemy, the Lord Lieutenant John Talbot. Furthermore, in 1415 Art also dispatched his son Gerald Kavanagh to England with Abbot John Doun of Graiguenamanagh, to take an oath of loyalty to Henry V. In the context of Art's alliance with the Butlers, the MacMurroughs' devastation in 1416 of the Wexford liberty belonging to Gilbert Talbot (d. 1419), the lord lieutenant's elder brother, makes some sense. The annals are divided on the date and circumstance of Art's death. One account records that he died in his bed during December 1416, while another tells that the greatest of the medieval kings of Leinster was fatally poisoned at New Ross in January 1417.

EMMETT O'BYRNE

References and Further Reading

Barry, Terry, Robin Frame,and, Katharine Simms, ed. *Colony and Frontier in Medieval Ireland*. London: Hambleton Press, 1995.

Bartlett, Thomas and Keith Jeffrey, ed. *A Military History of Ireland*. Cambridge: Cambridge University Press, 1996.

Byrne, Francis. *Irish Kings and High-Kings*. London: B. T. Batsford Ltd, 1973.

Colfer, William. *Arrogrant Trepass*. Enniscorthy: Duffry Press, 2002.

Hore, Herbert and James Graves, ed. *Social State of the Southern and Eastern Counties of Ireland in the Sixteenth Century*. Dublin, 1870.

Lydon, James. *The Lordship of Ireland in the Middle Ages*. Dublin: Four Courts Press, 2003.

Nicholls, Kenneth. "The Kavanaghs, 1400–1700." *Irish Genealogist*, v, no. 4 (1977): 435–436.

———. *Gaelic and Gaelicised Ireland*. Dublin: Lilliput Press, 2003.

O'Byrne, Emmett. *War, Politics and the Irish of Leinster 1156–1606*. Dublin: Four Courts Press, 2003.

O'Corrain, Donnchadh. "The Ui Cheinnselaig Kingship of Leinster 1072–1126." *Journal of Old Wexford Society* 6 (1976–1977): 48–52.

Otway-Ruthven, Jocelyn. *A History of Medieval Ireland*. London, 1969.

Simms, Katharine. "Warfare in Medieval Irish Lordships." *Irish Sword* 12 (1975–1976): 98–105.

———. *From Kings to Warlords*. Woodbridge: Boydell Press, 1987.

See also **Kings and Kingships; Leinster; Munster**

MACSWEENEY

The Mac Sweeneys (Mac Suib[h]ne) were a galloglass family (from *gallóglaigh*—warriors from the Innse Gall or Hebrides—Scottish mercenaries who fought as heavy armed foot in Ireland), which were to the forefront of the great migration of Scottish mercenary dynasties into the north of Ireland in the thirteenth and early fourteenth centuries. They became deeply established in the Ua Domnaill lordship of Tír Conaill, but spread to many other Irish lordships, making the Mac Sweeney family the second most important galloglass dynasty in Ireland after the Mac Domnaill family. Descended from an ancestor Suibne who flourished in Scotland circa 1200., the Mac Sweeneys originally possessed the fortress of Castlesween in Knapdale and were related to the Scottish families of Mac Sween and Mac Ewen. There is some controversy over the origin of the Mac Sweeneys. Medieval Irish genealogists gave the Mac Sweeney family an elaborate descent from Anradán son of Áed Athloman Ua Néill (d. 1033), who reputedly left Ulster and settled in Scotland. However, this genealogy is probably artificial, like that of the other major galloglass family of MacDomnaill. The Mac Sweeneys are more than likely of mixed Scots-Norse or Gall-Gaedheal descent (being a mixture of Irish settlers in Scotland and Scandinavian colonists in the Isles). The first Mac Sweeney mentioned in an Irish context was captured in western Connacht in 1267 and imprisoned by the earl of Ulster. The first of the family associated with the Ua Domnaill lordship of Tír Conaill was a daughter of Mac Sweeney, who married the chieftain Domnall Óg Ua Domnaill (lord of Tír Conaill 1258–81).

All the Mac Sweeneys in Ireland are descended from Murchad *Mear* (the crazy) Mac Sweeney whose son Murchad Óg left Scotland during the first Scottish war of Independence and settled in Ireland. Murchad Óg was an ally of Niall Garbh Ua Domnaill, (lord of Tír Conaill 1342–1343). His grandson was the first Mac Sweeney lord of Fanad to be recorded in the annals. These Mac Sweeneys agreed to supply their overlord, Toirrdelbach an Fhíona UaDomnaill, (lord of Tír Conaill 1380–1422), with two galloglass for each quarter of land they possessed. It was also around this time that the first Mac Sweeney Fanad underwent inauguration at Kilmacrennan by Ua Domnaill and Ua Firgill. Previous to this the Mac Sweeneys were inaugurated at Iona in the Scottish Isles.

From Fanad the Mac Sweeneys spread to the adjacent lordship of Doe in northwestern Tír Conaill, and many more branches of the Mac Sweeney family, all descended from Murchad Mear, spread throughout Ireland. The third Mac Sweeney family in Tír Conaill, Mac Sweeney Banagh, descended from an off-shoot of Mac Sweeney Connacht, settled in Banagh in the early fifteenth century and are first mentioned in the annals in 1496. The Mac Sweeneys of Connacht, were descended from Domnall na Madhmann Mac Sweeney of Rath Glas, County Galway (fl. 1420). Mac Sweeney Connacht had branches in Sligo, Roscommon (galloglass to O'Connor Don), Clanrickard (galloglass to Burke of Clanrickard), and Thomond (galloglass to O'Brien). Mac Sweeney of Ormond, galloglass to the earl of Ormond was descended from a Duinn Sléibe Mac Sweeney. The Mac Sweeneys of Desmond, galloglass to the Mac Carthys, were descended from Donnchad Mór Mac Sweeney Doe, and were noted for their seafaring galleys. Mac Sweeney Fanad was considered to be the most senior Mac Sweeney family, with the others in descending importance.

The Mac Sweeneys were established in Tír Conaill more deeply than galloglass in any other lordship in Ireland. There Mac Sweeney Fanad, Doe, and Banagh became important territorial lords, reminiscent of the *pomeshchiks* in contemporary Russia, the galloglass in other lordships only having scattered estates. The three Mac Sweeneys in Tír Conaill became very important in the sixteenth century as supporters of their lord, Ua Domnaill, and by the early seventeenth century Mac Sweeney Fanad and Mac Sweeney Doe each provided Ua Domnaill with 120 galloglasses, with Mac Sweeney Banagh supplying 60 and a man to carry the breastplate and stone of Colum Cille.

Prominent Mac Sweeneys were Máel Muire, lord of Fanad from 1461 to 1472, his son Ruaidrí, lord of Fanad from 1472 to 1518, and the last inaugurated lord of Fanad, Domnall, who was still alive in 1619. Toirrdelbach Mac Sweeney, lord of Fanad from 1529 to 1544 commissioned the compilation of the narrative text the *Craobhsgaoileadh Chlainne Suibhne*. Murchad Mall Mac Sweeney (d. 1570) was the most important lord of Doe, being prominent in Hugh Mac Manus O'Donnell's defeat of Shane O'Neill at the battle of Farsetmore in 1567. Eoghan Óg Mac Sweeney Doe (d. 1596) was foster-father to the famous Red Hugh O'Donnell, was also a noted patron of bardic poets and sheltered Spanish Armada survivors and the lord of Bréifne in his territory. Donough Mac Sweeney was the last lord of Banagh. Of the other Mac Sweeney septs, Connor, constable of Thomond and Edmond, constable of Clanrickard, were both killed at the battle of Spancel Hill in 1559, and Domhnall son of Owen of the Lake Mac Sweeney, constable of Muskerry, "a man who had good tillage, and kept a house of hospitality," died in 1589.

DARREN MCGETTIGAN

References and Further Reading

Hayes-McCoy, Gerard A. *Scots Mercenary Forces in Ireland (1565–1603)*. Dublin, 1937.

Knott, Eleanor, ed. and trans. *The Bardic Poems of Tadhg Dall hUiginn 1550–1591*. 2 vols. London, 1922/1926.

McGettigan, Darren. "The Renaissance and the Late Medieval Lordship of Tír Chonaill 1461–1555." In *Donegal History and Society*, edited by William Nolan, Liam Royayne, and Mairead Dunlevy, pp. 203–228. Dublin, 1995.

O'Donovan, John, ed. and trans. *Annala Rioghachta Eireann, Annals of the Kingdom of Ireland*. Dublin, 1856.
Simms, Katharine. *From Kings to Warlords: The Changing Political Structure of Gaelic Ireland in the Later Middle Ages*. Woodbridge: Suffolk, 1987.
Walsh, Paul. *Leabhar Chlainne Suibhne*. Dublin, 1920.

See also **Colum Cille; Connacht; Inauguration Sites; Mac Donnell; Ulster; Ua Domnaill; Ua Néill; Ulster, Earldom of**

MÁEL-ÍSU UA BROLCHÁIN (D. 1086)

Biography

Máel-Ísu Ua Brolcháin was a religious poet from Donegal who was a member of the Armagh community. His death in Lismore is mentioned in the Annals of Innisfallen in 1086. He is recognized as one of the primary poets of his age, and there is a full-page account of his life and family in the sixteenth century *Acta Sanctorum* by Colgan. He was educated in the monastery of Both Chonais, Gleneely, beside the present day Culdaff, County Donegal. W. Reeves suggests a site in the town land of Carrowmore outside Culdaff. His death is mentioned in all major annals, but the Annals of the Four Masters give a longer notice than others:

> The senior scholar of Ireland, learned in wisdom, in piety and in poetry in both languages. So great was his erudition and scholarship that he himself wrote books and compositions of wisdom and intellect. His spirit ascended into heaven on the 16th January, as is said: On the sixteenth of January/ on the night of fair Fursa's feast,/ Máel-Ísu Ó Brolcháin perished,/ Oh! Who lives to whom this not a great distress.

Máel-Ísu reveals no personal details in his poetry but genealogical sources give his father as Máel-Brígte and his three brothers as Áed, Diarmait, and Muirecán. The Uí Brolcháin descended from the Ulster king Suibne Mend and further from Niall Naí nGiallach.

His Work

The manuscript sources attribute eight poems to Máel-Ísu: *A Aingil, beir*, *A Choimdiu báid*, *A Choimdiu, nom-chomét*, *Buaid crábuid*, *Deus Meus, adiuva me*, *Dia hAíne ní longu*, *In Spirut Naem immunn*, and *Ocht n-éric na nDualach*. Many of these are published in the anthologies of lyrical poetry by Gerard Murphy, David Greene, and James Carney. A full collection of the poems are published by M. Ní Bhrolcháin. Scholars mention him as the possible author of four further compositions. Fr. F. Mac Donncha suggested that he may also be the author of the *Passions and Homilies* because he was well educated with a deep knowledge of the scriptures and of Latin and had access to an extensive library.

The content of his poems reflect the concerns of his age, the secularization of the church and the budding reform. He composed devotional, personal prayers as well as didactic poems that reflect the beliefs and the teaching of the *Céili Dé* (culdees) in preaching restraint, fasting, continence, and study as a way of life. He prays directly to the Trinity, to Saint Michael, and to God himself, using his poetry as a vehicle for religious teaching and for personal prayer. Some of the poetry may be directed at his students—*Dia hAíne ní longu* says: "You eat,/ as for me, I shall fast,/ on account of fire which water does not extinguish/ and cold which heat does not quench." He may have moved to Lismore in search of the reforming spirit that was absent in the secular world of Armagh.

The poetry appears in a wide range of manuscripts including Laud 610 and 615, the Yellow Book of Lecan, 23 N 10 and 23 Q 1 in the Royal Irish Academy. He utilizes a wide range of meter such as *Treochair*, *Rinnard*, *Aí Freisligi* and *Cró cummaisc etir casbairdni ocus lethrannaigecht móir*.

Three poems are attributed to him by Carney; *A Chrínóc*, *At-lochar duit*, *Mo chinaid i comláine*; and Kuno Meyer cites him as the author of *Rob soraid*. Carney argues cogently for *A Chrínóc*'s being a poem to a Psalter that the poet rediscovers in old age. Greene accepts the attribution with the caveat that nothing else of Máel-Ísu's work attains the same standard. The metaphoric style of *A Chrínóc* is not found in any other of Máel-Ísu's poetry. All three poems address the themes of old age, sickness, a sinful life and impending death. Carney's ascription is primarily based upon the poet's reference to his northern origin and the improbability that two northern poets, both ill and dying, should reside in Munster at the same time. The poem *At-lochar duit* also refers to the north of the country. Carney does not examine considerations such as material, style, and meter. *Rob soraid* was attributed to him by Meyer and in common with Máel-Ísu's poetry it pleads protection for a journey and shows similarities of phrases with some others of his verse. In the shorts prayers and invocations such as *In Spirut Naem immunn* and the prayer to St. Michael, *A Aingil, beir*, he begs protection against the vices of the world. The three syllable initial line of *A Aingil, beir* intensifies the emotion: "Do not delay!/ bring my exorbitant prayer/ to the King, to the High-King." The *lorica A choimdiu, nom-chomét* seeks protection from the eight deadly sins for eight parts of the body: eyes, ears, tongue, heart, stomach, male organ, hands, and feet. The sins associated with each are outlined, for example:

"Protect my ears so that I do not listen to scandal, so that I do not listen to the foolishness of the evil world," and he continues: "Do not allow me to fall into the principal sins of the eminent, reputed eight, Christ come to me, to hunt them, to defeat them." In this he follows the teachings of the Penitentials as he does in his longest poem *Ocht n-éric na nDúalach* that treats the eight vices. Some five or six stanzas are given over to each vice and to its cure, for example: "Greed—what it does is/ to force miserliness upon you;/ a craving for all things,/ pillage, plunder and robbery. The sole cure is/ contempt for the dark world,/ being in continual poverty/ without acquiring wealth."

The renowned bilingual *Deus Meus adiuva me* is still used as a hymn in the modern Irish church, a testament to Máel-Ísu's talent: "My God help me!/ Son of God give me your love/ Son of God give me your love/ My God help me!."

MUIREANN NÍ BHROLCHÁIN

References and Further Reading

Colgan, John. *Acta Sanctorum veteris et majoris Scotiae seu Hiberniae . . . Sanctorum Insulae*. Vol. 1. Louvain: 1645. Reprint, Dublin: 1947.

Reeves, William, ed. *The Life of Columba, Written by Adamnán*. Dublin: 1857.

Carney, J. "Old Ireland and Her Poetry." In *Old Ireland*, edited by Robert Mac Nally. Dublin: 1965.

Murphy, Gerard. *Early Irish Lyrics*. Oxford: Oxford University Press, 1965. Reprint, Dublin: Four Courts Press, 1998.

Greene, David, and Frank O'Connor. *A Golden Treasury of Irish Poetry: A.D. 600 to 1200*. London: Macmillan, 1967.

Mac Donncha, F. "Medieval Irish Homilies." In *Biblical Studies: the Medieval Irish Contribution*, edited by Martin MacNamara. Dublin: 1976.

Ní Bhrolcháin, Muireann. *Maol-Íosa Ó Brolcháin*. Maynooth: An Sagart, 1986.

——— "Maol-Íosa Ó Brolcháin: an Assessment." *Seanchas Ard Mhacha* 12 (1986): 43–67.

See also **Armagh; Bardic Schools, Learned Families; Lismore, Book of; Lyrics; Metrics; Penitentials; Poetry; Irish; Uí Néill**

MÁEL-MURA OTHNA (D. 887)

Distinguished poet and historian and an ecclesiastic of Othan (Fahan, Co. Donegal), **Máel-Mura Othna** probably belonged to one of the northern Uí Néill lineages. His personal name suggests a devotion to Mura, patron saint of the Cenél nÉogain. He joined a community that had an established tradition of learning; one of its members, Fothad na canóine (d. 819), was eminent in canon law. However, Máel-Mura's reputation as a scholar rests mainly on poetry with historical themes concerned, for the most part, with the ancestry of the Irish people and the kingship of Ireland.

Of the poems ascribed to Máel-Mura, "Can a mbunadus na nGáedel?" ("Whence the origin of the Gael?") tells of the mythical Gáedel Glas and his descendant Milesius, ultimate ancestors of the "Gaelic race," who supposedly brought their people from Egypt of the pharaohs via Spain to their "promised land" of Ireland. The storyline is clearly inspired, as McCone demonstrates, by the Book of Exodus and marks an important stage in the cross-fertilization of ecclesiastical and native learning. Máel-Mura is also credited with a poem, included in the thirteenth century Book of Lecan, addressed to the Uí Néill king Flann Sinna, which charts the kings of Tara from the (probably mythical) Tuathal Techtmar to Flann.

Máel-Mura died in 887 (AU; AFM). He is styled *rígfili Érenn* (chief poet of Ireland) in his obit, while an appended verse describes him as *senchaid* (historian). Certainly the surviving works credited to him represent a major contribution to the genre of pseudohistorical literature.

AILBHE MACSHAMHRÁIN

References and Further Reading

Best, R. I., O. Bergin, and M. A. O'Brien, ed. *The Book of Leinster*. 6 vols. Vol. 2, p. 333. Dublin: Dublin Institute for Advanced Studies, 1954—1983.

McCone, Kim R. *Pagan Past and Christian Present in Early Irish Literature*, pp. 24, 27, 68. Maynooth: An Sagart, 1990.

See also **Uí Néill, Northern; Lecan, Book of; Invasion Myth; Tara**

MÁEL-RUAIN (D. 792)

Máel-Ruain was the leading proponent of the *Céli Dé* movement and founder of its most important center, Tallaght. Contemporary information about Máel-Ruain's life is very scant. His name, máel (tonsured one), and ruain (of Rúadán), suggests that he may have come from Saint Rúadán's monastery of Lothra (North County Tipperary). Máel-Ruain founded the *Céli Dé* monastery of Tallaght (old Irish *Tamlacht*) most probably in the third quarter of the seventh century. O'Dwyer has suggested, based upon a line in the *Book of Leinster*, that Máel-Ruain was given the site at Tallaght by the Leinster over king Cellach mac Dunchada (sl. 776) in the year 774. This claim is not, unfortunately, corroborated by the annals.

As a leader of the *Céli Dé*, Máel-Ruain had a number of followers. A tract from the *Book of Leinster* known as *Oentu Máel-Ruain*, "Folk of the Unity of Máel-Ruain" lists twelve of the most prominent followers of his teachings. Although the list is not contemporary to Máel-Ruain, it does give some sense of the influence and range of his teaching. Among the most notable followers were Máel-Díthruib of *Tír dá glass* (Terryglass); Fedelmid mac Crimthainn, king of Cashel; Diarmait ua hÁedo Róin of *Dísert Diarmata* (Castledermot, County Kildare) and Óengus, author of the *Félire Óengusso*.

Many of his followers did not live with him at Tallaght, but were rather personal associates residing at their own houses either attached to older monastic foundations or in some cases newly founded centers. He seems to have been in continuous contact with followers outside his Tallaght community, advising *Céli Dé* leaders on how to guide their communities toward greater spiritual purity.

Although Máel-Ruain is credited with the authorship of *The Rule of the Céli Dé*, the surviving text has been changed from the original verse into prose. It still, however, offers some insight into the sort of ascetic practices advocated at Tallaght. Additionally, there are three other works associated with the *Céli Dé* movement that show evidence of his influence. From these texts, *The Monastery of Tallaght*, *The Teaching of Máel-Ruain* and *Félire Óengussa*, we gain some understanding of the sort of community he led at Tallaght, and advocated among his associates. Máel-Ruain appears to have run a very disciplined and strict community. His monks said the whole Psalter daily, and two monks remained in the church saying the psalms until matins (the night office), when they were then replaced by another pair who said the psalms from matins until lauds. He advised his monks not to ask for news from outside, to avoid becoming involved with worldly disputes, and to never plead for anyone in a law court or assembly. In general he was wary of contact with the outside world, and forbade pilgrimage outside of Ireland, fearing that such influences would distract monks' minds from God.

Máel-Ruain also saw physical pleasures of the body as a constant threat. Women in particular presented a serious danger to one's spiritual purity. A priest who broke his vow of chastity was no longer allowed to say mass. Married couples under spiritual direction were likewise held to a severe regimen, and expected to abstain from sex for four days and nights in every seven. He encouraged the practice of the cross-vigil—praying with one's arms outstretched—as well as vigils standing in water and flagellation performed by another monk. His community abstained entirely from the consumption of alcohol, even on feast days. Similarly, he did not allow the playing of music, and when the anchorite and piper Cornán asked if he might play for him, Máel-Ruain responded that "his ears were not lent to earthly music, that they may be lent to the music of heaven." He expected his monks to consult their *anamchara*, "confessor," no less than once a year.

The overall impression of the life of a monk of Máel-Ruain is one of severe discipline and rigorous self–denial; however, excessiveness was not encouraged. He did not wish for his monks to leave the monastery, particularly in permanent exile, as this brought only questionable benefit to the exile and deprived the monastery of an important member. Even occasional exile to a local wilderness could have negative effects, for without the control of the monastic environment an anchorite could be overzealous in his penance and become unfit for normal work or communal life. Ultimately, Máel-Ruain advocated that one live in a communal lifestyle, moderate in its ascetic practice, and devoted to God.

MICHAEL BYRNES

References and Further Reading

Stokes, W., ed. and trans. *The Martyrology of Oengus the Culdee: Félire Óengussa Céli Dé*. Henry Bradshaw Society 29. London, 1905, Reprint, Dublin, 1984.

Gwynn, E. J. and W. J. Purton, ed. and trans. "The Monastery of Tallaght." *Proceedings of the Royal Irish Academy* 29, C, no. 5 (1911): 115–179.

Gwynn, E. J., ed. and trans. "The Rule of Tallaght." *Hermathena* 44, second supplemental volume (1927).

Kenney, James F. *The Sources for the Early History of Ireland: Ecclesiastical*. New York: Columbia University Press, 1929.

Hughes, Kathleen. *The Church in Early Irish Society*. London: Methuen,1966.

O'Dwyer, Peter. *Céli Dé*. Dublin: Carmelite Publications, 1977.

Etchingham, Colmán. *Church Organisation in Ireland A.D. 650 to 1000*. Maynooth: Laigin Publications, 1999.

See also **Céle Dé; Devotional and Liturgical literature; Ecclesiastical Organization; Hagiography and Martyrologies; Metrics; Moral and Religious Instruction; Penitentials; Poetry, Irish; Scriptoria**

MÁEL-SECHNAILL I (D. 862)

Máel-Sechnaill I was the son of Máel-ruanaid and belonged to Clann Cholmáin of Mide, a southern Uí Néill dynasty. The first of the kings of Tara who received the submission of all the provincial kings and earned the title "king of all Ireland," his reign witnessed the appearance of the Vikings on the political stage as allies and mercenaries. In the 840s, Clann Cholmáin were divided between the descendants of Donnchad Midi (king of Tara 770–797). Máel-Sechnaill eliminated his rivals in 845. The same year he drowned the Viking leader Tuirgéis, who had been raiding the midlands from a base at Lough Ree. When Niall Caille of Cenél nÉogain died in 846, Máel-Sechnaill succeeded him as king of Tara. A year later Fedelmid mac Crimthainn died, and the Irish kings now turned their attention to the Vikings. Máel-Sechnaill launched a successful attack on Dublin in 849. In 851 he killed the king of North Brega, who had burned the churches and fortresses of South Brega in cooperation with the Vikings. In the same year Máel-Sechnaill organized a royal conference with the king of Ulaid at Armagh, but it remains uncertain whether this was related to the presence of the Vikings. In the following years he concentrated on the subjugation of Munster, which no king of Tara had attempted before. In 854 and 856, he took hostages of the province, and hence his authority nominally covered the whole island of Ireland. In the meantime the Vikings had gained in strength with the arrival of Amlaíb (Olaf) and Imar (Ivar) on the scene. In 856, Máel-Sechnaill attempted to curb their activities by hiring Gall-Goídil (Norse-Irish) as his mercenaries to fight for him. The situation rapidly escalated. In the south the Vikings teamed up with Cerball of Osraige, and in the north Áed Finnliath, king of Ailech, and Flann, king of North Brega, were also hostile toward the ambitious king of Tara. This did not stop Máel-Sechnaill from taking the hostages of Munster once again in 858. The next year saw the war being carried to Mide itself, and Máel-Sechnaill reacted by hosting a royal conference at the border between Mide and Munster. Supported by the Irish clergy, he forged an alliance with Cerball of Osraige, who may have married Máel-Sechnaill's daughter Ailbi on this occasion. The agreement was warranted by the king of Munster, who was killed by Vikings the next year. Having pacified the south, Máel-Sechnaill gathered the forces of the southern Uí Néill, Munster, Leinster, and Connacht, and marched to Armagh in 860. This unprecedented show of force was insufficient to bring Áed Finnliath and Flann to heel. In the following years, they were joined by the Vikings in their attacks on Mide. Máel-Sechnaill's power waned, and he died in 862. The contemporary *Annals of Ulster* style him "king of all Ireland," which reflects his nominal kingship over the island.

Although Domnall mac Áeda (d. 642) and his grandson Loingsech (d. 703) of Cenél Conaill are also called "king of Ireland" in early sources, it is uncertain whether they had received the submission of all the Irish provincial kings. In Máel-Sechnaill's case it is clear that he did so. Even if his authority over the island was temporary and disputed, he had shown that a king of Tara was capable of dominating all Ireland. Typically, there were other Uí Néill kings who resisted him, and who could not be won over by diplomacy or force. The Vikings played an important part in the overall struggles, but in the long run they remained mercenaries rather than political allies. After 862, they went back to raiding in Ireland and overseas, or teamed up with the enemies of Áed Finnliath and Flann Áed Finnliath (king of Tara 863–879) never matched the successes of Máel-Sechnaill in the south, unlike his successor Flann Sinna (king of Tara 879 916). Flann was born in 848 or 849, and was the only son of Máel-Sechnaill and Lann, the sister of Cerball, who later married Áed Finnliath. On his turn, Flann later married another wife of Áed Finnliath, the daughter of the king of Scotland. Such marriages were largely symbolic, but maintained the close bonds between the leading branches of the Uí Néill. The period around 850 to 940 witnessed a number of strong kings of Tara, but at the same time the Uí Néill were disintegrating. After the reign of Máel-Sechnaill II (d. 1022), the kingship of Tara lost its old meaning.

BART JASKI

References and Further Reading

Ó Corráin, Donncha[dh]. *Gill History of Ireland*. Vol. 2, *Ireland before the Normans*, pp. 99–101. Dublin: Gill and Macmillan, 1972.

Byrne, Francis John. *Irish kings and high kings*. London: Batsford, 1973. Reprint, Dublin: Four Courts Press, 2001.

Jaski, Bart. "The Vikings and the Kingship of Tara." *Peritia: Journal of the Medieval Academy of Ireland* 9 (1995): 310–351.

See also **Cerball mac Dúngaile; Mide; Munster; Osraige; Uí Néill, Southern; Vikings**

MÁEL-SECHNAILL II (949/950–1022)

Máel-Sechnaill II was the son of Domnall Donn and was the last great king of his dynasty, Clann Cholmáin of Mide, which had been a powerful force in Irish politics since the mid-eighth century. He was also the last of the "old-style" kings of Tara, who claimed to be over kings of both the southern and the northern Uí Néill and their traditional allies. His reign was marked by his control over Dublin, his struggle with Brian Boru of Munster, and further disintegration of the Uí Néill. After Brian's death in 1014, Máel-Sechnaill was the most powerful king in Ireland, and he temporarily managed to act as a king of Tara of old.

Career

With the death of Muirchertach "of the Leather Cloaks," king of Ailech, in 943, and of Máel-Sechnaill's grandfather, Donnchad Donn son of Flann Sinna, king of Tara, a year later, the Uí Néill were in disarray. There was no agreed successor to the kingship of Tara, which was contested by two outsiders. With the death of Ruaidri ua Canannáin of Cenél Conaill in 950, Congalach Cnogba of Brega came out as the winner. Until his death in 956, he tried to keep the kings of Mide under his authority. He was aided by the fact that the descendants of Flann Sinna were quarreling about the kingship of Mide. Domnall Donn, one of Donnchad Donn's sons, is recorded as king of one half of Mide at his death in 952. His wife was Dúnfhlaith, daughter of Muirchertach "of the Leather Cloaks," who had given birth to Máel-Sechnaill some years earlier. It was probably after Domnall's death that Dúnfhlaith became the wife of Amlaíb Cuarán, to whom she bore Glún Iarn (Iron Knee). Amlaíb and his allies of Leinster and Brega dominated affairs in eastern Ireland in the 960s and 970s. Muirchertach's son Domnall Ua Néill (king of Tara 956–980) led several ferocious campaigns in the area, but was ultimately unsuccessful in his efforts to subjugate the Dublin Norse and their allies, and was finally repulsed from Mide by Clann Cholmáin. Because of internal divisions, the kingship of Mide was in abeyance from 974 to 978. Finally, Máel-Sechnaill claimed it when he was still in his late twenties. Two years later, Domnall Ua Néill died, and according to the traditional rule of alternation, it was the turn of Clann Cholmáin to deliver the next king of Tara. Aided by allies from overseas, the ambitious Amlaíb attacked Máel-Sechnaill at Tara, perhaps at his inauguration. Together with the forces of Ulster and Leinster, Máel-Sechnaill defeated his enemies, and beleaguered Dublin for three days in a row until the inhabitants came to terms. The new king of Tara obtained a large tribute in cattle and jewelry, and freed the Irish hostages kept in Dublin. Some annals refer to this as the end of the Babylonian Captivity of Ireland. Máel-Sechnaill was now the new overlord of Dublin, which was henceforth ruled by his half-brother Glún Iarn. His marriage to Máel-Muire (d. 1021), daughter of Amlaíb, may also stem from this period. His control over Dublin provided Máel-Sechnaill with additional resources to check the progress of Brian Boru in Leinster and Connacht. Hence he immediately laid siege to Dublin after the killing of Glún Iarn by a slave in 989. Once again the Norse made their submission and paid a huge tax. Their new king, Sitriuc Silkenbeard, was the son of Amlaíb and Gormfhlaith, the daughter of the king of Leinster. Gormfhlaith also became Máel-Sechnaill's wife, perhaps to forge an alliance between the two parties. When Máel-Sechnaill was forced to come to terms with Brian in 997, he conceded his overlordship over Dublin to his rival. It marks the beginning of the situation in which control over Dublin was tantamount to the control over Ireland. Máel-Sechnaill failed to rally the northern Uí Néill to his banner, and submitted to Brian in 1002. He remained king of Tara, and as such attempted to revive the Fair of Tailtiu, the assembly of the Uí Néill and their allies, which had fallen into disuse in the early tenth century. The poem which Cúán Ua Lothcháin composed about the event in 1007 shows that the main kings of the northern Uí Néill and Connacht did not show up. Máel-Sechnaill's fortunes reversed when he pulled out before the battle of Clontarf in 1014. With the death of Brian and the Dál Cais weakened, he became the most powerful king in Ireland. In 1015, Máel-Sechnaill was joined by Flaithbertach Ua Néill, king of Cenél nÉogain; the kings of Cenél Conaill and Bréifne; and by the son of the king of Connacht in an attack on Dublin, which was burned. Afterward he took the hostages of Leinster and plundered Osraige. The next year the hostages of Osraige and Ulster were secured. Máel-Sechnaill was now at the height of his power and had almost matched the successes of Brian. However, the remarkable unity among the Uí Néill and their allies did not last. In 1018, the northern Uí Néill were at war with him, and Máel-Sechnaill received aid from the Éoganachta of Munster in his expedition to the north. Yet in 1020 the annals record that the king of Tara was joined by Flaithbertach Ua Néill, Art Ua Ruairc of Bréifne, and Donnchad mac Briain of Munster in an expedition to the Shannon, where they gave the hostages of Connacht to him. Máel-Sechnaill's power in this period earned him the epithet *Mór* (the Great), and the description "high king of Ireland and pillar of the dignity and nobility of the western world" at his death in 1022.

Legacy

The aftermath of Clontarf had given Máel-Sechnaill the opportunity to reunite the Uí Néill and reestablish the power of the king of Tara over Ireland. This was no mean achievement, and it testifies to his abilities as a leader, which were overshadowed only by those of Brian Boru. But the period of disintegration of the Uí Néill and the changes Brian Boru had caused could not be undone in a matter of years. After his death, Flaithbertach Ua Néill did not claim the kingship of Tara, and the title became the rather empty prerogative of the kings of Mide. The kingship of Tara, which had been a steady force in Irish politics for centuries, had thus been rendered ineffectual. Yet this situation actually dates from the 940s onward, when none of the kings of Tara had been able to make good of his claim to be over king of all the Uí Néill. While Máel-Sechnaill may have been recognized as such in the period from 1015 to 1022, this did not herald the return to the old political order. After the death of Flaithbertach Ua Néill in 1036, the kingship of Cenél nÉogain was taken by another branch, which would bring forth the Mac Lochlainn lineage. Clann Cholmáin were equally divided. Although Máel-Sechnaill had ruled Mide for almost forty-five years, he was succeeded by Máel-Sechnaill Got, a member from another branch. His own sons were either dead or not powerful enough to claim the kingship. After 1030, the kingship of Mide was held by the descendants of Máel-Sechnaill's son Domnall (d. 1019 as head of Clonard), but Clann Cholmáin were often too divided and the Ua Máelshechlainn (O Melaghlin) family never recovered their former glory. Henceforth Mide became the battleground for the more successful kings from Leinster, Munster, Connacht, and Cenél nÉogain.

BART JASKI

References and Further Reading

Ó Corráin, Donncha[dh]. *Gill History of Ireland*. Vol. 2, *Ireland before the Normans*, pp. 120, 121–24, 128–29, 131. Dublin: Gill and Macmillan, 1972.

MacShamhráin, A. "The Battle of Glenn Máma, Dublin and the High-Kingship of Ireland."

Duffy, Seán, ed. *Medieval Dublin*. Vol. 2, pp. 57–59, 62–63. Dublin: Four Courts Press, 2001.

See also **Amlaíb Cuarán; Brian Boru; Dublin; Gormlaith (d. 1030); Mide; Sitriuc Silkenbeard; Uí Néill, Northern; Uí Néill, Southern**

MAGUIRE

The Maguire (*Mag Uidhir*) family rose to prominence in the lordship of Fermanagh at the end of the thirteenth century to become kings of Lough Erne. Their rise was associated with the fall of the Mac Lochlainn dynasty in Tír nEógain, and was sponsored by the de Burgh, earls of Ulster. The Maguires replaced the ruling Fermanagh dynasties of Ua hEignigh (to whom the Maguires were related), Ua Duib Dara and Ua Máel Ruanaid, and maintained forty soldiers for the earl of Ulster and paid tribute of eighty cows. Following the collapse of the earldom in the mid-fourteenth century, the Maguire lordship of Fermanagh was noted for its peace and prosperity. There were a large number of learned families in the lordship, poets, historians, brehons, and physicians, and in the early seventeenth century, the inhabitants of Fermanagh were "reputed the worst swordsmen of the north, being rather inclined to be scholars or husbandmen than to be kerne or men of action."

Under two powerful lords, Tomás Mór Maguire (1395–1430) and his son and heir, Tomás Óg (1430–1471), abdicated. the lordship of Fermanagh and became very wealthy. It was stated of Tomás Mór that he was "a man of universal hospitality toward poor and mighty," noted as a founder of churches and monasteries and for "the goodness of his government." Tomás Óg's position as lord of Fermanagh was so secure that he was able to go on pilgrimage to Rome, and twice to Santiago Compostella in Spain. Both of these lords were also noted for their strong support for their overlord, Ua Néill of Tír nEógain.

Within the Maguire lordship, there was a continual spreading out of junior Maguire septs, who dispossessed other families of less importance. For example, the Clann Amlaimh Maguires established themselves in Muinntear Peodacháin, west of Lough Erne, dispossessing the chieftain, Mac Gille Fhinnéin. The Maguires also took Clankelly from the Mac Mahons. By the late fifteenth and early sixteenth centuries, Ua Domnaill interference in the Maguire lordship began to increase, and Fermanagh became part of the lord of Tír Chonaill's sphere of influence. The O'Donnells put fleets of ships on Lough Erne and even took over the castle of Enniskillen for long periods. However, Shane O'Neill, the lord of Tyrone from 1559 to 1567, won Fermanagh back to the Ua Néill fold.

In the late sixteenth century, the chieftain Cú Chonnacht Maguire, lord of Fermanagh from 1566 to 1589, was an important ruler. Known for "his munificence toward churches, ollaves, soldiers, and servants," and for his knowledge of Latin and Irish, Cú Chonnacht was a noted patron of bardic poets, and he commissioned the collection now known as *Duanaire Mhéig Uidhir*(Maguire's Poembook), which survives in a manuscript in Copenhagen. Cú Chonnacht's son, Hugh Maguire, lord of Fermanagh from 1589 to 1600 was a very important confederate cavalry commander during the Nine Years' War. He defeated an English force at the Battle of the Ford of the Biscuits in 1594 and commanded the Irish cavalry at the Battle of Clontibret in 1595 and the Battle of the Yellow Ford in 1598. He was killed while accompanying Hugh O'Neill's progress into Munster in March 1600. Hugh's half-brother, Cú Chonnacht Óg Maguire, succeeded him as lord of Fermanagh. He was a prominent participant in the Flight of the Earls in 1607 and died in Genoa in 1608.

DARREN MCGETTIGAN

References and Further Reading

O'Donovan, John, ed. and trans. *Annala Rioghachta Eireann, Annals of the Kingdom of Ireland*. Dublin, 1856.

Greene, David. *Duanaire Mhéig Uidhir, The Poembook of Cú Chonnacht Mág Uidhir, Lord of Fermanagh 1566–1589*. Dublin, 1991.

Simms, Katharine. "The Medieval Kingdom of Lough Erne." *Clogher Record* 9, no. 2 (1977): 126–141.

See also **Mac Lochlainn; Learned Classes; Pilgrimage; Ua Néill; Ua Domhnaill; Ulster, earldom of**

MALACHY (MÁEL-MÁEDÓIC)

Due to the fact that St. Bernard, the great abbot of Clairvaux, wrote the Life of St. Malachy, much more is known about him than anyone else involved in the twelfth-century Church Reform in Ireland. Known in

Irish sources as Máel-Máedóic Ua Morgair, Malachy was born in Armagh in 1095. He was the son of Mugrón Ua Morgair, a man of learning and head of the monastic school (*aird-fher leighind*) of Armagh, who died in Mungret in 1102. From a young age he came under the influence of the ascetic Ímar Ua hAedacáin, the founder of the monastery of St. Peter and St. Paul in Armagh and was just ten years old when the new coarb of Patrick (i. e., the abbot of Armagh), Cellach, joined the reform movement, just then gaining ground in the Irish church. His formation, therefore, took place at a very critical time for the church in Ireland, and he was in a position to observe it at close quarters. In fact, he soon came to the attention of Cellach, who ordained him deacon around 1118 and priest one year later. Obviously impressed by Malachy's qualities, Cellach appointed him as his vicar at Armagh while he (Cellach) was in Dublin for a period from 1121 pursuing the aims of the newly formed church hierarchy. During this vicariate he is said to have vigorously pursued reforms, reinstituting sacraments which had lapsed, and establishing, in particular, the customs of the Roman church. On Cellach's return, he went to Lismore to acquaint himself with the practices of the universal church; Máel Ísa Ua hAinmire, one of the main architects of reform whom Cellach had met at the synod of Raith Bressail, was bishop there and had previously been a Benedictine monk at Winchester in England. After his training, he was recalled to Armagh because his uncle, then coarb of Bangor, wished to retire and hand over that monastery to Malachy in order that he might restore it to its former glory. On the instructions of his mentor, Ímar, he took with him ten of Ímar's monks and began his task; one of these monks was Máel Ísa, brother of the first abbot of Mellifont and later papal legate, Gille Críst Ua Connairche. In 1124, Malachy was consecrated bishop but it is not clear which diocese he ruled; Connor, according to St. Bernard, but he continued to govern Bangor, which is in the diocese of Down—perhaps he ruled both. An outbreak of violence in 1127 saw him and a group of his monks flee to Lismore, a setback he turned to advantage by founding a new monastery, the location of which is disputed (*monasterium Ibracense*).

Malachy and the Primacy

Shortly before he died in 1129, Cellach nominated Malachy as his successor. Cellach would have been well aware that this nomination would meet resistance in Armagh because it represented a major break in tradition in that Malachy did not belong to the family that had controlled the abbacy of Armagh since the middle of the tenth century. For this reason he laid responsibility specifically on the two kings of Munster to help Malachy take up his new office. Immediately after Cellach's death, a member of the traditional ruling family, Muirchertach, was installed as the coarb of Patrick. As a result, Malachy was reluctant to take up his appointment, as he knew this would be violently resisted. For the reformers this was an intolerable situation, as it was essential that Armagh, the seat of the primate, should remain in the possession of a reformer. Because of this, those two stalwarts of reform, the papal legate Gille and Máel Ísa Ua hAinmire, both of whom, along with Cellach, had signed the decrees of Raith Bressail, strongly urged him to take up his position; he continued to be indecisive however. Finally, in 1132, after three years of unsuccessful persuasion, they called together an assembly of bishops and lay princes to add strength to their urging, saying that they were prepared to use force if necessary. Eventually Malachy relented, but he did not enter the city (Armagh) at this stage; for two years he carried out his episcopal duties from outside while Muirchertach remained within. When a new coarb succeeded Muirchertach in 1134, Malachy's supporters acted; they forcefully but successfully installed him in the city, despite the resistance met. Armagh was finally in the possession of the reformers, but it was an uneasy possession. Because of this, Malachy devised a strategy; he would resign in favor of a candidate who would be acceptable to both the reformers and the local power, the Cenél nEógain. The candidate chosen was Gilla Mac Liac. The office of bishop and coarb was now successfully merged and protection was assured. Malachy was free to pursue reform at a different level.

Malachy the Church Reformer

In 1139/1140, Malachy traveled to Rome to get papal approval for the decisions made earlier at Raith Bressail. The pope received him well and questioned him closely about the church in Ireland; as a result he decided that the time was not yet ripe for the granting of his approval. Appointing him papal legate, he advised Malachy to return home, call general council, get the agreement of all, and at that stage seek papal approval. Back in Ireland, Malachy set about getting the agreement of all. Because Dublin had remained outside the structure agreed at Raith Bressail, and because the interests of the king of Connacht, Tairrdelbach Ua Conchobair (now king of Ireland), had to be accommodated, some compromises had to be made. We know very little about his negotiations, but in 1148 a synod was held on Saint Patrick's island. It would appear that agreement was reached there; as a result, Malachy was sent by the synod to the pope to get his approval. However, he never got to meet the pope, but the request he was to make did; what is more, it was successful and four years later the papal legate, Cardinal John Paparon, gave the pope's formal approval to a revised church structure (which incorporated the compromises Malachy had negotiated) when he presided over the synod of Kells.

Cistercians and Augustinians

Malachy had failed to meet the pope because he died at Clairvaux on November 2, 1148; he was fifty-four years of age. He had died in the place that he had come to love. Eight years earlier he made what appears to have been his first contact with it and was so taken by what he found there that he wished to join it as a monk. However, pope Innocent II refused his request; Malachy, therefore, determined that he would introduce its form of monasticism—the Cistercian order—into Ireland. To this end he left four of his companions to be trained at Clairvaux when he returned to Ireland and then sent more out from Ireland to join them. Meanwhile he found a site at Mellifont, near Drogheda, which would become, in 1142, the first Cistercian foundation in Ireland; in it were Irish monks trained at Clairvaux and some French confreres. Although not without its difficulties, the Cistercian order in Ireland spread rapidly; by the time of Malachy's death, Mellifont had five daughter-houses. Thereafter the order continued to spread.

Malachy also introduced into Ireland the rule followed by the Canons Regular of St. Augustine who lived in Arrouaise in Flanders; he visited them in 1140, but he may have known about it from houses in England. He was directly associated with the establishment of a house of canons regular in Saul in County Down; he was also closely associated with the introduction of the Arrouaisian rule into Bangor and Down (one of which formed the chapter of the diocese of Down), Knock, Termonfeckin, Louth (probably the head of the Arrouaisian congregation in Ireland), and, most likely, St. Patrick's Purgatory in Lough Derg. Outside the province of Armagh little is known about canons there during Malachy's lifetime, but some of the monastic houses in Munster may have adopted the rule under his influence. He was also responsible for the establishment of houses of canonesses. The introduction of the canons was likely to have been of considerable help to bishops who had the task of setting up dioceses without any infrastructure; the canons became the cathedral chapter in some dioceses, the first step in the formation of a sub-diocesan administrative structure. This may have been the reason why Malachy introduced them in the first place.

At a personal level, it is clear that Malachy made a great impression on St. Bernard and on his community. This is clear from his sermons, letters, and the Life he wrote. But Bernard also wore the habit in which Malachy died whenever he said mass and was later buried in it. As well as that, Cistercians continued to honor Malachy, spreading his cult and being responsible for his canonization in 1190. He remains a saint in the Cistercian calendar to the present day.

MARTIN HOLLAND

References and Further Reading

Bernard of Clairvaux. *The Life and Death of Saint Malachy the Irishman*. Translated by R. T. Meyer. Kalamazoo: 1978.

Lawlor, H. J. *St. Bernard of Clairvaux's Life of St. Malachy of Armagh*. London: Macmillan Company, 1920.

Conway, C. *The Story of Mellifont*. Dublin: 1958.

Dunning, P. J. "The Arroasian Order in Medieval Ireland." *Irish Historical Studies* 16 (1945): 297–315.

Gwynn, A. *The Irish Church in the Eleventh and Twelfth Centuries.* Edited by Gerard O' Brien. Dublin: Four Courts Press, 1992.

Hughes, K. *The Church in Early Irish Society.* London: Methuen, 1966.

Gwynn, A. *A History of Irish Catholicism.* Vol. 2, *The Twelfth-Century Reform.* Dublin & Sydney: Gill and Son, 1968.

Watt, J. *The Church in Medieval Ireland.* 2nd ed. Dublin: University College Dublin Press, 1998.

Holland, Martin. "Dublin and the Reform of the Irish Church in the Eleventh and Twelfth Centuries." *Peritia: Journal of the Medieval Academy of Ireland* 14 (2000): 111–160.

See also **Church Reform, Twelfth Century; Gille (Gilbert) of Limerick; Kells, Synod of; Raith Bressail, Synod of**

MANORIALISM

Manorialism was the system by which both land tenure and political control were exercised throughout the Anglo-Norman lordship of Ireland. The original classic French system of feudal land holdings, modified by English custom, was introduced into Ireland from 1169–1170 onward. The greatest lords held their land from the king as great estates or lordships, all of which were divided into manors. In areas like the great Butler lordship in Tipperary and Kilkenny, there were large "caput" (chief) manors, which were then further divided into smaller manors, which could still be as large as 5,000 acres in extent. The major service owed by these great tenants-in-chief to the crown was military service, usually expressed in the number of knights they were to provide to the royal army. For instance, the important de Clare lordship of Leinster was held from the king by service of 100 knights, but it must be remembered that this service had been almost wholly commuted to a money payment, or scutage, (known as royal service in Ireland), by the later Middle Ages. Other feudal obligations included owing suit of court, giving counsel to the lord when requested, and to aid him with money grants on certain occasions. Some kinds of service were largely formalized by a symbolic gift, such as the hawk that the manor of Dalkey, an important outport of medieval Dublin, was expected to provide each year.

It is difficult to give a strict definition of what constituted a manor, which was the basic economic and juridical unit of this feudal system of land holding. Most manors would have some evidence of the following: a legal recognition of the landowner's rights of lordship over the people within the manor, the existence of demesne land (farmland directly cultivated by the landlord), land held by tenants, and, finally, some evidence of a dependent tenantry there. One of the major institutions of the manor was the manorial court that mainly decided on disputes over land tenure, although minor breaches of the law were also tried there. It is a pity that so few manorial court rolls have survived in Ireland, as they could give us a unique insight into the every day life of the majority of the population within the English lordship.

There appear to have been wide regional variations in the amount of land in any particular manor occupied by the demesne farm, especially in the marchlands of Connacht. In the more prosperous parts of Leinster, it varied from around 13 percent to over 25 percent of the entire lands of the manor. The free tenants obviously held most land within the manor, paying a money rent or even holding their land by military service. As well as being liable for certain feudal obligations, they had to attend the manorial court. A particularly Irish type of free tenure was that of the gavillers, who on top of paying rent were liable for some labor services. In Ireland, many of the unfree tenants were often known as betaghs, a corruption of the Irish word *biatach*, meaning "food-provider," a rank of semi-free tenants in pre-Norman society, owing some kinds of labor service but who owned their own land. They seem generally to have been better off in material terms than the other major class of servile tenants, known as the cottars.

In the eastern half of the country, where there was a denser Anglo-Norman settlement and where the new settlers felt more secure, these manors appeared to have been largely modeled on similar ones in England and Wales. But in the more westerly edges of the colony, where the "land of war," or less securely held areas predominated, it seems as though these manors really only existed in the documents of the Anglo-Norman lordship. One of the most significant differences on Irish manors was the existence of people who held their land by burgage tenure. These burgesses lived in what have been called "rural boroughs," which were little more than villages but which had this attractive form of tenure obviously designed to attract colonists from England, Wales, and possibly even Flanders to settle and develop the under-populated lands of Ireland. These tenants were a distinctive community within many Irish manors, each holding their burgage plots at the low rent of one shilling per annum, enjoying other privileges including having their own court, quite separate from the manor court, but still under the overall jurisdiction of the lord. Indeed, in the settlement of Kilmaclenine, County Cork, the burgesses there owed some labor services to their lord, the bishop of Cloyne.

The other difference between Ireland and Britain in this period was the fact that the lands of the church were often held in "free alms," or sometimes as fee farms, which meant that they were removed from the military obligations of the feudal system. More generally, grants to sub-tenants of fee farms, a form of hereditary tenure not liable to military service but to a fixed rent, were much more common in Ireland than in Britain. The system of agriculture most often associated with the manors in the eastern half of the country was the arable open fields. Here, all the plowed arable land was distributed over two or three large unenclosed fields (each one up to 500 acres in extent), one of which lay fallow for a year in order for it to recover its productivity. Each field was then subdivided into scattered holdings or strips of land that were parceled up among the tenants of the manor, with everyone sharing both good and poor land. It is, however, unclear how far this system extended in the westernmost manors, where pastoral farming arguably predominated.

Understanding the population of one of these manors is fraught with many difficulties, firstly because the extents or surveys that do survive often do not have a complete itemization of all the classes of people on a particular manor. In the second place, those listed are only the heads of households within the manor, so some estimate of the average medieval household has to be reached, probably in the region of four to five people. When these caveats are taken into account, it would appear that an average-sized manor probably only had a population of a few hundred people. One such manor was Knocktopher in County Kilkenny, where a fairly detailed extent for 1312 lists four farmers holding between 5 and 74 acres of arable land, then at least forty-five free tenants holding from as much as 2,520 acres of arable land all the way down to just one house plot. There are also ninety-seven burgesses who held 360 acres of arable land, and finally there is mention of a settlement of betaghs, but the record does not give any indication of their probable numbers, although they farm 120 acres of arable land. These figures reveal the great differences in wealth within each particular category of landholder. Finally, the extent also tells us about the buildings at the manorial center, which include a castle, with a farm and a cattle shed and other diverse farm buildings, a columbarium for doves, three gardens for fruit and vegetables, and two mills.

In the expansionary period of the thirteenth century, production on the demesne farms, particularly in the southeast, was so successful that large amounts of hides, wool, and grain, in particular, were exported from the region's major ports, such as Waterford to Britain and to Continental Europe. At an individual manorial level, this period of economic expansion is best illustrated in the few surviving manorial account rolls that were compiled for the greatest landholders, such as the Bigod, earls of Norfolk, for their extensive landholdings in Wexford and Carlow. These record annually, at the end of September, all items of manorial income and expenditure, whether in cash or kind. But the more severe times of the fourteenth and fifteenth centuries put pressure on this surplus production, so that by this latter period Ireland was again importing grain for its own use. This classic system of medieval manorialism declined in importance as the area controlled by the Dublin government shrank throughout the later Middle Ages. It probably came to an end in the seventeenth century, as the many wars of that century were followed by large-scale land redistribution that finally broke up many of the old feudal estates.

TERRY BARRY

References and Further Reading

Empey, Adrian. "Medieval Knocktopher: A Study in Manorial Settlement–Part 1." *Old Kilkenny Review* 2 (1982): 329–342.

Hennessy, Mark. "Manorial Economy in Early Thirteenth-Century Tipperary." *Irish Geography* 29 (1996): 116–125.

Lydon, James. *The Gill History of Ireland*. Vol. 6, *Ireland in the Later Middle Ages*. Dublin: Gill and Macmillan, 1973.

See also **Agriculture**

MANUFACTURING

See **Craftwork**

MANUSCRIPT ILLUMINATION

The production of manuscripts formed a significant activity in early Christian Ireland. The arts of calligraphy and decoration were widely practiced, with scribes holding a high position in society. Illuminated manuscripts from the period between the sixth and ninth centuries represent high points in Ireland's artistic history and have helped to define the country in a cultural-historical sense.

Folio 58v from the Book of Kells. *The Board of Trinity College Dublin.*

The travels of Irish missionaries abroad exerted wide influence on calligraphy and decorative techniques. Important Irish manuscripts survive from centers in Europe. Early seventh-century copies of works by Jerome (Milan, Biblioteca Ambrosiana, S.45.sup) and Orosius (Milan, Biblioteca Ambrosiana, D.23.sup) were probably produced at Bobbio, founded in 612 by St. Columbanus of Bangor, County Down. The latter manuscript contains the earliest surviving "carpet page" in insular art. (A carpet page is one composed entirely of ornament. "Insular" is commonly used as a broad and neutral term to describe the characteristics of the style in art and script.) A fragmentary gospel book, "Codex Usserianus Primus" (Dublin, Trinity College, 55), contains a cross monogram, set within a triple frame, between the gospels of Luke and Mark. The Greek letters *alpha* and *omega* are placed on either side of the cross. Generally believed to have been made early in the seventh century, "Usserianus Primus" has recently been ascribed to the fifth century and to a continental center (Dumville, 1999). Several important manuscripts, including a strikingly decorated Gospel book from the eighth century (St. Gallen, Stiftsbibliothek, 51), survive from St. Gallen in Switzerland, which was founded by one of Columbanus's disciples.

St. Colum Cille, or Columba (c. 521–597) is a key figure in any account of Irish illuminated manuscripts. Born in Donegal around 521 into the ruling Uí Néill dynasty, Colum Cille traveled to Scottish Dál Riata with twelve companions around 561. The monastery he founded on Iona, off Mull (Argyll), became the head of a wealthy monastic confederation stretching from Ireland through Scotland to the north of England, where Lindisfarne was its most significant foundation.

A life of Colum Cille written by his successor as abbot, Adomnán (c. 628–704), contains references to the copying of texts on Iona, and to Colum Cille's own prowess as a scribe. Three manuscripts written at intervals of roughly one hundred years—the Psalter of circa 600 known as the Cathach ("Battler") (Dublin, Royal Irish Academy, 12.R.33); the famous gospel manuscripts in the Book of Durrow, from circa 700 (Dublin, Trinity College, 57), and the Book of Kells, produced circa 800 (Dublin, Trinity College, 58)—all have strong associations with St. Colum Cille and serve as landmarks in the progression of insular styles of decoration. The Cathach, perhaps the earliest surviving manuscript with an unquestioned Irish origin, was traditionally believed to be the copy made by St. Columba of a Psalter lent to him by St. Finnian. A dispute about the ownership of the copy was resolved by King Diarmait mac Cerbhaill with the judgment "to every cow her calf and to every book its copy." This is frequently cited as an early instance of copyright law. It is not clear whether the Cathach was written and decorated by St. Colum Cille or was the work of a copyist undertaken some years after his death. Its artistic techniques include trumpet and spiral devices, the fish and the cross (symbols of Christ), as well as the calligraphic device of "diminuendo," in which the opening letters of a verse are formed in diminishing sizes. Its initials are frequently outlined in red dots. Red is used for rubrics, and there are some yellow and white pigments, but the damaged condition of the manuscript—a result of its having been kept in a shrine since the eleventh century—inevitably leads to a diminished appreciation of its artistry. On certain folios there are creatures that have been described as dolphin-like. Such uncertainty over the identity and meaning of particular devices, the purpose of which was presumably clear to the artist, is a feature of the study of insular art.

The Book of Durrow employed red dotting, not only around letters, but also, executed with remarkable delicacy, in places like the face of the Man, symbol of Matthew (folio 21v). Broad ribbon interlace, in red, green, and yellow, dominates the carpet pages and symbols pages preceding its Gospel texts, while trumpet and spiral devices and panels set into the carpet pages are strongly reminiscent of metalwork and jewelery. The Eagle, symbol of Mark (folio 84v), is derived from a Roman imperial model, while the Lion on folio 191v has joint features in common with Pictish representations of animals. Both Durrow and the Cathach are thought to be the work of single artist-scribes. This is not so with the Book of Kells, which has such a diversity of approach as to indicate that it was executed by several different practitioners, probably working discontinuously rather than together to a common plan. Its artists and scribes showed extraordinary assurance and a vivid sense of color in integrating native Irish art with animal and figure drawings derived from classical and other Mediterranean prototypes. The Book of Armagh (Dublin, Trinity College, 52) was produced around the same time. It contains the earliest extant New Testament copied in Ireland, along with a dossier of texts relating to St. Patrick and a life of St. Martin of Tours. Sections of it can be attributed to a known scribe, Ferdomnach, "a scholar and an excellent scribe," as he was termed by the Annals of Ulster. According to an inscription in the manuscript, Ferdomnach made it for Torbach, who was abbot of Armagh in 807. Ferdomnach's work resembles the Book of Kells in its style and virtuosity, though he used only pen and ink. The identities of other artist-scribes in the Book of Armagh are not known.

The Macregol Gospels (Oxford, Bodleian Library, Auct. D.2.19) can also be securely dated and localized. An inscription on the final page indicates that it was written and decorated by Macregol, abbot of Birr, County Offaly, who died in 822. A gloss added in Old English late in the tenth century indicates that it left Ireland at a relatively early date and demonstrates how difficult it is to anchor insular manuscripts in time and place. Had the final page of the Macregol Gospels, with its telling colophon, been lost, it might have been mistaken for one produced outside Ireland. Macregol's work has considerable vigor and impact, containing initials that characteristically have purple or yellow fillers and are surrounded by red dots. These remain gleaming on the page and in relief, in contrast to the Book of Kells, where nineteenth-century processing has flattened and reduced the impact of such effects. A few pages that were not glossed in the tenth century give an unsullied impression of Macregol's artistry.

It is likely that most major monasteries produced and cherished great manuscripts as relics of their founder and as status symbols. It is known from the comments of the thirteenth-century historian, Giraldus Cambrensis, that Kildare owned such a book, though one no longer extant:

> It contains the concordance of the four gospels according to Saint Jerome, with almost as many drawings as pages, and all of them in marvellous colours.... If you look at them carelessly and casually and not too closely, you may judge them to be mere daubs rather than careful compositions.... But if you take the trouble to look very closely, and penetrate with your eyes to the secrets of the artistry, you will notice such intricacies, so delicate and subtle, so close together, and well-knitted, so involved and bound together, and so fresh still in their colourings that you will not hesitate to declare that all these things must have been the result of the work, not of men, but of angels (Fox, 1999).

It does not seem fanciful to suggest that similar manuscripts were probably produced at monasteries of the stature of Clonmacnoise, County Offaly, or Terryglass, County Tipperary, where a "pocket copy" of St. John's Gospel and the Stowe Missal (bound together as Dublin, Royal Irish Academy, D.II.3) may have originated. Both manuscripts have high-grade decoration in the small format characteristic of the "pocket gospel" group of manuscripts. Examples from the group include the late eighth-century Book of Mulling (Dublin, Trinity College, 60), from St. Mullins, County Carlow, which has finely executed portraits of three evangelists, as well as a sadly damaged diagram, on its final page, of twelve crosses with inscriptions to accompany a sequence of prayers. The contemporary Book of Dimma (Dublin, Trinity College, 59), from Roscrea, County Tipperary, contains less naturalistic images.

The tradition and skills of insular decoration continued into the tenth century, seen in, for example, a fire-damaged Psalter from the Cotton library (London, British Library, Cotton Vitellius F.XI); the eleventh century, of which the *Liber Hymnorum* (Dublin, Trinity College, 1441) is a handsome example; and the twelfth century, where a Psalter signed by the scribe Cormac contains decoration that is assured and coherent (London, British Library, Add. 36929).

Styles imported from England after the Norman invasion of 1169 are reflected in a Psalter from Christ Church Cathedral, Dublin, produced in 1397 (Oxford, Bodleian Library, Rawl. C. 185), and an illustrated early-fifteenth-century missal (London, Lambeth Palace, 213). Both manuscripts probably originated in England. In the late fourteenth century, the charter roll of the city of Waterford was decorated in a lively manner, perhaps locally, while in the early fifteenth century a decorated copy of Ranulf Higden's chronicle (Oxford, Bodleian Library, Rawl. B.179) is probably a Dublin production.

BERNARD MEEHAN

References and Further Reading

Fox, Peter, ed. *The Book of Kells, MS 58, Trinity College Library Dublin: Commentary*, pp. 320–321. Faksimile Verlag Luzern, 1990.

Henry, F. *Irish Art in the Early Christian Period (to 800 A.D.)* London, 1965.

———. *Irish Art during the Viking Invasions 800–1020 A.D.* London, 1967.

———. *Irish Art in the Romanesque Period 1020–1170 A.D.* London, 1970.

Alexander, J. J. G. *Century, A Survey of Manuscripts Illuminated in the British Isles*. Vol. 1, *Insular Manuscripts, 6th to the 9th*. London, 1978.

Meehan, Bernard. "Aspects of Manuscript Production in the Middle Ages." In *The Illustrated Archaeology of Ireland*, edited by M. Ryan, pp. 139–145. Dublin, 1991.

Adomnán of Iona. *Life of St. Columba*. Edited by Richard Sharpe. Penguin, 1995.

Meehan, Bernard. "The Book of Kells and the Corbie Psalter (with a note on Harley 2788)." In *"A Miracle of Learning": Irish Manuscripts, Their Owners and Their Uses. Festschrift William O'Sullivan*, edited by T. Barnard, K. Simms, and D. Ó Cróinín, pp. 29–39. Scolar Press, 1998.

———. "'A Melody of Curves Across the Page': Art and Calligraphy in the Book of Armagh." *Irish Arts Review* 14 (1998): 91–101.

Herity, Michael and Aidan Breen. *The Cathach of Colum Cille*. An introduction and interactive CD-ROM. Dublin: Royal Irish Academy, 2002.

Dumville, David N. *A Palaeographer's Review: The Insular System of Scripts in the Early Middle Ages*, pp. 38–39. Kansai: Kansai University Press, 1999.

See also **Armagh, Book of; Columbanus; Colum Cille; Durrow, Book of; Early Christian Art; Kells, Book of; Scriptoria**

MARCH AREAS

The March or the marches were terms used by the central government at Dublin to describe the lands that lay between the "land of peace" (territory firmly under the control of the Anglo-Irish) and the "land of war" (territory under native control). Contemporaries probably understood what was meant when reference was made to marches, but this medieval shorthand has meant that march areas have remained ambiguous to historians; an ambiguity only added to by the distinctions between marches. For, despite the blanket terminology, not all marches were the same. Physically, they could vary in terms of height above sea level and land use. They also varied in terms of the customs or march law that operated within them. And there were further variations within given marches: for example, between those areas of a march where the agents of local government could operate effectively, and the "strong marches" where the will of the Dublin government through the person of the sheriff might be less easily enforced despite the government's jurisdictional competence. In contrast to the march of Wales, the king's writ did run throughout most march areas in Ireland where the marches were not synonymous with wide jurisdictional privilege.

What seems certain is that marches were highly militarized areas where defense was of the utmost importance. Some historians have focused on other aspects of the marches and have attempted to define them in cultural or economic terms, but it is in military terms that marches were primarily referred to in the administrative records of the Dublin government. Marches figured prominently in the statutes issued by the Dublin parliament of 1297, whose purpose was to bring order and peace to the lordship of Ireland. These statutes sought to counter the problems created by lords who, by residing in England or on their property in the land of peace, had abandoned their lands in the marches with the result that the marches were destroyed and laid waste. Other contemporary evidence is more positive, however. In particular, the agreements between de Geneville and his leading tenants concerning the division of prey taken in the marches show that at least some resident lords were active in the defense of their marches.

Those parts of the marches at the very edge of the English settlement in Ireland must always have been subject to depredations by the native Irish. However, the retreat of the English colony in the face of the Gaelic Revival, brought the edges of the settlement closer to Dublin, whose rhetoric concerning march areas became more insistent as the fourteenth century progressed. The Gaelic Revival not only affected the boundaries of the marches, but also their ethnic composition: the Gaelic population increased considerably, although Anglo-Irish continued to live in the difficult environment of the march. But this recovery of land was not met with an adoption of the Anglo-Irish terminology by the Irish: government records might record reference by an Irish lord to "his marches," but the terms remained absent from the Gaelic annals themselves.

BETH HARTLAND

References and Further Reading

Davies, R. R. "Frontier Arrangements in Fragmented Societies: Ireland and Wales." In *Medieval Frontier Societies*, edited by Robert Bartlett and Angus MacKay, pp. 77–100. Oxford, 1989.

Frame, Robin. "Military Service in the Lordship of Ireland, 1290–1360: Institutions and Society on the Anglo-Gaelic Frontier." In Robin Frame, *Ireland and Britain 1170–1450*, pp. 279–299. London, 1998.

Lydon, James, ed. *Law and Disorder in Thirteenth-Century Ireland: the Dublin Parliament of 1297*. Dublin, 1997.

Smith, Brendan. "The Concept of the March in Medieval Ireland: The Case of Uriel." *Proceedings of the Royal Irish Academy* 88, sec. C (1988): 258–269.

See also **Central Government; Gaelic Revival; Local Government; March Law**

MARCH LAW

"March law" was a term used in medieval Ireland to refer to the customs and practices of dispute settlement at a local level in the "marches," the shifting and undefined border areas between the more settled areas of English colonization, which continued to acknowledge the overall control of the English crown and to be governed by the Common Law, and those areas controlled by Irish rulers owing little or no allegiance to the English crown and which followed Brehon law. It was the subject of fierce condemnation by Richard fitz Ralph as archbishop of Armagh (1347–1369). He described it as the "law of the devil" in so far as it sanctioned the killing and despoiling of the native Irish by the Anglo-Irish of County Louth. The Irish parliament and Dublin administration were increasingly concerned in the middle years of the fourteenth century with the extension of the use of the "law of the March" to disputes within the Anglo-Irish community itself, which should have been determined in accordance with the Common Law. Legislation of 1351 imposed a penalty of imprisonment and ransom (fine) for its use in such disputes, and a further mandate of 1360 threatened the loss of life and members for the same offense. Chapter four of the Statute of Kilkenny (1366) increased the penalties to imprisonment and conviction as a traitor. However, none of these statutes attempted to proscribe its use in its proper context. The legislation suggests that the Irish "law of the March" possessed two particular features. One was the use of private, unauthorized, and unlimited distraint (seizure of animals and other movables and perhaps also persons) accompanied by the use or threat of force against those against whom the distrainor had a claim or grievance. This would then normally be met by a counter-distraint by the persons distrained who now had their own grievance. The second was the use of "parleys," discussions under a temporary truce between the two disputing parties (perhaps with the aid of mediators) in an attempt to resolve their disputes. Historians have sometimes written as though "March law" covers various other observable legal phenomena in the later medieval Irish lordship which may reflect Irish influence, such as the imposition of collective responsibility for criminal offenses on families, the practice of ransoming rather than hanging Irish offenders, and even changes in inheritance practice, but there seems to be no contemporary warrant for this extension of the term. The Irish usage of the term seems also to be very different from its usage in the Marches between England and Scotland or in the Marcher lordships of Wales. On the Scottish border it seems to have been used for the body of law that helped resolve disputes between inhabitants from either side of a relatively stable and fixed border and was enforced by the Wardens of the Scottish March. In Wales it was generally used to refer to the laws and customs of the individual Marcher lordships within Wales, which might differ considerably from Marcher lordship to Marcher lordship and were subject to the ultimate control of the Lord of that Marcher lordship.

PAUL BRAND

References and Further Reading

Frame, Robin. "Power and Society in the Lordship of Ireland, 1272–1377." *Past and Present* 76 (1977): 3–33.

MacNiocaill, Gearoid. "The Interaction of Laws." In *The English in Medieval Ireland*, edited by James Lydon, pp. 105–117. Dublin: Royal Irish Academy, 1984.

See also **Brehon Law; Common Law**

MARIANUS SCOTTUS

Marianus Scottus, whose real name was Máel Brigte, was born 1028 in the north of Ireland and died December 22, 1082 at Mainz. He composed a work called the *Chronicon ex chronicis*. This world history extended from Creation to the year of his death, and it was written probably at Mainz.

Asides in his chronicle provide what little is known of Marianus' life. He was a poet who identifies himself in an acrostic verse. Marianus entered religious life in 1052 at Moville (Co. Down), but was banished from Ireland in 1056 by Abbot Tigernach Bairrcech. The reason is unknown, although he claims that it was a minor offense. He went to the monastery of St. Martin at Cologne, arriving on August 1. In 1058 he migrated to Fulda, where he was walled up as an *inclusus* in 1059. Immediately prior to this, Marianus had been ordained a priest at the Church of St. Killian the Martyr at Würzburg. Ten years later, in 1069, Marianus went to the Church of St. Martin at Mainz, again as an *inclusus*, where he died in 1082.

The chronicle of Marianus is an important source of information about the contemporary Irish clergy on the continent as well as for Irish and Scottish affairs. The earliest copy is Vatican MS Codex Palatino-Vaticanus no. 830, written in part by Marianus and in part by his Irish amanuensis. The chronicle contains a prologue and three books, with the third, from the Ascension to Marianus' own day, having unique information such as the claim that Brian Boru was slain while at prayer during the battle of Clontarf. The first part of the manuscript provides a glimpse of the medieval historian at work, as Marianus collects materials and develops a chronology that is 21 or 22 years in advance of the actual date. He used a variety of materials, including a late-ninth-century king-list extending from the legendary Conn of the Hundred Battles to Flann "Sinna" mac Máele Sechnaill (d. 916), records from the continental Irish communities, and accounts from contemporary informants. Marianus' informants included his amanuensis, who dated one year by reference to the slaying of the Leinster king Diarmait mac Máele na mBó (1072) and whose journey across Scotland provided the exact dates for the deaths of kings Malcolm II, Duncan, Macbeth, and Lulach. He also copied verses on the creation of Adam by Airbertach Mac Cosse.

Marianus' work was widely respected and influential. His chronicle was used by Bishop Robert de Losinga of Hereford and the chroniclers Sigebert of Gembloux and John of Worcester, among others. Bishop Robert brought Marianus' chronicle to England and made a digest of it for his own use. Through Robert's friendship with Bishop Wulfstan II of Worcester, a copy was used at that church.

Marianus provided a point of contact between the Irish and continental writing centers. Efforts at chronological precision, respect for his continental predecessors, and interest in his homeland make Marianus' chronicle an important resource for historians. The fame of his work brought Irish historical writing into the view of Europe as a whole.

Benjamin Hudson

References and Further Reading

Waitz, G. *Monumenta Germaniae Historica, Scriptores*. Vol. 5, edition of book 3 of the chronicle, pp. 481–564. Hanover, 1884.

MacCarthy, B. *Todd Lecture Series*. Vol. 3, *The Codex Palatino-Vaticanus, no. 830*. Dublin: Royal Irish Academy, 1892.

Kenney, James. *Sources for the Early History of Ireland: Ecclesiastical*. Edited by L. Bieler, pp. 614–616. New York, 1966. Reprint, Dublin: Pádraic Ó Táilliúir, 1979.

See also **Annals and Chronicles; Poetry, Irish; Scriptoria**

MARRIAGE

Marriage in medieval Ireland was the result of a symbiotic relationship between native Brehon law and canon law. But the theory, as defined by canonists and jurists, was, as elsewhere in medieval Europe, very different from the social and economic conditions that impinged upon the practical realities. Recent studies have shown that even native law surviving in Old Irish tracts like *Cáin Lánamna*, "The law of marriage," was influenced by Roman law. Canon law, written in Latin, was, of course, influenced by the Bible and early church canons, but it was also influenced by Brehon law. The main sources of canon law were the text known as "The teachings of the Apostles," *Didascalia Apostolorum*, and the statements or formulations of the early church fathers. But the clerical jurists relied primarily upon the Bible, texts from which were quoted with great frequency.

Native Irish law was, like the Roman law of marriage, primarily concerned with property inheritance and the transference of property and thus most immediately affected the higher classes, those with property and status. Marriage among the lower classes counted for little or nothing—they had no birthright, little property, and lived outside the bounds of decent, respectable society. *Cáin Lánamna* is similarly concerned with property transference in the marital union and with the arrangements for its distribution in the event of separation. It emphatically does not deal with moral issues—although it states that marriage should not be entered into "with mutual deceit"—or with any area of the marital relationship falling within the sphere of canon law. Matters relating to morality in marriage and the family were more properly dealt with in canon law. The earliest piece of canonical legislation that has canons relating to marriage is the so-called "Second Synod of Patrick," a collection of 31 canons from the seventh century. It is the only synodal document to survive from the pre-Norman Irish Church in its entirety, and is therefore a valuable witness to the type of legislation being passed by the church at the same time as *Cáin Lánamna* was being drawn up. Several of its canons concern marriage, betrothal, adultery, and canonical incest by affinity or marriage.

Native lawyers distinguished three principal kinds of marriage: *lánamnas comthinchuir*, "a marriage of joint income," *lánamnas for ferthinchur*, "a marriage on a man's income," and *lánamnas for bantinchur*, "marriage on a woman's contribution." The first kind seems to have been the normative form of marriage. Marriage was publicly sealed by *arnaidm*, "binding, tying," marked by the formal exchange of property between the families of the bride and groom, and was witnessed and secured by guarantors of appropriate status. The giving of property or wealth from the man or his family to the bride's family, known in Roman law as *donatio ante nuptias*, was rooted in Germanic and Celtic law. The giving of such a gift may not have been essential to a marriage, but no good marriage was complete without it. Consequently most of the preliminaries about marriage were concerned with the size of the dowry, because that determined the character of the marriage, as is clear from the three principal forms of marriage discussed in *Cáin Lánamna*. The introduction of a betrothal contract, sealed with a vow, was made by the church. The right of the woman to choose her own partner was severely limited. As *Synodus II Patricii* puts it: "The maiden shall do what the father wishes, because 'the man is the head of the woman' (Eph 5:23). But the father must ascertain the will of the maiden, for 'He [God] left man in the hands of his own counsel' (Sir 15:14)" (can. 27). If the woman married without her parents' consent, or refused to accept the partner chosen for her, she could forfeit her right to the family inheritance. The law generally tried to protect women against gross abuse, such as being forced into marriage by someone who only had his own interests at heart. In comparison with the male-dominated prescriptions of Roman law, which upheld a double standard, one for men and quite another for women, Brehon law was quite liberal and compassionate, and permitted divorce to women for several reasons.

Some of the prescriptions on marriage in the *Collectio canonum Hibernensis*, Book 46 (*De ratione matrimonii*), are (1) that the bride should be virginal and the marriage ceremony properly and publicly conducted; (2) That she should remain faithful to her husband (46:2); (3) That either partner should put aside the other for adultery until she/he does penance, after which they may be reconciled, which marked a break with earlier ecclesiastical practice; (4) that neither partner may remarry while the other is alive (46:15); (5) that it is permissible to remarry after the death of one's partner, but it is more acceptable before God to remain chaste (46:13b), more especially for the woman; and (6) that a surviving brother should never marry his late brother's wife (46:35,: also *Synodus II Patricii*, can. 25), a prescription clearly not observed in medieval Irish society.

Essentially, the *Collectio Hibernensis* provides an idealized statement of Christian society, in which all eventualities or possibilities were provided for. The canonists knew better than we do just how little observed in fact these prescriptions were: polygamy, concubinage (among lay as well as clergy), arbitrary repudiation, incest, and multiple serial unions were common in Irish and other societies throughout the Middle Ages. The notion of the ubiquitous validity of legal norms, such as exists in the modern state, scarcely existed. It is precisely for this reason that some of the canons of the *Collectio* are inconsistent or repetitive, a feature it fully shared with earlier canon law and the barbarian law-codes: different circumstances and different schools of law gave rise to different judgments. In addition, no distinction was made between canon law or ethics and theology: the same group of people drew up penitential decrees, with their tariffs of penance, which included prayer, fasting, prostrations, almsgiving, and temporary sequestration with Scripture reading. The sanctions of Brehon law were largely monetary. Later Irish synods issued and reissued canons against simony, clerical marriage, the freedom of the church from rent and exaction, and incestuous marriage (a matter complained of by both Lanfranc and Anselm), but apparently without much effect.

A. Breen

References and Further Reading

Wasserschleben, F. W. *Die irische Kanonensammlung.* 2nd ed., lib. 45 ("De questionibus mulierum"), xlvi ("De ratione matrimonii"). Leipzig, 1885.

Thurneysen, R. "*Cáin lánamna*: die Regelung der Paare." In *Studies in Early Irish Law*, edited by R. Thurneysen and D. A. Binchy, pp. 1–80. Dublin, 1936.

Ó Corráin, D. "Irish Law and Canon Law." In *Irland und Europa: die Kirche im Frühmittelalter/ Ireland and Europe: The Early Church*, edited by P. Ní Chatháin and M. Richter, pp. 157–166. Stuttgart, 1984.

Ó Corráin, D. "Marriage in Early Ireland." In *Marriage in Ireland*, edited by A. Cosgrove, pp. 5–24. Dublin, 1985.

Kelly, F. *A Guide to Early Irish Law.* Dublin, 1988.

Reynolds, P. L. *Marriage in the Western Church: The Christianization of Marriage during the Patristic and Early Medieval Periods.* Leiden, 1994.

Ó Corráin, D. "Women and the Law in Early Ireland." In *Chattel, Servant or Citizen*, edited by M. O'Dowd and S. Wichert, pp. 46–57. 1995.

Jaski, B. "Marriage Laws in Ireland and on the Continent in the Early Middle Ages." In *The Fragility of her Sex?*, edited by C. Meek and K. Simms, pp. 14–42. Dublin, 1996.

Tatsuki, A. "The Early Irish Church and Marriage: An Analysis of the *Hibernensis*." *Peritia* 15 (2001): 195–207.

See also **Brehon Law; Canon Law; Law Tracts; Penitentials**

MARSHAL

Marshal family, earls of Pembroke and lords of Striguil and Leinster, were prominent in Ireland in the late twelfth and early thirteenth centuries after William I Marshal (d. 1219) married Strongbow's heiress, Isabel de Clare, in 1189.

Perhaps the most influential family in Ireland and England, the Marshals epitomize that type of Anglo-Norman noble—common in Ireland in this period—whose lands were scattered across the dominions of the king of England. As such, the Marshals brought the Irish lordship onto the main stage of English politics. William I Marshal, regent for Henry III, is reputed to have suggested using Leinster as a refuge for the king from the threat of civil war and French invasion. When William's eldest son, William II (d. 1231) countered Hugh II de Lacy in his attempt to regain the earldom of Ulster in 1223–1224, he was in fact tackling a confederation that embraced both sides of the Irish sea and included de Lacy's ally, the Welsh prince Llywelyn ab Iorwerth. The most notorious intrusion of English politics onto an Irish stage is the case of Richard Marshal (d. 1234), brother and heir of William II. Richard had opposed Henry III's foreign favoritism and fled to Leinster in 1234 where he was murdered, almost certainly with the king's connivance. The murder shocked the English political community, and Henry III had hastily to distance himself from any involvement.

Given their importance, the attention that the Marshals paid Ireland is remarkable. The family's interest there should perhaps be linked to the loss of Normandy in 1204, after which Leinster was seen as an alternative source of revenue. William I spent considerable periods in Ireland from 1207. He risked censure from King John in 1210 for harboring the king's enemy William de Braose, but was not deprived of his estates, and in 1212 Marshal declared his allegiance to the king, to whom he remained steadfastly loyal throughout the Magna Carta crisis.

It has been suggested that the Marshals were leaders in Ireland of a party that wished to make their estates profitable through the expropriation of the native Irish. The king's consent to this project has been interpreted as a reward for the declaration of loyalty of 1212. Such an interpretation requires some modification. Firstly, the Gaelic Irish had been undermined by royal charters as early as John's expedition to Ireland in 1185. The policy, therefore, was not an innovation. Moreover, the Marshals had close relations with the native Irish. William I's wife was the product of a mixed marriage between Strongbow and the daughter of Diarmait Mac Murchada. Moreover, in 1226, the Gaelic annals refer to William II as the "personal friend" of Áed Ua Conchobair, claimant to the kingship of Connacht. The family, therefore, was neither the originator nor the perpetuator of an aggressive racial policy against the native Irish.

That is not to say that they did not attempt to profit from involvement in Ireland. Marshal died in 1219 and was succeeded by his son, William II. Both father and son encouraged the economic development of Leinster through the towns of Kilkenny and New Ross, and they administered their lordship from their castles at Carlow, Dunamase, Kildare, Kilkenny, and Wexford. The profit accruing from Leinster had long concerned the crown. Prince John attempted to prevent William I from gaining possession of Leinster after 1189 but was undermined by King Richard I who interceded on the Marshals' behalf. Even so, Marshal was constantly harried in Ireland by John's justiciar, Meiler fitz Henry, until a settlement was reached in 1208. At William II's death in 1231, Henry III similarly attempted to prevent Richard Marshal from gaining his lands in Ireland.

Following Richard's murder in 1234, Henry III did not dare to deny his heir control of Leinster. Leinster passed to Richard's brothers, Gilbert (d. 1241), Walter (d. 1245), and Anselm (d. 1245), who each died childless. These deaths profoundly affected Ireland. Leinster was divided between the five daughters of William I, and some of the greatest noble houses in England became entangled through marriage with Ireland. Carlow descended to the Bigod earls of Norfolk, Kilkenny to the de Clare earls of Gloucester,

Dunamase to Mortimer of Wigmore, Kildare to the de Vescys, and Wexford to the de Valence family. This fragmentation weakened the colony, both because decentralization complicated Leinster's defense and because the new lords—whose English holdings were far superior in extent and profitability—paid little attention to their inherited interests in Ireland. Such absenteeism was to be a constant source of complaint in the colony for the remainder of the medieval period.

PETER CROOKS

References and Further Reading

Frame, Robin. "Aristocracies and the Political Configuration of the British Isles." In *Ireland and Britain: 1170–1450*, pp. 151–169. London: The Hambledon Press, 1998. First published in *The British Isles, 1100–1500: Comparisons, Contrasts and Connections,* edited by R. R. Davies, pp. 141–159. Edinburgh: Donald, 1988.

Walker, R. F. "The Supporters of Richard Marshal, Earl of Pembroke, in the Rebellion of 1233–34." *Welsh History Review* 17 (1994): 41–65.

Warren, W. L. "The Historian as 'Private Eye.'" In *Historical Studies: Papers Read before the Irish Conference of Historians*, edited by J. G. Barry, vol. 9, pp. 1–18. Belfast: 1974.

See also **Áed Ua Conchobair; Clare, de; Connacht; FitzHenry, Meiler; John; Kildare; Kilkenny; Lacy, Hugh de; Leinster; Mac Murchada, Diarmait; Mortimer; Strongbow; Ulster, Earldom of; Valence, de; Wexford**

MAYNOOTH

Although the town and castle of Maynooth, County Kildare, date from the aftermath of the Anglo-Norman settlement, the name *Mag Nuadat*, or Nuadu's Plain, is far older. Speculation links the area either to Nuadu Argatlám ("of the Silver Arm"), legendary leader of the Tuatha Dé Danann, or to Eógan Mór, also known as Mug Nuadat, after whom the southern half of Ireland became known as Leth-Moga (Moga's Half). The discovery of Gaelic round-houses at Maynooth shows the area was settled before the tenth century, but was overshadowed by its close neighbors, the monasteries of Donaghmore, Taghadoe and Laraghbryan. For unknown reasons the settlement was abandoned at some point before the twelfth century.

After the Anglo-Norman invasion, Maynooth formed the central manor of the lands granted to Maurice fitz Gerald. Construction of a castle began in the 1170s, no doubt at first an earth-and-timber building; at some date before the 1190s this was replaced with a stone keep. Architectural evidence points to three main phases of building. In 1248, the castle's chapel was made a prebendary of St. Patrick's cathedral, Dublin. The town of Maynooth grew alongside the castle, and together they formed the nucleus of a thriving manor. In 1286, a royal grant confirmed the town's commercial role with a weekly market and an annual fair (7–9 September). A list of the vill's tenants in 1328/1329 found in the Red Book of Kildare lists fifty-five tenants, including eighteen betaghs of Irish surname, paying a total annual rent of £35 7s. 6d.

The rising power of the FitzGerald earls of Kildare in the fourteenth and fifteenth centuries added to the town's prestige. Although the county court met at Kildare, Maynooth was the real center of power in the earldom, and one of the key border fortresses of the Pale. With the Kildare earls' near-monopoly over the office of chief deputy in the later fifteenth century, some historians have claimed that by 1500 Maynooth had become a virtual capital of Ireland. By the time of Gearóid Mór FitzGerald (the eighth earl of Kildare), the castle was a mansion worthy of the finest of Renaissance princes, complete with a library of books in Latin, Irish, French, and English (listed in a catalog of 1526). In 1518, the College of the Blessed Virgin was established in Maynooth.

During the rebellion of Silken Thomas (Lord Offaly, son of the ninth earl), the English were quick to identify Maynooth as a key strategic point, which, once captured, would remove the FitzGeralds' ability to resist. After seven days of fighting in March 1535, the castle was stormed, possibly due to the treachery of Thomas' foster-brother Christopher Parese. In the aftermath it was reported that the manor of Maynooth had been "made waste to the gates of the castle." The College was dissolved during the Reformation.

Maynooth was restored to the FitzGeralds in 1580, but the castle was destroyed and looted during the heavy fighting of the 1640s. Today the keep remains as a ruin.

JAMES MOYNES

References and Further Reading

Cullen, Mary. *Maynooth: A Short Historical Guide*. 1995.

FitzGerald, Walter. "Historical Notes on Maynooth Castle." *Journal of the Royal Society of Antiquaries of Ireland* 57 (1914): 81–99.

MEDICINE

The practice of medicine in medieval Ireland is a large topic that has attracted surprisingly little scholarly attention. Naturally sources are not plentiful, yet there is a body of diverse material, most of which has not been studied systematically and some of it not at all. It includes annals, law tracts, administrative records, leech books, translations into Irish of English and European medical texts, ecclesiastical documents, archaeological remains, place names, and folklore.

That medicine was widely practiced in Ireland both before and after the coming of the Anglo-Normans is clear from numerous references in the annals and other sources. The annals also regularly recorded serious epidemic, as well as endemic, diseases, many of which are hard to identify today. But it is probable that leprosy, smallpox, and typhus were major scourges. Bubonic plague raged in Ireland during the sixth and seventh centuries. The destruction of crops and animals, usually the result of bad weather or military conflict, inevitably produced famine, which in turn generated an array of diseases. The late thirteenth and early fourteenth centuries were especially disastrous in this regard, even before the reappearance of the Black Death in the late fourteenth and fifteenth centuries.

Physicians practiced in Ireland from an early period, as is clear from references in the sagas. In the *Táin Bó Cuailnge*, for example, King Conchobor's personal doctor demonstrated remarkable skill in the use of herbs to heal even the most grievously wounded warrior. Practitioners obviously enjoyed high status and widespread recognition: the *Annals of the Four Masters*, under the year 860, recorded the death of the "most learned physician of Ireland." But, despite their status, doctors were subject to close regulation. The laws made a number of references to the duties, entitlements, and qualifications of physicians. Fees for medical treatment were specified in some detail, and doctors whose treatments failed could be fined. What was entailed in medical education is unclear, but as medicine was largely a hereditary profession then presumably skills were passed on from one generation to the next by means of an apprenticeship system.

After the twelfth century it is possible to identity certain families of hereditary physicians who served ruling dynasties over many generations. By the fifteenth century, and probably earlier, some of these families were compiling leech books or medical manuals. Such works often described disorders and recommended remedies and, in doing so, demonstrated great practical knowledge. But, at the same time, they also showed considerable familiarity with English and European medical theories of the time and earlier. Irish translations of extracts from Bernard de Gordon's *Lilium Medicinae* (1303) and John Gaddesden's *Rosa Anglica* (1314) are to be found among the medical manuscripts held by the libraries of the Royal Irish Academy and Trinity College, Dublin. The medical manuscripts in the King's Inns Library contain extracts from classical authorities such as Hippocrates and Galen, and from Constantinus Africanus, who was instrumental in transmitting knowledge of Arabic medicine to Europe after the eleventh century. Irish medicine was clearly not practiced in isolation; Irish physicians were familiar with European traditions and developments. There is some scattered evidence of members of Irish medical families studying and working in England and Europe during the fifteenth, sixteenth, and seventeenth centuries, while the greatest Scottish medical family of the early modern period had emigrated from Ireland in the fourteenth century.

In both the Irish and Anglo-Norman parts of the country, medicine was by no means solely the preserve of males. The Brehon laws referred to *banliaig* (female physicians), who are usually presumed by scholars to have been midwives, although it is likely that their expertise was more extensive. The guilds of barbers and surgeons found in towns such as Dublin admitted women, who were often relatives of male members. It is certain then that Irish women practiced widely within their communities as healers, herbalists, nurses, and midwives, although evidence for their activities is scanty.

There is certainly evidence, however, of women working in hospitals. While figures are far from reliable, it has been estimated that there were some 211 hospitals in late medieval Ireland run by religious orders. Possibly around half of these were facilities intended to segregate lepers, whose breath was considered to be infectious. The rest were mainly almshouses, hospices for poor travelers, and pilgrims and institutions caring for the sick poor. Place names, such as Spiddal in County Meath, Spital in County Cork, Hospital in County Limerick, Cloonalour in County Antrim, and Leopardstown and Palmerstown in County Dublin, all suggest the existence of some sort of hospital during the medieval period—the latter three catering specifically for lepers.

Often attached to religious houses and staffed by brothers and nuns, these institutions generally offered food and warmth and, more particularly, religious consolation, rather than medical treatment; although the excavation during the 1920s and 1930s of trephined skulls at monastic sites in Cos Down and Meath testified to complex surgery being conducted, at least occasionally, and more recent excavations in urban sites have produced further evidence of cranial and other surgery.

St. Stephen was especially associated with lepers, and from the twelfth century Dublin boasted a St. Stephen's hospital for lepers, as did Cork from the thirteenth century. Perhaps the largest hospital in late medieval Ireland was also found in Dublin: the priory and hospital of St. John the Baptist without the New Gate, known as Palmer's Hospital, established during the 1180s and operated by the *Fratres Cruciferi* (or Crutched Friars). In 1334, the hospital had 155 beds, although when the priory was dissolved in the 1530s only fifty beds remained. Yet, like a number of medieval hospitals, St. John's appears to have continued to function under secular control into the seventeenth century. St. Stephen's in Dublin also remained a medical and charitable institution, being transformed in the 1730s into Mercer's Hospital, which did not close until the 1980s. Ireland's medieval medical heritage is thus a very long, if little understood, one.

ELIZABETH MALCOLM

References and Further Reading

Bannerman, John. *The Beatons: A Medical Kindred in the Classical Gaelic Tradition.* Edinburgh: John Donald Publishers, 1998.

Fleetwood, John F. *The History of Medicine in Ireland.* 2nd ed. Dublin: Skellig Press, 1983.

Gwynn, Aubrey and R. Neville Hadcock. *Medieval Religious Houses: Ireland.* Dublin: Irish Academic Press, 1970.

Dublin Institude of Advanced Studies. *Irish Script on Screen.* Reproductions of the Twenty-Eight Irish Medical Manuscripts held by Trinity College, Dublin, Library. 2003. http://www.isos.dcu.ie/english/index.html.

Kelly, Fergus. *A Guide to Early Irish Law.* Dublin: Dublin Institute of Advanced Studies, 1988.

Kelly, Maria. *A History of the Black Death in Ireland.* Stroud, Gloucestershire: Tempus, 2001.

Lee, Gerard A. *Leper Hospitals in Medieval Ireland.* Dublin: Four Courts Press, 1996.

McNeill, Charles. "Hospital of S. John Without the New Gate, Dublin." *Journal of the Royal Society of Antiquaries of Ireland*, ser. 6, no. 55 (1925): 58–64.

Wulff, Winifred, ed. *Rosa Anglica. Sev Rosa Medicinae. Johannis Anglici.* Irish Texts Society, Vol. 25. London: Simpkin, Marshall, 1929.

See also **Anglo-Norman Invasion; Annals and Chronicles; *Annals of the Four Masters*; Archaeology; Black Death; Brehon Law; Classical; Conflict; Education; English; Famine and Hunger; Fraternities and Guilds; Law Tracts; Manuscript; Nuns; Pilgrims and Pilgrimage; Placenames; Records Administrative; Records Ecclesiastical; Religious Orders; Scottish; Women**

MELLIFONT

The first Cistercian monastery in Ireland situated on the banks of the river Mattock (Co. Louth) approximately five miles northwest of Drogheda. The impetus for the foundation came from visits made in 1139 and 1140 by St. Malachy (Máel M'áedóic) to the Abbey of Clairvaux in Burgundy, then at the height of its influence under its charismatic abbot St. Bernard. Malachy left some of his entourage at Clairvaux to receive monastic formation, and they, with a number of French companions, returned to Ireland in 1142 to a site granted by Donnchad Ua Cerbaill, King of Airgialla. Between 1143 and 1153, seven new foundations were made from Mellifont, and its filiation or network of daughter houses eventually came to number twenty

The Lavabo, Mellifont, Co. Louth. © *Department of the Environment, Heritage and Local Government, Dublin.*

houses. In 1170, Mellifont itself contained one hundred monks and three hundred lay brothers. This rapid growth rested on an insecure foundation, for, unlike England and the Continent, Ireland had no significant tradition of Benedictine monasticism from which to draw seasoned recruits. A number of the French monks returned to Clairvaux and it proved necessary to recall some Irish monks for further formation. Despite this, the monastery and its abbots were closely associated with the latter stages of the twelfth-century reform movement in the Irish church. In 1151, Abbot Christian (Giolla Chríost Ó Connairche) was appointed Bishop of Lismore and Papal Legate, and in 1152 Cardinal John Paparo held a session of the Synod of Kells in the monastery.

With its continental contacts and extensive network of daughter houses Mellifont exercised a tremendous influence on Irish church architecture. Archaeological excavations have uncovered four different stylistic phases in the church and claustral buildings. As the largest stone structure of its day, it was known in native sources as *An Mhainistir mhór* (the great monastery).

Despite securing a number of English royal confirmations of their lands, rights, and privileges, the establishment of new monasteries by the Anglo-Normans created rival filiations to Mellifont and introduced an element of racial tension among the Cistercians in Ireland. This was exacerbated by the reluctance of Irish abbots to travel to the order's annual chapter at Citeaux, which meant that the Gaelic houses became increasingly isolated from the Order's disciplinary mechanisms. By 1216, it was evident that a general breakdown of discipline had occurred, and successive attempts by the order's central authorities met with stiff resistance from Mellifont and her daughter houses. This revolt, known in a contemporary phrase as the "conspiracy of Mellifont," was in large measure resolved by the Abbot of Stanley, Stephen of Lexington, who conducted a visitation of the Irish houses in 1128 and whose letter book detailing his labours survives. At Mellifont he presided over the election of Jocelyn of Bec as Abbot and broke up the Mellifont filiation placing her daughter houses under various English and Continental monasteries. No candidates were to be admitted unless they could confess in Latin or French, and the numbers at Mellifont were fixed at fifty monks and sixty lay brothers. The filiation was restored in 1274, but racial tension continued to affect the community. In 1321, Edward II complained to the Abbot of Citeaux that the house would only admit novices who swore that they were not of the English race. In 1380, the situation was reversed, and the monastery was under English control and so continued until the dissolution. Other difficulties arose however: in 1367, John Terrour was accused of murdering another monk, John White, but the case was never proved and Terrour became Abbot in 1371. The maladministration of Abbot John Waring (c. 1458-1471) almost ruined the community through alienation of resources and lands, though most of these were recovered by his successor Roger Boley (d. 1486). His successor, Abbot John Troy, was appointed visitator of the Irish houses by the general chapter around 1497, and his report paints a bleak picture of decline, abuses, and neglect in most of the Irish houses with only two houses, Mellifont and St. Mary's, Dublin, celebrating the Divine Office or wearing the religious habit. Abbot Richard Contour surrendered the monastery on July 23, 1539, the abbot and eighteen monks receiving pensions or annuities. In 1540, the property of the monastery, which included approximately 5,000 acres, 300 messuages and cottages, granges, mills, fisheries, and boats was valued at £352 3s. 10d. Though this represents a significant under valuation of the monastery's true worth, it places Mellifont in the same league as some of the major English houses. As part of the seventeenth century Irish Cistercian revival, a small community was reestablished in Drogheda under Abbot Patrick Barnewall in 1623.

COLMÁN N. Ó CLABAIGH, OSB

References and Further Reading

Carville, Geraldine. *The Impact of the Cistercians on the landscape of Ireland.* Ashford: K.B. Publications, 2002.

Gwynn, Aubrey, and Hadcock, R. Neville. *Medieval Religious Houses: Ireland.* London: Longman, 1970 [rept. Dublin: Irish Academic Press, 1988].

Mac Niocaill, Gearóid. *Na Manaigh Liatha in Éirinn, 1142-c.1600.* Dublin: Cló Morainn, 1959.

Ó Conbhuidhe, Colmcille. *Studies in Irish Cistercian history.* Dublin: Four Courts Press, 1998.

——— *The Story of Mellifont.* Dublin: M.H. Gill and Son, 1958.

O'Dwyer, Barry. "The problem of reform in the Irish Cistercian monasteries and the attempted solution of Stephen of Lexington in 1228." *Journal of Ecclesiastical History*, 15 (1964): 186-91

Stalley, Roger. *The Cistercian monasteries of Ireland.* London & New Haven: Yale University Press, 1987.

See also **Abbeys; Anglo-Norman invasion; Architecture; Church Reform, Twelfth Century; Ecclesiastical Settlements; Gaelic Revival; Kells, Synod of; Malachy (Máel M'áedoic); Military orders; Nuns; Papacy; Racial and Cultural Conflict; Religious Orders**

METALWORK

Little is known of the process of commissioning and paying for metal objects in early-medieval Ireland. Some have argued that craftsmen worked in highly controlled circumstances thus allowing potentates to control the supply of luxury goods and so help to perpetuate their power. In law, however, metalworkers were free and could rise to fairly high status. Metalworking evidence is widespread on Irish sites of the early medieval period—on ringforts, crannogs, and ecclesiastical foundations. Iron-working was ubiquitous and this may reflect the need, known to farmers of today, to attain some skill so as to keep agricultural equipment in repair—in other words it may be unspecialized metalwork or the jobbing work by an itinerant craftsman. The iron was probably obtained mostly from bog iron ore, but other sources may well have been exploited. Iron was sourced in sufficient quantity to make sword blades—although the characteristic sword in pre-Viking Ireland was small like the Roman gladius, and made of fairly soft metal at that. In the Viking period, blades of high quality were imported, and much larger iron objects, such as plow coulters, were fabricated.

Luxury objects were mostly made of bronze, and theoretically much of this could have been recycled scrap. There is some evidence of copper mining at this period at Ross Island, Killarney, County Kerry, and it is likely that native sources continued to supply the needs of the bronzesmith. Tin, essential for bronze, was probably imported. It is also likely, despite the presence of lead ore (galena), that lead also came from overseas. The lead-ores of Ireland are silver-rich and were extensively mined for silver from the seventeenth century onward, but there is no evidence that silver was produced in Ireland before the thirteenth century, when foreign expertise was required to make it possible. In the pre-Viking period, silver was clearly in short supply and was obviously adulterated with copper even on pieces of high status, and it is likely that Roman silver was constantly recycled. In the Viking Age, silver became abundant as a result of Scandinavian imports of coinage from the Islamic world, and metal objects made of very pure silver were manufactured in large quantity. Gold, once apparently abundant in Ireland, seems to have been very scarce, and it too was likely to have been recycled or imported—a text on kingship speaks of knowing gold by its foreign ornaments. At the height of the production of luxury metalwork in the eighth and ninth centuries, gold was always used sparingly as filigree, granulation, or gilding, and only one surviving complete object of substance, the ninth-century Loughan Brooch, is fabricated of gold (see Jewelery and Personal Ornaments).

Techniques of metalworking seem to have been somewhat conservative. Objects—examples include the Ardagh and Derrynaflan chalices and the Derrynaflan Paten—are often elaborate constructions where continental analogs are often structurally relatively simple. Casting of metal, especially for the making of brooches, pins, and other smaller objects, was remarkably competent with much ornament, to all appearances engraved or chased, but actually produced in the mold. Casting in bivalve clay molds formed on lead, wax, or wooden models was the preferred method, although there is evidence of the use of lost wax casting for complex pieces such as the components of the stem of the Ardagh Chalice. Casting was, however, limited in *scale*, and it was not until the production of fine hand bells, made entirely of bronze, that pieces of any great size were produced in the ninth or tenth centuries. The older bells—not unlike a cowbell in shape—were made of sheet iron folded to shape and dipped in bronze. These are the ones that were often enshrined—St. Patrick's Bell, provided with a reliquary at the end of the eleventh or beginning of the twelfth century is a good example.

Inventiveness within the conservative tradition was often remarkable—the extraordinary ornament of fine dark trumpet spirals on the reverse of the Tara Brooch, long thought to have been made of niello (a black sulphide of silver), is in fact a pattern raised on copper plates by stamping and then covering the area with a wash of silver solder and polishing it down until the copper shows through in a remarkably delicate fashion.

At the beginning of the period, craftsmen turned out brooches and pins for cloak fastening which were very like those being produced in Late Roman Britain (see Jewelery and Personal Ornaments). These were predominantly of bronze. The brooches were sometimes enriched with enamel, and the decorative repertoire was limited to stylized palmettes and later to spiral scrollwork. By the seventh century, more sophisticated products with a wider range of decorative techniques were appearing. Filigree, gilding, granulation, the occasional use of amber, were all adopted by workshops by about the year 700 A.D. A characteristic of the finest metalwork is the appearance of cast polychrome glass studs with angular inset metal grilles designed to mimic gem-set garnets so beloved of Germanic jewelers.

By this time also, many of the greater monasteries had become wealthy and powerful, and they not only were able to commission craftsmen but also to have craft workshops themselves. By the end of the seventh century, elaborate decorated house-shaped shrines to protect the relics of native saints were being produced (see Early Christian Art). In the eighth century, the production for the church clearly accelerated, and the numbers of reliquaries must have been significant to judge by the surviving corpus of complete examples and fragments of house-shaped shrines. Specialized reliquaries such as book shrines had made their appearance by the eighth century—the oldest known is the Lough Kinale shrine, the earliest in a series which continues into the high Middle Ages. The enshrinement of bells associated with native saints seems to have begun at an early date—what may be the crest of such a shrine was preserved at Killua Castle, County Westmeath until acquired by the National Museum in the early twentieth century. The obverse of the shrine crest shows in openwork an *orant* figure between two beasts—almost certainly a representation of Christ. This remarkable composition may be traced to early-medieval Merovingian and Burgundian belt-buckles. The same motif occurs twice on a recently reconstructed large altar or processional cross from Tully, County Roscommon.

The Tully Cross, made of wood covered with sheets of bronze and cast bronze decorative bosses and plates, introduces us to the manner in which almost interchangeable parts were created—square and round bronze bosses, binding strips, hinged tabs, animal-headed terminals—which could appear on objects of different type. It is often implied that the work of the period was so intricate that long periods of time were required to create some of the surviving objects, but it is clear that workshops were practical places where the techniques employed, such as casting and die-stamping, were designed for the efficient production of multiples, and by no means was every object a masterpiece.

Some pieces were exceptional, and the finest are the altar vessels from the hoards of Ardagh, County Limerick and Derrynaflan, County Tipperary. The Ardagh Hoard consisted of two chalices and four brooches. Probably deposited during the tenth century, the brooches, which may all have been made for ecclesiastical use, represent the major phases of personal ornament development from the eighth and ninth centuries (see Jewelry and Personal Ornaments). One of the chalices was a simple vessel of bronze, the other an elaborate construction of silver decorated with great brilliance using the full repertoire of filigree, polychrome glass, gilding, die-stamped plates, knitted silver- and copper-wire mesh, casting, and a limited series of amber settings. It is matched by the Derrynaflan Paten found in a hoard composed exclusively of altar vessels (chalice, paten, liturgical sieve, basin) on a monastic site in County Tipperary. This great silver communion plate carries twenty-four filigree panels, many of them bearing iconographical scenes; brilliant polychrome glass studs; and superb die-stamped panels. Both the bowl of the Ardagh Chalice and the plate of the paten were spun on lathes indicating that, in addition to hand tools, simple machinery was available to aid the processes of manufacture. The paten carried an engraved assembly code suggesting strongly that a literate person—presumably a cleric—was involved in its manufacture. Both pieces were made in the eighth century—at a guess in the later part of that century. The chalice from Derrynaflan is remarkable for its simpler, graphic filigree style, which reflects common and simple iconographical motifs of birds and beasts, including lions, probably associated with the Tree of Life and related iconographies. It was probably created in the ninth century. The chalices belong to a distinctive insular type, while the paten reflects an old tradition of large communion plates now missing from the surviving corpus of ecclesiastical metalwork from the west. It is difficult to escape the impression that the great altar vessels were created to reflect the traditions of the important metropolitan churches of Rome and Gaul. Their commissioning must have been a significant act of patronage either by a king or an important foundation anxious to provide their church with fitting plate to rival that seen abroad by clergy and pilgrims.

One of the more complex types of object from the workshops was the crosier. Often described as crosier-shrines, it is assumed that the metal casings had been designed to enshrine the wooden staff of an early saint. This is in most cases very unlikely, as the constructions of relatively thin bronze tubing required an armature internally to give support. Irish crosiers are comparatively short—although some may have lost sections in antiquity and may originally have been somewhat longer, many are now of walking stick proportions. The Irish type has a very distinctive crook that ends in a straight edge (the "drop") and a series of bulbous knops that join the tubular segments together. The tubes are generally plain, but the knops are often a field for the display of ornament. The majority of surviving crosiers were either made or restored in the eleventh and twelfth centuries, but the type is known to have achieved its traditional form by the eighth century. A very damaged early example in the National Museum is associated with Durrow, County Offaly. Particularly fine examples are the Crosier of Clonmacnoise—dating to the later eleventh century and the Lismore Crosier dated by an inscription to the early twelfth century. A thirteenth- or early-fourteenth-century grave slab at Kilfenora, County Clare shows a bishop holding an Irish-style crosier. Crosiers were certainly valued as relics, and one, the Bacall Íosa (Staff of Jesus), thought to have been St. Patrick's, was kept at Ballyboughal, County Dublin until in the sixteenth century it was seized and burned by the Lord Deputy. Irish clergy are almost invariably depicted with their bell and crosier on early medieval sculpture. A mounted ecclesiastic on the ninth century (?) Banagher shaft, now in the National Museum, is shown carrying his crosier over his shoulder. An odd figure on the little pillar at Killadeas, County Fermanagh carries both bell and staff.

The tradition of native metalwork production survived the Viking invasions and continued in a modified form until the twelfth century. Viking influences were absorbed—particularly the use of distinctive Scandinavian animal styles. Production of fine metalwork in rural habitations is not wellattested in the later period—it seems to have shifted mainly to monasteries and towns. Some of the very finest products of the early-medieval period were made in later eleventh and twelfth century in styles that show obvious Viking influence as well as a return to early inspiration. Inscriptions on metalwork objects clearly indicate that kings and other notables, as well as leading clerics, commissioned works of importance—Domhnall Ua Lochlainn was chief patron of the Shrine of St. Patrick's Bell, Tairrdelbach Ua Conchobair of the Cross of Cong. Craftsmen, too, are named in the inscriptions, but this emergence from anonymity was to be short-lived. The traditional pattern of patronage seems to have withered away during the later twelfth century, having been dealt a heavy blow by the Norman Invasion, and the products of native schools of metalwork were supplanted by continental imports and by manufactures of workshops in towns now dominated by the conquerors.

MICHAEL RYAN

References and Further Reading

Edwards, Nancy. *The Archaeology of Early Medieval Ireland*. London, 1990.

Harbison, Peter. *The Golden Age of Irish Art The Medieval Achievement 600—1200*. London, 1999.

Henry, Francoise. *Irish Art in the Early Christian Period to A.D. 800*. London, 1965.

———. *Irish Art during the Viking Invasions 800–1020 A.D.* London, 1967.

———. *Irish Art in the Romanesque Period 1020—1200.* London, 1970.

Ó Floinn, Raghnall. "Schools of Metal Working in Eleventh- and Twelfth-Century Ireland." In *Ireland and Insular Art A.D. 500–1200*, edited by M. Ryan, pp. 179–187. Dublin, 1987.

Organ, Robert M. "Examination of the Ardagh Chalice—A Case History." In *Application of Science in Examination of Works of Art*, edited by W. J. Young, pp. 238–271. Boston, 1973.

Ryan, Michael, ed. *Ireland and Insular Art.* Dublin, 1987.

———. *Studies in Medieval Irish Metalwork.* London, 2000.

Smith, Reginald A. "Irish Brooches of Five Centuries." *Archaeologia* 65 (1914): 223–250.

Stevenson, R. B. K. "The Hunterston Brooch and its Significance." *Medieval Archaeology* 18 (1974): 16–42.

Whitfield, Niamh. "The 'Tara' Brooch an Irish emblem of Status in its European Context." In *From Ireland Coming: Irish Art from the Early Christian to the Late Gothic Period and its European Context*, edited by Colum Hourihane, pp. 210–247. Princeton, 2001.

Youngs, S., ed. *The Work of Angels: Masterpieces of Celtic Metalwork 6th–9th Centuries A.D.* London, 1989.

See also **Jewelry and Personal Ornament; Weapons and Weaponry**

METRICS

Four major metrical systems are attested, which follow each other in a roughly chronological sequence from the sixth to the seventeenth century, though with some overlap. The first is found in an archaic stratum of Irish poetry containing legal aphorisms, gnomes, genealogies, and the heightened language of prophecy embedded in prose sagas. Metrically, this poetry was characterized by a fixed number of syllables per line (most commonly seven), loosely accentual in the first part but with a fixed end-of-line cadence following a caesura. Take, for example, *to-combacht selb soertellug*, "landed property has been recovered by means of high occupation," in which a first unit of four syllables with variable stress and marked with a caesura after *selb* (x`xx`xl) is followed by a unit of three syllables with a fixed cadence (`xxx). Close parallels with the meters of certain other languages, notably Sanskrit, Greek, and Slavic, argue for its ultimate origins in a shared Indo-European heritage. This meter was used by the *filid*, the preeminent learned class of early Ireland (see *Áes Dána*), no doubt long before the introduction of Christianity.

During the seventh century, a new metrical system appeared. Known collectively as the *nuachrutha* ("new forms, meters"), it held sway until the late twelfth century. It was characterized by a fixed number of syllables per line (commonly seven), by a caesura after the fourth syllable, by end-rhyme, and by a stanzaic structure. These features are generally attributed by scholars to a conscious imitation of similar features in late Latin poetry, especially Latin hymns, though some would argue that the *nuachrutha* were a natural development from the earliest meters. During the Old and Middle Irish periods, the *nuachrotha* were the sole verse medium of the poets, secular and ecclesiastical.

The third metrical system, the *dán díreach*, ("strict meter"), came to the fore in the late twelfth century. Basically, it was a reworking of the previous *nuachrotha*, which were severely reduced in number and adapted to the twelfth-century phonology of Irish. Its practitioners were the so-called bardic poets who dominated the literary scene for the next four centuries. Their primary concern was metrical ornaments such as alliteration and rhyme. These ornaments, which had been used sparingly by early practitioners of the *nuachrotha*, now became widespread, their use prescribed with elaborate rules. The collapse of the Gaelic order in the seventeenth century spelled the end of the bardic schools and the demise of *dán díreach*.

Already by the sixteenth century another type of meter was appearing, the *amhrán* or song poem, though it probably had a much earlier history among the common people. Metrically, it was characterized by strophic structure (quatrains in the earliest examples); regular rhythm based on the interplay of accented (usually four to six per line) and unaccented syllables both within and between lines; and by systematic use of ornamental assonance. It was also meant to be sung to a particular tune. This meter, foreign in origin and popular in usage, eventually became the dominant form for the next three centuries.

PÁDRAIG Ó NÉILL

References and Further Reading

De Brún, Pádraig et al. *Nua-Dhuanaire*, pt. 1. Dublin: School of Celtic Studies, Institute for Advanced Studies, 1971.

Greene, David and Fergus Kelly. *Irish Bardic Poetry.* Dublin: The Dublin Institute for Advanced Studies, 1970.

Murphy, Gerard. *Early Irish Metrics.* Dublin: Royal Irish Academy, 1961.

Watkins, Calvert. "Indo-European Metrics and Archaic Irish Verse." *Celtica* 6 (1963): 194–249.

See also **Áes Dána; Bardic Schools, Learned Families; Hiberno-Latin; Poetry, Irish**

MICHAEL OF KILDARE

See **Hiberbo-English Literature**

MIDE (MEATH)

Mide, meaning the "Middle Territory," was originally the district around the hill of Uisnech (Usnagh, County Westmeath); Uisnech was considered to be the center of Ireland. Twelfth-century king-lists styled the kings of Mide as *ríg Uisnig* (kings of Uisnech). This district was controlled by the Laigin up to the early sixth century. According to the annals, Fiachu, one of the sons of Niall Noígiallach, defeated the Uí Failgi at the battle of Druimm Derge in 516; thereafter Mag Midi (the Plain of Mide) was lost to the Laigin. Apart from the inclusion of Fiachu in the lists of the kings of Uisnech, there is no evidence that Cenél Fiachach ever adopted the style *rí Midi* ("King of Meath"). While some of the other annals accord the title *rí Midi* to certain seventh-century members of Clann Cholmáin Máir (see Uí Néill, Southern), the earliest evidence in the Annals of Ulster for a kingship of Mide relates to the mid-eighth century when Follaman mac Con-chongailt (d. 766), of the relatively insignificant Clann Cholmáin Bic, was appointed to that position, possibly by Donnchad Midi, of Clann Cholmáin Máir, as part of the latter's campaign to consolidate his control of the midlands. Thereafter, the kingship was confined to dynasts of Clann Cholmáin Máir. The blessing of antiquity was conferred upon Uí Néill control of Mide by later propagandists who claimed that Tuathal Techtmar, grandfather of Conn Cétchathach, and the common ancestor of the Uí Néill, the Connachta, and the Airgialla, had created the kingdom of Mide for himself by cutting off the neck (*méde*) of each surrounding province. According to another tradition, Fintan mac Bóchra, a wise man of phenomenal longevity, set up a five-cornered stone at Uisnech at the point where the five great provinces of Ireland were said to meet.

The rapid growth of Clann Cholmáin Máir during the eighth century and subsequently resulted in the name "Mide" being applied to the extensive territory over which that dynasty held sway, which included the modern county of Westmeath, together with parts of counties Longford and Offaly. It is unclear whether Uisnech remained as an inaugural site for the kings of Mide after the expansion of the kingdom.

The kingdom of Mide, under the rule of Clann Cholmáin, was at the forefront of Irish political life from the late eighth until the eleventh century. Many kings of Mide during this period also became kings of Tara and had not entirely unrealistic aspirations to rule over the entire country. Most prominent among these kings were the following: Donnchad Midi (d. 797), Máel-Sechnaill I mac Máele-ruanaid (d. 862), Flann Sinna mac Máele-Sechnaill (d. 916), and Máel-Sechnaill II mac Domnaill (d. 1022). The power and influence of Mide posed a serious threat to the ambitions of other rival Uí Néill kingdoms. On two occasions during the ninth century, kings of Cenél nÉogain (see Uí Néill, Northern) sought to avail themselves of the opportunity, presented by the kingships of less illustrious Clann Cholmáin lords, to curb this influence by dividing the kingdom of Mide between rival claimants. These arrangements were short-lived.

The status of Mide and its kingship began to fall into decline, along with that of the Uí Néill dynasty generally, during the latter years of the reign of Máel-Sechnaill II mac Domnaill, who died in 1022. Mide's eastern neighbor Brega (a territory extending from south County Louth to north Co. Dublin) suffered an even greater eclipse during this period to such an extent that the name Mide was extended, by the twelfth century, to include Brega as well.

The territory of Mide fell prey to internal feuding among the leading Clann Cholmáin family of Ua Máelsechlainn expansion from the Uí Briúin of Bréifne during the eleventh century. During the late eleventh and the twelfth centuries, the kingdom of Mide frequently assumed the status of a puppet state as warlords from the then powerful dynasties of Ua Briain of Munster, Mac Lochlainn of Cenél nÉogain (see Uí Néill, Northern), and Ua Conchobair of Connacht dismissed and appointed kings and divided the kingship, seemingly at will.

In 1172, following the Anglo-Norman invasion of Ireland, King Henry II granted the Liberty of Meath to Hugh de Lacy. The Liberty of Meath included the more extensive territory of Mide referred to above. The Irish Parliament of 1542 divided the "shire of Methe" into the present-day counties of Meath and Westmeath.

PAUL BYRNE

References and Further Reading

Byrne, Francis John. *Irish Kings and High-Kings*. London: Batsford, 1973. Reprint, Dublin: Four Courts Press, 2001.

Byrne, Paul. "Certain Southern Uí Néill Kingdoms (Sixth to Eleventh Century)." PhD diss., University College Dublin, 2000.

O'Rahilly, T. F. *Early Irish History and Mythology*. Dublin: Dublin Institute for Advanced Studies, 1946.

See also **Anglo-Norman Invasion; Laigin; Máel-Sechnaill I; Máel-Sechnaill II; Uí Néill; Uí Néill, Northern; Uí Néill, Southern**

MILITARY ORDERS

The conflict between Christendom and its Islamic and non-Christian neighbors that emerged in the eleventh century gave rise to a new form of religious life: the Military Order whose members combined monastic life with active military service. To support their activities in the Middle East, the various orders were granted properties and privileges throughout Europe. The principal orders were the Knights of the Temple of Solomon (Templars) and the Knights of the Hospital of St. John of Jerusalem (Hospitalers) both of which possessed extensive estates in Ireland. The Knights of St. Thomas of Acre also had Irish possessions.

The Templars were founded around 1119 by Hugh de Payens for the protection of pilgrims to the Holy Land. Initially guided by the Rule of St. Augustine they later adopted Cistercian practices under the influence of St. Bernard. After securing ecclesiastical approval at the Council of Troyes (1129), the order spread rapidly and increased in wealth, prestige, and influence.

The earliest reference to the Templars in Ireland occurs about 1180 when Matthew the Templar witnessed a deed whereby Henry II granted them the vill of Clontarf as their principal Irish foundation or preceptory. Five other preceptories were established by the end of the twelfth century as well as nine smaller houses (*Camerae*). Though more military than monastic in appearance, these preceptories functioned as religious houses in which the Divine Office was celebrated, novices were recruited and trained, and to which older members retired. Like the Hospitalers, the Templars recruited almost exclusively from the Anglo-Norman community and sided with the colony in its struggles against the native Irish population.

The wealth and influence of the Templars aroused the envy of other religious orders and secular rulers. Opposition was particularly strong in France where King Philip the Fair orchestrated a campaign that culminated in the suppression of the order by Pope Clement V in 1312. Their properties were to pass to the other military orders, principally the Hospitalers.

As part of the general campaign against the order, fifteen Irish Templars were tried in St. Patrick's cathedral, Dublin in 1310. The judges and accusers were for the most part mendicant friars, and, as elsewhere in Europe, the case against them was quite weak: one knight was regarded as suspect because he was observed not gazing on the host during the elevation at mass in Clontarf. In 1311, three preceptories were assigned to accommodate the Irish Knights for the rest of their lives while the rest passed to the Knights Hospitaler after 1312.

The Knights of the Hospital of St. John of Jerusalem emerged in the twelfth century as the military wing of an institution originally established to care for pilgrims and the sick in Jerusalem. Like the Templars, they came to Ireland in the wake of the Anglo-Normans and had established fifteen preceptories in every province except Connacht by the second decade of the thirteenth century. Their chief house was the priory of St. John the Baptist at Kilmainham, and they formed part of the English division or *langue* in the order's general structure. They took an active role in the defense of the colony: in 1274 the prior of Kilmainham, William Fitz Roger, was captured by the Irish but escaped and subsequently led a royal army into Connacht. Other priors held important posts in the colonial administration, including those of chief governor and chancellor.

The acquisition of six former preceptories of the Knights Templar in 1312 greatly augmented the Hospitalers' wealth. The shrewd administration of Prior Roger Outlaw between 1317 and 1341 consolidated these gains, and he also used his terms as deputy justiciar, chancellor, and justiciar to acquire further lands and rights for the order. In the forty years after Outlaw's death, English Knights Hospitaler took control of the Irish priory and its resources. This was greatly resented by the Anglo-Irish members who in 1384 elected an Anglo-Irish knight, Richard White, as prior. Exploiting the divisions caused by the Great Western Schism, they also transferred their allegiance from the Grand Master of the Avignon, obedience to whom, paradoxically, was recognized by the English Hospitallers, to the Grand Master of the Roman obedience, thereby confirming their independence.

Despite their great wealth, the number of knights in the Irish priory was miniscule so that at the dissolution only five had to be pensioned. Of these, the prior Sir John Rawson received 500 marks and an annuity of £10. In 1557, the Irish priory was restored but was finally dispersed the following year on the accession of Elizabeth I.

COLMÁN N Ó CLABAIGH

References and Further Reading

Falkiner, C. L. "The Hospital of St. John of Jerusalem in Ireland." *Proceedings of the Royal Irish Academy* 26 (1906–1907): 275–317.

Gwynn, Aubrey, and R. Neville Hadcock. *Medieval Religious Houses: Ireland*. London: Longman, 1970. Reprint, Dublin: Irish Academic Press, 1988.

McNeill, Charles, ed. *Registrum de Kilmainham*. Dublin: Irish Manuscripts Commission, 1943.

Tipton, C. L. "The Irish Hospitallers during the Great Schism." *Proceedings of the Royal Irish Academy* 69 (1970): 33–43.

Wood, Herbert. "The Templars in Ireland." *Proceedings of the Royal Irish Academy* 26 (1906–1907): 327–377.

See also **Anglo-Norman Invasion; Architecture; Church Reform, Twelfth Century; Henry II; Military Service, Anglo-Norman; Military Service, Gaelic; Racial and Cultural Conflict; Religious Orders**

MILITARY SERVICE, ANGLO-NORMAN

Anglo-Norman military service was employed by the English king and his government at Dublin for the defense of English lordship in Ireland. This form of military feudalism generally reflected the situation in England and the march of Wales. The most common unit of Anglo-Norman military service in Ireland was the knight's fee. But the knight's fee in Ireland seems to have differed considerably from that of England, being defined more precisely and based on territorial units. Thus the crown was able to assess military service due to it. However, the territory covered by a knight's fee could vary from region to region. For example, a knight's fee in County Dublin could amount to ten plowlands—while the average size of a knight's fee in Meath covered twenty plowlands. In Ireland, nearly half of the military service owed to the crown was due from the four great royal tenants in Leinster, Meath, and Cork. The great Marshal lordship of Leinster was divided into 180 knight's fees—but only owed to the crown the service of 100. Similarly, the neighboring liberty of Meath possessed 120 knight's fees—but only rendered the service of 50 knights. Only 60 knight's fees were due from the two grantees of Cork. All were obliged to serve when a royal service was proclaimed in Ireland. The king's tenants by knight service—both English and Irish—brought with them their own military sub-tenants to make up the feudal host. A feudal host was an assembly under arms of the royal tenants in chief, each with the quota of knights that his enfeoffment required. Essentially, the arms of the feudal host were made up of knights, men at arms, footmen, archers, and the hobelars—forces of lightly armed and mobile horsemen adapted to the conditions of Irish warfare. However, the nature and composition of the feudal host was changing. Even before 1100, it was clear that the feudal host was gradually becoming an obsolete form of military organization in England—but the nature of warfare in Ireland hastened its demise further. In Ireland, English settlements were often subject to raids; royal service could mean frequent absence on campaign—leading to increased settler vulnerability. Because of the incessant nature of frontier warfare, the royal government was careful not to deplete a country of its men of fighting age by strictly enforcing observance of a royal summons. Accordingly, the royal government from early in the thirteenth century introduced scutage to lighten the burden of feudal service on smaller military tenants. Scutage first appeared in England around 1100 and was adopted to ease the burden of military service upon frontier lords in Ireland, allowing them to render a money payment instead of royal service. Scutage is first mentioned in Ireland in 1222. Then the royal tenants in Munster (Tipperary and Limerick), Decies (Waterford), Desmond, and the vale of Dublin were ordered to pay scutage rather than join the justiciar on campaign. On occasion, though, the levy of scutage could be unpopular. During the 1280s, some tenants complained to Edward I that they preferred military service to scutage.

As time passed, Anglo-Norman military service evolved further—adapting to suit local conditions in Ireland. Clearly, English magnates in Ireland were adopting elements of Gaelic military service. In Ulster, the de Burgh earls of Ulster famously adopted the *buannacht* ("bonaght" – wages and provisions of a galloglass) of Ulster. The bonaght involved the quartering of galloglass throughout the earldom of Ulster, while the earls levied the *tuarastal* ("wages") of these elite soldiers upon the peasantry. Increasingly life on the frontier between the lands of the English and the Irish became even more hybrid—as demonstrated dramatically during the proceedings of the parliament of 1297. It emerged that English magnates often hired Irish troops, billeting them upon their English tenants. When the commons complained bitterly that English settlers were greatly impoverished by the imposition of these hired "kerne," this billeting was outlawed. However, the de Burgh earls of Ulster were not the only English magnates in Ireland to adopt this Irish practice of billeting troops upon their tenants. In the fifteenth century the earls of Ormond, Kildare, and Desmond adopted the practice. It was reputed that James Fitzgerald (d.1463), seventh earl of Desmond, first imposed *coinnmheadh* ("quartering or billeting" – better known as coyne and livery) upon the Desmond earldom. The activities of the Desmonds did not go without rebuke. In 1467, the abbot of Odorney in Kerry wrote to the pope,complaining about their exactions. According to the abbot, they were forcing the local population to maintain the *ceithearn tigh* ("kernety" or "household kerne"—a form of military police), while they extracted military service from both horsemen and footmen. If a horseman failed to answer a summons, he was compelled to pay a fine of three cows or 15s.—while a kerne was liable for one cow or 5s. as a penalty for non-attendance. Similarly, the Butler earls of Ormond imposed coyne and livery upon their lands. During the early decades of the fifteenth century, James Butler (d.1452), fourth earl of Ormond, imposed forces of kernety and galloglass throughout his patrimony in Tipperary and Kilkenny—granting them the right to take a *cuid oidhche* (anglicized as "cuddy"—meaning a night's portion of food, drink, and entertainment extracted by an Irish lord from a subject) from every freeholder's house. The evolution of Anglo-Norman military service by the fifteenth century is dramatically illustrated in this usage of Desmond and Ormond kernety.

But the development of these large private armies by the English magnates of Ireland was crucial to the survival of their power on the frontiers. The best example of this was in the rise of the Fitzgerald earls of Kildare from 1456 to 1534. In his parliament of 1474, Thomas Fitzgerald (d.1478), seventh earl of Kildare, established a permanent fighting force, the "Fraternity of St. George" compromising 160 archers and 63 spearmen, whose captains included Kildare's son. However, the Kildares' real military muscle was built up by their importation of MacDomnaill galloglass from the Western Isles. By the late fifteenth century they were able to do the unthinkable—billet their MacDomnaill galloglass upon the Pale before levying coyne and livery upon Englishmen for the maintenance of these troops. The Dublin government's first priorities following the defeat of the Kildare rebellion in 1535 was to shore up the frontiers of the Pale and extend royal jurisdiction throughout the country. This process demanded reform and the dissolution of the private armies maintained by the great Irish lords and English magnates in Ireland. Accordingly, the government demanded that common law be the only observed writ in the country. By seeking the abolition of coyne and livery, the Dublin government intended to destroy the military power that underpinned the power of both the Irish and English nobles of Ireland. With the reestablishment of the royal writ, successive chief governors backed by English forces strove to ensure that all military service was due alone to the monarch.

Emmett O'Byrne

References and Further Reading

Barry, Terry, Robin Frame, and Katharine Simms, ed. *Colony and Frontier in Medieval Ireland*. London: Hambleton Press, 1995.

Bartlett, Robert and Angus McKay, ed. *Medieval Frontiers Societies*. Oxford: Oxford University Press, 1989.

Bartlett, Thomas and Keith Jeffrey, ed. *A Military History of Ireland*. Cambridge: Cambridge University Press, 1996.

Lydon, James. "The Hobelar: An Irish Contribution to Medieval Warfare." *Irish Sword* 2 (1954–1956): 13–15.

———, ed. *Law and Disorder in the Thirteenth-Century Ireland*. Dublin: Four Courts Press, 1997.

———. *The Lordship of Ireland in the Middle Ages*. Dublin: Four Courts Press, 2003.

Morgan, Hiram. *Tyrone's Rebellion*. London: Boydell Press, 1999.

Nicholls, Kenneth. *Gaelic and Gaelicized Ireland*. Dublin: Lilliput Press, 2003.

O'Byrne, Emmett. *War, Politics and the Irish of Leinster 1156–1606*. Dublin: Four Courts Press, 2003.

Otway-Ruthven, Jocelyn. "Knight Service in Ireland." *Journal of the Royal Society of Antiquities* 79 (1959): 1–7.

———. "Royal Service in Ireland." *Journal of the Royal Society of Antiquities* 98 (1968): 37–39.

———. *A History of Medieval Ireland*. London: Ernest Benn, 1968.

Simms, Katharine. *From Kings to Warlords*. Woodbridge: Boydell Press, 1987.

See also **Chief Governors; Common Law; Feudalism; Military Service, Gaelic; Tuarastal**

MILITARY SERVICE, GAELIC

According to Brehon law, an Irish king at an *óenach* ("general assembly") could exact military service from his followers by issuing a legally binding summons to arms—provided that the *garmsluaigh* ("hosting, rising out") was just. The attendance at the rising out demonstrated the king's power over the people of his country, as his real wealth lay in the farmers living under his protection. Accordingly, these farmers owed military service in return for this protection. All the able-bodied population of a country—apart from the learned and the clergy—were eligible for service either as horsemen or footmen. To enforce a call to arms, Irish kings appointed officers to ensure their fighting men obeyed the summons. These officers were also entrusted with the levying of fines upon those who choose to ignore the call. For example, MacCarthy Mór of Kerry in 1598 expected the men of his country to answer his summons within three days—with victuals and sufficient weapons—while anyone who failed to serve was required to pay a fine of 20s. In the neighboring palatinate of Kerry,a horseman who failed to answer a summons paid a fine of three cows or 15s., while an absentee footman had a choice between a fine of a cow or 5s. The most important and lucrative office within the military hierarchy of an Irish kingdom was that of the *marasgal* ("marshal"). This prestigious office was hereditary and was confined to members of a noble family close to the king. The origins of the office of the marshal are probably to be found in the much older office of *dux luchta tige* ("the head of the king's household"). In the execution of his duties—particularly those of levying and billeting troops—the marshal was assisted by a team of submarshals. Billeting troops upon the people of the kingdom was the principal task of the marshal—this was known as *coinnmheadh* ("quartering, billeting") later more commonly known as "coyne."

Permanent standing forces were also a feature of Irish warfare. From at least the eleventh century, the Irish kings were maintaining small permanent fighting forces. These forces were known as *teclach* or more aptly *lucht tige* ("troops of the household"). These household troops were well-equipped footmen and *marcsluag* ("cavalry") skilled in the use of arms, living upon the king's mensal lands. The majority of these highly mobile and well-armed horsemen were drawn from the upper classes. A major development in Gaelic military service of the Middle Ages was the increasing dependence of Irish kings upon retained bands of mercenaries. The origins of the much-demonized sixteenth-century *cethern* ("woodkerne") are to be probably found among these *ceithirne congbála* ("retained bands") or in the large recruited companies of mercenaries known as *rúta* ("routes"). As early as the 1100s, Irish kings—particularly of Ulster—were recruiting among the large communities of Hebridean-Norse fighting men in the Western Isles of Scotland hiring large forces and fleets for service in Ireland. Irish kings also hired English or Welsh mercenaries, as Diarmait

Mac Murchada of Leinster (d. 1171) famously did between 1167 and 1170. Other Irish kings were not slow in following Mac Murchada's example, Domnall Mac Gille Pátraic (d. 1185) hired Maurice de Prendergast in 1169, while Cathal Crobderg Ua Conchobair in 1195 employed the services of Gilbert de Angulo. Hiring mercenaries, however, could be a risky business. In 1310, Áed Breifnech Ua Conchobair (sl. 1310), king of Connacht, was killed by his own mercenary captain, Johnock Mac Uigilin (Mac Quillan). Ideally, the preferred option of Irish kings of Ulster and Connacht was to hire directly from the Western Isles of Scotland. In 1259, Áed son of Feidlim Ua Conchobair (d. 1274), prince of Connacht, formed an alliance with these Hebridean-Norse communities. That year he married the daughter of King Dubgall mac Ruaidrí of the Hebrides, gaining as part of his bride's dowry 160 fighting men known as galloglass. Domnall Óc Ua Domnaill of Tír Conaill (sl. 1281) followed Áed's example—marrying two brides drawn from the great galloglass families of Mac Domnaill of the Western Isles and Mac Suibne of Argyll. These galloglass, led by their own nobility, were traditionally huge men and fearless—preferring often to fight to the end rather then surrender. These forces were to play a dominant role in the Irish wars of raid and counter raid. Often the galloglass formations were employed as defensive shields to protect the retreating horsemen from their pursuers. A galloglass wore a helmet and was clad from head to toe in a mail coat. His arms were the *tuagh* ("axe"), broad swords, and daggers, and he employed a manservant to tend to the care of his armor and weapons. The "cess" or quartering of galloglass on a country was called the *buannacht* ("bonaght"), while the Irish kings levied the *tuarastal* ("wages") and provisions due to these elite soldiers from the people of their countries. This Irish practice of billeting galloglass upon the peasantry was later copied by English magnates of Ireland such as the de Burgh earls of Ulster in the fourteenth century and later by the earls of Ormond, Kildare, and Desmond a century later.

The role of mercenaries in Irish warfare was to develop in importance as a feature of Gaelic military service. But the widespread use of foreign mercenary forces by Irish kings only became commonplace in the fifteenth century, heralding the rapid intensification and scale of Irish warfare. In 1428, Niall Garbh Ua Domnaill (d. 1439) imported a great force of Scots to besiege Carrickfergus Castle. This was the first recorded use of seasonal Scottish soldiers, or "redshanks" as they became more commonly known. Unlike the galloglass, these redshanks did not engage in long-term contracts, but were imported directly from the Western Isles in greater numbers for shorter periods. As this custom became more widespread—particularly in the latter half of the sixteenth century—it greatly increased the destructive scale of Irish warfare. In comparison to Ulster and Connacht, the Irish of Leinster had traditionally always hired native born soldiers of fortune from either Connacht or Ulster. But the rising power of the earls of Kildare in Leinster from the 1450s may have denied the Leinster Irish access to their traditional sources of mercenaries, forcing them to look elsewhere. Moreover, the Kildares were also importing large forces of Mac Domnaill galloglass from the Western Isles into Leinster from the 1460s, forcing the Leinster Irish to maintain themselves by recruiting galloglass forces of their own. As a document dated to about 1483 illustrates, there was a huge influx of galloglass into Leinster during the 1470s and the 1480s—recording that Mac Murchada, Ua Broin, Mac Gille Pátraic, Ua Conchobair Failge, and Ua Mórda each employed a "battle" of galloglass. And such was the hybrid nature of Irish warfare that by the 1500s the earls of Kildare were billeting their Mac Domnaill galloglass upon the Pale—levying "coyne and livery" upon Englishmen for the maintenance of these troops.

The introduction of firearms into Ireland in the 1470s further speeded Irish warfare along its increasingly destructive path. The later widespread usage of firearms among the Irish brought major innovations in warfare—including a raise in the status of the woodkerne. Traditionally, the kerne was a lightly armed and nimble footman, armed with a set of three javelins, a small shield, and a sword. In 1399, the kerne of Mac Murchada and Ua Broin displayed how effective they could be fighting in a naturally protecting environment of mountain and forest—harrying mercilessly the beleaguered army of Richard II of England (d. 1400). In the sixteenth century, many of these kerne became extremely proficient in the use of firearms, wounding three chief governors between 1510 and 1534. Toward the end of the sixteenth century, a series of Irish leaders such as Áed Ua Néill (d. 1616), second earl of Tyrone, and Fiach Ua Broin (sl. 1597) emerged to revolutionize Irish warfare by adopting foreign ideas, tactics, training, and formations—adapting them to suit the Irish landscape. Tyrone, a far-seeing and ruthless man, trained a red coated Ulster army to fight in the Spanish formation of the terico, using both pike and musket. With this army at his back, Tyrone won great victories at Clontribret in 1595 and at Yellow Ford three years later—but his defeat at Kinsale in 1601 and subsequent submission in 1603 effectively ended the Irish military establishment.

EMMETT O'BYRNE

References and Further Reading

Barry, Terry, Robin Frame, and Katharine Simms, ed. *Colony and Frontier in Medieval Ireland*. London: Hambleton Press, 1995.

Bartlett, Robert and Angus McKay, ed. *Medieval Frontiers Societies*. Oxford: Oxford University Press, 1989.

Bartlett, Thomas and Keith Jeffrey, ed. *A Military History of Ireland*. Cambridge: Cambridge University Press, 1996.

Byrne, Francis. *Irish Kings and High-Kings*. London: B. T. Batsford.

Harbison, Peter. "Native Irish Arms and Armour in Medieval Gaelic Literature, 1170–1600." *Irish Sword* 12 (1975–1976): 174–180.

Lydon, James. *The Lordship of Ireland in the Middle Ages*. Dublin: Four Courts Press, 2003.

Morgan, Hiram. *Tyrone's Rebellion*. London: Boydell Press, 1999.

Nicholls, Kenneth. "The Kavanaghs, 1400–1700." *Irish Genealogist*, vol. 5, no. 4 (1977): 435–436.

———. *Gaelic and Gaelicised Ireland*. Dublin: Lilliput Press, 2003.

O'Byrne, Emmett. *War, Politics and the Irish of Leinster 1156–1606*. Dublin: Four Courts Press, 2003.

Simms, Katharine. "Warfare in Medieval Irish Lordships." *Irish Sword* 12 (1975–1976): 98–105.

———. *From Kings to Warlords*. Woodbridge: Boydell Press, 1987.

See also **Brehon Law; Chief Governors; Kings and Kingship; Mac Murchada, Diarmait; Tuarastal**

MILLS AND MILLING

The technology of building horizontal and vertical watermills entered Ireland from the Roman world, although the manner and exact date of transmission are unclear. Both forms, dating to around 630, are known from Little Island in Cork Harbor, while Cogitosus's *Life of Brigid* (c. 650) describes a mill and gives an account of the cutting and fitting of a millstone. The horizontal watermill was the preferred form in early medieval Ireland, probably because it was better suited to small, fast-flowing steams and, also, because of the absence of gears, it was comparatively simple and cheap to build. Typically the horizontal mill was housed within a two-story, rectangular structure consisting of an upper and a lower room. The upper room contained the grinding stones and the hopper mechanism for the grain, while a vertical shaft connected the upper grinding stone with a horizontal waterwheel, composed of paddles, in the chamber below. Water was channeled by means of a millrace and a chute so that it fell onto the horizontal wheel causing it to turn. One revolution of the waterwheel produced one revolution of the upper rotary stone, which was usually no more than about three feet across.

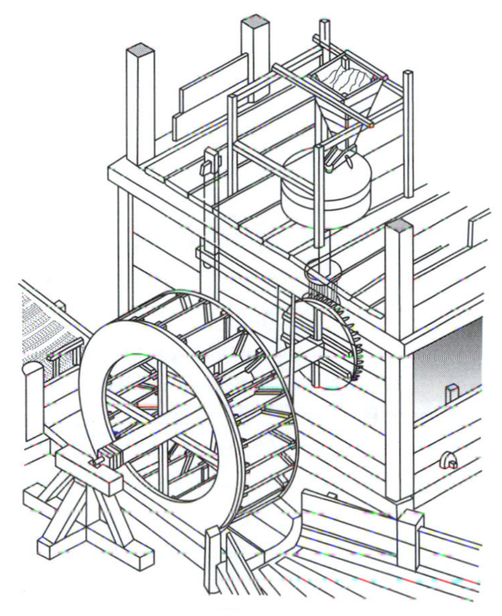

Vertical Watermill, circa 630 CE.

Vertical mills had an upright waterwheel with a horizontal axle that was geared to a vertical shaft, which was connected to the grinding stones. The gearing made it possible to adjust the rotation speed of the millstones, something that was impossible in the horizontal mill. Vertical waterwheels could be fed from above (overshot) or below (undershot) and both forms are evidenced in the Roman world. The fall of water from above gave the overshot mill greater power but it was more expensive and time-consuming to build. Accordingly, undershot mills are much more common. Apart from Little Island, another early example of a vertical undershot watermill, dating to about 710, is known from Morett, County Laois. Tide mills, in which a current was created by water descending from a pool where it had been trapped at high tide, are a feature of Atlantic Europe. The earliest Irish example is at Strangford Lough (619–621), while the Little Island mills, already mentioned, were also tidal. Early medieval mills are frequently found at ecclesiastical sites. Some have been found in isolation, but insufficient work has been done to determine whether they formed part of ecclesiastical and aristocratic estates or not. All of the known early examples are in rural locations but from the twelfth century onward, watermills are found in the Hiberno-Scandinavian port towns, where they tended to be located on feeder streams rather than tidal reaches.

Despite the ubiquity of watermills in early medieval Ireland, the grinding of grain by hand, using quern stones, remained commonplace. This changed after the Anglo-Norman invasion, when all grain had to be ground at designated mills. Such mills were a significant source of income for the ecclesiastical and territorial lords who monopolized the manufacture of flour until the close of the Middle Ages. A good example of a vertical undershot watermill of thirteenth- and fourteenth-century date was excavated at Patrick Street, Dublin. As is common with mills established at this time, the Patrick Street mill continued in use, periodically remodeled and rebuilt, into modern times.

Windmills are documented in Britain from 1137, but in Ireland they seem to be a feature of the thirteenth century. The earliest form was the post mill, a small wooden-framed building, which pivoted on a large upright beam (the post), and whose interior was accessed by means of a ladder. The entire building was turned so that the sails pointed into the wind. The structures were usually built on low mounds and survive today as circular earthworks with a distinctive internal cross pattern, which is all that survives of the foundations. There is a fine example at Shanid, County Limerick. The internal construction was much the same as that of watermills except that the vertical shaft fell downward to the stones. Tower windmills, consisting of a circular stone tower with a rotating cap that carried the sails, such as the example at Rindown, County Roscommon, are not evidenced in Ireland until the late sixteenth and seventeenth centuries. Bridge mills were built at Dublin despite the obvious dangers that they posed in times of flooding. Although beer mills, fulling mills, and iron mills, which used pounders rather than rotary stones, are known from continental Europe by the eleventh century, there is little evidence for them in Ireland prior to the sixteenth century. Similarly, no evidence for boat mills is known.

JOHN BRADLEY

References and Further Reading

Bennett, Richard and John Elton. *History of Cornmilling*. Vol. 4, *Some Feudal Mills*. London and Liverpool: Simpkin, Marshall & Co., 1904. Reprint, Wakefield: EP Publishing, 1975.

Lydon, J. F. "The Mills of Ardee in 1304." *Journal of the County Louth Archaeological and Historical Society* 19, no. 4 (1981): 259–263.

Rynne, Colin. "The Introduction of the Vertical Watermill into Ireland: Some Recent Archaeological Evidence." *Medieval Archaeology* 33 (1989): 21–31.

———. "The Craft of the Millwright in Early Medieval Munster." In *Early Medieval Munster, Archaeology, History and Society*, edited by M. A. Monk and John Sheehan, pp. 87–101. Cork: Cork University Press, 1998.

Walsh, Claire. *Archaeological Excavations at Patrick, Nicholas & Winetavern Streets, Dublin*. Dingle: Brandon, 1997.

See also **Agriculture; Diet and Food**

MINING

See **Metalwork**

MODUS TENENDI PARLIAMENTUM

Modus Tenendi Parliamentum is a later medieval treatise describing the workings of parliament. It exists in both an "English" and an "Irish" version. Both claim a spurious antiquity for their descriptions, perhaps in order to enhance their authority. The longer English version claims to be an account compiled for William the Conqueror of how parliament had functioned in the reign of Edward the Confessor; the shorter Irish version to contain instructions from Henry II to his Irish subjects on the holding of parliaments in Ireland. The earliest surviving manuscripts of the English version were probably written in the 1380s; the earliest surviving manuscript of the Irish version is contained in an official *inspeximus* dating from 1419, which now forms part of the Ellesmere manuscripts, at the Huntington Library in California. Historians used to believe that the English version itself belonged to the 1380s, but almost all modern historians (other than Richardson and Sayles) have accepted Maud Clarke's arguments for composition in the 1320s on the basis of similarities between the treatise's description of parliament's working and the workings of the English parliament in that period, though not her specific suggestion of a date of 1322. The Irish version probably belongs to the early fifteenth century. Clarke suggested a specific connection with Archbishop O'Hedigan of Cashel (1406–1440). Sayles argued for a lost Irish original treatise dating from shortly after 1381 behind both English and Irish versions, but this view has not met general acceptance. Historians have also disagreed about the nature of the original treatise. Some have argued for an underlying political purpose; others suggested that it provided a generally honest, if sometimes tendentious, description of parliament in the 1320s; still others that it was intended only to provide an ideal picture of how parliament ought to be run.

PAUL BRAND

References and Further Reading

Clarke, M. V. *Medieval Representation and Consent: A Study of Early Parliaments in England and Ireland, with Special Reference to the Modus Tenendi Parliamentum*. London and New York: Longmans Green, 1936.

Pronay, Nicholas, and John Taylor. *Parliamentary Texts of the Later Middle Ages*. Oxford: Clarendon Press and New York: Oxford University Press, 1980.

See also **Henry II; Parliament**

MO-NINNE (D. C. 517 OR 519)

She is the reputed founder and abbess of Cell Sléibhe Cuilinn or Killevy in County Armagh, a prominent monastery for nuns. Her original name was Darerca (or Sárbile, according to the Martyrology of Oengus), but she is better known by the hypocoristic Mo-Ninne or Monenna. Her feast day is July 6. Killevy was sacked by the Norse in 790 and again in 923; records indicate that it survived as female monastic house well into the twelfth century and afterward. By the sixteenth century, it had become a convent for Augustinian nuns, which was dissolved in 1542.

Almost all of Mo-Ninne's life and works are legendary; her Lives consist mainly of a series of miracles and wonder-working. Three versions of her Life remain extant; the earliest, by a monk named Conchubranus, dates only from the eleventh century and, like a twelfth-century Irish redaction, is probably based on an earlier Life. Two hymns in honor of Mo-Ninne, perhaps composed at Killevy, date from about the eighth century. A third Life, written by Geoffrey of Burtonin in the twelfth century, is based in part on Conchubranus but is in honor of St. Modwenna of Burton-on-Trent, whom he identified with Mo-Ninne.

Mo-Ninne's traditions make her a contemporary of Patrick and Brigit. According to her legends, she sought out Patrick for baptism and consecration, along with eight other virgins and a widow, all of whom became her disciples. She adopted the widow's son, Luger, as her foster-son and eventually saw him ordained as a bishop. For a time, she and her nuns lived under the rule of Ibar, a prominent bishop and teacher. She then visited Brigit and lived for a time at the monastery of Kildare before making her own foundation at Killevy. Mo-Ninne was famous for her rigorous asceticism: she frequently lived as a hermit in the wilderness, in prayer and fasting; she wore a garment of badger skins; she combed her hair only once a year, at Easter; she tilled the ground herself in order to grow her own food. She was compared to two famous biblical desert dwellers, John the Baptist and the prophet Elijah, and praised for her "manly spirit." After her death, her hoe, her comb, and her badger-skin dress were kept as relics at her monastery at Killevy. Her ascetic regime extended to her community; her legends relate how several of her nuns died of fasting and hunger until Mo-Ninne miraculously supplied them with food.

Mo-Ninne's cult spread to Scotland and to England. She is said to have sent one of her nuns to the monastery of St. Ninian at Whithorn in Scotland for further instruction, and her own legendary travels, as told by Conchubranus, took her to Scotland and England, where she founded several monasteries, and to Rome. After her death, her remains were translated to England.

DOROTHY ANN BRAY

References and Further Reading

Bray, Dorothy Ann. "The Manly Spirit of Saint Monnena." In *Celtic Connections: Proceedings of the Tenth International Congress of Celtic Studies*, edited by Ronald Black, William Gillies, and Roibeard Ó Maolalaigh, vol. 1, pp. 171–181. East Linton: Tuckwell Press, 1999.

Gwynn, Aubrey and R. Neville Hadcock. *Medieval Religious Houses: Ireland*. Dublin: Irish Academic Press, 1970.

Harrington, Christina. *Women in a Celtic Church, Ireland 450–1150*. Oxford: Oxford University Press, 2002.

Kenney, James F. *The Sources for the Early History of Ireland: Ecclesiastical*. New York: Columbia University Press, 1929.

de Paor, Liam, trans. "The Life of St. Darerca, or Moninna, the Abbess." In *Saint Patrick's World*, edited by Liam de Paor, pp. 281–294. Dublin: Four Courts Press, 1993.

Sharpe, Richard. *Medieval Irish Saints' Lives*. Oxford: Clarendon Press, 1991.

Ulster Society for Medieval Latin, ed. and trans. "The Life of Saint Monenna by Conchubranus." *Seanchas Ardmacha* 9 (1978–1979): 250–275; *Seanchas Ardmacha* 10.1 (1980–1982): 117–141; *Seanchas Ardmacha* 10.2 (1982): 426–454.

See also **Hagiography and Martyrologies; Nuns**

MONKS

See **Ecclesiastical Organization**

MORAL AND RELIGIOUS INSTRUCTION

By the time Christianity arrived in Ireland, assuming it had a strong foothold by the later fifth century, it had a firm self-perception (1) as a teaching religion with a body of doctrine to be transmitted and understood, and (2) that it demanded an ethical and religious discipline (but which varied with the different kinds of Christians: laypeople, clerics, monks, nuns). Thus a major part of the church's concerns, and a key part of ecclesiastical organization, was concerned with teaching that doctrine and discipline. We see this concern with instruction in a number of ways, but most obviously in saints' lives where part of the pattern of most lives is to portray their subject as one who was "illustrious" as a teacher and whose lifestyle was an example of discipline to others. We can also observe the importance attached to teaching in canon law: the *Collectio canonum hibernensis*, for instance, assumes that teaching is one of the duties attached to the senior grades of cleric (deacon, presbyter, and bishop), and has a special section devoted to teachers (*De doctoribus*: book 38); while the *Collectio* itself is a major repository of the various demands of the Christian life. However, most of our knowledge of moral and religious instruction must be derived from their extant writings, which can be grouped under four broad headings: biblical exegesis being the most important. Moreover, we must not expect the modern distinction between "moral" and "religious" texts to be always clearly made: for instance, a biblical commentary may be primarily doctrinal in its interests, but distinguish several kinds of exegesis in the same text and label one kind "spiritual" (roughly equivalent to "religious" meaning), and another kind "moral" in which case it is usually the shortest section; likewise, a homily may be devoted to a doctrinal subject, such as the

Nicene Creed, but contain much instruction on how Christians should live.

Biblical Exegesis

By the fifth century, Christianity's approach to scripture, both in content and form, were already fixed. The Latin West, found in the fourth and fifth century writers such as Ambrose, Jerome, and Augustine, masters it could revere and whose books would be the basis of all they wrote. Seeing themselves as disciples of these great "fathers," they believed it was their task to repeat this material, organize it systematically, and make it as accessible as possible to students. Hence their emphasis on the repetition of patristic themes, staccato questions and answers, the production of collections of facts, and manuals that survey complex questions through a series of quotes from authorities. The aim of all Irish exegesis was to provide textbooks and syntheses within an established intellectual paradigm. These scholastic repetitions were original works, and their innovation lay in the way they systematized the inherited tradition.

Irish exegesis has to be examined against this background for it is similar in content and quality with the work from Italy, Visigothic Spain, Merovingian Gaul, and, slightly later, the Anglo-Saxon and the Carolingian writers. However, it does present some striking qualities of its own. The productivity of Europe in the period between 500 and 800 is meager when compared with the ninth century, but in the earlier period the work of Irish scholars, in Ireland and abroad, is significant disproportionally to the country's size or background. Hence we can assert that there was a significant Irish input into the exegesis and theological life of the period, and we must view scholars such as Eriugena (John Scottus)—who saw himself as engaged in exegesis—not as a lone phenomenon, but as the most famous expression of a well-established exegetical culture.

One peculiarity of Irish exegesis is how much of it is anonymous or pseudonymous, for we have only a handful of names: Adomnán, Ailerán, and Laidcenn. While works of major importance such as the *De mirabilibus sacrae scripturae* presents itself as Augustine's, the *De ordine creaturarum* is attributed to Isidore, and Cummian's (?) *Commentarius in Marcum* to Jerome; and most of the exegetical material bears no name and is attributed to Irish writers only on the basis of modern comparative research. This has raised the question of "an Irish school" of exegesis, and has promoted the search for telltale "Irish symptoms" in such works. While there are features that figure prominently in Irish works, such as interest in grammar or computistics questions in the midst of exegesis, these cannot settle the question of origins as such features are not exclusively Irish. The presence of even several "characteristics" in a single work cannot be decisive, and must be viewed only as increasing the probability of Irish origins.

At present we are still in the period of discovery: finding the texts, providing editions, and making preliminary studies of their contents. Only when this process is complete, and the material compared with that from Gaul and Spain, will the true character of the Irish group emerge. Only then will the attribution of works to places of origin be possible on a secure footing and allow considered answers to be given to questions such as why so much Irish writing is anonymous.

Manuals

While many manuals produced in Ireland are linked to exegesis, as a form of instruction they deserve special attention; and because they were often works produced by teachers responding to their local situation they exhibit regional differences not found in works aimed at the larger church. These manuals range form single pages (e.g., the plan of the New Jerusalem in the Book of Armagh), to works to be committed to memory (e.g., Ailerán's *Kanon evangeliorum*), to textbooks that distill many of the major problems of Latin theology into a user-friendly system (e.g., the *De ordine creaturarum*). The notion of such manuals was seen as being sanctioned by their authorities, the task being to go through the materials they had in their libraries, abstract the relevant bits, and present it in an easily taught format. Adomnán's *De locis sanctis* is an example of this process where, on the basis of what could be found in his library, he produced the manual Augustine had said would be so useful for teachers—and the work was found useful throughout Europe. In a similar spirit, glossaries of Hebrew and Greek words were produced, synopses of major texts (e.g., of Augustine's *De Genesi ad litteram*), and key references assembled in convenient packets (e.g., the *Liber ex lege Moysi*). Lastly, some large gospel books may have been specially prepared with intention that they would be reference resources (e.g., the complexity of the marginal apparatus in the Book of Durrow, or the amount of material relating to text-division found in the Book of Armagh).

Monastic Instruction

Life-long instruction of the monks/nuns in a monastery is part of the very reason for the monastery's existence; and we know that in the West, and nowhere more than in Ireland, the writings of John Cassian not only formed the basis of instruction, but provided the theoretical basis for all on-going formation. This instruction took several forms, penitentials and rules (there was no dominant Western monastic rule before the ninth century) being the most obvious, and examples survive from Ireland in both Latin and Irish. It also took narrative forms—inspired by Gregory the Great's *Dialogi*—whereby an ideal monk is praised for his holy life (e.g., Adomnán's *Vita Columbae*) or an ideal monastery is envisaged (e.g., the *Navigatio sancti Brendani*). However, it also took the form of the "conference" (a lecture or sermon to the community), the outstanding example of which is the collection of *Instructiones* by Columbanus (whose authenticity was often doubted prior to 1997).

Sermons

Sermons were, in all likelihood, the means by which most instruction was delivered, and certainly the activity of preaching is one that is praised formally and offered as good example in our sources. However, sermons, as such, do not survive. So when we have a sermon text we are already removed from the actual instruction and seeing something that was either a model of a good sermon (does this mean that without such models the preaching was inept or simply that these were what an individual teacher thought a sermon should be?) or a skeleton around which an actual sermon could be composed: in either case the sort of person who would compose such a sermon is different from the average cleric delivering the sermon. We have extant examples of both full sermons and skeleton sermons in both Latin and Irish; and, on the whole, they are remarkably similar to sermons from the same period from elsewhere in the Latin world. With regard to this particular literary genre, we should note that while some texts are obviously sermons (e.g., the "Cambrai Homily"), and texts found in homily collections are equally obviously so (e.g., the so-called *Catechesis Celtica*), there are many other sermon texts that have been cataloged under other headings (e.g., Christmas sermons which contain apocryphal themes and so are studied under the heading "apocrypha" rather than as instructional materials), and a full listing of all such texts is desirable as a benchmark in advancing our understanding of preaching in medieval Ireland.

THOMAS O'LOUGHLIN

References and Further Reading

Adomnán, *De locis sanctis*. In *Scriptores Latini Hiberniae*, vol. 3, edited and translated by D. Meehan and L. Bieler. Dublin: Dublin Institute for Advanced Studies, 1958.

Ailerán, *Interpretatio mystica et moralis progenitorum Domini Iesu Christi*. Edited and translated by A. Breen. Dublin: Four Courts Press, 1995.

Columbanus. *Sancti Columbani Opera*. Edited and translated by G. S. M. Walker. Dublin: Dublin Institute for Advanced Studies, 1957. Reprint, 1970.

Fletcher, Alan J. "Preaching in Late-Medieval Ireland: The English and the Latin Tradition." In *Irish Preaching: 700–1700*, edited by Alan J. Fletcher and Raymond Gillespie, pp. 56–80. Dublin: Four Courts Press, 2001.

Kelly, Joseph F. "A Catalogue of Early Medieval Hiberno-Latin Biblical Commentaries." *Traditio* 44 (1988): 537–571; and 45 (1989): 393–434.

Kenney, James F. *The Sources for the Early History of Ireland: Ecclesiastical—An Introduction and Guide*. New York: Columbia University Press, 1929. New edition, Dublin: Pádraig Ó Táilliúir, 1979.

Lapidge, Michael, and Richard Sharpe. *A Bibliography of Celtic–Latin Literature 400–1200*. Dublin: Royal Irish Academy, 1985.

Murdock, Brian. "Preaching in Medieval Ireland: The Irish Tradition." In *Irish Preaching: 700–1700*, edited by Alan J. Fletcher and Raymond Gillespie, pp. 40–55. Dublin: Four Courts Press, 2001.

O'Loughlin, Thomas. *Celtic Theology: Humanity, World and God in Early Irish Writings*. London: Continuum, 2000.

O'Loughlin, Thomas. "Irish Preaching before the End of the Ninth Century: Assessing the Extent of our Evidence." In *Irish Preaching: 700–1700*, edited by Alan J. Fletcher and Raymond Gillespie, pp. 18–39. Dublin: Four Courts Press, 2001.

O'Loughlin, Thomas, ed. *The Scriptures and Early Medieval Ireland*. Turnhout: Brepols, 1979.

Stancliffe, Clare. "The Thirteen Sermons Attributed to Columbanus and the Question of their Authorship." In *Columbanus: Studies in the Latin Writings*, edited by Michael Lapidge, pp. 93–202. Woodbridge: Boydell and Brewer, 1997.

See also **Adomnán; Biblical and Church Fathers; Canon Law; Columbanus; Conversion to Christianity; Devotional and Liturgical Literature; Ériugena (John Scottus); Grammatical Treatises; Paschal Controversy; Penetentials; Scriptoria**

MORTIMER

The Mortimers were among the most influential absentee families in later medieval Ireland. Calculated marriage alliances, military endeavor, and personal service to the crown brought the Mortimer earls of March lordship across broad swathes of Ireland. Earls of Ulster and lords of Clare, Connacht, Kilkenny, and Meath (Mide), they gave frequent personal attention to their Irish lands at a time when English landholding in Ireland was waning. Such focus was required as the defense of their interests became increasingly problematic, three Mortimer earls, indeed, dying in Ireland, victims of the nature of their estates.

Inheritance

On September 24, 1301, Roger Mortimer (c. 1287–1330), lord of Wigmore in Herefordshire, married Joan de Geneville (1286–1356), heiress to the Irish liberty of Trim in eastern Meath, giving the Mortimers transmarine estates of real consequence for the first time. Roger had inherited his grandmother's portion of the Marshal lands at Dunamase in Laois, but his immediate forebears had not maintained them. Widely anglicized, valuable, and imbued with seigniorial privileges enjoyed in no other Irish liberty, including the four royal pleas of arson, forestalling, rape, and treasure trove, Trim was worth fighting for. In mid-November 1308, months after coming of age, Roger received the lordship from Joan's grandfather, Geoffrey de Geneville (c. 1226–1314). Geoffrey, a former chief governor of Ireland with long experience of Irish political and military affairs and of landholding across frontiers, instilled in Roger the desirability of personal lordship in Ireland. For six of the following twelve years (1308–09, 1310–13, 1315, 1317–18, 1319–20), Roger resided in Ireland, establishing his lordship against his wife's kin, the Lacys of Rathwire and the Scots under Edward Bruce, and cementing his family's position among the elite of Irish landholding society. Despite forfeiting his lands in rebellion against Edward II in 1321-22, his notorious subsequent relationship with Queen Isabella and leadership of the invasion that deposed the king gave Roger almost unfettered power. Elevated to the earldom of March in October 1328, he launched a spree of acquisition in Ireland, gaining custody of the western half of Meath, during the minority of the de Verdun heiresses, with liberty status. This reestablished the lordship of Meath, which had been divided after the death of Walter de Lacy in 1241. Roger also obtained custody of the heir to the earldom of Kildare and expanded into Louth, coming close to creating an "empire" on the threshold of Dublin.

This potential evaporated upon Roger's execution by Edward III on November 29, 1330. But, while his lands were forfeited to the crown, his legacy provided the springboard for his successors' ambitions. Roger, second earl of March (1329–1360), emulated his proximity to the crown, helping to found the Order of the Garter and becoming one of Edward III's most trusty generals in his continental wars. Consequently, upon restoration to the earldom of March in 1355, he regained Meath as a liberty. Roger's prestige, moreover, secured the marriage of his son, Edmund (1352–1381), to Phillippa, daughter of Lionel, duke of Clarence, and Elizabeth, granddaughter of William de Burgh, late Earl of Ulster, in May 1368. Edmund thus became earl of March and Ulster, lord of Clare, Connacht, Kilkenny and Meath.

Problems

Despite their wealth, the Mortimers faced intractable problems in Ireland. The importance of their estates made their defense imperative, but the attractions of English court life, prolonged minorities, and the unfortunate brevity of their forays into Ireland made it increasingly difficult to maintain a firm grip against nascent Gaelic Revival. In many ways, the fate of their lands reflected the decay experienced across Ireland in the fourteenth century.

It is noticeable that each Mortimer lord received livery of his inheritance while still a minor, for the value of their estates was only matched by their vulnerability. As early as 1323, reports claimed that the castle and manor of Dunamase were worthless, as no English tenants remained after the onslaught of the Laois Irish. Both the first and second earls became embroiled in disputes with the men of Carbry. In December 1309, the king pardoned men of Trim who had chased a raiding party back into Carbry, killing several of them. In 1355, the steward of Trim was captured and imprisoned at Carbry after levying rents at Rathwire. Inquisitions returned into the English chancery demonstrate that most of the lands pertaining to the earldom of Ulster had been rendered of little value by 1368 thanks to destruction wrought by native armies across Ireland, emboldening Edmund Mortimer to attempt repairs at his fortresses of Greencastle and Carrickfergus and the bridge at Coleraine. By the time of Edmund's son, Roger (1374–1398), the fourth earl, supremacy in Ulster and Connacht had passed to the Irish, his tenants in Ulster, English and Irish, performing homage to Niall Ua Néill.

One of the Mortimers' most tangible solutions to the problems caused by absenteeism involved the transmission of retainers from the Welsh marches to their estates in Ireland. In the aftermath of his defeat of the Lacys in 1317, Roger Mortimer granted escheats in counties Meath and Dublin to Herefordshire and Shropshire retainers, giving his English tenants a stake in the maintenance of his Irish lordship. It is noticeable that members of the Hakelut and Harley families, for example, returned to Ireland with successive Mortimer lords. Less tangible are the Mortimers' relations with native communities. The Wigmore chronicle boasts of the Mortimers' lineage from both Strongbow and Diarmait Mac Murchada. How and if they played upon this heritage is not known. In Laois, the O'Mores were a constant thorn in their flesh, but they formally recognized their liege status at least once. In 1350, Maurice Sionnach, "king" of Fartullagh and Fergal Mac Eochagáin, "duke" of Moycashel, agreed to serve the earl of Kildare against all men save de Geneville's heir, evidence perhaps of a longer association between the

Mortimers and their Irish tenantry. Whether militarily or in a social context, therefore, the Mortimers certainly had experience of native society.

Chief Governors

It was exactly this, combined with their position at court and in Ireland, that brought all but one of the earls to the chief governorship of Ireland, and with some success. On November 23, 1316, around the time of Robert Bruce's landing in Ulster, Roger Mortimer was named as the king's lieutenant in Ireland. During the ensuing eighteen months, he set about restoring peace, brokering compromises between disputing Anglo-Irish lineages, particularly in Cork and Waterford, where large fines were extracted from warring factions, and making war on the lordship's enemies. Having exiled the Lacys of Rathwire in June 1317, he devastated Irish communities in Connacht and what is now County Wicklow. On his return to Ireland as justiciar in June 1319, in the aftermath of the battle of Faughart, more importantly, he received native Irishmen into English common law, attempting to mitigate one of the grievances expressed in the Irish Remonstrance.

When Edmund Mortimer arrived in Ireland in May 1380 by popular acclaim, the administration was penniless. Nevertheless, if chronicle accounts are accurate, he was able temporarily to regain his lordship in Ulster and northern Connacht, taking the homage of many "nobles of the Gael" and Niall Óg Ua Néill, captain of the Irish of Ulster, before sweeping south, crossing the Shannon, and tackling recalcitrant Irish and English kin groups in Thomond.

Pneumonia, however, ended his life on December 27, 1381, and his successes evaporated. Experiments with Thomas Mortimer, Edmund's brother, as deputy for his infant nephew, Roger, provided no boon against Irish encroachment and forced Richard II into journeying to Ireland himself. He was accompanied in 1394 by Roger Mortimer, fourth earl of March, who had initially been made lieutenant in July 1392, but had been delayed by disputes over his inheritance. During Richard's stay, the king, who wished to conciliate some of the leading Irish kin groups, forced Roger into accepting the negotiated homage of the O'Neills. Upon Richard's departure, however, Roger was left as lieutenant in Ulster, Connacht, and Meath. After gathering an army including the earls of Ormond and Desmond and many other prominent members of the settler community under the king's banner, he ravaged modern counties Longford and Cavan in an attempt to regain control of his lordship of Meath. He then launched an attack on the position of his rivals, the O'Neills, in Armagh, bringing them temporarily to heel. In April 1397, Roger was granted the sole governorship of Ireland, and he appears to have attempted to wrest control of the country back for the king. Such ephemeral successes, however, were curtailed by his murder on a raid into Leinster in August 1398.

Roger would be the second of his name to die in Ireland. He would not be the last. His son, Edmund (1391-1425), fifth earl of March, died at Trim on January 18, 1425, while vainly trying to employ the resources of Ireland as lieutenant in the defense of his familial estates.

Throughout the fourteenth century, the Mortimer earls of March had accumulated the single most important patrimony in Ireland. Far from remaining permanent absentees, unlike many of their contemporaries, they made frequent, if fleeting visits to Ireland, where their skills as warlords, peacemakers, and figures of compromise ensured their place at the zenith of landholding society and made them essential agents in the maintenance and development of English lordship. Premature deaths and minorities, however, meant that, ultimately, their lands could not be adequately defended, and they too became the unfortunate personal victims of the failure of the English lordship in Ireland to make any temporary successes endure.

PAUL R. DRYBURGH

References and Further Reading

Cosgrove, Art. *Late Medieval Ireland, 1370–1541*. Dublin: Helicon, 1981.

———. *A New History of Ireland: Medieval Ireland, 1169—1534*. Edited by F. J. Byrne, Art Cosgrove, and T. W. Moody, 2nd ed. Oxford, 1993.

Curtis, Edmund. *Richard II in Ireland, 1394–5, and the Submissions of the Irish Chiefs*. Oxford: Clarendon, 1927.

Dryburgh, Paul R. "The Career of Roger Mortimer, First Earl of March (c.1287–1330)." PhD diss., University of Bristol, 2002.

Frame, Robin. *Colonial Ireland, 1169–1369*. Dublin: Helicon, 1981.

———. *English Lordship in Ireland, 1318–61*. Oxford: Clarendon, 1982.

———. *The Political Development of the British Isles, 1100–1400*. Oxford: Clarendon, 1990.

———. *Ireland and Britain, 1170–1450*. London: The Hambledon Press, 1998.

Hagger, Mark S. *The Fortunes of a Norman Family: The de Verduns in England and Wales, 1066–1316*. Dublin: Four Courts Press, 2001.

Johnston, Dorothy. "Chief Governors and Treasurers of Ireland in the Reign of Richard II." In *Colony and Frontier in Medieval Ireland. Essays Presented to J. F. Lydon*, edited by T. B. Barry, Robin Frame, and Katherine Simms, pp. 97–115. London: The Hambledon Press, 1995.

Mortimer, Ian. *The Greatest Traitor. The Life of Sir Roger Mortimer, 1st Earl of March, Ruler of England, 1327–1330*. London: Jonathan Cape, 2003.

Otway-Ruthven, A. J. *A History of Medieval Ireland*. London: Ernest Benn, 1968.

See also **Anglo-Irish Relations; Bruce, Edward; Connacht; Gaelic Revival; Geneville, Geoffrey de; Kilkenny; Lacy, de; Lionel of Clarence; Lordship of Ireland; March Areas; March Law; Mide; Richard II; Trim; Ulster, Earldom of; Verdon, de**

MOTTE-AND-BAILEYS

These are the archetypal earthwork and timber castles of the Anglo-Norman invasion of Ireland that were used to hold the country down during the military phase of the campaign. The "motte" is a Christmas-pudding-shaped mound of earth constructed from the upcast from the excavation of the fosse around it, usually ranging in height from as little as 3 m to over 10 m, whose circular perimeter was defended by a wooden palisade, often with a tower, initially built of timber, constructed in the center of this summit. There seems to have been two major methods of construction: the first has been well illustrated in the Bayeux Tapestry, an almost contemporary pictorial account of the events of 1066, where the motte at Hastings was shown to have a perfect "reverse" stratigraphy of different layers of earth as the mound was raised from the soil excavated directly from its perimeter fosse. But there is also evidence at some mottes, such as at Lorrha in County Tipperary, where an earthen ringbank was first constructed around its perimeter, and then its center was filled in until it reached its required height. There are also a few examples that have a "squared off" summit, such as at Aghaboe in County Laois, although this may be the result of the construction of stone walls at a later date. The "bailey" was a much lower and larger defensive earthwork delineated by an earthen bank and palisade with an external fosse, classically rectangular in layout but with other shapes as well, attached to the motte by a wooden "flying bridge." There are also some examples of mottes with more than one bailey, such as the impressive double bailey at Mannan Castle, Donaghmoyne, County Monaghan. There are also many mottes that lack baileys, especially in the earldom of Ulster, and this has led scholars to speculate as to whether they may have been built by native Irish lords. Also, in comparison with examples in England and Wales, Irish baileys are often very small, which makes it difficult to envisage them as containing the hall, as well as other domestic and farm buildings of the classic manorial center. Thus the motte functioned as the citadel to which the inhabitants of the castle would retreat if they came under sustained attack, while the bailey was the area usually inhabited by the occupants when they were at peace.

It is of great interest to scholars that motte castles were constructed in Ireland more than a century after their first use during the Norman conquest of England, at a point when most of the castles in Britain were being constructed of stone, This reveals much about their main function: that of campaign castles. Their success was due to the fact that they could be constructed quickly, probably in a few weeks, with materials such as earth and timber that can readily be found in most locations. The small circular defensive perimeter of the motte also has the great advantage of being defensible by a small force against more sizeable besieging armies, as was often the situation in Ireland. Other strengths included their height advantage that meant that it was an "uphill" battle for any attacker. Also, the earthen composition of the motte meant that it was almost totally immune to mining and attacks by fire, two of the most common methods of contemporary siege warfare. These were some of the main reasons that they were such a successful and necessary component to the Norman military system.

Many Irish examples are situated either on top of earlier settlements, especially ringforts, which were the most numerous settlement sites in pre-Norman Ireland, or utilized preexisting natural features in the landscape, such as the gravel esker ridges that are to be found all over the midlands. Examples of the former include Rathmullan in County Down and Dunsilly, County Antrim, both of which have been excavated. Many of these castles became manorial centers in the new Anglo-Norman lordship, while some of the most significant and strategically sited examples were soon converted into important stone castles, such as at Dublin and Kilkenny.

Mottes are, hardly surprisingly, most densely distributed in the eastern half of the island where Anglo-Norman settlement was strongest, and especially in the province of Leinster, which has well over half of the national total of around five hundred examples. The two northern counties of Antrim and Down make up the majority of most other surviving mottes. There are very few of them in the north and the west, much of which remained outside the Anglo-Norman lordship, although they are also few in number in the province of Munster in the south, which did experience a sizeable level of Anglo-Norman settlement. This has given rise to the idea that there was also another early type of castle that was being constructed in Ireland at about the same time as the mottes, the military ringworks, which were particularly concentrated in south Wales from which many of the early Anglo-Norman invaders originated. There are also a few examples of mottes in areas that remained under the control of the native Irish lords throughout the Middle Ages, especially in Ulster.

Chronologically, most Irish mottes probably date to the short period from the end of the twelfth century up until the first quarter of the thirteenth century, but with some, especially in Leinster, still being built in the second half of that century. Contemporary documentary sources relating to Irish mottes are limited, and the archaeological record is concentrated geographically, almost wholly to one county, that of Down. Additionally, most excavations have concentrated on the motte perimeter to the exclusion of the other important elements of these castles. Therefore, future archaeological research should examine the baileys as well as the external peripheries of these important sites in order to further progress our knowledge of this important component of Anglo-Norman military settlement.

TERRY BARRY

References and Further Reading

Barry, Terry. *The Archaeology of Medieval Ireland*. London: Routledge, 1999.

McNeill, Tom. *Castles in Ireland, Feudal Power in a Gaelic World*. London: Routledge, 1997.

O'Conor, Kieran. *The Archaeology of Medieval Rural Settlement in Ireland*. In *Discovery Programme Monographs*, vol. 3. Dublin: Royal Irish Academy, 1998.

Sweetman, David. *The Medieval Castles of Ireland*. Cork: The Collins Press, 1999.

See also **Anglo-Norman Invasion; Armies; Castles; Ringforts**

MUIRCHERTACH MAC LIACC

Described as *ardollamh Erenn* ("chief poet of Ireland") in the Annals of Ulster on his death in 1016, is primarily associated with Brian Boru in later tradition whose literary biography he is credited with composing, along with poetic works on various aspects of the career of the Munster king. These include the mournful elegy, *Anoir tánic tuitim Briain* ("Brian's downfall came from the east"), and *A Chinn Choraidh, caidi Brian* lamenting the neglected state of Brian's County Clare citadel, rendered into English by James Clarence Mangan as "O, where, Kincora, is Brian the great?," as well as versified genealogies of Brian's Dál Cais dynasty. Frequently associated with Mac Coisse, said to be chief poet of Brian's main rival, Máel Sechnaill mac Domnaill, he engages him in dramatic dialogue in one particular work. What these compositions have in common is that their varied late dates link them with the polished poetic persona Mac Liacc subsequently became, rather than with a historical personage of that name about whom we know next to nothing. The latter may have been the author of a *dinnshenchas* text explaining the origin of Carn Conaill in south County Galway, in which he describes himself, metaphorically, according to Edward Gwynn, as Mac Liacc *Linni na nÉces* ("of Linn na nÉces," literally "of the pool of the poets"). Of the same "pool" was his son, Cumara, who is also described as a poet on his death in 1030. However, to the later literary legend bearing his name, this elusive Mac Liacc bears scant resemblance.

MÁIRE NÍ MHAONAIGH

References and Further Reading

Gwynn, Edward, ed. and trans. *Todd Lecture Series*. 5 vols. Vols. 8–12, *The Metrical Dindshenchas*, pp. 440–449. Dublin: Royal Irish Academy, 1903–1935.

Ó Lochlainn, Colm. "Poets on the Battle of Clontarf." *Éigse: A Journal of Irish Studies*, vol. 3 (1941–42), pp. 208–218; vol. 4 (1943–44), 33–47.

See also **Brian Boru; Dál Cais**

MUIRCHÚ

Muirchú Moccu Macthéni (d. + 697) was the author of a *Life of Patrick*, written at the "dictation"—i.e., command—of his bishop Aéd of Sletty (d. 700). According to *Félire Óengusso*, Muirchú was of Leinster origin and went with bishop Áed to Armagh. The Additamenta in the Book of Armagh states that Áed had gone to Armagh during the abbacy of Segéne (661–688) and gave a bequest of his church and "kin" to Patrick. It is likely that Muirchú wrote his Life of Patrick shortly after Fland Feblae's succession to the abbacy following the death of Segéne, while Áed was still at Armagh, where he died, probably in retirement as an anchorite. Both Muirchú and Áed appear on the guarantor list of the *Cáin Adomnáin* drawn up at a synod of the ecclesiastics and nobles of Ireland held at Birr, County Offaly, in 697. The note in *Félire Óengusso* for his feast on June 8 says "Medron and Murchon, two brothers, in Cell Murchon among the Huí Ailella. Medron and Murchu, sons of Húa Machthéni . . . " (p. 145). Muirchú was probably the father of Colmán mac Murchón, abbot of Moville County Down and author of a Latin hymn to St. Michael ("In Trinitate spes mea"). Muirchú's death is not recorded in the annals, and nothing more is known of him.

The prologue to his Life is modeled on the prologue to Luke's Gospel ("Quoniam quidem mi domine Aido . . . ") and is also modeled on earlier hagiography, especially Sulpicius Severus' *Vita Martini* and the early Life of St. Samson of Dol. He claims Cogitosus, the author of an earlier Latin Life of Brigit, as his "father," that is, his predecessor in the new genre of Irish hagiography, but probably also at some time his abbot. His reference to Cogitosus would seem to confirm Muirchú's Leinster origins, though other sources place him among the Mochtheine, who resided in the district of Armagh. In either case, his portrayal of Patrick as a conquering hero and a patriarch in the Old Testament tradition is clearly a piece of propaganda for the church of Armagh. His Life played a crucial role in assisting Armagh's alliance with the Uí Néill dynasty, and by insisting upon Armagh's conformity on the Easter question helped establish her claims over the oldest churches in the country, namely those of Auxilius and Isserninus in the midlands and east (*Vita Patricii*, I 19:3–4, p. 92). Muirchú says that they were conferred with lower orders on the day Patrick was created bishop, in order to be sent as his helpers.

The principal MS witnesses to Muirchú's Life are the early-eighth-century Book of Armagh (A), the incomplete eleventh-century text in Brussels, Bibliothēque Royale (B), and the late-eighth-century fragments in Vienna, Österreichische Nationalbibliothek, ser.n. 3642. The Vienna fragment represents a better version of the "B-text," which is sometimes superior to A, showing that Muirchú's Life had a separate transmission outside of Armagh circles. Muirchú's Life falls into two main parts. The first section (I 1–12) tells of the story of his early years, his captivity, escape, and return home; his decision to go to Rome for training in the religious life; but how, on his way there, he decides to stay with bishop Germanus of Auxerre. Having remained with him for many years, Patrick is summoned in a vision to return to Ireland. On his way, he learns that Palladius, the archdeacon to Pope Celestine entrusted by him with a mission to the "wild Irish," had died. Patrick goes to bishop Amatorex nearby and is finally consecrated bishop by him, and he then sets out for Ireland. So far, the narrative is straightforward and unadorned. The first section ends with Patrick's encounter with Miliucc, his former slave-master, and his celebration of his first Easter in Ireland on the plain of Brega, close to the royal seat of Tara.

With Patrick's arrival in Ireland, Muirchú's account (I 10 ff.) takes on the character of a full narrative. The later chapters (I 13–22) deal with his confrontation with king Lóegaire of Tara, his contest of miracles with the king's druids, Lóegaire's subsequent conversion, and Patrick's setting forth from Tara to convert the Irish. Muirchú here seems to have followed and elaborated upon a more primitive narrative, which in some form was also known to Tírechán, but he gave it vividness and detail. He was certainly familiar with Patrick's *Confessio*, upon which he is dependent in the opening chapters. He evidently also drew upon some other vernacular material of great age, both oral and written, concerning Patrick. There then follows a number of brief stories in the order and content of which texts A and B differ. Armagh then has a colophon closing Book I and opening II. From II 4 onward, this deals with Patrick's final days, his imminent death in Saul, County Down, on March 17 at the patriarchal age of 120, and his burial at Dún Lethglaisse (Downpatrick).

Muirchú claimed at the outset to write "in a poor style" (Prologue 3), but that, as Bieler put it (1974), is "a hagiographical commonplace" and "is belied . . . . by the very context in which it is made." His style at times is full of rhetorical *colores*, and some of his episodes have literary merit. As a piece of hagiography, it falls in terms of literary accomplishment between his predecessor Cogitosus, to whom he is superior in Latinity and literary ability, and Adomnán's *Vita Columbae*. His principal source is the Vulgate Bible, but he seems also to have some limited knowledge, however it was acquired, of classical literature, because he quotes (II 8:1) one line of Virgil, *Aeneid* viii, 369, and one line of Caelius Sedulius, *Carmen paschale* iii, 221, as well phrases borrowed from Jerome and Sulpicius. His work also contains some allusions to biblical apocrypha, specifically an apocryphal text on St. Peter and Simon Magus, perhaps that known as *Actus Petri cum Simone*.

AIDAN BREEN

References and Further Reading

Fél. Óeng.; Kenney, *Sources* 269, 331–3.

Bieler, L. *The Life and Legend of St. Patrick*. Dublin, 1949.

———. "Studies on the Text of Muirchú." *P.R.I.A.* 52, sec. C (1950), 179–220; 59, sec. C (1959), 181–195.

———. *Bibliotheca Sanctorum* 9, 666–668.

———. "Muirchú's Life of St. Patrick as a Work of literature." *Medium Aevum* 43 (1974): 219–233.

———. *The Patrician Texts in the Book of Armagh*. (SLH X), (1979).

Sharpe, R. "Palaeographical Considerations in the Study of the Patrician Documents in the Book of Armagh." *Scriptorium* 36 (1982): 3–28.

C. Doherty. "The Cult of St. Patrick and the Politics of Armagh in the Seventh Century." In *Ireland and Northern France*, edited by J.-M. Picard. Dublin, 1991.

Sharpe R. and M. Lapidge. *A Bibliography of Celtic Latin Literature 400–1200*, pp. 84 ff. 1985.

See also **Armagh; Armagh, Book of; Cogitosus; Hagiography and Martyrologies; Patrick; Tírechán**

MUNSTER

Munster is the most southerly of Ireland's provinces. Its name, *Mumu (a quo* Munster), is of unknown origin. In the second century C.E., Ptolemy showed Munster being populated by the Érainn (Iverni). In the course of the fifth century, however, Munster came to be dominated by the Eóganachta, who may have been Irish returnees from Roman Britain. There was significant

Irish settlement in Wales and western England at that time, and Cashel, the capital of the Eóganachta in Munster, possibly took its name from the *castella* erected to protect western Britain from Irish raiders. Roman influences are also reflected by the number of pre-Patrician saints tradition claimed for the province, and by the distribution of ogham stones preserving the earliest examples of Irish writing.

The Eóganachta spawned a series of collateral branches that took possession of rich lands across Munster from the fifth to the ninth centuries. The Eóganachta of Cashel provided most of Munster's kings throughout that period, but failed to concentrate that royal power within narrow dynastic limits. The Eóganachta of Glanworth, Lough Leane and of Raithlinn regularly provided kings of Munster too. Even the relatively minor Eóganachta of Ainy provided a number of kings. The failure of individual Éoganachta dynasties to consolidate their hold on the provincial kingship meant that they were never in a position to challenge the novel claims put forward by the Uí Néill for the high kingship of Ireland. The Eóganachta's relative lack of martial prowess is reflected in the fact that the Vikings succeeded in establishing sizable towns at Waterford and Limerick, and a smaller port at Cork.

In 963 Mathgamain, king of the Dál Cais, seized the kingship of Munster away from the Éoganachta. In 978, Mathgamain's brother, Brian Boru, became king of Munster. He succeeded in harnessing the province's economic and demographic strength to make himself the king of Ireland. Brian's supremacy did not survive his death at the battle of Clontarf in 1014, but his dynasty, the Uí Briain, held the kingship of Munster tightly in their grip into the twelfth century. One of Brian Boru's grandsons, Tairrdelbach Ua Briain (+1086), seized the high kingship of Ireland, as did his son and successor, Muirchertach Ua Briain (+1119). Muirchertach seemed to be poised to transform the high kingship of Ireland into a true national monarchy. In 1101, he granted Cashel, the ancient symbolic capital of Munster, to the church while he established the Hiberno-Viking city of Limerick as his capital. Muirchertach sponsored the twelfth-century reform of the Irish church, a process that gradually remodeled the church along more conventional Roman lines.

However, in 1118, Tadc Mac Carthaig (MacCarthy), king of the Eóganachta of Cashel, led a rebellion of the various Eóganachta of southern Munster against the Uí Briain and succeeded, with support from Tairrdelbach Ua Conchobair, king of Connacht, in creating the kingdom of Desmond (from Irish *Desmuma*, literally "South Munster"). Tairrdelbach Ua Briain became the king of Thomond (from *Tuadmuma*, "North Munster"). A divided Munster was impotent in terms of national politics. In 1127, Cormac Mac Carthaig was made the king of Munster with Ua Briain support in order that he would undermine Tairrdelbach Ua Conchobair's hegemony. Mac Carthaig led a coalition of armies from Munster and other provinces to victory against Ua Conchobair in 1131–1133. However, once victory was achieved, the Ua Briain turned against Mac Carthaig and, in 1138, succeeded in having him assassinated. Tairrdelbach Ua Briain reigned over a reunited Munster until 1151, when a second rebellion, led this time by Cormac Mac Carthaig's son, Diarmait, and backed by Tairrdelbach Ua Conchobair, resulted in Munster being divided once more into the kingdoms of Thomond and Desmond. For the next two decades, the kings of Thomond and Desmond vied with each other in vain for supremacy in Munster.

In 1171, Henry II, king of England, came to Ireland to assert his authority over Strongbow (Richard de Clare, lord of Pembroke) and other Anglo-Norman adventurers who came to Ireland at the behest of Diarmait Mac Murchada, the erstwhile king of Leinster. Diarmait Mac Carthaig, king of Desmond, approached Henry II to seek an "alliance" against Domnall Ua Briain, king of Thomond—but Ua Briain approached Henry II too to neutralize the threat posed by Mac Carthaig. Unwittingly, the two Munster kings may have helped to persuade Henry II to maintain an English presence in Ireland by the alacrity with which they appeared to submit to him.

In 1177, Henry II granted the kingdoms of Desmond and Thomond to Anglo-Norman adventurers who had served him well. Robert fitz Stephen and Milo de Cogan succeeded in capturing the Hiberno-Viking port of Cork and establishing it as the basis for an English colony in the heart of Desmond. Philip de Braose, the grantee of Thomond, failed in his assault on Limerick. Nonetheless, when the Lord John came to Ireland in 1185, he made important grants of lands in Thomond to Englishmen, including Theobald Walter, ancestor of the future Butler earls of Ormond. Domnall Ua Briain proved to be a formidable adversary to the English, though, and it was only after his death in 1194 that English conquest and colonization gathered pace in Thomond. Diarmait Mac Carthaig's son, Domnall, held back the tide of English colonization in Desmond until internecine struggles following his death in 1206 facilitated massive English conquests in southern Munster.

The English transformed Munster, forcing the Irish aristocracy into ever-shrinking enclaves, to the defensible west of the Shannon in Thomond and into the mountainous southwest of Munster. The fertile lands elsewhere were extensively manorialized, and English peasants were settled over wide areas. There was an economic boom, with large agricultural surpluses supporting a dense network of villages and towns and forming the basis for high volumes of exports from Munster's ports.

The tide of English expansion was reversed spectacularly in southwestern Munster in 1261 when Fingín Mac Carthaig, king of Desmond, won a decisive victory at the Battle of Callan and followed it up by destroying a string of English castles as far east as the outskirts of Kinsale. English expansion into Thomond continued until 1317 when the local English magnate, Richard de Clare, was killed and his forces routed by the Ua Briain at the Battle of Dysert O'Dea.

The English colonies in Munster contracted over the course of the fourteenth century. The Bruce invasion helped to undermine English royal authority in distant regions in Ireland and facilitated the emergence of the "rebel English," lawless men who preyed on lesser folk. Climatic change, exacerbated by the Black Death later in the century, resulted in a massive reduction in the agricultural surpluses that had underpinned the manorial system in Munster. The Irish warlords were more effective in facing the colonists in war. As the colony contracted, power became more concentrated in the hands of Anglo-Irish magnates; particularly the Fitzgerald earls of Desmond, whose estates covered a discontinuous tract from north Kerry, through Limerick, to Imokilly in east Cork and Decies in County Waterford, and the Butler earls of Ormond whose estates formed a more consolidated block in Counties Tipperary and Kilkenny. In Cork there was a series of lesser lordships held by the Barretts, Barrys, and Roches and the Condons, Cogans, and Courceys, all of them subject to some degree of influence by the earl of Desmond, and most of them under pressure from the MacCarthys of Muskerry or the MacDonogh MacCarthys of Duhallow.

By the end of the Middle Ages, an uneasy equilibrium had been achieved between the various lordships in Munster. The province was ruled over by lords of Gaelic Irish or Anglo-Irish descent, with very little interference from England. English influences persisted, especially in the towns and in their more heavily colonized hinterlands. The Anglo-Irish lords of Munster generally observed the English custom of primogeniture and took care to secure recognition from the English crown for their titles. Yet, the Irish language, culture, and laws prevailed over most of Munster, even in Ormond. The extent of gaelicization was to pose a challenge to Tudor schemes for reform in the sixteenth century.

HENRY A. JEFFERIES

References and Further Reading

Byrne, Francis J. *Irish Kings and High-Kings*. London: Batsford, 1973.

Down, Kevin. "Colonial Economy and Society in the High Middle Ages." In *A New History of Ireland*, edited by A. Cosgrove, ii. Oxford: Oxford University Press, 1987.

Nicholls, Kenneth W. *Gaelic and Gaelicised Ireland in the Later Middle Ages*. Dublin: Gill & Macmillan, 1972.

Nicholls, Kenneth W. "The Development of Lordship in County Cork." In *Cork: History and Society*, edited by P. O'Flanagan and C. G. Buttimer. Dublin: Geography Publications, 1993.

Ó Corráin, Donnchadh. *Ireland before the Normans*. Dublin: Gill & Macmillan, 1972.

See also **Brian Boru; Cork; Dál Cais; Eóganachta; Érainn; Fitzgeralds of Desmond; Limerick; Mac Carthaig, Cormac; Uí Briain; Waterford**

MUSIC

Ireland's achievement in music has been noted by admirers as early as the twelfth-century chronicler Giraldus Cambrensis. Acclaiming little else in his twelfth-century first-hand account, Giraldus extols the range and talent of Irish musicians, commenting specifically on their distinctive melodies, harmonies, and composition. Renowned for both its vocal and instrumental skills, medieval Ireland boasts a rich history in music.

Stringed instruments dominate any study of music in medieval Ireland. The harp is Ireland's best-known instrument and one of its most enduring national symbols. The instrument appears in several different shapes and forms over the medieval period, suggesting an evolution from earlier four-sided instruments to the well-known triangular harp of today. While "harp" remains the standard interpretation of all stringed instruments mentioned in the sources, there

Irish musician playing the harp from *Topographia Hibernica*. © *The British Library.*

is little to suggest that the harp, as it is known today, corresponds to the earlier instruments. This problem of identification stems mainly from the often general and unspecific nature of the sources. The national and symbolic significance of the harp further compounds the difficulty.

Evidence for stringed instruments preceding the modern Irish harp is both abundant and diverse. Images of the instruments often appear on carved panels of high crosses and in illustrations within medieval manuscripts. In general the instruments seem to have been crafted entirely from wood, often willow. While modern harps use gut strings and are played with the pads of the fingers, medieval harps often used metal strings and were played with plectra. Early physical depictions of stringed instruments, particularly those found on high crosses, reveal smaller quadrangular instruments with relatively few strings. These smaller instruments are contemporaneous to the early Irish term *crott*, describing a popular instrument found throughout the sources.

From the earliest references, the *crott*, generally translated "harp," held a clear prominence above all musical instruments. An early text describes a tune on it as one of the three accomplishments of Ireland. Equally laudatory, a similar passage states that all music is holy until compared to that of the harp. The harp's unique sound is also often admired, once described as the sweetest and lowest of the musical instruments. This praise is echoed in several religious texts, where the harp is seen to enjoy particularly high standing. As in modern times, the instrument and its music were especially valued and endorsed by the church, at the exclusion of most other musical entertainment.

The *cláirseach* ("harp"), the triangular-framed instrument familiar today, known as the "modern harp" seems to have enjoyed prominence from the fourteenth century. The famous 'Brian Boru" harp, displayed in the Long Room of Trinity College Dublin, is a classic example of this harp. Dating from the fifteenth century, its likeness is used on Irish coinage and on the Guinness label. The triangular harp enjoyed the same prominence in Irish music as had its predecessors.

Music from the harp accompanied all manner of entertainment and ceremony. Harps and music played on harps can be found in descriptions of nearly all medieval gatherings, from festivals and royal banquets to wakes and ale houses. Early sources consistently mention three strains of music a skilled harper must be able to perform. The three are consistently described as ones that bring about sleep, laughter, and tears. The harp was clearly the most likely instrument at any gathering or assemblage. As a result of this the harpist was often permanently employed, by anyone who might afford one. There is likewise evidence for professional itinerant players. Professional musicians seemed to have enjoyed a fairly lucrative and in some cases celebrated career. A particularly skilled player might have attained the propitious status of king's musician, travelling and boarding with the king as part of his retinue.

While the harp is Ireland's best-known instrument, other stringed instruments were also played. The *timpán* was a small, handheld stringed instrument. Most often described as having three strings, the instrument was played by plucking and striking both the strings and frame. In several accounts the use of a bow is also mentioned. Other stringed instruments, including those resembling the psaltery and lyre, were also played. The deaths of three particularly well-known players of the *timpán* are recorded in the annals, attesting to their popularity and status.

Several types of horn and wind instruments were played in medieval Ireland. Numerous metal horns of Bronze Age provenance have been found in archaeological excavations. These horns, while having been tested and shown to emit sounds, were most likely used for military and decorative functions. Later archaeological evidence for horns provides several instruments fashioned from wood. Too little of the instruments remain for precise dating though they clearly date from the Christian era. Wooden horns are described in the early literature.

Despite the lack of descriptive evidence for musical horns, the horn blower is a common figure of the early sources. The horn blower is listed in a Law Tract as an entertainer expected in the banquet hall of a king, and in several similar descriptions he is depicted as a musician expected at festivals and feasts.

Wind instruments such as pipes and whistles also enjoy a rich history in Ireland. The *cuisle* ("bagpipe") is a commonly attested instrument in Ireland's early literature. While the *cuisle* was clearly a wind instrument, there is unfortunately no evidence to determine whether it was a straight tubular pipe, a bagpipe, or possibly a combination of the two. Multi-tuned pipes are often described in the literature. Today's *uilleann* pipes derive from the medieval instrument, so named as they are held and pressed between the ribs and elbow (*uillinn*).

The pipe is most often mentioned as accompanying another instrument or instruments. An early Law Tract states that proper musical arrangement consists of a harp accompanied by pipes. Countless references in the medieval literature mention the pipe player as a standard compliment to festivities and gatherings. Interestingly, his status seems to have been among the lowest of musicians and entertainers. *Fedán*, "a whistling or hissing sound," is also used occasionally to describe a musical instrument, undoubtedly a whistle or flute. Pipes, flutes, and whistles have been found in archaeological excavations, mainly in Viking contexts. They are generally crafted from the bones of birds and fowl.

Vocal performance and accompaniment are also well documented in medieval Ireland, showing an impressive variety in the use of the human voice. Two early terms in Irish for general musical entertainment derive from the Irish term, "to blow," or "blow out." This suggests that mere humming and whistling, as well as wind instruments, were principal among early musical entertainment. *Airfitech* ("minstrel") is the most common term used for any musician involved in vocal and instrumental performances. The *airfitech* is a regular feature of the saga literature, providing professional vocal and instrumental entertainment for gathered guests.

Humming and crooning as musical performances are well documented in medieval Ireland. Several different named types of humming and crooning are mentioned, each associated with different strains of music and activities. Keening, a mournful humming or moaning traditionally associated with Irish funerals, is an enduring example of these vocal styles. Additionally, an abundance of terms in the surviving literature attest to the popularity of formal humming or lilting, a recognized feature of Irish music that survives to the modern day.

The late medieval period witnessed significant change in Ireland's music tradition. With the breakup of the native ruling class in the late sixteenth century came permanent influences from Britain. Newly introduced instruments such as the standard flute and violin became popular, as did novel vocal arrangements and formal dance. This period also saw the arrival of sheet music for both instrumental and vocal compositions. Surviving sheet music is indeed abundant for Ireland's early modern period. Historically an oral music tradition with emphasis on improvisation, this innovation had a profound effect on Irish music. Formal dance in Ireland is not well attested before the sixteenth century. Doubtless a part of Ireland's musical culture, little is known of its expression in the medieval period.

ANGELA GLEASON

References and Further Reading

Buckley, Anne. "Musical Instruments in Ireland: Ninth–Fourteenth Centuries." In *Musicology in Ireland*, edited by Gerald Gillen and Harry White. Dublin, 1990.

Flood, W. H. Grattan. *A History of Irish Music*. Dublin, 1905.

O'Curry, Eugene. *The Manner and Customs of the Ancient Irish*. Vol. 3. Dublin, 1973.

See also **Entertainment**

MYTHOLOGICAL CYCLE

The Mythological Cycle is that body of medieval Irish narrative literature chiefly concerned with the deeds of the inhabitants of the Irish otherworld. Characters from this cycle frequently appear in the other three main cycles (the Ulster Cycle, the King Cycle, and the Finn Cycle) as well as in other categories of narrative that fall outside this classification such as the *dinnsenchas* ("place lore").

The inhabitants of the otherworld are collectively known as Túatha Dé Danann ("tribes of the goddess Danu/Danann"), a term that first appears around the turn of the eleventh century in poems associated with *Lebor Gabála Érenn* (see Invasion Myth). Earlier texts called them Túatha Dé ("tribes of gods"), but the more common designation at all periods was *áes síde* ("folk of the fairy-mound"). They inhabited the *síde* ("fairy-mounds"), which were natural hillocks or man-made mounds such as the ancient tomb complex at Brug na Bóinne (Newgrange, Co. Meath). Their divine origin is sometimes explicitly recognized, but Christian writers occasionally rationalize them as fallen angels or demons. Attempts to euhemerize them (to suppose that they were mortals who were worshipped after their deaths) are most successful in the influential *Lebor Gabála Érenn*, where they are usually portrayed as descendants of Noah.

The names of some of the principal characters suggest an origin in native mythology. One of their chiefs was the Dagda whose name means "good god" (Dago-dēvos). The king of the Túatha Dé Danann at the battle of Mag Tuired is Núadu Argatlám ("Núadu silver-arm") whose name is cognate with Lludd Llawereint ("Lludd silver-arm") of Welsh saga and is thought to be the same as the deity called Nodons who appears on Romano-British dedications. Lug Lámfhata ("Lug long-arm") is associated with the harvest festival (Lugnasad) and is thought to be a reflex of the Celtic god whose name may be preserved in various continental European place names such as Lyon and Leiden (Lugudunum).

Cath Maige Tuired

There are two distinct but related battles of Mag Tuired. The first, often known as *Cath Maige Tuired Conga* ("the battle of Mag Tuired of Cong") is recounted in *Lebor Gabála Érenn* and tells of the invasion of Ireland by the Túatha Dé Danann and the subsequent battle with the Fir Bolg. Although chronologically anterior to the second battle, it was composed at a later date. In the battle, Núadu, king of the Túatha Dé Danann, loses his arm. A replacement is fashioned from silver by the physician Dian Cécht and the smith Credne, and so Núadu is called *Arcatlám*, "silver arm." However, this physical blemish renders Núadu unsuitable for kingship and he is replaced by Bres, the son of the king of the Fomoiri ("demons from over the sea") and a woman of the Túatha Dé Danann. Bres proves himself in battle by overthrowing the Fir Bolg and exiling them to Connacht and distant coastal islands.

The second battle of Mag Tuired (*Cath Maige Tuired*) is probably the most important of the mythological tales because of its narrative sophistication and its inclusion of almost all the known Irish gods. The earliest versions are an eleventh- or twelfth-century composition based on Old Irish material, and a version that was incorporated into *Lebor Gabála Érenn* in the eleventh century. It was influenced by *Lebor Gabála Érenn*, from which it draws accounts of the first battle of Mag Tuired and the loss of Núadu's arm. This is an important prelude to the second battle, as it explains how Bres came to be king of the Túatha Dé Danann. Bres's rule is unjust, and he forces the champions of the Túatha Dé Danann to perform demeaning tasks. When the poet Coirpre is treated inhospitably, he composes a satire on Bres after which the Túatha Dé Danann expel him. Bres then seeks help from the Fomoiri. Meanwhile, Núadu's arm grows back and he is restored to the kingship. The Túatha Dé choose Lug as their leader in battle. In the subsequent conflict with the Fomoiri, Núadu is slain by Balor whose single eye, when opened, paralyses the opposing army. Lug and Balor are engaged in combat when Balor again opens his eye, but Lug slays him with a slingshot in a scene reminiscent of the biblical story of David and Goliath. The Túatha Dé Danann win the battle and drive their enemies into the sea. Bres, however, is spared when he promises to teach the victors the secrets of agriculture. Lug, the Dagda, and Ogma pursue the Fomoiri to Bres's banqueting hall where they retrieve their cattle and the Dagda's harp.

Various attempts have been made to distill ancient myth from the surviving texts. T. F. O'Rahilly's theory that the original myth dealt with the defeat of the sun-god Balor by the divine hero Lug has been discredited. Dumézil places the tale within the context of Indo-European myth according to which battle is waged by representatives of the first (sacred) and second (martial) functions against the third function (material), resulting in the integration of the three functions. This is represented here by the defeat of the Fomoiri by Núadu and Lug and the divulging of the secrets of agriculture by Bres to the victors. While this interpretation requires some modification to accommodate the extant narrative, it has been widely accepted. The story also functions as an exemplary myth embodying and validating in dramatic form the ideology of its originators. This is done here negatively through the portrayal of undesirable characters such as Bres, and affirmatively through the depiction of positive role models such as Lug. In doing so it explores the nature of kinship and kingship, intertribal and intratribal relationships, and the interplay between social and cosmic order.

Other interpretations have focused on the significance and meaning of the tale within its contemporary context, indicating that monastic writers had a profound influence on the form of the tale. The negative portrayal of the *cáinte* ("satirist") who overburdens the unfortunate Dagda may be clerically inspired. The threat from the Fomoiri is depicted as an alliance among Scandinavian forces intent upon the conquest of Ireland, and this may be a reflection of the threat from the Viking incursions of the ninth century when the tale was first written. The text may have had an overarching contemporary political message, namely, the importance of unity around the Tara kingship in order to repel foreign attacks. If so, it could be regarded as propaganda for the Uí Néill dynasty that controlled the kingship of Tara during this period.

Aislinge Óengusso (The Dream of Óengus)

This Old Irish story tells how Óengus, son of the Dagda and Bóann, fell ill after seeing a beautiful maiden in a dream. The forces of the otherworld are marshaled to reveal that the woman is Cáer Iborméith from Síd Úamain in Connacht, and the assistance of Ailill and Medb, king and queen of Connacht, is enlisted to procure her from her father. It is revealed that she takes the form of a swan every other year and her father reveals where she will be the following Samain. Óengus goes to her there, and, in the form of swans, they sleep together before going to Brug na Bóinne. The episode is used to explain Óengus's participation in the cattleraid of Cooley in the Ulster Cycle.

Tochmarc Étaíne (The Courtship of Étaín)

This is actually a sequence of three interrelated stories. In the first, Midir of Brí Léith obtains as compensation for an alleged injury Étaín, the fairest of all the maidens of Ireland and daughter of Ailill, king of northeast Ulster. However, in her jealousy Midir's first wife, Fúamnach, turns her consecutively into a pool of water, a worm, and a fly or butterfly. A wind conjured up by Fúamnach drives Étaín out to sea, and she wanders the coast for seven years. She eventually encounters Óengus, son of the Dagda, who carries her in a crystal cage until Fúamnach again drives her off. She falls into the cup of the wife of an Ulster king and is reborn 1,012 years after her first birth. In the second story, another thousand years have passed, and the descendants of Míl Espáine (the mortal Gaels) rule Ireland. She is married to Eochaid Airem, king of Tara, but Eochaid's brother is also in love with her and falls ill as a result. Étaín agrees to meet with him at a separate location so as not to bring shame on her husband. However, on her third rendezvous she discovers that she has been sleeping with her former husband, Midir. Nevertheless, Eochaid's brother is cured when she returns to the palace, and her virtue remains intact. In the third story, Midir abducts Étaín from Eochaid's house through trickery and carries her off. Eochaid destroys many fairy mounds in his pursuit of Midir before catching up with him at Brí Léith. Midir again tricks Eochaid, this time by passing the daughter of Étaín and Eochaid off as Étaín herself. Enraged, Eochaid destroys Brí Léith and retrieves his wife. But Eochaid's daughter had already borne him a daughter. She is abandoned to die but is found and reared by a herdsman. When she grows up, she marries Eterscél, king of Tara.

Later Tales

Interest in the Mythological Cycle continued into the Early Modern Irish period (c .1200–c. 1650), although the earlier tales were only rarely copied. *Cath Maige Tuired* was revised in the later Middle Ages, and several other tales were either revised or composed anew. *Altram Tige Dá Medar* ("The nourishment of the houses of the two milk vessels") opens with a description of the settlement of the fairy mounds by the Túatha Dé Danann after their defeat by the descendants of Míl Espáine. Brug na Bóinne was initially assigned to Elcmar but Óengus, son of the Dagda, expels him at the instigation of Manannán. A beautiful daughter by the name of Eithne is born to Óengus. When she is fully grown, a bawdy insult causes her to fast, after which she will only take milk from Óengus's marvelous cow which had been brought from India. When summoned to Manannán's palace she again refuses to eat, drinking only milk from Manannán's marvelous cow. Manannán reveals that the demon of the Túatha Dé Danann had left her when she was insulted to be replaced by an angel and that she is therefore unable to eat their food. Thereafter, Eithne refuses to eat food of the otherworld and consumes only milk from marvelous cows. Centuries later, she is baptized by St. Patrick and dies a fortnight later.

The tale of the death of the children of Lir (*Oidheadh Chloinne Lir*) was probably written in the fifteenth century. In later manuscripts it is enumerated among the three sorrows of storytelling (*trí truaighe na sgéalaigheachta*), although it was originally intended as an explication of the transient nature of temporal pleasure and the purgative effects of suffering. It tells how Aoife, wife of Lir of Síd Fionnachaidh, turned his four children into swans out of jealousy. After nine hundred years of exile, they settle on an island where the saint Mochaomhóg finds and comforts them. Aoife's spell is finally broken, and the four children are transformed into wizened old people whom Mochaomhóg baptizes before they die. The death of the children of Tuireann (*Oidheadh Chloinne Tuireann*), a tale of murder and revenge, is also numbered among the three sorrows of storytelling. Although the earliest surviving text was written in the later Middle Ages, a version existed as early as the eleventh century. The three sons of Tuireann (Brian, Iuchar, and Iucharbha) slay Cian, the father of Lugh (Lug) of the Túatha Dé Danann, and as compensation Lugh demands that they undertake various dangerous quests. They perish during the final quest, and the mortally wounded Brian carries his brothers home. Lugh refuses to save Brian with a healing pigskin and he dies. Tuireann buries his three sons in a single grave and dies himself soon after.

GREGORY TONER

References and Further Reading

Breatnach, Caoimhín. "The Religious Significance of *Oidheadh Chloinne Lir*." *Ériu* 50 (1999): 1–40.

Carey, John. "Myth and Mythography in *Cath Maige Tuired*." *Studia Celtica* 24–25 (1989-90): 53–69.

Dillon, Myles. *Early Irish Literature*. Chicago: University of Chicago, 1948. Reprint, Dublin & Portland: Four Courts, 1994.

———, ed. *Irish Sagas*. Dublin & Cork: Mercier, 1968.

Gantz, Jeffrey. *Early Irish Myths and Sagas*. London: Penguin, 1981.

Gray, Elizabeth. *Cath Maige Tuired, the Second Battle of Mag Tuired*. Irish Texts Society 52 (1982).

Mac Cana, Proinsias. *Celtic Mythology*. Feltham: Hamlyn, 1970.

McCone, Kim. *Pagan Past and Christian Present*. Maynooth: An Sagart, 1990.

Ó Cathasaigh, Tomás. "*Cath Maige Tuired* as exemplary myth." In *Folia Gadelica*, edited by Pádraig de Brún et al., pp. 1–19. Cork: Cork University Press, 1983.

O'Rahilly, T. F. *Early Irish History and Mythology*. Dublin: Dublin Institute for Advanced Studies, 1946.

Rees, Alwyn and Brinley Rees. *Celtic Heritage: Ancient Tradition in Ireland and Wales*. London: Thames & Hudson, 1961.

Sjoestedt, Marie-Louise. *Gods and Heroes of the Celts*. Translated by Myles Dillon. London: Methuen, 1949. Reprint, Dublin & Portland: Four Courts, 1994.

***See also* Dinnsenchas; Invasion Myth; *Eachtrai*; *Imrama*; Pre-Christian Ireland; Scriptoria**

N

NATIONAL IDENTITY

It is problematical to work with the concept of "nation" in the Middle Ages. What will be investigated here are basic features of the social system prevalent in Ireland, which can be studied in unusual depth as compared to other contemporary European societies.

The investigation of this topic depends on the availability of written sources, and these are the products of only a small segment of the population as a whole. It is not clear how representative they were. On the other hand, from these sources one can deduce certain results that transcend the individual author.

From the earliest written sources from within Ireland there emerges the concept of the society as a whole expressed in the terminology used for the island and its population. The Romans called the Irish *Scotti*, the island *Scotia*. In his *Confessio* written in the fifth century, St. Patrick calls the Irish *Hiveriones,* most likely the Latinization of a native term. The earliest attested Irish term for the island as a whole is *Ériu*, and we have thus from within Irish society the expression of territorial and social unity, no doubt helped by the fact of Ireland being an island.

The social knowledge and its value systems were from the earliest attested times the domain of a class of learned specialists, later known generally as *aes dána* (people of skill). They surface in the early Irish native law texts of ca. 700 in various manifestations, as highly respected and highly valued professionals. They occur in passing in earlier sources written in Latin. There is no unanimity among modern scholars as to the age of these professions, yet there are striking parallels to professionals among the continental Celts as described by Greek and Roman authors. Professionals of native learning did survive prominently in Christian Ireland, and indeed into the modern period, although their status would not have been unchanged throughout.

The fields of their expertise were primarily language (poetry in a variety of genres) and, also expressed in language, history (genealogies), law, healing skills, and so forth. According to the law tract *Uraicecht na ríar* (supported elsewhere in written material), the specialists had to undergo a rigorous training, and there were various levels of learning to be mastered until one reached the top. This law tract contains a generic term for these professionals—*filid*—which later became restricted to the poets in a more narrow sense.

In sources from Ireland, legal experts are the first to be referred to, namely in St. Patrick's *Confessio* (ref) written more than two centuries before the compilation of the Irish law tracts. In the highly fragmented political landscape of early and later medieval Ireland the professionals of knowledge enjoyed special status and protection throughout the island (as well as later in Scotland).

In view of their status it is not surprising, but remarkable nevertheless, that the language of the written sources in Irish from the Old Irish period (before ca. 900) appears as a standardized language with no regional variants. This would be contributory toward the sociocultural homogeneity of Irish society insofar as it is reflected in the written sources. It is likely that this language was accessible to the population in general, and thus expressed joint culture.

Law was one of the important domains of the professionals of learning. The surviving voluminous Irish law tracts (no comprehensive codification) are incomplete ('ein Trümmerhaufen', so Thurneysen in 1935): Two great collections, Senchas Mar and Bretha Nemed are associated by modern scholars with different regions (northern and southern). Nevertheless, the law was known genetically as *fénechas* (law of the *feni*, or law of the Irish free men). Irish law knew of no boundaries of the *túatha*.

In a similar manner the social values articulated in political poetry, even when applied to individual people, expressed not so much the personal features of the individuals in question but their living up to values and expectations held generally.

MICHAEL RICHTER

NAVAL WARFARE

The importance of ships and shipping to the early Irish finds eloquent expression in the hoard of gold objects that were deposited at Broighter, approximately 4k north of Limavady, County Derry. Along with gold collars, chains, torcs, and a model cauldron was a model boat of beaten sheet gold. It had benches, oars, rowlocks, steering oar, and a mast. It would appear to have been a nine-bench, wooden ocean-going ship. The deposit lay on the ancient shoreline midway along Lough Foyle. It is very likely that this was a votive offering to the sea god Manannán Mac Lir. It has been dated to the first century BC. With the Romanization of Britain, Ireland now had a wealthy neighbor. It is clear that through trade and maritime raiding, many goods, including slaves, were brought into Ireland. Ammianus Marcellinus records further raids in the 360s by the *Scotti* (Irish) and *Picti* (Picts), among others, after a treaty had been broken. It is clear that such raids were intense throughout the fourth and fifth centuries. Irish settlement began in Wales and Cornwall, and soon afterward in the southwest of Scotland and the Isle of Man. During this period the Irish dominated the Irish Sea. It was a dominance that was to continue until the coming of the Vikings. Marauding Britons challenged their control of the Isle of Man during this period.

While the currach was used from ancient times to the present, and is specifically referred to by Roman sources as the vessel of the Picts and Scots, it is likely that shipbuilding techniques were enhanced during the Roman period. The word *long* (ship, boat, vessel) was borrowed from the Latin (*navis*) *longa*. It was also referred to as a *long fata* (long-ship, galley) in the law tracts. A *long chennaig* was a merchant ship. Also borrowed from Latin were *barca*, *bárc* (ship) and *libern*, from *liberna* (merchant ship). In the late sixth century Irish kings went on naval expeditions as far as the Orkneys. As may be seen in Adamnán's "life" of Saint Colum Cille, Iona was a hub of seagoing activity. He tells of a pilot on Rathlin Island, County Antrim, who guided ships through the dangerous tides and currents. The abbot of Applecross drowned in 737, *cum suis nautis*, "with his sailors," twenty-two in number. The crew were likely to have been monks. Adamnán also refers to sailors, and it would seem that some of the monks of these communities were specialist seafarers. The document known as the *Senchus Fer nAlban* (History of the Men of Scotland) originates in a seventh-century Latin text. It provides a remarkable picture of the organization in Scottish Dál Riata for manning the fleet. Houses were grouped into twenties for the purpose of naval recruitment. Two seven-benchers were required from every twenty houses. It is clear from this work and that of Adamnán that navies were highly organized with bodies of professional sailors. Propulsion was by rowing, but there was also a single sail. Each oar would have had two men, so a warship would have had at least twenty-eight men on board. The importance of this fleet may be seen in the agreement reached at the Convention of Druim Cett (Co. Derry) in 575. The Irish portion of the kingdom of Dál Riata was to serve the high king with land forces. The Scottish portion was to be independent, except that it must serve the high king with its fleet when required. In 734, the last Cenél Conaill high king, Flaithbertach mac Loingsig, was defeated in a sea battle off the mouth of the river Bann, despite having the help of the Dál Riata fleet.

The arrival of the Vikings brought about profound changes in Irish society, or rather, accelerated the rate of change. As in the Roman period, words were borrowed, now from Old Norse. The main part of this vocabulary was in the area of ships, shipping, and fishing. The common Irish word for boat, *bát* (modern Irish *bád*), is a borrowing from ON *bátr*. The word for a wine ship, *fínbárc* (*vinum* + *barca*), now became *fíncharb*, where the ON *Karfi* has replaced *bárc*. *Ancaire*, from Latin *ancora*, now has a companion in *accaire*, from ON *akkeri* (anchor). *Stiúir*, from ON *styri* (rudder, helm), and *stiúrusmann* (helmsman), are again from Old Norse. This is merely a sample of a large vocabulary and emphasizes the technical superiority of Norse shipping. The shallow-draft, clinker-built ships of the Norse were revolutionary and allowed for new strategies in warfare. The Vikings beached their ships and threw up a protective bank around them for protection. In the ninth century the compound *longphort* appeared to describe these encampments, made up of *long* (ship) and *portus* (harbour), from Latin *portus*. The *longphort* came to mean a military encampment.

The first large fleets of sixty ships came to the mouth of the Liffey in 837. There is uncertainty about the size of the crews of these ships, but they would seem to have consisted of forty to fifty men. Some of the fleets were large. In 871, Amlaíb and Ímar returned to Dublin from Scotland with 200 ships. Warfare was now on a scale not seen before. By the tenth century the Irish had assimilated this new technology. Much of the wealth upon which it was based was generated by the slave trade.

The Irish were now increasingly making use of fleets in their battles with the Norse and among themselves. The lakes and rivers of the interior were being exploited militarily, as well as the usual activity on the sea. In 955, the northern Uí Néill king, Domnall Ua Néill, brought a fleet from the mouth of the Bann into Lough Neagh, along the river Blackwater into Lough Erne, and from there to Lough Owel to force the submission of Fergal ua Ruairc, king of Bréifne. In 963, he brought a fleet along the Blackwater across Slíabh Fuait (Fews, Co. Armagh) to Lough Ennell, where the main branch of the southern Uí Néill had their headquarters. Brian Bóruma put 300 vessels on Lough Ree in 988, from where he harried the midlands and Connacht. His opponent Máel-Sechnaill maintained a fleet on Lough Ree, and in 1016 Brian's son plundered the main churches on the islands of the lake and captured his ships.

One type of ship is called a *serrcend* (galley). For example, in 1035 the men of Bréifne came down the Shannon with fourteen galleys and plundered Clonfert. Brian's son pursued them with the crew of one ship and slaughtered them at the confluence of the Suck and the Shannon.

Fleets then became an indispensable part of military strategy and must have consumed considerable resources in building and maintenance. From twelfth-century texts it would seem that each *tricha cét* (cantred) was to support ten ships. The crews were filled by a levy of men who must take wages and accompany their lords and be provisioned by their own families while they were on service. By the twelfth century local kings had become the officers of the major kings. Ó Flaithbheartaigh (O'Flaherty), for example, was the *taísech nócoblach* (admiral) of Ó Conchobair's (O'Connor's) fleet based at the mouth of the Corrib, at Galway.

The largest ship known from the Viking world was built in Dublin about the year 1060. This was a long warship, perhaps of the type known as a *skei*, and was built of Irish oak. It was c. 30 m long and had a crew of up to 100 men, with about sixty on the oars. By the mid-eleventh century the Dublin fleet was formidable and was frequently for hire. It was part of an attack on England in 1058 by Gruffydd ap Llywelyn and Magnus, son of Harald Hardrada, king of Norway. Gruffydd invaded Gwynedd with the fleet in 1075. Diarmait Ua Briain plundered Wales with the fleet in 1080. In 1137, Conchobar Ua Briain and Diarmait Mac Murchada, with 200 ships from Dublin and Wexford, laid siege to Waterford. Malcolm IV, king of Scotland, hired the Dublin fleet in 1164, and in the following year none other than King Henry II of England hired the fleet for six months.

A major industry at Dublin must have been the building, repairing, and provisioning of ships. From eleventh- and twelfth-century texts we have a vocabulary that is in support of this. The word *longboth* means "shipyard" or "boat-shed," perhaps the temporary structure placed over a ship under repair; *longthech* (boat-house) may represent a more permanent structure. These words may correspond to the *hrof*, a less-substantial structure, on the one hand and the *naust*, a proper building mentioned in Icelandic literature, on the other.

Given the nature of the evidence it is not possible to determine how battles were organized or what strategies were used by opposing fleets. What is clear, however, is the scale of naval warfare and the quality of the ships and the sailors who manned them.

Charles Doherty

References and Further Reading

Bannerman, John. *Studies in the History of Dalriada*. Edinburgh: Scottish Academic Press, 1974.

Doherty, Charles. "The Vikings in Ireland: a Review." In *Ireland and Scandinavia in the Early Viking Age*, edited by Howard B. Clarke, Máire Ní Mhaonaigh, and Raghnall Ó Floinn. Dublin: Four Courts Press, 1998, pp. 288–330.

Smith, Brendan. *Britain and Ireland 900–1300: Insular Responses to Medieval European Change*. Cambridge: Cambridge University Press, 1999.

See also **Dublin; Military Service, Gaelic; Ships and Shipping; Viking Incursions**

NIALL GLUNDUB

See **Uí Néill, Northern**

NIALL NOÍGIALLACH

Niall was the eponymous ancestor of the Uí Néill dynasty, which originated in north-eastern Connacht and was dominant in Ireland until the end of the tenth century. His real name was probably *Nél* (cloud); the change to Niall may be due to the influence of his epithet *Noígiallach* (of the nine hostages), referring to nine tributary peoples owing allegiance to him. According to some of the later annals, Niall died in the early fifth century; his actual floruit may have been in the fourth century. There is no firm evidence about Niall's life, as he predates the period of written history. However, he features prominently in myth and legend.

Niall is said to have been the son of Eochaid Mugmedón and Cairenn, who may have been of British origin. The *Echtra mac n-Echach Muigmedóin* (which is intended to establish the relative political prestige of the Uí Néill and their Connacht cousins), recounts that Eochaid and his other wife, Mongfind, had four sons—Brian, Fiacha, Ailill, and Fergus—three of whom were the progenitors of the Connacht dynasties of Uí Briúin, Uí Fiachrach, and Uí Ailello. Mongfind resented Niall, and she asked Eochaid to judge between all his sons to determine who was to succeed him. The sons were sent on a hunting expedition, during which each one, in turn, went to a well guarded by an ugly, old woman. She demanded a kiss from each before she would permit him to draw water. Fergus and Brian refused; Fiacha kissed her and the woman foretold that he would visit Tara (two of his descendants, Ailill Molt and Nath Í, took the kingship). Niall, however, said that he would lie with her as well as kiss her, whereupon she was transformed into a beautiful woman. She identified herself as "sovereignty" and promised Niall that sovereignty would be his and his children's forever, save for the two (aforementioned) descendants of Fiacha and Brian Boru. Niall is acknowledged as a king of Tara in all of the extant king lists dating from early eighth century onward.

Niall is assigned two wives: Indiu, daughter of Lugaid mac Óengusa Finn of Dál Fiatach of Ulster, and Rígnach, daughter of Meda mac Rosa, also of Dál Fiatach. The main tradition identifies Indiu as the mother of Niall's sons; an alternative account relates that their mother was Rígnach. Niall is credited with between three and fourteen sons, in sources of differing dates, some of whom were recognized as the progenitors of dynasties. The accretion of additional sons would have followed the absorption of various dynastic groupings into the Uí Néill.

According to the saga of Niall's death, he was slain, while on an expedition in Scotland, by Eochu mac Énnai Cheinnselaig. Cináed Ua hArtacáin, the tenth-century poet, relates that Niall raided Britain seven times and that he was slain on the last of these raids by Eochu, acting in conjunction with the Saxons. (The reference to the Saxons is almost certainly anachronistic.) Niall is said to be have been buried at Ochan (Faughan Hill in Co. Meath).

References and Further Reading

Byrne, Francis John. *Irish Kings and High-Kings*. London: Batsford, 1973.

Byrne, Paul Francis. "Certain Southern Uí Néill Kingdoms (Sixth to Eleventh Century)." Ph.D. dissertation, University College Dublin, 2000.

Connon, Anne. "Prosopography II: A Prosopography of the Early Queens of Tara". In *Tara: A Study of an Exceptional Kingship and Landscape* (*Discovery Programme, Monograph X*), edited by Edel Bhreathnach. Dublin: Royal Irish Academy/Discovery Programme (forthcoming).

Dillon, Myles. *The Cycles of the Kings*. London: Oxford University Press, 1946.

Ó Corráin, Donnchadh. "Legend as Critic." In *The Writer as Witness*, edited by Tom Dunne. Cork: Cork University Press, 1987, pp. 23–38.

O'Rahilly, T. F. *Early Irish History and Mythology*. Dublin: Dublin Institute for Advanced Studies, 1946.

See also **Uí Néill; Cináed Ua hArtacáin; Tara; Connacht**

NICHOLAS MAC MÁEL-SU

Archbishop of Armagh from 1272 to 1303, Nicholas mac Máel-su was the last representative of the old Irish ecclesiastical tradition to serve in that position in medieval Ireland. Little is known about his background, except that he was a native of the diocese of Ardagh (Cos. Longford and Roscommon). He came of Irish (as opposed to Anglo-Norman) stock, probably from a prominent local family, some of whose members were later charged with killing the king's knights. His forename is likely to have been assumed upon taking holy orders, rather than testifying to an accommodation with Anglo-Norman culture in Ireland. He may have received a university education abroad, as suggested by references in a contemporary obituary to his secular eloquence and his title of *magister*.

Elected in 1270 and consecrated by the Cardinal Archbishop of Tusculum in 1272 (Pope Gregory X was then in the Holy Land), Nicholas then rendered homage to Henry III in England. He may have attended the Council of Lyons (1274). Certainly he made its main concerns—excessive secular interference in episcopal elections and the proper administration of lands owned by the church—the central issues of his own career. To further his cause, Nicholas cultivated good relations with the English administration in Ireland, especially with Stephen Fulbourne, bishop of Waterford and justiciar of Ireland. He tried to have Stephen's brother, Walter, appointed bishop of Meath against the wishes of the local diocesan chapter. The attempt backfired and Nicholas became the subject of a royal investigation, being summoned to answer charges at Drogheda in 1284.

More broadly, he fought the crown on the issue of the king's right to the temporalities of a diocese during a vacancy. Edward I rejected the claim on the grounds that English common law gave him this right and that the same law was deemed to apply in all Ireland, both Anglo-Norman and Gaelic. In reality, the English administration in Ireland was in no position to enforce common law in native-held areas, so Nicholas was able to retain control of temporalities in most of the disputed dioceses.

In another famous exchange, when Edward tried to levy a special tax on the Irish church to finance his wars, Nicholas reacted by summoning a council of the Armagh province at Trim in 1291. From this meeting emerged a united front of Irish and Anglo-Norman bishops, who vowed to defend each other's rights against any lay power trying to hinder them in the exercise of their episcopal duties. Although nothing more is heard of this movement, it demonstrated Nicholas's ability to marshal support from traditional enemies in defense of basic ecclesiastical rights.

Yet Nicholas was no less diligent in protecting those same rights against the native Irish rulers of the small kingdoms that formed much of his province. When Boniface VIII published *Clericis laicos*, a papal bull forbidding secular rulers such as the kings of England and France from levying taxes on the church without first obtaining Rome's permission, Nicholas deftly appropriated it for his own purposes. Armed with the bull and the relics of Ireland's three greatest saints, Patrick, Colum Cille, and Brigit (their location at Saulpatrick had been revealed to him in 1293) Nicholas did a circuit of the neighboring Gaelic kingdoms. He persuaded Domnall Ua Néill of Tír nEógain, Brian Mac Mathgamna of Airgialla (Oriel), and Donn Mag Uidhir of Fermanagh to put their names to a document protecting the church from various secular infringements. Thus, the deed made provision for fines of cattle (an Irish custom) for injuries done to ecclesiastical property and persons (including damage done by hired mercenaries, the Irish kern and Scots gallowglass); stipulated penalties for the followers of those lords who injured clerks going to Rome, and nuns or widows; and upheld the church's right to goods arising from intestacy.

As indicated by these cases, Nicholas does not fit the stereotypical characterization of the church in Ireland of the thirteenth century as divided into two perpetually hostile camps, Irish and Anglo-Norman. Certainly, as a native Irish archbishop of a predominantly Irish province he knew how to use the traditional weapons of his culture. Witness his exploitation of the relics of Ireland's three great saints and his recourse to the customary fines of cattle as exacted in native Irish law. Yet he seems to have cultivated good relations with the Anglo-Norman bishops. In addition to plotting with Stephen of Fulbourne to have the latter's brother made bishop of Meath, he maintained good relations with the Anglo-Norman archbishop of Dublin, avoiding the potentially explosive topic of which ecclesiastical province held primacy of all Ireland. And although he fought with the crown, he was fully prepared to cooperate with it as long as ecclesiastical rights were honored.

PÁDRAIG Ó NÉILL

References and Further Reading

Gwynn, Aubrey. "Nicholas Mac Maol Íosa, Archbishop of Armagh, 1272–1303." In *Féilsgríbhinn Éoin Mhic Néill*, edited by John Ryan, 394–405. Dublin: The Sign of the Three Candles, 1940.

Otway-Ruthven, A. J. *A History of Medieval Ireland*. New York: Barnes and Noble, 1993.

Watt, John. *The Church in Medieval Ireland*. Dublin: Gill and Macmillan, 1972.

Watt, John. "English Law and the Irish Church: The Reign of Edward I." In *Medieval Studies presented to Aubrey Gwynn, S.J.*, edited by J. A. Watt et al., 133–167. Dublin: The Sign of the Three Candles, 1961.

NUNS

There were nuns in Ireland from the arrival of St. Patrick in the fifth century until the dissolution of the monasteries in the 1540s, although they left fewer traces of their lives and spiritual interests than their brothers did.

Fifth to Twelfth Century

The earliest Christian writings from Ireland, those of St. Patrick, speak of the large numbers of women who were living under religious vows. In the earliest days these women must have lived privately, as it was not until the sixth century that there is evidence of women's communities in the records. After this, religious women began to live together on land often set aside for their use by their families. Some of the communities were short-lived, while others flourished and have been remembered in place names, stories about saints, and the surviving buildings.

Bridget of Kildare (sixth century) is the best known of the nuns from early Ireland, although there is little certainty about the events of her life. Hagiography and other texts indicate that the nunnery at Kildare was a large and very important community from at least the seventh century. Its political importance is underlined by the fact that all the recorded abbesses of Kildare were from the families of the kings of Leinster, such as the Uí Dúnlainge. From the hagiography of prominent women saints such as Bridget of Kildare, Íte of Cell Íte (Killeedy, Co. Limerick), Mo-Ninne of Cell Shléibe Cuilinn (Killeevy, Co. Armagh), and Samthann of Cluain Brónaig (Clonbroney Co. Longford), it is clear that professed nuns were involved in fostering and educating children, pastoral care, negotiating for the release of hostages, and caring for the sick. Hagiography also records nuns reading and writing, teaching psalms, and lending manuscripts. These well-educated and privileged women lived under religious rules that were probably devised by the founders of their nunneries. There were other religious women who lived under less formal vows. Some of these were widows living either in small groups or around churches, where they prayed and undertook some pastoral care.

Twelfth Century

Many early Irish nunneries survived until the twelfth century, when church reforms by Irish churchmen included the introduction of continental religious orders for both men and women. Secular leaders and reforming bishops were instrumental in founding convents to house the women who wanted to join these revitalized nunneries in the twelfth century. One of the first women's communities to be founded under the Arroasian observance of the Augustinian rule was St. Mary's at Clonard, County Meath, founded by Murchad Ua Maél Sechlainn, ruler of Meath in association with Malachy of Down. St. Mary's had monks and nuns living in separate buildings and worshipping in the same church. Diamait Mac Murchadha also founded important nunneries in his territories in Dublin (St. Mary del Hogges) and near Waterford (Kilculliheen). These nunneries and their smaller dependencies formed loose federations that maintained some connections throughout the medieval period. After the Anglo-Norman invasion, religious houses were often refounded or reinvigorated with donations of money and land from their newly conquered territories. Nunneries were included in this pattern of monastic foundations, and there were important, well-endowed convents founded in the late twelfth and early thirteenth centuries in County Dublin (Grace Dieu), Meath (Lismullin), and Kildare (Graney and Timolin).

Thirteenth to Sixteenth Century

In the late twelfth and early thirteenth century there was renewed interest by Gaelic Irish kings and bishops in including nunneries in their own reorganization of local monastic houses. Two major foundations dating from this time are the Ua Briain convent of Killone (Co. Clare) and the Ua Conchubair convent at Kilcreevanty (Co. Galway) and its dependencies. Both of these houses were well endowed with land and buildings and were staffed with women from the founders' families.

There were some small convents for nuns following other rules. There were Cistercian convents at Ballymore (Co. Westmeath), Derry, Downpatrick and St John's in Cork followed the Benedictine rule. The majority of nunneries in the later medieval period were Augustinian, usually using the Arroasian observance of that rule, either from the time of their foundation or not long afterward. These nuns were under the care of the local bishop and were subject to visitation and correction of any lapses of adherence to their rule. Most of the recorded lapses were for breaking enclosure or neglecting monastic property, though there were also nuns who broke their vows of chastity.

Nuns in later medieval Ireland were usually enclosed, at least officially. That is, they were not permitted to leave the convent walls nor were lay people permitted to enter. However, for these nuns, as with their sisters in the rest of medieval Europe, enclosure was often not closely followed, as it proved difficult for nuns to manage their properties and negotiate with secular leaders if they remained inside their walls.

These nuns were involved in their local lay communities by providing prayers for their founders and lay patrons, educating children who were destined for the church, and giving hospitality and alms to travellers and those in need. Some nunneries prospered throughout the later medieval period, retaining the support of the local laity, either the Gaelic kings or Anglo-Norman landholders. When nuns did not have this lay support they were more vulnerable to diminution of their income and, ultimately, to closure. Nuns who lived under formal vows were mostly from relatively wealthy families, and there were probably never more than 12 at any one time. By the time of the dissolution of the monasteries in the 1540s, most nunneries had only a handful of nuns. There were also many vowed women who lived privately, either in their family homes or beside churches, throughout Gaelic and Anglo-Norman Ireland. These women have left few traces of their existence.

Although some nunneries such as Grace Dieu tried to survive the tide of change at the dissolution, by the end of the sixteenth century the medieval nunneries of Ireland had faded away, the nuns themselves had died out, and their estates and buildings were sold or given away. There are physical remains of some of the nunneries, particularly in the west of Ireland, however none have been excavated to date.

Dianne Hall

References and Further Reading

Bitel, L. *Land of Women: Tales of Sex and Gender from Early Ireland*. Ithaca and London: Cornell University Press, 1996.

Hall, D. *Women and the Church in Medieval Ireland, c. 1140–1540*. Dublin: Four Courts Press, 2003.

Harrington, C. *Women in a Celtic Church : Ireland 450–1150*. Oxford: Oxford University Press, 2002.

Gwynn, A., and R. N. Hadcock. *Medieval Religious Houses: Ireland*. Harlow: Longmans, 1970.

Bitel, L. "Women's Monastic Encloussre in Early Ireland: A Study of Female Spirituality and Male Monastic Mentalities." *Journal of Medieval History* 12 (1986): 15–36.

OIREACHT

See **Society, Functioning of Gaelic**

OLD AGE

See **Society, Functioning of**

ÓENACH

The word derives from *óen* (one) and has a primary meaning of "coming together," a "reunion" for the purpose of burial at the traditional tribal burial ground. It is the normal word for a popular assembly or gathering. It also occurs in place-names, meaning "a place of assembly."

The more elaborate burial mounds at an *óenach* were associated with kings and kingship and were a focal point for the expression of identity of king and people. The references to social intercourse on such occasions, including horse racing, is reminiscent of the funeral games of the ancient Mediterranean world. By the seventh century, burial was for the most part in Christian graveyards, but the *óenach* had become the assembly (still at the traditional site) of king and people on set occasions for the transaction of public business, with games, music, social interaction, and trade also part of the activities. The early Latin glosses on *óenach—theatrum* and *agon regale*—are a further indication of this. The Latin *circio* (Hiberno-Latin for "circus") is also used for *óenach*, as in the entry of the *Annals of Ulster* for the year 800 A.D. when the death of the local king is recorded at the fair, on the feast-day of St. MacCuilinn of Lusk (Co. Meath). The implication is that the *óenach* was held at the local monastery. As local territories were incorporated into more powerful kingdoms, the political significance of these local *óenach*s declined, leaving only the social and commercial aspect intact. In this way the *óenach* (Modern Irish *áonach*) survived as a "fair" into modern times.

There are references (dating between the eighth and twelfth centuries) to local fairs being held at churches such as Lusk, Armagh, Kildare, Glendalough, Lynally, Roscrea, Cashel, and Kells. The marketplace was marked with a cross in some of these sites. The word *margad* (market), from Old Norse *markadr*, itself from Latin *mercatus*, was probably borrowed into Irish during the tenth century. It glossed Latin *nundinae*, the market held every ninth day. It sometimes seems interchangeable with *óenach*, and there are references to markets being held at the great provincial *óenach*s held by the great kings, in which the political function was still of prime importance. Such was the *Óenach Tailten*, at *Tailtiu* (Telltown) situated on a loop of the river Blackwater in County Meath. This was the *óenach* of the Uí Néill dynasty and was closely associated with Tara. In Leinster, *Óenach Carmain* was the main assembly of the Leinstermen. Like *Óenach Tailten*, it was situated on a loop of the river Liffey in the parish of Carnalway, east of Kilcullen in County Kildare. Fairs such as these were sometimes not held or were disrupted for political reasons. The king who presided demonstrated his right to rule. A poem celebrating the *Óenach Carmain* provides most of what is known about these fairs. It was held on the feast of Lugnasad (August) every third year.

CHARLES DOHERTY

References and Further Reading

Gwynn, Edward, ed. *The Metrical Dindshenchas* (Royal Irish Academy,

Todd Lecture Series X), *Part III*. Dublin, 1913, pp. 2–24.

Ó Murchadha, Diarmuid. "Carman, Site of Óenach Carmain: A Proposed Location," *Éigse. A Journal of Irish Studies* 33 (2002): 57–70.

Doherty, Charles. "Exchange and Trade in Early Medieval Ireland," *Journal of the Royal Society of Antiquaries of Ireland* 110 (1980): 67–87.

See also **Burials; Fairs; Games; Kings and Kingship; Trade**

OSRAIGE

Osraige is the name of a kingdom (and people) located in the southeast quadrant of Ireland in an area roughly coterminous with the present diocese of Ossory (Cos. Kilkenny and Leix), which preserves the name. Originally, the kingdom may have extended westward to the river Suir and eastward to Gowran (Co. Kilkenny), near the river Barrow. Osraige is probably a compound of the name of an eponymous ancestor and the common suffix, *–raige* (a kingdom).

Its strategic location gave Osraige an importance belied by its relatively modest size and status. It controlled the main route into Munster for armies coming down from the north, which had to cross Belach Gabráin in south Osraige. (Several famous battles were fought in the kingdom.) More importantly, it served as a buffer between the provinces of Munster and Leinster, a position that ensured for it a role in the politics of both. Indeed, so prominent was its role in Leinster affairs that the genealogists falsely traced the Osraige back to Leinster stock. Osraige belonged to Munster in the sixth to eighth century, and probably long before then. In addition, early Irish genealogical and hagiographical traditions link Osraige with another Munster people, the Corcu Loígde, who were located in maritime southwest County Cork. Thus, St. Ciarán of Saigir (Seirkieran), patron saint of the Osraige, came from the Corcu Loígde. It appears that the Osraige had broken away from dependence on the Corcu Loígde by the seventh century. Both peoples may have been much more prominent at an earlier, prehistoric period, as suggested by the special status they enjoyed in relation to the provincial overking of Munster at Cashel. Thus, the Osraige were a free people who did not have to pay tribute, because (it was said) they had once been the rulers of Munster. Yet they seem to have been subject to Uí Néill kings of Tara in the late sixth century, perhaps because the new ruling family of Munster, the Eóganacht, owed allegiance to the latter.

The second half of the ninth century witnessed a period of marked Osraige influence in both Munster and Leinster. The Osraige king, Cerball mac Dúngaile (847–888), made a name for himself in Leinster by defending the waterways of the Barrow and Nore against Viking attacks. He played off Norse forces against each other and forged marriage alliances with the Viking rulers of Dublin. (He is remembered in Icelandic genealogies as Kjarvalr Írakonungur, Kjarvall the Irish king.) But he was no match for the political ambitions of Máel-Sechnaill I, the Uí Néill king of the northern half of Ireland, who invaded Munster. As a result, Munster was forced to alienate Osraige to him in 859. Thereafter, Osraige drifted toward a Leinster sphere of influence as Cerball's successors laid claim to the kingship of Leinster—unsuccessfully for the most part. The Norman invasion of 1170, which was directed at the southeast, meant that Osraige was one of the first Gaelic kingdoms to fall. Within a decade the Normans separated it from Leinster, making it part of the royal demesne lands of Waterford under Robert le Poer.

PÁDRAIG Ó NÉILL

References and Further Reading

Byrne, Francis John. *Irish Kings and High-Kings*, 2nd ed. Dublin: Four Courts Press, 2001, esp. 162–163, 180–181, 202, 262–263.

Charles-Edwards, Thomas. *Early Christian Ireland*. Cambridge: Cambridge University Press, 2000, esp. 476, 488–489, 541–542.

Radnor, Joan Newlon. *Fragmentary Annals of Ireland*. Dublin: Dublin Institute for Advanced Studies, 1978.

See also **Cerball mac Dúngaile; Eóganachta; Leinster; Munster; Uí Néill**

P

PALE, THE

The idea of the Pale has become one of the defining features of late-medieval Ireland, symbolizing the political and cultural differences that divided the Gaelic Irish from the English settlers. The word derives from the Latin *pallus*, meaning a stake, and by extension a defensive wall built from stakes; the derivation is identical to that of the word "palisade." But in the Irish context the word came to refer, not to a defensive perimeter, but to the area enclosed by such a notional perimeter; the area in which English culture and English law was observed. The Pale (roughly comprising the four "loyal" counties of Louth, Meath, Dublin, and Kildare) corresponded to "the land of peace," as opposed to "the land of war" where Gaelic rule held sway.

Defense of these counties from the Gaelic Irish became an increasingly precarious matter throughout the fifteenth century, a fact reflected by the widespread construction of defended tower houses. However, the first mention of an enclosing perimeter around the four counties comes from a statute of Poynings in 1494, urging that defensive ditches be constructed around the Pale. Before his appointment to Ireland, Poynings had served as governor of Calais, where the territory around the port was referred to as "the English Pale of Calais;" the term probably came to Ireland with the new deputy lieutenant.

Stretches of earthworks matching the description in Poynings' statute can still be seen, for example at Kilteel in County Kildare, but the construction of a continuous barrier encircling the four counties was never attempted. The Palesmen never had the resources to garrison or maintain such a fortification. Rather, given the prevailing Irish strategy of cattle raiding, such earthworks as were constructed served to impede the movement of herds through open land into Irish-held territories.

Far from comprising a continuous defensive rampart, the frontier between the Irish and the settlers was an ill-defined and fluid affair, which fluctuated over the years according to the fortunes of war. Although Louth was counted as one of the loyal counties of the Pale, by the late-fifteenth century the community was paying "black rents" (or protection money) to the Uí Néill (O'Neills) on an almost annual basis. Since the Pale was neither a solid defensive line nor a strictly-defined territory, perhaps the best definition of "the Pale" is as the name of a community of people; that is, those people of English descent, settled in the counties around Dublin, whose political loyalty remained strongly with the English crown.

Historians have often discussed the Pale and the Palesmen in this sense, frequently in relation to events almost a century before the first historical appearance of the term in 1495. This makes sense, since the defining characteristics of the Pale emerged long before the word itself became common currency. The 1366 Statutes of Kilkenny show that fears about the erosion of English culture and customs was widespread in the fourteenth century, and in the mid-fifteenth century commentators were already lamenting that English rule had been restricted to an area along the east coast scarcely thirty miles long and twenty miles deep. This siege mentality shaped the emerging consciousness of the Pale community, and expressed itself in their frequent appeals to the king to provide them with strong leadership and military aid to crush, or at least stem the advance of, the Gaelic Irish.

In reality, the English kings were too distant to provide the strong leadership required to bolster the Pale, and such English deputies as were sent from time to time often found themselves overwhelmed by the complexities of Irish factionalism. The great magnates, such as the earls of Desmond, Ormond, and Kildare, who should have been the natural leaders of the Pale gentry, were so tainted by Gaelic alliances and customs as to arouse the suspicion of the Palesmen; their attempts to impose coyne and livery within the Pale were a frequent cause of discontent. Only the Anglo-Irish magnates could ensure the protection of the Pale, while in return the support of the Palesmen was a prize the magnates could not afford to ignore. But for all that, relations between the two groups were often uncertain.

The renewed vigor of Tudor policies led to the breakdown of the medieval concept of the frontier in Ireland, although in the 1540s commentators were writing of "the English Pale in Scotland," and the phrase "beyond the Pale" remains part of the English language to this day.

JAMES MOYNES

References and Further Reading

Cosgrove, A., ed. *A New History of Ireland. II Medieval Ireland.* Oxford: Clarendon, 1987.

Devitt, M. "The Ramparts of the Pale at Clongowes Wood." *Journal of the Kildare Archaeological Society* 3 (1899–1902): 284–288.

Ellis, S.G. *Ireland in the Age of the Tudors 1447–1603: English Expansion and the End of Gaelic Rule.* London: Longman, 1998.

——— *Tudor Frontiers and Noble Power: The Making of the British State.* Oxford: Clarendon, 1995.

——— *Reform and Revival: English Government in Ireland, 1470–1534.* London: The Royal Historical Society, 1986.

Empey, C.A. and Simms, K. "The Ordinances of the White Earl and the Problem of Coign in the Later Middle Ages." *Proceedings of the Royal Irish Academy* 75 (1975): sect. C.

Gorman, V. "Richard, Duke of York, and the Development of an Irish Faction." *Proceedings of the Royal Irish Academy* 85 (1985): sect. C, 169–179.

Lydon, J.F. *The Lordship of Ireland in the Middle Ages.* 2d ed. Dublin: Four Courts Press, 2003.

Lyons, M.A. Church and Society in County Kildare, c. 1480–1547. Dublin: Four Courts Press, 2000.

Manning, C. "Excavations at Kilteel Church, Co. Kildare." *Journal of the Kildare Archaeological Society* 16 (1981–1982):173–229.

Matthew, E. "The Financing of the Lordship of Ireland under Henry V and Henry VI." In *Property and Politics: Essays in Later Medieval English History,* edited by A.J. Pollard, 97–115. Gloucester: Alan Sutton, 1984.

Otway-Ruthven, A.J. *A History of Medieval Ireland.* 2d ed. London: Benn, 1980.

See also **Dublin; Kildare; Mide; Tower Houses**

PALLADIUS

In 431 (according to Prosper of Aquitaine, *Chronicle,* S.A.) Pope Celestine I dispatched the newly-ordained Palladius as "first bishop to the Irish believing in Christ" (*primus episcopus ad Scottos in Christum credentes*). Neither Palladius nor his mission is mentioned in official Roman sources, and references to Palladius in later Irish documents derive either from the *Chronicle* or from Book I cap. 13 of Bede's *Ecclesiastical History of the English People* (731), who also found the information in Prosper. Prosper appears to allude again to the mission of Palladius in his polemical tract *Contra Collatorem* (written in the later 430s). He refers to Celestine's having made Britain ("the Roman island") Catholic, while making Ireland ("the barbarous island") Christian. This was in reference, in the first instance, to an earlier episode, in 429, when Celestine dispatched Germanus, bishop of Auxerre, to Britain in order to combat a recent recrudescence of the heresy known as Pelagianism. That mission (again according to Prosper) had been undertaken at the instigation of a deacon named Palladius, who is undoubtedly identical with the man of that name sent to Ireland in 431. It is generally assumed that the mission to Ireland in 431 followed on from the one to Britain in 429, on the basis that the ecclesiastical authorities in Rome would probably have feared for the orthodoxy of any fledgling Christian community in Ireland because of its geographical proximity to the compromised Christians of Britain. It is assumed also that Palladius was, by whatever means, familiar with the situation in Ireland.

Nothing more was known about Palladius himself until a recent discovery that casts new light on his youthful years, especially those apparently spent in Rome studying law circa 417, following which he made a "conversion" to radical Christianity. According to this new theory, Palladius has been proposed as the previously unidentified author of a group of radical Christian-socialist tracts known to scholars as the "Caspari Letters" (after their first editor, Carl Paul Caspari), a collection with strong links to Pelagius and his circle and composed probably around 417. The fierce denunciation of wealth and property in the letters suggests that their author was a recent (and relatively youthful) convert to Pelagian views. We can only speculate as to whether or not those views find a reflection in later Irish Christianity.

Though some Irish writers of the late-seventh century maintained that Palladius's mission was either not successful, or else that he abandoned the missionary effort, there is a general consensus amongst historians of today that he did reach Ireland, presumably with a party of helpers, and established his mission probably in the area around the present-day County Meath. The place names Dunshaughlin and Killossy/Killashee are understood to derive from the Irish *dún* "fort" + Secundinus and cell "church" + Auxilius respectively (in their Irish forms Sechnall and Ausille), denoting early foundations by continental missionaries probably associated with Palladius. No church dedicated to Palladius, however, has survived.

At just this point, however, Palladius disappears entirely from view, his role and that of his followers completely submerged by the all-conquering legend surrounding the great Saint Patrick. Native tradition associates the beginnings of Irish Christianity with Patrick, not Palladius, who was written out of history in the seventh century. Because of Palladius's "disappearance" after 431, Irish historians filled the void by dating Patrick's arrival in 432. No document from the Palladian mission has survived, whereas Patrick's two writings became the foundation for a body of legends that turned the humble Briton into an all-powerful, conquering Christian hero. In the process, however, the true character of the man was sacrificed for the purpose of creating a mythological figure whose "heroic" deeds formed the basis for outlandish claims made in the centuries after him.

When the Irish churches emerge fully into the light of history at the beginning of the seventh century, the famous Paschal letter of Cummian (632) refers only to "the holy Patrick" (*sanctus Patricius*) as *papa noster* ("our father")—the earliest indication we have that Patrick, and not Palladius, enjoyed a special status as the "Father" of the Irish Church. Historians have been troubled, however, by the fact that Patrick nowhere in his writings makes mention of Palladius or anyone else involved in missionary activity in Ireland, but constantly reiterates the claim that he has gone "where no man has gone before." It is not at all impossible, therefore, that Patrick came to Ireland *before* Palladius, rather than after him, perhaps in the late fourth century, or in the generation before Palladius was dispatched by Pope Celestine to the "Irish believing in Christ." Whatever his eventual fate, Palladius made nothing like the same impression on the Irish historical mind as Patrick did, and is now a forgotten figure in Irish history.

Dáibhí Ó Cróinín

See also **Christianity, conversion to; Patrick**

PAPACY

The earliest reference to papal contact with Ireland occurs in 431 when the *Chronicle* of Prosper of Aquitaine recorded the sending of Palladius as bishop to "the Irish believing in Christ" by Pope Celestine I (d. 432) as part of a wider papal mission to the church in the British Isles. The Irish church developed distinctive structures and practices and the Irish method of calculating Easter was a particular cause of controversy. The Venerable Bede (d. 735) makes reference to two seventh- century papal letters to Irish ecclesiastics concerning this paschal controversy. The Irish *peregrinus Columbanus* (d. 615) corresponded with Pope Gregory the Great (d. 604) and forcefully reminded Pope Boniface IV (d. 615) of his responsibility to exercise the Petrine ministry to stamp out error. A similar respect for papal primacy is evident in some eighth-century Brehon law texts.

Increased contacts between Ireland, England, and the Continent from the mid-eleventh century brought the Irish Church into contact with the Gregorian reform movement. There were a number of Irish royal pilgrimages to Rome during this period and an Irish monastery was established on the Celian hill. Pope Gregory VII (d. 1085) corresponded with King Tairrdelbach Ua Briain encouraging his efforts at church reform. This momentum culminated in a series of synods; Cashel I (1101), Rath Breassail (1111), and Kells-Mellifont (1152), presided over by papal legates, in which a diocesan structure was established, sacramental and liturgical life renewed, and attempts made to reform sexual mores and ensure the payment of tithes. Irish prelates were well represented at both the great reform councils of the Middle Ages: Lateran III (1179) and Lateran IV (1215).

The Anglo-Norman presence in Ireland from 1169 was a complicating factor in Irish-papal relations. Much academic controversy has been generated over the significance of the 1155 bull *Laudabiliter* of the Pope Alexander IV by which Ireland was granted to Henry II of England. While its authenticity is now generally accepted, in many respects the attitude of the papacy after the invasion is more significant as the English right to the lordship of Ireland was never challenged by the popes before the Reformation, particularly after King John agreed to hold Ireland as a papal fief from Innocent III in 1213.

The decline in the English colony that became evident in the latter half of the thirteenth century continued into the fourteenth. Tension between the two nations became particularly pronounced in the wake of the Bruce invasion (1315–1317) and King Edward II enlisted papal support for the correction of clergy and religious who sided with the rebels. The grievances of the Irish population found expression in the 1317 *Remonstrance* addressed to Pope John XXII by Domnall Ua Néill in which he claimed that as the English had failed to fulfill the conditions of *Laudabiliter*, they should be deprived of their Lordship. Though John XXII did not concede this he did urge Edward II and Edward III to attend more carefully to the rights of their Irish subjects.

The transfer of the papal court to Avignon in 1315 brought the papacy closer to Ireland and there is a corresponding rise in the volume of Irish material preserved in the various series of papal records. Unlike England, where recourse to the papacy in legal matters was strictly controlled by statute, Gaelic Ireland was under no such strictures and the papal records abound in references to disputes relating to appointments to benefices, elections to bishoprics, matrimonial cases and dispensations from illegitimacy, and other canonical impediments to ordination. While papal intervention in the twelfth and thirteenth centuries was generally on the side of church reform these later involvements were less edifying and more mercenary as an impoverished papal curia exploited every avenue of financial opportunity. This was particularly pronounced during the Great Schism (1378–1418). Like England, the Irish church sided with Urban VI and his successors in the Roman obedience, though there is some evidence for support for Clement VII and the Avignon line in Connacht and amongst the friars in the early stages of the controversy.

Ireland remained largely untouched by the conciliar movement and any impetus towards reform came through the Observant movement among the mendicant friars, which emerged at the end of the fourteenth century. The Observance brought the friars into close contact with the papacy and they emerged as its chief champions when the challenge to papal authority arose after 1536 when the Irish parliament recognized Henry VIII as the supreme head of the Church in Ireland.

Colmán N. Ó Clabaigh, OSB

References and Further Reading

Bracken, Damian. "Authority and Duty: Columbanus and the Primacy of Rome." *Peritia: Journal of the Medieval Academy of Ireland* 16 (2002): 168–213.

Flanagan, Marie Therese. "Hiberno-papal Relations in the Late Twelfth Century." *Archivium Hibernicum* 34 (1976–1977): 55–70.

Gwynn, Aubrey. *The Irish Church in the 11th and 12th Centuries.* Dublin: Four Courts Press, 1992.

Walsh, Katherine. "Ireland, the Papal Curia and the Schism: a Border Case." *Genèse et Débuts du Grand Schisme d'Occident. Colloques Internationaux du Centre Nationale de la Recherché Scientifique* 586 (1980) 561–74.

Watt, John. *The Church in Medieval Ireland.* 2d ed. Dublin: University College Dublin Press, 1998.

———. "The Papacy and Ireland in the Fifteenth Century." In *The Church, Politics and Patronage in the Fifteenth Century,* edited by R. B. Dobson, 133–45. Gloucester: 1984.

See also **Anglo-Norman Invasion; Black Death; Bruce, Edward; Cashel, Synod of I (1101); Cashel, Synod of II (1172); Christ Church Cathedral; Church Reform, Twelfth Century; Ecclesiastical Organization; Ecclesiastical Settlements; Fitzralph, Richard; Gaelic Revival; Henry II; Kells, Synod of; Malachy (Máel M'áedoic); Paschal Controversy; Racial and Cultural Conflict; Scotti/Scots**

PARISH CHURCHES, CATHEDRALS

Parish Churches

Parish churches and cathedrals are a product of the church reforms of the twelfth century in Ireland. While the country was dotted with churches and ecclesiastical sites in the early medieval period, there was a lack of overall organization and the degree of pastoral care available is a matter of debate and certainly varied greatly from place to place. There has also been debate about the extent to which parishes were well established and tithes levied before the Anglo-Normans arrived. Certainly in the areas settled by the Anglo-Normans parishes became well established and were often coextensive with the local manor. Parish churches and manorial centers are commonly situated in close proximity in these areas. On the other hand many of these churches were older ecclesiastical sites and some of the manors were older political units or centers. However, in areas not settled by the Anglo-Normans, parishes appear to have been established more gradually and haphazardly.

The parish system was based on tithes, a tax amounting to one tenth (a tithe) of farm produce payable to the parochial clergy for their maintenance. In many cases the lord granted the tithes to a monastic establishment, who would supply a priest from the community to serve the parish or more usually pay a priest to do so. A small amount of land known as the glebe, usually situated close to the parish church, was set aside for the priest's residence and for grazing and tillage on a small scale.

Medieval parish churches are normally divided into a nave, the main body of the church where the congregation worshipped, and the smaller chancel, where

Kildare Cathedral. © *Department of the Environment, Heritage and Local Government, Dublin.*

the priest performed the ceremonies at the altar. The upkeep of the nave was the responsibility of the parish, while that of the chancel fell to the priest. The chancel was usually a separately roofed, lower and smaller section of the building and was entered and visible from the nave through a chancel arch. Usually set immediately west of the chancel arch was a wooden rood screen, which takes its name from a large wooden crucifix (rood) suspended above it. The screen, which had doors in the center giving access to the chancel, normally had a narrow loft above it, accessed by a stairs within the screen or in the thickness of the wall on one side. Not a single medieval wooden rood screen survives from Ireland but evidence for their former existence can be seen in many churches in the form of corbels or beam-holes in the walls and windows or stairs that served them. The rood screen would have further emphasized the division between nave and chancel and in plain rectangular churches would have been the main demarcation between these two areas of the church.

Irish medieval parish churches are mostly in ruin and are generally considerably smaller than contemporary examples in England. Some of the largest examples from the thirteenth century were built in towns and good examples can be seen at New Ross, County Wexford and Gowran, County Kilkenny. At New Ross only the chancel and transepts survive from an ambitious early thirteenth-century church. At Gowran most of the nave survives with its aisles, all built around 1270.

Parish churches were frequently altered and added to and it is rare to find an example that is of one period only. Some incorporate remains of older churches from the tenth to twelfth centuries, such as Tullaherin, County Kilkenny and Fore, County Westmeath. The sequence of alterations and rebuildings can in some cases be very complicated as at St. Audoen's in Dublin or the larger church at Liathmore, County Tipperary.

The fifteenth century saw a great boom in building in Ireland and many churches were built anew or older ones altered. Many of the small ruined medieval churches in graveyards around the country date from this period. Often they have opposing doorways, with pointed two-centered heads, in the north and south walls of the nave. The most popular form of window in late medieval times was that with an ogee head and these are found also in the contemporary tower houses. Some of the finest churches of this period were those built in towns or in the more settled lands of the Pale. Larger examples have fine traceried windows such as those seen at St. Nicholas's in Galway or Dunsany, County Meath.

A feature of churches of this period is the provision of accommodation for the ministering priest in the west end of the church or in an attached tower. In many cases the area of the nave to the west of the doorways was walled off from the rest of the church and provided with a first floor for extra accommodation. A tower at the west end or attached to one side or incorporating the chancel probably served as a castellated presbytery. A common feature of the period was the addition of crenellated parapets on churches, rendering them suitable for defense on a small scale. Churches were attacked and damaged in raids and warfare and the provision of defenses was a serious consideration. Also churches were used to store property and valuables, making them a lucrative target for raiders and thieves.

Chantry chapels were sometimes formed within or added to parish churches, especially those in towns. These and collegiate churches were sometimes endowed under the terms of a will, when money or property was bequeathed for this purpose and to support the priest or priests to say mass there for the souls of the deceased. Some chantry chapels were controlled by guilds, such as the guild of St Anne at St Audoen's in Dublin.

Many parish churches, especially the rural ones, became ruined during the wars of the sixteenth and seventeenth centuries or as a result of the Reformation, which made them the property of the established Protestant church while the vast majority of the population remained Catholic. Also, as a consequence of the dissolution of the monasteries, many parishes became impropriate to lay people, whose main concern was not the cure of souls.

Cathedrals

Though bishops were important figures in the early medieval Irish church, churches associated with them were not referred to as cathedrals until the reforms of the twelfth century divided the entire country into dioceses. Most of the buildings then designated as cathedrals would have been older principal churches on ecclesiastical sites. Some of these survive as ruins such as those at Clonmacnoise and Glendalough, while others were incorporated into buildings still in use by the Church of Ireland such as Clonfert, County Galway and Kilfenora, County Clare. Other churches had short-lived claims to cathedral status such as those at Ardmore, County Waterford and Scattery Island, County Clare. The newly appointed bishops in the twelfth century succeeded in acquiring for their dioceses some of the old termon lands of early medieval monastic/ecclesiastical sites and, where these were successfully held and well managed, they became episcopal manors, which helped to support the bishop and other diocesan dignitaries. If extensive, these see lands could help in supplying funds for building projects, such as a new cathedral or additions to an old one.

Many Irish cathedrals incorporate Romanesque fabric from the twelfth century and indeed the Romanesque style is often seen as closely associated with the reform movement. A Romanesque church appears to have an essential prerequisite in order to make a serious claim for episcopal status in the twelfth century. Cathedrals with surviving Romanesque include Tuam, County Galway; Ardfert, County Kerry; and Clonfert, County Galway.

The largest cathedral in Ireland is St. Patrick's in Dublin, and both it and the other Dublin cathedral, Christ Church, were influential in introducing the Gothic style from the west of England in the thirteenth century. Other medieval cathedrals in Ireland are smaller in scale, while some are much smaller than many English parish churches. Well-preserved examples still in use by the Church of Ireland include St. Canice's in Kilkenny, Cloyne, County Cork, Killaloe, County Clare, and Limerick. All four are mainly of thirteenth-century date and Kilkenny and Limerick have both nave aisles and transepts. The cathedral at Cashel, County Tipperary, is also a large, mainly thirteenth-century structure, but is unroofed, having been abandoned in the eighteenth century.

The only medieval cathedral, which was also monastic, was Christ Church in Dublin, which was a monastery of Augustinian canons. It is also the only cathedral with the remains of a formal chapter house; the lower parts of it survive to the south of the south transept. At other sites the diocesan chapter presumably met within part of the cathedral or in an attached or associated building or chapel such as the Romanesque Cormac's Chapel at Cashel, which is known to have been used as a chapter house in post-medieval times. In Ireland there is little evidence for major associated developments around cathedrals and even medieval episcopal palaces are very rare. The present palace at Kilkenny incorporates the medieval palace, while in Dublin parts of the medieval Palace of St. Sepulchre survive within later buildings. At Cashel the hall and dormitory of the vicar's choral, endowed in the fifteenth century, have been reroofed as visitor facilities.

CONLETH MANNING

References and Further Reading

Fitzpatrick, Elizabeth and O'Brien, Caimin. *The Medieval Churches of County Offaly*. Dublin: Government of Ireland, 1998.

Leask, H.G. *Irish Churches and Monastic Buildings*. 3 vols. Dundalk: Dundalgan Press, 1955–60.

Manning, Conleth. "Clonmacnoise Cathedral." *In Clonmacnoise studies Volume 1: Seminar Papers 1994*, edited by Heather A. King, 56–86. Dublin: Dúchas The Heritage Service, 1998.

———. *Rock of Cashel, Co. Tipperary*. Dublin: Dúchas The Heritage Service, 2000.

Milne, Kenneth, ed. *Christ Church Cathedral, Dublin: A History*. Dublin: Four Courts Press, 2000.

Ní Ghabhláin, Sinéad. "Church and Community in Medieval Ireland: The Diocese of Kilfenora." *Journal of the Royal Society of Antiquaries of Ireland* 125 (1995), 61–84.

See also **Abbeys; Altar-Tombs; Architecture; Burials**

PARLIAMENT

From the early days of the Irish lordship the chief governor was expected to act with the advice of the feudal tenants-in-chief, who were, in turn, required to proffer this advice as part of their feudal obligations. As the central government expanded, the King's Council in Ireland included more and more permanent salaried officials. The Great Councils, when the king's officers were joined by the chief magnates of the land, gradually evolved in the course of the thirteenth century into parliamentary sessions, as in England.

Beginnings of Parliament in Ireland

Up until the nineteenth century, historians believed that parliament in Ireland originated when Henry II was in Dublin in 1171–1172. However, regular parliaments were only instituted in England after 1258 and it appears that a similar system was introduced into Ireland at around the same time. Certainly, the first documentary reference to an Irish parliament comes from 1264 and was used to describe a gathering in Castledermot, County Kildare. Little is known about the business transacted at this parliament beyond the fact that an inquisition was taken before the justiciar and council. In the first century of its existence it remains difficult to distinguish which assemblies were indeed Irish parliaments as the term *parliamentum* was used ambiguously and sometimes implied no more than a parley with the Irish or others at war against the king.

Composition of Parliament

The nucleus of the parliament was the council, the permanently-constituted body of ministers who advised the justiciar. This group was augmented by the chief magnates and higher clergy of the land who were summoned to attend in person but were frequently represented by their stewards and bailiffs. There is no certain evidence of the presence of elected representatives in parliament before 1297 when two knights were summoned from each of ten counties and five liberties. On this occasion the sheriffs of counties and stewards of liberties were also present.

The towns were not requested to send representatives in 1297, but cities and boroughs to which foreign merchants came were required to send two citizens or burgesses to the parliament of 1299 and one year later, all cities and boroughs were required to send representatives. At the start of the fourteenth century, therefore, representatives of all the principal local communities were being summoned to attend parliament. This did not imply that the Irish parliament had become a democratic institution. Only the knights of the countryside and the burgesses of the towns were allowed to vote for representatives who were themselves drawn from this narrow class. Where the names of the representatives are known, it is clear that the local communities chose men of experience to assent to the legislation which they would ultimately be required to implement.

The makeup of any single parliament depended on the precise nature of the business to be discussed and the justiciar would therefore tailor the summonses to parliament to meet specific purposes. It is thought that during the early period of the institution's existence, there was a distinction between "general parliaments" and simple "parliaments," with only the former requiring the attendance of local representatives. Popular representatives—also referred to as "the commons"—cannot be seen as playing an essential part in parliament before the middle of the fourteenth century when their summons to Irish parliaments became invariable. Until this date parliaments are certainly viewed by contemporaries as primarily aristocratic occasions and it would appear that as an institution the Irish parliament made little or no impact on the popular consciousness.

By the end of the reign of Edward III (1377) the commons had an established place in the Irish parliament and had been joined by the clerical proctors, representatives of the lower clergy. This group was summoned for the first time in 1371 and thereafter became a regular part of the Irish parliament.

By the reign of Richard II, the idea of a defined and limited class of peers of parliament had appeared in Ireland. This combined with other factors to result in the steady decline in numbers summoned to parliament. In the early fourteenth century as many as ninety laymen might be summoned by individual writ, but by the end of the fifteenth century there were only fifteen temporal peers. The numbers of heads of religious houses summoned shrank from around twenty to six, often because of claims for exemption from the abbots and priors themselves. The majority of the thirty-two Irish bishops seem to have been summoned but not all attended. Fourteen counties and liberties and twelve towns were entitled to send representatives but this number similarly shrank as the area controlled by the central government became more constricted and centred on what was to become the Pale.

The 1297 Parliament

The parliament held in Dublin in 1297 is noteworthy for a number of reasons. Although some fragmentary memoranda survive for the 1278 parliament, the 1297 one is the first from which substantial legislation survives. Furthermore, 1297 is regarded by many as the first real parliament to meet in Ireland as it was the first time, as far as is known, that the writs of summons contained a *plena potestas* clause. This meant that the representatives sent by the shires and provinces were to be given full powers to speak and decide for all and that the legislation of the parliament was to be fully binding. The legislation passed by the 1297 parliament had much significance for the localities and this was no doubt why the presence of their representatives was deemed important. The parliament considered the vital question of how best to establish and maintain peace in Ireland, and the legislation placed the burden of defense and peacekeeping firmly on communities and individuals. It identified the twin evils of absenteeism on the part of the great lords and what was termed "degeneracy" on the part of some resident subjects who were not preserving their distinctive English customs and traits.

These problems and how to deal with them occupied Irish parliaments for many decades to come.

Parliamentary Legislation

Parliaments had a judicial and a legislative role and were used to deal with petitions and settlement of disputes as well as to formulate legislation and obtain consent to various measures. One of the most important pieces of business regularly transacted in parliaments in the later Middle Ages was the granting of subsidies to the chief governor. The amounts to be levied from these subsidies varied considerably until they became standardized in the fifteenth century. With the growing threat to the lordship in the fourteenth century came increased demands for taxation. The representatives of the local communities, which had to bear the main burden of taxation, played an important part in the granting of subsidies. During the 1370s William of Windsor's heavy fiscal demands provoked opposition in parliament and when the king tried to bring representatives of the Irish counties and towns to England in order to ply them with demands for money, their electors reacted by denying them the authority to make grants.

It would appear that parliamentary rolls on the English model were not kept in medieval Ireland and prior to the fifteenth century our knowledge of parliamentary legislation comes from a variety of documentary sources. In 1427 the statute rolls, which contain enactments of the Irish parliament, began as a series and while all those for the medieval period were destroyed in 1922, a great deal of the legislation survives in transcripts. It has been said, however, that much of the legislation was trivial or ephemeral, intended to deal with immediate and often personal problems. There were exceptions, of course, and perhaps the most famous legislation of an Irish parliament is the body of statutes passed by the assembly summoned by Lionel of Clarence to meet in Kilkenny in 1366.

No other corpus of legislation passed in Ireland during the Middle Ages is accorded a status equivalent to that of the Statutes of Kilkenny. The preamble to the legislation identifies a special preoccupation with the problem of "degeneracy" and one group of statutes is concerned with regularizing relations between the Irish and English communities. In particular, the English are forbidden to adopt Irish language, dress, and culture or to form alliances with the Irish through marriage or fosterage, while the Irish are to be excluded from appointment to certain church offices. Other statutes deal with economic matters, such as price and wage fixing, with the reform of central government, and with the relationship between church and state. It is now generally acknowledged that while some of this legislation was new, most can be traced back to 1350 and some to 1297. However, in the Statutes can be seen an attempt to combine a whole series of enactments whose purpose was to deal with the problems of the lordship.

The Statutes of Kilkenny demonstrate the extent to which Irish parliaments and great councils had became concerned with issues caused by the growing problems of absenteeism and "degeneracy." As time went on parliamentary assemblies were increasingly used by the Anglo-Irish as occasions to formulate appeals to the king and send messengers to the English court. For example, in 1385, meetings in Dublin and Kilkenny were used to draw up arguments to convince Richard II to personally intervene in the colony.

Parliament in the Fifteenth Century

The role of the parliament as a high court where legislation particular to Ireland was enacted and where petitions from the lords, gentry, and communities of the lordship were dealt with continued in the fifteenth century. However during this century there also emerged an enhanced sense of parliament's status. This was exploited in 1460 by Richard, duke of York, when the Drogheda parliament (in its own way exploiting Richard's vulnerability) issued its declaration asserting Ireland's jurisdictional identity under the crown, and denying the validity of English statutes unless these were accepted "by the lords spiritual and temporal and the commons of Ireland in a great council or parliament." The legislation of this parliament was truly revolutionary in nature but its real significance has been much debated. Some have characterized it as an aberration, a reaction to the particular circumstances of the time which had no historical precedent. Another school of thought sees the declaration as having a certain amount of historical justification and interprets it as an important expression of Anglo-Irish separatism.

The support in Ireland for Yorkish pretenders to Henry VII's throne helped prompt the enactment of Poynings's Law by the parliament that met in Drogheda in 1494–1495. Henceforth, no parliament could be summoned to meet in Ireland without the king's explicit license and no legislation could be enacted until it had been inspected and approved by the king and his council in England. Thus the English government gained unprecedented control over the business of the Irish parliament and the medieval phase of Irish parliamentary history was brought to a close.

The similarities between the parliaments of medieval Ireland and England have been correctly acknowledged by parliamentary historians but so too have the differences. All English institutions were to some extent modified and adapted to the Irish situation and parliament was no exception. Among the distinctive characteristics of the Irish parliament was the regular inclusion from the late fourteenth century of the proctors of the lower clergy who in England met in separate convocations. Another respect in which the Irish parliament differed from its English counterpart was the practice of financially penalizing those who were absent from parliament when summoned. However, the greatest difference between the two institutions was surely that whereas in England those summoned to parliament were expected to represent all parts of the country, if not all sections of society, in medieval Ireland the parliament was representative of only one of the two nations within the country—the English nation. In 1395 Richard II made an attempt to include Gaelic Irish leaders within parliament. This attempt failed and it was not until the reign of Henry VIII that some Gaelic lords were allowed to take their place in the ranks of the peerage. The medieval Irish parliament therefore was an institution that represented only the colonial section of the Irish population.

MARGARET MURPHY

References and Further Reading

Cosgrove, Art. "Parliament and the Anglo-Irish Community: the Declaration of 1460." In. *Parliament and Community. Historical Studies XIV*, edited by Art Cosgrove and J.I. McGuire. Dublin: Appletree Press, 1983.

Duffy, Seán. *Ireland in the Middle Ages*. Dublin: MacMillan, 1997.

Ellis, Stephen. *Ireland in the Age of the Tudors 1447–1603*. London: Longman, 1998.

Frame, Robin. *Colonial Ireland, 1169–1369*. Dublin: Helicon, 1981.

Lydon, James. *Ireland in the Later Middle Ages*. Dublin: Gill and MacMillan, 2003.

Lydon, James, ed. *Law and Disorder in Thirteenth-century Ireland: The Dublin Parliament of 1297*. Dublin: Four Courts Press, 1997.

Richardson, H.G. and Sayles, G.O. *The Irish Parliament in the Middle Ages*. Philadelphia: University of Pennsylvania Press, 1952.

Otway-Ruthven, A.J. A *History of Medieval Ireland*. 3d ed. New York: Barnes and Noble, 1993.

See also **Central Government; Chief Governors**

PASCHAL CONTROVERSY

The earliest Christians, being converted Jews, followed the Jewish practice of observing the day on which the Paschal lamb was slaughtered at Passover, which was the 14th day of the Jewish lunar First Month, Nisan. By the second century Christians of Asia Minor kept Nisan 14, irrespective of what day of the week it fell on, thereby acquiring the name of "Quartodecimans," later to become a term of anathema. The paradox of a Jewish-Christian practice, traced by its adherents back to the Apostle John and acknowledged even by its opponents to be the older practice in the Church, subsequently being declared a heresy was not lost on the later Irish and English churches, in which controversialists (including the 7th–*c*. Roman curia) professed to detect residual traces of the practice.

By the time Christianity reached Ireland in 431–432, Christians everywhere had agreed that Easter should be kept on a Sunday, following the 14th day of the First Month (*luna XIV*), after the spring equinox. These early Christians, following the practices of Greek-speaking gentiles, also supposed that *pascha* derived from *paschein* "to suffer," and therefore concluded that Easter denoted the Passion, rather than the Resurrection. The combination of these uncertainties with the fact that churches in Rome and Alexandria (and elsewhere too) differed in their methods of calculating *luna XIV* and the First Month, as well as the correct date of the equinox, led to the situation in which churches in different parts of the Christian world celebrated Easter Sunday on different dates.

Stark divergences in 444 and again in 455 between the Roman and Alexandrian cycles led Pope Leo to ask his archdeacon, Hilarus, to commission a new table, which was drawn up by the Aquitanian mathematician Victorius and published in 457. Victorius's incompetence made an awkward situation impossible, and the result was chaos. Two advantages of his table, however, enabled it to secure widespread adherence: it was a perpetual cycle of 532 years (running from C.E. 28 to C.E. 559), and it followed familiar Roman practice by starting the year on January 1. Although Pope John I commissioned another study in 529 with a view to solving the persisting problems, Victorius's faulty tables were declared the official tables for the Gallican church in 541 (significantly enough, 84 years after their first appearance). While Pope John's attempted reform resulted in the publication in 525 of the 19-year tables of Dionysius Exiguus (which ran for 95 years from 532–626) based on (the correct) Alexandrian principles, Victorius's tables continued to be used for several centuries afterward.

It is not certain which Easter cycles were introduced into the fifth-century Irish church, but it may be assumed that Palladius introduced whatever cycle was prevalent in Gaul in the 430s (either an 84-year cycle or an early version of the Alexandrian 19-year "Metonic" cycle championed in the 390s by Ambrose of Milan), while St. Patrick (supposedly active in Ireland at around the same time as Palladius, or perhaps a generation or so later), would, in all likelihood, have introduced a form of the 84-year Easter cycle then in use by the British church. Victorius's tables were certainly known in Ireland by the sixth century, and when Columbanus of Bangor traveled from Ireland to Gaul circa 590 his realization that they were the standard tables there, apparently sanctioned by Rome (because archdeacon Hilarus had since succeeded Leo as Pope), occasioned his famous first letter addressed to Pope Gregory I—"a letter equally remarkable for baroque Latinity and studied insolence"—in which he damned the Aquitainian's tables by declaring that they had been dismissed by Irish computists and scholars as being "more worthy of ridicule and pity than of authority." Columbanus told Gregory that his fellow countrymen used an 84-year Easter table and a related tract *De ratione paschali* attributed to Anatolius, Bishop of Laodicea, in modern Syria († *c*. 282). The table was apparently used by all churches in Ireland by that time, and it remained in use by the community of Iona (off the western coast of Scotland) until 716, after which it disappeared without trace. Anatolius, for his part, was cited by controversialists throughout the sixth and seventh centuries (though some suspected the text was a forgery). The rediscovery and publication of the long-lost "Irish 84" in 1985 has for the first time allowed a correct reconstruction of the historical Irish Easter dates, and thereby cleared up many misconceptions about the controversy.

Papal suspicions of Irish Easter practices came to a head in 628–629, when Pope Honorius I addressed a letter to the Irish clergy, admonishing them for their erroneous ways and urging conformity with Gallican and Roman usages. A synod of southern Irish clerics, which was convened in response to the papal letter, is described in the famous Paschal Letter of Cummian (632–633), without doubt the most important Irish document surviving from this period. Cummian and his colleagues appear to have adopted the Victorian tables (though not before a delegation sent to Rome reported back on their findings), which provoked a letter of response from Ségéne, abbot of Iona, accusing them of heresy. The fierceness of the language in Cummian's Letter indicates how heated the debate had become in Irish circles. Not all Irish churches followed Cummian's party, however. In 640 a group of northern Irish churches (headed by Armagh) wrote to Rome seeking papal advice on how to reckon the Easter date for 641. As it happened, both Victorius and the Irish 84-year Easter table gave Easter Sunday on April 1 of that year, whereas the Dionysiac table gave April 1 as *luna XIV*, a date on which Easter Sunday was not allowed to fall in the Alexandrian reckoning (which had Easter Sunday therefore on April 8 that year). The same problem arose also in Visigothic Spain, and a letter of Bishop Braulio of Saragossa in response to an unknown enquirer may possibly have been addressed to an Irish correspondent.

It is not known whether Armagh and the northern churches changed their observances after 641, but such evidence as exists suggests that they did not. Certainly, the community of Iona, which from its foundation in 563 was the dominant church in Scotland, and which from 634–635 was in control also of all the newly established churches in the north of England, and whose paruchia included important houses in both Ireland and Britain, held fast to the 84-year Easter tables of its founder, Colm Cille. This in turn led to difficulties, first in the north of England and subsequently in Scotland, which came to a head at the famous synod of Whitby (Northumbria) in 664. The exact nature of the conflict is unclear, but our principal source of information, the Venerable Bede (*Ecclesiastical History of the English Nation, III 25*) presents the debate as one between Irish traditionalists, partisans of the old 84-year tables, and Romanist "reformers" led by Wilfrid of Hexham and York, who advocated adoption of the Dionysiac tables. The decision of the presiding king, Osuiu, was against the Irish, whose leader, Bishop Colman of Lindisfarne, decided to withdraw from Northumbria and retire back to Iona (and eventually Ireland) along with those of his community (both Irish and Anglo-Saxon) who wished to remain loyal to the old ways.

Only Iona itself amongst Irish foundations appears to have held out in the struggle. Even here, however, after many years' effort by the Englishman Ecgberct, the island community finally relented and in 716 Easter was celebrated there in accordance with the new (Dionysiac/Alexandrian) ways. After the expulsion of Iona monks from the Pictish kingdom in 717, the only insular churches to resist change after that were the British, some of whom, perhaps in the kingdom of Strathclyde, had come into line already in 703–704 (Bede, *Ecclesiastical History of the English Nation* V 15) while the last of them, led by Bishop Elfoddw of Bangor, conformed in 768.

DÁIBHÍ Ó CRÓINÍN

References and Further Reading

Blackburn, Bonnie, & Leofranc Holford-Strevens. *The Oxford Companion to the Year*, 791–812. (Oxford: 1999).

Edwards, T.M. Charles. *Early Christian Ireland*, 391–415. Cambridge: 2000.

Jones, C. W. *Bedae Opera de Temporibus. Cambridge*, MA: 1943.

Ó Cróinín, D. *Early Irish History and Chronology.* Dublin: 2003.

Wallis, F. *Bede: The Reckoning of Time*, xxxiv-lxxxv. Liverpool: 1999.

Walsh, M., & Ó Cróinín, D. *Cummian's Letter 'De Controversia Paschali'* (Toronto 1988).

PATRICK

Two of Patrick's Letters, the *Epistola ad Milites Corotici*, written to soldiers of a petty king who had killed some of his catechumens and enslaved others, and the *Confessio*, written to explain and justify his mission to converts and critics alike, tell what he wanted us to know of his life as Apostle of the Irish.

St Patrick asleep on a knoll from *La Vie des Sains*. © *The British Library.*

The Letters survive in eight manuscripts: the earliest giving an abridged text copied at the beginning of the ninth century in Armagh; a second during the tenth century, perhaps in the diocese of Soissons; a third about the year 1000 at Worcester; a fourth during the eleventh century, owned if not written at Jumièges; three during the twelfth century in northern France and in England; and one during the seventeenth century. As among these eight manuscripts seven are independent of each other, the text of the Letters is fairly secure.

Patrick wrote the name of his grandfather as *Potitus*, which means "empowered man," and the name of his father as *Calpornius*, associated with the name of the Roman plebeian gens *Calpurnius* and the name of Julius Caesar's wife *Calpurnia*, derived from καλπη + *urna* + *-ius*, designating one who bears a "pitcher" or "urn" in religious ceremonies. Patrick's own name, derived from *pater* + *-icius*, meaning "like a father," designates "a man noble in rank." Because of the meaning of his name and because of the status of his family Patrick explicitly claimed *nobilitatem* "nobility" for himself. He described his grandfather as a *presbyter* "priest" and his father as both a *diaconus* "deacon" and a decurio "*decurion*," a member of a municipal senate, an official responsible for the rendering of taxes. His father owned slaves of both sexes and land, a *villula* "little villa" near *Bannavem Taburniae*, perhaps *Bannaventa Berniae* "market town at the rock promontory of Bernia." Although some would identify *Bernia* with the territory of people described in Old Welsh as *Berneich* and in Anglo-Latin as *Bernicii*, inhabitants of the northern province of the kingdom of Northumbria, and one later Life states that Patrick was born in Strathclyde, the northern part of Britain was a military zone. Patrick's Roman Christian family of landowning and slave-holding clerics and imperial civil servants are likelier to have lived somewhere in the civilian zone of southwestern Britain, along the Severn estuary, where Ordnance Survey maps show many villas. As Patrick describes his fatherland in the plural as *Brittanniae* "the Britains" and neighboring regions as *Galliae* "the Gauls," he must have been born while these regions were still divided into multiple provinces in which Roman ecclesiastical and civil administration still functioned, perhaps about A.D. 390, certainly before 410. As he refers incidentally to coinage, *solidi* and *scriptulae*, and contrasts the behavior of the presumably post-Roman *tyrannu Coroticus* (a name related to Old Welsh *Ceredigion*, modern Cardigan) with that of Romano-Gaulish Christians dealing with pagan Franks, his mission probably preceded the conversion of the Franks, perhaps in 496, certainly before 511. This is consistent with dates of *The Annals of Ulster*, which record Patrick's arrival as a missionary in Ireland in 432, his foundation of Armagh in 444, and his death in 461, when he would have fulfilled the Biblical span of 70 years, alternately 491, when he would have been about 100.

Captured as a 15-year-old *adolescens* "adolescent" (15–21) on his father's *villula*, Patrick worked as a slave for six years near the Forest of Foclut in Ireland, where began a series of seven dreams that informed his career. After learning in the first that he would escape to his fatherland he journeyed 200 Roman miles (188 of ours), presumably from northwest to southeast across Ireland, whence he sailed for three days. On landing, his company wandered for 28 days through wilderness, nearly starving until discovery of food after Patrick's prayer for help. There followed a great temptation by Satan in a second dream, escape from perils which lasted one month, an account of a later dream which foretold accurately a captivity of two months, then return to his family in Britain, and a fourth dream *in iuuentute* "in youth" (22–42), in which a man named *Victoricius*, sometimes identified with Victricius bishop of Rouen (*c.* 330–*c.* 407), bore a letter with the *Vox Hiberionacum* "the voice of the Irish" summoning him to evangelize them. In the fifth dream Christ spoke within him. In the sixth he saw and heard the Holy Spirit praying inside his body, *super me, hoc est super interiorem hominem* "above me, that is above my inner man." In the triumphant seventh vision, following his degradation, he was joined to the Trinity as closely as to the pupil of an eye.

After the raid by Coroticus Patrick sent a letter seeking redress with a priest *quem ego ex infantia docui* "whom I have taught from infancy" (implying, since infancy ended at 7 and ordination to the priesthood occurred at 30, that he had been in Ireland more than 23 years). Rejection of that letter elicited the letter of excommunication we know as the *Epistola*. As Patrick states in it *Non usurpo* "I am not claiming too much," one infers that his critics believed he was exceeding the limits of his authority. His attempt to excommunicate from Ireland a tyrant in Britain may have provoked the attack that he relates at the thematic crux of the *Confessio*, an attack on his status as bishop when he was at least 51, in his senectus "old age" (which began after 42) by ecclesiastical *seniores* "elders" in Britain who tried him during his absence. They charged him with a sin, committed when he was 14, confessed at least 7 years later, after escaping from Ireland, before becoming a deacon. The sin was revealed by the *amicissimus* "dearest friend" to whom he had confessed it, the man whose statement *Ecce dandus es tu ad gradum episcopatus* "Behold, you are bound to be appointed to the grade of bishop" stands at the symmetrical center of the *Confessio*.

Although modern scholars have supposed that Patrick was poorly educated, a barely literate rustic who struggled to express himself in a language he could not master, his two extant Letters are, not by Ciceronian standards, but certainly by Biblical standards, masterpieces. If Patrick was a *homo unius libri* "a man of one book," that book was the Latin Bible, which he quoted both economically and brilliantly, using its phrases to claim identity of his vocation and mission with those of the Lawgiver Moses and the Apostle Paul, relying upon readers' knowledge of the unquoted contexts of his quotations and allusions to clarify his explicit meanings, to suggest implicit overtones and undertones, and to attack his critics. To establish his literary credentials he composed in Ciceronian clausular rhythms, which he arranged by type, only in the paragraph of the *Confessio* in which, addressing *domini cati rethorici* "lords, skilled rhetoricians," he appears to proclaim his ignorance. Elsewhere his cursus rhythms, like his Biblical orthography, diction, and syntax, are faultless. His prose, arranged *per cola et commata* "by clauses and phrases," exhibits varied forms of complex word play. Every paragraph is both internally coherent and bound in larger patterns within comprehensively architectonic compositions, in which every line, every word, every letter has been arithmetically fixed. His prose consistently evokes Biblical typology, an effective means of linking the events of his personal life with sacred and universal history.

Patrick nowhere states that he brought any ecclesiastical assistants with him from Britain, but he affirms repeatedly that he is a bishop in Ireland, referring often to those converted, baptized, confirmed, ordained as clerics, and admitted to the religious life as both monks and nuns in Ireland. He never describes his education, nor does he name any authoritative teacher or ecclesiastical patron. In stating at the beginning of the *Epistola* that he is *indoctus* he does not lament that he is "unlearned;" rather he proclaims that he is "untaught" by men, and he continues directly *Hiberione constitutus episcopum me esse fateor. Certissime reor a Deo accepi id quod sum* "established in Ireland I confess myself to be a bishop. Most certainly I think I have received from God what I am." He mentions his dealings with Irish kings, *praemia dabam regibus* "I habitually gave rewards to kings," with the sons of kings in his retinue, *dabam mercedem filiis ipsorum qui mecum ambulant* "I habitually gave a fee to the sons of the same [kings] who walk with me," with the lawyers or brehons *qui iudicabant* "who customarily judged," to whom he distributed *non minimum quam pretium quindecim hominum* "not less than the price of fifteen men," with noble women, *una benedicta Scotta genetiua nobilis pulcherrima adulta erat quam ego baptizaui* "there was one blessed Irish woman, born noble, very beautiful, an adult whom I baptized," and with others *quae mihi ultronea munuscula donabant et super altare iactabant ex ornamentis suis, et iterum reddebam illis* "who habitually gave to me voluntary little gifts and hurled them upon the altar from among their own ornaments, and I habitually gave them back again to them."

Though Patrick mentions no absolute date, he makes it abundantly clear that the milieu in which he lived and worked was late-Roman and post-Roman Britain and Ireland of the fifth century. From at least the sixth century onward Patrick has been revered as the effective founder of the Church in Ireland, celebrated in the panegyric "Saint Sechnall's Hymn" *Audite Omnes Amantes Deum* composed probably during the fourth quarter of the sixth century and quoted during the seventh. Patrick is cited as *papa noster* "our father" in Cummian's Letter about the Paschal controversy written in the year 633 to Ségéne, Abbot of Iona, and the Béccán the Hermit. There are references to three lost Lives of Patrick written by Bishop Columba of Iona, Bishop Ultán moccu Conchobuir of Ardbraccan, and Ailerán the Wise, Lector of Clonard by the middle of the seventh century. From the end of the seventh century a hagiographic dossier in support of the metropolitan claims of the church at Armagh includes memoranda, *Collectanea*, by Tírechán, a pupil of Bishop Ultán, and a *Vita* by Muirchú moccu Machthéni, a pupil of Cogitosus of Kildare. By the end of the eleventh century or the beginning of the twelfth there were four additional *Vitae*.

The *Synodus Episcoporum* or "First Synod of Saint Patrick," extant in a single manuscript copied from an Insular exemplar and written at the end of the ninth century or the beginning of the tenth in a scriptorium under the influence of Tours, may have issued from a synod between 447 and 459 by the missionaries Palladius, Auxilius, and Isserninus, the former sent in 431 by Pope Celestine *ad Scotos in Xpistum credentes* "to the Scots [i.e., Irish] believing in Christ," the text attracted to the Patrician dossier by propagandists at Armagh in the seventh century.

Patrick is commemorated on March 17.

David Robert Howlett

References and Further Reading

Bieler, Ludwig, ed. *Libri Epistolarum Sancti Patricii Episcopi*, introduction, text, and commentary. In *Classica et Mediaevalia* XI (1950), XII (1951). Irish Manuscripts Commission. Dublin: Stationery Office, 1952. Reprinted as *Clavis Patricii* II, Royal Irish Academy, *Dictionary of Medieval Latin from Celtic Sources, Ancillary Publications IV* Dublin: RIA, 1993.

———"The Hymn of St. Secundinus." *Proceedings of the Royal Irish Academy LV C*. (1952–1953), pp. 117–127.

———. *The Works of St. Patrick, St. Secundinus Hymn on St. Patrick, Ancient Christian Writers XVII*. London: Longmans, Green & Co. and Westminster MD: Newman Press, 1953.

———. *Four Latin Lives of St. Patrick, Scriptores Latini Hiberniae VIII*. Dublin: The Institute for Advanced Studies, 1971.
———. *The Irish Penitentials: Scriptores Latini Hiberniae V*, 54–59. Dublin: The Institute for Advanced Studies, 1975.
———. *The Patrician Texts in the Book of Armagh, Scriptores Latini Hiberniae X*. Dublin: The Institute for Advanced Studies, 1979. Devine, Kieran. *A Computer-Generated Concordance to the Libri Epistolarum of Saint Patrick, Clavis Patricii I*, Ancillary Publications III. Dublin: RIA DMLCS, 1989.
Gwynn, John, ed. *Liber Ardmachanus. The Book of Armagh*. Dublin: 1913.
Howlett, David Robert, ed. *Liber Epistolarum Sancti Patricii Episcopi, The Book of Letters of Saint Patrick the Bishop*. Dublin: Four Courts Press, 1994.
———. "*Synodus Prima Sancti Patricii*: An Exercise in Textual Reconstruction." *Peritia XII* (1998), 238–253.
The Bishops' Synod, ARCA Classical and Medieval Texts Papers and Monographs I. Liverpool: Francis Cairns, 1976, pp. 1–8, facsimile pp. 65–75.
Lapidge, Michael, and Richard Sharpe. *A Bibliography of Celtic-Latin Literature 400—1200*, Ancillary Publications I. Dublin: RIA DMLCS, 1985, nos. 25-26, pp. 9-11.
Orchard, Andy, ed. *Audite Omnes Amantes*: A Hymn in Patrick's Praise. In *Saint Patrick, A.D. 493–1993*. Woodbridge: Boydell, 1993, pp. 153–173.

See also **Christianity, Conversion to; St. Patrick's Cathedral**

PENITENTIALS

Strictly speaking, a "penitential" is a *libellus* (small book) designed for pastoral use covering every kind of misbehavior that Christians consider "sinful" (i.e., offensive to God in contrast to a breach of legal requirements), arranged within a specific theological framework, specifying detailed amounts of penance as remedies. In this sense few penitentials with Irish links have survived: four in Latin (those of Finnian (6th century); Columbanus (6th–7th century); Cummean (7th century); and the Bigotian (8th century)) and one in Irish (before late 8th century). It is clear from surviving texts that these are only a fraction of the number that were compiled or used in Ireland. The term is, however, applied more widely to cover a range of early medieval legal texts which make prescriptions, regarding sinful acts, using the pattern found in penitential *libelli* (e.g., the *Canones Hibernenses*). The term is also used more loosely for the system of Christian penance, usually with the gloss that it emerged in Ireland, which was used in the West between the disappearance of "public penance" and the appearance of individual "confession."

By the fifth century, Latin Christianity had developed a practice with regard to "sins committed after baptism" known as "public penance" (admission of the faults to the bishop followed by public separation within the community) which applied to the "greater sins:" murder, apostasy, and fornication. This practice was a failure. And, that failure was compounded by the theological justifications made in its defense (e.g., by Jerome and Augustine) that it was a "laborious baptism" available only once in the Christian's life and was to be truly difficult. The practice invoked a notion of sin as a crime deserving divine retribution where "doing the penance" was simply the sinner applying this punishment to themself. While several writers (e.g., Caesarius of Arles (*c.* 470–542)) pointed out that the whole system was a pastoral disaster, such voices went unheeded for fear of breaking faith with the past and its eminent supporters. Moreover, the system did not take account of the everyday sins, nor link the notion of penance for sins with the "doing of penitence" (*cf.* Mt 3:2 as found in Latin) preached as a basic part of Christian living.

Where in the British Isles the break with that practice was made is not clear (some of the earliest penitential-like texts have titles that link them to sub-Roman Britain: e.g., the "Synod of the Grove of Victory," and such legislation supposes the theoretical understanding that only a full penitential could supply), but the oldest extant formal penitential is by Finnian. The penitentials present a new view of (1) a sinful offense's nature, (2) of the purpose of doing penance, and, (3) a new understanding of religious culpability. In contrast to the notion of a crime demanding a punishment—an assumption in Roman law—they adopt a notion of crime that closely resembles the system of debts found in Brehon Law whereby a crime, for example, homicide, produced a debt for the murderer to the dead person's family which had to be repaid, and the size of the fine varied with the gravity of the action, the status of the offender and the offended, and the intention of the offender. Thus any sinful act's penitential "loading" depended on the action (e.g., homicide is worse than theft), the actor (e.g., a cleric is more culpable), the one offended—if this is applicable (e.g., stealing from a church is worse than other thefts), and with what intention (e.g., by accident or neglect, or in hot temper, or cold-bloodedly). So just as a crime against another person produced a debt, so a crime against God produced a debt that could be worked-off (the system inherently allowed for the repetition) with suitable religious payments of prayer, fasting, and alms (*cf.* Mt 6:2–18). The other key element in the penitentials' understanding of sinfulness is that penance is not seen as retribution, but therapy; while sin is viewed as a symptom of sickness rather than a manifestation of evil. This derives from John Cassian (*c.* 360–435) whose writings form the basis of western monasticism. Cassian saw sinful acts as expressions of eight underlying vices (called "principal" as they are the *principia* (sources) from which sins flow) imagined as chronic illnesses deep within individuals and requiring suitable medicines prescribed by a physician. The monastery was the place where these received chronic therapy following a dominant assumption of late antique medicine: "contraries heal contraries"—just as, for instance, the physical illness of fever needs cold, so the spiritual illness of gluttony requires fasting. Thus in extant *libelli* even when Cassian is not quoted, medical language is applied to the reconciliation process, and the sins are arranged systematically under the vices which produce them. Since the actual practice depends on this theological underpinning, legal texts with penitential-like materials should not be seen as proto-penitentials, as often happens, but as legal supplements to an established system of penitence based in the use of penitentials.

The penitentials' originality lay in extending to everyone and every action, however serious, a method for helping monks overcome ongoing imperfections. This avoided the problems of public penance in being repeatable and linking penance for sins with everyday penitential practice. There is no evidence that new discipline met resistance in Ireland, but the system did encounter some resistance among Anglo-Saxon clergy, and later much sterner opposition among Frankish clergy. However, the practice gradually gained ground probably due to its pastoral practicality, and left a complex legacy to the Western church: it generated an increasing awareness of the place of internal contrition and conscience in sin, it led directly to the development of indulgences whereby one penance was replaced by another of equal worth but with less physical demands, and it provided the practical—and some of the theological—background to all later Western systems of penance (e.g., what the twelfth-century canonists called the "sacrament of penance").

THOMAS O'LOUGHLIN

References and Further Reading

Bieler, Ludwig, ed. *The Irish Penitentials: Scriptores Latini Hiberniae V.* Dublin: The Dublin Institute for Advanced Studies, 1975.

McNeill, John T., and Gamer, Helena M. *Medieval Handbooks of Penance*. New York: Columbia University press, 1938.

O'Loughlin, Thomas. *Celtic Theology: Humanity, World and God in Early Irish Writings*. London: Continuum, 2000, pp. 48–67.

O'Loughlin, Thomas. "Penitentials and Pastoral Care." In *A History of Pastoral Care*, edited by Gillian R. Evans, 93-111. London: Cassell, 2000.

O'Loughlin, Thomas, and Conrad-O'Briain, Helen. "The 'Baptism of Tears' in Early Anglo-Saxon Sources." *Anglo-Saxon England* 22 (1993): 65-83.

See also **Adomnán; Biblical and Church Fathers Scholarship; Brehon Law; Canon Law; Columbanus; Christianity, Conversion to**

PEREGRINATIO

The classical meaning of *peregrinus* is "stranger"; in the Middle Ages the term was used to express the concept of "pilgrim." While the stranger was by definition a person without legal standing, the pilgrim, in principle at least, was to enjoy a privileged status. In Irish society *peregrinatio* stood for an ascetic Christian ideal in the form of a self-imposed, life-long exile in pursuit of the personal salvation in a life lived according to Christ's commands.

This *peregrinatio* comes into view first with Columbanus who left Ireland in 590 and lived for the rest of his life in Gaul and Italy. He died November 23, 615, in Bobbio (diocese of Piacenza). In his own writings the term *peregrinus* does occur without being further defined, but the concept of *peregrinatio pro Christo* can be deducted from the corpus of his writings as a whole, and in particular from his *Instructiones*. However, the concept is clearly articulated in the *Life of Columbanus*, written by Jonas of Bobbio one generation after the saint's death. This shows that the basis of Columbanus's Christian concept was cherished in Bobbio beyond his death. According to Jonas, Columbanus was confronted with the concept of two degrees of *peregrinatio*, a lesser one practiced in the form of a self-chosen exile within Ireland and a stronger one by leaving Ireland. (I, 3). Jonas also reports that Columbanus successfully resisted plans to have him brought back to Ireland when he was forced to leave his monasteries in Burgundy circa 610.

Contrary to widespread views the Irish *peregrinatio pro Christo* was not connected with missionary intentions but remained the pursuit of personal salvation. This emerges clearly in Columbanus's letter to Frankish bishops of circa 602 when he refused to attend a synod and instead expressed his wish for his community to be left alone to mourn their dead brethren in the wilderness (Ep. 2 p. 16: *mihi liceat cum vestra pace et caritate in his silvis silere et vivere iuxta ossa nostrorum fratrum decem et septem defunctorum* "that I may be allowed with your peace and charity to enjoy the silence of these woods and to live beside the bones of our seventeen dead brethren"). In Lombard, Italy, Columbanus did preach Catholicism against the Arians, and he wrote a treatise on the subject (which has not survived). He did this at the request of the king, and it would appear that this was the price he had to pay for the permission to settle there.

It is most likely that the Irish concept of *pererinatio pro Christo* was developed under the inspiration of Irish secular law (not yet written) which knew two groups of foreigners, one within Ireland and one from overseas. The two classes of "foreigners" implied different status.

After Columbanus there were some more Irish Christians who pursued the *peregrinatio pro Christo*, such as Fursa or Cellach who settled in Peronne in Picardy. (It is not clear whether the *Schottenklöster* from the eleventh century onwards on the continent can be taken as expressions of this ideal.) Contrary to the widespread view which finds some apparent support in Adomnán's Vita Columbae, Colum Cille, abbot of Iona (d. 597), was not a representative of this ideal because he visited Ireland after he had settled in Iona. While the number of Irish *peregrini pro Christo* was small, their exemplary lifestyle proved to be inspiring on a large scale and was responsible for the enormous influence of Irish spirituality on early continental Christianity, including the system of penance. The question remains whether it seemed impossible for radical Irish Christians to live such a radical Christian life at home.

MICHAEL RICHTER

References and Further Reading

Charles-Edwards, T. M. *Early Christian Ireland.* Cambridge: 2000.

Richter, M. *Ireland and Her Neighbours in the Seventh Century.* Dublin: 1999. (both with further references)

See also **Columbanus**

PERSONAL NAMES

The earliest personal names are attested in inscriptions carved on stone in the Ogam alphabet and dating from the fifth to the seventh century. They may have served to memorialize or to mark boundaries. Some four hundred such names (normally in the possessive case) have been found; (e.g., CATABAR MOCO VIRICORB) ("[of] Cathbarr descendant of Fer Corb"). Their evidence, however, is not always trustworthy either because of difficulties in deciphering the readings or problems with the phonology of the forms. Additionally, since the geographical spread of the Ogam inscriptions is quite narrow its names may not be representative of the whole country.

The primary source of names are works from the period, notably genealogies, annals, secular tales, martyrologies and lives of Irish saints, and the poetry of the Bardic Schools. These works, however, tend to reflect names of personages belonging to the privileged classes, secular and ecclesiastical; very little is known about name-giving among the lower orders of society. In the secular genealogies of the early Irish ruling families over 12,000 names of people are listed, providing some 3,500 separate names. Yet even this abundance of evidence has its problems: many of the names occur only once, raising the possibility that some of them may have been invented by professional genealogists. Distribution is also limited: over 4,000 of the persons listed share a mere 100 of the names, the five most common being *Aéd, Eochaid, Fiachnae, Ailill*, and *Fergus*.

Given the patriarchal character of early Irish society and the ecclesiastical provenance of its written records, it is hardly surprising that women's names are poorly represented: some 100 in the genealogies and 300 more in a twelfth-century poem on famous women. Most of the names are rare and confined to the earliest period. Among the most common, well attested in tenth- to twelfth-century sources, were: *Aífe*, *Ailbhe*, *Áine*, *Cacht*, *Eithne*, *Mór*, *Gormlaith*, and *Órlaith*. A few names could serve for either men or women, such as *Ailbhe*, *Cellach*, *Columb*, *Flann*, and *Medb.*

Morphologically, in both men's and women's names four types of formation are evident: (1) simple, uncompounded names, many of which are identifiable with nouns or adjectives of known meaning, for example *Áed* ("fire"), *Art* ("a bear"), *Donn* ("brown"); (2) derived names, formed by adding to an existing word a diminutive, adjectival, or agency ending, such as, *Aéd-án* ("little Aéd"), *Dún-amail* ("like a fort"), and *Mucc-aid* ("a keeper of pigs"), respectively; (3) close compounds consisting of combinations of noun and adjective elements; for example, *Fer+gus* (noun+noun), *Barr+fhinn* (noun+adjective), *Cóemgen* (adjective+noun), *Findchaem* (adjective+adjective); and (4) loose compounds, consisting of noun/substantival adjective+adjective (*Cú Buide*) or proper name (*Cú Chulainn*) or noun (*Donn Bó*). The fourth type, although apparently of non-Indo-European origin, became the most dynamic source of new names from the seventh century on, vastly outnumbering the other three.

Informal varieties of personal names abound, especially pet (hypocoristic) names and nicknames. The rules for hypocoristic formations have not yet been fully elucidated though some patterns are evident. The most common involves the shortening of the normal form to a single root syllable ending in a doubled consonant followed by *a*/*e*; for example *Diarmait* becomes *Dímme* and *Colmán* becomes *Conna* (compare Modern English "Samuel" and "Sammy"). In another type the shortened form is preceded by *mo* ("my") or *to* ("your") in the vocative, a type of affectionate naming used for monastic saints; such as *Mo Chumma* (<*Colmán*) and *To Lua* (<*Lucaill*). Also of monastic provenance are pet names in *–óc*, a formation borrowed from British; for instance *Mo Chíaróc* or *Tu Medóc*. Nicknames are more problematic since it is not always possible to tell in individual cases whether a specific emphasis on the name's meaning was intended. Nevertheless, certain names that denote prominent physical or psychological features suggest ultimate origins as nicknames; such as *Becc* ("the small one"), *Baíthéne* ("the little simpleton"), and *Lonngargán* ("the fierce and eager one"). Such nicknames could also be added to a personal name as a soubriquet; for example, *Conán Maol* ("C. the bald") or *Domnall Rua* ("Domnall with the red hair").

The presence of other languages and cultures in medieval Ireland gave rise to the borrowing of foreign names (see Languages). As already noted, British (Welsh) influence is evident in the formation of pet names. Conversion to Christianity led to borrowing from early missionaries of such names as *Pátraic* (<Lat. *Patricius*) and *Sechnall* (<Lat. *Secundinus*), while the ecclesiastical culture that they introduced provided biblical names and names of foreign saints; such as *Aindrias*, *Martan*, and *Petar*. From Anglo-Saxon derived names such as *Conaing* (<OE *cyning*) and *Éamonn* (<OE *Eadmund*); and from Scandinavian *Amlaíb* (<ON *Óláfr*) and *Ímar* (<ON *Ívarr*). But the most prominent group of borrowed names came from Norman French, many of which were probably introduced as a result of the Norman practice of conferring saints' names at baptism. Popular women's names from French were *Caitilín* (<*Cateline*), *Máire* (<*Marie*), and *Nóra* (<*Honora*); and among male names, *Seaán* (<*Jehan*), and *Séamus* (<*Jacobus*).

The most significant innovation in personal names was the introduction of surnames. In the early medieval period an individual of note, say *Colmán*, could be further identified by reference to his sept or his father; thus, *Colmán mac Rímedo* (C. son of Rímed). But by the tenth century the formula "X son of Y" was undergoing a change in function whereby its "son of Y" element now indicated a surname. The litmus test for such surnames is that the *mac* element no longer has its literal meaning of "son." Thus the obit of Dermot mac Murrough (+1171), identifies him as *Diarmait Mac Murchada*, where the "Mac M." element is not literally true since Dermot's father was called Donnchad. Likewise the other common surname prefix, *Ua* (later *Ó*), "grandson," introduces a surname in the name *Comaltán Ua Cléirig* (+980) whose actual grandfather was Máel Fábaill. Of the two prefixes, *Ua* is older, the *Mac* element becoming popular in the late twelfth century, a development which may be connected with the breakdown of Gaelic family structures following the Anglo-Norman invasion. Much work remains to be done on archiving and classifying the personal names of medieval Ireland.

PÁDRAIG Ó NÉILL

References and Further Reading

O'Brien, M.A. *Corpus genealogiarum Hiberniae*. Dublin: Dublin Institute for Advanced Studies, 1962.

——— (ed. R. Baumgarten). "Old Irish Personal Names." *Celtica* 10 (1973): 211–36.

Ó Corrain, Donnchadh, and Maguire, Fidelma. *Gaelic Personal Names*. Dublin: The Academy Press, 1981.

Ó Cuív, Brian. "Aspects of Irish Personal Names." *Celtica* 18 (1986): 151–84.

Russell, Paul. "Patterns of Hypocorism in Early Irish Hagiography." *In Studies in Irish Hagiography: Saints and Scholars*, edited by John Carey et al., 237–49. Dublin: Four Courts Press, 2001.

See also **Annals and Chronicles; Christianity, Conversion to; Genealogies; Hagiography and Martyrologies; Languages**

PILGRIMS AND PILGRIMAGE

Because of minimal documentation, we often know little more about medieval Irish pilgrims than their names and where they died, though their activity must have played an important role in the religious life of Ireland during the Middle Ages. St. Colum Cille was among the first of the Irish *peregrini* to leave their native country and go on pilgrimage abroad, and he was followed by many others who traveled to the European continent, spreading the faith and seeking a higher place in heaven. Their numbers declined after bishops ordained in Ireland were banished from the Carolingian Empire at the Council of Châlons-sur-Saone in 813, after which the Irish Church appears to have encouraged more pilgrimage at home. Nevertheless, individual pilgrims continued to go to Rome throughout the medieval period, the reasons for doing so summed up by the words of Celedabhaill, abbot of Bangor, quoted in the Annals of the Four Masters under the year 956. There were doubtless more pilgrims to Rome than the Irish kings and clerics listed in the Annals, as evidenced by the thirteenth-century pewter pilgrim badge bearing images of SS. Peter and Paul found in the Old Dublin excavations, which also produced an inscribed ampulla of the same period probably brought back as a souvenir from the shrine of St. Wulfstan in Worcester.

Historical sources record Irish pilgrims journeying to Santiago de Compostela from the thirteenth to the fifteenth century, but art historical evidence could push this back to the twelfth. There are fifteenth-century Irish carvings representing St. James, with his stick, wallet, and shell. A pilgrim shell was found in a medieval burial in Athenry, County Galway, and other pilgrimage souvenirs are also known.

By far the most important place of pilgrimage in Ireland during the Middle Ages was St. Patrick's Purgatory in Lough Derg, where a Welsh knight Owein claimed to have seen the torments and joys of the Otherworld in an island cave there around 1140. His story spread quickly and brought pilgrims to the small Donegal lake from many parts of Europe, from Hungary to Catalonia, many of whom wrote down their experiences—their texts usefully assembled by Shane Leslie in 1932. The cave was closed in 1632, but not before many had tried to follow in Owein's footsteps. Pilgrimage to the island continues today, as it does to another Patrician site on Croagh Patrick, County Mayo. In Lough Derg, Patrick's cult overshadowed that of a local saint Daveoc and, in a similar fashion, Brendan seems to have eclipsed St. Malkedar in the pilgrimage to Mount Brandon. This latter seems to have been sea-based, with pilgrims probably traversing the western Atlantic seaboard, as mirrored in the *Navigatio Brendani*.

Medieval Irish pilgrimage must have been undertaken for various reasons (e.g., penance, fulfilment of vows, to cure sickness, and save one's soul), and have taken a variety of forms. Patterns to holy wells—still practiced today—would have been the small-scale manifestation. But more important were opportunities to venerate saints at the larger monasteries they had founded, where their relics were enshrined in metal reliquaries, not all as large as the sarcophagi containing the remains of SS. Brigid and Conlaed in front of the high altar at Kildare, as described by Cogitosus. Island sanctuaries, both sea and inland, were evidently popular pilgrimage sites (e.g., Skellig Michael, the Aran Islands, Inishmurray, Co. Sligo, Inishcealtra, Co. Clare, and Monaincha, Co. Tipperary).

The first of a number of pilgrim deaths in Ireland was recorded at Clonmacnoise in 606 and, to judge by annalistic entries, they—and pilgrimage activity in Ireland generally—would seem to have reached their peak in the twelfth century. But many of the pilgrimage sites, particularly in Gaelic Ireland, continued in use for hundreds of years, as we can see from the places visited by Heneas Mac Nichaill in 1543 to expiate the sin of having murdered his son.

Small boulders with man-made holes for water, called bullauns, are often found on pilgrimage sites, and beehive huts may have served as pilgrim shelters. Some old pilgrim paths are known (e.g., Cosán na Naomh in the Dingle Peninsula, St. Kevin's Way from Hollywood to Glendalough, Ballintubber Abbey to Croagh Patrick, Lemonaghan, Co. Offaly and Saints Island, Lough Derg), many of which have recently been revitalized for walking. Figures on cross-decorated stones at Carndonagh, County Donegal and Ballyvourney, County Cork, may represent early medieval Irish pilgrims.

PETER HARBISON

References and Further Reading

Harbison, Peter. *Pilgrimage in Ireland. The Monuments and the People*. London and Syracuse, N.Y.: 1991.

Hughes, Kathleen. "The Changing Theory and Practice of Irish Pilgrimage." *Journal of Ecclesiastical History* 11, 1960, 143–51.

Leslie, Shane. *Saint Patrick's Purgatory, A Record from History and Literature*. London: 1932.

Stalley, Roger. "Maritime Pilgrimage from Ireland and its Artistic Repercussions." In *Actas del II Congreso Internacional de Estudios Jacobeos, Ferrol, Septiembre 1996*, Vol. II, 255–75. Santiago: 1998.

PLACENAMES

The majority of Irish placenames have their origins in the Irish language and accordingly constitute a very considerable body of valuable linguistic data. Most names in an English language context are written in an Anglicized orthography that is often a fairly thinly-disguised version of the original Irish form. Other languages represented in the body of Irish toponymy include Latin, Norse, Norman French, and, of course, English, while there is a handful of names that may arguably be of pre-Celtic origin. Placenames of such diverse origins reflect the complexity of Ireland's linguistic history. Some names that are at least partly Latin in origin date from the early Christian centuries, while from the Viking era, the ninth and tenth centuries, we have (in contrast to the Highlands and Islands of northern and western Scotland) a remarkably small number of Norse names, almost all of them on or near the eastern and southern coasts. Comparatively few names of indisputably Norman-French origin survive. As one might expect, English has had a significant impact on Irish nomenclature, although purely English names are considerably less numerous than Anglicized or hybrid English-Gaelic name-forms.

The body of Irish toponymy may be likened to a pyramid. Starting at the top, there is the name of the island (from the prehistoric form *Iuerne* through *Ériu* in Old Irish to *Éire* in Modern Irish); below this are, in turn, the early bifold division of the island into Leth Cuinn and Leth Moga ("Conn's half" and "Mug's half," respectively); the "fifths" or provinces whose roots go back to prehistoric times; the counties, 32 of which were established between the early post-Norman period (c. A.D. 1200) and the beginning of the seventeenth century; the baronies, of which there are 324, some of them representing ancient Gaelic divisions; the civil parishes, some 2420, reflecting medieval ecclesiastical parishes that may have been established at the same time as the dioceses in the twelfth century, although many have roots going back much further; and finally, at the bottom of the pyramid, is the smallest administrative unit, that peculiarly Irish division, the townland, of which there are more than 60,000, ranging in area from less than one acre to more than 7,000 acres. Below this level is a vast body of microtoponymy, much of which has never been recorded, let alone printed on maps, and is therefore in imminent danger of being lost forever; such names are most numerous in areas in which Irish is, or was until recently, spoken.

A few dozen Irish placenames can be traced back almost two millennia to the work of the Alexandrian geographer Ptolemy, although identifying many of these with certainty is problematical. Many others occur in a wide range of documents from early to late medieval times, but the majority of names are attested in documents from the later sixteenth century onwards. The corpus of Irish placenames reflects such things as natural features, flora and fauna, land divisions, settlement patterns (secular and ecclesiastical), ways of life and death, land clearance and cultivation, historical events, and religious and mythological matters, as well as land ownership (through inclusion of personal and family names).

The scholarly study of Irish placenames has its roots in the work of the Ordnance Survey, established in Ireland in 1824. The Survey's first superintendent, Thomas Larcom, had the foresight to employ the young Kilkenny scholar John O'Donovan to assist with processing the thousands of placenames that any scheme to map the entire country would encounter. O'Donovan—later joined in the Survey's Topographical Department by George Petrie, Eugene O'Curry and others—went on to become the greatest Irish scholar of the nineteenth century. Most of his early scholarly work was related to the collection and interpretation of tens of thousands of placenames, many of them to be found only on the lips of native speakers of Irish, never having previously been written down. O'Donovan's pioneering work on a great range of sources, most of them lying unpublished in medieval manuscripts, led to his masterly editions of various Irish texts, most notably the *Annals of the Four Masters*.

Further valuable work on Irish placenames—though largely based on O'Donovan's researches—was done by Patrick Weston Joyce, while early in the twentieth century Edmund Hogan, SJ, produced his remarkable *Onomasticon Goedelicum*, a dictionary of placenames culled mainly from medieval sources (many of them still in manuscript). Other noted twentieth-century contributors to Irish toponymical studies included Canon Patrick Power on the placenames of Decies, County Waterford, and of east Cork, Pádraig Ó Siochfhradha ("An Seabhac") on those of the barony of Corkaguiny, County Kerry, Fr. Paul Walsh on County Westmeath, and Liam Price on the placenames of County Wicklow.

In 1946 the Irish government established the Irish Placenames Commission to further scholarly research into Irish placenames, in order to furnish authoritative Irish forms for public use. Work commenced on the names of more than 3,000 postal towns throughout Ireland, resulting, eventually, in the book *Ainmneacha Gaeilge na mBailte Poist* (1969)—later incorporated in the *Gazetteer of Ireland/Gasaitéar na hÉireann* (1989). In 1955, the Commission's research functions were transferred to the newly established Placenames' Branch of the Ordnance Survey, while the Commission retained an overseeing and advisory role in relation to the Branch's activity. The Branch commenced work in the early 1970s on the names of townlands, on a county-by-county basis. This led to a book on the placenames of County Limerick (1991) and bilingual lists of the townland and other names of six counties (Limerick, Louth, Waterford, Kilkenny, Monaghan, and Offaly—with Dublin, Tipperary, and Galway soon to follow). In 1999 the Placenames' Branch was detached from the Ordnance Survey; it now operates under the Department of Community, Rural and Gaeltacht Affairs. Meanwhile, in 1987, with British government funding, the Northern Ireland Place-Name Project was established in Queen's University Belfast to undertake similar work to that of the Placenames' Branch in relation to the six counties of Northern Ireland. Part of the fruits of the Project's researches was published in seven substantial volumes before official funding ceased in 1997. Since then, with alternative (but much reduced) funding, a smaller staff has continued the research and produced three further volumes. Valuable work is also being done on the corpus of earlier Irish placenames by the LOCUS Project which was established in 1996, with a grant from Toyota Ireland, Ltd., in the Department of Early and Medieval Irish, University College, Cork. Representing the initial instalment of a detailed revision of Hogan's *Onomasticon*, the first fascicle of the new *Historical Dictionary of Gaelic Placenames* appeared in 2003.

NOLLAIG Ó MURAÍLE

References and Further Reading

Andrews, John H. *A Paper Landscape: The Ordnance Survey in Nineteenth-Century Ireland*. Dublin: Four Courts Press, 2002. [Reprint of original Oxford University Press edition, 1975.]

Brainse Logainmneacha na Suirbhéireachta Ordanáis. *Gasaitéar na hÉireann / Gazetteer of Ireland*. Dublin: Oifig an tSoláthair, 1989.

———. *Liostaí Logainmneacha: Lú/Louth*. Dublin: Oifig an tSoláthair, 1991.

———. *Liostaí Logainmneacha: Luimneach/Limerick*. Dublin: Oifig an tSoláthair, 1991.

———. *Liostaí Logainmneacha: Port Láirge / Waterford*. Dublin: Oifig an tSoláthair, 1991.

———. *Liostaí Logainmneacha: Cill Chainnigh/Kilkenny*. Dublin: Oifig an tSoláthair, 1993.

———. *Liostaí Logainmneacha:Uíbh Fhailí/Offaly*. Dublin: Oifig an tSoláthair, 1994.

———. *Liostaí Logainmneacha: Muineachán/Monaghan*. Dublin: Oifig an tSoláthair, 1996.

Flanagan, Deirdre, and Flanagan, Laurence. *Irish Place Names*. Dublin: Gill and Macmillan, 1994.

General Alphabetical Index to the Townlands and Towns, Parishes and Baronies of Ireland—Based on the Census of Ireland for the Year 1851 [recte 1861]. Baltimore, Md: Genealogical Publishing Company, Inc., 1986. [Reprint of work published in 1861.]

Hogan, Edmund. *Onomasticon Goedelicum Locorum et Tribuum Hiberniae et Scotiae: An Index, with Identifications, to the Gaelic Names of Tribes and Places*. Dublin, 1910; reprint Dublin, Four Courts Press: 1993.

Hughes, A.J, and Hannan, R.J. *The Place-Names of Northern Ireland: Vol. 2: County Down II—The Ards*. Belfast: Institute of Irish Studies, 1992.

Joyce, Patrick Weston. *The Origin and History of Irish Names of Places*, I-III. Dublin: 1869, 1875, 1913. [Reprinted by Éamonn de Búrca, Dublin, 1995.]

Mac Gabhann, Fiachra. *The Place-Names of Northern Ireland: Vol. 7: County Antrim II—Ballycastle and North-East Antrim*. Belfast: Institute of Irish Studies, 1997.

McKay, Patrick. *The Place-Names of Northern Ireland: Vol. 4: County Antrim I—The Baronies of Toome*. Belfast: Institute of Irish Studies, 1995.

Muhr, Kay. *The Place-Names of Northern Ireland: Vol. 6: County Down IV—North-West Down/Iveagh*. Belfast: Institute of Irish Studies, 1996.

Ó Mainnín, Mícheál B. *The Place-Names of Northern Ireland: Vol. 3: County Down III—The Mournes*. Belfast: Institute of Irish Studies, 1993.

Ó Maolfabhail, Art, ed. *Logainmneacha na hÉireann, Iml. I: Contae Luimnigh*. Dublin: Oifig an tSoláthair, 1990.

———. *The Placenames of Ireland in the Third Millennium/ Logainmneacha na hÉireann sa Tríú Mílaois*. Dublin: Ordnance Survey/Placenames Commission, 1992.

Ó Muraíle, Nollaig. *Mayo Places: Their Names and Origins*. Dublin, 1985.

———. "The Place-names of Clare Island." In *New Survey of Clare Island. Volume I: History and Cultural Landscape*, edited by Críostóir Mac Cárthaigh and Kevin Whelan, 99–141. Dublin: Royal Irish Academy, 1999.

———. "Seán Ó Donnabháin, An Cúigiú Máistir." In *Scoláirí Gaeilge (Léachtaí Cholm Cille* XXVII), edited by Ruairí Ó hUiginn, 11–82. Maynooth: An Sagart, 1999.

———. "Settlement and Place-names." In *Gaelic Ireland c. 1350–1600: Land, Lordship and Settlement*, edited by Patrick Duffy, David Edwards, and Elizabeth FitzPatrick, 223–45. Dublin: Four Courts Press, 2001.

Ó Murchadha, Diarmuid, and Murray, Kevin. "Place-names." In *The Heritage of Ireland: Natural, Man-made and Cultural Heritage: Conservation and Interpretation, Business and Administration*, edited by Neil Buttimer, Colin Rynne, and Helen Guerin, 146–155. Cork: The Collins Press, 2001.

Ó Riain, Pádraig, Ó Murchadha, Diarmuid, and Murray, Kevin, ed. *Historical Dictionary of Gaelic Placenames. Fascicle I (Names in A-)*. London, Dublin: Irish Texts Society, 2003.

Oftedal, Magne. "Scandinavian place-names in Ireland." *In Proceedings of the Seventh Viking Congress, Dublin . . . 1973*, edited by Bo Almqvist and David Greene, 125–133. Dublin, 1976.

Price, Liam. *The Place-Names of County Wicklow*, I-VII. Dublin: The Institute for Advanced Studies, 1945–1967.

Toner, Gregory, and Ó Mainnín, Mícheál B. *The Place-Names of Northern Ireland: Vol. 1: County Down I—Newry and South-West Down*. Belfast: Institute of Irish Studies, 1992.

Toner, Gregory. *The Place-Names of Northern Ireland: Vol. 5: County Derry I—The Moyola Valley*. Belfast: Institute of Irish Studies, 1996.

See also **Annals of the Four Masters; Languages; Leth Cuinn and Leth Moga**

PLUNKETT

The Plunkett family arrived in Ireland with the Anglo-Norman conquest, and rose from relatively obscure beginnings to become one of the leading families of the Pale. A legend arose that they were of Danish origin and arrived in Ireland in the eleventh century, but the name is almost certainly derived from the French *blanchet* (from *blanc*, white). The Plunketts came to hold lands in Dublin, Meath, and Louth, but probably first settled near Dublin where a Walter Plunkett held property before his death circa 1270. Walter had a son John who joined the Franciscans in Dublin, and a grandson, also called John, who in 1336 was seized of holdings in Greenoge, County Meath and the manor of Clonaghlis in County Kildare.

For ambitious families of the fourteenth century, the route to riches lay through law, commerce, property acquisitions, and fortuitous marriages, and all these avenues were exploited by the Plunketts. A Thomas Plunkett of Louth was a chief justice of the common pleas in 1316, and his contemporary John Plunkett, the real founder of the family's future greatness, was a professional sergeant-at-law, representing litigants in the Dublin courts. John married Alice, granddaughter of Henry of Trim, who had been mayor of Drogheda in 1272. Through this marriage John inherited the manors of Redmore, Stachliban and Beaulieu, all held of de Verdon. John selected Beaulieu as his principal residence and had a new parish created for his church there.

As John's descendents prospered they put down roots throughout the Pale, giving rise to cadet branches; ultimately the family came to hold no less than three peerages. A number of fortunate marriages linked the Plunketts to the other chief families of the Pale, as well as leading to the acquisition of further lands. In 1432 Sir Christopher Plunkett was appointed deputy to the lord lieutenant, John Stanley, on his recall to England. Sir Christopher married Lady Joan Cusack, and a splendid (though badly-damaged) fifteenth-century tomb in Killeen church is probably their final resting place; in 1449 the Killeen Plunketts were ennobled. A grandson of Christopher inherited the Barony of Dunsany through his Cusack relatives. One of Christopher's younger sons, Edward Plunkett of Balrath, County Meath, was implicated in the quarrel between John Tiptoft and the earl of Desmond in 1468. On Tiptoft's orders he was arrested and flogged through the streets of Drogheda, but he avoided the executioner's axe, which dispatched the earl ten days later, and lived to serve as seneschal of Meath in 1472.

In 1484 Bishop Thomas Barrett, as envoy of Richard III, singled out Sir Oliver and Sir Alexander Plunkett for their valour in repelling the king's Irish enemies. But during the Lambert Simnel affair one of the Killeen Plunketts served in the rebel army, and Thomas Plunkett, chief justice of the common bench, only received pardon after lengthy pleading. After the Reformation the Plunketts remained one of the leading Catholic families of the Pale, counting the martyr Oliver Plunkett among their number.

James Moynes

References and Further Reading

Brand, P. "The Formation of a Parish; The Case of Beaulieu, Co. Louth." In *Settlement and Society in Medieval Ireland. Essays presented to F.X. Martin*, edited by J. Bradley, 261–76. 1988.

O'Reilly, M. "The Plunkett Family of Loughcrew." *Ríocht na Midhe* vol 1 no 4 (1958): 49–53.

See also **Pale, The**

POER

The surname Poer (also Poher, Puher: the earliest references omit the prefix *le*. The modern form is, of course, Power) seems to have denoted a native of Picardy in northern France, although a connection with the district of Poher in Brittany has also been suggested and a multiple origin is possible. A number of bearers of the name, members of a family associated with the de Courcys in Somerset and Devon, figured prominently in the invasion of Ireland from 1170 on. Robert Puher, a member of the household of King Henry II, whom he accompanied to Ireland in 1171, was appointed governor of Waterford in 1177. It is probably the same man who had acquired Dunshaughlin and Ratoath in County Meath before 1191. Others were Roger (killed 1188), William and Reginald Poer, while Simon le Poer was briefly (1185-90) lord of half the "kingdom of Cork" as husband of Milo de Cogan's daughter and heiress Margaret.

Descendants of some of these Poers certainly endured in County Kilkenny, and it is possible that Henry Poher, to whom King John granted the great barony of Dunoil (Dunhill) in County Waterford belonged to this family, but the fact that a charter of his begins with the formula "to all my men: French, English, Welsh and Irish" suggests rather an origin in the Welsh marches, where the surname also occurs. Henry's grandson, John fitz Robert le Poer of Dunoil (dead by 1242), acquired also lands in Limerick and Connacht, but the lineage remained overwhelmingly connected with County Waterford and the surrounding areas. John's son, another John, produced King John's charter in court in 1262, but unfortunately the text does not survive. By 1300 the Poers were one of the most numerous of the "Anglo-Norman" lineages and included a substantial criminal element. One branch bore the strange epithet of the "blackman" Poers, while conversely Sir John fitz William le Poer (died 1295) was known as "the white Poer." He founded the County Cork branch of the family. In the 1320s the family were involved in a bitter feud with Maurice fitz Thomas, first earl of Desmond. The direct line of the barons of Dunoil comes to an end shortly before 1360, an event followed by bitter internal feuds within the lineage.

Sir Eustace fitz Benedict le Poer (died 1311) the younger son of a junior branch, was a remarkable self-made man who, having married a rich widow, built up an enormous landed estate, including the great barony of Kells in County Kilkenny. He died childless, having divided up his lands among his kinsmen. His nephew, Sir Arnold fitz Robert, seneschal of Kilkenny, died in 1328 in Dublin Castle, where he had been imprisoned on a charge of heresy through his involvement in the famous Kilkenny witchcraft case. His son, another Sir Eustace, having taken part in the rebellion of his family's former enemy, the earl of Desmond, was captured in County Kerry in 1346 by the chief governor, Sir Ralph Ufford, and executed, his lands being confiscated. Nicholas fitz John le Poer of Kilmeaden (died after 1393), the largest landowner of the lineage in his day, was a nephew or grandnephew of Sir Eustace fitz Benedict, some of whose lands he inherited. He was the ancestor of the later Kilmeaden line, who in the fifteenth century also obtained possession of Dunoil itself. His rivals for the leadership of the lineage were Richard fitz John le Poer (died. 1376) and his son David Rothe (*Ruad*, "the red"). Around this time commenced the long and bitter feud between the Poers and their neighbors, the citizens of Waterford, which was to continue for a century and a half.

David Rothe's son Nicholas, known patronymically as *Mac Daibhid Ruaid*, was appointed in 1425 to the sheriffship of County Waterford, an office that he and his descendants were, uniquely in Ireland, to convert into a hereditary lordship. He was a man of sufficient note for his death in 1446 to be picked up by an annalist in far-away Fermanagh. His son Richard (died 1483) succeeded him as sheriff, surviving a parliamentary attempt in 1476 by the citizens of Waterford to have him removed as a Gaelicized rebel, who used only "brehon" law. The sheriffship, converted into a local lordship over the eastern half of County Waterford, passed in turn to his son Piers and his grandson, another Richard (died 1539), who was raised to the peerage in 1535–1536 as Lord Power of Curraghmore, a title which remained with his descendants. His remarkable widow, Katherine Butler, was the last to exercise autonomous authority over "the Power Country."

Kenneth Nicholls

References and Further Reading

Mac Cotter, Paul, and Nicholls, Kenneth. "Sir John 'The White Poer.'" *The Pipe Roll of Cloyne*, 144–148., Cloyne: 1996.

See also **Anglo-Norman Invasion; Bermingham; Gaelicisation; Desmond, Fitzgeralds of; Mac Carthaig; Munster; Ormond; Waterford**

POETRY, HIBERNO-LATIN

Hiberno-Latin Poetry is one of the most extensive genres of Latin poetry to emerge from the West in the early Middle Ages. Some of it is of an incidental, spontaneous nature, but much of it is religious, devotional, or hagiographic. It can best be dealt with through its authors, since, unlike the bulk of Hiberno-Latin literature, much of it can be attributed (perhaps coincidentally) to named persons. The writing of Latin verse in Ireland extended into the late Middle Ages.

Columbanus is the earliest composer of Latin poetry from Ireland. The authorship of his works is controverted, especially a group of six metrical poems, which are disputed primarily because of their implications for a knowledge of Classical literature in the early Irish schools. The poems *In mulieres* and *Monosticha* are unlikely to be his, but the internal textual evidence of *De mundi transitu* clearly shows it to be the work of Columbanus. Some of the reasons recently adduced for eliminating the others or attributing them to Columbanus of St. Trond were subsequently refuted, but their authorship still remains in doubt.

Colmán moccu Cluasaig (d. *c.* 665) was abbot and *fer léigind* of the monastery of Cork. His most important composition, *Sén Dé don de for don te* ("God's blessing, bear us, succour us") was composed, according to the *Liber Hymnorum*, to avert the "Yellow Plague" of 664–665. It is one of the earliest pieces of macaronic verse in any western European vernacular, interspersing Latin phrases into an Irish adaptation of an early liturgical *ordo* for the dead. The list of Old Testament saints invoked, Abel, Elias, and so forth, betrays Eastern liturgical influence: nothing like it exists elsewhere in Western Europe at this early date.

Colmán is an otherwise unknown ninth-century author of two well-constructed poems in Vergilian hexameters on (1) a miracle of Brigit and (2) a farewell salutation to a younger namesake of his, another Colmán, on the eve of his return to Ireland (*Colmano versus in Colmanum perheriles*). One manuscript attribution names the author Colmanus "nepos Cracavist," a corruption of "ep(is)c(opu)s craxavit"—meaning. "Colmanus the bishop wrote (this)." The poem on St. Brigit (*Quodam forte die caelo dum turbidus imber*) relates a version of the story of her hanging her cloak on a sunbeam to dry, found also in a slightly variant version in Vita I of Brigit. It seems to have been written for someone who may have been composing a Life of Brigit. The second is an *envoi* to a younger compatriot returning home (*Dum subito properas dulces invisere terras*). It describes the dangers of the sea voyage ahead of his companion and the sorrow of their parting, and asks him to remember him, an old man. Both poems are full of classical reminiscences from Virgil and are good examples of Hiberno-Latin versecraft. There are some striking similarities between the poem to Colmán and the *Versus ad Sethum* attributed to Columbanus of Bobbio. The names of both writers are almost identical, and therefore easily confused, so that it is certainly possible that the ninth-century Colmán was the author of *Ad Sethum*, formerly attributed to Columbanus.

Donatus, bishop of Fiesole, is the author of an epic Latin poem in hexameters on St. Brigit, which drew on earlier lives of that saint by Ultán, Ailerán, and Cogitosus (qqv), as well as one "Animosus." It has been suggested that it was dedicated to the famous Dúngal of St. Denis/Pavia. The poem, of which over 2000 lines survive, is replete with classical references, which would indicate that his school had the facilities to teach classical poetry.

Sedulius Scottus, Irish scholar and poet at the court of Charles the Bald, was the most prolific and best Latin versifier of the mid-ninth century. He is perhaps best known for the 83 poems that he composed in a variety of Classical Latin meters for his patrons, friends, and colleagues, which rank him as the most skillful poet of his day. His poetry can still be appreciated for its inventive freshness, delicacy of sentiment and humor. Among the addressees are his patron, Hartgar of Liège, bishop Hilduin of Cologne, Eberhard of Friaul, and Hatto of Fulda.

"Hibernicus exul" is an anonymous late eighth- and early ninth-century Irish poet of the Carolingian Renaissance. His two main pieces are a panegyric on Charlemagne's victory in 787 over Tassilo, duke of Bavaria. The second is a poem in two parts, of praise for and admonition to his students (*Discite nunc, pueri*). He also wrote a piece for the imperial coronation of Charlemagne in 800. The relative freshness of his verse typifies the literary revival, which took place under Charlemagne, though his poetic craft is not otherwise of remarkable quality. His total output of 38 poems survives in a unique manuscript, Vatican, Bibl. Apostolica, Reg. lat. 2078 (saec. ix in).

After the Anglo-Norman invasion, new schools of Hiberno-Latin poetry emerged. Michael of Kildare, a Franciscan friar, is author of some poems in British Library, Harleian MS 913, written in the early fourteenth century. The manuscript is of Irish origin and contains a collection of poems in Latin (31) and English (17), written in an Irish Franciscan milieu. The collection also contains what has been described as the first Christmas carol in English.

Some of the earliest known English songs written by Richard Ledrede, Bishop of Ossory (1317–1360), are preserved in the *Red Book of Ossory*, where there are sixty Latin verses. The verses were written in about 1324 "for the Vicars Choral of Kilkenny Cathedral, his priests, and clerics, to be sung on great festivals and other occasions." The sixty pieces are in honor of Our Lord, the Holy Ghost, and the Blessed Virgin Mary, and the first of them is entitled: *Cantilena de Nativitate Domini*, a sort of Christmas Carol, followed by three others "*de eodem festo*."

AIDAN BREEN

References and Further Reading

Bernard, J.H. and Atkinson, R. ed. *The Irish Liber Hymnorum* (HBS 13, 14), 2 vols. London, 1898.

Colledge, Edmund, ed. *The Latin Poems of Richard Ledrede, O.F.M., Bishop of Ossory, 1317–1360*. In *The Red Book of Ossory, Studies and Texts 30*. PIMS, 1974.

Esposito, M. "The Poems of Colmanus 'nepos Cracavist'; and Dungalus 'praecipuus Scottorum.'" *Journal of Theological Studies* 33 (1932): 113–119;

Godman, P. *Poetry of the Carolingian Renaissance*. London, 1985.

Heuser, W. *Die Kildare-Gedichte: die ältesten mittelenglischen Denkmäler in Anglo-irischer Überlieferung*, xiv. Bonner Beiträge zur Anglistik, 1904.

Jacobsen, P.C. "Carmina Columbani." In *Die Iren und Europa im Früheren Mittelalter* I, edited by H. Löwe, 434–467, at 465–467. Stuttgart, 1982.

Kenney, J.F. *The Sources for the Early History of Ireland: 1 Ecclesiastical*. New York, 1929; reprinted Dublin, 1993.

Lapidge, M. & Sharpe, R. *A Bibliography of Celtic-Latin literature 400–1200*. Dublin, 1985.

Löwe, H., ed. *Die Iren und Europa im früheren Mittelalter*. Stuttgart, 1982. (esp. articles by K. Schäferdiek, P.-C. Jacobsen, and D.Schaller).

Manitius, M. *Geschichte der lateinischen Literatur des Mittelalters* I (1911, repr. 1973), 315–323.

Meyer, K. "Colman's Farewell to Colman." *Ériu* 3 (1907), 186–189.

Murphy, G. "Scotti Peregrini: The Irish on the Continent in the Time of Charles the Bald," *Studies xvii* (1928): 39–50, 229–244. Seymour, St John D. *Anglo-Irish literature 1200–1582*, 52–57.- Cambridge, 1929.

Stemmler, Theo, ed. *The Latin Hymns of Sir Richard Ledrede. Gnomon* 52 (1980): 58–59.

Traube, L. "O Roma nobilis: Philologische Untersuchungen aus dem Mittelalter," *Abhandlungen der phil.-philol. Classe der königlich bayerischen Akademie der Wissenschaften* xix, 297–395. 1892.

Waddell, H. *Medieval Latin Lyrics*. London, 1929. Walker, G.S.M., ed. *Sancti Columbani Opera*. Dublin, 1957.

See also **Columbanus; Lyrics; Metrics; Sedulius Scotus**

POETRY, IRISH

Before Approximately 1200

The earliest datable poetry in Irish Gaelic are poems of praise and of lament, themes continuing throughout Gaelic literature. The earliest of these, the *Amra Coluim Cille* ("The Eulogy of Colum Cille") (d. 597), is a lament and praise poem in a meter employing lines of irregular length, with extensive alliteration. Among the earliest purely praise poems extant, from the 700s, concerns Aed, a chief of North Leinster. It is referred to as "In Praise of Aed," and employs one of the *dán díreach* meters.

A dividing line can be made for Irish poetry, and for Irish history, at circa 1200. It was then that the effects of the Anglo-Norman Invasion and Church Reforms were being felt in Ireland. The poetry before this was different from what came after. Among the characteristic works of this time are nature poetry, religious poetry, and poetry of personal comment (including the Viking Incursions), produced largely, if not exclusively, by monks, and largely found as Glosses in early manuscripts. It is mainly lyrics, contemplative, spontaneous, graceful, which often appeal to modern taste. This poetry perhaps originated with the Irish hermits of the 500s and 600s, perhaps in the songs of Pre-Christian Ireland, perhaps under the influence of some Latin verse—but it certainly became a distinctive medieval Irish style—and is perhaps the best known today.

In addition to the clerical poets there were other classes, professional poets of the *Áes Dána*, and amateurs, with education provided by ecclesiastical schools, native Irish schools (in schools of poetry, law Schools, and schools of history), and by tutors, in both Irish and Latin languages for many.

While much of this poetry is anonymous, or of doubtful authorship, not all of it is. We have such names as Dallán Forgaill, a professional poet, to whom the *Amra Coluim Cille* is ascribed; Bláthmac Mac Con Brettan (*fl.* mid-700s); Feidilmid mac Crimthainn (*fl.* early 900s); Cormac mac Cuileannáin (d. 908); Mac Liag (*fl.* ca. 1014), and his lamentation for Kincora (upon the death of Brian Boru); and Máel Ísu Ó Brolcháin (d. 1086).

There are also other genres using verse: larger religious works, for example, on biblical history, theological poetry, rules for monastic life, praises of saints, and a body of Hagiography. There are didactic works, such as grammatical treatises, for instruction in language and poetry, and the *Félire Oengusso* ("The Martyrology of Oengus the Culdee") by a member of the *Céile Dé*, and other such martyrologies and calendrical works. There are humorous works, and humor within tales, such as has been claimed for the *Táin Bó Cuailgne* ("The Cattle Raid of Cooley") (the earliest version dated to the 600s or 700s), and there are works of satire. There is love poetry, and love *aislings* (visions). There are historical works, such toponomical works as the *Dinnsenchus* ("Lore of Place Names"), works of genealogy, and law tracts expressing the Brehon Law. There are adaptations of foreign, including classical, works; and there are prophecies, and magical spells. Almost all genres are represented in verse, or partly so, largely as an aid to memory.

There is narrative literature, often in a mixture of prose and verse, historical, fictional, or a mixture of the two. Among these are the tales of the Ulster Cycle, the most famous of which is the *Táin Bó Cualgne* (see above). There is the Mythological Cycle, and the Historical Tales, though the medieval Irish based their classification on the first word of a story's title, for instance, *Aided* (violent deaths), *Eachtrai* (adventures), *Comperta* (conceptions and births), and *Imrama* (voyages).

Finally, there is the Fionn Cycle, the stories associated with Finn Mac Cumhaill, which became prominent in the 1100s, with the appearance of the *Acallam na Senórach* ("The Colloquy of the Old Men"). It finds its origins in folk literature, and comes to replace the Ulster Cycle in popularity. It is more fantastic, magical, romantic (perhaps receiving Norman French influence in this), and more humorous. It is also more often in verse, and in ballad form, something new to Ireland, and to Europe.

After Approximately 1200

Changes took place, from around 1200, as a result of the Anglo-Norman Invasion and Church Reforms: the end of the composition of nature poetry, a growing dominance of the professional poets, the *fili* (pl. *filid*), and a change in popular literature. From this point on, bardic praise poetry was the dominant poetic composition, and the Fionn Cycle was the dominant narrative literature.

There were *filid* before this, members of the ancient *Áes Dána*, but they now regained something of their former dominance. Their presence can be seen in praise poetry recorded before circa 1200, in earlier literature, and in their conflicts with the Church. The *filid*, now members of bardic families, attained their position by inheritance and ability, after instruction in Bardic schools, and, unlike many of earlier times, are no longer anonymous. In addition to the *filid*, there are poets and performers of other sorts, such as the bards, who performed the compositions of the *fili* (known to the English as "bards" because of the *bard's* higher visibility), and the musicians, notably the harpers, who accompanied these performances.

The *fili* and *bard* were not always so divided. Earlier, the *fili* was a guardian and narrator of traditional knowledge, and a person of supernatural powers, who occasionally composed and performed praise poetry, while the latter function principally fell to the *bard*. But this changed with time, and after approximately 1200 the bard was subsidiary to the fili, the *bard* performing his works. Also, poets had earlier performed at the *Óenach* (pl. *Óenaigh*), in addition to royal courts, but now the *Óenaigh* were gone.

In composing their poetry, the *filid* used traditional Irish materials, promises and threats. The poems were composed in dark rooms and later committed to writing, their works preserved in part in the *Duanairí*. But they also used ever more foreign material, such as from Classical tradition, from French Romance, and to some extent from Welsh/British tradition—though the conservatism of the Irish resisted this. The poets were also trained in language, rhetoric, and metrics, employing the complex rules of *dán díreach*, and the bardic dialect. The bardic meters, and the bardic language, achieved a standard during this period, with little change, and no dialectical differences, and they were helped in maintaining these standards by grammatical treatises.

Their principal productions were, as stated above, praise poetry, but there were also inaugural odes, satire, religious poetry, homiletic poems, laments, appeals, complaints, poetical instruction, and personal commentary—but especially praise—for both the Gaelic aristocracy, and the Anglo-Norman, as they underwent Gaelicization. These poets served both groups, and moved freely within a politically divided Ireland, both before and after 1200.

This poetry has been criticized as being too formal and stylized, too pragmatic, lacking in feeling, and not appealing to modern taste. But, this is not always the case. Among the more famous poets we might list from among the bardic families (e.g., the Ua Dálaigh, Mac Con Midhe, Ua hUiginn, Mac an Bhaird, Ua Gnímh, Ua hEoghusa) are Muireadhach Albanach Ua Dálaigh (*fl.* early 1200s), Giolla Brighde Mac Con Midhe (*fl.* mid 1200s), Donnchadh Mór Ua Dálaigh (d. 1244), Gofraidh Fionn Ua Dálaigh (d. 1387), Tadhg Óg Ua hUiginn (d. 1448), and Tadhg Dall Ua hUiginn (d. 1593), all considered among the best of the *filid*.

As early as the 1300s, there was a new genre, a type of love poetry different from the earlier sort. Some claim it developed under French influence, others claim English, and others that it is of purely Irish derivation. Whatever the case, it had become uniquely Irish. They are generally by amateurs, and use the *dán díreach* meters, though usually of the simpler *ógláchas* type. The earliest extant of these is by Gearóid Iarla Mac Gearailt, Gerald (the Earl) Fitzgerald, fourth Earl of Desmond (d. 1398). This Gerald also produced other poetry, for instance concerning his imprisonment by Brian Ó Briain, and, though an amateur, who had been a student of Gofraidh Finn Ó Dálaigh, he became even more famous than his teacher.

Finally, we have the end of this era, and those poets who, while continuing the poetry of praise, also produced poetry of lament. In the 1600s came the collapse of the Gaelic aristocracy, and of the *fili*. We have Laoiseach Mac an Bhaird (*fl.* late 1500s), Ferghal Mhac an Bhaird (*fl.* late 1500s), Fear Flatha Ó Gnímh (*fl.* late 1500s), and Eochaidh Ó hEoghusa (*fl. c.* 1600), and in the end, we have the verse contest known as the *Iomarbháigh na bhfileadh* ("Contention of the Bards,") (early 1600s), but it is the sad act of a "a dog fighting over an empty dish."

There was a new era dawning, using *amhrán* meters, for a different audience, and complained of by such as Eochaidh Ó hEoghusa. It is perhaps attested to as early as the 1300s, and growing in importance from the 1500s on, but this goes beyond the limits of this entry.

MICHAEL TERRY

References and Further Reading

Bergin, Osborn. *Irish Bardic Poetry.* Dublin: Institute for Advanced Studies, 1970.

Carney, James. *The Irish Bardic Poet.* Dublin: Dolmen Press, 1967.

Greene, David H. *An Anthology of Irish Literature*, Vol. I. New York: New York University Press, 1971 (c1954).

Kinsella, Thomas, ed. *The New Oxford Book of Irish Verse.* Oxford, New York: Oxford University Press, 1986.

Knott, Eleanor. *Irish Classical Poetry.* Cork: Published for the Cultural Relations Committee of Ireland by Mercier Press, 1966, 1973 printing.

Murphy, Gerard. T*he Ossianic Lore and Romantic Tales of Medieval Ireland.* Cork: Published for the Cultural Relations Committee by Mercier Press, 1971.

O'Connor, Frank. *A Short History of Irish Literature.* Putnam, 1967.

Welch, Robert, ed. *The Oxford Companion to Irish Literature.* Oxford: Clarendon Press and New York: Oxford University Press, 1996.

Williams, J.E. Caerwyn. *The Court Poet in Medieval Ireland.* London: Oxford University Press, 1972.

Williams, J.E. Caerwyn, and Ford, Patrick K. *The Irish Literary Tradition.* Cardiff: University of Wales Press, and Belmont, Massachusetts: Ford & Bailie, 1992.

POETS/MEN OF LEARNING

Poets constitute the most important and prolific literary group of Medieval Ireland. Poets were highly trained professionals of high status and social eminence. An early Irish text lists the three broad types of poetry expected of a poet as white, black, and speckled. The text expounds the classification, white representing praise poetry, such as eulogies, black representing satire, and speckled representing poetry concerning legal issues. As shown by the surviving literature, however, Ireland's medieval poets were not limited to these three types. Poetic compositions included elegies, legal formulae, ancestral tables, historical recitation, prophetic visions, grammatical tracts, religious verse, and so forth. A poet was much more than simply a composer of verse; he was among other things an historian, a man of letters, a public official, a legal expert, a satirist, and a genealogist. Poets composed in both Irish and Latin. Interestingly, many poems of the medieval period written in Latin, while resembling more closely Latin style and composition, contain manifold examples of idiomatic Irish. While names and dates of individual authors are relatively rare in the early medieval period, the poems of a few well-known poets survive, including Colmán, Dallán, Ninine, and Senchán. More is known about the lives and reputations of the later poets, particularly Bardic poets, as authorship is usually given.

Attesting to their preeminent status, poets were the only professionals who retained personal rights and privileges of custom beyond the confines of their territory. Poets traveled freely between borders, even in times of conflict. The auspices for such travel usually included praise poetry for a distant king or lord, or to demand a cross-border claim. Poets were paid highly for their compositions. Compensation for poems usually consisted of a payment or tribute, comprised of various forms including cattle, horses, jewellery, weaponry, and the like. Payment was demanded by the poet himself, a fee determined by his grade or rank, the difficulty of composition, and in all likelihood, the relative wealth of the patron. The threat of satire upon non-payment for a poem seems to have guaranteed prompt payment in full. Satire was a heavy blow to the rank and status of its victim. Short satirical poems survive, sometimes including a patron's name or family, publicizing paltry and ungenerous payments.

Early traditional accounts specify seven distinct grades of poet, modeled on the seven ecclesiastical grades within the church. The highest grade, that of the *ollam* or "master poet," was attainable only through bloodline, that is if the poet's own father and grandfather were also poets. The remaining six grades were hierarchical, demanding longer study and knowledge of proportionately more verse compositions per grade. Poets studied and trained in schools down to the seventeenth century. Standard instruction for a poet lasted seven years, dominated by the study of countless compositional forms, each consisting of different metrical and rhyme schemes. Poetic composition was bound by strict rules of form and content.

The term bard has come to denote any Celtic poet. While the terms poet and bard are often synonymous in modern contexts, an exact and important medieval distinction existed between the two. Until the thirteenth century the poet was distinct from the bard through his professional status and technical training. The bard of this period and earlier was a low ranking poet of modest social rank and skill. Poetic schools of the early period are most closely associated with legal and monastic institutions. In later years such schools became primarily concerned with the study and preservation of Gaelic literature, grammar, and instruction.

Bardic poetry dominates Irish literature from the Anglo-Norman invasion until the late medieval period and is responsible for the vast majority of surviving poetic material. Bardic poetry differs considerably from earlier poetry, most remarkably in its greater length through a characteristically elaborate, embellished style. Interestingly, for nearly half a millennium the lexicon of Bardic poetry remained largely unchanged. Bardic poets composed their poems in a standard, fossilized literary dialect.

Bardic poetry, consisting predominantly of lengthy elegies and praise poems, recorded and immortalized the heroic exploits and largesse of its patrons. These poems were addressed almost exclusively to members of the ruling and educated classes. Bardic poets enjoyed special status within the household of their patrons, status that usually terminated upon the patron's death. A poet in a favorable relationship with his patron would often write compositions in a role consistent with a lover or spouse. Freedom of travel and an itinerant profession allowed poets to advance from patron to patron. The fall of Gaelic society brought with it the demise of Bardic poetry as professional poets were no longer supported and maintained by their patrons.

ANGELA GLEASON

References and Further Reading

Breatnach, Liam, ed. *Uraicecht na Ríar; The Poetic Grades in Early Irish Law.* Dublin: Dublin Institute for Advanced Studies, 1987.

Breatnach, Liam. "Poets and Poetry.". *In Progress in Medieval Irish Studies*, edited by K. McCone & K. Simms, 65–77. Maynooth, 1996.

Carney, James. *The Irish Bardic Poet.* Dublin, 1967.

Murphy, Gerard. *Early Irish Lyrics.* Oxford, 1956.

See also **Bardic Schools/Learned Families; Duanairí; Poetry Irish; Poetry Hiberno-Latin; Satire; Society, Grades of, Gaelic**

POPULATION

The central role of the human population of any socio-political unit has long been recognized by historians: the number of people, along with such variables as rates of fertility and mortality, help to determine certain economic processes and social policies, and are in turn influenced by other variables such as the efficacy of medical practices and the incidence of warfare. Even when census-derived head counts become available in the early nineteenth century, the precise nature and significance of population movements are not easy to evaluate. These preliminary remarks underline the one stark, unalterable fact about the population of medieval Ireland: there is no evidence on which to base a scientifically respectable figure. Nevertheless two numerical counts, one for the early Middle Ages and one for the late Middle Ages, more certainly in the future and more tentatively here provide grounds for arguments by which generally agreed estimates might be attained.

The Early Middle Ages

The first numerical count is of over forty-five thousand ringforts—earthen raths and stone-built cashels to defend against cattle raids—in the period centered on the seventh to ninth centuries. Crucial to this argument is that these enclosed farmsteads were the dominant settlement form and that they were built and occupied contemporaneously. These were the homes of the great majority of lords and farmers (divided by brehon lawyers into numerous subcategories). A generous multiplier of ten to allow for dependent relatives and servile personnel would produce a base figure of 450,000. In addition there were at least two larger settlement forms: the secular dún, many of which may no longer have served as permanent residential complexes, and ecclesiastical sites. The latter, often visible in the modern landscape as ovoid enclosures, numbered several hundred. To judge by the well-known description of Kildare in circa 630 by Cogitosus in his *Vita Brigitae*, the greatest monasteries were populous places, with outer zones (here called "suburbs") that provided accommodation for resident craftworkers and visiting pilgrims. On this basis a minimum figure of half a million inhabitants would be a reasonable estimate.

Whatever "guestimate" is favored, the population of early medieval Ireland was not stable but, on the contrary, was subject to both upward and downward pressures. The wealth of Roman Britain had attracted Irish as well as Germanic raiders and settlers. In the late fourth and the fifth century there was migration from Ireland to western parts of Britain: to Dyfed and Gwynedd in Wales and to Cornwall in England. The best indication of the scale and scope of this migration is the distribution of ogam inscriptions, which are concentrated in southwest Wales and which suggest that the majority of migrants came from Munster and Leinster. At the opposite end of the island an ultimately more significant migration of the Ulaid to southwestern

Scotland began around 500 C.E. when the royal family of Dál Riata, under Fergus son of Erc, abandoned Dunseverick and resettled across the North Channel. This population movement took on a well-known religious dimension with the settlement of Colum Cille and his followers on Iona in 563. Like Colum Cille's mission, these migrations may have been purely opportunistic; they may on the other hand be indicative of localized overpopulation in parts of Ireland.

Then came the first historically recorded plague pandemic of the mid-sixth century, whose demographic consequences cannot be measured scientifically but which may have been severe. The introduction of the mouldboard plough around 600 C.E. is believed by some scholars to have led to dietary improvements and to a steady growth of population, coinciding with the great age of ringfort and monastic construction. The introduction of water mills and legal texts dealing with mill construction, milling, and the status of millwrights point in the same direction. The subsequent influx of Scandinavian settlers (Vikings) in the ninth and tenth centuries was probably limited, to judge by the lack of rural placenames when compared with those of England, Scotland, and Normandy. Most of the newcomers would have been males from Norway or from Norwegian colonies in western Scotland; their influence is likely to have been cultural rather than numerical.

The Late Middle Ages

The second numerical count is that of parishes, of which there were about twenty-four hundred. Medieval parishes varied enormously in size: at one extreme were the extensive parishes of the most mountainous districts; at the other the smallest parishes of inner Dublin amounting to a few acres of ground, though with a population of several hundred. In the year 1300 Dublin had sixteen parish churches, seven within the walls and nine outside. If the estimated total population of the city at that time of eleven thousand can be accepted, the average number of parishioners per parish would have been a little under seven hundred. This sort of calculation gives some indication of maximum density of inhabitants, but until a wide range of local studies of medieval parishes and their population has been undertaken, we cannot go beyond mere guesswork. Thus, for example, average densities across the whole island of three to four hundred parishioners yield crude minimum and maximum totals of 720,000 and 960,000. Since it is generally supposed that in the great age of population growth in the twelfth and thirteenth centuries the number of inhabitants doubled or even trebled, the higher of these estimates is probably nearer the mark. One critical factor here is the comparatively low level of urbanization in medieval Ireland, where only about one-fifth of the island would have had regular access to town life even at the height of the Anglo-Norman colony around 1270. Accordingly the population increase experienced in Ireland is unlikely to have matched that achieved in more urbanized parts of Britain and the Continent.

In the late Middle Ages the great demographic enforcer was the second recorded plague pandemic commonly known as the Black Death, which first struck the country in 1348. Before then, there was significant immigration into Ireland, mainly from England and Wales, of people who congregated on rural manors and in towns. In addition aristocratic households established themselves in castles. The size of this influx of "new foreigners" is unknown, but their administrative and cultural impact is likely to have far outweighed their numerical strength. Epidemiological observations based on modern incidences of plague, together with a small amount of contemporary evidence, suggest that the colonists (as distinct from the natives) may have suffered a 40 percent reduction of population through a combination of mortality and emigration. An inquisition at Youghal, for instance, implies a mortality rate of around 45 percent in the case of burgess households. To all appearances and for a variety of reasons, the Gaelic Irish experienced lower death rates and indeed some of them migrated to the towns, including Dublin. Large herds of livestock (Irish *caoraigheacht*) are a manifestation of widespread internal migration from the late fourteenth century onwards. The demographic low may have reached the half million mark and the country remained underpopulated during the sixteenth century.

H.B. CLARKE

References and Further Reading

Barry, Terry, ed. *A History of Settlement in Ireland*. London and New York: Routledge, 2000.

Clarke, H.B. "Decolonization and the Dynamics of Urban Decline in Ireland, 1300–1550." In *Towns in Decline, A.D. 100–1600*, edited by T.R. Slater, 157–192. Aldershot: Ashgate, 2000.

Down, Kevin. "Colonial Society and Economy in the High Middle Ages." In *A New History of Ireland*, Vol 2, *Medieval Ireland 1169–1534*, edited by Art Cosgrove, 439491. Oxford: Clarendon Press, 1987.

Kelly, Maria. *A History of the Black Death in Ireland*. Stroud: Tempus Publishing, 2001.

Russell, J.C. "Late-thirteenth-century Ireland as a Region." *Demography* 3 (1966): 500–512.

Stout, Matthew. The Irish Ringfort (*Irish Settlement Studies, no. 5*). Dublin: Four Courts Press, 1997.

See also **Black Death; Diet and Food; Famine and Hunger; Slaves; Tribes; Túatha, Vikings**

PORTS

Although saints' lives and other monastic sources have occasional references to ships and trade and certain of the larger monasteries such as Armagh and Kildare may have functioned as proto-towns, it was the Vikings who established the first towns and ports in Ireland. Beginning in the ninth century with their *longphoirt*, "fortified enclosures protecting their ships," in County Louth and at Dublin, they expanded to become permanent settlements and centres of trade in the tenth century. Carrickfergus, Carlingford, Drogheda, Dublin, Wicklow, Arklow, and Wexford were ideally placed along the east coast of the Irish Sea to benefit from the traffic on the trade route from Scandinavia and the northern isles to England and the continent. The excavations of Viking Dublin show a thriving city of merchants and craftsmen in wood, metal, bone, and cloth who manufactured goods which they traded locally and abroad. Evidence of finds also suggests that food supplies in particular came from the local Irish, but perhaps the smaller Viking settlements to the north and south also shipped food and fuel supplies to Dublin.

When the Anglo-Normans came in the twelfth century they immediately recognized the importance of the Viking settlements, and these were quickly taken over, with Dublin and the larger ports coming under the king's direct control. In other places local magnates founded a port as a gateway to trade with their lands. John de Courcy tried to develop Down Patrick, a long-established religious and dynastic center, as a port as well as the main town of his lands. He chose Carrickfergus as a strategic site and constructed a strong keep there possibly as early as 1178. A settlement grew up beside the castle, which had a parish church by 1205 and was described as a "vill" in 1226. After the earldom of Ulster reverted to the king in 1333 the port in the shadow of the castle functioned as a government outpost useful for trade with the Gaelic hinterland in a region beyond the jurisdiction of the crown throughout the later Middle Ages. However, archaeological finds of imported pottery indicate that the citizens maintained foreign contacts as well.

Bertram de Verdon may be regarded as the founder of Dundalk after he was granted most of north County Louth by the future King John. The early settlement was close to the motte and bailey at Castletown, but the town was to develop a little over a mile downstream close to the estuary and to take advantage of the Irish Sea trade in the early thirteenth century. Indications are that the port was operational before the official customs were established in the late 1270s, and Dundalk was reckoned as one of the "ports of Ulster." Such references to the trade of the port as survive indicate the usual commodities: wine, salt, iron, and cloth imported and corn, fish, and hides exported, but compared with Drogheda and Dublin, Dundalk remained a minor port in the Middle Ages.

Hugh de Lacy fortified the site of Drogheda on the river Boyne five miles from the open sea. The earliest surviving charter of the town was granted in 1194 by Walter de Lacy. To attract citizens from England it offered attractive privileges to the burgesses, large plots within the town, three acres in the countryside close by and free access to the river Boyne. The walls enclosed 113 acres (45 hectares), making it comparable in size to Bristol, Oxford, New Ross, Kilkenny, and Dublin. It had at least four gates and seven towers, and the barbican of St Laurence's gate is the finest surviving in Ireland. A series of at least thirteen murage grants, levying tolls on goods coming into the town for the construction and maintenance of the walls, are extant between 1234 and 1424.

Drogheda flourished as a port despite difficulties with silting and sandbars, a problem it shared with Dublin. The port records of Chester and Bristol suggest that the bulk of its trade was across the Irish Sea, but Drogheda merchants such as the Symcocks and the Prestons did venture to France, particularly to Bordeaux, for wine, and in the fifteenth century there was traffic with Brittany, the Baltic, and Iceland. In the early years there was a significant export trade in corn and victuals as supplies to the royal armies campaigning in Scotland and Wales. For most of the later Middle Ages the archbishops of Armagh resided in the manor of Termonfechin close by, and this added to the town's status and prosperity. Occasionally the archbishops provided the townsmen with safe conducts to travel and trade with the Ulster Irish.

Wicklow and Arklow declined in importance in the later Middle Ages due largely to their hinterland being dominated by the O'Byrnes and O' Tooles who had little interest in trade. Dalkey, however, functioned as the deepwater port of Dublin. Due to the shallowness of the Liffey estuary, large ships could not berth at Dublin's quays, and wine ships in particular had to anchor at Dalkey and unload on to lighters which carried the wine tuns up the Liffey. Remains of tower houses and castles suggest that the little port profited from its deep anchorage.

Wexford (Veigsfjorthr) was an important Viking settlement by the end of the ninth century as archaeological excavations have shown. It had trading connections with Bristol, and the links continued after the town came under the control of Diarmait mac Máel-na-mBó in the mid-eleventh century. In 1169 it was the first town to be taken by Diarmait mac Murchada and his Anglo Norman allies. It was walled in the fourteenth century, but little is recorded of its ships or trade before the seventeenth century.

Around 1210 William Marshal, earl of Pembroke, established New Ross, which was to become the chief port of the lordship of Leinster, but despite its powerful lord it never managed to break the monopoly of the royal port of Waterford. Throughout the later Middle Ages all ships entering Waterford Harbour were by law obliged to disembark first at Waterford and, having paid customs there, were free to proceed to New Ross.

Ports of the South and West Coast

On the south coast, the mouth of the river Blackwater where Youghal now stands appears to have been settled by Vikings in the ninth century. The town's founder may have been Maurice fitz Gerald in the early thirteenth century. Youghal was walled after 1275 and remained under Geraldine influence throughout the medieval period. At the other end of their huge earldom was the port of Dingle, which appears to have been walled during the medieval period, but a murage grant and decree of incorporation was only issued in 1585. The town was an embarkation point for the pilgrimage to Santiago de Compostela, but the citizens and the fitz Geralds benefited most from the revival of the herring fisheries in the late fifteenth and sixteenth centuries.

The port of Kinsale, County Cork developed from a Viking trading post to being settled by the Anglo-Normans around 1200. Its first charter dates to 1333, and there is a murage grant of 1348. In the fifteenth century it grew to be a prosperous port, and its ships are recorded as trading with Bristol and with France. However, its relatively isolated position and excellent harbor made it attractive to pirates and freebooters in the later Middle Ages.

The town of Galway grew around a castle built by Richard de Burgh in the thirteenth century. The prosperity of the families who controlled the town, later known as the "tribes of Galway" (Athy, Blake, Bodkin, Browne, D'Arcy, Deane, Flont, Joyce, Kirwin, Lynch, Martin, Morris, and Skerret), is evident in the building of St Nicholas's church in 1320. Galway's loyalty to the language and traditions of England made it increasingly isolated as the influence of the Crown waned in the west of Ireland. By 1396 it attained the status of a royal borough, relatively free from the control of the de Burghs. In the same year St Nicholas's church was granted collegiate status, separating it from the local Irish bishop of Annaghadown and empowering the citizens to elect a warden responsible for ecclesiastical affairs in the city. Galway was destroyed by fire in 1473 and again in 1500, but continuing prosperity enabled rebuilding in stone, and pictorial maps of the early seventeenth century show a city of elegant buildings of unified style.

The medieval port had links not only with England, from where Bristol merchants leased the Corrib salmon and eel fisheries, but also with Flanders, France, Spain, and Portugal importing wine in exchange for cattle hides and fish procured from local magnates in exchange for salt and luxury goods. Wills of members of the Blake family dated 1420 and 1468 indicate that a barter system was in operation with numbers of hides being owed for wine, cloth, and salt. This contrasts with Limerick where surviving wills of the Arthur merchants record money transactions with the local Irish. The opening up of the Atlantic seaways in the fifteenth century benefited Galway. Henry the Navigator had an agent in the city, and according to a letter in the Portuguese Archives dated 1447, he promised to send a lion on board his next ship to Galway as he thought the people of that city had never seen a lion! Some of the more adventurous merchants such as Germanus Lynch (*fl.* 1441-1483) sailed frequently to England, Spain and, like an enterprising consortium from Drogheda at the end of the fifteenth century, made the hazardous trip to Iceland to service the developing fishery there.

Merchants from Bristol also sailed as far as Sligo having secured permits to bring wine, salt, and cloth to trade with the king's lieges there for salmon. Sligo, like Galway, had come to prominence during the de Burgh invasion of Connacht in the thirteenth century. It was originally granted to Maurice fitz Gerald, ancestor of the earls of Kildare, who built a castle there in 1245 and founded a Dominican friary close by in 1253. Richard de Burgh built a new castle and laid out a town in 1310, but the town passed into the control of the O'Conors of Sligo and remained in Irish hands until the end of the sixteenth century.

References and Further Reading

Bradley, John. "The Topography and Layout of Medieval Drogheda." *Journal of the County Louth Archaeological and Historical Society.* 19, no. 2 (1978): 98–127.

Mac Niocaill, Gearóid. *Na Buirgéisí, XII-XV aois* 1 vol. In 2. Dublin: Cló Morainn, 1964.

O' Neill, Timothy. *Merchants and Mariners in Medieval Ireland.* Dublin: Irish Academic Press, 1987.

See also **Cork; Dublin; Limerick; Trade; Urban Administration; Waterford; Walled Towns**

PRE-CHRISTIAN IRELAND

Sources

Because literacy arrived only with conversion to Christianity from the fifth century C.E. onward, there is no contemporary native record for pre-Christian Ireland, while its remote location on the western edge of Europe meant that it attracted little attention from Classical commentators. The archaeological record attests to ritual practice, but not to religious belief; a limited insight in this connection may be obtained from historical sources (including certain early ecclesiastical texts in the *Collectio Canonum Hibernensis*, some Middle Irish antiquarian tracts, and occasional stray references in later medieval accounts), from hagiography, and from Old and Middle Irish literature. The extent to which such evidence reflects practices or beliefs comparable to those of the Continental Celts is open to debate; scholars have noted the limited quantity and range of archaeological data and its qualified resemblance to the style of La Tène, a site in Switzerland the material heritage of which is widely viewed as the definitive characteristic of Iron Age Celtic culture. Concerns have also been expressed about the lateness of literary references to pagan custom, and about acceptance of the apparently La Tène type settings of Old Irish stories (especially the Ulster Cycle), from the eighth century C.E. onwards, as representing a "window on the Iron Age." McCone, in particular, stresses the ecclesiastical environment in which Old Irish literature was produced, and the Christian and Classical influences affecting it, while not denying the possible survival of some influence from oral tradition with pagan roots. Some have argued for a recasting, in this ecclesiastical environment, of earlier sources to present a more supernatural view of the poet's profession. In Carey's view, the late ninth-century glossary of Cormac mac Cuilennáin represents an important step in reconstructing a "pagan heritage" for Ireland.

Ritual Sites

Two prominent earthwork sites, which apparently served ritual functions, bear traces of fire-ceremonies—while at another site there are indications that horses were perhaps ceremonially killed. At Emain Macha ("Navan Fort"), near Armagh, excavated by Waterman in the 1960s–70s, one phase of later Iron-Age activity (with dendrochronological date of 95 B.C.E.) involved the construction of what looks to be a shrine comprised of concentric circles of wooden posts. There is evidence of intense burning, seemingly deliberate. Parallels have been drawn with the Dún Ailinne earthwork, County Kildare (radiocarbon dates ranging from 390 B.C.E. to 320 C.E.), where Wailes found a circle of wooden posts inside an enclosure—also destroyed by fire. Cosmological interpretations of these structures focus on an apparent resemblance to the sky wheel (a symbol elsewhere associated with the Celtic deity Taranis, who had solar connotations) and see particular significance in the destruction of these sites by fire. At Tara, the discovery by Roche of animal remains—especially horse—suggests ritual activity and prompts comparisons with Danesbury in England, and perhaps with Belgic or north-Gaulish sites like Gournay or Ribemont. The find is especially curious in view of a colorful account in the twelfth-century topography of Giraldus Cambrensis, which purportedly describes a regnal inauguration, whereby the new king engaged in ritual mating with a white horse before it was slaughtered and eaten.

Priesthoods

The separation of embankment from interior by a fosse at such sites, possibly intended to distance observers from proceedings within the enclosure, has led some archeologists to infer the existence of a priestly class. Historical evidence from the Early Christian period in the form of ecclesiastical legislation (particularly the so-called "First Synod of St. Patrick," which may reflect a sixth-century C.E. reality) refers to seers before whom pagans swore solemn oaths. Hagiographical works from the seventh century onward—including the Latin Lives of St. Brigit—commonly refer to druids; such references, and the term *driú* in Old Irish, may mean that this priesthood (described in a Gaulish context by Caesar and by earlier commentators) historically did exist in Ireland. The late medieval description by Giraldus Cambrensis of nine women who guarded an eternal flame at Kildare has been viewed as testimony to a priestly role for females in connection with a fire-cult—and comparisons drawn with a flame at Bath and with Classical accounts of all-female sanctuaries in Gaul. Several episodes in the Life of Brigit, including the description of her veiling, when a column of fire was seen to rise from her head—otherwise open to interpretation as Christian symbolism—have been cited as possible reflections of a fire-priestess role. However, recent opinion, as represented by Harrington, is more skeptical.

Sacrifices

The archaeological record includes several discoveries of La Tène artifacts, hoards, and single finds, which seemingly represent ritual deposits. In the late nineteenth century, a collection of swords, scabbard plates, and spear fitments of La Tène style was discovered at Lisnacrogher, County Antrim. Although now a bog, the site was probably, as Raftery suggests, a votive lake in which valuables were deposited—like Llyn Cerrig Bach in Wales or, indeed, La Tène itself. At another almost-dry lake, Loughnashade (*Loch na Séad*; "Lake of the Valuables"), near Emain Macha, workers in the eighteenth century found four elaborate bronze ceremonial horns featuring late La Tène ornament. Isolated finds of weapons and ornamental objects have been made in rivers, especially in the Bann and Shannon, a trend that is paralleled in Britain and on the Continent. Water cults are certainly well attested among the Celts. Disposal of valuable items in sacred places presumably represents vicarious sacrifice—although there are indications that animals, and sometimes humans, were ritually deposited.

Loughnashade also produced animal remains and several human skulls. In the 1970s, Lynn's excavation at "The King's Stables" (Co. Armagh)—the site of an artificial pond, now dry—yielded an impressive collection of animal bones including cattle, deer, dog, pig, and sheep. This was plainly not an occupation site, and it seemed unlikely that these were food remains. The case for ritual deposit here was greatly strengthened by the discovery of the facial portion of a human skull. Similarly, an enclosure at Raffin, County Meath, produced a skull burial. These finds clearly point to the ritual deposit of bodily remains, and prompt questions in relation to human sacrifice. The nineteenth-century unearthing at Gallagh, County Galway, of the body of a young man preserved in a bog provides a rather compelling case for ritual killing. The finders' accounts are emphatic about a rope or ligature around the individual's neck, and he was evidently immersed in water. Mistreatment of the remains at the time of discovery and afterwards render it difficult to now ascertain whether or not the body, when found, also bore evidence of wounding. The closest parallels are provided by bog bodies of the first century B.C.E. from Lindow Moss, in England, and from Tollund and other locations in Denmark, which lie beyond what is generally construed as the "Celtic Zone." In these instances the individuals concerned, also young men, were wounded (the first struck on the head with a stone, the other stabbed in the throat), hanged, or garroted, and immersed in water. Finds of this order are significant in the light of the widely discussed "threefold death" motif, found in Old Irish and Welsh literature. Tales such as that of Áed Dub, king of Ulaid, or Diarmait mac Cerbaill, king of Tara (composed, as Borsje points out, in an explicitly Christian context), concern an anti-hero who inescapably perishes amidst prophesies of doom. The individual in question is generally stabbed, and falls from wood (possibly an image for hanging) into water, or is burned and then "drowned."

Burials and Afterlife

Ireland's record of general Iron Age burials is limited, in terms of quantity and quality. Of the small number of examples found to date, there is nothing comparable to the impressive earthen barrows associated with the Hallstatt and La Tène cultures of the Continent, characterized by the presence of a "wagon" or "chariot" and featuring a range of grave-goods, including weaponry, ornaments, and abundant indicators of a funerary feast. Often taking the form of modest "ring-barrows," which suggest continuity with the Bronze Age, Irish burials display a mixture of cremation and inhumation rites, with paltry grave-goods and no clear evidence of food or drink to send off the deceased. Typical of the ring-barrows so far excavated is Grannagh, County Galway, which produced just a bronze *fibula* brooch, some bone pins, and glass beads; only in one grave at Tara were animal bones found accompanying a burial, and it is uncertain whether or not they are primary. Nonetheless, even such sparse grave-goods still indicate belief in an afterlife. Old Irish literature features tales of Donn, viewed by some as a counterpart to the Continental deity Dis Pater, who ruled over a realm of the dead. Donn appears as an isolated figure who has little association with other gods; moreover, it is clear that he was host only to the "glorious dead"—the warrior elite. Ordinary folk are accorded little attention in Early Irish literature, whether in relation to this life or the next.

Deities

Archaeology tells us little regarding the deities worshiped in pre-Christian Ireland. The absence of inscriptions, prior to the introduction of *ogham* in an Early Christian context, makes it difficult to identify figural representations—even if they could be confidently dated, which is another issue. Stone sculptures found, which arguably belong to the Iron Age, include a head from Corleck (Co. Cavan), and the "Tanderagee idol," reportedly from Armagh. The tricephalic character of the Corleck head, paralleled on the Continent, has led some to view this as a representation of the god Lug—with whom this characteristic is associated. The way in which the Tanderagee figure holds its right arm has prompted identification with the deity Nuadu (or Nodens); the latter features in the Irish Mythological Cycle in company with In Dagda—the "good god"—as one of the Tuatha Dé Danann, or divine people. As king, Nuadu loses his arm and, although given a silver replacement, ultimately abdicates in favor of Lug.

Various placenames seem to commemorate these deities; for instance, Lugmad (Lugmoth? = Lug's penis (?) = Louth), or Magh Nuadat (Nuadu's Plain = Maynooth). There is record of the population-group Luigne—"descendants of Lug"—whose name is left on the Barony of Leyney, County Sligo. Problems arise, however, concerning the identity of the eponyms in question and the probable date at which the placenames were coined.

In addition to deities accorded prominent roles in mythological tales, others, it has been argued, are reflected in hagiography. Debate concerning apparent solar symbolism in the lives of Brigit notwithstanding, a goddess of that name was known to the Continental Celts and is discussed in Cormac's glossary. The story of Ailbhe bishop of Emly and his rearing by wolves, nowadays interpreted as a borrowing from heroic literature, was formerly viewed as a reflection of older traditions—associating the saint with a sacred animal. There are representations in the archaeological record of what may be divine animals or animal deities; these include stone bear figures from Armagh—if Iron Age in date—and a recently discovered janiform figure with "human" and "animal" (wolf?) sides. There are further hints regarding cults of inanimate nature; the notion of the bile, or sacred tree, persisted well into the historical period—while some claim that magical properties assigned to certain "holy wells" in modern times point to pagan origins.

Festivals

Feast-days known from the Continental Celtic calendar, including Imbolc (February 1), Beltene (May 1), Lugnasa (August 1), and Samain (November 1)—all representing turning-points of the year, are noted in Cormac's Glossary. Samain, in particular, features prominently in Old Irish literature. Many tales, including some of the *echtrai* genre, are set at this feast of the dead, which saw the suspension of barriers between earth and otherworld, permitting reciprocal access. It was the appropriate time for the demise of heroes, and a suitable backdrop for "threefold death" tales. Stories of these festivals and of deity-figures associated with them were carried into later tradition, and customs relating to them—including the lighting of bonfires—long survived in modern folk practice.

AILBHE MACSHAMHRÁIN

References and Further Reading

Bieler, Ludwig, ed. "The First Synod of St. Patrick." *The Irish Penetentials*, 56–57. Dublin: Dublin Institute for Advanced Studies, 1963.

Borsje, Jacqueline. "Fate in Early Irish Texts.' *Peritia* 16 (2002): 214–231.

———. "Het mensenoffer als literair moteif in het middeleeuwse Ierland," *Nederlands Theologisch Tydschrift* 58 (2004)-46–60.

Byrne, Francis John. *Irish kings and High-kings*, 97–102, 155. New ed. Dublin: Four Courts Press, 2001.

Carey, John. "The Three Things Required of a Poet." *Ériu* 48 (1997): 41–58.

Connolly, Seán, trans. "Vita Prima Sanctae Brigitae." *Jnl. Roy. Soc. Antiq. Ire.* 119 (1989): esp. 14–17, 18, 41.

Harrington, Christina. *Women in a Celtic Church: Ireland 450–1150*, 63–67. Oxford: Oxford University Press, 2002.

Lynn, C.J. "Trial Excavations at the Kings Stables, Co. Armagh." *Ulster Jnl. Arch.* 40 (1977): 42–62.

———. "Navan Fort: A Draft Summary of D.M. Waterman's Excavations." *Emania* 1 (1986): 11–19.

MacCana, Proinsias. *Celtic Mythology*. London: Hamlyn, 1970, passim.

McCone, Kim R. *Pagan Past and Christian Present*. Maynooth: An Sagart, 1990, esp. chapters 1, 7, 8.

MacGiolla-Easpaig, Dónall. "Noun-noun Compounds in Irish Placenames." *Études Celtiques* 18 (1981): 151–163, esp. 162–163.

MacShamhráin, Ailbhe. "Iarsmaí 'Ceilteacha' na Danmhairge ón Iarannaois." In *Bliainiris 2001*, edited by Ruairí Ó hUiginn & Liam MacCóil, 181–202. Rath Cairn, Co. Meath: Carbad, 2001.

Muhr, Kay. "The Early Place-names of County Armagh." *Seanchas Ardmhacha* (2002): 1–54, esp. 17–18.

Ó hÓgáin, Daithi. *The Sacred Isle: Belief and Religion in Pre-Christian Ireland*. Cork: Collins Press, 1999.

O'Meara, J.J., trans. Giraldus Cambrensis (Gerald of Wales), *The History and Topography of Ireland* (Topographia Hiberniae), 81–82, 88, 109–110. Revised edition. Mountrath, Co. Laois: Dolmen Press, 1982.

Raftery, Barry. *Pagan Celtic Ireland*. London: Thames & Hudson, 1994, chapters 4, 8.

Roche, Helen. "Late Iron Age Activity at Tara, Co. Meath." *Ríocht na Mídhe* 10 (1999): 18–30.

Ross, Ann. "Lindow Man and the Celtic tradition." *In Lindow Man: The Body in the Bog*, edited by Ian Stead, J. B. Bourke, and Don Brothwell, 162–169. London: British Museum, 1986.

Wailes, Bernard. "Dún Ailinne: A Summary Excavation Report." *Emania* 7 (1990): 10–21.

Warner, Richard. "Two Pagan Idols: Remarkable New Discoveries." *Archaeology Ireland* 17 no. 1 (Spring 2003): 24–27.

See also **Brigit; Burials; Christianity, Conversion to; Cormac mac Cuilennáin; Eachtrai; Emain Macha; Giraldus Cambrensis; Hagiography; Inauguration Sites; Inscriptions; Invasion Myth; Kings and Kingship; Mythological Cycle; Tara; Ulster Cycle; Witchcraft and Magic**

PROMONTORY FORTS

A promontory fort is a fortified coastal headland or sea-girt promontory of land. The seaward sides are naturally defended by a cliff while one or more straight or curved ramparts of earth or stone, with accompanying ditches, protect the landward side. The main purpose in using a headland for fortification was to take advantage of the natural defense provided by a vertical cliff face. The location of these forts predicated engagement with the sea and maritime activity for their occupants. Many of them incorporate the Irish word *dún* (fort) in their name. Over 350 promontory forts have been identified on the Irish coast of which just nine have been the subject of archaeological excavation. The first scientific excavation of a promontory fort was carried out at Larribane, County Antrim in 1936 with a subsequent season of excavation in 1962. This was followed by excavations at Dunbalor on Tory Island, County Donegal in 1949, at Dalkey Island, County Dublin between 1956 and 1959 and by three excavations at the promontory forts of Carrigillihy, Dooneendermotmore, and Portadoona, County Cork in 1952. Dunbeg, County Kerry was excavated in 1981 and Doonagappul and Doonamo, County Mayo in 1999. Much of what is known about promontory forts is based on the pioneering work of Thomas Johnson Westropp who, between 1898 and 1922, visited and recorded 195 sites primarily in the west and southwest of Ireland and published twenty papers dealing with his findings. As late as the end of the twentieth century archaeologists tended to classify promontory forts as a sub-class of the less numerous inland hillforts.

Promontory forts are attributed various functions. Among the suggestions are that they may have been used as landing places for seagoing invaders and temporary refuges during inland attack. They have also been proposed as trading bases, ceremonial enclosures, observation posts, and livestock pounds. In several cases the interiors of promontory forts show no visible sign of occupation, which favors the idea that some may have served as temporary refuges.

Although there are, as yet, no firm dates for the construction of this monument type, archaeologists have tended to view promontory forts as primarily Iron Age in origin. However, evidence from excavations, Tudor maps, historical documents, and upstanding structures within promontory forts clearly indicate that occupation also took place within some of them in the early (fifth century to *c.* 1100) and later (*c.* 1100 to *c.* 1600) medieval periods. Three of the nine promontory forts scientifically excavated have produced substantial evidence of medieval occupation. Barry's excavation at Dunbeg, County Kerry revealed that the first phase of occupation provided a radiocarbon date spanning the period from the end of the ninth century to the late sixth century B.C.E. However, a radiocarbon date from the innermost ditch of the promontory fort ranged from the late seventh to the early eleventh centuries C.E. proving that the fort was in use in the early medieval period. In addition, an early medieval souterrain ran outward from the entrance to the fort, and the earlier of two occupation layers within a large *clochán* or circular stone hut in the interior of the fort was dateable to the period from the late ninth century to the mid-thirteenth century C.E. Childe's excavation at Larribane, County Antrim and subsequent excavations there by Proudfoot and Wilson suggested that the fort had been built and occupied around 800 C.E. No evidence was produced however, to prove that the occupation of the headland coincided with the actual construction of the stone wall and external ditch that defended the landward side of the site. Liversages excavations on Dalkey Island, County Dublin demonstrated that a midden, datable to the fifth or sixth century C.E. on the basis of imported pottery found within it, represented the first early medieval occupation of the promontory before it was actually fortified. A second phase of occupation in the seventh century, constituting a hearth, a midden, and a possible house site, which postdated the construction of the rampart of the promontory fort, was identified in the interior of the fort. Apart from the evidence for early medieval activity found during scientific excavations, the Irish chronicles also allude to the occupation of some promontory forts during the Viking Age. For instance, Dunseverick, County Antrim was the target of a Viking raid in the ninth century, which suggests that it was a substantial settlement of some wealth in that period.

Several promontory forts enjoyed periods of occupation in later medieval times. In the late twelfth and early thirteenth centuries Anglo-Norman colonizers saw the immediate advantages of adopting and enhancing promontory forts that had previously been used as strongholds by Gaelic lords. Dun Contreathain or Donaghintraine on the Atlantic coastline of County Sligo is, for instance, mentioned in the Irish chronicles as a base for the activities of the Anglo-Norman magnate, de Bermingham, in 1249. By 1297 the de Bermingham family had built themselves a manorial hall house within a promontory fort at Castleconnor, overlooking the estuary of the River Moy in County Mayo. At the commencement of the Anglo-Norman colonisation of Ireland, the Anglo-Norman knight, Raymond le Gros, greatly augmented the defences of an existing promontory fort called Dundonuil at Baginbun, County Wexford in order to secure an initial base for his army in May 1170 and for that which followed under the command of Strongbow in August of that year.

The results of O'Kelly's excavation at Dooneendermotmore, County Cork perhaps best exemplifies the enduring nature of the promontory fort as a form of defended settlement. The defences of the fort were constructed in two phases. No date was confirmed for the first phase but the rock-cut ditch was modified during the later medieval period and crossed by means of a drawbridge. The parapet wall of the fort and a large two-roomed house site in the interior were also constructed in that period. In fact, no occupation levels earlier than the sixteenth century were identified during excavation.

The remarkably late use of promontory forts can also be seen on a sixteenth-century map-picture of Portrush, County Antrim made by a Tudor cartographer and at the impressive Dooncarton on Broadhaven Bay in County Mayo where a series of stone buildings in the interior of the fort constituted the homestead of a local Gaelic family as late as the seventeenth century.

ELIZABETH FITZPATRICK

References and Further Reading

Barry, Terry. "Archaeological Excavations at Dunbeg Promontory Fort, County Kerry." *Proceedings of the Royal Irish Academy* 81 (1977): 295–329.

Edwards, Nancy. *The Archaeology of Early Medieval Ireland*. London: Batsford, 1990.

O'Conor, Kieran. "A Reinterpretation of the Earthworks at Baginbun, Co. Wexford." In *The Medieval Castle in Ireland and Wales*, edited by John R. Kenyon and Kieran O'Conor, 17–31. Dublin: Four Courts Press, 2003.

O'Kelly, Michael J. "Three Promontory Forts in Cork." *Proceedings of the Royal Irish Academy* 55 (1952): 25–59.

Westropp, Thomas J. "The Promontory Forts and the Early Remains on the Islands of Connacht." *Journal of the Royal Society of Antiquaries of Ireland* 44 (1914): 297–337.

———. "The Promontory Forts of the Three Southern provinces of Ireland." *Journal of the Galway Archaeological and Historical Society* 11 (1922): 112–131.

———. "The Promontory Forts and Adjoining Remains in Leinster." *Journal of the Royal Society of Antiquaries of Ireland* 52 (1922): 52–76.

PROPHESIES AND VATICINAL LITERATURE

Prophesies and Vaticinal literature were important elements in medieval Ireland. Fore-knowledge is claimed in St. Patrick's *Confessions*, while Adomnán devotes a third of his *Life of Columba* to the saint's prophesies. Prophesy was used to justify political conditions. The tenth-century *Tripartite Life of Patrick* has the saint predicting Ireland's political history. Prophesies were a convenient medium for commentary or dissension. The Prophecy of Berchán was begun in the ninth century with verses on the Vikings, and continued in the eleventh century with a critical recitation of Irish and Scots high kings. A contemporary prophesy is attributed to Bec mac Dé, and it condemns father to son succession in the headship of Armagh. Opinion on important clergy is found in the Prophecy of Bricín, composed around 1000.

Prophesies were also attributed to legendary individuals. An early eighth-century recitation of princes is the Prophecy of Conn of the Hundred Battles. This work was the model for an eleventh-century composition known as the "Phantom's Frenzy," written by Dubdá-Leithe of Armagh. A phantom and a lady who represents the sovereignty of Ireland tell Conn who will rule Ireland. Conn's son Art is made the author of a prophesy that foretells his death at the battle of Mag Mucruimhé and the arrival of St. Patrick.

A group of prophesies from the tenth and eleventh centuries has the theme of the Last Days. Works such as the "Fifteen Signs of Doomsday" and the "[Day of] Judgment" describe the end of the world. After the mid-tenth century are prophesies about destruction associated with the feast of John the Baptist. This culminated in a panic throughout Ireland in 1096 when certain chronological conditions, described in the "Second Vision of Adomnán," were believed to herald this disaster.

The twelfth century saw a reaction to prophetic works. The "Vision of Mac Con Glinne" mocks the "Phantom's Frenzy," and instead of a list of rulers there is a list of delicacies for a feast. Nevertheless, prophetic texts continued to be produced. The "Poem of Prophecies," about the evils of the age, is a continuation of an eleventh-century text. Contemporary is a prophesy attributed to St. Moling, concerned mainly with Leinster affairs.

The Anglo-Norman invasions inspired a new wave of original composition. A prophesy attributed to St. Columba, addressing his friend Baitín, called "Harken O Baitín," places the invaders in the general context of Irish history. Ironically, the Anglo-Normans were enthusiastic students of Irish prophesies, and Gerald Cambrensis' *Conquest of Ireland* originally was called the Vatican History. He claims that the adventurer John de Courcy had a volume of Irish prophesies, which he believed had foretold his conquests. In later medieval Ireland prophesy increasingly became subordinated to contemporary affairs. Odes to princes usually claimed that their reigns had been foretold by an ancient prophet. At the end of the Middle Ages, prophesy had become a cliché in Irish society.

BENJAMIN HUDSON

References and Further Reading

Adomnan. *Life of Columba*, edited by A.O. & M.O. Anderson. Oxford: Clarendon Press, 1991.

Hudson, B., *Prophecy of Berchán*. Westport: Greenwood, 1996.

———. "Time is Short." In *Last Things*, edited by C. Bynum and P. Freedman, 101–123. Philadelphia: University of Pennsylvania, 2000.

Knott, Eleanor. "A Poem of Prophecies." *Ériu* 18 (1958): 55–84.

O'Curry, Eugene. *Lectures on the Manuscript Materials of Ancient Ireland*. Dublin: Wm. A. Hinch & Patrick Traynor, 1878.

O'Kearney, S. *The Prophecies of Saints Colum-Cille, Maeltamlacht, Ultan, Senan, Bearcan, and Malachy.* Dublin, repr. 1925.

See also **Hagiography and Martyrologies; Historical Tales; National Identity; and Poets, Men of Learning**

PROVINCES

See **Connacht; Leinster; Munster; Ulster, Earldom of**

QUEENS

Just as there were several different levels of kingship in early Ireland, so are there instances where the title "queen" is applied to women at all levels of the royal hierarchy: the wives of petty kings, of provincial kings, of the Uí Néill kings of Tara, and of the later high kings of Ireland. The title usually bestowed is "queen of a king," although there are some instances where the outright title of "queen" is used. Most of the latter instances involve women who fall into the rare category of wives who predeceased their husbands, a pattern that may indicate Irish queenship denoted a distinct office rather than merely being the king's wife. Possibly, there could have been only one queen of a kingdom at any given time, so that when a king died his widow became simply "queen of a king," with the wife of her husband's successor becoming *the* queen. Alternatively, since many of the instances involving the outright title of "queen" are later than those involving "queen of a king," it has been suggested that the switch may indicate an elevation in the status of Irish queenship over time.

Other evidence indicating that Irish queenship may have been a distinct office include the existence of specific mensal lands that seem to have been assigned to queens, and the phenomenon that no matter how many wives a king was known to have had, the *Annals of Ulster* mostly record obits for only one queen per king, and never accord the title of either "queen" or "queen of a king" to more than one wife of the same king. Perhaps, then, a king could have only one designated queen per reign, and any other spouse was regarded simply as the king's wife.

While Irish queenship may have been a distinct office, historical Irish queens did not—despite depictions of the legendary pre-Christian queens Medb and Macha as strong monarchs in their own right—rule independently of their husbands. This is not to say, however, that the women were powerless. Both historical and literary accounts attest to their involvement in politics, acknowledging that the counsel of royal women, both solicited and not, could influence their husbands' and sons' actions considerably. The frequently noted presence of queens on their spouses' royal circuits and military hostings would have facilitated greatly their direct involvement in royal affairs.

Back at home, an important function of the queen appears to have been the provision of hospitality at her husband's court. Other dimensions of the queen's role included patronage of the church, as both a benefactor of religious institutions—for many queens seem to have had considerable personal wealth—and an intercessor between the church and her husband. If the evidence of later bardic compositions may be applied to pre-Norman queens, the queen's role included patronage of poets as well.

Perhaps most important of all was the queen's function as a partner in interdynastic alliance. Her participation in a royal marriage was intended to cement or inaugurate a political alliance between her family and that of her husband, both in the generation of the union itself and in that of any children resulting from the marriage, for motherhood constituted another major element of Irish queenship. Multiple marriages were very common among both kings and queens, with divorce and death contributing to marital careers encompassing what seems to have been an average of three partners per spouse.

A particularly fascinating aspect of the involvement of queens in multiple marriages is that in situations involving a switch in dynastic power within a kingship, the new king not infrequently married the widow of the old. It has been suggested that in these cases the queen was seen to symbolize the sovereignty of the land and that marriage to her constituted a claiming of the kingship. It should be noted, however, that in virtually every one of these instances, great animosity had existed between the new king and his predecessor. While the sovereignty interpretation may have some validity, the marriage should likely also be seen as an act of hostile triumph. A large proportion of queens ended their days in religious life within a convent, some of whom may have retired there, voluntarily or otherwise, after having fulfilled such symbolic requirements. Others may have retired there precisely in order to avoid being treated as a pawn in this way. That many chose to retire to a monastery associated with their birth family underlines that despite multiple marriages, the queens still had strong bonds to their native dynasties.

ANNE CONNON

References and Further Reading

Connon, Anne. "A Prosopography of the Early Queens of Tara." In *Tara: A Study of An Exceptional Kingship and Landscape*. Dublin: Discovery Programme Monographs, edited by Edel Bhreathnach. 2004, forthcoming.

Edel, Doris. "Early Irish Queens and Royal Power: A First Reconnaissance." In *Ogma: Essays in Celtic Studies in honour of Próinséas Ní Chatháin,* edited by Michael Richter and Jean-Michel Picard, Dublin: Four Courts Press, 2002.

Ní Dhonnchadha, Máirín, ed. "Medieval to Modern." In *The Field Day Anthology of Irish Writing, vol. iv: Irish Women's Writing and Traditions*, edited by Angela Bourke, Siobhán Kilfeather, Maria Luddy, Margaret Mac Curtain, Geraldine Meaney, Máirín Ní Dhonnchadha, Mary O'Dowd, and Clair Wills. Cork: Cork University Press, 2002.

Ó Corráin, Donnchadh. "Women in Early Irish Society." In *Women in Irish Society: The Historical Dimension*, edited by Margaret MacCurtain and Donnchad Ó Corráin. Dublin: The Womens' Press, 1978.

See also **Aífe; Derborgaill; Gormfhlaith (d. 948); Gormfhlaith (d. 1030); Kings and Kingship; Women**

R

RACIAL AND CULTURAL CONFLICT

Racial and cultural conflict in medieval Ireland is most famously described in a document written to the Pope, John XXII, in 1317 known as the *Remonstrance of the Irish Princes*. Composed as a justification of the Bruce invasion, it describes the fallout that resulted from English attempts to "extirpate" the native population: "Whence . . . relentless hatred and incessant wars have arisen between us and them [the Irish and the English], from which have resulted mutual slaughter, continual plundering, endless rapine, detestable and too frequent deceits and perfidies."

English policy never included anything approaching a "final solution" to the Irish problem during the Middle Ages, yet the description in the *Remonstrance* of a turbulent relationship between the two nations was not a fiction. Racial and cultural conflict was real and sprang from multifarious factors: economic disadvantage, legal disability, cultural suppression, violence, and fear of expropriation. At its simplest, it stemmed from an invasion that put two different cultures in competition for the same resources.

The Anglo-Norman invasion of the late 1160s was in fact not the first cultural clash Ireland had experienced. The first Viking incursion came in 795 C.E., with permanent settlements appearing in the mid-ninth century. The English invasion was a more thorough affair and was also more thoroughly documented, but there were definite similarities. In both cases, the invaders met a Gaelic race that was not politically centralized but that had a profound awareness of national identity. One result of this was that both sets of invaders were immediately identified as something different. A distinction emerged between the *Góidil* (the native inhabitants) and the *Gaill* (the foreign invaders), and although both Viking and English underwent Gaelicization over time, the terminology endured. Already by approximately1100, the propaganda work *Cogad Gáedel re Gallaib* (the war of the Irish with the foreigners)—written for the aspirant to the high kingship of Ireland, Muirchertach Ua Briain (d. 1119)—represented an historical tradition that celebrated the conflict between native and foreigner, irrespective of how important the Vikings had become to the Irish polity. This distinction, which was transferred seamlessly from Viking to Englishman, is now a commonplace of Irish history, but it remains important because perceived differences were the building blocks of racial and cultural conflict.

The perception of difference was equally strong on the part of the invader. Both the Viking and English invasions had sprung from economic imperatives: the problems of overpopulation and lack of land were to be solved by conquest and the opportunity to gain plunder, power, and political preeminence. Overlying the base motive for conquest was an ideology that saw the invaded as inferior and the invasion as justified. The most famous exponent of this view is Giraldus Cambrensis (1146–1223), the first "foreigner" to describe in detail Ireland and the Irish. His description was not flattering. He saw the Irish as a barbarous people, economically backward, morally and sexually debased, lazy, and wicked. They may have been Christian in name, but in reality they were a pagan and "fifthy people, wallowing in vice." For centuries these racial condemnations have been pounced upon with either delight or disgust, and historians have long labored to show how misguided Giraldus was. The various marriages between the early settlers and native Irish prove that the situation was indeed more complex than one of total racial segregation. Undoubtedly, however, the contrast between the mainstream "Frankish" culture of the Anglo-Normans and that of Gaelic Ireland provoked in many of the invaders a reaction similar to Giraldus's.

For all the cultural differences, the language of racial conflict could be remarkably similar. Both communities, for instance, charged the other with treachery. Giraldus was adamant that the Irish were "constant only in their fickleness" and should be feared "more for their wile than their war . . . their honey than their hemlock," but the Irish similarly saw treachery as a key characteristic of the foreigner. The *Annals of Inisfallen* in 1233 relate a story of one Tadc Duibfedha Mac Carthaig, who after being blinded was given a prod with a knife by one of his Gaelic captors. The anecdote continues: "He [Tadc] enquired who that was, and he was told it was Domnall *Gall* (i.e., foreign Domnall). 'That is true, indeed,' said he. 'He did that like a foreigner.'" The *Remonstrance* of 1317, referred to above, similarly details the treachery practiced by the English upon the Irish population.

Racial and cultural conflict was, however, more than just rhetoric. Some newcomers to Ireland—like Stephen of Lexington, who was sent in 1228 to reform the Cistercian monasteries of Ireland—strove to avoid the charge of racial discrimination against the Irish, but many were less sensitive. Broadly speaking, colonial policy toward the native Irish came to be one of exclusion. The Irish in general had no access to the colony's English-style justice. Attempts were made to exclude natives from positions in the colony's cities and towns and to prohibit the promotion of Irish clergy to church offices, and the Irish nobility was not represented in the colony's parliament. More fundamentally still, the native population was driven from the most fertile land and many of them were compelled to survive by raiding and plundering the colonists.

Of course, there were always exceptions to the rule. The king, for instance, qualified the exclusion of native Irish clergy by saying that it should not apply to the Irish who lived faithfully within the territory controlled by the royal government. Moreover, some of the exclusionist policies should be related to general European attitudes in the later Middle Ages. The prohibition on admitting native Irishmen to municipal office was coincidental with similar attempts in cities on the German-Slav frontier to exclude from guilds those who were not of German origin. Then there is the fact that much of the hysterical rhetoric of both sides flies in the face of what was the reality of colonial warfare. In nearly every engagement between colonist and native, there was a native fighting on the colonial side (and vice versa in many instances).

These are important qualifications, but they should not disguise the fact that national antagonisms were real. It would be impossible to plot precisely the growth or decline of racial and cultural conflict, yet some general trends may be discerned. It seems clear that England's attempts to lord it over the whole British Isles heightened hostility in Ireland. There was considerable sympathy in Ireland for the Welsh and Scottish struggles against English dominance from the mid-thirteenth century, and when Edward Bruce invaded Ireland in 1315 he was a focus for anti-English sentiment. Equally, England's adventures against France in the Hundred Years War heightened its insecurity about all things foreign.

This insecurity became particularly important in the last century of the Middle Ages when the lordship of Ireland became increasingly culturally alienated from England (*see* Anglo-Irish relations). It is now generally accepted that the famous Statutes of Kilkenny of 1366 were not a direct attack on Gaelic culture. They were an attempt to curb Gaelicization by prohibiting, among other things, the English community from marrying or fostering children with the Gaelic Irish or even using the Irish language. The statutes reveal a deep insecurity about the fact that, to survive in frontier conditions, the character of the lordship of Ireland had for the most part departed from English norms. In the fifteenth century, those from the lordship of Ireland, whether Irish or English, were classified as aliens in England. By the time of the later Tudors, a reversion to a policy of reconquest and plantation was deemed necessary to deal with the Irish problem. The result was a rekindling of racial and cultural antagonisms, but with the added spice of religious conflict. One remarkable fact about sixteenth century commentators on Ireland was how little their ideas had advanced on the racial pronouncements of Giraldus Cambrensis in the twelfth century.

PETER CROOKS

References and Further Reading

Bartlett, Robert. *Gerald of Wales 1146–1223.* Oxford: Clarendon Press, 1982.

———. *The Making of Europe: Conquest, Colonization and Cultural Change, 950–1350.* London: Penguin, 1994.

Davies, R. R. *Domination and Conquest: The Experience of Ireland, Scotland and Wales 1100–1300.* Cambridge: University Press, 1990.

———. *The First English Empire: Power and Identities in the British Isles, 1093–1343.* Oxford: University Press, 2000.

Gillingham, John. "The English Invasion of Ireland." In *The English in the Twelfth Century: Imperialism, National Identity and Political Values.* Woodbridge: Boydell and Brewer, 2000.

Giraldus Cambrensis. *The History and Topography of Ireland.* London: Penguin, 1982.

———. *Expugnatio Hibernica: The Conquest of Ireland,* edited with translation by A. B. Scott and F. X. Martin. Dublin: Royal Irish Academy, 1978.

Ó Corráin, Donnchadh. "Nationality and Kingship in Pre-Norman Ireland." In *Nationality and the Pursuit of National Independence*, edited by T. W. Moody. Belfast: Appleton Press, 1978.

Remonstrance of the Irish Princes. In *Irish Historical Documents, 1172–1922,* edited by Edmund Curtis and R. B. McDowell. London: Menthuen and Co., 1943; Reprinted in Seán Duffy, *Robert the Bruce's Irish Wars: The Invasions of Ireland 1306–1329.* Stroud: Tempus Publishing, 2002.

See also **Anglo-Norman Invasion; Bruce, Edward; Gaelicization; Lordship of Ireland; National Identity; Viking Incursions**

RÁITH BRESSAIL, SYNOD OF

The synod of Ráith Bressail, which met near Borrisoleigh (County Tipperary) in the year 1111, is, by far, the most important of all the synods associated with the twelfth-century Church Reform movement in Ireland. While the synod of Cashel, which preceded it by ten years, has been seen as introducing reform, Ráith Bressail has been perceived to be revolutionary. It sought nothing less than to bring about a complete change in the way the church in Ireland was administered. Up to the time it was convened the church did not have the administrative structure (a few Hiberno-Norse cities excepted) that existed in most of the rest of the Western church—a hierarchical, territorially based system of dioceses under the control of bishops. It was precisely that system that the synod would now set about introducing.

Before this happened, some preparations were made. A bishop was chosen for a new diocese that was established in Limerick, the headquarters of Muirchertach Ua Briain, then the most powerful king in Ireland. This bishop, Gille, set about preparing the clergy of Ireland for the changes that were about to be implemented. He prepared a tract on the constitution of the church, *De statu ecclesiae*; this explained the organization of the overall church within which the structure, about to be introduced, fitted. He sent this to the bishops and priests of the whole country accompanied by a letter that deplored what he saw as the fault of the contemporary organization: lack of uniformity of religious practice. In this he also urged them to be zealous in striving for unity of practice in conformity with the rules of the Roman church. At some time prior to the actual meeting of the synod, Gille was appointed papal legate by Pope Paschal II and it was in this capacity that he presided over it.

The Synod Meets

The synod is widely reported in the annals; all report the presence of the king, Muirchertach Ua Briain; the coarb of Patrick (i.e., the abbot of Armagh); the important Munster cleric, bishop Máel Muire Ua Dúnáin; and varying numbers of other unnamed clerics and laymen. None report Gille's presence despite its importance. Nor do they report on what it decreed, being content with only rather formulaic references to it. For this we have to depend upon a chance survival. As part of his great work on the history of Ireland, *Foras Feasa ar Éirinn*, Geoffrey Keating, the seventeenth-century historian, transcribed some details about the synod from an old book, now lost, that he found in Clonenagh. It is here that we also discover the important role played by Gille. Keating reports one of the more important decisions when he summarizes what he has read in the old book: "It was at this synod that the churches of Ireland were given up entirely to the bishops free for ever from the authority and rent of lay princes." The property of the existing church was thus to be handed over to the bishops and they were to hold it free of any charge that may have been exercised against it by laymen—a major transfer that must have been very difficult to implement in practice. Even more important, however, was its decision to divide Ireland into dioceses and to nominate their sees.

The Diocesan Structure

Based apparently upon what was originally planned (but never put into practice) for the English church as described by the Venerable Bede, it was decided that there would be two ecclesiastical provinces in Ireland, one for the northern half with its archiepiscopal see in Armagh and the other for the southern half with its see in Cashel. This corresponded with the long-established tradition of two political divisions in Ireland, Leth Cuinn (the northern half) and Leth Moga (the southern half). Lesser political divisions had to be taken into account when decisions were made about individual dioceses. This was likely to have been difficult, exacerbated as it was by the number of existing ecclesiastical establishments often associated with these political divisions that would have aspired to become diocesan sees. In this the chosen English model proved to be of considerable help in that it provided what was likely to have been an acceptable precedent for the number of dioceses to be established: thirteen, including the diocese of the archbishop, in each ecclesiastic province. Given that Bede's work was known and respected in Ireland, it would have provided a bulwark against pressures to establish a multiplicity of dioceses, and there is strong evidence that a major concern was that a cap be placed on the number of dioceses to be established.

For each province the synod specified not just the sees but the boundaries of each diocese, except in the case of Limerick, which, reflecting the role played in the synod by its bishop, Gille, is described in considerable detail: the diocesan boundaries are delimited by four named topographical points such as a mountain, a river, or the sea. The diocesan sees chosen for the northern province were Armagh (the primatial and metropolitan see), Clogher, Ardstraw, Derry or Raphoe, Connor, and Down (all six in Ulster); Duleek and

Clonard (both in Meath); and Clonfert, Tuam, Cong, Killala, and Ardcarn or Ardagh (all five in Connacht). Those chosen for the southern province were Cashel (the metropolitan see), Lismore or Waterford, Cork, Ratass, Killaloe, Limerick, and Emly (all seven in Munster); Kilkenny, Leighlin, Kildare, Glendalough, and Ferns or Wexford (all five in Leinster).

Armagh and Munster Predominant

After outlining the sees and diocesan boundaries that were to be located in Connacht and Leinster, the synod added a rider that had the same purpose in both cases. This is more explicit in the case of Connacht. It states: "If the Connacht clergy agree to this division, we desire it, and if they do not, let them divide it as they choose, and we approve of the division that will please them, provided there be only five bishops in Connacht." Here we see clearly that the synod is more concerned with the number of dioceses that are being created in Connacht and Leinster than with the actual boundaries that are being delimited. Something else is apparent from these riders; the clergy of these political provinces were unlikely to have attended the synod—otherwise, the riders would not have been necessary or would have been written differently. It seems unlikely also that the clergy from Meath were present or, if they were, they were not representative of the whole of the province of Mide since, later in the same year as the synod met, the Meath clergy held their own synod at Uisnech and redivided their territory differently, with Clonmacnoise and Clonard as the agreed sees.

The absence of these clergy suggests that the synod was predominantly a Munster and Armagh synod. This is confirmed by the names of those who subscribed to the report of the synod as found in Keating's transcription: Gille, papal legate and bishop of Limerick; Cellach, coarb of Patrick and primate of Ireland; and Máel Ísu Ua hAinmire, archbishop of Cashel. It is further confirmed by the names that we have already seen in the reports of the synod that appeared in the annals. At this stage of the reform process, therefore, the main thrust behind the move to introduce a new administrative structure into the Irish church was to be found in both Munster and Armagh.

It will have been noticed that when the dioceses of the southern province were specified by the synod, only twelve dioceses (including that of the metropolitan at Cashel) were given—that meant that it was left with one diocese short, if it were to follow the original English model. It will also have been noticed that there is no mention of the diocese of Dublin, despite the fact that it had at that time been a canonical diocese for nearly a hundred years. The reason for its omission seems to be reasonably clear; Dublin had aspirations to be the metropolitan see for the whole of Ireland under the primacy of Canterbury. What was happening at Ráith Bressail was an effort to counteract this by setting up an ecclesiastical structure within Ireland itself, independent from Canterbury. However, such a structure would obviously have to accommodate Dublin at some stage and, to allow for this, room was now left for it to join in ultimately. Had the synod wished to exclude Dublin completely it would simply have allocated twelve suffragans to Cashel as the model it was following suggested. It did not do that and it would appear, therefore, that it was its intention that Dublin would be encouraged to join and would be expected to do so at some point in the future.

Although there would be difficulties experienced in the implementation of its decisions and many adjustments would be made subsequently, the synod's work was, nevertheless, momentous and what it set out to do can still be recognized in the church structure that has endured until the present day.

MARTIN HOLLAND

References and Further Reading

Gwynn, A. *The Twelfth-century Reform, A History of Irish Catholicism II*. Dublin and Sydney: Gill and Son, 1968.

Gwynn, A. *The Irish Church in the Eleventh and Twelfth Centuries*, edited by Gerard O' Brien. Dublin: Four Courts Press, 1992.

Holland, Martin. "Dublin and the Reform of the Irish Church in the Eleventh and Twelfth Centuries." *Peritia: Journal of the Medieval Academy of Ireland* 14 (2000): 111–60.

Hughes, K. *The Church in Early Irish Society*, London: Methuen 1966.

Mac Erlean, John. "Synod of Ráith Breasail: Boundaries of the Dioceses of Ireland." *Archivium Hibernicum* 3 (1914): 1–33.

Watt, J. *The Church in Medieval Ireland*, (2nd ed.) Dublin: University College Dublin Press, 1998.

See also **Cashel, Synod of I (1101); Church Reform, Twelfth Century; *Foras Feasa ar Éirinn*; Gille (Gilbert) of Limerick**

RAWLINSON B 502

Description

Rawlinson B 502 is a vellum and paper composite manuscript now housed in the Bodleian Library, Oxford, and known after its pressmark there. It is one of the manuscripts received by the Bodleian in 1756 as part of the bequest of Richard Rawlinson (1690–1755), a graduate of the University of Oxford (St. John's College). In 1909 a collotype facsimile edition, with introduction and indexes, was published by Kuno Meyer. The manuscript is described in great detail by Brian Ó Cuív (2001) and his description has been drawn on heavily below. There are a total of 175 folios, which also includes binder's leaves. There are two vellum sections that were originally independent of one another. Their combination in this volume together with the paper leaves is due to Sir James Ware (1594–1666), auditor general under the English administration in Ireland. During his lifetime, Ware was very active in collecting manuscripts, both in Irish and in Latin, and in using them in his historical researches. Rawlinson B 502 is one of thirteen manuscripts in the bequest of Richard Rawlinson that were part of the collection of Irish manuscript material built up over a number of years by Ware.

First Vellum Section

There are twelve folios in the first vellum section, foliated as folios one through twelve that would seem to date from the late eleventh or early twelfth centuries. Writing is in two columns. Incomplete text at beginning and end suggests the original manuscript is likely to have been much more extensive. It contains a fragment of the "Irish World-Chronicle," a Latin-Irish chronicle of ancient world history based on Latin sources, mainly Eusebius, Orosius, and Bede. The text is in a fairly large and careful minuscule script. Various features of the text have led some scholars to conclude that one scribe wrote the first four folios and a second scribe wrote the remaining eight. The second scribe also added many textual glosses in both sections. Glosses were also added by at least two other scribes. It would seem to be the case that this section of the manuscript was glossed, and possibly written, in a scriptorium attached to the church of Clonmacnoise.

Second Vellum Section

The second vellum section, foliated as folios nineteen through eighty-nine, consists of seventy-one folios and consists mainly of material in the Irish language. Ó Cuív has postulated eight gatherings for this section. A number of leaves have been lost. The writing is generally in two columns, but there are exceptions. Genealogical material, for example, is in many instances set out over the full page divided into more than two columns. Care taken with the preparation of the vellum, neatness of presentation of the various texts, and the structured order and quality of its decoration indicate that this section of Rawlinson B 502 is the surviving part of one of the finest medieval Irish illuminated manuscripts and the most magnificent of the extant manuscripts containing for the most part material in the Irish language. Its decoration indicates that this manuscript was carefully planned from the outset. Splendid examples of script and decorative features from this section of the manuscript may be seen in the *Catalogue of the Irish Language Manuscripts in the Bodleian Library* (*cf.* Brian Ó Cuív [2003], plates 15–21). Apart from the decoration, the whole manuscript seems to be the work of one scribe who wrote a very neat and regular hand using minuscule letters for the most part, but with majuscule or larger letters in conjunction with minuscule in some texts.

Date and Place of Composition of Second Vellum Section

There are no scribal notes to indicate even approximately the date or place of writing of this manuscript. Ó Cuív observes that comparison with similar manuscripts of known date on the basis of script, layout, and illumination as well as linguistic features would suggest a date about the end of the eleventh century or the beginning of the twelfth. He also points out, however, that the evidence of some of the versified king lists points to a date well into the twelfth century. The last king of Connacht listed, for example, is Tairdelbach Ua Conchobair, who reigned from 1106 to 1156. Many of the texts in this section reflect a special interest in Leinster history and prehistory, which would point to compilation in a Leinster monastery. Places suggested are Glendalough in County Wicklow and Killeshin in County Laois. It has been argued by Pádraig Ó Riain that these seventy-one folios are what remain of a manuscript known as *Lebar Glinne Dá Locha* (The Book of Glendalough), a source quoted in later manuscript sources. It has been argued by Caoimhín Breatnach and Brian Ó Cuív, however, that this is not the case. As part of his argument Breatnach has pointed out that the title *Saltair na Rann* (The Psalter of Verses), the title of the first text in the extant manuscript, was applied to the manuscript by seventeenth-century scholars, one of whom was James Ware. Breatnach and Ó Cuív also pointed out that some of these scholars also cite *Lebar Glinne Dá Locha* as one of their manuscript sources. It is most unlikely that two different titles would have been applied to the one manuscript contemporaneously by these scholars.

Contents of Second Vellum Section

At the beginning of this section (folios nineteen through forty-six) is a medley of prose and verse texts mostly on historical topics, either directly or indirectly inspired by the Old Testament. Within this biblico-historical unit is the only complete copy of *Saltair na Rann* (The Psalter of Verses), a verse account of Biblical history divided into 150 sections and consisting of 7,788 lines. This important work has been dated to approximately 988. It has been argued, but not widely accepted, that its author was Airbertach mac Cosse. Also within this unit is the tract known as *Sex Aetates Mundi*, which treats of various topics of world history in the framework of the six periods of time from Adam to the end of the world. The unit ends with four poems generally attributed to Airbertach mac Cosse. Among the miscellaneous contents of the rest of this section of Rawlinson B 502 is a copy of *Amra Choluim Chille* (The Eulogy of Colum Cille), attributed to a sixth-century poet, Dallán Forgaill. The canonical text of the latter is written in majuscule letters and is accompanied by a wealth of glosses and commentary that are in minuscule script. There are also important collections of genealogies. These include secular genealogies that are arranged with a Leinster dynasty bias and genealogies of Irish saints. There are a number of historical and literary texts in prose and verse also reflecting a special interest in Leinster. Among the prose tales are *Esnada Tige Buchet* (The Melodies of Buchet's House), *Orgain Denna Ríg* (The Destruction of Dinn Ríg), *Gein Branduib maic Echach ocus Áedáin maic Gabráin* (The Birth of Brandub mac Echach and Áedán mac Gabráin), and *Orgun Trí Mac nDiarmata meic Cerbaill* (The Slaying of the Three Sons of Diarmait mac Cerbaill). The manuscript also contains the earliest extant copies of law tracts, namely *Gúbretha Caratniad* (The False Judgments of Caratnia), a text concerned with some exceptions to basic principles of Irish law, and *Cóic Conara Fugill* (The Five Paths of Judgment), a text on procedure. The format of the law tracts is main text with gloss and commentary. In the case of *Gúbretha Caratniad*, the main text is in larger letters and the glosses and commentary are in minuscule. The text and glosses of *Cóic Conara Fugill* are all in minuscule.

Paper Leaves

The paper leaves belong to the seventeenth century. Those foliated as folios thirteen through eighteen and ninety through 103 contain copies of various items relating to Irish history, mostly in Latin. According to Ó Cuív, it can be argued that the inclusion of the paper leaves implied an intention on the part of Sir James Ware to record in the bound volume, in a continuing project, copies of documents relating to Irish history. The large number of blank leaves after folio 103 implies that he abandoned the project.

Caoimhín Breatnach

References and Further Reading

Breatnach, Caoimhín. "Rawlinson B 502, Lebar Glinne Dá Locha and Saltair na Rann." *Éigse* 30 (1997): 109–132.

———. "Manuscript Sources and Methodology: Rawlinson B 502 and *Lebar Glinne Dá Locha*." *Celtica* 24 (2003): 40–54.

Ó Cuív, Brian. *Catalogue of the Irish Language Manuscripts in the Bodleian Library at Oxford and Oxford College Libraries*. 2 vols: Part 1—Descriptions. Dublin, 2001; Part 2—Plates and Indexes. Dublin: 2003.

Ó Néill, Pádraig. "Airbertach mac Cosse's Poem on the Psalter." *Éigse* 17 (1977–1979): 19–46.

Ó Riain, Pádraig. "The Book of Glendalough or Rawlinson B 502." *Éigse* 18 (1981): 161–176.

———. "Rawlinson B 502 alias Lebar Glinne Dá Locha: A Restatement of the Case." *Zeitschrift für celtische Philologie* 51 (1999): 130–147.

See also **Annals and Chronicles; Manuscript Illumination; Scriptoria**

RECORDS, ADMINISTRATIVE

No administrative records survive from any of the native lordships of medieval Ireland. The only records that do survive come from the English lordship of Ireland, which was administered partly from England and partly from within the lordship itself. This system of dual control produced three distinct but related categories of administrative records: records created by the Irish administration that remained in Ireland; records created by the Irish administration that were sent to England for administrative reasons and retained there; and records of the English administration relating to Ireland that were produced in England and kept there.

Irish Administrative Records as Originally Produced

The main constituent parts of the Irish administrative machinery to produce and keep records were the Irish chancery and the Irish (or Dublin) exchequer. Both were modeled on the corresponding English institutions and produced some of the same record series. The English chancery produced and kept multiple series of rolls on which were recorded (in slightly abbreviated form) some of the voluminous writs and other documents issued by chancery in the king's name. Each roll took the form of multiple individual membranes of parchment written on both sides and sewn together, with the bottom of one membrane attached to the top of the next. The earliest material was on the inside or top of the roll and the latest on the outside or bottom in roughly chronological order. In most series of rolls there was a separate roll for each regnal year. The Irish chancery seems to have followed this general model from some time in the thirteenth century but to have produced and kept only two main series of rolls, the Close Rolls and Patent Rolls. On the Close Rolls were recorded many, but certainly not all, of the letters close issued by the Irish chancery. These were documents issued in the king's name with a wax impression of the king's seal attached in such a way as to damage the wax when they were opened to be read. They generally took the form of instructions or authorizations to particular individuals or groups to take specific actions. These included writs of *liberate*, which authorized the Irish exchequer to make payments to particular individuals, whose counterparts were enrolled on a separate series of Liberate Rolls in England, but that were an important constituent element of the Irish Close Roll. As in England, the Close Rolls were also used for recording private acknowledgments of debt and private deeds. On the Patent Rolls were recorded letters patent issued in the king's name, to which an impression of the king's seal had been attached in such a way as to allow the document to be read on multiple occasions without damage to the wax. Appointments to offices, grants of land or privileges, pardons, and protections all took this general form. Both sets of rolls were kept in Ireland. The Irish chancery probably also, like its English counterpart, from the thirteenth century onward kept files of the writs that it had sent out with instructions to take action or collect certain information once these had been returned with a report on the action taken or the information required. It also kept on file other written warrants for other action that it took. All these record series were retained in Ireland.

The English exchequer also compiled and kept various series of rolls relating to Ireland. The oldest of these were the Pipe Rolls, annual rolls recording the accounting of local sheriffs (and later, others as well) at the exchequer for the sums of money they and others owed the king. These rolls took the form of multiple membranes of parchment mainly sewn together at the top (although some individual membranes were lengthened by adding membranes at the bottom). There were also from the thirteenth century two overlapping (but not identical) annual series of Memoranda Rolls, whose membranes were sewn together at the top, that recorded a variety of different materials relating to the exchequer's functions in collecting money due to the king and disbursing moneys as required. From the early thirteenth century onward the Irish exchequer produced Pipe Rolls and by the end of the thirteenth century, if not before, what seems to have been a single series of Memoranda Rolls that resembled their English counterparts. Both series of rolls were retained in Ireland. The Irish exchequer, like its English counterpart, also produced three copies of its annual Receipt Rolls, recording on a daily basis moneys paid into the treasury of the exchequer, and of its annual Issue Rolls, recording moneys paid out of the treasury of the exchequer. Two copies of both series of rolls were taken to England when the treasurer of Ireland was required to present his accounts at the Westminster exchequer and they were then retained there permanently. The treasurer also took with him to Westminster proof of proper authorization of payments he had made in the form of writs of *liberate* and receipts for those payments. These too were retained among the records of the English exchequer.

The accounts of the treasurer of Ireland rendered at the English exchequer and enrolled on the English Pipe Roll (and later on the related roll of Foreign Accounts) are among the more important of the records relating to Ireland produced in England. Treasurers of Ireland were required to account at Westminster from 1293 onward, although the practice died out in the mid-fifteenth century, and there was also a period in the late fourteenth and early fifteenth centuries when it seems to have been in suspension. There also survive some slightly earlier enrolled accounts that were audited at Westminster because of allegations of misconduct made against specific treasurers and a justiciar of Ireland. A considerable quantity of material relating to Ireland is also enrolled on the rolls of the English chancery, reflecting the ultimate control of the Irish administration by the king in England. There was no separate set of rolls for Irish material like the series of Gascon, Welsh, and Scotch Rolls. Instead, Irish enrolments are to be found interspersed with material of purely English relevance in the main English series of enrollments. Most judicial and administrative appointments in Ireland (including appointments of the justiciar and chancellor) are recorded on the Patent and Close Rolls. Various kinds of license (including most licenses to grant land "in mortmain" to the church prior to 1380) are recorded on the Patent Rolls. Some grants and confirmations of lands in Ireland are to be found on the Charter Rolls. The files of the English chancery also include relevant material. This includes copies of returned inquisitions *post mortem* relating to lands in Ireland held by tenants-in-chief of the crown and copies of inquisitions *ad quod damnum* into proposed mortmain alienations of property in Ireland.

Surviving Irish Administrative Records

Irish administrative records retained in Ireland have suffered badly from neglect and destruction over the centuries. By the early nineteenth century the earliest

Irish chancery rolls to survive were a Patent Roll for 1302–1303 and a Close Roll for 1308–1309. The Irish Record Commission set to work to produce a calendar (in Latin) of the medieval rolls then surviving, which was published under the editorship of Edward Tresham in 1828. Transcripts of some entries on those rolls were also made by individual scholars both in the nineteenth century and earlier and survive in manuscript. All the surviving original rolls, however, were subsequently destroyed in the fire at the Four Courts in Dublin in 1922 during the Civil War. Fate had been kinder to the Pipe Rolls of the Irish Exchequer, at least prior to their wholesale destruction. Many more of these survived, the earliest being for 14 John (1211–1212). Before their wholesale destruction in 1922, a full transcript had been made of the earliest Pipe Roll (although this was not published till 1941), and a full transcript of the Pipe Roll for 45 Henry III (1260–1261) and of much the Roll for the following year survives in the Royal Irish Academy. Later rolls down to 1348 survive only in the form of the summaries printed in appendices to the *Reports of the Deputy Keeper of the Public Records in Ireland* published between 1903 and 1927. There is a further unpublished calendar of the Pipe Roll for 1356–1357 in the National Archives of Ireland in Dublin. Transcripts and calendars of other material from the Pipe Rolls survive in other manuscript collections. By the early nineteenth century the earliest surviving memoranda roll belonged to 22 Edward I (12931294), but they survived thereafter in relatively large quantity. In 1922, all but two of them were destroyed, the sole survivors being the rolls for 3 Edward II (1309–1310) and 13–14 Edward II (1319–1320). The destroyed rolls are, however, calendared at some length in forty-three Record Commission calendars made prior to their destruction, now available in the National Archives of Ireland. There are also other transcripts and calendars made by private scholars. None of the series of Receipt and Issue Rolls of the Irish Exchequer retained in Ireland survives.

The records produced in Ireland by the Irish exchequer but sent to England for administrative purposes have fared much better. Irish Receipt and Issue Rolls survive with a few gaps for most of the period during which treasurers of Ireland found themselves accounting at the English exchequer and are now available at The National Archives (formerly the Public Record Office) at Kew in London. They are an essential source for Irish medieval administrative and political historians. The British National Archives is also the location for the surviving records of the English exchequer and chancery. The latter are also mainly available in calendared form as well.

PAUL BRAND

References and Further Reading

Brand, Paul. "The Licensing of Mortmain Alienations in the Medieval Lordship of Ireland." In *The Making of the Common Law*, London and Rio Grande: The Hambledon Press, 1992.

Connolly, Philomena. *Irish Exchequer Payments, 1270–1446*. Dublin: Irish Manuscripts Commission, 1998

———. *Medieval Record Sources (Maynooth Research Guides for Irish Local History, No. 4)*. Dublin: Four Courts Press, 2002.

Lydon, J. F. "Survey of the Memoranda Rolls of the Irish Exchequer, 1294–1509." *Analecta Hibernica*, 23 (1966): 49–134.

Otway-Ruthven, A. J. "The Medieval Irish Chancery." In *Album Helen Maud Cam*, vol. II, *Etudes presénteés a la Commission Internationale pur l'histoire des assemblées d'Etats*, vol. 24. Louvain: Publications Universitaires de Louvain, 1961.

Richardson, H. G., and G. O. Sayles. *The Administration of Medieval Ireland, 1172–1377*. Dublin: Irish Manuscripts Commission, 1963.

RECORDS, ECCLESIASTICAL

Although the great days of Hiberno-Latin composition were past, the tradition of compiling annals, martyrologies, and hagiographical composition continued apace in the post-Norman church. There seems to have been a distinct impetus to the redaction of older materials and the composition of new ones in the centuries following the Anglo-Norman invasion, a spurt of assertive cultural creativity not seen since the early Christian period. But in the post-conquest age we are dealing with a dual tradition of compiling ecclesiastical records. The churches *inter Anglicos* and *inter Hibernicos* were run and organized on quite different lines. For most of the Middle Ages, ten sees, the wealthier ones, were in Anglo-Norman hands, thirteen in Gaelic hands, and the remaining nine fluctuated between both communities or were held by absentees. Among the Irish, tenure of church lands, religious houses, and the custody of sacred relics were concentrated in the hands of hereditary ecclesiastical estate managers, erenaghs (from Old Irish *airchinnech*), and coarbs (from Old Irish *comarba*). Marriage within the native clergy was thus essential to the maintenance of the system, since ecclesiastical families ruled the church. Within the colonial enclaves, the clergy operated within defined territorial limits and were controlled by the state and by a carefully regulated system of ecclesiastical courts. Senior clergy were royal officers, episcopal *temporalia* were controlled by the crown, and the clergy at all levels were subject to tax. Diocesan synods and episcopal visitations were much more regular within the Pale. In consequence, we are much better provided with documentation from the colonial church, but we can be sure that the functioning of the Gaelic church in most fundamental respects was not entirely independent of or unlike that of the Anglo-Norman church.

The major sets of annals, monastic in origin and compiled and r-combined from various sources, were continued through the Anglo-Norman period and written for the most part in Irish. Other Anglo-Irish annals, written in Latin, were compiled in the new monastic centers established by the continental Orders, the Cistercians, Franciscans, and so forth, for example, the so-called *Annals of St Mary's, Dublin* and the *Annals of Multyfarnham*, compiled by Stephen Dexter, O.F.M., the so-called Kilkenny Chronicle, and the annals of Friar John Clyn, which ceased in the Great Plague of 1348–1349. They certainly continued to be written by men in clerical orders, and as such, although strictly a secular source, they are a valuable complement to the often scanty material from ecclesiastical sources.

From the eleventh century onward, new impetus was given to the composition of hagiographical material in both Irish and Latin, and an Irish *homiliarium* was composed. The manuscripts of the saints' lives survive in three large medieval collections put together in the fourteenth century. They include lives of Irish saints as well as imported lives of continental saints, all of them redacted in the period following the diocesan reorganization of the Irish Church at the synod of Kells-Mellifont (1152) from sources now lost. These lives were intended for an ecclesiastical audience, but the great vernacular religious compilations such as *Leabhar Breac* and the *Book of Lismore* were written for educated lay patrons but by clerical scholars.

The earlier martyrologies were adumbrated and new ones composed. They took external sources like the ninth-century martyrology of Bishop Ado of Vienne and merged it with native material, especially the commentary on the Martyrology of Óengus compiled between 1170 and 1174 at Armagh. At least four new martyrologies—those of Drummond, Turin, Cashel, and York—were compiled in the immediate post-conquest period, perhaps in response to Anglo-Norman accusations of cultural backwardness.

Documents detailing transfer of ecclesiastical property or rights existed in the pre-Norman Church, and indeed in the early Christian period. However, they are not, strictly speaking, charters, since Ireland had no central administration and no chancery. What might be described as primitive charters defining ecclesiastical property rights and prerogatives can be found in the eighth-century *Book of Armagh*, but they are best defined as *notitiae* (records of legal transactions or proceedings). They are only records of transactions between donor and recipient and lack the disposition and witness list found in later charters. The Irish charters of the eleventh and twelfth centuries differ in character and in form but derive their authenticity and sanction from transcription in Gospel books, a tradition found all over Europe. They are bilingual, written partly in Irish and partly in Latin. They were not produced by royal scribes, writing in a distinctive notarial hand, and carry no royal seal, as do later charters; they were produced for new church foundations, particularly of continental origin, by their own scribes. Diplomas, writs, and deeds are a product of the Anglo-Norman administration. A probable link has been suggested between the reform of the Irish church in the eleventh and twelfth centuries and the introduction of "full" charters (Flanagan 1998).

The native Irish church in the twelfth and thirteenth centuries continued to hold synods and issue decrees, quite separate for the most part from those that came from Anglo-Norman centers like Dublin. The Synods of Cashel (1101) issued canons against simony, clerical marriage, the exemption of the Church from rent and exaction, incestuous marriage, and the joint administration of contiguous monastic dioceses. The synods of Ráith Bressail (1111) and Kells-Mellifont (1152) reorganized the native diocesan system and issued similar canons against the recurring issues of simony and concubinage as well as inheritance of benefices, taxation of churches, and abuse of sanctuary. Like all synodalia of the medieval period, however, they are badly preserved and scattered among sources and manuscripts of later date. We know little of the provincial constitutions or councils held within the Gaelic dioceses. The earliest record of a synod held within the Pale is that convened in 1186 by Archbishop Cumin of Dublin, preserved in a confirmation of Pope Urban III. From the later thirteenth century, it can be seen that English synodal statutes were commandeered into service in modified form in many Irish foundations and that, consequently, little original formulation or legislation was undertaken in Ireland. The undated group of canons in the *Crede Mihi*, the register of the see of Dublin, probably belong to the episcopate of Archbishop Fulk de Sanford (1256–1271) and, perhaps, specifically to his visitation of his diocese in 1256–1257. They relate chiefly to the education of diocesan clergy. They derive ultimately from statutes promulgated at York between 1241 and 1255 and were adapted, without acknowledgment of source, for use in Dublin by Fulk or his predecessor Luke. Similarly, the surviving statutes of the diocese of Ferns derive from English legislation. Cross-miscegenation and repromulgation to suit local circumstances are universal in early canon law.

The Vatican archives are an invaluable source of information on the affairs of the medieval Irish Church, but many are as yet unpublished. The *Calendar of Papal Letters* contains breviates of letters from the papacy in response to individual queries relating to a great range of issues: dispensations from impediments to the holding of benefices on the grounds of illegitimacy, provisions to benefices or disputes over the possession of them, dispensations relating to marriage within the forbidden degrees of consanguinity or affinity, hereditary succession to an ecclesiastical office, and many others. They are of equal value for both the Gaelic and Anglo-Norman churches. The *obligationes pro annatis* relate to the payment of the first year's income from a benefice to the papal *camera* and are a rich source of information about local nomenclature. Most papal bulls relating to Ireland are lost, destroyed as symbols of papal government during the Reformation. Valuations of Irish dioceses for the purpose of taxation by the crown survive in some numbers for most dioceses from the early fourteenth century.

The surviving records of episcopal secretariats relate to everything from the upkeep of buildings and the management of temporalities to the chastisement of clergy and laity, but they are not contemporary and are preserved for the great part—and then only imperfectly—in later compilations. Most of these functions in each diocese and their recording lay in the hands of the archdeacon, a functionary unknown to the preconquest Irish church but indispensable to the church *inter Anglicos*. Among the most important of episcopal registers is the *Liber Niger Alani*, a compilation made for John Alen, archbishop of Dublin, in the early sixteenth century. It contains copies of documents dating from as early as the twelfth century. Some of them are found in the thirteenth-century collection of grants, charters, and letters entitled *Crede Mihi*. For the archdiocese of Armagh, we have a series of registers from 1361 to 1535, from Milo Sweetman, who became archbishop in 1361, down to George Cromer in the mid-sixteenth century. They give us a fairly complete picture of the metropolitan jurisdiction of Armagh, but they are an inchoate source in which entries are made with no obvious chronological order or diplomatic principle, composed of miscellaneous notarial notes and drafts. No doubt, later rearrangements of material and regatherings of the manuscripts or fragments of manuscripts have contributed to that state of affairs. The famous *Red Book of Ossory*, compiled in the fourteenth century, contains much nonecclesiastical material, such as hymns and poems in Latin and Norman-French. There are only two surviving original rent rolls of episcopal estates, but later copies of the rentals of Dublin and Ossory survive in the *Liber Niger Alani* and in the *Red Book of Ossory*. Episcopal deeds and cathedral registers survive in very few numbers. Most parish records of the medieval period have also perished. Collections of deeds dating back to the thirteenth century survive for only two Dublin parishes, St. Catherine and St. James. The only surviving parish account is that of St. Werburgh, Dublin.

The separate communities with the Irish Church patronized their monastic foundations with grants of land, rentals, tithes, endowments, and other privileges. Monastic records show that such grants, in addition to the original foundation charter, where such survives, were scattered over several counties. Many English and Welsh monasteries also held possessions in Ireland. Cartularies of these survive for the Augustinian priories of Llanthony Prima in Wales, Llanthony Secunda near Gloucester, and St. Nicholas's priory in Exeter. Records also survive of the possessions of each house at the time of their dissolution, 1540–1541. Each house would certainly have had some record of its possessions and their administration, including copies of charters, grants, deeds, and leases made to and by the mother house. Deeds of several monasteries survive, including the Cistercian abbeys of Duiske, Kells, and Jerpoint in County Kilkenny and of Holy Cross in Tipperary, the properties of which came into the Butler family of Ormond and are preserved among the family papers in the National Library in Dublin. The original cartularies of the abbeys of St. Mary and St. Thomas in Dublin survive, and copies of others now lost, were made by the seventeenth-century antiquarian James Ware. They are not solely cartularies but contain copies of episcopal and papal instruments and of miscellaneous grants and confirmations. The only obituary books—naming those to be commemorated in the prayers of the community—to survive are for Holy Trinity and later Christ Church, Dublin, and extracts made by Ware of the Franciscan monastery at Galway. Only two other books, miscellanies of administrative and literary material, survive from the entire medieval library of Christ Church, the so-called *Liber Niger* and *Liber Albus*, compilations of the fourteenth and sixteenth centuries. Almost the entirety of medieval Irish monastic and cathedral libraries has long since disappeared.

AIDAN BREEN

References and Further Reading

Broun, D. "The Writing of Charters in Scotland and Ireland in the Twelfth century." In *Charters and the Use of the Written Word in Medieval society*, edited by K. Heidecker. Turnhout, 2000.

Cheney, C. R. "A Group of Related Synodal Statutes of the Thirteenth century." In *Medieval Studies presented to Aubrey Gwynn, S. J.* edited by J. A. Watt, J. B. Morrall, and F. X. Martin. Dublin: 1961.

Connolly, P. *Medieval Record Sources*. Dublin: 2002, c.2 "Ecclesiastical Records."

Cosgrove, A., ed. *Medieval Ireland 1169–1534*. Dublin: 1987.

Davies, W. "The Latin Charter-tradition in Western Britain, Brittany and Ireland in the Early Mediaeval Period." D. Whitelock et al. (ed.), In *Ireland in Early Mediaeval Europe: Studies in Memory of Kathleen Hughes*, edited by D. Whitelock et al. Cambridge: 1982.

Flanagan, M-Th. "The Context and Uses of the Latin Charter in Twelfth-century Ireland. H. Pryce (ed.), In *Literacy in Medieval Celtic societies*, edited by H. Pryce. Cambridge: 1998.
Grabowski, K., and D. Dumville. *Chronicles and Annals of Mediaeval Ireland and Wales*. 1984.
Gwynn, A. "Provincial and Diocesan Decrees of the Diocese of Dublin during the Anglo-Norman period." *Archivium Hibernicum* xi (1944): 31–117.
Gwynn, A., and R. N. Hadcock. *Medieval Religious Houses: Ireland*. Dublin: 1970.
Mac Niocaill, G. *Notitiae as Leabhar Cheanannais 1033–1161*. Dublin: 1961.
MacNiocaill, G. *Medieval Irish Annals*. 1975.
Milne, K. *Christ Church Cathedral Dublin: A History*. Dublin: 2000.
Mooney, C. *The Church in Gaelic Ireland: Thirteenth to Fifteenth Centuries*. Dublin: 1969.
Ó Riain, P., ed. *Four Irish Martyrologies: Drummond, Turin, Cashel, York*. London: 2002.
Quigley, W. G. H., and E. F. D. Roberts, ed. *Registrum Johannis Mey: The Register of John Mey, Archbishop of Armagh, 1443–1456*. Belfast: 1972.
Sharpe, R. *Medieval Irish Saints' Lives: An Introduction to Vitae Sanctorum Hiberniae*. Oxford: 1991.
Sheehy, M. P. *Pontificia Hibernica: Medieval Ppapal Chancery Documents concerning Ireland, 640–1261*. Dublin: 1962–1965.
Stevenson, J. "Literacy in Ireland: The Evidence of the Patrick Dossier in the *Book of Armagh*." In *The Uses of Literacy in Early Medieval Europe*, edited by R. McKitterick. Cambridge: 1990.
Watt, J. A. "The Church and the Two Nations in Late Medieval Armagh." In *The Churches, Ireland and the Irish*, edited by W. J. Sheils and D. Wood. Oxford: 1989.
White, N. B. *Irish Monastic and Episcopal Records A.D. 1200–1600*. Dublin: 1936.

See also **Annals and Chronicles; Charters and Chartularies; Ecclesiastical Organization; Hagiography and Martyrologies**

RELIGIOUS ORDERS

Early Developments

Early Irish monasteries were largely unaffected by Benedictine monasticism, although Irish foundations on the continent played a key role in transmitting the Benedictine rule. In the ninth century the liturgical practices of the *Céle Dé* movement showed some slight Benedictine influences, and the appointment of a number of Irish Benedictine monks from English monasteries as bishops of the Norse-Irish sees of Dublin, Waterford, and Limerick in the eleventh and twelfth centuries brought Ireland into contact with monastic reformers in England. A Benedictine priory at Dublin existed from approximately 1085 to 1096.

In 1076, Muiredach Mac Robartaig (d. 1088), an Irish pilgrim and anchorite, settled in Regensburg in Germany and in 1090 his disciples established the Benedictine monastery of St. James. This became the mother house of an Irish Benedictine congregation (*Schottenklöster*) in German-speaking lands that numbered ten monasteries at its peak. The congregation established two priories in approximately 1134 at Cashel and Roscarbery for recruitment and fundraising purposes. The congregation went into decline in the fourteenth and fifteenth centuries and in 1515 the house at Regensburg was taken over by Scottish monks.

The increasing number of Irish pilgrims to Rome in the eleventh century led to the establishment of a Benedictine monastery, Holy Trinity of the Scots, on the Celian Hill.

Fore Abbey, Co. Westmeath. © *Department of the Environment, Heritage and Local Government, Dublin.*

New Orders and the Twelfth Century Reform

The orders most favored by the twelfth-century reformers of the Irish church were the Augustinian canons and Canonesses and the Cistercian monks. The Canons combined monastic observance with pastoral work and over 120 foundations were established by the mid-thirteenth century. Houses belonging to the congregations of Arrouaise and St. Victor were the most numerous, but contacts between the Irish houses and the Orders' central authorities were poor. The Premonstratensian canons founded approximately six abbeys and five smaller houses in Ireland between 1182 and 1260.

The early progress of the Cistercian monks in Ireland can be traced in great detail from the writings of St. Bernard and Irish references in the order's general statutes. In 1139, while en route to Rome, St. Malachy (Máel Máedóic) visited the Cistercian monastery at

Clairvaux, then at the height of its influence. Leaving a number of his entourage to be trained as monks he procured a site near Drogheda for the first Irish foundation, Mellifont, which was colonized in 1142 by French and Irish monks. Differences over observance soon led to the return of the French brethren to Clairvaux. Despite this, the monastery flourished, eventually numbering twenty daughter houses in its filiation.

The Anglo-Norman invasion in 1169 had a profound effect on monastic and ecclesiastical life. Although initially welcomed by Irish churchmen as promoters of church reform, racial tension soon emerged and the issue of the "two nations" in the Irish church became a dominant and divisive one for the rest of the Middle Ages. The colonists' establishment of new Cistercian houses created rival filiations to Mellifont. These foundations were staffed by English or French personnel and generally maintained a higher standard of monastic discipline so that racial animosity became fused with issues of religious observance. This contributed to a breakdown in relations between the order's general chapter and the Gaelic houses between 1217 and 1230. Known as the "Mellifont conspiracy" the dispute was largely resolved by the visitation of Abbot Stephen of Lexington in 1228. He disbanded the Mellifont filiation, imposed French and English abbots on a number of houses, dismissed nuns from the vicinity of the monasteries, and insisted that all monks be able to confess in either Latin or French. The Mellifont filiation was restored in 1274.

The Anglo-Normans also introduced the Hospitaller and military orders to Ireland. Of these, the Knights Templar with their principal preceptory at Clontarf and the Knights Hospitaller at Kilmainham were the most important. They were granted extensive lands and recruited their members almost exclusively from the ranks of the colonists. A monastery for the Trinitarians, a group dedicated to the redemption of Christian slaves, was established at Adare in Limerickin approximately 1226 and the Order of the Holy Cross (Crutched Friars) had established seventeen priory hospitals by the early thirteenth century.

The twelfth century also saw the emergence of anchorites or religious recluses attached to parochial and monastic churches in many of the towns and cities of the colony. The 1306 will of John de Wynchedon, a wealthy Cork merchant, lists four such recluses at various churches in the city, and there is contemporary evidence for their presence at sites in Dublin, Waterford, Fore, and Cashel.

The small number of early nunneries that survived into the later Middle Ages generally adopted the Augustinian rule during the twelfth century. A number of new houses for Augustinian Canonesses were also made, of which Clonard in Meath (*c.* 1144) was initially the most important. In 1195, it was listed as having thirteen daughter houses. The Ua Conchobair foundation at Kilcreevanty, County Galway, although originally Benedictine, had become Augustinian by 1223 when it was recognized as the mother house of the Canonesses in Connacht. Other important Augustinian nunneries were Killone, County Clare, St. Mary de Hogges (*c.* 1146), and Grace Dieu (*c.* 1190) in Dublin city and county, respectively. There are also references to Cistercian nuns at Derry and Ballymore, County Westmeath.

The Mendicant Friars

The mendicant friars experienced rapid growth in Ireland in the thirteenth century. The Dominicans arrived in 1224, and an independent Irish Franciscan province was erected in 1230. The Carmelites are first mentioned in 1271, and the Augustinians made their first foundation in 1282. All these Orders were founded from England, and the Irish Dominicans, Carmelites, and Augustinians formed part of the English provinces for most of the pre-Reformation period. The friars initially gravitated to the towns and boroughs of the colony, although a number of important early Gaelic foundations were also made: Franciscan and Dominican houses were established at Ennis, Armagh, and Roscommon while the Dominican foundation at Athenry, through a de Bermingham foundation, enjoyed the patronage of local native Irish lords.

The Irish friars promoted pastoral renewal through preaching and hearing confessions. Each order developed a network of *studia* or schools in which young friars were instructed. More promising students were sent for higher studies to the respective orders' *studia* at Oxford, Cambridge, Paris, Strasbourg, Bologna, Milan, and Padua. The mendicants also provided the teaching staff for the short-lived University of Dublin in the 1320s.

The friars' success brought them into conflict with the Anglo-Irish secular clergy, and Archbishop Richard Fitz Ralph of Armagh (d. 1360) proved a formidable and influential opponent.

Despite their initial fervor the mendicants were also divided by racial tension. The most frequently cited example was the death of sixteen friars as a result of a dispute between Gaelic and Anglo-Irish Franciscans in Cork in 1291. The campaign of Edward Bruce in Ireland between 1315 and 1317 further polarized the friars, and the pope strongly condemned the native Irish friars for supporting Bruce. Irish grievances found expression in theapproximately 1317 *Remonstrance* of Domnall Ua Néill to Pope John XXII, who denounced the Cistercian monks of Granard and other

Anglo-Irish religious for hunting and killing the Gaelic population without compunction. Separatist tendencies on the part of the Anglo-Irish Augustinians, Dominicans, and Knights Hospitaller were the cause of tension with their English confreres. In 1380, the attempts of an English Dominican, Friar John of Leicester, to assert his authority as head of the Irish Dominican vicariate occasioned an armed riot in Dublin during which the friars on both sides were found to be wearing chain mail under their habits.

The general decline in the fortunes of the colony from the end of the thirteenth century in the face of war, famine, and Gaelic revival also affected the religious life. No new houses of Cistercian or Augustinian Canons were founded after 1272, and by 1300 the first wave of mendicant expansion had peaked, with only a small number of foundations made after that date. The Black Death (1348–1349) had a devastating effect on religious and monastic life in Ireland as elsewhere in Europe. The Kilkenny chronicler, Friar John Clyn, records the death of twenty-five Franciscans in Drogheda and twenty-three in Dublin before Christmas 1348. As well as devastating the monasteries numerically the plague exacerbated the decline in recruitment and morale that characterized fourteenth-century Irish monasticism. Conventual life all but collapsed in many Cistercian and Canons' monasteries. The disappearance of the lay brother from Cistercian houses deprived them of their labor force and meant that the land was rented out, while speculation on the wool trade led some monasteries into financial difficulties.

The Observant Reform and the Dissolution of the Monasteries

The emergence of the Observant movement among the mendicant friars at the end of the fourteenth century brought the Irish friars into contact with one of the most vibrant reform currents in the late medieval church. Within each Order the Observants promoted rigorous discipline and strict adherence to the rule and constitutions as antidotes to the lax observance known as "Conventualism." To facilitate this, the continental Observants received papal and conciliar permission to elect their own superiors thus, forming a hierarchical structure within each order, nominally subordinate to the Conventual or unreformed authorities. In the Irish context this mechanism proved politically attractive to Gaelic friars who, by becoming Observants, could withdraw from the jurisdiction of the Anglo-Irish and English friars who had dominated each order since the thirteenth century. Although this may have contributed to the initial success of the reform in Gaelic areas, the genuine religious zeal of the reformers was recognized and many of the older foundations also adopted the reform. The movement first emerged in Ireland in 1390 among the Dominicans of Drogheda and increased in influence throughout the fifteenth century, with a distinct Observant congregation emerging by 1503. Franciscan reformers were active by 1417, establishing an Irish Observant vicariate in 1460. The Augustinian Observants made their first foundation at Banada, County Sligo, in 1423 and by 1517 numbered eight houses.

The Observants were highly regarded as confessors, preachers, and moral authorities and attracted widespread and influential patronage. The Franciscans in particular were keen promoters of the "Third Order" among their lay followers. Initially intended for zealous lay people who continued in their normal secular occupations, the Third Order or Tertiary Rule also provided the canonical basis for communities of professed religious and between 1426 and 1540 forty-nine communities of Franciscan tertiaries and one of Dominicans were founded. These Third Order houses were concentrated in the Gaelic areas of Connacht and Ulster, and their members engaged in educational and pastoral work.

Only the Franciscan movement had any impact on women's religious life, with six houses of the Order of St. Clare being listed in 1316. A later list gives three foundations tentatively identified as Carrick-on-Suir (County Tipperary), Youghal (County Cork), and Fooran (County Westmeath). The Franciscan nunnery recorded in Galway in 1511 was probably a Third Order house.

The late fifteenth century saw the establishment of colleges of secular priests at Youghal (1464), Athenry (1484), Galway (1484), and Kildare (1494) and the re-emergence of the anchoritic vocation in parts of Gaelic Ireland. Attempts at reform on the part of the Cistercians in the same period met with little success: in approximately 1497 Abbot John Troy of Mellifont asked to be excused from acting as visitator of the Irish houses because of the difficulties this entailed. Another report asserted that in only two of the monasteries, Mellifont and Dublin, was the religious habit worn or the Divine Office celebrated.

Owing to the incomplete nature of the Tudor conquest, the dissolution policy was administered unevenly in Ireland. In areas under crown control most religious houses were officially suppressed between 1536 and 1543. The earls of Thomond and Desmond were allowed to run the suppression campaigns in their own territories and their connivance ensured that some monastic houses and many friaries remained unmolested. In Gaelic Ireland the policy had little effect, allowing the friars in particular to regroup and, through their well-established Continental links, ally themselves with the forces of Counter Reformation Catholicism.

COLMÁN N. Ó CLABAIGH

References and Further Reading

Bradshaw, Brendan. *The Dissolution of the Religious Orders in Ireland under Henry VIII*. Cambridge: Cambridge University Press, 1974.

Cotter, Francis. *The Friars Minor in Ireland from their Arrival to 1400*. New York: Franciscan Institute Publications, 1994.

Flynn, Thomas S. *The Irish Dominicans, 1536-1641*. Dublin: Four Courts Press, 1994.

Gwynn, Aubrey, and R. Neville Hadcock. *Medieval Religious Houses: Ireland*. London: Longman, 1970 [reprint, Dublin: Irish Academic Press, 1988].

Hall, Dianne. *Women and the Church in Medieval Ireland, c. 1140-1540*. Dublin: Four Courts Press, 2003.

Kinsella, Stuart, ed. *Augustinians at Christ Church: The Canons Regular of the Cathedral Priory of the Holy Trinity, Dublin*. Dublin: Christ Church Cathedral Publications, 2000.

Mac Niocaill, Gearóid. *Na Manaigh Liatha in Éirinn, 1142–c. 1600*. Dublin: Cló Morainn, 1959.

Martin, Francis Xavier. "The Augustinian Friaries in pre-Reformation Ireland." *Augustiniana*, 6 (1956): 346–384.

———. "The Irish Augustinian Reform Movement in the Fifteenth Century." In *Medieval Studies Presented to Aubrey Gwynn, S. J.*, edited by J. A. Watt, J. B. Morall, and F. X. Martin. Dublin: The Three Candles, 1961.

Ó Clabaigh, Colmán N. *The Franciscans in Ireland, 1400–1534*. Dublin: Four Courts Press, 2002.

———. "Preaching in Late-medieval Ireland: The Franciscan contribution." In *Irish preaching: 700-1700*, edited by Alan J. Fletcher and Raymond Gillespie. Dublin: Four Courts Press, 2001.

Ó Conbhuidhe, Colmcille. *Studies in Irish Cistercian History*. Dublin: Four Courts Press, 1998.

———. *The Story of Mellifont*. Dublin: M. H. Gill and Son, 1958.

O'Dwyer, Barry. "The Problem of Reform in the Irish Cistercian Monasteries and the Attempted Solution of Stephen of Lexington in 1228." *Journal of Ecclesiastical History*, 15 (1964): 186–191.

O'Dwyer, Peter. *The Irish Carmelites*. Dublin: Carmelite Publications, 1988.

Stalley, Roger. *The Cistercian Monasteries of Ireland*. London and New Haven: Yale University Press, 1987.

Watt, John. *The Church in Medieval Ireland*. Dublin: University College Dublin Press, 1998.

See also **Abbeys; Anglo-Norman Invasion; Annals and Chronicles; *Annals of the Four Masters*; Architecture; Bermingham; Black Death; Bruce, Edward; Christ Church Cathedral; Church Reform, Twelfth Century; Devotional and Liturgical; Ecclesiastical Organization; Ecclesiastical Settlements; Education; Fitzralph, Richard; Fraternities and Guilds; Gaelic Revival; Hagiography and Martyrologies; Henry II; Kells, Synod of; Malachy (Máel M'áedoic); Military Orders; Moral and Religious Instruction; Nuns; Papacy; Racial and Cultural Conflict; Scots, Scotti**

RELIQUARIES

Relics, physical tokens of sanctity, were essential to the medieval Church and representative of Christ and of holy people and holy places. Their containers, termed reliquaries (or shrines), are a conspicuous manifestation, although most relics were never formally enshrined. Relics, whether primary (corporeal remains) or secondary (material things hallowed by contact or association), when subject to enshrinement were simultaneously protected, honored, and enhanced, and many shrines were readily portable. Reliquaries themselves have secondary status, having continued to be revered, in some cases to the present day, by a transference of sanctity from their original contents. The latter for certain categories of shrine are characteristically lost, notably in the case of Irish tomb-shaped shrines of seventh- to ninth-century date. These are, plausibly, the reflex of a native cult of corporeal relics, although the imported (presumably secondary) relics mentioned in early sources may have inspired their initial manufacture. The oldest, the late seventh-century, all-metal shrine from Clonmore, County Armagh, and its Irish-made analogue in Bobbio imitate Continental (and ultimately Classical) forms, but the only probable import, a stone box with sliding lid from Dromiskin, County Louth, is unique in Ireland. The Clonmore and Bobbio shrines, and their eighth- and ninth-century descendants (one of which, preserved in Tuscany, contains some human bone), appear to be miniature versions of larger containers such as the *sargifagum martyrum* in seventh-century Armagh or the *monumenta* of Brigid and Conláed that flanked the altar in contemporary Kildare.

Just as a small relic was representative of the complete, if disarticulated, skeleton of the saint, so the portable, wearable shrine represented its larger container. The latter shrines have not survived, although betokened by such remnants as the metal finials long preserved at St. Germain-en-Laye. While Christian altars incorporated relics from the sixth century, this usage in Ireland is scantily attested: a cavity within the altar of *Teach Molaisse* on Inishmurray was recognized in the nineteenth century, and the building was itself Molaisse's shrine, but the *mionna* of the altar of Clonmacnoise, noted in the annals at 1143, might have been accessories and need not have been an integral endowment. The burial, or enshrinement, of the holy dead in proximity to altars, as at Kildare, was seemingly an Insular alternative. However, not all monumental tombs were necessarily housed indoors, and those of stone surviving in the open include both box- and tent-shaped forms; one of the latter, at Killabuonia, County Kerry, is pierced at one end to allow repeated access and the creation of relics secondarily. If whole bodies were enshrined in the seventh century, parts thereof were probably so treated, as in the case of Oswald in contemporary Anglo-Saxon England. The shrine of St. Lachtaín's arm, of approximately 1120, is among the earliest of its kind; the shrine of St. Patrick's hand (another arm-reliquary) is the only example of Gothic form in Ireland.

Ireland shared in the cult of the True Cross, a relic of which was enshrined (in the Cross of Cong) in the 1120s, while another gave its name and fame to Holy Cross, County Tipperary, a twelfth-century Cistercian foundation. An early hint of native diversification is Tírechán's allusion in the seventh century to containers, *bibliothicae*, made for patens in honor of Patrick, and books (which themselves enshrined the Word), belts, and bells were afterward enshrined, typically with the effect of removing the relic from its primary functional context. The earliest book shrine, from Lough Kinale, County Longford, dates to the eighth century; the latest, that of St. Caillín of Fenagh, County Leitrim, was made in 1536. Neither retains its original contents, but in the case of the *Cathach*, a psalter, both manuscript and reliquary survive. Of belt shrines one is extant, from Moylough, County Sligo, and encloses a relic in four pieces. Bell shrines—for iron bells—are a numerous group, a reflection of the ubiquity of hand bells as cult accessories and of their creative adaptation when superannuated by physical deterioration. Croziers are likewise numerous; although unlikely to be the staffs of saints encased in metalwork, they are shrines to the extent that the crook can be designed to house relics.

The reliquaries referred to were symbols of ecclesiastical succession and sometimes made under royal patronage but were doubtless once outnumbered by the personal and smaller sort, albeit that these rarely survive: a twelfth-century silver box just 2 centimeters square from "Straidcayle," County Antrim, although of English or Continental origin, contains a ring of plaited rush wrapped in linen that is conceivably a Brigidine relic; tin-lead ampullae that held water have been found in Dublin and were brought from Canterbury and Worcester in the thirteenth century as pilgrims' souvenirs; and adventitious in the Irish record is a gold, book-shaped reliquary of the sixteenth century from the *Girona*, a ship of the Spanish Armada that sank off the County Antrim coast.

Portability is characteristic of Irish reliquaries, many being designed to be carried on a strap around the neck, a quality that allowed their use in processions, in oath-taking, in healing, in levying tribute, in solemnizing treaties, and as battle talismans. However, this is not to state their whole function or application. The tombs of the saints in dedicated churches or parts of churches were a focus of pilgrimage in which it seems that portable insignia, including croziers, were combined with statues or otherwise displayed. The Reformation dissolved this nexus or conjunction and reliquaries, by their very portability, were at once sundered from their settings and enabled to survive. Their custodianship, often hereditary, was a social institution, bound up with status and the tenure of land. They were dignified, and familiarized, with proper names: the *Breac* was the "speckled" shrine of Máedóc; the *Ballach* was the "spotted" shrine of Damnat. The nineteenth century was an age of antiquarian acquisition, but some reliquaries are still in Church hands.

CORMAC BOURKE

RENAISSANCE

In the absence of a native university before 1592 and a developed printing industry until after 1600, it is perhaps understandable that sixteenth-century Ireland did not foster an indigenous intellectual and cultural ferment as did the Renaissance elsewhere. Yet even before the Reformation and Counter-Reformation brought fresh thinking, humanist and religious strains of the European revival of learning were influencing sections of the island's population. Resort by Irish students to the universities of Oxford, Cambridge, and other academies in Europe as well as the Inns of Court in London may have given rise to the strand of civic humanism detected in their engagement with political and social reform in the earlier Tudor period. Signs of religious renewal inspired by northern Renaissance spirituality may be seen in pre-Reformation lay devotional practice among the older English community, while the upsurge of observant mendicantism in the Gaelic regions from about 1450 bore the hallmarks of the reformed piety and culture of the late medieval Italian city-states.

Among the socio-political leadership, the Anglo-Norman aristocratic families such as the Fitzgeralds of Kildare and Butlers of Ormond showed signs of being animated by the English courtly Renaissance, most notably in the case of the former's library of humanistic and other works at Maynooth and the stately architecture of the latter in Kilkenny and Carrick-on-Suir. The Gaelic lords too manifested an openness to trends in contemporary power politics in their more professional administrations, their patronage of innovative poetry and biographical projects, and their adroitness at diplomatic maneuvring, as during the Geraldine League, for example.

The Reformation brought not only an upsurge of theological debate but also a boost for the vernacular languages. The printing press, established in Dublin in 1551, produced an English book of common prayer, and a Gaelic typeface for Irish printing sponsored by Queen Elizabeth was used briefly as a tool of evangelization from 1571, although the venture was not seriously persevered with. Both Protestant and Catholic Reformations gave momentum to new pedagogical developments in later Tudor Ireland. The former pioneered diocesan grammar schools as well as the university at Dublin, while the latter fostered its own education, as at Limerick under the Louvain-educated Richard Creagh and at Kilkenny under the Oxford-trained Peter White. Graduates of these academies proceeded to participate in the foundation of Irish colleges on the continent from the end of the century.

By that time, Irish scholars were participating in a range of later Renaissance learned pursuits. The doyen was the Dublin-born Richard Stanihurst, who became the first Irish humanist in print with his London publication on Aristotelian dialectics in 1570 and whose later scholarship included Irish history and topography, a translation of the *Aeneid* into English hexameters, a treatise on alchemy, and devotional tracts. The generation produced by the Renaissance schoolmen of the earlier Elizabethan period flourished in the *fin de siècle* years, mostly in exile, and included William Bathe, Stephen White, Peter Lombard, and David Rothe, while James Ussher was the most notable product of the Anglican educational system in Ireland.

COLM LENNON

References and Further Reading

Bradshaw, Brendan. "Manus 'the Magnificent': O'Donnell as Renaissance Prince." In *Studies in Irish History Presented to R. Dudley Edwards*, edited by Art Cosgrove and Donal McCartney. Dublin: University College Press, 1979.

Lennon, Colm. *Richard Stanihurst the Dubliner, 1547–1618*. Dublin: Irish Academic Press, 1981.

Silke, John J. "Irish Scholarship and the Renaissance, 1580–1673." In *Studies in the Renaissance* 20 (1973): 169–206.

Walsh, Katherine. "In the Wake of Reformation and Counter-Reformation: Ireland's Belated Reception of Renaissance Humanism?" In *Die Renaissance im Blick der Nationen Europas*, edited by Georg Kauffmann. Wiesbaden: Otto Harrassowitz, 1991.

See also **Architecture**

RHETORIC

The principles of rhetoric, the second of the three *artes liberales*, were certainly known to the Irish, although not as much explicit evidence of its study, as opposed to practice, survives as does for Anglo-Saxon England. The Irish were well aware of the use of rhetorical embellishments in the early Christian tradition: the best of their surviving Latin compositions show that they understood the Augustinian doctrine that Christian rhetoric must reveal the truth of Scripture, make it pleasing, and move the reader/hearer. Almost certainly they had access to handbooks of rhetoric, not just Augustine *De doctrina Christiana*, the standard manual for teachers and preachers throughout the Middle Ages, and Cassiodorus' *Institutiones* but also Martianus's *De nuptiis Philologiae et Mercurii*, one of the standard textbooks on the liberal arts in the Middle Ages, and perhaps the compositions of Victorinus and the Latin panegyrists. However, the only tract on rhetoric of Irish composition that we know by name is the *Rethorica Alerani,* written probably by Ailerán, lector of Clonard (d. 665). Although no trace of it survives, it was still in the monastic library of St. Florian, near Linz, up to the twelfth century.

The epistles and sermons of Columbanus (d. 615) are our earliest evidence of a developed form of Latin rhetoric. They are marked by a complex clausular structure, prose rhythm and rhyme, alliteration, and other rhetorical devicesthat presuppose an education in quite advanced rhetoric, which he would have received at home. The sustained use of rhyming prose is very evident in the seventh-century moral-theological treatise *De XII abusiuis* and in other homiletic and exegetical pieces of the seventh and eighth centuries. The style of extravagant Latin composition known as *hisperic* contains a great many rhetorical devices and may have been inspired by the rhetorical style of Gaulish Latin authors of the fifth and sixth centuries. The Irish were also addicted to learned word play and the interweaving of their sources, biblical and patristic, into complex mosaic patternsthat can be seen in abundance in seventh-century exegetical and homiletic material.

There are rhetorical forms in Old Irish, used in saga texts especially, known as *roscada* that contain passages in rhyming prose with obscure vocabulary and strings of alliterative nouns or adjectives and nouns. Evidence of rhetorical practice in the epistolary style in Irish does not survive before the twelfth century. The earliest example is a letter written to Áed mac Crimthaind in about 1150 by Finn mac Gormáin, bishop of Kildare (d. 1160), which has the usual parts of a rhetorical epistle: the *salutatio* greeting him, the *captatio benevolentiae* praising him for his learning as "chief historian of Leinster in wisdom and knowledge." The *petitio* (request) asks that the tale *Cath Maige* being dictated to his scribe by Finn be completed by Áed, who apparently had access to a better or fuller copy, a quatrain in praise of Áed, and a request that a copy of the *duanaire* (poem book) of Mac Lonáin be sent him. It concludes with a pious subsalutation. This is rhetoric in the Latin mode, very effectively transposed into Irish idiom.

A. BREEN

References and Futher Reading

Best, R. I., O. Bergin, and M. A. O'Brien. *The Book of Leinster, formerly Lebar na Núachongbála*, vol. 1. Dublin: 1954.

Breatnach, L. "Zur Frage der 'Roscada' im Irischen." In *Metrik und Medienwechse—Metrics and Media*, edited by Hildegard L. C. Tristram. Tübingen: 1991. ScriptOralia 35.

Forste-Grupp, S. L. "The Earliest Irish Personal Letter." *Proceedings of the Harvard Celtic Colloquium* 15 (1995): 1–11.

Mac Cana, P. "On the Use of the Term *Retoiric*." *Celtica* 7 (1966): 65–90.

Oberhelman, S. M. *Rhetoric and Homiletics in Fourth-century Christian Literature: Prose Rhythm, Oratorical Style, and Preaching in the Works of Ambrose, Jerome and Augustine*. Atlanta: Scholars Press, 1991.

Tristram, H. L. C. "Early Insular Preaching: Verbal Artistry and Method of Composition." Österreichische Akademie der Wissenschaften, Phil.-hist. Kl. Sitzungsberichte, 623. Band

See also **Columbanus; Áed mac Crimthainn; Hiberno-latin**

RICHARD II

Richard II was born on January 6, 1367, at Bordeaux in the duchy of Aquitaine. With the death in 1376 of his father, Edward of Woodstock, the Black Prince, Richard became heir to his grandfather, Edward III of England, whom he succeeded in 1377 at the age of ten. His reign of twenty-two years saw a number of domestic crises, from the Peasants' Revolt (1381) to later conflicts with a disaffected nobility, culminating in his usurpation by his cousin, Henry Bolingbroke (crowned Henry IV). Richard was deposed in September 1399 and died in captivity at Pontefract Castle in February 1400, events made familiar through their dramatization in Shakespeare's *Richard II*.

Richard II goes to Ireland from *Froissart's Chronicles* (Volume IV, part 2). © *The British Library*.

Richard was the first English king to visit the lordship of Ireland since John in 1210 and the only reigning monarch to make two such expeditions. This unusual degree of involvement has prompted historical speculation about his motives and degree of success, for although recurring crises may have justified royal intervention no other monarch responded to the lordship's needs with such a commitment. In recent biographies drawing on modern studies of Richard's reign, English historians have set his aspirations in a wider British context, linking his policies in Ireland with those in other peripheral regions and with his views on royal authority and the obedience of subjects.

By the time Richard became king, the Irish lordship was regularly requiring support from England to meet its critical military and financial needs. In 1385, a council in Dublin asked for personal royal intervention. Sustainable recovery could not be effected by passing the burden to chief governors, whether the king conferred authority on them by military indenture or, as in 1385 with Robert de Vere, Marquess of Dublin, by the grant of extensive powers commonly reserved to the crown. In these circumstances, amid fears of the lordship's complete collapse, Richard's first Irish expedition (1394–1395) was an extraordinary success.

Richard arrived in Ireland in October 1394 with a substantial force. His primary objectives were military and political, but he also intended administrative and financial reforms. The presence of the king in Ireland provided an unprecedented opportunity to establish peace between the different interests in the country. A combination of diplomacy and of overwhelming force, demonstrated in the defeat of Art Cáeánach Mac Murchada in Leinster, won the submission of Gaelic Ireland. Correspondence from the rebel Irish lords records their willingness to accept the English crown and their desire that Richard arbitrate in their disputes with the English of Ireland. Richard accepted this position as the basis for a common approach to all the submitting Irish. Rather than focus on their punishment for past rebellion, he welcomed the Irish lords as his lieges, requiring them to make oaths of allegiance that effectively recognized their status as his subjects. Special sensitivities within some areas called for additional arrangements. In Leinster, an area of particular difficulty because of Mac Murchada's authority and the lordship's vulnerability, he attempted to revitalize the English interest, requiring the Irish to yield lands they had seized and making grants to English knights. In Ulster, however, there had been no resolution of the differences between Ua Néill and Roger Mortimer, earl of Ulster, when Richard left Ireland in April 1395.

Although little evidence survives about the lordship's government and administration in the 1390s, occasional references show that in the following years

Richard attempted, for a time, to maintain his expeditionary settlement. It was from the start under great pressure, protected by a greatly reduced military capacity. The collapse of the fragile peace was hastened by many factors. These included the unresolved difficulties between Gaelic Ireland and the lordship, the ambitions of local lords, the crown's dependence on Mortimer as lieutenant, the conflict within the Irish administration, and the financial and political problems in England that demanded Richard's attention.

In late 1397, with the Irish lords of both Ulster and Leinster once more at war, Richard was already planning to return. His second expedition, in June and July 1399, again brought a significant army to Ireland, but in less favorable circumstances. The campaign in Leinster was already in difficulty when the venture was cut short by news of Bolingbroke's return to England in arms against Richard. The bulk of the expeditionary forces withdrew in haste and disorder from Ireland, leaving behind a vulnerable lordship and, for both Gaelic Ireland and the Anglo-Irish community, a message of royal impotence. Richard's Irish aspirations ended in failure, both for himself and for the English interest in Ireland.

Dorothy B. Johnston

References and Further Reading

Curtis, E. *Richard II in Ireland, 1394–5, and Submissions of the Irish Chiefs*. Oxford: Clarendon Press, 1927.

Johnston, Dorothy. "Richard II and the Submissions of Gaelic Ireland." *Irish Historical Studies*, 12 (1980): 1–20.

———. "The Interim Years: Richard II and Ireland, 1395–99." In *England and Ireland in the Later Middle Ages: Essays in Honour of Jocelyn Otway-Ruthven*, edited by J. F. Lydon. Dublin: Irish Academic Press, 1981.

Lydon, J. F. "Richard II's Expeditions to Ireland." *Journal of the Royal Society of Antiquaries of Ireland*, 93 (1963): 135–149.

Saul, Nigel. *Richard II*. New Haven and London: Yale University Press, 1997.

See also **Anglo-Irish Relations; Chief Governors; O'Néill; Ulster, Earldom of**

RINGFORTS

Ringforts are the most ubiquitous, abundant and, ironically, among the less-studied monuments on the Irish landscape. Over 45,000 have been identified, the densest concentrations occurring in north Connacht, north Munster, part of south Leinster, and in a band extending from south Antrim, through Monaghan and Cavan in Ulster, southwest into the Leinster county of Longford. They tend to be located on sloping ground in the well-drained soils of lowland areas.

A ringfort is essentially a circular or near circular space, which in some instances is raised above ground

Ballyconran Ringfort, Co. Wexford. © *Department of the Environment, Heritage and Local Government, Dublin.*

rounded in turn by a ditch. The interior was reached via a causeway across the ditch and a gate entrance in the bank. In the manner of a property boundary, the "ring" generally defined the perimeter of a homestead and encompassed and protected a dwelling or group of dwellings. Where the surrounding bank or banks of a ringfort are built of earth the term *rath* is given to the construction, while the words *caiseal* and *cathair* are generally used to describe ringforts made of stone. Unlike their earthen counterparts, stone forts tend not to have an external ditch. The word *lios*, which is frequently embraced in Irish place names, refers to the interior of a ringfort, while *urlann* is the term given to an open space in front of a ringfort. Most ringforts have one bank, but there are some with two or three banks and intervening ditches. The bank would have been augmented by a timber palisade, a quick-set hedge (hawthorn), or a sturdy growth of bushes and trees. Ringforts vary greatly in size from approximately 27 meters to 75 meters in diameter internally, but the average *rath* tends to be 27–30 meters and, in general, stone forts tend to have smaller diameters. It is argued that the size of ringforts and the complexity of their enclosing banks suggest something about the status of their occupants. Larger ringforts appear to have accommodated the highest grades of society and to have attracted a clustering of smaller ringforts around them. A feature often found in ringforts is a souterrain, or underground passage, generally constructed of stone but also created by tunneling into natural rock or compact clay. Souterrains provided refuge for the inhabitants of ringforts and were also possibly used as storage facilities.

Ringforts constructed in the early medieval period essentially enclosed single farmsteads engaged in pastoral farming. This interpretation of their primary function is especially borne out by the results of Lynn's excavations at Deer Park Farms, County Antrim, where a *rath* was found to enclose a group of five contemporary wicker houses dating to approximately 700 C.E. The layout of this farmstead, the design of its houses, and the artifacts recovered are considered consistent with the material attributes of lower grade ringfort occupants noted in the seventh-century Irish law tract called *Crith Gabhlach*. Typical early medieval finds from ringforts include bronze, iron, and bone pins; glass beads; crude handmade pottery called souterrain ware; and wheel-thrown pottery termed E ware. Cattle bones are also particularly numerous, showing the dominance of cattle meat in the diet of ringfort inhabitants and an emphasis on dairying.

Stout's seminal study of ringforts proposes that the majority were constructed in a three hundred-year period from the beginning of the seventh century to the end of the ninth century C.E. This conclusion is based on the predominantly early medieval radiocarbon and tree-ring dates obtained from just forty-seven excavated ringforts. At present there is insufficient evidence from archaeological excavations to support any claim that ringforts continued to be constructed after 1200. However, structural features in the fabric of some upstanding stone forts, significant information derived from a few excavations, depictions of ringforts as "living sites" on Tudor maps of the late sixteenth and early seventeenth centuries, and historical references and analysis of distribution patterns combine to suggest that ringfort settlement did not become obsolete at the end of the early medieval period. The well-known map picture of Tulach Óg in Tyrone drawn by Richard Bartlett in 1602 shows the dwelling of the Uí Ágáin family as a single-banked ringfort containing a large house and a cabin. As late as 1619, the Lindsey family who had received the lands at Tulach Óg during the Plantation of Ulster occupied the ringfort.

Some ringforts appear to have enjoyed a measure of continuity of use from the early into the later medieval period. Excavations by Jope at the *rath* of Ballymacash, County Antrim, revealed that the ringfort had been built in two phases. A radiocarbon date of 1020–1250 was obtained from the remains of an oak post positioned in the clay floor of one of three houses associated with the second phase of occupation of the ringfort, and stratified shards of everted-rim ware dating to the late thirteenth or early fourteenth century were found in association with another of the three houses. Excavations by Raftery at Rathgall, County Wicklow, have proved that the *caiseal* enjoyed substantial occupation in the late thirteenth and fourteenth century, probably as the *caput* of the Gaelic sept of Uí Bhróin. Over two thousand shards of medieval pottery including locally produced glazed ware and Leinster cooking ware were found within the *caiseal* in association with coins of late thirteenth- and fourteenth-century date.

Some of the well-preserved stone forts of the Burren, County Clare, contain structural features indicative of modification and occupation by leading Gaelic families in the later medieval period. Cathair Mhór and Cathair Mhic Neachtain both retain the remains of late medieval two-story gate-tower entrances. Within Cathair Mhór there is a large masonry dwelling with rounded quoins and a stout batter of the same period. Cathair Mhic Neachtain, which was the residence of the Uí Dhubhdábhoireann legal family, contains the foundations of several buildings, among them a large dwelling and a kitchen house, described in a seventeenth-century document. Stone forts occupied in the later medieval period served much the same purpose as the bawn wall of tower houses—they afforded a measure of defense and defined a courtyard in which domestic buildings were situated.

The longevity of ringforts and the modifications made to them over time are not yet fully appreciated and are likely to dominate future scholarly investigations of this most commonplace, versatile, and complex of Irish medieval monuments.

ELIZABETH FITZPATRICK

References and Further Reading

Edwards, Nancy. *The Archaeology of Early Medieval Ireland.* London: Batsford, 1990.

O'Conor, Kieran D. *The Archaeology of Medieval Rural Settlement in Ireland.* Dublin: Discovery Programmme/Royal Irish Academy, 1998.

Stout, Matthew. *The Irish Ringfort.* Dublin: Four Courts Press, 1997.

ROADS AND ROUTES

Irish place names preserve a variety of words for roads of different status, construction, and quality. Among these are *bóthar* (a cattle track), *bealach* (a passage, gap, or road), and *tóchar* (a causeway), which was usually of timber. The word *casán* is used of a path, while *ceis* specifically refers to a path made of wattles. The term *slighe* is given to a high road or a more significant national route. Apart from roads, major rivers such as the Shannon, the Erne, the Liffey, the Barrow, and the Boyne were important highways in early medieval Ireland and not surprisingly they attracted both ecclesiastical and secular settlements such as the monastery of Clonmacnoise on the River Shannon and St. Mullins on the Barrow, which was both a monastic site and an Anglo-Norman stronghold during the twelfth century. The Norse Vikings made particularly good use of Ireland's navigable rivers during their ninth-century inland raids and for the purpose of establishing over-wintering encampments or *longphort* settlements. Norse *longphoirt* tended to be located either at the mouth of a river, like Athlunkard on the River Shannon near Limerick City, or at the confluence of rivers such as those at Dubh Linn and Áth Cliath set up in the Liffey estuary during the 840s and the alleged stronghold of the Viking leader Rodolph at Dunrally, County Laois, which was sacked in the year 862.

Five high roads reputedly emanated from Tara in ancient Mide. The Slighe Mór followed the east–west route of the gravel ridge known as the Eiscir Riada between Dublin and Clarinbridge in County Galway. The Slighe Dhála or Slighe Dhála Meic Umhóir was the road from west Munster to Tara and formed part of the boundary of north Munster on its southwest course, which took it as far as Tarbert in County Kerry. It passed through what are now the modern counties of Dublin, Kildare, Laois, Tipperary, and Limerick. Its route in the Laois area took in Ballyroan, Abbeyleix, Shanahoe, Aghaboe, Borris-in-Ossory, and Ballaghmore. Créa, after whom the Tipperary town of Roscrea is named and through which the Slighe Dhála passed, was the wife of Dála. The road is also known as Belach Muighe Dála, a name that partly survives in the name Ballaghmore (Great Road) given to the two townlands of Ballaghmore Upper and Lower that lie just a mile and a half west of the early medieval church site of Confert-Molua. Slighe Assail connected ancient Mide with Connacht, Slighe Mhidhluachra linked Tara to Emain Macha in Armagh, and Slighe Chualann led from Tara to the southeast. All five roads are attributed supernatural origins in early medieval literature. Various legendary heroes are attributed with the discovery of the roads as they traveled to Tara to celebrate the birth of king Conn Céadcathach at the Feast of Tara in the first century C.E. Surprisingly little research has been done on this road system. In the only comprehensive enquiry into the subject by Colm Ó Lochlainn in 1940, he suggested that Cormac, son of Art and grandson of Conn Céadcathach, could have been the king who instigated the road-building program. Perhaps what Cormac undertook was the construction of link roads connecting Tara with a long-established countrywide road system.

Irish archaeology has made a significant contribution to our understanding of the nature of roads and routes and, in particular, their association with early medieval monasteries. Proximity to the great roads was apparently an important influence in the choice of location for the establishment and development of societal monasteries. The monasteries of Glendalough, County Wicklow; Clonard, County Meath; Clonmacnoise, Durrow, Gallen, Lemanaghan, Rahan, and Tihilly, County Offaly; and Clonfert, County Galway were sited on the route of the Slighe Mór. Likewise, several monastic communities established their foundations close to the Slighe Dhála. Among those that developed into important ecclesiastical centers on that route were Aghaboe and Clonfert Molua, County Laois; Monaincha and Roscrea, County Tipperary, and Killaloe, County Clare. Evidence from excavations suggests that the construction of networks of lesser roads connecting monastic sites with the great roads and other landmarks also took place in the early medieval period. Many of these lesser routes are named "Pilgrim's Road" or "Monk's Path" in local tradition. Some of the more impressive remains of *tóchars* have been found in association with the monastery of Clonmacnoise. A gravel and flagstone road constructed and used between 566 and 770 C.E. and then intermittently until the second half of the thirteenth century was uncovered during drainage works in Bloomhill Bog, a short distance northeast of Clomacnoise. It formed part of a larger network of routes focusing on the monastery. A similar gravel road found in Coolumber Bog appears to have been the principal western approach to Clonmacnoise and may have connected with a wooden bridge over the Shannon downstream from the monastery. Substantial evidence for the bridge was recovered during survey and excavations in the late 1990s. It ran for a distance of approximately 160 meters and constituted about fifty oak posts each up to 15 meters long driven 3 to 4 meters down into the riverbed. Dendrochronologically dated oak trunks, used as vertical timbers, suggest a date of 804 for the construction of the bridge.

While roads and routes feature on sixteenth- and seventeenth-century regional maps of Ireland, their mapping was highly selective and generally related to military activity. Roads were often marked only in the context of routes followed by military forces which, for instance, is the case with the roads depicted on both John Grafton's map of Mayo and Sligo (1586) and on Richard Bartlett's map of southeast Ulster (1602). More commonplace roads in everyday use tended not to be included in Elizabethan regional cartography. Routes might be noted in circumstances where they traversed difficult terrain such as bogland or where they afforded a pass through forest or a fording point on a river. A map of Leix-Offaly made approxmately 1560 marks twenty-five passes through bogland and woodland. It was not until the eighteenth century that the network of Irish roads and routes was more comprehensively addressed in cartography.

ELIZABETH FITZPATRICK

References and Further Reading

Andrews, John H. "The apping of Ireland's Cultural Landscape, 1550–1630." In *Gaelic Ireland c. 1250—c. 1650: Land, Lordship and Settlement*, edited by Patrick J. Duffy, David Edwards, and Elizabeth FitzPatrick. Dublin: Four Courts Press, 2001.

Moloney, Aonghus. "From East and West, Crossing the Bogs at Clonmacnoise." In *Clonmacnoise Studies* (Vol. 1, Seminar Papers 1994), edited by Heather A. King. Dublin: Dúchas, 2000.

Ó Lochlainn, Colm. "Roadways in Ancient Ireland." In *Essays and Studies Presented to Professor Eoin MacNeill*, edited by John Ryan. Dublin: Three Candles, 1940.

O'Sullivan, Aidan, and Donal Boland, Donal. *The Clonmacnoise Bridge: An Early Medieval River Crossing in County Offaly* (Archaeology Ireland Heritage Guide No. 11). Dublin: Wordwell, 2000.

ROMANCE

The study of romance in medieval Ireland has been hindered by doubts about its literary merits, a certain prejudice against its foreign origins, and by the sheer number of such works and the lack of printed editions. Thus, it is not yet possible to offer a full assessment of the influences that shaped Irish romance or of its place in Irish literary history. The period of written romance in medieval Ireland broadly coincides with Early Modern Irish (1200–1600).

The earliest evidence for romance occurs in an Irish manuscript of the late twelfth century that contains a catalog of traditional tales, among them one called *Aigidecht Artúir* (The entertaining of Arthur). Several trends were converging about this time to make Ireland more receptive to Continental literature. Ecclesiastical reforms brought Continental religious orders and a realignment of the Irish learned classes. The new learned class, the bardic families, concentrated on eulogistic poetry designed to flatter the numerous petty kings that emerged after the Anglo-Norman invasion. One casualty was the traditional repertoire of Irish prose tales, which lost much of their *raison d'être* in the changed literary and political order. Moreover, the new Anglo-Norman ruling class favored French literature, especially romance.

The influence of foreign romance at first manifested itself indirectly. The traditional Irish tales continued to circulate, but they were now altered in quantity and quality under the influence of foreign romance. Thus, the number of tales that enjoyed currency became much smaller. A principal casualty was the cycle of Historical Tales (or King Tales). Even the other two major groups of tales from Old-Irish literature, the Mythological Cycle and the Ulster Cycle, underwent a process of severe selection. Most of the newly selected tales dealt with love or the marvelous or lent themselves to ready expansion with numerous incidents of the marvelous—all characteristic features of romance. Moreover, the tone was transformed. It is telling to compare the ninth-century tale *Longes Mac n-Uislenn* with its fourteenth-century revised counterpart, *Oidheadh Chloinne hUisneach*. The former is marked by a heroic ethos, austere style, and tight narrative in contrast with the romantic ethos, verbose style, and loose structure of the latter. A fourth group of traditional tales, the Fenian Cycle, which related the deeds of Finn Mac Cumaill and his *fiana* (band of warriors), had the advantage over the older cycles of being still in formation after the Anglo-Norman invasion. Thus, it could accommodate itself more easily to contemporary literary tastes. It also had a wealth of romantic matter: elements of the marvelous (from folk tradition), heroes, and the ill-starred lovers Diarmait and Gráinne (compare Tristan and Iseult). The trend of tailoring native tales to romantic tastes continued as late as the fifteenth century.

By the fifteenth century the influence of foreign romance became overt. The Norman-English settlers had become so wedded to Gaelic society that they were in a position to influence it directly, a process helped by the fact that the English administration was preoccupied with the political turmoil in England that culminated in the Wars of the Roses. Romantic tales of English and French origin, a central part of the Anglo-Norman literary heritage, were now translated into Irish. Representative works of the well-known cycles of romance appear in translation: from "The Matter of Britain," "The Quest of the Holy Grail," from "The Matter of France" tales from the Charlemagne Cycle; from "the Matter of Greece and Rome," *Stair Ercuil* (based on Caxton's English translation of the French); and from Middle English (fourteenth century) such romances as *William of Palerne*, *Guy of Warwick*, and *Bevis of Hampton*. In addition, translations were made of certain works that, although not romance, contained elements of the marvelous; notably, *The Travels of Marco Polo*, *The Travels of Sir John Mandeville*, the *Letter of Prester John*, and the *Book of Alexander.*

Such were the foreign influences at work in the shaping of the Irish Romantic Cycle (Ir. *rómánsaíocht*). Its individual tales (in prose) are very difficult to date since most of them are anonymous and written in a standardized literary language. Although appearing mostly in manuscripts of the seventeenth and eighteenth century, they were composed well before that time as suggested by the presence of a few of them in fifteenth-century manuscripts. Their authors were professional men of letters, knowledgeable not only in the native repertoire of stories and folk motifs but also in foreign romances. In origin literary (rather than oral) productions, they were intended for oral delivery to an aristocratic audience.

For the most part their subject matter is drawn from the exploits (real or imagined) of characters from Ireland's literary history: the court of Conchobar of Ulster and his opponents, as portrayed in the Ulster Cycle; Finn Mac Cumaill and his *fiana* as found in the

Fenian Cycle; the Tuatha Dé Danann, descendants of a pre-Celtic people, who inhabited the Otherword—although their role is more as catalysts than protagonists—and occasional "historical" provincial or high kings. At least one foreign cycle is evident, the Arthurian. Although Irish in subject matter, these romances share the same themes as their Continental counterparts, adventure and love, narratively framed as a quest conducted by an individualized hero. More specifically, they often have as a theme a conflict between an Irish hero and an enemy from outside, whether foreigner or someone from the Irish Otherworld. The foreigner, alone or with an army, invades Ireland and after a long struggle is defeated. Elsewhere, the foreigner is imbued with magical powers (often because he is of the Otherworld) and can only be defeated with the help of friendly supernatural agents. The most popular theme involves the hero in an external adventure, either outside Ireland or in the Otherworld, a genre known as *Eachtraí*. As with Continental romance, the adventure involves a quest: for a woman who has been carried off; for someone bound by enchantment; on behalf of a foreign woman in trouble; or to satisfy a requirement imposed by a wicked character. Likewise, the Irish romance is often composed structurally of a chain of loosely connected episodes. Unlike most other types of medieval Irish literature, the romance survived the collapse of the Gaelic order in the sixteenth century, finding a new home in the folk tale.

Pádraig Ó Néill

References and Further Reading

Baumgarten, Rolf. *Bibliography of Irish Linguistics and Literature 1942–71*. Dublin: Dublin Institute for Advanced Studies, 1986.

Best, Richard I. *Bibliography of Irish Philology and of Printed Irish Literature*. Dublin: Stationery Office, 1913.

———. *Bibliography of Irish Philology and Manuscript Literature, Publications 1913–1941*. Dublin: Hodges and Figgis, 1942.

Bruford, Alan. *Gaelic Folk-Tales and Mediæval Romances*. Dublin: the Folklore of Ireland Society, 1969.

Murphy, Gerard. *The Ossianic Lore and Romantic Tales of Medieval Ireland*. Dublin: At the Sign of the Three Candles, 1955.

———. "Irish Storytelling after the Coming of the Normans." In *Seven Centuries of Irish Learning, 1000–1700*, edited by Brian Ó Cuív. Dublin: The Stationery Office, 1961.

See also **Bardic Schools/Learned Families; Echtrae; Historical Tales; Mythological Cycle; Ulster Cycle**

S

SATIRE

Satire has a long-standing role in Ireland's history and literature. Still a modern characteristic feature of its prose, satire is best attested in the verse literature of medieval Ireland.

Several Irish terms exist for satire, all deriving from the basic notion "to cut" or "strike." Satire was levied through verse compositions, artfully crafted within a set of accepted technical rules. As a skill of the trained poet, the proper composition of satire required lengthy training, memorization, and study of the traditional styles, meters, and rhymes. The professional poet, and in the later period the bard, were the master craftsmen of satire, paid highly for their art. As expected, the more skillful the poet, the higher his rank, and the higher his rank, the more expensive his satire. Interestingly, while female poets were uncommon, female satirists seem to have been relatively familiar and accepted.

In the early period, satire seems to have been fairly short and concise, consisting of blunt sarcasm or ridicule. Compositions lengthened in the later medieval period, often with a less specific victim or objective. Satire was believed to cause facial blemishes and blisters, and in extreme cases, even death. Early annals relate the deaths of notable figures, deaths brought on by particularly potent satirical verse. Literature of the Middle Ages, English and Gaelic, including works by Shakespeare and Spenser, mention Irish satire employed to kill men and animals, mainly rats. Throughout the later period, and up to the present day, satire developed a less specific and individual nature, evolving into the general lampooning so characteristic in modern Irish literature.

Satire was employed for various means, and not simply public defamation of character. At its basic level, it was used to threaten and insult a targeted individual. Common topics of satire included moral and intellectual faults such as cowardice, stinginess, inhospitality, ignorance, treachery, and conceit. Everyday devices of satire included sarcasm, innuendo, and creating nicknames that stuck.

From the early period, satire was also a sanction used to enforce and ensure legal remedy. The threat of satire could prompt payment of claims, fines, and penalties. It could also force a high-ranking member of society to submit to legal arbitration. The formal procedure of the latter made it illegal to ignore satire, ensuring the cooperation of the higher-ranking defendant. Attesting to its pervasive role in society, the Church itself was not immune from the power of satire and its poets. Saint Columba (Colum-Cille) himself is once described in a cold sweat, fearing satire from a poet he cannot remunerate. Columba is saved when his sweat turns to gold, which he generously and immediately offers in compensation.

As a powerful legal tool, Irish law maintained strict regulation for the proper use of satire. Illegal satire was anathema to both society and its legal system. The illegal satirist was punished heavily, usually stripped of all social rank and standing. Illegal satire included publicizing a false story, mocking a disability or deformity, wrongful accusations, and technically or metrically imperfect satire. It was also illegal to satirize someone after his death.

ANGELA GLEASON

References and Further Reading

Kelly, Fergus, *A Guide to Early Irish Law.* Dublin, 1988, 43–44, 137–138

Mercier, Vivian, *The Irish Comic Tradition.* Oxford, 1962

See also **Brehon Law; Colum-Cille; Poetry, Irish; Poets/Men of Learning**

SAVAGE

The Savage family was one of the principal English families of medieval Ulster. William Savage witnessed one of John de Courcy's charters and his son was among those taken hostage by King John in 1204 as sureties for their lord. In 1276 Robert Savage held a large estate in north Antrim, probably granted by Hugh de Lacy between 1227 and 1242. Under the de Burgh Earls, in the later thirteenth century, the family of de Mandeville in particular overshadowed the Savages, although Richard Savage attended parliament in 1310 and held the north Antrim manor in 1333. The family prospered from the eclipse of its main rivals during the serious political crisis of the 1330s. The Bissets were deeply involved in the Bruce wars; the De Mandevilles were involved, through Henry, in the conspiracy of the earl of Desmond in 1331. More importantly, the earl of Ulster was murdered in 1333 by a combination of De Mandevilles and Logans. Robert Savage was the chief juror in the subsequent inquisition into the Earl's lands and, with Henry de Mandeville, took charge of the earldom on behalf of the earl's widow and baby daughter. Along with John Savage, he was rewarded in 1342 with grants of land in the Six Mile Water valley of modern south County Antrim. Together with the Savage estate in north Antrim, his lands straddled the client Irish kingdom of the O'Flynns of Uí Tuirtre. Robert's position was such that he was perceived by the annalist of St. Mary's Dublin as the mainstay of English power in Ulster in 1358 and that his death in 1360 marked a serious blow to that cause.

By the mid-fifteenth century, lands of the Earldom of Ulster had been seized or granted to the O'Neills of Clandeboy, displacing English and Irish families alike. The Savages established themselves in an estate in the south of the Ards peninsula of modern County Down. The Savages appear to have acquired (either by grant or force) former lands of the Earldom there: Ballyphilip (or Portaferry) and Ardkeen, both sites of Savage tower houses. To the north of their lands in seventeenth century Inquisitions it was recorded that the area around Kircubbin was held by O'Flynns "of the Turtars;" the two families' fortunes seem to have been linked together. In the sixteenth century, the towers of Quintin, Ballygalget, and Kirkistown castles were added to the family holdings. In the fourteenth century Henry Savage was summoned to Parliament, while through the later fourteenth and fifteenth centuries, Savages were frequently appointed seneschals of Ulster: They were established as the senior family of the English interest in the region. The family had established a landed estate that, in economy, politics, and social connections was closely parallel to the gentry estates of the north Pale. During the seventeenth century, they made a successful transition to the new order, intermarrying with the Montgomeries, Viscount Ards and turning Protestant, although Patrick Savage of Portaferry was severely indebted and needed to be rescued by Montgomery.

T. E. McNeill

References and Further Reading

Flanagan, D.E. "Three settlement names in Co. Down." in *Dinnseanchas*, 5, 1973, 65–71

McNeill, T.E. *Anglo-Norman Ulster.* Edinburgh: John Donald, 1980

Savage-Armstrong, G.F. *Ancient and noble family of the Savages of the Ards.* London: M. Ward, 1888

See also **Ulster, Earldom of**

SCANDINAVIAN INFLUENCE

Scandinavian influence on Medieval Ireland is evident in the considerable number of place-names of Norse origin that dot the Irish landscape, including Larne in the north which has been equated with *Ulfrecksfjörðr*, Wicklow (*Vikingaló*), Wexford (*Weigsfjörðr*), and Waterford (*Vetrafjörðr*) on the east coast, and Limerick (*Hlymrekr*) in the southwest. In addition, Old Norse has bequeathed to Irish a substantial number of loanwords, most notably in the fields of seafaring and trade where the impact of the Scandinavians was most pronounced, including perhaps *bád* (boat), Old Norse *bátr*, and *margad* (market), Old Norse *markaðr*. As well as adopting their words, Irish rulers also eagerly followed many of the practices of the Norsemen, as is evident in their extensive use of naval forces from the tenth century onward. Those same native leaders were also quick to appreciate the benefits that Norse trading activities brought in their wake. Archaeological evidence suggests that much Viking Age silver found its way into Irish hands, whereby it was used as currency. Michael Kenny, for example, has interpreted the important hoards from Lough Ennell County Meath, as indicative of a close economic relationship between the Clann Cholmáin kings of Mide and their Dublin neighbors based in part on their mutual interest in slaves. In the same way, John Sheehan has speculated that one of the most characteristic Hiberno-Norse silver objects, the broad-band arm ring, may have served as tribute between ever more interdependent Irish and Norse rulers. Silver was also exchanged for everyday commodities, as coins were from the tenth century onward. Moreover, by the time a mint was set up in Dublin about 995, the more powerful Irish kings sought actively to gain a foothold in what were becoming increasingly urban trading centres achieving some success in this respect during the course of the eleventh and twelfth centuries.

While the beneficial effects of Norse-driven commercial activity have long been recognized, assessment of their influence in other areas has been more divided. In general, however, their overall impact is now viewed as far less catastrophic than was once maintained. In the first place, the intense period of hit-and-run type raiding that marked the first part of the ninth century soon gave way to a series of attacks mounted from fixed bases against which the Irish had considerable success. Moreover, their sustained presence in the country from the 840s onwards meant that the Scandinavians were gradually incorporated into the political landscape, forming significant alliances with a wide range of rulers. In the 850s, for example, they were aligned for a time with Áed Finnlíath, king of the Northern Uí Néill, whose daughter may well have been married to the Norse king of Dublin. Nor was this mixed marriage unusual, as indicated by the increasing number of Irishmen bearing Scandinavian names and vice versa. Among the latter were the children of a tenth-century king of Dublin, Amlaíb Cúarán, who in addition to acquiring an Irish nickname (*cúarán*, "sandal") commissioned poetry from one of the leading poets of the day, Cináed úa hArtacáin.

Elsewhere in the literature, Vikings also appear in a range of guises reflecting the complex, ever-changing relationship between them and their Irish neighbors. A ninth-century ecclesiastic may extol the stormy sea preventing the arrival *dond láechraid lainn úa Lothlind* (of the fierce warriors from Scandinavia), yet little more than a century later a Welsh poet presumed to count on the support of both *gynhon Dulyn* (the foreigners of Dublin) and *Gwydyl Iwerdon* (the Irish of Ireland) in battle against the Saxons. Similarly, while the pagan Viking invader assumed a literary life of his own in a variety of texts whose authors conveniently ascribed to him a plethora of ills, his more sober, real-life descendant was long since enmeshed in the cultural milieu of his adopted homeland. In artwork too his presence is felt, as demonstrated by the only hog-back tomb known in Ireland at Castledermot, Co. Kildare, as well as by an eleventh-century book shrine made by one whose Hiberno-Norse origin is revealed in his name, Sitric mac Meic Áeda. Nor was influence solely one-way: the Irish silver thistle brooch became popular in Norway in the tenth century. In addition, a number of Irish kings feature as ancestor figures in the account of the settlement of Iceland recorded in *Landnámabók* (The Book of Land-taking). Whether their intellectual interdependency stimulated the Norse to first produce sagas, however, as has sometimes been claimed, is questionable. Nonetheless, intensive interaction on many levels over a long period has certainly left its mark in Scandinavia, but particularly in Ireland.

MÁIRE NÍ MHAONAIGH

References and Further Reading

Kenny, Michael. "The Geographical Distribution of Irish Viking-age Coin Hoards." *Proceedings of the Royal Irish Academy,* 87 C (1987): 507–525.

Ní Mhaonaigh, Máire. "Friend and Foe: Vikings in Ninth- and Tenth-century Irish Literature." In *Ireland and Scandinavia in the Early Viking Age*. Edited by Howard B. Clarke, Máire Ní Mhaonaigh, and Raghnall Ó Floinn, Dublin: Four Courts Press, (1998): 381–402.

Ó Cuív, Brian. "Personal Names as an Indicator of Relations between Native Irish and Settlers in the Viking Period." In *Settlement and Society in Medieval Ireland: Studies Presented to F.X. Martin o.s.a.* Edited by John Bradley, 79–88. Kilkenny: Boethius Press, 1988.

Ó Floinn, Raghnall. "Irish and Scandinavian Art in the Early Medieval Period." In *The Vikings in Ireland*. Edited by Anne-Christine Larsen, 87–97. Roskilde: The Viking Ship Museum, 2001.

Oftedal, Magne. "Scandinavian Place-names in Ireland." In *Proceedings of the Seventh Viking Congress, Dublin, 15–21 August 1973*. Edited by Bø Almqvist and David Greene, 125–133. Dublin: Royal Irish Academy, 1976.

Sheehan, John. "Early Viking Age Silver Hoards from Ireland and their Scandinavian Elements." In *Ireland and Scandinavia in the Early Viking Age*. Edited by Howard B. Clarke, Máire Ní Mhaonaigh, and Raghnall Ó Floinn, 166–202. Dublin: Four Courts Press, 1998.

See also **Amlaíb Cúarán; Battle of Clontarf; Brian Boru; Cináed úa hArtacáin; Coinage; Dícuil; Hoards; Metalwork; Personal Names; Sculpture; Viking Incursions; Weapons and Weaponry**

SCIENCES

Natural Science

The earliest dateable text, composed in southern Ireland in 655, is *De mirabilibus Sacrae Scripturae*, by an author using the pseudonym of Augustinus. It gives exceptionally rationalistic explanations of natural phenomena, though within the pre-scientific spirit of the Middle Ages. These physical explanations for the miracles in the Bible and strange natural phenomena–(e.g., why the Sun stood still at Joshua's command, how Lot's wife could turn into a pillar of salt, why certain animals are not found in Ireland) are, although naïve, strikingly original in conception by the standards of the age. His analysis of tidal flow is one the best in medieval literature.

Computus and Astronomy

Computistical literature must have been well established in Ireland by the sixth century. The basis of their reckoning of the paschal term was the fourth-century Latin translation of Anatolius of Laodicea's, *De ratione paschali* ("On the reckoning of the Pasch"). Cumian's paschal letter to Ségéne in 632/3 refers to ten different paschal cycles, including Anatolius. The anonymous seventh-century treatise *De ratione computandi* shows the author to have had a very competent knowledge of calendrical science and computistics, Bede and the Anglo-Saxons were deeply indebted to the Irish for their knowledge of the computus.

The astronomical entries in the Irish annals, recorded from 627 to 1133, constitute one of the most extensive and careful series of records in Western Europe of observations of eclipses, comets, auroras, volcanic dust clouds, and the supernova of 1054. Such phenomena were recorded either from Ireland itself or from its offshore monastic communities, such as Iona, and are accurate in their chronological and observational details. These observations and recordings of astronomical and atmospheric phenomena in Irish monasteries must be considered as evidence of scientific enquiry, even if they ultimately served the eschatological purpose of looking for the "signs of Doomsday." They show that the Irish had a deep interest in the observation of natural phenomena and of the measurement of the passage of time, and were capable of doing so with considerable accuracy. Their astronomical observations are therefore part of the same intellectual phenomena that engendered their study of the computus.

One of the earliest astronomical treatises penned by an Irishman is that by Dungal of Pavia (d. *c.* 830), "instructor in the imperial court," *magister palatinus*, under Charlemagne and his son Louis. In 811 he wrote a letter to Charlemagne, giving an account of a double solar eclipse in the previous year. It is based mainly upon the astronomical system of Macrobius and shows Dungal to have possessed a rich learning in the Liberal Arts and astronomy. He complains of the lack of certain works of classical astronomy that would have helped him to answer the question more fully.

Cosmology

Pseudo-Isidore, *De ordine creaturarum*, a cosmological-theological treatise on the structure and order of divine creation, was written toward the end of the seventh century. It used the *De mirabilibus* and other Hiberno-Latin material. It has some striking cosmological theories and was one of the earliest texts to expound the medieval doctrine of Purgatory in some detail. Many of the problems connected with the sources of this text remain unsolved. Later Irish Scriptural commentaries on the Hexameron, the six days of creation, have some cosmological material, but most of it is derived from the Fathers.

The greatest Irish thinker by far was John Scottus, author of the massive *Periphyseon* (*De natura rerum*), a complete system of cosmology, dealing with the fourfold nature of God and creation, set in the form of a dialog between master and pupil. It is the most complete synthesis of medieval cosmology, and its profound philosophical analysis of being and creation is a tour-de-force unequalled in early medieval Europe.

An Irish origin has been claimed for Honorius of Regensburg (*fl.* ante 1156), one of the most enigmatic and prolific authors of the Middle Ages, wrote a number of works of cosmology and natural science, the *Clauis physicae* and *Imago Mundi.* He disseminated the theology and cosmology of Anselm and Johannes Scottus in Germany before anyone else. His admission into the canon of Hiberno-Latin literature would make a considerable addition to the corpus of scientific work by the Irish.

Geography

Dícuil (*c.* 760-post 825), Irish scholar-exile at the courts of Charlemagne and Louis the Pious and an important author of several works on geography, computus, grammar, and astronomy. The only details known of his life are what can be garnered from incidental references in his works. He was teacher at the Palace School of Louis the Pious in about 815. His first work, *Liber de astronomia*, is a verse-computus written between 814–816 in four books, to which a fifth book was later added. In 818 he wrote the *Epistula censuum*, a verse treatise on weights and measures. In the same year he also wrote two other works, of which the most important is his treatise on geography, *Liber de mensura orbis terrae*. Dícuil used a wide range of sources for this treatise, some of them now lost or only partly preserved, such as the *Cosmographia* of Julius Caesar, in the recension of Julius Honorius, as well as some derivative of the emperor Agrippa's map of the world, probably that known as the *Diuisio* or *Mensuratio orbis*. Among his other sources are Pliny, Solinus, and Isidore of Seville. He had spent some time in the islands north of Britain and Ireland, and had made a note of their size and location. Dícuil seems to have acquired his geographical knowledge of the islands around Britain from some time spent as a monk on Iona, and he is a witness to the settlement in the eighth century by Irishmen of the northernmost islands of Britain and of Iceland. His descriptions of Egypt and

Palestine are largely derived from written sources, though he also refers to oral information communicated from a traveler to those parts, a "brother Fidelis," from whom he also got one of the earliest descriptions in Western vernacular literature of a Nile crocodile.

References and Further Reading

Further bibliography in M. Lapidge, R. Sharpe, *A Bibliography of Celtic-Latin Literature 400–1200* (Dublin, 1985).

Kenney, J. F. *The Sources for the Early History of Ireland: 1 Ecclesiastical* (New York, 1966 [1929], reprinted Dublin, 1993).

Monumenta Germaniae Historia Epistolae IV, 570–583.

M. Esposito, "An unpublished astronomical treatise by an Irish monk Dicuil," *R.I.A. Proc.* xxvi, sect. C (1907), 378–446 (with addenda and corrigenda by the editor in *Z.C.P.* viii, 1910: 506–507).

van de Vyver, A. "Dicuil et Micon," *Revue Belge de philologie et d'histoire* 14 (1935), 25–47 (with an edition of Liber censuum).

Tierney, J. J., editor. with contribution by L. Bieler, *Dicuili Liber de mensura orbis terrae* (SLH 6, 1967).

Reynolds, R. "Further evidence for the Irish origin of Honorius Augustodunensis." *Vivarium* 7 (1969), 1–7.

Ferrari, M. "In Papia conveniant ad Dungalum." *Italia mediovale e humanistica* 15 (1972), 1–52.

Flint, V. "The career of Honorius Augustodunensis: some fresh evidence." *Revue Benedictine* 82 (1972), 63–86.

Gautier-Dalché, Patrick. "Tradition et renouvellement dans la représentation d'espace geeographie au IX siecle" *Studi Medievali* ser. 3, 24.1 (1983), 121–165.

Garrigues, M.O. "L'oeuvre d'Honorius Augustodunensis: Inventaire critique." *Abhandlungen der Braunschweigischen Wissentschaftlichen Gesellschaft* 38 (1986), 7–136; 39 (1987), 123–228.

Bermann, W. "Dicuils De mensura orbis terrae." Edited by P. L. Butzer and D. Lohrmann, *Science in Western and Eastern Civilization in Carolingian Times* (Basel, 1993), 527–537.

Stansfield Eastwood, B. "The astronomy of Macrobius in Carolingian Europe: Dungal's letter of 811 to Charles the Great." *Early Medieval Europe* 3.2 (1992), 117–134.

Ó Cróinín, D. *Early Medieval Ireland 400–1200,* (London, New York, 1995).

Smyth, Marina. *Understanding the Universe in Seventh-Century Ireland.* Studies in Celtic History 15 (Woodbridge, 1996).

McCarthy, D., and A. Breen. "An evaluation of the astronomical observations in the Irish annals." *Vistas in Astronomy* 40.1 (1997), 117–138.

McCarthy, D., and A. Breen. "Astronomical observations in the Irish annals and their motivation." *Peritia* 11 (1997), 1–43.

See also: **Dícuil; John Scottus Ériugena; Paschal controversy**

SCOTTI/SCOTS

Scotti is a Latin word which initially was applied to all Gaels, but which later came to be used in a more confined way to refer only to Gaels in northern Britain. It is first attested in the late third century A.D. in Roman texts, and gradually came to replace *Hiberni* as the people-term for the Gaels. St. Patrick used it in his fifth-century epistles—sometimes with a pejorative sense, as when he refers to the followers of Coroticus as "socii Scottorum atque Pictorum apostatarumque" (companions of Scots and Picts and apostates; Epistola, I, ii). Its etymology is uncertain, because it is unrelated to any known Gaelic word, but it may have meant "raider," describing Gaels who attacked Britain in the Late-Roman period.

In the early medieval period it was also adopted by both the Gaels in Ireland and Britain to describe themselves. Adomnán's *Life of St. Columba*, written about 700 A.D., uses *Scoti* for both Gaels in Ireland and Britain, *Scoti Brittaniae* for Dál Riata in Britain, and the derivative term *Scotia* denotes Ireland, rather than part of Britain, while Muirchú's Life of St Patrick describes Tara as "capital of the *Scotti*." A reflex of this is that a woman called *Scotta* was created as one of the eponymous ancestors of the Gaels in their origin legends.

From the ninth century onward the use of these and cognate terms began to change. David Dumville has recently pointed out that Old Norse *Skotar* and Old English *Scottas* (from about 900 onward) were used for Gaels in northern Britain only and that *Scotland* in Old English came to be confined in meaning to Gaelic parts of northern Britain, while Old Norse *Írar* and Old English *Iras* were terms for the Irish. The exact explanation of these changes is unclear, but the effects of Scandinavian raids and the transition from Pictish to Gaelic identity in much of Scotland north of the Firth of Forth may have increased the need for separate designations for the Gaels in Ireland and Britain.

Equivalent developments did not occur in Gaelic nomenclature until much later. In the late tenth century *Scotti* and *Scotia* were first used for part of northern Britain in contrast to the Irish and Ireland, but the change was more gradual. By the twelfth century *Scotia* had come to denote the core territory of the Gaelic kingdom of Alba, rather than areas of later conquest, such as Lothian and Moray, but in the early thirteenth century the term gradually came to mean all the territories of the kingdom of Alba, whereas *Scoti* was used by chroniclers in Melrose to describe themselves by the 1290s. As the kingdom of Alba expanded, so did the geographical area and number of people covered by these terms increased to their modern extent. However, the usage of *Scotti* in Alba was never as clearcut as for *Scotia*, because of the remaining connections between the Gaels in Ireland and Scotland and a continuing belief in a common origin as late as the fourteenth century. In Ireland the wider meaning of these terms was still used in the early fourteenth century, when the "Irish Remonstrance" employed

Scotia minor for Scotland and *Scotia maior* for Ireland, to show support for Edward Bruce's invasion of Ireland.

NICHOLAS EVANS

References and Further Reading

Anderson, Alan Orr, editor, and Marjorie Ogilvie Anderson, translator. *Adomnán's Life of Columba* (Oxford: Clarendon Press, revised edition, 1991).

Hood, A. B. E., editor and translator. *St. Patrick: His Writings and Muirchú's Life* (London and Chichester: Phillimore, 1978).

Howlett, D. R., editor and translator. *The Book of Letters of Saint Patrick the Bishop* (Dublin: Four Courts Press, 1994).

Broun, Dauvit. "Gaelic literacy in eastern Scotland, 1124–1249." In *Literacy in Medieval Celtic Societies.* Edited by Pryce. (Cambridge 1998).

Dumville, David N. "Ireland and Britain in *Táin Bó Fraích*," *Études Celtiques.* 32 (1996), 175–187.

Rivet, A. L. F., and Colin Smith, *The Place-Names of Roman Britain* (London: B.T. Batsford Ltd, 1979).

See also **Invasion Myth; National Identity; Patrick**

SCOTTISH INFLUENCE

Throughout the medieval period there were close connections between Ireland and Scotland, largely because many areas of Scotland shared a common Gaelic culture with Ireland. This led to a considerable degree of cultural interaction, although in general Irish culture was more influential in Scotland than vice versa. The peoples of medieval Scotland held a place in the geographical consciousness of the Irish, as the frequent journeys in tales of Irish heroes to Dál Riata, Pictland, and Alba indicate, but there are fewer clear examples where the Irish were influenced by those in Scotland.

In the Early Medieval period much of the influence probably came through Dál Riata in western Scotland, which had important ecclesiastical establishments among the Picts and Northumbrians. It is likely that Dál Riata was a crucial intermediary in the seventh-century development of the Hiberno-Saxon art style shared by Dál Riata, Pictland, Northumbria, and Ireland, combining artistic attributes from each region. Iona, founded by the Irish Columba, seems to have been particularly significant. The *Iona Chronicle*, kept from at least 660 to 740, forms a significant element in all the surviving Irish chronicles of the period, and Adomnán's *Life of St Columba* was probably a model for later Irish saints' lives, such as the late eighth-century *Life of St Cainnech.* The descriptions of rulers as "kings of Ireland" in both of these Iona texts are the first explicit references to the concept of a high-kingship of Ireland.

Later in the Middle Ages a number of historical texts from Scotland were transmitted to Ireland, then copied and sometimes adapted in Ireland so that they became part of the Irish historical tradition. The clearest case of this is *Lebor Bretnach*, a Gaelic version of the Latin *Historia Brittonum*, or "The History of the Britons." This text was produced in lowland Scotland in the late eleventh century specifically for an Irish audience, and survives in a number of Irish manuscripts. Other Scottish texts are found in Irish manuscripts, including the tenth-century *Míniugud Senchasa Fher nAlban*, whereas versions of the Pictish king-list occur in *Lebor Gabála* or accompany *Lebor Bretnach*. In some cases, such as the late-eleventh century *Duan Albanach*, the genealogy of the rulers of Moray and the Alba king-lists, Scottish materials were adapted for the creation of texts in Ireland. The effect of these Scottish texts on Irish historical thinking is at present unclear, but they may reflect an increased interest in Scottish affairs caused by the dominance of Gaelic culture in formerly Pictish areas and the prestige of the expansionist Gaelic kings of Alba.

From the twelfth century onward lowland Scotland became increasingly English-speaking, so that there was less cultural interaction with Ireland. However, the Highlands and the Isles of Scotland maintained a high degree of contact with Ireland, with bards often travelling between Ireland and Scotland and maintaining a single poetic culture, but the degree to which Ireland was influenced by Scotland in terms of literature during this time is uncertain.

NICHOLAS EVANS

References and Further Reading

Bannerman, John. *Studies in the History of Dalriata* (Edinburgh: Scottish Academic Press, 1974).

Thomas Clancy. "Scotland, the 'Nennian' recension of the *Historia Brittonum*, and the *Lebor Bretnach*." In *Kings, Clerics and Chronicles in Scotland, 500–1297: Essays in Honour of Marjorie Ogilvie Anderson on the Occasion of Her Ninetieth Birthday.* Edited by Simon Taylor. (Dublin: Four Courts Press, 2000).

Henry, Françoise. *Irish Art in the Early Christian Period (to 800 A.D.)* (London: Methuen and co., revised edition, 1965).

Herbert, Máire. *Iona, Kells and Derry: The History and Hagiography of the Monastic Familia of Columba.* (Oxford: Oxford University Press, 1988; reprinted Dublin: Four Courts Press, 1996).

Herbert, Máire. 'The *Vita Columbae* and Irish Hagiography: A Study of *Vita Cainnechi*." In *Studies in Irish Hagiography: Saints and Scholars.* Edited by John Carey, Máire Herbert, and Pádraig Ó Riain. (Dublin: Four Courts Press, 2001).

See also **Annals and Chronicles; Early Christian Art; Hagiography and Martyrologies**

SCRIPTORIA

Every medieval monastery needed a stock of working copies of bibles, Psalters, and missals for the liturgical life of the community; they also needed copies of the Rule, penitentials, and, indeed, secular documents such as deeds and letters in their dealings with the world. For religious study multiple copies of commentaries on the scriptures, lives of the saints, and the writings of the early Church Fathers were made. The monastery library and school needed texts and scholarly glosses. Psalters were the primers for the novices. Copies had to be handwritten in the scriptorium or medieval monastic secretariat. The first call on the scribes in the scriptorium, usually situated adjacent to the library, was the reproduction of books for the services in choir and the readings in the refectory. Some scriptoria (like that at Armagh—which, according to the Annals of Tigernach, escaped destruction in the great fire of 1020) acquired a reputation for excellent calligraphy and beautiful illuminations and royal patrons paid handsomely for *de luxe* products.

The work of scribes, *copisti*, and illuminators to produce legible and correct copies required not only a scholarly mastery of reading and interpreting of the texts called *exemplars* (borrowed from neighbouring monasteries) but also great skills in the art and craft of writing in the distinctive insular majuscule and minuscule Irish hands. Animal skins for vellum and parchment were expensive in the seventh and eighth centuries, hence the use of minuscule to get more writing onto a page and likewise the use of palimpsests, that is, previously used parchments, showing over-writing, marginal notes, and the interlinear glosses much favoured by Irish scribes. Scraping, curing, and dyeing of vellum was an important activity in the preparation of writing materials in the scriptorium, as was the making of inks—mainly black, red, yellow, and purple; this last, procured from a particular seashell. So too was the cutting of quills sliced off in a chisel edge to produce the distinctive thick down strokes and thin horizontal line of insular minuscule script. Perhaps the finest early example of this script in Irish is that of Ferdomnach who penned the *Book of Armagh* (807). The development of minusculeor lower case letterforms was arguably the most important Irish contribution to written language in Western Europe. Until the rise of the university stationers in the thirteenth century monastic scriptoria had a monopoly of book production.

Often a copyist, pure and simple, made errors and copied errors of syntax and grammar and of fact—but a scribe with better latin and sounder scholarship corrected copy. Irish trained scribes are to be found not just in Northumbrian scriptoria but also in those of Scotland, Anglo-Saxon England, and naturally in the original Columban monasteries of Europe. The technicalities of distinguishing Northumbrian, Anglo-Saxon, and Irish scripts are complex and have led to much controversy, for example, over the date, place, and origin of the celebrated Book of Kells. While the majority of monastic scribes could copy texts and documents for practical usage, few could have been skilled enough artists to execute the illuminated art treasures which the Irish monasteries gave the world in such plenitude of artistry as in the Books of Durrow, Kells, Armagh, and Lindisfarne, and indeed, the Latin Gospel Books of Echternach, (in the Bibliotheque Nationale in Paris-latin Ms 9389), Durham, and Lichfield, all of which display a marriage of calligraphy and illumination reaching its full maturity in the case of Kells.

After the twelfth-century manuscripts were no longer produced solely for monastic usage, books were executed for lay patrons by secular professional scribes from the learned families. The "Golden Age" of the Gospel manuscripts, the illuminated service books and the book shrines of the Celtic saints, was largely the product of Irish monastic scriptoria, ensuring them a unique place in the history of western European civilization.

J. J. N. McGurk

References and Further Reading

Bieler, Ludwig. *Ireland, Harbinger of the Middle Ages* (1963).

Flower, Robin. *The Irish Tradition.* (Oxford, 1947).

Henry, Francoise. *Irish Art in the Early Christian Period* (to A.D. 800) (1957).

Henry, Francoise, and M. Marsh. "A century of Irish illumination, 1070–1170." *Proceedings of the Royal Irish Academy,* 62C (1962), 101.

Kenney, J. F. *The Sources for the Early History of Ireland.* I, Ecclesiastical, (NewYork: Columbia Univesity Press, 1929; reprint Dublin 1966).

McGurk, J. J. N. "The Origins of our letters." *History Today,* xviii no.10 (1968).

McGurk, Patrick. "The Irish Pocket Gospel Book." *Sacris Erudiri* no. 8 (1956), 249-.

Moody,T. W., F. X. Martin, and F. J. Byrne, editors. *New History of Ireland: Medieval Ireland.* (Oxford: Oxford University Press, 1987).

Ó Cuív, Brian, editor. *Seven Centuries of Irish Learning.* (Dublin, 1961).

O'Neill, Timothy. "Script." *Encyclopaedia of Ireland.* Edited by B. Lalor (Dublin: Gill & Macmillan, 2003).

The Ordnance Survey Map of Monastic Ireland. (Dublin 1964).

Ryan, John. *Irish Monasticism: Origins and Early Development.* (Dublin: Talbot Press, 1931).

Thompson, E. M. "Calligraphy in the middle ages." *Bibliographica,* iii. (1897).

See also **Armagh; Armagh, Book of; Manuscript Illumination**

SCULPTURE

Apart from cross-inscribed stones that may have started as early as the seventh century in Ireland, sculpture should be seen to begin with High Crosses and related monuments. With the exception of the crosses in the Barrow Valley, including the tall cross at Moone, which has wonderfully graphic stylized figures in flat false relief, the ninth- and tenth-century High Crosses often have squat figures carved in high false relief which are more naturalistically represented than in Celtic art, suggesting that they copy sculpture in a classical tradition, probably influenced by the Carolingian empire, with stucco having been suggested as a possible medium of transmission. A gap in the eleventh century is followed by a style of cross with Christ triumphant standing out in high relief (partially in the style of the Volto Santo in Lucca), often accompanied by an Episcopal figure also in high relief.

At the same time, architectural sculpture emerged on Romanesque churches, particularly on doorways and chancel arches, though occasionally also on windows. Often of a high quality, inspired probably by English and French models, the sculpture includes bearded masks and capitals, well modeled in relief, sometimes semi-naturalistic, at other times (e.g., Tuam) wonderfully stylized. Often the carved ornament is geometric, more a superficial veneer than an integral part of the architecture it ornaments. Through the so-called School of the West, the style continued confidently in Connacht until about 1230. By that time, the rest of the country had adopted a Gothic style, with the early-thirteenth century plant ornament on the capitals at Corcomroe being prematurely naturalistic before stiff-leaf foliage becomes the norm by the middle of the century. The spread of Anglo-Norman hegemony through much of the country between 1169 and 1235 introduced a new trend, particularly noticeable in tomb-sculpture.

O'Brien/Butler tomb (1626) from St. Mary's Church, Iniscealtra, Co. Clare. © *Department of the Environment, Heritage and Local Government, Dublin.*

Starting around 1200, we find effigies of ecclesiastics, knights, and civilians, carved in high relief, and dependent on inspiration from England, from whence many masons must have come to create them and ornament Gothic churches such as St. Patrick's Cathedral in Dublin. The carving of such effigies continued into the seventeenth century, but those of the fifteenth and sixteenth centuries rested on tomb chests ornamented with Weepers under arcades—at first in Meath, and later practiced by the Ormond and O'Tunney schools in the Kilkenny area. North Leinster also had well-carved baptismal fonts and wayside crosses in the later medieval period. But the West, too, had a strong tradition of later Gothic sculpture (e.g., Strade) and, in Clare, panels in Ennis imitate English alabasters, and stone figures of the *Pieta* copied wooden models imported into Ireland at the time. But, even in the thirteenth century local sculptors had been carving their own versions of Romanesque Madonnas, and continued to carve statues of varying quality for ornamenting churches. These are doubtless rare survivals of impressive woodcarving schools, whose quality can be measured on the only Irish wooden misericords that fortunately survive in St. Mary's Cathedral in Limerick.

PETER HARBISON

References and Further Reading

Garton, Tessa. "Masks and monsters: Some recurring themes in Irish Romanesque sculpture." In *From Ireland Coming: Irish Art from the Early Christian to the Late Gothic Period andIts European Context.* Edited by Colum Hourihane. Princeton, 2001, 121–140.

Harbison, Peter. *The High Crosses of Ireland.* 3 vols. Bonn, 1992.

Hunt, John. "The Limerick Cathedral misericords." *Ireland of the Welcomes.* September-October 1971, 12–16.

Hunt, John. *Irish Medieval Figure Sculpture.* 2 vols. London/Dublin, 1974.

King, Heather. "Late medieval Irish crosses and their European background." *From Ireland Coming: Irish Art from the Early Christian to the Late Gothic Period and Its European Context.* Edited by Colum Hourihane. Princeton, 2001.

MacLeod, Catriona. "Some mediaeval wooden figure sculptures in Ireland." *Journal of the Royal Society of Antiquaries of Ireland.* 76, 1946, 155–170.

MacLeod, Catriona. "Some late mediaeval wood sculptures in Ireland." *Journal of the Royal Society of Antiquaries of Ireland.* 77, 1947, 53–62.

McNab, Susanne. "Celtic antecedents to the treatment of the human figure in early Irish art." In *From Ireland Coming: Irish Art from the early Christian to the Late Gothic Period and Its European Context.* Edited by Colum Hourihane. Princeton, 2001, 161–82.

Nelson, E. Charles, and Roger Stailey. "Medieval naturalism and the botanical carvings at Corcomroe Abbey (County Clare)." *Gesta* XXVIII/2, 1989, 165–174.
Roe, Helen M. *Medieval Fonts of Meath*, 1968.

See also **Altar-tombs; Architecture; Carolingian Influence; High Crosses; Iconography**

SEDULIUS SCOTTUS

Few dates can be assigned with any reliability to this important figure, who, during the Renaissance and even afterwards was conflated with the late Roman poet (Caelius) Sedulius, author of the *Carmen paschale*. Sedulius Scottus flourished in the middle of the ninth century at Liège, where he was the nominal head of a circle of Irish scholars. His special patron was Bishop Hartgar of Liège to whom Sedulius dedicated a number of poems, including a lament on his death. He also addressed poems to King Louis, King Charles (the Bald), and to Emperor Lothar and Empress Ermingard. Beyond these he mentions a number of his Irish friends, some of whose names appear in ninth-century manuscripts as contributors of glosses or scholia, notably Fergus and (Bishop?) Marcus.

As with John Scottus Eriugena, we do not know a great deal about Sedulius's early education in Ireland. Traces of Irish influence can be seen in his use of material found also in the *Collectio canonum hibernensis*, and the use of Pelagius. It is also possible that Sedulius gained a rudimentary knowledge of Greek in his homeland. However, his command of a variety of classical metres was almost certainly not acquired at home, nor was his exceptional knowledge of classical Latin literature (which far exceeded that of John Scottus).

Sedulius's activities as a writer and scholar were richly diverse. He wrote biblical commentaries, commentaries on three grammarians (Priscian, Eutyches, and Donatus's *Ars minor* and *maior*), numerous poems, and a long treatise entitled *De rectoribus christianis* ("On Christian Rulers"). He was also responsible for the *Collectaneum*, a large anthology of selections from classical and patristic writers that affords a glimpse into the state of learning during the third generation of the Carolingians. Sedulius also wrote scholia on classical poets.

Two major works of scriptural scholarship are the work of Sedulius: *Collectaneum in omnes beati Pauli epistolas*, based prominently but not exclusively on Pelagius and Jerome, and the *Collectaneum in Mattheum*, a miscellany of commentary on the Gospel of Matthew. Shorter works include a commentary on the Eusebian Canons, *Expositio in epistolam Hieronimi ad Damasam papam* and the *Expositio argumenti Hieronimi in decem canones,* as well as the *Explanatiuncula in argumentum secundum Matthaeum, Marcum, Lucam.* Another significant contribution to biblical studies is his autograph copy of the Greek psalter (not glossed or translated, Paris, Bibliothéque de l'Arsénal, MS 8407).

Sedulius's *Collectaneum* on Paul has been especially valuable to modern scholars interested in reconstructing Pelagius's commentary on the Apostle's writings; it further demonstrates the enduring interest of the Irish in Pelagius's work. Sedulius's text of Paul was probably not Pelagius's own version, as once thought, but a variant of the type of biblical text transmitted in Ireland. In the prologue he outlines the subject of the Epistles according to the seven *circumstantiae* (*persona*, *res*, *causa*, *tempus, locus*, *modus*, *materia sive facultas*), a type of *accessus* found in several other ninth-century Irish commentaries.

The grammatical commentaries, now in modern editions, tell us much about Sedulius's reading and erudition. In commenting on Donatus's *Ars maior*, for example, Sedulius cites the opinions of other grammarians on the use of grammatical terms, or other grammatical questions. He often seizes the opportunity to explain a lemma with a Greek equivalent, or to expand a mythological reference, even to give full *argumenta* of literary works cited. Biblical passages are cited alongside secular ones. The commentary on Priscian's *Institutiones* is another testament to ninth-century Irish interest in that grammarian. In addition to the commentary by John Scottus Eriugena (as yet unprinted), there are several glossed manuscripts of the *Institutiones* from the ninth century written in Irish hands, of which the most famous is St. Gall (904, recently edited). An interesting feature of the commentary on Eutyches is evidence for the continuing use of Virgilius Maro Grammaticus, whose text tradition originated in Ireland.

The *De rectoribus christianis* is an example of a *Fürstenspiegel* ("mirror of princes," i.e., treatises that give advice to rulers). Several of these were written in the Carolingian period (by Smaragdus of St. Mihiel, Jonas of Orléans, Dhuoda, and Hincmar of Reims). Sedulius's *De rectoribus* is unusual in that it is a *prosimetrum*, the metrical forms being almost as diverse as those found in Boethius's *Consolatio*, which doubtless served as its formal model. The work is more a showpiece of erudition than a political tract. Many of the texts cited in the work were also used in Sedulius's other *Collectaneum* (to be distinguished from the biblical commentaries with the same title). This last is a collection of excerpts from classical and patristic writers. The selection of works seems clearly to have been based on their relevance to moral questions and practical wisdom; some of the excerpts also appear in the *Collectio canonum hibernensis*. Examples of secular works include the so-called *Proverbia Graecorum* and the *Sententia Ciceronis de virtutibus et vitiis* (drawn from *De inventione*), excerpts from Cicero's orations and his works on rhetoric, and passages from Vegetius, Valerius Maximus, Frontinus, and the *Scriptores Historiae Augustae*. Patristic works include Augustine, Ambrose, Jerome, Lactantius, and Cassiodorus.

Sedulius's poems, more than any other of his writings, have attracted the most attention in modern times. His mastery of a variety of meters set him apart from his Irish contemporaries, the Sapphic being particularly favoured next to the heroic and elegiac; examples of rhythmical verses are few. Most of the poems are panegyrics to rulers and influential clergy, especially Hartgar, but two poems have attracted attention for their charm and wit: the *De certamine liliae et rosae* ("Debate between the Lily and the Rose") and an amusing poem, *De quodam verbece a cane discerpto* ("On a Wether Mangled by a Dog"). Sedulius's glosses on classical poets in the miscellany found in Bern 363 are identifiable—a hint that they may have been drawn from lost full commentaries, or at least scholia collections.

MICHAEL W. HERREN

References and Further Reading

Bischoff, Bernhard. "Irische Schreiber im Karolingerreich." In *Mittelalterliche Studien: Ausgewählte Aufsätze zur Schriftkunde und Literaturgeschichte*. 3 vols. Stuttgart: Anton Hiersmann, 1967–1981, 3: 39–54.

Contreni, John. "The Irish in the Western Carolingian Empire (According to James F. Kenney and Bern, Burgerbibliothek, 363)." In *Die Iren und Europa im früheren Mittelalter*, edited by Heinz Löwe. Stuttgart: Klett-Cotta, 1982, [Teilband 2] 758–798.

Dictionnaire des lettres françaises: le moyen âge, edited by Robert Bossuat, Louis Pichard, and Guy Raynaud de Lage. 2nd edition. Paris: Fayard, [1991], 1371–1372.

Düchting, Reinhard. *Sedulius Scottus, Seine Dichtungen*. Munich: Wilhelm Fink Verlag, 1968.

Frede, Hermann Josef. *Pelagius, Der irische Paulustext, Sedulius Scottus*. Freiburg: Verlag Herder, 1961.

Godman, Peter. *Poets and Emperors: Frankish Politics and Carolingian Poetry*. Oxford: Clarendon Press, 1987.

Hellmann, Sigmund. *Sedulius Scottus*. Quellen und Untersuchungen zur lateinischen Philologie des Mittelalters 1. Munich: C. H. Beck'sche Verlagsbuchhandlung, 1906.

Kenney, J. F. *The Sources for the Early History of Ireland. Ecclesiastical: An Introduction and Guide*. Records of Civilization 11. New York: Columbia University Press, 1929; revised by L. Bieler, 1966, 553–569.

Lapidge, Michael, and Richard Sharpe. *A Bibliography of Celtic-Latin Literature 400–1200*. Dublin: Royal Irish Academy, 1985.

Löfstedt, Bengt, editor. *Sedulius Scottus. Kommentar zum Evangelium nach Matthaus*. Aus der Geschichte der lateinischen Bible 14. 2 vols. Freiburg: Verlag Herder, 1989.

Meyer, Jean, editor. *Sedulii Scotti Carmina*. Corpus Christianorum Continuatio Mediaevalis 117. Turnhout: Brepols, 1991.

Simpson, D. W., editor. *Sedulii Scotti Collectaneum miscellaneum*. Corpus Christianorym Continuatio Mediaevalis 67. Turnhout: Brepols, 1988.

See also **Biblical and Patristic Scholarship; Canon Law; Classical Influence; Devotional and Liturgical Literature; Ériugena, John Scottus; Glosses; Grammatical Treatises; Hiberno-Latin; Metrics; Moral and Religious Instruction; Poetry, Latin**

SERFS

See **Society, Anglo-Norman Grades of**

SHEELA-NA-GIG

A *sheela-na-gig* (which may translate as "Síle of the breasts") is a medieval female exhibitionist figure posed in a manner that displays and emphasizes the genitalia. Their background is to be found in European Romanesque churches, particularly those located along pilgrimage routes in France and Spain, where a range of male and female exhibitionist figures and related carvings are found that served to alert the faithful to the dangers of the sin of lust. Emphasis on the genitalia, which are usually enlarged, may relate to the church's teaching that sinners were punished in hell through the bodily organs by which they had offended.

By contrast with the Continental carvings, most of the Irish *sheela-na-gigs* are isolated figures located on buildings that are otherwise sculpturally plain. Carved in stone, they were placed on churches, castles, and town walls, located usually near a door or window or on quoins. Some variations in pose exist but it is by no means certain whether these have any real significance. In general, *sheela-na-gigs* appear to be evenly divided between those which seem to be standing and those that may be seated. The legs may be splayed widely or, alternatively, the thighs may be splayed but with the heels together. In some cases the legs appear not to have been represented at all. The commonest position of the arms is that whereby the hands are placed in front with a gesture towards the abdomen or, more explicitly, towards the pudenda. The hands may join in front of the genitalia or may be shown gripping the pudenda. In some instances the arms are placed behind the thighs. Irish *sheela-na-gigs* occur predominantly within or adjoining areas of heavy Anglo-Norman settlement in north Munster, Ossory (the Co. Kilkenny area), and the midlands, being virtually absent from the far west and north of the island. One of the most westerly examples, that from Aghagower, County Mayo, is located along a pilgrim's road to Croagh Patrick. The earliest Irish examples are located on churches where they fulfilled a purpose similar to the continental carvings. Later examples found in secular contexts on town walls and tower houses may have functioned as protective carvings. The change in meaning, from figure of lust to protective icon may have arisen as a result of Gaelic cultural resurgence during the later Middle Ages. Elements of ancient pre-Christian beliefs were embedded within Irish concepts of lordship. The land was a female entity to which the lord was wedded metaphorically and therefore responsible for its protection, wealth, and fecundity. These beliefs found reinforcement in the Irish literary tradition that included ancient epic mythological tales featuring female characters such as Queen Medb, who are literary versions of the ancient earth goddess. *Sheela-na-gig* supplied a readymade visual image that could be expropriated and displayed on a lord's residence to provide validation of his role and status. Medieval lords who displayed *sheela-na-gigs* in secular contexts included Ua Briain (O'Brien) and Ua Máelsechnaill (O'Melaghlin), who were the descendants of Irish high-kings, while others such as Butler and Fitzmaurice were of Anglo-Norman stock.

EAMON P. KELLY

References and Further Reading

Guest, E. "Irish Sheela-na-gigs in 1935." T*he Journal of the Royal Society of Antiquaries of Ireland*, Dublin, 66 (1936), 107–129.

Kelly, Eamonn P., *Sheela-na-gigs: Origins and Functions* (Dublin, 1996).

McMahon, Joanne, and Jack Roberts. *The Sheela-na-gigs of Ireland & Britain: The Divine Hag of the Christian Celts: An Illustrated Guide.* Cork: Mercier Press, 2001.

Weir, Anthony. *Images of Lust: Sexual Carvings on Medieval churches*. (London :Routledge, 1999).

See also **Architecture; Castles; Ecclesiastical Sites; Iconography; Parish Churches and Cathedrals; Pre-Christian Ireland; Sculpture; Tower Houses; Wall Paintings; Women**

SHIPS AND SHIPPING

As an island, Ireland has a history of boats stretching back into prehistory. The National Museum of Ireland's dug-out canoe from Lurgan, County Galway dating to about 2200 B.C.E. is one of the oldest specimens known and the Museum's beautiful gold model boat found at Broighter, County Derry, has been assigned to the first century B.C.E. It appears to be a model of a wooden ocean-going vessel rather than a skin-covered one and has eight seats, sixteen oars complete with rowlocks, a central mast with crossbeam, and a steering oar.

The Irish geographer Dicuil, who recorded in the early ninth century voyages made a century earlier by Irish monks to the northern islands, described long sea voyages in ships out of sight of land. Several types of boats are described in Adomnán's *Life of St. Columba* written in the island monastery of Iona about 700, including the curach (*curucus*) made of hides sewn together over a wooden framework. These were light, portable, and seaworthy. The larger of them were fitted with a sail that hung from a yard, which was raised and lowered using ropes. Adomnán also mentions building a "long ship" (*longa navis)* from pine and oak timbers that were partly prepared before being brought to Iona by land and sea. He reports numerous boat trips made from Derry to Britain, and from Gaul to Iona, and from Gaul to Britain. Generally the voyages were under sail but sometimes the sailors were forced to row. The relationship of the early medieval monks with the sea is best reflected in the *Voyage of St. Brendan* (*Nauigatio Sancti Brendani*), which is Ireland's most popular contribution to medieval literature, composed about a century after Adomnán's *Life of Columba*.

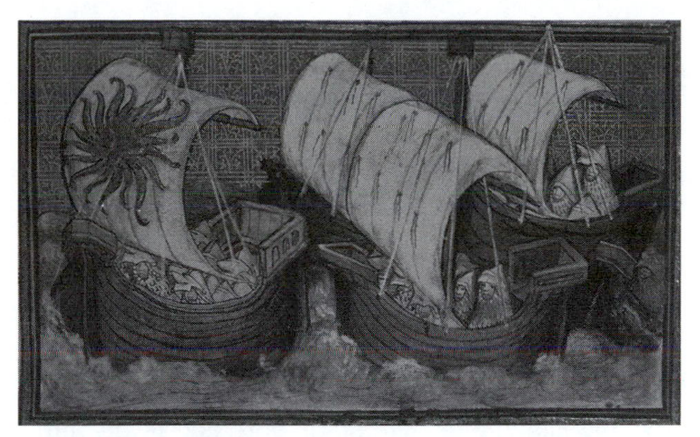

Richard II sails from Ireland from *Histoire du Roy d'Angleterre Richard II. © The British Library*

According to Scandinavian archaeologists the first Viking attacks were probably launched in ships of the Oseberg type. This restored ship, originally built about 800, is seventy-five feet long with a mast and sail and could be rowed by thirty men. A century later, when the Gokstad ship was used as a burial, Viking vessels were more seaworthy for longer voyages under sail or oars. They were clinker built in a shell of overlapping planks nailed to the keel and stems and strengthened and stabilised by ribs and crossbeams. The sailing rig was one square sail on a central mast, and the ship was guided by a steering oar or side rudder. Ship timbers found in the Dublin excavations—keels, stems, planking, and framing—are examples close to the Gokstad ship, indicating that shipbuilding in Dublin, up to the middle of the twelfth century, was in the mainstream Viking tradition. These ships were the key to the Viking success; fast on the water, they were easily beached and could be taken up rivers and carried by the crew if necessary, in effect "the only ocean-going landing craft ever devised"(Bertil Almgren). Ships of this type (with a stern rudder) were in use throughout the middle ages and were particularly favored by the Scots of the Isles (and most likely the O'Malleys of Connacht) as pirate galleys.

Trading and Fishing Vessels

Later medieval records mentioning ships sometimes give the name, type, and home port, and if this information is accompanied by a description of the amount of cargo and number of crewmen, then a reasonably accurate picture of the vessel in question can be deduced. The principle types of trading boats in Irish ports during the period were: cogs, hulks, barges, balingers, caravels, and, more rarely, carracks. Crayers and pickards were most numerous and appear to have been used mainly for fishing.

By the thirteenth century, *cog* is the term used to describe the typical seagoing sailing vessel of northern Europe. It was a sturdily built, single-masted, square-sailed, rather tubby ship with a length to breadth proportion of about 3:1. Cogs varied in size. Two transporting wine were in Southampton in 1326; the *Johan* of 160 tons and a crew of 47, and *La Seintemari* of 60 tons and a crew of 27. In 1338 Maurice, son of the earl of Desmond, hired *La Rodecogge* of Limerick to sail to Gascony. The *hulk* was originally crescent—shaped, made of strong curved boards fastened together by external pieces, and bound together at each end of the ship without stem or stern posts. It was driven by a large square sail and steered by side rudders. A hulk from Waterford, the *Blessed Mary*, carried corn to Gascony in 1297, and a stylized representation of a hulk appears on the municipal seal of Youghal, 1527.

Barges appear around the beginning of the fifteenth century. They may have carried oars as well as sail and have been about 50 to 100 tons capacity. *La barge de Saint John* left Ireland with a cargo of 8,400 hides in 1413, and barges are mentioned in 1392–1393; one leaving Drogheda to trade with Irish enemies and another being purchased from Gerald le Byrne, "captain of his nation." *Balingers* appear to have been two-masted sailing vessels between 20 and 50 tons used for fishing and trade. The *Katherine*, a 50-ton balinger of Waterford, was requisitioned for the king's service in 1414.

Merchant ships increased in size during the fifteenth century; the average size of Bristol ships using the port of Bordeaux rose from 88 tons (*c.* 1400) to around 150 tons (in 1450). The huge, rounded, two- and three-decked Carracks with high aftercastles and three masts are noted in the later fifteenth century. It is not known whether they regularly visited Irish ports, although in 1431 a Venetian carrack chartered by Italian and Aragonese merchants was captured while on a voyage from Brittany to Ireland, and the 320-ton London ship bound for Santiago de Compostela, which picked up four hundred pilgrims in New Ross in 1477 was in all likelihood a carrack. With minor additions to the sail plan this vessel represents the full-rigged ship of the sixteenth century. In 1567, John Goghe's map of Ireland has a drawing of two of these ships in the Irish Sea. The map also shows a *caravel* off the south coast. Developed by the Portuguese, it was long and low in the water and by the sixteenth century carried square sails as well as lateen (triangular) sails. Because of its speed and maneuverability it was a favorite of Breton pirates.

Crayers and Pickards seem to be smaller versions of the balinger. A number of ships coming to Ireland from Bristol around 1400 are described as crayers. Between 20 and 30 tons, they were crewed by up to sixteen men. The earl of Ormond hired a crayer in Waterford several times in the 1380s to travel to England. Crayers were probably decked and with a cabin, whereas pickards were open. Customs tolls at Ardglass, County Down, in 1515 charged five shillings for boats "with a top" and three shillings and four pence for every pickard or ship "without a top."

Shipbuilding

Shipbuilding took place in the Irish ports in the later Middle Ages. In 1234 the king ordered two sixty-oared and four forty-oared galleys to be built in Ireland, and Drogheda was ordered to make a second galley in 1241, Waterford to make two, and Cork and Limerick one each. Some idea of the numbers of ships in the Irish ports may be formed from the requisitioning that took place in 1301 and in 1303 to transport soldiers to Scotland. In 1301, 46 out of 74 arrested were Irish and in 1303, 37 out of 173. A survey (by W. A. Childs) of ships using Bristol 1480–1489 showed that at least seventy but possibly ninety ships and ninety-three working shipmasters were from Ireland. Add about thirty more operating in and out of Chester and it would appear that Ireland was well provided with shipping in the medieval period.

TIMOTHY O' NEILL

References and Further Reading

O'Neill, Timothy. *Merchants and Mariners in Medieval Ireland*. Dublin: Irish Academic Press, 1987.

Unger, R. W. *The Ship in the Medieval Economy, 600–1600.* London: Croom Helm and Montreal: Mc Gill-Queens University Press. 1980.

See also **Adomnán; Dicuil; Naval Warfare; Trade; Ports**

SICK, MAINTENANCE OF THE

See **Brehon Law**

SITRIUC SILKENBEARD

(Sigytryggr Silkiskeggi), son of Amlaíb (Óláfr) Cuarán, was king of Dublin for a period of almost fifty years between 989 and 1036 During this period, the great growth in trade and economic prosperity that had begun under the reign of his father continued to flourish, and the first coinage to be minted in Ireland appeared in Dublin around 997, with the inscription *SIHTRIC REX DYFLIN* and variants thereof.

Several short interruptions punctuated the long reign of this Hiberno-Scandinavian king. In the early 990s, an ongoing rivalry between the descendants of Amlaíb Cuarán and those of Ímar (Ivarr) of Waterford was manifested in a struggle for control of Dublin. Sitriuc managed to expel Ímar and three ships of Ímar's men from Dublin in 993, but was himself expelled in 994, the kingship going to Ragnall mac Ímair. Upon Ragnall's death in 995, Sitriuc returned to power, only to be expelled again in 999 by Brian Boru. Sitriuc's expulsion followed Brian's successful siege of Dublin in the aftermath of the Battle of Glenn Máma; however, after submitting to Brian and giving him hostages, Sitriuc was allowed to return to his kingdom the subsequent year. Sitriuc's third departure—a pilgrimage to Rome in 1028—was voluntary. During this third absence, Sitriuc's son Amlaíb (Óláfr) appears to have become king of Dublin, ceding authority back to his father at some point before 1031. Sitriuc's pilgrimage is a reflection of the Christian devotion characteristic of his family since the conversion of his father, the most marked legacy of which was Sitriuc's founding of Christ Church cathedral (*c.* 1030).

The family's conversion to Christianity no doubt helped to further facilitate the marital alliances that had been taking place between Norse and Irish dynasties since the mid-ninth century. Not only was Sitriuc himself the product of such a union—his mother was Gormfhlaith, daughter of the Uí Fáeláin king of Leinster Murchad mac Finn—but he was also a partner in one through his marriage to Sláine, daughter of Brian Boru. Sitriuc's sister Máel Muire, meanwhile, was married to the king of Tara Máel Sechnaill II, whereas another sister, Ragnall (Ragnhildr), was the wife of Domnall mac Congalaig, king of Brega. The potential benefit of such marriage alliances is seen to best effect in the case of the mutual support afforded between Sitriuc and his mother's brother, Máel Mórda, king of Leinster from 1003 to 1014. It was Máel Mórda who killed Ragnall mac Ímair in 994, thereby enabling Sitriuc's return to power, whereas Sitriuc returned the favor in 995 by helping his uncle seize the current king of Leinster, Donnchad mac Domnaill of the Uí Dúnchada, thus clearing the way for Máel Mórda's accession. Thereafter, the two supported each other consistently, most notably in their unsuccessful stands against Máel Sechnaill II and Brian Boru at Glenn Máma in 999, and against Brian in the battle of Clontarf in 1014. Clearly, Sitriuc's bonds to Máel Sechnaill and Brian by marriage were in no way as effective as his bonds by blood to the Uí Fáeláin.

While Sitriuc's relations with Máel Sechnaill were generally strained across the board, those with Brian were more mixed. After Brian reinstated Sitriuc following the events of 999, the Dublin king, in conjunction with the Leinstermen, regularly accompanied Brian on his hostings throughout the country. In 1013, however, this cozy relationship changed when Sitriuc joined with Máel Mórda against Brian in the hostilities that eventually led to Clontarf. According to contemporary sources, Sitriuc and the Leinstermen were assisted at Clontarf by contingents of Scandinavians from the Orkneys and Hebrides; in later sources this alliance becomes exaggerated to include representatives from almost every corner of the Viking world. The later sources also relate that Sitriuc stayed inside the battlements of the city throughout the conflict, defending his fortress from within, while his Scandinavian allies fought alongside the Leinstermen from without. Given that Sitriuc survived to rule Dublin for more than twenty more years after the battle, despite huge casualties on the parts of his Leinster and Scandinavian allies as well as of Brian's forces, this account of his whereabouts may indeed reflect the truth.

Ultimately, it was Sitriuc's fellow Hiberno-Scandinavians rather than any Irish ruler who put an end to his reign as king of Dublin. In 1035, Sitriuc murdered Ragnall Ua Ímair, king of Waterford, enflaming the feud with the descendants of Ímar. The following year, one such descendant, Echmarcach mac Ragnaill, took over control as king of Dublin and Sitriuc went into exile "across the sea," possibly Wales, and died in 1042.

Anne Connon

References and Further Reading

Clarke, Howard B., Máire Ní Mhaonaigh, and Raghnall Ó Floinn, editors. *Ireland and Scandinavia in the Early Viking Age.* Dublin: Four Courts Press, 1998.

Duffy, Seán. "Irishmen and Islesmen in the Kingdoms of Dublin and Man, 1052–1171." *Ériu*, 43, (1992), 93–133.
Kinsella, Stuart. "From Hiberno-Norse to Anglo-Norman, 1030–1300." In *Christ Church Cathedral Dublin: A History*, edited by Kenneth Milne, 25–52. Dublin: Four Courts Press, 2000.
Mac Shamhráin, Ailbhe. "The battle of Glenn Máma, Dublin, and the high-kingship of Ireland: a millennial commemoration." In *Medieval Dublin II. Proceedings of the Friends of Medieval Dublin Symposium 2000.* Edited by Seán Duffy (Dublin: Four Courts Press, 2001), 53–64.
Ryan, Fr. John S. J. "The battle of Clontarf." *Journal of the Royal Society of Antiquaries of Ireland*, 68 (1938), 1–50.

See also **Amlaíb Cuarán; Christ Church Cathedral; Pilgrims and Pilgrimage; Waterford**

SLAVES

Slavery may be characterized negatively by an absence of judicial status, meaning that the slave was considered by law to be an object in a slave-owner's possession rather than as a person in his or her own right. Status as a slave might be temporary or permanent, and although the actual procedures are not known, a slave could obtain status as a free person. There were three sources of recruitment for slaves: (1) prisoners of war, (2) debt slaves, and (3) children born of slaves. Medieval Irish sources for slavery are abundant but often treat the subject cursorily and non-systematically. A full picture of Irish slavery must therefore remain impressionistic in character.

The most common names for slaves in Irish were *mug* for male and *cumal* for female slaves. *Cumal* was also widely used as a unit of value for cattle and land. "Martyrologies" often refer to slave labor as an image of personal debasement. The vita of St. Senan tells of the men of Corcu Baiscind who were admonished to obey St. Senan to not suffer such hunger that "a man would sell his son and daughter in distant territories for nourishment." A vita of the ninth century relates that St. Ciaran, a slave to the king, had to grind the grain every day. Slaves are never associated with husbandry but mainly with heavy agricultural labor such as sowing, harrowing, thrashing, and grinding.

Ship raids on Britain in the fifth century after the collapse of the Roman Empire provided prisoners of wars who were treated as slaves. These raids seem to have ceased as a result of the stabilization of Britain in the seventh century. Children born of these prisoners continued to be a source of an Irish slave population, although they are rarely mentioned in the idealistic status system depicted by early medieval Irish laws. Recurring mention of the sale of children in hunger years attests to the existence of "debat-slavery" as an institution throughout the early Middle Ages.

The effect of the Viking attacks and subsequent settlements was to accentuate slavery as a social institution. Viking warfare did not respect the sanctity of monasteries and brought about a change in the norms of warfare, which included an acceptance to reduce prisoners of war to slave status. The Irish annals record 23 instances when Vikings took prisoners *en masse*, which must be taken as an indication of slaving operations. While hostages for tribute were termed *géill*, the annals refer to these prisoners as *brat* (captives). The early instances of Viking slave raids do not, however, indicate large-scale operations for a full-blown slave market, but rather seem to be spectacular acts of defiance and humiliation against the enemy. After the battle of Tara in 980, the king of Meath is reputed to have freed all the Irish slaves of Dublin, an act that was repeated by Brian Boru's and Máel Sechnaill's joint action on Dublin after the victory over Leinster and the mercenaries of Dublin in 999. During the first half of the tenth century, slaves were still a by-product of a particular kind of war, namely retaliatory actions and military adventures designed to vaunt the capabilities of the would-be successor. Overall, there is no evidence to suggest that the institution of slavery in Ireland and even in Dublin was anything more than a marginal phenomenon of luxury for the nobles.

However, in connection with the heavy expenditures caused by the struggle for the kingship of Ireland in the early eleventh century, Irish kings began taking captives in large numbers. In the decisive campaigns up to 1014, punitive actions seem to have greatly increased. The Cenél nEógain king Flaithbertach was the leading slaver in a number of actions on neighboring territories. In 1011, he united with the son of Brian Boru and allegedly took many cows and 300 captives (*brait*, a word hitherto restricted to Viking assaults) from the Cenél Conaill, and the following year he is credited with the largest booty any king had taken of captives and kine from the Ulaid. In later years, the annals repeatedly note the massive taking of prisoners, and more mundane events were changed by new attitudes to the defeated. Irish warfare had traditionally seen many plain raids that were not part of a larger political scheme but rather must be seen as seasonal traditional manifestations of the bravado of young warriors. This long-established custom was called *crech*, a prey or a raid for cattle. By the mid-eleventh century, the taking of captives also became part of these heroics. The rising power of the Northern over kings is marked in the annals by heavy exactions upon neighboring kingdoms. The Ua Conchobair kings of Connaught, who were at times near achieving total supremacy over Ireland, also practiced the new kind of warfare in their campaigns. The climax came in 1109, when Muirchertach of the Dál Cais mustered a large force against the Uí Briúin of Connaught and took many captives from the islands of Loch Oughter. The Uí Briúin took revenge upon Meath, the ally of Muirchertach, by burning, killing, and leading off many captives. The final blows to Dál Cais supremacy were accompanied by great predatory expeditions in 1115 and 1116, but the prisoners of the last campaign were released afterward as an homage to God and to St. Flannán of Killaloe—the patron saint of the Dál Cais.

What were the driving motives behind the massive taking of prisoners by Irish kings of the eleventh and early twelfth centuries? First, there was a striking similarity between the warfare of Dublin and Irish kings. Simple lessons of the humiliating function of massive imprisonment to the prestige of any king were well learned by the Irish, and it seems plausible that once learned, they put it to their own use. Further, Irish cattle raids and petty warfare between minor kings took on a far larger and more devastating character when the Irish invited Viking warriors for wars of conquest and paid them in kind by the wealth of the enemy, including prisoners of war. We know that the great struggles of the over kings for supremacy were largely decided by the use of naval fleets. These fleets were either indirectly controlled by the over king as a consequence of their control of Norse cities or they were hired from Norse settlements in Ireland or the Scottish Isles. The decline of Dublin's political power forced many warriors either to settle or to take up freebooting, more or less out of control of the Dublin king. These half-independent warriors may have supplied the Dublin slave market with captives that had not been taken because of political complications, but simply for profit. From the middle of the twelfth century, we know that the Dublin fleet was hired for thousands of cattle that were driven to the city in payment. It is also conceivable that payment in the eleventh century was in slaves.

As the evidence stands, there is, however, no way to substantiate the hypothesis that the Irish captives of war were in fact sold to Norse slave dealers. What exists is a relatively clear-cut case that slavery became more widespread during the course of the eleventh century. We have much circumstantial evidence of the importance of slavery to Irish kings in eleventh century writings such as *Lebor na Cert* (Book of Rights) and *Cogadh Gaedhel re Gallaibh*. The distinction in these texts between male and female slaves reveals some functions of slavery. Female slaves are referred to as "full-grown," "swarthy," "fair," "graceful," and "valuable;" and the Leinster king is obliged to give "eight women whom he has not dishonored." Male slaves are described as "lads," "hard working," "strong-fisted," "willing," "expensive," and "spirited." If we may deduce anything from these descriptions, the slaves seem primarily to have been intended for the household: as servants, concubines, mountebanks, and the *drabants* of the court. The old use of *cumal* for a female slave was evidently obsolete by 1100, and instead *mná (daera)* or the crude *banmog* were used. In the Leinster list, Dublin is entitled to "thirty women with large families"—an indication perhaps of the furnishing of Dublin warriors with concubines. Further, Lebor na Cert draws a clear distinction between native and foreign slaves ("foreigners who do not know Irish," "women from over the great sea"), an indication that not only were slaves recruited by internal warfare but some were also supplied by foreign trade.

Ireland has no mineral wealth, and foreign luxury goods could be bought by Irish kings mainly for two export goods, cattle and people. Labor and concubines were in demand wherever a new elite had established itself, and hides for parchment were in strong demand. Tenth and eleventh century wars and not least the Norman conquest of Britain must have generated a strong market for the Irish commodities. Very little is known about the actual trade mechanisms and balances, but one indicator is the growing number of instances recorded in the annals of the taking of slaves by the Irish. In the eleventh century, Dublin was probably the prime slave market of western Europe, furnishing customers in the British Isles, Anglo-Saxon as well as Norse, and the Scandinavian countries. In 1102, however, Dublin's slave trade to Bristol was prohibited on religious grounds, while the trade also seems to have been despised for its antisocial character. Demand in Scandinavia declined for the same religious and social reasons as it did in Britain. The trade and Irish slave raids seem therefore to have petered out in the early twelfth century. However, some trade must have continued, as indeed the Irish synod of 1170 welcomed the Norman Conquest as just punishment for the abuses of the slave trade. Slavery as such was not put to an end overnight, as we are well reminded by the synod of Armagh of 1170. Even as late as 1235, the mark of slavery was still felt by some people; in Waterford a man was known as Philippus Leysing, Philip the manumitted, or freed slave.

Poul Holm

SOCIETY, FUNCTIONING OF ANGLO-NORMAN

Anglo-Norman, or English society, in medieval Ireland was shaped by two distinct but related forces: one was aristocratic, the other royal. The conquest was initially effected by aristocratic adventurers who wished to improve their fortunes by carving out lordships for themselves in Ireland. From the royal expedition in 1171–72 of King Henry II (1154–89), however, royal authority was stamped upon Ireland. As the trappings of royal power in England grew during the thirteenth century, they were systematically extended to Ireland. By the turn of the fourteenth century, almost all the principal English institutions had a miniature colonial counterpart. With royal power came assumptions about feudal customs, the use of the common law, and English inheritance practice. Despite all this, Anglo-Norman lords were frequently left to fend for themselves, and the social reality was that independent aristocratic forces remained powerful. For those operating on the frontier with the native Irish, survival sometimes depended on departing from the central government's legalistic ideal of how society should function. It was, perhaps, the principal lesson of the fourteenth century that royal power was better served by working with the nobles of Ireland rather than by reacting against them.

Royal Government

It is dangerous to press the distinction between royal and aristocratic interests too hard. Their intentions were fundamentally identical: Both saw Ireland as a source of land; both hoped to make that land profitable by colonizing and cultivating it on the model of the English manor. Nonetheless, the imposition of royal authority was of the greatest importance. Pointing to Scotland's experience, R.R. Davies has reminded Irish historians that Anglo-Norman colonization could have progressed without Ireland becoming an English colony. Henry II's intervention removed that possiblity by ending the entrepreneurial nature of the conquest. The Anglo-Norman adventurers now held their land of the king of England by feudal tenure, with all the obligations that entailed. Any new projects they engaged in—for instance the de Burgh conquest of Connacht in the 1230s—were initiated at the nod, or rather by charter, of the king.

The feudal relationship was central to the functioning of colonial society. In return for grants of land, the nobles owed the lord of Ireland—who, for most of the period, was king of England—military service, which in practice was frequently commuted to a cash payment called "scutage" (royal service [see *Feudalism*]). This relationship was replicated at each level of society so that theoretically everyone was engaged in a bond with some social superior, ultimately leading to the king. The king, in return, was obliged to protect his subjects, to lead them in war, and to provide justice. In fact, the king was usually absent from Ireland, but his obligations of lordship were exercised by proxy through his chief governor. Just as royal authority in England was bolstered by its precocious administrative institutions—for instance the chancery, the exchequer, and the court system—so all these organs of royal power were provided for the lordship of Ireland. Dependence on England carried with it the assumption that the colony would be governed by English law. This assumption was sometimes made explicit, as in the famous case of *Magna Carta*, which was sent to Ireland as the *Magna Carta Hiberniae* (Great Charter of Ireland) in 1217. To the end of the medieval period, many other English statutes were transmitted to Ireland and there promulgated by the central government or later ratified in the Irish parliament. Royal authority was also strong in local government because of the shire system. By the turn of the fourteenth century, royal sheriffs executed the king's writs across most of Ireland.

In a society in which land was the source of wealth and power, the most far-reaching effects of English law were in connection with property. The king as feudal overlord exercised considerable rights in this respect. By the late twelfth century, primogeniture—the descent of lands to the eldest son—had become the normal practice for dealing with inheritance in England. Its application in Ireland was in stark contrast to the Gaelic system of partible inheritance. In the colony, all land theoretically reverted to the lord of Ireland at the death of one of his tenants. The heir then had to pay a large sum of money—known as a relief—to gain possession of his inheritance. If the heir to a great estate was a minor (usually meaning under 18-years-old), the king took him into his custody as a ward of the crown, and all his property was placed under royal protection. This practice was of considerable profit to the king, both because of the revenues he received from the lands during the minority of the heir, and because it was source of patronage. The wardship did not have to be held by the crown personally, but could be granted out to another great noble. A grant of wardship was much sought after, but could have disastrous consequences for the ward's lands. In 1344, the earldom of Ormond was granted in wardship to the first earl of Desmond (d. 1356) during the minority of James Butler (d. 1382). Rather than protecting the lands, the earl of Desmond ravaged them.

The king also reserved the right to consent to the marriages of his vassals' daughters. Marriage, in this period, cannot be divorced from politics. A good marriage was intended to secure inheritance, forge alliances, and patch up former disputes. One of the more spectacular marriage networks in Ireland was devised by Richard de Burgh, the "Red Earl" of Ulster (d. 1326). Among the husbands of his female progeny numbered some of the most exalted nobles in Ireland, England, and Scotland, including Robert Bruce, king of Scotland (d. 1329); Gilbert de Clare, earl of Gloucester (d. 1314); the second earl of Kildare (d. 1328); the first earl of Desmond (d. 1356); and John de Bermingham, earl of Louth (d. 1329). The marriages of minor heirs were also arranged by the Crown, and in the later Middle Ages, the king employed this right to bind the nobility of England and Ireland together. The young fourth earl of Kildare (d. 1390) was present at the siege of Calais in 1347, and Edward III took the opportunity to marry him to the daughter of one of his knights, Sir Bartholomew de Burghersh.

Although primogeniture had advantages in terms of settling succession disputes, in an Irish context it sometimes had serious drawbacks. Besides the vulnerability caused by minorities, there was a risk that an estate could be fractured if a male line of heirs failed. This happened dramatically in the mid-thirteenth century. Five successive sons of William Marshal (d. 1219) died childless with the result that the great lordship of Leinster was divided between his five daughters. Meath was similarly divided at the death of the heirless Walter de Lacy in 1241. There were further subdivisions in the thirteenth and fourteenth centuries, and the splintered estates that resulted were of little interest to the absentee English nobles to whom they descended. Their neglect was a source of considerable weakness in the lordship. It was the risk of division that led to the practice of granting estates in "tail male." This practice—employed in the cases of the earldoms created in the early fourteenth century and also used extensively among the lesser nobility—meant that an estate always descended to the nearest male relative and so could not suffer fracture between heiresses.

Magnate Power

Irish historians assessing the lordship of Ireland have long tracked the growth of royal lordship to about the year 1300 and have thereafter bemoaned the general decline that saw the contraction of royal authority to the Pale by the later fifteenth century. However, equating royal lordship and an efficient administration with a successful society can be perilous in a frontier region like Ireland. The work of Robin Frame is particularly important in this regard. He has reminded us that although royal authority was theoretically extensive by the end of the thirteenth century, "the areas with which the administration was closely involved were probably less securely held and less prosperous than they had been sixty years earlier, when royal government was simpler and less intrusive, and the large, undivided lordships of Leinster and Meath still existed" (Frame 1981).

For all the crown's complicated administrative machinery, the resident magnates themselves remained of prime importance. How they exercised power, both officially and unofficially, reveals some important points of divergence from England. Although an English earl might take his title from a region in which he held lands, those lands were not usually geographically consolidated. He would usually hold manors scattered across several counties, which ensured that the county or shire court remained the primary focus of local jurisdiction. Ireland was different. Lands were held in large blocks, which gave their lords a territorial dominance highly unusual in England. Moreover, from the start of the conquest, grants were often made of liberty jurisdiction. This meant that the king delegated royal authority to the lord within the bounds of the liberty or franchise, except for four pleas reserved to the crown: arson, rape, forestall (highway robbery), and treasure trove. Leinster, Meath, and Ulster were liberties in the thirteenth century, and in the fourteenth century the earls of Kildare, Ormond, and Desmond each held a former county as a liberty (Kildare, Tipperary, and Kerry respectively). Even when the liberty of Kildare was suppressed in 1345, never to be restored, the county of Kildare effectively remained under the control of the local earl.

This official policy—although often decried by later historians as spelling the ruin of royal government (see *Feudalism*)—in fact ensured the strength of English lordship in much of Ireland at moments when direct royal aid could not be forthcoming because of preoccupations elsewhere. This is true both in the early stages of the conquest and in the fifteenth century when the three resident earls dominated much of Ireland. Particularly in the later middle ages, however, less official methods of maintaining power became common. In England, since the time of the early Norman kings, private war had been prohibited. For the nobility of Ireland defending territories on the march (or frontier) with the hostile native Irish, that prohibition was impracticable. Although it horrified English administrators, private armies were common and were often billeted on the countryside (see *Coyne and Livery*). A famous case is the force controlled by the first earl of Desmond known as "MacThomas's rout." It should not be imagined that these private forces were perpetually inimical to the native Irish. In fact, since the earliest points of the conquest when the native Irish recruited Anglo-Norman knights, there had been alliances between the two nations. By the fourteenth century, these arrangements were often standardized, such as the "bonnaght" of Ulster (the 345 Gaelic *satellites*, or troops, that the northern chieftains owed the earl of Ulster), or the agreements of retinue made between the earls of Ormond and his neighboring Gaelic chiefs in the 1350s.

In the first half of the fourteenth century, royal administrators endeavored on several occasions to bring the Irish magnates under control and curb their "unofficial" practices. Such attempts bred hostility between the "English of Ireland" and the "English of England," and it soon became clear that it was easier to rule through the nobility rather than against them. By the late fifteenth century, the magnates were strikingly independent, but they were also buttresses of English lordship. The fourth earl of Ormond (d. 1452), who served several times as the king's lieutenant in Ireland, promulgated private ordinances for the governance of his lordship. These ordinances intertwined different legal traditions—common, march, and brehon law—and are an indication of his extraordinary independence and self-confidence. A famous and possibly apocryphal story about Gerald FitzGerald, the Great Earl of Kildare (d. 1513), like all great clichés, conveys a basic truth: in 1496, on being told up that all Ireland could not control the Great Earl, King Henry VII (1485–1509) reputedly replied: "Then, in good faith, shall this Earl rule all Ireland."

The practices described above are often ascribed to "Gaelicization," the process common to border societies in which the settlers adopted many of the customs of the indigenous population. In Ireland this involved abandoning common law and the English inheritance system. But points of divergence ought not to be stressed alone. Although it is a less familiar concept to Irish historians, both Gaelic and colonial societies were operating in ways similar to "bastard feudal" England, where each lord had a private affinity of retainers who served him in peace and war. Even the custom of designating the leaders of "Gaelicized" lineages as "chiefs of their nations" and allowing them to discipline their own extended families is reminiscent of the claim by English lords that they should be allowed to discipline their own retainers. What is more, in England and Ireland alike, "bastard feudal" practices were condemned by royal administrators. Later medieval Ireland, in other words, was not a totally alien society. When viewed solely from a royal standpoint, it is revealed in too negative a light. The shifting politics may have been difficult to follow and some social practices unconventional, yet others must have been familiar. Although royal authority retreated in the later middle ages, the magnate power that took its place was not negative. Indeed it was the resident nobility's resilience that maintained nominal English control over much of Ireland into the early modern period.

PETER CROOKS

References and Further Reading

Curtis, Edmund. "The Clan System among the English Settlers in Ireland." *English Historical Review* 15 (1910): 116–120.

———. "The 'Bonnaght' of Ulster" *Hermathena: A Series of Papers on Literature, Science and Philosophy by Members of Trinity College, Dublin* 46 (1931): 87–105.

Davies, R.R. "Lordship or Colony?" In *The English in Medieval Ireland*, edited by James Lydon, 142–60. Dublin, Royal Irish Academy, 1984.

Empey, C.A., and Katherine Simms. "The Ordinances of the White Earl and the Problem of Coign in the Later Middle Ages," *Proceedings of the Royal Irish Academy* 75 section C (1975), 161–187.

Frame, Robin. "Power and Society in the Lordship of Ireland, 1272—1377." *Past and Present: A Journal of Historical Studies* 76 (August 1977): 3–33.

———. *Colonial Ireland, 1170–1370.* Dublin: Helicon Ltd, 1981.

———. *English Lordship in Ireland, 1318–1361.* Oxford: Clarendon Press, 1982.

———. "War and Peace in the Medieval Lordship of Ireland." In *The English in Medieval Ireland*, edited by James Lydon, 118–41. Dublin: Royal Irish Academy, 1984.

Hicks, Michael. *Bastard Feudalism.* Essex: Longman, 1995.

Otway-Ruthven, A.J. "Knight Service in Ireland." *Journal of the Royal Society of Antiquaries* 89 (1959): 1–15.

———. "Royal Service in Ireland." *Journal of the Royal Society of Antiquaries* 98 (1968): 37–46.

Reynolds, Susan. *Fiefs and Vassals: The Medieval Evidence Reconsidered.* Oxford: University Press, 1994.

Simms, Katharine. *The Changing Political Structure of Gaelic Ireland in the Later Middle Ages.* Woodbridge: The Boydell Press, 1987; reprinted 2000.

See also **Brehon Law; Central Government; Chief Governors; Common Law; Courts; Feudalism; Gaelicization; Henry II; Local Government; Lordship of Ireland; Manorialism; March Law; Pale, The; Parliament; Society, Grades of Anglo-Norman**

SOCIETY, FUNCTIONING OF GAELIC

The study of Gaelic Irish society has suffered in the past from an exaggerated belief in the degree of continuity in its laws and institutions, a belief which has been aided by the fact that—thanks to the survival of the early Irish law-tracts—we know more of the details of its functioning in the seventh and eighth centuries than in the fourteenth and fifteenth. Even here there is a danger of mistaking the ideals of the jurists who compiled the law-tracts for the actual condition of the society in which they lived. An extreme form of this belief in continuity has imagined Gaelic Irish society as imprisoned in a web of immutable law and custom. However, quite apart from the constant changes in local power structures caused by the proliferation of the ruling lineages, medieval Irish society experienced two periods of traumatic shock separated by one of intense political change. Following the devastating Viking slave-raids of the ninth and tenth centuries, the old polity, based on the autonomous petty kingdom (*tuath*) and on the increasingly powerful and equally autonomous ecclesiastical establishments, gave way to one of the regional dynasties whose kings imposed their rule on the local kinglet (*rí tuaithe*) and introduced new ideas of royal authority. Like their contemporaries elsewhere in Europe, they claimed paramount rights over the land and its inhabitants, including the right to transfer these rights to monasteries and, probably, individuals. Again, the later Gaelic world which came into existence after the second, and even greater, shock of the Anglo-Norman invasion and the partial conquest which followed it had been followed in turn by the Gaelic Revival, which involved the passing into the Gaelic cultural sphere in a greater or lesser degree of most of the lordships of Anglo-Norman origin and was very different from the earlier one. The title of king (*rí*) gradually passed into desuetude. In the legal sphere native Irish law absorbed elements of English law and, in its last two centuries, like that of every other European country except England, was moving towards an acceptance of the "common law" (*Jus Commune*) of Europe, the system composed of the Roman and Canon Laws and favored by the church. An example of this was the replacement of the earlier marriage system based on the *coibhche* or bride price by one that required the woman to bring a dowry to her husband. Nevertheless, by a manipulation of the Canon Law rules governing marriage, the Irish upper classes were able to continue their practice of serial marriage.

Nevertheless, certain predominant features of Gaelic Irish society persist throughout the medieval period and down to its final destruction in the early seventeenth century. The most notable is the constant proliferation of the dynastic and other dominant lineages, so that there is a recurrent process of replacement from the top downwards as the offshoots of the ruling lineage take the place of former locally dominant groups and are in their turn supplanted by the most recent offshoots of the lineage, the immediate offspring of ruling lords. The displaced elements in turn replace their former inferiors, who descend - if they survive – into the propertyless bottom layer of the population. This proliferation, found also in the other Celtic countries, Scotland and Wales, contrasts starkly with the usual Western European situation where ruling lineages tend to die out and be replaced by others ascending from below. In a lineage-based society, where power, property, and status are conferred by membership of a high-ranking descent group, a woman maximizes the opportunities for her children by having them fathered by a man belonging to such a group, so that in such societies there is in fact a competition for women that favors the dominant lineages. In Gaelic Ireland, an inclusive rule of legitimacy and an easy process of affiliation produced the same effects as, for example, chiefly polygyny in Bantu Africa. This proliferation from the top characterizes Gaelic Irish society throughout its history, being as true of dynastic lineages such as the Uí Dúnlainge and Uí Chennselaig in Leinster in the early medieval period as of Mac Murchada (MacMurrogh), Mág Uidir (Maguire) and Ua Néill (O'Neill), as well as those of Anglo-Norman descent, such as Burkes, Butlers, and FitzGeralds, in the later. The process ensured that the structures of power and landholding in Gaelic society were dynamic rather than static, and coupled with the mechanisms of succession to power (theoretically by seniority but in practice, as was recognized from the earliest period, to "the person who possesses most clients and power" within the ruling lineage), ensured continuing political instability and frequent bloody struggles for succession (see *Tánaiste*). Out of these uncertain mechanisms of succession grew the principle of "segmentary opposition," by which those sections of the ruling lineage that were out of power would automatically oppose the ruling chief (drawn from a rival section) and ally themselves with his enemies, that is to say, the external enemies of their own people. This was as true of sixteenth-century Ireland as it was of the eighth.

Another feature of Gaelic society in all periods was the importance of the corporate lineage-group (in anthropological terminology, the "clan," a term derived through Scotland from the Gaelic *clann*). The Gaelic term in the earlier period was *fine*, a term which (although it continued in use in Gaelic Scotland) was in Ireland replaced in the later Middle Ages by *sliocht* (literally "section," rendered in English as "sept"). Although the law-tracts lay down precise definitions of the extent of the *fine*, it has been plausibly suggested that these were in fact paradigms, and certainly in the latest period, the boundaries of the *sliocht* were determined by individual practice. The corporate lineage-group was the landowning unit, its land being divided between its members on a shifting basis, either every year or on the death of an individual member. When it became too extended over generations, it could or would break up into separate groups, each operating on the same basis. Bloodshed within the clan arising out of landholding was almost a norm. Because the clan itself was the unit for prosecuting the homicide of a member, killing within the clan presented legal difficulties and could in practice lead to retaliation in kind. In theory, crimes such as homicide and theft were matters for private suit and, following an award by a brehon (*breitheamh*), to the payment of compensation. The blood price (*éraic*) for homicide was determined by the status of the victim: the compensation for theft was twice the value of the stolen object. In the later period, however, local rulers were imposing fines and penalties far in excess of the compensation awarded to the victim and taking the larger part of the *éraic* for themselves, while certain crimes that outraged public opinion, such as sacrilege, could involve the death penalty. In the latest period, the same fate could befall a thief whose clan could not or would not pay the compensation.

By the fifteenth century, the characteristic forms of Gaelic social and political organization—lineage expansion, collective landownership by the *sliocht*, irregular forms of succession to power, and Gaelic criminal law—had become established in many of the former colonial areas, most especially in Connacht and Westmeath, while in Munster and Kilkenny mixed systems had come into existence, with criminal law being entirely Gaelic.

Kenneth Nicholls

References and Further Reading

Fergus Kelly. *A Guide to Early Irish Law.* Dublin, 1988.

Nicholls, Kenneth. "Gaelic Society and Economy in the High Middle Ages." in *A New History of Ireland, II,* edited by Art Cosgrove. Oxford, 1987.

Nicholls, Kenneth. *Gaelic and Gaelicized Ireland in the Middle Ages.* 2nd edition, Dublin, 2003.

Simms, Katherine. *From Kings to Warlords: The Changing Political Structure of Gaelic Ireland in the Later Middle Ages.* Woodbridge, 1987; reprinted, Dublin, 2000.

See also **Common Law; Gaelic Revival; Gaelicization; Law Texts; Lordship of Ireland; Society, Grades of Gaelic**

SOCIETY, GRADES OF ANGLO-NORMAN

Anglo-Norman Ireland was a colony of Anglo-Norman England, and its social hierarchy closely imitated prevailing English fashions. England, of all the kingdoms in Western Europe, corresponded most closely to the historian's ideal of a "feudal pyramid." At the apex was the king, of whom all land was technically held. In Ireland, the supreme landholder was the "lord of Ireland." This was, in practice, the king of England. In 1177, shortly after the initial conquest, King Henry II (1154–89) granted the lordship of Ireland to his son John (d. 1216) in a plan to make Ireland a kingdom that would descend in the cadet line of the royal house. This was the only occasion on which Ireland seemed as if it might be separated from the English crown. The putative kingdom of Ireland did not materialize. When John became king of England in 1199, he brought the lordship of Ireland back to the English crown. This situation was confirmed forever by King Henry III (1216–72) in 1254: by a charter granting Ireland to his eldest son, the future Edward I, he stipulated that the king of England must also henceforth be lord of Ireland.

The Titled Nobility

Beneath the lord of Ireland came the landholding class—the nobility. The nobility was essentially a military caste that also exercised very important administrative functions. In Ireland, because the king was almost permanently absent, these roles were particularly significant. The king frequently chose the chief governor—his representative and head of the Irish administration—from the ranks of the Irish nobility; and given the almost permanent state of war on the marches of Ireland, the military function of the nobility was constantly being tested.

In the early phases of the Anglo-Norman colony in Ireland, hereditary noble titles were extremely rare. In fact, the only such title was earl (Latin: *comes*), and, from 1205 until 1316, Ulster was the only earldom in Ireland. This reflected the situation in England, where the earls were a tiny elite within the nobility. It was not until the fourteenth and fifteenth centuries that titles proliferated and the nobility became more strictly stratified. The earldom of Ulster was created in 1205 by King John for Hugh de Lacy II. It lapsed at de Lacy's death in 1242. In 1263, the earldom was revived for Walter de Burgh (d. 1271), whose descendant, Richard de Burgh, the "Red Earl" (d. 1326) was the premier noble in Ireland—both in terms of dignity and wealth—in the early fourteenth century. After his grandson, William, was murdered in 1333, the earldom came into the hands of absentees and ultimately became an appanage of the English crown.

Meanwhile, in reward for service during the Bruce invasion of Ireland (1315–18), several new creations added to the ranks of the comital elite. In 1316, John fitz Thomas (d. 1316), baron of Offaly, was created earl of Kildare, and for defeating and killing Edward Bruce at the battle of Faughart in 1318, John de Bermingham was made earl of Louth. This latter earldom expired along with de Bermingham when he was murdered in the Braganstown massacre of 1329. Two further earldoms were created during the disturbed minority of King Edward III (1327–77). In 1328, James Butler (d. 1338) was created earl of Ormond, and the following year Maurice fitz Thomas (d. 1356) of the Munster Geraldines was created earl of Desmond (see *Factionalism*). The new earl of Ormond's father, Edmund Butler (d. 1321), had been granted lands to hold under the name "earl of Carrick" in 1315, but he is rarely described as such in official documents and does not appear to have been officially "belted" as earl. His son's earldom of Ormond was, by contrast, highly successful. These elevations were of the greatest importance. The three comital houses of Kildare, Ormond, and Desmond became central to the history of the lordship of Ireland for the rest of the MiddleAges.

Their endurance was guaranteed because the grants were made in "tail male," meaning that should the male line fail, the earldom could not suffer partition between heiresses but would descend intact to the nearest male relative.

On only one occasion was a dignity higher than earl introduced to Ireland. In 1383, Robert de Vere, earl of Oxford and an intimate of King Richard II (1377–99), was created marquis of Dublin. A "marquis" was intended to be superior to an earl but inferior to the dignity of "duke." The latter title had not existed in England until 1337 and thereafter was usually reserved for members of the royal family. Yet, in 1386, de Vere was further promoted to be duke of Ireland. The experiment proved ephemeral and personally disastrous for de Vere. He never came to Ireland, and resentment against him in England, culminating in the Appellant crisis of 1385–86, forced him to flee the kingdom, never to return.

The Parliamentary Peerage and the Lesser Nobility

Below the earls were all the other nobles of Ireland. This non-comital class was extremely fluid. The fortunes of each family depended just as much on luck and biological accident as on landed wealth and military ability. Status could not be guaranteed to survive into the next generation. This truism also held for the English nobility, but it was particularly obvious in Ireland. The families that were granted earldoms in the fourteenth century descended from the mere adventurers who had arrived at the earliest stages of the conquest. Their elevation allowed new families to occupy the second rank of the nobility, for instance the le Poers, Roches, and Cauntetons. The heads of these families usually held *in capite* (i.e., "in chief," or directly) of the crown, but they were also leaders of great lineages. By the fourteenth century, due to the frontier conditions in which they had survived, these lineages had often departed from English norms (see *Gaelicization*). Problems arose if the supply of available land began to dry up. Landless "idlemen" (Latin: *ociosi*; French: *gentz udyves*) from the junior branches of these families were a persistent source of disorder, and legislation was enacted to make the heads of lineages responsible for disciplining the men who bore their family names.

Just as the development of parliament in England was instrumental in defining the nobility more rigorously, so a similar development can be discerned in Ireland where, from the late thirteenth century, parliament became increasingly important. The lists of those who were summoned to parliament became customary over time. A son was summoned because his father had been, and in this manner a parliamentary peerage crystallized by the late fourteenth century. In particular, the title "baron" began to be used with some consistency. This vague term had been used loosely during the thirteenth and fourteenth centuries to denote a major landholder, for instance the FitzGerald barons of Offaly, who became earls of Kildare in 1316. It was a means of identification rather than a hereditary honorific. But families such as the Flemings of Slane, the Prestons of Gormanstown, and others who received regular summonses to parliament, now began to refer to themselves as barons. In fact, the first formally created baronies date from the fifteenth century. In 1462, the baronies of Portlester and Trimlestown were created by writ for the FitzEustace and Barnewall families respectively, and were followed in 1468 by the short-lived barony of Ratoath created for Robert Bold (d. 1479). Because the fourteenth-century baronies were not created formally, but rather emerged over time, the incumbents frequently disputed their antiquity, which dictated seniority and precedence in parliament. The prolonged competition for precedence between the barons of Slane and Gormanstown was finally won when the Prestons of Gormanstown were made the first viscounts in Ireland in 1478. A viscountcy should have conveyed a rank between baron and earl. In practice, however, the struggle had not ended; in the summons to parliament of 1489, the new viscount Gormanstown was not accorded his proper title and was ranked below the baron of Slane.

Beneath the emerging parliamentary peerage was the lesser nobility or gentry. This grade of society played an important part in filling the retinues of the greater nobles and acted in various administrative capacities, for instance as seneschals, or stewards, in the great liberties of Ireland or sheriffs in the royal shires. Among these men were found the knights who came to make up the commons in parliament. Knighthood was a personal, not a hereditary, honor. The mounted knight had once been the elite of the Anglo-Norman military machine. In England, where by the later Middle Ages knightly families had gained land and ceased to exercise a purely military function, the expense and responsibilities of knighthood came to be seen as something to avoid. In Ireland it is not hard to imagine that the Anglo-Irish—isolated in other respects from the English court—were pleased to enjoy the honor. There are only fleeting glimpses of its ceremonial aspect. One such occasion is the report of the Dublin annalist which states that when Lionel of Clarence first arrived in Ireland in 1361, he knighted many residents of the colony. Among them was one Robert Preston, who had acted for some time as chief justice of the Dublin bench. The Preston family had begun in the early fourteenth century as a merchant family. They then progressed to careers in law. That the descendants of the new Sir Robert Preston of 1361 went on to become viscounts of Gormanstown shows the extent to which an upwardly mobile family could succeed over a few generations.

The Non-Noble Class

Most colonists, of course, were not noble. They were tenants on the great lordships, and, together with the unfree laborers, they kept the manorial economy of Anglo-Norman Ireland operating. The pioneering study of Anglo-Norman agriculture by Professor Otway-Ruthven revealed several grades within tenant society. Free tenants held their lands perpetually either by feudal tenure, meaning that they owed military service to their lords, or else by payment, a fixed annual money rent. Farmers also leased their lands but for a finite term of years, and they often owed their lords labor services in addition to rent. Gavillers and cottiers were less secure again, being tenants at will who owed heavier labor services and money rents. They appear, however, to have been personally free, and in this respect are distinct from the betaghs. The term "betagh" comes from the Irish *biatach*, referring to a food provider in Gaelic Ireland. The Gaelic *biatach* was not servile, but on the Anglo-Norman manors, the betagh seems to have become bound to the land and to have had a status equivalent to the villein or serf of medieval England. They owed labor service rather than money rent and, being unfree, had no legal recourse other than the manor court of their immediate lord. The various services such as plowing, reaping, and carrying crops that were owed to Anglo-Norman lords in Ireland were considerably less onerous than was normal in England. This no doubt acted as an incentive for the first peasants to migrate to Ireland. The prospect of burgage tenure also made Ireland attractive to prospective settlers. Burgesses—the tenants of a medieval borough—were granted liberties often based on the Law of Breteuil (from the small town of Breteuil-sur-Iton in Normandy), such as the right to their own court, the right to sell their burgage lands, and the right to marry freely. As a result, they were extremely common in Ireland. The boroughs they lived in sometimes grew into towns, but it was not uncommon to find the residents of tiny rural settlements exercising the rights of burgesses. In practice, these various grades were as fluid as were those of the nobility. In the later Middle Ages, as labor services came to be commuted in favor of rent, the distinction between servile and free tenants became increasingly blurred. By the end of the medieval period, serfdom had disappeared.

Peter Crooks

References and Further Reading

Cokayne, G.E. *Complete Peerage of England, Scotland, Ireland, Great Britain and the United Kingdom*. edited by V. Gibbs, *et al.* 12 volumes. London: St. Catherine Press, 1910–59.

Frame, Robin. *English Lordship in Ireland, 1318–1361*. Oxford: Clarendon Press, 1982.

Given-Wilson, Chris. *The English Nobility in the Later Middle Ages*. London and New York: Routledge and Kegan Paul, 1987.

MacNiocaill, Gearóid. "The Origins of the Betagh." *The Irish Jurist*, new series 1 (1966): 292–98.

Otway-Ruthven, A.J. "The Organization of Anglo-Irish Agriculture in the Middle Ages." *Journal of the Royal Society of Antiquaries of Ireland* 81 (1951): 1–13.

See also **Agriculture; Anglo-Irish Relations; Chief Governors; Feudalism; John; Lordship of Ireland; Manorialism; Ulster, Earldom of**

SOCIETY, GRADES OF GAELIC

Gaelic Ireland was divided into a number of social classes. Each class was based on qualifications in terms of property, learning, or skill. The members of these classes were subdivided into distinct grades, each with its own legal attributes.

Some of these classes were considered to be *nemed*, "noble" or "privileged." According to the law text *Bretha Nemed toísech* (the first collection of the Judgements of Privileged Persons), these were the poets, the clergy, the men of ecclesiastical learning, and the lordly grades. The lordly grades (which included the kings) derived their wealth and political power by advancing cattle to clients.

A person's grade affected their legal rights. For example, no one could offer legal protection to a person of a higher grade. Their grade also determined the size of the retinue they were entitled to have with them when exercising their rights to hospitality. Most importantly, each grade had its own honor price (*eneclann* or *lóg n-enech*). A person's honor price determined part of the payment they received in the case of legal offenses against them (for these were considered to be a slight against their honor). It was paid to them if they agreed to submit themselves to a lord and thereby become his "base client." A person's right to act independently in making a contract, or giving a gift of property, was usually restricted to the level of his own honor price. Likewise, a person could act as a surety or guarantor to the contracts of others only up to the level of his own honor price.

Secular Grades in the Early Medieval Period

Free persons were divided into two main classes: the noble freemen (*grád flatha*) and the common freemen (*grád Féne*). The members of both classes owned their own land. The principle distinction between these two classes was that the common freemen served as the base clients of the noble freemen. A base client owed his *flaith* (lord) payments of labor and food. By the later medieval period, the regular name for such a client in nonlegal Gaelic sources was *biatach*, "food provider." (The Anglo-Irish lawyers, however, naturally associated the providers of such services with the feudal serf. The Anglicized term "betagh" was therefore applied to dependent tenants and farm-laborers working the land of their masters.)

The lowest grade of common freeman was that of the *fer midboth* (a man between huts). This grade's name indicated that he was in transition between living as a dependent in the house of his father (or foster father) and establishing a house of his own. All young freemen between the ages of fourteen and twenty belonged to this grade. Furthermore, those over twenty years of age continued to belong to this grade until they had inherited sufficient land to advance to a higher one.

Once a young man had inherited sufficient land to support a small herd of about seven cows, he rose to the grade of *ócaire* (young freeman). The minimum landholdings of the more prosperous commoner grade, the *bóaire* (cow freemen), were twice those of the *ócaire*.

Above these common grades were the noble grades of lordship. The lowest ranking noble was the *aire désa*. His name (freeman of lordship) indicated that he had received the submission of clients. The *aire désa* may well have had an entire kin-group serving him as clients.

The next highest grade of lord, the *aire tuíse* (freeman of leadership), appears to have had authority over a number of related kin-groups, for he is described as the leader of a *cenél* ("extended family" or "people"). He may have attained this position by obtaining several lords of the *aire désa* grade as "noble clients." (Noble clients were quite different from base clients; in particular the noble client was not paid his honor price by his lord because he did not subordinate himself to him.)

Next came the *aire ard* (high freeman), whose authority in the whole kingdom was such that he appears to have been called upon to act on its behalf as an ambassador in interkingdom relations. Above him was the *aire forgaill* (commanding freeman), a man of such political power that he was considered a likely candidate for the kingship itself. Finally came the *rí* (king) of the *túath* (petty kingdom).

Elaboration of Grades in the Legal Texts

This basic structure was considerably elaborated in those early Irish law texts that dealt specifically with status. These elaborations vary from text to text. For example, in the text called *Críth Gablach* ("the Bifurcated Purchase"), most of the grades are split into two subdivisions. These early status texts also differentiated between the king of a single petty kingdom and various types of over king.

Out of these early elaborations there later developed a new hierarchy of grades. By the twelfth century, the legal commentaries were using a grade structure based on seven sets of three, as follows: (1) the king of Ireland (without opposition), the king of Ireland with opposition, and the king of a province; (2) the king of several kingdoms, the king of a great kingdom, and the king of a petty kingdom; (3) three subdivisions of the *aire forgaill* grade; (4) the *aire ard*, the *aire tuíse*, and the *aire désa*; (5) three subdivisions of *bóaire*; (6) three subdivisions of *ócaire*; and (7) three subdivisions of *fer midboth*.

Beneath the grades of independent freemen, the status tract *Uraicecht Becc* (the Small Primer) lists three subdivisions of young boys not yet of sufficient age for independent legal status. (These match the three subdivisions of apprentice poets and novice ecclesiastics mentioned in the subsequent paragraphs.)

Women and Children

Children under the age of fourteen were dependent on their fathers. Most women also remained dependent on a man (usually either their father or their husband). Such dependents had an honor price equal to half of that of the person on whom they were dependent. However, a woman could inherit land if she had no brothers. In such cases, she was entitled to the independent honor price appropriate to her property.

The Semi-Free and Unfree Classes

Beneath the class of freemen came a class of semi-free men. These men lacked the normal qualifications of property or the support of a kin-group from whom they might expect to inherit such property and so placed themselves in a position of dependency upon a landed freeman. As a result, their status, and their honor price, was diminished. Beneath them came the slaves, who had no status at all.

The Poets

The grade structure of the poets was somewhat different from the normal secular grade structure. Most texts recognize seven grades of poets as follows: *ollam* (the master poet, whose honor price was equal to that of the king), *ánruth*, *clí*, *cano*, *dos* (who had the same honor price as the common *bóaire*), *macfhuirmid*, and *fochloc*.

These grades set out the career path of the professional poet. Each grade marked his mastery of increasingly advanced poetical meters. (Beneath the lowest grade of independent poet were three subdivisions of apprentice poets, still dependent on their masters.)

The Church

Like the poets, those in holy orders followed a sevenfold grade structure. This structure is typically given as bishop, priest, deacon, subdeacon, lector, exorcist, and usher. (Beneath these came three subdivisions of novice; young trainees not yet of independent status.) Like the master poet, the bishop had an honor price equal to that of the king. Unlike the poets, each of the other Church grades also claimed an honor price equal to one of the other grades of lord. This claim was hard to accommodate to the original structure of five grades of lords in all. It may well explain the subsequent splitting of the *aire forgaill* grade into three subdivisions.

There was a separate grade structure for those not in orders but trained in Christian learning (the *gráda ecnai*). The highest grade, often called the *fer léigind* (man of learning) or *suí* (sage), again had an honor price equal to that of the king.

The Professions

According to *Uraicecht Becc*, some craftsmen had an honor price equal to that of the lowest rank of lord (i.e., equal to that of the *aire désa*). This was true of the blacksmith (who worked in iron), the metal-wright (who worked in copper, bronze, and precious metals), and of physicians. It was also the case for a wood-wright proficient in the construction of mills, churches, or boats, and the wood-wright who produced cups and bowls of yew wood. A wood-wright who combined all four of these skills could rise in status as high as the *aire tuíse*. Indeed, if he added yet other professions to this, he could rise as high as the *aire ard*.

The calculation of the fees appropriate to these professions was set out in specific law texts. A judge (or "brehon") who was competent to sit in judgment on each of these professions likewise had an honor price equal to that of their exponents (i.e., equal to the *aire désa*). A judge proficient in all areas of Brehon Law rose to an honor price equal to that of the *aire tuíse*. One who was expert not only in Brehon Law but also in the rights and obligations of poets and in canon law, reached the honor price of the *aire ard*.

Among musicians, only the harpist (who accompanied recitations of poetry) had free status on account of his craft. He ranked with the highest of the commoners (the *bóaire*).

Mere carpenters (such as the builder of carts or the maker of shields) had a modest status equal to that of the mid-ranking commoner (the *ócaire*). This was also true of decorative craftsmen such as relief-carvers and cloth figurers.

Lesser craftsmen ranked only with the lowest grade of freemen (the *fer midboth*). This was true of fishermen, leather-workers, and comb-makers.

Other occupations had no status at all. Their exponents were semi-free, rating merely as the dependents of those for whom they worked. This was true of musicians (other than the harpist previously mentioned), jesters, and chariot drivers.

Movement Between Classes

A person could advance within his own class by increasing his property or skill. However, movement between classes was not as straightforward. Full membership of a particular class was usually restricted to those whose father and grandfather had been members of it.

So, for example, a common freeman might become so prosperous that he was able to acquire clients. This did not, however, entitle him to the rank of a lord. Instead, he became a member of a transitional class. He required twice as many clients as the *aire désa* before he could claim the same honor price.

Among the semi-free, those who had remained so for a number of generations lost all hope of inheriting sufficient land to regain their independence. As a result, they were locked into servitude. They were no longer able to legally separate from the freemen on whom they had become dependent.

Professions, including that of the poet, were also usually hereditary.

Necessity to Make Productive Use of Wealth or Skills

Legal independence went hand in hand with economic independence. It was not enough to possess the property or acquire the learning appropriate to a particular grade. It was necessary to put that land or learning to productive use. A person who did not work their land, or a professional who turned away work from members of the public, had no claim to the honor price that they might otherwise have deserved.

Worthy Behavior

A person's status was also dependent on their good name. In effect, a person who had the property or learning necessary for a particular grade lost half their honor price if they behaved unworthily. This was true, for example, of those who failed three times to properly fulfill their obligations as a surety. Half of a person's honor price was also lost if they were guilty of crimes such as intentional wounding or arson. In cases of very serious criminality, (such as killing one's own kinsman or covert homicide), their entire honor price was lost.

NEIL MCLEOD

References and Further Reading

Binchy, Daniel. editor. *Críth Gablach.* Dublin: Dublin Institute for Advanced Studies, 1941.

Breatnach, Liam. *Uraicecht na Ríar: The Poetic Grades in Early Irish Law.* Dublin: Dublin Institute for Advanced Studies, 1987.

Kelly, Fergus. *A Guide to Early Irish Law.* Dublin: Dublin Institute for Advanced Studies, 1988.

MacNeill, Eoin. "Ancient Irish Law: The Law of Status or Franchise." *Proceedings of the Royal Irish Academy* 36 C (1923): 265–316.

McLeod, Neil. "Interpreting Early Irish Law: Status and Currency." (Part 1) *Zeitschrift für celtische Philologie* 41 (1986): 46–65; (Part 2) ibid. 42 (1987): 41–115.

See also **Áes dána; Brehon Law; Canon Law; Children; Craftwork; Ecclesiastical Organization; Entertainment; Kings and Kingship; Law Tracts; Music; Slaves; Poets/Men of Learning; Society, Functioning of Gaelic; Women**

ST. PATRICK'S CATHEDRAL

The earliest known reference to a St. Patrick's church in Dublin occurs circa 1121. Often described as "in insula," between two branches of the Poddle river, it was one of several churches south of the Liffey with Irish dedications. Six cross-slabs in the cathedral, including the "St. Patrick's well" slab, date to the Hiberno-Norse tenth and eleventh centuries, demonstrating that, from the twelfth century onwards, the cult of St. Patrick was intricately bound up with competition between Dublin and Armagh. "Edanus, priest of St. Patrick's" is the earliest known individual, listed in a grant by Archbishop Lorcán Ua Tuathail to his canons of Holy Trinity (Christ Church) cathedral in 1178.

By 1191, the first Anglo-Norman archbishop of Dublin, John Cumin, established St. Patrick's beyond the city walls as a collegiate church, dedicated on St. Patrick's Day to "God, our Blessed Lady Mary and St. Patrick." His successor, Henry of London, unhappy with Holy Trinity as a priory of Augustinian canons regular (subject to monastic rule), began to elevate St. Patrick's to cathedral status. In 1214, he instituted the secular offices of precentor, chancellor, and treasurer, followed by the adoption of the Sarum rite, and by 1221, St. Patrick's had a charter which, completing the 4-square secular organization, included the office of dean.

The late foundation of St. Patrick's explains the paucity of its landed property around Dublin, some 1,800 acres, mainly consisting of Clondalkin, Saggart, and Shanganagh. Thirteen prebendaries were established from archiepiscopal lands, the richest being the "Golden prebend" of Swords. Liberties of St. Patrick's and of St. Sepulchre's respectively surrounded the cathedral and the nearby archbishop's palace.

In 1225, Henry III granted protection to preachers begging for alms for the fabric, suggesting a new cathedral was underway, probably modeled on Old Sarum. By 1235, a chapel was dedicated to St. Mary, for which Archbishop Luke provided lights and vicars attending the mass. By 1254, the cathedral was consecrated and a lady chapel was added in the 1260s, attributed to Fulk de Sandford, first archbishop of Dublin to be buried in the cathedral. Meanwhile, the emergence of two diocesan chapters of Holy Trinity and St. Patrick's caused serious difficulties in the appointment of an archbishop. Matters came to a head in 1300 when the two institutions agreed to sign a pact known as the *composicio pacis*, which recognized both of them as diocesan cathedrals, but the seniority of Holy Trinity. Unsuccessful attempts at establishing a university at St. Patrick's were made in 1311 by Archbishop John Lech and again in 1318 by his successor, Alexander Bicknor.

In 1316, St. Patrick's suffered a storm which destroyed the spire; a fire set by the citizens to thwart the progress of Edward Bruce; and the resultant loss of many treasures. Bicknor, appointed a year later, perhaps then allowed the north transept to be used for the parish of St. Nicholas without the walls. An accidental fire caused by John the sexton in 1362 led to the destruction of the tower and bells, as recorded the following year in a petition to the pope by Archbishop Minot. The Minot tower, completed circa 1370, suffered further in late fourteenth century. The ensuing building campaigns shed much light on late gothic architecture in the Pale.

Services at St. Patrick's were probably daily, often choral, and occurred in the aisle chapels of St. Paul, St. Michael, St. Peter or St. Stephen, as well as the nave, which probably filled for such prestigious occasions as episcopal consecrations or governmental ceremonies such as the creation of knights or barons. Unlike the strong civic links of Holy Trinity, St. Patrick's links were often to the state. A number of deans were chancellors of Ireland, masters of the rolls, and Dean John Colton was lord deputy of Ireland.

St. Patrick's and Holy Trinity were destinations in 1434 for the mayor and citizens' barefoot penitential walk through Dublin for taking the earl of Ormond prisoner and other offenses. The earls of Ormond and Kildare were also reconciled there in 1492 following a feud concluded by Fitzgerald, who thrust his hand through a hole in the chapter house door (exhibited in the cathedral) to shake that of Ormond's and thus restore trust.

The Reformation at St. Patrick's saw statues in the choir destroyed in 1537 and the cathedral itself dissolved in 1546, but not before Christ Church based its new secular constitution on that of St. Patrick's. The nave vault collapsed during this period, yet in 1548, it was used as "a common hall for the Four Courts of Judicature." The cathedral was restored under Mary in 1555, but its walls were being painted and scripture passages erected in 1559, under the Protestant reforms of Elizabeth.

STUART KINSELLA

References and Further Reading

Mason, W.M. "*The History and Antiquities of the Collegiate and Cathedral Church of St Patrick near Dublin.*" Dublin, Ireland, 1820.

See also **Architecture; Christchurch Cathedral; Cumin, John; Dublin; Patrick**

STRONGBOW (RICHARD FITZ GILBERT)

Richard fitz Gilbert, also known as Strongbow, a sobriquet first accorded to his father, was the son of Gilbert fitz Gilbert, a member of a cadet branch of a prominent family, arrived in England with William the Conqueror and acquired the lordship of Clare in Suffolk, whence he is also sometimes termed Richard de Clare. His father held lands in the duchy of Normandy, in nine different counties in England, and in the Welsh borders. His castle at Strigoil (modern Chepstow in Monmouthshire) was deemed to be his principal residence among his assemblage of estates. In 1138, during the civil war of King Stephen's reign (1135–54), when the king was eager to win support to his side, Gilbert was created earl of Pembroke, thereby acquiring not just a title but additional lands in south Wales. He died in 1148. His son, Richard, was probably born about 1130 because he occurs as a witness to a number of his father's charters before the latter's death, suggesting that he had come of age by that date. In 1148, Richard inherited his father's estates, including the earldom of Pembroke. However, when Henry II succeeded Stephen as king of England in 1154, he refused to acknowledge Strongbow as earl and took the lordship of Pembroke into his own hands. That Henry's disfavor derived from the circumstances of the civil war of Stephen's reign is indicated by the fact that in 1153, even before his accession as king of England, Henry's supporters had already seized Strongbow's lordships of Orbec and Bienfaite when they took control of the duchy of Normandy. From the time of Henry's accession, Strongbow was out of favor at the royal court. In 1164, he should have succeeded to additional lands in right of his mother from the partition of the Giffard inheritance, but Henry withheld them from him. The king also did not provide him with a wife commensurate with his status, for as a tenant who held lands directly of the king, Strongbow would have expected to marry an heiress whose estates would augment his own, for which, however, he required both the king's patronage and consent. According to the Anglo-Norman apologist, Gerald of Wales, when in the autumn of 1167, Diarmait Mac Murchada, the exiled king of Leinster, encountered Strongbow, "he had succeeded to a title rather than possessions." There is other evidence that Strongbow was in straitened circumstances, had been obliged to mortgage some of his landed property, and was in debt to Aaron, a Jew of Lincoln. Diarmait Mac Murchada's offer of rewards to be gained in Ireland would therefore have been attractive to Strongbow, especially if Diarmait was promising him marriage with his daughter, Aífe, together with succession to the kingdom of Leinster after Diarmait's death, in other words marriage to an heiress. As a tenant who held his lands directly of the king, Strongbow required the king's permission to marry, and he may have sought such permission in 1168 when his presence was recorded at the English royal court for the first time since 1155. Whether Henry consented—and contemporary sources are ambiguous on this point—Strongbow departed for Ireland. In August 1170, he landed at Waterford, captured the city, and his wedding to Aífe was celebrated almost immediately. Together with the forces of Diarmait Mac Murchada, Strongbow then set out to take the city of Dublin and, having done so, embarked on expansionist raids into Meath. In May 1171, Diarmait Mac Murchada died at Ferns, but Strongbow's control of Leinster was secured by the fact that he had the support of Diarmait's son, Domnall Caemánach. Ruaidrí Ua Conchobair, king of Connacht, and claimant to the high-kingship of Ireland, in a bid to assert control over Strongbow, besieged Dublin from June to August 1171, but failed to dislodge the Anglo-Norman garrison. Towards the end of September 1171, Henry II returned from his continental dominions to England and, delaying only long enough to gather together an army and provisions, made ready a large expedition to Ireland. In the meantime, the king had sequestrated Strongbow's estates in England and south Wales. This constituted a serious threat to Strongbow who now risked losing those lands that he held in Henry's dominions for a potential lordship in Leinster, over which he had not yet asserted secure control. Strongbow had first sent messengers in July to negotiate on his behalf with the king at Argentan in Normandy, and, in advance of Henry's arrival in Ireland, he himself now traveled in person to meet Henry at Newnham in Gloucester. His anxiety to confer with the king strongly suggests that he conceived Henry's expedition to Ireland as being aimed primarily at him. Henry, however, was not deflected from his intention to go to Ireland himself, where he remained from October 1171 until April 1172, during which time bargaining between the two must have been protracted. Henry obliged Strongbow to acknowledge him as his overlord for Leinster, while Strongbow persuaded Henry II to restore to him his dignity as an earl, although the king refused to concede him the earldom of Pembroke. From 1172 onwards, Strongbow was titled "earl of Strigoil," which, however, brought him no additional lands. Henry II also removed control of Dublin, Waterford, and Wexford from Strongbow, retaining them for his own use. In 1173, Henry II tested Strongbow's loyalty by summoning him to fight in Normandy, where he played a critical role in defending the castle of Gisors and recapturing Verneuil for the king. As a reward for faithful service, Henry then returned the port town of Wexford to Strongbow. During Strongbow's absence from Leinster, a number of Irish had seized the opportunity to attack Anglo-Norman garrisons, but, on his return to Ireland, Strongbow regained the upper hand militarily and set about planting Leinster with his own tenants, probably using Irish disloyalty as a justification for at least some of his land-grants. Strongbow was now also appointed Henry's principal agent in Ireland, and, in that capacity, he issued charters on behalf of the king relating to the now royal city of Dublin. He died unexpectedly in April 1176 from an injury to his foot and was buried in Christ Church Cathedral (the tomb there that is traditionally associated with him is of later date), his funeral presided over by Lorcán Ua tuathail (Laurence O'Toole), archbishop of Dublin. He left as his heir a three-year-old son, Gilbert, and a daughter, Isabella, both children having been given stock forenames in the de Clare family. His death meant that responsibility for the lordship of Leinster fell on his overlord, Henry II, for the duration of the minority of his heir. This undoubtedly slowed down Anglo-Norman settlement in Leinster. Gilbert died sometime between 1185 and 1189, and, in 1189, Isabella, who was now Strongbow's sole heir, was given in marriage by Henry II to William Marshal, who became lord of Leinster in right of his wife. William systematically set about reconstituting his father-in-law's landholdings both in Ireland and elsewhere, culminating in his recovery of the earldom of Pembroke in 1199. In the period between Strongbow's death in 1176 and the marriage in 1189 of his daughter, Isabella, Strongbow's widow, Aífe, is mentioned in English royal records as enjoying her widow's dower from his estates in England and may have resided for a period at his castle in Strigoil. In 1184, she may even have been responsible for the organization of defenses against the Welsh. The date of her death is not known, but she was buried in Tintern Abbey in Monmouthshire, alongside her father-in-law, Gilbert, who had founded that monastery. As settlers in Leinster, Strongbow chose tenants chiefly from his English landholdings rather than promoting Cambro-Normans from south Wales, such as the Geraldines, which largely accounts for the hostile portrayal of him in Gerald of Wales's *Expugnatio Hibernica*. The English chronicler, William of Newburgh, commented that as a result of his acquisitions in Ireland, Strongbow, who had little fortune previously, became celebrated in England and Wales for his great wealth and prosperity. His intervention in Ireland returned him to the favor of the English royal court, augmented his landed resources, secured him a wife commensurate with his status, restored to him the dignity of an earl, and afforded the means by which his son-in-law, William Marshal, was to recover the earldom of Pembroke. From Henry II's perspective, Strongbow's strategically located acquisitions in Ireland had raised the prospect that a disaffected subject, who deemed himself to have been arbitrarily deprived of the earldom of Pembroke, might destabilize the king's control of South Wales. This undoubtedly was a key consideration in Henry's decision to intervene personally in Ireland and to reach an accommodation with Strongbow. Strongbow's marriage to Aífe is the subject of a noted historical painting by Daniel Maclise (1806–1870), which depicts a downcast Aífe, symbolically representing Ireland, being reluctantly wed to an overbearing conqueror under the authority of her father, Diarmait. References to her in English royal sources as "the Irish countess" and the "countess of Strigoil," and charters which she issued in her own name, including one in which she styled herself "Countess Eva, heir of King Diarmait," coupled with the fact that she never remarried, suggest that she retained a notable degree of independence over her own career.

M. T. Flanagan

References and Further Reading

Duffy, Seán. *Ireland in the Middle Ages*. London: Macmillan, 1997.

Flanagan, Marie Therese. *Irish Society, Anglo-Norman Settlers, Angevin Kingship: Interactions in Ireland in the Late Twelfth Century*. Oxford: Oxford University Press, 1989; reprinted, 1998.

See also **Diarmait Mac Murchada; Henry II; Waterford**

T

TÁNAISTE

The word *tánaiste*, anglicized as "tanist," refers to the candidate who, by the Gaelic method of succession, was recognized as next in line to rule a lordship or kingdom, and was so designated during the ruler's lifetime: literally, "second to the chief." By choosing a tanist, Gaelic lineages were able to avoid the turmoil that would accompany succession by a minor. In its original Old Irish form it meant "the expected one," which, within a few centuries, merged in meaning with the Latin *secundus*. Recent scholarship has disputed the suggestion that it was equivalent to the early medieval terms *ádbar ríg* and *rígdamna* (the former meaning worthy and eligible, the latter referring to the head of the main dynastic segment not in actual power). However, the fact that during later medieval times annalists sometimes separately described certain known *tánaistigh* as *ádhbhar ríogh* or *ríoghdhamhna* shows that the terms became interchangeable to some extent.

The nomination of a tanist usually took place at the assembly gathered for the election of a new chieftain. According to a late sixteenth-century description by the English writer Spenser, the tanist was entitled to place one foot on the chief's inauguation stone, and the chief had to swear, among other things, to deliver the succession peaceably to him. Apart from high status, being tánaiste brought other benefits, principally the right to a share of the profits and revenues of the lordship, and also the right to rule part of the dynastic territory more or less independently, as a sublordship. The English crown abolished the office of tanist and the practice of tanistry early in the seventeenth century.

David Edwards

References and Further Reading

Binchy, D. A. "Some Celtic Legal Terms." *Celtica* 3 (1956): 221–231.

Charles-Edwards, Thomas. "The Heir-Apparent in Irish and Welsh Law." *Celtica* 9 (1971): 180–190.

Mac Niocaill, Gearoid. "The 'Heir Designate' in Early Medieval Ireland." *The Irish Jurist* 3 (1968): 326–329.

Nicholls, Kenneth. *Gaelic and Gaelicised Ireland in the Middle Ages*, 2nd ed. Dublin: 2003.

Ó Corráin, Donnchadh. "Irish Regnal Succession: A Reappraisal." *Studia Hibernica* 11 (1971): 7–39.

Simms, Katharine. *From Kings to Warlords: The Changing Political Structure of Gaelic Ireland in the Later Middle Ages*. Woodbridge: Boydell, 1987, 2000.

See also **Inauguration Sites; Kings and Kingship**

TAPESTRIES

See **Wall Paintings**

TARA

Tara is a prehistoric sacred site in County Meath that held a powerful place in the early medieval Irish imagination and acquired national significance as a symbolic center of sovereignty and over-kingship. Tara's ritual importance to ancient peoples rested in its situation, which provided commanding views over an agriculturally rich landscape. It is a ridge 2 km long, rising to a height of 155 m, unimposing from the east but affording extensive views over a great part of the central plain from the west, while further afield Mount Leinster, the Slieve Blooms, and the Mourne Mountains are to be seen. Taken together these features place Tara in visual contact with one-fifth of the surface area of Ireland. To early farmers it was an ideal venue at which to intercede with the gods for the fertility of the lands below.

Little is known of the earliest monument on the hill—a large, possibly palisaded enclosure dating to the Neolithic—but comparable sites in Britain were used for seasonal gatherings. This was replaced around

3000 B.C. by a passage tomb, known today by its medieval name as the "mound of the hostages." The tomb is oriented to the east, and alignments have been observed with the full moon of Lughnasa (August) and the rising sun of the festivals of Samhain (November) and Imbolc (February). One of the side stones of the passage is decorated with concentric circles and zigzags, typical of passage-tomb art. The tomb was used for communal burial, and some 1000 pounds of cremated bone were recovered, estimated as representing about 200 people. The artifacts included passage-tomb pottery, decorated stone pendants, stone balls, and mushroom-headed bone pins, the latter two of which are thought to have a fertility significance. Aligned onto the mound of the hostages is a linear earthwork known by its medieval name as *Tech Midchúarta* (banqueting hall). It is unexcavated, but it is thought to be a ceremonial avenue or cursus of Neolithic date, although some scholars have expressed the view that it may belong to the Iron Age modifications of the hill. Some forty burials of Early/Middle Bronze Age date (*c.* 2400–1400 B.C.) were inserted into the mound of the hostages, showing that it remained an important monument, while dozens of small barrows (earthen burial mounds) were also erected across the ridge at this time. Little is known about the burial customs of the Late Bronze Age, but Tara evidently remained a sacred site, as is shown by the discovery there of two great gold torcs dating to around 1200 B.C., which were deposited as a votive offering.

During the first century B.C., the hilltop was rearranged and the summit was enclosed by a great ditch with an external rampart. This monument is known by its medieval name as *Ráith na Ríg* (fort of the kings). In fact it was not a fort, but rather a ritual enclosure that included within it the mound of the hostages as well as the *Forrad* and *Tech Cormaic*. The *Forrad* is a flat-topped mound, enclosed by two banks and ditches, built over earlier Bronze Age barrows, and which probably played a role in inauguration rituals. A granite pillar in the centre of the *Forrad* is supposed to be the *Lia Fáil* (stone of destiny); its phallic shape indicates that it is a fertility symbol. This accords well with one of the traditional attributes of kingship and with the inauguration ceremony, with which it was linked according to medieval lore. In medieval Irish mythology Tara was connected with the god Lug, who was the divine manifestation of kingship, and with the goddess Medb, the embodiment of fertility.

Tech Cormaic is a ringfort adjoining the *Forrad* that may have been inhabited in the early middle ages. Definite evidence of habitation on the hill during the early centuries A.D. was uncovered when the ringfort known as the Rath of the Synods was excavated. This revealed four major phases of activity, during which the use of the site changed from a cemetery to a ceremonial enclosure, then back to a cemetery before finally becoming a ringfort. Several of the finds were high-status, imported objects from Roman Britain, dating mainly from the second to the fourth centuries A.D.

There are no descriptions of actual inaugurations at Tara, and it is thought that the *Feis Temro* (assembly at Tara) was a celebration held at the height of a king's reign. The last assembly was held by Diarmait mac Cerbaill in 558/560, and celebration of the event seems to have declined as conversion to Christianity increased. When Tara is mentioned by Muirchú around 680, it was already an abandoned, legendary place—the *caput Scottorum* (capital of the Irish) associated with a powerful pre-Christian kingship. From the seventh century onward, medieval historians developed the theme of Tara as the seat of the high kings of Ireland, a concept that was intimately connected to the contemporary ambitions of the Uí Néill and that provided them with the legitimacy of tradition, albeit an invented one. The title of *rí Temrach* (king of Tara) was applied to an over king, although from the time of Máel-Sechnaill I it was gradually replaced by that of *rí Érenn* (king of Ireland). In 980, Tara was the setting for an important battle in which Máel-Sechnaill II defeated the Scandinavians of Dublin, and it was during his reign that the *Dinnsenchas Érenn* was compiled. Tara comes first in the account, and the detailed description of the hill is effectively a survey of the earthworks that were visible at the time.

After the coming of the Anglo-Normans Tara fell into the hands of the de Repentini family, and a church is first mentioned there in 1212, when it belonged to the House of the Knights Hospitallers at Kilmainham. It functioned as a parish church until the sixteenth century, when it fell into disrepair. The iconic status enjoyed by Tara in recent centuries rests largely on the literary skill of Geoffrey Keating's *Foras Feasa ar Éirinn* (written *c.* 1634–1636), in which he formulated and popularized the idea of Tara functioning continuously as a national institution from prehistoric times into the middle ages.

JOHN BRADLEY

References and Further Reading

Bhreathnach, Edel. *Tara, a Select Bibliography.* Dublin: The Discovery Programme, 1995.

Bhreathnach, Edel, and Conor Newman. *Tara.* Dublin: The Stationery Office, 1995.

Newman, Conor. *Tara, an Archaeological Survey.* Dublin: The Discovery Programme, 1997.

Roche, Helen. "Excavations at Ráith na Ríg, Tara, Co. Meath, 1997." *Discovery Programme Reports* 6 (2002): 19–165.

See also **Burials; Dinnshenchas; Earthworks; Feis; Forus Feasa ar Éirinn; Kings and Kingship; Máel Sechnaill I; Máel-Sechnaill II; Pre-Christian Ireland; Ringforts; Uí Néill**

TAXES AND TITHES

See **Coyne and Livery; Central Government; Ecclesiastical Organization; Kings and Kingship; Local Government**

TÍRECHÁN

A bishop, and author of a Latin memoir of St. Patrick, Tírechán flourished in the second half of the seventh century. The nature of his episcopacy is not known, although he may have ruled a diocese in his native territory. A native of Tirawley, a region of northwest Mayo in the province of Connacht, he belonged to its ruling family, the Uí Amolngado, a collateral branch of the Uí Fiachrach, the most influential dynastic family in Connacht during his time. He was also a descendant of Amolgnid—a son of Nath Í, King of Connacht and putatively king of Tara, in succession to the famous Niall Noígiallach, eponymous ancestor of the Uí Néill. His connections with this dynasty, whose seat was in Meath, may explain how he became the fosterling and pupil of Bishop Ultán of Ardbraccan (Co. Meath) in the middle years of the seventh century.

Tírechán's memoir is preserved with other documents relating to St. Patrick in a single manuscript, the *Book of Armagh*. Although the memoir cannot be closely dated, its reference to recent plagues (*nouissimas plagas*)—if it means those of 664–666, 680, and 683—would suggest a date around 690. Lacking a formal title, it has been variously labeled by modern scholars as a *Collectanea*, *a Memoranda*, and an *Itinerarium*. The first term recognizes that it was compiled from a variety of sources, oral and written; the second that it may have been intended as a record of events relating to St. Patrick's mission; and the third that its narrative is framed as a circular journey which takes Patrick from Meath westward to Connacht and back. In its present form the work is structurally awkward, perhaps even defective: It is preceded and followed by supplementary notes; its division into two books may be the work of a later redactor rather than Tírechán; and it ends abruptly with the mention of a visit by Patrick to Cashel, in the southern province of Munster.

Nor should it be regarded as in any sense a Life of Patrick. Following a brief synopsis of Patrick's early career, the work focuses entirely on events that allegedly took place during the year or two after his arrival in Ireland, and it presents Meath (rather than Armagh) as Patrick's headquarters. Tírechán evidently compressed several journeys into a single one, as evidenced by inconsistencies in his narrative that imply at least three forays by the saint into Connacht. While it is possible that some of Tírechán's information reflects genuine traditions about Patrick's activities and foundations in the fifth century, most of his material belongs to the seventh century.

Despite these deficiencies, his work is one of the most important historical sources for early medieval Ireland, specifically for the seventh century, and for the west of Ireland, which is otherwise poorly represented in contemporary sources. Tírechán provides invaluable (and detailed) information about the political and ecclesiastical landscape of northern Connacht (including a host of personal and place names), though he is not so good in regard to central and south Connacht. These latter areas presumably did not have traditions about Patrick, or else did not welcome inquiries that might make them liable to claims of allegiance from the Patrician community.

Tírechán says that he was inspired to compose the memoir by his mentor, Bishop Ultán, who provided him with a book in his possession about Patrick. Another source that he mentions is a *plana historia* (straightforward narrative) of the saint's life, perhaps a primitive Life based on Patrick's own autobiography, the *Confessio*. Tírechán drew on oral sources, notably information relayed to him by Ultán and by the senior clergy of Meath. He also visited quite a number of the foundations that he attributes to Patrick, such as Armagh, Tara, Baslick, and Cruachu. He seems to assume familiarity on the part of his readers with the general story of Patrick's life, as for example in his reference to the inhabitants of the Wood of Fochluth (§15.5) who appealed to Patrick in a dream to come and convert them, an episode described in the *Confessio*.

Tírechán's work seems to have been assembled without much concern for literary style or structure. Indeed, it has been characterized as "the raw material on which the hagiographer could, later, work." While using the narrative framework of the circuit—borrowed from secular Irish tales—it relies for the substance of its story on a tedious pattern whereby Patrick arrives in a certain district, converts the local magnates, and receives from them land for a foundation, which his successors maintain to the present day. Syntactically, the narrative is sustained by means of the connective *et*, perhaps under the influence of a similar device in Old-Irish storytelling. Indeed, Tírechán's Latin is strongly influenced by Irish syntax and idiom. Yet his work contains occasional passages of striking beauty, the most notable being the story of the conversion of Loíguire's daughters (§26), with its rhythmic, quasi-poetical language. However, some scholars would attribute these characteristics to his dependence on an Irish poetical source.

The traditional view holds that Tírechán wrote his memoir to establish a list of the ecclesiastical foundations in Meath and Connacht, which by virtue of having been founded by Patrick owed allegiance to the Patrician community of his own time, centered at Armagh. Certainly, Tírechán was a partisan of that cause, as evident from his lament that ecclesiastical enemies were encroaching on such foundations, especially the powerful communities of Colum Cille and Clonmacnoise, Armagh's main rivals in the second half of the seventh century. But an alternative theory proposes that Tírechán's concern was not Armagh but two other churches, one in his native Tirawley and the other (more important) in Meath, at the site of Patrick's first celebration of Easter. According to this interpretation, Tírechán addressed his work to the Uí Néill kings of Meath (a dynasty that had long supported the Patrician churches) in an effort to curry favor by supporting their claim to lands in Connacht.

References and Further Reading

Bieler, Ludwig. *The Patrician Texts in the Book of Armagh*. Dublin: The Dublin Institute for Advanced Studies, 1979.

Kenney, James F. *The Sources for the Early History of Ireland: Ecclesiastical*. New York: Columbia University Press, 1929, pp. 329–331.

Swift, Catherine. "Tírechán's Motives in Compiling the *Collectanea*: An Alternative Interpretation." *Ériu* 45 (1994): 53–82.

Swift, Catherine. "Patrick, Skerries and the Earliest Evidence for Local Church Organization in Ireland." In A. MacShamhráin, ed., *The Island of St. Patrick*. Dublin: Four Courts Press, 2004, pp. 61–78.

PÁDRAIG Ó NÉILL

See also **Armagh, Book of; Connacht; Ecclesiastical Organization; Hagiography and Martyrologies; Hiberno-Latin; Mide; Patrick; Uí Néill**

TOWER HOUSES

Thousands of small castles (known to contemporaries as such), which consisted primarily of towers, were built in late medieval Ireland. The normal form is of a square tower, with a vaulted ground floor and at least two (usually three) upper floors. Each floor has one main room; the first floor provided the principal one, with a good chamber normally above, and then lesser rooms in the attics. Some towers have attached turrets which housed stairs, latrines, or small chambers; a few towers are circular in plan. They provided accom-

Knockane Tower House, Co. Tipperary. © *Department of the Environment, Heritage and Local Government, Dublin.*

modation for the lord and family alone, without the trappings of English lordship, law, and administration, which needed a hall. Many had attached enclosures (in Ireland called bawns; in Scotland, barmkins) which may have held halls, farm buildings, and so forth made of perishable materials, but the core was the stone tower. The great majority date from the late fourteenth to the sixteenth centuries, sharing some architectural details with the friaries of the time. Their origin lies in the chamber towers of halls of the thirteenth century. Ireland shares the common use of the tower in late medieval castles with other European countries, but principally with Scotland and northern England. They overlap with the large residential towers for magnates (e.g., Askeaton), but seem to be a separate type.

Because tower houses are thick-walled and equipped with battlements, narrow loops, machicolations or "murder holes" (to drop rocks on people below), and other defensive features, they are often seen as primarily defensive in purpose, sometimes falsely contrasted with the more peaceful manor houses of lowland England. The defensive features are often impractical, such as rooftop battlements where it is impossible to move around the wall-walk unimpeded by gables, chimneys, and so forth. Doors are at ground level (in earlier, more defensive, castles tower doors were raised to first floor level) and wells are universally lacking. Side towers are common in the Pale area but do not flank the doors with archery or gun loops. The defensive features are part of display, not evidence that even low-intensity war was common in fifteenth-century Ireland.

Tower houses are most common in a swathe across Ireland from Galway, Clare, and Limerick in the West through the Midlands toward Wexford or south Dublin in the southeast. Ulster and the extreme Southwest have relatively few; the north Pale or Wicklow also have fewer than might be expected. Tower houses proliferated particularly in areas where lordship was disturbed in the fourteenth century. Within Ireland there are clear regional types of tower houses: attached towers in the Pale and southeast Ulster, elaborate roof-level display in Tipperary and Kilkenny, and plan type in Limerick. Within each region, however, the towers conform closely to type, unlike most castles, which emphasize individual design apparently to stress the owner's membership of a particular class. They are often found on sites that are not known to have been manorial centers in the thirteenth century, and they appear to be most common where new lordships were set up, often as a result of the decline of the earlier thirteenth-century magnates and the consequent division of their lands, in some parts. They appear to be sited to profit from the new economy of pasture and seaborne trade established after the Black Death. They give a general impression of a break with the earlier period.

The main builders were of the gentry class; although a few were built on lesser manors of major lords, they are not the castles of the great. They provided for a lifestyle without the trappings of lordship and administration through a public hall. As such, they were the first type of castle to have been built widely by Gaelic Irish lords. Their use in stabilizing control of land and resources is seen in north Donegal, where the incoming gallowglass lords, the MacSweeneys, built a number of towers. It is characteristic that these were castles of a second-rank Gaelic lord; Maguire, O'Cahan, or Clandeboy O'Neills built castles before the great lords, Ua Néill of Tyrone or Ua Domnaill. This is also seen in Scotland, where the lords of the Isles built few, if any, castles, but their chief men—such as Campbell, MacLean, or MacLeod—did. Below the level of gentry, tower houses were also a feature of towns, where they were the houses of the richer merchants and wealthier rural priests. Their numbers and cost are evidence of some real level of prosperity in late medieval Ireland, both in the countryside, profiting from the shift to cattle economy, and in towns.

T. E. McNeill

References and Further Reading

Bradley, J. "The Interpretation of Scandinavian Settlement in Ireland." In *Settlement and Society in Medieval Ireland*, edited by J. Bradley, 49–78. Kilkenny: Boethius Press, 1998.

Duffy, P., D. Edwards, and E. Fitzpatrick. *Gaelic Ireland c. 1350–c. 1600*. Dublin: Four Courts Press, 2001.

Edwards, N. *The Archaeology of Early Medieval Ireland*. London: Batsford, 1990.

Hurley, M., and O. Scully. *Late Viking Age and Medieval Waterford: Excavations 1986–1992*. Waterford: Waterford City Council, 1997.

Leask, H. G. *Irish Churches and Monastic Buildings* (3 vols.) Dundalk: Dundealgan Press, 1955–1960.

McCorry, M. *The Medieval Pottery Kiln at Downpatrick, Co. Down*. London: British Archaeological Reports no. 326, 2001.

McNeill, T. E. *Anglo-Norman Ulster*. Edinburgh: John Donald, 1980.

McNeill, T. E. *Castles in Ireland*. London: Routledge, 1997.

Mytum, H. *The Origins of Early Christian Ireland*. London: Routledge, 1992.

O'Conor, K. D. *The Archaeology of Medieval Rural Settlement*. Dublin: Discovery Programme/Royal Irish Academy, 1998.

O'Keeffe, T. *Medieval Ireland, an Archaeology*. Stroud: Tempus, 2000.

Stalley, R. A. *Cistercian Monasteries in Ireland*. London & New Haven: Yale University Press, 1987.

Sweetman, P. D. *Medieval Castles of Ireland*. Woodbridge: Boydell, 2000.

See also **Architecture; Castles**

TOWNS

See **Villages; Wall Towns**

TOYS

See **Children; Games**

TRADE

Substantive records of trade showing the imports and exports of medieval Ireland coincide with the establishment of municipal organization in the later twelfth and early thirteenth centuries. Archaeology and literary sources give some indication of the nature of trade in earlier times. St. Patrick's Confessio suggests the export of dogs from Ireland, and wine ships must have brought that essential commodity to the numerous monasteries. In the ninth and early tenth century the Hiberno-Norse craftsmen of the developing port towns imported huge amounts of silver, which was minted into coins for trade and fashioned into a variety of brooches and personal ornaments. The approximately 150 silver hoards unearthed in the countryside suggest that trade with local Irish lords may have involved more than food and fuel. Viking Dublin was an important link in the international trade of the period, which involved slaves as well as the more exotic walrus ivory, amber, and oriental silks.

After the establishment of the Anglo-Norman colony and the setting up of manorial farms in the east and southeast of the country, there was an export trade in the products of these farms, principally grain (oats and wheat) and wool. The economy of the colony was greatly stimulated in the thirteenth century by the activities of Italian merchant bankers. They were responsible for the collection of all royal and papal taxes and the organization of food supplies for shipment to the king's armies in Wales, Scotland, and France. Their share of the vast sums of money collected was used to buy and export wool or was loaned to enterprising merchants in the towns to expand their business, or to lords developing their estates.

Fish from the Irish Sea and hides made up the balance of the thirteenth-century exports. Imports, then and throughout the period, were headed by wine, along with salt, fine cloth, spices, and other luxuries. Building stone and iron were important imports in the first phase of church and castle building before 1250. The ports and newly established inland towns were situated, like many wool-producing Cistercian monasteries, on the major rivers, which were vital for the transport of bulky commodities. Merchants and craftsmen—stonemasons, smiths and carpenters, tanners, leatherworkers, weavers, butchers, and bakers—lived in the towns. They served their fellow citizens and traded with the inhabitants of the hinterland at the regular markets and fairs. The Suir, Nore, and Barrow rivers in particular were important avenues of trade, and there are references to wool being bought and loaded in Clonmel and leaving Waterford for Bordeaux. Large quantities of hides were stored in Inistioge and Thomastown, County Kilkenny, to be shipped downriver to Waterford for export. Firewood was another essential commodity that was transported by water whenever possible, and shippers, particularly on the Liffey, sometimes complained about newly constructed fishing weirs that impeded navigation. Some idea of the scale of exports from Ireland of wool, woolfells (sheepskins), and hides may be seen in the total of the customs receipts of the Irish exchequer for these commodities (tabulated by Gearóid Mac Niocaill) for various years between 1275 and 1345. When the trade was at its peak in the 1270s and 1280s, an average export of 4,000 to 4,500 sacks of wool or 400,000 to 450,000 hides a year was a possible maximum. At the same time, the amount of corn being shipped out to supply the royal armies abroad would have required the tillage of more than 30,000 statute acres (according to James Lydon).

As the fourteenth century progressed there was a decline in manorial farming in Munster and the southeast and a shift in trading emphasis to cattle hides and fish, commodities less affected by wars than tillage and wool production. However, between 1300 and 1550 there appears to have been little basic change in the economy of Dublin, Drogheda, and the ports and towns of the east, mainly because the economy of their hinterlands, in spite of political and social upheavals like the Bruce invasion and the Black Death, largely remained based on agricultural production.

The Fish Trade

Fish, particularly herrings, were caught and sold in the ports of the Irish Sea coast and exported, sometimes to the detriment of the home market. It was noted in 1515 that "merchants convey out of this land into France, Brittaine and other strange parts, salmonds, herrings, dry lings, haaks and other fish, so abundantly that they leve none within the land to vitall the King's subjects." Huge numbers of men were employed, 6,000 on one occasion in 1535, with a fleet of 600 English boats fishing off Carlingford, County Louth.

A similar number of Spanish boats was noted in west Cork in 1572, when there was a proposal to fortify an island in Baltimore harbor to collect customs from them. The previous century had witnessed a migration of shoals and a consequent huge growth in the herring fishery off the south and west coasts, where, to the annoyance of the government in 1465, the "Kings Irish enemies . . . were much advanced and strengthened as well in victuals and harness [and by] great tributes of money." The foreign fishermen had to anchor in the havens and land on the beaches to process their catch, and paid the local Irish lords for the privilege. The local economy benefited enormously, and the extra wealth is one explanation for the building boom that saw the construction and refurbishment of so many friaries and tower houses in the south and west of the country. No doubt that Ua Domnaill of Donegal, who was known as "King of the fish" in the sixteenth century, used his income from the herring fishery based around Arranmore Island, County Donegal, to hire, equip, and train the substantial army he fielded in the 1590s.

The Organization and Hazards of Trade

Apart from the haven-based herring fishery of the west, the imports and exports of the country came to be managed by a relatively small group of wealthy families in the ports, whose prosperity depended completely on the hides, wool, fish, flax, and furs procured from the local lords in exchange for wine, salt, iron, and fine cloth supplied by the merchants. This interdependence transcended the divisive politics of the period in Ireland. Through family connections and the guild structure, Irish and foreign merchants operated within codes of practice recognized all over western Europe. Binding agreements governed the chartering of ships and crews, freight charges, destinations, and turnaround times, as well as arrangements between the individual merchants who often formed a consortium for a particular venture.

Round trips were economically the most viable. One such was the voyage of the *Julian* of Bristol. In 1454, an agreement was drawn up between a group of Bristol merchants that included John May, a member of a prosperous Irish family settled in that city, and their ship was loaded with merchandise in Bristol. It sailed for Lisbon, where the English products were sold and wine, honey, and salt were taken on board. The ship next went to "legge de Breon" (unidentified) and Galway, where the wine, honey, and salt were exchanged for hides. From there they went to Plymouth for the final decision on whether to go and sell the hides in Normandy, Brittany, or Flanders.

Apart from the attentions of over-zealous customs inspectors in English ports, shipwreck and attacks by pirates were the main problems faced by merchants trading to and from Ireland. Navigation out of sight of land was a science that developed very slowly. The prototype of the mariners' compass was in use during the thirteenth century, and the following century saw the first Italian *portulan* maps of Ireland. The earliest, dated 1339, names over fifty places and islands. These maps were intended for use with a book of sailing directions containing bearings and distances. By the mid-fifteenth century the maps had about thirty new Irish names, which included Malahide (Co. Dublin), Ardglass (Co. Down), Baltimore (Co. Cork), Galway, and Sligo, indicating the development of the herring fisheries and interest in Atlantic seaways. But the directions given were very general and the outline of the coast inaccurate, so that ships rarely ventured into unfamiliar territory without a pilot.

Irish pirates were sailing the Irish Sea in the fifth century, sometimes directing their attention to the coastal communities of Britain. After the Viking raiders became settled traders in Ireland, shipping continued to be menaced by their relatives operating out of the Norse kingdom of Man and the Isles. Throughout the 1300s there is a continual series of complaints about the depredations of the Scots in the Irish Sea north of Holyhead. Merchant ships were commandeered and brought to Scotland; hostages were taken and were not freed until a ransom, generally in the form of victuals, was paid. After 1400, increasing numbers of Breton and Spanish pirates appeared in the southern Irish Sea. The Bretons demanded money ransoms for prisoners captured at sea. According to the Law Merchant, towns were responsible for the conduct of their sailors abroad, and as a result, the goods of many legitimate Breton merchants were confiscated and used to compensate victims. Nicholas Arthur of Limerick lost goods worth 700 marks in 1425 and spent two years imprisoned in Mont Saint Michel, until a ransom of 400 marks was paid. On his release he got "letters of reprisal" to the value of 8,000 marks against any Bretons "within the dominions of the king of England whether by land or sea," which, we are told, he levied "to the last farthing."

The modest trade and prosperity of Ireland increased for a time in the sixteenth century. In the Elizabethan period, however, systematic and ruthless warfare and displacement of people crushed the independence of lords and merchants, not least by wasting the countryside on which the prosperity of so many depended.

TIMOTHY O'NEILL

References and Further Reading

Longfield, A. K. *Anglo-Irish Trade in the Sixteenth Century.* London: George Routledge, 1929.

O' Neill, Timothy. *Merchants and Mariners in Medieval Ireland.* Dublin: Irish Academic Press, 1987.

Childs, Wendy, and Timothy O' Neill. "Overseas Trade." In *A New History of Ireland: Vol. 2, Medieval Ireland 1169–1534*, edited by Art Cosgrove. Oxford: Oxford University Press, 1987.

See also **Cork; Dublin; Fishing; Fraternities and Guilds; Limerick; Ports; Roads, Routes; Ships and Shipping; Walled Towns; Waterford**

TRIADS

A predilection for triadic arrangements—groupings of three items that share a characteristic feature—is already discernible in ancient Celtic imagery and ornamentation. As a method of systematizing and preserving mytho-historical and legal lore, as a vehicle for proverbial and ethical statements, and finally as a literary genre with some affinities to wisdom texts, triads are known from the literatures of Ireland and, even more prominently, of Wales. The triad is just one possible mnemonic device for grouping information. Catalogs based on other numbers such as 2, 5, 7, or 9 also occur throughout medieval Irish literature. An entire law tract is called *Heptads*. The Christian doctrine of the Trinity must have reinforced the special popularity of triadic arrangements, but an outright derivation of the device from Old Testament models (e.g., Prov. 30: 18–31) as suggested by Meyer (1906: xii) is unlikely because of the great stylistic differences between the biblical and the Irish patterns.

Medieval Irish tales abound in triplets, for example, the three female personifications of Ireland (*Banba*,

Fódla, and *Ériu*), numerous triple adversaries of Cú Chulainn, the motif of the threefold death, and so on. *Trefocul* (three-words) is the name in Irish metrics for a collection of precepts that identify and correct faults in versification. Descendants of mythic or divine figures are frequently encountered in groups of threes, for example, the Dagda's three daughters called *Brigit*, or the three sons of Bres son of Elathu (*Brian, Iuchar*, and *Iucharba*), with the long triadic list of their belongings and relations in LL 3902–3915. Triads in a narrow literary sense denote short, gnomic, occasionally aphoristic statements that comprise three items (persons, places, events, human activity, etc.) with a characteristic and memorable trait in common. They are normally in prose and consist of an introductory reference to the common characteristic, followed by the three items, sometimes accompanied by an explanation. The earliest examples of triads in the Irish written tradition are found in the late-seventh century *Cambrai Homily*, for example,"There are three kinds of martyrdom which count as a cross for man, that is, white martyrdom, green martyrdom and red martyrdom" (*Thesaurus Palaeohibernicus, vol. II*, 246.27–31).

Trecheng Breth Féine

The central collection of Irish triads, constituting only a fraction of the triadic sayings scattered throughout Irish literature, is known under the name *Trecheng Breth Féine* (TrBF) or "The Triads of Ireland" (literally "a threefold grouping [?] of the judgments of the Irish"), a title found only in a single manuscript. The rare OIr. term *trecheng*, also *trethenc* (triad), not used elsewhere in the collection, cannot be separated from *decheng* (a pair of persons), but the further etymology is unclear. The common OIr. word for "three things" is the numeral abstract *tréide*. From the custom of naming characteristic traits in triads, the plural of *tréide*, sometimes spelled *tréithe*, later developed the meaning "accomplishments, qualities, trait." TrBF, which survives in nine manuscripts from the fourteenth to the nineteenth century, can be dated on linguistic grounds to the second half of the ninth century. In all manuscripts in which it appears, TrBF belongs to a body of wisdom texts comprising *Tecosca Cormaic*, *Audacht Morainn*, and *Bríathra Flainn Fína maic Ossu*. That TrBF had a separate origin, however, and came to be associated with the other three texts only later is shown by the fact that it is not included in the oldest manuscript containing the three, the *Book of Leinster*. Despite its title, only 214 of the 256 entries in TrBF are enumerations of threes. Single items (§§ 1–7, 9–31), pairs (§§ 8, 124, 133–134), tetrads (§§ 223, 230, 234, 244, 248, 251–252), a heptad (§ 235), and an ennead (§ 231) are also included. The entries are loosely arranged in groups according to their contents. The first sixty-one contain mere topographical enumerations. A large portion of the entries comment in a moralistic, sometimes misogynistic, tone on human nature and on social conduct. Triads 149–186 are almost entirely legal, some of them direct quotes from the law tracts. But a general legal orientation, also apparent from the title of the collection, underlies many more entries. Only a few (e.g., §§ 62, 236–237) display the mytho-historical focus that is so prominent in the Welsh triads. Stylistically, the best examples are distinguished by climactic or anticlimactic conclusions and by paradoxical formulations, for example, "Three rejoicings with sorrow afterwards: a man wooing, a man stealing, a man giving testimony" (§ 67). Entries are often paired with antithetical formulations, for example, "Three things that make a fool wise . . . " vs. "Three things that make a wise man foolish . . . " (§§ 193–194).

It is better not to speak of an original author of TrBF but rather of a compiler, who was not overly concerned with giving his anthology a homogeneous appearance. Mention has already been made of entries excerpted from law texts. TrBF's compilatory character also emerges from linguistic and stylistic features. In the majority of triads the numeral *trí* is used both attributively (three X) and as a substantivized neuter (three things). Only in a group of thematically distinct triads concerned with the prerequisites of professional classes (§§ 116–123, 202) taken from the law tract *Bretha Nemed Toísech*, and a few others (e.g., §§ 77–78, 80–81) does the numeral abstract *tréide* appear substantivally. One triad (§ 239) is formulated as a question: "What are the three wealths of fortunate people? . . . " Two entries (§ 230, a tetrad; § 231, an ennead) make no reference to the number of items in their introductions at all, but start with *cenéle* (types of . . .).

Occasional triads are encountered in almost all wisdom texts, such as *Audacht Morainn*, *Tecosca Cormaic*, and *Apgitir Chrábaid*, and the *Prouerbia Grecorum* includes a number of triads, pentads, heptads, and octads. *Bríathra Flainn Fína maic Ossu* consists mainly of three-word sayings, for example, *Ad·cota cíall cainchruth* (Good sense results in fair form, § 1.3) and *Ferr dán orbu* (A skill is better than an inheritance, § 6.1). Proverbial sayings in the form of triads were still popular in twentieth-century Ireland, for example, "Three kinds of men who fail to understand women: young men, old men, and middle-aged men."

DAVID STIFTER

References and Further Reading

Bromwich, Rachel. *Trioedd Ynys Prydein. The Welsh Triads.* Cardiff: University of Wales Press, 1961.

Green, Miranda J. "6. Triplism and multiple images." In *Symbol and Image in Celtic Religious Art*, 169–205. London – New York: Routledge, 1989.

Kelly, Fergus. *A Guide to Early Irish Law (Early Irish Law Series, vol. 3).* Dublin: Dublin Institute for Advanced Studies, 1988, pp. 284–285.

Meyer, Kuno. *The Triads of Ireland (Todd Lecture Series, vol. 13).* Dublin: Hodges, Figgis & Co.; London: Williams & Norgate, 1906.

Sims-Williams, Patrick. "Thought, Word and Deed: An Irish Triad." *Ériu* 29 (1978): 78–111.

See also **Law Tracts; Moral and Religious Instruction; Wisdom Texts**

TRIBES

Early Irish society is often referred to as tribal. The term was and still is used by writers to describe the early Irish socio-political organization based on the *túath*. As a term to characterize Irish society, "tribe" has passed in and out of favor among scholars— namely due to the vagaries of its definition, but also because of its perceived derogatory connotations.

In the nineteenth century, under Marxist and Darwinian theories of social evolution, the term "tribe" was used to place medieval Irish society within an evolutionary model, identifying Ireland as prefeudal, and thus inferior in its development. For early scholars of Irish history, such as Orpen, Ireland did appear to be a fragmented and tribal island, by which was meant a system organized under local rule, with each aggregate of people united by ties of blood or belief in descent from a common ancestor. This does bear some of the hallmarks of the *túath* social system, but it could equally be said to reflect some of the qualities of the smaller nonpolitical group, the *fine*, which is a group of people of the same family.

Mac Néill, writing in 1919, believed that the term "tribal" was inaccurate, even derogatory, and consciously avoided using it. He felt that the term "tribe" as a translation for *túath* suggested that Ireland was divided into numerous groups or clans, with a chief at the head of each group and with all members of the clan considering themselves to be of one blood. He felt that this description of the *túath* was at odds with the evidence of the source material, which clearly demonstrated a *túath* to be a territorial unit, brought together under the rule of a *rí* (king), without a common ancestral bond or common ownership of land.

Binchy reintroduced "tribe" to early Irish historiography in 1970 when he suggested that despite the derogatory nature and vagueness attached to the term, "tribe" could usefully be employed as a description of Early Irish society, as long as one was specific in one's definition. For Binchy this meant "a primary aggregate of people in a primitive or barbarous condition under a headman or chief." The tribal character of Irish society was for Binchy embodied in the society portrayed by the law tracts, which presented a fossilized picture of Irish society corresponding to a period just prior to the shift toward a dynastic political structure. He saw the new dynastic families as having superimposed "a whole series of 'mesne' kingdoms ruled by scions of their own kindred" on the old tribal pattern. However, despite this change he felt that at the most basic level "the old tribal substratum still remained."

Similarly, Byrne also employed "tribe" to describe ancient Irish society, which for him was also exemplified in the law tracts. He believed "tribe" was particularly useful for defining the process of change whereby ancient tribal rule was being replaced by the rise of dynastic rule. Thus, for Byrne, the Uí Néill represented a new political principle, a change from local tribal identity to a dynastic hierarchy, which was embodied in the very form of their name. Names in *Uí*, *Cenél*, *Clann*, and *Síl* all denoted descent from a living ancestor within the historical period and belonged to dynastic families, whereas names in *Maccu*, *Dercu*, *Corcu*, *Dál*, and *–rige* denoted older tribal names.

This distinction has been followed by historians, more recently by Charles-Edwards, who identifies the disappearance of the Latin term *gens* (OI *muccu*) from use in the sources with the change from a tribal system to a dynastic system. His concept of tribe is similar to that of Binchy and Byrne, where by "tribe" he means a small political unit without a bureaucratic government, but governed instead by existing social bonds.

Opposition remained, however, and Scott has rejected both Binchy's and Byrne's use of "tribe" for what he sees as the imprecision of their use and the confusion and misinterpretation to which their definitions can lead. Furthermore, although there can be little doubt that dynastic and territorial kingship replaced tribal kingship in medieval Ireland, the dates suggested by historians for this transformation have ranged from the late fifth to the eleventh century. The term "tribe" thus implies the social structures prevalent in Ireland, prior to the introduction of dynastic kingship, that were based around the local and territorial rule of the *túath*, which was not derived from ties of blood and was administered through communal tradition rather than a fixed bureaucratic structure.

MICHAEL BYRNES

References and Further Reading

Binchy, D. A. *Celtic and Anglo-Saxon Kingship*. Oxford: Clarendon Press, 1970.

Byrne, F. J. "Tribes and Tribalism in Early Ireland." *Ériu* 22 (1971): 129–166.

Charles-Edwards, T. M. *Early Christian Ireland*. Cambridge: Cambridge University Press, 2000, pp. 97–98, 100.

Ó Corráin, D. "Nationality and Kingship in Pre-Norman Ireland." In *Nationality and the Pursuit of National Independence* (*Historical Studies 12*) 1–35, edited by T. W. Moody. Belfast: Blackstaff Press, 1978.

Patterson, N. T. *Cattle-Lords and Clansmen. The Social Structure of Early Ireland*, 2nd ed. Notre Dame, Ind.: University of Notre Dame Press, 1994.

Scott, B. G. "Tribes and Tribalism in Early Ireland." *Ogam* 22–25 (1970–1973): 197– 206.

See also **Feudalism; Kings and Kingship; National Identity; Túatha**

TRIM

A market town and liberty (1244–1494) in County Meath, Trim is situated at a fording point on the River Boyne that has been important since prehistoric times. The place name is derived from the Irish *Áth Truim* (ford of the elder tree), first mentioned in an account of around 700, which was later transcribed into the *Book of Armagh*. Trim's earliest traditions hint at the presence of a fifth-century British Christian community connected with Lommán, who was subsequently remembered as the patron of Trim. With the rise of the Uí Néill to prominence in the sixth century, the older ruling dynasty, the Uí Lóegaire, was displaced but managed to hold onto Trim and its immediate environs. Under increasing pressure in the late seventh and early eighth centuries the Uí Lóegaire affiliated Trim with the Armagh *paruchia*, thereby introducing an association with St. Patrick. In the eighth and ninth centuries, the Uí Cholmáin, a branch of the Uí Lóegaire, achieved significance as ecclesiastics at both Trim and Clonard, while another member, Rumann mac Colmáin (d. 747) was a poet of distinction. Little is known of Trim in the tenth and eleventh centuries. In 1128, the settlement and its churches were burnt, while an Augustinian priory was established by 1150. Despite the paucity of sources, Trim evidently remained important, otherwise it would not have been chosen by Hugh de Lacy in 1172 as the capital of his lordship of Meath.

Immediately after his arrival, de Lacy commenced the construction of one of the largest and most impressive castles in Ireland. With its distinctive cruciform plan, the tall keep rose above the surrounding curtain wall to enclose an area of over two acres. The castle functioned as the administrative center of the lordship of Meath and liberty of Trim, which passed by marriage to Geoffrey de Geneville in 1252 and to Roger Mortimer, first earl of March, in 1308. With the extinction of the Mortimer male line in 1425, the castle and liberty passed to Richard Plantagenet, duke of York, and after his death at the Battle of Wakefield in 1460, it was absorbed by the crown.

The thirteenth century was a period of urban expansion reflected in the foundation of Newtown Trim, with its substantial cathedral and hospital, two miles downriver, as well as a Dominican friary in 1263 and a Franciscan house at about the same time. The town was fortified with walls and developed trading connections along the Boyne, with Drogheda as its port. Unlike most Irish towns Trim seems to have survived the vicissitudes of the fourteenth century fairly well, and it was the setting for twelve meetings of parliament between 1392 and 1492. Nonetheless, urban decline commenced in the fifteenth century, when reduced revenues made it difficult for the lordship of Meath to function. As the lordship suffered so did the town, and with the declining fortunes and final abolition of the liberty, the wealth and importance of the town faded. Trim has never since regained the centrality that it enjoyed in the Middle Ages.

JOHN BRADLEY

References and Further Reading

Potterton, Michael. "The Archaeology and History of Medieval Trim, County Meath." Ph.D. dissertation, The National University of Ireland, Maynooth, 2003.

See also **Armagh; Geneville, Geoffrey de; Lacy, Hugh de; Mortimer; Parliament; Patrick; Uí Néill**

TUARASTAL

Tuarastal (OI [n]: stipend, retaining fee, wages, reward, remuneration) was the fee or stipend paid by an over king to subordinates for their services in his army. In the Lebor na Cert we find horses, swords, shields, horns, ships, hounds, rings, bridles, bracelets, cups, and other types of military or luxury items being given as *tuarastal*. This system of reward was not found in the Early Medieval period, and in fact the word *tuarastal* in that period refers to eyewitness testimony. In the Early Medieval period the word used for surety payment was *Ráth*. There was in this period no need to remunerate an army, as they were not expected to remain away from home for any length of time.

The changing practice of warfare in Ireland in the post-Viking period brought with it a change in terminology. In the eleventh and twelfth centuries over kings needed armies that could be sent far from home for extended periods of time. This need did not fit with the earlier practice whereby the over king's leading vassals might expect to be in his service for two or three weeks in any given year. To compensate the vassals for their extended service, the king rewarded them with *tuarastal*.

Over kings also bestowed vassals with *tuarastal* at the time of their submission. Later still in the Anglo-Norman period *tuarastal* came to mean a purely monetary transaction, with Anglo-Norman armies hiring themselves to Irish kings without any sense of subordination or political vassalage.

MICHAEL BYRNES

References and Further Reading

Dillon, Myles, ed. and transl. *Lebor na Cert: The Book of Rights*, vol.46. Dublin: Irish Texts Society, 1962.

Doherty, Charles. "Exchange and Trade in Early Medieval Ireland." *Journal of the Royal Society of Antiquaries of Ireland* 110 (1980): 67–89.

Ó Corráin, Donncha. *Ireland Before the Normans*. Dublin: Gill and Macmillan, 1972.

Simms, Katharine. *From Kings to Warlords: The Changing Political Structure of Gaelic Ireland in the Later Middle Ages*. Woodbridge: The Boydell Press, 1987.

See also **Armies; Feudalism; Kings and Kingship; Military Service, Gaelic; Military Service, Anglo-Norman**

TÚATH

In Early Irish society the *túath* (plural *túatha*; people, tribe; country, territory; the state, as opposed to the church) was the basic territorial unit to which every individual belonged. It could also refer to the laity residing there. It is usually translated as "petty kingdom" or "tribe," although neither is a completely accurate translation. The *túath* was a small political unit—much smaller than the contemporary kingdoms of Mercia or Kent in England. It has been variously estimated that there were as many as one hundred *túatha* in Ireland between the fifth and twelfth centuries. Provinces were made up of several such *túatha*, each with its own king. With an estimated population in Ireland of roughly 500,000, an average *túath* would have had around five thousand members. Despite the fact that the *túath* was a rather small unit, it was the center of almost all social, political, and religious interaction for its members.

The boundaries of a *túath* were often defined by bog, woodland, or mountains, but it would usually also have had an area of well-cultivated land, that is a *mag* (plain), contained within it. Some *túatha* were named after the *mag*, for example, Mag Cerai from the territory of the Fir Cherai. Each *túath* had its own sacred site where its kings were inaugurated. Such sites were often signified by a sacred flagstone (*lecc*) or sometimes by a sacred tree (*bile*). To remove or destroy these was one of the most grave violations that could befall a *túath*.

According to the law tracts a *túath* must have a church, an ecclesiastical scholar, a poet, and a king. Of these, the king was perhaps the most important to the *túath*. All free men of the *túath* owed him loyalty, and he in turn acted as their appointed representative in all matters concerning the *túath* and its neighbors. It was through the king that the *túath* made *cairde* (pacts) with other *túatha*. The king was also responsible for summoning a *slógad* (hosting) to defend the *túath* or to launch an attack as required. The king might also become a client to another, more powerful king, thus placing the *túath* in a network of *túatha* led by a regional or even a provincial ruler.

Although *túath* is often translated as "petty kingdom" it also refers to the people of that kingdom; thus *túath* has the additional meaning of "people" or "tribe." This sense is particularly important with regard to the functioning of the Early Irish law codes. An individual had very limited rights outside his or her own *túath*, and in fact to leave one's *túath* permanently was considered dishonorable. The exceptions to this were members of the church (priests, monks, abbots, and bishops who fell under the ecclesiastical laws), poets (members of the learned class), and kings, as well as a woman marrying into a family outside her own *túath*. For ordinary people, leaving the boundary of the *túath* could be very dangerous, unless a treaty had been agreed guaranteeing the protection of the individual while in another *túath*. Without such a treaty a stranger could legally be killed, because once outside his own *túath* he no longer had any *éraic* (honor price), thus leaving his kin-group with no means of legal redress. Thus, the Early Irish legal system, which was essentially a system of civil law, was tied to the *túath*, although the laws themselves were uniform across all *túatha*.

It is partly due to this system of law, as well as to the segmentation of ruling dynasties, that the *túath* continued to have a place in Irish society. The *túatha* enabled the great dynasties to maintain a degree of cohesion, and to stem competition for the overkingship by allowing the heads of subordinate branches to be kings of their own *túatha*. This practice helped to preserve the system of petty kingdoms into the post-Viking period.

However, the king never had executive command over the *túath*. Rather, the Law worked through a system of pledges and bindings using the individual's honor price. These pledges were administered by lawyers. In addition the *túath* did not provide the king with any real tax base, nor did he have control of a standing army. Thus the *túath* never evolved beyond a very simple state, and under ever-increasing pressure from the provincial kingships, the *túath* eventually lost its independent status. By the eleventh and twelfth centuries *túath* had come to mean a small, semiautonomous territory ruled by a *taoiseach* (leader).

MICHAEL BYRNES

References and Further Reading

Binchy, D, A., ed. *Corpus Iuris Hibernici* (6 vols.) Dublin: Dublin Institute for Advanced Studies, 1978, pp.1111–1138, esp.1123–1132.

Binchy, D. A. *Celtic and Anglo-Saxon Kinship*. Oxford: Clarendon Press, 1970.

Byrne, F. J. *Irish Kings and High-Kings*. London: Batsford, 1973.

Charles-Edwards, Thomas. *Early Christian Ireland*. Cambridge: Cambridge University Press, 2000.

Gwynn, E. J., ed. "An Old-Irish Tract on the Privileges and Responsibilities of Poets." *Ériu* 13 (1942): 1–60, 220–236.

Kelly, Fergus. *A Guide to Early Irish Law*. Dublin: Dublin Institute for Advanced Studies, 1988.

O'Keefe, J. G., ed. "The Ancient Territory of Fermoy." *Ériu* 10 (1926–1928): 170–189.

Power, P., ed. *Crichad an Chailli: Being the Topography of Ancient Fermoy*. Cork: Cork University Press, 1932.

See also **Ecclesiastical Settlements; Feis, Inauguration Sites; Kings and Kingship; Óenach; Tribes**

U

UA BRIAIN (UÍ BRIAIN, O'BRIEN)

The O'Briens (Uí Briain) are descended from Brian Boru, king of the Dál Cais from 976. The Dál Cais, previously known as the Déisi Tuaiscirt until around 934, rose to prominence in the course of the ninth century. They expanded across modern County Clare and, in the second half of the tenth century, imposed their authority over the Viking town of Limerick. Compared with other dynasties in Munster, the Dál Cais were more dynamic and ruthless and tended to take direct possession of their neighbors' lands rather than content themselves with mere overlordship.

High Kingship

Brian Boru displaced the Eóganachta to become king of Munster, with support from the Vikings of Waterford, and he asserted his authority over Leinster. He forced Máel Sechnaill II of the Uí Néill to relinquish his claim to overlordship of the south of Ireland. Brian Boru showed determination in harnessing the wealth of Munster and in projecting the inherent demographic and economic power of the province over the rest of Ireland. In 1001, Brian Boru attacked the Uí Néill and subsequently established himself as the high king of Ireland. On Good Friday 1014, Brian and his allies, including Vikings from the southern towns, fought a rebellious king of Leinster, Máel Mórda and his Viking allies from Dublin, at Clontarf. Brian Boru was killed in battle, as was the king of Leinster. Brian was buried at Armagh, Ireland's premier church. The *Book of Armagh* called him the *imperator Scottorum* ("emperor of the Irish"). That title was a little grandiose, but Brian Boru had succeeded in breaking the Uí Néill monopoly of the high kingship of Ireland and made himself the king of Ireland. It is not surprising that his descendants called themselves Ua Briain ("descendant of," literally "grandson of") after him.

Brian Boru's son and successor, Donnchadh (*c.* 1064), was unable to maintain his father's hold on the high kingship. However, Tairrdelbach Ua Briain, a nephew of Donnchadh's, succeeded in making himself "king of Ireland with opposition" between 1072 and 1086, meaning that his authority was recognized throughout most but not quite all of Ireland. Tairrdelbach's son, Muirchertach Ua Briain, went further, ruling as king of Ireland from 1088 until 1118 and coming close to establishing a true Irish monarchy. Muirchertach is strongly associated with the twelfth century church reforms, and particularly with the synod of Cashel I (1101) and the more important synod at Ráith Bressail (1111), which sought to transform the Irish church along Roman lines. Muirchertach may also have commissioned one of the most effective pieces of propaganda produced in medieval Ireland—*Cogadh Gaedheal re Gallaibh* ("The war between the Irish and the Foreigners"). The *Cogadh*, written circa 1109–1118, glorified Brian Boru as the national savior of the Irish from the Viking onslaught at the Battle of Clontarf. It made Brian Boru a legend and created a myth of national resistance to foreign oppression that resonated among Irish nationalists well into the twentieth century.

Partition of Munster

A rebellion by the MacCarthys in southern Munster in 1118, backed by Tairrdelbach Ua Conchobair, king of Connacht, resulted in the division of Munster in two: an Uí Briain kingdom of Thomond in northern Munster and a Meic Cárthaig kingdom of Desmond in southern Munster. Tairrdelbach Ua Conchobair kept Munster divided to undermine the Uí Briain and preempt any challenge they might attempt to his aspirations to becoming the king of Ireland. It seems that in the late 1120s Ua Conchobair annexed the Uí Briain heartland of Clare to Connacht, driving them to accept Cormac Mac Cárthaig, king of Desmond, as the king of Munster. Mac Cárthaig led a coalition of forces from across the south of Ireland in a prolonged and savage war to overthrow Ua Conchobair's hegemony. However, once the threat from Connacht was ended in 1133 the Uí Briain ended their alliance with Mac Cárthaig, and Munster was split in two again. In 1138, Tairrdelbach Ua Briain, king of Thomond, succeeded in having Cormac Mac Cárthaig assassinated. Thereafter he ruled all of Munster until 1151 when another rebellion assisted by Tairrdelbach Ua Conchobair, king of Connacht, this one led by Cormac Mac Cárthaig's son Diarmait, resulted in the province being partitioned again. The kings of Thomond and Desmond each looked upon the other as a deadly rival for the kingship of Munster.

In 1168, Diarmait Mac Cárthaig, king of Desmond, had Tairrdelbach Ua Briain assassinated in an attempt to unite Munster under his authority. His ambitions, however, were thwarted by Ruaidrí Ua Conchobair, king of Connacht and high king of Ireland, who preferred to see Munster remain divided and easier to overawe. Ua Conchobair obliged Mac Cárthaig to pay an *éraic* ("compensation") for the killing. Mac Cárthaig's assault on Limerick, the capital of Thomond, early in 1171 failed to advance his aspirations for the kingship of Munster for the same reason. Indeed, Domnall Mór Ua Briain, king of Thomond, had aspirations of his own for the kingship of Munster, and he formed an alliance with Ua Mathgamna (O'Mahony), one of Mac Cárthaig's chief subordinates, in a plan to invade Desmond in October 1171. Coincidentally, Henry II, king of England, had just landed in Ireland in order to assert his authority over some Anglo-Norman adventurers (they knew themselves simply as "English") led by Richard de Clare, earl of Pembroke, better known as Strongbow. Mac Cárthaig submitted to Henry II in a bid to avoid the planned Ua Briain invasion, prompting Domnall Ua Briain and Ua Mathgamna to do likewise. Impressed by the alacrity with which the kings in Ireland were willing to submit to him, Henry II decided to revive the papal grant of Ireland to him in the privilege known as *Laudabiliter.*

Anglo-Norman Invasion

Limerick was captured by Anglo-Norman or English adventurers in 1175, but its occupation was short-lived. In 1177, Henry II granted the Ua Briain kingdom of Thomond and the Mac Cárthaig kingdom of Desmond to three of his leading knights. Domnall Ua Brian managed to repulse Philip de Braose, the grantee of Thomond, at the walls of Limerick and save Thomond from invasion. By contrast, the unwalled town of Cork fell easy prey to its grantees and became the center of a new English colony in southern Ireland. Ua Briain took advantage of Mac Cárthaig's discomfiture by invading Desmond late in 1177. Ua Briain's efforts failed, but they were instrumental in forcing Mac Cárthaig to temporize with the English and concede a large part of his kingdom to them.

Domnall Ua Briain's strategy in dealing with the English thereafter was to continue to offer robust resistance while showing a willingness, nonetheless, to reach an accommodation with the English crown in order to safeguard as much as possible of his kingdom. When the Lord John came to Ireland in 1185, Ua Briain, together with Ruaidrí Ua Conchobair of Connacht and Diarmait Mac Cárthaig of Desmond, submitted to the prince, but the Irish kings were treated with open contempt and derision. John granted northeastern Thomond to Theobald fitz Walter, ancestor of the future Butler earls of Ormond, and he granted much of southeastern Thomond to Philip of Worcester and William de Burgh. Ua Briain offered stout resistance to the invaders, and it was only following his death in 1194 that the English were able to consolidate their control of Ormond and capture the capital of Thomond, Limerick. John granted Limerick the status of a royal borough around 1197, and he gave away much of the Ua Briain lands in Limerick diocese to the sons of Maurice fitz Gerald, ancestor of the future earls of Desmond.

Donnchad Cairbrech Ua Briain (*c.* 1242), the next strong king of Thomond, concentrated his efforts on safeguarding the Uí Briain heartland in Clare from English incursions. He reached an accommodation with King John wherein he accepted a knighthood and committed himself to paying a substantial rent to the English crown for his diminished kingdom. It was a strategy that largely succeeded, though there were lesser Uí Briain who dissented from his policy of collaboration with the English.

Conchobar, Donnchad's son and successor, continued his father's strategy, but from 1248 Henry III, king of England, made grants resulting in English colonization in and around Bunratty. When the English sought to make further inroads into Thomond, Conchobar fought back hard and routed an English force sent against him in 1257. In the following year his son Tadc Ua Briain met with Áed Ua Conchobair, king of Connacht, and Brian Ua Néill, king of the Irish of Ulster, at Cáel Uisce to agree to the formation of a national confederacy against the English. The confederation proved to be short-lived, however. Tadc died in 1259, and Ua Néill was killed soon afterward. The Irish were too divided among themselves, and too weak in any case, to mount an effective nationwide resistance to the English. Nonetheless, Ua Briain hostility toward the colonists in Thomond continued unabated, and there was widespread devastation in the borderlands around Bunratty and Limerick.

In a bid to overcome the determined opposition of Ua Briain, Edward I granted all of the Ua Briain lordship to Thomas de Clare, brother to the king's chief adviser, the earl of Gloucester. De Clare made some kind of agreement with Brian Ruad Ua Briain, the king of Thomond, but treacherously killed Ua Briain in 1277. Brian Ruad's sons prevented de Clare from taking immediate advantage of their father's murder. However, Thomond was without a strong leader, and internecine rivalry among the Uí Briain allowed de Clare to force the next king of Thomond, Tairrdelbach Ua Briain, into an arrangement whereby he agreed to hold all of Thomond beyond Bunratty for an annual rent of £120. Following his death, there was further internecine strife among the Uí Briain which was exploited by Richard de Clare. De Clare sought to extend his sway across County Clare, but at the battle at Dysert O'Dea in 1317 de Clare was killed by forces led by Muirchertach Ua Briain, a son of Tairrdelbach Ua Briain. The threat from the de Clares was ended and the Uí Briain's control of Thomond was assured. The battle at Dysert proved to be a decisive encounter.

The Uí Briain Revival

After Dysert O'Dea, the Uí Briain went on the offensive against the English colonists. The colony at Bunratty was put under sustained pressure and fell to the Irish in the 1350s. The frontiers of Ó Briain power were pushed right up to the walls of Limerick. There were repeated raids east of the Shannon to harass the English lordships in Ormond. In 1370, Brian Sreamhach Ua Briain, king of Thomond, won a great victory against the earl of Desmond south of the Shannon. His subordinates captured and sacked the city of Limerick. Such audacity, however, prompted the intervention of the English chief governor, Sir William de Windsor, and an uneasy modus vivendi was established between Ua Briain and the embattled English colonists. Limerick was restored to English control, but Ua Briain was in the ascendant.

In 1466, Tadc Ua Briain, lord of Thomond, led an army across the Shannon and imposed his overlordship over the MacBriens of Coonagh and Aherlow and the Clanwilliam Burkes. He imposed a "black rent" on the inhabitants of Limerick city and the east of County Limerick, a financial tribute reflecting his military power. At the close of the Middle Ages, the Uí Briain were again one of the most powerful dynasties in Ireland.

Conclusion

It has to be conceded that the Uí Briain were fortunate in that their heartlands, in modern County Clare, were relatively remote from England and accordingly less attractive to English colonists. Also, their core territory had the advantage of geographical cohesion, bounded as it was by the Shannon river and estuary, the Atlantic Ocean, and the lordship of the Gaelicized Clanrickard Burkes with whom they maintained good relations. Nonetheless, the survival of the Ó Briain lordship in the later Middle Ages was due primarily to the tenacity and resourcefulness of its leaders. Their grasp of realpolitik, and their readiness to seek and maintain accommodations with the English while offering stiff resistance to incursions west of the Shannon, helped them to come through the most threatening phase of English colonization up to 1317. Their firm governance and control of their subordinates within Thomond gave them the strength to take advantage of English weaknesses subsequently so that they became a force to be reckoned with across much of the territory of the former kingdom of Thomond. That strength and adaptability was demonstrated again in 1542 when Muirchertach Ua Briain became the first earl of Thomond under the auspices of Henry VIII's policy known as "surrender and regrant." The earldom was a recognition of Ua Briain's stature as one of the most important lords in Ireland.

Henry A. Jefferies

References and Further Reading

Byrne, Francis. *Irish Kings and High-Kings*. London, 1973.
Duffy, Seán. *Ireland in the Middle Ages*. Dublin, 1997.
Nicholls, Kenneth. *Gaelic and Gaelicised Ireland in the Later Middle Ages*. Dublin, 1972.

See also **Brian Boru; Dál Cais; Eóganachta; Munster; Ua Briain, Muirchertach; Ua Briain, Tairrdelbach**

UA BRIAIN, MUIRCHERTACH (1050–1119)

Muirchertach Ua Briain (1050–1119) was the son of Tairrdelbach (d. 1086), son of Tadc (d. 1023), son of Brian Boru, the latter's most successful successor as king of Dál Cais, Munster, and Ireland. The Annals of Tigernach (perhaps anachronistically) record his birth in 1050, and he otherwise appears on record only in 1075 when his father led the armies of Munster, Leinster,

Osraige, Mide, Connacht, and Dublin to Ardee, County Louth, to gain submission from the Airgialla and Ulaid; the former stood their ground at Ard Monann (unidentified) where they slaughtered the Munstermen under Muirchertach, described as *rígdamna Muman*, "makings of a king of Munster" (AFM). In the same year, his father appointed him king of Dublin with which he was sporadically associated for the rest of his life. He then disappears without a trace until October 29, 1084 when he and the forces of Dublin, Leinster, Osraige, and Munster defeated Donnchad Ua Ruairc of Uí Briúin Bréifne at Móin Cruinnióce, near Leixlip, County Kildare (a Norse settlement which he may have ruled from Dublin). In this major battle 4,000 were killed, Muirchertach cutting off Ua Ruairc's head and bringing it back to his father's palace at Limerick.

When Tairrdelbach died in 1086, Munster was divided between his three sons, Tadc, Diarmait, and Muirchertach. Tadc died within a month, and Muirchertach banished Diarmait and seized the whole province. In 1087, he fought a battle against Diarmait and the king of Leinster at Ráith Étair (possibly the promontory fort at the tip of Howth Head), and Muirchertach's forces triumphed. In 1088, he sent one fleet up the Shannon as far as Incherky, south of Clonfert, where the king of Connacht, Ruaidrí na Saide Buide Ua Conchobair, father of the great Tairrdelbach, slaughtered Muirchertach's men. Another Munster fleet sent around the west coast was also slaughtered, the Connacht army then invading Corco Mruad (the Burren, Co. Clare).

Domnall Mac Lochlainn, king of the Northern Uí Néill, now emerged as Ua Briain's rival for supremacy, allying with the sons of Muirchertach's uncle Donnchad mac Briain (d. 1064) by marrying the daughter of his grandson Cennétig (d. 1084), who had been king of Telach Óc in Tír nEógain with Mac Lochlainn's support. In 1088, the latter attacked Connacht and forced its king to submit, and then together they marched on Munster in Ua Briain's absence in Leinster; Limerick, Kincora, and Emly were burned and 160 hostages seized, whom Muirchertach later bought back with cattle, horses, gold, silver, and meat, victory being symbolically sealed when the severed head of Donnchad Ua Ruairc was brought back to Connacht. A Munster source claims that Ua Briain avenged himself in 1089 by invading Mide and Leinster, whose king he killed, making himself king of Leinster and Dublin, before proceeding to Connacht to cut down the sacred inaugural tree of the Connachtmen. Other annals suggest Muirchertach did not gain the kingship of Leinster and Dublin and had no role in killing the Leinster king (apparently assassinated by kinsmen), a version of events borne out by the fact that his men burned Lusk in Fine Gall, killing 160 church occupants, suggesting that it (and presumably Dublin) was in enemy hands.

Also in 1089, Muirchertach sailed to Lough Ree and looted its islands, but Ua Conchobair blocked the Shannon near Clonmacnoise, denying the Munster ships a route home. Driven back to Athlone, they were forced to surrender both ships and supplies to the king of Mide, Domnall Ua Máel Sechnaill, returning home under safe conduct overland while the Connacht and Mide armies sailed the confiscated vessels southward, purportedly reducing the Plain of Cashel to a desert. In 1090 Ua Briain, Mac Lochlainn, Ua Conchobair, and Ua Máel Sechnaill held a conference that resulted in the three other kings giving hostages (presumably in submission) to Mac Lochlainn, before departing in peace. But Muirchertach was on the march again in Mide that same year, though he was defeated by Ua Máel Sechnaill who invaded Munster, as did Ua Conchobair at about the same time. Ua Briain then marched into Connacht, raided Leinster along with the men of Dublin (which he had obviously retaken), and marched to Athboy in Brega, where Mac Lochlainn apparently aided him against Ua Máel Sechnaill.

Ua Conchobair invaded Munster again in 1091 but was blinded in the following year, whereupon Muirchertach led an army into the province, took its hostages, and, according to the Inisfallen Annals, assumed the high kingship of Connacht. In 1092, Ua Briain expelled (temporarily) the ruling dynasty of Connacht into Tír nEógain, and the king of Mide came to Limerick to submit to him, while in 1094 he unprecedentedly partitioned Mide between two rival members of the Uí Máel Sechnaill. His contemporary power is evident from the request by the nobility of Man and the Isles to provide a ruler following their king's death in 1095, whereupon Muirchertach apparently found an outlet for the wayward energies of his nephew, Domnall mac Taidc, by dispatching him to Man. Two ominous expeditions by the Norse king Magnus III ("Barelegs") in 1098 and 1102–1103, were skillfully handled by Ua Briain, who bought off Norse aggression in 1102 by marrying his daughter to Magnus's younger son (though the threat dissipated following Magnus's killing by the Ulaid in 1103). He likewise improved relations with the Normans of South Wales when, in 1101, another daughter was married off to Arnulf de Montgomery, then in rebellion against the new English king, Henry I: when the latter responded with a trade blockade, Muirchertach apparently relented, but Arnulf is said to have fled to Ireland hoping to succeed him as king, and Muirchertach later wrote to Anselm of Canterbury thanking him for interceding with Henry on his son-in-law's behalf.

Now the dominant figure in Ireland, as king of Munster and usually overlord of Osraige, Leinster, Dublin, Mide, and Connacht, Muirchertach nevertheless discovered (like his father and his great-grandfather, Brian Boru) that his authority remained incomplete. Each year his vast interprovincial army went north to Assaroe on the Erne or the Sliab Fuaid/Mag Muirthemhne area on the southeastern frontier of Ulster, only to be forced back, often following the intervention of the *comarbae Pátraic* (the abbot of Armagh) to secure a year's truce. In 1100, for example, he led "the men of Ireland" to Assaroe to force Cenél Conaill to submit, simultaneously sending the Dublin fleet around the coast to Inishowen, but was forced to retreat, and the fleet was massacred. In 1101, he was appropriately called for the first time "king of Ireland" in the Annals of Tigernach, and made his most spectacular campaign yet, called *An Slógadh Timcheall* (the Circular Hosting) by the other annals. The six-week expedition again involved the armies of all the provinces, save those of the north, marching to the Erne at Assaroe, then on to Inishowen, burning en route Ardstraw and Fahan, and culminating in the demolition of Grianán of Ailech in revenge for Mac Lochlainn's earlier destruction of Kincora. Muirchertach's men were ordered to bring back to Limerick one stone for every sack of provisions they had, and his forces returned home along the ancient Slige Midlúachra. For the first time, Ulaid was successfully invaded and its submission won, making Ua Briain master of all Ireland except for the northwestern corner which, though undermined, had not submitted and never did. Muirchertach regularly returned, sometimes with disastrous consequences, as in 1103 when (following his possibly conscious emulation of Brian Boru in making a donation of gold to Armagh) his allies were severely routed by Mac Lochlainn in the battle of Mag Coba; but usually the annual expedition ended in stalemate in what is now South Armagh.

In 1101, following his father's example, he presided over the Synod of Cashel, which attempted to reduce lay interference in church affairs, saw Cashel being handed to the church in perpetuity, and prohibited marriage within specified degrees of consanguinity. It was probably at Muirchertach's behest about this juncture that the propaganda tract known as *Cogad Gáedel re Gallaib* ("The war of the Irish with the foreigners") was composed. Combining annalistic data with romantic embellishments allegedly recounting the exploits of Brian Boru against his Norse and other enemies, in fact it uses Brian as a paradigm for Muirchertach and seeks to glorify his reign. The culmination of the latter was perhaps his institution of another reform synod at Ráith Bressail in 1111, in which a formal territorial diocesan structure was established for the country, and two ecclesiastical provinces established at Armagh and Cashel, the former having primacy.

But political opposition remained. In 1112, Domnall Mac Lochlainn defied him by marching to Dubgaill's Bridge in Dublin, raiding Fine Gall, and carrying off livestock and prisoners. In the same year, Domnall invaded Ulaid, annexing part of it, which he intended to rule in person. Muirchertach came to Ulaid in response, Mac Lochlainn moved his armies to Mag Coba ready for battle, and the *comarbae Pátraic* intervened to secure a truce. But the Munster army remained encamped for a month in Brega, Mac Lochlainn's forces observing from the lands of Fir Rois, County Louth, both prepared for war. Domnall's strength is apparent from his refusal to negotiate, although the crisis was again resolved by intervention from Armagh.

Muirchertach's position was weakened further by opposition from his brother Diarmait and the sons of Tadc. His own intended heir, his son Domnall, was proving a disappointment (his nickname *gerrlámhach* ("short-armed") may indicate a disability). The convention of apprenticing the heir to Dublin had been followed, and Domnall managed one major success in battle there in 1115, but subsequently vanished from view and ended his days in monastic obscurity. Muirchertach fell dangerously ill in 1114 whereupon the kingship was seized by his brother Diarmait. When Muirchertach recovered in 1115 and set about regaining his kingdom, his principal ally was Brian (d. 1118) son of Murchad (d. 1068) son of Donnchad mac Briain, his own father's archenemy. He needed all the support he could get, since Domnall Mac Lochlainn reacted to news of the high king's illness by forcing the submissions of Ulaid, Mide, and Bréifne. But, being of similar age to Muirchertach, his day had passed.

Tairrdelbach Ua Conchobair was the new aspirant to national power and had invaded Thomond in 1115. Ua Briain's problems were eased by the death in that year of his troublesome nephew Domnall mac Taidc, and he took to the campaign trail again in Osraige, Leinster, and Brega, but pressure from Ua Conchobair's repeated invasions of Munster finally forced Muirchertach's resignation of the kingship of Munster to his brother Diarmait and his own retirement to Lismore. When Diarmait died in 1118, Ua Conchobair partitioned Munster, giving Desmond to Tadc Mac Carthaig and Thomond, not to Muirchertach's sons, but to those of Diarmait. Muirchertach was dead within a year, the Inisfallen annalist tersely recording his passing in the words "*Murchertach Ua Briain, rí Érend, fo buaid aithirgi quieuit* [Muirchertach Ua Briain, king of Ireland, rested after a victory of repentance]," the

Annals of Ulster calling him "the tower of the honour and dignity of the Western World." All future kings of Thomond were descended from his brother Diarmait, his son son Mathgamain being ancestors of the MacMahons of Corco Baiscind.

SEÁN DUFFY

References and Further Reading

Candon, Anthony. "Muirchertach Ua Briain, Politics, and Naval Activity in the Irish Sea, 1075 to 1119." In *Keimelia: Papers in Memory of Tom Delaney*, edited by Gearóid Mac Niocaill and P. F. Wallace, pp. 397–415. Galway, 1988.

Duffy, Seán. "Irishmen and Islesmen in the Kingdoms of Dublin and Man. 1052–1171," *Ériu* 43 (1992): 93–133.

Duffy, Seán. "The Career of Muirchertach Ua Briain in Context." In *Ireland and Europe in the Twelfth Century: Reform and Renewal*, edited by Damian Bracken and Dagmar Ó Riain-Raedal. Dublin, 2004.

Ó Corráin, Donncha. *Ireland before the Normans*, pp. 142–150. Dublin, 1972.

Ryan, John. "The O'Briens in Munster after Clontarf." *North Munster Antiquarian Journal* 2 (1941) 141–152; 3 (1942–43) 1–52, 189–202.

See also **Brian Boru; Cashel, Synod of; Dál Cais; Munster**

UA BRIAIN, TAIRRDELBACH, (*c.* 1009–JULY 14, 1086 AT KINCORA)

Tairrdelbach Ua Briain was king of Munster (1063–1086) and high king of Ireland (1072–1086). His parents were Tadc, son of Brian Boru, (d. 1023) and Mor, daughter of Gilla Brigte hUí Máel Muaid of Cenél Fhiachach.

Tairrdelbach's early career was dominated by a feud with his uncle Donnchad mac Briain (d. 1065) who had incited the Éile to kill Tairrdelbach's father. Tairrdelbach's marriage to an Éile princess named Gormfhlaith was probably a later peace effort. Tairrdelbach raided his uncle's lands in upper County Clare in 1053, and attacked Donnchad's son Murchad in 1055. The feud intensified when Tairrdelbach received support from his foster-father, the Leinster king Diarmait mac Máele na mbó. They attacked Donnchad in 1058, who was forced to burn Limerick to prevent them from capturing it. In 1062, Tairrdelbach and Diarmait returned to County Limerick, destroying Donnchad's lands and followers. Tairrdelbach was recognized as king of Munster in 1063.

Tairrdelbach could be ruthless. He preempted rebellion in southwest Ireland by a massive raid in 1064. Three years later, Tairrdelbach and Diarmait paid Áed Ua Conchobhair 30 ounces of gold to kill the heir-designate of Teffa. The following year Tairrdelbach, Diarmait, and Domnall Ua Gillai Patraic of Ossory invaded Connacht and provoked a civil war in which Áed and his brother Conchobar were slain. These unsettled conditions sparked a crime wave, and Tairrdelbach had to proclaim legislation in 1068 forbidding the concealment of livestock.

Tairrdelbach always maintained good relations with Diarmait. In 1068, Tairrdelbach received gifts from Diarmait that included his grandfather's sword and the standard of the English king Edward the Confessor. He returned to Leinster again in 1070, receiving valuables and taking custody of Diarmait's troublesome nephew Donnchad mac Domnaill. At the same time, he imposed his lordship over Ossory. The death of Diarmait's sons Glúniairn and Murchad in 1070 led to a power struggle in Leinster. In 1071, Tairrdelbach intervened in the conflict between Diarmait's grandson Domnall mac Murchada and his nephew Donnchad. Tairrdelbach afterward imposed his lordship over Mide, giving the hostages to Diarmait, and had two bridges built across the Shannon at Thomond Bridge and Killaloe.

When Diarmait was slain in 1072, Tairrdelbach promptly asserted his control over Leinster, Mide, and Dublin. He removed Diarmait's nephew Donnchad from Dublin and installed Godred son of Olaf, the uncle of his daughter-in-law Mór. The next year, two of his kinsmen joined Godred's brother Sitric for a raid on the Isle of Man. Family ties were, however, no guarantee of safety. For reasons that are never stated, Tairrdelbach sent Godred into exile in 1075. Domnall mac Murchada briefly seized the kingship, but died before Tairrdelbach returned and installed his son Muirchertach. The five Jews who brought gifts to Tairrdelbach in 1079 may have met him at Dublin.

The final extension of Tairrdelbach's authority began in 1073. He led a massive assault on eastern Mide after his client king Conchobar was slain by his brother Murchad. He also raided the coast, despoiling the Gailenga and killing Máelmorda Ua Cathusaigh of Brega. Tairrdelbach then went into Connacht and received hostages from Uí Conchobhair and Breifné. The north was a special concern to Tairrdelbach, in part because his cousins Conchobar and Cennetig, sons of Donnchad, had found sanctuary with Cenél nEógain of Tuloch Óg. When the king of Ulaid, Donn Sléibe Ua hEochada, was deposed by his kinsman Áed "the furious" in 1078, he submitted to Tairrdelbach at Kincora. Donn Sléibe was reinstated as king by 1080. The unfortunate Áed was in Tairrdelbach's custody when he drowned at Limerick in 1083. During all this, Tairrdelbach's son Diarmait raided Wales.

Good relations with the church were maintained by Tairrdelbach. In 1068 he allowed the comarb of

Armagh to make a circuit of Munster and to take away a stipend and gifts. Tairrdelbach was among those who petitioned Archbishop Lanfranc of Canterbury to consecrate Gilla Patraic as bishop of Dublin in 1074. Gilla Patraic praised Tairrdelbach's good government to Lanfranc. In 1082, a bishop from Tairrdelbach's home of Dál Cais corresponded with Lanfranc. Pope Gregory VII wrote to Tairrdelbach and urged him to support church reform.

Tairrdelbach's empire began to show signs of strain two years before his death. In 1084, Donn Sléibe began his own empire building when Donnchad Ua Ruairc of Bréifne submitted to him at Drogheda. In response, Tairrdelbach led his forces against Ua Ruairc and, with his sons Tadc and Muirchertach, ravaged Bréifne. Ua Ruairc moved south—destroying churches in Dál Cais—and east—raiding the lands around Dublin. At this time the men of Mide rose in revolt, and the northern Uí Néill raided Ulaid. Tairrdelbach's troops put down the rebellion in Meath. On October 19, Tadc and Muirchertach defeated an invading Connacht army at Monecronock; among the slain was Donnchad Ua Ruairc.

In 1085, Tairrdelbach had the first attack of his fatal illness when his hair fell out. He died at the age of 77. Historians have not regarded his career as favorably as that of his illustrious grandsire, but in some respects it was more notable. Tairrdelbach dominated Irish affairs for a longer period, and he died peacefully, with his enemies cowed. Able to adapt to change, Tairrdelbach's hold on the important Viking commercial centers reflected the increasing financial demands that princes throughout Europe were facing. This sophistication is reflected in the *Book of Rights*, a treatise on stipends and dues. Tairrdelbach was more than just a political being. While not hesitating to bribe one day and attack the next, as with Aed Ua Conchobhair, he was unfailingly loyal to his foster-father Diarmait, even when he had become the more powerful party. Tairrdelbach's career shows the problems and opportunities faced by princes throughout late-eleventh-century Europe.

BENJAMIN HUDSON

References and Further Reading

Mac Airt, Sean. *The Annals of Inisfallen*. Dublin: Dublin Institute for Advanced Studies, 1951.

Ó Corráin, Donncha. *Ireland before the Normans*. Dublin: Gill & Macmillan, 1972.

Ryan, John. "The O'Briens in Munster after Clontarf." *North Munster Antiquarian Journal* 2 (1941), 141–153; 3 (1942–43), 1–52 and 189–202.

See also **Dal Cais; Jews in Ireland; Kings and Kingship; O Brien**

UA CATHÁIN

The Ua Catháin (later Anglicized O Cahan, eventually O'Kane) lineage (like the Uí Neill, a branch of the Cenél nEógain) first appear in the annals in 1138, when they were already rulers of the territories of Fir na Craíbe, Fir Lí, and Ciannachta, forming most of the northern part of the present County Derry. The southward shift of the center of power in Tír nEógain following the final replacement of the Meic Lochlainn by the Uí Neill as kings after 1242 was to favor their rise to independent status, as was their cooperation with elements within the Ulster colony. Although Magnus Ua Catháin and fourteen others of his lineage fell fighting against the colonists with Brian Ua Neill at the battle of Down in 1260, his son Cú Muige (later called from this circumstance Cú Muige "na ngall," "of the foreigners") was immediately made chief by Sir Henry de Mandeville, seneschal of Ulster, against the claims of a rival. Thereafter he remained Sir Henry's ally in his struggle against his fellow colonists in Ulster. Subsequently the Uí Chatháin seem to have cooperated with Richard de Burgh, the "Red" earl of Ulster (to whom they paid an annual tribute of forty cows), and in 1312 Cú Muige's son, Diarmait Ua Catháin, styling himself "king of Fir na Craíbe" surrendered to the earl the territory of Glenconkeen (in the southeast corner of County Derry), which the earl immediately regranted to Henry Ua Néill, ancestor of the Clann Áeda Buide. After the collapse of the earldom, the Uí Chatháin became again vassals of the O'Neills, the most powerful and most refractory, with intervals (in the early fifteenth and early sixteenth centuries) when they were forced to submit to the control of the aggressively expansionist Uí Domnaill of Tír Conaill. The uneasy relationship between the Uí Neill and the Uí Chatháin, the former seeking to maximize their control, the latter to minimize it, was to reach its climax after 1603, in the few years before both were destroyed by the Plantation of Ulster.

KENNETH NICHOLLS

References and Further Reading

McCall, Timothy. "The Gaelic Background to the Settlement of Antrim and Down, 1580–1641." Unpublished MA thesis, Queen's University, Belfast, 1983.

MacNeill, T. E. *Anglo-Norman Ulster: The History and Archaeology of a Medieval Barony*. Edinburgh, 1980.

Nicholls, Kenneth. *Gaelic and Gaelicised Ireland in the Middle Ages*. 2nd ed. Dublin, 2003.

Simms, Katharine. "Late Medieval Donegal." In *Donegal History and Society*, edited by William Nolan, Liam Ronayne, and Mairead Dunlevy, pp. 183–201. Dublin, 1995.

Simms, Katharine. "Tír Eoghain 'North of the Mountain.'" In *Derry and Londonderry: History and Society*, edited by Gerard O'Brien and William Nolan, pp. 149–174. Dublin, 1999.

Ua Ceallaigh, Séamus. *Gleanings from Ulster History.* Cork: Cork University Press, 1951. Reprint, Draperstown: Ballinascreen Historical Society, 1994.

UA CONCHOBAIR (UÍ CONCHOBAIR, Ó CONCHOBAIR)

The Connacht dynasty of Ua Conchobair claimed descent from the legendary king Conn Cétchathach. Conn was the purported ancestor of Brión, whose descendants, the Uí Briúin, gained ascendancy over their dynastic rivals in Connacht, and from the ninth century the kingship of the province was reserved for those of Uí Briúin extraction. The Uí Briúin split into the Uí Briúin Seola, the Uí Briúin Bréifne, and the Uí Briúin Aíi. From the latter came the Síl Muiredaig (named after their ancestor Muiredach Muillethan (696–702)), from whom sprang the Uí Chonchobair.

Early Ua Conchobair Kings

Within the Uí Briúin, there was contention over the kingship of Connacht as well. One of the dynasties contending for the kingship was of Uí Briúin Bréifne stock, the Uí Ruairc. After a period of Ua Ruairc sovereignty the kingship came into Uí Briúin Aí hands and it was Conchobar (966–973) son of Tadg "an Túir" ("of the Tower," sometimes called "of the Three Towers") who took on the kingship of Connacht. Conchobar became the progenitor of the Uí Chonchobair; his grandson Tadg "in Eich Gil" ("of the White Steed") (1010–1030) being the first to take the surname. Among the early Uí Chonchobair kings of some influence was Cathal (973–1010), Conchobar's son. He built a stone bridge over the Shannon at Athlone and was a patron of the monastery of Clonmacnoise to which he retired in 1003 and where he died in 1010. His daughter married the high king Brian Boru.

Until the end of the eleventh century the Uí Chonchobair claim to the kingship of Connacht was often heavily contested by the Uí Ruairc, as for instance during the reign of Áed "in Gaí Bernaig" ("of the Broken Spear"). Throughout Áed's reign, the Uí Ruairc disputed his claim to supremacy. After being killed in battle fighting the Uí Ruairc in 1067, Áed was succeeded by Áed Ua Ruairc, the kingship effectively alternating between the families. Subsequent to Áed Ua Ruairc's reign, Ruaidrí Ua Conchobair (d. 1118) ruled until he was blinded by Ua Flaithbertaig of Uí Briúin Seola. This event marked the end of his reign as, according to Brehon law, it made him unfit to govern. After an interval of several years, one of his sons, Tairrdelbach, was inaugurated as king of Connacht at Áth an Termoinn (probably to be identified with Áth Carpait in County Roscommon). However, Carnfree was the traditional inauguration site for the kings of Connacht, and it was here that they were made king up to the fifteenth century.

The Uí Chonchobair's Heyday

Ruaidrí's son Tairrdelbach Mór Ua Conchobair (d. 1156), in his spectacular career, managed to overcome the powerful ruler of the northwest of Ireland, Muirchertach Mac Lochlainn and Tairrdelbach Ua Briain (king of Thomond) and introduce the Uí Chonchobair to the high kingship of Ireland. He built several bridges and castles and was patron of the church and supporter of church reform. At the strategically important location of Athlone, Tairrdelbach built a new bridge in 1124, a castle in 1129 and, according to tradition, a monastery in about the mid-twelfth century. One of the monasteries under Ua Conchobair patronage was Cloontuskert, which was possibly refounded as an Augustinian (Arroasian) priory by Tairrdelbach, and his grandson Áed son of Ruaidrí (d. 1244) was buried there. Tairrdelbach founded an Augustinian priory or hospital in Tuam in 1140. He made his son Conchobar king of Dublin and, in theory at least, Leinster, and later king of Mide; however, Conchobar was assassinated in 1144. Tairrdelbach was considered to be *ardrí co fressabra*, high king with opposition, as his supremacy was contested.

Tairrdelbach fathered nearly twenty sons, one of them being Brian Luignech (d. 1181) the progenitor of the Clann Briain Luignig, from which came the branch called Ua Conchobair Sligig. Another son of Tairrdelbach, Muirchertach Muimnech, was the progenitor of the Clann Muirchertaig of Bréifne. Members of this branch of the dynasty contended and provided kings for the kingship of Connacht during the second half of the thirteenth and the first half of the fourteenth century. However, they subsequently lost their lands and led an almost nomadic life until the fifteenth century, when they disappear from the records.

But it was Ruaidrí (d. 1198) who assumed kingship when his father Tairrdelbach died in 1156. After some struggles he was inaugurated king of Ireland in Dublin in 1166. Like his father before him, he celebrated the Óenach Tailten (the fair of Tailtiu, in 1168, the last occasion it was held). He was a patron of learning and endowed Armagh with an annual income for the teaching of scholars from both Ireland and Scotland. During his reign, Diarmait Mac Murchada went into exile and sought protection from overseas, thus triggering the Anglo-Norman invasion of Ireland. When King Henry II came to Ireland in 1171, Ruaidrí refused to submit. He defeated Strongbow at Thurles (1174), invaded Meath and ravaged Munster, and was in a very strong position in 1175 when he signed the Treaty of Windsor with Henry. In the treaty Ireland was divided into a zone—comprising the lands already overrun by the invaders—under the direct control of the English king's delegate in Ireland, and one under the rule of Ruaidrí. The Treaty proved impossible to maintain and was soon abandoned, however. Ruaidrí suffered from several rebellions by his sons. His son Conchobar Máenmaige (1183–1189) made a bid for the kingship, and Ruaidrí finally retired to the monastery of Cong in 1183, leaving his son king of Connacht. When the latter died, Ruaidrí left the monastery, but was unable to resume the kingship, which was taken by his grandson, Cathal Carrach son of Conchobar Máenmaige (1189–1202). Ruaidrí died in 1198 and was buried at Cong, which had been refounded by his father Tairrdelbach and where Ruaidrí himself had built a new monastery. In 1207 his bones were disinterred and he was reburied in Clonmacnoise, where his father, who had made rich gifts to the monastery, was also buried.

Cathal Carrach's claim was disputed by his kinsman Cathal Crobderg ("of the Red Hand") (1189–1224). The latter managed to get ascendancy by obtaining the support of the leading Anglo-Norman baron in Connacht, William de Burgh, who formerly had supported Cathal Carrach. Crobderg slew Cathal Carrach and was subsequently inaugurated at Carnfree. Like his father Tairrdelbach, he was a great patron of the church and among the many abbeys he founded were Knockmoy (for Cistercian monks) and Ballintober (for Augustinian canons). He was married only once, namely to Mór Muman, daughter of Domnall Mór Ua Briain, and the annals explicitly say he was faithful to her, whereas his father and grandfather were notorious for their polygamous unions. Cathal undertook to maintain good personal ties with the English king. In 1205, he resigned two thirds of Connacht to King John and agreed to pay 100 marks annually for the remaining one third. When John visited Ireland in 1210, Cathal joined forces with him. They parted on bad terms due to the fact that Ua Conchobair refused to hand over his eldest son Áed as a hostage. This notwithstanding, Cathal's diplomatic efforts to obtain a formal grant of Connacht did meet with success five years later.

Cathal Crobderg died in 1224 and was succeeded by his eldest son, Áed (d. 1228). Cathal had attempted to obtain a grant of Connacht for him from the king of England. However, Áed forfeited the land through rebellion. Connacht, with the exception of the so-called "King's Five Cantreds," an area roughly consisting of County Roscommon and parts of counties Sligo and Galway, was then granted to Richard de Burgh. When Áed was murdered, Ruaidrí's sons Tairrdelbach and Áed contested each other for the kingship, with the latter finally emerging victorious. But when Áed rebelled against the Anglo-Norman leaders in Connacht, he was deposed by them, and Feidlim (d. 1265), son of Cathal Crobderg, was set up as king in his stead.

By around 1235, de Burgh had conquered much of Connacht, and the Anglo-Normans proceeded to build castles in the province. Feidlim was left with the "King's Five Cantreds." Like his father and grandfather before him, he was a patron of the church, and he founded Roscommon priory for Dominican friars in 1253. He attempted to maintain a good personal relationship with the English king, and he visited him in England and joined the king's military campaign in Wales in 1245. But by the end of his reign, the territory he officially ruled had been reduced by the granting of lands to royal favorites. His son Áed na nGall (d. 1274) was involved in the attempt at reviving the high kingship of Ireland, and what was later termed the Gaelic revival. In 1260, he fought alongside Brian Ua Néill in the Battle of Down against the settlers in east Ulster where the Irish troops, including Brian, were slaughtered. Through his marriage to a daughter of Dubgall, Mac Ruaidrí from the Hebrides, Áed had at his disposal a band of galloglass, Scottish mercenaries. In 1270, he defeated the English led by Walter de Burgh, earl of Ulster, in the Battle of Áth an Chip. When he died, the kingship was fought over by the Clann Muirchertaig and the line of Áed son of Cathal Crobderg for decades. The latter were able to cling to the kingship in spite of intervals in which members of the Clann Muirchertaig ruled. By the end of the thirteenth century the area officially under Ua Conchobair rule was reduced to a mere three cantreds. Very rarely did any one claimant manage to hold on to the kingship of Connacht for more than a few years. An exception to this was Áed son of Eógan, who ruled from 1293 to 1309, and whose fortress at Cloonfree in County Roscommon is referred to in bardic poetry.

In the early fourteenth century an attempt was made at reintroducing traditional inaugurations of kings at Carnfree. In a series of succession disputes, Feidlim (descendant of Áed son of Cathal Crobderg) was set up by his foster-father, Mac Diarmata. To give this young contender more credibility, Feidlim was inaugurated as king at Carnfree in a grand style reminiscent of ancient traditional customs. According to an inauguration tract, Ó Conchobair was inaugurated by his *ollamh*, Ua Máel Chonaire. Mac Diarmata of Moylurg (north Co. Roscommon) occupied another important place at the inauguration. The Meic Diarmata were among the most prominent subjects of the Uí Chonchobair and provided them with their hereditary chief marshal.

During the Bruce invasion, Feidlim initially joined the earl of Ulster, Richard de Burgh, but subsequently accepted Edward Bruce's offer of the kingship of Connacht and left de Burgh's army. Feidlim fell in 1316 in the Battle of Athenry, fighting against the earl's cousin William de Burgh.

In the subsequent decades, succession disputes were rife and contestants found it near impossible to obtain and maintain ascendancy for any length of time. Not only the descendants of Áed son of Cathal Crobderg, and members of the Clann Muirchertaig sought the kingship, but also members of other branches, as for instance Cathal of the Clann Briain Luignig, who managed to become king (1318–1324). Tairrdelbach, brother of Feidlim Ua Conchobair, also had to contend with Walter son of William de Burgh, who attempted to have himself made king of Connacht. However, de Burgh was defeated and starved to death by his kinsman the earl of Ulster in 1332. Tairrdelbach was killed in 1345, and was succeeded by his son Áed. His reign was brief, and Áed son of Feidlim managed to become king without opposition. He was succeeded by Ruaidrí son of Tairrdelbach. After Ruaidrí's death in 1384, the descendants of Áed son of Cathal Crobderg split in two branches, the Uí Chonchobair Ruad ("Red") and the Uí Chonchobair Donn ("Brown"). The former, with their territory in the north, were supported by Mac Diarmata and the branch of the de Burghs called MacWilliam Burke of Mayo, whereas the latter allied with Ua Ceallaigh and the de Burgh branch of ClanRickard.

The Uí Chonchobair Sligig, descended from Tairrdelbach Mór's son Brian Luignech, were lords of Carbury. In 1420, the leading member of this branch of the family, Brian son of Domnall son of Muirchertach, built Bundrowes Castle, County Donegal. In 1425, Cathal son of Ruaidrí Ua Conchobair Donn became king over all Connacht. Feidlim Geancach Ua Conchobair Donn was the last king of Connacht, and after the 1460s the title "king of Connacht" disappeared.

FREYA VERSTRATEN

References and Further Reading

Dillon, Myles. "The Inauguration of O'Conor." In *Medieval Studies. Presented to Aubrey Gwynn*, edited by J. A. Watt, J. B. Morrall, and F. X. Martin, pp. 186–202. Dublin: Printed by Colm Ó Lochlainn at the Three Candles, 1961.

Fitzpatrick, Elizabeth. "The Inauguration of Tairdelbach Ó Conchobair at Áth an Termoinn." *Peritia: Journal of the Medieval Academy of Ireland* 12 (1998): 351–357.

O'Flaherty, Roderic. *A Chorographical Description of West or h-Iar Connaught.* Edited by James Hardiman. Dublin: Irish Archaeological Society, 1846.

O'Donovan, J., ed. *The Annals of the Four Masters*. Vols. 2–5.

Charles, Owen and Don O'Conor. *The O'Conors of Connaught: An Historical Memoir.* Compiled from a MS of the late John O'Donovan, LL.D. With additions from the State Papers and Public Records. Dublin: Hodges, Figgis, and Co., 1891.

Orpen, Goddard Henry. *Ireland under the Normans 1216–1333.* 4 vols. Oxford: Oxford University Press, 1911 and 1920.

Simms, Katharine. "A Lost Tribe—the Clan Murtagh O'Conors." *Journal of the Galway Archaeological and Historical Society* 53 (2001): 1–22.

Simms, Katharine. "'Gabh umad a Fheidhlimidh'—a Fifteenth-Century Inauguration Ode?" *Ériu* 31 (1980): 132–145.

Walsh, Paul. "Christian Kings of Connacht." *Journal of the Galway Archaeological and Historical Society* 17 (1937): 124–143.

Walton, Helen. "The English in Connacht, 1171–1333." PhD diss., Trinity College Dublin, 1980.

See also **Connacht; King and Kingship; Uí Briúin; Uí Ruairc**

UA CONCHOBAIR, RUAIDRÍ (*c.* 1116–1198)

Ruaidrí Ua Conchobair, son of Tairrdelbach Ua Conchobair (d. 1156) and his third wife Caillech Dé, daughter of Ua hEidin, was the last undisputed high king of Ireland. Ruaidrí was born about 1116, he and his sister Mór being the only fruit of their parents' brief union. During the late 1110s, Tairrdelbach married Mór (d. 1122), daughter of Domnall Mac Lochlainn (d. 1121), allowing Caillech Dé to later successively marry Tairrdelbach Ua Briain (d. 1167) and Murchad Ua Briain. From the events of Ruaidrí's early life it would appear that his father did not favor him. During 1136, Tairrdelbach's fortunes were at an all-time low, encouraging the ambitions of Ruaidrí and some of his brothers. Although under the protection of Bishop Muiredach Ua Dubthaig (d. 1150) and Ua Domnalláin, Tairrdelbach arrested Ruaidrí with Uada Ua Concheanainn (d. 1168), ordering his intended heir Conchobar Ua Conchobair (sl. 1144) to blind another son, Áed Ua Conchobair. How long Ruaidrí spent in confinement is uncertain, but he again incurred his father's wrath in 1143. Then Tairrdelbach ordered Conchobar with Tigernán Ua Ruairc of Bréifne (sl. 1172) to arrest Ruaidrí, breaking sureties given by Archbishop Ua Dubthaig, Tadg Ua Briain (d. 1154), and Murchad Ua Ferghail. Ruaidrí's arrest led the clergy and the nobility to fast at Rathbrendan for his release and, on hearing their petitions, Tairrdelbach outwardly relented, promising to release Ruaidrí in April 1144. Tairrdelbach had little intention of releasing Ruaidrí, but the killing of Conchobar in Mide (Meath) during 1144 forced a rethink. Under pressure from Archbishop Gilla mac Liac (Gelasius) of Armagh (d. 1173) with the clergy and nobility of Connacht, the high king finally released Ruaidrí along with Domnall Ua Flaithbertaig and Cathal Ua Conchobair.

Tairrdelbach now favored another son Domnall Midech Ua Conchobair (d. 1176) but, with typical determination, Ruaidrí slowly rose in his father's estimation, punishing Ua Ruairc with a raid into Dartry during 1146. During 1146 and 1150, he improved his status, capturing and later killing Domnall Ua Conchobair, Tairrdelbach's nephew and enemy. In 1147, Domnall Midech was defeated in Mide, and his fall was completed by his arrest in 1151, allowing Ruaidrí to stake his claim as his father's heir. This role gave him new confidence, attacking Thomond successfully during 1151. After Tairrdelbach's great victory over the army of Tairrdelbach Ua Briain (d. 1167) at Móin Mór, Muirchertach Mac Lochlainn (sl. 1166) compelled the high king to give hostages. Mac Lochlainn's intervention did not prevent Ruaidrí from pursuing his hard line against Thomond, burning Croome in Limerick. And during 1153 Tairrdelbach expelled Ua Briain into the north, causing Mac Lochlainn to come south. In Mide, Tairrdelbach, Ruaidrí, Diarmait Mac Murchada (d. 1171), and Tadg Ua Briain attempted to halt Mac Lochlainn. But Tairrdelbach retreated to Connacht after his allies had suffered heavy losses, leaving Ruaidrí exposed. At Fardrum in west Mide, Mac Lochlainn pounced, routing Ruaidrí and his west-Connacht troops. However, Tairrdelbach and Ruaidrí were determined not to allow Mac Lochlainn to get the better of them, defeating him at sea during 1154. But Mac Lochlainn proved stronger on land, plundering east Connacht and Bréifne that year. Upon the death of Máelsechlainn Ua Máelsechlainn of Mide in 1155, Mac Lochlainn came south again. Tairrdelbach and Ruaidrí vehemently resisted the enforcement of Muirchertach's settlement of Mide, building a bridge at Athlone and sacking Cullentragh castle.

Ruaidrí finally became king of Connacht upon the high king's death in May 1156, inheriting the struggle with Mac Lochlainn. He quickly stamped his authority on his familial rivals, arresting three brothers, blinding one. On learning of Tairrdelbach's death and Ruaidrí's accession, Mac Lochlainn took hostages from Mac Murchada before plundering Osraige (Ossory). Ruaidrí competed with Mac Lochlainn for control of Leinster and the midlands, transforming the whole region into an arena where their respective clients struggled for its kingships. That winter, Ruaidrí strengthened Connacht's midland frontier, positioning a fleet on the Shannon in anticipation of Mac Lochlainn's next move. In 1157, Mac Lochlainn deposed Donnchad Ua Máelsechlainn of Mide (sl. 1160), giving the kingship of Mide to Diarmait Ua Máelsechlainn (sl. 1169) before taking hostages of Mac Murchada. He then attacked Ruaidrí's client kings in Uí Failge and Loígis (partly reflected in the modern counties Offaly and Laois) and Osraige, forcing them to flee to Connacht before subduing Munster and Ostman Limerick. Ruaidrí had to hit back or lose face. Taking advantage of Mac Lochlainn's absence in Munster, Ruaidrí attacked the north, burning Incheny near Strabane and plundered parts of Derry. As Mac Lochlainn hurried home, Ruaidrí then doubled back and appeared in Munster to overturn the high king's settlement of Munster, dividing it between Tairrdelbach Ua Briain and Diarmait Mac Carthaig (sl. 1185).

During 1158 Ruaidrí proved even more crafty. With Mac Lochlainn putting down a rebellion in Tír Conaill (Co. Donegal), Ruaidrí invaded Leinster and reversed the high king's settlement of Loígis and Osraige, carrying Macraith Ua Mórda of Loígis in chains to Connacht. As king of Connacht, he proved equally ruthless, blinding the two sons of the rebel Murchad Ua Ceallaig. Avoiding Mac Lochlainn on land, Ruaidrí used Connacht's maritime superiority to hit the high king and his clients, plundering Inishowen before ransacking Tethbae in Mide. But the greatest weapon in Ruaidrí's arsenal was political skill. As an intriguer, he was unequalled, winning over Donnchad Ua Máelsechlainn in 1158. In 1159, he dumped Ua Máelsechlainn to take advantage of Ua Ruairc's discontent at Mac Lochlainn's deposition of Diarmait Ua Máelsechlainn (sl. 1169) as king of Mide. Ruaidrí now decided to make his move on Mac Lochlainn, attacking Mide and capturing the Ua Máelsechlainn caput on Lough Sewdy before invading Airgialla, home of Donnchad Ua Cearbaill (sl. 1167), Mac Lochlainn's ally. Close to Ardee, Mac Lochlainn annihilated the Connacht army and its allies and pursued Ruaidrí to the Shannon and wasted Bréifne. Later Mac Lochlainn invaded Connacht and humiliated Ruaidrí, sacking his capital at Dunmore and several other forts. Ruaidrí stubbornly refused to submit and left the high king no alternative but to withdraw. While a chastened Ruaidrí licked his wounds, he watched Mac Lochlainn win back Ua Ruairc, strengthen Donnchad Ua Máelsechlainn's kingship of Mide, confirm Leinster to Mac Murchada, and expel Fáelán Mac Fáeláin (d. 1203) from Leinster to Connacht.

In 1160, Ruaidrí, displaying considerable tenacity, exploited Mac Lochlainn's difficulties in putting down another northern rebellion. And the killing of Donnchad Ua Máelsechlainn allowed him to sweep into Mide, take its hostages, and make Diarmait Ua Máelsechlainn king. At Assaroe near Ballyshannon, Mac Lochlainn met him, but they failed to agree to a peace. Mac Lochlainn then marched into Mide to take its hostages and those of Bréifne, prompting Ruaidrí to come to the aid of Diarmait Ua Máelsechlainn and Ua Ruairc. Mac Lochlainn backed off and returned home, allowing Ruaidrí to erode his hold on Munster, placing a fleet on the Shannon to take hostages of Tairrdelbach Ua Briain. Ruaidrí continued to erode the high king's authority in 1161. He and Ua Ruairc went into Mide and Leinster and took the hostages of

Uí Fáeláin and Uí Failge, leaving Fáelán Mac Fáeláin and Muirchertach Ua Conchobair Failge (sl. 1166) there as kings. This forced Mac Lochlainn to reaffirm his high kingship, plundering Bréifne and west Mide and taking the submissions of the Dublin Ostmen and Mac Murchada. Faced by Mac Lochlainn's might, Ruaidrí pragmatically negotiated with Mac Lochlainn on the plain of Tethbae. Although Ruaidrí gave Mac Lochlainn four hostages for Uí Briúin, Conmaicne, Munster, and Mide, the Clonmacnoise annals say he gave twelve. It is clear, though, that his submission was far from unconditional given that the high king did not depose Connacht's allies in Uí Fáeláin and Uí Failge. In return, Ruaidrí received Connacht and was granted west Mide by Mac Lochlainn, while east Mide was confirmed to Diarmait Ua Máelsechlainn. At this time, Mac Lochlainn was king of Ireland *cen fressabra* ("without opposition"), presiding over a convention of the laity and the clergy of Ireland at Dervor in Mide. While Ruaidrí presumably attended, he was biding his time. Afterwards he returned to Connacht and executed Domnall Ua Laeghacháin despite the sureties of the bishop of Clonmacnoise.

A sign of Ruaidrí's acknowledgment of Mac Lochlainn's superiority was the presence of Connacht troops at the high king's siege of Dublin during 1162. He also drew Diarmait Ua Máelsechlainn closer to him, returning west Mide to him for five score ounces of gold in 1162. During 1163, he sent a timely message to Mac Lochlainn, allowing his heir Conchobar Máenmaige Ua Conchobair (sl. 1189) to capture the high king's son in Connacht before sending him home. By 1164, Ruaidrí was up to his old tricks, resuming his favorite ploy of exploiting Mac Lochlainn's difficulties in the north. In Thomond, Ruaidrí's half-brother Muirchertach Ua Briain (sl. 1168) seized the kingship from his father Tairrdelbach Ua Briain. Ruaidrí and Muirchertach tried to expel Tairrdelbach, but failed. Undeterred, Ruaidrí campaigned with Ua Ruairc to the borders of Dublin before crowning the year by transferring his capital from Dunmore to his new fortress of Tuam. With Mac Lochlainn distracted by troubles in the north, Ruaidrí consolidated his grip on the midlands and parts of Leinster. In 1165 he punished the Leinster subkingdom of Cairpre (Carbury) for its participation in the killing of Sitriuc Ua Ruairc. That year he restored Diarmait Ua Máelsechlainn to the kingship of west Mide, subdued the latter's enemies in the Mide subkingdoms of Brega and Saithne, campaigned into Leinster, and asserted his superiority over Diarmait Mac Carthaig.

Ruaidrí's great moment came in 1166 when Mac Lochlainn's high kingship unwound amid a serious rebellion in the north. Around April 24, Mac Lochlainn blinded Eochaid Mac Duinnsléibe of Ulaid, prompting the outraged Donnchad Ua Cearbaill of Airgialla (sl.1167) (Eochaid's foster-father) to repair to Connacht to Ruaidrí. Realizing his time had come, Ruaidrí marched on Dublin and was acknowledged as high king. Deep inside Mac Lochlainn's sphere at Drogheda, Ruaidrí took Ua Cearbaill's submission, but instead of attacking Mac Lochlainn, he first drummed up support against Mac Murchada in Uí Fáeláin and Uí Failge. In May, Ruaidrí invaded Mac Murchada's home kingdom of Uí Chennselaig, defeating him at Fid Dorcha. With Mac Murchada subdued, Ruaidrí marched to Tír Conaill to take the submissions of its lords, ensuring they did not go to Mac Lochlainn's aid. Thereafter, the collapse of Mac Lochlainn in the north was rapid; Ruaidrí's allies killed him in the Fews of Armagh. He began a circuit of Ireland in Tír nEógain (Tyrone), dividing it between Niall Mac Lochlainn (sl. 1176) and Áed Ua Néill (sl. 1177) and took the submission of the king of Ulaid (east Ulster). He then entered Leinster, took the submissions of the king of Osraige before traveling to Munster to take the allegiance of its kings. In Ruaidrí's absence, Mac Murchada attempted to reassert himself, leading to the August invasion of Uí Chennselaig by Ua Ruairc, Diarmait Ua Máelsechlainn, the Leinster princes, and the Dublin Ostmen. Mac Murchada fled in search of Henry II, resulting in the division of Uí Chennselaig between the Mac Gilla Pátraic dynasty of Osraige and Murchad Mac Murchada (sl. 1172). Before the close of the year, Ruaidrí at Athlone rewarded all his clients who had played decisive roles in his capture of the high kingship.

As high king, he was determined to rule the disparate kingdoms of Ireland, taking steps towards the achievement of effective royal government by presiding over an almost national assembly at Athboy in Mide during 1167. But he was compelled to campaign against Niall Mac Lochlainn, marching with the kings of Mide, Thomond, Desmond, and Ulaid to Armagh before catching the fleet to attack Derry. After forcing Mac Lochlainn's submission, he redivided Tír nEógain between him and the Uí Néill. In August, Mac Murchada returned with English troops and reconquered Uí Chennselaig. Ruaidrí reacted quickly and brought him to heel after two clashes at Kellistown. Feeling secure, Ruaidrí celebrated the Óenach Tailten ("fair of Teltown"), an act proclaiming his dominance over the island. Although the most powerful man in Ireland, Ruaidrí had difficulties in welding his kingdom together, particularly in Mide and Thomond. But he coped competently, extracting a fine of 800 cows from Diarmait Ua Máelsechlainn for killing a client. But the trouble did not end there, for angered by Ua Máelsechlainn's payment of the fine, the Meathmen deposed him; and Ruaidrí's troops sent to restore him were routed. In Munster Ruaidrí's power was also threatened by the killing of Muirchertach Ua Briain (his half-brother), but again he dealt easily with the crisis. He divided the province between Domnall Mór Ua Briain (d. 1194) and Mac Carthaig before levying a fine of 720 cows for Muirchertach's killing. At Athlone he received the fealty of Mac Gilla Pátraic, and such was his power that later the kings of Tír nEógain made their submission there too.

In 1169, Ruaidrí's confidence in his high kingship was plain, granting the lector of Armagh ten cows in perpetuity to lecture Irish and Scottish students in literature. Alarmingly, Mac Murchada's second wave of English troops landed in May, and Diarmait Ua Máelsechlainn was killed by Domnall Ua Máelsechlainn (sl. 1173), who established himself as king of east Mide. Ruaidrí quickly shored up his position, expelling Domnall, kept west Mide for himself, and gave the east to Ua Ruairc. But as he plugged one leak, others appeared. Mac Murchada now attacked the high-king's clients in Osraige and West Leinster. A concerned Ruaidrí summoned the men of Ireland, meeting them probably at Tara. With Murchad Ua Cearbaill of Airgialla (d.1189) and Magnus Mac Duinnsléibe of Ulaid (sl.1171), he went to Dublin to confer with its ruler before returning to Connacht. To counter Mac Murchada's successes in Osraige, Ruaidrí began a circuit through Munster, Leinster, and Osraige to reassure his allies. With an army of Irish and Ostmen, he entered Uí Chennselaig to confront Mac Murchada. Although Ruaidrí proved militarily superior, he lost confidence in his ability to impose a military solution. Characteristically, he changed tactics, opting for politics. Messengers were dispatched to tempt Robert fitz Stephen (d. 1210) to desert Mac Murchada. When fitz Stephen refused, Ruaidrí switched to Mac Murchada himself, offering an alliance if he would turn on the English. Mac Murchada turned him down, forcing Ruaidrí to review his options and dispatch clerics to treat with the Leinster king. They found him receptive and struck a deal that confirmed Mac Murchada as king of Leinster in return for his recognition of Ruaidrí's high kingship; the English were to be sent home; Mac Murchada's last legitimate son, Conchobar, was taken by Ruaidrí as a hostage and to him was promised one of Ruaidrí's daughters. Satisfied with these arrangements, Ruaidrí departed.

Matters worsened considerably in 1170. Domnall Ua Briain revolted against Ruaidrí, distracting the high king's attention from Leinster. During the summer, Ruaidrí's problems mounted when Mac Murchada attacked Osraige, Leinster, and Mide and dispatched English troops to aid Ua Briain. Ruaidrí was forced to retreat from Thomond and had to content himself with wasting Ormond. On August 23, Mac Murchada took Waterford and marched on Dublin. Ruaidrí hastened to the aid of the Ostmen, positioning his army at Clondalkin to block Mac Murchada. The Leinster king, however, cut through the Wicklow mountains to reach Dublin. The Ostman king, judging that Mac Murchada had bested Ruaidrí, entered into negotiations with him. Aware of this treachery, Ruaidrí prudently withdrew, leaving the Ostmen to be repaid in their own faithless coin when the English seized the city on September 21. Ruaidrí's withdrawal left Mac Murchada in complete control of east Leinster and exposed his clients in west Leinster and Mide to Mac Murchada's revenge. The crisis gripping Ruaidrí's high kingship was graphically illustrated when Ua Ruairc forced him to execute Mac Murchada's hostages for his continued fealty. Ruaidrí and the Irish kings in general were so alarmed that they may have dispatched a delegation to Henry II of England seeking protection.

In spite of these terrible reverses, Ruaidrí recovered in 1171, forcing Ua Briain to submit before mid-year and was boosted when Mac Murchada died in May. Ruaidrí now planned a major campaign to support the Leinstermen fighting the English. With their help and fleets from the Western Isles and Man, Ruaidrí besieged Dublin through August and September 1171, reducing the English to desperate straits. With success within his grasp, Ruaidrí dictated a peace: the English could retain the Ostman towns of Dublin, Wexford, and Waterford, but nothing more. His confidence was such that he now divided his army. According to the Irish annals, he left a large contingent at Castleknock to contain the English at Dublin, and moved off to rendezvous with his Leinster allies, who were maintaining the blockade south of the city. He also led an expedition inland to punish those still loyal to the family of Diarmait Mac Murchada and the English, while he dispatched that cavalry of Bréifne and Airgialla to burn the cornfields of the English near Dublin. The weakening of the Irish grip around Dublin was Ruaidrí's undoing. Twilight was falling as the English descended on the unprepared camp at Castleknock, slaughtering hundreds. Ruaidrí's presence at the rout is disputed. The Irish sources uniformly say that he was still away campaigning in Leinster, while the near-contemporary source known as the *Song of Dermot and the Earl* makes no mention of his presence during the attack. Only Giraldus Cambrensis (Gerald of Wales) has that Ruaidrí was there; he claimed that Ruaidrí was having a bath when the English attacked and that he escaped through the slaughter naked. Be that as it may, Ruaidrí's high kingship had suffered an irreversible shock, forcing him to withdraw from Dublin. On October 18 Henry II landed at Waterford before proceeding to Dublin, taking the submissions of Ruaidrí's allies such as Ua Briain, Mac Carthaig, Domnall Mac Gilla Pátraic (d. 1185), Ua Máelsechlainn, and even Ua Ruairc. In November, Henry II sent emissaries to Ruaidrí demanding his submission. All English sources with the exception of Giraldus tell of how the high king refused, informing the emissaries that Ireland was his by right and that he owed the English king no fealty. That Henry II considered leading an expedition against Ruaidrí also indicates that the high king's army was largely still intact.

In any event, Henry returned to England in March 1172 to deal with the rebellion of his sons. In Leinster, Ruaidrí's allies still resisted the English, but he suffered a major blow when his father-in-law Ua Ruairc was killed. That year Ruaidrí confined himself to Connacht, presiding over a convention of laity and clergy at Tuam. In 1173, he aided the Irish fighting the English advance, allowing Conchobar Máenmaige and the men of west Connacht to join Domnall Ua Briain to sack Kilkenny. He also had the hand of Domnall Ua Ruairc, his father-in-law's nemesis, nailed to the top of Tuam castle. During 1174, he took the field himself against the English, blocking their advance into Ormond, forcing them to send for reinforcements. He then dispatched Ua Briain and Conchobar Máenmaige to attack the reinforcements, defeating them with great loss at Thurles, forcing the English to retreat to Waterford and abandon Kilkenny to the Irish. Jubilant, Ruaidrí returned to Connacht and assembled an army largely drawn from that province, Ulster, Mide, and west Leinster. Taking advantage of the absence of Hugh de Lacy (sl. 1186), he invaded Mide, sacking its castles and penetrated as far as Dublin. But he was unable to strike the fatal blow, and Raymond le Gros forced him to retreat to Connacht, leaving his supporters in Leinster and Mide with no option but to take refuge in Connacht. As a result of Ruaidrí's inability to press home his advantage, Ua Briain revolted in 1175. Ruaidrí duly deposed Ua Briain, raising his own half-brother, the son of Murchad Ua Briain, to the kingship of Thomond. Ua Briain, though, continued to resist, leading Ruaidrí to resort to a game of divide and rule. Before October 1, he invited the English and Domnall Mac Gilla Pátraic to aid him, intending to use them to administer a decisive defeat upon his sometime enemy. At the same time, Ruaidrí dispatched a delegation, consisting of Archbishop Laurence O'Toole of Dublin, Archbishop Cadhla Ua Dubthaig of Tuam, and his chancellor Master Laurence, to negotiate a treaty with Henry II at Windsor. By its terms, Ruaidrí on October 6, 1175 acknowledged Henry II as his overlord and agreed to stay out of much of Leinster and part of east Munster, while Henry would leave the rest of the island to Ruaidrí. Around this time, Raymond le Gros held separate conferences with Ruaidrí and Ua Briain and received pledges of loyalty. Sensing Ua Briain's weakness, Ruaidrí pounced, forcing him to give up seven hostages.

By 1177, the treaty of Windsor had become unworkable due to continuing English encroachments into Connacht. And like his father, Ruaidrí had a troubled relationship with his sons. In 1177, Ruaidrí's son Murchad Ua Conchobair (d. 1216) guided Milo de Cogan's invasion of Connacht. Even though they sacked Tuam, the invaders fled before Ruaidrí's forces. As an example to others, Ruaidrí blinded the captive Murchad for his treachery. The invasion of Connacht now caused Ruaidrí to question the loyalty of his other sons, arresting the able Conchobar Máenmaige before the close of the year. Even though Conchobar Máenmaige escaped in 1178, father and son were reconciled and drove de Lacy's forces away from Clonmacnoise that year. Indeed, Ruaidrí's hand may even be detected in the attacks of dispossessed Leinster princes upon English forces in 1179.

While Conchobar Máenmaige put down an Ua Ceallaig rebellion in Connacht during 1180, Ruaidrí resumed his political machinations, seeking to divide his enemies. He dispatched Archbishop Loréan and a son to negotiate a new peace with Henry, while at the same time he formed an alliance with de Lacy. The alliance was sealed with the marriage of de Lacy to Ruaidrí's daughter Róis, angering King Henry who thought de Lacy too powerful in Ireland. In allying with de Lacy, Ruaidrí hoped his son-in-law would stem the colonial flood. But in Ruaidrí's struggle to keep Connacht afloat, the colonists were not his only challenge. Ever since the death in 1176 of Domnall Midech, lord of north Connacht, Ruaidrí's rule over north Connacht was precarious. In 1181, Domnall Midech's sons joined Ruaidrí's former son-in-law Flaithbertach Ua Máeldoraid (d. 1198), king of Tír Conaill, against Ruaidrí. The high king dispatched an army to crush their rebellion, but they wiped it out at Cairpre (the Carbury area of Co. Sligo) on May 23. If this was not bad enough, Ruaidrí's miserable year was capped off by the death of his wife Dubchoblach, daughter of Ua Ruairc. In 1182, Ruaidrí ordered Conchobar Máenmaige to gather his army. And when they cornered the rebels in Sligo, they slaughtered them in a complete rout. With Ruaidrí now approaching 70, Conchobar Máenmaige was eager to succeed him as king of Connacht, leading to considerable tension between father and son. In 1183, the situation seemed to have been resolved when Ruaidrí renounced the world and entered the monastery of Cong in Mayo, allowing Conchobar Máenmaige to assume the Connacht kingship. After such a life, it was natural that Ruaidrí would find a life of contemplation tedious. And so in 1185, he decided to reclaim his kingship from Conchobar Máenmaige. This selfish decision was disastrous for the recovering Ua Conchobair kingdom of Connacht, heralding decades of civil war. After enlisting Domnall Ua Briain and the English, Ruaidrí pillaged throughout west Connacht in 1185, burning churches and terrorizing the population. Even though Conchobar Máenmaige ravaged Thomond in revenge, the slaughter appalled him, leading him to agree to share Connacht with his father. In 1186, Conchobar Máenmaige thought better of this accommodation and exiled Ruaidrí to Munster, but later recalled him and allotted him new lands. Despite this, Ruaidrí refused to relinquish his dream of taking the kingship back and continued to plot against Conchobar Máenmaige. His dream became a reality after Conchobar Máenmaige was assassinated by Ruaidrí's supporters, causing the Connacht nobility to recall him. Upon his return, Ruaidrí was triumphantly welcomed by the nobility and received their hostages. This was to prove his last triumph. In reality, he was too old for the rigors of Connacht politics and was deposed before the end of the year. In 1191, he tried yet again to reclaim his kingship, traveling to Ulster, Mide, and Munster to gather troops. His decline was evident as, everywhere he went, none would help him. Finally, the Connacht nobility prevailed upon him to return home, telling him that lands had been put aside for him in southwest Galway. The old man returned, but soon entered the monastery of Cong to begin his penance. Ruaidrí, last of the high kings, died at Cong in 1198 and was buried with his father in the church of Clonmacnoise. Even in death, Ruaidrí proved turbulent, as his remains were disinterred and placed in a stone shrine in 1207.

EMMET O'BYRNE

References and Further Reading

Giraldus Cambrensis, *History and Topography of Ireland*. Dublin: 1952.

———. *Expugnatio Hibernica*. Dublin: Royal Irish Academy, 1978.

Hennessy, W. and B. MacCarthy, ed. *The Annals of Ulster*. 4 vols. Dublin: 1887–1901.

———. *The Annals of Loch Cé*. 2 vols. London, 1871.

Luard, H., ed. *Matthiaei Parisiensis Chronica Majora*. Vol. 2. London, 1884.

O'Donovan, John, ed. *The Annals of the Four Masters*. 7 vols. Dublin, 1851.

Ó hInnse, Séamus, ed. *Miscellaneous Irish Annals*. Dublin: Irish Manuscript Commission, 1947.

Orpen, Godfred, ed. *The Song of Dermot and the Earl*. Felinfach, 1994.

Mac Airt, Seán. *The Annals of Inisfallen*. Dublin: Dublin Institute for Advanced Studies, 1951.

Murphy, Denis. *The Annals of Clonmacnoise*. Dublin: Royal Irish Academy, 1896.

Stokes, Whitley, ed. "The Annals of Tigernach." *Rev. Celt* 16–18 (1895–1897).

Stubbs, W., ed. *The Historical Works of Gervase of Canterbury*. Vol. 1. London: 1879.

———. *Chronica Magistri Rogeri de Houedene*. Vol. 2. London: 1869.

See also **Henry II; Lacy, Hugh de; Mac Lochlainn, Muirchertach; Mac Murchada, Diarmait; Ua Conchobair, Tairrdelbach; Uí Néill**

UA CONCHOBAIR, TAIRRDELBACH (1088–1156)

Tairrdelbach Ua Conchobair, son of Ruaidrí Ua Conchobair (d. 1118), king of Connacht, and Mór (d. 1088), daughter of Tairrdelbach Ua Briain (d. 1086), high king of Ireland. Tairrdelbach's early life was troubled. According to the Annals of Tigernach, Tairrdelbach's mother died the year he was born, suggesting his birth was arduous, and in 1092 his father Ruaidrí was blinded by Flaithbertach Ua Flaithbertaig (blinded, in turn, in 1098). Thereafter Connacht fell largely under the sway of Tairrdelbach's maternal uncle Muirchertach Ua Briain (d. 1119), high king of Ireland. Ua Briain possibly took Tairrdelbach into his household to groom him for the day when he would be king of Connacht. In 1106, that day came when Ua Briain replaced Domnall Ua Conchobair (d. 1118), Tairrdelbach's elder half-brother, with his protégé.

Tairrdelbach carefully maintained his alliance with Ua Briain, sending troops to aid the high king against the Uí Ruairc of Bréifne in 1109. But he was also determined to defend his kingdom against predators such as Domnall Mac Lochlainn (d. 1121), king of the north of Ireland. In 1110, Mac Lochlainn attacked Connacht, carrying captives and cattle back to Ulster. The raid rattled Tairrdelbach, leading him to attack Conmaicne and Bréifne with mixed fortunes. He beat the former at Mag nAí, but the latter defeated his troops at Mag Brenair. During 1111, he raided north, plundering Termonn Dabeoc in Tír Conaill and ravaged Fermanagh to Lough Erne. By 1114, Tairrdelbach was undisputed master of Connacht, having banished Domnall into Munster as well as expelling the Conmaicne from Mag nAí. Recognition came in a prestigious marriage to his second wife Orlaith (d. 1115), daughter of Murchad Ua Máelsechlainn of Mide (d. 1152), sometime enemy of Ua Briain. After Ua Briain fell ill in 1114, Toirdelbach's greater ambitions became evident. He turned to Ua Briain's enemies, reaching an agreement with Mac Lochlainn and Ua Máelsechlainn. Pooling their forces, they attacked Munster, forcing the Uí Bhriain to sue for peace. Such was Tairrdelbach's new strength that in 1115 he gave the kingship of Thomond to Domnall son of Tadg Ua Briain. The latter proved no puppet and revolted against Connacht, prompting an outraged Tairrdelbach to devastate Thomond and dispatch his former protégé.

During 1115, Tairrdelbach's rule led to great disquiet among some of his own vassals, leading to an unsuccessful attempt to kill him at Áth bó. Moreover, the death of his wife Orlaith that year ended Tairrdelbach's Ua Máelsechlainn alliance, granting him a pretext to attack Mide (Meath), inflicting defeat on Domnall Ua Fergail's fleet before forcing Ua Máelsechlainn's submission. At the close of 1115, Tairrdelbach gave thanks, bestowing gifts of a drinking horn inlaid with gold and a golden cup and patina for a chalice upon the monastery of Clonmacnoise. He then married the Connacht noblewoman Caillech Dé, daughter of Ua hEidin, mother of Ruaidrí Ua Conchobair (d. 1198). Their union was brief as Tairrdelbach soon married Mór (d. 1122), daughter of Mac Lochlainn. Throughout 1116 and 1117, Tairrdelbach was opposed on the political front by Diarmait Ua Briain (d. 1118). But during 1118 Tairrdelbach, Ua Máelsechlainn, and Áed Ua Ruairc (sl. 1122) joined a recovered Muirchertach Ua Briain to attack Tadg Mac Carthaig of Desmond (d. 1124). However, they turned on Ua Briain at Glanmire near Cork, allying with Mac Carthaig to depose the high king for good. Tairrdelbach then broke Ua Briain's hold over Leinster, Osraige, and Ostman Dublin, expelling Domnall Ua Briain (d. 1135) from that city. And he even invaded Thomond itself, demolishing the Ua Briain fortress at Kincora, hurling it into the Shannon. During 1119, Tairrdelbach demonstrated his power, compelling Leinster, Osraige, and Ostman Dublin to campaign against the Uí Bhriain. But his exiling of Ua Máelsechlainn to Ulster in 1120 and his celebration of Óenach Tailten ("the fair of Teltown"), an act proclaiming his highkingship, attracted Mac Lochlainn's unwelcome attentions. Mac Lochlainn reinstated Ua Máelsechlainn in Mide, compelling Tairrdelbach to back off and make "false peace" with them at Athlone. Luck, though, was on Tairrdelbach's side, for Mac Lochlainn died during 1121, leaving him the most powerful man in Ireland. And he made the most of it, subduing Munster, causing "the people to cry aloud."

In 1122, the Munster question was briefly settled when Tadg Mac Carthaig submitted. As his political fortunes soared, Tairrdelbach suffered a blow when his fourth wife Mór died that year. Despite his grief, Tairrdelbach did not remain single, taking a fifth wife in Tailltin (d. 1128), daughter of Ua Máelsechlainn. Moreover, he threw himself into his campaigns with zest, capturing Tairrdelbach Ua Briain (d. 1167), forcing the submission of Énna Mac Murchada of Leinster (d. 1126), and probing the north to Lough Erne. Although primarily a soldier king, Tairrdelbach pragmatically cultivated church support through generous patronage. In 1123, he capitalized upon the visitation of a relic of the true cross to Ireland, enshrining a piece of it at Roscommon and commissioned the later processional Cross of Cong to hold it. Indeed, Tairrdelbach displayed traits of contemporary European kings, making land grants to both clerical and lay supporters, levying taxation, and possibly issuing a form of coinage. He was also a builder, erecting abbeys, as well as improving his communication and defensive abilities by building bridges and Irish castles.

In 1123, the Munster problem reappeared with Cormac Mac Carthaig (sl. 1138) determined to fight. Although Tairrdelbach forced the Munstermen to submit, they rose up again during 1124. The Munster troubles then spread to Leinster and the midlands, culminating in an alliance between Munster, Mide, Osraige, the Conmaicne, and Leinster. While Tairrdelbach routed the Conmaicne, Desmond, Leinster, and Mide invaded west Mide and moved to attack him at Athlone. Contemptuously, the high king executed the hostages of Desmond, causing the alliance to splinter for fear of more executions. Tairrdelbach now taught his enemies a lesson, beginning with Mide in 1124. In 1125 he took the hostages of Osraige, and forced the submission of Tigernán Ua Ruairc of Bréifne (sl. 1172) and banished Ua Máelsechlainn to the north. In spite of considerable opposition and the loss of his bridges at Athlone and Áth Croich, he divided Ua Máelsechlainn's kingdom among three family rivals and Ua Ruairc, before confirming the Leinster kingship of Énna Mac Murchada (d. 1126). Tairrdelbach's dominance was such that the Annals of Tigernach record in 1126 that he assumed the Leinster kingship after the death of Mac Murchada and installed his son Conchobar (sl. 1144) as king of Dublin. He then routed Mac Carthaig, Osraige, and the Meic Murchada of Uí Chennselaig before transferring the Leinster kingship to Conchobar.

In 1127, he proved his superiority over the Munstermen, routing their armies and fleets before dividing the province. But while Tairrdelbach was in Munster, the Leinstermen and the Dublin Ostmen deposed Conchobar as king of Leinster and Dublin. This brought Tairrdelbach back into Leinster, but even he was forced to concede that Conchobar was unsuitable as provincial king, turning instead to Domnall Mac Fáeláin of Uí Fáeláin (sl. 1141). Again he gave thanks for his success, granting lands to the archiepiscopal see of Tuam (though it did not obtain metropolitan status until 1152). After the death of Tailltin in 1128, Tairrdelbach married Mac Lochlainn's daughter, Derbforgaill (d. 1151). In Leinster there was trouble, leading to Tairrdelbach's campaign against the Meic Murchada of Uí Chennselaig. Before the end of the year Tairrdelbach again forced Munster to sue for peace and devastated Tír Conaill in 1130, leading to a truce with Conchobar Mac Lochlainn (sl. 1136).

Munster remained tempestuous. And the tide was turning. After defeating Desmond during 1131, Tairrdelbach was confronted by the armies of Munster and the north. While Tairrdelbach dealt masterfully with them, defeating the Ulstermen first before scattering the Munster army, their audacity was unsettling. The year 1132 proved that Tairrdelbach was just not strong enough to defeat his rivals decisively. The year began well with victories over Munster and a fresh division of Mide, but the balance tipped against him when Ua Ruairc and the Conmaicne joined Conchobar Ua Briain of Thomond (d. 1142). A Munster fleet burnt Galway, and Ua Briain and Mac Carthaig were to sack Tairrdelbach's capital at Dunmore in 1133, while Ua Máelsechlainn destroyed his bridge at Athlone, compelling him to conclude a year's peace with Ua Briain. For all his brilliance, Tairrdelbach was on the ropes when his enemies closed for the kill in 1134, leading him to finally acknowledge reality and dispatch Bishop Muiredach Ua Dubthaig (d. 1150) to Mac Carthaig to sue for peace. Tairrdelbach's defeat encouraged Ua Ruairc, the Conmaicne, and Ua Briain to test the territorial integrity of Connacht in 1135, fanning also the ambitions of some sons. During 1136, an ill Tairrdelbach arrested his son Ruaidrí Ua Conchobair (d. 1198) and authorized his intended heir Conchobar to blind another son, Áed Ua Conchobair. However, Connacht's fortunes remained in the doldrums. During 1137, Connacht was "laid waste from Assaroe to the Shannon and to Echtach of Munster." The first sign of a Connacht recovery came after Mac Carthaig's killing, as evidenced by Tairrdelbach's Mide campaign of 1138. During 1139, he worked hard to revitalize his forces, employing them to improve Connacht's natural defenses by diverting the Suck to form a flood plain. He also dealt with rebels, blinding Donnchad Ua Máelruanaid of Mag Luirg (d. 1144).

The clearest sign of Tairrdelbach's return to form came in 1140. Then Archbishop Gilla mac Liag (Gelasius) of Armagh (d. 1173) visited Connacht and received tribute as primate of all Ireland. Reinvigorated, Tairrdelbach then made a false peace with Ua Máelsechlainn and threw a new bridge over the Shannon at Athlone. He swept into the Mide subkingdom of Tethbae, plundering it mercilessly. Not content with that, he banished Ua Máelsechlainn. Although 1141 began badly with Ua Briain burning much of west Connacht, Tairrdelbach recovered and consolidated his midland hegemony through the restoration of Ua Máelsechlainn to Mide and the taking of Ua Ruairc's hostages. Ua Briain's death in 1142 bolstered Tairrdelbach's fortunes, allowing him again to claim the high kingship. In 1143, he consolidated his resurgent power by defeating Tairrdelbach Ua Briain of Thomond at Roevehagh and exiling Ua Máelsechlainn to Munster. To emphasize his resurgence, Tairrdelbach granted new lands to the church, but controversially replaced Ua Máelsechlainn as king of Mide with his heir Conchobar. On the home front, Tairrdelbach faced considerable pressure from Bishop Muiredach Ua Dubthaig about his continued imprisonment of his son Ruaidrí. Although Tairrdelbach promised to release Ruaidrí, the assassination of Conchobar in Mide during 1144 forced him to do it earlier than expected. In Mide he hunted the assassins down, and divided that kingdom before making peace with Tairrdelbach Ua Briain, whereupon he proceeded to subdue Leinster.

Peace was short-lived as Mide again rebelled in 1145, leading Tairrdelbach to dispatch his trusted son Domnall Midech Ua Conchobair (d. 1176) to subdue it. Anxiously, Tairrdelbach Ua Briain watched as Connacht tightened its grip upon the Shannon and the adjoining midlands. Ua Briain then challenged Connacht's overlordship in the midlands, but was forced to retreat and content himself with a raid into Connacht. More seriously, Ua Briain gathered a combined Munster and Ostman fleet to break Connacht's grip on the Shannon. But Tairrdelbach hit first, sinking Ua Briain's fleet at the mouth of the Shannon, ensuring Connacht's dominance into 1146 despite the emergence of an alliance between Mide, Munster, Ua Ruairc, and the Conmaicne. Even though Tairrdelbach easily defeated Munster, Ua Ruairc's defection was embarrassing as it stoked trouble in Mide, contributing to Domnall Midech's defeat in Tethbae during 1147. Tairrdelbach attempted to make peace with Ua Ruairc during 1148, but failed due to the determination of Domnall Ua Fergail to kill the Bréifne king. But in 1149 a bigger threat emerged to Connacht in the person of Muirchertach Mac Lochlainn (sl. 1166).

The aging Tairrdelbach reluctantly recognized Mac Lochlainn's dominance in 1150, sending him hostages. Even so he was still a force to be reckoned with on the battlefield, plundering Munster that year. Early in 1151, Tairrdelbach welcomed Archbishop Gilla of Armagh to Connacht, presenting him with a golden ring. But it was on the battlefield that Tairrdelbach had his greatest success in 1151. Ua Briain invaded Desmond, forcing Diarmait Mac Carthaig (sl. 1185) to ask Tairrdelbach for help. Secretly, Tairrdelbach and Diarmait Mac Murchada of Leinster (d. 1171) marched into Desmond and met Mac Carthaig before tracking Ua Briain. Using mist as cover, Tairrdelbach attacked Ua Briain's rearguard, throwing his army into confusion. The annals record that 7,000 men fell and Ua Briain fled. Even though this victory must have been the pinnacle of Tairrdelbach's military career, Muirchertach Mac Lochlainn invaded Connacht through the Curlew Mountains. Prudently, the old king decided not to risk all on a wager of battle and gave Mac Lochlainn hostages. Tairrdelbach's long marriage to Derbforgaill, daughter of Mac Lochlainn, ended with her death on pilgrimage to Armagh that year, prompting him to wed Dubchoblach, daughter of Ua Máelruanaid (d. 1168).

Yet Tairrdelbach remained supreme over much of southern Ireland. In 1152, he met Mac Lochlainn again near Ballyshannon and renewed their peace. But peace was the last thing on his mind, for he banished Ua Briain into the north before dividing Munster again. With Mac Murchada, he evened scores with Ua Ruairc, briefly giving his kingdom to a rival before restoring him. In 1153, he compelled Mac Murchada to return Ua Ruairc's wife before marching against Ua Briain. But his banishment of Ua Briain to the north brought Mac Lochlainn south. On the approach of the northern army into Mide, Tairrdelbach ordered the retreat to Connacht, but Mac Lochlainn mauled the rearguard under Ruaidrí. Tairrdelbach's reluctance to challenge Mac Lochlainn may have been due to poor health, as the annals record a serious illness late that year. In 1154, he recovered enough to resume sparring with his northern rival, joining his fleet to plunder Tír Conaill and Inishowen, enjoying a major naval victory over Mac Lochlainn's hired Hebridean fleets. On land, Mac Lochlainn was stronger, invading east Connacht that year. And to Tairrdelbach's chagrin, he divided Mide in 1155 despite Ruaidrí's resistance. During early 1156 Tairrdelbach obtained some redress, undermining Mac Lochlainn's support in southern Ireland. Then Ua Briain submitted and Ua Ruairc agreed to a peace until May. The old man did not get the chance to break the peace, as this great king died aged 68 at his capital of Dunmore and was buried beside the altar in the church of Clonmacnoise. During Tairrdelbach's life, he married seven times, fathering a recorded three daughters and ten sons. He was survived by his seventh wife Dubchoblach and was succeeded as king of Connacht by his son Ruaidrí.

EMMET O'BYRNE

References and Further Reading

Hennessy, W. and B. Mac Carthy, ed. *The Annals of Ulster.* 4 vols. Dublin, 1887–1901.

———. *The Annals of Loch Cé.* 2 vols. London, 1871.

O'Donovan, John, ed. *The Annals of the Four Masters.* 7 vols. Dublin, 1851.

Ó hInnse, Séamus, ed. *Miscellaneous Irish Annals.* Dublin Irish Manuscript Commission, 1947.

Mac Airt, Seán. *The Annals of Inisfallen.* Dublin: Dublin Institute for Advanced Studies, 1951.

Murphy, Denis. *The Annals of Clonmacnoise.* Dublin: Royal Irish Academy, 1896.

Stokes, Whitley, ed. "The Annals of Tigernach." *Rev. Celt* 16–18 (1895–1897).

See also **Mac Lochlainn, Muirchertach; Mac Murchada, Diarmait; Ua Briain, Muirchertach; Ua Briain, Tairrdelbach**

UA CONCHOBHAIR-FÁILGE

This Irish lordship comprised eastern Co. Offaly and northern Co. Laois. At the time of the Anglo-Norman invasion, the Ua Conchobhair-Fáilge quickly came to an agreement with the invaders. Little is known of this initial arrangement, but it probably reflected the pattern of loose overlordship which had governed relations between Irish kings and their subkings prior to the Anglo-Norman invasion. The Ua Conchobhair-Fáilge retained much of their lordship after the invasion because of the wooded and boggy character of the region, however the land lost to the Anglo-Irish consisted of the best agricultural land. The 1270s saw a general increase in hostility between the surviving Irish lordships in Leinster and the Anglo-Irish of that province, and the initial agreement between the Ua Conchobhair-Fáilge and the Anglo-Irish seems to have collapsed around that time. It has been suggested that this change in relations was due to Anglo-Irish efforts to transform their loose overlordship into more formal tenurial lordship during the thirteenth century, but further factors, such as the absence of the Archbishop of Dublin (a major landholder in Leinster) and the minority of the lord of Offaly (an important local magnate), probably explain the timing of this increased hostility.

By the end of the thirteenth century, the royal government in Ireland had given two local magnates—John fitz Thomas, lord of Offaly, and Piers Bermingham—responsibility for pacifying the region, but with little success. A key event during this period was Piers Bermingham's murder of the ruler of the Ua Conchobhair-Fáilge and some of his men at a feast in 1305. These murders were the subject of considerable comment and condemnation in contemporary Irish sources, and special note was made of them in the *Remonstrance of the Irish Princes* (a condemnation of English rule in Ireland sent to Pope John XXII around 1317, during Edward Bruce's invasion of Ireland). Bermingham's aim may have been to render the Ua Conchobhair-Fáilge leaderless, making them easier to bring to peace, but the murders led to increased warfare in the region. Throughout the fourteenth century, the relationship between the Anglo-Irish and the Ua Conchobhair-Fáilge continued to be unstable. Sporadic warfare and endemic raiding continued throughout the century.

During the late fourteenth and early fifteenth centuries a succession of strong Ua Conchobhair-Fáilge leaders were able to gain the advantage against the Anglo-Irish of Meath and Kildare through a series of successful raids and campaigns, beginning a period of minor expansion. Their submission to Richard II in 1395 and the short peace that followed proved to be only a brief interruption of this growth. During the first half of the fifteenth century, the Ua Conchobhair-Fáilge made territorial gains in the region and exacted black rent (protection money) from the Anglo-Irish of western Meath (modern Co. Westmeath). However, by the 1470s the lordship of the Ua Conchobhair-Fáilge was in decline and came increasingly under the lordship and control of the earls of Kildare. The English plantation of Laois and Offaly in the mid-sixteenth century saw the final collapse of the Ua Conchobhair-Fáilge lordship.

KEITH A. WATERS

References and Further Reading

Ó Cléirigh, Cormac. "The Problem of Defence: A Regional Case-Study." In *Law and Disorder in Thirteenth-Century Ireland*, edited by James Lydon, 25-56. Dublin: Four Courts Press, 1997.

Ó Cléirigh, Cormac. "The O'Connor Faly Lordship of Offaly, 1395–1513." *Proceedings of the Royal Irish Academy* (C) 96 (1996): 87–102.

Otway-Ruthven, A. J. *A History of Medieval Ireland*. New York: Barnes & Noble, 1980.

O'Byrne, Emmett. *War, Politics and the Irish of Leinster, 1156–1660*. Dublin: Four Courts Press, 2003.

See also **Anglo-Norman Invasion; Bermingham; Bruce, Edward; Richard II**

UA DÁLAIGH

The Ua Dálaigh family of bardic poets traced their ancestry to the legendary Dálach, a pupil of the famous Colmán mac Lénéni, founder of the church of Cloyne, County Cork, who died in 604. His descendants served as bardic poets for aristocratic courts and monasteries throughout Ireland and Scotland during the Middle Ages and as late as the eighteenth century (see Doan 1985b). These include the infamous Muiredach Albanach ("the Scotsman"), who killed a taxman and fled to various Irish, Anglo-Norman, and Scottish lords, becoming the ancestor of the MacMhuirich family of Scottish Gaelic poets (see Ó Cuív and Thomson). Another Ua Dálaigh poet, famous for his religious verse, was Donnchad Mór, alleged to have become abbot of Boyle in County Roscommon in later life (see McKenna 1922 and 1938 for editions of his poetry).

Among the most famous of the name were various thirteenth-century individuals called Cerball Ua Dálaigh, including Cerball Buide of Connacht (d. 1245), his brother Cerball Fionn, and his nephew Cerball Bréifnech of Bréifne (modern Co. Cavan). However, four other poets of the name contributed to the development of Cerball's composite persona in Irish literary and folk tradition: Cerball, *ollamh* ("chief poet") of Corcomroe abbey (d. 1404); two County Wexford men called Cerball (probably father and son) who flourished between the 1590s and 1640s and whose names are found in Elizabethan *fiants* ("legal pardons") dating from 1597 and 1601; and possibly a late seventeenth-century Cerball who worked in Ulster circa 1680–1690.

This literary tradition begins with *Bás Cerbaill agus Ferbhlaide* ("The Death of Cerball and Ferbhlaid"), a late medieval romance concerning the tragic love and death of Cerball "son of Donnchad Mór" Ua Dálaigh, presumably the *ollamh* of Corcomroe, and Ferbhlaid, daughter of King Séamas "son of Turcall" of Scotland, based on a fifteenth-century Scottish King James. The tale exists in some twenty manuscripts, dating from 1600 to 1800, as well as in a later adaptation, *Eachtra Abhlaighe . . . agus Chearbhaill . . . .* ("The Adventure of Abhlach . . . and Cearbhall . . . "), probably composed in the mid- or late-seventeenth century. Both versions have been edited and translated (see Doan 1985a and 1990, and Ní Laoire 1986). Two of the poems attributed to Cerball in the original version of the romance are written in *dán díreach* ("strict meter"), as one would expect from a professional poet, or *file*, during this period. However, three remaining poems attributed to Cerball are in *ógláchas* (a looser metrical form). The poetry ascribed to Ferbhlaid is also in *ógláchas*, appropriate for a medieval aristocratic Gaelic-speaking woman who, while educated, would not be expected to compose poetry in as strict a form as a professional male poet.

At least five poems attributed to the Wexford Cerballs survive in Irish manuscripts, as well as the popular Irish *amhrán* or folksong, "Eibhlín (or Eilíonóir), a rúin" ("Eileen [or Eleanor], my love"), which purports to be an exchange between one of these poets, probably the younger, with Eleanor, daughter of Sir Murchadh Caomhánach (Morgan Kavanagh, d. 1643), inviting her to elope with him, which she accepts. This song and the tale accompanying it are among the best known in the Irish tradition, though the extant melody and text probably date from the late seventeenth, or early eighteenth, century (see Doan 1985c).

Like the poems found in *Bás Cerbaill agus Ferbhlaide*, the poems ascribed to the Wexford Cerballs show considerable skill, although these are composed in *ógláchas* rather than in *dán díreach* meters. Many of these poems fall within the *dánta grádha* ("love poetry") tradition, with the poet often suffering lovesickness, as in "Ní truagh galar acht grádh falaigh" ("There is no disease so pitiful as hidden love"). Another deals with the pleasures of the scholarly life ("Aoibhinn beatha an scoláire" – "Delightful is the life of a scholar"), including backgammon, harping, and making love to a beautiful woman. Probably the most famous is "The Quarrel of Echo and Cearbhall Ua Dálaigh" ("A mhac-alla dheas" – "Oh, fair echo"), a debate over the poet's love for a woman named Cáit, which he compares to Echo's love for Narcissus, finally making Echo agree that Cáit surpasses Narcissus in beauty. This is also the most technically proficient of the syllabic poems attributed to Cerball (see Doan 1990, 147–172, for editions and translations of these poems).

Already in the poetry of Pádraigín Haicéad (*fl. c.* 1620–1630), we find references to a "Cerball Ua Dálaigh" who has earned a reputation as a figure renowned in poetry and wisdom; famed for speech, music, and feats; cognizant of spells; and highly attractive to women (see Ní Cheallacháin, 6–9). Pádraigín's perception anticipates the modern view of Cerball as lover, craftsman, trickster, and archetypal poet, which continues to this day in Irish folk tradition (see Doan 1981, 1982, 1983).

James E. Doan

References and Further Reading

Doan, James E. "Cearbhall Ó Dálaigh as Archetypal Poet in Irish Folk Tradition." *Proceedings of the Harvard Celtic Colloquium* 1 (1981): 77–83.

———. "Cearbhall Ó Dálaigh as Craftsman and Trickster." *Béaloideas* 50 (1982): 54–89.

———. "Cearbhall Ó Dálaigh as Lover and Tragic Hero." *Béaloideas* 51 (1983): 11–30.

———, ed. and trans. *The Romance of Cearbhall and Fearbhlaidh.* Mountrath: The Dolmen Press, 1985a.

———. "The Ó Dálaigh Family of Bardic Poets, 1139–1691." *Éire-Ireland* 20 (1985b): 19–31.

———. "The Folksong Tradition of Cearbhall Ó Dálaigh." *Folklore* 96 (1985c): 67–86.

———, ed. and trans. *Cearbhall Ó Dálaigh: An Irish Poet in Romance and Oral Tradition.* New York: Garland Press, 1990.

Harrison, Alan. *An Chrosántacht.* Dublin: An Clóchomhar Teoranta, 1979.

McKenna, Lambert, ed. *Dán Dé: Poems of Donnchadh Mór Ó Dálaigh.* Dublin, 1922.

———, ed. *Dioghluim Dána.* Dublin, 1938.

Ní Cheallacháin, Máire. *Filíocht Phádraigín Haicéad.* Dublin: An Clóchomhar Teoranta, 1962.

Ní Laoire, Siobhán, ed. *Bás Cearbhaill agus Farbhlaidhe.* Dublin: An Clóchomhar Teoranta, 1986.

Ó Cuív, Brian. "Eachtra Mhuireadhaigh Í Dhálaigh." *Studia Hibernica* 1 (1961): 56–69.

Thomson, Derick S. "The MacMhuirich Bardic Family." *Transactions of the Gaelic Society of Inverness* 12 (1966): 281.

Watson, Seosamh, ed. *Mac na Míchomhairle.* Dublin: An Clóchomhar Teoranta, 1979.

See also **Bardic Schools, Learned Families; Education; Entertainment; Games; Music; Poetry, Irish; Poets, Men of Learning; Romance**

UA DOMNAILL (O'DONNELL)

The Ua Domnaill (O'Donnell) dynasty, were a leading family of the northern Uí Néill, and became the rulers of the lordship of Tír Conaill in the late medieval period. They came to prominence about the year 1200 C.E., when the first Ua Domnaill ruler, Éigneachán (*c.* 1201–1207) came to power. Previous to this, the O'Donnells were local kings of Cenél Lugdach in northern Tír Conaill, with a crannog at Lough Gartan and an inauguration site at Kilmacrennan. The Ua Domnaills dispossessed the previous ruling dynasties of Ua Máel Doraid, who may have died out around 1197, and Ua Canannáin, the last lord of whom, Ruaidrí, was deposed and slain in 1248. The Ua Domnaill lords proved themselves to be an innovative and talented, but very violent, family. In the mid-thirteenth century, they became the first Irish dynasty to employ galloglass mercenaries, in their case the Mac Suibne (Mac Sweeney) family, who became deeply established in three separate branches in Tír Conaill. The Ua Domnaill chieftains, Máel Sechlainn (1241–1247), Gofraid (1248–1258), and Domnall Óc (1258–1281), were prominent fighters against English colonialism in the northwest. Gofraid in particular defeated the chief governor Maurice Fitzgerald, in "a brave battle . . . in defense of his country," at Credrán in Cairbre (Carbury) in 1257, a battle that succeeded in keeping the English out of Tír Conaill. The O'Donnaills also violently resisted all attempts by the Ua Néill lords of Tír nEógain to establish provincial hegemony in Ulster. However, they also fought fierce internal civil wars throughout the fourteenth century until the powerful Ua Domnaill chieftain, Tairrdelbach an Fhíona (1380–1422) established himself as lord of Tír Conaill.

The ruling Ua Domnaill dynasty enjoyed crucial support from the Ua Gallchobhair (Gallagher) family, who commanded their household troops, and Ua Baoigill of Boylagh and Ua Dochartaig, lord of Inishowen, who were Ua Domnaill's two most important subchieftains. The ruling Ua Domnaills were also great patrons of the Gaelic learned classes, endowing their chief practitioners, such as Ua Cléirigh, *ollamh* in history, and Mac an Bháird, *ollamh* in poetry, with much land. At the same time, the Ua Domnaills amassed great wealth through the exploitation of salmon and herring fisheries, becoming known as "king of fish" on the continent. The Ua Domnaill lords had a particularly good relationship with merchants from the city of Bristol in England, who traded wine, firearms, and luxury goods for the fish, tallow, and hides exported from Tír Conaill.

Tairrdelbach an Fhíona's son Niall Garbh, lord of Tír Conaill from 1422 to 1439, was an innovative ruler. He joined with Eógan Ua Néill, the lord of Tír nEógain, to raid the English Pale, but was captured in 1434 and imprisoned, first in London, and then in the Isle of Man. A long civil war followed his imprisonment, fought between Niall Garbh's sons and his brother Neachtan, lord of Tír Conaill from 1439–1452. This war only ended in 1497 with the assassination of Neachtan's brother, Éigneachán Mór Ua Domnaill.

From 1461 to 1555 Tír Conaill was ruled by a series of three very successful Ua Domnaill warlords, Áed Ruad (1461–1505), Áed Dub (1505–1537) and Maghnus (1537–1555), who were father, son, and grandson. Shrewd and religious, these three rulers expanded Ua Domnaill power into the neighboring lordships of Fermanagh and Lower Connacht, and they called themselves "Prince of Ulster," in direct opposition to the claims of Ua Néill of Tír nEógain. Áed Ruad, aided by Máel Muire Mac Suibne, seized power in Tír Conaill in 1461. In 1481, he inflicted a severe defeat on Mac William Burke in Tirawley. A deeply religious man, Áed Ruad introduced the Franciscan Observant order into Tír Conaill, establishing a monastery at Donegal in 1474. Áed Dub Ua Domnaill was in an unusually secure position in Tír Conaill, so much so that he went on a two-year pilgrimage to Rome from 1510 to 1512, during which he stopped off for thirty-two weeks at the court of Henry VIII, who knighted him. The highlight of his career was his defeat of Conn Bacach Ua Néill at the battle of Knockavoe, fought near Strabane in 1522. Áed Dub's son, Maghnus Ua Domnaill, was a canny ruler. Involved in the Geraldine League, he also made attempts to be made earl of Sligo. Maghnus was a noted Gaelic scholar, composing poetry and commissioning a biography of Colm Cille, the *Betha Cholaim Cille*, in 1532. All three pioneered the hiring of "redshank" mercenaries (Scottish Highland soldiers). They also utilized firearms, guns being mentioned in Tír Conaill from 1487. These Ua Domnaill chieftains also had close links with the Stuart kings of Scotland. In 1495, Áed Ruad visited James IV when he "went to the house of the king of Scotland." In 1513, Áed Dub also visited James IV, where Ua Domnaill received a suit of clothes, £40 in plate, £160 in cash, in addition to the promise of a cannon and a culverin. Artillery arrived in Tír Conaill in 1516 when a French knight brought over an artillery piece sent by the earl of Albany. From 1534 to 1537, Maghnus Ua Domnaill was in contact with king James V.

Following the deposition of Maghnus in 1555, his two sons, Calbach (1555–1566) and Áed (1566–1592) were weak rulers, in whose time Tír Conaill descended into anarchy as Shane O'Neill terrorized the lordship. However, the famous Red Hugh Ua Domnaill, lord of Tír Conaill from 1592 to 1602 reestablished Ua Domnaill power when he joined in the great Gaelic confederacy, which fought the Nine Years' War. Red Hugh participated in the major Irish victory at the Yellow Ford in 1598 and won a spectacular success in his own right in the Curlew mountains in 1599. However, following defeat at the battle of Kinsale (1602), a loss for which Ua Domnaill was blamed, Red Hugh left for Spain where he subsequently died. His brother, Rury, was created earl of Tír Conaill by King James I, but fled Ireland in the Flight of the Earls in 1607. Rury died in Rome in 1608, thus ending Ua Domnaill power.

Darren McGettigan

References and Further Reading

Bradshaw, Brendan. "Manus 'The Magnificent': O'Donnell as Renaissance Prince." In *Studies in Irish History*, edited by Art Cosgrove and Donal McCartney, pp. 15–36. Naas, 1979.

Mac Carthy, B. ed. and trans. *Annals of Ulster, 1379–1541*. Vol. 3. Dublin, 1895.

McGettigan, Darren. "The Renaissance and the Great Ua Domnaill Mór, Aodh Dubh Ua Domnaill, Prince of Tír Conaill, 1505–37." M.A. diss. University College Dublin, 1995.

———. "The Renaissance and the Late Medieval Lordship of Tír Conaill, 1461–1555." In *Donegal History and Society*, edited by William Nolan, Liam Ronayne, and Mairead Dunlevy, pp. 203–228. Dublin, 1995.

O'Donovan, John, ed. and trans. *Annala Rioghachta Eireann, Annals of the Kingdom of Ireland*. Dublin, 1856.

Ua Riain, Pádraig, ed. *Beatha Aodha Ruaidh: The Life of Red Hugh O'Donnell Historical and Literary Contexts*. Irish Texts Society, 2002.

Simms, Katharine. "Late Medieval Donegal." In *Donegal History and Society*, edited by William Nolan, Liam Ronayne, and Mairead Dunlevy, pp. 183–201. Dublin, 1995.

———. "Niall Garbh II O'Donnell, King of Tír Conaill, 1422–39." *Donegal Annual* 12 (1977): 7–21.

See also **Burke; Colm Cille; Fitzgerald; Inauguration Sites; Mac Sweeney; Pilgrimage; Ua Néill; Uí Néill, Northern; Ulster, Earldom of**

UA DUBHAGÁIN, SEÁN MÓR

See **Dinnshenchas**

UA NÉILL (Ó NÉILL)

Origin of the Surname

The Ua Néill family were the first Irish dynasty to develop a surname, literally "grandson of Niall." This derived from Niall Glúndub, or "Black-knee," king of the northern Uí Néill territory of Cenél nEógain, then comprising the area covered by the modern counties of Derry, Tyrone, and north Armagh. In 916, Niall Glúndub succeeded Flann Sinna, king of the southern Uí Néill, as high king of Tara by agreement of both northern, and southern Uí Néill, ending a long period

Silver signet ring with the arms of O'Neill attached to a silver chain. © *The Trustees of the National Museums of Scotland.*

of interdynastic rivalry. He led their united forces against renewed Viking incursions along the southern and eastern coasts of Ireland, but was slain in battle by the Dublin Norse in 919.

Niall was succeeded as high king by Flann Sinna's son, Donnchad, and as king of Cenél nEógain by his own son, Muirchertach na Cochall Craicinn, "of the Leather Cloaks." Muirchertach continued to battle against the Norse invaders. Although initially opposing the rule of the high king Donnchad, he eventually made common cause with him His nickname "of the Leather Cloaks" was traditionally said to be earned by a winter campaign in 941, when he brought his leather-clad soldiers on a circuit of southern Ireland, capturing Cellachán, "king of Cashel" (over king of Munster), and forcing him to submit to Donnchad. Before he could succeed Donnchad as high king, Muirchertach was killed by a Norse army near Clonkeen, County Louth, in 943.

The surname Ua Néill first emerged with Muirchertach's son, Domnall of Armagh, otherwise "Domnall Ua Néill." Domnall succeeded his father as king of Cenél nEógain, but only won recognition as high king of Ireland in 956, after an interregnum, 944–956, when two long-excluded branches of the northern and southern Uí Néill, the Cenél Conaill and the Síl nÁedo Sláine respectively, fought unsuccessfully for supremacy. Domnall (d. 980) was the last of his line to hold high kingship. He was succeeded in Cenél nEógain by two sons. The first, Áed Craeibe Telcha, or "of Crewe Hill, County Antrim," was named after the battle in which he was killed in 1004 while attempting to assert lordship over the Ulaid. His brother, Flaithbertach an Trostáin Ua Néill (1004–1036), "of the pilgrim's staff," was so called because he transferred his kingship to his son Áed in 1030 and went on pilgrimage to Rome. Áed died in 1033, and although the aged Flaithbertach resumed kingship for a further three years, his death in 1036 was followed by a succession struggle among different branches of the Cenél nEógain dynasty, leading to the rise of a collateral kindred, the Mac Lochlainn kings of Cenél nEógain, two of whom successively claimed to be high kings of Ireland "with opposition."

The Medieval Lords of Tír nEógain (Tyrone)

The Ua Néill family went through a period of obscurity until the fall of the high king Muirchertach Mac Lochlainn in 1166. The next high king, Ruaidrí Ua Conchobair of Connacht, divided Cenél nEógain in two in 1167, giving the northern half to Niall, son of Muirchertach Mac Lochlainn, and the southern half to Áed Ua Néill, an Macaem Tóinlesc, "the lazy-rumped lad," traditionally so called because as a boy he had failed to stand up respectfully when Muirchertach Mac Lochlainn entered the house where he was staying. Áed Ua Néill was opposed and eventually killed by the sons of Muirchertach Mac Lochlainn near Armagh in 1177. His brief reign had nevertheless restored the claims of the Ua Néill line, and in 1199 his son Áed Méith, "Áed of Omeath" (County Louth), began a long and militarily successful career as ruler of the Cenél nEógain.

The Anglo-Norman invasion of Ireland began in 1169, leading to the conquest and settlement of Ulaid by John de Courcy in 1177. Áed Méith's first recorded exploit was an attack on the Anglo-Norman port of Larne, County Antrim, which forced de Courcy to retreat from his own invasion of Tír nEógain, the land ruled by the Cenél nEógain. Thereafter, Áed headed an alliance of Tír nEógain with Tír Conaill (most of modern County Donegal) under Ua Domnaill, and Fir Manach (County Fermanagh) under Ua hÉignig. He came to terms with John de Courcy, lord of Ulster, and afterward with Hugh de Lacy, who replaced de Courcy and was created first earl of Ulster in 1205. When de Lacy rebelled and his castle of Carrickfergus was besieged by King John in 1210, Áed refused to yield hostages to the English king, and succeeded in destroying the Anglo-Norman castles with which John's chief governor ringed Ulster subsequently between 1211 and 1214. However the pipe roll of John for 1211 and 1212 shows Ua Néill paid a fine of at least 293 cows to obtain pardon for rebellion, and a further 321 cows or more as rent for the kingship of Tír nEógain. When the exiled Earl Hugh returned from 1222 to 1224 to win back his earldom by force, Áed Ua Néill supported him against the justiciar's army, enabling de Lacy to negotiate the restoration of his title. Ua Néill was less successful in 1225 when he invaded Connacht in support of one side in a succession dispute among the Ua Conchobair kings. At his death in 1230, he was described in the annals as "king of Conchobar's Province," that is, of the whole of Ulster and not just Tír nEógain, "a prince eligible *de jure* for the kingship of Ireland."

Áed's son and heir Domnall was killed within a few years by Domnall Mac Lochlainn, last of his line to hold the kingship of Tír nEógain. In 1241, Áed's nephew Brian Ua Néill allied with Máel Sechlainn Ua Domnaill, king of Tír Conaill, to defeat and kill Domnall Mac Lochlainn and ten of his closest kinsmen. After Hugh de Lacy's death in 1243, the earldom of Ulster was taken into the hands of royal administrators, and Brian Ua Néill began raiding to reconquer eastern Ulster from the Anglo-Normans. At a meeting in 1258 at Cáel Uisce near Belleek, County Fermanagh, attended by Áed Ua Conchobair, son of the king of Connacht, and Tadc Ua Briain, son of the king of Thomond (north Munster), Brian was acknowledged "king of the Irish of Ireland." In 1260, Áed Ua Conchobair brought a force to join Ua Néill in an allied attack on the Ulster colonists, but they were defeated outside Downpatrick. Ua Néill was killed, his head cut off and sent to King Henry III in England.

The next three kings of Tír nEógain were descendants of Áed Méith, who had come to an arrangement with the new earls of Ulster, Walter de Burgh or Burke (d. 1271), created earl in 1263, and his son Richard de Burgh, "the Red Earl" (d. 1326). Áed Buide ("the Yellow-haired"), son of Domnall son of Áed Méith, married Earl Walter's kinswoman, Eleanor de Nangle, in 1263, and allied with the Anglo-Normans to defeat and kill Domnall Óc Ua Domnaill, king of Tír Conaill, who invaded Tír nEógain in 1281. After Áed Buide's death in 1283, Brian Ua Néill's son Domnall seized the kingship, but was deposed by the earl in 1286, in favor of Áed Buide's brother Niall Cúlánach ("of the long back hair"). Domnall persisted, killing Niall Cúlánach in 1291 and killing the earl's next appointee, Brian son of Áed Buide, in 1296, after which the earl left Domnall in the kingship, perhaps because Henry son of Brian son of Áed Buide was still too young to be king. A grant of land by the earl to Henry Ua Néill in 1312 may signal the rising power of his potential rival that induced Domnall Ua Néill to associate himself with King Robert the Bruce of Scotland and his brother Edward just after their victory against the English at Bannockburn in 1314. From the first landing of Edward Bruce with an invading army of Scots at Larne, County Antrim, in 1315, to his eventual defeat and death in 1318, Domnall Ua Néill was his closest ally, and the ravaging of eastern Ulster by the Scottish army during those three years significantly undermined the wealth and power of Earl Richard de Burgh.

De Burgh expelled Domnall in 1319 in favor of the descendants of Áed Buide, led by Henry Ua Néill, but Domnall had recovered power at least partially before his death in 1325. When the next de Burgh earl of Ulster, William "the Brown Earl," was assassinated by his own Anglo-Norman vassals in 1333, inquisitions record that Tír nEógain was shared between Henry Ua Néill and Domnall's son Áed Remar ("the Fat"), who were jointly responsible for paying rent for the kingship of Tír nEógain and supporting a quota of the earl's mercenary soldiers billeted on their territory. In practice, we are told, Henry supported his share of the soldiers and paid the whole of the rent hoping to be acknowledged as sole lord of Tír nEógain.

However, Henry had joined the Anglo-Norman rebellion against Earl William, and in 1344 the justiciar, Ralph d'Ufford, deposed him in favor of Áed Remar, Domnall's son, who adopted the title "King of the Irish of Ulster." By a peace treaty in 1338, Henry and his descendants were granted a stretch of war-ravaged land in south county Antrim, where they established a separate lordship as the Clann Áeda Buide, "the descendants of Áed Buide," later known as the O'Neills of Clandeboye.

The earldom of Ulster passed through Earl William's daughter, Elizabeth, to her husband, Prince Lionel of Clarence, and then to his son-in-law, Edmund Mortimer, all absentees. The resulting power vacuum in the north was filled by the rise of Áed Remar (1325–1364), his son Niall Mór ("the Elder," 1364–1397), and his grandson Niall Óc ("the Younger," 1397–1403). They not only won the submission of the other chiefs in their province, but took over the Ulster earls' custom of billeting a quota of mercenary soldiers, in their case the Mac Domnaill galloglass (heavy-armored foot soldiers imported from the Western Isles of Scotland), on each of the territories subject to them, a custom known as the "bonaght of Ulster" (from *buannacht*, "military billeting"). Their attempts to overrun remaining English settlements on the coast of County Down were, however, unsuccessful.

Alliance with the Earls of Kildare

The fifteenth century began with a civil war between Niall Óc's son Eogan and his nephew Domnall (reigned 1404–1432), the first lord to be called by the English "the Great O'Neill," to distinguish the ruler of Tír nEógain from the Ua Néill Buide, or lord of Clandeboye. This war allowed Ua Domnaill of Tír Conaill and Ua Néill Buide to build up significant overlordships of their own on either side of Tír nEógain. To counter the threat they posed, Eógan Ua Néill (reigned 1432–1455) and his son Henry (1455–1489) increasingly used alliances with the Anglo-Irish earls of Ormond and Kildare, who sent troops from the Pale area to help quell the rebellions of Ua Domnaill and of junior members of the Ua Néill dynasty inside Tír nEógain itself. Thus reinforced, Eógan and his son Henry managed intermittently to dominate an area equivalent to the nine counties of modern Ulster, including their newly acquired overlordship of Ua Raigillig's territory of East Bréifne (County Cavan). This close association with the earls gradually developed into dependency. Henry's son Conn Mór (1483–1493) and the latter's son Conn Bacach ("the Lame," 1519–1559) respectively married Eleanor the sister, and Alice the daughter, of Gerald Mór Fitzgerald, eighth earl of Kildare (d. 1513). Ua Néill of Tír nEógain lent important political support to the house of Kildare both before and after the rebellion of Silken Thomas, the tenth earl, in 1534. It was to win Ua Néill back to the government's side that Conn Bacach was created first earl of Tyrone in 1542.

KATHARINE SIMMS

References and Further Reading

Davies, Oliver and David B. Quinn, ed. "The Irish Pipe Roll of 14 John, 1211—1212." *Ulster Journal of Archaeology*, ser. 3, vol. 4. (July 1941): supplement.

Freeman, A. Martin, ed. *Annala Connacht: The Annals of Connacht (A.D. 1224–1544)*. Dublin: The Dublin Institute for Advanced Studies, 1944.

Hayes-McCoy, Gerard A. "The Making of an O'Neill." *Ulster Journal of Archaeology*, ser. 3, vol. 33 (1970): 89–92.

Hogan, James. "The Irish Law of Kingship, with Special Reference to Ailech and Cenél nEógain." *Proceedings of the Royal Irish Academy* 40 sec. C (1932): 186–254.

Ua Ceallaigh, Séamus, *Gleanings from Ulster History*. Cork: Cork University Press, 1951. Reprint, Draperstown: Ballinascreen Historical Society, 1994.

Simms, Katharine. "The Archbishops of Armagh and the O'Neills, 1347–1461." *Irish Historical Studies* 19 (1974): 38–55.

Simms, Katharine. "'The King's Friend': O'Neill, the Crown and the Earldom of Ulster." In *England and Ireland in the Later Middle Ages*, edited by James Lydon, pp. 214–236. Dublin: Irish Academic Press, 1981.

Simms, Katharine. "Tír nEógain North of the Mountain." In *Derry & Londonderry: History and Society*, edited by Gerard O'Brien, pp. 149–174. Irish County History Series. Dublin: Geography Publications, 1999.

Simms, Katharine. "Late Medieval Tír nEógain: The Kingdom of 'the Great Ua Néill.'" In *Tyrone: History and Society*, edited by Charles Dillon and Henry A. Jefferies, pp. 127–162. Irish County History Series. Dublin: Geography Publications, 2000.

See also **Armagh; Fitzgerald; Fitzgerald, Gerald 8th Earl; Gaelic Revival; Mortimer; Ua Domnaill; Ua Néill, Domnall; Clandeboye; Uí Néill; Uí Néill, Northern; Viking Incursions**

UA NÉILL, DOMNALL (ANTE 1260–1325)

Domnall Ua Néill was the son of Brian Chatha an Dúna (d. 1260), son of Niall Ruad (d. 1223), son of Áed "In Macáem Tóinlesc" (d. 1177), and was king of the Cenél nEógain line of the Northern Uí Néill. After his father's death in battle at Downpatrick trying to overthrow the earldom of Ulster, the kingship was wrested by Domnall's second cousin, Áed Buide (d. 1283), progenitor of the Clandeboye O'Neills, who later took an Anglo-Norman wife related to the de Burgh earls, on whose support he could rely. At Áed's death, Domnall seized the kingship but was deposed in 1286 by Earl Richard de Burgh, who instated Áed's brother Niall Cúlánach as king. Domnall ousted him in 1290 with support from his brother-in-law Tairrdelbach Ua Domnaill of Cenél Conaill, and possibly the latter's Clann Domnaill galloglass relatives from Islay (a late source cites Domnall as the first to introduce galloglass to Cenél nEógain). In 1291, de Burgh again deposed Domnall in favor of Niall Cúlánach, and when he killed Niall in 1291 Domnall still found himself deposed again, this time by Brian son of Áed Buide, aided by the earl's Mandeville and Bisset vassals. It was only by killing Brian and his Anglo-Norman supporters at Maidm na Craibe in 1295 that Domnall obtained an unchallenged grip on power.

Although he appears on record with other Ulster kings in 1297 agreeing to the archbishop of Armagh's request that he control the excesses of his Irish and Scottish troops (*satellites et Scoticos nostros*), the sources are then silent on Domnall's activities for many years, which suggests some accommodation with de Burgh. Domnall, it has been suggested, built the first castle at Dungannon, but was probably not pleased with de Burgh's grant of Roe Castle and lands (near Limavady) to his new brother-in-law James the Steward of Scotland, nor with his construction of Northburgh Castle in Inishowen in 1305, and certainly not with his continued (or revived) support for the line of Áed Buide, whose grandson Énrí was granted hitherto Ua Catháin lands at Glenconkeen.

Domnall may have responded favorably to Robert Bruce's request for military aid in 1306 and 1307, and he certainly ignored that of Edward II in March 1314 asking him and many other Irish lords to serve against the Scots under (ironically) Richard de Burgh. After Bannockburn, negotiations probably commenced on the proposal to have Edward Bruce installed as king of Ireland. When the latter arrived in Ulster in May 1315, Domnall joined forces with him and was consistently by his side thereafter, although he apparently did not participate in the battle of Faughart in which Bruce was killed in 1318 (perhaps being preoccupied consolidating his succession following the violent death of his son Seán that same year, protecting Derry from Cenél Conaill encroachment).

While Bruce was in Ireland, Domnall produced an extraordinarily emotive letter (and possibly two, if that to Mac Carthaig urging a national alliance in favor of Bruce and in opposition to the English is not, as has been suggested, a forgery). Addressed to the pope circa 1317, and preserved in Scottish manuscripts, it is generally known as the "Remonstrance of the Irish Princes" and is a remarkable statement of Irish discontent under English rule. Domnall, asserting his entitlement to the high kingship of Ireland established by his Uí Néill ancestors, declares that he has invited Edward Bruce to Ireland and renounced his claim in favor of him. He states that Edward is descended from "our noblest ancestors," which may point to a family marriage alliance (it is possible that Bruce's maternal grandfather Niall, earl of Carrick in Galloway, is named after Domnall's grandfather of the same name), and one contemporary thought that Bruce had been "educated" with the man who invited him to Ireland, which may suggest fosterage by one in the other's household.

The backlash that followed the collapse of the Bruce regime at Faughart saw Domnall temporarily expelled in 1319 by the forces of de Burgh and Clann Áeda Buide, and the slaying of his son and *tánaiste*, Brian. Domnall died in 1325 at Loch Lóegaire on the Cenél Conaill frontier, having failed to counter the threat from the line of Áed Buide, whose grandson Énrí succeeded as king. But the de Burghs were the real losers, their earldom passing after 1333 from the family to absentees. Domnall had used the title "king of Ulster" in his 1317 Remonstrance, and heirs lived up to it, his son Áed Remar emerging unopposed from 1345 to found the great O'Neill line, never again challenged for supremacy by their Clandeboy kinsmen.

SEÁN DUFFY

References and Further Reading

Phillips, J. R. S. "The Irish Remonstrance of 1317: An International Perspective." *Irish Historical Studies* 27 (1990): 62–85.

Simms, Katharine. "Tír nEógain North of the Mountain." In *Derry & Londonderry: History and Society*, edited by Gerard O'Brien, pp. 149–174. Irish County History Series. Dublin: Geography Publications, 1999.

Simms, Katharine. "Late Medieval Tír nEógain: The Kingdom of 'the Great Ua Néill.'" In *Tyrone: History and Society*, edited by Charles Dillon and Henry A. Jefferies, pp. 127–162. Irish County History Series. Dublin: Geography Publications, 2000.

See also **Ua Néill of Clandeboye; Uí Néill; Uí Néill, Northern; Ulster, Earldom of**

UA NÉILL OF CLANDEBOYE

The extensive territory of *Clann Áeda Buide*—Clandeboye—constituted what is now south County Antrim, north and east County Down, and southeast County Derry. It had formed part of the Anglo-Norman earldom of Ulster before the demise of the de Burgh earls in the early fourteenth century. By about 1350, the area had been seized and settled by a branch of the Uí Néill (O'Neills) descended from Áed Buide (Hugh "the yellow-haired"). These were breakaway members of the lineage with aspirations to the kingship, forced to the margins of Tí nEógain (Tyrone) in the decades after Aéd's death in 1283. Within a century the Clandeboye O'Neills had established themselves as one of the most successful *uirríthe*, or under kings, to emerge in later medieval Ireland. Theoretically vassals of Ua Néill of Tyrone, in reality they were largely autonomous, acknowledging Ua Néill's claims to overlordship and paying him tribute only by compulsion. By 1450, their power encompassed most of Antrim and Down, and their chieftain, Ua Néill Buide, was reputed a man of great wealth. According to a later English estimate, probably derived from local native sources, Clandeboye was cattle country, its extensive grazing lands capable of feeding many thousands of cows.

Despite the frequent enmity between them, the Clandeboye O'Neills owed their successful settlement of East Ulster to the actions of the O'Neills of Tyrone—particularly to Niall Mór Ua Néill (d. 1398). Initially, by driving out many Anglo-Norman settlers, Niall created the vacuum that the Clandeboye sept was able to exploit. Subsequently, by also waging war on the Scottish MacDonnells, he provided the lineage with ready-made allies willing to help them sustain their struggle against him and his successors. The O'Donnells also supported them, as did, occasionally, the English colonial government in Dublin. Thus, when Eóghan, the Great O'Neill, invaded the territory in 1444, the Clandeboye forces were strong enough to defeat him. A similar attempt by Eóghan's son, Henry (Énrí) Ua Néill, suffered the same fate in 1476. Efforts to rejuvenate the English colony in 1481 collapsed when Conn Ua Néill of Clandeboye had the government-appointed seneschal of Ulster blinded and castrated.

Clandeboye remained strong until the mid-sixteenth century, when a series of successional disputes weakened it internally, and externally it was menaced first by the territorial ambitions of their erstwhile allies, the MacDonnells, and later by the dramatic reemergence of English military power in East Ulster. After 1584, the English government split the lordship between rival claimants, dividing it into North and South Clandeboye, a development that hastened the family's decline. Their autonomous status disappeared after the Nine Years'

War, and early in the seventeenth century they lost large parts of their territory.

The valuable manuscript book, the *Leabhar Cloinne Aodh Buidhe* (Royal Irish Academy, MS 24 p. 33), was composed circa 1680 for the then head of the family, Cormac Ua Néill. It contains some uniquely valuable material of late-medieval provenance, most notably the "Ceart Uí Néill," a list of tributes claimed by the O'Neills throughout Ulster, and a *duanaire*, or bardic poem book, containing poems by members of the Ó Gnímh, Mac An Bhaird, Ó Maolchonaire, and Mac Mhuireadhaigh families.

David Edwards

References and Further Reading

Chart, D. A., "The Break-up of the Estate of Conn O'Neill." *PRIA* 48 C (1942–1943).

Frazer, William. "The Clandeboy O'Neills' Stone Inauguration Chair." *JRSAI*, ser. 5, vol. 8 (1898).

McCall, Timothy. "The Gaelic Background to the Settlement of Antrim and Down, 1580–1641." Unpublished MA thesis, Queen's University, Belfast, 1983.

McNeill, T. E. "County Down in the Later Middle Ages." In *Down: History & Society*, edited by L. Proudfoot and W. Nolan. Dublin, 1997.

Morgan, Hiram. *Tyrone's Rebellion: The Outbreak of the Nine Years' War in Tudor Ireland.* Woodbridge, 1993.

Nicholls, Kenneth. *Gaelic and Gaelicised Ireland in the Middle Ages*. 2nd ed. Dublin, 2003.

O Donnchadha, Tadhg, ed. *Leabhar Cloinne Aodh Buidhe*. Dublin: Irish Manuscripts Commission, 1931.

Simms, Katharine. "'The King's Friend': O'Neill, the Crown and the Earldom of Ulster." In *England and Ireland in the Later Middle Ages*, edited by James Lydon, pp. 214–233. Dublin, 1981.

UA RUAIRC (O'ROURKE)

The Ua Ruairc (O'Rourke) family were the rulers of the Gaelic Irish lordship of West Bréifne, or Bréifne O'Rourke (modern Co. Leitrim), throughout the late medieval period. Descended from a leading segment of the Uí Briúin Bréifne dynasty, who had provided a number of kings of Connacht in the late tenth and early twelfth centuries, the O'Rourkes came to national prominence under an exceptional ruler—Tigernán Ua Ruairc, rí Bréifne (sometimes called Tigernán Mór)who ruled from approximately 1128 to 1172. After Tigernán's reign, the O'Rourkes remained lords of Bréifne until the early seventeenth century. Although remaining a prominent Gaelic family, and more or less independent, the family did not produce many notable leaders throughout the late medieval period. However, the dynasty became a Gaelic power again in Connacht in the latter half of the sixteenth century under two very able rulers who played a prominent role in the events of that time.

Fergal (d. 966 or 967), Art Uallach (d. 1046), Aodh (d. 1087), and Domhnall (d. 1102) were all Ua Ruairc kings of Connacht. However, the O'Rourkes lost the kingship of that province to the related but more strategically located Ua Conchobhair dynasty of Uí Briúin Aí. To compensate for the loss of the kingship of Connacht, the O'Rourkes turned to the fertile plains of Mide. The great leader of this encroachment was Tigernán Mór Ó Ruairc. Tigernán was an exceptionally powerful ruler during his time as king of Bréifne. He is mentioned in the annals almost every year from 1124 to 1172. He both supported and opposed the powerful Tairrdelbhach Ua Conchobair, king of Connacht from 1118 to 1156, and pursued the same policy with O'Connor's son, the high king of Ireland, Ruaidrí Ua Conchobair. However, it is Tigernán's actions in Mide that warrant the most attention.

In 1130, Ua Ruairc defeated the men of Mide at the battle of Sliabh Guaire, Co. Cavan. In 1138 he invaded Mide again in the company of king Tairrdelbhach Ua Conchobair. In 1144, Tigernán received a grant of half of east Mide from king Tairrdelbhach. Tigernán was a ruthless ruler. In 1137, he had Domhnall Ua Caindealbhain, lord of Cinel Laeghaire in Mide, executed, and in 1139 he inflicted the same punishment on Fearghal Mac Raghnaill, lord of Muintir Eolais. However, it is the abduction in 1852 of Tigernán's wife, Dearbhforgaill, daughter of Ua Maeleachlainn, king of Mide, by the king of Leinster, Diarmait Mac Murchada, for which Ua Ruairc is most famous. It led to deep and lasting enmity between Ua Ruairc and Mac Murchada, although Tairrdelbhach Ua Conchobair took Dearbhforgaill back for Ua Ruairc in 1153. Mac Murchada added to the insult when he defeated Tigernán at the battle of Cuasan, near Tara in Mide, in 1156. As a result Tigernán insisted that the new high king, Ruaidrí Ua Conchobair, banish Mac Murchada from Ireland entirely, and in 1166 Ua Ruairc invaded Mac Murchada's kingdom of Uí Cheinnselaig and destroyed the Leinster king's castle at Ferns. In 1167, when Mac Murchada returned with his first Anglo-Norman reinforcements, Ua Ruairc took the lead in opposing him. Tigernán defeated his rival and exacted 100 ounces of gold as *eineach* (atonement) for the abduction of his wife in 1152.

As the Anglo-Norman invasion of Ireland gathered pace, Tigernán was a staunch supporter of the high king, Ruaidhrí Ua Conchobair. Ua Ruairc accompanied Ua Conchobair against Strongbow, and to maintain his authority over east Mide he executed their hostages in 1170. Tigernán raided east Mide and Dublin in 1171, and his son Aodh was killed. In 1172, he was lured to a parley at Tlachtgha (the hill of Ward, Co. Meath) and treacherously killed by Hugh de Lacy. Tigernán, who apparently had only one eye, was called in his annalistic obit "a man of great power for a long time." Indeed, the

Ua Ruairc dynasty never enjoyed such strength and prominence after him.

After the death of Tigernán Ua Ruairc, the family retreated into its Bréifne heartland. Throughout the thirteenth and early fourteenth centuries Bréifne was something of a backwater. However, even there the family was challenged by the O'Reilly (Ua Raghailligh) dynasty, who rose to prominence in east Bréifne (Co. Cavan) at that time. From the mid-fourteenth century onward, the lords of Bréifne also had to conduct a major war with the nomadic Clann Murtough O'Connors, who were encroaching on Ua Ruairc territory. In 1343, Ualgharg Ua Ruairc, lord of Bréifne from 1316 to 1346, drove the Clann Murtough out of Bréifne, although they killed him at the battle of Calry in 1346. The Clann Murtough killed Ualgharg's wife, Dearbháil, in 1367. Tighearnán Mór Ua Ruairc, lord of Bréifne from 1376 to 1418, managed to defeat the Clann Murtough in 1391, and this chieftain also waged "great war" with the neighboring O'Reillys. Tadhg Ua Ruairc, lord of Bréifne from 1419 to 1435, was also at war with the O'Reillys. In 1429, when supported by Ua Néill, the O'Reillys defeated Tadhg at the battle of Achadh Chille Moire.

It was only in the late sixteenth century that the O'Rourkes of Bréifne again became a Gaelic power to be reckoned with. Brian Ballagh Ua Ruairc, chieftain from 1536 to 1562, with some interruptions, capitalized on the turmoil among the Ua Domhnaill dynasty to become very powerful in Connacht. His son, Brian na Múrtha Ua Ruairc, lord of Bréifne from 1566 to 1591, was also a successful chieftain. Noted for his proud nature, he sheltered coiners and Spanish Armada survivors in his lordship, for which he was attacked in 1589 and 1590 by the president of Connacht, Sir Richard Bingham. Brian na Múrtha fled to Tír Chonaill and then to Scotland. However, he was arrested at Glasgow by King James VI and extradited to England, where Queen Elizabeth had him executed at Tyburn in 1591. Brian na Múrtha's son, Brian Óg, succeeded him as Ua Ruairc and became an important figure in the Gaelic confederacy that fought the Nine Years' War. Brian Óg participated in the Irish victory in the Curlew mountains in 1599, and marched with Red Hugh O'Donnell to the battle of Kinsale. Following that defeat, it was O'Rourke's advice that every chieftain should return to defend his own lordship that was followed by most Irish leaders. Brian Óg Ua Ruairc died in Galway city in 1604.

DARREN MCGETTIGAN

References and Further Reading

Byrne, Francis John. *Irish Kings and High-Kings.* London, 1973.

Gallogly, Rev. Daniel. "Brian of the Ramparts O'Rourke (1566–1591)." Breifne vol. 11, 5 (1962): 50–79.

Morgan, Hiram. "Extradition and Treason-Trial of a Gaelic Lord: The Case of Brian O'Rourke." *The Irish Jurist* 22 (1987): 285–301.

O'Donovan, John, ed. and transl. *Annala Rioghachta Eireann, Annals of the Kingdom of Ireland.* Dublin, 1856.

Simms, Katharine. "The Norman Invasion and Gaelic Recovery." In *The Oxford Illustrated History of Ireland*, edited by R. F. Foster, 3–103.

See also **Uí Briúin; Connacht; Ó Conchobhair: Mide; Tairrdelbhach Ua Conchobair; Ruairí Ua Conchobair; Leinster; Diarmait Mac Murchada; Anglo-Norman Invasion; Strongbow; Hugh de Lacy; Ó Néill; Ó Domhnaill**

UA TUATHAIL (O'TOOLE), ST. LAWRENCE (d. 1180)

The first Irishman canonized as a saint of the Roman Catholic Church (the second was Máel-Máedóc—St. Malachy), Lorcán Ua Tuathail achieved distinction as prelate, church reformer, and diplomat. He belonged to the Uí Muiredaig dynasty; his father, Muirchertach Ua Tuathail, was principal subking of Diarmait Mac Murchada, over king of Leinster, and his mother a daughter of an Uí Fáeláin dynast, Cerball mac meic Bricc. One of seven siblings, his half-sister Mór later married Mac Murchada. Lorcán was born around 1128, tradition placing his birth at Mullach Roírenn (Mullaghreelion Hillfort, Co. Kildare).

According to his Latin "Life," the young Lorcán was held hostage by Mac Murchada—probably after the Leinster purge of 1141. As relationships with the over king improved, he was placed in fosterage at Glendalough. He received his education there, later joining the religious community. By the early 1150s, differences with Mac Murchada had apparently been settled; the Synod of Kells (1152) confirmed diocesan boundaries for Glendalough, which encompassed the regional kingdom of Uí Muiredaig, and shortly afterward his sister Mór wed Mac Murchada. It seems reasonable, as Flanagan considers, that the over king's influence lay behind the appointment of Lorcán (aged only twenty-five) as abbot of Glendalough in 1153, and as archbishop of Dublin in 1162—when the Hiberno-Norse kingdom came under Mac Murchada's sway.

Whether or not dynastic influences underlay his preferment, Lorcán's selection as papal legate probably acknowledged his efforts for church reform. With diocesan reorganization already well advanced, he pursued behavioral and attitudinal change. Contending that continental religious orders offered a suitable model for community discipline, he introduced the Augustinian Canons to Glendalough in the 1150s and later to Dublin, where they formed the chapter of Holy Trinity (Christ Church) Cathedral. His concern with extending behavioral reform to the laity is evidenced by his leading role, with the political support of high king Ruaidrí Ua Conchobair, at the synods of Athboy (1167) and Clonfert (1179).

Presumably, Lorcán realized that initiatives to enforce clerical celibacy and to end hereditary ecclesiastical succession would conflict with dynastic agendas—including those of Uí Muiredaig. The appointment of his nephew Thomas as abbot of Glendalough may represent a compromise in this regard. The latter, whether or not in priests orders, was apparently a noncelibate cleric, and had a son—and later a grandson—who witnessed early-thirteenth-century charters. Furthermore, the succession of Thomas, sometime in the mid 1160s (the record is unclear), apparently took place against a background of dynastic intervention. The assertion of Lorcán's hagiographer that Thomas was chosen not because of his lineage, but on account of his worthiness, hints at some controversy.

After the Anglo-Norman invasion, Lorcán found himself increasingly drawn into the sphere of politics—torn between obedience to the new political order, loyalty to his dynasty, and commitment to reform. In the summer of 1170, when MacMurchada and Strongbow advanced on Dublin, Lorcán was implored by the leading citizens to negotiate on their behalf. Existing difficulties, posed by his in-law relationship to the over king, were exacerbated when, during the negotiations, an English contingent seized control of the town. In the event, the archbishop apparently salvaged his integrity. To some, Lorcán was clearly an Irish partisan; Giraldus Cambrensis alleges that he incited resistance to Strongbow when the latter claimed sovereignty of Leinster following MacMurchada's death in May 1171. Yet, that autumn, when Dublin was besieged by Ua Conchobair, Lorcán was chosen as negotiator by Strongbow—now married to the archbishop's niece, Aífe. Throughout the crisis of 1171, as his hagiographer emphasizes, Lorcán exerted himself, at great personal risk, ministering to the hard-pressed populace.

If Lorcán's submission to the English King Henry II in December 1171 implied recognition of political realities, he apparently trusted in Henry's support for church reform—which perhaps explains his support for the Synod of Cashel, which the king summoned in 1172. Quite likely, Lorcán nurtured expectations that Henry, as Lord of Ireland, would bring political stability. However, the situation continued to deteriorate. Offensives by Irish regional kings in 1173, and retaliatory attacks the following year, increased anxieties. Lorcán was persuaded by Irish interests to mediate with Henry in an effort to restrain English colonial expansion. As an ambassador of Ua Conchobair, he attended the Council of Windsor in 1175, although he did not lead the negotiations.

Concern for Glendalough properties, threatened by colonial expansionism, perhaps motivated Lorcán to seek confirmation of the abbatial possessions; he is the principal witness to Strongbow's charter. However, the death of Strongbow (May 1176) prompted rapid expansion of the colony, with the effective abandonment of the Windsor agreement. Following the dispossession of Uí Muiredaig (1178) from ancestral territories in County Kildare, Lorcán perhaps collaborated in resettling remnants of his dynasty on Glendalough lands; the archbishop, it appears, conveyed several holdings to his nephew Thomas around this time. Certainly, the closing years of his life saw relationships between Lorcán and Henry II deteriorate dramatically.

On his way to the Lateran Council in 1179, Lorcán was warned by King Henry (who no longer trusted him) not to prejudice English interests. However, he persuaded the papacy of the threat to Ireland's ecclesiastical and political establishment from English expansionism, securing papal protection for the dioceses of Dublin and of Glendalough—which the English administration wanted to suppress. Returning to Ireland as papal legate, he consecrated Tommaltach, nephew of Ua Conchobair, as archbishop of Armagh. Although greatly incensed, Henry, already responsible for murdering Thomas Becket, archbishop of Canterbury, could not risk confrontation with a papal emissary. When, the following year, Lorcán undertook a diplomatic mission to England on behalf of Ua Conchobair, the king refused to meet him. Realizing that Henry had departed for France, Lorcán followed but collapsed with fever and died at the priory of Eu, Normandy on November 14, 1180. The Life commissioned by the community of Eu, stressing his spiritual qualities—with a persuasive account of his asceticism, charity, and dedication to pastoral responsibilities—helped to expedite his canonization by Pope Honorius III in 1226.

AILBHE MACSHAMHRÁIN

References and Further Reading

Plummer, Charles, ed. "Vie et miracles de S Laurent, archeveque de Dublin." *Analecta Bollandiana* 33 (1914): 121–186.

McNeill, Charles, ed. *Calendar of Archbishop Alen's Register*, pp. 2, 8, 9. Dublin: Royal Society of Antiquaries of Ireland, 1950.

Giraldus, Cambrensis. *Expugnatio Hibernica*. Edited by A. B. Scott and F. X. Martin, pp. 67, 79, 99, 167, 197, 306n116, 342n281. Dublin: Royal Irish Academy, 1978.

Flanagan, Marie-Therese. *Irish Society, Anglo-Norman Settlers, Angevin Kingship*, pp. 101–102. Oxford: Oxford University Press, 1989.

Gwynn, Aubrey. *The Irish Church in the Eleventh and Twelfth Centuries*. Edited by G. O'Brien, pp. 66, 135–143. Dublin: Four Courts Press, 1992.

MacShamhráin, Ailbhe. *Church and Polity in Pre-Norman Ireland: The Case of Glendalough*, pp. 103, 104, 154–155, 157–159, 161. Maynooth: An Sagart, 1996.

MacShamhráin, Ailbhe. "The Emergence of the Metropolitan See: Dublin 1111–1216." In *History of the Catholic Diocese of Dublin*, edited by J. Kelly and D. Keogh, pp. 58–62. Dublin: Four Courts Press, 2000.

Martin, F. X. "St. Bernard, St. Malachy, St. Laurence O'Toole." *Seanchas Ardmhacha* 15, no. 1 (1992): 19, 25, 28, 30–32.

See also **Church Reform, Twelfth Century; Hagiography**

UÍ BRIÚIN

Origins

From the late eighth century until the Anglo-Norman invasion, the Uí Briúin were the most powerful dynasty in Connacht. Their eponym—Brión son of Eochu Mugmedón, king of Ireland, by Mongfhind, daughter of the Munster king Feradach son of Dáre Cerbba—is depicted by the genealogies and saga literature as the elder brother of Fiachrach and Ailill, ancestors of the early Connacht dynasties of Uí Fiachrach and Uí Ailella. Their half-brother, the son of Eochu by the British slave girl Cairenn Casdub, was said to be Niall Noígiallach ("of the Nine Hostages"), ancestor of the Uí Néill. It is uncertain whether the depiction of the eponyms as brothers reflects actual bonds of kinship or was simply a biological metaphor for political relationships between the dynasties concerned.

Dynasts alleged by the genealogists to have been members of the Uí Briúin appear in the annals by the early sixth century. The earliest references to Uí Briúin specifically as a dynasty, however, are mid-seventh century, occurring both within a series of annal entries and in Tírechán's life of Patrick. Tírechán relates that the saint traveled to Duma Selchae in Mag nAí, where the "halls of the sons of Brión" were located. Tírechán neither enumerates nor names these sons, but the equivalent passage in the *Vita Tripartita*, a possibly ninth-century life of Patrick, names six sons of Brión. A series of later sources dating from the eleventh century onward, meanwhile, enumerates Brión's progeny as no less than twenty-four. No doubt the increasing power of the Uí Briúin was responsible for this dramatic swelling of the ranks, as tribes and dynasties newly coming under Uí Briúin sway were furnished with ancestries that would link them genealogically to their overlords. Into this category fall the Uí Briúin Umaill and likely also the Uí Briúin Ratha and Uí Briúin Sinna.

Uí Briúin Aí

According to the later sources, Brión's youngest son, Duí Galach, was the ancestor of the three most important branches of the dynasty: Uí Briúin Aí, Uí Briúin Bréifne, and Uí Briúin Seóla. Of the three, by far the most powerful branch of the dynasty was Uí Briúin Aí, based in Mag nAí in north-central County Roscommon. Throughout the seventh and first half of the eighth centuries, Uí Briúin Aí struggled with Uí Fiachrach to control the kingship of Connacht. By the end of the eighth century, they had managed to squeeze out their Uí Fiachrach rivals to gain a near monopoly on the provincial kingship. At this point, the controlling branch of the Uí Briúin was a sept known as Síl Muiredaig whose ruling family were to become the Ua Conchobair dynasty. Spreading out from their Mag nAí homeland, Síl Muiredaig took direct control over most of modern County Roscommon and much of east Galway, in addition to their overlordship of the rest of the province.

From the late eighth century until the coming of the Anglo-Normans, Uí Briúin Aí provided all but six kings of Connacht. They lost their grasp on the kingship once in the mid-tenth century as a consequence of intense rivalry from within Síl Muiredaig, and five times in the eleventh century when they faced fierce opposition from Uí Briúin Bréifne and Uí Briúin Seóla within Connacht, and from the Uí Briain of Munster without. After the vicissitudes of the eleventh century, however, the Uí Briúin Aí made a remarkable recovery in the twelfth, not only firmly recovering the kingship of Connacht but also gaining the kingship of Ireland in the personages of Tairrdelbach Ua Conchobair and his son Ruaidrí.

Uí Briúin Bréifne

On those occasions when Uí Briúin Aí lost control of the Connacht kingship, it was predominantly dynasts from Uí Briúin Bréifne, specifically the ruling family of Ua Ruairc, who seized the kingship. Having crossed east of the Shannon by the late eighth century into present day counties Leitrim and Cavan, Uí Briúin Bréifne gradually expanded in a diagonal direction so that at the peak of their power in the late twelfth century they controlled a diagonal band of territory stretching from Leitrim and northeast Sligo down to Kells and Drogheda. Although under the suzerainty of the king of Connacht, most of this kingdom lay beyond the technical limits of the province, defined as west of the Shannon.

Uí Briúin Seóla

Least powerful of the three main Uí Briúin dynasties, but by no means inconsequential, were Uí Briúin Seóla. Also known as the kings of "Uí Briúin In Déisceirt" ("Uí Briúin of the south") and of "Iar-Chonnacht" ("west Connacht"), the Uí Briúin Seóla were based in the area around Moyola, County Galway, east of Lough Corrib. Their two main divisions were the Clann

Coscraig, whose ruling family were the Meic Áeda, and the Muinter Murchada, whose ruling family were the Uí Flaithbertaig. By the end of the eleventh century, the Ua Flaithbertaig family were dominant within the Uí Briúin Seóla and indeed managed to very briefly take the provincial kingship in 1098 during the height of dynastic instability within Connacht. In the mid-thirteenth century, the Uí Flaithbertaig were deprived of their possessions east of the Corrib by the de Burgh family and moved west into present-day Connemara.

Common Uí Briúin Identity

Until the first half of the eleventh century, the various Uí Briúin dynasties seemed to actively maintain some sort of common Uí Briúin identity. Up until at least the 1030s, there existed a title "king of Uí Briúin" that was mostly bestowed upon the rulers of Uí Briúin Seóla, functioning as subkings under the Ua Conchobair kings of Connacht. Likely due to the dissension of the eleventh century, however, the names of the constituent elements of the dynasty thereafter proclaimed themselves to be much more discrete entities. Uí Briúin Aí became known exclusively as Síl Muiredaig, while Uí Briúin Seóla became known solely by the territorial designation of Iar-Chonnacht. Only Uí Briúin Bréifne retained the "Uí Briúin" element of their name, so much so that by at least the beginning of the twelfth century, the term Uí Briúin denoted Uí Briúin Bréifne alone.

ANNE CONNON

References and Further Reading

Byrne, Francis John. *Irish Kings and High-Kings*. London: Batsford, 1973.

Duignan, M. V. "The Kingdom of Bréifne." *Royal Society of Antiquaries of Ireland Journal* 65 (1935): 113–140.

Mac Niocaill, Gearóid. *Ireland before the Vikings*. Dublin: Gill & Macmillan, 1972.

Ó Corráin, Donnchad. *Ireland before the Normans*. Dublin: Gill & Macmillan, 1972.

Ó Mórdha, Eoghan. "The Uí Briúin Bréifni Genealogies and the Origins of Bréifne." *Peritia* 16 (2002): 444–450.

Walsh, Rev. Paul. "Christian Kings of Connacht." *Journal of Galway Archaelogical Society* 17 (1935): 124–142.

See also **Burke; Connacht; Connachta; Clonmacnoise; Leth Cuinn; Mide (Meath); Niall Noígiallach; Tírechaán; Ua Conchobair; Ua Conchobair, Ruaidrí; Ua Conchobair, Tairrdelbach; Ua Ruairc; Uí Néill**

UÍ CHENNSELAIG

Uí Chennselaig was one of the most important population groups and dynasties in early medieval Leinster. Uí Chennselaig considered themselves to be of the Laigin and traced their descent to one Énna Cennselach ("Énna the Dominant"), a grandson of Bressal Bélach, the ancestor also of Uí Dúnlainge. It is probable that Uí Chennselaig originally came from the Barrow valley and moved eastward into central Leinster, where their early royal center was Ráth Bilech (Rathvilly, Co. Carlow); later they expanded southward across the Blackstairs Mountains into the fertile plains of County Wexford and won supremacy in the south of the province, so that their rulers were occasionally styled *rí Laigin Desgabair* ("king of south Leinster") in the annals.

The first significant Uí Chennselaig king was Brandub mac Echach (d. 605) who defended Leinster against Uí Néill encroachments. Despite this success, Brandub's branch of the dynasty, Uí Felmeda of Ráth Bilech, were unable to compete with Uí Dúnlainge of north Leinster, and their power was eclipsed. In the eighth century, members of Síl Cormaic and Síl Máeluidir competed for the Uí Chennselaig kingship. Some of them also became kings of Leinster, but after 738 Uí Dúnlainge excluded the other Laigin dynasties from the provincial kingship for over three hundred years. Accordingly, Uí Chennselaig set about consolidating their hold on south Leinster: Síl Cormaic expanded into the territories of Uí Dróna (south Co. Carlow) and the area around the church of St. Mullins on the Barrow with which Uí Chennselaig had long-standing associations; Síl Máeluidir had meanwhile taken control of the lower Slaney and the area adjacent to Wexford harbor, in the process isolating earlier Leinster peoples, Uí Bairrche and Fothairt.

In the ninth century, Uí Chennselaig power was centered on the church of Ferna Mór Máedóc (Ferns, Co. Wexford). Uí Chennselaig had close associations with the church, and several members of the dynasty also held ecclesiastical office there, sometimes in combination with the kingship; Cathal mac Dúnlainge is titled *rex nepotum Cennselaig et secnap Fernann* ("king of Uí Chennselaig and prior of Ferns") in the annals at his death in 819. During this period, Viking incursions in south Leinster led other churches to look increasingly to the protection and patronage of Uí Chennselaig kings, and so the annals report Cairpre mac Cathail fighting with the *muinter* ("community") of the church of Tech Munnu (Taghmon, Co. Wexford) against Vikings in 828. Viking activities had an increasing effect on the intradynastic struggles of Uí Chennselaig, particularly after the establishment of Viking settlements at Wexford and Waterford. From the late ninth century, branches of the dynasty including Síl Cormaic, Síl nÉladaig, and Síl nOnchon were all contenders for the overall kingship. Síl nOnchon, originally an obscure group, provided two kings of Uí Chennselaig in the late ninth century, Tadg mac Diarmata (d. 865) and his brother Cairpre (d. 876). Although both these kings were killed feuding with their own relatives, most of Cairpre's descendants in the tenth century were kings of Uí Chennselaig, and were gradually able to increase their power despite internecine strife and intrusions from outside south Leinster. Diarmait mac Máele-na-mBó of Síl nOnchon inaugurated a new period of success for the dynasty, firstly by consolidating his hold on Ferns and the rich royal demesne thereabout, and then eliminating rivals for the kingship of Uí Chennselaig. The power of the Uí Dúnlainge kings had declined as a result of attacks by Uí Néill, Vikings, and the kings of Osraige, and Diarmait mac Máele-na-mBó was able to take the kingship of all Leinster in 1042. While he was engaged on his campaigns, his son Murchad acted as regent in Leinster, maintained Uí Chennselaig dominance over Uí Dúnlainge, conducted border raids against Mide, and focused his attention particularly on the control of Dublin, which was increasingly integrated into the Leinster province. Murchad's obit in 1070 calls him *rí Laigen & Gall* ("king of Leinster and the Foreigners"), and he was buried at Dublin. His descendants took the family name of Mac Murchada (Mac Murrough), and though they retained the Leinster kingship they were unable to challenge for the high kingship of Ireland, though Diarmait Mac Murchada played a pivotal role in the politics of the twelfth century.

MARK ZUMBUHL

References and Further Reading

Byrne, Francis J. *Irish Kings and High-Kings*. London: B. T. Batsford, 1973; Reprint, Dublin: Four Courts Press, 2001.

O'Byrne, Emmett. *War, Politics and the Irish of Leinster, 1156–1606*. Dublin: Four Courts Press, 2003.

Smyth, Alfred P. *Celtic Leinster. Towards an Historical Geography of Early Irish Civilization A.D. 500–1600*. Dublin: Irish Academic Press, 1982.

See also **Diarmait mac Máele-na-mbó; Dublin; Laigin; Leinster; Mac Murchada, Diarmait; Mac Murchada; Uí Dúnlainge; Uí Néill; Viking Incursions**

UÍ DÚNLAINGE

Uí Dúnlainge was one of the most important population groups and dynasties in early medieval Leinster. Uí Dúnlainge considered themselves to be of the Laigin and to be descended from Dúnlaing, grandson of Bressal Bélach, the ancestor also of Uí Chennselaig. Uí Dúnlainge occupied fertile land in north Leinster, including the Liffey valley and Kildare plains. Uí Dúnlainge initially owed their rise to the decline of earlier Leinster dynasties in the face of the expansion of the southern Uí Néill, and in the seventh century they drove the Uí Garrchon and Uí Enechglaiss across the Wicklow mountains. The first significant king of Uí Dúnlainge was Fáelán mac Colmáin (d. 666 or earlier) who defeated competitors from the rival dynasties of Uí Máil and Uí Chennselaig to attain the kingship of Leinster; he was also allied with the southern Uí Néill.

From the mid-eighth century, Uí Dúnlainge monopolized the kingship of Leinster. Three of the sons of Murchad mac Brain (d. 727), Dúnchad, Fáelán, and Muiredach, reigned in turn after him as kings of Leinster. These kings were progenitors of the most powerful branches of Uí Dúnlainge in the following three centuries: Uí Dúnchada, Uí Fáeláin, and Uí Muiredaig. Uí Dúnchada were settled between the lower Liffey and the Wicklow mountains, their territory later extending to the outskirts of Dublin. Their center was at Liamain (Lyon's Hill, Co. Kildare), and several members of the family enjoyed the abbacy of Kildare. Uí Fáeláin settled to the southwest, with their center at Naas. Further south dwelt Uí Muiredaig, with their base at Maistiu (Mullaghmast, Co. Kildare); they had links with the church of Glendalough.

The kingship of Leinster rotated between these groups, and this pattern of a "circuit among the branches" of an Irish dynasty has been used as one of the models of Irish kingship. For all the apparent neatness, the succession was often accompanied by feud and fratricide. Additionally, though later king-lists call many of these dynasts kings of Leinster, the contemporary evidence of the annals sometimes gives them lesser titles such as "king of Uí Dúnlainge," *rí Iarthir Liphi* ("king of western Liffey"), or even "king of Naas." This may be due to the dominance that the Uí Neill periodically asserted over them, forcing them to submit and give hostages. The southern Leinster kings must often have rejected Uí Dúnlainge overlordship, and as time went on the authority of the Uí Dúnlainge kingship was eroded.

From the ninth century, domination of Leinster became a main point of contention between the kings of Uí Néill and the kings of Munster. A highly significant development in this period was the establishment of a Viking base at Dublin in 841. Soon the Dublin Vikings controlled a substantial hinterland north and south of the Liffey estuary (Fine Gall) and were a considerable threat to their neighbors. However, internal dissensions among the Dublin Norse facilitated an attack in 902 by Cerball mac Muirecáin, Uí Fáeláin king of Leinster, in alliance with the king of Brega, which drove the Vikings from Dublin for fifteen years. Cerball Mac Muirecáin also fought against Cormac mac Cuillenáin in 908 and married Gormfhlaith (d. 948), said to have been the latter's widow.

Uí Dúnlainge also had to contend with the kings of Osraige, whose power grew steadily in the late ninth and tenth centuries. Cerball mac Dúngaile and his son Diarmait mac Cerbaill, both kings of Osraige, allied with Vikings and attempted to dominate Leinster. The Uí Dúnlainge themselves quickly learned the advantages of allying with Vikings; in 956 the Leinstermen and Dublin Vikings killed the king of Tara. In the late tenth century Dublin and its hinterland, though politically independent of Leinster, was seen as an important center of wealth and power, and Irish kings, including those of Uí Dúnlainge, attempted to assert control of the settlement.

In the 980s and 990s, Uí Dúnlainge lost out as Brian Boru attempted to control Leinster and Dublin as part of his struggles with Máel-Sechnaill II for the high kingship of Ireland. In 997, Máel-Sechnaill gave to him the hostages of Leinster and Dublin he had previously held. The Laigin and Dublin Vikings were no more amenable to the overlordship of Brian than to that of the Uí Neill kings, and they rebelled in 999, to be crushed by Brian. His success in Leinster was in part due to the divisions between Uí Dúnlainge and Uí Chennselaig, and between the branches of Uí Dúnlainge themselves. In 1003, Brian deposed Donnchad mac Domnaill of Uí Dúnchada as king of Leinster and installed Máelmórda mac Murchada of Uí Fáeláin. Brian was in fact married to Máelmórda's sister, Gormfhlaith (d. 1030), who had previously been married both to Amlaíb Cuarán, king of Dublin, and to Máel-Sechnaill. However, relations deteriorated over the following years, and Máelmórda rebelled against Brian. This culminated in the Battle of Clontarf in which both were killed. The power of Osraige subsequently grew again. Donnchad mac Gilla-Pátraic, king of Osraige, intervened in Leinster several times, and in 1036 he took the kingship of the province. A fatal blow had been dealt to declining Uí Dúnlainge power, and in 1042 the kingship of Leinster passed to Uí Chennselaig in the person of Diarmait mac Máele-na-mBó.

Uí Dúnlainge retained considerable power in their own districts into the twelfth century. Uí Dúnchada had suffered territorially at the hands of the Dublin Vikings, although they retained land at Lyon's Hill and in the area of the Dublin-Wicklow border. Their family name at the time of the Anglo-Norman invasion was Mac Gilla Mo-Cholmóc, and their descendants survived under the new regime as the Fitzdermots. The English invaders forced the Uí Fáeláin and Uí Muiredaig dynasties, at that time represented by the families of Ua Brain (O'Byrne) and Ua Tuathail (O'Toole) eastward into the less fertile lands of the Wicklow Mountains. However, they were able to survive and even prosper as Gaelic lordships until the end of the Middle Ages.

Mark Zumbuhl

References and Further Reading

Byrne, Francis J. *Irish Kings and High-Kings.* London: B. T. Batsford, 1973.

Mac Shamhráin, Ailbe. *Church and Polity in Pre-Norman Ireland: The Case of Glendalough.* Maynooth: An Sagart, 1996.

O'Byrne, Emmett. *War, Politics and the Irish of Leinster, 1156–1606.* Dublin: Four Courts Press, 2003.

Smyth, Alfred P. *Celtic Leinster. Towards an Historical Geography of Early Irish Civilization* A.D. *500–1600.* Dublin: Irish Academic Press, 1982.

See also **Anglo-Norman Invasion; Brian Boru; Cerball mac Muireccáin; Diarmait mac Máele-na-mbó; Dublin; Glendalough; Gormlaith (d. 1030); Kildare; Laigin; Leinster; Máel-Sechnaill; Uí Néill; Viking Incursions**

UÍ MAINE, BOOK OF

The Book of Uí Maine is a fourteenth-century vernacular Irish manuscript, part of which at least was written for a member of the ruling family of Uí Maine, Muirchertach Ua Ceallaig, bishop of Clonfert (1378–1392), and later archbishop of Tuam (1392–1407), as a colophon in the hand of one of its scribes, Fáelán mac a' Ghabann, makes clear. Fáelán is one of only two scribes named in the codex, the second being the principal redactor, Adhamh Cúisín, whose hand has been identified on 99 of the Book's 161 folios. As only approximately 40 percent of the original codex has survived, it is impossible to determine whether this Anglo-Norman scribe was responsible for the greater part of the manuscript in its pristine state, as William O'Sullivan has remarked. As he was in Ua Ceallaig's employ when the latter was archbishop of Tuam, however, his proximity to the patron of the work is not in doubt. Moreover, as O'Sullivan's analysis of the various scripts has shown, he undertook his writing later than all but one of the Book's eight or so other scribes, and thus may have overseen completion of the work. Indicative of this perhaps is the fact that he was also responsible for the preparation of the manuscript for its first binding. Be that as it may, Cúisín and his predecessors were followed by a number of secondary scribes, the two principal hands of which were at work in the sixteenth century. In the seventeenth century, as Nollaig Ó Muraíle has demonstrated, our manuscript was known by the alternative title, *Leabhar Uí Dhubhagáin* ("The Book of Ua Dubagáin"), because of an association with the Uí Dhubagáin who supplied hereditary poets to the Uí Cheallaig. A number of texts attributed to the most famous *ollam* ("chief poet") of this family, Seáan Mór Ua Dubagáin (d. 1372), are preserved in the manuscript, and it has been speculated that a direct descendant of his, Seáan son of Corbmac Ua Dubagáin

(d. 1440), may have been one of the anonymous scribes. In any event, some of the material contained therein is indeed aimed at an Ua Ceallaig patron, in particular the Uí Maine genealogies, which take pride of place in the manuscript's considerable genealogical corpus. On the whole, however, it comprises a varied collection of texts of different types that would have had widespread appeal. Saints' pedigrees take their place alongside secular ones. In addition, it preserves important versions of two Middle Irish compilations, *Dinnshenchas Érenn* ("The Place-Name Lore of Ireland") and *Banshenchas* ("Women Lore"). Other pseudohistorical matter includes a copy of the early-twelfth-century Munster text, *Lebor na Cert* ("The Book of Rights"), as well as a body of dynastic poetry. An interest in the workings of verse is underlined by a metrical treatise contained therein, and grammar is represented by the pivotal seventh-century tract, *Auraicept na nÉces* ("The Scholars' Primer"). Notwithstanding its heterogeneity, it remained in the possession of the Uí Cheallaig for a considerable period. It was acquired by the Royal Irish Academy in the nineteenth century where it still remains, apart from four folios that became separated and now form British Library Manuscript Egerton 90.

MÁIRE NÍ MHAONAIGH

References and Further Reading

Macalister, R. A. S. *The Book of Uí Maine*. Dublin: Irish Manuscripts Commission, 1941.

Mulchrone, Kathleen, ed. *Catalogue of the Irish Manuscripts in the Royal Irish Academy*. 28 fascicles, fasc. 26, 3314–3356.

Ó Muraíle, Nollaig. "Leabhar Ua Maine *alias* Leabhar Uí Dhubhagáin." *Éigse: A Journal of Irish Studies* 23 (1989): 167–195.

O'Sullivan, William. "The Book of Uí Maine formerly the Book of Ó Dubhagáin: Scripts and Structure." *Éigse: A Journal of Irish Studies* 23 (1989): 151–166.

See also **Dinnshenchas; Gaelic Revival; Genealogy; Grammatical Treatises; Leabhar Breac; Lecan, Book of; Lecan, Yellow Book of; Manuscript Illumination; Metrics**

UÍ NÉILL

The Uí Néill were the most prominent political dynasty in Ireland from the seventh to the late tenth century. The annals for this period contain copious references to the dynasty that allow historians to reconstruct their story in reasonable detail. The Uí Néill claimed descent from Niall Nóigiallach who lived in the fourth or fifth century. Further back, the Uí Néill were said to be a branch of the Connachta, descended from Conn Cétchathach ("Conn of the Hundred Battles"). The origins of the dynasty pre-date recorded history and are shrouded in obscurity. T. F. O'Rahilly believed that they emerged from the east midlands. However, it is much more likely that they originated in northeastern Connacht.

The Uí Néill comprised a number of distinct dynastic groupings each of which claimed descent from a different son of Niall. J. V. Kelleher argued that the Uí Néill were a federation of tribes who contrived a common descent from Niall Noígiallach. It was certainly the case that the number of sons attributed to Niall grew over the centuries as other tribal groupings fell under Uí Néill control and assumed Uí Néill identity by crediting their founding ancestor figure as a son of Niall. However, it seems fairly certain that there was an early core Uí Néill grouping onto which these later accretions were grafted. This core group probably included the descendants of Lóegaire, Coirpre, Fiachu, and Conall (Ailbhe Mac Shamhráin has argued that the two putative sons of Niall named Conall, namely Conall Gulban and Conall Cremthainne, were one and the same person). The full flowering of Uí Néill expansion saw no fewer than fourteen sons being claimed for Niall.

By the sixth century, the basic dynastic structure of the Uí Néill, which was to characterize the history of the dynasty for centuries to come, had begun to take shape. Cenél Conaill, Cenél nÉogain, and the relatively obscure Cenél nÉnnai were settled in northwest Ulster (see Uí Néill, Northern), Cenél Coirpri to the south in northeastern Connacht, Cenél Lóegaire in various locations from Loch Erne to the Slieve Bloom mountains, the descendants of Conall Cremthainne (later to emerge as Clann Cholmáin and Síl nÁedo Sláine) in the east midlands, Cenél Fiachach in the center of the country near Uisnech (see Uí Néill, Nouthern). Uí Néill dominance of the northern half of Ireland gave rise to that area's being known as Leth Cuinn ("Conn's Half").

From about the mid-seventh century onward, the Uí Néill had assumed proprietorial rights to the ancient sacral kingship of Tara, and indeed the term *rí Temro* ("king of Tara") was to become synonymous with the over kingship of the Uí Néill dynasty. The balance of power between the northern and southern branches of the Uí Néill was maintained from the mid-eighth to the late tenth century by means of an arrangement whereby, with only one exception, the kingship of Tara alternated between a member of Cenél nÉogain (who had emerged as the dominant grouping among the Northern Uí Néill) and the Clann Cholmáin (the strongest branch of the Southern Uí Néill).

Various Uí Néill kings styled themselves "king of Ireland" from at least the time of Domnall mac Áedo of Cenél Conaill (d. 642) onward. However, it was not until the ninth and tenth centuries, during the reigns of Máel-Sechnaill I mac Máele-ruanaid (d. 862), and a number of his successors, that the Uí Néill achieved anything approaching dominance over the entire country.

The rise of the Uí Néill to prominence, and the subsequent maintenance of that dominance, owed much to their relationship with the church. From the seventh century onward, they were closely allied to Armagh, which was at that time asserting its own claims to primacy over the church in Ireland. Branches of the Uí Néill also had close associations with the various churches of the Columban federation; St. Colum Cille (d. 597), belonged to Cenél Conaill of the Northern Uí Néill. Later, the Uí Néill, and especially the Clann Cholmáin, were to forge strong links to the major monastic foundation of Clonmacnoise, where a number of the antiquities from the tenth and eleventh centuries bear testimony to the close association between church and dynasty.

The decline of the Uí Néill dynasty in the late tenth century was due in large measure to the emergence of the Dál Cais dynasty in Munster—a dynasty which under Brian Bóruma (d. 1014), his grandson Tairrdelbach (d. 1086), and his great-grandson, Muirchertach Ua Briain (d. 1119) demonstrated the ambition and vigor to lay claim to the overlordship of Ireland.

PAUL BYRNE

References and Further Reading

Byrne, Francis John. "The Rise of the Uí Néill and the High-Kingship of Ireland." O'Donnell Lecture. Maynooth: National University of Ireland, 1969.

Byrne, Francis John. *Irish Kings and High-Kings*. London: Batsford, 1973. Reprint, Dublin: Four Courts Press, 2001.

Byrne, Paul. "Certain Southern Uí Néill Kingdoms (Sixth to Eleventh Century)." PhD diss., University College Dublin, 2000.

Kelleher, J. V. "The Rise of the Dál Cais." In *North Munster Studies: Essays in Commemoration of Monsignor Michael Moloney*, edited by Etienne Rynne, pp. 230–241. Limerick: The Thomond Archaeological Society, 1967.

Mac Shamhráin, Ailbhe. "Nebulae discutiuntur? The Emergence of Clann Cholmáin, Sixth–Eighth Centuries." In *Seanchas: Studies in Early and Medieval Irish Archaeology, History and Literature in Honour of Francis J. Byrne*, edited by Alfred P. Smyth, pp. 83–97. Dublin: Four Courts Press, 2000.

O'Rahilly, T. F. *Early Irish History and Mythology*. Dublin: Dublin Institute for Advanced Studies, 1946.

See also **Armagh; Clonmacnois; Colum Cille; Connachta; Dál Cais; Diarmait mac Cerbaill; Leth Cuinn; Lóegaire mac Néill; Máel Sechnaill I; Máel Sechnaill II; Mide; Niall Nóigiallach; Tara; Uí Néill, Northern; Uí Néill, Southern**

UÍ NÉILL, NORTHERN

The Early Period—Cenél Conaill Dominance

"The Northern Uí Néill" is the collective name for the dynasties established in northwest Ulster by Eógan, Conall, and Éndae, three putative sons of Niall Noígiallach, known respectively as Cenél nEógain, Cenél Conaill, and Cenél nÉnnai. If Cenél nÉnnai ever held a prominent position, it has left no trace in the historical records, and the history of the Northern Uí Néill is, essentially, the history of Cenél Conaill and Cenél nEógain.

From the sixth century, when the Northern Uí Néill emerged onto the pages of history, until the latter half of the eighth century, Cenél Conaill were the dominant grouping among the Northern Uí Néill. However, the standing of Cenél nÉogain was by no means insignificant during this period, in the course of which several members of that dynasty attained the over kingship of the Uí Néill. At the battle of Mag Roth (Moira, Co. Down) in 637, Domnall mac Áedo of Cenél Conaill, who styled himself *rex Hibernie* ("king of Ireland"), defeated an alliance of the king of Ulaid and the Dál Riata from Scotland. After this battle, the Uí Néill were established as the dominant power in the north of Ireland. The base of Cenél Conaill and Cenél nEógain lay in the relatively poor lands of Donegal in the northwest of Ireland (the massive stone fortress of Ailech, in the Inishowen peninsula, was to remain synonymous with the kingship of Cenél nEógain long after the center of Cenél nEógain power had moved to the east). Between the Northern Uí Néill and the Ulaid (largely confined to counties Antrim and Down) lay the extensive lands of the Airgialla ("those who give hostages"), a group that increasingly fell under the sway of the Northern Uí Néill.

The Emergence of Cenél nEógain

Flaithbertach mac Loingsech, who abdicated the over kingship of the Uí Néill in 734, having been challenged by Áed Allán of Cenél nEógain, was the last Cenél Conaill dynast to attain that status. The reign of Áed witnessed the beginning of the arrangement, which was maintained for two and a half centuries (with only one exception), whereby the kingship of the Uí Néill alternated between the Cenél nEógain and the Clann Cholmáin of the Southern Uí Néill. The battle of Clóitech in 789, in which the Cenél Conaill were roundly defeated by Cenél nEógain, seems to have consolidated the already superior status of the latter dynasty.

Cenél nEógain expansion from their northeast Donegal base into Derry, Tyrone, and Fermanagh was at the expense of the Airgialla. So extensive was this expansion that, by the ninth century, Airgialla had been restricted to the southeast of the modern province of Ulster. Sometime during this period of expansion, Cenél nEógain established their inauguration site at Tulach Óc (Tullaghogue) in county Tyrone. The descendants of the Cenél nEógain kings, the O Néill family of Tyrone, were inaugurated standing on the "Stone of the Kings" at Tullaghogue until 1602 when the stone was smashed by Lord Mountjoy.

Following their defeat at the hands of Niall Caille mac Áedo of Cenél nEógain at the battle of Leth Cam near Armagh in 827, the Airgialla became a subject people of the Northern Uí Néill. Of even greater significance was the fact that, after Leth Cam, the abbacy of the church of Armagh also fell under the effective control of Cenél nEógain—a control that they exercised through the Airthir, a branch of Airgialla settled near Armagh. So close were the ties between Cenél nEógain and Armagh that Áed Findliath, the Cenél nEógain king of Tara, had his own house there in 870. A century later, Domnall Ua Néill, a great-grandson of Áed, whom the annalists styled *ardrí Érenn* ("High King of Ireland"), died after penance in Armagh.

The tenth century in Ireland witnessed renewed Viking incursions. Among the most effective Irish kings at dealing with the enhanced Norse presence was the Cenél nEógain king, Muirchertach na Cochall Craicinn ("of the Leather Cloaks") mac Néill, who defeated the Vikings near Armagh in 921 and at Carlingford and Annaghassan in 926. He slew the jarl Torulb in 932, defeated the combined forces of the king of the Ulaid and the foreigners in 933, and launched a successful onslaught, in conjunction with Donnchad Donn, the Clann Cholmáin, over king of the Uí Néill, against the Dublin Norse in 938. Not surprisingly, he met his death at the hands of the Norse in 943.

The Northern Uí Néill During the Eleventh and Twelfth Centuries

Domnall ua Néill (d. 980), son of Muirchertach of the Leather Cloaks, was the last king of Cenél nEógain for over a century to assume a position of national dominance. In the period following Domnall's reign, the Ua Briain dynasty of Dál Cais in Munster succeeded in excluding all branches of the Uí Néill from political power on the national stage. However, the emergence of Domnall Ua Lochlainn in the late eleventh century represented something of a revival in the fortunes of Cenél nEógain. At various stages in his career, he achieved dominance over, among others, the Ulaid, Cenél Conaill, and Uí Máelsechlainn of Clann Cholmáin. The submission of Muirchertach Ua Briain to Domnall in 1090 was short-lived, and he was to continue to oppose Domnall. Domnall's successes, such as they were, were due in no small part to the support that he received from the church of Armagh. The annalists, recording Domnall's death in 1121, accord him the title *ardrí Érenn*—a designation scarcely matched by the evidence. The achievements of Domnall's grandson, Muirchertach Mac Lochlainn, between the middle 1140s and his death in 1166, were more substantial. Following a series of successful military campaigns, Muirchertach had become the foremost dynast in the country by the early 1150s. His position was, however, never secure and was continually challenged by Tairrdelbach Ua Conchobair of Connacht and his son Ruaidrí. However, in 1161, Mac Lochlainn secured the submission of all the principal kings of Ireland, including Ruaidrí Ua Conchobair. By 1166, Muirchertach was facing a rebellion of Ruaidrí Ua Conchobair, who had the support of most of the major Irish kings, and the backing of Muirchertach's enemies within Cenél nEógain. Muirchertach died in 1166 as he prepared to confront Ua Conchobair's invading army. The annalists called him *rí Érenn*—a dignity that was certainly not universally recognized at the time of his death.

PAUL BYRNE

References and Further Reading

Byrne, Francis John. *Irish Kings and High-Kings.* London: Batsford, 1973. Reprint, Dublin: Four Courts Press, 2001.

MacShamhráin, Ailbhe. "Cenél nEógain and the Airgialla from the Sixth to the Eleventh Centuries." In *Tyrone: History and Society*, edited by C. Dillon and H. J. Jefferies, pp. 55–84. Dublin: Geography Publications, 2000.

Ó Corráin, Donnchadh. *Ireland before the Normans.* Dublin: Gill and MacMillan, 1972.

Ó Fiaich, An tAth Tomás. "The Church of Armagh Under Lay Control." *Seanchas Ard Mhacha: Journal of the Armagh Diocesan Historical Society* 5, no. 1 (1969): 75–127.

Simms, Katharine. "Late Medieval Tír Eoghain: The Kingdom of the Great Ó Néill." In *Tyrone: History and Society*, edited by C. Dillon and H. J. Jefferies, pp. 127–162. Dublin: Geography Publications, 2000.

See also **Ailech; Airgialla; Armagh; Dál Cais; Inauguration Sites; Mac Lochlainn; Niall Noígiallach; Ua Néill family; Uí Néill; Uí Néill, Southern; Ulaid; Viking Incursions**

UÍ NÉILL, SOUTHERN

Background and Early Period

"The Southern Uí Néill" is little more than a convenient designation for the several branches of the Uí Néill that occupied territories stretching across the center of Ireland from north County Dublin to northeastern Connacht.

During the earliest period of recorded history, the fifth and sixth centuries, the dominant branches of the Southern Uí Néill were the descendants of three putative sons of Niall Noígiallach: Lóegaire (Cenél Lóeguiri), Coirpre (Cenél Coirpri) and Fiachu (Cenél Fiachach). Cenél Lóeguiri had their base near the church of Trim in County Meath and controlled large tracts of land from Lough Erne in the north to Rathlihen, near the Slieve Bloom mountains, in the south. Many historians have concluded that, from the earliest historical period, Cenél Coirpri extended from its base in northeast Connacht to the northern borders of Leinster. It is more probable that the primary Coirpre kingdom was in northeastern Connacht and that the incursion by one branch into northern Tethbae (County Longford), where a separate kingdom was established, occurred as late as the eighth century, while Cenél Coirpri settlement on the Mide–Leinster borders only occurred in the mid-twelfth century, although they had made incursions into Mide before that time. Cenél Fiachach were located in the territory around Uisnech (the traditional center point of Ireland). The earliest list of the kings of Tara, *Baile Chuind*, includes the names of Lóegaire and his son Lugaid, as well as Coirpre and his grandson Tuathal Máelgarb. The history of these early groups is closely associated with the struggle to wrest the northern midlands from the Laigin.

By the mid-sixth century, a branch of Uí Néill claiming descent from Conall Cremthainne mac Néill had assumed a dominant position. Diarmait mac Cerbaill (d. 565) was the first of that line to become over king of the Uí Néill. Two sons of Diarmait, Áed Sláine and Colmán, were the ancestors of the two major dynasties that dominated the Southern Uí Néill from the seventh century onward, viz. Síl nÁedo Sláine and Clann Cholmáin. Originally these groupings were based in Brega (modern County Meath along with north County Dublin and south County Louth). During the seventh and early eighth centuries, the Síl nÁedo Sláine were in the ascendant, and several of their number became over kings of the Uí Néill during this period. Among these, Fínsnechta Fledach (d. 695) was perhaps the most notable. He is remembered in later tradition as the king who remitted the *Bóruma* ("the Cattle Tribute") that the Laigin had to pay to the Uí Néill. By the middle of the eighth century, Síl nÁedo Sláine had divided into northern and southern branches. The northern branch, which assumed the name *Ciannachta*—after the people whose lands they had appropriated—were centered on Knowth; the southern branch was based at Lagore.

The Emergence of Clann Cholmáin

Clann Cholmáin, which had hitherto been largely subordinate to their Síl nÁedo Sláine cousins, began, during the first half of the eighth century, to assume a position of power in the territory to the west of Brega that was to become the kingdom of Mide. From 728 onward, Clann Cholmáin excluded Síl nÁedo Sláine from the over kingship of the Uí Néill for some two hundred years.

It was during the reign of Donnchad Midi that Clann Cholmáin became a dynasty of national importance. Donnchad succeeded in subduing Leinster and the Northern Uí Néill and sought to control Munster as well. Much of his reign was, however, taken up with attempting to subdue the Ciannachta branch of Síl nÁedo Sláine. Donnchad's death, in 797, was followed by a period of limited achievement for Clann Cholmáin.

Ninth and Tenth Centuries – The Era of Clann Cholmáin Dominance

Máel Sechnaill I mac Máele-Ruanaid advanced the cause of Clann Cholmáin further in the mid-ninth century. His obituary in the annals, for the year 862, describes him as *rí Hérenn uile* ("king of all Ireland"). The claim is somewhat inflated; while he achieved a significant level of dominance over most of the major dynasties and some notable victories over the Vikings, he faced persistent opposition from the Ciannachta and from Áed Findliath of Cenél nEógain. The reign of Máel Sechnaill's son, Flann Sinna ("Flann of the Shannon"), enjoyed a remarkably lengthy reign as king of Tara; when he died in 916, he had held the kingship for thirty-seven years. Flann achieved a reasonable measure of military dominance throughout this career, but, not too surprisingly, he faced the growing unrest of his ambitious sons who challenged him on a number of occasions. However, it was as a patron of the church—and, in particular, the monastery of Clonmacnoise—that Flann is best remembered. He built *Tempul na Ríg* at Clonmacnoise in 909, and his patronage of the monastery is commemorated by an inscription on the Cross of the Scriptures.

The emergence of Congalach mac Máel-mithig of Síl nÁedo Sláine as over king of the Uí Néill in 944 represented a break in the two-hundred-year-old convention whereby the kingship alternated between Clann Cholmáin and Cenél nEógain (see Uí Néill). The accession of Congalach, who was something of a compromise candidate, was made possible by the complex political circumstances within the Uí Néill dynasty arising from the reemergence of Cenél Conaill as a significant force within the Northern Uí Néill.

Congalach's reign was relatively unremarkable, and, following his death in 956, the alternating succession between Cenél nEógain and Clann Cholmáin was restored.

Máel Sechnaill mac Domnaill (d. 1022) was the last major dynast from the Southern Uí Néill; in fact, his power was already in decline by the turn of the eleventh century, as the balance of power had shifted to Brian Bóruma. The eleventh and twelfth centuries witnessed the eclipse of the Uí Néill by the Uí Briain of Dál Cais and the Uí Chonchobhair of Connacht. The power of the Southern Uí Néill was further diminished during this period as a result of the seizure of large tracts of their territories by, among others, the Uí Briúin Bréifne.

PAUL BYRNE

References and Further Reading

Byrne, Francis John. *Irish Kings and High-Kings*. London: Batsford, 1973.

Byrne, Paul. "Certain Southern Uí Néill Kingdoms (Sixth to Eleventh Century)." PhD diss., University College Dublin, 2000.

Ó Corráin, Donnchadh. *Ireland before the Normans*. Dublin: Gill and MacMillan, 1972.

Ryan, Fr. John. *Clonmacnois: A Historical Summary*. Dublin: The Stationery Office, 1973.

See also **Uí Néill; Dál Cais; Ua Briain; Ua Conchobhair; Mide; Brian Boru; Diarmait mac Cerbaill; Lóegaire mac Néill; Máel Sechnaill I; Máel Sechnaill II**

ULAID

In 1177 the Anglo-Norman adventurer John de Courcy conquered the kingdom of the Ulaid and established his own lordship based upon its historical center in and around Downpatrick. This action brought to an end one of the most enduring polities in Irish history. In his "Life of Patrick," the late-seventh-century writer Muirchú described the territory of the Ulothi as lying between the Boyne and the Lagan. The same territory seems to be ascribed to the Uoluntii by the Greco-Roman geographer Ptolemy in the second century A.D. In the Ulster Cycle, surviving in literary texts from the eighth century onward, the territory of the Ulaid was said, in pre-Patrician times, to have extended over the whole of Ireland north of the Boyne, but the coincidence of Ptolemy's and Muirchú's location of the tribe makes this seem unlikely. It should also be noted that Ptolemy's Isamnion (O. I. *Emain*) is a coastal promontory in County Down and not the site near Armagh city with which the medieval authors identified it.

In medieval times, the Ulaid were dominated by the Dál Fiatach dynasty with their royal center at Dún Lethglaise (Downpatrick), who extended their sway over most of modern Down and Louth and parts of Armagh. The tradition of a greater sway than this may have some basis in fact even if it was not as extensive as legends suggest. The Ulaid were counted, along with the Laigin and the Féni, as one of the three major peoples of Ireland, and their claims to hegemony in the northeast were only gradually eroded by the Uí Néill. The Dál Fiatach seem to have been finally marginalized after the battle of Leth Cam in 827 when they seem to have been attempting to detach the Airgialla from Uí Néill overlordship. Indeed, the aspirations leading up to this battle may have inspired the Ulster Cycle vision of a greater Ulaid. Earlier Ulidian kings, such as Báetán mac Cairill (*c.* 581) and his nephew Fiachnae mac Demmáin (626), had been able to operate as major players on the Irish scene.

By the late ninth century, the Dál Fiatach had fallen on hard times and were forced to accept the overlordship of their northern neighbors, the Dál nAraide of Moylinny (Antrim). The propaganda produced to legitimize the dominance of this Cruthni dynasty over the Ulaid has obscured the original distinctiveness of the two peoples. Dál Fiatach fortunes were restored by Eochaid mac Ardgail in 972, but inevitably this led to renewed conflict with the Uí Néill culminating in Eochaid's death in battle in 1004. The period of Dál nAraide dominance coincided broadly with the presence of at least one Viking base on Strangford Loch, which must have discomfited the Ulaid, but in the long run no enduring Hiberno-Norse settlements were established in their territory. The Ulaid even developed their own fleet, which was active in the Irish Sea and even along the British coast.

The dynasty had a final revival under Donn Sléibe macEchdacha (sl. 1091) and his descendants, and developed close relations with the kingdom of the Isles. In the course of the twelfth century Mac Lochlainn kings were twice able to divide the Dál Fiatach into four *tigernae* ("lordships"), and Mourne was lost to Donnchad Ua Cerbaill, king of the Airgialla, who negotiated the restoration of Ulidian kingship with Muirchertach Mac Lochlainn. After de Courcy's conquest, the MacDúinnshléibe retained the title "king of Irish Ulster," but they had lost any real independence.

ALEX WOOLF

References and Further Reading

Byrne, Frances John. *Irish Kings and High Kings*, pp. 106–129. London: Batsford, 1973.

Mallory, J. P. and T. E. MacNeill. *The Archaeology of Ulster*, pp. 143–248. Belfast: The Institute of Irish Studies, 1991.

ULSTER CYCLE

The large body of stories and poems that constitute the Ulster Cycle concern the exploits of the Ulaid, a group of peoples that in the early Middle Ages are confined to northeastern Ireland but in the tales stretch across the whole of the North. Their king is Conchobor mac Nessa who has his royal palace at Emain Macha (Navan Fort near Armagh City). Among its best-known warriors are Fergus mac Róich, Conall Cernach, and Cú Chulainn. There is a state of almost constant warfare between the Ulaid and the Connachta whose capital is at Crúachu or Crúachain (Rathcroghan, Co. Roscommon). They are led by their king, Ailill mac Máta, and his queen, Medb, the daughter of Eochaid Feidlech, king of Tara. The events of the Ulster Cycle are traditionally dated to around the time of Christ by medieval scholars who largely believed in the historicity of the main events and characters of the Cycle.

The earliest accounts of the deeds of the Ulstermen were written in the seventh century. It has been suggested that the interest in the Ulster Cycle tales was first cultivated in the great monastery of Bangor, County Down, in the district where the Ulaid were located in the early Middle Ages, but some of the earliest references to the events of the Cycle are contained in the work of the seventh-century poet Luccreth moccu Chérai who is associated with Munster. The Cycle was very popular throughout Ireland until the twelfth century. The earliest manuscript copies were written at Clonmacnoise, County Offaly and Terryglass, County Tipperary in the eleventh and twelfth centuries. Several tales of the Ulster Cycle were reworked in the later Middle Ages, but it no longer dominated the literary scene as it had done up to the twelfth century.

The Cattle Raid of Cooley

The central tale of the Cycle is the Cattle Raid of Cooley (*Táin Bó Cúailnge*), which tells of the heroic single-handed defense of Ulster by the young Cú Chulainn. The men of Ireland, led by Ailill and Medb, attack Ulster in order to obtain the Brown Bull of Cúailnge (Cooley peninsula, Co. Louth). Cú Chulainn fends them off by engaging them in single combat, tragically slaying his beloved foster-brother Fer Diad in the process. The Bull is carried off and dies fighting against the White-Horned Bull (*Finnbennach*) of Connacht. A number of other tales, called foretales (*remscéla*), purport to explain the events that lead up to the Cattle Raid, although the connection between the foretales and the Cattle Raid is often tenuous. The reason for the inability of the Ulaid to defend themselves is given in the tale *Ces Ulad* "the debility of the Ulaid." The otherworld woman Macha is forced to race against the king's horses while heavily pregnant. She gives birth to twins on winning the race, and as she lies dying she curses the Ulstermen so that they will suffer the pangs of childbirth at times of greatest danger. The origin of the two bulls is explained in *De Chophur in dá Muccida* ("Of the generation of the two swineherds"). The swineherds of the title transform themselves into various animals to demonstrate their magical powers. When they take on the form of worms, they are swallowed by two cows that subsequently give birth to the two bulls. Another important prefatory tale is "The Exile of the Sons of Uisnech" (*Longas macc nUisnig*), which explains how various Ulster warriors, most notably Fergus mac Róich, went into exile in Connacht and so accompany Ailill and Medb on the Cattle Raid.

The earliest surviving version of the tale was compiled in the eleventh century from ninth-century material, and the earliest copy is preserved in *Lebor na hUidre*. This version has been heavily criticized for the lack of unity that results from the presence of different linguistic strata, doublets, variants, inconsistencies, and interpolations. However, the aim of the redactor was scholarly rather literary, and it has been suggested that he deliberately juxtaposed contradictory versions in an attempt to establish the historical facts. In the twelfth century, the tale was revised to produce a more consistent narrative, and this version is found in the Book of Leinster. The story was clearly known long before this, as it is referred to in three seventh-century poems: one attributed to the Morrígain, which is preserved in the Cattle Raid, *Verba Scáthaige* ("Scáthach's words"), and a poem by Luccreth moccu Chérai. A later tradition attributes the "finding" of the story to the son of the seventh-century poet, Senchán Torpéist, who supposedly obtained it directly from Fergus mac Róich.

Historicity

The surviving texts postdate the period in which the Cycle is set by some six hundred years, and so they cannot be viewed as historically reliable. Nevertheless, many scholars have been struck by parallels between the Ulster Cycle and classical accounts of the Gauls and Britons of the second and first centuries B.C.E., which appear to suggest that the tales were remarkably conservative. More recent scholarship has shown that Christian monks had a far greater creative influence on these tales than formerly believed. Many of the surviving tales are fresh compositions, while others may have adapted traditional material to suit a contemporary context. Studies of the material culture depicted in the Cycle have shown that it broadly reflects post-Viking society, and analyses of the tales themselves show a concern for contemporary matters, often of directly Christian interest. According to this approach, the convincingly archaic nature of the tales was deliberately cultivated by the writers of the tales. Nevertheless, it is likely that a small number of elements, such as the enmity between Ulster and its southern neighbors, do preserve genuine memories. Emain Macha itself was an important site at the time in which the Cycle is set, although archaeological investigations have shown that it was a religious structure rather than a habitation site.

Some scholars have sought the origins of the Cattle Raid of Cooley and associated tales in pagan myth. The conflict between the two bulls at the end of the Cattle Raid resulting in the reshaping of the physical landscape is widely thought to reflect a cosmogonic myth. Medb, whose name may mean "drunken one" or "she who makes drunk," is seen by many as a reflex of the goddess of sovereignty, but she has also been interpreted as a vague memory of a once-powerful queen such as Boudicca of the Iceni of Roman Britain. Conchobor is described as an earthly god (*día talmaide*) of the Ulstermen in *Lebor na hUidre*, but this may merely be an expression of his exulted status rather than a belief in his divinity. Conall Cernach has been compared to Cernunnos, who is usually depicted sporting stag's antlers in continental European art and is often accompanied by other animals, but the evidence is inconclusive.

Heroic Conduct

The Ulster Cycle is a heterogeneous body of material written at different times and locations with diverse aims. Nevertheless, most of the texts are concerned with the fundamentals of heroic behavior: valor, loyalty, martial prowess, and adherence to the martial code of honor. Cú Chulainn, for instance, preferred fame to long life, and the Cattle Raid of Cooley is a celebration of his bravery and skill against superior odds. Warriors were morally bound by a heroic code of conduct that guaranteed fair play in battle. In *Breslech Mór Maige Muirthemne* ("The great rout of Mag Muirthemne"), Conall Cernach ties one of his arms behind his back before fighting Lugaid who has lost an arm fighting against Cú Chulainn. Warriors eschewed any semblance of cowardice and were given to vaunting their own bravery. The originally eighth-century tale of Mac Dathó's pig (*Scél Mucce Meic Dathó*) shows the Ulstermen and Connachtmen engaged in a series of boasts about their conquests in a bid to win the right to the champion's portion (*curadmír*). However, these acts of bravado end in a devastating battle and humiliation for the kings of both Ulster and Connacht. The posturing of warriors is further parodied in *Fled Bricrenn* ("The feast of Bricriu").

Women are often portrayed negatively. Medb is by far the most prominent woman in the Cycle, even rivaling her husband Ailill in some tales. In the Cattle Raid of Cooley, she usurps the role of the king, and she is portrayed as foolhardy, manipulative, and immodest, offering sexual favors to warriors who will fight against Cú Chulainn. The positive traits of women are frequently depicted as virtue, modesty, fidelity, wisdom, beauty, and skillfulness. When portrayed in such a light, they often act as a counterpoint to their menfolk. In *Aided Óenfhir Aífe* ("the death of Aífe's only son"), Cú Chulainn's wife Emer attempts to prevent him from engaging his own son in mortal combat. In *Scél Mucce Meic Dathó*, the Leinster hosteller Dathó falls ill with worry when both the Connachta and the Ulaid ask him for his famous hound. His wife determines the cause of his illness and devises a clever ruse that results in a battle between the Connachta and the Ulaid.

GREGORY TONER

References and Further Reading

Aitchison, N. B. "The Ulster Cycle: Heroic Image and Historical Reality." *Journal of Medieval History* 13, no. 2 (1987): 87–116.

Bhreathnach, Edel. "Tales of Connacht: *Cath Airtig, Táin Bó Flidhais, Cath Leitreach Ruibhe*, and *Cath Cumair*." *Cambrian Medieval Celtic Studies* 45 (Summer 2003): 21–42.

Dillon, Myles, ed. *Irish Sagas*. Dublin & Cork: Mercier, 1968.

Gantz, Jeffrey. *Early Irish Myths and Sagas*. London: Penguin, 1981.

Kelleher, J. V. "The Táin and the Annals." *Ériu* 22 (1971): 107–127.

Kinsella, Thomas. *The Tain*. Oxford: Oxford University Press, 1969.

Mallory, J. P., ed. *Aspects of the Táin*. Belfast: December Publications, 1992.

O'Leary, Philip. "*Fír fer*: An Internalized Ethical Concept in Early Irish Literature?" *Éigse* 22 (1987): 1–14.

O'Rahilly, Cecile, ed. *Táin Bó Cúalnge from the Book of Leinster*. Dublin: Dublin Institute for Advanced Studies, 1967.

Radner, J. N. "'Fury Destroys the World': Historical Strategy in Ireland's Ulster Epic." *Mankind Quarterly* 23 (1982): 41–60.

Thurneysen, Rudolf. *Die irische Helden—und Königsage bis zum siebzehnten Jahrhundert*. Halle (Saale): Max Niemeyer, 1921.

Toner, Gregory. "The Ulster Cycle: Historiography or Fiction." *Cambrian Medieval Celtic Studies* 40 (Winter 2000): 1–20.

See also ***Aided*; *Comperta*; Connachta; Emain Macha; Historical Tales; Mythological Cycle; Scriptoria; Ulaid**

ULSTER, EARLDOM OF

The Earldom of Ulster grew to be the most powerful territorial unit in Anglo-Norman Ireland by the end of the thirteenth century; yet from the mid-fourteenth century onward, the earldom went into decline, suffering a dramatic reduction in its territorial extent, and the original Anglo-Norman settlers—who became increasingly Gaelicized—lost virtually all contact with the central government in Dublin.

Control of the territory that eventually became the earldom of Ulster was first won by John de Courcy in 1177 from the ancient Mac Duinn Sléibe dynasty of Ulaid. De Courcy's conquest, which roughly comprised the modern counties Down and Antrim, was consolidated with settlement from England, notably from Cumbria—just over seventy miles away across the Irish Sea—where de Courcy had family connections. De Courcy ruled Ulster with exceptional independence for over twenty-five years until, in 1205, he was expelled by Hugh II de Lacy (d. 1242). De Lacy was rewarded by King John with de Courcy's lands, and it was at this point that Ulster became the colony's first earldom. De Lacy was himself expelled by King John during his expedition to Ireland of 1210, for association with the king's enemies. De Lacy was not restored until, in the 1220s, he resorted to open war with the government. In 1227, he was confirmed as earl of Ulster for life, so that when he died in 1242, the earldom reverted to the crown.

By this time, the earldom had developed into the five administrative areas of Down, Antrim, Carrickfergus, Newtown Blathewyc (Newtownards), and Coleraine. The borders of these regions were not precisely defined and fluctuated periodically, but the earldom protected itself by densely covering the landscape with mottes, although rarely with the accompanying bailey. There were important castles at Dundrum, Greencastle (Co. Down), and Coleraine. The impressive fortress at Carrickfergus was a royal castle and remained garrisoned to the end of the Middle Ages. Ulster held liberty jurisdiction, meaning that, with only minor exceptions, the earls ruled independently of the crown and kept separate administrative records. These records were stolen from Trim Castle in the 1490s, and their loss may partly explain why the earldom has been so neglected by historians of the Irish lordship. Fortunately, royal records afford a glimpse into its workings during the periods when the earldom lapsed or the earl was a minor.

Ulster remained in the king's hands from the death of de Lacy until, in 1263, it was granted to the lord of Connacht, Walter de Burgh. Walter died in 1271 and was succeeded by a minor, Richard de Burgh (d. 1326), the "Red Earl." Richard gained control of Ulster in 1281, and during his tenure the earldom reached the height of its territorial extent and influence. Richard's control extended west of the river Bann to Derry, and he built Northburgh Castle on the Inishowen peninsula. He gained the submission of all the native Ulster lords except Ua Domnaill, claiming from them military service known as the "bonnaght" of Ulster (from *buana*, a hired soldier).

The career of the "Red Earl" illustrates how Ulster, far from being a peripheral region, was part of a wider political community linked by the Irish Sea. As noted, de Courcy colonized Ulster from the north of England. Following de Lacy's forfeiture in 1210, large tracks of the earldom's coastline were granted to the Scottish earls of Athol, Carrick, and the lord of Galloway. This interconnection was perpetuated under de Burgh, who captured the Isle of Man for Edward I in 1290 and served with his Gaelic retinue in Scotland in 1296 and 1303. He was, moreover, linked by marriage to the Bruces—earls of Carrick and future kings of Scotland—whose claims to land in the earldom cannot have been forgotten when between 1315 and 1318 Ulster was the base for Edward Bruce's attempt to claim the kingship of Ireland. When, in 1328, Richard de Burgh's grandson William (the "Brown Earl") attempted to take control of his earldom, which had been disturbed since the death of his grandfather in 1326, he was supported by a now ailing King Robert Bruce of Scotland.

The Bruce invasion had a devastating effect on Ulster, and the earldom suffered a further blow in 1333 when the "Brown Earl" was assassinated by his own vassals. Thereafter, the earldom fell into the hands of absentees. It descended by marriage to Lionel of Clarence, and thence to the Mortimer earls of March.

Without the influence of a resident earl, the Gaelic Irish and mercenary Scots—notably the Clandeboye O'Neills and the Mac Donnells of the Glens of Antrim—encroached on the earldom, pushing the principal families into south County Down and the Ards peninsula. These families increasingly adopted Irish customs. We should, however, be careful not to exaggerate this development; it had begun long before 1333. Since de Courcy's time, there had been veneration for Irish saints and alliances with native Ulster lords. At his death in 1326, the "Red Earl" was the subject of a Gaelic praise poem. Moreover, acculturation also moved in the other direction, as is shown by the appearance of the name "Henry" among the Ua Néills of Clandeboye. Nor in practical terms did English rule in Ulster end immediately with William de Burgh's murder in 1333. It was a gradual process, and the earldom was still providing revenue, under its hereditary seneschals the Savages, in the 1350s. Moreover, for the rest of the medieval period, successive earls of Ulster—chosen for their connection with Ireland to serve as chief governor—attempted to regain the "bonnaght" of Ulster, which had been appropriated by the Ua Néills. The Gaelic chiefs repeatedly promised to render service. These promises were not made a reality, but this was in part due to chance rather than impotence. For instance, both the sixth (d. 1381) and eighth (d. 1425) Mortimer earls died within two years of their first successes in Ulster, with the result that the submissions they had taken could not be given practical effect.

In 1425, at the death of Edmund Mortimer, Ulster passed to Richard, duke of York. When the house of York came to the throne in 1461, Ulster became a permanent appanage of the English crown. Proposals for a reconquest appear in the accounts of the early Tudor period but were not implemented. Nonetheless, in 1541 when Conn Bacach Ua Néill suggested that he be made earl of Ulster, Henry VIII strongly rebuked him, reputedly calling Ulster one of the great earldoms of Christendom and an ornament of the crown. The Ua Néills were only made earls of Tyrone, and the royal claim to Ulster continued to be a factor in crown policy into the early modern period.

PETER CROOKS

References and Further Reading

Duffy, Seán. "The First Ulster Plantation: John de Courcy and the Men of Cumbria." In *Colony and Frontier in Medieval Ireland: Essays Presented to J. F. Lydon*, edited by Terry Barry, Robin Frame, and Katharine Simms, pp. 1–27. London: The Hambledon Press, 1995.

MacNeill, T. E. *Anglo-Norman Ulster: The History and Archaeology of a Medieval Barony.* Edinburgh, 1980.

Orpen, Goddard Henry. "The Earldom of Ulster." *Journal of the Royal Society of Antiquaries of Ireland* 43 (1913), 30–46, 133–143; 44 (1914), 51–66; 45 (1915), 123–142; 50 (1920), 166–177; 51 (1921), 68–76.

See also **Bruce, Edward; Courcy, John de; John; Lacy, Hugh de; Lionel of Clarence; Mac Donnell; Mortimer; Ua Néill of Clandeboye**

URBAN ADMINISTRATION

The practice of urban self-government in Europe long predated the historical evidence for such activity. The sense of community that made townspeople feel different from country people came in the first instance from living together in even closer proximity and in larger numbers, from the need to import most of their food, and from a desire to protect wealth accumulated by means of craftworking and trading. The urban "community" (Latin *communitas*) included everyone in theory, or at least all adult males. This is why open-air assemblies were the norm in the early Middle Ages; only later did urban administration become the preserve of more exclusive groupings. Before the Anglo-Norman invasion, the only real towns in Ireland were the Hiberno-Norse trading settlements—few in number and scattered around the coastline of the southern half of the island. Being of Scandinavian origin, they may have had an assembly (Norse *thing*) at which decisions were reached collectively in accordance with local custom. Only Dublin provides convincing evidence: outside the town to the east there was an assembly place or Thingmót, where warrior-merchants met presumably under the presidency of the king, and possibly also of the bishop (later the archbishop) after about 1030. That the townspeople came to think of themselves as burgesses (Latin *burgenses*) is indicated by a letter sent to the archbishop of Canterbury in 1121. Eventually it may have become customary to meet in a large hall, referred to by Giraldus Cambrensis as the "court" (Latin *curia*). Modern archaeologists have seen in the regularity of the house plots of Hiberno-Norse Dublin a sign of some kind of regulatory authority in the period before the Anglo-Norman takeover in 1170. The mint that was operating in Dublin from 997 down to the 1120s must have had a designated and publicly accessible location, possibly in the precinct of Christ Church Cathedral. Collective decision making, therefore, was a tradition rather than a novelty by the late twelfth century, if only on a limited scale.

The Formalization of Municipal Self-Government

Many existing towns in western Europe came to acquire more complete independence and more formal recognition of that independence during the twelfth and early thirteenth century. These developments—more or less universal—happened to coincide with the colonial process in Ireland. In addition, more towns were founded in parts of the country, and, as elsewhere, townspeople petitioned their rulers for charters as expressions and guarantees of urban "liberty." There were two types of town: those (generally larger towns) whose lord was the king of England and those (generally smaller towns) whose lord was a lay or an ecclesiastical aristocrat. For the former, the legal model was Bristol in England; for the latter, the small town of Breteuil-sur-Iton in Normandy. In a general sense, however, Dublin acted as the chief role model in Ireland, and its progress toward self-government is a classic demonstration of the stage-by-stage process whereby rulers made considerable sums of money by granting concessions in a piecemeal fashion. Having been selected by King Henry II as the main focus of loyalty to the English crown in Ireland, Dublin was handed over to merchants from Bristol for a whole generation. Only in 1192 was the city granted its first charter of urban liberties as an independent entity. At this stage the essence of municipal administration was the hundred court (named after a subdivision of Anglo-Saxon shires), which met weekly and enjoyed a wide range of administrative competence. An important feature of its procedures was trial by fellow burgesses. A further liberty with wide-ranging administrative implications was granted in 1215—the right to assess and to collect the fee farm or city rent of two hundred marks (a mark was two-thirds of one pound sterling) payable to the exchequer in two annual installments. The municipal authorities now had a direct financial relationship with all householders in the city, though no detailed records have survived. An even more decisive advance toward autonomy in administrative matters was made in 1229, when King Henry III granted the citizens permission to elect a mayor (Latin *maior*). After 1229, the city administration was headed by the mayor and two provosts, called "bailiffs," from 1292. A council of twenty-four members also gained official recognition, and the commonalty's status as a corporate body (though not yet legally incorporated) was confirmed by a common seal for authenticating documents.

Other towns in Ireland obtained privileges that would form the basis of self-government at different times and with varying results. The earliest example of a purely Anglo-Norman-chartered town is Drogheda-in-Meath, whose burgesses were accorded a version of the laws of Breteuil by Walter de Lacy in 1194. A slightly later seigneurial creation is Kilkenny, whose first documented privileges date from around 1200 and whose *Liber Primus* begins about thirty years later with the election of a town council of twelve members, together with a town administrator called the sovereign. As at Dublin, Kilkenny's hundred court met every week. To judge by the size of its fee farm, one hundred marks, Waterford was the second most important town in medieval Ireland; its burgesses were allowed to collect this money themselves from 1232 and to elect their own mayor from about 1254. In yet other towns, the chief administrative officer might be called the portreeve or the seneschal. All of these officers acted both as figureheads and as intermediaries with the overlord, whether king or nobleman. This relationship was mediated through oaths of loyalty. A major responsibility of urban administrators was the construction and maintenance of military defenses, in the shape of walls, mural towers, gates, and ditches. To that end, English kings granted special murage charters to royal and non-royal townspeople to enable them to raise funds. These operations would have placed an enormous financial and logistical burden on administrators and on those who were administered by them. The most dramatic expression of this is the Anglo-French poem describing the excavation of the town ditch at New Ross in 1265. A weekly rota was drawn up, in order that different socioeconomic groups would perform their share of the labor; even the town's priests and womenfolk were recruited. This is a fine illustration of that collective sense of responsibility that lay at the heart of medieval self-government.

Late Medieval Developments and Difficulties

Formal charters granting urban privileges were expressed as a rule in terms of general principles; it was left to their recipients to work out the details of urban administrative procedures. Broadly speaking, we have more evidence about these matters from the fourteenth and fifteenth centuries than from the great age of town growth itself, the records of Dublin being the most informative. One positive development there was the initiation of a new municipal book, known as the Chain Book because it was secured by a chain for public consultation in the tholsel, or city hall. The main item is a long list of by-laws ("laws and usages") drawn up in French, still the official language of legal enactments, both central and local, and a widely known vernacular. By the early fourteenth century the mayor and two bailiffs headed a complex structure of twenty-four *jures* (making up the regular city council), forty-eight *demi-jures*, and a body called the ninety-six (together forming the common council). The latter met four times a year, and from 1447 onward its minutes or assembly rolls have survived. The principal functions of the mayor were to preside over the hundred court, to execute decisions reached by the city council and the common council, and to represent the citizens vis-à-vis the outside world. The bailiffs assisted the mayor, enrolling contracts, supervising the seizure of goods, confiscating stray animals, and performing other tasks with quasi-legal connotations. The chief officers in turn were assisted by a host of functionaries ranging from the recorder, treasurer, and auditors at the top, to sergeants, jailers, the keeper of the dockside crane, and the water bailiff at the bottom. The annual appointment of these men ensured a degree of control over their activities. There are signs that the administrative burden was onerous: men elected as mayor or bailiff were fined for refusal to serve their term in office. Depopulation caused by the Black Death may have made the pool of eligible men too small: the surviving franchise roll for the years 1468–1512 implies that the majority of new admissions came from lower social levels (by apprenticeship) and from outside the city.

Waterford had a similar set of by-laws modeled on those of Dublin, as well as a tripartite structure of councilors. In the city's records, there are comparable signs of the difficulties experienced in persuading leading citizens to serve in high office. A serious deterrent was the need for mayors of Waterford to be proactive militarily from circa 1320, as the position of the colonists steadily worsened; the unfortunate John Malpas was actually killed while on campaign in 1368. For its citizens' heroic services to the English crown, Waterford was conferred with a civic sword for ceremonial purposes in 1462, as Dublin already had been in 1403. Apart from these two cities, however, the mechanisms of urban administration are not well recorded. One unusual survival is a mid-fifteenth-century land-gavel ("ground rent") roll for Cork. It is unfortunately incomplete, but the fact that nine individuals held between them 40 percent of the recorded properties may point toward a sharp reduction in the city's population. Indeed, Cork's fee farm payments to the central administration declined from the 1340s and ceased altogether after 1416. Over forty functioning cities and towns survived in Ireland into the sixteenth century, but the degree of competence and diligence with which most of them were administered is virtually impenetrable.

H. B. CLARKE

References and Further Reading

Bateson, Mary, ed. *Seldon Society*, Vols. 18, 21, *Borough Customs.* London: Bernard Quaritch, 1904–1906.

Clark, Mary, and Gráinne Doran. *Serving the City: The Dublin City Managers and Town Clerks, 1230–1996.* Dublin: Dublin Public Libraries, 1996.

Clarke, H. B. "The 1192 Charter of Liberties and the Beginnings of Dublin's Municipal Life." *Dublin Historical Record* 46 (1993): 5–14.

Edwards, R. D. "The Beginnings of Municipal Government in Dublin." *Dublin Historical Record* 1 (1938–1939): 2–10.

Gilbert, J. T. "Archives of the Municipal Corporation of Waterford." In *Historical Manuscripts Commission, Tenth Report*, app., pt. 5, pp. 265–339. London: Her Majesty's Stationery Office, 1885.

Lennon, Colm, and James Murray, ed. *The Dublin City Franchise Roll, 1468–1512.* Dublin: Dublin Corporation, 1998.

McEneaney, Eamonn, ed. *A History of Waterford and its Mayors from the 12th Century to the 20th Century.* Waterford: Waterford Corporation, 1995.

Mac Niocaill, Gearóid. *Na Buirgéisí XII-XV Aois.* 2 vols. Dublin: Cló Morainn, 1964.

———. "Socio-economic Problems in the Late Medieval Irish Town." In *Historical Studies*, vol. 13, *The Town in Ireland*, edited by David Harkness and Mary O'Dowd, pp. 7–21. Belfast: Appletree Press, 1981.

Thomas, Avril. *The Walled Towns of Ireland.* 2 vols. Blackrock: Irish Academic Press, 1992.

Webb, J. J. *Municipal Government in Ireland, Mediaeval and Modern.* Dublin: Talbot Press, 1918.

See also **Charters and Chartularies; Fraternities and Guilds; Local Government; Ports; Records, Administrative; Walled Towns**

V

VALENCE, DE

The connection of the de Valence family with Ireland began in 1247 when William de Valence (d. 1296), the Poitevin half-brother of Henry III, married Joan, daughter of Joan de Munchensy and heiress to one-fifth of the Marshal lordship of Leinster. By this marriage William, lord of Montignac and other lands in France in his own right, became lord of Wexford in Ireland and gained the lands of the earldom of Pembroke in Wales, as well as lands in England. William's career was focused on the English court; and he served Henry III and Edward I as a counselor and envoy, roles that his son, Aymer, was to fulfill for Edward II. William was particularly interested in increasing his lands and rights in the earldom of Pembroke, the official title to which he coveted.

William was, nevertheless, also interested in his lands in Wexford. These constituted between six and thirteen percent of William de Valence's total annual income of about £2,500. It was perhaps in order to acquire a local ally to facilitate the maintenance of his Irish interests that William married his daughter, Agnes, to Gerald Fitz Maurice, one of the Geraldine lords of Offaly and a member of one of the most important settler families in Ireland. Although he did not make the trip, William considered traveling to Ireland in early 1272 regarding the purchase of the custody of the lands, with marriage of the heirs, of Gerald Fitz Maurice, a potential rival to the claims of his daughter, Agnes. Thirty years later, the absentee Agnes was still tenacious in her pursuit of her rights in Ireland.

On the death of William in 1296 the management of Wexford fell to Joan, countess of Pembroke, who was succeeded by her son, Aymer de Valence, the earl of Pembroke, in 1307. Aymer does not appear to have been very interested in his lordship of Wexford. Transcripts of legal records show that he (and Joan) were interested in maintaining their rights in Ireland, but that neither were particularly litigious in this respect. Wexford contributed just over ten percent of Aymer's total annual landed income of £3,160, but it was not only financial considerations that decided where the earl's focus lay. Aymer, as one of the few magnates who demonstrated continuous loyalty to Edward II, was heavily involved in English politics. His lands in France made him particularly valuable to Edward II as a diplomat on the continent. Indeed, it was his service as an envoy to Avignon in 1316 that accounted for his absence from the list of absentees summoned to the defense of the lordship of Ireland, in response to the Bruce invasion.

On his childless death in 1324, Aymer's estates were divided between his nephew and two nieces. His wife, Mary de Sancto Paulo, Countess of Pembroke, held dower lands in Ireland, which she was summoned to defend in 1331 and, again, in 1361. The interest of the de Valence family proper in Ireland, however, had already ended in 1324.

BETH HARTLAND

References and Further Reading

Dictionary of National Biography. Oxford: 1917–.

Ó Cléirigh, Cormac. "The Absentee Landlady and the Sturdy Robbers: Agnes de Valence." In *"The Fragility of Her Sex?" Medieval Irishwomen in Their European Context*, edited by Christine Meek and Katherine Simms, 101–118. Dublin: Four Courts Press, 1996.

Phillips, J. R. S. *Aymer de Valence, Earl of Pembroke 1307–1324: Baronial Politics in the Reign of Edward II.* Oxford: 1972.

Ridgeway, Huw. "William de Valence and His *Familiares*, 1247–72." *Historical Research* 65 (1992): 239–257.

See also **Leinster; Marshal**

VERDON, DE

In 1185, Bertram III de Verdon (d. 1194) was sent to Ireland as seneschal of the Lord John. Bertram established himself in Louth, and his service was rewarded with the grant of the lordship of Dundalk in 1189. This led to a reorientation of the family's landed interests, which had previously been concentrated in the English midlands and Normandy. The de Verdons sought to establish themselves within Anglo-Irish society through marriage. For example, Thomas (d. 1199) married his sister, Leselina, to Hugh de Lacy, endowing them with half of his Irish lands. This did not lead to cordial relations in the short term, as Nicholas (d. 1231) sought to regain these lands, resulting in a period of sustained conflict with Hugh. It was not until 1235, when Rohesia (d. 1247) recovered part of the Dundalk lands, that this dispute was settled.

The marriage of John de Verdon (d. 1274) to Margery de Lacy may have been intended to smooth over relations between the two families, but it also boosted the de Verdon family within the social hierarchy. Following the failure of the de Lacys in the male line in 1241, John became lord of the western half of Meath in Ireland and lord of Ewyas Lacy, in the Welsh March, in right of his wife. The other half of the de Lacy inheritance passed to Geoffrey de Geneville.

In 1266, John began his attempt to regain the full judicial liberties once held by Walter in Meath. This legal battle was continued by Theobald I (d. 1309), but the de Verdons were unsuccessful, although this privilege had been granted to de Geneville. The failure to secure these rights may have contributed to the reorientation of the main de Verdon line away from Ireland and toward the Welsh March, where their franchise remained wide. Nevertheless, de Verdon authority over their tenants in Louth remained strong, such that Robert (a younger brother of Theobald II) was able to lead much of the county in the still unexplained "de Verdon rebellion" in 1312.

The Gaelic Revival may also have played a part in the reorientation of the main de Verdon line. During John's absence on crusade in 1271, his sons Nicholas and John were killed defending the family lands in Louth. Another John, the eldest son of Theobald I, was also killed by the Irish in 1297. Theobald II divided his time between England and Ireland far less equally than his immediate forbears, and he appointed his younger brother Milo as chief guardian of his lands and fees in Ireland in 1309. Arguably this did not represent a lack of interest in Ireland, but rather a sensible approach to the problem of cross–Irish Sea land-holding in a period of political uncertainty.

Theobald de Verdon had no sons. His lands were divided between his widow, Elizabeth de Clare (d. 1360), and his four daughters. The final partition was effected in 1332, custody of the lands in the interim being granted to Theobald's brothers, Milo and Nicholas, who actively defended them during the Bruce invasion. The absentee de Verdon co-heirs eventually sold their lands in Ireland between 1366 and around 1378.

Beth Hartland

References and Further Reading

Hagger, Mark S. *The Fortunes of a Norman Family: The De Verduns in England, Ireland and Wales, 1066–1316.* Dublin: 2001.

Hartland, Beth. "Vaucouleurs, Ludlow and Trim: The Role of Ireland in the Career of Geoffrey de Geneville (c. 1226–1314)." *Irish Historical Studies* 33 (2001): 457–477.

Otway-Ruthven, A. J. "The Partition of the de Verdon Lands in Ireland in 1332." *Proceedings of the Royal Irish Academy,* 66C (1968): 401–455.

Smith, Brendan. *Colonization and Conquest in Medieval Ireland: The English in Louth 1170–1330.* Cambridge: 1999.

See also **Anglo-Norman Invasion; Bruce, Edward; Clare, de; Courcy, John de; Gaelic Revival; Geneville, Geoffrey de; Lacy, de; Lacy, Hugh de; March Areas; Mide (Meath); Mortimer; Pale; Plunkett; Trim**

VIKING INCURSIONS

Viking incursions are first recorded in Ireland in A.D. 795. The earliest targets were churches and communities located on islands or near the coast. While surprise was an essential feature of these early hit-and-run

Viking settlements.

attacks, Irish armies began to intercept the raiders, with varying degrees of success, from the 810s. By the 820s, Viking fleets had circumnavigated Ireland (for example, raiding Skellig Michael in 823). The long distance of some of these campaigns from Scandinavia has led some commentators to suggest that the first raiders came from colonies in the Northern and Western Isles of Scotland.

Initially, Vikings seem to have been motivated by desire for portable wealth in the form of tribute, stolen goods (including reliquaries—the contents were sometimes discarded), and slaves. Some of the loot made its way back to Scandinavia, as demonstrated by finds of insular metalwork in Norwegian graves.

In the 830s and 840s, Vikings made more strenuous efforts to establish a foothold in Ireland. Numerous bases, sometimes called *longphoirt* (ship ports), were founded. The earliest recorded examples are Arklow (836), Lough Neagh (839), and Dublin and Annagassan (841). At the same time Viking campaigns extended further inland, exploiting the major river systems of Ireland. The bases enabled booty to be ransomed or traded locally, and they fostered closer interaction between Vikings and the Irish. Many bases were temporary, but others, notably Dublin, Limerick, and Waterford, have been occupied ever since.

A number of Irish kings used this new turn of events to their advantage by recruiting Viking support against their enemies. Alliances between Vikings and Irish, such as that involving the Osraige king Cerball mac Dúngaile, are well attested from the mid-ninth century. In consequence, the reasons behind Viking incursions became more sophisticated, combining desire for booty with political strategy. Of the native rulers who sought to block Viking expansion in Ireland, Máel-Sechnaill, son of Máel-ruanaid, achieved particular success.

It is evident that a number of different Viking armies operated in Ireland. During the 850s, three major groups jostled for power: *Finngaill* (Fair Foreigners), *Dubgaill* (Dark Foreigners), and *Gall-Goídel* (Foreign-Gaels). The Dark Foreigners were ultimately successful under the leadership of Ívarr (who died in 873) and his descendants, who were based in Dublin.

During the late ninth century there was a decrease in recorded attacks in Ireland, which may be linked with the activities of Ívarr and his associates in Britain. In 866 and 867, Ívarr's absence encouraged Irish rulers to destroy a number of Viking bases. After a resurgence of Viking attacks in Ireland in the late 870s and 880s, the power of Ívarr's descendants was compromised by dynastic infighting. This led to the expulsion of leading Vikings from Dublin in 902 by a coalition of troops from Leinster and Brega. The exile of the dynasty of Ívarr lasted until 914. In the interim, there is scant record of Viking activity in Ireland, while Cerball mac Muirecáin, who had ousted the Viking leaders from Dublin, died in 909.

After the restoration of the dynasty of Ívarr, there was a period of vigorous Viking activity that lasted until the 940s. These years perhaps mark the zenith of Viking power in Ireland. Recurrent attacks were led against Irish power centers such as Armagh and Clonmacnoise. There was also fierce competition between the Viking settlements of Dublin and Limerick. Numerous Viking bases were established across Ireland in these years as the rival groups sought to extend their sphere of influence. These incursions were curtailed in 937, when the Vikings of Limerick were crushed by their Dublin rivals, and in the 940s by defeats inflicted on Dublin by Congalach, over king of Brega.

In the late tenth century, Viking settlements increasingly fell under the influence of Irish rulers. Not only did Viking incursions decrease in number, but their actions became less autonomous. Brian Boru brought Waterford, Limerick, and (temporarily) Dublin under his control before his death at the battle of Clontarf in 1014.

From that time, ambitious Irish over kings vied for control of the wealth and military resources of Dublin, which was the premier town of Ireland. Viking armies increasingly acted under the direction of Irish leaders. Nevertheless, the Viking dynasty of Ívarr still remained influential in Dublin. A branch of the dynasty, which ruled the Hebrides and Man, also continued to intervene in Irish affairs. In 1091, 1142, and perhaps in 1162, Viking kings of the Isles seized control of Dublin. After Magnús, king of Norway, took control of the Isles in 1098, he also intruded in Irish politics. He was killed on a raid in Ulster in 1103, and his alleged son from an Irish or Hebridean lover went on to rule Norway. Thus, for an extended period, the kingdom of the Isles was closely linked with Viking activity in Ireland.

The Viking Age in Ireland ended in the 1170s when the English seized control of the Hiberno-Scandinavian towns. In a final gasp for power, Ascall, the deposed ruler of Dublin, led a contingent from Man and the Isles against the town in 1171, but he was captured and beheaded. The term *gall* (foreigner), most frequently applied by Irish chroniclers to the Vikings, was soon after transferred to the English.

Viking incursions had a significant impact on Irish history. To them, past scholars have attributed both the decline of Uí Néill as over kings of Ireland and the increasing lay control of churches—although such views have since been modified. In the economic sphere, Vikings stimulated trade through their network of external contacts. They founded towns and introduced coinage. Irish rulers adopted Viking military techniques, and Vikings also made their mark on Irish art and literature. In turn, the Irish exercised a profound influence on Viking settlers. As Vikings from Ireland made incursions elsewhere, this influence extended to other colonies. Thus, Hiberno-Viking impact can be traced in diverse sources such as place names, saints cults, or medieval literature in Normandy, Iceland, and Western Britain.

CLARE DOWNHAM

References and Further Reading

O'Donovan, John, ed. and trans. *Annala Rioghachta Eireann: Annals of the Kingdom of Ireland by the Four Masters, from the Earliest Times to the Year 1616*, 2nd ed. Dublin: 1856.

Clarke, Howard B. et al., eds. *Ireland and Scandinavia in the Early Viking Age*. Dublin: Four Courts Press, 1998.

Duffy, Seán. "Irishmen and Islesmen in the Kingdoms of Dublin and Man, 1052–1171." *Ériu* 43 (1992): 93–133.

Etchingham, Colmán. *Viking Raids on Irish Church Settlements in the Ninth Century: A Reconsideration of the Annals* (Maynooth Monographs, Series Minor 1). Maynooth: 1996.

Lucas, A. T. "Irish-Norse Relations: Time for a Reappraisal?" *Journal of the Cork Historical and Archaeological Association* 71 (1966): 62–75.

See also **Brian Boru; Cerball mac Deungaile; Cerball mac Muirecáin; Dublin; Limerick; Máel-Sechnaill I; Waterford**

VILLAGES

Very little is known about village life in medieval Ireland. Archaeological work has been limited, and the poor survival of records makes it difficult for the historian, but fortunately it has been a subject of research by historical geographers. The lack of interdisciplinary study is further exacerbated by the fact that from about the tenth century onward most English and continental peasants lived in villages, while in Ireland dispersed settlements appear to have remained the norm.

A village is a settlement intermediate in size between a hamlet and a town, but in practice the borderlines are vague and undefined. Commonly, a medieval village consisted not only of the built-up area of houses, outbuildings, gardens, haggards, and orchards, but also the surrounding fields from which the inhabitants derived their livelihoods. It has often been remarked that the Latin word *villa* should really be translated as "township," rather than "village." Furthermore, the medieval village was more than a settlement form. It was also a community and, indeed, a special type of community, in that it was one defined by common residence and a shared economic and social interest, rather than one bound together purely by the ties of kinship. Medieval Irish villages varied in size, physical form, function, and population. These differences suggest that there was a hierarchical ordering in the landscape as well as an economic and social complexity, but insufficient work has been done to establish the patterns.

Village life prior to the Anglo-Norman invasion remains nebulous, but during the eleventh and twelfth centuries the evidence for nucleation increases. This is indicated by the appearance of new words such as *baile* and *sráidbaile* as well as the archaeological and documentary evidence for the concentration of craftsmen and artisans at ecclesiastical sites. The phrase "monastic town" has been coined for larger settlements such as Armagh, Clonmacnoise, Kells, and Kildare, but there were also places of intermediate size that could be called monastic villages. Seventy or eighty houses, for instance, are recorded as being burnt at Duleek in 1123, while eighty houses were destroyed in the remodeling of Derry in 1162. Houses are also recorded at Ardagh (Co. Longford), Ardpatrick (Co. Limerick), Ardstraw, Cloyne, Devenish, Emly, Louth, Ratass, Roscommon, and Slane, among others, and nucleation was not confined to ecclesiastical sites alone. Excavations at Knowth, County Meath, have uncovered six or seven houses clustered below the royal site during the eleventh and twelfth centuries, while by the mid-twelfth century, there was a mixed community of clerics and burgesses at Killaloe, the settlement at the foot of the Ua Briain royal site of Kincora. Similarly, there were intermediate-sized settlements in the Hiberno-Scandinavian world, such as Arklow and Wicklow, which probably functioned primarily as fishing villages. Nonetheless, in view of the kin-based structure of early Irish society and the somewhat tantalizing nature of the evidence, the degree to which these settlements were villages in the accepted sense rather than kin-based agglomerations remains open to question.

After the Anglo-Norman invasion, the migration of English peasants led to the foundation of many new villages. The manorial lords frequently offered burgess rights to the colonists, leading to the establishment of what scholars have called rural boroughs: settlements with an agricultural economy, but in which property holders had the status of townspeople. This permitted the development of an organized village community led by a reeve, who may have been appointed by the lord or elected by the burgesses, and whose responsibility was to oversee the annual performance of obligations to the lord and the collection of rents and dues.

There has been a debate about the extent to which the colonists introduced the English village system and the degree to which the traditional Irish pattern of dispersed settlement was adopted. English-syle villages are found in the densely settled parts of the Anglo-Norman colony, such as Dublin, Kildare, Kilkenny, Meath, and Tipperary. Typically these would have had an arable infield set in strips while the outfield was grazed in common, and in Meath the ridges dividing the strips are referred to by the Middle English word selion. In the less densely colonized areas, such as Ulster and Connacht, villages are largely absent, and it has been argued that the pre-existing townland scheme militated against the formation of large, English-style nucleations. It has also been suggested, however, that the absence of villages in the landscape may be the result of a historical phenomenon—the movement of English tenants to the periphery of the manorial lands when villages were abandoned or a phase of secondary colonization in the thirteenth century.

Deserted village earthworks are rare in Ireland, with the greatest number occurring in south Tipperary. Here one frequently finds a church and a manorial center (typically a motte), surrounded by peasant houses of the English-speaking settlers. Farther away lived the Gaelic Irish-speaking betaghs who farmed their land in common and owed labor services and rents to the lord of the manor. In her study of deserted villages in Westmeath, Meenan found that village earthworks were predominantly associated with churches, with or without the presence of a motte. The church was an indicator of centrality and drew allegiance to the village. The remains typically consist of three to five houses with their associated garths. The numbers of houses do not indicate the original figures, but rather were the last ones to be deserted. Settlements that were abandoned early tend to have few earthworks, and the more prominent the archaeological features the later the date of desertion tends to have been. Peasant long-houses have been excavated at Caherguillamore (Co. Limerick) and Jerpointchurch (Co. Kilkenny). These were rectangular buildings divided into two rooms, one of which was the dwelling room and the other a byre.

Many villages were abandoned in the late thirteenth and early fourteenth centuries. Dunamase, for instance, had 127 burgesses in 1283 but only 40 in 1324. The reasons for desertion were varied: economic decline; famine; the Bruce wars (1315–1318) and the lawless nature of the countryside for twenty years after; the Black Death (1348–1349); increasing Gaelicization; and simply because the inhabitants thought that they might find better opportunities elsewhere.

JOHN BRADLEY

References and Further Reading

Glasscock, R. A. "The Study of Deserted Medieval Settlements in Ireland (to 1968)" and "Gazeteer of Deserted Towns, Rural-Boroughs and Nucleated Settlements in Ireland." In *Deserted Medieval Villages, Studies*, edited by Maurice Beresford and J. G. Hurst, 279–301. London: Lutterworth Press, 1971.

Meenan, Rosanne. "Deserted Medieval Villages of County Westmeath." M. Litt. dissertation, Trinity College Dublin, 1985.

O'Conor, K. D. *The Archaeology of Medieval Rural Settlement in Ireland*. Dublin: The Discovery Programme, 1998.

Simms, Anngret, and Patricia Fagan. "Villages in County Dublin: Their Origins and Inheritance." In *Dublin City and County: From Prehistory to Present, Studies in Honour of J. H. Andrews*, edited by F. H. A. Aalen and Kevin Whelan, 79–119. Dublin: Geography Publications, 1992.

See also **Armagh; Clonmacnoise; Ecclesiastical Sites; Houses; Kildare; Motte-and-Baileys**

WALL PAINTINGS

Medieval interiors were far more ornate than one might envisage from what survives of them today. One aspect of this decoration, wall painting, was used to ornament not just the walls and ceilings but also the carved details of the interior, such as capitals and tomb surrounds.

Approximately sixty-five medieval sites both ecclesiastical and secular, principally abbeys and castles, have extant and/or recorded wall paintings. Most of these buildings retain just a few small traces of the original decoration surviving in sheltered positions, as in the transept niches at Muckross Friary, County Kerry, and the double sedilia in the chancel at Fore Abbey, County Westmeath. Fragments from archaeological excavation also contribute to the number of surviving examples. A few wall paintings are known only from earlier records. Nothing survives today of the Trinity recorded in 1886–1887 by drawings and photographs at St. Audeon's, Dublin.

Lime-wash or plaster layers, accidental accretions, and microbial growth often conceal wall paintings. In addition to stabilization, conservation at a number of sites has revealed details of the imagery and subject matter, and aspects of materials and technology and dating. Information on patronage and ecclesiastical matters and details of weaponry and dress are gleaned, contributing greatly to the multidisciplinary study of the medieval period.

Most wall paintings are applied to one or more lime-plaster layers. This is tapered to a thin lime-wash preparation for painting of finely carved features such as tomb surrounds or window details. A guide or preliminary drawing was often mapped out into the still-damp plaster with a sharp implement. The pigments were applied onto either wet plaster (*fresco*) or a lime-wash layer overlying the plaster. Further colors could be added using a binding medium to the dry surface (*secco* additions). Yellow, red, and brown ochres, lime white, and bone or charcoal black colors have been identified visually. More costly pigments of cinnabar, *lapis lazuli*, and gold have been identified by analysis at a few sites.

The imagery includes imitation masonry patterns, consecration crosses, boats, and at a number of sites quite extensive figurative narratives. Associated with the O'Kelly burial monument (*c.* 1401–1403) at Abbeyknockmoy, County Galway, is the popular morality theme of the Three Living and the Three Dead Kings. With the message "we have been as you are, you shall be as we are," the skeletons admonish the kings for their vanity and encourage them to consider their own end. Below is a damaged Trinity alongside the Martyrdom of St. Sebastian. St. Sebastian (also found at Ballyportry Castle, Co. Clare, with an archbishop) is one of the patron saints of the plague, and here he reaffirms the theme of death and is in keeping with the commemorative role of the O'Kelly tomb.

St. Michael Weighing the Souls is found at Ardamullivan Castle with a bishop and scenes from the Passion cycle, and at Clare Island Abbey (Phase Two paintings) with diverse imagery. Set between painted imitation ribs, apparently secular and some aristocratic activities of musicians, fishing, hunting, and cattle raiding occur with fabulous beasts, dragons, and serpents alongside scenes of obvious religious meaning. Stag hunting, occurring with a Gaelic horseman, is also found on the earlier painting at Clare Island Abbey, and is recorded on the paintings at Urlan More Castle, County Clare (now collapsed) and Holycross Abbey, County Tipperary.

Karena Morton

Wall painting from Knockmoy Abbey. © *Department of the Environment, Heritage and Local Government, Dublin.*

References and Further Reading

Cochrane, R. "Abbey Knockmoy, County Galway: Notes on the Buildings and 'Frescoes'." *JRSAI* 34 (1904): 244–253.

Cochrane, R. "The Frescoes, Abbey Knockmoy. County Galway." *JRSAI* 34 (1905): 419–420.

Crawford, H. S. "Mural Paintings and Inscriptions at Knockmoy Abbey." *JRSAI* 49 (1919): 25–34.

Morton, K. "Medieval Wall Paintings at Ardamullivan." In *Irish Arts Review Yearbook 2002*, vol. 18 (2001): 104–113.

Morton, K. "Irish Medieval Wall Painting." In *Medieval Ireland: The Barryscourt Lectures I-X*. Kinsale: Gandon Editions, 2004, pp. 312–349.

Westropp, T. J. 1911. "Clare Island Survey: history and archaeology." *Proceedings of the Royal Irish Academy* 31 (1911–15), section 1, part 2, 1–78.

See also **Abbeys; Castles; Iconography**

WALLED TOWNS

There were four major phases of town foundation in medieval Ireland, and these follow one another in roughly chronological order: first, "monastic" towns; second, Scandinavian towns; third, Anglo-Norman towns; and fourth, Gaelic towns of the late middle ages.

"Monastic" towns and Scandinavian towns developed in the tenth century, although both had earlier origins. From the late seventh century onward some ecclesiastical settlements performed the urban functions of harbors, trading places, and centers of iron-working and craft production, while in the ninth century the Viking invaders established permanent settlements at sites such as Dublin, which are described in the annals as *longphoirt* (ship fortresses). Nonetheless,

Town wall, Waterford City. © *Department of the Environment, Heritage and Local Government, Dublin.*

little is known about these early settlements, and scholars are now agreed that towns in the sense of nucleated, densely populated centers, whose inhabitants were not engaged in primary production, are a feature of the tenth century and later.

The group of five Scandinavian port towns (Dublin, Wexford, Waterford, Cork, and Limerick) established, or in some cases re-established, between 914 and 922 are important in this regard. Of these Dublin is the best known, and excavations at Fishamble Street revealed an organized urban layout from around 925, when the settlement was first enclosed by an earthen bank. About the middle of the tenth century the embankment was raised and an external ditch added, while around the year 1000 the earthen defenses were enlarged and crowned by a post-and-wattle fence, later replaced by a stave palisade. These were the defenses that witnessed the battle of Clontarf in 1014. A stone wall was built around 1100 and endured until the Anglo-Norman invasion, although the town had acquired extramural suburbs by that time. Within the defenses virtually all of the buildings were of wood and were constructed of post-and-wattle. The remains of over 200 houses have been excavated, and the town was essentially the home of craftsmen and traders. Dublin's trading connections were extensive, and imported goods included silks from Byzantium and silver from the Arab world. The increasing status of the Dubliners and their identity as townspeople, distinct from others, is evidenced by a reference to them in 1127 as burgesses (*burgenses*). The archaeological evidence from Waterford is second only to that of Dublin. The same house types are evidenced, and they have also been discovered in Wexford and Cork, leading to the recognition that Hiberno-Scandinavian towns had a distinctive physical identity. Three houses of mid-twelfth century date have been excavated on the site of King John's Castle, Limerick.

Ecclesiastical settlements were enclosed by ramparts from at least the seventh century, but it is not until the eleventh and twelfth centuries that these can be described as defenses. In 1103, Armagh resisted a week-long siege, while the surviving twelfth-century gatehouse at Glendalough suggests that it was also defended.

The Anglo-Normans founded some fifty new towns and established the urban network that still endures over much of eastern and southern Ireland. Although chequer plans, such as at Drogheda and Galway, are occasionally found, the predominant street plan was linear, with the marketplace located in the center of the street and with houses positioned so that the gable was on the street frontage. Access to the house was often by means of a side lane, thus giving rise to the laneways that still characterize towns such as Clonmel, Drogheda, and Kilkenny. The houses themselves were positioned on long narrow properties, known as burgage plots, which frequently stretched from the main street to the town wall. These plots, combined with an acreage of arable land outside the walls and common of pasture, were granted by the lord of the town to the incoming colonial heads of household, who were given the status of burgesses in return for the payment of an annual rent, generally set at one shilling. The earliest town defenses were earthen, such as the example from the 1190s found in the course of archaeological excavation at Drogheda. Other towns, such as Duleek, retained earthen defenses throughout the Middle Ages. Defenses of earth and timber could be every bit as strong and difficult to capture as stone walls, but from about the 1220s onward the larger towns began to replace earthen ramparts with mortared stone. Stone defenses were more expensive, but they were also more prestigious, and in medieval art and cartography they were depicted as the symbol of a town. Town walls served not merely as barriers to attack; they also enabled the control of movement to and from the town, and the town gates were important points for gathering tolls. Among the tolls collected was murage, a tax on all goods coming into the town for sale, which was levied in order to raise monies to pay for the construction of the town walls. At first the grants were short and simple, but by the mid-fifteenth century the lists of taxable commodities had become long and elaborate. Although town defenses fell out of use by 1700, some towns, such as Cashel, continued to collect murage until the 1960s. The new Anglo-Norman towns are usually characterized by having one parish church, by the location of the lord's castle on the edge of the town, and by having religious houses and hospitals situated either just inside the town wall or outside the town completely.

In general terms the thirteenth century was a period of urban expansion and population increase, with extramural suburbs being a feature of many towns. By contrast, the fourteenth century was one of decline, brought on for much the same reasons as the contemporary desertion of villages. Some towns, such as Athlone, Rindown, and Roscommon, were abandoned completely. The fifteenth century was a period of consolidation, and it is not until the dissolution of the monasteries in the sixteenth century that expansion is again evidenced, when urban land once more became available for redevelopment.

The final phase of medieval urbanization, the development of towns such as Cavan and Longford in areas controlled by the Gaelic Irish, is still little understood. The towns copied their form and layout from the neighboring late-medieval towns within the Pale, and the townspeople seem to have profited particularly from the sale of horses and livestock to Anglo-Irish merchants. There is no evidence, however, that any of these Gaelic-Irish towns were walled.

JOHN BRADLEY

References and Further Reading

Bradley, John. *Walled Towns in Ireland.* Dublin: Town House, 1995.

Doherty, Charles. "The Monastic Town in Early Medieval Ireland." In *The Comparative History of Urban Origins in Non-Roman Europe*, edited by H. B. Clarke and Anngret Simms, 45–75. Oxford: British Archaeological Reports, 1985.

Thomas, Avril. *The Walled Towns of Ireland.* Dublin: Irish Academic Press, 1992.

Wallace, P. F. "The Archaeological Identity of the Hiberno-Norse Town." *Journal of the Royal Society of Antiquaries of Ireland* 122 (1992): 33–66.

See also **Anglo-Norman Invasion; Battle of Clontarf; Ecclesiastical Settlements; Houses; Viking Incursions; Villages**

WATERFORD

Waterford is one of the major medieval ports along the east coast of Ireland, originally founded as a Hiberno-Norse urban center in the tenth century. Its original Old Norse name was *Vedrarfjordr*, which probably means "windy fjord," where this inlet of the River Suir offered a safe haven for their ships. Indeed, it has been calculated that the quays of the city in the later Middle Ages could berth around 60 cargo ships. Thus it is hardly surprising that although Dublin remained the governmental and administrative capital of Ireland throughout the Middle Ages, it was ports such as Waterford in the Southeast that dominated her international trade. Throughout much of this period Waterford remained the largest exporter of wool, wool products, and hides, as well as the biggest importer of wines. It also became a significant entrepôt for French

Ardmore Round Tower, Co. Waterford. © *Department of the Environment, Heritage and Local Government, Dublin.*

wine, re-exporting much of it in the late thirteenth century to supply the armies of the English Crown fighting in Scotland. Its status as a royal borough strategically located on the important river system of the Suir-Nore-Barrow helped it to dominate both the political and the economic life of much of South Leinster and North Munster. Waterford also fought a bitter economic war with its near neighbor, New Ross, for control of this rich hinterland.

In the Hiberno-Norse period the original town defenses were first constructed, and its principal streets were also laid out. The first phase of its defensive perimeter, which was an external fosse and an internal earthen rampart surmounted by a wooden palisade, has been dated by dendrochronology to the last quarter of the eleventh century. A stone town wall was first erected in Waterford in the mid-1130s, at a time of growing tension between the rival kingdoms of Leinster and Desmond. This coincided with the great animal murrain of 1133, which would have put much pressure on the food supply of the region. After the city was captured in 1170 by an Anglo-Norman army that successfully breached its walls, there were many references to the grant of "murage" by the crown to the burgesses of the city, from as early as 1207, to help defray the cost of building and maintaining their walls. It is a testimony to their industry that there were some fifteen gates and twenty-three mural towers in the circuit at its height, of which six towers still survive to this day.

In the latter part of twentieth century the large-scale redevelopment of the city center allowed an impressive series of archaeological excavations to go ahead between 1988 and 1992 in the center of the Viking-Age core of the city. These excavations are doubly important, in the first place because they covered 20 percent of the Hiberno-Norse walled area of the city—the largest proportion of any Irish city that has been archaeologically investigated. In addition, the city's archaeological horizons run uninterruptedly from the tenth century to the post-medieval period, a hitherto unparalleled sequence of survival in an Irish urban environment.

These excavations concentrated upon the area around two of the principal streets of Waterford—High Street and Peter Street—that run in an east-west direction along the top of a natural promontory of land. The excavation of this complete block of the city center bounded by four streets produced a large quantity of structural evidence for its original housing from as early as the eleventh century, with the majority of these houses fronting onto the street. Up until the middle of the twelfth century the houses were single-storied rectangular structures with wattle walls, very similar to those found in Hiberno-Norse Dublin. From the period just before the Anglo-Norman invasion (1169–1170), there were the beginnings of a new tradition in building with the introduction of sill-beam houses, where rectangular-shaped oak beams were sited as opposing pairs on the long walls of these structures. It was also in this period that four sunken buildings and two stone-lined entrance passages to other structures, all of late eleventh-century date, were constructed, which represents the greatest number yet found in any Irish urban settlement. In the following century stone houses started to be constructed, with three extensive stone undercrofts dating from the middle of the thirteenth until the fifteenth century being excavated.

The most significant major building that was excavated was the complete ground plan of St. Peter's parish church, along with its associated burial ground. Six major building phases were identified up to the seventeenth century; the earliest was represented by a possible wooden church dating from the middle of the eleventh to the early twelfth century. Later in the twelfth century it was replaced by a stone church with an apse, a unique feature in a medieval Irish parish church that might be associated with the continental influences on the reform of the church at that time. The excavation of its burial ground also produced much useful evidence about death rates, nutritional deficiencies, degenerative joint disease, and dental attrition in its medieval urban population.

The more than 200,000 artifacts recovered in these excavations illustrated the importance of trade to this urban community, mainly evidenced by the many pottery shards, although very few medieval coins were located, surprisingly. Many examples of fine tablewares from western France, especially in the form of jugs, were located. The excavations also revealed the extent and importance of locally made pottery production.

The finds revealed that Waterford had strong trading links with Western England, Northwest France, and the Low Countries. Some of the more important crafts that have left an archaeological trace were bone combmaking and the production of other antler objects in Peter Street, woodworking, and leather production. There was also extensive metalworking, producing everyday objects such as locks all the way to a very rare and beautiful early twelfth-century kite brooch, or a thirteenth-century gold ring brooch.

Although in European terms the walled area of the medieval city was quite modest in size, it was still the port chosen by English kings such as Henry II, John, and Richard II for their landfall in Ireland because of the security afforded by its walls and its location close to the major ports of western Britain. Because of its status as a royal port it prospered throughout the Middle Ages, even in the difficult years of the fourteenth and fifteenth centuries, and its prosperity continued well into the sixteenth century.

TERRY BARRY

References and Further Reading

Clarke, Helen, and Bjorn Ambrosiani. *Towns in the Viking Age.* Leicester and London, Leicester University Press, 1991.

Hurley, Maurice et al. *Late Viking Age and Medieval Waterford Excavations, 1986–1992.* Waterford: Waterford Corporation, 1997.

Thomas, Avril. *The Walled Towns of Ireland,* 2 vols. Dublin, Irish Academic Press, 1992.

Walton, Julian. *The Royal Charters of Waterford.* Waterford: Waterford Corporation, 1992.

See also **Ports; Ships and Shipping; Walled Towns**

WEAPONS AND WEAPONRY

Early Medieval Period (*c.* 400–800)

Early medieval Ireland was not highly militarized, and without significant external threats there was little pressure to improve military technology. The quality of pre-Viking Irish weapons has been questioned, although metallographic study of weapons of this period has found that while some were technologically poor, others were quite effective. Irish sources of the ninth to twelfth centuries depict the ideal weaponry of a warrior as a shield, a sword, and one or more spears. Significantly, these are the only weapons represented in the historical or archaeological record for the pre-Viking period and, indeed, for the preceding Iron Age. The shield/sword/spear combination seems to have been the ideal throughout the Iron Age and early medieval periods.

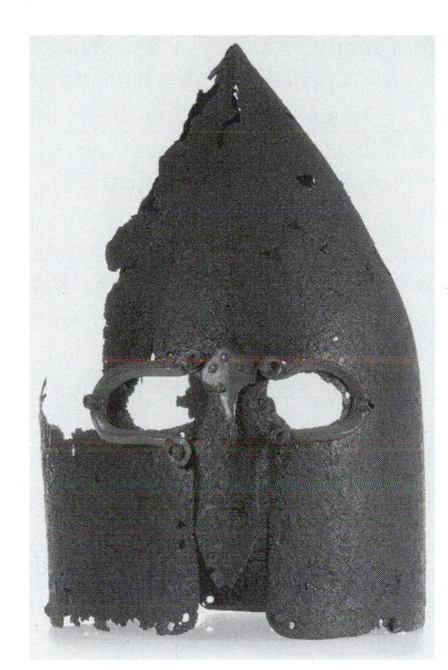

Iron helmet, Lough Henney, Co. Down. *Photograph reproduced with the kind permission of the Trustees of the National Museums and Galleries of Northern Ireland.*

The spear was the most common, and in that sense the most important of these weapons. Used by all races, classes, and types of warrior, it could be thrown as a missile or retained in the hand to thrust and parry in close combat. At least twelve different terms for spears are found in Irish sources—testimony to its ubiquity. Archaeological study of medieval spearheads is hampered, however, both by the scarcity of examples from dateable contexts and by the recurrence of similar forms over long periods. Early medieval sources contain two terms for "sword"—*claideb* and *colg*. Mallory suggested that *colg* is the earlier, originally applied to small Iron Age thrusting swords, whereas *claideb* is a fifth- or sixth-century introduction, denoting longer swords designed for slashing or cutting. Rynne suggested that swords of "sub-Roman" type developed during the fourth to seventh centuries, followed by other forms ("expanded-ended" and "crannog" swords), which may have remained in use until the ninth century. In view of the scarcity of good contextual information, however, Rynne's classification and chronology must be regarded as provisional.

Viking/Hiberno-Norse Period (*c.* 800–1170)

The normal range of Viking military equipment is well known; their main weapons were spears, swords, axes, and bows and arrows. Initial Viking technological superiority could have been made up fairly readily, however, by the greater Irish kings. Military technology always responds rapidly to new influences, and

Ballinderry crannog, County Westmeath, may illustrate how far this process had advanced by the tenth century. This Irish site produced Ireland's finest "Viking" sword, a bow that must be ultimately of Viking background, and other typical "Viking" weapons. Borrowing of weaponry is not easily detected in the historical record because, on paper, few new weapon types were introduced by the Vikings (with the exceptions of bows and axes). Spears and swords continued to be the main offensive weapons, and shields the main means of defense. Undoubtedly, the form and technology of spears, swords, and shields developed, but this must be investigated through sustained archaeological research (such as works by Walsh and Pierce), rather than documentary sources.

Swords were always expensive and only available to the relatively wealthy. The Viking Age saw the introduction of finer but even more expensive swords, and among the Irish, swords were largely replaced by the cheaper axe. Introduced by the Vikings, axes were so widely adopted by the Irish that in the late twelfth century Giraldus Cambrensis portrays them as a veritable national weapon. Axes feature throughout Giraldus' *Expugnatio Hibernica*, culminating in his parting advice that the English "must never grow careless of the axes of the Irish." Such major figures as Hugh de Lacy, Miles de Cogan, and Ralph FitzStephen met their deaths by the dreaded Irish axe, while Meiler FitzHenry is described as having three axes stuck in his horse and two more in his shield during an Irish attack in 1173. Giraldus knew the Irish had borrowed the axe from the Norse, and the earlier twelfth-century *Cogad Gáedel re Gallaib* also refers to the Irish using "Norwegian axes." This is confirmed by archaeology, since all known battle axes of this period are derived from a Scandinavian type, Petersen's Type M.

Archery was apparently unknown since prehistory in Ireland, until reintroduced by the Vikings in the ninth century. Indeed, there is little evidence for Gaelic Irish use of archery before the thirteenth century. Archaeological evidence is largely confined to the Hiberno-Norse towns, where surviving bow fragments indicate an established bowmaking tradition, largely anticipating the better-known late-medieval English tradition. The sheer volume of archaeological evidence (mainly arrowheads) testifies to widespread Hiberno-Norse use of archery, but its military significance is less clear. Archery was apparently used mainly in preliminary missile exchanges prior to battle. There is no evidence for its exploitation to anything like the same extent as in the later Middle Ages, nor is there evidence for specialist archers—the bow was simply one of the weapons used by Viking warriors.

In contrast to weaponry, the use of armor clearly distinguishes the Irish from the Vikings. Irish sources indicate that the Irish did not wear armor, while the Norse are consistently described as doing so. Armor of this period rarely survives, and is discussed largely on the basis of representational evidence. The main body armor was a mail shirt reaching usually to the knees (the hauberk or byrnie), worn over a padded undergarment (the aketon or gambeson). Helmets were typically simple and conical, either of single-piece construction or formed of triangular plates riveted to a framework of iron bands—the *Spangenhelm*. Circular shields were replaced by triangular or kite-shaped forms in the eleventh century, but it is unclear when this happened in Ireland. Both forms were constructed of wooden boards covered with leather or other material, with a central iron boss and, probably, an iron binding strip around the edge. Clothing worn in battle by the Irish, even the nobility, did not differ significantly from civilian dress. This probably explains Giraldus's statement that the Irish went "naked and unarmed into battle." The contemporary *Song of Dermot and the Earl* describes the Irish as quite naked," with neither hauberks (*haubers*) nor byrnies (*bruines*). Undoubtedly, some Irish warriors could have obtained Norse armor through trade or combat, but such borrowing clearly did not happen to the same extent as with weaponry.

Anglo-Norman Period (*c.* 1170–1300)

Anglo-Norman weapons and armor were little different from those of Hiberno-Norse warriors; their military success must be explained in terms of organization and tactics, rather than technology. Hauberks remained the main body armor, supplemented from the later twelfth century by separate mail chausses, mufflers, and coifs, worn over the legs, hands, and head and neck, respectively. Conical helmets continued in use alongside hemispherical and cylindrical forms, which developed by around 1200 into the "great helm," fully enclosing the head. Triangular shields tended to become broader and shorter in the thirteenth century. Knights used spears and swords, much like those of the Hiberno-Norse period; long, double-edged blades, designed for cutting blows either from horseback or on foot, predominated. A series of surviving Irish swords, of twelfth- to fourteenth-century date, are typical of what would have been used by the first Anglo-Normans and their successors. Maces with spiked heads of bronze were also used, but the bow and arrow remains the most common weapon in the archaeological record. Archers made up the bulk of Anglo-Norman forces, and in the thirteenth century there is the first clear evidence for the use of archery by the Gaelic Irish. References to the capture of English armor imply that the Irish were also using armor in the thirteenth century. This is difficult to quantify, but undoubtedly the English conquest markedly increased the amount of armor circulating in Ireland.

Late Medieval Period (*c.* 1300–1550)

The lack of dramatic changes in military technology continued through the late Middle Ages, which is characterized by the use of apparently antiquated armor and weapons. There is a logical explanation for this, first expressed by Giraldus Cambrensis. Even in the twelfth century it was clear that the trend toward increasingly strong—and heavy—armor must be constrained, in Ireland, by the overriding requirement of mobility, dictated by the physical environment and prevailing tactics of warfare. Despite initial military superiority, the Anglo-Irish largely adopted Gaelic warfare tactics, based on raiding rather than large campaigns or battles. Late-medieval European developments in plate armor came at too high a price, in terms of increased weight and rigidity, for Irish combatants. Fifteenth- and sixteenth-century Anglo-Irish effigies depict armor consisting of a "pair of plates" (early plate armor for the torso) over a mail shirt, with separate plate defenses for arms and legs and a visored helmet worn over a mail mantle, covering throat and shoulders. Elsewhere such armor might be dated around the late-fourteenth century, but it appears on Irish effigies into the later-sixteenth century. Sculptural and documentary evidence suggests that Gaelic nobles and their gallowglass mercenaries routinely wore armor that differed only slightly from the Anglo-Irish pattern. Mail shirts (and the aketons underneath) tend to be longer; separate plate defenses for torsos, legs and arms are rare; and helmets are usually not visored. Even these distinctions are not rigid, however. Early sixteenth-century effigies of an O'More at Abbeyleix, County Laois, and a MacGillapatrick at Fertagh, County Kilkenny, display armors close to Anglo-Irish style, while the finest example of the "Gaelic" pattern, the Glinsk (Co. Galway) effigy, apparently represents an Anglo-Irish de Burgh. Archaeology confirms the sculptural evidence for armor forms. Although no surviving armor is known, late medieval Irish swords and arrowheads retain blade forms designed for use against mail armor; forms developed for use against plate armor—common elsewhere in Europe—are absent.

The aristocratic figures represented on the effigies would have fought as horsemen, and their principal weapon would have been the lance or spear. Gaelic horsemen did not use their spear in the couched position—the lack of saddle and stirrups would not have permitted this—but rather held it overarm, for throwing or thrusting. Their other main weapon, the sword, is depicted on effigies as a single-hand weapon, as would be expected for horsemen. Swords on Anglo-Irish effigies conform to common European styles, but no surviving examples are known. Swords on Gaelic effigies, however, display characteristic hilt forms found on a substantial group of surviving swords. This form is also found in Scotland as early as the fourteenth century, and its presence in Ireland probably reflects gallowglass activity. In the sixteenth century these may have been replaced by a distinctively Irish sword form, characterized by open-ring pommels. Surviving battleaxes, including fine ceremonial weapons inlaid with silver, are also often attributed to gallowglass. However, they were also used by the Gaelic Irish and clearly developed from the Viking battleaxe tradition, which predates gallowglass activity in Ireland. Common (non-noble) soldiers rarely wore armor and used a range of weapons. Archery was hugely important in late medieval English warfare, and from the mid-fourteenth century the Anglo-Irish government almost invariably employed English retinues composed mainly of archers. Deliberate efforts were made to foster archery among the Anglo-Irish commoners, but were ultimately unsuccessful outside of the core of the Pale (essentially Dublin and Meath). Besides bows and arrows, Anglo-Irish archers also used swords and bucklers (small shields). The poorest commoners used staff weapons such as bills and glaives. Gaelic common footsoldiers, or kern, might be armed with a sword, axe, or long knife; perhaps a bow and arrows; or a number of spears and a shield.

Artillery—although sporadically used from as early as 1361—was extremely rare until the late fifteenth century, when there is evidence for artillery and handguns being used by the Anglo-Irish and even by some Gaelic Irish. Artillery was first used effectively in government campaigns of the 1520s and 1530s (notably the Fitzgerald revolt), but it was not until the later sixteenth century that English armies decisively abandoned the longbow in favor of the musket. The attraction of firearms was not based on superior range or penetrative power, but simply on the fact that their use required little or no training, whereas archery, to be effective, demanded large numbers of highly trained men. Firearms ultimately revolutionized every aspect of warfare, but in Ireland this was a post-medieval development.

ANDY HALPIN

References and Further Reading

Bartlett, T., and K. Jeffery, eds. *A Military History of Ireland.* Cambridge: University Press, 1996.

De hÓir, S. "Guns in Medieval and Tudor Ireland." *Irish Sword* 15 (1982–1983): 76–88.

Halpin, A. "Irish Medieval Swords, *c.*1170–1600." *Proceedings of the Royal Irish Academy* 86C (1986): 183–230.

Halpin, A. "Irish Medieval Bronze Maceheads." In *Keimelia: Studies in Medieval Archaeology and History in Memory of Tom Delaney,* edited by G. MacNiocaill and P. F. Wallace, 168–92. Galway: Galway University Press, 1988.

Harbison, P. "Native Arms and Armour in Medieval Gaelic Literature, 1170–1600." *Irish Sword* 12 (1975–1976): 173–199, 270–284.

Hayes-McCoy, G. A. "The *Galloglach* Axe." *Journal of the Galway Archaeological and Historical Society* 17 (1937): 101–121.

Hayes-McCoy, G. A. "The Early History of Guns in Ireland." *Journal of the Galway Archaeological and Historical Society* 18 (1938–1939): 43–65.

Hayes-McCoy, G. A. *Sixteenth-Century Irish Swords in the National Museum of Ireland.* Dublin: National Museum of Ireland, 1977.

Hunt, J. *Irish Medieval Figure Sculpture 1200–1600.* Dublin and London: Irish University Press/Sotheby Parke Bernet, 1974.

Mahr, A. "The *Galloglach* Axe." *Journal of the Galway Archaeological and Historical Society* 18 (1938): 66–68.

Mallory, J. P. "The Sword of the Ulster Cycle." In *Studies on Early Ireland: Essays in Honour of M. V. Duignan,* edited by B.G. Scott, 99–114. Belfast: Association of Young Irish Archaeologists, 1982.

Petersen, J. *De Norske Vikingesverd: En Typologisk-Kronologisk Studie Over Vikingetidens Vaaben.* Kristiana/Oslo, 1919.

Peirce, I. *Swords of the Viking Age.* Woodbridge, Suffolk: The Boydell Press, 2002.

Rynne, E. "A Classification of Pre-Viking Irish Iron Swords." In *Studies on Early Ireland: Essays in Honour of M. V. Duignan*, edited by B.G. Scott, 93–97. Belfast: Association of Young Irish Archaeologists, 1982.

Scott, B. G. *Early Irish Ironworking.* Belfast: Ulster Museum, 1990.

Walsh, A. "A Summary Classification of Viking Age Swords in Ireland." In *Ireland and Scandinavia in the Early Viking Age, edited by H. B.* Clarke, M. Ní Mhaonaigh, and R. Ó Floinn, 222–235. Dublin: Four Courts Press, 1998.

See also **Armies; Military Service, Gaelic; Military Service, Anglo-Norman; Naval Warfare**

WELSH INFLUENCE

Welsh influence on various aspects of medieval Irish life can be glimpsed occasionally in the extant sources, its precise nature, however, is difficult to assess. It can be detected in the formation of Ireland's particular brand of Christianity, in which British ecclesiastics, exemplified by Patrick, played a primary role. Moreover, communities termed Gailinne *na mBretan* and Dermag *Britonum* (of the Britons) may point to religious establishments founded and perhaps run by British monks, though how long these were likely to have remained British in any real sense is a matter for debate. In any event, individual Britons continued to occupy pivotal positions in the Irish Church for a considerable period if the designation *Britt* (the Briton) applied to Aedgen, bishop of Kildare who died in 864, is to be believed. In addition, an interest in the Welsh Church is revealed by the inclusion of St. David in *Félire Óengusso* (The Calendar of Óengus), a ninth-century metrical list of mainly Irish saints. That the interest was mutual is underlined by the extended stay in Ireland of Sulien, an eleventh-century bishop of the foundation to which St. David gave his name, whose thirteen-year study trip abroad was motivated by the wondrous wisdom of the Irish, according to his son, Ieuan. In actual fact, such scholarly sabbaticals on both sides of the Irish Sea may not have been unusual. The eighth-/ninth-century Juvencus manuscript with its mixture of Old Irish and Welsh glosses bears witness to active cooperation between Irishmen and Britons in one particular scriptorium. Irish scholars also formed part of the group of intellectuals patronized by the successive kings of Gwynedd, Merfyn Frych and his son, Rhodri Mawr, and we may suspect that manuscripts regularly found their way to and fro across the Irish Sea. Thus may the author of *Sanas Cormaic* (Cormac's Glossary), possibly the ninth-century king-bishop of Cashel, Cormac mac Cuilennáin, have acquired his Welsh, considerable use of which is made in his Glossary. This degree of knowledge of the neighboring culture, however, is likely to have been the exception rather than the rule.

Ecclesiastical and cultural connections of this nature are mirrored in the political sphere. The presence of a considerable body of Irish settlers in Dyfed as early as the sixth century, as manifested most tangibly in the ogam inscriptions they left behind, provided their kinsmen at home with a gateway through which all manner of ideas and influences might emerge. Indeed it was via this channel that the Irish acquired a name for themselves and for their language, *Goídel* (Irishman) and *Goídelc* (Irish) being borrowings from Welsh *Gwyddel* and *Gwyddeleg*, respectively. This Irish power base in South Wales did not survive; later centuries, however, saw both Irish and Ostmen kings seeking to involve themselves in the affairs of their nearest neighbor. The eleventh-century king, Díarmait mac Maíl na mBó, is described as king of Wales, no less, in his death notice in the *Annals of Tigernach*, and while the claim may have no basis in fact it points to intensive involvement with Wales on the part of the Leinster ruler. Nor was such political trafficking all one-way. Among the Welsh rulers to have a close association with Ireland was Cynan ab Iago, whose son Gruffudd, by the daughter of the Hiberno-Norse king of Dublin, grew up in Swords and drew extensively on Irish assistance in his attempt to regain his Gwynedd patrimony. His name and those of other leading Welshmen are recorded in the Irish chronicles, ample testimony that they formed a significant presence in the Irish political scene.

It need not surprise us that literary reflexes of these links have also survived. The Welsh prose tale, *Branwen uerch Lyr* (Branwen, daughter of Llyr) has as its starting point a marriage alliance between Branwen, sister of the Welsh king Bendigeidfran, and his Irish counterpart, Matholwch. An alliance of a different kind with the war leader Ingcél, mac ríg Bretan (son of the king of the Britons) forms the core of the Irish narrative *Togail Bruidne Da Derga* (The Destruction of Da Derga's Hostel). Ecclesiastical intercourse is a commonplace motif in the Lives of a number of Welsh and Irish saints, and while textual borrowing may explain the resemblances in some instances, both hagiographical cultures could well reflect independently contacts taking place on the ground. In the same way, we would do well to assess carefully the perceived similarities between both literatures, recognizing that while ample opportunity for borrowing may have arisen, the two textual traditions are ultimately the product of two distinct societies, however intertwined.

Máire Ní Mhaonaigh

References and Further Reading

Charles-Edwards, Thomas. "Language and Society among the Insular Celts A.D. 400–1000." In *The Celtic World*, edited by Miranda J. Green, 703–736. London and New York: Routledge, 1995.

Duffy, Seán. "Ostmen, Irish and Welsh in the Eleventh Century." *Peritia: Journal of the Medieval Academy of Ireland* 9 (1995): 378–396.

Lapidge, Michael. "Latin Learning in Dark Age Wales: Some Prolegomena." In *Proceedings of the Seventh International Congress of Celtic Studies, Oxford 1983*, edited by D. Ellis Evans, John G. Griffith, and E. M. Jope, 91–107. Oxford: Cranham Press, 1986.

O'Rahilly, Cecile. *Ireland and Wales: Their Historical and Literary Relations*. London: Longmans, Green and Co., 1924.

Sims-Williams, Patrick. "The Evidence for Vernacular Irish Literary Influence on Early Mediaeval Welsh Literature." In *Ireland in Early Mediaeval Europe: Studies in Memory of Kathleen Hughes*, edited by Dorothy Whitelock, Rosamond McKitterick, and David Dumville, 235–257. Cambridge: Cambridge University Press, 1982.

See also **Cormac mac Cuilennáin; Hagiography and Martyrologies; Scriptoria**

WILLIAM OF WINDSOR (*c.* 1330–1386)

William of Windsor was born sometime around 1330 in Westmoreland. He was chief governor of Ireland from1369 to 1372 and from 1373 to 1376. Windsor's appointment as king's lieutenant in 1369 was a continuation of the policy of large-scale military intervention, funded from England, that began with Lionel of Clarence in 1361. Unlike Lionel, Windsor was not a great prince of royal blood, but a mere knight. Yet his experience of frontier conditions in the north of England, where he had repeatedly demonstrated military and administrative skill, and his service in Ireland under Lionel from 1363, made him an obvious choice to govern the lordship of Ireland.

Windsor's tenure as chief governor was hampered by acrimonious relations with the Anglo-Irish community over the question of taxation. Whereas Lionel had been appointed in the wake of a peace treaty of 1360 with France, Windsor's appointment coincided with the renewal of hostilities. It was feared that France would attempt an invasion of Ireland as a "back door" into England, and part of Windsor's mandate was to secure the southern coast. Windsor was heavily subsidized from England, receiving some £22,300. This figure was, however, modest compared to the sums invested against France, and it was inadequate to maintain Windsor's large army. Furthermore, the strain that the Anglo-French war put on the English exchequer meant that money was often slow in arriving. Windsor therefore repeatedly summoned the Irish parliament and demanded that it contribute to the cost of the lordship's defense.

The frequency and extent of Windsor's demands were unprecedented, and he resorted to coercion to gain the funds he required. This breached the principle that taxation had to be voted by parliament of its free will and caused great antagonism. Lists of grievances against him were sent to England, and in 1372 he was recalled. He was reappointed, however, in 1373 and continued as before. The policy reached a climax when Irish representatives were summoned to England, presumably in the hope that they could be browbeaten into voting funds. Representatives were duly elected, but the communities specifically withheld the power to grant a subsidy.

The opposition may in part have stemmed from discontentment with Windsor's record. He did succeed in gaining some submissions from Irish lords, notably capturing and executing the king of Leinster, Diarmait Láimhderg Mac Murchada. But these submissions lasted only as long as they were directly enforced. This dissatisfaction was mirrored in England, where parliament found it difficult to justify continued investment in Ireland when there was little sign of a return to self-sufficiency, let alone profitability. Renewed complaints sent to England found a ready audience in the growing opposition to Edward III. Windsor was vulnerable because he was the husband of Alice Perrers, Edward III's influential and despised mistress. Windsor was recalled to England in 1376 to coincide with the "Good Parliament," where the accusations against his administration in Ireland may have been used to attack Perrers.

Several of Windsor's appointees in the Irish administration were dismissed, but the charges directed at Windsor himself were subsequently dropped. Modern historians have depicted him as a victim of circumstance and his own excessive zeal, rather than a corrupt official. He went on to have a highly successful career and died at Haversham in 1386. The Anglo-Irish community continued to demand aid from England while simultaneously resenting the intrusion of English officials, and the policy of military intervention was to culminate in the two expeditions under King Richard II of 1394 and 1399.

References and Further Reading

Clarke, Maude V. "William of Windsor in Ireland, 1369–1376." In *Fourteenth-Century Studies by M. V. Clarke*, edited by L. S. Sutherland and M. McKisack, 146–241. Oxford: Clarendon Press, 1937. Reprint 1968.

Connolly, Philomena. "The Financing of English Expeditions to Ireland, 1361–1376." In *England and Ireland in the Later Middle Ages: Essays in Honour of Jocelyn Otway-Ruthven*, edited by James Lydon, 104–121. Dublin: Irish Academic Press, 1981.

Harbison, S. "William of Windsor, the Court Party and the Administration of Ireland." In *England and Ireland in the Later Middle Ages: Essays in Honour of Jocelyn Otway-Ruthven,* edited by James Lydon, 53–74. Dublin: Irish Academic Press, 1981.

Lydon, James F. "William of Windsor and the Irish Parliament." In *English Historical Review* 80 (1965): 252–267.

See also **Chief Governors; Lordship of Ireland**

WILLS AND TESTAMENTS

Testaments, in later medieval Ireland, were the sets of instructions, left on their death by testators, for the burial of their body and for the disposition of such of their property as was at their free disposal, and appointing executors to ensure that these wishes were carried out.

Testaments of this general type were a common feature of general Western European legal practice, but seem only to have been introduced into Ireland by the Synod of Cashel of 1172. Testaments seem most commonly to have been made in writing. They could also be made orally, provided there were witnesses to prove what the testator had said. They were normally made on the testator's deathbed. Indeed, the Synod of Cashel required them to be made in the testator's last sickness in the presence of his confessor and neighbors. The executors were required to probate the testament, normally in the local bishop's court—that is, to produce the testament and prove it was genuine—before they would be authorized by sealed letters of administration to carry out the testator's last instructions. For testators who had property in towns, a second probate in the town court was often required. The executors could then proceed to pay the debts of the deceased, collect moneys owing to him, and then distribute his or her estate. When they had finished doing so they were required to provide written accounts of their administration to the bishop's court. In general principle, lands, houses, and other similar kinds of property could not be left by testament and were supposed to pass by the general rules of intestate inheritance, except in towns where local town or city custom authorized this. Married men, on their deaths, had at their free disposal only one third (if they had a wife and children) or one-half (if they had only a wife) of their money and other goods. Under secular law married women could only make testaments with the consent of their husbands, and of such property as the husband assigned to them. The term "last will" or "will" was often used as a synonym for testament, but it was also used more specifically from the later fourteenth century onward for the instructions left to trustees (feoffees to uses), who held the nominal legal title to lands, to execute the wishes of the beneficial owner of the lands after their death. Last wills of this kind effectively gave landowners a power of testation over their lands and allowed them to determine to whom they passed. Wills did not require probate, but were often included in testaments and probated with them. Some original testaments and wills survive in collections of medieval deeds, and others as copied into cartularies. There is also a single surviving (and published) register of enrolled copies of testaments and wills, plus accompanying inventories of the possessions of the deceased testator submitted by the executors at the time of probate, for the diocese of Dublin for the period from1457 to 1483.

PAUL BRAND

References and Further Reading

Berry, H. F. *Register of Wills and Inventories of the Diocese of Dublin, 1457–1483*. Dublin: Royal Society of Antiquaries of Ireland, 1898.

Connolly, Philomena. *Medieval Record Sources (Maynooth Research Guides for Irish Local History, No. 4)*. Dublin: Four Courts Press, 2002.

Murphy, Margaret. "The High Cost of Dying: An Analysis of *pro anima* Bequests in Medieval Dublin." In *The Church and Wealth (Studies in Church History vol. 25)*, edited by W. J. Sheils and Diana Wood. Oxford: Blackwell, 1987.

See also **Burials**

WISDOM TEXTS

Literature that in a pithy, sententious style comments on the nature of humankind and the world or that gives aphoristic, moralizing precepts is known as gnomic, or wisdom, literature. The wide variety of wisdom texts and the large number of gnomic statements scattered throughout tales and poetry give evidence of the high esteem that the genre enjoyed in Irish literature.

Tecosca

A distinct subgroup of Irish wisdom texts is called *tecosc* (instruction), plural *tecosca*, consisting of advice on the moral qualities appropriate to kings that was usually attributed to legendary figures of the past and directed at their pupils or foster sons. *Tecosca* can be equated with the later continental genre of *specula principum* (Mirrors of Princes). These texts emphasize moderate and considerate social conduct and encourage defense and maintenance of the traditional law. A distinct lack of heroic, warriorlike ethics is noticeable. The central theme of the *tecosca* is that of *fír flathemon* (the ruler's truth). By acting in accord with this concept, the ruler secures peace, stability, and prosperity for himself and for his people, since his justice and righteousness correlates with the welfare of his country. The idea of *fír flathemon* has been compared with similar concepts in the Indo-Iranian world, such as Vedic *rtá-* (right order) and Avestan *a2a-* (rightness). Keating claims that *tecosca* were recited at the inauguration of kings until the Norman period.

The oldest of these in the Irish language is *Audacht Morainn* (AM) (The Testament of Morann), which on linguistic grounds can be dated in its oldest recension to the late seventh century. AM, which consists of 164 lines in sixty-three paragraphs, expresses the ideas of *fír flathemon* in its most concise form of all *tecosca*, cp. the catalogue of its beneficial effects introduced by the phrase *Is tre fír flathemon* . . . "It is through the ruler's truth that . . . " (§§ 12–28). AM shows affinities in form and in substance with law tracts of the *Bretha Nemed* school. The probably ninth-century *Tecosca Cormaic* (TCor) (The Instructions of Cormac), advice given by the wise king Cormac in reply to questions of his son and successor Cairpre, is the longest Irish wisdom text. In some manuscripts it consists of thirty-seven sections (746 lines), but only the first eighteen are believed to make up the original part of the work. Other paragraphs may have been borrowed from *Senbríathra Fíthail* (SF) (The Old Sayings of Fíthal (king Cormac's judge)) (245 lines in thirteen sections). The focus of SF, composed around 800, lies less on political instruction than on statements of a general nature (§§ 1–6, 9), which are cast to a large extent into three-word maxims such as *Dligid fír fortacht* (Truth should be supported) (§ 5.2) and *Tosach éolais imchomarc* (Inquiry is the beginning of knowledge) (§ 1.4). Sections 7–8 and 10–12 are, like TCor, in the form of a dialogue between Fíthal and his son. AM, TCor, and SF are frequently found combined in the manuscripts.

The short *Tecosc Cuscraid* (TCus) (The Instruction to Cuscraid (son of king Conchobar)) (26 lines), attributed to the Ulster hero Conall Cernach, forms part of the tale *Cath Airtig* (The Battle of Airtech). *Bríatharthecosc Con Culainn* (BrCC) (The Precept-Instruction of Cú Chulainn) to the future king Lugaid Réoderg (40 lines), is included in the tale *Serglige Con Culainn* (The Sick-Bed of Cú Chulainn). Middle-Irish compositions such as *Diambad Messe bad Rí Réil* (If I were an Illustrious King) (37 quatrains) and *Cert Cech Ríg co Réil* (The Tribute of Every King is Clearly Due) (72 quatrains), which draw on the older tradition, are cast into metrical form. Close in sentiment to the *tecosca* is what the seventh-century Latin tract *De Duodecim Abusivis Saeculi* (The Twelve Evils of the World) has to say in chapter 9 on the *rex iniquus* (unjust king) and the *rex bonus* (good king). This chapter was also included in the *Collectio Canonum Hibernensis* (XXV, 3–4) and contributed strongly to the concept of the continental *specula principum*. Sedulius Scottus, an Irish scholar who lived in the Frankish empire in the ninth century, drew on classical and Christian tradition for his Latin *Liber de Rectoribus Christianis* (Book on Christian Rulers) for king Lothair II.

Gnomic Texts

Other wisdom texts contain general advice, not aimed at specific social classes. *Aibidil Luigne maic Éremóin* (ALE) (The Alphabet of Luigne Son of Éremón) (157 lines) is a miscellaneous collection of legal and proverbial maxims in three distinct sections brought together from various sources such as law tracts and *tecosca*. *Bríathra Flainn Fína maic Ossu* (BFF) (The Sayings of Flann Fína Son of Oswiu) is in its core identical with sections 1–5 of SF, but has expanded the number of maxims from 139 to 261. The ascription of the authorship to Flann Fína, the Irish name for king Aldfrith of Northumbria (d. 705), is doubtful since the language of the collection is that of the eighth or ninth century. Another collection with the title *Roscada Flainn Fína maic Ossu Ríg Sacsan* (RFF) (The Maxims of Flann Fína Son of Oswiu, King of the English) is close in content to BFF, but the sections and maxims have a different order. Colin Ireland, their most recent editor, treats SF, BFF, and RFF as different recensions of one original gnomic collection. For *Trecheng Breth Féine* (TrBF) (The Triads of Ireland) see Triads. The *Prouerbia Grecorum* (PG) (Proverbs of the Greek), which may go back to the sixth or seventh century, purports to be a Latin translation of some eighty proverbs from Greek, but its Western manuscript tradition and its ideas about the just king, similar to those expressed in the *tecosca*, speak for an Irish provenance.

Though having a different bias, religious texts such as *Apgitir Chrábaid* (The Alphabet of Piety) or *The Rule of Ailbe of Imlech*, and legal texts such as *The Advice to Doidin* show affinities in style, structure, and expression with wisdom texts. Irish law tracts contain many didactic passages, and stylistically many legal axioms are expressed in a manner similar to that of wisdom literature. Medieval Irish tales are also interspersed with nuggets of wisdom, such as *Gel nech nua* (Any new thing is bright) (TCor § 14.23), which is used in *Serglige Con Culainn* of 720.

The earlier texts (AM, TCus, BrCC, parts of TCor and SF) are composed in a rhythmical prose whose prime stylistic features are repetition, alliteration, and sometimes unusual syntax. BFF, ALE, parts of SF and TCor display a monotonous, formulaic style with terseness of expression bordering on obscurity. A strong legal interest is apparent in all wisdom texts. Women are usually depicted in an unfavourable way. Although pre-Christian origins are frequently assumed for Irish wisdom literature, stylistic parallels with biblical models such as the *Book of Proverbs* are observable and may have influenced, if not engendered, the Irish texts. Due to the nature of the genre, its formulaic style, and the compilatory character of many of the texts, a great amount of mutual borrowing has taken place and the collections could easily have been added to in the process of transmission, so that it is now largely impossible to get a clear picture of the original shapes of the texts.

David Stifter

References and Further Reading

Best, Richard I. "The Battle of Airtech." *Ériu* 8/2 (1916): 170–190. [= TCus]

Breen, Aidan. "De XII Abusivis: Text and Transmission." In *Ireland and Europe in the Early Middle Ages: Texts and Transmission*, edited by P. Ní Chatháin and M. Richter. Dublin: Four Courts Press, 2002.

Ireland, Colin. *Old Irish Wisdom Attributed to Aldfrith of Northumbria: An Edition of Bríathra Flainn Fhína maic Ossu (Medieval and Renaissance Texts and Studies, Vol. 205)*. Tempe, Arizona: Arizona Center for Medieval and Renaissance Studies, 1999.

Kelly, Fergus. *Audacht Morainn*. Dublin: Dublin Institute for Advanced Studies, 1976.

Meyer, Kuno. *The Triads of Ireland, (Todd Lecture Series, Vol. 13)*. Dublin: Hodges, Figgis & Co.; London: Williams & Norgate, 1906. [= TrBF]

——— *The Instructions of King Cormac mac Airt (Todd Lecture Series, Vol. 15)*. Dublin: Hodges, Figgis & Co.; London: Williams & Norgate, 1909. [= TCor]

O'Donoghue, Tadhg. "Cert Cech Ríg co Réil." In *Miscellany Presented to Kuno Meyer*, edited by Osborn Bergin and Carl Marstrander, 258–277. Halle an der Saale: Niemeyer, 1912.

——— "Advice to a Prince." *Ériu* 9 (1921–1923): 43–54.

O'Rahilly, Thomas F., ed. *A Miscellany of Irish Proverbs*. Dublin: Talbot Press, 1922.

Simpson, Dean. "The 'Prouerbia Grecorum'." *Traditio* 43 (1987): 1–22.

Smith, Roland. "On the Briatharthecosc Conculaind." *Zeitschrift für Celtische Philologie* 15 (1925): 187–192.

——— "The *Speculum Principum* in Early Irish Literature." *Speculum* 2 (1927): 411–445.

——— "The Alphabet of Cuigne mac Emoin." *Zeitschrift für Celtische Philologie* 17 (1928): 45–72. [= ALE]

——— "The Senbriathra Fithail and Related Texts." *Revue Celtique* 45 (1928): 1–92. [= SF & BFF]

See also **Kings and Kingship; Law Tracts; Moral and Religious Instruction; Sedulius Scottus; Triads**

WITCHCRAFT AND MAGIC

Terminology

By "magic" we understand words and acts performed by human beings, which are believed to bring about changes in the empirical world or to produce knowledge of hidden things in a supernatural way. The term "supernatural" refers to the nonempirical dimension of life, which is central to religious belief systems. The difference between the categories "natural" or "empirical" and the "supernatural" becomes clear when applied, for instance, to the human sense of "seeing." If someone looks at a cow in a field in a natural way, the empirical information about the cow's location and form is passed on to the brain of this person. If a person is believed to look at this beast in a supernatural way, the cow may be said to have been affected by the look, because of which it stops yielding milk. This way of "supernatural" looking is known as "casting the evil eye" or "bewitching." The relation between cause and effect in magic is not dictated by laws of science but is part of belief systems; hence, magic is a religious concept. "Witchcraft" is magic performed by witches—people believed to be professionals in magic.

A study of the semantic history of the term "magic" would reveal that the word has often been used in a polemic context. It has been seen as a "wrong" kind of religion. Originally, Magoi were the priests of the ancient Zoroastrian religion of Iran, but in the course of the fifth century B.C.E the Greeks started to use the term for those engaged in occult arts and private rituals (see Bremmer 2002a). Modern scholars such as James Frazer (1854–1941) defined magic in opposition with religion: By magic, people believed to bring about changes in an automatic, supernatural way or by commanding supernatural beings (often demons), whereas in religion, these changes are believed to be brought about by the supplication and veneration of supernatural beings (usually God or gods). This opposition is, however, not medieval but stems from Victorian middle-class elitist thinking (see Bremmer 2002b).

The polemic view of magic is also found in Christianity. Medieval Irish literature, composed in monasteries, is no exception to this rule. It is, therefore, not surprising that Irish equivalents for the term "magic" are *díabuldánacht* (diabolic art) and *gentliucht* (pagan art). Other general terms are *druídecht* (druids' art) and *ammaitecht* (witchcraft). Words such as *corrguinecht* and *fithnasacht* may have referred to a specific type of sorcery.

Magic in Early Irish Literature

In conformity with general Christian doctrine, magic is associated with pre-Christian or non-Christian religion in early Irish literature. In hagiography, druids and magic are described in antithesis with saints and miracles; the former representing evil and the latter good. Supernatural acts performed by druids and saints may be similar, but their evaluation differs. A good example is the contest between Saint Patrick and the druids as described in Muirchú's *Life of Patrick* (see O'Loughlin 2003). The aim of magic in hagiography is always destructive, hence the art of *magi* (magicians, i.e., druids) is designated in Hiberno-Latin, for example, *ars diabolica* (devilish art) or *maleficia* (evil deeds).

In non-hagiographic narrative literature (see Ulster Cycle, Mythological Cycle), the negative image of magic is less pervasive. Divination—the supernatural art to acquire knowledge about hidden or future things plays an important role in portrayals of pre-Christian society. As in hagiography, the source of knowledge or power with regard to such magical practices is sometimes explicitly identified as "demons," but at other times such indications are absent, and in this way, a more neutral description is given. We do not know whether divination and other rituals as described in this literature have ever taken place. Some descriptions may just as well reflect Christian assumptions about the pre-Christian past, influenced by Biblical and/or Classical literature. Certain portrayals of magic may be influenced by a Middle-Irish (*c*. 900–1200) trend to romanticize the pre-Christian past (Carey 1997).

Another difference between this kind of literature and hagiography is that magic is also associated with non-human inhabitants of Ireland: the supernatural beings of early Irish literature. Thus, the *áes síde* (people of the hollow hills or "fairies;" see Mythological Cycle) are believed to possess knowledge of magic. The so-called Túatha Dé Danann (see Invasion Myth, Mythological Cycle) have been said to have acquired supernatural knowledge in northern islands before their settlement in Ireland. Several magical practices are described in *Cath Maige Tuired* (The Battle of Mag Tuired; Gray 1983) as supernatural weapons in a war between the Túatha Dé Danann and their enemies, the Fomoire. The association of magic with the left, the north and evil is a recurring theme in early Irish literature (see Borsje 2002).

Magic in Daily Practice

As magic was considered to be useful in criminal acts, it is also mentioned in early Irish law (Kelly 1997: 174–175). Not only professional witches but also ordinary people were believed to harm others with magic, for example, by casting the evil eye (Borsje and Kelly 2003).

Magic was, however, also seen as useful for good and neutral purposes: for example, healing, protective, and divination charms that were written in Christian manuscripts. The supernatural entities referred to are both non-Christian and Christian. These, often complicated, texts are still largely ignored in Celtic and Medieval Studies (Carey 2000).

Witch Persecution in Medieval Ireland

In general, Christian doctrine condemned magic and witchcraft. In this spirit, belief in a *lamia* or *striga* "a dangerous supernatural female associated with witchcraft" was forbidden at the *First Synod of Saint Patrick* (Bieler 1963: 56–57). In later medieval Ireland, the general condemnation did not lead to witch hunts on the large scale as have taken place on the European Continent during the fifteenth to seventeenth centuries. A famous, and probably the first, witch trial was that of Alice Kyteler and her associates in Kilkenny (1324). Bishop Richard de Ledrede, a British cleric schooled in France, played a crucial role in the trial and wrote a contemporary narrative of the events. He seems to have tried to introduce continental ideas about witchcraft to Ireland. The few trials that did take place in Ireland have, however, never led to a "witch craze."

JACQUELINE BORSJE

References and Further Reading

Bieler, Ludwig. *The Irish Penitentials. Scriptores Latini Hiberniae V.* Dublin: Dublin Institute for Advanced Studies, 1963.

Borsje, Jacqueline. "The Meaning of *túathcháech* in Early Irish Texts." *Cambrian Medieval Celtic Studies* 43 (2002): 1–24.

——— and Fergus Kelly. "Examples of 'the Evil Eye' in Early Irish Literature and Law." *Celtica* 24 (2003): 1–39.

Bremmer, Jan N. "The Birth of the Term 'Magic'." In *The Metamorphosis of Magic from Late Antiquity to the Early Modern Period*, edited by Jan N. Bremmer and Jan R. Veenstra, 1–11. Leuven: Peeters, 2002a.

———. "Appendix: Magic *and* Religion." In *The Metamorphosis of Magic from Late Antiquity to the Early Modern Period*, edited by Jan N. Bremmer and Jan R. Veenstra, 267–271. Leuven: Peeters, 2002b.

Carey, John. "The Three Things Required of a Poet." *Ériu* 48 (1997): 41–58.

———. "Téacsanna Draíochta in Éirinn sa Mheánaois" (Magical Texts in Early Medieval Ireland). *Breis faoinár nDúchas Spioradálta: Léachtaí Cholm Cille* 30 (Maigh Nuad: An Sagart, 2000): 98–117.

Gray, Elizabeth A. *Cath Maige Tuired. The Second Battle of Mag Tuired (Irish Texts Society LII)*. London: The Irish Texts Society, 1983.

Kelly, Fergus. *Early Irish Farming: A Study Based Mainly on the Law-Texts of the 7th and 8th centuries* A.D. Dublin: Dublin Institute for Advanced Studies, 1997.

Mackey, James P. "Magic and Celtic Primal Religion." *Zeitschrift für Celtische Philologie* 45 (1992): 66–84.

Neary, Anne. "The Origins and Character of the Kilkenny Witchcraft Case of 1324." *Proceedings of the Royal Irish Academy* 83 C (1983): 333–350.

O'Loughlin, Thomas. "Reading Muirchú's Tara-Event Within its Background as a Biblical 'Trial of Divinities'." In *Celtic Hagiography and Saints' Cults*, edited by Jane Cartwright, 123–135. Cardiff: University of Wales Press, 2003.

Seymour, St. John D. *Irish Witchcraft and Demonology*. Dublin: Hodges, Figgis & Co. and London: Humphrey Milford, 1913.

See also **Hagiography and Martyrologies; Invasion Myth; Mythological Cycle; Patrick; Pre-Christian Ireland; Satire; Ulster Cycle**

WOMEN

Women in Sagas

Irish sagas set in the pre-Christian period feature some very masterful heroines, notably Medb, queen of Connacht, who has equal property and power with her husband, King Ailell, and leads a great army to invade the province of Ulster in the famous saga *Táin Bó Cúailnge* (the Cattle-raid of Cooley), from the Ulster cycle. This can give people the impression that women had greater freedom and control in pagan Ireland before the norms of Christianity redefined their role in society. However, there are two problems with this interpretation. First, most sagas were actually written between the ninth and the twelfth centuries or later, by Christian scribes adapting their rich inheritance of old traditions to suit the taste of their own times. Second, a number of their female protagonists, Queen Medb in particular, were based on goddesses or female symbols of sovereignty, whose extensive powers reflect their own supernatural attributes rather than the role of ordinary women at any date.

Women in Saints' Lives

Female saints also had supernatural attributes, in the sense that the Latin or Irish accounts of their lives credit them with many miracles. Otherwise they are shown as respected abbesses running communities of nuns, and the Lives may give us clues about the life of female religious communities in the early period. They show the nuns employing men to plow the lands attached to their communities, entertaining visiting bishops and abbots to hospitable meals that might include home-brewed beer, fostering young boys ultimately destined for the priesthood, and giving them their early education. Certain saints, like Lasair of Kilronan, are reputed to have pursued academic studies under the instruction of male saints and to have become qualified to instruct male clerics themselves, but the Life of St. Lasair is a late text written in a secular school of hereditary male historians, and it is uncertain if this feature of the Life is based on very early tradition. The fact is, we have no Latin works from early Ireland attributed to female authors, though we may have some Irish poems, such as "St. Íte's Lullaby to the Baby Jesus" or "The Lament of the Hag (or Nun) of Beare." Another feature of the Lives of Irish saints, male and female, is the saint's tendency to wander through the countryside from church to church, founding new communities, prescribing the tribute to be paid to the mother church, and blessing future generations of local families as long as they continue to be obedient to the saint's "heir," or successor, the head of the prinicipal church dedicated to that saint. This is clearly a literary device by the writer of a saint's Life to cast an aura of sanctity over territorial and financial rights claimed by the principal church in later generations, so again it is uncertain whether this reflects a real tendency of early nuns to leave their convents to wander on extensive tours of affiliated churches. However, as "heir" to the lands and authority endowing her nunnery, any abbess qualified as a female landowner, and this was the one class of female who did enjoy a degree of independence and power in early Irish law.

Landownership in the Laws

Old Irish law tracts discuss property rights, forms of marriage, and legal capacity. Full status as a free citizen in early Ireland depended on landownership, and family lands could only be transmitted through male heirs. If a man had no sons, his daughter might inherit his share of the family estate for her lifetime. Such an heiress would have the legal rights of a property owner, and the same public liability for tax and services as a male landowner. According to commentaries added to the law tracts around the eleventh or twelfth centuries, a female heiress, if she wished to hold all instead of only half of her father's land, must undertake to provide military service at the local king's summons, by paying and arming a kinsman to fight on her behalf. However, she could not pass on her estate to her children. After her death it would revert to her father's kindred, unless she married her first cousin on her father's side or another close relative, allowing her children to inherit the land through their father.

Legal Capacity

Apart from these exceptional heiresses, women received only movable property—cows, household goods, or silver—from their fathers, normally as marriage goods. They were thus "second-class citizens," legally dependent on their fathers or brothers if they were single, or on their husbands or grown-up sons if they were married. These male guardians were responsible for seeing that compensation was received for injuries inflicted on their womenfolk, or that fines were paid for crimes or damage committed by the women.

However, women were not completely without rights. Honor price (*lóg n-enech*) was a graded value applied to different classes in society, and used by lawyers to calculate the amount of compensation a freeman or noble could claim for insults or injuries. A wife's honor price was set at half the value of her husband's. This gave an officially married wife the same status as an adult son still living under his father's roof. If the male head of the household struck a bad bargain involving an overpayment that might result in financial loss to his family, the wife or son could object and dissolve the husband's contract within a period of ten days after the initial agreement. The husband had an even greater right to object to his wife's contracts, for a period of fifteen or twenty days after she agreed to a bargain. Secondary wives or concubines with children had lesser rights, and concubines with no children had even less control. However, the looser the tie between a woman and her partner, the stronger the connection she retained to her own kindred, and this could provide protection against wife-beating, for example.

Marriage

Although Old Irish treatises on customary law bear all the signs of having been written by or for clerics, surprisingly they recognize many more types of union between man and woman than a monogamous Christian marriage. They were compiled between the seventh and the ninth century C.E., before Carolingian church reforms gave Continental clergy a greater role in regulating marriage laws, and at a time when Christian Merovingian and Anglo-Saxon kings publicly kept concubines and sometimes passed on their thrones to the sons of those concubines. Old Irish law tracts give pride of place to a man's one official wife, the "first in the household" (*cétmuinter*), who normally contributed movable property of her own to the joint housekeeping and was entitled to receive it back, with any accumulated profits, if the couple divorced later. Divorce could be initiated by either the husband or the wife, on a number of grounds. A wife, for example could cite her husband's impotence or sterility, beating her severely enough to leave a scar, homosexuality causing him to neglect her marriage bed, failure to provide for her support, discussing her sexual performance in public, spreading rumors about her, his having tricked her into marriage by using magic arts, or his having abandoned her for another woman. In this last case, however, the first wife had the right to remain in the marriage if she wished, and was then entitled to continued maintenance from her husband.

A man could only marry another *cétmuinter* if his first wife was a permanent invalid unable to fulfill her marital duty, but it was not uncommon for husbands to acquire one or more secondary wives or concubines, known in the Old Irish tracts as *airech*, but significantly described in the later commentaries as *adaltrach* (adulteress). Irish marital customs attracted severe criticism from church reformers in the late eleventh century. Archbishop Lanfranc of Canterbury referred to Irishmen arbitrarily divorcing one wife in exchange for another "by the law of marriage or rather the law of fornication," and Pope Gregory VII heard it rumoured that many Irish "not only desert their lawful wives, but even sell them."

The Later Middle Ages

Officially all this was changed after the twelfth century church reform. Roman canon law was enforced through the decisions of church courts in each diocese. Following the Anglo-Norman invasion, feudalism was introduced into Ireland, along with English common law, which was particularly rigid in its insistence that a landowner's son could only inherit his father's estate if he was born after the canonically legitimate marriage of his parents.

However, it soon became obvious that English common law would apply in Ireland only to the settlers of English descent. An attempt by Irish church leaders to bribe King Edward I to extend common law to all native Irishmen living south of Ulster was blocked by the Anglo-Irish barons, and the Irish continued to be ruled by their own customary law, or "brehon law." Since this allowed illegitimate sons to inherit land along with those born of a church marriage, there was no economic incentive for Irish nobles to reform their marital habits. Arbitrary divorce followed by a remarriage that was invalid in the eyes of the church continued to be common, together with legally recognized contracts of concubinage, sealed by a bride-price paid by the man to the girl's family. It was open to a divorced wife to appeal to a church court to have her marriage declared still valid, but aristocratic erring husbands were normally able to demonstrate, through the arguments of their advocates, that the marriage in question had never been valid because they were too closely related to their wife, or they had already been married to a former repudiated wife who was still living when the second marriage took place.

The medieval Irish women who were most likely to sue their husbands to have their marriage reinstated were the wives of chieftains, because as the local queen, the chief's wife received certain lands and taxes, and occupied a seat on the council of nobles who represented her husband's territory. Queens' dowries formed an important source of movable wealth that could be drawn on for the ransoming of hostages, and this gave them a role in negotiating peace treaties and the release of captives. The most influential of all the queens were those who brought not wealth, but a regiment of soldiers to their husband as their dowry. Some of these retained considerable control over these military forces after their marriage, the best-known being the Scottish princess Iníon Dubh, warlike mother of the famous Red Hugh O'Donnell, and Gráinne O'Malley, the "pirate queen," both in the late sixteenth century.

Ordinary Irishwomen are first described by foreigners, medieval pilgrims to St. Patrick's Purgatory, or the bureaucrats of the Tudor reconquest. All report a generally relaxed attitude toward nudity and sex, which may relate to the failure of the Gregorian drive for clerical celibacy to make much headway in rural Ireland. Christina Harrington has noted that Irish churchmen, often themselves married, did not normally demonize woman in their writings or project her as a temptress responsible for man's sins. Young girls in Cork were seen by Fynes Moryson grinding corn stark naked, presumably to preserve their clothes from flour. The rural prostitutes of sixteenth century Gaelic Ireland, described by Edmund Spenser as *monashul* (*mná siúl*: wandering women), in default of urban centers wandered from place to place and fair to fair, and were seen as just one of the lower-class entertainers like gamesters or jugglers, suitable recipients of a great lord's fringe hospitality. Moryson noted as unusual that gentlewomen and Irish chieftains' wives stayed drinking "health after health" with the men at banquets, though unmarried maidens might be sent away after the first few rounds. Modern Irish Puritanism originated in the seventeenth century, promoted by the Counter-Reformation missionaries and the extension of English common law to Gaelic Ireland under James I.

KATHARINE SIMMS

References and Further Reading

Binchy, Daniel A., ed. *Studies in Early Irish Law.* Dublin: Royal Irish Academy, 1936.

Cosgrove, Art, ed. *Marriage in Ireland.* Dublin: College Press,1985.

Hadfield, Andrew, and Willy Maley, eds. *Edmund Spenser: A View of the State of Ireland.* Oxford: Blackwell, 1997.

Hall, Dianne Patricia. *Women and the Church in Medieval Ireland.* Dublin: Four Courts Press, 2003.

Harrington, Christina. *Women in a Celtic Church: Ireland 450–1150.* Oxford: Oxford University Press, 2002.

Mac Cana, Proinsias. "Aspects of the Theme of King and Goddess in Irish Literature." *Études Celtiques* 7 (1955): 76–114 and 356–413; 8 (1958): 59–65.

MacCurtain, Margaret, and Donnchadh Ó Corráin, eds. *Women in Irish Society: The Historical Dimension.* Dublin: Arlen House Press, 1978.

MacCurtain, Margaret, and Mary O'Dowd, eds. *Women in Early Modern Ireland.* Edinburgh: Edinburgh University Press, 1991.

Meek, Christine Elizabeth, and Katharine Simms, eds. *"The Fragility of Her Sex?"—Medieval Irish Women in Their European Context.* Blackrock, County Dublin: Four Courts Press, 1996.

Moryson, Fynes. "A Description of Ireland." In *Ireland Under Elizabeth and James I,* edited by Henry Morley, 413–445. London: George Routledge and Sons, 1890.

Nicholls, Kenneth, *Gaelic and Gaelicised Ireland in the Middle Ages.* Dublin: Gill and Macmillan, 1972. Second ed. Dublin: Lilliput Press, 2003.

Simms, Katharine. "The Legal Position of Irishwomen in the Later Middle Ages." *The Irish Jurist* 10 (1975): 96–111.

O'Dowd, Mary, and Sabine Wichert, eds. *Chattel, Servant or Citizen: Women's Status in Church State and Society* (Historical Studies XIX). Belfast: The Institute of Irish Studies, 1995.

See also **Brehon Law; Canon Law; Children; Church Reform, Twelfth Century; Common Law; Hostages; Íte; Law Tracts; Nuns; Papacy; Pre-Christian Ireland; Queens; Religious Orders; Slaves; Society, Grades of: Gaelic; Ulster Cycle; Witchcraft and Magic.**

WOODLANDS

In medieval Ireland, woodlands were a significant source of raw materials, fuel, and livelihood. They were often seen as a significant part of the landscape, bounded with fences and walls and protected by law and custom. There is a range of archaeological, historical, and paleoecological evidence that can be used to reconstruct the character of woodlands in this medieval landscape. Palynological studies, macrofossil plant studies, and beetle analyses all can indicate the range and relative quantities of tree species. The archaeological, technological and dendrochronological analyses of wooden structures (houses, waterfronts, mills) and artifacts (wooden bowls, spoons, tools, and equipment) is also revealing on species selection, age and growth patterns, trunk and branch morphology, and woodcrafts. Historical sources, particularly for the latter part of the Middle Ages, also reveal the presence, extent, ownership, and use of named woodlands in the landscape.

In the early medieval period (A.D. 400–1100), woodland was already a distinct, valued zone in a generally open agricultural landscape. Pollen analysis indicates that woodlands that had regenerated during the Iron Age were now being cleared for agricultural purposes from at least the fifth century A.D. (and particularly in the ninth century A.D.), but undoubtedly discrete areas were maintained. Early Irish laws, saints' Lives, and wood-specialist studies on archaeological structures suggest at least some measure of woodland management, with large quantities of immature hazel, ash, and willow underwood required for the building of post-and-wattle houses such as those uncovered at Deer Park Farms ringfort, County Antrim. Oak timber was especially valued for building churches, horizontal mills, and bridges. The houses of Hiberno-Norse and Anglo-Norman Dublin, Waterford, and Cork were also constructed of vast amounts of hazel and ash underwood. Artifact studies suggest that the other woodland products to be used there included oak, ash, alder, willow, and yew wood for lathe-turned bowls, buckets, and other domestic equipment. Environmental analyses of urban deposits reveal the use of woodland mosses for latrine purposes; apples, hazelnuts, sloes, elderberries, cherries, and plums were probably gathered from woodlands around the towns for food.

In the manorial economy of Anglo-Norman Ireland, woodland was seen as a valuable source of income, as rights within woodlands encompassed a wide range of activities, including the harvesting of underwood and timber, deer and boar hunting, cattle pasturage, and the foddering of pigs. Anglo-Norman documents, such as the Pipe Roll of Cloyne, indicate that a distinction was made between timber woods (*silva*), woodlands for underwood (*boscus*), and scrubby woodlands used for fuel (*bruaria*). By the fourteenth century, analysis of land use in manors around Dublin suggests that about 8 percent of land was held in woodland. It is also likely that the native Irish were involved in trading woodland products into the town. It is also possible that there were different cultural perceptions and understanding of Irish woodlands, they perhaps being seen by Anglo-Norman colonists as the fearful retreats of the Gaelic Irish lordships. There may also have been periods of woodland regeneration, particularly after the Gaelic revival and the Black Death. It is also evident that extensive areas of medieval woodlands remained intact up until the sixteenth century, particularly in the southwest and west of the island. However, major woodland clearances in the seventeenth century, related to new agricultural practices and for iron working, led to the destruction of much woodland cover.

References and Further Reading

Hall, V. A. "The Documentary and Pollen Analytical Records of the Vegetational History of the Irish Landscape, A.D. 200–1650." *Peritia* 14 (2000): 342–371.

Nicholls, K. "Woodland Cover in Pre-Modern Ireland." In *Gaelic Ireland: Land, Lordship and Settlement, c. 1250–c.1650,* edited by Patrick J. Duffy, David Edwards, and Elizabeth Fitzpatrick, 181–206. Dublin: Four Courts Press, 2004.

O'Sullivan, A. "Trees, Woodland and Woodsmanship in Early Mediaeval Ireland." In *Plants and People: Economic Botany in Northern Europe, A.D. 800–180*, edited by J. H. Dickson and R. Mill, 674–681. Edinburgh: Royal Botanical Society of Scotland,1994.

O'Sullivan, A. "Woodmanship and the Supply of Timber and Underwood to Anglo-Norman Dublin." In *Dublin and Beyond the Pale: Studies in Honour of Patrick Healy*, edited by C. Manning, 63–73. Dublin: Wordwell, 1998.

O'Sullivan, A. "The Wooden Waterfronts: A Study of their Construction, Carpentry and Use of Trees and Woodlands." In *The Port of Medieval Dublin: Archaeological Excavations at the Civic Offices, Winetavern Street, Dublin 1993*, edited by A. Halpin, 62–92. Dublin: Four Courts Press, 2000.

Tierney, J. "Woods and Woodlands in Early Medieval Munster." In *Early Medieval Munster: Archaeology, History and Society,* edited by M. A. Monk and J. Sheehan, 53–58. Cork: Cork University Press, 1998.

AIDAN O'SULLIVAN

See also **Agriculture; Craftwork; Diet and Food; Gaelic Revival; Houses; Manorialism**

WOODWORK

See **Craftwork; Houses**

Index

Note on Gaelic names and terms. Early medieval spelling has been preferred. Variants of a term differing from it mainly in the use of markers for lenited consonants (b/bh, d/dh, g/gh, m/mh, p/ph, s/sh, t/th) have not been indicated; thus, *Cogadh Gaedheal re Gallaibh* appears under *Cogad Gáedel re Gallaib* without cross-referencing. Patronymic personal names are alphabetized under the given name with no cross-reference from the byname (i.e., under the noninverted form only).

C

F

G

M

N

Q

R

Y

Z